Bark Beetles
Biology and Ecology of Native and Invasive Species

Bark Beetles
Biology and Ecology of Native and Invasive Species

Edited by

Fernando E. Vega

Sustainable Perennial Crops Laboratory, United States Department of Agriculture,
Agricultural Research Service, Beltsville, MD, USA

Richard W. Hofstetter

School of Forestry, Northern Arizona University, Flagstaff, AZ, USA

ELSEVIER

AMSTERDAM • BOSTON • HEIDELBERG • LONDON
NEW YORK • OXFORD • PARIS • SAN DIEGO
SAN FRANCISCO • SINGAPORE • SYDNEY • TOKYO
Academic Press is an imprint of Elsevier

Academic Press is an imprint of Elsevier
32 Jamestown Road, London NW1 7BY, UK
525 B Street, Suite 1800, San Diego, CA 92101-4495, USA
225 Wyman Street, Waltham, MA 02451, USA
The Boulevard, Langford Lane, Kidlington, Oxford OX5 1GB, UK

First published 2015

British Library Cataloguing in Publication Data
A catalogue record for this book is available from the British Library

Library of Congress Cataloging-in-Publication Data
A catalog record for this book is available from the Library of Congress

ISBN: 978-0-12-417156-5

For information on all Academic Press publications
visit our website at **store.elsevier.com**

14 15 16 17 10 9 8 7 6 5 4 3 2 1

Working together
to grow libraries in
developing countries

www.elsevier.com • www.bookaid.org

Dedication

We dedicate this book to Dr. Donald E. Bright, in recognition of his outstanding contributions to systematics, biology, zoogeography, and the evolution of bark and ambrosia beetles. Don was born in 1934 in Columbus, Ohio, and received his Bachelor of Science degree from Colorado State University in 1957. He served as an entomologist in the U. S. Army from 1957-1959, and in 1961 received his Master of Science degree from Brigham Young University in Utah, where he worked with Stephen L. Wood. In 1965 he was awarded a Doctor of Philosophy degree from the University of California at Berkeley and in 1966 he started working as a Research Scientist at the Canadian National Collection of Insects, in Ottawa. Don retired in 2003 and moved to Fort Collins, Colorado, in 2006, where he joined Colorado State University as a Faculty Affiliate at the C. P. Gillette Museum of Arthropod Diversity in the Department of Bioagricultural Sciences and Pest Management.

Don has published nearly 100 bark and ambrosia beetle-related publications (below), including "A Catalog of Scolytidae and Platypodidae" with Steve Wood (Wood and Bright, 1987, 1992), as well as three supplements to the catalog (Bright and Skidmore, 1997, 2002, Bright, 2014). Don's contributions have been instrumental in gaining a better understanding of bark and ambrosia beetles.

Publications of D. E. Bright (in chronological order):

Bright, D.E., 1963. Bark beetles of the genus *Dryocoetes* Eichhoff (Coleoptera: Scolytidae) in North America. Ann. Entomol. Soc. Am. 56, 103–115.

Bright, D.E., 1964. Descriptions of three new species and new distribution records of California bark beetles. Pan-Pac. Entomol. 40, 165–170.

Bushing, R.W., Bright, D.E., 1965. New records of hymenopterous parasites from California Scolytidae. Can. Entomol. 97, 199–204.

Bright, D.E., 1966. New species of bark beetles from California with notes on synonymy (Coleoptera: Scolytidae). Pan-Pac. Entomol. 42, 295–306.

Bright, D.E., 1966. Support for suppression of *Xyleborus* Bowdich, 1825. Bull. Zool. Nom. 23, 132.

Bright, D.E., 1967. Catalogue of the Swaine types of Scolytidae (Coleoptera) with designations of lectotypes. Can. Entomol. 99, 673–681.

Bright, D.E., 1967. Lectotype designations for *Cryphalus amabilis* and *C. grandis* (Coleoptera: Scolytidae). Can. Entomol. 99, 681.

Bright, D.E., 1967. A review of the genus *Cactopinus*, with descriptions of two new species and a new genus (Coleoptera: Scolytidae). Can. Entomol. 99, 917–925.

Bright, D.E., 1968. Review of the genus *Leiparthrum* Wollaston in North America, with a description of one new species (Coleoptera: Scolytidae). Can. Entomol. 100, 636–639.

Bright, D.E., 1968. Three new species of *Pityophthorus* from Canada (Coleoptera: Scolytidae). Can. Entomol. 100, 604–608.

Bright, D.E., 1968. Review of the tribe Xyleborini in America north of Mexico (Coleoptera: Scolytidae). Can. Entomol. 100, 1288–1323.

Bright, D.E., 1969. Biology and taxonomy of bark beetle species in the genus *Pseudohylesinus* Swaine (Coleoptera: Scolytidae). University of California Publications in Entomology 54, 1–46.

Thomas, J.B., Bright, D.E., 1970. A new species of *Dendroctonus* (Coleoptera: Scolytidae) from Mexico. Can. Entomol. 102, 479–483.

Bright, D.E., 1970. A note concerning *Pseudohylesinus sericeus* (Mannerheim) (Coleoptera: Scolytidae). Can. Entomol. 102, 499–500.

Bright, D.E., 1971. New species, new synonymies and new records of bark beetles from Arizona and California (Coleoptera: Scolytidae). Pan-Pac. Entomol. 47, 63–70.

Bright, D.E., 1971. Bark beetles from Newfoundland (Coleoptera: Scolytidae). Ann. Soc. Entomol. Que. 16, 124–127.

Bright, D.E., 1972. The Scolytidae and Platypodidae of Jamaica (Coleoptera). Bulletin of the Institute of Jamaica 21, 1–108.

Bright, D.E., 1972. New species of Scolytidae (Coleoptera) from Mexico, with additional notes. 1. Tribes Xyleborini and Corthylini. Can. Entomol. 104, 1369–1385.

Bright, D.E., 1972. New species of Scolytidae (Coleoptera) from Mexico, with additional notes. II. Subfamilies Scolytinae and Hylesininae. Can. Entomol. 104, 1489–1497.

Bright, D., 1972. New species of Scolytidae (Coleoptera) from Mexico, with additional notes. III. Tribe Pityophthorini (except Pityophthorus). Can. Entomol. 104, 1665–1679.

Bright, D.E., 1973. *Xyleborus howdenae*, new name, and some corrections to "The Scolytidae and Platypodidae of Jamaica". Coleopterists Bull. 27, 18.

Bright, D.E., Stark, R.W., 1973. The bark and ambrosia beetles of California (Coleoptera: Scolytidae). California Insect Survey Bulletin 16, 1–169.

Bright, D.E., 1975. Comments on the proposed conservation of four generic names of Scolytidae (Insecta: Coleoptera). Z.N.(S.) 2069-2072. Bulletin of Zoological Nomenclature 32, 135.

Bright, D.E., 1976. The Insects and Arachnids of Canada, Part 2. The Bark Beetles of Canada and Alaska (Coleoptera: Scolytidae). Agriculture Canada Publication No. 1576, pp. 1–241.

Bright, D.E., 1976. Biological notes and new localities for three rare species of North American Trogositidae (Coleoptera). Coleopterists Bull. 30, 169–170.

Bright, D.E., 1976. Lectotype designations for various species of North American *Pityophthorus* Eichhoff (Coleoptera: Scolytidae). Coleopterists Bull. 30, 183–188.

Bright, D.E., 1976. New synonymy, new combinations, and new species of North American *Pityophthorus*(Coleoptera: Scolytidae). Part II. Great Basin Nat. 36, 425–444.

Bright, D.E., 1977. New synonymy, new combinations, and new species of North American *Pityophthorus*(Coleoptera: Scolytidae). I. Can. Entomol. 109, 511–532.

Bright, D.E., 1978. New synonomy, new species, and taxonomic notes of North American *Pityophthorus* (Coleoptera: Scolytidae). Part III. Great Basin Nat. 38, 71–84.

Bright, D.E., 1978. International voucher specimen collection of Scolytidae. Entomol. Soc. Can. Bull. 10, 42.

Campbell, J.M., Ball, G.E., Becker, E.C., Bright, D.E., Helava, J., Howden, H.F., Parry, R.H., Peck, S.B., Smetana, A., 1979. Coleoptera. In: Danks, H.V. (Ed.), Canada and its Insect Fauna.In: Memoirs of the Entomological Society of Canada, 108, pp. 357–387.

Bright, D.E., 1981. Studies on West Indian Scolytidae (Coleoptera) I. New species, new distribution records and taxonomic notes. Studies on Neotropical Fauna and Environment 16, 151–164.

Bright, D.E., 1981. *Afrotrypetus*, a new genus of bark beetles from Africa (Coleoptera: Scolytidae). Coleopterists Bull. 35, 113–116.

Bright, D.E., 1981. Taxonomic monograph of the genus *Pityophthorus* Eichhoff in North and Central America (Coleoptera: Scolytidae). Mem. Entomol. Soc. Can. 118, 1–378.

Bright, D.E., 1980. Studies on the Xyleborini 1. Three new species of *Schedlia* from New Guinea (Coleoptera: Scolytidae). Coleopterists Bull. 34, 369–372.

Bright, D.E., 1981. Eye reduction in a cavernicolous population of *Coccotrypes dactyliperda* Fabricius (Coleoptera: Scolytidae). Coleopterists Bull. 35, 117–120.

Bright, D.E., 1981. A new synonym of *Agrilus* sayi (Coleoptera: Buprestidae). Can. Entomol. 113, 871.

Bright, D.E., 1982. Studies on West Indian Scolytidae (Coleoptera) 2. New distribution records and descriptions of a new genus and species. Studies on Neotropical Fauna and Environment 17, 163–186.

Bright, D.E., 1982. Scolytidae (Coleoptera) from the Cocos Islands, Costa Rica, with description of one new species. Coleopterists Bull. 36, 127–130.

Bright, D.E., Stock, M.W., 1982. Taxonomy and geographic variation. In: Mitton, J.B., Sturgeon, K.B. (Eds.), Bark Beetles in North American Conifers. A System for the Study of Evolutionary Biology. University of Texas Press, Austin, pp. 46–73.

Stewart, W.E., Bright, D.E., 1982. Notes on *Pissodes fiskei* (Coleoptera: Curculionidae) with a redescription of the species. Coleopterists Bull. 36, 445–452.

Bright, D.E., 1985. New species and records of North American *Pityophthorus* (Coleoptera: Scolytidae), Part IV: The Scriptor Group. Great Basin Nat. 45, 467–475.

Bright, D.E., 1985. New species and new records of North America *Pityophthorus* (Coleoptera: Scolytidae), Part V: The Juglandis Group. Great Basin Nat. 45, 476–482.

Bright, D.E., 1985. Studies on West Indian Scolytidae (Coleoptera) 3. Checklist of Scolytidae of the West Indies, with descriptions of new species and taxonomic notes. Entomologische Arbeiten aus dem Museum G. Frey 33 (34), 169–187.

Bright, D.E., 1986. A Catalog of the Coleoptera of America North of Mexico: Family Mordellidae. United States Department of Agriculture, Agriculture Handbook Number 529–125.

Bright, D.E., 1987. New species and new records of North American *Pityophthorus* (Coleoptera: Scolytidae), Part VI. The Lautus group. Great Basin Nat. 46, 641–645.

Bright, D.E., 1987. New species and new records of North American *Pityophthorus* (Coleoptera: Scolytidae), Part VII. Great Basin Nat. 46, 679–684.

Bright, D.E., 1987. The metallic wood-boring beetles of Canada and Alaska (Coleoptera: Buprestidae). The Insects and Arachnids of Canada, Part 15. Agriculture Canada Publication 1810, pp. 1–335.

Bright, D.E., 1987. A review of the Scolytidae (Coleoptera) of the Azores with description of a new species of *Phloeosinus*. Bocagiana 107, 1–5.

Flores, J.L., Bright, D.E., 1987. A new species of *Conophthorus* from Mexico: descriptions and biological notes (Coleoptera: Scolytidae). Coleopterists Bull. 41, 181–184.

Wood, S.L., Bright, D.E., 1987. A Catalog of Scolytidae and Platypodidae (Coleoptera), Part 1: Bibliography. Great Basin Nat. Mem. 11, 1–685.

Bright, D.E., 1988. Notes on the occurrence of *Xyleborinus gracilis* (Eichhoff) in the United States. Coleopterists Bull. 41, 338.

Bright, D.E., 1988. *Polydrusus cervinus* (Linnaeus), a weevil new to Canada (Coleoptera: Curculionidae). Coleopterists Bull. 42, 337.

Bright, D.E., 1989. Two new species of *Phloeosinus* Chapuis from Mount Kinabalu, Borneo, with taxonomic notes (Coleoptera: Scolytidae). Coleopterists Bull. 43, 79–82.

Bright, D.E., 1989. New synonymy in North American *Sitona* (Coleoptera: Curculionidae). Coleopterists Bull. 43, 77–78.

Bright, D.E., 1989. Additions to the Scolytidae fauna of the Azores (Coleoptera). Bocagiana 129, 1–2.

Bright, D.E., 1989. 1) Scolytidae; 2) Platypodidae. In: Stehr, F.W. (Ed.), An Introduction to Immature Insects of North America. Kendall/Hunt Publ. Co., Dubuque, pp. 613–616

Bright, D.E., 1990. A new species of *Liparthrum* from Borneo with notes on its generic placement (Coleoptera: Scolytidae). Coleopterists Bull. 44, 485–488.

Atkinson, T.H., Rabaglia, R.L., Bright, D.E., 1990. Newly detected exotic species of *Xyleborus* (Coleoptera:Scolytidae) with a revised key to species in eastern North America. Can. Entomol. 122, 93–104.

Bright, D.E., 1991. A note concerning *Sitona tibialis* (Herbst) in North America (Coleoptera: Curculionidae). Coleopterists Bull. 45, 198–199.

Bright, D.E., 1991. Studies in Xyleborini 2. Review of the genus *Sampsonius* Eggers (Coleoptera: Scolytidae). Studies on Neotropical Fauna and Environment 26, 11–28.

Bright, D.E., 1991. Family Derodontidae. In: Bousquet, Y. (Ed.), Checklist of Beetles of Canada and Alaska. Research Branch, Agriculture Canada Publication 1861/E, pp. 195–196.

Bright, D.E., 1991. Family Melyridae. In: Bousquet, Y. (Ed.), Checklist of Beetles of Canada and Alaska. Research Branch, Agriculture Canada Publication 1861/E, pp. 211–213.

Bright, D.E., Skidmore, R.E., 1991. Two new records of Scolytidae (Coleoptera) from Canada. Coleopterists Bull. 45, 368.

Bright, D.E., 1992. Synopsis of the genus *Hemicryphalus* Schedl with descriptions of four new species from Borneo (Coleoptera: Scolytidae). Koleopterologische Rundschau 62, 183–190.

Bright, D.E., 1992. The Insects and Arachnids of Canada. Part 21. The Weevils of Canada and Alaska. Vol. 1. (Coleoptera: Curculionoidea, excluding Scolytidae and Curculionidae). Agriculture Canada Publication 1882, pp. 1–217.

Bright, D.E., 1992. Systematics research. In: Hayes, J.L., Robertson, J.L. (Eds.), Proceedings of a Workshop on Bark Beetle Genetics: Current Status of Research. U.S. Department of Agriculture, Forest Service, Pacific Southwest Research Station, p. 25, General Technical Report PSW–GTR–135.

Bright, D.E., Skidmore, R.E., Thompson, R.T., 1992. *Euophryum confine* (Broun), a new weevil record for Canada and the New World (Coleoptera: Curculionidae). Coleopterists Bull. 46, 143–144.

Wood, S.L., Bright, D.E., 1992. A Catalog of Scolytidae and Platypodidae (Coleoptera), Part 2: Taxonomic Index, Vols. A and B. Gt. Basin Nat. Mem. 13, 1–1553.

Bright, D.E., 1993. Systematics of bark beetles. In: Schowalter, T.D., Filip, G.M. (Eds.), Beetle-Pathogen Interactions in Conifer Forests. Academic Press, London, pp. 23–36.

Peschken, D.P., Sawchyn, K.C., Bright, D.E., 1993. First record of *Apion hookeri* Kirby (Coleoptera: Curculionidae) in North America. Can. Entomol. 125, 629–631.

Bright, D.E., Skidmore, R.E., Dunster, K.E., 1994. Scolytidae (Coleoptera) associated with the dwarf hackberry, *Celtis tenuifolia*, in Ontario, Canada. Coleopterists Bull. 48, 93–94.

Bright, D.E., 1994. New records and new species of Scolytidae (Coleoptera) from Borneo. Koleopterologische Rundschau 64, 257–274.

Bright, D.E., Poinar Jr., G.O., 1994. Scolytidae and Platypodidae (Coleoptera) from Dominican Republic amber. Ann. Entomol. Soc. Am. 87, 170–194.

Bright, D.E., 1994. A revision of the genera *Sitona* Germar (Coleoptera: Curculionidae) of North America. Ann. Entomol. Soc. Am. 87, 277–306.

Hobson, K.E., Bright, D.E., 1994. A key to *Xyleborus* of California, with faunal comments (Coleoptera: Scolytidae). Pan–Pac. Entomol. 70, 267–268.

Côté, S., Bright, D.E., 1995. Premières mentions Canadiennes de *Phyllobius intrusus* Kôno (Coleoptera: Curculionidae) et tableaux de détermination des espèces de *Phyllobius* et *Polydrusus* au Canada. Fabreries 20, 81–89.

Bright, D.E., 1996. Notes on native parasites and predators of the European pine shoot beetle, *Tomicus piniperda* (Linnaeus) in Canada (Coleoptera: Scolytidae). Proc. Entomol. Soc. Ontario 127, 57–62.

Cognato, A.I., Bright, D.E., 1996. New records of bark beetles (Coleoptera: Scolytidae) from Dominica, West Indies. Coleopterists Bull. 50, 72.

Bright, D.E., 1997. *Xyleborus fornicatus* Eichhoff. Crop Protection Compendium for Southeast Asia, Data Sheet. CABI Electronic Database, 20 p.

Bright, D.E., 1997. *Xylosandrus compactus* Eichhoff. Crop Protection Compendium for Southeast Asia, Data Sheet. CABI Electronic Database, 20 p.

Bright, D.E., 1997. *Xyleborus* spp. and related genera (Southeast Asia). Crop Protection Compendiumfor Southeast Asia, Data Sheet. CABI Electronic Database, 13 p.

Bright, D.E., Skidmore, R.E., 1997. A Catalog of Scolytidae and Platypodidae (Coleoptera), Supplement 1 (1990–1994). NRC Research Press, Ottawa, 368 p.

Bright, D.E., Peck, S.B., 1998. Scolytidae from the Galápagos Islands, Ecuador, with descriptions of four new species, new distributional records, and a key to species. Koleopterologische Rundschau 68, 223–252.

Bright, D.E., Rabaglia, R.J., 1999. *Dryoxylon*, a new genus for *Xyleborus onoharaensis* Murayama, recently established in the southeastern United States (Coleoptera: Scolytidae). Coleopterists Bull. 53, 333–337.

Bright, D.E., 2000. Scolytidae (Coleoptera) of Gunung Mulu National Park, Sarawak, Malaysia, with ecological notes and descriptions of six new species. Serangga 5, 41–85.

Vandenberg, N.J., Rabaglia, R.J., Bright, D.E., 2000. New records of two *Xyleborus* (Coleoptera: Scolytidae) species in North America. Proc. Entomol. Soc. Wash. 102, 62–68.

Bright, D.E., Skidmore, R.E., 2002. A Catalog of Scolytidae and Platypodidae (Coleoptera), Supplement 2 (1995-1999). NRC Research Press, Ottawa, 523 p.

Bright, D.E., 2004. Scolytinae. In: Cordo, H.A., Logarzo, G., Braun, K., Di Iorio, O. (Eds.), Catálogo de Insectos Fitófagos de la Argentina y sus Plantas Asociadas". South American Biological Control Laboratory and Sociedad Entomológica Argentina, Buenos Aires, pp. 155–162.

Schiefer, T.L., Bright, D.E., 2004. *Xylosandrus mutilatus* (Blandford), an exotic ambrosia beetle (Coleoptera: Curculionidae: Scolytinae: Xyleborini) in North America. Coleopterists Bull. 58, 431–438.

Bright, D.E., Torres, J.A., 2006. Studies on West Indian Scolytidae (Coleoptera) 4 A review of the Scolytidae of Puerto Rico, U. S. A. with descriptions of one new genus, fourteen new species and notes on new synonymy (Coleoptera: Scolytidae). Koleopterologische Rundschau 76, 389–428.

Bright, D.E., Bouchard, P., 2008. The Insects and Arachnids of Canada. Part 25. The Weevils of Canada and Alaska. Vol. 2. (Coleoptera: Curculionidae: Entiminae). NRC Research Press, Ottawa, 327 p.

Bright, D.E., 2010. *Stevewoodia minutum*, a new genus and species of Scolytidae (Coleoptera) from the West Indies. Studies on West Indian Scolytidae (Coleoptera) 6. ZooKeys 56, 45–48.

Burbano, E., Wright, M., Bright, D.E., Vega, F.E., 2011. New record for the coffee berry borer, *Hypothenemus hampei*, in Hawaii. J. Insect Sci. 11, 117.

Bright, D.E., Kondratieff, B.C., Norton, A.P., 2013. First record of the "Splendid Tamarisk Weevil", *Coniatus splendidulus* (F.) (Coleoptera: Curculionidae: Hyperinae), in Colorado, USA. Coleopterists Bull. 67, 302–303.

Goldarazena, A., Bright, D.E., Hishinuma, S.M., López, S., Seybold, S.J., 2014. First record of *Pityophthorus solus* (Blackman, (1928) in Europe. Bulletin OEPP/EPPO 44, 65–69.

Bright, D.E., 2014. A Catalog of Scolytidae and Platypodidae (Coleoptera), Supplement 3 (2000–2010), with notes on subfamily and tribal reclassifications. Insecta Mundi 0356, 1–336.

Contents

Contributors

Numbers in parentheses indicate the pages on which the authors' contributions begin.

Thomas H. Atkinson (41), University of Texas Insect Collection, University of Texas at Austin, Austin, TX, USA

Matthew P. Ayres (157), Department of Biological Sciences, Dartmouth College, Hanover, NH, USA

Barbara J. Bentz (157, 533), USDA Forest Service, Rocky Mountain Research Station, Logan, UT, USA

Peter H.W. Biedermann (85), Research Group Insect Symbiosis, Max Planck Institute for Chemical Ecology, Jena, Germany

Ryan Bracewell (305), Department of Ecosystem and Conservation Sciences, University of Montana, Missoula, MT, USA

Anthony I. Cognato (41, 351), Department of Entomology, Michigan State University, East Lansing, MI, USA

Thomas S. Davis (209), Plant, Soil and Entomological Sciences, University of Idaho, Moscow, ID, USA

Jamie Dinkins-Bookwalter (209), United States Department of Agriculture Forest Service Station, Southern Research Station, Asheville, NC, USA

Massimo Faccoli (371), Department of Agronomy, Food, Natural Resources, Animals and Environment (DAFNAE), Agripolis, Legnaro (PD), Italy

Christopher J. Fettig (555), Invasives and Threats Team, Pacific Southwest Research Station, USDA Forest Service, Davis, CA, USA

Jean-Claude Grégoire (1, 585), Biological Control and Spatial Ecology Laboratory, Université Libre de Bruxelles, Bruxelles, Belgium

Matthias Herrmann (247), Max Planck Institute for Developmental Biology, Department of Evolutionary Biology, Tuebingen, Germany

Jacek Hilszczański (555), Department of Forest Protection, Forest Research Institute, Sękocin Stary, Raszyn, Poland

Richard W. Hofstetter (209), School of Forestry, Northern Arizona University, Flagstaff, AZ, USA

Jiri Hulcr (41, 495), School of Forest Resources and Conservation and Department of Entomology, University of Florida, Gainesville, FL, USA

Francisco Infante (427), El Colegio de la Frontera Sur (ECOSUR), Carretera Antiguo Aeropuerto Km. 2.5, Tapachula, Chiapas, Mexico

Andrew J. Johnson (427), School of Forest Resources and Conservation, University of Florida, Gainesville, FL, USA

Anna Maria Jönsson (533), Department of Physical Geography and Ecosystem Science, Lund University, Lund, Sweden

Bjarte H. Jordal (41, 85), University Museum of Bergen, University of Bergen, Bergen, Norway

Lawrence R. Kirkendall (85), Department of Biology, University of Bergen, Bergen, Norway

Kier D. Klepzig (209), United States Department of Agriculture Forest Service Station, Southern Research Station, Asheville, NC, USA

Paal Krokene (177), Norwegian Forest and Landscape Institute, Ås, Norway

Bo Långström (371), Swedish University of Agricultural Sciences, Department of Ecology, Uppsala, Sweden

François Lieutier (371), Laboratoire de Biologie des Ligneux et des Grandes Cultures, Université d'Orléans, Orléans, France

B. Staffan Lindgren (1, 585), Natural Resources and Environmental Studies Institute, University of Northern British Columbia, Prince George, BC, Canada

Duane D. McKenna (41), Department of Biological Sciences, University of Memphis, Memphis, TN, USA

Kenneth F. Raffa (1, 585), Department of Entomology, University of Wisconsin-Madison, Madison, WI, USA

Diana L. Six (305), Department of Ecosystem and Conservation Sciences, University of Montana, Missoula, MT, USA

Sarah M. Smith (495), Department of Entomology, Michigan State University, East Lansing, MI, USA

Fernando E. Vega (427), Sustainable Perennial Crops Laboratory, United States Department of Agriculture, Agricultural Research Service, Beltsville, MD, USA

Aaron S. Weed (157), Department of Biological Sciences, Dartmouth College, Hanover, NH, USA

Rudolf Wegensteiner (247), University of Natural Resources and Life Sciences, BOKU–Vienna, Department of Forest and Soil Sciences, Institute of Forest Entomology, Forest Pathology and Forest Protection, Vienna, Austria

Beat Wermelinger (247), Swiss Federal Institute for Forest, Snow and Landscape Research WSL, Forest Dynamics, Birmensdorf, Switzerland

Preface

This is the first book broadly dedicated to the ecology, phylogeny, and management of bark beetles (Coleoptera: Curculionidae: Scolytinae) on a global scale. The ecological and economic impact of bark beetles on trees is a global issue that often surpasses all other disturbances including fire and storms. Bark beetles are of economic importance in forests, orchards, and urban areas as well as agricultural crops, and wood commodities. Despite these powerful impacts, most of the approximately 6,000 described species colonize stressed or dead tree tissues only. Most bark beetles feed on the phloem or fungi in the inner bark, but a minority specializes on other plant tissues such as cones and seeds.

The association of bark beetles with microbes has led to a variety of symbioses, and these relationships have led to the great success and diversification of bark beetles. Such symbioses may account for the majority of the world's most recent invasive tree and crop pests. Climate change and human-facilitated movement of bark beetles has also contributed to the explosive increase and range expansion of bark beetles, as is the case with the red turpentine beetle in China, the mountain pine beetle in Canada, and the coffee berry borer in tropical regions. Recent genomic data on bark beetles and associated microbes have increased our knowledge of the evolution and ecology of these complex communities.

The present volume includes chapters on ecology, morphology, taxonomy, phylogenetics, evolution, population dynamics, tree defense, symbioses, natural enemies, climate change, management strategies, and the economy and politics of bark beetles. In addition, individual chapters are dedicated to bark beetles in the genera *Dendroctonus*, *Ips*, *Tomicus*, *Hypothenemus*, and *Scolytus*. The editors have brought together an international team of authors, in an effort to combine the vast amount of literature and a diversity of viewpoints into one volume. We thank all the authors for their excellent contributions. We hope that this book's information and illustrations are valuable to entomologists, ecologists, foresters, land managers, and students interested in bark beetles.

We thank Pat Gonzalez and Kristi A.S. Gomez at Academic Press for their help and support throughout this project. Ann Simpkins cross-checked the references in many chapters, for which we are grateful. We appreciate the patience and support of Wendy S. Higgins, Ian G. Vega, Karen B. London, Brian J. Hofstetter and Evan M. Hofstetter during the creation of this book.

Fernando E. Vega and Richard W. Hofstetter

About the Editors

Fernando E. Vega is a Research Entomologist with the Agricultural Research Service of the United States Department of Agriculture in Beltsville, Maryland. He received a BS degree in Agriculture from the University of Puerto Rico, an MS degree in Horticulture from the University of Maryland, and a PhD in Entomology also from the University of Maryland. He has published extensively on the coffee berry borer (*Hypothenemus hampei*) and on fungal entomopathogens as fungal endophytes. Dr. Vega is a Fellow of the Linnean Society of London and of the Royal Entomological Society. He has co-edited *Insect-Fungal Associations: Ecology and Evolution* (Oxford University Press, 2005), *The Ecology of Fungal Entomopathogens* (Springer, 2010), and *Insect Pathology, Second Edition* (Academic Press, 2012).

Richard W. Hofstetter is Associate Professor of Forest Entomology in the School of Forestry, Northern Arizona University. He has a BS degree (1992) in Population Biology, and a MS degree (1996) in Entomology from University of Wisconsin-Madison, and a PhD (2004) in Ecology and Evolution from Dartmouth College. In 2005, he was hired as a Research Faculty in the School of Forestry at Northern Arizona University. In 2008, he started his tenure-track faculty position in the School of Forestry at Northern Arizona University. He teaches undergraduate and graduate courses in Forest Entomology, Tropical Forest Ecology, Symbioses, and Forest Health. His research focuses on bark beetle ecology related to plant-insect interactions, predator-prey dynamics, biological control, bioacoustics, and interactions between fungi and mites associated with bark beetles. He has contributed over 150 presentations and 60 peer-reviewed articles. He has offered a short course on bark beetle ecology and management open to both students and professionals in the fields of ecology, entomology and forestry. He is past-President of the Western Forest Insect Work Conference and the Symbiosis subject editor for *Environmental Entomology*.

Chapter 1

Natural History and Ecology of Bark Beetles

Kenneth F. Raffa[1], Jean-Claude Grégoire[2], and B. Staffan Lindgren[3]

[1]*Department of Entomology, University of Wisconsin-Madison, Madison, WI, USA,* [2]*Biological Control and Spatial Ecology Laboratory, Université Libre de Bruxelles, Bruxelles, Belgium,* [3]*Natural Resources and Environmental Studies Institute, University of Northern British Columbia, Prince George, BC, Canada*

1. INTRODUCTION

Bark beetles (Coleoptera: Curculionidae: Scolytinae) are a highly diverse subfamily of weevils that spend most of their life histories within plants. They occur in all regions of the world, and are associated with most major groups of terrestrial plants, almost all plant parts, and a broad array of invertebrate and microbial symbionts. Bark beetles have served as some of the most prominent model systems for studies of chemical ecology, symbiosis, sexual selection, population dynamics, disturbance ecology, and coevolution.

Bark beetles play key roles in the structure of natural plant communities and large-scale biomes. They contribute to nutrient cycling, canopy thinning, gap dynamics, biodiversity, soil structure, hydrology, disturbance regimes, and successional pathways. Several species in particular can genuinely be designated "landscape engineers," in that they exert stand-replacing cross-scale interactions.

In addition to their ecological roles, some bark beetles compete with humans for valued plants and plant products, and so are significant forest and agricultural pests. These species cause substantial socioeconomic losses, and at times necessitate management responses. Bark beetles and humans are both in the business of converting trees into homes, so our overlapping economies make some conflict of interest inevitable.

Anthropogenic activities are altering the environmental and genetic background on which bark beetles, their host plants, and symbionts interact. Factors that have already been shown to alter these relationships include transport of bark beetles and/or microbial associates, habitat manipulations in ways that homogenize or fragment plant communities, and climate change that raises temperatures and increases drought. These factors often lead to higher plant mortality or injury.

This chapter is intended to introduce, summarize, and highlight the major elements of bark beetle life history and ecology, for subsequent in-depth development in the following chapters. The enormous diversity of Scolytinae makes it impossible to address each of these elements for all permutations of their life histories. Only relatively few species (1) exert documented selective pressures on their host species and have major roles in landscape-scale processes, (2) pose significant challenges to natural resource management, and (3) provide the majority of our basic biological knowledge. These are disproportionately concentrated within species that colonize the main stems of conifers. We therefore place particular emphasis on that guild.

2. DIVERSITY OF LIFESTYLES AND ECOLOGICAL RELATIONSHIPS

The Scolytinae have a long evolutionary history (Cognato and Grimaldi, 2009). They are a subfamily within the Curculionidae, the weevils or snout beetles. They are distinct in having reduced snouts as an adaptation to spending much of their adult life within host plant tissues. These beetles are roughly cylindrical in shape, with short legs and antennae, suitable for a life of tunneling. The head is armed with stout mandibles and many scolytines have morphological adaptations to their elytral declivity (e.g., *Ips* spp.), head (e.g., male *Trypodendron* spp.), or legs for removing plant fragments from their breeding galleries, packing wood shavings in older parts of their gallery (e.g., some *Dendroctonus* spp.), or blocking unwanted conspecifics, competitors or natural enemies from galleries (S. L. Wood, 1982). Beyond those general traits, scolytine beetles are highly variable. While the common name "bark beetle" is sometimes applied to the entire subfamily, many are not associated with bark at all, but rather utilize a variety of plant tissues, both for reproduction and feeding. Many scolytine species are ambrosia beetles, which establish breeding galleries in wood, but feed on symbiotic fungi rather than directly on plant tissues. In this chapter, we focus on bark beetles *sensu stricto*, i.e., those species that breed in the inner bark of their host, but where appropriate we will reference other feeding guilds as well.

Bark Beetles. http://dx.doi.org/10.1016/B978-0-12-417156-5.00001-0

2.1 General Life Cycle

For the purpose of this chapter, we will emphasize well-studied species to illustrate a general bark beetle life cycle. There are many variations, but most species emerge from their brood galleries in spring or summer, and seek a mate and a new host. The effective dispersal flight is often no more than a few hundred meters (Salom and McLean, 1989; Zumr, 1992) where most successful attacks tend to occur (Wichmann and Ravn, 2001), but the potential to actively fly many kilometers has been demonstrated in laboratory flight mill (Forsse and Solbreck, 1985; Jactel, 1993) and field (Jactel, 1991; Yan et al., 2005) studies. Dispersal distances vary markedly among species, and within species with beetle condition, distribution of susceptible hosts, and environment (Franklin et al., 1999, 2000). Long-range, wind-aided dispersal can extend for hundreds of kilometers (Nilssen, 1984; Jackson et al., 2008; Ainslie and Jackson, 2011; de la Giroday et al., 2011; Samarasekera et al., 2012). Prior to colonizing new hosts, beetles may engage in maturation feeding, often in their brood gallery prior to dispersal. Some species disperse to a specific maturation feeding site, usually a live tree, prior to seeking a breeding site (Stoszek and Rudinsky, 1967; Långström, 1983; McNee et al., 2000). In several species, this behavior can result in vectoring of important pathogens, such as *Verticicladiella wageneri* W. B. Kendr. (Witcosky et al., 1986b) and *Ophiostoma novo-ulmi* Brasier (Webber, 1990).

Bark beetle reproductive strategies can be roughly divided into three types, depending on when and where mating occurs, and the gender initiating gallery construction. In *monogamous* species, females initiate the attack and are joined by a single male. Mating normally takes place on the bark or in the gallery, depending on species, but a small percentage of females may arrive at a host already mated (Bleiker et al., 2013). In *polygamous* species, the male initiates attacks, generally by excavating a nuptial chamber where he mates with several females. A few species are *solitary*, with mated females attacking weakened but living hosts. These species are parasitic on trees, and rarely kill their host, which would perhaps be maladaptive because of the protection host resin provides from predators and parasites. Females of solitary beetles often mate in their brood gallery, with either a brother or, possibly, an unrelated male.

Eggs are laid singly in niches excavated along a narrow gallery (tunnel), in groups on alternating sides, or sometimes grouped along one side of the gallery. In some species, a chamber is excavated in which the eggs are deposited. After hatching, larvae feed on phloem tissue in individual niches or galleries radiating away from the maternal gallery. The lengths of larval mines vary widely among species, ranging from an expansion of the original egg niche to accommodate growing ambrosia beetle larvae,

to extensive galleries 10–15 cm long in species that derive most of their nutrients directly from host tissue (S. L. Wood, 1982). In some species, larvae spend only a brief time in the inner bark, after which they migrate to the nutrient-poor outer bark. This is possibly an avoidance mechanism, as cerambycid larvae may both destroy the phloem and consume bark beetle larvae (Flamm et al., 1993; Schroeder and Weslien, 1994; Dodds et al., 2001). Larvae develop through 3–5 instars, after which they pupate. Metamorphosis is completed in 5–10 days in many species, and the adult beetle ecloses as a callow or teneral adult. These young adults are lightly colored due to incomplete sclerotization of the exoskeleton. After maturation feeding, adults exit through an emergence hole, which they excavate through the bark or were formed by an earlier emerging beetle, or in the case of ambrosia beetles through the entrance hole to the maternal gallery.

2.2 Variations to the Generalized Life Cycle

2.2.1 Feeding Substrate

Bark and wood are relatively poor nutritional substrates, so most bark beetles feed on the slightly more nutritious inner bark, or phloem. A considerable number of species exploit the ability of fungi to concentrate nitrogen, by consuming either fungus-infected phloem, or fungi (Ayres et al., 2000; Bleiker and Six, 2007). Associations between bark beetles and fungi range from facultative to obligatory symbioses. Bark beetles inoculate their fungal associates by carrying spores either on their exoskeleton, or by actively transporting and nurturing them. In evolutionarily advanced associations, the complexity and variety of specialized pockets (*mycangia*) that harbor symbionts suggest these symbioses have evolved independently multiple times (Six and Klepzig, 2004; Harrington, 2005). Ambrosia beetles represent the most advanced of such associations, and this specialization has allowed them to escape to the three-dimensional xylem from the essentially two-dimensional inner bark niche, where competition with other phloeophagous organisms may be fierce (Lindgren and Raffa, 2013). Thus, scolytine ambrosia beetles can occur at extremely high densities relative to bark beetles. Not surprisingly, ambrosia beetles have been extremely successful, particularly in the tropics, and another subfamily of the Curculionidae, the Platypodinae, have evolved to occupy a similar niche.

Many scolytine beetles have a relatively narrow host range, ranging from mono- to oligophagous. Some species may be associated with only one species of host tree, whereas others may be able to utilize most species within a genus, and on occasion other genera (Huber et al., 2009). Among bark beetles *sensu stricto* that colonize live trees, most have evolved adaptations to exploit Pinaceae despite formidable defenses that these trees can mount (Franceschi *et al.*, 2005). Many scolytines breed in angiosperms (Wood and Bright, 1992), but most of those are saprophages (Ohmart, 1989). Ambrosia beetles are often less constrained in their host range than phloem feeding bark beetles, and some are known to colonize many tree species (Hulcr *et al.*, 2007). This may be because the deciding factor is whether the tree can support the ambrosia fungus. For example, *Trypodendron lineatum* (Olivier), an economically important species in western North America and Europe, breeds in numerous Pinaceae genera, but also in at least four genera from three families of angiosperms (Lindgren, 1986).

In addition to species that utilize the trunk, a number of species breed in roots, twigs, or branches. Many scolytine beetles also utilize other plant parts, e.g., cone beetles in the genus *Conophthorus* breed in the cone axis of several species of conifers (Chapter 12), and *Hypothenemus hampei* (Ferrari), breeds in the seeds of two *Coffea* species and possibly other species in the family Rubiaceae (Chapter 11). Similarly, *Coccotrypes dactyliperda* F., breeds in the stone of green, unripened date fruits (Blumberg and Kehat, 1982), and a number of scolytine species breed in the woody petioles of *Cecropia* spp. (Jordal and Kirkendall, 1998). Furthermore, some species conduct maturation feeding outside the maternal gallery, e.g., in shoots of their host tree, such as several species of *Tomicus* (Långström, 1983; Kirkendall *et al.*, 2008; Chapter 10) and *Pseudohylesinus* (Stoszek and Rudinsky, 1967; Chapter 12).

2.2.2 Gender Roles

Host selection and gallery initiation are typically performed by females in monogamous (one male with one female) species, and males in polygamous (one male with several females) species, a distinction that holds at the genus level. In monogamous species, females arrive at a tree, and initiate a gallery while releasing pheromones. Males arrive at and attempt to enter the gallery. Male entrance, and hence mate choice, in these genera is typically dictated by female assessment of their suitability (Ryker and Rudinsky, 1976). A small percentage of females mate before they emerge (Bleiker *et al.*, 2013), and may arrive at the host already fertile, allowing them to construct a gallery and produce offspring without another male. This is assumed to occur either with a male that entered a natal gallery, or a sibling. In some parasitic species, e.g., *Dendroctonus micans* (Kugelann) and *Dendroctonus punctatus* LeConte, females attack by themselves, so mating occurs pre-emergence, or at least pre-attack (Grégoire, 1988; Furniss, 1995), or possibly both, as there is evidence that multiple mating can occur in *D. micans* (Fraser *et al.*, 2014). Exceptions to females being the pioneering sex

among monogamous species occur in some genera, such as the ambrosia beetle genus *Gnathotrichus*, in which the male initiates attack and is joined by one female. This may indicate that monogamy is a derived state in these genera. In polygamous species, the male initiates gallery construction in the form of a nuptial chamber. Females will attempt to join the male, who may resist entrance, i.e., in polygamous species the male controls mate selection (Wilkinson *et al.*, 1967; Løyning and Kirkendall, 1996). Subsequent females encounter increasing resistance by the male. In some cases, a late-arriving female may enter a gallery by excavating her own entrance, i.e., thus circumventing male mate selection.

Some polygamous species include pseudogynous females, i.e., females that require mating, but produce offspring parthenogenetically without the use of male gametes (Stenseth *et al.*, 1985; Løyning and Kirkendall, 1996). In some scolytine beetles, notably a few genera in the bark beetle tribe Dryocoetini and all species of the ambrosia beetle tribe Xyleborini, sex determination is by haplodiploidy, with unmated diploid females producing haploid dwarf males with which they may later mate (Normark *et al.*, 1999; Jordal *et al.*, 2000). Sib mating and fungal symbiosis are closely associated with this evolutionary path (Jordal *et al.*, 2000). A fascinating special case of sib mating occurs in the genus *Ozopemon*, where neoteny (sexual maturation of larvae) has evolved in males (Jordal *et al.*, 2002). Beetles with this haplo-diploid sex determination system are eminently well adapted for invading novel habitats, because even a single female is theoretically sufficient for establishment in a novel habitat (Jordal *et al.*, 2001; Zayed *et al.*, 2007; Hulcr and Dunn, 2011). Ambrosia beetles are particularly advantaged because host specificity is primarily determined by the ability of the ambrosia fungus to thrive in novel hosts. Consequently, ambrosia beetles are easily transported in dunnage or wood products, and many, e.g., *Xyleborinus saxeseni* (Ratzeburg), now have an almost worldwide distribution.

2.2.3 Symbiotic Associations

A wide diversity of symbionts has contributed to the success of bark and ambrosia beetles. Because parasitoids exert a form of delayed predation, we will not treat them as symbionts here. For most species, one or several symbionts play important roles. In many cases, the roles of symbionts are poorly understood, but recent findings have begun to shed light on the importance of some associations.

Most scolytine beetles appear to be closely associated with symbiotic fungi (Kirisits, 2004; Harrington, 2005). There are few exceptions, with *D. micans* currently considered an example (Lieutier *et al.*, 1992). The roles of fungi vary widely (Six, 2012). For some groups, i.e., ambrosia beetles, fungi serve as the sole source of nutrition for both adults and larvae. These species typically have a close association with one or two specialized symbiont fungi (Batra, 1966). For other groups, the relationship between the host and symbiont is less clear. For example, the mountain pine beetle, *Dendroctonus ponderosae* Hopkins, normally is associated with two or three species of fungi, but at least 12 (including yeasts) have been identified (Lee *et al.*, 2006). The roles of these symbionts can range from beneficial, e.g., as a source of food (Ayres *et al.*, 2000; Bleiker and Six, 2007) to detrimental (Harrington, 2005). A pervasive paradigm has been that fungi are necessary for, or at least contribute to, killing the host tree, as evidenced by inoculation experiments (Krokene and Solheim, 1998). Six and Wingfield (2011) argue against this premise, however. More recent studies suggest these fungi can contribute to overcoming tree defenses by metabolizing conifer phenolics and terpenes (see Section 3.4). Species associations vary markedly, with some relationships are facultative or even casual, rather than obligatory (Six, 2012).

Phoretic mites are frequently found on bark beetles (Hofstetter *et al.*, 2013; Knee *et al.*, 2013; Chapter 6). Large numbers of mite species from several families have been associated with the galleries of many bark beetle species (Lindquist, 1970). For example, *Dendroctonus frontalis* Zimmermann has at least 57 species of phoretic mites (Moser and Roton, 1971; Moser *et al.*, 1974; Moser and Macías-Sámano, 2000). Similarly, 38 species of mites are associated with *Ips typographus* L. captured in pheromone-baited traps in Europe (Moser *et al.*, 1989), and an additional three species were found on *I. typographus japonicus* Niijima (Moser *et al.*, 1997). The roles and impacts of mites are not well understood, but vary from detrimental (predatory on bark beetle larvae, parasitic on eggs) to beneficial (predators on nematodes, mycophagous) (Klepzig *et al.*, 2001; Lombardero *et al.*, 2003;

Kenis *et al.*, 2004). Some mites also contribute to the fungal diversity in bark beetle galleries by transporting spores in specialized sporothecae (Moser, 1985). Host specificity also varies, depending on the ecological role of the mite species (Lindquist, 1969, 1970).

Bark beetles are commonly associated with nematodes, most of which appear to be parasitic, phoretic, or commensal (Thong and Webster, 1983; Grucmanová and Holuša, 2013). Massey (1966) found 27 species of nematodes associated with *Dendroctonus adjunctus* (Blandford), and Grucmanová and Holuša (2013) list 11 phoretic, 12 endoparasitic, and eight species associated with frass of *Ips* spp. in central Europe. Cardoza *et al.* (2006b) found nematodes associated with special pockets, *nematangia*, on the hind wings of *Dendroctonus rufipennis* (Kirby). Nematangia have since been found on *Pityogenes bidentatus* (Herbst), containing the tree parasite *Bursaphelenchus pinophilus* Brzeski and Baujard (Nematoda: Parasitaphelenchinae) (Čermák *et al.*, 2012) and *Dryocoetes uniseriatus* Eggers, containing the insect parasite and nematode predator *Devibursaphelenchus* cf. *eproctatus* (Shimizu *et al.*, 2013).

Bacteria may play important roles in ensuring that the host environment remains hospitable, i.e., during initial attack when defense compounds may be high, and during later phases when contaminant antagonistic microorganisms could potentially harm the food supply or offspring. Scott *et al.* (2008) found that actinomycete bacteria associated with *D. frontalis* produce antibiotic compounds, a function similar to that of actinomycetes on leaf cutter ants, *Atta* spp. (Hymenoptera: Formicidae) (Currie *et al.*, 1999). Different bacteria vary in their tolerance of host terpenoids, and in particular bacteria associated with bark beetle species that breed in live resinous hosts are more tolerant than those that kill trees by mass attack (Figure 1.1A) (Adams *et al.*, 2011).

FIGURE 1.1 **Sample illustrations of bark beetle interactions with host plants.** (A) *Dendroctonus micans* tunneling through resin. (B) Extensive competition to *D. ponderosae* (note vertical ovipositional gallery in center) caused by *Ips* (note extensive network of surrounding galleries) in a windthrown *P. contorta* in Wyoming. *Reproduced with permission from Lindgren and Raffa (2013). Photos by (A): J.-C. Grégoire; (B): K. Raffa.*

2.3 Variation in Ecological Impacts of Bark Beetles: from Decomposers to Landscape Engineers, and from Saprophages to Major Selective Agents on Tree Survival

Various bark beetle species are prominent members among the succession of organisms that occupy tree tissues from initial decline to decay (Lindgren and Raffa, 2013). The vast majority of bark beetles are saprophagous, strictly breeding in dead trees or tree parts. The primary ecological role of such species is to initiate or contribute to the breakdown of wood by feeding, vectoring symbiotic microorganisms, or providing access for decay microorganisms. Lindgren and Raffa (2013) subdivided this guild into late succession saprophages, which occupy the resource once most or all of the defensive compounds have been detoxified, and early succession saprophages, which can tolerate some defense compounds. In some cases, the latter species may serve as thinning agents by attacking and killing moribund or severely weakened host trees (Safranyik and Carroll, 2006). They may also facilitate the facultative predatory beetles (Smith *et al.*, 2011). Tree-killing bark beetles, while relatively few in number, can have profound ecological effects, including impacts on species composition, age structure, density, woody debris inputs, and even global carbon balance (Kurz *et al.*, 2008; Lindgren and Raffa, 2013).

2.4 Major Groups

Based on phylogenetic analyses, the bark and ambrosia beetles have recently been reassigned from the family Scolytidae to the subfamily Scolytinae within Curculionidae. Wood (1986) used morphological characteristics to divide Scolytidae into two subfamilies and 25 tribes. Alonso-Zarazaga and Lyal (2009) divided the Scolytinae into 29 tribes. Of the more than 6000 species of Scolytinae described to date, the vast majority are tropical or subtropical (Knížek and Beaver, 2004). Yet, most of our knowledge of bark beetles is based on a large number of studies on a relatively small number of environmentally and economically important species across a few tribes, and within the temperate regions of the northern hemisphere. In particular, studies have emphasized tree-killing species in the Hylesinini, Hylastini, Ipini, Scolytini, and Dryocoetini. Additional focus has been centered on the large tribe of haplo-diploid ambrosia beetles comprising the Xyleborini, because of both their interesting reproductive biology and their prominence as invasive species (Cognato *et al.*, 2010). Another ambrosia beetle tribe, the holarctic Xyloterini, is also relatively well studied, particularly the genus *Trypodendron*, and specifically *T. lineatum* because of its economic importance in northern Europe and western North America (Borden, 1988).

3. INTERACTIONS WITH HOST PLANTS

3.1 Host location and Selection

Most bark beetles deposit all or most of their clutch within a single tree, so the ability to locate and select suitable hosts is crucial for their reproductive success. Many species can only utilize a host for one, or perhaps a few, beetle generations, so each cohort must locate a suitable host to reproduce. The choice of a host tree is laden with trade-offs (Raffa, 2001; Lindgren and Raffa, 2013): trees that are already dead or whose defenses have been severely compromised by environmental or endogenous stress pose little risk during colonization. However, such trees are relatively rare, ephemeral in space and time, are occupied by a diversity of interspecific competitors (Figure 1.1B), and often provide a lower quality nutritional substrate. At the other end of this continuum, relatively unstressed trees are consistently plentiful, in some cases nutritionally superior because of the thicker phloem accompanying their vigorous growth, and only become available to competitors after the primary beetle kills them. However, these trees possess vigorous defenses that can kill potential colonizers that enter them. Making this decision even more daunting is the fact that bark beetle adults normally survive for only a few days (Pope *et al.*, 1980) to a few weeks (Byers and Löfqvist, 1989) outside the tree, and they are subject to rapid energy depletion and predation. Furthermore, the more time a beetle takes to find a tree that elicits its entry behavior, the more trees are eliminated from the available pool by competing conspecifics.

Adult bark beetles employ multiple and integrated modalities, including visual, olfactory, tactile, and gustatory, to perform the difficult tasks of host location and selection (Borg and Norris, 1969; Wood, 1972; Raffa and Berryman, 1982; Pureswaran *et al.*, 2006). Their responses to these signals are influenced by external cues, internal physiology, heredity, and gene by environment interactions (Wallin and Raffa, 2000, 2004; Wallin *et al.*, 2002).

Initial landing is mediated by both visual and chemical cues (Saint-Germain *et al.*, 2007). Some species, such as *D. ponderosae*, show strong orientation to vertical silhouettes, which can be enhanced by coloration that provides greater contrast (Shepherd, 1966; Strom *et al.*, 1999). Chemical cues that elicit directed movement and landing can include host secondary compounds such as some monoterpenes, compounds indicative of stress such as ethanol, and compounds indicative of microbial infection or decay such as acetaldehyde. The extent to which initial attraction and landing in response to these compounds relates to ultimate host selection varies among species. In general, species that are solely associated with dead or highly stressed trees tend to respond to the latter groups of compounds and readily enter the hosts emitting these signals (Rudinsky, 1962). In contrast, species associated with less stressed or healthy trees tend to land initially in response to visual cues and some monoterpenes, and then make subsequent decisions post-landing (Wood, 1972; Moeck *et al.*, 1981). In these species, landing rates tend to be much higher than entry rates (Hynum and Berryman, 1980; Raffa and Berryman, 1980; Anderbrant *et al.*, 1988; Paynter *et al.*, 1990).

Following landing, beetles locate potential entry sites under bark crevices, in response to microclimatic and thigmotactic stimuli. Borg and Norris (1969) demonstrated that pronotal stimulation is required for host compounds to elicit tunneling behavior, a finding routinely incorporated into subsequent bioassays evaluating chemical signals (Elkinton *et al.*, 1981; Raffa and Berryman, 1982; Sallé and Raffa, 2007). In nature, this scales up to beetle within-tree orientation toward bark crevices, so physical texture can play an important role in microsite selection (Paynter *et al.*, 1990). A beetle's decision on whether or not to enter is largely driven by concentrations of host compounds, especially monoterpenes. In addition to the variable attractiveness vs. repellency of different compounds, a common pattern is for low concentrations of a particular monoterpene to elicit tunneling behavior while higher concentrations deter entry or continued tunneling (Figure 1.2). Often, the eliciting concentrations are typical of those which occur in constitutive tissues, but the concentrations present in induced tissue are adequate to deter continued tunneling (Wallin and Raffa, 2000, 2004). The concentrations that elicit entry versus rejection behavior vary among beetle species.

Many bark beetle species can detect cues associated with stress physiology of their host plants. Higher entry rates have been observed in response to root infection, defoliation, fire injury, and other stresses (Lindgren and Raffa, 2013). In addition to external cues, internal cues can affect beetles' responses to host chemicals. For example, as their lipids are depleted, as occurs during flight and extended host searching, some beetles become more responsive to host cues (Kinn *et al.*, 1994). Other internal cues that can increase a beetle's likelihood of entering a tree include age and the number of times it has already rejected putative

hosts, which likely relate to its dwindling likelihood of reproducing if it did not accept some host before dying (Wallin and Raffa, 2002). Beetle responses to host stimuli also show heritable variation. Laboratory experiments have demonstrated selection for "more discriminating" and "less discriminating" lines, based on the maximum monoterpene concentrations beetles will accept in amended media (Wallin *et al.*, 2002). In the field, *D. rufipennis* shows correlations between mothers and daughters in monoterpene concentrations that elicit entry, and these relationships persist for several generations. There are also differences among *D. rufipennis* from endemic vs. eruptive populations, with the latter showing a higher likelihood of entering high-terpene media when other beetles are present (Wallin and Raffa, 2004). Overall, there appears to be substantial plasticity in host selection among tree-killing bark beetles.

3.2 Host Defenses

Because phloem is essential to tree survival, conifers have evolved sophisticated defenses against bark beetle–microbial complexes. Five features of these defenses are particularly pertinent. First, they involve multiple modalities, including physical, histological, and biochemical components (Figure 1.3), and these modalities function in a highly integrated fashion (Raffa, 2001). Within each of these broad categories, there is further complexity and overlap. Chemical defense, for example, includes a variety of classes, and each class includes many different moieties of varying structure and chirality, and some chemical groups such as lignins and terpenoids contribute to physical barriers. Second, delivery of these toxins is augmented by physical structures such as resin canals and glands. Third, each of these physical, histological, and chemical components of defense include both constitutive defenses and rapidly induced defenses in response to attack (Reid *et al.*, 1967; Raffa and Berryman, 1983; Franceschi *et al.*, 2005; Bohlmann and Gershenson, 2009). Fourth, these components of tree defense inhibit multiple aspects of both beetles' and microbial symbionts' life histories, such as host entry, pheromone signaling, survival, growth, and sporulation. Fifth, in addition to heritable features of host resistance, fully integrated functioning of these mechanisms is associated with vigorous whole-plant physiology, so a variety of acute and chronic stresses can impair the extent and rate of these defenses (Raffa *et al.*, 2005; Kane and Kolb, 2010).

Outer bark provides a tough physical barrier, which screens out all but those relatively few herbivores adapted for penetrating it with powerful mandibles, specially modified legs, and other body morphologies. As soon as a tunneling beetle encounters live tissue, trees exude a rapid flow of resin (Chapter 5). The quantity and importance of this resin vary greatly among conifer genera (Berryman, 1972), and even among species within a genus,

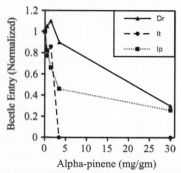

FIGURE 1.2 Effects of host chemistry on entry or aversion behavior in bark beetles. Artificial media amended with synthetic α-pinene for three species, *Dendroctonus rufipennis* (Dr), *Ips tridens* (It), and *Ips pini* (Ip). *From Wallin and Raffa (2000, 2004).*

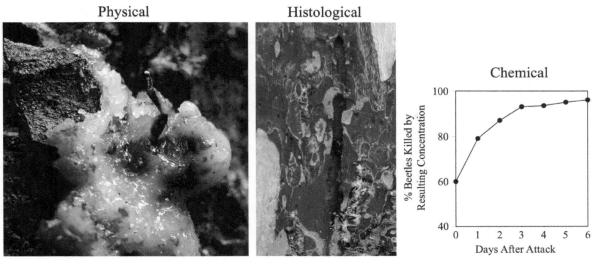

FIGURE 1.3 Integrated physical, histological, and chemical defenses of conifers. (A) *D. micans* killed in resin during attempted colonization of *P. abies*; (B) *D. ponderosae* killed in hypersensitive response during attempted colonization of *P. contorta*; (C) profile of toxicity to *I. pini* adults (48 hours *in vitro* assays) of concentrations of α-pinene present in constitutive and induced responses of *P. resinosa*. *From Raffa and Smalley (1995). Photos by (A): J.-C. Grégoire; (B): K. Raffa.*

as in *Pinus* (Matson and Hain, 1985). There is also substantial variation within species, and ontogenetic, phenological, and stress-mediated variation within individuals. Resin is stored in a variety of structures, such as specialized ducts and glands. This resin poses a significant physical barrier, and can entomb some beetles (Figure 1.3) (Raffa *et al.*, 2008). It also delays beetle progress, which can provide more time for histological and biochemical processes to achieve effective levels. Resin also contains various allelochemicals that can exert repellent and/or toxic effects. Tree-killing bark beetles, however, are often able to physiologically tolerate the concentrations present in constitutive resin.

As a bark beetle tunnels into a tree, inducible reactions begin rapidly. These include induced resinosis and traumatic resin duct formation, autonecrosis and associated alterations in polyphenolic parenchyma and stone cells, and biosynthesis of various compounds via a combination of mevalonic acid, 1-deoxy-D-xylulose-5-phosphate, and shikimic acid pathways (Safranyik *et al.*, 1975; Raffa and Berryman, 1983; Popp *et al.*, 1991; Martin *et al.*, 2003; Franceschi *et al.*, 2005; Keeling and Bohlmann, 2006; Boone *et al.*, 2011). These inducible responses are mediated by signaling compounds such as jasmonic acid and salicylic acid, which are ubiquitous among plants.

As the name induced resinosis implies, resin flow from a wound, especially one accompanied with a biotic inciter such as a beetle or its fungal symbionts, increases rapidly. In addition to delaying the beetle's progress, a copious flow of resin can inhibit a beetle's ability to elicit the arrival of other beetles with pheromones. This likely occurs through a combination of gummy resins physically blocking the

emission of volatile pheromones from the entry site, and high ratios of volatile host terpenes to pheromones that either mask perception or inhibit attraction (Zhao *et al.*, 2011; Schiebe *et al.*, 2012). If mass attack is not elicited relatively quickly, the ratio of monoterpenes to pheromones rises to such high levels that the likelihood of a tunneling beetle being joined by conspecifics becomes very low (Erbilgin *et al.*, 2003, 2006). While this is under way, the tree initiates an autonecrotic or "hypersensitive" reaction (Figure 1.3), in which rapidly progressing cell death forms a lesion that confines the attacking beetle and its symbionts. The nutritional value of this tissue is lost, and this reaction zone becomes the site of pronounced and rapid biochemical accumulation, apparently through both biosynthesis and translocation. Chemical changes include vastly increased concentrations of constitutive compounds, often from non-repellent to repellent, and non-lethal to lethal, doses, altered proportions of compounds present in constitutive synthesis, often with the more bioactive compounds undergoing disproportionately high increases, and production of new compounds that are not present (or at below detectable levels) in constitutive tissue (Raffa *et al.*, 2005). These abilities vary quantitatively among trees within a population. That is, all (or nearly all) trees are capable of this response, and fungal inoculation combined with mechanical wounding almost always elicits induced defenses. However, those trees that respond more extensively and rapidly are more likely to survive. For example, dose-dependent relationships between induced terpene accumulation and resistance to bark beetle attack in the field have been demonstrated in *Pinus* (Raffa and Berryman, 1983; Boone *et al.*, 2011), *Abies* (Raffa and Berryman, 1982), and *Picea* (Zhao *et al.*, 2011).

The major chemical groups that contribute to protection from bark beetle–microbial complexes include monoterpenes, diterpene acids, stilbene phenolics, and phenylpropanoids. These tend to have complementary activities (Raffa et al., 2005). In general, high concentrations of monoterpenes are repellent, ovicidal, larvicidal, and adulticidal toward the beetles. Tolerance appears somewhat higher among the solitary-parasitic than gregarious-tree killing species. Monoterpenes moderately inhibit fungal germination and growth, and are likewise highly toxic to a broad range of bacteria associated with beetles. Phenolics tend to have relatively low activity against bark beetles, but have moderate activity against their fungal associates. Their activity against beetle-associated bacteria is unknown. Diterpene acids are the most toxic group to beetle-associated fungi, greatly inhibiting mycelial growth, conidiophore production, and germination (Boone et al., 2013). In contrast, some beetle-associated bacteria are quite tolerant of diterpene acids. To date, no direct effects of diterpene acids against the beetles have been demonstrated. One phenylpropanoid, 4-allylanisole, is known to mediate conifer–bark beetle interactions by inhibiting the attraction of flying beetles to the pheromones emitted by a tunneling beetle (Hayes and Strom, 1994; Strom et al., 1999; Kelsey et al., 2001; Emerick et al., 2008). This compound occurs in both subcortical and foliar tissues. We currently have little information on how various defense compounds interact, but a variety of effects, including synergism, seem likely. The overall pattern, however, is that no single chemical influences all components of beetle–microbial systems, but all components of beetle–microbial systems are influenced by one or more compounds.

In addition to the very rapid synthesis of defense compounds and autonecrosis of utilizable substrate at the attack site, there may also be some longer-term effects. For example, Picea abies (L.) H. Karst., that had been inoculated with the root fungus Heterobadisium annosum (Fr.) Bref. or the bark beetle-vectored fungus Ceratocystis polonica (Siemaszko) C. Moreau showed reduced symptoms to inoculation with C. polonica 4 weeks later (Krokene et al., 2001). This appears to be primarily attributable to the induced formation of traumatic resin ducts, and the swelling and proliferation of polyphenolic parenchyma cells, a non-systemic response within the pretreated area (Krokene et al., 2003, 2008). Similarly, persistent elevated terpene levels induced by application of methyl jasmonate were localized, i.e., within the treated but not untreated stem sections. This agrees with work on Pinus resinosa Aiton, in which inoculation with Ophiostoma ips (Rumbold) Nannf. did not cause systemic alterations in lesion formation or monoterpene accumulation in response to subsequent inoculations (Raffa and Smalley, 1988; Wallin and Raffa, 1999). The extent to which prior beetle attacks influence susceptibility to subsequent attacks under natural conditions requires additional study. In the cases of the solitary D. micans and Dendroctonus valens LeConte, and the gregarious Dendroctonus rufipennis LeConte, previously attacked trees (P. abies, P. resinosa, and Picea glauca (Moench) Voss and Picea engelmannii Parry ex Engelm., respectively) were more likely to be attacked than unattacked trees (Gilbert et al., 2001; Wallin and Raffa, 2004; Aukema et al., 2010). These patterns are not consistent with priming or induced acquired resistance. However, they do not necessarily prove insect-induced susceptibility either, because subsequent cohorts of beetles could be responding to the same predisposing condition. Similar trends emerge from between-species, temporally spaced interactions. Prior sublethal infestation of the lower stems of Pinus ponderosa Douglas ex C. Lawson by D. valens and associated Leptographium is associated with increased subsequent attacks by Dendroctonus brevicomis LeConte (Owen et al., 2005), prior colonization of Pinus contorta Douglas ex Loudon by Pseudips mexicanus (Hopkins) is associated with increased subsequent attacks by D. ponderosae (Safranyik and Carroll, 2006; Smith et al., 2009, 2011; Boone et al., 2011), and prior infestation of the lower stems of Pinus resinosa by D. valens, Hylobius radicis Buchanan, and associated Leptographium is associated with increased likelihood of subsequent attacks by Ips pini (Say) (Aukema et al., 2010).

A fourth category of defenses, about which little is known, involves symbiotic associations. For example, endophytic bacteria in P. contorta can inhibit the growth of Grosmannia clavigera (Rob.-Jeffr. and R.W. Davidson), an important fungal symbiont of D. ponderosae (Adams et al., 2008). We do not yet know what roles these relationships play in nature. In some plant–herbivore interactions, symbioses involving mycorrhizae and endophytes can be quite important, so this area requires more investigation.

3.3 Host Substrate Quality

The quality, or suitability, of a host as a substrate for developing brood is distinct from its susceptibility, i.e., the relative ease or difficulty with which it can be killed. Stem-colonizing bark beetles consume a resource that is spatially limited, and of relatively poor quality. The phloem is a relatively thin subcortical layer, and different beetle species are confined by different minimal requirements of phloem thickness, which in turn limits the sizes of trees and heights along the bole they can colonize. This limitation creates a true "carrying capacity," in which the available resource per individual declines as the number of colonizing individuals increases (Coulson, 1979; Anderbrant, 1990). Hence, there are often direct relationships between phloem thickness and total beetle reproductive output, and between tree diameter and total beetle reproductive output (Amman, 1972).

Phloem tissue tends to be particularly low in nitrogen, which is often limiting to herbivorous insects (Mattson, 1980). Low nitrogen availability lengthens insect development times and reduces their fecundity. Bark beetles compensate for this resource deficiency with close associations with microbial symbionts, especially fungi and bacteria (Ayres *et al.*, 2000; Bleiker and Six, 2007; Morales-Jiménez *et al.*, 2009). The phloem resource also contains cellulose, but not in levels comparable to those in sapwood with which wood borers must contend, and so the cellulolytic capabilities of bacteria associated with bark beetles appear generally less than those of bacteria associated with cerambycids and siricids (Delalibera *et al.*, 2005; Adams *et al.*, 2011). Phloem tissue appears to have adequate concentrations of carbohydrates, sterols, and micronutrients for bark beetles, and there are no particular limitations in their availabilities.

Phenological changes in trees in temperate zones limit bark beetles to a relatively narrow window of resource availability. As the season proceeds, this tissue begins to harden, desiccate, and export resources. Once these changes begin, host quality declines. In multivoltine species such as *I. pini*, later-season development can be less productive, even though it may open periods of escape from predators (Redmer *et al.*, 2001). Host quality also deteriorates due to microbial exploitation following beetle colonization. The physical and chemical defenses that trees use against bark beetles also render this habitat unavailable to a diversity of saprophytic and antagonistic fungi. However, once the beetles have exhausted those defenses, the environment becomes available to competing organisms, which can exert substantial costs on beetle fitness.

Not much information is available on variation among host tree species in their resource quality for bark beetles, other than differences attributable to phloem thickness. In general, it appears that interspecific variation in substrate quality is mostly attributable to tree size and phloem thickness.

3.4   Roles of Symbionts in Host Plant Utilization

Symbionts play crucial roles in the life histories of bark beetles, especially in overcoming tree defense, utilizing host plant substrates, and protecting their resource. Numerous microbial taxa are associated with scolytines, and all scolytine species are associated with microorganisms.

Early work often depicted bark beetle-vectored fungi as virulent pathogens that killed the tree and thereby rendered it available for brood development. However, instances in which ophiostomatoid fungi directly kill trees appear limited to invasive species (such as *Ophiostoma ulmi* (Buisman) Nannf. and *O. novo-ulmi* in European and North

American *Ulmus* and *Leptographium procerum* (W. B. Kendr.) M. J. Wingf. in Chinese *Pinus*) (Gibbs, 1978; Brasier, 1991; Sun *et al.*, 2013), and a few species such as *C. polonica* (Krokene and Solheim, 1998) and certain strains of *G. clavigera* (Lee *et al.*, 2006; Plattner *et al.*, 2008; Alamouti *et al.*, 2011). Similarly, early researchers often envisioned these fungi as blocking the flow of resin to the point of attack, but subsequent experiments indicate that fungi probably do not grow quickly enough into tracheids to exert this effect (Hobson *et al.*, 1994).

More recent work indicates that microbial symbionts of bark beetles can metabolize host toxins. Specifically, *C. polonica* reduces concentrations of stilbene phenolics present in *Picea*, at least *in vitro* (Hammerbacher *et al.*, 2013). The fungus *G. clavigera* has genes that encode for terpene metabolism (DiGuistini *et al.*, 2011). Likewise, bacteria associated with *D. ponderosae* and their host trees have multiple genes encoding for detoxification of many terpenoids, and also greatly reduce concentrations on monoterpenes and diterpene acids *in vitro* (Adams *et al.*, 2013; Boone *et al.*, 2013). Furthermore, various bacteria species appear to have complementary metabolic activities, with different community members degrading specific compounds, but collectively all host chemicals being degraded by at least one bacterium. These relationships are dose dependent, as high concentrations of terpenes become toxic and negate bacterial activity. The tolerance of bacterial associates to host tree terpenes appears to vary with beetle life history strategy, with communities associated with species such as *D. valens* that often reproduce in live hosts being more tolerant than community members associated with mass-attacking species such as *D. ponderosae* (Adams *et al.*, 2011). In addition, yeasts can influence the composition of monoterpenes. When *Ogataea pini* (Holst) Y. Yamada, M. Matsuda, K. Maeda and Mikata from *D. brevicomis* mycangia was added to phloem disks of *P. ponderosae*, total monoterpenes were not reduced, but several individual components were higher or lower, relative to controls (Davis and Hofstetter, 2011). Overall, it appears that microbial associates function in concert with bark beetles to jointly overcome tree defenses, i.e., as cofactors (Klepzig *et al.*, 2009; Lieutier *et al.*, 2009). Further, microorganisms appear to detoxify tree chemicals in conjunction with, not in place of, detoxification by the beetles themselves, which are equipped with P-450 enzymes (Sandstrom *et al.*, 2006).

Microorganisms may also assist beetles in overcoming tree defense by contributing to biosynthesis of aggregation pheromones. For example, the bacterium *Bacillus cereus* converts α-pinene into verbenol *in vitro* (Brand *et al.*, 1975). However, it is not clear whether this plays an important role in nature. There are also instances in which fungi reduce tree defenses indirectly and with a time lag. For example, vectoring of *Leptographium* fungi into roots and lower stems by various *Hylastes* and solitary

Dendroctonus species impairs defenses against subsequent lethal stem-colonizing attack bark beetles (Witcosky *et al.*, 1986a; Klepzig *et al.*, 1991; Eckhardt *et al.*, 2007). There may be important interactions among microorganisms in overcoming tree defenses. For example, *G. clavigera* and other ophiostomatoid fungi are highly susceptible to diterpene acids, but the bacteria associated with *D. ponderosae* greatly reduce concentrations of these compounds. Likewise, bacteria associated with *D. ponderosae* and *D. valens* can enhance mycelial growth and spore germination of various fungal symbionts. These interactions can be either enhanced or inhibited by host tree terpenes (Adams *et al.*, 2009).

Fungi play crucial roles in nutrient acquisition by bark beetles (Six, 2012). Almost all bark beetle species show close associations with fungi, and benefit both from fungal metabolism of the substrate into utilizable nutrients, and by directly consuming fungi. Basidiomycetes can be particularly important in this capacity. In addition, symbiotic bacteria may assist beetle larvae in obtaining nitrogen, through nitrogen fixation in the gut (Morales-Jiménez *et al.*, 2012).

The specific composition of various symbiotic species on or in a beetle can have strong effects on bark beetle success, and can be influenced by a number of environmental factors. For example, temperature affects the relative abundance of *G. clavigera* and *Ophiostoma montium* (Rumbold) Arx in galleries of *D. ponderosae* (Addison *et al.*, 2013). This has important ramifications to the beetle's population dynamics in different parts of its range, in different habitats, and implication in response to climate change. In *D. frontalis*, the relative abundances of the mycangial nutritional mutualists, and the antagonist *O. minus*, are strongly influenced by phoretic mites (Hofstetter *et al.*, 2006). The mites, in turn, have variable relationships with these fungi. The outcomes of these interactions are mediated by tree chemistry and temperature (Hofstetter *et al.*, 2007; Evans *et al.*, 2011; Hofstetter and Moser, 2014). The relative composition of various fungal symbionts can also vary spatially and temporally with beetle population density, as in *D. rufipennis* (Aukema *et al.*, 2005a). Likewise, bacterial communities can vary regionally within a beetle species (Adams *et al.*, 2010).

One of the challenges to the lifestyle of bark beetles that colonize live trees is that their mode of overcoming defense (i.e., mass attack) renders the host environment suitable to a broad array of competitors. This can be highly deleterious to developing brood. Bacterial symbionts can play important roles in reducing these losses. As female *D. rufipennis* excavate ovipositional galleries, they egest oral secretions that contain several species of bacteria (Cardoza *et al.*, 2006a). These bacteria are highly toxic to antagonistic fungi such as *Aspergillus* and *Trichoderma*. They are also partially selective, showing less toxicity to the symbiont *Leptographium abietinum* (Peck) Wingf. Likewise, *D. frontalis* carry symbiotic Actinomycetes that produce mycangimicin, which

selectively inhibits the antagonist *O. minus* but not the mutualistic *Entomocorticium* sp. (Scott *et al.*, 2008). Competitors also include conspecific beetles that arrive after a tree's defenses have been overcome. Many bark beetles reduce this form of exploitation by producing anti-aggregation pheromones during the later stages of host colonization, and some fungi, including yeasts, appear to contribute to production of these masking compounds (Brand *et al.*, 1976; Hunt and Borden, 1990).

The degree of association between bark beetles and microbes that contribute to host utilization varies extensively. Closely linked mutualists, such as some Basidiomycete fungi, are transported in specialized mycangia (Six and Klepzig, 2004). Other fungi reside on the exoskeleton. Some of the bacteria that degrade host compounds may be both conifer and beetle associates, such that the ability to degrade terpenes is a requirement for inhabiting phloem, and the attacking beetles become the indirect beneficiaries of that association when they enter (Adams *et al.*, 2013).

3.5 Resource Partitioning

Although conifer bark beetles compete for a common resource, phloem tissue, they have several mechanisms for partitioning this resource and thereby reducing direct competition. The first level of separation is geographic range, and several species with similar life histories and host ranges occupy distinct or at least partially distinct zones. Some examples include *D. ponderosae* and *Dendroctonus adjunctus* Blandford in the northern and southern ranges of *P. ponderosae*, and *Dendroctonus murryanae* Hopkins and *D. valens* in the higher and lower elevations of *P. contorta*, respectively. A second level of resource partitioning occurs within a region, based on host range. This usually functions at the level of plant genus. Different species of bark beetles tend to be associated with a corresponding conifer genus, but can often colonize all the species within that genus within their geographic range (D. L. Wood, 1982). Some exceptions include *Dendroctonus jeffreyi* Hopkins that is closely associated with *Pinus jeffreyi* A. Murray, which in turn has unusual chemistry and is not attacked by most other scolytines. Also, *Pinus strobus* L. and *Pinus palustris* Mill. are not commonly attacked by *D. frontalis*, despite the high overlap with that insect's range. Although the most aggressive outbreak species are typically specialists on one genus, several of the moderately aggressive species, such as *Dendroctonus pseudotsugae* Hopkins, *Dryocoetes affaber* Mannerheim, and sometimes *I. pini*, utilize two genera, and the non-aggressive, often secondary, species such as *Orthotomicus caelatus* Eichhoff and *Dryocoetes autographus* (Ratzeburg) often colonize three or more genera.

Beyond the coarser levels of geographic region and host genus or species, various bark beetle species partition the phloem resource at several finer scales. First, different species are associated with different parts of a tree's stem (Coulson, 1979; Grünwald, 1986; Schlyter and Anderbrant, 1993; Flamm *et al.*, 1993). An example is the guild associated with southern pine beetle, in which the solitary or semi-solitary *Dendroctonus terebrans* Olivier colonizes the base, *D. frontalis* mass attacks the lower portion of the stem, *Ips grandicollis* Eichhoff often colonizes the portion above that, and *Ips calligraphus* (Germar) and *Ips avulsus* Eichhoff colonize both the main stem and lateral branches of the crown (Paine *et al.*, 1981). There are parallels within most systems. The degree of partitioning is typically partial rather than absolute when multiple species colonize a tree, and it is typically opportunistic rather than obligate, in that when one species is missing the others will extend into the zone the absent species normally occupies. Another level of partitioning can arise from seasonality, whereby one species tends to fly earlier, for a different length of time, or have different voltinism, than other species occupying the same host within a region. Finally, different species partition the resource based on host physiological condition (Rankin and Borden, 1991; Flechtmann *et al.*, 1999; Saint-Germain *et al.*, 2009). Many species only colonize dead trees or dead parts of trees. Others can colonize live trees, but only highly stressed individuals. Still other species can colonize healthy trees, but only during outbreaks. As with tree morphology, these relationships tend to be relative rather than strict. For example, beetle species that colonize healthy trees during outbreaks commonly rely on dead trees during lengthy endemic periods. Perhaps the species that comes most closely to relying solely on live trees is *D. frontalis*, which cannot be reared through its entire life cycle in dead logs. In general, those species that only colonize dead or severely stressed trees tend to be the most fit at competition, both when tree-killing species are limited to severely stressed trees, or when secondary beetles follow tree-killing species into healthy trees they overcome (Raffa and Berryman, 1987; Lindgren and Raffa, 2013).

In some cases, there is no apparent higher-level structuring to resource partitioning, but instead there initially appears to be scramble competition. However, in these cases there is often a secondary structuring mediated by pheromones (Lanier and Wood, 1975). That is, the first beetle to locate a susceptible stand or tree within a stand produces a species-specific pheromone that greatly biases local subsequent population ratios. For example, *I. pini* and *I. grandicollis* appear to interact much in this manner in the Great Lakes region of North America.

4. COMMUNICATION

Scolytine bark beetles are generally regarded as being largely subsocial (Wilson, 1971; Kirkendall *et al.*, 1997; Costa, 2006). Many species breed in aggregations on their host plants, and most species provide some care for their offspring (Jordal *et al.*, 2011). Even in some solitary species, larvae often exhibit aggregation behavior (Grégoire *et al.*, 1982). The ambrosia beetle *X. saxeseni* exhibits high levels of sociality, including gallery, fungus, and brood care by both the adult and larval offspring of a single foundress (Biedermann and Taborsky, 2011). Aggregation behavior and other social interactions require efficient means of communication, and scolytine beetles have evolved several means by which they influence the behavior of conspecifics, including physiological and anatomical adaptations for the production, emission, and reception of chemical signals (Dickens and Payne, 1977; Blomquist *et al.*, 2010). However, there is a high noise to signal ratio in the complex environments where these beetles generally dwell, so their communication systems need to be flexible in order to convey a correct message that varies with context.

Bark beetles attacking live hosts have evolved behavioral and physiological traits to contend with the dynamic defenses of their hosts. Once the primary physical defense of the bark is breached, a plant will flood the area with a blend of defensive compounds in a more or less viscous liquid, e.g., terpene-rich oleoresins in conifers and latex or sap in angiosperms. A major function of this liquid is to physically flush the wound and thereby remove invading organisms. An additional function is to repel attackers by toxins, and thus many of the constituent compounds in these defensive liquids are general or specific toxins, the potency of which may depend on dose (Raffa, 2014). In the Pinaceae, these compounds are also volatile, which may partially explain why bark beetles are particularly prominent in this family of plants (Franceschi *et al.*, 2005; Lindgren and Raffa, 2013). Volatile toxins constitute a very effective defense, but a drawback is that they broadcast a distress signal, which is subject to interception by additional enemies that can then orient to a plant that is injured or under attack (Dixon and Payne, 1980; Erbilgin and Raffa, 2001; Raffa, 2001).

4.1 Functions and Roles

In order to reproduce successfully, a bark beetle must locate the resource, quickly occupy it, attract a mate, and ward off both inter- and intraspecific competitors (Lindgren and Raffa, 2013). Throughout this sequence of events, both inter- and intraspecific communication play important roles, first as a means of locating the host, then to attract conspecifics, including a mate, and finally to prevent overcrowding (Nilssen, 1978; Byers, 1984, 1992b). The predominant modality of communication is through chemicals via olfaction and gustation (Raffa, 2014), although acoustic communication is also important.

The function of a specific semiochemical is context dependent, having different functions depending on the circumstance (Table 1.1). So-called pioneer beetles, the first to arrive at a resource, must use a variety of host cues (Borden *et al.*, 1986). Pioneer beetles attacking live hosts may first use visually directed landing that is random relative to host susceptibility, and make subsequent selection decisions on the bark (Hynum and Berryman, 1980). Beetles joining an attack in progress are aided by both host volatiles and semiochemicals emitted by conspecifics in the process of occupying the host. Once they have successfully occupied and acquired a resource, bark beetles can benefit from preventing additional beetles from arriving. This is accomplished by increased or decreased emissions of specific compounds, by special anti-aggregation or spacing pheromones, or by changes to the bouquet of host volatiles emitted because of cumulative biological and physical processes (Flechtmann *et al.*, 1999). Furthermore, bark beetles may use different cues for long-range and short-range orientation to a host (Saint-Germain *et al.*, 2007). Saprophages searching for dying, injured or fallen trees are guided by volatile emissions from the host (kairomones), such as monoterpenes and/or

TABLE 1.1 Functional Terminology of Semiochemicals (Nordlund, 1981) with Examples Relevant to Bark Beetles. Note that the Same Compound can be Assigned Different Functions Depending on the Context

Functional Term	Effect		Intra- or Interspecific	Description	Examples	Selected References
	Emitter	Receiver				
Pheromone	+	+	Intra	*Aggregation pheromones*, attracts both male and female conspecifics to a breeding resource. *Epideictic (spacing) pheromones*, produced by breeding pair to prevent crowding detrimental to their offspring. *Anti-aggregation pheromones*, a type of epideictic pheromone that interrupts aggregation (and hence crowding) on a resource.	trans-Verbenol Ipsdienol Frontalin exo-Brevicomin Verbenone MCH	Pitman *et al.*, 1969 Young *et al.*, 1973 Pitman and Vité, 1970 Rudinsky *et al.*, 1974 Shore *et al.*, 1992 Lindgren and Miller, 2002a Furniss *et al.*, 1974 D. L. Wood, 1982
Allomone	+	−	Inter	Semiochemical emitted by a bark beetle that prevents occupation by other species of an already occupied resource, thus preventing detrimental effects for the emitter.	Ipsdienol	Birch *et al.*, 1980 D. L. Wood, 1982
Kairomone	−	+	Inter	Host volatiles emitted by a live host tree that attracts bark beetles. Semiochemicals that attract potential natural enemies.	Monoterpenes Ipsdienol	Byers, 1992a Sun *et al.*, 2004 Dahlsten *et al.*, 2003 Hulcr *et al.*, 2005
Synomone	+	+	Inter	Semiochemical emitted by a bark beetle that prevents aggregation of a second bark beetle to an occupied resource, therefore reducing competition.	Ipsenol Verbenone	Borden *et al.*, 1992 Hulcr *et al.*, 2005
Apneumone	0	+	Inter	Volatiles emitted from a dead organism that attracts a predator or parasite even in the absence of their host insect.	Ethanol	Schroeder and Weslien, 1994

ethanol (Byers, 1992a; Miller and Rabaglia, 2009). In all cases, predators and competitors eavesdrop on these signals, using them to orient to the same resource.

Communal feeding by larvae occurs in a number of scolytine species, particularly in parasitic species like *D. micans* and *D. punctatus* (Grégoire, 1988; Furniss and Johnson, 1989) where solitary, mated females establish their brood gallery on a live tree, as well as in a few other *Dendroctonus* species attacking trees that tend to have high levels of oleoresin (Pajares and Lanier, 1990). Larval aggregation in *D. micans* is mediated by chemical communication (Grégoire *et al.*, 1982).

4.2 Chemicals

Beetles that attack live trees must be able to avoid, tolerate or detoxify tree defense chemicals, or they and/or their offspring will be killed by the plant (Lindgren and Raffa, 2013). Metabolism of host compounds by beetles, such as hydroxylation of terpenes, can substantially reduce toxicity, and some of the resulting alcohols and ketones may be exploited by the insect for communication (D. L. Wood, 1982; Raffa and Berryman, 1983; Sandstrom *et al.*, 2006). For example, trans-verbenol, a female-produced aggregation pheromone of *D. ponderosae*, is derived through simple hydroxylation of the host monoterpene α-pinene (Blomquist *et al.*, 2010). However, some bark beetles synthesize isoprenoid and monoterpenoid pheromones *de novo* (Ivarsson *et al.*, 1993; Seybold *et al.*, 1995; Blomquist *et al.*, 2010) through the mevalonate pathway, with specialized enzymes converting intermediates to pheromone components of the required stereochemistry (Blomquist *et al.*, 2010). Thus, *de novo* synthesis might be a predominant mode of pheromone production, at least among the Ipini, and for some semiochemicals used by members of Hylesinini. Lineatin, a complex tricyclic acetal that is an important aggregation pheromone or attractant for many *Trypodendron* species (Borden *et al.*, 1979; Schurig *et al.*, 1982; Lindgren *et al.*, 2000), is also synthesized *de novo*, as are exo- and endo-brevicomin, non-isoprenoid semiochemicals occurring widely in *Dendroctonus* (Blomquist *et al.*, 2010). In *Ips* and *Dendroctonus*, the most likely site of *de novo* pheromone production is the anterior midgut (Blomquist *et al.*, 2010).

Many bark beetle semiochemicals occur in more than one species and often in several tribes (Table 1.2). This supports the hypothesis that chemical communication has evolved primarily by exploitation of compounds that are naturally derived through commonly occurring, evolutionarily preserved biosynthetic processes. Significant overlap in aggregation pheromone blend components among species is common, e.g., frontalin is a primary component of the aggregation pheromone in a number of species in the genus *Dendroctonus* (Renwick and Vité, 1969;

Pitman and Vité, 1970; Dyer, 1975; Browne *et al.*, 1979), and ipsdienol and/or ipsenol are ubiquitous in the clade Ipini (Vité *et al.*, 1972; Phillips *et al.*, 1989) and also occur widely in the Dryocoetini (Klimetzek *et al.*, 1989). Many of these semiochemicals have also been found in non-insect taxa. For example, the aggregation pheromone of *Gnathotrichus sulcatus* (LeConte), sulcatol (Byrne *et al.*, 1974), has been identified in volatile extracts from various fungi (Vanhaelen *et al.*, 1978), and plants (Hüsnü Can Başer *et al.*, 2001), and frontalin, a common aggregation pheromone in the genus *Dendroctonus*, has been found in Asian and African elephants (Rasmussen and Greenwood, 2003; Goodwin *et al.*, 2006) and in the bark of angiosperms (Huber *et al.*, 1999). Sulcatol and frontalin are both produced through the mevalonic pathway with sulcatone as an intermediate product (Blomquist *et al.*, 2010).

The relative ubiquity of specific semiochemicals across many species, genera, and tribes (Table 1.2) suggests that reproductive isolation is achieved through multiple, not single, modalities. Host species fidelity, within-host niche separation, temporal and geographic isolation, as well as behavioral and physiological incompatibility reduces the likelihood of hybridization (Flamm *et al.*, 1987; Schlyter and Anderbrant, 1993; Kelley and Farrell, 1998; Pureswaran and Borden, 2003). In addition, receptor specificity for different enantiomers, enantiomeric ratios, and semiochemical blends prevents cross attraction (Pitman *et al.*, 1969; Birch *et al.*, 1980; Borden *et al.*, 1980; Schlyter *et al.*, 1992).

4.3 Acoustics

Volatile semiochemicals constitute an efficient means of communication, but many bark beetles also use acoustic signaling in intraspecific communication on the host (Rudinsky and Michael, 1973). Males and/or females of many species have specialized stridulatory organs (Barr, 1969), which appear to be significant for mate choice and male competition (Wilkinson *et al.*, 1967; Ryker and Rudinsky, 1976). The location and structure of these stridulatory organs vary widely among Scolytinae. The functions of acoustic communication, and how they integrate with chemical, visual, and tactile signals, are just becoming more fully understood.

4.4 Intraspecific Variation

Bark beetle semiochemical blends may be highly variable, both quantitatively and qualitatively (Schlyter and Birgersson, 1989). The context in which a pheromone is produced and emitted affects how the receiver responds to it. A number of studies have established geographic variation in response to host volatiles and/or pheromones (Lanier *et al.*, 1972; Borden *et al.*, 1982; Miller *et al.*, 1989, 1997). The response by *I. pini* to pheromones has

TABLE 1.2 Examples of Relative Ubiquity of Semiochemicals of the Scolytinae. Data from PheroBase (El-Sayed, 2012)

Semiochemical	Common Name	Presence in Tribes	No. of Species	Function*
2-methyl-6-methylene-7-octen-4-ol	Ipsenol	Dryocoetini	1	P
		Ipini	19	A2, P18
		Pityophthorini	1	A1
2-methyl-6-methylene-2,7-octadien-4-ol	Ipsdienol	Hylesinini	11	A10, P1
		Ipini	28	A16, P20
		Pityohthorini	1	A1
		Xyloterini	1	A1
1, 5-dimethyl-6,8-dioxabicyclo[3.2.1]octane	Frontalin	Hylesinini	11	A10, P7
		Cryphalini	1	A1
		Ipini	1	A1
		Pityophthorini	4	A4
		Scolytini	1	A1
		Xyloterini	1	K1
exo-7-ethyl-5-methyl-6,8-dioxabicyclo[3.2.1] octane	exo-Brevicomin	Hylesinini	10	A3, K1, P7
		Dryocoetinini	3	P3
endo-7-ethyl-5-methyl-6,8-dioxabicyclo[3.2.1] octane		Hylesinini	3	P3
		Dryocoetinini	3	P3
3,3,7-trimethyl-2,9-dioxatricyclo-[3.3. 1.0 4,7] nonane	Lineatin	Hylesinini	7	A7
		Cryphalini	1	A1
		Dryocoetini	1	A1
		Ipini	1	A1
		Xyleborini	1	A1
		Xyloterini	6†	A5, P1
6-methyl-5-hepten-2-ol	Sulcatol, Retusol‡	Corthylini	3	A1, P2
6-methyl-5-hepten-2-one	Sulcatone	Hylesinini	1	P1
2-(1-hydroxy-l-methylethyl)-5-methyltetrahydrofuran	Pityol	Pityophthorini	10	A3, K1, P7
(E)-2-methyl-6-methylene-octa-2,7-dienol	E-Myrcenol	Ipini	2	P2
cis-3-hydroxy-2,2,6-trimethyltetrahydropyran	Vittatol	Hylesinini	1	P1
2-methyl-3-buten-1-ol		Hylesinini	3	A2, P1
		Corthylini	1	A1
		Ipini	9	A7, P4
		Xyloterini	1	A1
2-methyl-3-buten-1-ol		Hylesinini	1	P1
		Ipini	1	P1
2-ethyl-1,6-dioxaspiro[4.4]nonane	Chalcogran	Ipini	3	A1, P3

Continued

TABLE 1.2 Examples of Relative Ubiquity of Semiochemicals of the Scolytinae. Data from PheroBase (El-Sayed, 2012)—cont'd

Semiochemical	Common Name	Presence in Tribes	No. of Species	Function*
7-methyl-1,6-dioxaspiro[4.5]decane	Conophthorin	Hylesinini	1	A1
		Ipini	4	A4
		Pityophthorini	4	P4
		Scolytini	5	A5
5-ethyl-2,4-dimethyl-6,8-dioxabicyclo[3.2.1] octane	α-multistriatin	Scolytini	3	A1, P3
trans-4,6,6-trimethylbicyclo[3.1.1]hept-3-en-2-ol	trans-Verbenol	Hylesinini	11	A4, 1Al, K1, P10
		Ipini	12	P12
cis-4,6,6-trimethylbicyclo[3.1.1]hept-3-en-2-ol	cis-Verbenol	Hylesinini	9	A3, Al1, P5
		Corthylini	1	A1
		Ipini	24	A10, P15
		Xyloterini	1	A1
4, 6, 6-trimethylbicyclo[3.1. l]-hept-3-en-2-one	Verbenone	Hylesinini	11	A3, P10
		Ipini	6	A2, P4
4-methylene-6,6-dimethylbicyclo[3.1.1] hept-2-ene	Verbenene	Hylesinini	1	P1
		Ipini	1	P1
3-methylcyclohex-2-en-1-ol	Seudenol	Hylesinini	4	A3, P2
		Polygraphini	1	A1
		Ipini	2	A2
1-methylcyclohex-2-en-1-ol	MCOL	Hylesinini	2	A1, P1
3-methylcyclohex-2-en-1-one	MCH	Hylesinini	4	Al1, P3

*Functions: A = Attractant, Al = Allomone, K = Kairomone, P = Pheromone
†Including two species from Lindgren et al. (2000)
‡Combined as Retusol is the S-(+)-enantiomer of Sulcatol

been found to vary seasonally (Teale and Lanier, 1991; Steed and Wagner, 2008). Geographic variation in both chemical species and enantiomer composition, as well as seasonal variation in response to pheromones, may be due to a number of factors, such as interspecific competition (Lanier *et al.*, 1972) and predator selection pressure (Raffa and Klepzig, 1989; Aukema and Raffa, 2000).

5. TRITROPHIC INTERACTIONS

Bark beetle-attacked trees provide abundant, though temporary, resources for hundreds of species of associated organisms. When attacked trees die, they provide a succession of spatio-temporal niches exploited by various guilds of natural enemies, competitors, and inquilines living off other components of these resources (see Section 2.2.3). The links between bark beetles and their associates vary from clear predator/prey or parasitoid/host relationships to more complex interactions that may vary according to circumstances (see Boone *et al.*, 2008a).

5.1 Major Predators, Parasitoids, Pathogens and their Life Histories

Natural enemies are discussed in detail in Chapter 7. Flies, beetles, wasps, mites, nematodes, and vertebrates (mostly birds) are either predators, parasitoids, or true parasites of Scolytinae, which are also affected by a variety of pathogens that include viruses, bacteria, entomopathogenic fungi, protozoa, apicomplexa, and microsporidia. Several general reviews are available (Dahlsten, 1982; Mills, 1983; Kenis et al., 2004; Wegensteiner, 2004). More specific reviews concern particular natural enemy taxa such as nematodes (Rühm, 1956), mites (Lindquist, 1964), chalcidoid parasitoids (Hedqvist, 1963), braconid parasitoids (Hedqvist, 1998), or microsporidia (Weiser, 1961). Other reviews focus on particular bark beetle species, e.g., *D. frontalis* (Berisford, 1980), *D. simplex* (Langor, 1991), or *Tomicus piniperda* (L.) (Hérard and Mercadier, 1996). In this section, we emphasize how the life history traits of these natural enemies are interwoven with those of the bark beetles

5.1.1 General Relationships with Bark Beetles

Some natural enemies arrive at the same time as the bark beetles on newly attacked trees. This is the case for phoretic mites, and for those nematodes and microorganisms that are attached to the beetles, either externally or internally. Lombardero et al. (2003) reported that more than 50 mite species and 40 species of fungi and bacteria are transported on *D. frontalis*; Knee et al. (2013) collected 33 mite species belonging to seven families on 18 bark beetle species in pheromone traps in Canada. The roles of these mites are highly diverse, including activities as predatory on various bark beetle stages, predatory on other associates, fungivorous, and saprophagous. Often these roles overlap, and in many cases they are unknown. Some coleopteran, dipteran, and hymenopteran predators and parasitoids also arrive early, in response to bark beetle pheromones. Clerid (*Thanasimus* spp. and *Enoclerus* spp.) and trogositid beetles (*Temnochila* spp.) feed on bark beetles landing on a new host (Vité and Williamson, 1970; Schroeder, 1999a, b; Zhou et al., 2001), oviposit in bark cracks (Gauss, 1954; Schroeder, 1999a), and their larvae enter the prey galleries and feed on any organism they encounter inside, including their conspecifics. Colydiid beetles (*Lasconotus* spp.) and ostomids (*Nemosoma* spp.) enter the galleries and oviposit therein (Hackwell, 1973). Histerid adults (e.g., *Platysoma*) enter the galleries where they prey on adults and eggs, and oviposit. Their larvae feed on bark beetle larvae and pupae (Aukema et al., 2004b). Staphylinidae also land early on attacked trees (Kennedy and McCullough, 2002). Dolichopodid predatory flies (e.g., *Medetera aldrichii* Wheeler) arrive early on attacked trees and oviposit near the prey galleries' entrance and ventilation holes (Fitzgerald and Nagel, 1972); the young larvae then enter the galleries. The larvae of *Medetera bistriata* Parent appear to paralyze their prey larvae with venom injected through their tentorial rods (Aukema et al., 2004b). Egg-larval endoparasitoid

Hymenoptera (e.g., Eulophidae: *Entedon* spp.) oviposit in the eggs of the hosts and their larvae develop in the host larvae. They arrive early enough on the trees to enter the galleries of their hosts and parasitize their eggs, with *Entedon ergias* (Ratzeburg) attacking *Scolytus scolytus* F. (Beaver, 1966a). Finely tuned timing is also important for endoparasitic wasps attacking adult bark beetles (e.g., the pteromalids *Tomicobia* spp. and *Mesopolobus* spp.; the braconids *Cosmophorus* spp. and *Cryptoxilos* spp.), which land on attacked trees at the same time as their hosts (Faccoli, 2000).

A large group of natural enemies, such as the hymenopteran ectoparasitoids of bark beetle larvae (Braconidae, Pteromalidae, some Ichneumonidae) land on attacked trees after bark beetle aggregation has ceased (Stephen and Dahlsten, 1976), when at least some host larvae have already reached some degree of maturity. They either enter the galleries to paralyze their hosts directly and oviposit on their bodies ("cryptoparasitoids," e.g., *Roptrocerus xylophagorum* (Ratzeburg) (Samson, 1984)) or locate hosts through the bark, drilling with their ovipositor to paralyze the host and oviposit (e.g., *Coeloides* spp.; Ryan and Rudinsky, 1962; Hougardy and Grégoire, 2003). Among the monotomid beetles, *Rhizophagus grandis* Gyll. colonizes prey broods at any stage from eggs to pre-emergent adults (Grégoire, 1988; Grégoire et al., 1992) and *Rhizophagus depressus* F. feeds mainly on the eggs of *T. piniperda* (Hérard and Mercadier, 1996), suggesting it arrives early in the tree colonization process.

Vertebrate predators show responses that are more diffuse. In a 15-year study in British Columbia, population densities of six woodpecker species increased in response to *D. ponderosae* epidemics, even though individual fecundity was not affected (Edworthy et al., 2011). Another study in South Dakota (Bonnot et al., 2009) focused on the black-backed woodpecker (*Picoides arcticus* (Swainson)), and showed that within 250 m of nests, nest location was best explained by densities of current *D. ponderosae*-infested trees. For those bark beetles that overwinter at least partly in the forest litter, insectivore mammals and rodents have probably some impact but, to our knowledge, this has never been measured.

5.1.2 Monoterpene Toxicity

In conifers, freshly attacked trees retain at least partly their own chemical and physical defenses, particularly when they survive attacks by parasitic bark beetles. After tree death, however, much larger communities are able to settle with the bark beetles with little or no exposure to toxic monoterpenes. *Rhizophagus grandis*, a specific predator of the parasitic bark beetle *D. micans*, has developed relatively high tolerance to monoterpene toxicity, which allows it to follow its prey in living, still fully defended, host trees. Tolerance to monoterpenes also provides an almost exclusive niche to this predator, as potential competitors, such as *Rhizophagus dispar* (Paykull), do not possess the same level of resistance (Everaerts *et al.*, 1988).

5.1.3 Limited Resources for Associates of Bark Beetles in a Confined Environment

Many of the bark beetle-associated organisms (e.g., insect larvae, mites, nematodes) cannot leave the trees on their own, and must therefore optimize use of available food. For example, predators can adjust oviposition to the available resources. *Rhizophagus grandis* regulates its egg production according to both the presence of conspecific females (Baisier and Grégoire, 1988) and the larval density of *D. micans* as perceived through oviposition stimuli present in the frass (Baisier *et al.*, 1988; Grégoire *et al.*, 1991). The density of *Thanasimus formicarius* L. larvae in bolts infested with *I. typographus* seemed to stabilize, by either egg-laying regulation or cannibalism, whether four or eight pairs of predators had been enclosed with the bolts (Weslien and Regnander, 1992).

Natural enemies may also develop various opportunistic strategies, or strategies for reducing intra- or interspecific competition to compensate for prey or host scarcity. They can attack a flexible range of host developmental stages, as does *Cephalonomia stephanoderis* Betrem parasitizing *H. hampei* (Lauzière *et al.*, 2000), or they can turn to alternative prey. The mite *Pyemotes parviscolyti* Cross and Moser is phoretic and predaceous on *Pityophthorus bisulcatus* Eichhoff, but preys on other scolytine larvae when their galleries cross those of *P. bisulcatus* (Moser *et al.*, 1971). Occasional fungivory (Hackwell, 1973; Hérard and Mercadier, 1996; Merlin *et al.*, 1986), facultative hyperparasitism, and intraguild predation are sometimes compensatory solutions to local host scarcity. For example, the primary parasitoid of *I. typographus*, *Dinotiscus eupterus* (Walk.), has been observed hyperparasitizing the other primary parasitoid *Dendrosoter middendorffii* Ratzeburg (Sachtleben, 1952); *T. formicarius* larvae were reported feeding on *Medetera* larvae (Nuorteva, 1959); xylophagous larvae of the longhorn *Monochamus* spp. prey on bark beetles (Dodds *et al.*, 2001; Schoeller *et al.*, 2012); and larvae of *Temnochila chlorodia* (Mannerheim) attack larvae of *Enoclerus lecontei* (Wolcott) (Boone *et al.*, 2008a).

Facultative cleptoparasitism may be a response to interspecific competition. Mills (1991) reported female *Cheiropachus quadrum* (F.) and *Eurytoma morio* Boheman (primary parasitoids of various bark beetles) stealing *Leperisinus varius* (F.) larvae from *Coeloides filiformis* Ratzeburg; Hougardy and Grégoire (2003) observed a similar behavior in *Rhopalicus tutela* (Walker) displacing *Coeloides bostrichorum* Giraud after the latter located

I. typographus larvae through the bark. Finally, when prey density is low relative to the predator population, contest competition in the form of cannibalism is regularly observed, for example among *Medetera* sp. larvae in galleries of *S. scolytus* (Beaver, 1966b), *R. grandis* larvae in brood systems of *D. micans* (Baisier *et al.*, 1984), larvae of *T. formicarius* in galleries of *I. typographus* (Hui and Bakke, 1997), and *T. dubius* in galleries of *I. pini* (Aukema and Raffa, 2002).

5.1.4 Shifting Prey: an Adaptation to Long Life Cycles or to Fluctuating Prey?

Many natural enemies have life cycles shorter than, or adjusted to, that of their prey or host. Some species, however, live longer than the bark beetles they exploit, a feature that could generate a shortage of resources. *Thanasimus formicarius* has a 2-year generation time (Schroeder, 1999b) and has a long flight period of more than 4 months, which begins at the same time as the flights of the first bark beetles in the season, i.e., *T. piniperda*, *T. lineatum*, and *Hylurgops palliatus* Gyll. (Gauss, 1954). Likewise, *T. dubius* can develop over 2 years (Reeve, 2000). Attacking many different prey may benefit predators that are partially asynchronous with prey. *Thanasimus formicarius* is recorded to attack at least 27 different prey species (Gauss, 1954; Mills, 1983; Tømmerås, 1988), with overlapping phenologies during a season. *Thanasimus dubius* is also described as a generalist (Costa and Reeve, 2011). One of its major prey, *D. frontalis*, has three to nine overlapping generations per year in the southern portion of its range (Wagner *et al.*, 1984), but its major prey in northern regions, *I. pini*, *I. grandicollis*, and *D. rufipennis*, are univoltine. Costa and Reeve (2011) also show that a predator could be conditioned by a previous exposure to respond preferentially to a particular prey.

5.1.5 Habitat Characteristics and Natural Enemies

At the tree level, height and orientation on the trees, as well as bark thickness, are important factors influencing the performances of larval parasitoids that oviposit through the bark (Dahlsten, 1982). Goyer and Finger (1980) found that all parasitoids of *D. frontalis* were negatively influenced by bark thickness, except for *Roptrocerus eccoptogastri* Ratzeburg, which enters the galleries. However, Gargiullo and Berisford (1981) found that *Roptrocerus xylophagorum* Ratzeburg was influenced by bark thickness. Understanding such relationships is confounded by underlying relationships of host beetle density with bark thickness. Studying the natural enemies of *Scolytus multistriatus* (Marsham), *S. scolytus*, and *Scolytus pygmaeus* (F.), Merlin (1984) found general trends very similar to those of Goyer and Finger's (1980), with bark thickness influencing all parasitoids except the cryptoparasitoid *Entedon leuco-gramma* (Ratzeburg). Bark thickness is also a limiting factor for *T. formicarius*. When the outer bark is too thin for the last instar larvae to create a pupal niche, they exit the tree (Warzée *et al.*, 2006). The relatively thicker outer bark of pine may explain why *T. formicarius* often has higher reproductive success on pine than spruce.

At the landscape scale, natural enemy performances are linked to several factors, such as stand composition and management history, and insect dispersal. Schroeder (1999a) found that *R. depressus* populations were higher in stands with high *T. piniperda* populations due to stumps and slash left after a thinning operation, than in unthinned stands, but *T. formicarius* showed little difference, suggesting that these predators had not moved preferentially into stands with high bark beetle densities. Schroeder (2007) confirmed the low mobility of natural enemies (including *T. formicarius*, *Medetera* spp., and parasitoids) between stands with high (left unmanaged since the 1995 storm) and low (windthrows removed) *I. typographus* densities, finding higher impact of the natural enemies in the managed stands two summers after the storm. Similarly, Ryall and Fahrig (2005) showed that ratios of predators (*T. dubius*, *Enoclerus nigripes* Say, and *Platysoma* sp.) to prey (*I. pini*) were significantly lower in isolated stands of *P. resinosa* than in contiguous forests, suggesting that the predators are less likely to exit habitat patches. This was further confirmed by Costa *et al.* (2013), who found that although *T. dubius* had a dispersal capacity 12 times higher than *I. grandicollis* (median: 1.54 km), it was less likely to disperse across fragmented landscapes. Their estimated dispersal distances strongly paralleled those of Cronin *et al.* (2000), who observed a median dispersal of 1.25 km in *T. dubius*, with 5% dispersing further than 5 km, and lower dispersal by the prey *D. frontalis*, with 95% of the predators flying as far as 5.1 km, and 95% of the prey reaching a maximum of 2.3 km.

Forest composition can significantly influence the abundance of polyphagous predators. Warzée *et al.* (2006) found that ratios of *T. formicarius* to *I. typographus* were higher in mixed spruce-pine stands than in pure spruce stands, presumably because the predators were more successful in pupating in pine. Abundant prey in the vicinity may also arrest dispersal of natural enemies. One year after releases of *R. grandis* for biological control of *D. micans* in France and England, this predator was recovered up to 200 m from the release sites (Grégoire *et al.*, 1985; Fielding *et al.*, 1991), although field observations suggest dispersal capacities up to at least 4 km (Fielding *et al.*, 1991). An additional but poorly understood aspect linked to stand composition is the need for synovigenic adult parasitoids to feed in order to reconstitute their egg load. However, plants producing pollen and nectar are frequent, including in even-aged, monospecific forest stands, and aphids in the tree crowns produce fair amounts of honeydew (Hougardy and Grégoire, 2000; VanLaerhoven and Stephen, 2008).

5.2 Relative Importance of Natural Enemies to Bark Beetle Ecology

5.2.1 Impact of Natural Enemies on Bark Beetles

The impacts of natural enemies on bark beetles have been measured through a variety of approaches, including laboratory assays, field sampling, and modeling. In the laboratory, direct observations (Aukema and Raffa, 2004b) and experiments (Barson, 1977; Senger and Roitberg, 1992; Schroeder, 1996; Reeve, 1997; Aukema and Raffa, 2002) typically focus on singular cases under controlled conditions, with both the advantages and disadvantages of omitting the more complex influences operating at the landscape level. Short-term field observations (Mills, 1985; Schroeder, 1996, 1999a; Erbilgin *et al.*, 2002; Wermelinger *et al.*, 2013) and experiments (Weslien and Regnander, 1992; Schroeder, 2007) provide further information on how different local conditions mediate the impact of natural enemies. However, a more complete picture may appear at a larger scale (Raffa *et al.*, 2008; Kausrud *et al.*, 2011a) and in this respect, recent modeling approaches shed a particularly interesting light on global relationships. In particular, they can suggest how different global bottom-up or top-down influences characterize different bark beetle systems, and delineate how and when natural enemies exert significant influences on the dynamics of these systems. Marini *et al.* (2013) analyzed demographic time series data of *I. typographus* and *T. formicarius* in Sweden from 1995 to 2011, and showed that the provision of breeding material by storms was the principal trigger of outbreaks, with intraspecific competition as a density-dependent negative feedback. There was no clear influence of *T. formicarius* on the bark beetles' demography. In contrast, Turchin *et al.* (1999) provided a time series analysis of fluctuations in *D. frontalis*, which suggested that a delayed density-dependent factor dominates beetle dynamics. With a long-term predator-exclusion experiment, they explored the hypothesis that *T. dubius* could act as such a delayed density-dependent factor, and detected a delayed impact (possibly due to the often longer life cycles of predators than bark beetles), suggesting a significant role of *T. dubius* in the population dynamics of *D. frontalis*. The different results between these studies are striking, and may highlight two systems with quite different drivers. That is, *I. typographus* is strongly driven by a bottom-up force, i.e., the availability of suitable hosts (windfelled trees) when at an endemic level. In contrast, it is less clear whether bottom-up forces other than lightning-struck trees exert significant influences on

D. frontalis (Hodges and Pickard, 1971; Coulson *et al.*, 1983). However, Friedenberg *et al.* (2008) and Martinson *et al.* (2013) have questioned Turchin *et al.*'s (1999) model. Another large-scale study is provided by Aukema *et al.* (2005b), who modeled populations of *T. dubius*, *Platysoma cylindrica* (Paykull), and *I. pini* in *P. resinosa* plantations during 2 years, and found evidence that predation exerts some density-dependent feedback.

5.2.2 Bark Beetle Behavior and Impact Mitigation of Natural Enemies

Several behavioral aspects of bark beetles may reduce the impact of natural enemies. Prolonged male residence in the galleries of *I. pini* can partly protect the eggs from predation (Reid and Roitberg, 1994). Increased *I. pini* densities can reduce the proportional impact of predation by *T. dubius* and *P. cylindrica*, suggesting that aggregation dilutes predation (Aukema and Raffa, 2004a). These findings indicate that predator dilution may be a viable benefit to aggregation. Additionally, *T. dubius* may attack disproportionately more responding males pioneer than responding males, and more males than females, suggesting that predators may stabilize bark beetle communication systems by selecting against cheating (only responding to pheromones rather than engaging in host searching) (Aukema and Raffa, 2004c).

5.2.3 Applications

Natural enemies have been used with limited or sometimes no success in a number of classical, augmentative or conservative biological control programs (reviewed in Kenis *et al.*, 2004). European natural enemies were introduced in North America against *S. multistriatus*, vector of Dutch elm disease, in New Zealand against *Hylastes ater* Payk., and in South Africa against *Orthotomicus erosus* (Wollaston). The only example of an entirely successful classical biological control program involves the mass production and release of *R. grandis* against *D. micans* in the Caucasus mountains of Georgia (Kobakhidze, 1965), France (Grégoire *et al.*, 1985), Great Britain (Fielding *et al.*, 1991), and Turkey (Alkan and Aksu, 1990)

5.3 Competitors

Many organisms compete with bark beetles for resources. Cerambycids such as *Monochamus* spp. also act as intraguild predators (Dodds *et al.*, 2001; Schoeller *et al.*, 2012), and exploit bark beetle semiochemicals as kairomones (Allison *et al.*, 2001). Sometimes, interspecific competitors can exert multiple effects. For example, *Ips* spp. are attracted to trees attacked by *Dendroctonus* and can both compete for resources and benefit predators that consume *Dendroctonus* (Boone *et al.*, 2008a; Martinson *et al.*, 2013).

5.4 Tritrophic Signaling

Natural enemies that arrive early in the colonization process of a newly attacked tree, including many predators, egg-larval parasitoids, and adult parasitoids, exploit bark beetle aggregation pheromones. This can exert a strong pressure on the bark beetles, which may sometimes modify their own communication system to obtain partial escape. For example, *I. pini* prefers stereospecific ratios of ipsdienol that differ from stereospecific preferences of local predators attracted to ipsdienol (Raffa and Dahlsten, 1995). Furthermore, *I. pini* produces and responds to lanierone in the Great Lakes region, to which the predators are non-responsive, even though predators in California, where *I. pini* does not produce lanierone, respond to this compound (Raffa *et al.*, 2007). These patterns suggest highly dynamic interactions.

Predators attacking multiple bark beetle prey have antennal receptor cells keyed to many pheromones produced by different prey. For example, *T. formicarius* has sensillae keyed to 22 bark beetle pheromone and conifer volatiles: (+) and (−)-ipsdienol; *(S)* and *(R)-cis-verbenol*; 2-methyl-3-buten-2-ol; (−) and (+)-ipsenol; (−) and (+)-verbenone; *(−)* and *(+)-trans-verbenol*; amitinol; exo- and endo-brevicomin; frontalin; (+)-lineatin; phenylethanol; (−) and (+)-α-pinene; myrcene; camphor; and pino-camphone (Tømmerås, 1985). Although receptive to many signals, some predators can learn to respond to one particular signal after exposure (Costa and Reeve, 2011). Early arrivers are also sensitive to signals indicating that mass attack has reached its end. *Thanasimus undatulus* (Say), *Enoclerus sphegeus* F., *E. lecontei*, and *Lasconotus* sp. are repelled by verbenone (Lindgren and Miller, 2002b). Late arrivers (larval ectoparasitoids) respond to odors produced by microbial symbionts (Sullivan and Berisford, 2004; Boone *et al.*, 2008b).

6. POPULATION DYNAMICS

6.1 Diversity in Bark Beetle Population Dynamics

Like most insects, bark beetles have high reproductive potentials that provide the capability to undergo rapid, exponential population increase (Coulson, 1979; Økland and Bjørnstad, 2006; Marini *et al.*, 2013). However, also like most insects, their realized rates of reproductive increase are usually far below that potential. Despite the enormous diversity in bark beetle population dynamics among species, three overlapping groups can be distinguished (Raffa *et al.*, 1993; Lindgren and Raffa, 2013). In the first group, most species exhibit relatively stable population dynamics, with local densities rising and falling with resource availability, temperature, and other features of environmental quality. The range of these numerical fluctuations can be orders of magnitude, but the populations do not become self-driving. A second smaller group can exert some positive feedback, such that once populations have risen in response to a resource pulse or more favorable environment, positive density dependence can contribute to numerical increases while conditions remain highly favorable, Finally, a third and much smaller group undergoes dramatic shifts in its relationships with host plants after a critical stand-level population threshold has been surpassed. Once this threshold, below which population growth is constrained by host defenses, is surpassed, populations become "eruptive" and enter a new reactive norm. Populations only return to endemic dynamics after resource depletions, intolerable temperatures, or some combination thereof, reduces beetle numbers below the critical density threshold.

The first group is highly diverse, and includes species that feed on dead plants, dead parts of live plants, reproductive organs, roots, and lateral branches, among others. They also include insects that colonize the main stem, both gregarious species that are always associated with severely stressed plants, and solitary/parasitic species that colonize live but usually less vigorous hosts, most commonly on the basal stem. Solitary species can play important roles in maintaining populations of semieruptive and eruptive species (groups 2 and 3) during their endemic phases (Aukema *et al.*, 2010; Smith *et al.*, 2011). The second group includes gregarious species that colonize the main stems of both gymnosperms and angiosperms, and can kill stressed trees. These insects are broadly distributed worldwide, but overall, they show less diversity than the first group. This group includes a number of species that can be economically important pests when habitats are managed in fashions that stress or concentrate host trees. The third group, eruptive species, exerts the strongest effects on ecosystem processes. These insects can be considered true "ecosystem engineers" in that they exert major effects on forest structure, biodiversity, successional pathways, nutrient cycling, and geophysical processes (Romme *et al.*, 1986; Kurz *et al.*, 2008; Griffin *et al.*, 2011; Kaiser *et al.*, 2012). The widespread tree mortality over large spatial scales caused by eruptive bark beetles also exerts major feedbacks to other bark beetle species (i.e., groups 1 and 2), by providing large resource pulses that facilitate their reproduction (Flamm *et al.*, 1989). Eruptive bark beetles show the lowest diversity, being gregarious colonizers of the main stems of conifers, restricted to the northern hemisphere, and mostly concentrated within North America. An important consideration in evaluating their population dynamics is that both positive and negative sources of feedback are always present, and thus net feedbacks are crucial.

6.2 Factors Affecting Survival, Development, and Reproduction: Sources of Positive and Negative Feedback behind Bimodality

Because the majority of research on bark beetles has been conducted during the eruptive phase of species capable of undergoing spatially synchronized outbreaks, it is easy to visualize a forest as one big salad bar. In fact, nothing could be further from the truth. Individual trees within a species show enormous variation in their resistance levels, even within a single age category and local population (Safranyik *et al.*, 1975; Ruel *et al.*, 1998; Rosner and Hanrup, 2004). Heterogeneity in tree defensive capacity arises from genetic, environmental, gene by environment, phenological, and ontogenetic contributions (Safranyik *et al.*, 1975; Sturgeon and Mitton, 1986; Raffa *et al.*, 2005; Roberds and Strom, 2006; Ott *et al.*, 2011). The importance of this diversity becomes obscured during outbreaks (Boone *et al.*, 2011), so models that are heavily informed by those relatively rare events must pool host type into a relatively homogeneous construct. Thus, generalized treatments of host availability and suitability have limited utility for understanding the more persistent condition of endemic population dynamics, or for understanding mechanisms by which populations transition from endemic to eruptive dynamics (Raffa *et al.*, 2008; Björklund *et al.*, 2009; Bleiker *et al.*, 2014).

For purposes of analysis, it is common to compartmentalize the different factors affecting an insect's replacement rate. But in the case of bark beetles, some of the most important drivers, such as food availability, plant defense, intraspecific competition, and interspecific competition, are so tightly interwoven that it is more useful to emphasize their interactions and linkages (Lindgren and Raffa, 2013; Marini *et al.*, 2013). A conceptual illustration of how these factors interact at the tree level is presented in Figure 1.4.

Beetle populations are most commonly at low, endemic population densities (Figure 1.4 top). Trees that are highly defended pose a high risk of attack failure to host-seeking beetles (solid line), due to the multifaceted defense mechanisms described above (Section 3.2). Trees with low defense level pose little risk, so the likelihood of successful colonization is high. In trees that have already died from some other cause, host defenses become nearly zero. Such undefended trees, however, are also available to a wide diversity of other phloeophagous and xylophagous species, including other Scolytinae (Figure 1.1B), woodborers, and microorganisms (Stephen and Dahlsten, 1976; Safranyik *et al.*, 1996; Wermelinger *et al.*, 2002; Saint-Germain *et al.*, 2007). Saprophytic Scolytinae are typically better competitors than tree-killing species (Poland and Borden, 1998a, b; Smith *et al.*, 2011, Lindgren and Raffa, 2013)

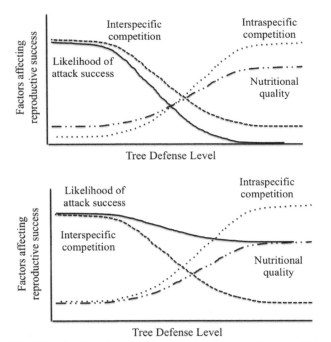

FIGURE 1.4 **Opposing effects of tree defense level on multiple selective pressures on bark beetles.** Top: Low population densities, which typify most generations of a species within an area. Trees with high defense pose a high risk of attack failure. However, trees with compromised defenses are available to other subcortical species that are superior competitors to tree-killing species. Stressed trees often have thinner phloem, and so are nutritionally inferior. When beetles colonize well-defended trees, they not only incur greater risk, but they experience greater intraspecific competition because the requisite mass attacks create crowding within a limited resource. Bottom: High population densities. The same general shapes of these relationships remain. However, the likelihood that attacks on well-defended trees will succeed is high. The within-tree intraspecific competition curve rises somewhat, but only partially because the beetles' antiaggregation system limits the number of attackers. The optimal choice for a beetle depends on stand-level population density.

and some of the cerambycid woodborers are both competitors and predators of bark beetles (Dodds *et al.*, 2001), while many of the saprogenic microorganisms are antagonistic to bark beetles (Paine *et al.*, 1997; Six and Klepzig, 2004; Cardoza *et al.*, 2006a). Consequently, interspecific competition tends to be very high in poorly defended trees, and comparatively lower when beetles attack well-defended trees (Figure 1.4 top, dashed line). Interspecific competition never totally disappears, because saprophages also exploit previously healthy hosts that are overcome by tree-killing bark beetles. In addition to harboring more interspecific competitors, highly stressed trees also tend to be less nutritionally suitable for the beetles' developing brood (Figure 1.4 top, dashed and dotted line). Trees experiencing drought, defoliation, age-related senescence, etc., often produce thinner phloem, i.e., the larval food base (Boone *et al.*, 2011; Powell *et al.*, 2012; Creeden *et al.*, 2014). When beetles attempt to attack well-defended trees,

on the other hand, they not only incur higher risk, but their mechanism of overcoming resistance, mass attack, incurs higher intraspecific competition (Figure 1.4 top, dotted line). A tree's phloem is a finite resource, so each additional attacker depletes the resource available for brood development (Coulson, 1979; Raffa and Berryman, 1983; Anderbrandt, 1990; Robins and Reid, 1997). Beetles can partially limit this cost by producing anti-aggregation pheromones and ceasing production of aggregation pheromones, once tree defenses are exhausted (D.L. Wood, 1982; Borden, 1985; Keeling *et al.*, 2013).

When beetle populations are high (Figure 1.4 bottom), the same qualitative relationships hold, but the coefficients change. Most importantly, the likelihood of successful attack becomes much higher, and much less sensitive to host defense. Because enough beetles are available to elicit (Erbilgin *et al.*, 2006) and conduct (Raffa and Berryman, 1983) mass attacks, colonization attempts are likely to succeed regardless of a tree's defensive capacity. The intraspecific competition curve can rise somewhat with increases in tree defense level, but this is again constrained by the beetle's sophisticated anti-aggregation pheromone system that nearly limits the number of attackers to that required to overcome defense. At some point, a tree can be so well defended that the number of beetles required to kill it is larger than the number of brood beetles that can develop in it. Beetles usually avoid such trees, but during the peak of intensive outbreaks and when these start to collapse, this relationship becomes apparent when beetles kill younger, smaller trees (Lindgren and Björklund, unpubl.).

The trade-offs between colonizing weak versus vigorous trees also have substantial higher-scale inputs. Trees undergoing acute stress due to lightning, root disease, and windthrow, for example, are relatively sparse in number across a landscape, yet are generally concentrated at highly localized spatial and temporal scales (Atkins, 1966; Coulson *et al.*, 1983; Smith *et al.*, 2011), which concentrates competition (Marini *et al.*, 2013). Further, tree-killing bark beetles can only utilize a host for one generation. Therefore, exploitable food resources are removed at each successful colonization event, which potentially leaves only healthier trees as an available, but not usually accessible, resource.

The importance of tree defense in bark beetle reproduction is somewhat of an enigma, in that its signature is mostly indirect. Although trees sometimes kill bark beetles directly (Figure 1.3), life tables usually show low within-tree mortality attributable to host resistance (Berryman, 1973; Coulson, 1979; Amman, 1984). This is somewhat expected, of course, because herbivores evolve sophisticated sensory apparati and neuromuscular sequences for avoiding plants that would kill them and their brood. Additionally, pheromone-mediated mass attacks are not analogous to human-wave assaults in which the first lines are slaughtered: such behavior would be highly maladaptive at the level of individual selection, and would require levels of kin selection that are highly unlikely to operate in the field (Raffa, 2001). When high populations succeed, the early arrivers experience only relatively minor costs (Raffa and Berryman, 1983; Pureswaran *et al.*, 2006; Latty and Reid, 2010). A better estimate of tree defense should incorporate the proportion of adults that emerge from brood trees, but do not encounter trees that elicit their entry behavior, before dying of other causes. When beetles are caged onto randomly selected trees within their host species and age range, for example, a sizable proportion will remain in their cage and die rather than enter (Raffa, 1988). Indeed, losses during host searching are high among bark beetles (Berryman, 1973, 1979; Pope *et al.*, 1980; Safranyik *et al.*, 2010). For example, Pope *et al.* (1980) estimated that even in the artificially homogeneous habitat structure of pine plantations, and even during outbreak conditions, 57% of emerging *D. frontalis* adults did not subsequently enter a new host, i.e., could not be accounted for by either new galleries or failed attacks. The proximate cause of death during host searching is not tree defense *per se* though, but rather is mostly energy depletion and predation (Rudinsky, 1962; Berryman, 1979; Gilbert and Grégoire, 2003).

Temperature is a major constraint and releaser on beetle populations (Régnière and Bentz, 2007; Trần *et al.*, 2007; Jönsson *et al.*, 2007; Powell and Bentz, 2009). Low winter temperatures can cause high mortality, and temperature is a major driver of beetle development rates (Bentz *et al.*, 1991). In multivoltine species such as *D. frontalis*, temperature strongly influences the number of generations per year, and in species such as *D. rufipennis* and *D. ponderosae*, temperature regimes determine whether local populations are univoltine or semivoltine. Temperature-driven survival and development rates translate directly into how closely bark beetles can approach their reproductive potential. Interactions between bark beetles and temperature are highly complex, and include multiple developmental and survival thresholds, often facultative diapause, and variable patterns of cold hardening (Bentz and Mullins, 1999; Lombardero *et al.*, 2000b; Hansen *et al.*, 2001; Hansen and Bentz, 2003; Koštál *et al.*, 2011; Inward *et al.*, 2012.). These multiple reaction norms overlay regionally genotypic variation (Mock *et al.*, 2007; Bentz *et al.*, 2011). The result of this complexity is that beetle responses are highly plastic, and a high diversity of outcomes can arise from variable inputs. A key feature of this plasticity is the linkage between bark beetle development and the need to overcome tree defense. Some bark beetle species exhibit relatively synchronous emergence despite a broad range of initial and developmental conditions, a relationship termed "adaptive seasonality" (Bentz *et al.*, 1991; Logan and Bentz, 1999). In addition to affecting beetle development rates, temperature can influence the relative abundance of symbionts such as mites and fungi (Hofstetter *et al.*, 2007; Addison *et al.*, 2013), which feed back to beetle reproduction.

The importance of predators as mortality and potential regulating agents appears to vary among bark beetles. Predacious beetles and flies that feed on multiple life stages can be particularly important sources of mortality. These predators often exploit bark beetle aggregation pheromones, again creating linkages between bark beetles' need to overcome tree defenses and other sources of mortality. Perhaps the strongest case for predatory regulating tree-killing bark beetle populations has been made for *D. frontalis* (Turchin *et al.*, 1999), but subsequent analyses have not supported such a role (Friedenberg *et al.*, 2008; Martinson *et al.*, 2013). Birds, especially woodpeckers, are likewise important and ubiquitous mortality agents (Fayt *et al.*, 2005). However, their roles are particularly difficult to quantify. A diversity of parasitoid species attack all stages of bark beetles (Linit and Stephen, 1983; Mills, 1991), and show sophisticated host location mechanisms, including responding to pheromones emitted by adults (Kudon and Berisford, 1981; Raffa *et al.*, 2007) and microbes associated with larvae (Sullivan and Berisford, 2004; Boone *et al.*, 2008b). Parasitoids can occasionally exert high mortality, but in general subcortical herbivores experience less parasitism than other insect guilds (Connor and Taverner, 1997). This presumably arises from the protection provided by the bark, and the energy and risk required to access hosts. Parasitism rates may be even further lowered in highly managed systems where nectar sources are reduced (Stephen and Browne, 2000). There is some evidence of density-dependent parasitism of some bark beetle species (Amman, 1984). However, to our knowledge, there are no bark beetles for which parasitoids have proven to be major population regulating agents.

Antagonistic microorganisms, including both pathogens and competitors, can likewise impose significant constraints on bark beetle reproduction. Gregarines and microsporidia can be among the most common pathogens, and can cause either mortality or sublethal effects such as reduced fecundity or dispersal ability. However, there is little evidence that they naturally exert enough mortality to be important regulating agents (Wegensteiner *et al.*, 2010). The fungus *O. minus* can be highly detrimental to several species of bark beetles, and substantially reduce brood survival. The mechanisms are not entirely understood, but appear to include competition for saccharides (Wang *et al.*, 2013) and reduction of immunocompetence (Shi *et al.*, 2012). Opportunistic fungi such as *Trichoderma* and *Aspergillus* can also reduce brood production (Fox *et al.*, 1992; Cardoza *et al.*, 2006b).

6.3   Transitions from Endemic to Eruptive Dynamics

The major consequence of the above positive and negative feedbacks is that some bark beetle species exhibit bimodal population dynamics. That is, their within-tree and within-stand replacement rates show strong relationships to stand-level population density, but these relationships vary both quantitatively and qualitatively between different population phases (Figure 1.5). Within the endemic phase, replacement rates can be represented using standard density-dependence curves: populations increase until they reach a stable, or endemic, equilibrium (EnEq), and when they exceed that density, negative feedbacks, such as depletion of the stressed-tree pool, prevail and the population declines. Over time, populations fluctuate with increasing or decreasing habitat favorability. In general, the beetles' within-tree reproductive gains are offset by within-tree and within-stand losses. In bimodal systems, however, if a population somehow reaches a critical eruptive threshold density (ErT), it then enters a new regime in which positive density-dependent feedback again prevails, and above that density, the population increases exponentially. Bimodal dynamics have been observed in diverse types of organisms, including locusts, Lepidoptera, sawflies, and fish among others, but the underlying bottom-up, lateral, and top-down mechanisms vary (Ricker, 1954; Southwood and Comins, 1976; Campbell and Sloan, 1977; Simpson *et al.*, 1999; Larsson *et al.*, 2000; Despland and Simpson, 2000, 2005; Dussutour *et al.*, 2008).

For a bimodal model to be both credible and useful to our understanding, it must satisfy two conditions. First, there must be some mechanism by which a population can increase from EnEq to ErT. The paradox is that, by definition, a population higher than EnEq will decline. Second, there must be a validated mechanism that drives continuous positive feedback above ErT. In the case of bark beetles, there is a substantial body of research informing both questions: (1) Bark beetle populations can rise quickly due to increased winter or summer temperatures, which improve overwintering survival and reduce development time (Bentz *et al.*, 1991; Safranyik and Carroll, 2006; Aukema *et al.*, 2008; Powell and Bentz, 2009; Preisler *et al.*, 2012; Régnière *et al.*, 2012), area-wide stresses that increase resource availability and within-stand replacement rates (Hicke *et al.*, 2006; Breshears *et al.*, 2009; McDowell *et al.*, 2011; Creeden *et al.*, 2014; Hart *et al.*, 2014), immigration that directly increases populations (Jackson *et al.*, 2008; Aukema *et al.*, 2008; Samarasekera *et al.*, 2012; Simard *et al.*, 2012). (2) Bark beetles have plastic host selection behaviors, which coupled with their ability to coordinate mass attacks, functionally expand their food supply in response to increasing beetle population density (Rudinsky, 1962; Berryman, 1981; Wallin and

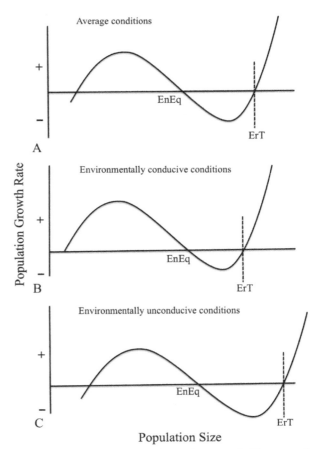

FIGURE 1.5   Conceptual model illustrating onset of self-perpetuating landscape-scale outbreaks based on underlying feedback structure. A high intrinsic rate of increase, coupled with negative density-dependence feedback, generates a classic parabolic relationship between population growth rate and population size. Past the endemic equilibrium (EnEq), any population increase results in population decline. However, the cooperative host procurement behavior of bark beetles generates a second zone of net positive feedback. If a population rises to the eruptive threshold (ErT), the beetle's relationship with its host changes, and defense is no longer a significant constraint. Under most conditions, populations do not bridge this gap. However, increased temperatures that reduce mortality and shorten development time, immigration, or a sudden widespread pulse of stressed trees such as during drought can raise the level to ErT. Once populations breach this threshold, a return of external drivers to their initial condition may not halt an outbreak. (A) Average stand conditions; (B) conditions favorable to bark beetles, such as high temperature or homogeneous mature stands narrow this gap, making transitions more likely; (C) suboptimal environmental conditions, such as cold temperature or stand heterogeneity, make transitions unlikely. For most bark beetle species, the gap from EnEq to ErT is essentially infinite, because they show relatively little density-dependent plasticity in host selection behavior. *Redrawn from Raffa and Berryman (1986).*

Raffa, 2004; Kausrud *et al.*, 2011a). The selection pressures on choices made by individual beetles change between low- and high-density conditions. Moreover, the initial conditions required to increase the population from EnEq to ErT are not necessarily required to maintain the population above ErT once that threshold has been breached (Raffa *et al.*, 2008; Creeden *et al.*, 2014)

Despite the potential of several single drivers to raise populations from EnEq to ErT, a combination of factors is more commonly required (Raffa *et al.*, in press). There are also many cases where outbreaks did not develop even when one or more of the above drivers were pronounced. The conditions most conducive to the development of outbreaks are a combination of: (1) amenable forest structure, specifically large, contiguous, relatively homogeneous, mature stands species; (2) high temperatures, which both benefit beetles directly and add to evapotranspiration stresses on trees; (3) region-wide stresses that reduce host defenses and cause a large number of trees to become abruptly and simultaneously accessible; and (4) reduced numbers of natural enemies, and an abundance of beneficial symbionts. An important feature of these constraints and drivers is how they interact. For example, a severe stress on trees can both make hosts available to an "eruptive" species, and also render them available to "secondary beetles" that can outcompete (conditions 3 and 4) the tree killers. Likewise, the rate at which populations increase in response to a temperature elevation (condition 2) or an environmental stress (condition 3) will be steeper in a more homogeneously than heterogeneously structured forest, and with larger than smaller trees. Similarly, the combination of symbiotic fungi (condition 4) varies with temperature (condition 2), and the local abundance of natural enemies (condition 4) varies with forest structure (condition 1). Overall, a rather uncommon set of conditions is required to facilitate the development of an outbreak at any particular time and place. However, at any one time, outbreaks are occurring somewhere in areas with suitable forest structure, composition, and weather.

The combination of these four conditions can be conceptualized as an "eruptive window" (Raffa *et al.*, in press), in which each coordinate expands or contracts, such that the area determines the distance from EnEq to ErT in Figure 1.5. The relative importance of these four coordinates in determining that area varies from system to system. The strength of the various underlying constraints that typically constrain bark beetle populations, relative to their population growth potential, can be also seen in the responses of historically non-eruptive species when anthropogenic activities relax or remove their actions. Some of the most dramatic examples have occurred when beetle–fungal complexes were introduced into areas in which native trees had not coevolved, and thus the important natural constraint of host resistance was lacking. These include instances of transport from Eurasia to North America and vice versa. Other examples include habitat manipulations that homogenize species and age composition, thus facilitating host finding, or that fragment patches in ways that reduce tracking by predators. More globally, elevated temperatures caused by climate change have resulted in outbreaks.

Several empirical examples demonstrate the importance of interactions among key drivers. Outbreaks by *D. rufipennis* in coastal Alaska can arise from warm temperatures that convert typically semivoltine to univoltine populations (Werner and Holsten, 1985; Berg *et al.*, 2006). However, in central Alaska, univoltine populations are the norm, yet populations rarely undergo outbreaks. This is probably due in part to drier conditions in central Alaska that yield drier phloem, which favors competitors (Werner *et al.*, 2006). Likewise, windstorms in Europe can release outbreaks by *I. typographus* (Marini *et al.*, 2013), but large windstorms in the midwestern US do not typically release outbreaks by *D. rufipennis*, probably because of the high tree species diversity and associated high predator and competitor abundances there (Gandhi *et al.*, 2009; Raffa *et al.*, in press). Furthermore, the effects of stress on tree defense can be complex, with severe water deficit reducing resistance, but moderate water stress resulting in increased constitutive but decreased induced levels (Lewinsohn *et al.*, 1993; Lombardero *et al.*, 2000a).

Comparisons among systems, locations, and seasons can be further developed within the conceptual model shown in Figure 1.5. When conditions for beetle reproduction are conducive (Figure 1.5A), the distance between the population size thresholds, EnEq and ErT, decreases (Figure 1.5B), and when they are less conducive this distance increases (Figure 1.5C). For example, at latitudes or elevations where temperatures are low, in forests that are sufficiently diverse, or in stands where the trees are too young to support high population densities, this gap becomes insurmountable. Similarly, many bark beetle species do not exhibit a high degree of plasticity in host selection behavior relative to defense physiology, so the distance between EnEq and ErT is essentially infinite. Some examples in which high initial populations increase in response to severe environmental stress but are not followed by sustained positive feedback include *Ips confusus* (LeConte) and *I. pini* following drought (Raffa *et al.*, 2008; Aukema *et al.*, 2010), and *I. grandicollis* following defoliation (Wallin and Raffa, 2001).

The support for bimodality of bark beetle populations arises from five complementary sources. The first is observation. Records by a number of early forest entomologists depict outbreak populations as being not only numerically but also dynamically different from endemic populations (Keen, 1938; Beal, 1939; Evenden *et al.*, 1943; Schwerdfeger, 1955; Zwölfer, 1957; Thalenhorst, 1958; Rudinsky, 1962; Atkins, 1966). Perhaps the most explicit statements of a link between individual- and population-level behavior are those of Keen (1938): "Endemic populations select weaker, less vigorous trees for attack, but no such selection is apparent during epidemic conditions," and Beal (1939): "When the Black Hills beetle is not numerous it breeds in weakened trees or those injured by lightning or in some other way...During outbreaks this insects attacks vigorous healthy trees...and shows a slight preference for the more vigorous, rapidly growing trees." A second line of support comes from theoretical, first-principles arguments (Berryman, 1979, 1981), which adapted similar relationships from other outbreak species (Southwood and Comins, 1976; Campbell and Sloan, 1977), extinction principles (Allee, 1949), and sustainable resource models (Ricker, 1954). Third, simulation models founded on the underlying assumptions of threshold-based bimodality have generated behaviors consistent with field observations (Økland and Bjørnstad, 2006; Kausrud et al., 2011b). Fourth, analyses of stand-level (Mawby et al., 1989; MacQuarrie and Cooke, 2011; Martinson et al., 2013), and within-tree (Berryman, 1974; Robins and Reid, 1997; Raffa, 2001) population replacement curves have demonstrated zones dominated by negative and positive density dependence, in agreement with historical data. Fifth, manipulative experiments testing the underlying mechanisms of positive feedback, the efficacy and individual benefit of cooperative attack, and the adaptive plasticity of host selection behavior have validated these processes (Raffa and Berryman, 1983; Lewis and Lindgren, 2002; Wallin and Raffa, 2004).

Two other features of the dynamic relationships illustrated in Figure 1.5 occur at the extremes. At very low population sizes, the within-tree Allee effect (Raffa and Berryman, 1983) can extend to stand-level extinction (Friedenberg et al., 2007). At the other extreme, populations can collapse due to resource depletion, cold temperatures, or both (not shown).

Threshold-based relationships pose special challenges to research on bark beetles. First, the intervals between EnEq and ErT are transient, highly unstable states, so populations rapidly jump to either condition, and thus are difficult to measure. One might conceptualize them similarly as we envision short-lived free radicals in chemical reactions. Second, the key transitions occur at very low stand-level densities, which are below detection by aerial tree-mortality surveys (Boone et al., 2011). Unfortunately, the available methods for studying such low-density populations are limited, costly, statistically challenging, and professionally risky. Third, key mechanisms that restrict populations at low densities may be unimportant at high densities. In systems characterized by thresholds at multiple levels of scale, there is often no correlation between key drivers and emergent patterns (Raffa et al., 2008), which impedes *post-hoc* analyses.

7. ROLES IN ECOLOGICAL PROCESSES AND SOCIOECONOMIC IMPACTS

7.1 Ecological Processes

Progar et al. (2009) and Müller et al. (2008) describe bark beetle activity as vital at many scales. Bark beetles influence forest regeneration by killing mature trees, thus creating gaps in the forest, which are beneficial to many species of wildlife. They also promote variability in tree sizes and ages, which increases forest and stand resiliency (Harvey et al., 2013). In some cases, bark beetle outbreaks were found to increase forest fire risk (Bigler et al., 2005), while in others they appear to lower fire risk (DeRose and Long, 2009) or have no measurable effect. Each of these processes is scale dependent.

The D. ponderosae outbreak in western North America provides a useful illustration of how profoundly bark beetles can affect forest ecosystem services. Costanza et al. (1997) list 17 ecosystem services provided by forests that could be affected by bark beetle outbreaks: gas regulation, climate regulation, disturbance regulation, water regulation, water supply, erosion control and sediment retention, soil formation, nutrient cycling, waste treatment, pollination, biological control, refugia, food production, raw materials, genetic resources, recreation and cultural services. Embrey et al. (2012) present a comprehensive review of ecosystem services that have been affected in the US and Canada by the D. ponderosae outbreak. Among these, the regulating and supporting services have been damaged, lowering the capacity of affected forests to regulate air and water quality and water flows, with increased water yields in the spring and shortage in the summer, because of a change in the capacity to receive snowmelt. Increased water runoff alters nutrient cycling and increases erosion, and water quality is threatened. The forest has also turned from a carbon sink to a source, at least prior to understory regeneration. Kurz et al. (2008) estimate that "the cumulative impact of the beetle outbreak in the affected region during 2000–2020 will be 270 megatonnes (Mt) carbon (or $36 \, \mathrm{g}$ carbon $\mathrm{m}^{-2} \, \mathrm{yr}^{-1}$ on average over $374{,}000 \, \mathrm{km}^2$ of forest)." In addition, the provisioning and cultural services, the commodities and immaterial services obtained from the forest are also jeopardized by insect damage. Products such as timber, firewood and pulp, and additional services such as cultural, aesthetic, and touristic values are being diminished. In the area of Davos, Switzerland, Bebi et al. (2012) examined the potential impact of natural disturbances such as fire or insect pests, and identified five ecosystem services that could be severely impacted: avalanche protection, recreation, CO_2 sequestration and storage, habitats of an endangered grouse (*Tetrao urogallus* L.), and timber production.

7.2 Socioeconomic Impacts

Socioeconomic impacts beyond lost timber values are difficult to calculate and require further attention. We provide below a brief summary of analyses addressing this topic. Costanza et al. (1997) estimated the value of annual ecosystem services offered by different biomes, including the temperate/boreal forests (Table 1.3). The global yearly

TABLE 1.3 Ecosystem Services Provided by the Temperate/Boreal Forests (Data from Costanza *et al.*, 1997)

Ecosystem Services	Cost 1994 (US$ ha^{-1} yr^{-1})
Climate regulation	88
Water regulation	0
Soil formation	10
Waste treatment	87
Biological control	4
Food production	50
Raw materials	25
Recreation	36
Cultural	2
Total value (ha^{-1} yr^{-1})	**302**

value of these services provided by temperate/boreal forests amounts to US$63.6 billion. These figures were calculated based on the "willingness-to-pay" for each service. Krieger (2001) proposed an analysis of ecosystem services provided by the forests in each of the US regions. He proposed a summary description of the various local indicators used to calculate values, when possible and relevant, for watershed services (water quantity; water quality; soil stabilization; air quality; climate regulation and carbon sequestration; biological diversity); recreation (economic impact; wilderness recreation; hunting and fishing; nontimber products); and cultural values (aesthetic and passive use; endangered species; cultural heritage).

The economic impact of timber loss in British Columbia directly resulted in the loss of 27,000 jobs (Abbott *et al.*, 2008). Price *et al.* (2010) applied hedonic analysis to the assessment of changes in property value after the mountain pine beetle epidemics. They found that property values declined by $648, $43, and $17, respectively, for every tree killed by mountain pine beetle infestations within a 0.1, 0.5, and 1.0 km buffer. Embrey *et al.* (2012) discussed the health impacts of the North American Rocky Mountain pine beetle outbreak. They included multiple factors, such as the direct and indirect effects of forest fire (although the extent to which bark beetle outbreaks predispose forests to fire is a matter of contention), quality losses in water supplies (also with possible long-range impact), consequences of property losses, and unemployment. Rittmaster *et al.* (2006) constructed an empirical air dispersion model to estimate the concentration of fine smoke particulate matter produced by a fire in Chisholm, Alberta, and used benefit transfer methods to estimate health impacts. The economic impacts

were high, second only to timber losses. Similarly, water quality is likely to be affected by insect damage, as deforestation-generated runoffs can translate into increased water turbidity, contamination with heavy metals, etc. Recent explosions in two sawmills in north-central British Columbia, which caused the death of four sawmill workers and severe injuries to many more, as well as the prolonged loss of work for other employees due to the destruction of the mills, was blamed in part on extremely dry sawdust generated from processing beetle-killed timber (Franck, 2012).

The management challenge to addressing bark beetles is essentially a matter of favoring the natural processes that promote their ecological services, while judiciously minimizing the socioeconomic costs they exert. For those species that can be locally damaging, either acutely or chronically, but do not undergo self-perpetuating outbreaks driven by positive feedback, desired results can often be attained by minimizing predisposing factors or reducing populations after environmental stresses raised them. For those species capable of landscape-scale outbreaks, management should emphasize keeping populations from surpassing the eruptive threshold. In all cases, however, it is essential to have clear and consistent management objectives. This poses a paradox, however: judicious human interactions with complex, large-scale, persistent systems such as forest biomes require consistency and integration over large scales of time and space, yet our sociopolitical institutions do the opposite (Chapter 15).

7.3 Invasive Species

The importance of effective plant defense can be readily seen during the initial stages of interaction between exotic organisms and novel hosts, i.e., new interactions where no co-evolutionary selection has been acting on the respective genomes. In such situations, large numbers of host trees may be killed by insects not known to cause mortality in their native range (Yan *et al.*, 2005; Poland and McCullough, 2006). For example, 5 years after *D. valens* was first detected in China, the beetle had spread over half a million hectares, killing 10 million *Pinus tabuliformis* Carrière (Yan *et al.*, 2005). Earlier, even more dramatic examples include *S. multistriatus* in North America. In some cases, introduced bark beetles may establish new associations with native phytopathogens, or introduced phytopathogens may establish new associations with native bark beetles. Similarly, high levels of mortality may result when native organisms encounter naïve hosts because of range expansion (Cudmore *et al.*, 2010). Extremely high populations of *D. ponderosae* in British Columbia have resulted in almost 100% mortality of lodgepole pine of susceptible host size class in many stands. In these examples, the population of the attacking insect increases rapidly, with dire consequences for the host plant populations and hence to local economies.

8   CONCLUSIONS

1. Bark beetles show a high diversity in their life histories. However, they also show some generalities arising from their reproduction within plant parts. These generalities include sophisticated host location systems, morphological adaptations facilitating tunneling, advanced communication systems, and close associations with microbial symbionts.

2. Life history strategies can be monogamous or polygamous with regard to mating, and solitary or gregarious with regard to intraspecific interactions. The gender responsible for host plant selection relates to mating strategy. The levels of parental care range from gallery maintenance in many tribes to eusociality in some ambrosia beetles.

3. Interactions with host plants vary markedly in terms of host species range, plant part, and physiological condition. Bark beetles are associated with a broad taxonomic range of plants, especially woody, perennial species. The tissues colonized by bark beetles are relatively poor in nutritional content for species that develop in roots, branches, and stems, and higher for species that develop in reproductive parts. The highest overall diversity of bark beetles is in tropical biomes.

4. Bark beetle species vary in the physiological condition of trees they colonize. Most species colonize dead plants or dead parts of plants. Some colonize trees that have been stressed by biotic or abiotic agents. A very few can colonize healthy trees. The last group exerts the strongest influences on ecosystem dynamics, and likewise poses the most serious challenges to natural resource management.

5. Bark beetles engage in sophisticated chemical signaling. Their pheromones serve to attract mates, and additional functions vary with their host–plant relationships. Some species use aggregation pheromones for cooperative resource procurement, by jointly overcoming tree defenses. Living trees pose formidable defenses that are multimodal, integrated, inducible, and capable of resisting attacks by individual or small numbers of beetles. Scolytine pheromones also incorporate plant chemicals, as precursors and/or synergists, into multicomponent signals.

6. A broad array of microbial symbionts, particularly fungi and bacteria, are associated with bark beetles. There is enormous diversity in their functional roles, including casual associations, antagonists, mycangial fungi transported in specialized structures, and ambrosia fungi that are actively gardened by the beetles. The fidelity of association ranges from incidental to obligatory, and in some cases there may be functional redundancy or substitutability. The benefits to beetles include assistance in procuring nutrients from phloem, a direct nutritional substrate, and assistance in overcoming tree defenses through detoxification of phytoalexins, among others. There are pronounced interactions among these symbionts, with outcomes mediated by host–plant chemistry, other phoretic organisms such as mites and nematodes, and temperature.

7. A wide diversity of predators, competitors, and parasites exploit bark beetles. Many of these exploit chemical signals associated with the beetles. Insect predators include several families of beetles and flies, and these, along with parasitoids of adult and egg stages, are often attracted to scolytine aggregation pheromones. Parasitoids of later stages are often attracted to volatiles emanating from the beetles' symbionts and deteriorating host plants. Despite the high diversity of predators and parasites affecting bark beetles, their habitat poses substantial physical protection that exerts substantial costs and challenges to organisms that exploit them. Thus, the effects of these natural enemies on bark beetle population dynamics are often limited.

8. Populations of most bark beetles in their native habitat tend to rise and fall with resource availability and weather. Because most species occupy a resource that is highly ephemeral in space and time, and can be utilized for only a limited duration, population increases incur substantial negative feedback. A few species, however, undergo intermittent landscape-scale population eruptions interspaced between much more extensive endemic periods. Populations of these species experience both the negative feedbacks of resource depletion, and also the positive feedbacks of increased resource availability driven by larger population size. The underlying mechanisms are driven by the cooperative behavior of mass attack and flexible, density-dependent host selection strategies. Critical thresholds, that operate at multiple levels of scale, and cross-scale interactions, govern these dynamics. Factors such as elevated temperature, drought, widespread environmental disturbance or immigration are needed for populations to surpass an eruptive threshold, and often multiple factors are required.

9. Bark beetles play important roles in ecosystem processing. These include nutrient cycling, decomposition, enhancing both animal and vegetative biodiversity, stand thinning, gap formation, and stand replacement depending on species. The eruptive species also have substantial influences on soil, hydrology, albedo, carbon sequestration, and other disturbance regimes. The landscape-transforming species are primarily associated with conifer biomes in northern temperate regions. These species exert substantial socioeconomic impacts.

10. Human activities have greatly magnified the reproductive success, and hence socioeconomic impacts of some bark beetles. These activities include transport of beetles and their symbionts to new regions, in which local trees have not evolved effective defenses; manipulation of the landscape in manners that reduce the heterogeneity of forest habitats or inhibit the success or dispersal of natural enemies; and climatic changes, specifically elevated temperatures that reduce overwintering mortality, accelerate beetle development and add to evapotranspiration stress on trees, and severe drought, which compromises tree defenses. Climate changes are resulting in both increased frequency and magnitude of outbreaks within historical ranges, and range expansions into new areas where trees lack coevolutionary adaptation.

ACKNOWLEDGMENTS

Helpful reviews of various portions of this manuscript were provided by Kier Klepzig, USDA Forest Service, and Patrick Tobin, USDA Forest Service.

REFERENCES

Abbott, B., Stennes, B., van Kooten, C.G., 2008. An economic analysis of mountain pine beetle impacts in a global context. Resource Economics and Policy Analysis (REPA) Research Group, Dept of Economics, University of Victoria, Victoria, BC, Canada.

Adams, A.S., Six, D.L., Adams, S.M., Holben, W.E., 2008. In vitro interactions between yeasts and bacteria and the fungal symbionts of the mountain pine beetle (Dendroctonus ponderosae). Microb. Ecol. 56, 460–466.

Adams, A.S., Currie, C.R., Cardoza, Y., Klepzig, K.D., Raffa, K.F., 2009. Effects of symbiotic bacteria and tree chemistry on the growth and reproduction of bark beetle fungal symbionts. Can. J. For. Res. 39, 1133–1147.

Adams, A.S., Adams, S.M., Currie, C.R., Gillette, N.E., Raffa, K.F., 2010. Geographic variation in bacterial communities associated with the red turpentine beetle (Coleoptera, Curculionidae). Environ. Entomol. 39, 406–414.

Adams, A.S., Boone, C.K., Bohlmann, J., Raffa, K.F., 2011. Responses of bark beetle-associated bacteria to host monoterpenes and their relationship to insect life histories. J. Chem. Ecol. 37, 808–817.

Adams, A.S., Aylward, F., Adams, S.M., Erbilgin, N., Aukema, B.H., Currie, C.R., et al., 2013. Metagenome of bacterial communities associated with mountain pine beetles and their host trees. Appl. Environ. Microbiol. 79, 3468–3475.

Addison, A.L., Powell, J.A., Six, D.L., Moore, M., Bentz, B.J., 2013. The role of temperature variability in stabilizing the mountain pine beetle-fungus mutualism. J. Theor. Biol. 335, 40–50.

Ainslie, B., Jackson, P.L., 2011. Investigation into mountain pine beetle above-canopy dispersion using weather radar and an atmospheric dispersion model. Aerobiology 27, 51–65.

Alamouti, S.M., Wang, V., DiGuistini, S., Six, D.L., Bohlmann, J., Hamelin, R.C., Feau, N., Breuil, C., 2011. Gene genealogies reveal cryptic species and host preferences for the pine fungal pathogen Grosmannia clavigera. Mol. Ecol. 20, 2581–2602.

Alkan, S., Aksu, Y., 1990. Research on rearing techniques for Rhizophagus grandis Gyll. (Coleoptera, Rhizophagidae). Proceedings of the Second Turkish National Congress of Biological Control, 173–179.

Allee, W.C., 1949. Group survival value for Philodina roseola, a rotifer. Ecology 30, 395–397.

Allison, J.D., Borden, J.H., McIntosh, R.L., deGroot, P., Gries, R., 2001. Kairomonal response by four Monochamus species (Coleoptera: Cerambycidae) to bark beetle pheromones. J. Chem. Ecol. 27, 633–646.

Alonso-Zarazaga, M.A., Lyal, C.H.C., 2009. A catalogue of family and genus group names in Scolytinae and Platypodinae with nomenclatural remarks (Coleoptera: Curculionidae). Zootaxa 2258, 1–134.

Amman, G.D., 1972. Mountain pine beetle brood production in relation to thickness of lodgepole pine phloem. Econ. Entomol. 65, 138–140.

Amman, G.D., 1984. Mountain pine beetle (Coleoptera: Scolytidae) mortality in three types of infestations. Environ. Entomol. 13, 184–191.

Anderbrant, O., 1990. Gallery construction and oviposition of the bark beetle Ips typographus (Coleoptera: Scolytidae) at different breeding densities. Econ. Entomol. 15, 1–8.

Anderbrant, O., Schlyter, F., Löfqvist, J., 1988. Dynamics of tree attack in the bark beetle Ips typographus under semi-epidemic conditions. In: Payne, T.L., Saarenmaa, H. (Eds.), Integrated Control of Scolytid Bark Beetles. Proceedings of the IUFRO Working Party and 17th International Congress of Entomology Symposium, Vancouver, BC, Canada, pp. 35–51, July 4, 1988.

Atkins, M.D., 1966. Behavioural variation among scolytids in relation to habitat. Can. Entomol. 98, 285–288.

Aukema, B.H., Raffa, K.F., 2000. Chemically mediated predator-free space: herbivores can synergize intraspecific communication without increasing risk of predation. J. Chem. Ecol. 26, 1923–1939.

Aukema, B.H., Raffa, K.F., 2002. Relative effects of exophytic predation, endophytic predation, and intraspecific competition on a subcortical herbivore: consequences to the reproduction of Ips pini and Thanasimus dubius. Oecologia 133, 483–491.

Aukema, B.H., Raffa, K.F., 2004a. Does aggregation benefit bark beetles by diluting predation? Links between a group-colonisation strategy and the absence of emergent multiple predator effects. Ecol. Entomol. 29, 129–138.

Aukema, B.H., Raffa, K.F., 2004b. Behavior of adult and larval Platysoma cylindrica (Coleoptera: Histeridae) and larval Medetera bistriata (Diptera: Dolichopodidae) during subcortical predation of Ips pini (Coleoptera: Scolytidae). J. Insect Behav. 17, 115–128.

Aukema, B.H., Raffa, K.F., 2004c. Gender- and sequence-dependent predation within group colonizers of defended plants: a constraint on cheating among bark beetles? Oecologia 138, 253–258.

Aukema, B.H., Werner, R.A., Haberkern, K.E., Illman, B.L., Clayton, M.K., Raffa, K.F., 2005a. Quantifying sources of variation in the frequency of fungi associated with spruce beetles: implications for hypothesis testing and sampling methodology in bark beetle–symbiont relationships. For. Ecol. Manage. 217, 187–202.

Aukema, B.H., Clayton, M.K., Raffa, K.F., 2005b. Modeling flight activity and population dynamics of the pine engraver, Ips pini, in the Great Lakes region: effects of weather and predators over short time scales. Pop. Ecol. 47, 61–69.

Aukema, B.H., Carroll, A.L., Zheng, Y., Zhu, J., Raffa, K.F., Moore, R.D., et al., 2008. Movement of outbreak populations of mountain pine beetle: influence of spatiotemporal patterns and climate. Ecography 31, 348–358.

Aukema, B.H., Zhu, J., Moeller, J., Rasmussen, J., Raffa, K.F., 2010. Interactions between below- and above-ground herbivores drive a forest decline and gap-forming syndrome. For. Ecol. Manage. 259, 374–382.

Ayres, M.P., Wilkens, R.T., Ruel, J.J., Vallery, E., 2000. Fungal relationships and the nitrogen budget of phloem-feeding bark beetles (Coleoptera: Scolytidae). Ecology 81, 2198–2210.

Baisier, M., Grégoire, J.-C., 1988. Factors influencing oviposition in *Rhizophagus grandis* (Coleoptera: Rhizophagidae), specific predator of the bark beetle *Dendroctonus micans* (Coleoptera: Scolytidae. Med. Fac. Landbouww. Rijksuniv. Gent 53 (3a), 1159–1167.

Baisier, M., Deneubourg, J.-L., Grégoire, J.-C., 1984. Death due to interaction between *Rhizophagus grandis* larvae. A theoretical and experimental evaluation. In: Grégoire, J.-C., Pasteels, J.M. (Eds.), Proceedings of the EEC Seminar on the Biological Control of Bark Beetles (*Dendroctonus micans*). Brussels, Belgium, pp. 134–139.

Baisier, M., Grégoire, J.-C., Delinte, K., Bonnard, O., 1988. Oviposition stimuli for *Rhizophagus grandis*, predator of *Dendroctonus micans*: the role of spruce monoterpene derivatives. In: Mattson, W.J., Lévieux, J., Bernard-Dagan, C. (Eds.), Mechanisms of Woody Plant Defenses Against Insects. Springer Verlag, New York, pp. 355–364.

Barr, B.A., 1969. Sound production in Scolytidae (Coleoptera) with emphasis on the genus *Ips*. Can. Entomol. 101, 636–672.

Barson, G., 1977. Laboratory evaluation of *Beauveria bassiana* as a pathogen of the larval stage of the large elm bark beetle, *Scolytus scolytus*. J. Invertebr. Pathol. 29, 361–366.

Batra, L.R., 1966. Ambrosia fungi: extent of specificity to ambrosia beetles. Science 153, 193–195.

Beal, J.A., 1939. The Black Hills beetle: a serious enemy of rocky mountain pines. USDA Farmer's Bulletin 1864.

Beaver, R.A., 1966a. The biology and immature stages of *Entedon leucogramma* (Ratzeburg) (Hymenoptera: Eulophidae), a parasite of bark beetles. Proceedings of the Royal Entomological Society, Series A 41, 37–41.

Beaver, R.A., 1966b. The biology and immature stages of two species of *Medetera* (Diptera: Dolichopodidae) associated with the bark beetle *Scolytus scolytus* (F.). Proceedings of the Royal Entomological Society, Series A 41, 145–154.

Bebi, P., Teich, M., Hagedorn, F., Zurbriggen, N., Brunner, S.H., Grêt-Regamey, A., 2012. Veränderung von Wald und Waldleistungen in der Landschaft Davos im Zuge des Klimawandels. Schweiz. Z. Forstwes. 163, 493–501.

Bentz, B.J., Mullins, D.E., 1999. Ecology of mountain pine beetle (Coleoptera: Scolytidae) cold hardening in the Intermountain West. Environ. Entomol. 28, 577–587.

Bentz, B.J., Logan, J.A., Amman, G.D., 1991. Temperature-dependent development of the mountain pine beetle (Coleoptera: Scolytidae) and simulation of its phenology. Can. Entomol. 123, 1083–1094.

Bentz, B.J., Bracewell, R.R., Mock, K.E., Pfrender, M.E., 2011. Genetic architecture and phenotypic plasticity of thermally-regulated traits in an eruptive species, *Dendroctonus ponderosae*. Evol. Ecol. 25, 1269–1288.

Berg, E.E., Henry, J.D., Fastie, C.L., De Volder, A.D., Matsuoka, S.M., 2006. Spruce beetle outbreaks on the Kenai Peninsula, Alaska, and Kluane National Park and Reserve, Yukon Territory: relationship to summer temperatures and regional differences in disturbance regimes. For. Ecol. Manage. 227, 219–232.

Berisford, C.W., 1980. Natural enemies and associated organisms. In: Thatcher, R.C., Searl, J.L., Hertel, G.D., Coster, J.E. (Eds.), The Southern Pine Beetle, pp. 31–52, USDA, Forest Service, Science and Education Administration Technical Bulletin 1631.

Berryman, A.A., 1972. Resistance of conifers to invasion by bark beetle fungus associations. BioScience 22, 598–602.

Berryman, A.A., 1973. Population dynamics of the fir engraver, *Scolytus ventralis* (Coleoptera: Scolytidae). I. Analysis of population behavior and survival from 1964 to 1971. Can. Entomol. 105, 1465–1488.

Berryman, A.A., 1974. Dynamics of bark beetle populations: towards a general productivity model. Environ. Entomol. 3, 579–585.

Berryman, A.A., 1979. Dynamics of bark beetle populations: analysis of dispersal and redistribution. Bull. Swiss Entomol. Soc. 52, 227–234.

Berryman, A.A., 1981. Population Systems: A General Introduction. Plenum Press, New York.

Biedermann, P.H.W., Taborsky, M., 2011. Larval helpers and age polyethism in ambrosia beetles. Proc. Natl. Acad. Sci. U. S. A. 108, 17064–17069.

Bigler, C., Kulakowski, D., Veblen, T.T., 2005. Multiple disturbance interactions and drought influence fire severity in Rocky Mountain subalpine forests. Ecology 86, 3018–3029.

Birch, M.C., Light, D.M., Wood, D.L., Browne, L.E., Silverstein, R.M., Bergot, B.J., et al., 1980. Pheromonal attraction and allomonal interruption of *Ips pini* in California by the two enantiomers of ipsdienol. J. Chem. Ecol. 6, 703–717.

Björklund, N., Lindgren, B.S., Shore, T.L., Cudmore, T., 2009. Can predicted mountain pine beetle net production be used to improve stand prioritization for management? For. Ecol. Manage. 257, 233–237.

Bleiker, K.P., Six, D.L., 2007. Dietary benefits of fungal associates to an eruptive herbivore: potential implications of multiple associates on host population dynamics. Environ. Entomol. 36, 1384–1396.

Bleiker, K.P., Heron, R.J., Braithwaite, E.C., Smith, G.D., 2013. Preemergence mating in the mass-attacking bark beetle, *Dendroctonus ponderosae* (Coleoptera: Curculionidae). Can. Entomol. 145, 12–19.

Bleiker, K.P., O'Brien, M.R., Smith, G.D., Carroll, A.L., 2014. Characterization of attacks made by the mountain pine beetle (Coleoptera: Curculionidae) during its endemic population phase. Can. Entomol. 146, 271–284.

Blomquist, G.J., Figueroa-Teran, R., Aw, M., Song, M., Gorzalski, A., Abbott, N.L., et al., 2010. Pheromone production in bark beetles. Insect Biochem. Mol. Biol. 40, 699–712.

Blumberg, D., Kehat, M., 1982. Biological studies of the date stone beetle, *Coccotrypes dactyliperda*. Phytoparasitica 10, 73–78.

Bohlmann, J., Gershenzon, J., 2009. Old substrates for new enzymes of terpenoid biosynthesis. Proc. Natl. Acad. Sci. U. S. A. 106, 10402–10403.

Bonnot, T.W., Millspaugh, J.J., Rumble, M.A., 2009. Multi-scale nest-site selection by black-backed woodpeckers in outbreaks of mountain pine beetles. For. Ecol. Manage. 259, 220–228.

Boone, C.K., Six, D.L., Raffa, K.F., 2008a. The enemy of my enemy is still my enemy: competitors add to predator load of a tree-killing bark beetle. Agric. For. Entomol. 10, 411–421.

Boone, C.K., Six, D.L., Zheng, Y., Raffa, K.F., 2008b. Exploitation of microbial symbionts of bark beetles by parasitoids and dipteran predators. Environ. Entomol. 37, 150–161.

Boone, C.K., Aukema, B.H., Bohlmann, J., Carroll, A.L., Raffa, K.F., 2011. Efficacy of tree defense physiology varies with bark beetle population density: a basis for positive feedback in eruptive species. Can. J. For. Res. 41, 1174–1188.

Boone, C.K., Adams, A.S., Bohlmann, J., Keefover-Ring, K., Mapes, A.C., Raffa, K.F., 2013. Bacteria associated with a tree-killing insect reduce concentrations of plant defense compounds. J. Chem. Ecol. 39, 1003–1006.

Borden, J.H., 1985. Aggregation pheromones. In: Kerkut, G.A., Gilbert, L. I. (Eds.), Comprehensive Insect Physiology, Biochemistry, and Pharmacology, Vol. 9. Pergamon Press, Oxford, pp. 257–285.

Borden, J.H., 1988. The striped ambrosia beetle. In: Berryman, A.A. (Ed.), Population Dynamics of Forest Insects. Plenum, New York, pp. 579–596.

Borden, J.H., Handley, J.R., Johnston, B.D., MacConnell, J.G., Silverstein, R.M., Slessor, K.N., et al., 1979. Synthesis and field testing of 4,6,6,-lineatin, the aggregation pheromone of *Trypodendron lineatum* (Coleoptera: Scolytidae). J. Chem. Ecol. 5, 681–689.

Borden, J.H., Handley, J.R., McLean, J.A., Silverstein, R.M., Chong, L., Slessor, K.N., et al., 1980. Enantiomer-based specificity in pheromone communication by two sympatric *Gnathotrichus* species (Coleoptera: Scolytidae). J. Chem. Ecol. 6, 445–456.

Borden, J.H., King, C.J., Lindgren, B.S., Chong, L., Gray, D.R., Oehlschlager, A.C., et al., 1982. Variation in response of *Trypodendron lineatum* from two continents to semiochemicals and trap form. Environ. Entomol. 11, 403–408.

Borden, J.H., Hunt, D.W.A., Miller, D.R., Slessor, K.N., 1986. Orientation in forest Coleoptera: An uncertain outcome of responses by individual beetles to variable stimuli. In: Payne, T.L., Birch, M.C., Kennedy, C.E.J. (Eds.), Mechanisms in Insect Olfaction. Clarendon Press, Oxford, pp. 97–109.

Borden, J.H., Devlin, D.R., Miller, D.R., 1992. Synomones of two sympatric species deter attack by the pine engraver, *Ips pini* (Coleoptera: Scolytidae). Can. J. For. Res. 22, 381–387.

Borg, T.K., Norris, D.M., 1969. Feeding responses by *Hylurgopinus rufipes* to combined chemical and physical stimuli. Ann. Entomol. Soc. Am. 62, 730–733.

Brand, J.M., Bracke, J.W., Markovetz, A., 1975. Production of verbenol pheromone by a bacterium isolated from bark beetles. Nature 254, 136–137.

Brand, J.M., Bracke, J.W., Britton, L.N., Markovetz, A.J., Barras, S.J., 1976. Bark beetle pheromones: production of verbenone by a mycangial fungus of *Dendroctonus frontalis*. J. Chem. Ecol. 2, 195–199.

Brasier, C.M., 1991. *Ophiostoma novo-ulmi* sp. nov., causative agent of current Dutch elm disease pandemics. Mycopathologia 115, 151–161.

Breshears, D.D., Myers, O.B., Meyer, C.W., Barnes, F.J., Zou, C.B., Allen, C.D., et al., 2009. Tree die-off in response to global change-type drought: mortality insights from a decade of plant water-potential measurements. Front. Ecol. Environ. 7, 185–189.

Browne, L.E., Wood, D.L., Bedard, W.D., Silverstein, R.M., West, J.R., 1979. Quantitative estimates of the western pine beetle attractive pheromone components, *exo*-brevicomin, frontalin, and myrcene in nature. J. Chem. Ecol. 5, 397–414.

Byers, J.A., 1984. Nearest neighbor analysis and simulation of distribution patterns indicates an attack spacing mechanism in the bark beetle, *Ips typographus* (Coleoptera: Scolytidae). Environ. Entomol. 13, 1191–1200.

Byers, J.A., 1992a. Attraction of bark beetles, *Tomicus piniperda*, *Hylurgops palliatus*, and *Trypodendron domesticum* and other insects to short-chain alcohols and monoterpenes. J. Chem. Ecol. 18, 2385–2402.

Byers, J.A., 1992b. Dirichlet tessellation of bark beetle spatial attack points. J. Anim. Ecol. 61, 759–768.

Byers, J.A., Löfqvist, J., 1989. Flight initiation and survival in the bark beetle *Ips typographus* (Coleoptera: Scolytidae) during the spring dispersal. Holarct. Ecol. 12, 432–440.

Byrne, K.J., Swigar, A.A., Silverstein, R.M., Borden, J.H., Stokkink, E., 1974. Sulcatol: population aggregation pheromone in the scolytid beetle. Gnathotrichus sulcatus. J. Insect Physiol. 20, 1895–1900.

Campbell, R.W., Sloan, R.J., 1977. Forest stand responses to defoliation by the gypsy moth. For. Sci. Monogr. 19, 1–34.

Cardoza, Y.J., Klepzig, K.D., Raffa, K.F., 2006a. Bacteria in oral secretions of an endophytic insect inhibit antagonistic fungi. Ecol. Entomol. 31, 636–645.

Cardoza, Y.J., Paskewitz, S., Raffa, K.F., 2006b. Travelling through time and space on wings of beetles: a tripartite insect-fungi-nematode association. Symbiosis 41, 71–79.

Čermák, V., Vieira, P., Gaar, V., Čudejková, M., Foit, J., Široká, K., Mota, M., 2012. *Bursaphelenchus pinophilus* Brzeski and Baujard, 1997 (Nematoda: Parasitaphelenchinae) associated with nematangia on *Pityogenes bidentatus* (Herbst, 1783) (Coleoptera: Scolytinae), from the Czech Republic. Nematology 14, 385–387.

Cognato, A.I., Grimaldi, D., 2009. 100 million years of morphological conservation in bark beetles (Coleoptera: Curculionidae: Scolytinae). Syst. Entomol. 34, 93–100.

Cognato, A.I., Hulcr, J., Dole, S.A., Jordal, B.H., 2010. Phylogeny of haplo-diploid, fungus-growing ambrosia beetles (Curculionidae: Scolytinae: Xyleborini) inferred from molecular and morphological data. Zool. Scr. 40, 174–186.

Connor, E.F., Taverner, M.P., 1997. The evolution and adaptive significance of the leaf-mining habit. Oikos 79, 6–25.

Costa, A., Reeve, J.D., 2011. Olfactory experience modifies semiochemical responses in a bark beetle predator. J. Chem. Ecol. 37, 1166–1176.

Costa, A., Min, A., Boone, C.K., Kendrick, A.P., Murphy, R.J., Sharpee, W.C., et al., 2013. Dispersal and edge behaviour of bark beetles and predators inhabiting red pine plantations. Agric. For. Entomol. 15, 1–11.

Costa, J.T., 2006. The Other Insect Societies. Harvard University Press, Cambridge.

Costanza, R., d'Arge, R., de Groot, R., Farber, S., Grasso, M., Hannon, B., et al., 1997. The value of the world's ecosystem services and natural capital. Nature 387, 253–260.

Coulson, R.N., 1979. Population dynamics of bark beetles. Annu. Rev. Entomol. 24, 417–447.

Coulson, R.N., Hennier, P.B., Flamm, R.O., Rykiel, E.J., Hu, L.C., Payne, T.L., 1983. The role of lightning in the epidemiology of the southern pine beetle. Z. angew. Entomol. 96, 182–193.

Creeden, E.P., Hicke, J.A., Buotte, P.C., 2014. Climate, weather, and recent mountain pine beetle outbreaks in the western United States. For. Ecol. Manage. 312, 239–251.

Cronin, J.T., Reeve, J.D., Wilkens, R., Turchin, P., 2000. The pattern and range of movement of a checkered beetle predator relative to its bark beetle prey. Oikos 90, 127–138.

Cudmore, T.J., Björklund, N., Carroll, A.L., Lindgren, B.S., 2010. Climate change and range expansion of an aggressive bark beetle: evidence of higher reproductive success in naïve host tree populations. J. Appl. Ecol. 47, 1036–1043.

Currie, C.R., Scott, J.A., Summerbell, R.C., Malloch, D., 1999. Fungus-growing ants use antibiotic-producing bacteria to control garden parasites. Nature 398, 701–704.

Dahlsten, D.L., 1982. Relationships between bark beetles and their natural enemies. In: Mitton, J.B., Sturgeon, K.B. (Eds.), Bark Beetles in North American Conifers: A System for the Study of Evolutionary Biology. University of Texas Press, Austin, pp. 140–182.

Dahlsten, D.L., Six, D.L., Erbilgin, N., Raffa, K.F., Lawson, A.B., Rowney, D.L., 2003. Attraction of *Ips pini* (Coleoptera: Scolytidae) and its predators to various enantiomeric ratios of ipsdienol and lanierone in California: implications for the augmentation and conservation of natural enemies. Environ. Entomol. 32, 1115–1122.

Davis, T.S., Hofstetter, R.W., 2011. Reciprocal interactions between the bark beetle-associated yeast *Ogataea pini* and host plant phytochemistry. Mycologia 103, 1201–1207.

de la Giroday, H.-M., Carroll, A.L., Lindgren, B.S., Aukema, B.H., 2011. Incoming! Association of landscape features with dispersing mountain pine beetle populations during a range expansion event in western Canada. Landscape Ecol. 26, 1097–1110.

Delalibera Jr., I., Handelsman, J., Raffa, K.F., 2005. Cellulolytic activity of microorganisms isolated from the guts of *Saperda vestita* (Coleoptera: Cerambycidae), *Ips pini*, and *Dendroctonus frontalis* (Coleoptera: Scolytidae). Environ. Entomol. 34, 541–547.

DeRose, J., Long, J., 2009. Wildfire and spruce beetle outbreak: simulation of interacting disturbances in the central Rocky Mountains. Ecoscience 16, 28–38.

Despland, E., Simpson, S.J., 2000. The role of food distribution and nutritional quality in behavioural phase change in the desert locust. Anim. Behav. 59, 643–652.

Despland, E., Simpson, S.J., 2005. Food choices of solitary and gregarious locusts reflect cryptic and aposmatic antipredator strategies. Anim. Behav. 69, 471–479.

Dickens, J.C., Payne, T.L., 1977. Bark beetle olfaction: pheromone receptor system in *Dendroctonus frontalis*. J. Insect Physiol. 23, 481–489.

DiGuistini, S., Wang, Y., Liao, N.Y., Taylor, G., Tanguay, P., Feau, N., et al., 2011. Genome and transcriptome analyses of the mountain pine beetle-fungal symbiont *Grosmannia clavigera*, a lodgepole pine pathogen. Proc. Natl. Acad. Sci. U. S. A. 108, 2504–2509.

Dixon, W.N., Payne, T.L., 1980. Attraction of entomophagous and associate insects of the southern pine beetle to beetle- and host tree-produced volatiles. J. Georgia Entomol. Soc. 15, 378–389.

Dodds, K.J., Graber, C., Stephen, F.M., 2001. Facultative intraguild predation by larval Cerambycidae (Coleoptera) on bark beetle larvae (Coleoptera: Scolytidae). Environ. Entomol. 30, 17–22.

Dussutour, A., Nicolis, S.C., Despland, E., Simpson, S.J., 2008. Individual differences influence collective behavior in social caterpillars. Anim. Behav. 76, 5–16.

Dyer, E.D.A., 1975. Frontalin attractant in stands infested by the spruce beetle, *Dendroctonus rufipennis* (Coleoptera: Scolytidae). Can. Entomol. 107, 979–988.

Eckhardt, L.G., Weber, A.M., Menard, R.D., Jones, J.P., Hess, N.L., 2007. Insect-fungal complex associated with loblolly pine decline in central Alabama. For. Sci. 53, 84–92.

Edworthy, A.B., Drever, M.C., Martin, K., 2011. Woodpeckers increase in abundance but maintain fecundity in response to an outbreak of mountain pine bark beetles. For. Ecol. Manage. 261, 203–210.

Elkinton, J.W., Wood, D.L., Browne, L.E., 1981. Feeding and boring behavior of the bark beetle, *Ips paraconfusus*, in extracts of ponderosa pine phloem. J. Chem. Ecol. 7, 209–220.

El-Sayed, A. M. (2012). The Pherobase: Database of Pheromones and Semiochemicals. Available online: http://www.pherobase.com. Last (accessed: 16.02.14.).

Embrey, S., Remais, J.V., Hess, J., 2012. Climate change and ecosystem disruption: the health impacts of the North American Rocky Mountain pine beetle infestation. Am. J. Publ. Health 102, 818–827.

Emerick, J.J., Snyder, A.I., Bower, N.W., Snyder, M.A., 2008. Mountain pine beetle attack associated with low levels of 4-allylanisole in ponderosa pine. Environ. Entomol. 37, 871–875.

Erbilgin, N., Raffa, K.F., 2001. Modulation of predator attraction to pheromones of two prey species by stereochemistry of plant volatiles. Oecologia 127, 444–453.

Erbilgin, N., Nordheim, E.V., Aukema, B.H., Raffa, K.F., 2002. Population dynamics of *Ips pini* and *Ips grandicollis* in red pine plantations in Wisconsin: within- and between-year associations with predators, competitors, and habitat quality. Environ. Entomol. 31, 1043–1051.

Erbilgin, N., Powell, J.S., Raffa, K.F., 2003. Effect of varying monoterpene concentrations on the response of *Ips pini* (Coleoptera: Scolytidae) to its aggregation pheromone: implications to the pest management and ecology of bark beetles. Agric. For. Entomol. 5, 269–274.

Erbilgin, N., Krokene, P., Christiansen, E., Zeneli, G., Gershenzon, J., 2006. Exogenous application of methyl jasmonate elicits defenses in Norway spruce (*Picea abies*) and reduces host colonization by the bark beetle *Ips typographus*. Oecologia 148, 426–436.

Evans, L.M., Hofstetter, R.W., Ayres, M.P., Klepzig, K.D., 2011. Temperature alters the relative abundance and population growth rates of species within the *Dendroctonus frontalis* (Coleoptera: Curculionidae) community. Environ. Entomol. 40, 824–834.

Evenden, J.C., Bedard, W.E., Struble, G.R., 1943. The mountain pine beetle, an important enemy of western pines. USDA Circular 664, 1–25.

Everaerts, C., Grégoire, J.-C., Merlin, J., 1988. The toxicity of spruce monoterpenes against bark beetles and their associates. In: Mattson, W.J., Lévieux, J., Bernard-Dagan, C. (Eds.), Mechanisms of Woody Plant Defenses Against Insects. Springer Verlag, New York, pp. 331–340.

Faccoli, M., 2000. Notes on the biology and ecology of *Tomicobia seitneri* (Ruschka) (Hymenoptera: Pteromalidae), a parasitoid of *Ips typographus* (L.) (Coleoptera: Scolytidae). Frustula Entomologica 23, 47–55.

Fayt, P., Machmer, M.M., Steeger, C., 2005. Regulation of spruce bark beetles by woodpeckers—a literature review. For. Ecol. Manage. 206, 1–14.

Fielding, N.J., O'Keefe, T., King, C.J., 1991. Dispersal and host-finding capability of the predatory beetle *Rhizophagus grandis* Gyll. (Col., Rhizophagidae). J. Appl. Entomol. 112, 89–98.

Fitzgerald, T.D., Nagel, W.P., 1972. Oviposition and larval bark-surface orientation of *Medetera aldrichii* (Diptera: Dolichopodidae): response to a prey-liberated plant terpene. Ann. Entomol. Soc. Am. 65, 328–330.

Flamm, R.O., Wagner, T.L., Cook, S.P., Pulley, P.E., Coulson, R.N., McArdle, T.M., 1987. Host colonization by cohabiting *Dendroctonus frontalis*, *Ips avulsus*, and *Ips calligraphus* (Coleoptera: Scolytidae). Environ. Entomol. 16, 390–399.

Flamm, R.O., Coulson, R.N., Beckley, P., Pulley, P.E., Wagner, T.L., 1989. Maintenance of a phloem-inhabiting guild. Environ. Entomol. 18, 381–387.

Flamm, R.O., Pulley, P.E., Coulson, R.N., 1993. Colonization of disturbed trees by the southern pine bark beetle guild (Coleoptera, Scolytidae). Environ. Entomol. 22, 62–70.

Flechtmann, C.A.H., Dalusky, M.J., Berisford, C.W., 1999. Bark and ambrosia beetle (Coleoptera: Scolytidae) responses to volatiles from aging loblolly pine billets. Environ. Entomol. 28, 638–648.

Forsse, E., Solbreck, C., 1985. Migration in the bark beetle *Ips typographus* L.: duration, timing and height of flight. Z. angew. Entomol. 100, 47–57.

Fox, J.W., Wood, D.L., Akers, R.P., Parmeter, J.R., 1992. Survival and development of *Ips paraconfusus* Lanier (Coleoptera: Scolytidae) reared axenically and with tree-pathogenic fungi vectored by cohabiting *Dendroctonus* species. Can. Entomol. 124, 1157–1167.

Franck, A. (2012). Wood processing in British Columbia: under the beetle's shadow. Hazardex, September 13, 2012. Available online: http://www.hazardexonthenet.net/article/53119/Wood-processing-in-British-Columbia–Under-the-beetle-s-shadow.aspx. (Last accessed: 10.06.14).

Franceschi, V.R., Krokene, P., Christiansen, E., Krekling, T., 2005. Anatomical and chemical defenses of conifer bark against bark beetles and other pests. New Phytol. 167, 353–375.

Franklin, A.J., Grégoire, J.-C., 1999. Flight behaviour of *Ips typographus* L. (Col., Scolytidae) in an environment without pheromones. Ann. For. Sci. 56, 591–598.

Franklin, A.J., Debruyne, C., Grégoire, J.-C., 2000. Recapture of *Ips typographus* L. (Col., Scolytidae) with attractants of low release rates: localized dispersion and environmental influences. Agric. For. Entomol. 2, 259–270.

Fraser, C.I., Brahy, O., Mardulyn, P., Dohet, L., Mayer, F., Grégoire, J.-C., 2014. Flying the nest: male dispersal and multiple paternity enables extrafamilial matings for the invasive bark beetle *Dendroctonus micans*. Heredity, in press.

Friedenberg, N.A., Sarkar, S., Kouchoukos, N., Billings, R.F., Ayres, M.P., 2008. Temperature extremes, density dependence, and southern pine beetle (Coleoptera: Curculionidae) population dynamics in east Texas. Environ. Entomol. 37, 650–659.

Furniss, M.M., 1995. Biology of *Dendroctonus punctatus* (Coleoptera, Scolytidae). Ann. Entomol. Soc. Am. 88, 173–182.

Furniss, M.M., Johnson, J.B., 1989. Description of the gallery and larva of *Dendroctonus punctatus* LeConte (Coleoptera, Scolytidae). Can. Entomol. 121, 757–762.

Furniss, M.M., Daterman, G.E., Kline, L.N., McGregor, M.D., Trostle, G.C., Pettinger, L.F., Rudinsky, J.A., 1974. Effectiveness of the Douglas-fir beetle antiaggregative pheromone methylcyclohexenone at three concentrations and spacings around felled host trees. Can. Entomol. 106, 381–392.

Gandhi, K.J.K., Gilmore, D.W., Haack, R.A., Katovich, S.A., Krauth, S.J., Mattson, W.J., et al., 2009. Application of semiochemicals to assess the biodiversity of subcortical insects following an ecosystem disturbance in a sub-boreal forest. J. Chem. Ecol. 35, 1384–1410.

Gargiullo, P.M., Berisford, C.W., 1981. Effects of host density and bark thickness on the densities of parasites of the southern pine beetle. Environ. Entomol. 10, 392–399.

Gauss, R., 1954. Der Ameisenbuntkäfer *Thanasimus* (*Clerus*) *formicarius* Latr. als Borkenkäferfeind. In: Wellenstein, G. (Ed.), Die grosse Borkenkäferkalamität in Südwest-Deutschland 1944–1951. Forstschutzstelle Südwest/Ringingen, pp. 417–429.

Gibbs, J.N., 1978. Intercontinental epidemiology of Dutch elm disease. Annu. Rev. Phytopathol. 16, 287–307.

Gilbert, M., Grégoire, J.-C., 2003. Site condition and predation influence a bark beetle's success: a spatially realistic approach. Agric. For. Entomol. 5, 87–96.

Gilbert, M., Vouland, G., Grégoire, J.-C., 2001. Past attacks influence host selection by the solitary bark beetle, *Dendroctonus micans*. Ecol. Entomol. 26, 133–142.

Goodwin, T.E., Eggert, M.S., House, S.J., Weddell, M.E., Schulte, B.A., Rasmussen, L.E.L., 2006. Insect pheromones and precursors in female African elephant urine. J. Chem. Ecol. 32, 1849–1853.

Goyer, R.A., Finger, C.K., 1980. Relative abundance and seasonal distribution of the major hymenopterous parasites of the southern pine beetle, *Dendroctonus frontalis* Zimmermann, on loblolly pine. Environ. Entomol. 9, 97–100.

Grégoire, J.-C., 1988. The greater European spruce beetle. In: Berryman, A.A. (Ed.), Dynamics of Forest Insect Populations: Patterns, Causes, and Implications. Plenum Press, New York, pp. 456–478.

Grégoire, J.-C., Braekman, J.-C., Tondeur, A., 1982. Chemical communication between the larvae of *Dendroctonus micans* Kug. Les Colloques de l'INRA, 7, Les Médiateurs chimiques, 253–257.

Grégoire, J.-C., Merlin, J., Pasteels, J.M., Jaffuel, R., Vouland, G., Schvester, D., 1985. Biocontrol of *Dendroctonus micans* by *Rhizophagus grandis* Gyll. (Col., Rhizophagidae) in the Massif Central (France). Z. angew. Entomol. 99, 182–190.

Grégoire, J.-C., Baisier, M., Drumont, A., Dahlsten, D.L., Meyer, H., Francke, W., 1991. Volatile compounds in the larval frass of *Dendroctonus valens* and *Dendroctonus micans* (Coleoptera: Scolytidae) in relation to oviposition by the predator, *Rhizophagus grandis* (Coleoptera: Rhizophagidae). J. Chem. Ecol. 17, 2003–2019.

Grégoire, J.-C., Couillien, D., Krebber, R., König, W.A., Meyer, H., Francke, W., 1992. Orientation of *Rhizophagus grandis* (Coleoptera: Rhizophagidae) to oxygenated monoterpenes in a species-specific predator-prey relationship. Chemoecology 3, 14–18.

Griffin, J.M., Turner, M.G., Simard, M., 2011. Nitrogen cycling following mountain pine beetle disturbance in lodgepole pine forests of Greater Yellowstone. For. Ecol. Manage. 261, 1077–1089.

Grucmanová, Š., Holuša, J., 2013. Nematodes associated with bark beetles, with focus on the genus *Ips* (Coleoptera: Scolytinae) in Central Europe. Acta Zool. Bulgarica 65, 547–556.

Grünwald, M., 1986. Ecological segregation of bark beetles (Coleoptera, Scolytidae) of spruce. Z. ang. Entomol. 101, 176–187.

Hackwell, G.A., 1973. Biology of *Lasconotus subcostulatus* (Coleoptera: Colydiidae) with special reference to feeding behavior. Ann. Entomol. Soc. Am. 66, 62–65.

Hammerbacher, A., Schmidt, A., Wadke, N., Wright, L.P., Schenider, B., Bohlmann, J., et al., 2013. A common fungal associate of the spruce bark beetle metabolizes the stilbene defenses of Norway spruce. Plant Physiol. 162, 1324–1336.

Hansen, E.M., Bentz, B.J., 2003. Comparison of reproductive capacity among univoltine, semivoltine, and re-emerged parent spruce beetles (Coleoptera: Scolytidae). Can. Entomol. 135, 697–712.

Hansen, M.E., Bentz, B.J., Turner, D.L., 2001. Temperature-based model for predicting univoltine brood proportions in spruce beetle (Coleoptera: Scolytidae). Can. Entomol. 133, 827–841.

Harrington, T.C., 2005. Ecology and evolution of mycetophagous beetles and their fungal partners. In: Vega, F.E., Blackwell, M. (Eds.), Insect Fungal Associations, Ecology and Evolution. Oxford University Press, Oxford, pp. 257–291.

Hart, S.J., Veblen, T.T., Eisenhart, K.S., Jarvis, D., Kulakowski, D., 2014. Drought induces spruce beetle (*Dendroctonus rufipennis*) outbreaks across northwestern Colorado. Ecology 95, 930–939.

Harvey, B.J., Donato, D.C., Romme, W.H., Turner, M.G., 2013. Influence of recent bark beetle outbreak on fire severity and postfire tree regeneration in montane Douglas-fir forests. Ecology 94, 2475–2486.

Hayes, J.L., Strom, B.L., 1994. 4-Allylanisol as an inhibitor of bark beetle (Coleoptera, Scolytidae) aggregation. J. Econ. Entomol. 87, 1586–1594.

Hedqvist, K.J., 1963. Die Feinde der Borkenkäfer in Schweden, 1. Erzwespen (Chalcidoidea). Studia Forestalia Suecica 11, 1–176.

Hedqvist, K.J., 1998. Bark beetle enemies in Sweden 2. Braconidae (Hymenoptera). Entomologica Scandinavica, Supplement 52, 1–86.

Hérard, F., Mercadier, G., 1996. Natural enemies of *Tomicus piniperda* and *Ips acuminatus* (Col., Scolytidae) on *Pinus sylvestris* near Orléans, France: temporal occurrence and relative abundance, and notes on eight predatory species. Entomophaga 41, 183–210.

Hicke, J.A., Logan, J.A., Powell, J., Ojima, D.S., 2006. Changing temperatures influence suitability for modeled mountain pine beetle (*Dendroctonus ponderosae*) outbreaks in the western United States. J. Geophys. Res.: Biogeosci. 111, G02019.

Hobson, K., Parmeter Jr., J.R., Wood, D.L., 1994. The role of fungi vectored by *Dendroctonus brevicomis* LeConte (Coleoptera: Scolytidae) in occlusion of ponderosa pine xylem. Can. Entomol. 126, 277–282.

Hodges, J.D., Pickard, L.S., 1971. Lightning in the ecology of the southern pine beetle, *Dendroctonus frontalis* (Coleoptera: Scolytidae). Can. Entomol. 103, 44–51.

Hofstetter, R.W., Moser, J.C., 2014. The role of mites in insect-fungus associations. Annu. Rev. Entomol. 59, 537–557.

Hofstetter, R.W., Cronin, J.T., Klepzig, K.D., Moser, J.C., Ayres, M.P., 2006. Antagonisms, mutualisms and commensalisms affect outbreak dynamics of the southern pine beetle. Oecologia 147, 679–691.

Hofstetter, R., Dempsey, T., Klepzig, K., Ayres, M., 2007. Temperature-dependent effects on mutualistic, antagonistic, and commensalistic interactions among insects, fungi and mites. Community Ecol. 8, 47–56.

Hofstetter, R.W., Moser, J.C., Blomquist, S.R., 2013. Mites associated with bark beetles and their hyperphoretic ophiostomatoid fungi. In: Seifert, K.A., de Beer, Z.W., Wingfield, M.J. (Eds.), Ophiostomatoid Fungi: Expanding Frontiers. In: CBS Biodiversity Series, 12, pp. 165–176.

Hougardy, E., Grégoire, J.-C., 2000. Spruce stands provide natural food sources to adult hymenopteran parasitoids of bark beetles. Entomol. Exp. Appl. 96, 253–263.

Hougardy, E., Grégoire, J.-C., 2003. Cleptoparasitism increases the host finding ability of a polyphagous parasitoid species, *Rhopalicus tutela* (Hymenoptera: Pteromalidae). Behav. Ecol. Sociobiol. 55, 184–189.

Huber, D.P.W., Gries, R., Borden, J.H., Pierce Jr., H.D., 1999. Two pheromones of coniferophagous bark beetles found in the bark of nonhost angiosperms. J. Chem. Ecol. 25, 805–816.

Huber, D.P.W., Aukema, B.H., Hodgkinson, R.S., Lindgren, B.S., 2009. Successful colonization, reproduction, and new generation emergence in live interior hybrid spruce, *Picea engelmannii* x *glauca*, by mountain pine beetle, *Dendroctonus ponderosae*. Agric. For. Entomol. 11, 83–89.

Hui, Y., Bakke, A., 1997. Development and reproduction of *Thanasimus formicarius* (L.) (Coleoptera: Cleridae) at three constant temperatures. Can. Entomol. 129, 579–583.

Hulcr, J., Dunn, R.R., 2011. The sudden emergence of pathogenicity in insect–fungus symbioses threatens naive forest ecosystems. Proc. R. Soc. B 278, 2866–2873.

Hulcr, J., Pollet, M., Ubik, K., Vrkoč, J., 2005. Exploitation of kairomones and synomones by *Medetera* spp. (Diptera: Dolichopodidae), predators of spruce bark beetles. Eur. J. Entomol. 102, 655–662.

Hulcr, J., Mogia, M., Isua, B., Novotny, V., 2007. Host specificity of ambrosia and bark beetles (Col., Curculionidae: Scolytinae and Platypodinae) in a New Guinea rainforest. Ecol. Entomol. 32, 762–772.

Hunt, D.W.A., Borden, J.H., 1990. Conversion of verbenols to verbenone by yeasts isolated from *Dendroctonus ponderosae* (Coleoptera: Scolytidae). J. Chem. Ecol. 16, 1385–1397.

Hüsnü Can Başer, K., Demirci, B., Tabanca, N., Özek, T., Gören, N., 2001. Composition of the essential oils of *Tanacetum armenum* (DC.) Schultz Bip., *T. balsamita* L., *T. chiliophyllum* (Fisch. & Mey.) Schultz Bip. var. *chiliophyllum* and *T. haradjani* (Rech. fil.) Grierson and the enantiomeric distribution of camphor and carvone. Flavour Fragrance J. 16, 195–200.

Hynum, B.G., Berryman, A.A., 1980. *Dendroctonus ponderosae* (Coleoptera: Scolytidae): pre-aggregation landing and gallery initiation on lodgepole pine. Can. Entomol. 112, 185–191.

Inward, D.J.G., Wainhouse, D., Peace, A., 2012. The effect of temperature on the development and life cycle regulation of the pine weevil *Hylobius abietis* and the potential impacts of climate change. Agric. For. Entomol. 14, 348–357.

Ivarsson, P., Schlyter, F., Birgersson, G., 1993. Demonstration of *de novo* pheromone biosynthesis in *Ips duplicatus* (Coleoptera: Scolytidae): inhibition of ipsdienol and *E*-myrcenol production by compactin. Insect Biochem. Mol. Biol. 23, 655–662.

Jackson, P.L., Straussfogel, D., Lindgren, B.S., Mitchell, S., Murphy, B.D., 2008. Radar observation and aerial capture of mountain pine beetle, *Dendroctonus ponderosae* Hopk. (Coleoptera: Scolytidae) in flight above the forest canopy. Can. J. For. Res. 38, 2313–2327.

Jactel, H., 1991. Dispersal and flight behaviour of *Ips sexdentatus* (Coleoptera: Scolytidae) in pine forest. Ann. For. Sci. 48, 417–428.

Jactel, H., 1993. Individual variability of the flight potential of *Ips sexdentatus* Boern. (Coleoptera: Scolytidae) in relation to day of emergence, sex, size, and lipid content. Can. Entomol. 125, 919–930.

Jönsson, A.M., Harding, S., Bärring, L., Ravn, H.P., 2007. Impact of climate change on the population dynamics of *Ips typographus* in southern Sweden. Agric. For. Meteorol. 146, 70–81.

Jordal, B.J., Kirkendall, L.R., 1998. Ecological relationships of a guild of tropical beetles breeding in *Cecropia* petioles in Costa Rica. J. Trop. Ecol. 14, 153–176.

Jordal, B., Normark, B.B., Farrell, B.D., 2000. Evolutionary radiation of an inbreeding haplodiploid beetle lineage (Curculionidae, Scolytinae). Biol. J. Linn. Soc. 71, 483–499.

Jordal, B.H., Beaver, R.A., Kirkendall, L.R., 2001. Breaking taboos in the tropics: inbreeding promotes colonization by wood-boring beetles. Glob. Ecol. Biogeogr. 10, 345–358.

Jordal, B.H., Beaver, R.A., Normark, B.B., Farrell, B.D., 2002. Extraordinary sex ratios and the evolution of male neoteny in sib-mating *Ozopemon* beetles. Biol. J. Linn. Soc. 75, 353–360.

Jordal, B.H., Sequeira, A.S., Cognato, A.I., 2011. The age and phylogeny of wood boring weevils and the origin of subsociality. Mol. Phylogen. Evol. 59, 708–724.

Kaiser, K.E., McGlynn, B.L., Emanuel, R.E., 2012. Ecohydrology of an outbreak: mountain pine beetle impacts trees in drier landscape positions first. Ecohydrology 6, 444–454.

Kane, J.M., Kolb, T.E., 2010. Importance of resin ducts in reducing ponderosa pine mortality following bark beetle attack. Oecologia 164, 601–609.

Kausrud, K., Økland, B., Skarpaas, O., Grégoire, J.-C., Erbilgin, N., Stenseth, N.C., 2011a. Population dynamics in changing environments: the case of an eruptive forest pest species. Biol. Rev. 87, 34–51.

Kausrud, K.L., Grégoire, J.-C., Skarpaas, O., Erbilgin, N., Gilbert, M., Økland, B., Stenseth, N.C., 2011b. Trees wanted—dead or alive! Host selection and population dynamics in tree-killing bark beetles. PLoS One 6, e18274.

Kelley, S.T., Farrell, B.D., 1998. Is specialization a dead end? The phylogeny of host use in *Dendroctonus* bark beetles (Scolytidae). Evolution 52, 1731–1743.

Keeling, C.I., Bohlmann, J., 2006. Genes, enzymes and chemicals of terpenoid diversity in the constitutive and induced defence of conifers against insects and pathogens. New Phytol. 170, 657–675.

Keeling, C.I., Chiu, C.C., Aw, T., Li, M., Henderson, H., Tittiger, C., et al., 2013. Frontalin pheromone biosynthesis in the mountain pine beetle, *Dendroctonus ponderosae*, and the role of isoprenyl diphosphate synthases *Proc*. Proc. Natl. Acad. Sci. U.S.A. 110, 18838–18843.

Keen, F.P., 1938. Insect enemies of western forests. USDA Misc, Pub. No, 273.

Kelsey, J.G., Peck, R.G., Niwa, C.G., 2001. Response of some scolytids and their predators to ethanol and 4-allylanisole in pine forests of central Oregon. J. Chem. Ecol. 27, 697–715.

Kenis, M., Wermelinger, B., Grégoire, J.-C., 2004. Research on parasitoids and predators of Scolytidae—a review. In: Lieutier, F., Day, K.R., Battisti, A., Grégoire, J.-C., Evans, H.F. (Eds.), Bark and Wood Boring Insects in Living Trees in Europe, a Synthesis. Kluwer, Dordrecht, pp. 237–290.

Kennedy, A.A., McCullough, D.G., 2002. Phenology of the larger European pine shoot beetle *Tomicus piniperda* (L.) (Coleoptera: Scolytidae) in relation to native bark beetles and natural enemies in pine stands. Environ. Entomol. 31, 261–272.

Kinn, D.N., Perry, T.J., Guinn, F.H., Strom, B.L., Woodring, J., 1994. Energy reserves of individual southern pine beetles (Coleoptera, Scolytidae) as determined by a modified phosphovanillin spectrophotometric method. J. Entomol. Sci. 29, 152–163.

Kirisits, T., 2004. Fungal associates of European bark beetles with special emphasis on ophiostomatoid fungi. In: Lieutier, F., Day, K.R., Battisti, A., Grégoire, J.-C., Evans, H.F. (Eds.), Bark and Wood Boring Insects in Living Trees in Europe, a Synthesis. Kluwer, Dordrecht, pp. 181–236.

Kirkendall, L.R., Kent, D.S., Raffa, K.F., 1997. Interactions among males, females and offspring in bark and ambrosia beetles: the significance of living in tunnels for the evolution of social behavior. In: Choe, J.C., Crespi, B.J. (Eds.), The Evolution of Social Behavior in Insects and Arachnids. Cambridge University Press, Cambridge, pp. 181–215.

Kirkendall, L.R., Faccoli, M., Ye, H., 2008. Description of the Yunnan shoot borer, *Tomicus yunnanensis* Kirkendall and Faccoli sp. n. (Curculionidae: Scolytinae), an unusually aggressive pine shoot beetle from southern China, with a key to the species of *Tomicus*. Zootaxa 1819, 25–39.

Klepzig, K.D., Raffa, K.F., Smalley, E.B., 1991. Association of insect-fungal complexes with red pine decline in Wisconsin. For. Sci. 37, 1119–1139.

Klepzig, K.D., Moser, J.C., Lombardero, F.J., Hofstetter, R.W., Ayres, M.P., 2001. Symbiosis and competition: complex interactions among beetles, fungi and mites. Symbiosis 30, 83–96.

Klepzig, K.D., Adams, A.S., Handelsman, J., Raffa, K.F., 2009. Symbioses, a key driver of insect physiological processes, ecological interactions, evolutionary diversification, and impacts on humans. Environ. Entomol. 38, 67–77.

Klimetzek, D., Köhler, J., Krohn, S., Francke, W., 1989. Das pheromonsystem des waldreben-borkenkäfers, *Xylocleptes bispinus* Duft. (Col., Scolytidae). J. Appl. Entomol. 107, 304–309.

Knee, W., Forbes, M.R., Beaulieu, F., 2013. Diversity and host use of mites (Acari: Mesostigmata, Oribatida) phoretic on bark beetles (Coleoptera: Scolytinae): global generalists, local specialists? Ann. Entomol. Soc. Am. 106, 339–350.

Knížek, M., Beaver, R., 2004. Taxonomy and systematics of bark and ambrosia beetles. In: Lieutier, F., Day, K.R., Battisti, A., Grégoire, J.-C., Evans, H.F. (Eds.), Bark and Wood Boring Insects in Living Trees in Europe, a Synthesis. Kluwer, Dordrecht, pp. 41–54.

Kobakhidze, D.N., 1965. Some results and prospects of the utilization of beneficial entomophagous insects in the control of insect pests in Georgian SSR (USSR). Entomophaga 10, 323–330.

Koštál, V., Doležal, P., Rozsypal, J., Moravcová, M., Zahradníčková, H., Šimek, P., 2011. Physiological and biochemical analysis of overwintering and cold tolerance in two Central European populations of the spruce bark beetle. Ips typographus. J. Insect Physiol. 57, 1136–1146.

Krieger, D.J., 2001. The economic value of forest ecosystem services: a review. Wilderness Society, Washington, DC.

Krokene, P., Solheim, H., 1998. Pathogenicity of four blue-stain fungi associated with aggressive and nonaggressive bark beetles. Phytopathology 88, 39–44.

Krokene, P., Solheim, H., Christiansen, E., 2001. Induction of disease resistance in Norway spruce (*Picea abies*) by necrotizing fungi. Plant Pathol. 50, 230–233.

Krokene, P., Solheim, H., Krekling, T., Christiansen, E., 2003. Inducible anatomical defense responses in Norway spruce stems and their possible role in induced resistance. Tree Physiol. 23, 191–197.

Krokene, P., Nagy, N.E., Solheim, H., 2008. Methyl jasmonate and oxalic acid treatment of Norway spruce: anatomically based defense responses and increase resistance against fungal infection. Tree Physiol. 28, 29–35.

Kudon, L.H., Berisford, C.W., 1981. An olfactometer for bark beetle parasites. J. Chem. Ecol. 7, 359–366.

Kurz, W.A., Dymond, C.C., Stinson, G., Rampley, G.J., Neilson, E.T., Carroll, A.L., et al., 2008. Mountain pine beetle and forest carbon: feedback to climate change. Nature 454, 987–990.

Langor, D.W., 1991. Arthropods and nematodes co-occurring with the eastern larch beetle *Dendroctonus simplex* (Col.: Scolytidae) in Newfoundland. Entomophaga 36, 303–313.

Långström, B., 1983. Life cycles and shoot-feeding of the pine shoot beetles. Studia Forestalia Suecica 163, 1–29.

Lanier, G.N., Wood, D.L., 1975. Specificity of response to pheromones in the genus *Ips* (Coleoptera: Scolytidae). J. Chem. Ecol. 1, 9–23.

Lanier, G.N., Birch, M.C., Schmitz, R.F., Furniss, M.M., 1972. Pheromones of *Ips pini* (Coleoptera: Scolytidae): variation in response among three populations. Can. Entomol. 104, 1917–1923.

Larsson, S., Ekbom, B., Björkman, C., 2000. Influence of plant quality on pine sawfly population dynamics. Oikos 89, 440–450.

Latty, T.M., Reid, M.L., 2010. Who goes first? Condition and danger dependent pioneering in a group-living bark beetle (*Dendroctonus ponderosae*). Behav. Ecol. Sociobiol. 64, 639–646.

Lauzière, I., Pérez-Lachaud, G., Brodeur, J., 2000. Behavior and activity pattern of *Cephalonomia stephanoderis* (Hymenoptera: Bethylidae) attacking the coffee berry borer, *Hypothenemus hampei* (Coleoptera: Scolytidae). J. Insect. Behav. 13, 375–395.

Lee, S., Kim, J.J., Breuil, C., 2006. Diversity of fungi associated with the mountain pine beetle, *Dendroctonus ponderosae* and infested lodgepole pines in British Columbia. Fungal Divers. 22, 91–105.

Lewinsohn, E., Gijzen, M., Muzika, R.M., Barton, K., Croteau, R., 1993. Oleoresinosis in grand fir (*Abies grandis*) saplings and mature trees. Modulation of this wound response by light and water stresses. Plant Physiol. 101, 1021–1028.

Lewis, K.J., Lindgren, B.S., 2002. Relationship between spruce beetle and *Tomentosus* root disease: two natural disturbance agents of spruce. Can. J. For. Res. 32, 31–37.

Lieutier, F., Vouland, G., Pettinetti, M., Garcia, J., Romary, P., Yart, A., 1992. Defence reactions of Norway spruce (*Picea abies* Karst.) to artificial insertion of *Dendroctonus micans* Kug. (Col., Scolytidae). J. Appl. Entomol. 114, 1–5.

Lieutier, F., Yart, A., Sallé, A., 2009. Stimulation of tree defenses by Ophiostomatoid fungi can explain attack success of bark beetles on conifers. Ann. For. Sci. 66, 801–823.

Lindgren, B.S., 1986. *Trypodendron lineatum* (Olivier) (Coleoptera: Scolytidae) breeding in big leaf maple. Acer macrophyllum. J. Entomol. Soc. B.C. 83, 44.

Lindgren, B.S., Miller, D.R., 2002a. Effect of verbenone on five species of bark beetles (Coleoptera: Scolytidae) in lodgepole pine forests. Environ. Entomol. 31, 759–765.

Lindgren, B.S., Miller, D.R., 2002b. Effect of verbenone on attraction of predatory and woodboring beetles (Coleoptera) to kairomones in lodgepole pine forests. Environ. Entomol. 31, 766–773.

Lindgren, B.S., Raffa, K.F., 2013. The evolution of tree-killing by bark beetles: trade-offs between the maddening crowds and a sticky situation. Can. Entomol. 145, 471–495.

Lindgren, B.S., Hoover, S.E.R., MacIsaac, A.M., Keeling, C.I., Slessor, K.N., 2000. Lineatin enantiomer preference, flight periods, and effect of pheromone concentration and trap length on three sympatric species of *Trypodendron* (Coleoptera: Scolytidae). Can. Entomol. 132, 877–887.

Lindquist, E.E., 1964. Mites parasitizing eggs of bark beetles of the genus *Ips*. Can. Entomol. 96, 125–126.

Lindquist, E.E., 1969. Review of holarctic tarsonemid mites (Acarina: Prostigmata) parasitizing eggs of ipine bark beetles. Mem. Entomol. Soc. Can. 60, 1–111.

Lindquist, E.E., 1970. Relationships between mites and insects in forest habitats. Can. Entomol. 102, 984–987.

Linit, M.J., Stephen, F.M., 1983. Parasite and predator component of within-tree southern pine-beetle (Coleoptera, Scolytidae) mortality. Can. Entomol. 115, 679–688.

Logan, J.A., Bentz, B.J., 1999. Model analysis of mountain pine beetle (Coleoptera: Scolytidae) seasonality. Environ. Entomol. 28, 924–934.

Lombardero, M.J., Ayres, M.P., Ayres, B.D., Reeve, J.D., 2000a. Cold tolerance of four species of bark beetle (Coleoptera: Scolytidae) in North America. Environ. Entomol. 29, 421–432.

Lombardero, M.J., Ayres, M.P., Lorio, P.L., Ruel, J.J., 2000b. Environmental effects on constitutive and inducible resin defences of *Pinus taeda*. Ecol. Lett. 3, 329–339.

Lombardero, M.J., Ayres, M.P., Hofstetter, R.W., Moser, J.C., Klepzig, K.D., 2003. Strong indirect interactions of *Tarsonemus* mites (Acarina: Tarsonemidae) and *Dendroctonus frontalis* (Coleoptera: Scolytidae). Oikos 102, 243–252.

Løyning, M.K., Kirkendall, L.R., 1996. Mate discrimination in a pseudogamous bark beetle (Coleoptera: Scolytidae): male *Ips acuminatus* prefers sexual to clonal females. Oikos 77, 336–344.

MacQuarrie, C.J.K., Cooke, B.J., 2011. Density-dependent population dynamics of mountain pine beetle in thinned and unthinned stands. Can. J. For. Res. 41, 1031–1046.

Marini, L.L., Lindelöw, Å., Jönsson, A.M., Wulff, S.S., Schroeder, L.M., 2013. Population dynamics of the spruce bark beetle: a long-term study. Oikos 122, 1768–1776.

Martin, D.M., Gershenzon, J., Bohlmann, J., 2003. Induction of volatile terpene biosynthesis and diurnal emission by methyl jasmonate in foliage of Norway spruce. Plant Physiol. 132, 1586–1599.

Martinson, S.J., Tlioja, T., Sullivan, B.T., Billings, R.F., Ayres, M.P., 2013. Alternate attractors in the population dynamics of a tree-killing bark beetle. Population Ecology 55, 95–106.

Massey, C.L., 1966. The nematode parasites and associates of *Dendroctonus adjunctus* (Coleoptera: Scolytidae) in New Mexico. Ann. Entomol. Soc. Am. 59, 424–440.

Matson, P.A., Hain, F.P., 1985. Host conifer defense strategies: a hypothesis. In: Safranyik, L. (Ed.), The Role of the Host in the Population Dynamics of Forest Insects. Canadian Forestry Service, Pacific Forest Research Centre, Victoria, BC, Canada, pp. 33–42.

Mattson, W.J., 1980. Herbivory in relation to plant nitrogen content. Annu. Rev. Ecol. Syst. 11, 119–161.

Mawby, W.D., Hain, F.P., Doggett, C.A., 1989. Endemic and epidemic populations of southern pine beetle: implications of the two-phase model for forest managers. For. Sci. 35, 1075–1087.

McDowell, N.G., Amthor, J.S., Beerling, D.J., Fisher, R.A., Raffa, K.F., Stitt, M., 2011. The interdependence of mechanisms underlying vegetation mortality. Trends Ecol. Evol. 26, 523–532.

McNee, W.R., Wood, D.L., Storer, A.J., 2000. Pre-emergence feeding in bark beetles (Coleoptera: Scolytidae). Environ. Entomol. 29, 495–501.

Merlin, J., 1984. Elm bark beetles and their main parasitoids in Belgium: emergence and some aspects of their ecological relations. Med. Fac. Landbouww. Rijksuniv. Gent 49, 857–865.

Merlin, J., Parmentier, C., Grégoire, J.-C., 1986. The feeding habits of *Rhizophagus dispar* (Col., Rhizophagidae), an associate of bark beetles. Med. Fac. Landbouww. Rijksuniv. Gent 51, 915–923.

Miller, D.R., Rabaglia, R.J., 2009. Ethanol and (−)-α-pinene: attractant kairomones for bark and ambrosia beetles in the southeastern US. J. Chem. Ecol. 35, 435–448.

Miller, D.R., Borden, J.H., Slessor, K.N., 1989. Inter- and intrapopulation variation of the pheromone, ipsdienol, produced by male pine engravers, *Ips pini* (Say) (Coleoptera: Scolytidae). J. Chem. Ecol. 15, 233–247.

Miller, D.R., Gibson, K.E., Raffa, K.F., Seybold, S.J., Teale, S.A., Wood, D.L., 1997. Geographic variation in response of pine engraver, *Ips pini*, and associated species to pheromone, lanierone. J. Chem. Ecol. 23, 2013–2031.

Mills, N.J., 1983. The natural enemies of scolytids infesting conifer bark in Europe in relation to the biological control of *Dendroctonus* spp. in Canada. Biocontrol News and Information 4, 303–326.

Mills, N.J., 1985. Some observations on the role of predation in the natural regulation of *Ips typographus* populations. Z. angew. Entomol. 99, 209–215.

Mills, N.J., 1991. Searching strategies and attack rates of parasitoids of the ash bark beetle (*Leperisinus varius*) and its relevance to biological control. Ecol. Entomol. 16, 461–470.

Mock, K.E., Bentz, B., O'Neill, E., Chong, J., Orwin, J., Pfrender, M., 2007. Landscape-scale genetic variation in a forest outbreak species, the mountain pine beetle (*Dendroctonus ponderosae*). Mol. Ecol. 16, 553–568.

Moeck, H.A., Wood, D.L., Lindahl Jr., K.Q., 1981. Host selection behavior of bark beetles (Coleoptera: Scolytidae) attacking *Pinus ponderosa*, with special emphasis on the western pine beetle, *Dendroctonus brevicomis*. J. Chem. Ecol. 7, 49–83.

Morales-Jiménez, J., Zúñiga, G., Villa-Tanaca, L., Hernández-Rodríguez, C., 2009. Bacterial community and nitrogen fixation in the red turpentine beetle, *Dendroctonus valens* LeConte (Coleoptera: Curculionidae: Scolytinae). Microb. Ecol. 58, 879–891.

Morales-Jiménez, J., Zúñiga, G., Ramírez-Saad, H.C., Hernández-Rodríguez, C., 2012. Gut-associated bacteria throughout the life cycle of the bark beetle *Dendroctonus rhizophagus* Thomas and Bright (Curculionidae, Scolytinae) and their cellulolytic activities. Microb. Ecol. 64, 268–278.

Moser, J.C., 1985. Use of sporothecae by phoretic *Tarsonemus* mites to transport ascospores of coniferous bluestain fungi. Trans. Br. Mycol. Soc. 84, 750–753.

Moser, J.C., Roton, L.M., 1971. Mites associated with southern pine bark beetles in Allen Parish, Louisiana. Can. Entomol. 103, 1775–1798.

Moser, J.C., Macías-Sámano, J.E., 2000. Tarsonemid mite associates of *Dendroctonus frontalis* (Coleoptera: Scolytidae): implications for the historical biogeography of *D. frontalis*. Can. Entomol. 132, 765–771.

Moser, J.C., Cross, E.A., Roton, L.M., 1971. Biology of *Pyemotes parviscolyti* (Acarina: Pyemotidae). Entomophaga 16, 367–379.

Moser, J.C., Wilkinson, R.C., Clark, E.W., 1974. Mites associated with *Dendroctonus frontalis* Zimmermann (Scolytidae: Coleoptera) in Central America and Mexico. Turrialba 24, 379–381.

Moser, J.C., Perry, T.J., Solheim, H., 1989. Ascospores hyperphoretic on mites associated with *Ips typographus*. Mycol. Res. 93, 513–517.

Moser, J.C., Perry, T.J., Furuta, K., 1997. Phoretic mites and their hyperphoretic fungi associated with flying *Ips typographus japonicus* Niijima (Col., Scolytidae) in Japan. J. Appl. Entomol. 121, 425–428.

Müller, J., Bussler, H., Gossner, M., Rettelbach, T., Duelli, P., 2008. The European spruce bark beetle *Ips typographus* in a national park: from pest to keystone species. Biodivers. Conserv. 17, 2979–3001.

Nilssen, A.C., 1978. Spatial attack pattern of the bark beetle *Tomicus piniperda* L. (Col., Scolytidae). Norwegian J. Entomol. 25, 171–175.

Nilssen, A.C., 1984. Long-range aerial dispersal of bark beetles and bark weevils (Coleoptera, Scolytidae and Curculionidae) in northern Finland. Ann. Entomol. Fenn. 50, 37–42.

Nordlund, D.A., 1981. Semiochemicals: a review of the terminology. In: Nordlund, D.A., Jones, R.L., Lewis, W.J. (Eds.), Semiochemicals: Their Role in Pest Control. John Wiley and Sons, New York, pp. 13–23.

Normark, B.B., Jordal, B.H., Farrell, B.D., 1999. Origin of a haplodiploid beetle lineage. *Proc. R. Soc. Lond., Ser.* Bull. Am. Meteorol. Soc. 266, 2253–2259.

Nuorteva, M., 1959. Untersuchungen über einige in den Frassbildern der Borkenkäfer lebende *Medetera*-Artern (Dipt., Dolichopodidae). Ann. Entomol. Fenn. 25, 192–210.

Ohmart, C.P., 1989. Why are there so few tree-killing bark beetles associated with angiosperms? Oikos 54, 242–245.

Økland, B., Bjørnstad, O.N., 2006. A resource-depletion model of forest insect outbreaks. Ecology 87, 283–290.

Ott, D.S., Yanchuk, A.D., Huber, D.P.W., Wallin, K.F., 2011. Genetic variation of lodgepole pine, *Pinus contorta* var. *latifolia*, chemical and physical defenses that affect mountain pine beetle, *Dendroctonus ponderosae*, attack and tree mortality. J. Chem. Ecol. 37, 1002–1012.

Owen, D.R., Wood, D.L., Parmeter, J.R., 2005. Association between *Dendroctonus valens* and black stain root disease on ponderosa pine in the Sierra Nevada of California. Can. Entomol. 137, 367–375.

Paine, T.D., Birch, M.C., Svihra, P., 1981. Niche breadth and resource partitioning by 4 sympatric species of bark beetles (Coleoptera, Scolytidae) Oecologia 48, 1–6.

Paine, T.D., Raffa, K.F., Harrington, T.C., 1997. Interactions among scolytid bark beetles, their associated fungi, and live host conifers. Annu. Rev. Entomol. 42, 179–206.

Pajares, J.A., Lanier, G.N., 1990. Biosystematics of the turpentine beetles *Dendroctonus terebrans* and *D. valens* (Coleoptera: Scolytidae). Ann. Entomol. Soc. Am. 83, 171–188.

Paynter, Q.E., Anderbrant, O., Schlyter, F., 1990. Behavior of male and female spruce bark beetles, *Ips typographus*, on the bark of host trees during mass attack. J. Insect. Behav. 3, 529–543.

Phillips, T.W., Atkinson, T.H., Foltz, J.L., 1989. Pheromone-based aggregation in *Orthotomicus caelatus* Eichhoff (Coleoptera: Scolytidae). Can. Entomol. 121, 933–940.

Pitman, G.B., Vité, J.P., 1970. Field response of *Dendroctonus pseudotsugae* (Coleoptera: Scolytidae) to synthetic frontalin. Ann. Entomol. Soc. Am. 63, 661–664.

Pitman, G.B., Vité, J.P., Kinzer, G.W., Fentiman, A.F., 1969. Specificity of population-aggregating pheromones in *Dendroctonous*. J. Insect Physiol. 15, 363–366.

Plattner, A., Kim, J.-J., DiGuistini, S., Breuil, C., 2008. Variation in pathogenicity of a mountain pine beetle-associated blue-stain fungus, *Grosmannia clavigera*, on young lodgepole pine in British Columbia. Can. J. Plant Pathol. 30, 1–10.

Poland, T.M., Borden, J.H., 1998a. Competitive exclusion of *Dendroctonus rufipennis* induced by pheromones of *Ips tridens* and *Dryocoetes affaber* (Coleoptera: Scolytidae). J. Econ. Entomol. 91, 1150–1161.

Poland, T.M., Borden, J.H., 1998b. Disruption of secondary attraction of the spruce beetle, *Dendroctonus rufipennis*, by pheromones of two sympatric species. J. Chem. Ecol. 24, 151–166.

Poland, T.M., McCullough, D.G., 2006. Emerald ash borer: invasion of the urban forest and the threat to North America's ash resource. J. For. 104, 118–124.

Pope, D.N., Coulson, R.N., Fargo, W.S., Gagne, J.A., Kelly, C.W., 1980. The allocation process and between-tree survival probabilities in *Dendroctonus frontalis* infestations. Res. Pop. Ecol. 22, 197–210.

Popp, M.P., Johnson, J.D., Massey, T.L., 1991. Stimulation of resin flow in slash and loblolly pine by bark beetle vectored fungi. Can. J. For. Res. 21, 1124–1126.

Powell, E.N., Townsend, P.A., Raffa, K.F., 2012. Wildfire provides refuge from local extinction but is an unlikely driver of outbreaks by mountain pine beetle. Ecol Monogr. 82, 69–84.

Powell, J.A., Bentz, B.J., 2009. Connecting phenological predictions with population growth rates for mountain pine beetle, an outbreak insect. Landscape Ecol. 24, 657–672.

Preisler, H.K., Hicke, J.A., Ager, A.A., Hayes, J.L., 2012. Climate and weather influences on spatial temporal patterns of mountain pine beetle populations in Washington and Oregon. Ecology 93, 2421–2434.

Price, J.I., McCollum, D.W., Berrens, R.P., 2010. Insect infestation and residential property values: a hedonic analysis of the mountain pine beetle epidemic. Forest Pol. and Econ. 12, 415–422.

Progar, R. A., Eglitis, A., and Lundquist, J. E. (2009). Some ecological, economic, and social consequences of bark beetle infestations.

The Western Bark Beetle Research Group: A Unique Collaboration With Forest Health Protection, 71. Proceedings of a Symposium at the 2007 Society of American Foresters Conference, October 23–28, 2007, Portland, Oregon. United States Department of Agriculture Forest Service Pacific Northwest Research Station General Technical Report PNW-GTR-784 April 2009.

Pureswaran, D.S., Borden, J.H., 2003. Is bigger better? Size and pheromone production in the mountain pine beetle, *Dendroctonus ponderosae* Hopkins (Coleoptera: Scolytidae). J. Insect. Behav. 16, 765–782.

Pureswaran, D.S., Sullivan, B.T., Ayres, M.P., 2006. Fitness consequences of pheromone production and host selection strategies in a tree-killing bark beetle (Coleoptera: Curculionidae: Scolytinae). Oecologia 148, 720–728.

Raffa, K.F., 1988. Host orientation behavior of *Dendroctonus ponderosae*: integration of token stimuli and host defenses. In: Mattson, W.J., Levieux, J., Bernard-Dagan, C. (Eds.), Mechanisms of Woody Plant Resistance to Insects and Pathogens. Springer-Verlag, New York, pp. 369–390.

Raffa, K.F., 2001. Mixed messages across multiple trophic levels: the ecology of bark beetle chemical communication systems. Chemoecology 11, 49–65.

Raffa, K.F., 2014. Terpenes tell different tales at different scales: glimpses into the chemical ecology of conifer–bark beetle–microbial interactions. J. Chem. Ecol. 40, 1–20.

Raffa, K.F., Berryman, A.A., 1980. Flight responses and host selection by bark beetles. In: Berryman, A.A., Safranyik, L. (Eds.), Dispersal of Forest Insects: Evaluation, Theory, and Management Implications. Conference Office, Cooperative Extension Service, Washington State University, Pullman WA, pp. 213–233.

Raffa, K.F., Berryman, A.A., 1982. Gustatory cues in the orientation of *Dendroctonus ponderosae* (Coleoptera: Scolytidae) to host trees. Can. Entomol. 114, 97–104.

Raffa, K.F., Berryman, A.A., 1983. The role of host plant resistance in the colonization behavior and ecology of bark beetles (Coleoptera: Scolytidae). Ecol. Monogr. 53, 27–49.

Raffa, K.F., Berryman, A.A., 1986. A mechanistic computer model of mountain pine beetle populations interacting with lodgepole pine stands and its implications for forest managers. For. Sci. 32, 789–805.

Raffa, K.F., Berryman, A.A., 1987. Interacting selective pressures in conifer-bark beetle systems: a basis for reciprocal adaptations? Am. Nat. 129, 234–262.

Raffa, K.F., Klepzig, K.D., 1989. Chiral escape of bark beetles from predators responding to a bark beetle pheromone. Oecologia 80, 566–569.

Raffa, K.F., Dahlsten, D.L., 1995. Differential responses among natural enemies and prey to bark beetle pheromones. Oecologia 102, 17–23.

Raffa, K.F., Smalley, E.B., 1988. Response of red and jack pines to inoculations with microbial associates of the pine engraver, *Ips pini* (Coleoptera: Scolytidae). J. For. Res. 18, 581–586.

Raffa, K.F., Smalley, E.B., 1995. Interaction of pre-attack and induced monoterpene concentrations in conifer defense against bark beetle-microbial complexes. Oecologia 102, 285–295.

Raffa, K.F., Phillips, T.W., Salom, S.M., 1993. Strategies and mechanisms of host colonization by bark beetles. In: Schowalter, T.D., Filip, G.M. (Eds.), Beetle-Pathogen Interactions in Conifer Forests. Academic Press, San Diego, pp. 103–128.

Raffa, K.F., Aukema, B.H., Erbilgin, N., Klepzig, K.D., Wallin, K.F., 2005. Interactions among conifer terpenoids and bark beetles across multiple levels of scale, an attempt to understand links between population patterns and physiological processes. Recent Adv. Phytochem. 39, 80–118.

Raffa, K.F., Hobson, K.R., LaFontaine, S., Aukema, B.H., 2007. Can chemical communication be cryptic? Adaptations by herbivores to natural enemies exploiting prey semiochemistry. Oecologia 153, 1009–1019.

Raffa, K.F., Aukema, B.H., Bentz, B.J., Carroll, A.L., Hicke, J.A., Turner, M.G., Romme, W.H., 2008. Cross-scale drivers of natural disturbances prone to anthropogenic amplification: the dynamics of bark beetle eruptions. BioScience 58, 501–517.

Raffa, K. F., Aukema, B. H., Bentz, B. J., Carroll A. L., Hicke J. A., and Kolb, T. E. Responses of tree-killing bark beetles to a changing climate. In "Climate Change and Insect Pests" (C. Björkman and P. Niemelä, Eds.). CABI, Wallingford, in press.

Rankin, L.J., Borden, J.H., 1991. Competitive interactions between the mountain pine beetle and the pine engraver in lodgepole pine. Can. J. For. Res. 21, 1029–1036.

Rasmussen, L.E., Greenwood, D.R., 2003. Frontalin: a chemical message of musth in Asian elephants (*Elephas maximus*). Chem. Senses 28, 433–446.

Redmer, J.S., Wallin, K.F., Raffa, K.F., 2001. Effect of host tree seasonal phenology on substrate suitability for the pine engraver, *Ips pini* (Coleoptera: Scolytidae): implications to population dynamics and enemy free space. J. Econ. Entomol. 94, 844–849.

Reeve, J.D., 1997. Predation and bark beetle dynamics. Oecologia 112, 48–54.

Reeve, J.D., 2000. Complex emergence patterns in a bark beetle predator. Agric. For. Entomol. 2, 233–240.

Régnière, J., Bentz, B., 2007. Modeling cold tolerance in the mountain pine beetle, *Dendroctonus ponderosae*. J. Insect Physiol 53, 559–572.

Régnière, J., Powell, J., Bentz, B., Nealis, V., 2012. Effects of temperature on development, survival and reproduction of insects: Experimental design, data analysis and modeling. J. Insect Physiol. 58, 634–647.

Reid, M.L., Roitberg, B.D., 1994. Benefits of prolonged male residence with mates and brood in pine engravers (Coleoptera: Scolytidae). Oikos 70, 140–148.

Reid, R.W., Whitney, H.S., Watson, J.A., 1967. Reactions of lodgepole pine to attack by *Dendroctonus ponderosae* Hopkins and blue stain fungi. Can. J. Bot. 45, 1115–1126.

Renwick, J.A.A., Vité, J.P., 1969. Bark beetle attractants: mechanism of colonization by *Dendroctonus frontalis*. Nature 224, 1222–1223.

Ricker, W.E., 1954. Stock and recruitment. J. Fish. Res. Board Can. 11, 559–623.

Rittmaster, R., Adamowicz, W.L., Amiro, B., Pelletier, R.T., 2006. Economic analysis of health effects from forest fires. Can. J. For. Res. 36, 868–877.

Roberds, J.H., Strom, B.L., 2006. Repeatability estimates for oleoresin yield measurements in three species of the southern pines. For. Ecol. Manage. 228, 215–224.

Robins, G.L., Reid, M.L., 1997. Effects of density on the reproductive success of pine engravers: is aggregation in dead trees beneficial? Ecol. Entomol. 22, 329–334.

Romme, W.H., Knight, D.H., Yavitt, J.B., 1986. Mountain pine beetle outbreaks in the Rocky Mountains: regulators of primary productivity? Am. Nat. 127, 484–494.

Rosner, S., Hannrup, B., 2004. Resin canal traits relevant for constitutive resistance of Norway spruce against bark beetles: environmental and genetic variability. For. Ecol. Manage. 200, 77–87.

Rudinsky, J.A., 1962. Ecology of Scolytidae. Annu. Rev. Entomol. 7, 327–348.

Rudinsky, J.A., Michael, R.R., 1973. Sound production in Scolytidae: stridulation by female *Dendroctonus* beetles. J. Insect Physiol. 19, 689–705.

Rudinsky, J.A., Morgan, M.E., Libbey, L.M., Putnam, T.B., 1974. Anti-aggregative rivalry pheromone of the mountain pine beetle, and a new arrestant of the southern pine beetle. Environ. Entomol. 3, 90–98.

Ruel, J.J., Ayres, M.P., Lorio, P.L., 1998. Loblolly pine responds to mechanical wounding with increased resin flow. Can. J. For. Res. 28, 596–602.

Rühm, W., 1956. Die Nematoden der Ipiden. Parasitologische Schriftenreihe 6, Veb Gustav Fischer Verlag, Jena.

Ryall, K.L., Fahrig, L., 2005. Habitat loss decreases predator-prey ratios in a pine-bark beetle system. Oikos 110, 265–270.

Ryan, R.B., Rudinsky, J.A., 1962. Biology and habits of the Douglas-fir beetle parasite, *Coeloides brunneri* Viereck (Hymenoptera: Braconidae), in western Oregon. Can. Entomol. 94, 748–763.

Ryker, L.C., Rudinsky, J.A., 1976. Sound production in Scolytidae: aggressive and mating behavior of the mountain pine beetle. Ann. Entomol. Soc. Am. 69, 677–680.

Sachtleben, H., 1952. Die parasitischen Hymenopteren des Fichtenborkenkäfers. Ips typographus L. Beitr. Entomol. 2, 137–189.

Safranyik, L., Carroll, A.L., 2006. The biology and epidemiology of the mountain pine beetle in lodgepole pine forests. In: Safranyik, L., Wilson, W.R. (Eds.), The Mountain Pine Beetle: A Synthesis of Biology, Management, and Impacts on Lodgepole Pine. Natural Resources Canada, Canadian Forest Service, Pacific Forestry Centre, Victoria, B.C., Canada, pp. 3–66.

Safranyik, L., Shrimpton, D.M., Whitney, H.S., 1975. An interpretation of the interaction between lodgepole pine, the mountain pine beetle and its associated blue stain fungi in western Canada. In: Baumgartner, M. (Ed.), Management of Lodgepole Pine Ecosystems, Washington State University Cooperative Extension Service, Pullman, WA, pp. 406–428.

Safranyik, L., Shore, T.L., Linton, D.A., 1996. Ipsdienol and lanierone increase *Ips pini* Say (Coleoptera: Scolytidae) attack and brood density in lodgepole pine infested by mountain pine beetle. Can. Entomol. 128, 199–207.

Safranyik, L., Carroll, A.L., Régnière, J., Langor, D.W., Riel, W.G., Shore, T.L., et al., 2010. Potential for range expansion of mountain pine beetle into the boreal forest of North America. Can. Entomol. 142, 415–442.

Saint-Germain, M., Buddle, C.M., Drapeau, P., 2007. Primary attraction and random landing in host-selection by wood-feeding insects: a matter of scale? Agric. For. Entomol. 9, 227–235.

Saint-Germain, M., Drapeau, P., Buddle, C.M., 2009. Landing patterns of phloem- and wood-feeding Coleoptera on black spruce of different physiological and decay states. Environ. Entomol. 38, 797–802.

Sallé, A., Raffa, K.F., 2007. Interactions among intraspecific competition, emergence patterns, and host selection behaviour in *Ips pini* (Coleoptera: Scolytinae). Ecol. Entomol. 32, 162–171.

Salom, S.M., McLean, J.A., 1989. Influence of wind on the spring flight of *Trypodendron lineatum* (Olivier) (Coleoptera: Scolytidae) in a second-growth coniferous forest. Can. Entomol. 121, 109–119.

Samarasekera, G.D., Bartell, N.V., Lindgren, B.S., Cooke, J.E., Davis, C. S., James, P.M., et al., 2012. Spatial genetic structure of the mountain pine beetle (*Dendroctonus ponderosae*) outbreak in western Canada: historical patterns and contemporary dispersal. Mol. Ecol. 21, 2931–2948.

Samson, P.R., 1984. The biology of *Roptrocerus xylophagorum* (Hym.: Torymidae), with a note on its taxonomic status. Entomophaga 29, 287–298.

Sandstrom, P., Welch, W.H., Blomquist, G.J., Tittiger, C., 2006. Functional expression of a bark beetle cytochrome P450 that hydroxylates myrcene to ipsdienol. Insect Biochem. Mol. Biol. 36, 835–845.

Schiebe, C., Hammerbacher, A., Birgersson, G., Witzell, J., Brodelius, P. E., Gershenzon, J., et al., 2012. Inducibility of chemical defenses in Norway spruce bark is correlated with unsuccessful mass attacks by the spruce bark beetle. Oecologia 170, 183–198.

Schlyter, F., Birgersson, G., 1989. Individual variation in bark beetle and moth pheromones: a comparison and an evolutionary background. Holarctic Ecol. 12, 457–465.

Schlyter, F., Anderbrant, O., 1993. Competition and niche separation between two bark beetles: existence and mechanisms. Oikos 68, 437–447.

Schlyter, F., Birgersson, G., Byers, J.A., Bakke, A., 1992. The aggregation pheromone of *Ips duplicatus*, and its role in competitive interactions with *I. typographus*. Chemoecology 3, 103–112.

Schoeller, E.N., Husseneder, C., Allison, J.D., 2012. Molecular evidence of facultative intraguild predation by *Monochamus titillator* larvae (Coleoptera: Cerambycidae) on members of the southern pine beetle guild. Naturwissenschaften 99, 913–924.

Schroeder, L.M., 1996. Interactions between the predators *Thanasimus formicarius* (Col.: Cleridae) and *Rhizophagus depressus* (Col: Rhizophagidae), and the bark beetle *Tomicus piniperda* (Col.: Scolytidae). Entomophaga 41, 63–75.

Schroeder, L.M., 1999a. Population levels and flight phenology of bark beetle predators in stands with and without previous infestations of the bark beetle *Tomicus piniperda*. For. Ecol. Manage. 123, 31–40.

Schroeder, L.M., 1999b. Prolonged development time of the bark beetle predator *Thanasimus formicarius* (Col.: Cleridae) in relation to its prey species *Tomicus piniperda* (L.) and *Ips typographus* (L.) (Col.: Scolytidae). Agric. For. Entomol. 1, 127–135.

Schroeder, L.M., 2007. Escape in space from enemies: a comparison between stands with and without enhanced densities of the spruce bark beetle. Agric. For. Entomol. 9, 85–91.

Schroeder, L.M., Weslien, J., 1994. Reduced offspring production in the bark beetle *Tomicus piniperda* in pine bolts baited with ethanol and α-pinene which attract antagonistic insects. J. Chem. Ecol. 20, 1429–1444.

Schurig, V., Weber, R., Klimetzek, D., Kohnle, U., Mori, K., 1982. Enantiomeric composition of "lineatin" in three sympatric ambrosia beetles. Naturwissenschaften 69, 602–603.

Schwerdfeger, F. (1955). Pathogenese der Borkenkäfer-Epidemie. *Nordwestdeutschland. Schriftenreiheder Forstlichen Fakultät der Universität Göttingen*, 13/14, 135 p.

Scott, J.J., Oh, D.-C., Yuceer, M.C., Klepzig, K.D., Clardy, J., Currie, C.R., 2008. Bacterial protection of beetle-fungus mutualism. Science 322, 63.

Senger, S.E., Roitberg, B.D., 1992. Effects of parasitism by *Tomicobia tibialis* Ashmead (Hymenoptera: Pteromalidae) on reproductive parameters of female pine engravers, *Ips pini* (Say). Can. Entomol. 124, 509–513.

Seybold, S.J., Quilici, D.R., Tillman, J.A., Vanderwel, D., Wood, D.L., Blomquist, G.J., 1995. *De novo* biosynthesis of the aggregation pheromone components ipsenol and ipsdienol by the pine bark beetles *Ips paraconfusus* Lanier and *Ips pini* (Say) (Coleoptera: Scolytidae). Proc. Natl. Acad. Sci. U. S. A. 92, 8393–8397.

Shepherd, R.F., 1966. Factors influencing the orientation and rates of activity of *Dendroctonus ponderosae* (Coleoptera: Scolytidae). Can. Entomol. 98, 507–518.

Shi, Z.-H., Wang, B., Clarke, S.R., Sun, J.H., 2012. Effect of associated fungi on the immunocompetence of red turpentine beetle larvae, *Dendroctonus valens* (Coleoptera: Curculionidae: Scolytinae). Insect Sci. 19, 579–584.

Shimizu, A., Tanaka, R., Akiba, M., Masuya, H., Iwata, R., Fukuda, K., Kanzaki, N., 2013. Nematodes associated with *Dryocoetes uniseriatus* (Coleoptera: Scolytidae). Environ. Entomol. 42, 79–88.

Shore, T.L., Safranyik, L., Lindgren, B.S., 1992. The response of mountain pine beetle (*Dendroctonus ponderosae*) to lodgepole pine trees baited with verbenone and *exo*-brevicomin. J. Chem. Ecol. 18, 533–541.

Simard, M., Powell, E.N., Raffa, K.F., Turner, M.G., 2012. What explains landscape patterns of tree mortality caused by bark beetle outbreaks in Greater Yellowstone? Glob. Ecol. Biogeogr. 21, 556–567.

Simpson, S.J., McCaffery, A.R., Haegele, B.F., 1999. A behavioural analysis of phase change in the desert locust. Biol. Rev. 74, 461–480.

Six, D.L., 2012. Ecological and evolutionary determinants of bark beetle-fungus symbioses. Insects 3, 339–366.

Six, D.L., Klepzig, K.D., 2004. *Dendroctonus* bark beetles as model systems for the study of symbiosis. Symbiosis 37, 207–232.

Six, D.L., Wingfield, M.J., 2011. The role of phytopathogenicity in bark beetle–fungus symbioses: a challenge to the classic paradigm. Annu. Rev. Entomol. 56, 255–272.

Smith, G.D., Carroll, A.L., Lindgren, B.S., 2009. The life history of a secondary bark beetle, *Pseudips mexicanus* (Coleoptera: Curculionidae: Scolytinae), in lodgepole pine. Can. Entomol. 141, 56–69.

Smith, G.D., Carroll, A.L., Lindgren, B.S., 2011. Facilitation in bark beetles: endemic mountain pine beetle gets a helping hand (Coleoptera: Curculionidae: Scolytinae). Agric. For. Entomol. 13, 37–43.

Southwood, T.R.E., Comins, H.N., 1976. A synoptic population model. J. Anim. Ecol. 45, 949–965.

Steed, B.E., Wagner, M.R., 2008. Seasonal pheromone response by *Ips pini* in northern Arizona and western Montana, USA. Agric. For. Entomol. 10, 189–203.

Stenseth, N.C., Kirkendall, L.R., Moran, N., 1985. On the evolution of pseudogamy. Evolution 39, 294–307.

Stephen, F.M., Browne, L.E., 2000. Application of Eliminade™ parasitoid food to boles and crowns of pines (Pinaceae) infested with *Dendroctonus frontalis* (Coleoptera: Scolytidae). Can. Entomol. 132, 983–985.

Stephen, F.M., Dahlsten, D.L., 1976. The arrival sequence of the arthropod complex following attack by *Dendroctonus brevicomis* (Coleoptera: Scolytidae) in ponderosa pine. Can. Entomol. 108, 283–304.

Stoszek, K.J., Rudinsky, J.A., 1967. Injury of Douglas-fir trees by maturation feeding of the Douglas-fir hylesinus, *Pseudohylesinus nebulosus* (Coleoptera: Scolytidae). Can. Entomol. 99, 310–311.

Strom, B.L., Roton, L.M., Goyer, R.A., Meeker, J.R., 1999. Visual and semiochemical disruption of host finding in the southern pine beetle. Ecol. Appl. 9, 1028–1038.

Sturgeon, K.B., Mitton, J.B., 1986. Allozyme and morphological differentiation of mountain pine beetles *Dendroctonus ponderosae* Hopkins associated with host tree. Evolution 40, 290–302.

Sullivan, B.T., Berisford, C.W., 2004. Semiochemicals from fungal associates of bark beetles may mediate host location behavior of parasitoids. J. Chem. Ecol. 30, 703–717.

Sun, J., Miao, Z., Zhang, Z., Zhang, Z., Gillette, N.E., 2004. Red turpentine beetle, *Dendroctonus valens* LeConte (Coleoptera: Scolytidae), response to host semiochemicals in China. Environ. Entomol. 33, 206–212.

Sun, J., Lu, M., Gillette, N.E., Wingfield, M.J., 2013. Red turpentine beetle, innocuous native becomes invasive tree killer in China. Annu. Rev. Entomol. 58, 293–311.

Teale, S.N., Lanier, G.N., 1991. Seasonal variability in response of *Ips pini* (Coleoptera: Scolytidae) to ipsdienol in New York. J. Chem. Ecol. 17, 1145–1158.

Thalenhorst, W., 1958. Grundzüge der Populations dynamik des grossen Fichtenborkenkäfers *Ips typographus* L. Schriftenreiheder Forstlichen Fakultät der Universität Göttingen 21, 1–126.

Thong, C.H.S., Webster, J.M., 1983. Nematode parasites and associates of *Dendroctonus* spp. and *Trypodendron lineatum* (Coleoptera, Scolytidae), with a description of *Bursaphelenchus varicauda* n. sp. J. Nematol. 15, 312–318.

Tømmerås, B.Å., 1985. Specialization of the olfactory receptor cells in the bark beetle *Ips typographus* and its predator *Thanasimus formicarius* to bark beetle pheromones and host tree volatiles. J. Comp. Physiol. A 157, 335–341.

Tømmerås, B.Å., 1988. The clerid beetle, *Thanasimus formicarius*, is attracted to the pheromone of the ambrosia beetle, *Trypodendron lineatum*. Experientia 44, 536–537.

Trần, J.K.I., Ylioja, T., Billings, R.F., Régnière, J., Ayres, M.P., 2007. Impact of minimum winter temperatures on the population dynamics of *Dendroctonus frontalis*. Ecol. Appl. 17, 882–899.

Turchin, P., Taylor, A.D., Reeve, J.D., 1999. Dynamical role of predators in population cycles of a forest insect: an experimental test. Science 285, 1068–1071.

Vanhaelen, M., Vanhaelen-Fastré, R., Geeraerts, J., 1978. Volatile constituents of *Trichothecium roseum*. Sabouraudia 16, 141–150.

VanLaerhoven, S.L., Stephen, F.M., 2008. Incidence of honeydew in southern pine-hardwood forests: implications for adult parasitoids of the southern pine beetle, *Dendroctonus frontalis* (Coleoptera: Scolytidae). Biocontrol Sci. Technol. 18, 957–965.

Vité, J.P., Williamson, D.L., 1970. *Thanasimus dubius*: prey perception. J. Insect Physiol. 16, 233–239.

Vité, J.P., Bakke, A., Renwick, J.A.A., 1972. Pheromones in *Ips* (Coleoptera: Scolytidae): occurrence and production. Can. Entomol. 104, 1967–1975.

Wagner, T.L., Gagne, J.A., Sharpe, P.J.H., Coulson, R.N., 1984. A biophysical model of southern pine beetle Dendroctonus frontalis (Coleoptera: Scolytidae) development. Ecol. Model. 21, 125–147.

Wallin, K.F., Raffa, K.F., 1999. Altered constitutive and inducible phloem monoterpenes following natural defoliation of jack pine: implications to host mediated inter-guild interactions and plant defense theories. J. Chem. Ecol. 25, 861–880.

Wallin, K.F., Raffa, K.F., 2000. Influences of external chemical cues and internal physiological parameters on the multiple steps of post-landing host selection behavior of *Ips pini* (Coleoptera: Scolytidae). Environ. Entomol. 29, 442–453.

Wallin, K.F., Raffa, K.F., 2001. Effects of folivory on subcortical plant defenses: can defense theories predict interguild processes? Ecology 82, 1387–1400.

Wallin, K.F., Raffa, K.F., 2002. Prior encounters modulate subsequent choices in host acceptance behavior by the bark beetle *Ips pini*. Entomol. Exp. Applic. 103, 205–218.

Wallin, K.F., Raffa, K.F., 2004. Feedback between individual host selection behavior and population dynamics in an eruptive insect herbivore. Ecol. Monogr. 74, 101–116.

Wallin, K.F., Rutledge, J., Raffa, K.F., 2002. Heritability of host acceptance and gallery construction behaviors of the bark beetle *Ips pini* (Coleoptera: Scolytidae). Environ. Entomol. 31, 1276–1281.

Wang, B., Lu, M., Cheng, C., Salcedo, C., Sun, J., 2013. Saccharide-mediated antagonistic effects of bark beetle fungal associates on larvae. Biol. Lett. 9, 20120787.

Warzée, N., Gilbert, M., Grégoire, J.-C., 2006. Predator/prey ratios: a measure of bark-beetle population status influenced by stand composition in different French stands after the 1999 storms. Ann. For. Sci. 63, 301–308.

Webber, J.F., 1990. The relative effectiveness of *Scolytus scolytus, S. multistriatus and S. kirschii* as vectors of Dutch elm disease. Eur. J. For. Pathol. 20, 184–192.

Wegensteiner, R., 2004. Pathogens in bark beetles. In: Lieutier, F., Day, K. R., Battisti, A., Grégoire, J.-C., Evans, H.F. (Eds.), Bark and Wood Boring Insects in Living Trees in Europe, a Synthesis. Kluwer, Dordrecht, pp. 291–313.

Wegensteiner, R., Dedryver, C.-A., Pierre, J.-S., 2010. The comparative prevalence and demographic impact of two pathogens in swarming *Ips typographus* adults: a quantitative analysis of long term trapping data. Agric. For. Entomol. 12, 49–57.

Weiser, J., 1961. Die Mikrosporidien als Parasiten der Insekten. Monographien zur angewandte Entomologie 17. Hamburg and Berlin, Paul Parey.

Wermelinger, B., Duelli, P., Obrist, M.K., 2002. Dynamics of saproxylic beetles (Coleoptera) in windthrow areas in alpine spruce forests. Forest Snow and Landscape Research 77, 133–148.

Wermelinger, B., Obrist, M.K., Baur, H., Jakoby, O., Duelli, P., 2013. Synchronous rise and fall of bark beetle and parasitoid populations in windthrow areas. Agric. For. Entomol. 15, 301–309.

Werner, R.A., Holsten, E.H., 1985. Factors influencing generation times of spruce beetles in Alaska. Can. J. For. Res. 15, 438–443.

Werner, R.A., Raffa, K.F., Illman, B.L., 2006. Insect and pathogen dynamics. In: Chapin III, F.S., Oswood, M., Van Cleve, K., Viereck, L.A., Verbyla, D. (Eds.), Alaska's Changing Boreal Forest. Oxford University Press, Oxford, pp. 133–146.

Weslien, J., Regnander, J., 1992. The influence of natural enemies on brood production in *Ips typographus* (Col. Scolytidae) with special reference to egg-laying and predation by *Thanasimus formicarius* (Col.: Cleridae). Entomophaga 37, 333–342.

Wichmann, L., Ravn, H.P., 2001. The spread of *Ips typographus* (L.) (Coleoptera, Scolytidae) attacks following heavy windthrow in Denmark, analysed using GIS. For. Ecol. Manage. 148, 31–39.

Wilkinson, R.C., McClelland, W.T., Murillo, R.M., Ostmark, E.O., 1967. Stridulation and behavior in two southeastern *Ips* bark beetles (Coleoptera: Scolytidae). Fla. Entomol. 50, 185–195.

Wilson, E.O., 1971. The Insect Societies. Harvard University Press, Cambridge.

Witcosky, J.J., Schowalter, T.D., Hansen, E.M., 1986a. *Hylastes nigrinus* (Coleoptera: Scolytidae), *Pissodes fasciatus*, and *Steremnius carinatus* (Coleoptera: Curculionidae) as vectors of black-stain root disease of Douglas-fir. Environ. Entomol. 15, 1090–1095.

Witcosky, J.J., Schowalter, T.D., Hansen, E.M., 1986b. The influence of time of precommercial thinning on the colonization of Douglas-fir by three species of root-colonizing insects. Can. J. For. Res. 16, 745–749.

Wood, D.L., 1972. Selection and colonization of ponderosa pine by bark beetles. In "Insect–Plant Relationships" (H. F. van Emden, Ed.). Symp. R. Entomol. Soc. Lond. 6, 101–107.

Wood, D.L., 1982. The role of pheromones, kairomones, and allomones in the host selection and colonization behavior of bark beetles. Annu. Rev. Entomol. 27, 411–446.

Wood, S.L., 1982. The bark and ambrosia beetles of North and Central America (Coleoptera: Scolytidae), a taxonomic monograph. Great Basin Nat. Mem. 6, 1–1359.

Wood, S.L., 1986. A reclassification of the genera of Scolytidae (Coleoptera). Great Basin Nat. Mem. 10, 1–126.

Wood, S.L., Bright, D.E., 1992. A catalog of Scolytidae and Platypodidae (Coleoptera), part 2: Taxonomic index. Great Basin Nat. Mem. 13, 1–1553.

Yan, Z.-G., Sun, J., Owen, D., Zhang, Z., 2005. The red turpentine beetle, *Dendroctonus valens* LeConte (Scolytidae): an exotic invasive pest of pine in China. Biodivers. Conserv. 14, 1735–1760.

Young, J.C., Silverstein, R.M., Birch, M.C., 1973. Aggregation pheromone of the beetle *Ips confusus*: isolation and identification. J. Insect Physiol. 19, 2273–2277.

Zayed, A., Constantin, S.A., Packer, L., 2007. Successful biological invasion despite a severe genetic load. PLoS One 2, e868.

Zhao, T., Krokene, P., Hu, J., Christiansen, E., Björklund, N., Långström, B., Solheim, H., Borg-Karlson, A.-K., 2011. Induced terpene accumulation in Norway spruce inhibits bark beetle colonization in a dose-dependent manner. PLoS One 6, e26649.

Zhou, J., Ross, D.W., Niwa, C.G., 2001. Kairomonal response of *Thanasimus undatulus, Enoclerus sphegeus* (Coleoptera: Cleridae), and *Temnochila chlorodia* (Coleoptera: Trogositidae) to bark beetle semiochemicals in Eastern Oregon. Environ. Entomol. 30, 993–998.

Zumr, V., 1992. Dispersal of the spruce bark beetle *Ips typographus* (L.) (Col., Scolytidae) in spruce woods. J. Appl. Entomol. 114, 348–352.

Zwölfer, W., 1957. Ein Jahrzehnt forstentomologischer Forschung. 1946–1956. Z. angew. Entomol. 40, 422–432.

Chapter 2

Morphology, Taxonomy, and Phylogenetics of Bark Beetles

Jiri Hulcr[1], Thomas H. Atkinson[2], Anthony I. Cognato[3], Bjarte H. Jordal[4], and Duane D. McKenna[5]

[1] School of Forest Resources and Conservation and Department of Entomology, University of Florida, Gainesville, FL, USA, [2] University of Texas Insect Collection, University of Texas at Austin, Austin, TX, USA, [3] Department of Entomology, Michigan State University, East Lansing, MI, USA, [4] University Museum of Bergen, University of Bergen, Bergen, Norway, [5] Department of Biological Sciences, University of Memphis, Memphis, TN, USA

1. INTRODUCTION

1.1 Historical Development and Current Status of the Field

Wood and Bright (1992a) list several hundred authors that have contributed to scolytine and platypodine classification and taxonomy. Many of these were forest entomologists or general biologists who contributed one or two papers, but several authors contributed extensively and were essential to the development of the field. These key historical figures are highlighted in this section.

Although most classical authors worked on classification as well as taxonomy (see Box 2.1 for definitions), these two fields are increasingly distinct. We separately describe the development and the current status of each field.

1.2 Development of Taxonomy

1.2.1 Taxonomists

The first bark beetles officially described as zoological species were four common European species listed in the 10th edition of Linné's *Systema Naturae* (Linnaeus, 1758): *Ips typographus*, *Pityophthorus micrographus*, *Polygraphus poligraphus*, and *Tomicus piniperda*. All were included in the genus *Dermestes*, which is currently classified in the beetle family Dermestidae. Subsequently, the accumulation of described genera and species of bark beetles progressed in a relatively steady manner, until a sudden decline in the 1970s (Figure 2.1). The most productive scolytine taxonomists came from Germany, Japan, the UK, and the USA (Table 2.1). Although very productive in terms of species descriptions, most of the foundational taxonomists of the late 19th and the early 20th centuries were trained in classical typological taxonomy, and were preoccupied by alpha-taxonomy (descriptions of individual

species), rather than by creating a classification. Only a few, such as Eggers, Hopkins, Schedl, and Wood, can be credited with attempts to create a comprehensive scolytine classification, including such higher-level taxonomic groups as tribes and genera.

Table 2.1 lists the major taxonomists that have worked on bark beetles, ordered by the number of species they described (minimum 50 species described). Only currently accepted names are included here, not the number of originally proposed names. This is an important distinction because, as a rough estimate, approximately 30% of species' names proposed to date are now considered synonyms of other species (i.e., the same species described more than once, often by the same author). The top three authors, Karl Schedl, Stephen L. Wood, and Hans Eggers, collectively published more than 50% of all described species of bark beetles.

By far the greatest number of species was described by the Austrian taxonomist Karl Schedl, whose collection, currently housed in the Naturhistorisches Museum in Vienna, is one of the greatest depositories of scolytine specimens in the world. Schedl was a prolific alpha-taxonomist. For example, he was one of the few authors to tackle the African fauna. However, he also appears to have been a slightly uncooperative figure, and his legacy includes hundreds of species that were described out of failure to consult other taxonomists' work, resulting in a high proportion of synonymy. Further inflation of the number of species described by Schedl probably resulted from his frequent habit of describing species based on minute details of little biological relevance. This is contrary to his early publications where he clearly appreciated intraspecific morphological variation. Wood and Bright (1992a) give an insightful account of the "Schedl factor in scolytine taxonomy."

The greatest synthesizer of scolytine taxonomy and the father of modern bark beetle classification was undoubtedly Stephen L. Wood (1924–2009) (Bright, 2010). His

Bark Beetles. http://dx.doi.org/10.1016/B978-0-12-417156-5.00002-2

Ambrosia beetle—a species in either of the weevil sub-families Scolytinae or Platypodinae that is obligately associated with nutritional fungal symbionts. Obligate symbiosis with fungi is present in at least 11 independent scolytine and platypodine groups. Ambrosia beetles are therefore not monophyletic, and the name is not a taxonomic designation.

Bark beetle—"bark beetle" is both a taxonomic and ecological designation. In the taxonomic sense, bark beetles are all species in the weevil subfamily Scolytinae, including species that do not consume bark. In the ecological sense, bark beetles are species of Scolytinae whose larvae and adults live in and consume phloem of trees and other woody plants.

Character—a feature that can be used to designate and compare species. Characters can be morphological (e.g., number of antennal segments, color), anatomical (e.g., type of proventriculus), molecular (most commonly nucleotides in DNA), behavioral (type of a gallery), and ecological (e.g., host plant). Typically, characters are shared among related species, but the state of the character varies, allowing for species or group recognition.

Character state—a particular variant of a character, usually discrete. For example, a spine on an elytron may have two states: present or absent. For a character state to be useful in phylogenetics, it should be variable between groups but conserved within a group.

Classification—hierarchical classification of organisms into named groups, such as species or genera.

Phylogenetics—the science of inferring evidence-based groupings of organisms based on shared ancestry.

Phylogeny—a representation of evolutionary relationships between organisms, typically a phylogram (with nodes and branch lengths proportional to the evolutionary distance between taxa) or cladogram (with nodes and branch lengths constant, i.e., not proportional to evolutionary distances).

Taxonomy—the science of, and a set of rules for, describing and naming organisms and their relationships.

landmark contributions to the science of bark and ambrosia beetles include several synthetic works, some co-authored by Donald E. Bright. These works brought order to what could be accurately described as former chaos. That by no means implies that S. L. Wood's publications and the Wood and Bright catalogs (Wood and Bright, 1992a, b) are free of mistakes and controversies, but such shortcomings are inevitable when synthesizing hundreds of papers by hundreds of authors, corroborating those with thousands of specimens deposited in museums, and describing over 1000 new species and many new genera. As a result, the Wood and Bright catalogs and their supplements (Bright and Skidmore, 1997, 2002; Bright, 2014) serve as the foundation for essentially all current taxonomic work on bark and ambrosia beetles.

Interestingly, several of the foundational taxonomists, and specifically S. L. Wood, introduced a large amount of new terminology into scolytine morphology. This makes their taxonomic texts a challenge to read even for entomologists; not all these terms are well known, and some of the terms appear to have been newly invented (e.g., "subvulcanate"). Contemporary literature on bark beetles is moving away from use of this terminology.

Of the many other foundational taxonomists, two deserve further mention here, since both are still very productive. Dr. Roger Beaver has single-handedly reviewed and redescribed a large portion of the scolytine fauna from the Paleotropics and from portions of South America, including treatment of the natural history of these taxa. Dr. Donald E. Bright, a close collaborator and former student of the late S. L. Wood, is playing a similar role, describing and synthesizing the North American and Caribbean fauna.

1.2.2 Accumulation of Described Species

The cumulative number of species described and the rate of species description are shown in Figure 2.1. Looking only at the cumulative numbers, it is tempting to conclude that the number of species of Scolytinae has leveled off and that the current total of 6056 species accounts for nearly all extant species. However, a closer look at the rate of species description (bottom graph) reveals that the steep rise in the rate of new species described reflects the overlapping periods of activity (dates of publications in Table 2.1) of Eggers, Schedl, and Wood, and also the activity of F. G. Browne and D. E. Bright. The peak in the decade of 1970–1979 actually coincides with the massive collection effort by Bright and Wood in the southwestern USA, Mexico, and Central America that culminated in monographs for the enormous genus *Pityophthorus* (Bright, 1981) and the entire subfamily Scolytinae (Wood, 1982) in North and Central America. Since then, the rate of discovery has fallen precipitously, with a slight increase between 2000 and 2009 caused by the large number of species described by Wood in his monograph of the South American fauna (Wood, 2007). This pattern is therefore best interpreted as a reflection of reduced collection effort and reduced taxonomic activity rather than an indication that most species have been described.

While taxonomic work on bark beetles has slowed, activity has by no means come to a halt. In an ongoing tabulation of around 100 publications since 1992 of nomenclatorial significance (T. H. Atkinson, unpubl.) the following preliminary statistics have been generated. Since the publication of the world catalog in 1992, 508 new species have been described from around the world. One of the inevitable effects of the publication of new catalogs and monographs is that it makes it easier for taxonomists to spot omissions,

FIGURE 2.1 Historical patterns of species descriptions and accumulations. The totals are based on date of description of species currently considered valid. For a given decade, totals include the years 0–9 (e.g., 1980–1989).

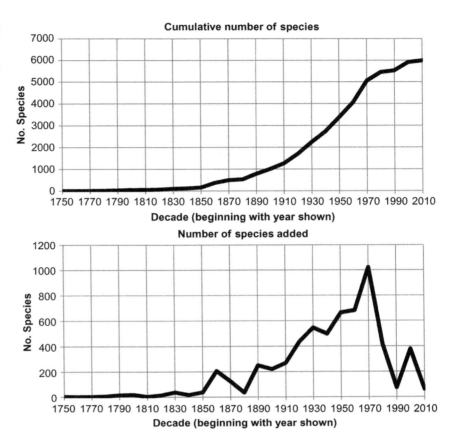

TABLE 2.1 Numbers of Species Described by Different Authors, including Year of First and Last Publication Describing a New Species

Author	Species	1st year	Last year
Schedl	1607	1931	1982
Wood	1267	1951	2007
Eggers	732	1908	1951
Blandford	263	1893	1905
Browne	221	1949	1997
Bright	189	1964	2010
Blackman	173	1920	1943
Eichhoff	146	1864	1886
Hopkins	78	1895	1916
Hagedorn	76	1903	1912
LeConte	65	1857	1885
Murayama	53	1929	1963
Swaine	53	1910	1925
Chapuis	52	1869	1875

These numbers reflect only the number of species names still in use, not the number of names originally proposed by these authors and later synonymized.

deletions, and errors. Consequently, since 1992, 403 species have been moved to different genera and approximately 400 new synonymies have been detected (at least 400 species recognized in the world catalog have been removed from the list). On the other hand, 59 species that were previously considered synonyms have been restored to species status. Eleven replacement names have been proposed in cases where generic reassignment has resulted in two species with the same name within a genus.

Perhaps the most important outcome from the current revival is the emergence of a new generation of bark beetle systematists. The global community of bark beetle systematics enthusiasts now includes several dozen colleagues, and as of the time of this writing, we estimate that several hundred species of scolytine beetles are being described. This promises to increase the rate of species discovery, which has been relatively low since the middle 1970s.

The degree of exploration in different parts of the world is uneven. North America, Western Europe, Russia, Japan, Korea, Australia, and New Zealand have been well explored and the faunas of these regions are known to a high degree. Even so, new species are still being encountered. The bark beetle fauna in specific countries where knowledgeable collectors have lived and worked, such as Nigeria, Ghana, parts of the Democratic Republic of Congo (formerly known as Zaire), Zambia, Thailand, Malaysia, Taiwan, and Venezuela, is relatively well known. Meanwhile, most of South America, much of Africa, and large parts of Asia has not been well collected by competent collectors. While many species are abundant, most are not routinely encountered by general collectors or with methods such as light traps.

1.3 Development of Classification

1.3.1 Traditional

Most of the foundational bark beetle taxonomists in Table 2.1 designated "species groups." These were the first attempts to bring order to the bewildering diversity of morphologically uniform scolytine beetles. Schedl, for example, used the beetle's elytral declivity for his classification, often as a sole character. Although those typological species groups lack the rigor of genera and tribes based on shared ancestry and fall outside of the International Code of Zoological Nomenclature, they were nevertheless useful in organizing collections and suggesting patterns of similarity.

1.3.2 Evolutionary, Pre-analysis

One of the first and most productive classifiers of bark beetles in the pre-phylogenetic era was S. L. Wood. He is notable for being the first to attempt to use evolutionary and biogeographical data to support his groupings. By working with many characters simultaneously, he departed from the typological tradition of his predecessors. His classification was also based on consideration of all species and genera at a comprehensive, global scale. Wood did not employ a discrete analytical approach. In most cases, he preferred gestalt, and handpicked some characters over others. He was also one of the most outspoken proponents of treating bark beetles as a separate family (Scolytidae), rather than a subfamily of weevils (the convention), contrary to most phylogenetic analyses. Yet, despite a few shortcomings, many of his taxonomic groups have been corroborated by modern methods of phylogenetic inference.

1.3.3 Phylogenetic—Pre-DNA Sequence

A phylogenetic approach to scolytine classification was pioneered by entomologists who were not bark beetle taxonomists. For example, G. N. Lanier conducted several pre-phylogenetic, biosystematics studies. He collected karyological and mating data, which he used to test the biological species concept for closely related *Ips* and *Dendroctonus* species. His studies revealed several cryptic sibling species diagnosable by microscopic morphology (Lanier, 1970). These species were later confirmed through phylogenetic analysis of molecular phylogenetic data (Kelley and Farrell, 1998; Cognato and Sun, 2007).

Perhaps the first true phylogeny of bark beetles was that by Bentz and Stock (1986). They reconstructed a phylogeny of 10 *Dendroctonus* species using allozyme frequency data. The phenograms they produced agreed, in part, with previously proposed species groups (Wood, 1963; Lanier, 1981). Their work was still based on the traditional "species groups" instead of clades, and its main goal was to order species on an artificial scale from "primitive" to "advanced." Nonetheless, their work served as a basis for subsequent phylogenetic projects. It could be said that the work by Six and Paine (1999) on the coevolution between *Dendroctonus* and their fungal associates was ahead of its time. The authors used characters with limited phylogenetic power (isozymes electrophoresis), but the analytical and conceptual framework of an empirical test of cospeciation events was cutting-edge.

Allozyme and DNA fragment data are useful in elucidating the relationships of closely related species. Cane *et al.* (1990) created a phenogram of the *Ips* species in the *grandicollis* group using allozyme frequency data. Cognato *et al.* (1995) re-examined the phylogeny of the *grandicollis* species group with a larger sample of species and using DNA fragment data (RAPDs) in a cladistics analysis that produced similar results. de Groot *et al.* (1992) used allozymes to test the validity of three *Conophthorus* species.

Phylogenetic analyses using DNA sequence data supported many of the species relationships found in the above studies (Kelley and Farrell, 1998; Cognato and Sperling, 2000; Cognato et al., 2005a). However, technical advances in PCR and DNA sequencing offered a much larger source of phylogenetic characters, and contributed to the general disuse of allozyme and DNA fragment data near the turn of the 20th century.

1.3.4 Phylogenetic—Molecular and Morphological

The early 21st century saw a revival of bark beetle classification with a strong phylogenetic emphasis. Some studies (e.g., Farrell et al., 2001) introduced phylogenetics and DNA sequence data into the study of bark beetles, and with it new optimism for reconstructing bark beetle relationships. The results were influential, and were used for the classification of bark beetles in major textbooks such as *American Beetles* (Arnett et al., 2002). Additional studies from the same research group also had a significant molecular phylogenetic component, and were important for the development of new DNA characters/markers. These studies (Farrell et al., 2001) also highlighted for the first time failures in morphology-based classifications, such as the placement of *Premnobius* in Xyleborini, the separate position of Xyleborina from Dryocoetini, and the separation of Hylastini and Hylurgini (Hylastini are nested in Hylurgini).

Several of the authors of this chapter are using increasingly greater numbers of markers, morphological characters, and taxa to create a true phylogeny-based taxonomy of bark and ambrosia beetles, setting the stage for the next big reclassification since S. L. Wood (Wood, 1986) defined most of the tribes more than 30 years ago.

1.3.5 Scolytidae or Scolytinae?

An interesting debate regarding evolutionary classification of scolytine beetles concerns their collective taxonomic status (Bright, 2014; Jordal et al., 2014). While there has never been any serious question about whether the bark and ambrosia beetles are closely related to (or derived from within) the true weevils (family Curculionidae) within the larger superfamily Curculionoidea, their specific phylogenetic placement has been debated for years. Since the bark and ambrosia beetles collectively form a large, ecologically cohesive group, most specialists were fully engaged within that group. This led to a certain "taxonomic inertia." Wood strenuously defended the position that the Scolytidae deserved family status and were not closely related to the true weevils (Wood, 1978, 1982, 1986). Recent studies, however, are consistent with the notion that Scolytinae as a group is best treated as a subfamily within the family Curculionidae (e.g., Thompson, 1992; Kuschel, 1995; Lyal, 1995; Marvaldi, 1997; Marvaldi et al., 2002; McKenna et al., 2009; McKenna, 2011; Haran et. al., 2013; Gillett et al., 2014).

The traditional classification of scolytines as a distinct family was based on the magnitude of their differences from other Curculionidae, as well as the uniqueness of scolytine morphological characters. While the combination of some scolytine morphological characters is specific to the group, the individual characters are found in other weevil groups and are not specific to scolytines (Jordal et al., 2014). To reflect natural groupings, classifications need to recognize the branching structure of species relationships, regardless of the magnitude of morphological or molecular differentiation or other unique features of its evolutionary history. If a group of species is nested within another group, it is best treated as a subgroup of the group encompassing it. In this regard, most modern phylogenetic analyses support the nested position of scolytine beetles within weevils, specifically within the family Curculionidae (true weevils).

2. CURRENT APPROACHES AND STATUS OF THE FIELD

2.1 Morphological Approaches

Although molecular phylogenetics has grown to play an important role in bark beetle classification, the majority of the group's taxonomy is based on external morphological characters. Most characters can be readily observed through a binocular microscope. However, it is worth emphasizing that there are three character systems commonly used in classification of insects that are only rarely used in bark beetle taxonomy: internal structures, mouthparts, and genitalia.

Internal morphology of bark beetles can be a rich source of taxonomic and phylogenetic characters. The typical approach to observing internal structures is to dissolve soft tissues of the beetle in diluted KOH and fix the preparation in Euparal on a glass slide (Peterson, 1964). The organs that have yielded valuable taxonomic information include the partially sclerotized proventriculus, spermatheca, and the parts of head that are typically concealed under the pronotum. The proventriculus is the chitinized part of the gut and contains many morphological features, including spines and chitinous plates, and can be divided into several segments (Díaz et al., 2003; Totani and Sugimoto, 1987). The overall extent of its sclerotization also appears to reflect the ecological strategy of the species: the anterior plate is reduced in obligately fungus-feeding ambrosia beetles, while it is extensively sclerotized in beetle species with increased dependence on woody tissues (Nobuchi, 1988). Where the anterior plate is well developed, it may contain characters useful for both species-level and higher-level classification. A good example is the peculiar form of the anterior plate in all Ipini genera, supporting their monophyly.

Mouthparts similarly reflect different feeding modes, but also show several features that are useful in classification. The best example given so far is the shape of the first labial palp segment, which is round and swollen with an oblique row of setae in inbreeding dryocoetine genera and all xyleborines, the so-called haplodiploid clade reflecting the genetic system of males in these taxa (Jordal *et al.*, 2002a). Although various features of mouthparts have been illustrated in taxonomic treatments of various scolytine groups, they have not been used in classification.

Genitalia have been used rarely in bark beetle systematics and phylogenetics at a higher taxonomic level partly because of their very small size and the associated technical challenges of studying them. Within specific genera, they have proven useful for distinguishing species. Male genitalia and female abdominal plates have been used as characters for distinguishing cryptic species, including economically important ones such as the southern pine beetle *Dendroctonus frontalis* (Zimmermann) and its sister species *Dendroctonus mexicanus* (Hopkins) (Rios-Reyes *et al.*, 2008). They also contain reliable characters for higher-level studies such as those linking *Dryocoetes* with the haplodiploid clade of dryocoetines and xyleborines (Jordal *et al.*, 2002a).

2.1.1 Morphological Approaches: Improvements in Imaging

The availability of detailed electronic illustrations of scolytine beetles and their characters has revolutionized the field. Even though only a small fraction of the global diversity of Scolytinae has been illustrated, in some regions (e.g., North America and Europe), nearly all species have been photographed, and most of the images are freely available online.

This increase in bark beetle photographs stems primarily from developments in microphotography. Typical approaches involve taking many high-resolution, low depth-of-field photos through a camera coupled with a microscope, and assembling the images into a so-called "stacked" image with more or less complete depth of field. This technology is increasingly available to most taxonomists.

2.1.2 Current Electronic Image Depositories

The following list offers a selection of sources of scolytine photographs that are well curated and of high quality as of April 2014. It is not an exhaustive list.

T. H. Atkinson's Bark and Ambrosia Beetles site (http://barkbeetles.info/) focuses on Scolytinae and Platypodinae of the Americas. With 4485 images, it is currently the most information-rich electronic source of information, images, and metadata on Scolytinae and Platypodinae in the world.

Xyleborini Ambrosia Beetle information resource is a curated, taxonomically up-to-date database of information and images (1051 digital photos): http://xyleborini.myspecies.info/. It is a community-driven resource, compiled by six taxonomists from four different countries and curated by J. Hulcr.

Forestry Images (http://www.forestryimages.org) is a community sourcing approach to assembling images of forests pests, including several thousand bark and ambrosia beetle images. It is managed by the University of Georgia, and it is mostly up to date taxonomically.

Bark Beetle Genera of the United States (http://idtools.org/id/wbb/bbgus/) is a comprehensive identification tool that includes illustrated fact sheets and an interactive key. It is a product of collaboration between the USDA-APHIS-PPQ and J. Mercado. Each genus is illustrated by a single species.

Xyleborini of Papua New Guinea is a printed monograph (Hulcr and Cognato, 2013), but it is accompanied by a databank of more than 500 images freely accessible at http://www.ambrosiasymbiosis.org/PNG_Xyleborini/.

Coléoptères Scolytidae D'Alsace is an exhaustive series of high quality photographs of scolytine and platypodine species from the French province of Alsace by C. Schott: http://claude.schott.free.fr/Scolytidae/Scolytidae-liste-PL.html.

Atlas of Beetles of Russia, a project led by the Zoological Institute of the Russian Academy of Sciences, contains high-quality scolytine photographs: http://www.zin.ru/animalia/coleoptera/eng/atl_sl.htm. It is dedicated to the 100th anniversary of G. G. Jacobson's book *Beetles of Russia* (Jacobson, 1913).

The Pest and Disease Image Library (PaDIL), administered by the Australian government, is a collection of figures and information on species with invasion potential. It includes many high-quality scolytine photographs: http://www.padil.gov.au.

2.2 Molecular Approaches: DNA Sequences in Phylogenetic Analyses

With the advent of molecular methods in ecology and systematics, new tools were introduced in the late 1980s and 1990s. The past two decades have thus hosted a major leap forward in the development of genetic markers that can be reliably amplified and sequenced across scolytine taxa, providing phylogenetic resolution across taxonomic levels (Table 2.2).

The first studies on taxonomic relationships in Scolytinae using molecular data included a data set for the

TABLE 2.2 Characteristics of Molecular Markers used in Phylogenetic Studies of Scolytinae

Gene	Symbol	Genome	Length[a]	Intron	Rate nuc[b]	aa var[c]	Paralogs[d]
Cytochrome oxidase I	COI	mt	1200	0	high	yes	yes[e]
Large subunit ribosomal RNA	16S	mt	500	–	high	n/a	no
Small subunit ribosomal RNA	12S	mt	400	–	high	n/a	no
Mitogenome[f]	mtDNA	mt	13,792	0	high	yes	no
Arginine kinase	ArgK	nuc	1120	1	low	no	no
Multifunctional, incl. carbamyl-phosphate synthetase	CAD	nuc	900	1	high	yes	no
Elongation factor 1α, C1 copy	EF-1α, C1	nuc	927	0–1	low	no	yes
Elongation factor 1α, C2 copy	EF-1α, C2	nuc	1020	1–3	moderate	no	yes
Enolase, no intron copy	Eno, ni	nuc	687	0	high	yes	yes
Enolase, intron copy	Eno, 2i	nuc	900	1–6	high	yes	yes
Histone H3	H3	nuc	328	0	low	no	yes
Large subunit ribosomal RNA	28S	nuc	800	–	low	n/a	no
RNA polymerase II	Pol II	nuc	822	0	low	no	yes
Small subunit ribosomal RNA	18S	nuc	1900	–	low	n/a	no
Sodium–potassium pump	NaK	nuc	713	1	high	yes	no

[a]Approximate longest fragment used in phylogenetic analyses of Scolytinae (bp).
[b]Substitution rates in nucleotide sequences across tribes and genera.
[c]Amino acid sequence variation across tribes and genera.
[d]In Scolytinae.
[e]Mitochondrial copies inserted into the nuclear genome ("numts": Nuclear MiTochondrial Sequences, pseudogenes).
[f]Mitogenomic bulk sequencing (Haran et al., 2013).

mitochondrial gene cytochrome oxidase I (COI) from a small sample of European *Ips* (Stauffer *et al.*, 1997), and the northern hemisphere boreal genus *Dendroctonus* (Kelley and Farrell, 1998). The COI gene has now become a standard marker in studies on bark beetle relationships, ranging from population structure to higher-level relationships (Normark *et al.*, 1999; Cognato and Sperling, 2000; Farrell *et al.*, 2001; Sequeira *et al.*, 2001; Cognato *et al.*, 2005b; McKenna *et al.*, 2009; Jordal *et al.*, 2011; McKenna, 2011; Jordal and Cognato, 2012).

Genes in the mitochondrial genome are easy to amplify because they appear in many copies per cell. They were therefore preferred markers in early PCR-based molecular studies. In addition to COI, there have also been several published studies on Scolytinae using other mitochondrial gene fragments from the small (12S) and large (16S) ribosomal units (Jordal *et al.*, 2000; Cognato and Vogler, 2001; Jordal *et al.*, 2002a, b; Jordal and Hewitt, 2004; Jordal *et al.*, 2006). The performance of mitochondrial

markers was not always optimal and resulted in a search for additional molecular markers. The main problem with mitochondrial markers is the high evolutionary rates that characterize such genes, especially at deeper phylogenetic levels. Saturation of substitutions is therefore common, leaving little to no phylogenetically informative signal due to homoplasy.

Parallel to the development of mitochondrial markers, the development of nuclear markers focused on ribosomal RNA genes, which included the small (18S) and large (28S) subunits. These genes occur in multiple copies and are therefore easier to amplify than single-copy protein coding nuclear genes. They are therefore widely used in molecular phylogenetic studies of insects in general, but due to a generally low substitution rate they are mainly used to resolve relationships at higher taxonomic levels (e.g., above the genus level). The small subunit rRNA (18S) is less commonly used within Scolytinae, but was nonetheless one of the main sources of evidence for placing Scolytinae in the context of superfamily Curculionoidea (Farrell, 1998; Marvaldi *et al.*, 2002; McKenna *et al.*, 2009; McKenna, 2011). While the smaller subunit is best suited for distinguishing families and superfamilies of beetles, the larger subunit is more variable and hence informative within subfamilies, and sometimes even within genera of Scolytinae (Sequeira *et al.*, 2000; Farrell *et al.*, 2001; Jordal *et al.*, 2008, 2011; Dole *et al.*, 2010; Cognato *et al.*, 2011; Jordal and Cognato, 2012; Cognato, 2013a). Ribosomal RNA genes are difficult to align due to long extension segments and loops of variable length. Alignments guided by secondary structure may therefore be helpful to improve homology, but this is a very labor-intensive procedure and efforts to do this across the superfamily Curculionoidea have not recovered significant additional resolution (McKenna, unpubl.). Nonetheless, trees produced via alignment software such as Muscle or Mafft are often reasonably well resolved, especially after masking ambiguously aligned regions using programs designed for this purpose such as Gblocks (Jordal *et al.*, 2008; Cognato *et al.*, 2011; Cognato, 2013a).

Due to the many methodological pitfalls associated with sequencing and analyzing data from mitochondrial genes and nuclear ribosomal RNA genes, new markers from protein-coding nuclear genes have been developed. Except for occasional studies on functional genes, such as dieldrin resistance in the coffee berry borer (*Hypothenemus hampei* (Ferrari); Andreev *et al.*, 1998), the first gene used for scolytine phylogenetics was elongation factor 1-alpha (EF-1α) (Normark *et al.*, 1999). This is a highly conserved gene that has provided valuable information in bark beetle phylogenies, especially at the generic level (Cognato and Vogler, 2001). Although multiple copies exist for this gene in bark beetles, the different copies are usually highly divergent, typically with different intron structures, and hence readily distinguished (Jordal, 2002; McKenna, unpubl.).

Primers and amplification conditions for other promising genes were optimized, such as enolase and histone H3 (Farrell *et al.*, 2001; Sequeira *et al.*, 2001; Jordal and Hewitt, 2004; Jordal *et al.*, 2006), but their utility has been limited to a small minority of taxa and they therefore have not yet been used broadly across Scolytinae. Enolase occurs as two copies in Scolytinae (Farrell *et al.*, 2001) and each copy has generally low amplification success. Histone H3 has three copies in all animals, which tend to co-amplify (Jordal, 2007). Multiple copies also exist for RNA polymerase II (Pol II) based on indirect phylogenetic evidence. Copy-specific primers must be developed before any of these genes can be used as reliable phylogenetic markers for bark beetles. Arginine kinase (ArgK) and carbamoyl-phosphate synthetase 2 (CAD) are two recently developed genes that show perhaps the greatest potential for robust

phylogenetic resolution in scolytine beetles, similar to EF-1α (Jordal, 2007; Dole *et al.*, 2010; Jordal *et al.*, 2011). Arginine kinase occurs in multiple copies in some groups of beetles, including near-relatives of weevils, but not in true weevils (McKenna, unpubl.), while CAD is apparently a single-copy gene in all weevils. Both genes have only a single short intron present in the most widely sequenced fragment, providing for easy amplification and making them useful for broad-scale phylogenetic analyses.

Current phylogenetic efforts on bark and ambrosia beetles thus utilize nucleotide sequence data mainly from five genes: COI, EF-1α, CAD, ArgK, and 28S. These genes have helped resolve some phylogenetic relationships in Scolytinae, but a large proportion of the scolytine phylogeny is still unresolved. Large-scale screening of additional nuclear genes is under way to obtain additional data for phylogenetic reconstruction.

2.2.1 Limitations of Marker-based Phylogenetics

Optimism in the early days of PCR and sequencing was considerable and many expected that important taxonomic and phylogenetic issues could be solved in a short time span. This has not been the case, and much research remains to resolve many controversial or unresolved relationships. The reasons for limited progress are many, but fall into roughly two categories: biased taxon sampling, and the application of few markers with limited phylogenetic signal. While the first problem can be solved with expanded and more thoughtful taxon sampling, the second is more technical and dependent on selecting genes with appropriate levels of variation that can be properly/effectively modeled in phylogenetic analyses. A practical solution is the implementation of numerous molecular markers that have the potential for complementary information at various phylogenetic depths. Screening of new molecular markers is therefore being conducted across a broad assemblage of bark beetles and other weevil taxa, with some recent progress made for this group (Table 2.2). Another approach pursued by the McKenna laboratory is to sequence hundreds of orthologous nuclear loci using phylogenomic approaches, such as anchored hybrid enrichment (see below).

2.2.2 The Biggest Picture: Phylogenomic Data

Considering the limited resolution and nodal support in phylogenetic trees resulting from the analysis of traditional molecular phylogenetic data sets (Farrell, 1998; Farrell *et al.*, 2001; Marvaldi *et al.*, 2002; Hundsdoerfer *et al.*, 2009; McKenna *et al.*, 2009; Jordal *et al.*, 2011; McKenna, 2011), it is clear that more data are needed to

BOX 2.2

Although DNA barcoding has frequently been questioned as a tool for inferring evolutionary relationships, its utility in cryptic species identification is generally undisputed. An example of the use of DNA barcoding in bark beetles is the collaboration between taxonomists and regulatory agencies in the protection of coffee from the coffee berry borer (*Hypothenemus hampei*) in Papua New Guinea.

Hypothenemus hampei is the most destructive pest of coffee worldwide (see Chapter 11). In Papua New Guinea, coffee is one of the most important commodities, and the first step towards protecting the country from introduction of the pest is to develop the capacity to identify it. Unfortunately, there are many native *Hypothenemus* in the country, all with very similar morphology. International quarantine standards are strict about movement of coffee beans potentially infested with *H. hampei*. Misidentification of a local harmless beetle with the coffee berry borer can thus cause huge losses to the coffee industry in Papua New Guinea.

DNA barcoding based on a single marker (COI) is a powerful technique in distinguishing the coffee berry borer from native *Hypothenemus*. Hulcr *et al.* (unpubl.) compared the DNA barcode of *H. hampei* to the DNA barcodes of 22 species of very similar, unidentified species in the tribe Cryphalini that are native to Papua New Guinea. The diagram shows several key results:

1. The COI "barcode" of *H. hampei* can be easily amplified and sequenced.
2. DNA barcodes of all other beetle species and genera included in the test, with which the coffee berry borer may be easily confused, can be also easily sequenced.
3. The barcode of *H. hampei* is distinctly different from similar native beetles in Papua New Guinea.

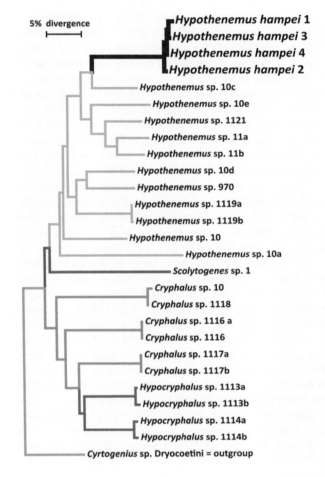

comprehensively resolve both the interrelationships of the subfamilies of Curculionidae and the internal relationships of Scolytinae, Platypodinae, and the other curculionid subfamilies. Fortunately, phylogenomic approaches (the use of genome scale data to infer phylogenetic trees) are now feasible, allowing for the sequencing and analysis of thousands to millions of base pairs of DNA sequence data (Niehuis *et al.*, 2012; Lemmon *et al.*, 2012).

Published phylogenomic studies of weevils are currently limited to a single paper (Haran *et al.*, 2013) based on a relatively small data set (by phylogenomic standards) comprised of DNA sequences from 12 of the 13 mitochondrial protein-coding genes (≈ 10 kbp total data) for each of 29 species of Curculionoidea (27 new partial mitochondrial genome sequences), including 22 species of Curculionidae. Among these were the scolytines *Ips cembrae* (Heer) (GenBank accession JN163961) and *Scolytus* sp. (GenBank accession JN163962) and the platypodine *Platypus cylindrus* Fab. (GenBank accession JN163963). The phylogenetic trees resulting from analyses of these data were well resolved, and contained moderate to high bootstrap support for most nodes, though relationships among early-divergent Curculionidae (including Platypodinae) were not all well supported (Haran *et al.*, 2013).

Beetle genomes are quite variable in size, with a mean of 974 megabases (MB) and range of 154–2578 MB for the 66 species of Coleoptera in 23 families sampled by Hanrahan and Johnston (2011). In the few weevils sequenced to date, the mitochondrial genome is ≈ 15 kbp in length (Song *et al.*, 2010), whereas the nuclear genome ranges from ≈ 170 Mb to ≈ 3 Gb in the few weevils that have been studied, with a likely average of slightly less than 1 Gb (Hanrahan and Johnston, 2011; Normark, 1996). Genome size estimates are available for four scolytines: *Dendroctonus ponderosae*

Hopkins (208 Mb; Gregory *et al.*, 2013), *H. hampei* (170–180 Mb; Nuñez *et al.*, 2012), *Xyleborus* sp. (230 Mb; Hanrahan and Johnston, 2011), and *Ips pini* (Say) (509–587 Mb; Cognato and Johnson, unpubl.), suggesting a small average genome size for the subfamily compared to other beetles. There are no published genome size estimates for Platypodinae. Gene content in weevil genomes is poorly known. However, based on the number of gene models (13,088) reported by Keeling *et al.* (2013) for the *D. ponderosae* genome, gene content might be expected to be similar, perhaps at least on average, to *Tribolium castaneum* (Herbst) (Tenebrionoidea: Tenebrionidae: 16,404 gene models) reported by Richards *et al.* (2008).

While the only published phylogenomic study of weevils to date uses data from the mitochondrial genome, phylogenomic studies of weevils employing nuclear DNA sequence data and genomic structural information are now under way. These studies have been facilitated by recent advances in phylogenetic methods and the increasing availability of weevil transcriptome and genome sequence data, both published and unpublished. For example, anchored hybrid enrichment probes (*sensu* Lemmon *et al.*, 2012) that target nearly 1000 nuclear loci (known 1:1 orthologs) have been designed for weevils (McKenna *et al.*, unpubl.). These probes have now been used successfully across all families and subfamilies of weevils, including multiple exemplars of the subfamilies Scolytinae and Platypodinae (McKenna *et al.*, unpubl.).

Only one weevil genome has been published to date, that of the scolytine *D. ponderosae* (male: NCBI BioProject PRJNA162621 and female: NCBI BioProject PRJNA179493; Keeling *et al.*, 2013). *Hypothenemus hampei* is the only other weevil for which a genome project has been publicly reported, though the genome assembly is not yet publicly available. However, a first draft of the genome was reported by Nuñez *et al.* (2012), who also mention that the genome of both males and females had been sequenced, and that sequence-based physical mapping from a BAC library is ongoing (Nuñez *et al.*, 2012). Global or tissue-specific transcriptome assemblies of variable completeness, EST libraries, and/or other similar genomic data are available from NCBI (as of April 2014) for the scolytines *D. ponderosae, D. frontalis, I. pini, Tomicus yunnanensis* Kirkendall and Faccoli, and *Ips typographus* (L.). A global transcriptome has been completed for *I. typographus* as part of the 1000 Insect Transcriptome Evolution project (1KITE, 2013), and a global transcriptome is under way for the platypodine *Oxoplatypus quadridentatus* Olivier, as part of the 1KITE Project. However, so far there are no publicly available genomes or transcriptomes from the subfamily Platypodinae. Thus, with the first weevil genomes being representatives of Scolytinae, scientists are well positioned for pursuing comparative genomic and phylogenomic studies.

2.3 Species Delimitation, Population Genetics, and Phylogeography

High substitution rates in mitochondrial genes pose problems for higher-level phylogenetic studies, but are essential in population genetics and phylogeography, along with the study of recently diverged species. However, the use of mitochondrial markers is not problem free. One obvious challenge is the maternal inheritance of such loci, which have certain idiosyncratic properties, especially if dispersal is sex biased. Others are related to selective sweeps such as those associated with *Wolbachia* infection. Additionally, serious problems are caused by the existence of pseudogenes, which are gene-like sequences with lost function. For mitochondrial genes, a pseudogene develops from a mitochondrial copy, which is transferred to the nuclear genome (NUMT: NUclear MiTochondrial sequence) and may co-amplify with the true mitochondrial gene sequence, or occasionally it might amplify instead of the targeted mitochondrial gene. The most extreme case of NUMTs in bark beetles is found in *I. typographus*, with several dozen COI NUMTs lacking indels and stop codons, only some 1–3 base pairs different from the mitochondrial gene (Bertheau *et al.*, 2011). NUMTs of COI have been reported from many other scolytine taxa, including *Pityogenes chalcographus* (L.), *Ips acuminatus* (Gyllenhal), *Orthotomicus laricis* (Fab.), *Polygraphus poligraphus* (L.), *P. punctifrons* Thomson, *Dryocoetes alni* (Georg), *Hylurgops glabratus* (Zetterstedt), and *Hylastes attenuates* Erichson (Jordal and Kambestad, 2014), but the problem is anticipated to be more universal. Mitochondrial ribosomal RNAs are particularly vulnerable to errors because stop codons are not present and indels cannot be readily detected in such genes. Solutions to these problems are being developed and include the use of pure mitochondrial extracts or next generation sequencing to detect the numbers and kinds of nuclear pseudogene copies of mtDNA (see below).

To obtain a more complete picture of the genetic variation in populations, nuclear markers (Table 2.3) are needed to complement those from the mitochondrial genome. Several studies on species complexes have used nucleotide sequence data to assess the deep mitochondrial divergence in bark beetles. Among the four nuclear genes regularly used in higher-level phylogenetics, only CAD (multifunctional protein that includes carbamyl-phosphate synthetase) seems to provide more than a few variable sites within a bark beetle species (Dole *et al.*, 2010; Andersen *et al.*, 2012). Several other nuclear genes were screened for population genetic variation in *Araptus attenuatus* Wood and related corthyline beetles (Garrick *et al.*, 2009). Gene fragments such as ATP synthetase subunit α (ATPS-α) and muscle protein 20 (MP20) include introns that are particularly useful as they contain indels

TABLE 2.3 Molecular Markers Typically used in Population Genetic Studies of Scolytinae

Gene	Symbol	Genome	Length[a]	Intron	Rate nuc[b]	Taxa
Microsatellites, various	Mic.sat.	nuc	variable	–	high	*Coccotrypes* spp., *Hypothenemus hampei*, *Pityogenes chalcographus*, *Ips typographus*, *Hypocryphalus mangiferae*, *Xyleborus affinis*, *Xylosandrus germanus*, *Dendroctonus* spp.
Anonymous microsatellite	AML	nuc	135–166	–	high	*Araptus attenuatus*
ATP synthetase subunit α	ATPSα	nuc	196–210	1	high	*A. attenuatus*
Calmodulin	Cal	nuc	?	1	?	*Dendroctonus* spp. (Kelley, pers. comm.)
Carbamoyl-phosphate synthetase 2	CAD	nuc	900	1	moderate	*Xylosandrus morigerus*, *Thamnurgus petzi*
Cytochrome Oxidase I	COI	mt	1200	0	high	*Dendroctonus* spp., *Ips* spp., *H. hampei*, + many others
Dieldrin resistence gene	rdl	nuc	800	1	low	*H. hampei*
Elongation Factor 1α, C1 copy	EF-1α, C1	nuc	927	0–1	low	*Thamnurgus petzi*, *A. attenuatus*, *Aphanarthrum glabrum*
Enolase, no intron	Eno-ni	nuc	168	0	moderate	*A. attenuatus*, *A. glabrum*
Histone H3	H3	nuc	328	0	low	*A. glabrum*
Internal transcribed spacer 2	ITS-2	nuc	747	–	low	*H. hampei*
Kuzbanian	Kuz	nuc	266	0	moderate	*A. attenuatus*
Large subunit ribosomal RNA	16S	mt	500	–	low	*A. glabrum*
Lysidyl aminoacyl transfer RNA synthetase	LTRS	nuc	217	0	low	*A. attenuatus*
Muscle protein 20	MP20	nuc	303–323	1	high	*A. attenuatus*
Wingless	wnt	nuc	229	0	moderate	*A. attenuatus*

[a]*Approximate longest fragment used in genetic analyses of Scolytinae (bp).*
[b]*Substitution rates in nucleotide sequences within species and species complexes.*

for fragment length polymorphism but also contain generally higher nucleotide substitution rates.

The most frequently applied data in population genetics are microsatellite data. These short repetitive nucleotide segments may vary in the number of repeated segments and hence provide polymorphic allele data. The first microsatellites for bark beetles were developed for the seed beetles *Coccotrypes dactyliperda* F. and *C. carpophagus* Hornung (Berg *et al.*, 2003; Holzman *et al.*, 2009). Although generally recognized as particularly difficult to develop in bark beetles, other studies soon followed, focusing on *I. typographus* (Sallé *et al.*, 2003), *H. hampei* (Gauthier and Rasplus, 2004), and various *Dendroctonus* spp. (Schrey *et al.*, 2007). Amplification across different species and genera has also enabled population genetic analyses in other species, such as *Hypocryphalus mangiferae* Stebbing and *Xyleborus affinis* Eichhoff (Masood *et al.*, 2011) and *Xylosandrus germanus* Blandford (Keller *et al.*, 2011). Most obstacles in developing microsatellites have now been solved with the application of next generation sequencing. Massively parallel sequencing with subsequent filtering of sequences with nucleotide repeats was recently used to rapidly develop 18 polymorphic loci in *I. typographus* (Stoeckle and Kuehn, 2011).

Studies on bark beetle populations quite often reveal significant geographical structure. This is by no means unique to bark beetles, but divergences within species can be very deep, suggesting long-term geographical separation. Life in concealed niches under bark tends to homogenize morphology, which may increase the number of cryptic species. Molecular methods are therefore crucial in taxonomic work of such beetles, particularly those that lack selection for secondary sexual characters, such as the frons or declivity.

2.3.1 The Most Detailed Resolution: Genotyping-by-Sequencing

Population structure of bark and ambrosia beetle species is not only a key component of our knowledge of the species evolution and ecology, but also a critical variable in applied decision-making, particularly with regards to invasive species designation and management. The most common approach in molecular studies of population structure in scolytine beetles has been single-marker sequencing and microsatellite genotyping, as described above. These markers are often species specific, only represent a few genomic regions, and can be costly to use on a large number of specimens. These limitations can be overcome by using high-throughput sequencing technologies, where hundreds of new loci can be genotyped for hundreds of specimens without any marker development and for a cost that does not scale up with additional genotypes.

A popular emerging technique that combines high-throughput marker discovery with genotyping for populations is double digest restriction-site associated DNA sequencing (ddRADseq; Peterson *et al.*, 2012). This genotype-by-sequencing (GBS) method uses two restriction enzymes to reduce the genome to particular size-selected fragments that can be recovered from >80% of the sequenced individuals. Library construction includes ligating unique barcoded adapters to fragments from each individual. Ligated fragments are then amplified in PCR, selected for size, pooled, and sequenced on a next-generation sequencing platform, typically Illumina.

RADseq has received increased attention for its utility in the study of non-model organisms (Narum *et al.*, 2013) and it appears to be just as suitable for scolytine beetles, including populations of widespread and highly inbred species. In preliminary work by Storer and Hulcr (unpubl.), ddRADseq was used to identify populations of introduced *Xylosandrus crassiusculus* Motschulsky in the southeastern USA. In 16 individuals collected from six locations, 2947 loci were genotyped. The population appears to be very homogeneous, lacking any obvious structure. This may be a result of multiple introductions of the pest by several clones that spread independently and do not mix with each other. In addition to examining population structure, the level of inbreeding was also measured at all loci. While high inbreeding (FIS >0.8) was detected at the majority of loci, some outbreeding was detected at 107 loci. This indicates that some proportion of mating occurs between families, contrary to the prevailing assumption of complete inbreeding in Xyleborini. While further investigations are ongoing, it is already clear that ddRADseq will prove to be a useful new tool for studying bark and ambrosia beetle populations.

2.3.2 DNA Barcoding for Species Identification

Species identification has traditionally been based on morphological data and implemented in dichotomous identification keys. With easy access to increasingly affordable DNA sequencing, specimens can also be identified through sequence similarity in taxonomically curated sequence databases. Even a very short stretch of DNA can be sufficiently informative to enable clustering of conspecific species. A single molecular marker is therefore often sufficient for DNA "barcoding," where a unique sequence of a particular marker is referred to as a species barcode. The 5′ end of the mitochondrial COI gene has become the standard marker in DNA barcoding of animals, including insects. International collaborative effort with the base at the University of Guelph (Hebert *et al.*, 2002, 2003; Hajibabaei *et al.*, 2006) has led to standardized protocols and databases, the so-called Barcoding of Life database system (BoLDsystems, www.boldsystems.org). Since its implementation in 2002, DNA barcoding has grown rapidly in popularity and by the end of 2013, the BoLD database included almost three million barcode sequences from more than 200,000 species.

DNA barcoding is based on the principle that intraspecific variation is less than and not overlapping with variation between closely related species. This is often the case, especially in temperate regions, which harbor limited diversity compared to the tropics. DNA barcoding of 70 Palearctic bark beetle species resulted in only three cases of paraphyletic sequence clusters (Jordal and Kambestad, 2014). Two of these were in fact cryptic species supported by nuclear sequence data, concluding that only a single case of DNA barcoding error occurred. This exception involved mitochondrial introgression through hybridization in *Pityophthorus micrographus* L. and *P. pityographus* Ratzeburg and could therefore be resolved with additional nuclear sequence data. This pattern is also observed in the *Aphanarthrum glabrum* (Wollaston) complex in the Canary Islands, where mitochondrial genes are shared between species (Jordal *et al.*, 2006). Cone beetles in the genus *Conophthorus* are similarly problematic with rampant polyphyly observed for *C. ponderosae* Hopkins (Cognato *et al.*, 2005a). Their analyses suggest the occurrence of geographically isolated morphologically cryptic species, which await further investigation via multi-gene phylogenetic analysis.

There is, however, no universal threshold for species-delimiting sequence divergence usable across taxa (Cognato, 2006; Jordal and Kambestad, 2014). In scolytine identification, the use of a standard percent sequence difference to delimit species should be restricted to an advisory role (Cognato, 2006). This standard is only useful for comparison among closely related species (e.g., within a genus) and after thorough sampling of species and multiple individuals of each species. For example, Cognato and Sun (2007) reconstructed a COI phylogeny for a near-complete sample of *Ips* species of which 67% were represented by multiple individuals. They demonstrated a mean 6.6% interspecific nucleotide difference between sister species and a mean 1.0% intraspecific difference for *Ips* species. Using this information, along with a multiple gene phylogeny and morphological diagnostic characters, they justified the designation of *Ips shangrila* Cognato and Sun as a valid species. Furthermore, they used mean inter- and intraspecific differences as a guide to identify other clades that were in need of taxonomic attention.

Another issue concerning DNA barcoding is the unintentional amplification of non-homologous mtDNA that originate as copies of the mitochondrial gene and are incorporated in the nuclear genome (NUMTs). The most misleading type of NUMTs are those not readily detected by indels or stop codons. Nuclear copies of the mitochondrial genes can be anything from near identical to quite diverged without signs of non-functionality and hence misguide interpretation of genetic variation in a species (Bertheau *et al.*, 2011). Fortunately, in bark beetle barcoding there have been relatively few cases where NUMTs were a problem. In eight bark beetle species where NUMTs have been detected, only one of these contained paraphyletic gene clusters when NUMTs were included, but could be detected by indels (Jordal and Kambestad, 2014).

2.4 Pheromones and Ecology Corroborate Species Limits

Closely related species of bark beetles are often difficult to distinguish using only morphological characters. In some cases, behavioral or ecological features can be more informative. Since species recognition in bark beetles is generally facilitated by pheromonal communication, it is no surprise that pheromonal compounds can diverge rapidly between sister species and consequently can be used to distinguish those. A study by Sullivan *et al.* (2012) may illustrate this type of species divergence. They studied a tree-killing species of *Dendroctonus* in Mexico. The pest was initially assumed to be the southern pine beetle *D. frontalis*; however, it was noted that the population consists of two morphotypes. These differed only slightly in their surface sculpturing, but more significantly in their production of *endo*-brevicomin, a pheromone blend compound. Subsequent analyses showed that the observed differences in surface sculpturing, pheromone production, and cuticular hydrocarbons support the delimitation of two different species (Sullivan *et al.*, 2012).

2.5 Fossil Bark Beetles

Fossilized remains of scolytines are known for 56 species as impressions in sedimentary rock or inclusions in amber (Table 2.4). The fossils date from 120 to 5 millions of years ago (Ma) (Wood and Bright, 1992a; Bright and Poinar, 1994; Cognato and Grimaldi, 2009; Kirejtshuk *et al.*, 2009; Cognato, 2013b).

The preservation of known scolytine fossil impressions is poor and diagnostic scolytine characters are sometimes difficult to discern (Wickham, 1913). Ten species from Florissant and Green River compressions are known (Eocene-Oligocene ≈ 34 Ma) (Wood and Bright, 1992a). Generic placement of some specimens is dubious and at best only tribal affiliations can be inferred. This fauna is likely similar to the Baltic amber fauna described below.

Amber, on the other hand, provides excellent preservation of many inclusions, allowing for more informed taxonomic determinations and a better understanding of the diversity of the ancient fauna. Inclusions in Baltic amber (mid-Eocene ≈ 40–47 Ma) represent 24 species among seven extant and two extinct genera of bark beetles (Table 2.4) (Wood and Bright, 1992a). These species existed in a subtropical to warm temperate climate with periodic cooling (Wolfe *et al.*, 2009). Conifers, mostly Pinaceae and Sciadopityaceae, were abundant. The extinct pine *Pinites succinifera* Göppert (Pinaceae) may have contributed to the production of Baltic amber; however, recent evidence implicates Sciadopityaceae as the more likely source (Wolfe *et al.*, 2009). It is unknown if extinct

TABLE 2.4 Fossilized Scolytinae Taxa. Asterisk (*) Indicates an Extinct Genus

Dominican Amber, ≈20 Ma	
Corthylini	*Corthylites* bicolor* Bright and Poinar
Corthylini	*Gnathotrichus fosser* Bright and Poinar
Corthylini	*Microcorthylus antiquarius* Bright and Poinar
Corthylini	*Paleophthorus* bispinus* Bright and Poinar
Corthylini	*Pityophthorus antiquarius* Bright and Poinar
Corthylini	*Pityophthorus aphelofacies* Bright and Poinar
Corthylini	*Pityophthorus temporarius* Bright and Poinar
Cryphalini	*Hypothenemus avitus* Bright and Poinar
Dryocoetini	*Dryomites* incognitus* Bright and Poinar
Hexacolini	*Pycnarthrum senectum* Bright and Poinar
Hexacolini	*Scolytodes electrosinus* Bright and Poinar
Hexacolini	*Scolytodes neoschwarzi* Bright and Poinar
Hylesinini	*Electroborus* brighti* Cognato
Micracidini	*Micracites* squamifera* Bright and Poinar
Phloeosinini	*Cladoctonus angustostriatus* Bright and Poinar
Phloeosinini	*Cladoctonus ruber* Bright and Poinar
Phloeosinini	*Paleosinus* fossulatus* Bright and Poinar
Phloeosinini	*Protosinus* hispaniolensis* Bright and Poinar
Phloeotribini	*Phloeotribus antiguus* Bright and Poinar
Scolytini	*Cnemonyx priscus* Bright and Poinar
Scolytini	*Scolytus poinari* Bright
French Oligocene, Sediment Impression, ≈30 Ma	
Hylesinini	*Hylesinus neli* Petrov
Rock Impressions, ≈34 Ma	
Dryocoetini	*Dryocoetes carbonarius* Scudder
Dryocoetini	*Dryocoetes diluvialis* Wickham
Hylastini	*Hylastes americanus* Wickham
Hylastini	*Hylurgops piger* Wickham
Hylesinini	*Hylesinus dromiscens* Scudder
Hylesinini	*Hylesinus extractus* Scudder
Hylesinini	*Hylesinus hydropicus* Wickham
Phloeosinini	*Phloeosinus arcessitus* (Scudder)
Xyloterini	*Trypodendron impressum* Scudder
Incertae Sedis	*Xyleborites* longipennis* Wickham
Rovno Amber, Late Eocene, ≈38 Ma	
Dryocoetini	*Taphramites rovnoensis* Petrov and Perkovsky
Hylurgini	*Xylechinus mozolevskae* Petrov and Perkovsky

TABLE 2.4 Fossilized Scolytinae Taxa. Asterisk (*) Indicates an Extinct Genus—cont'd

Baltic Amber, ≈45 Ma	
Dryocoetini	*Taphrorychus immaturus* Schedl
Hylastini	*Hylastes aterites* (Schedl)
Hylastini	*Hylurgops corpulentus* Schedl
Hylastini	*Hylurgops dubius* (Hagedorn)
Hylastini	*Hylurgops electrinus* (Germar)
Hylastini	*Hylurgops granulatus* (Schedl)
Hylastini	*Hylurgops pilosellus* Schedl
Hylastini	*Hylurgops schellwieni* (Hagedorn)
Hylastini	*Hylurgops tuberculifer* Wood
Hylesinini	*Hylesinus facilis* Heer
Hylesinini	*Hylesinus lineatus* Foster
Hylurgini	*Xylechinites anceps* Hagedorn
Phloeosinini	*Phloeosinus assimilis* (Schedl)
Phloeosinini	*Phloeosinus brunni* (Hagedorn)
Phloeosinini	*Phloeosinus regimontanus* (Hagedorn)
Phloeosinini	*Phloeosinus rehi* (Hagedorn)
Phloeosinini	*Phloeosinus robustus* (Schedl)
Phloeosinini	*Phloeosinus sexspinosus* (Schedl)
Phloeosinini	*Phloeosinus tuberculifer* (Schedl)
Phloeosinini	*Phloeosinus wolffi* (Schedl)
Incertae Sedis	*Carphoborites* keilbachi* Schedl
Incertae Sedis	*Carphoborites* posticus* Schedl
Incertae Sedis	*Taphramites* gnathotrichus* Schedl
Burmese Amber, ≈100 Ma	
Hexacolini	*Microborus inertus* Cognato and Grimaldi
Lebanese Amber, ≈120 Ma	
Cylindrobrotini	*Cylindrobrotus* pectinatus* Kirejtshuk, Azar, Beaver, Mandelshtam, and Nel

scolytine genera fed upon Sciadopityaceae, though none have been recorded thus far from the sole extant species, *Sciadopitys verticillata* (Thunb.) Siebold and Zucc. (Wood and Bright, 1992a). If indeed Sciadopityaceae was the host, then these genera may have experienced a host switch from the dwindling numbers of Sciadopityaceae to the emerging dominant Pinaceae in Europe during the Late Eocene.

Our knowledge of fossil scolytines and platypodines occurring in New World amber has grown significantly in the past 20 years. Before the 1990s only four platypodine species were known from Mexican and Dominican

Republic amber (mid-Miocene ≈15–20 Ma) (Schedl, 1962; Schwaller, 1981). Bright and Poinar (1994) published the first comprehensive treatment of the Dominican Republic amber scolytine and platypodine fauna. They described 27 species in 12 extant and six new genera, representing seven tribes. Cognato (2013b) described a new genus for a previously unrecorded tribe (Hylesinini). The faunal assemblage in Dominican amber is very similar to the Recent Neotropical diversity except for the absence of Xyleborini, which likely radiated ≈20 Ma in Asia after the deposition of the Dominican amber (Cognato *et al.*, 2011; Jordal and Cognato, 2012). Although Earth's climate was cooling during the Miocene, these species existed during a relatively brief period ($\approx$5–10 Ma) of global warming when tropical climates expanded into present day temperate regions (Zachos *et al.*, 2001). *Hymenaea protera* Poinar (Fabaceae) produced the resin that resulted in Dominican amber. It is likely that many of the scolytine genera included in this amber used *H. protera* as their food since several extant species utilize *Hymenaea courbaril* L. (Wood and Bright, 1992a).

Although the faunas have similar species diversities, the taxonomic composition of Scolytinae are quite different in Baltic and Dominican ambers. These faunas only overlap with representatives of Platypodinae, Hylesinini, Phloeosinini, and Dryocoetini (Cognato, 2013b; Wood and Bright, 1992a; Bright and Poinar, 1994). Though Platypodines are few in Baltic amber, a greater diversity of Phloeosinini are preserved as compared to the Dominican amber. The fauna included in the Dominican amber exhibits a greater diversity of *Cenocephalus* species (Platypodinae) and Scolytinae genera (19) (Cognato, 2013b; Bright and Poinar, 1994). These differences are likely better explained by habitat/climate than geologic age, given that recent scolytine assemblages differ spatially in relation to habitat/climate (Wood, 1986).

Two amber inclusions dated from the Cretaceous are perhaps the most significant to the understanding of Scolytinae evolution. *Cylindrobrotus pectinatus* Kirejtshuk, Azar, Beaver, Mandelshtam and Nel was described from Lebanese amber ($\approx$120 Ma), likely originated from Araucariaceae resin (Kirejtshuk *et al.*, 2009). Although this species resembled Dryocoetini, the unique combination of characters found in other tribes resulted in the erection of a new tribe to accommodate the genus. The phylogenetic placement of this species is ambiguous, and it may represent a stem lineage of Scolytinae.

The other Cretaceous amber inclusion, *Microborus inertus* Cognato and Grimaldi, was described from Burmese amber ($\approx$100 Ma), which likely originated from Taxodiaceae resin (Cognato and Grimaldi, 2009). Its significance lies in the conservation of its morphology. The generic placement of this species is undisputable given the presence of the generic characters that define *Microborus* (Blandford, 1897). The phylogenetic placement of the genus is less certain, but molecular phylogenetic analyses suggest that it may be at the base of all extant Scolytinae (Jordal *et al.*, 2011) or nearly so (Jordal and Cognato, 2012). The fossil also provides insights into Scolytinae evolution. Extant *Microborus* species feed on angiosperms, thus it is reasonable to assume that *M. inertus* also fed on angiosperms. The occurrence of this lone specimen among 4000 Burmese amber inclusions also suggests that the conifer that produced the amber was not the preferred food of Cretaceous scolytines. These observations suggest angiosperm feeding occurred early in scolytine evolution, relatively soon after the Cretaceous diversification of flowering plants (Cognato and Grimaldi, 2009).

Microborus inertus is unexpectedly old, particularly for a representative of crown-group Scolytinae, and for an extant genus. The age given is relatively consistent with existing phylogenetic studies when considering confidence intervals around stem age estimates from molecular studies. While the age of Burmese amber was until recently contentious, Shi *et al.* (2012) document quite convincingly a maximum age of 98.79 $\pm$ 0.62 Ma (Cenomanian), consistent with the estimate (100 Ma) reported by Cognato and Grimaldi (2009). Ideally, additional fossils representing crown-group Scolytinae will be discovered in Cretaceous ambers, lending further support for the unexpectedly old age of *Microborus inertus* and insight into the early evolution of the subfamily.

Trace fossils often attributed to bark beetles are preserved galleries in fossil wood remains. The oldest one currently known is from the Early Cretaceous (Jarzembowski, 1990), although its attribution to Scolytinae is not entirely certain (Petrov, 2013). Because shapes of galleries are often highly conserved within scolytine groups, the remains can sometimes be attributable to extant genera and described as new ichnotaxa (Petrov, 2013).

2.6   Timing of Bark Beetle Origin and Evolution

To date there are only three published molecular time trees (chronograms) for weevils that sample sufficient Scolytinae and Platypodinae to shed light on timing and patterns of diversification in these groups (McKenna *et al.*, 2009; Jordal *et al.*, 2011; Jordal and Cognato, 2012). McKenna *et al.* (2009) were concerned with higher-level relationships across the superfamily Curculionoidea, but their taxon sample also included 22 genera of Scolytinae representing 16 tribes. Based on the BEAST analyses they reported, stem-group Scolytinae are proposed to have originated $\approx$90–115 Ma. Among the several curculionoid fossils used to inform the application of prior constraints on node ages in the BEAST analysis was a 55 Ma fossil scolytine from the London Clay (Britton, 1960), and a 25 Ma fossil platypodine in Apenninian amber (Skalski and Veggiani, 1990; Kohring and Schlüter, 1989). Even though these fossils are considerably younger than the recently discovered 100 Ma *Microborus* fossil, the age estimate in McKenna *et al.* (2009) fit well with a more recent estimate based on older fossils (Jordal *et al.*, 2011). The oldest known curculionid fossil is only 116 Ma, which makes these time estimates based on tree topology and fossil ages congruent.

2.7   Methods in Bark Beetle Identification

2.7.1   Traditional Dichotomous Keys

Bark and ambrosia beetle identification methodologies have followed a very similar path as that of other insect groups. The first generation of identification keys was dichotomous printed keys. The only authoritative key to the scolytine tribes and genera with a truly global scope is that of S. L. Wood (1986). Other keys typically deal with regional fauna: Maiti and Saha (2004, 2009) for South Asia, Pfeffer (1994) for Europe, Wood (1982) for North and Central America, Wood (2007) for South America, etc. Unfortunately, some of these publications are out of print and difficult to obtain. Other regional keys may be available and sufficient for the user. The publications referenced below are not intended to be an exhaustive list, but rather the first step for anyone attempting to identify bark and ambrosia beetles. Higher-level identification, for example to genera, is often possible with general guides to beetles, such as Arnett *et al.* (2002) for North America.

2.7.2   Computer-based Identification Keys

In the recent past, the availability of software for creating custom identification keys has greatly improved identification possibilities in some insect groups. Typically, computer-based interactive tools for insect identification are based on a matrix of taxa and characters. Compared to traditional dichotomous keys, matrix-based keys have the advantage of easy update and easy electronic production. A particularly prominent disadvantage in scolytine computer-based identification is that the tool only works well if all taxa are scored for all characters. In bark beetles, this creates two problems. First, the number of characters should scale up approximately with the number of taxa, which is nearly impossible in a hyperdiverse group of species with a limited number of variable features. Second, keys are not an analog of cladistics scoring and cannot distinguish between truly shared characters and homoplasy (convergently evolved apparent similarity). In scolytine beetles, some of the most variable and easy-to-observe characters are completely homoplastic (i.e., they are only convergently similar, not due to relatedness) at higher taxonomic levels, which make them of limited use. For example, while elytral declivity or antennal features are, in principle, useful for species-level identification across all Scolytinae, most of their morphological variation is repeated in many tribes, which renders them unsuitable for genus- or tribe-level matrix-based identification.

Due to the combination of advantages and disadvantages of matrix-based interactive keys in scolytine identification, some publicly available ones are hybrids of matrix-based and dichotomous approaches (Baker *et al.*, 2009).

2.7.3   Digital Catalogs

According to our estimates, no more than 25% of the scolytine diversity has ever been included in any identification key. This leaves three quarters of the world species unidentifiable without access to comparative material in museum collections. This fact highlights both the importance of curated insect collections as well as the need to accelerate the process of documenting the bark and ambrosia beetle diversity digitally, ultimately making the information and photographs available online. It is important that specimens are safeguarded in collections worldwide for taxonomic work, but such arrangement is not suitable for all applications. It results in limited access for most users, and precludes rapid identification of potential pests, which is increasingly important around the world as various scolytine beetles invade new territories.

Online digital catalogs offer a more flexible alternative. Table 2.5 lists some of the more comprehensive online resources for Scolytine taxonomy. It is important to realize that in the majority of cases, taxonomic data are not collected *de novo*, but rather transcribed from older printed catalogs or checklists, typically Wood and Bright (1992a). Only in exceptional cases, the resource represents a truly new dataset "ground-checked" against primary literature and museum specimens.

Some of these resources include images, typically photographs. This would not be too significant if not for the fact that comprehensive collections of photographs are in many cases superseding more traditional identification tools. In an insect group where morphological characters are minute and rife with homoplasy, character-based keys are difficult to create and use (see above). It is often more efficient to compare a large number of full-body images for at least preliminary identification. This "virtual collection" approach to identification of insects, including scolytines, is gaining popularity, and is providing justification for the production of illustrated catalogs.

3.   BARK BEETLE MORPHOLOGY

3.1   Morphological Characteristics and Variation

Scolytines share the following general characteristics:

1. The body is cylindrical in cross-section.
2. The head is enlarged to accommodate large mandibles and chewing muscles. In most groups, the rear portion of the head with muscle attachments fills up most of the space within the pronotum.
3. Legs and antennae are short with respect to the length of the body and can be retracted or flattened against the body. In most groups, tibiae perfectly accommodate tarsi in a folded position.

TABLE 2.5 Public Databases on Bark Beetle Taxonomy and Images

Resource	URL	Contains Images?	Ground Checked? (see text)
HISL Xyleborini taxonomic database	http://xyleborini.speciesfile.org/public/site/scolytinae/home	No	Partially
Scolytinae and Platypodinae literature database	http://xyleborini.speciesfile.org/public/site/scolytinae/home/db_intro	No	Partially
Bark beetles of North America (including Mexico)	http://www.barkbeetles.info/about.php	Yes	Yes
Cryphalini Life Desks	http://cryphalini.lifedesks.org/	Few	No
Xyleborini of North America	http://www.ambrosiasymbiosis.org/northamericanxyleborini/	Yes	Yes
Xyleborini image database	http://xyleborini.myspecies.info/	Yes	Yes
Bark beetle genera of the United States	http://idtools.org/id/wbb/bbgus/index.html	Yes	Partially
Xyleborini of New Guinea	http://www.ambrosiasymbiosis.org/PNG_Xyleborini/	Yes	Yes

4. Together, 1, 2, and 3 are important adaptations for constructing and moving inside tunnels in the woody tissues of plants.
5. All tarsi apparently consist of four visible segments. In fact, the true fourth segment is very much reduced and generally not visible, except in larger species.
6. The antenna is elbowed (first segment much longer than any others) and clubbed (three terminal segments more or less fused and abruptly wider than previous segments; Figures 2.2 and 2.3).
7. Unlike most members of the Curculionidae, the mouthparts are not extended anteriorly into a snout or rostrum.

Compared to other insect groups of equal diversity, Scolytinae and Platypodinae are morphologically uniform. The uniformity is undoubtedly the result of the lifestyle that all species share: tunneling through plant tissues. Consequently, most morphological variation mainly occurs on the two unconstrained ends of the insect: the anterior end (head, pronotum) and the posterior end (mostly elytral declivity). The highly restricted "morphospace" resulted in an incredible amount of homoplasy, or the convergent evolution of character states. For example, many tribes have independently evolved most of the same variations of the elytral declivity, from rounded to sharply truncated, excavated, armed with spines, or attenuated (Figure 2.8). Because the elytral declivity is one of the most prominent features of many species, "typological" taxonomists have occasionally attempted to classify Scolytinae according to elytral characters. However, empirical analyses of the phylogenetic information content of characters suggest that this is one of the least reliable character sets for higher-level classification (e.g., Xyleborini: Hulcr *et al.*, 2007). Rather, any morphology-based classification of Scolytinae has to rely on *combinations* of characters in multiple character classes in order to mitigate the effect of homoplasy.

The following section focuses on features that are used in identification of tribes and genera. It is not intended to serve as a complete review of scolytine external morphology.

3.1.1 Head

Many bark beetle tribes differ in whether or not the head is visible from above (Figure 2.5). In most cases, this is a complementary character to whether or not the anterior margin of the pronotum projects forward in lateral view, covering the head as a hood. Less common alternative combinations include a strongly curved and short pronotum with the head not visible from above due to the downward directed foramen (e.g., *Chramesus*), or a straight and long pronotum with invisible head (e.g., *Hylastes*).

3.1.2 Eyes

Compound eyes of Scolytines may be coarsely or finely faceted. The eyes are always flush with the surrounding levels of the head and do not bulge or protrude. This is useful in distinguishing them from superficially similar groups such as the Bostrichidae. Eyes may be entire (oval in shape), emarginate (especially around the antennal base),

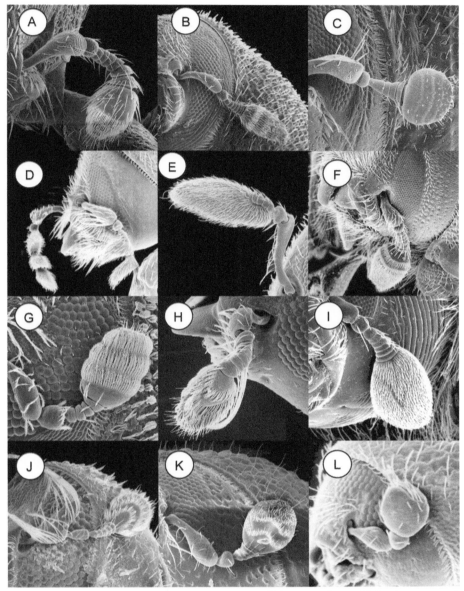

FIGURE 2.2 **Antennae of bark and ambrosia beetles.** (A) *Hylastes tenuis*; (B) *Hylesinus aculeatus*; (C) *Dendroctonus terebrans*; (D) *Phloeotribus texanus*; (E) *Chramesus chapuisii*; (F) *Dendrosinus bourreriae*; (G) *Carphoborus bifurcus*; (H) *Loganius vagabundus*; (I) *Scolytus dimidiatus*; (J) *Thysanoes fimbricornis*; (K) *Ips avulsus*; (L) *Crypturgus alutaceus*.

or completely divided into dorsal and ventral sections (Phrixosomatini, Hyorrhynchini, Xyloterini, etc.).

3.1.3 Antennae

Antennal characters are very important in identifying genera and tribes (Figures 2.2 and 2.3). The antenna is elbowed. The first segment (scape) is elongated, often as long as all subsequent segments combined. It may be variously ornamented with setae (Figure 2.2D, J), is generally slightly curved, and sometimes enlarged distally or with projections (Figures 2.2J and 2.3F, G).

The segments between the scape and the terminal club are much smaller, flexible and often considered collectively as the funicle. The number of funicular segments is very important for identification, generally consistent within a genus, and varies from seven (Hylastini, Scolytini), down to one (some Corthylini). Within a given genus, the number of segments in the funicle may be lower in smaller species. The definition of funicle in scolytine taxonomy typically includes the larger second segment, also known as the pedicel. This convention is at odds with the definition of the funicle in other insect taxa, which excludes the pedicel, and used in morphological ontologies (relational

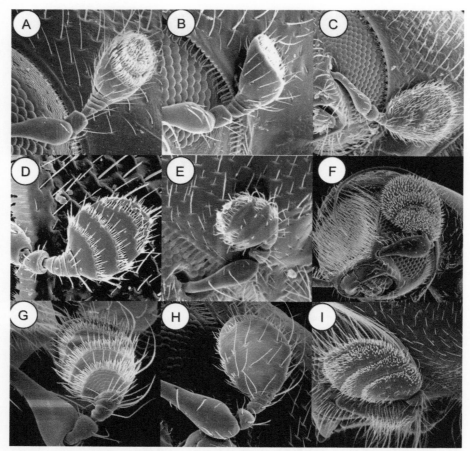

FIGURE 2.3 **Antennae of bark and ambrosia beetles.** (A) *Ambrosiodmus lecontei*; (B) *Xyleborus pubescens*; (C) *Scolytogenes jalapae*; (D) *Conophthorus coniperda*; (E) *Pityophthorus confusus*; (F) *Corthylus papulans*; (G) *Gnathotrichus materiarius*, anterior face; (H) *Gnathotrichus materiarius*, posterior face; (I) *Monarthrum fasciatum*.

terminology) based on the insect developmental morphology. For more information on this uncertainty, see Section 5.3.

The final three segments of the antenna, known as the club, are partially or totally fused. Sometimes the club appears to have more than three segments, which are not true segments but superficial lines formed by setae. A notable exception is the tribe Phloeotribini (Figure 2.2D) where the segments of the club are joined along one side and freely movable. The club is conspicuously wider than the segments of the funicle. The degree of flattening of the antennal club varies widely. In Hylastini, Hylurgini, and some Hylesinini (Figures 2.2A, B) the segments are nearly circular in section, and the club is cone shaped or nearly so. In most cases however, the club itself is strongly flattened

The phylogenetically basal plesiomorphic state of the club consists of independent segments that have been fused to varying degrees and are no longer movable with respect to each other. In most cases the suture lines between adjacent segments are still visible as external grooves

(Figures 2.2A–C, F, G and 2.3D, G). In many cases, these sutures may also be marked with lines of setae. In other cases, the surface of segments between sutures may be more sclerotized than the segmental areas and the sutural lines hidden among the abundant setae. In more derived cases, the external sutures themselves may no longer be visible under optical microscopy, but are still marked either by bands of setae, or by more polished, sclerotized surface of the segments between sutures (Figure 2.2I). Another variation, common in the Xyleborini, many Dryocoetini and some Ipini, is the strongly sclerotized basal segment of the club with terminal segments reduced and less well defined (Figures 2.2L and 2.3A, B).

The sutures between club segments vary from horizontal (i.e., perpendicular to the axis of the club, Figure 2.2A, B) to prominently curved in different fashions (Figures 2.2J, K and 2.3G). In some groups of Pityophthorini, the sutures are only visible along the outside edges of the club (Figure 2.3E). In some groups it is important to observe whether the external suture lines are matched by a corresponding internal barrier, or septum.

In practice, this can be difficult to see in pinned specimens or specimens preserved in alcohol, and depend strongly on the condition of the specimen. In other genera, there may be a partial septum, even in the absence of external sutures or markings of setae or sclerotized areas (Figure 2.3F).

Other useful characters include the symmetry of the club, both in lateral aspect and anterior–posterior aspect. When looking at the anterior face of the antennal club, the basal segment is generally attached at its midpoint to the last segment of the pedicel and the club is symmetrical along its centerline. There are notable exceptions where the club connection is more unusual (Figure 2.2E) or its shape is not laterally symmetrical (Figure 2.3F, I). In some groups, the posterior face of the antenna is similar to the anterior face. When the insect is on an open surface, the antennae are typically held perpendicular to the body with the anterior face forwards. In other poses, especially within galleries, the antennae are folded back against the body with the posterior face flush against the prothorax. It is generally believed that the setae of the club are chemoreceptors. In genera with flattened antennal clubs, the setae on the posterior face are generally less abundant, even when the sutures are similar in shape and position to those of the anterior face (Figure 2.3G, H). In the Xyleborini and some Dryocoetini there has been a distal displacement of the sutures on the posterior face (Figure 2.3A, B). In some cases, the posterior sutures have been entirely shifted to the anterior face (*Xyleborus, Xylosandrus*).

3.1.4 Thorax

As with most beetles, only the prothorax is visible in dorsal view, with the meso- and meta-thorax hidden by the elytra. In most of the groups formerly placed in the subfamily Hylesininae (Wood and Bright, 1992a), the prothorax curves in a smooth arc from the base of the elytra towards the head. In most cases, part of the head is visible from above (Figure 2.5). In many genera formerly placed in the subfamily Scolytinae the prothorax extends forwards and partially encloses the head, completely hiding it (Figure 2.6). In many, but not all, of these genera there is a marked change in curvature from the pronotal disc, which is flat with respect to the elytral surface to the anterior slope. The point or area of this abrupt change is referred to as the pronotal summit. In both cases, there is often a pronounced change in surface texture and vestiture (hair or setae) between the pronotal disc and its anterior slope. In many genera, there are small, backwards-pointing projections (asperities) on the anterior slope. These may be few in number and not arranged in any particular order (e.g., *Hylesinus, Chaetophloeus*), or they may be very numerous, or even arranged in concentric patterns (e.g., *Pseudothysanoes, Pityophthorus*).

The only mesothoracic structure typically visible in dorsal view is the scutellum. In many hylesinines, it is rounded and set back in a notch between the elytral bases (Figure 2.5). In most scolytines, it is flush with the surface of the elytra and its anterior margin is in line with the margins of the adjacent elytra (Figure 2.6).

3.1.5 Legs

Characteristics of the legs, especially the protibiae, are important in distinguishing certain tribes and genera. Tibiae are generally flattened. The interior margin refers to the margin that would contact the femur when folded inwards. The anterior face is the portion that faces forward when the insect is walking. Generally, the protibia is wider at its distal end (Figure 2.4G–J) and appears somewhat triangular, or the sides may be parallel (Scolytini, Figure 2.4D; Micracidini, Figure 2.4F), or the protibia can be conspicuously slender (Figure 2.4K) or widened (Figure 2.4C).

Most beetles have various projections on their tibiae. These include socketed denticles, teeth, spines, spurs, etc., and the terms are used interchangeably. Socketed denticles are typical for most Scolytinae, but these features are occasionally found in other wood-boring weevil groups (e.g., Conoderinae, Molytinae, and Cossoninae). At lower magnifications, these simply appear to be small spines on the sides and apex of the tibiae. At higher magnification, it is clear that these are actually setae that are jointed and set within sockets (Figure 2.4). Socketed denticles are not present in all species. To complicate matters, many genera have a large spine at the apex of the tibia that is not socketed. In the Scolytini, the protibiae have a single, curved spine at the apex with no socketed denticles along its margins (Figure 2.4D). The terminal apical mucro or spine is generally on the interior margin (Figure 2.4A, E–I).

3.1.6 Elytra

The anterior margin of the elytral bases and their position with respect to the scutellum is a major landmark. In many tribes formerly placed in Wood's Hylesininae (Wood, 1986), the scutellum is set back into a notch and the anterior margins of the elytra are conspicuously curved (Figure 2.5). Often there are elevated, curved projections (crenulations) along this margin. In most other cases, the anterior margins of the elytra are smooth, without elevations, and straight (Figure 2.6).

One of the major features of the elytral surface is the striae and interstriae. Striae are marked by longitudinal rows of surface punctures. These can be very prominent (Figure 2.5) or nearly obsolete (Figure 2.6). Interstriae are the spaces between the striae. There is a great deal of variation in the degree of prominence of the striae, diameter and spacing of punctures, surface asperity, and vestiture

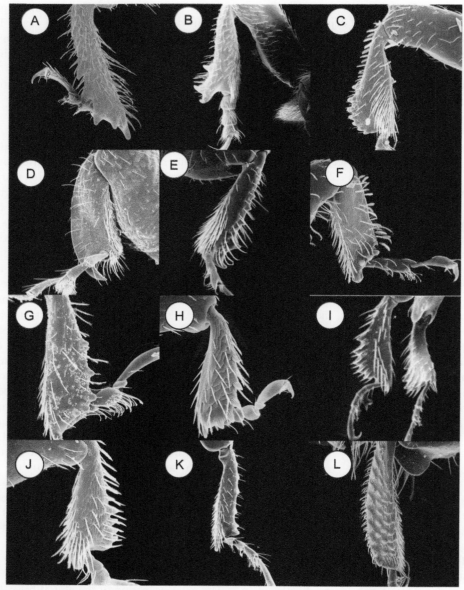

FIGURE 2.4 **Protibiae of bark and ambrosia beetles.** (A) *Dendroctonus terebrans*, posterior; (B) *Cnesinus strigicollis*, posterior; (C) *Chramesus chapuisii*, anterior; (D) *Scolytus muticus*, posterior; (E) *Scolytodes schwarzi*, anterior; (F) *Thysanoes fimbricornis*, anterior; (G) *Lymantor decipiens*, posterior; (H) *Crypturgus alutaceus*, posterior; (I) *Xyleborus ferrugineus*; (J) *Hypothenemus crudiae*, anterior; (K) *Gnathotrichus materiariu*, anterior; (L) *Monarthrum fasciatum*, posterior.

(hair or setae). These characters are typically more important in distinguishing species than genera.

The apical (rear) portion of the elytra is the declivity. The change in curvature or profile between the elytral disc and the declivity can be gradual or abrupt. In most bark and ambrosia beetles, the ventral surface of the thorax and abdomen is more or less flat and the elytra curve down to meet that level. Differences in the curvature in lateral view of both the elytra and abdomen may be useful in distinguishing genera. In the genus *Scolytus*, the dorsal surface of the elytra is nearly flat with little or no declivity while the ventral surface of the abdomen rises to meet it. In other Scolytini and some Cryphalini, Xyloctonini, and Hexacolini, the ventral surface of the abdomen also curves upwards in a characteristic way.

There is extreme variation among species in the form, sculpture, armature (spines and projections), and vestiture (hairs and setae) of the declivity. The most common form is for the declivity to be gradual and for the surface to be similar to that of the disc, but variations of this are found within almost all tribes. In many cases, elytra may be markedly different between sexes. Elytral sculpturing can

FIGURE 2.5 External anatomy of *Cnesinus strigicollis* (LeConte) (Bothrosternini).

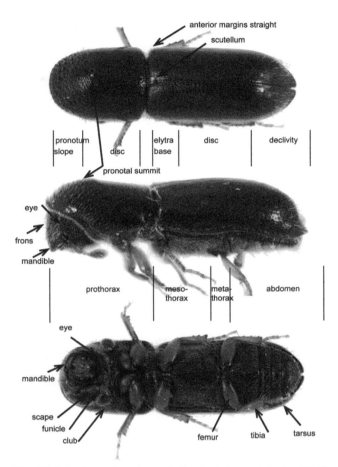

FIGURE 2.6 External anatomy of *Gnathotrichus materiarius* (Fitch) (Corthylini).

be very useful for separating species, but is typically too variable for separation at a generic or tribal level. Many of these modifications are convergent in totally unrelated groups, and/or sexually dimorphic: characters that are typical of females in one group may be found on males in other (different) groups.

3.1.7 Mycangia and their Role in Classification

While there is no shortage of characters that vary within Scolytinae, there are very few characters that vary in a synapomorphic fashion, i.e., correlated with phylogeny. One of the few exceptions—character systems where character states are generally synapomorphic—is mycangia (Hulcr *et al.*, 2007; Smith and Cognato, 2010). A mycangium is an organ for fungus transport. Various kinds of mycangia evolved many times within Scolytinae, reflecting the richness of the beetles' relationships with fungi. From a phylogenetic and taxonomic perspective, mycangia appear to be important characters since they are highly evolutionarily conserved in many scolytine groups, and since they presumably correlated

with the evolution of fungus feeding. As such, they can potentially tell the story of beetle–fungus coevolution and of the rise of some groups to dominance better than any other character. However, the complete picture of evolution of mycangia is still precluded by two factors.

First, there is no clear definition of mycangium. Instead, beetles have been shown to carry fungal spores in various pits, grooves, and sacs, with or without glands, and with varying degrees of fidelity (see Chapter 6 for more discussion on the topic). As many as six different classes of mycangia have been defined (Six, 2003).

The second problematic factor is that many scolytine groups living in obvious symbiosis with fungi remain poorly studied, and the presence, type, and position of mycangia is unknown. These groupd include, for example, *Sueus* spp., the *Stylotentus* group of *Hypothenemus* (Schedl, 1962), *Scolytodes unipunctatus* Wood and Bright (Hulcr *et al.*, 2007), *Cnesinus lecontei* Blandford and *Eupagiocerus dentipes* Blandford (Kolarik and Kirkendall, 2010), among others.

For the purpose of illustrating the phylogenetic value of mycangium, we restrict the definition to the most highly evolved type, the "glandular sac mycangium" *sensu*

TABLE 2.6 Scolytine Clades with Independently Evolved Sac Mycangia

Clade	Position of the Glandular Sac Mycangium
Xyleborini	mandibular; mesonotal; elytral
Dendroctonus frontalis group	thorax
Scolytoplatypodini	dorsal or large lateral pit on thorax
Xyloterini	tubular structure opening near procoxae
Corthylini	procoxal cavity

Six (2003). These invaginated, often complex and large structures are typically present in beetle groups obligately dependent on their fungal symbionts (Table 2.6). These include all ambrosia beetles and several bark beetles whose larval development depends on the presence of specific symbionts. Based on the preliminary consensus phylogeny (Figure 2.9), glandular sac mycangia appear to have evolved at least seven times in Scolytinae. In groups that evolved a glandular sac mycangium, the organ is usually present in all species in the group, and is either absent from other groups or present in a different form (Hulcr *et al.*, 2007; Smith and Cognato, 2010). There has been at least one and possibly more losses: *Diuncus*, a genus of Xyleborini ambrosia beetles evolved mycocleptism, or fungus stealing, and consequently appear to have lost their mycangium (Hulcr and Cognato, 2010). Similarly, *Dryoxylon*, an enigmatic xyleborine genus, does not appear to carry any fungi in any part of its body (Bateman and Hulcr, unpubl.).

In the other large group of ambrosia weevils, the Platypodinae, mycangia seem to be a phylogenetically unstable character. The presence, absence, and gender-specific presence of mycangia often varies, even between closely related species of Platypodinae.

3.1.8 Character Variation

There are several character classes that appear variable and thus taxonomically informative, but which reveal significant intraspecific variation. These include body size, vestiture, and color. The variation in body size in some scolytine groups appears to be a result of two different factors: nutrition and altitudinal gradient. Nutrient content of diet, whether tree tissue or fungus, can have significant effects on the body dimensions and weight of a scolytine beetle. Bark beetles are generally much more variable in body size than ambrosia beetles, which may be because tree tissue is much more variable in nutrition than fungal matter. In some groups (such as Xyleborini) individuals of the same

species increase in size with elevation as much as 1.5 times (Beaver, 1976; Hulcr and Cognato, 2013). The mechanism behind this variation is not clear, but experimental data for insects indicate that longer development in colder environments results in larger bodies.

Vestiture density ("hairiness") and color can be conserved and reliable characters. However, both can also vary dramatically with age of an individual beetle, and because of collecting techniques. Many museum collections contain specimens collected in glue traps. These specimens have usually lost all or most of their setae, and are darker than normal specimens. Likewise, specimens collected from their galleries before they had a chance to mature and fully melanize are usually pale, while mature specimens of most species have a typically dark color.

3.2 Images of Morphology

The overall morphology of Scolytinae is perhaps somewhat more constrained than that of many other weevils, primarily because of the subcortical wood-boring lifestyle. However, fine-scale morphological variation is enormous. This section presents a glossary of morphological characters and an illustration of morphological variation across Scolytinae. We chose representatives of five tribes, either common ones or those that contain economically important species. The goals are to illustrate variation in scolytine morphology, and to annotate different characters and character states as they vary between groups.

3.2.1 Bothrosternini

Cnesinus strigicollis is a common bothrosternine species (Figure 2.5) from North America. Its morphology is a good example of Wood's former subfamily Hylesininae, which contains many important pests, such as the hylurgine genera *Tomicus* and *Dendroctonus*. While Wood's subfamilies are no longer recognized, some of the characters used to distinguish the two groups are still generally useful.

3.2.2 Corthylini

The tribe Corthylini (Figure 2.6) represents one of the largest radiations of fungus farming ambrosia beetles. They are largely confined to the New World tropics. *Gnathotrichus materiarius* is a common North American species, now also introduced to Europe.

3.2.3 Cryphalini

A comprehensive morphological sketch for the tribe Cryphalini is included in Chapter 11.

3.2.4 Scolytini

Scolytini (Figure 2.7) is a large tribe containing several genera that are morphologically rather distinct from most

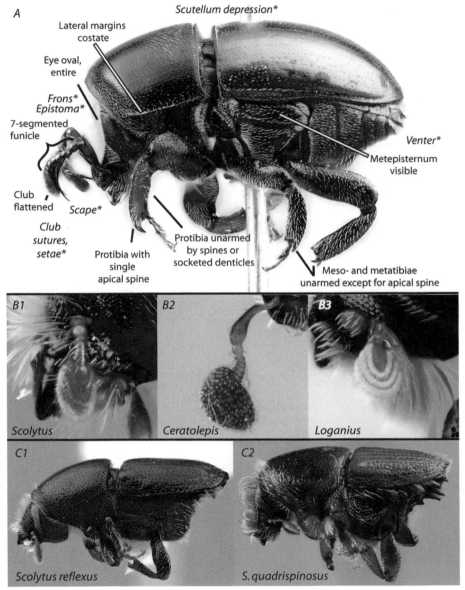

FIGURE 2.7 **Morphology of Scolytini.** (A) Overall morphology of *Camptocerus noel*. (B1–B3) Variation in antennal clubs shown on *Scolytus*, *Ceratolepis*, and *Loganius*. (C1, C2) Deviations from the ancestral venter (refer to A) in *Scolytus*. Characters marked with asterisks are important for the identification of species within Scolytinae.

other Scolytinae. The group's phylogenetic position is not clear, but they may be sister to most other Scolytinae (Figure 2.9). Scolytini include many economically and ecologically important species, such as several *Scolytus* spp. that are vectors of the Dutch elm disease.

3.2.5 Xyleborini

The characters and character systems most important for the internal classification of Xyleborini (Figure 2.8) include the antennal club, protibiae and their spines, shape and surface of pronotum, and the elytral declivity (the sloped end of the elytra). The elytral declivity is by far the most

variable feature within xyleborine beetles. However, despite its variation, its utility for classification is limited because it also displays considerable homoplasy. Many xyleborine genera include species that independently evolved nearly identical shapes of the elytral declivity.

4. CURRENT SCOLYTINE AND PLATYPODINE CLASSIFICATION

4.1 Where do Bark Beetles Belong?

Evidence in support of the classification of Scolytinae as a subfamily within the weevil family Curculionidae comes

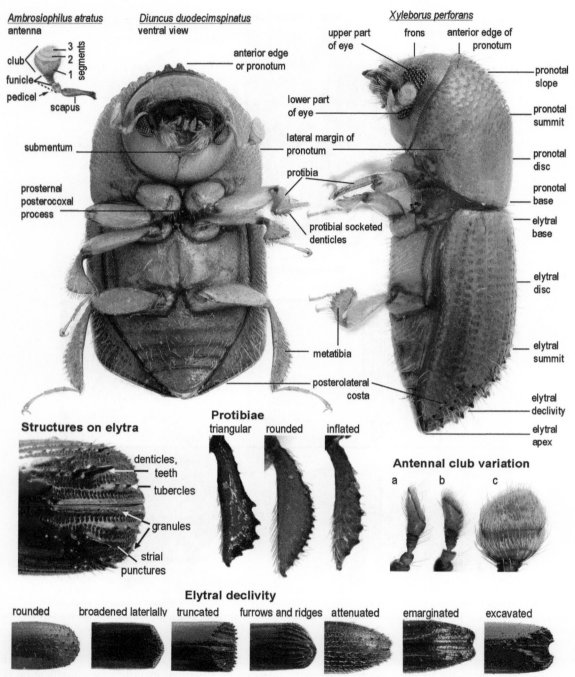

FIGURE 2.8 **Xyleborini morphology.** Overall structure of the body; nomenclature of elytral armature; variation in protibiae; variation in antennal club (a: first segment covers all of the rear face, b: second and third segments are also visible on the rear face, c: first segment is not dominant); and variation in elytral declivity.

from studies of both adult (Kuschel, 1995) and larval (Marvaldi, 1997) morphology, and from molecular and morphological phylogenetics. The first phylogenetic study that placed Scolytinae as a derived curculionid subfamily close to Cossoninae was the pioneering work by Kuschel (1995) based on adult and larval morphological characters.

Several subsequent publications based on morphological and molecular data have confirmed the derived position among the "higher Curculionidae." The most extensively gene- and taxon-sampled molecular phylogenetic study of the weevil superfamily Curculionoidea to date (135 Curculionoidea; 100 Curculionidae; 22 genera of Scolytinae; six

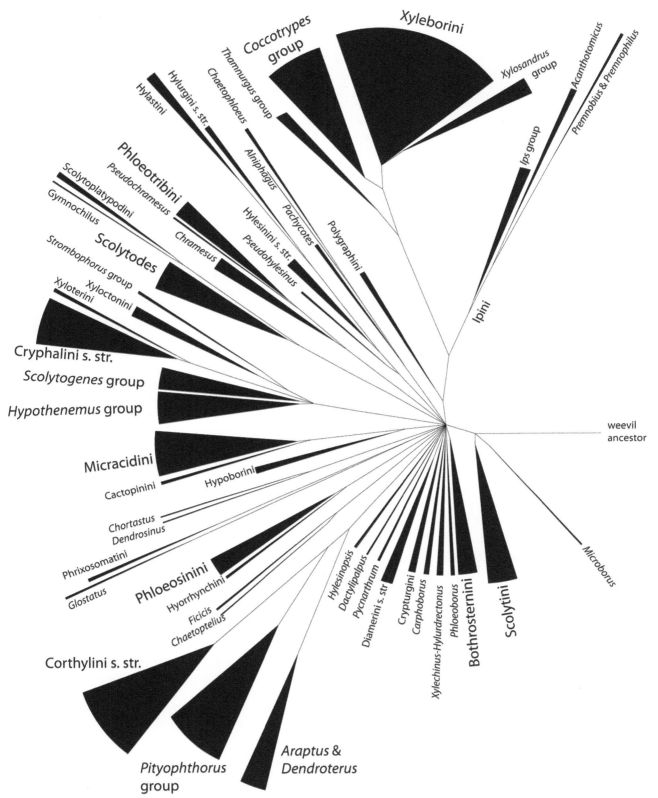

FIGURE 2.9 Bark and ambrosia beetle evolution and diversity (Scolytinae only).

genes) was published by McKenna *et al.* (2009). In this study, Scolytinae were recovered as sister to a clade comprised of Molytinae, Cossoninae, Conoderinae, and most Curculioninae, but without strong nodal support. Haran *et al.* (2013) compared partial mitochondrial genome sequences for 29 species of Curculionoidea, including 22 species of Curculionidae (12 of the 13 mitochondrial protein-coding genes, ≈ 10 kbp total data). The study recovered strong support for the aforementioned placement of Scolytinae. Jordal *et al.* (2011) also recovered Scolytinae within "higher Curculionidae," but in varying positions. Thus, these studies are consistent in placing Scolytinae within the weevil family Curculionidae, but there remains no clear consensus about which curculionid group is the closest relative of bark beetles. It is nonetheless clear that the Entiminae, Dryophthorinae, Bagoinae, Hyperinae, Cyclominae, and Brachycerinae are not likely near-relatives.

The phylogeny of Curculionidae and its constituent subfamilies (including Scolytinae and Platypodinae) remains uncertain because of limited taxon sampling, limited and/or inconsistent resolution, and low nodal support in most studies to date. Consequently, other evolutionary inferences—the timing of lineage divergences, patterns of host-use evolution, taxonomic diversification, etc., in Curculionidae and its constituent subfamilies—remain unsettled. There is much need for a comprehensive tribal-level phylogeny of higher Curculionidae and associated estimates for the timing of lineage divergences. Within such a framework it will be much easier to establish, for example, the timing and taxonomic location of major evolutionary transitions in host taxon associations, shifts (increases or decreases) in diversification rate, patterns in the evolution of host taxon and tissue specialization, and overall patterns in the evolution of larval feeding habits.

4.2 Internal Phylogenetic Relationships

Figure 2.9 aims to visualize the stunning radiation of scolytine beetles, and the large differences in species diversity between clades. The largely unresolved deeper structure is likely a result of the rapid radiation of the group, beginning 120 Ma, resulting in short internal branches in the phylogeny, and the relatively long terminal branches, which together contribute to the difficulty of phylogeny reconstruction. The branching structure follows the latest molecular phylogenies of Jordal and Cognato (2012) and Cognato *et al.* (2011). The width of terminal branches reflects the number of species in the branches (see Table 2.8 for a comprehensive list). Only monophyletic clades are shown; note that some clades are not congruent with the tribal classification of Wood (Wood, 1978) (revised in Alonso-Zarazaga and Lyal, 2009).

The *Coccotrypes* group contains the dryocoetine genera *Taphtorychus*, *Cyrtogenius*, *Dryocoetes*, *Ozopemon*, *Dryocoetiops*, and *Coccotrypes*. Cryphalini *sensu stricto*

contains *Cryphalus* and *Hypocryphalus*. The *Hypothenemus* group contains the cryphaline genera *Allernoporus*, *Ptilopodius*, *Hypothenemus*, and *Trypophloeus*. The *Scolytogenes* group contains another group of cryphaline genera: *Ernoporus*, *Ernoporicus*, *Procryphalus*, *Cosmoderes*, and *Scolytogenes*. These two cryphaline clades are not monophyletic with the Cryphalini *sensu stricto* group because of Xyloterini, and are not reciprocally monophyletic because of Xyloctonini and the *Strombophorus* group. The *Strombophorus* group contains the hylesinine genera *Hapalogenius* and *Rhopalopselion*, and formerly diamerine genera *Acacicis* and *Strombophorus*. The *Ips* group contains the ipine genera *Pityokteines*, *Pityogenes*, *Orthotomicus* and *Ips*. Hylesinini *sensu stricto* contains *Pteleobius*, *Hylesinus*, and *Hylurgopinus*. The *Thamnurgus* group contains the dryocoetine genera *Thamnurgus*, *Xyllocleptes*, *Dactylotrypes*, *Triotemnus*, and *Lymantor*. The *Xylosandrus* group contains xyleborine genera with mesonotal mycangium: *Anisandrus*, *Hadrodemius*, *Eccoptopterus*, *Xylosandrus*, *Cnestus*, and *Diuncus*. The *Pachycotes* group contains the hylurgine genera *Pachycotes*, *Sinophloeus*, *Hylurgonotus*, and *Xylechinosomus*. Hylurgini *sensu stricto* contains *Tomicus*, *Dendroctonus*, and *Hylurgus*. Hypoborini *sensu stricto* contains the genera *Liparthrum*, *Hypoborus*, and *Styracoptinus*. Micracidini contains the genera *Pseudothysanoes*, *Thysanoes*, *Hylocurus*, *Micracis*, *Lanurgus*, and *Miocryphalus*. Phloeosinini *sensu stricto* contains the genera *Phloeosinus* and *Hyledius*. Polygraphini *sensu stricto* contains the genera *Dolurgocleptes*, *Polygraphus*, and *Serrastus*. Genera absent in this phylogeny are missing either because they contained fewer than 10 species (arbitrary cutoff for the sake of resolution), or they have not been included in the underlying molecular phylogenies and thus their position is unconfirmed.

4.3 A Checklist of all Currently Recognized Genera

Table 2.7 lists all currently recognized tribes of Scolytinae. While most of the public and non-specialist researchers are aware of several, usually economically important species, there are almost 6000 species in this hyper-diverse weevil subfamily.

The most recent comprehensive classification of bark and ambrosia beetles is that of Wood (1986, 1993), which is entirely based on morphological characters. Earlier classification schemes were covered by Wood (1986, 1993) and are not reviewed here. Although Wood's classification was accompanied by many phylogenetic hypotheses, it was not tested using phylogenetic methods. Wood's classification included 25 tribes in two subfamilies, Hylesininae and Scolytinae. An additional tribe, Amphiscolytini was added by Mandelshtam and Beaver (2003) to include a single species

TABLE 2.7 Current Tribal Classification of the Scolytinae based on Wood (1986), Wood and Bright (1992), Alonso-Zarazaga and Lyal (2009), and Bright (2014)

Tribe	Afrotropical	Oriental	Austro-Pacific	Palearctic	Neotropical	Nearctic	Other	Total
Amphiscolytini	1							1
Bothrosternini					131			131
Cactopinini					17	4		21
Carphodicticini		3			2			5
Corthylini	50	8	2	30	863	258		1211
Cryphalini	145	184	145	114	89	15	10	702
Crypturgini	25	3		24		3		55
Diamerini	62	55	5	10				132
Dryocoetini	115	137	75	81	49	10	7	474
Hexacolini					242			242
Hylastini				28		27		55
Hylesinini	69	13	11	31	30	10		164
Hylurgini	4	12	23	18	39	34		130
Hyorrhynchini	1	15		3				19
Hypoborini	10	4	4	20	24	12		74
Ipini	87	25	6	52	11	49		230
Micracidini	53			1	187	57		298
Phloeosinini	6	37	10	26	113	35		227
Phloeotribini			4	17	81	8		110
Phrixosomatini	10				15			25
Polygraphini	59	18		50		27		154
Scolytini				58	124	27		209
Scolytoplatypodini	20	27		6				53
Xyleborini	199	484	160	65	243	10	7	1168
Xyloctonini	64	11	3					78
Xyloterini		5		12		5		22
Grand Total	980	1041	448	646	2260	591	24	5990

For convenience, tribes placed by Wood in his subfamily Hylesininae are shaded. Numbers for different biogeographic regions are based on extant species only. Species classified as "other" are mostly widely distributed tropical species whose region of origin is not known.

that did not fit into any of the previously recognized tribes. The genus *Coptonotus* was placed by Wood in the tribe Coptonotini within the Platypodidae. Later scientists (Thompson, 1992; Kuschel *et al.*, 2000) concluded that this genus belonged within the Scolytinae and treated it as a separate tribe related to the Hylesinini. More recent phylogenetic analyses based on combined molecular and morphological data (Jordal *et al.*, 2011) and mitochondrial genome sequences (Gillett *et al.*, 2014) rejected this

hypothesis and demonstrated that *Coptonotus* is not related to Scolytinae or Platypodinae.

Since the publication of Wood's classification (1986) and the world catalog (Wood and Bright, 1992a) several phylogenetic studies involving molecular and morphological characters have noted problems and inconsistencies with the arrangement of genera and subtribes (Kuschel, 1995; Marvaldi, 1997; Farrell *et al.*, 2001; Marvaldi *et al.*, 2002; Jordal, 2007; Jordal *et al.*, 2008; Jordal and Cognato, 2012).

Notably Wood's "Scolytinae" is definitely not monophyletic. His tribes Hylesinini and Hylurgini (with Hylastini as a nested group) are not defensible in their current form. Likewise, the tribe Dryocoetini is paraphyletic with respect to Xyleborini. There have been numerous inconsistencies between molecular studies and no clear consensus has been reached among students of the group on how to construct a new classification that incorporates these new data. At the same time, these studies have helped elucidate the evolution of major biological features such as host associations (Sequeira *et al.*, 2000, 2001; Farrell *et al.*, 2001), inbreeding (Normark *et al.*, 1999; Jordal *et al.*, 2002b), and the development of the ambrosia habit (Jordal and Cognato, 2012). Eventually, a new classification that synthesizes this new information will emerge.

In the meantime, some sort of taxonomic structure is needed to keep track of taxa at all levels. Noting the discrepancies between recent phylogenetic studies and the published classification, Alonso-Zarazaga and Lyal (2009) suggested keeping all of Wood's tribes at a tribal level while dispensing with his subfamilies altogether. Their proposal was put forward strictly as an interim measure until higher-level relationships could be resolved. Their system is followed here. The classification is summarized in Tables 2.7 and 2.8. Presently, 5990 species are included in 282 genera (Table 2.8).

Species are tabulated in Tables 2.7 and 2.8 according to broad biogeographical regions. This inevitably introduces some subjectivity and error into the tabulations in the many cases where a species is known only from the single locality or narrow region from which it was described, or in other cases where the distribution of a species may cross biogeographical boundaries. On the other hand, use of political boundaries introduces other errors since these generally do not correspond to geographic or biological reality. In the absence of a fully resolved generic- and tribal-level phylogeny for Scolytinae, it is premature to draw broad conclusions about the evolutionary history of major taxonomic groupings. Even so, examination of Tables 2.7 and 2.8 shows that the distribution of genera and tribes is highly uneven, especially if one considers relative abundance rather than absolute presence or absence.

5. CONCLUSION: UNRESOLVED ISSUES

5.1 Scolytinae are Definitely a Subfamily of Weevils, but What about Platypodinae?

Because of their ecological similarity, Scolytinae and Platypodinae have often been studied by the same taxonomists. Consequently, their taxonomic history has been largely parallel, including the traditional status of both as unique families, and the recent shift from treatment as separate families of beetles to the status of curculionid weevil subfamilies.

Nonetheless, the phylogenetic placement of Platypodinae is still debated. Various analyses, morphological as well as molecular, support one of four hypotheses:

1. Platypodinae are best treated as a family (Platypodidae), distinct from, but closely related to, Curculionidae. They were treated as a family until the first comprehensive morphological cladistic analyses suggested their subfamily-level status within Curculionidae.

2. Platypodinae are a subfamily of Curculionidae closely related to Scolytinae. The strongest evidence for the subfamilial classification of Platypodinae within the advanced weevils comes from morphological studies of adults (Kuschel, 1995; Marvaldi *et al.*, 2002) as well as larvae (Marvaldi, 1997) and from most molecular–phylogenetic analyses using one or a few genes. Nonetheless, a close relationship to Scolytinae is not comprehensively supported. The molecular–phylogenetic study of Jordal *et al.* (2011) included a range of weevil taxa and morphological and molecular data, and showed that the position of Platypodinae may be close to Scolytinae, but also showed that this was highly dependent on taxon sampling. Unfortunately, molecular data for these highly divergent and diverse groups are still weakly developed. The close relationship with Scolytinae could therefore also be a result of "long branch attraction," a phylogenetic phenomenon whereby highly divergent taxa are erroneously recovered together in certain phylogenetic analyses. The DNA sequences of Platypodinae are particularly notorious for being highly divergent from other weevil groups (McKenna *et al.*, 2009; Jordal *et al.*, 2011; Haran *et al.*, 2013; Gillett *et al.*, 2014)

3. Platypodinae are a distinct subfamily within weevils that is not closely related to Scolytinae but rather originated separately. Specifically, some studies of weevil larval morphology (Marvaldi, 1997; Oberprieler *et al.*, 2007) have suggested a close relationship between Platypodinae and Dryophthorinae. McKenna *et al.* (2009) included five platypodine genera into their pan-weevil molecular phylogenetic study, and recovered Platypodinae in an early-divergent position within Curculionidae, where they were the sister group of most Dryophthorinae. These results are compatible with the results of recent studies by Haran *et al.* (2013) and Gillett *et al.* (2014). Interestingly, even the comprehensive study of Jordal *et al.* (2011) that proposed Platypodinae as closely related to Scolytinae could not reject a close relationship between Dryophthorinae and Platypodinae. In their analysis, Platypodinae grouped with Dryophthorinae when *Mecopelmus*, *Schedlarius*, and *Coptonotus* are excluded, and these results are congruent with the molecular phylogenies above. However, when these taxa are included, Platypodinae

TABLE 2.8 Current Genera of the Scolytinae based on Wood (1986), Wood and Bright (1992), Alonso-Zarazaga and Lyal (2009), and Bright (2014) and Taxonomic Literature from 2000 to Present

Genus	Tribe	Afro-tropical	Oriental	Austro Pacific	Pale-arctic	Neo-tropical	Nearctic	Wide-spread	Total
Acacicis	Diamerini	4	5	2					11
Acanthotomicus	Ipini	53	23	6	1	11			94
Acorthylus	Cryphalini					6			6
Afromicracis	Micracidini	17							17
Akrobothrus	Bothrosternini					1			1
Allernoporus	Cryphalini				1				1
Allothenemus	Cryphalini					1			1
Alniphagus	Hylesinini				1		2		3
Amasa	Xyleborini	2	25	13	1				41
Ambrosiodmus	Xyleborini	31	25	4	2	17	1		80
Ambrosiophilus	Xyleborini		6	2					8
Amphicranus	Corthylini					66			66
Amphiscolytus	Amphiscolytini	1							1
Ancipitus	Xyleborini		1						1
Anisandrus	Xyleborini	1	7	1	3		2		14
Aphanarthrum	Crypturgini	24	3		2				29
Araptus	Corthylini					172			172
Aricerus	Phloeotribini			3					3
Arixyleborus	Xyleborini		28	4					32
Asiophilus	Phloeosinini		2						2
Beaverium	Xyleborini		3	4					7
Bothinodroctonus	Polygraphini		3						3
Bothrosternoides	Diamerini		1						1
Bothrosternus	Bothrosternini					11			11
Brachyspartus	Corthylini					1			1
Cactopinus	Cactopinini					17	4		21

Continued

TABLE 2.8 Current Genera of the Scolytinae based on Wood (1986), Wood and Bright (1992), Alonso-Zarazaga and Lyal (2009), and Bright (2014) and Taxonomic Literature from 2000 to Present—cont'd

Genus	Tribe	Afro-tropical	Oriental	Austro Pacific	Pale-arctic	Neo-tropical	Nearctic	Wide-spread	Total
Camptocerus	Scolytini					31			31
Cardroctonus	Polygraphini	3							3
Carphobius	Polygraphini						3		3
Carphoborus	Polygraphini	1	2		10		21		34
Carphodicticus	Carphodicticini					1			1
Carphotoreus	Phloeosinini					1			1
Catenophorus	Phloeosinini	1							1
Ceratolepsis	Scolytini					7			7
Chaetophloeus	Hypoborini					14	10		24
Chaetoptelius	Hylurgini			6	1				7
Chiloxylon	Dryocoetini					1			1
Chortastus	Polygraphini	5							5
Chramesus	Phloeosinini					86	6		92
Cisurgus	Crypturgini	1			8				9
Cladoctonus	Phloeosinini	5		1		7			14
Cnemonyx	Scolytini					23			23
Cnesinus	Bothrosternini					95			95
Cnestus	Xyleborini		22	5	1	4			32
Coccotrypes	Dryocoetini	30	57	24	5	7		6	129
Conophthorus	Corthylini						13		13
Coptoborus	Xyleborini		1			22			23
Coptodryas	Xyleborini	1	32	2	1				36
Coriacephilus	Cryphalini		5						5
Corthylocurus	Corthylini					15			15
Corthyloxiphus	Corthylini					21			21
Corthylus	Corthylini					150	9		159

Genus	Tribe							
Cortisinus	Phloeosinini					1		1
Cosmoderes	Cryphalini	5	6	7	2			20
Craniodicticus	Carphodicticini		3					3
Cryphalogenes	Cryphalini		2					2
Cryphalomimus	Xyloctonini	3						3
Cryphalus	Cryphalini	16	55	55	60	1	3	190
Cryphyophthorus	Hypoborini	1	1					2
Cryptocarenus	Cryphalini	1				15		16
Cryptocurus	Hylesinini	1						1
Cryptoxyleborus	Xyleborini	15		3				18
Crypturgus	Crypturgini				13		2	15
Ctonoxylon	Xyloctonini	28						28
Cyclorhipidion	Xyleborini	27	49	9	1			86
Cynanchophagus	Dryocoetini				1			1
Cyrtogenius	Dryocoetini	24	40	41	1			106
Dacnophthorus	Corthylini					5		5
Dacryostactus	Hypoborini	1						1
Dactylipalpus	Hylesinini	9	2					11
Dactylotrypes	Dryocoetini	1						1
Debus	Xyleborini		14	2				16
Dendrochilus	Ipini	9						9
Dendrocranulus	Dryocoetini					41	2	43
Dendroctonus	Hylurgini				2		18	20
Dendrodicticus	Carphodicticini					1		1
Dendrosinus	Phloeosinini					7		7
Dendroterus	Corthylini					15		15
Dendrotrupes	Hylurgini			2				2
Deropria	Crypturgini				1			1
Diamerus	Diamerini	11	20	3				34
Diuncus	Xyleborini	2	10	5				17

Continued

TABLE 2.8 Current Genera of the Scolytinae based on Wood (1986), Wood and Bright (1992), Alonso-Zarazaga and Lyal (2009), and Bright (2014) and Taxonomic Literature from 2000 to Present—cont'd

Genus	Tribe	Afro-tropical	Oriental	Austro Pacific	Pale-arctic	Neo-tropical	Nearctic	Wide-spread	Total
Dolurgocleptes	Polygraphini	2							2
Dolurgus	Crypturgini						1		1
Dryocoetes	Dryocoetini		2	1	27		6	1	37
Dryocoetiops	Dryocoetini		13	2	3				18
Dryocoetoides	Xyleborini					25			25
Dryotomicus	Phloeotribini					4			4
Dryoxylon	Xyleborini				1				1
Eccoptopterus	Xyleborini		5					1	6
Eidophelus	Cryphalini		3	1	1				5
Ernocladius	Cryphalini	1	1						2
Ernoporicus	Cryphalini	1	3		10		1		15
Ernoporus	Cryphalini		9	3	4				16
Eupagiocerus	Bothrosternini					3			3
Euwallacea	Xyleborini	5	26	10	3			1	45
Ficicis	Hylesinini		6	8					14
Fortiborus	Xyleborini		3	3					6
Glochinocerus	Corthylini					2			2
Glochiphorus	Hypoborini	2							2
Glostatus	Xyloctonini	18							18
Gnatharus	Corthylini				1				1
Gnatholeptus	Corthylini					4			4
Gnathotrichus	Corthylini					2	14		16
Gnathotrupes	Corthylini					30			30
Gymnochilus	Hexacolini					9			9
Hadrodemius	Xyleborini		3						3
Halystus	Polygraphini	2							2

Genus	Tribe								Total
Hapalogenius	Hylesinini	32							32
Hemicryphalus	Cryphalini		4	3					7
Hylastes	Hylastini				16		16		32
Hylastinus	Hylesinini				4				4
Hyledius	Phloeosinini		22	2					24
Hyleops	Phloeosinini			1					1
Hylesinopsis	Hylesinini	16							16
Hylesinus	Hylesinini		5	3	19	2	8		37
Hylocurus	Micracidini					62	16		78
Hylurdrectonus	Hylurgini			4					4
Hylurgonotus	Hylurgini					4			4
Hylurgopinus	Hylurgini						1		1
Hylurgops	Hylastini				12		9		21
Hylurgus	Hylurgini				3				3
Hyorhynchus	Hyorhynchini		9		2				11
Hypoborus	Hypoborini	1			1				2
Hypocryphalus	Cryphalini	4	28	20					52
Hypothenemus	Cryphalini	82	21	10	12	45	3	10	183
Immanus	Xyleborini			2					2
Indocryphalus	Xyloterini		5		3				8
Ips	Ipini				14		31		45
Kissophagus	Hylesinini				3				3
Lanurgus	Micracidini	22							22
Leptoxyleborus	Xyleborini		6						6
Liparthrum	Hypoborini	1	3	3	19	9	2		37
Loganius	Scolytini					16			16
Longulus	Hylesinini				1				1
Lymantor	Dryocoetini				2		2		4
Margadillius	Cryphalini	1	7	5					13
Metacorthylus	Corthylini					13			13

Continued

TABLE 2.8 Current Genera of the Scolytinae based on Wood (1986), Wood and Bright (1992), Alonso-Zarazaga and Lyal (2009), and Bright (2014) and Taxonomic Literature from 2000 to Present—cont'd

Genus	Tribe	Afro-tropical	Oriental	Austro Pacific	Pale-arctic	Neo-tropical	Nearctic	Wide-spread	Total
Micracis	Micracidini					22	4		26
Micracisella	Micracidini					16	4		20
Microborus	Hexacolini					8			8
Microcorthylus	Corthylini					38			38
Microdictica	Phloeosinini		1						1
Microperus	Xyleborini		9	5	2				16
Mimiocurus	Corthylini	11	1	2	1				15
Monarthrum	Corthylini				1	125	14		140
Neocryphus	Cryphalini					2			2
Neopteleobius	Hylesinini				1				1
Orthotomicus	Ipini		1		16		3		20
Ozopemon	Dryocoetini		16	5					21
Pachycotes	Hylurgini			9					9
Pagiocerus	Bothrosternini					5			5
Peridryocoetes	Dryocoetini		4	2					6
Periocryphalus	Cryphalini					2			2
Peronophorus	Diamerini	5							5
Phelloterus	Corthylini					3			3
Phloeoborus	Hylesinini					27			27
Phloeocleptus	Micracidini					11			11
Phloeocranus	Phloeosinini		1						1
Phloeocurus	Micracidini	1							1
Phloeoditica	Phloeosinini		2						2
Phloeosinopsioides	Phloeosinini		1	2					3
Phloeosinus	Phloeosinini		7	4	26		29		66
Phloeoterus	Corthylini					1			1

Phloeotribus	Phloeotribini			1	17	77	8	103
Phrixosoma	Phrixosomatini	10				15		25
Pityoborus	Corthylini						7	7
Pityodendron	Corthylini	1						1
Pityogenes	Ipini		1		16		7	24
Pityokteines	Ipini				4		6	10
Pityophthorus	Corthylini	37	6		25	143	174	385
Pityotrichus	Corthylini				1		2	3
Planiculus	Xyleborini		4	3				7
Polygraphus	Polygraphini	44	13		40		3	100
Premnobius	Ipini	23						23
Premnophilus	Ipini	2						2
Procryphalus	Cryphalini				1		2	3
Pseudips	Ipini				1		2	3
Pseudochramesus	Phloeosinini					11		11
Pseudodiamerus	Diamerini	3						3
Pseudohylesinus	Hylurgini						13	13
Pseudohyorrhynchus	Hyorhynchini		2		1			3
Pseudomicracis	Micracidini	8						8
Pseudopityophthorus	Corthylini				1	1	25	27
Pseudothamnurgus	Dryocoetini	1			4			5
Pseudothysanoes	Micracidini				1	66	25	92
Pseudowebbia	Xyleborini		3	3				6
Pseudoxylechinus	Hylurgini		7	2				9
Pteleobius	Hylesinini				2			2
Ptilopodius	Cryphalini	2	8	7				17
Pycnarthrum	Hexacolini					18		18
Remansus	Scolytoplatypodini	4						4
Rhopalopselion	Hylesinini	11						11
Sampsonius	Xyleborini					22		22

Continued

TABLE 2.8 Current Genera of the Scolytinae based on Wood (1986), Wood and Bright (1992), Alonso-Zarazaga and Lyal (2009), and Bright (2014) and Taxonomic Literature from 2000 to Present—cont'd

Genus	Tribe	Afro-tropical	Oriental	Austro Pacific	Pale-arctic	Neo-tropical	Nearctic	Wide-spread	Total
Sauroptilus	Corthylini	1							1
Saurotocis	Micracidini	2							2
Schedlia	Xyleborini		4	2					6
Scierus	Hylastini						2		2
Scolytodes	Hexacolini					207			207
Scolytogenes	Cryphalini	25	32	34	10	6			107
Scolytomimus	Xyloctonini		11	3					14
Scolytoplatypus	Scolytoplatypodini	16	27		6				49
Scolytopsis	Scolytini					6			6
Scolytus	Scolytini				58	41	27		126
Serrastus	Polygraphini	2							2
Sinophloeus	Hylurgini					1			1
Spermophthorus	Corthylini					2			2
Sphaerotrypes	Diamerini	8	29		10				47
Stegomerus	Cryphalini					7			7
Stenoclyptus	Micracidini						2		2
Stephanopodius	Cryphalini	6							6
Sternobothrus	Bothrosternini					16			16
Stevewoodia	Micracidini					1			1
Streptocranus	Xyleborini	2	8	1					11
Strictodex	Xyleborini		1	1					2
Strombophorus	Diamerini	31							31
Styphlosoma	Corthylini					4			4
Styracoptinus	Hypoborini	4							4
Sueus	Hyorhhynchini	1	4						5
Taphronurgus	Dryocoetini				1				1

Genus	Tribe								Grand Total
Taphrorychus	Dryocoetini	1			18				19
Taurodemus	Xyleborini					15			15
Thamnurgus	Dryocoetini	22			11				33
Theoborus	Xyleborini					11			11
Thysanoes	Micracidini					9	6		15
Tiarophorus	Dryocoetini	7		1					8
Tomicus	Hylurgini				7				7
Traglostus	Micracidini	3							3
Tricolus	Corthylini					50			50
Triotemnus	Dryocoetini	8		2	5				15
Trischidias	Cryphalini	1				4	2		7
Truncaudum	Xyleborini	1		5		1			7
Trypanophellos	Hypoborini					1			1
Trypodendron	Xyloterini				9		4		13
Trypophloeus	Cryphalini				13		4		17
Urocorthylus	Corthylini			1					1
Wallacellus	Xyleborini			1	2				3
Webbia	Xyleborini			35	3				38
Xyleborinus	Xyleborini	36		14	5	18	2	1	76
Xyleborus	Xyleborini	90	99	56	43	108	7	1	404
Xylechinosomus	Hylurgini					11			11
Xylechinus	Hylurgini	4		5	3	24	2	2	40
Xylocleptes	Dryocoetini	22		1	3				26
Xyloctonus	Xyloctonini	15							15
Xylosandrus	Xyleborini	1		24	8	2	1	3	39
Xyloterinus	Xyloterini						1		1
Zygophloeus	Hypoborini			1					1
Grand Total		980	1041	448	646	2260	591	24	5990

Numbers of species for different biogeographic regions are based on extant species only. Species classified as "widespread" are widely distributed tropical species whose region of origin is not known. Bufonus Eggers is excluded because it was known only from a unique type, since destroyed, and cannot be reliably placed.

group with Scolytinae. This uncertainty is likely attributable to the character sets (mostly morphology) that these transient taxa contribute to the phylogeny.

4. Platypodine beetles are actually a subgroup of Scolytinae. Earlier molecular studies, notably Farrell *et al.* (2001), recovered Platypodinae as apomorphic derivatives of Scolytinae. However, the study was not designed to test this relationship (e.g., the taxon sample does not include other curculionids needed to comprehensively evaluate this placement), and as such does not provide strong evidence in favor or against a close Platypodinae–Scolytinae relationship.

5.2 How many Species are There?

Undoubtedly, there remain many undescribed bark and ambrosia species in the forests of the world. An estimate of the unknown portion of the fauna can be "not very many" or "large numbers" depending on the group and the region in question.

Some regions are insufficiently explored and undocumented. For example, tropical South America appears to be the last frontier of scolytine alpha-taxonomy. Wood (2007) noted that over half of the South American bark beetle species are known from a single collection. He also estimated that only about one-third of the Neotropical diversity has been described.

In some groups, large amounts of species diversity may have escaped morphological classification because species differences are apparently cryptic, and are mainly distinguishable using molecular tools. This appears to be the case in some Cryphalini. A preliminary analysis of molecular diversity in *H. eruditus* at a single locality in Central America revealed up to 29 potential cryptic species supported by the COI and 28S gene sequences. Interestingly, a retroactive examination of the morphology of these specimens in light of the molecular phylogeny revealed previously obscure morphological differences not mentioned in any previous identification keys or descriptions (Kambestad, 2011).

On the other hand, scolytine taxonomy also undoubtedly suffers from significant inflation caused by incorrectly described species. Invalid species descriptions are often the result of the failure to consult other taxonomists' work, which was frequent in the first half of the 20th century and in some cases later. An analysis of the xyleborine fauna of Papua New Guinea failed to uncover as many new species as were expected (Hulcr and Cognato, 2013); only 10 new species were discovered after a significant collecting effort, while 59 species were synonymized. Particularly problematic is Browne's work on the fauna of South-East Asia and Oceania published in several issues of Kontyu. Our analysis of his collection in the British Museum of Natural History revealed that most of his species appear to be

synonyms (Hulcr and Cognato, 2013). His years of employment as a bark beetle taxonomist in Japan were particularly "productive" in unconsulted descriptions of species that had been intercepted once in Japanese ports on exotic timber shipments.

5.3 Unsettled Terminology

In the wake of a recent communal effort to homogenize beetle terminology (Leschen and Beutel, 2014), the current terminology of bark beetle features does not always conform to the general beetle system. Most taxonomic research on bark beetle morphology applied Hopkins' (1909) terminology, e.g., Wood (1982, 1986). The following traditional terms (left) are now changed (term shown on right) to accommodate the aforementioned uniform beetle ontology:

- Pregula → submentum (see also Lyal, 1995)
- Episternum → anepisternum
- Metepisternum → metanepisternum
- Sternite 3–7 → ventrite 1–5
- Fore wing vein R1 → apical stripe
- Fore wing vein R2 → RP1
- Fore wing vein M1 → RP2
- Fore wing vein M → MP1+2
- Fore wing vein M2 → medial spur (for more wing vention details see Jordal, 2009)

In the scolytine literature, and in much of the literature on the morphology of Coleoptera in general, the funicle (=funiculus) is usually reported as containing the pedicel. On the other hand, morphological terminology of other groups usually excludes pedicel from the funiculus. For example, the Hymenoptera taxonomy, the fly ontology (relational classification of morphological characters in Diptera), as well as the emerging beetle ontology, all exclude pedicel from funicle (A. D. Smith and N. Franz, unpubl., pers. comm.). This definition is based on the different ontogenesis, musculature, and articulation of the pedicel compared to the rest of the funicular segments. The issue is far from settled, and taxonomists are encouraged to explicitly state how they define the number of funicular segments in their publications on Scolytinae.

We encourage taxonomists to include informative characters even if they have been rarely used before, especially in studies above the species level. These include features of the notum such as the shape of the scutoscutellar suture, the length of the scutellar groove, and the degree of fusion between the metanotum and postnotum, which are otherwise separated by a membrane. It also includes aspects of the legs such as the absence or presence of a true corbel on the meso- and metatibiae, sclerolepidia on the metanepisternal suture, the shape of maxillary and labial palpi, and the shape of the proventriculus and male genitalia.

REFERENCES

1KITE (2013). Available online: http://www.1kite.org/. Last accessed: April 28, 2014.

Alonso-Zarazaga, M., Lyal, C.H.C., 2009. A catalogue of family and genus group names in Scolytinae and Platypodinae with nomenclatural remarks (Coleoptera: Curculionidae). Zootaxa 2258, 1–134.

Andersen, H.F., Jordal, B.H., Kambestad, M., Kirkendall, L.R., 2012. Improbable but true: the invasive inbreeding ambrosia beetle *Xylosandrus morigerus* has generalist genotypes. Ecology and Evolution 2, 247–257.

Andreev, D., Breilid, H., Kirkendall, L., Brun, L.O., ffrench-Constant, R. H., 1998. Lack of nucleotide variability in a beetle pest with extreme inbreeding. Insect Mol. Biol. 7, 197–200.

Arnett Jr., R.H., Thomas, M.C., Skelley, P.E., Frank, J.H., 2002. American Beetles, Polyphaga: Scarabeoidea through Curculionoidea. CRC Press, Boca Raton.

Baker, J.R., LaBonte, J., Atkinson, T., Bambara, S., 2009. An Identification Tool for Bark Beetles of the Southeastern United States. North Carolina State University, Raleigh, NC. http://keys.lucidcentral.org/keys/v3/bark_beetles/.

Beaver, R.A., 1976. Biology of Samoan bark and ambrosia beetles (Coleoptera, Scolytidae and Platypodidae). Bull. Entomol. Res. 65, 531–548.

Bentz, B.J., Stock, M.W., 1986. Phenetic and phylogenetic relationships among ten species of *Dendroctonus* bark beetles (Coleoptera: Scolytidae). Ann. Entomol. Soc. Am. 79, 527–534.

Berg, P.R., Dawson, D.A., Pandhal, J., Kirkendall, L.R., Burke, T., 2003. Isolation and characterization of microsatellite loci from two inbreeding bark beetle species (*Coccotrypes*). Mol. Ecol. Notes 3, 270–273.

Bertheau, C., Schuler, H., Krumböck, S., Arthofer, W., Stauffer, C., 2011. Hit or miss in phylogeographic analyses: the case of the cryptic NUMTs. Mol. Ecol. Resour. 11, 1056–1059.

Blandford, W.F.H., 1897. Scolytidae. In: Godman, F.D., Salvin, O. (Eds.), Biologia Centrali-Americana. Insecta. Coleoptera, Vol. 4.

Bright, D.E., 1981. A taxonomic monograph of the genus *Pityophthorus* Eichhoff in North and Central America (Coleoptera: Scolytidae). Entomol. Soc.Can. Mem. 118, 1–378.

Bright, D.E., 2010. Stephen Lane Wood. ZooKeys 56, 7–16.

Bright, D.E., 2014. A Catalog of Scolytidae and Platypodidae (Coleoptera), Supplement 3 (2000–2010), with notes on subfamily and tribal reclassifications. Insecta Mundi 0356, 1–336.

Bright, D.E., Poinar Jr., G.O., 1994. Scolytidae and Platypodidae (Coleoptera) from Dominican Republic amber. Ann. Entomol. Soc. Am. 87, 170–194.

Bright, D.E., Skidmore, R.E., 1997. A Catalog of Scolytidae and Platypodidae (Coleoptera), Supplement 1 (1990–1994). NRC Research Press, Ottawa, Ontario.

Bright, D.E., Skidmore, R.E., 2002. A Catalog of Scolytidae and Platypodidae (Coleoptera), Supplement 2 (1995–1999). NRC Research Press, Ottawa, Ontario.

Britton, E.B., 1960. Beetles from the London Clay (Eocene) of Bognor Regis, Sussex. Bull. Br. Mus. (Nat. Hist.) Geol. 4, 27–50.

Cane, J.H., Stock, M.W., Wood, D.L., Gast, S.J., 1990. Phylogenetic relationships of *Ips* bark beetles (Coleoptera: Scolytidae): electrophoretic and morphometric analysis of the grandicollis group. Biochem. Syst. Ecol. 19, 359–368.

Cognato, A.I., 2006. Standard percent DNA sequence difference for insects does not predict species boundaries. J. Econ. Entomol. 99, 1037–1045.

Cognato, A.I., 2013a. Molecular phylogeny and taxonomic review of Premnobiini Browne 1962 (Coleoptera: Curculionidae: Scolytinae). Front. Ecol. Evol. 1, 1–12.

Cognato, A.I., 2013b. *Electroborus brighti*: the first Hylesinini bark beetle described from Dominican amber (Curculionidae: Curculionidae: Scolytinae). Can. Entomol. 145, 501–508.

Cognato, A.I., Sperling, F.A.H., 2000. Phylogeny of *Ips* DeGeer (Coleoptera: Scolytidae) inferred from mitochondrial cytochrome oxidase I sequence. Mol. Phylogenet. Evol. 14, 445–460.

Cognato, A.I., Vogler, A.P., 2001. Exploring data interaction and nucleotide alignment in a multiple gene analysis of *Ips* (Coleoptera: Scolytinae). Syst. Biol. 50, 758–780.

Cognato, A.I., Sun, J.H., 2007. DNA based cladograms augment the discovery of a new *Ips* species from China (Coleoptera: Curculionidae: Scolytinae). Cladistics 23, 539–551.

Cognato, A.I., Grimaldi, D., 2009. 100 million years of morphological conservation in bark beetles (Coleoptera: Curculionidae: Scolytinae). Syst. Entomol. 34, 93–100.

Cognato, A.I., Rogers, S.O., Teale, S.A., 1995. Species diagnosis and phylogeny of the *Ips grandicollis* group (Coleoptera: Scolytidae) using random amplified polymorphic DNA. Ann. Entomol. Soc. Am. 88, 397–405.

Cognato, A.I., Gillette, N.E., Bolaños, R.C., Sperling, F.A.H., 2005a. Mitochondrial phylogeny of pine cone beetles (Scolytinae, *Conophthorus*) and their affiliation with geographic area and host. Mol. Phylogenet. Evol. 36, 494–508.

Cognato, A.I., Sun, J.H., Anducho-Reyes, M.A., Owen, D.A., 2005b. Genetic variation and origin of red turpentine beetle (*Dendroctonus valens* LeConte) introduced to the People's Republic of China. Agric. For. Entomol. 7, 87–94.

Cognato, A.I., Hulcr, J., Dole, S.A., Jordal, B.H., 2011. Phylogeny of haplo-diploid, fungus-growing ambrosia beetles (Coleoptera: Curculionidae: Scolytinae: Xyleborini) inferred from molecular and morphological data. Zool. Scr. 40, 174–186.

de Groot, P., Harvey, P.T., Roden, P.M., 1992. Genetic divergence among eastern North American cone beetles, *Conophthorus* (Coleoptera: Scolytidae). Can. Entomol. 124, 189–199.

Díaz, E., Arciniega, O., Sánchez, L., Cisneros, R., Zuñiga, G., 2003. Anatomical and histological comparison of the alimentary canal of *Dendroctonus micans*, *D. ponderosae*, *D. pseudotsugae*, *D. rufipennis*, and *D. terebrans* (Coleoptera: Scolytidae). Ann. Entomol. Soc. Am. 96, 144–152.

Dole, S.A., Jordal, B.H., Cognato, A.I., 2010. Polyphyly of *Xylosandrus* Reitter inferred from nuclear and mitochondrial genes (Coleoptera: Curculionidae: Scolytinae: Xyleborina). Mol. Phylogenet. Evol. 54, 773–782.

Farrell, B.D., 1998. "Inordinate fondness" explained: why are there so many beetles? Science 281, 555–559.

Farrell, B.D., Sequeira, A.S., O'Meara, B.C., Normark, B.B., Chung, J.H., Jordal, B.H., 2001. The evolution of agriculture in beetles (Curculionidae: Scolytinae and Platypodinae). Evolution 55, 2011–2027.

Garrick, R.C., Meadows, C.A., Nason, J.D., Cognato, A.I., Dyer, R.J., 2009. Variable nuclear markers for a Sonoran Desert bark beetle, *Araptus attenuatus* Wood (Curculionidae: Scolytinae), with applications to related genera. Conserv. Genet. 10, 1177–1179.

Gauthier, N., Rasplus, J.Y., 2004. Polymorphic microsatellite loci in the Coffee Berry Borer, *Hypothenemus hampei* (Coleoptera, Scolytidae). Mol. Ecol. Notes 4, 294–296.

Gillett, C.P.D.T., Crampton-Platt, A., Timmermans, M.J.T.N., Jordal, B. H., Emerson, B.C., Vogler, A.P., 2014. Bulk *de novo* mitogenome assembly from pooled total DNA elucidates the phylogeny of weevils (Coleoptera: Curculionoidea). Mol. Biol. Evol. http://dx.doi.org/10.1093/molbev/msu154.

Gregory, T.R., Nathwani, P., Bonnett, T.R., Dezene, P.W., 2013. Sizing up arthropod genomes: an evaluation of the impact of environmental variation on genome size estimates by flow cytometry and the use of qPCR as a method of estimation. Genome 56, 505–510.

Hajibabaei, M., Janzen, D.H., Burns, J.M., Hallwachs, W., Hebert, P.D.N., 2006. DNA barcodes distinguish species of tropical Lepidoptera. Proc. Natl. Acad. Sci. U. S. A. 103, 968–971.

Hanrahan, S.J., Johnston, J.S., 2011. New genome size estimates of 134 species of arthropods. Chromosome Res. 19, 809–823.

Haran, J., Timmermans, M.J., Vogler, A.P., 2013. Mitogenome sequences stabilize the phylogenetics of weevils (Curculionoidea) and establish the monophyly of larval ectophagy. Mol. Phylogenet. Evol. 67, 156–166.

Hebert, P.D.N., Cywinska, A., Ball, S.L., deWard, J.R., 2002. Biological identifications through DNA barcodes. Proc. R. Soc. London, Ser B 270, 313–321.

Hebert, P.D.N., Ratnasingham, S., deWaard, J.R., 2003. Barcoding animal life: cytochrome *c* oxidase subunit 1 divergences among closely related species. Proc. R. Soc. London, Ser. B 270 (Suppl), S96–S99.

Holzman, J.P., Bohonak, A.J., Kirkendall, L.R., Gottlieb, D., Harari, A.R., Kelley, S.T., 2009. Inbreeding variability and population structure in the invasive haplodiploid palm-seed borer (*Coccotrypes dactyliperda*). J. Evol. Biol. 22, 1076–1087.

Hopkins, A.D., 1909. Contributions toward a monograph of the scolytid beetles. I. The genus Dendroctonus. USDA Bureau of Entomology Technical Bulletin 17, 1–164.

Hulcr, J., Cognato, A.I., 2010. Repeated evolution of theft in fungus farming ambrosia beetles. Evolution 64, 3205–3212.

Hulcr, J., Cognato, A.I., 2013. Xyleborini of New Guinea, a Taxonomic Monograph (Coleoptera: Curculionidae: Scolytinae). Thomas Say Publications in Entomology: Monographs, Entomological Society of America, Annapolis.

Hulcr, J., Beaver, R., Dole, S., Cognato, A.I., 2007. Cladistic review of xyleborine generic taxonomic characters (Coleoptera: Curculionidae: Scolytinae). Syst. Entomol. 32, 568–584.

Hundsdoerfer, A.K., Rheinheimer, J., Wink, M., 2009. Towards the phylogeny of the Curculionoidea (Coleoptera): reconstructions from mitochondrial and nuclear ribosomal DNA sequences. Zool. Anz. 248, 9–31.

Jacobson, G.G., 1913. [Beetles of Russia and West Europa. Manual to the determination of beetles], Izdanie A. Ph. Devriena, Sankt Peterburg [In Russian].

Jarzembowski, E.A., 1990. A boring beetle from the Wealden of the Weald. In: Boucot, A.J. (Ed.), Evolutionary Paleobiology of Behavior and Coevolution. Elsevier, Amsterdam, pp. 373–376.

Jordal, B.H., 2002. Elongation Factor 1a resolves the monophyly of the haplodiploid ambrosia beetles Xyleborini (Coleoptera: Curculionidae). Insect Mol. Biol. 11, 453–465.

Jordal, B.H., 2007. Reconstructing the phylogeny of Scolytinae and close allies: major obstacles and prospects for a solution. Proceedings from the Third Workshop on Genetics of Bark Beetles and Associated Microorganisms. US Forest Service RMRS P-e 45, 3–8.

Jordal, B.H., 2009. The Madagascan genus *Dolurgocleptes* Schedl (Coleoptera: Curculionidae, Scolytinae): description of a new species and transfer to the tribe Polygraphini. Zootaxa 2014, 41–50.

Jordal, B.H., Hewitt, G.M., 2004. The origin and radiation of Macaronesian beetles breeding in *Euphorbia*: the relative importance of multiple data partitions and population sampling. Syst. Biol. 53, 711–734.

Jordal, B.H., Cognato, A.I., 2012. Molecular phylogeny of bark and ambrosia beetles reveals multiple origins of fungus farming during periods of global warming. BMC Evol. Biol. 12, 133.

Jordal, B.H., Kambestad, M., 2014. DNA barcoding of bark and ambrosia beetles reveals excessive NUMTs and consistent east–west divergence across Palearctic forests. Mol. Ecol. Resour. 14, 7–17.

Jordal, B.H., Normark, B.B., Farrell, B.D., 2000. Evolutionary radiation of an inbreeding haplodiploid beetle lineage (Curculionidae, Scolytinae). Biol. J. Linn. Soc. 71, 483–499.

Jordal, B.H., Beaver, R.A., Normark, B.B., Farrell, B.D., 2002a. Extraordinary sex ratios and the evolution of male neoteny in sib-mating *Ozopemon* beetles. Biol. J. Linn. Soc. 75, 353–360.

Jordal, B.H., Normark, B.B., Farrell, B.D., Kirkendall, L.R., 2002b. Extraordinary haplotype diversity in haplodiploid inbreeders: Phylogenetics and evolution of the sib-mating bark beetle genus *Coccotrypes*. Mol. Phylogenet. Evol. 23, 171–188.

Jordal, B.H., Emerson, B.C., Hewitt, G.M., 2006. Apparent "sympatric" speciation in ecologically similar herbivorous beetles facilitated by multiple colonizations of an island. Mol. Ecol. 15, 2935–2947.

Jordal, B.H., Gillespie, J.J., Cognato, A.I., 2008. Secondary structure alignment and direct optimization of 28S rDNA sequences provide limited phylogenetic resolution in bark and ambrosia beetles (Curculionidae: Scolytinae). Zool. Scr. 37, 1–14.

Jordal, B.H., Sequeira, A.S., Cognato, A.I., 2011. The age and phylogeny of wood boring weevils and the origin of subsociality. Mol. Phylogenet. Evol. 59, 708–724.

Jordal, B.J., Smith, S.M., Cognato, A.I., 2014. Classification of weevils as a data-driven science: leaving opinion behind. Zookeys, in press.

Kambestad, M., 2011. Coexistence of habitat generalists in Neotropical petiole-breeding bark beetles: molecular evidence reveals cryptic diversity, but no niche segregation. MSc thesis, University of Bergen, Norway.

Keeling, C.I., Yuen, M.M.S., Liao, N.Y., Docking, T.R., Chan, S.K., Taylor, G.A., et al., 2013. Draft genome of the mountain pine beetle, *Dendroctonus ponderosae* Hopkins, a major forest pest. Genome Biol. 14, R27.

Keller, L., Peer, K., Bernasconi, C., Taborsky, M., Shuker, D., 2011. Inbreeding and selection on sex ratio in the bark beetle *Xylosandrus germanus*. BMC Evol. Biol. 11, 359.

Kelley, S.T., Farrell, B.D., 1998. Is specialization a dead end? The phylogeny of host use in *Dendroctonus* bark beetles (Scolytidae). Evolution 52, 1731–1743.

Kirejtshuk, A.G., Azar, D., Beaver, R.A., Mandelshtam, M.Y., Nel, A., 2009. The most ancient bark beetle known: a new tribe, genus and species from Lebanese amber (Coleoptera, Curculionidae, Scolytinae). Syst. Entomol. 34, 101–112.

Kohring, R., Schlüter, T., 1989. Historische und paläontologische Bestandsaufnahme des Simetits, eines fossilen Harzes mutmaßlich mio/pliozänen Alters aus Sizilien. Documenta Naturae 56, 33–58.

Kolarik, M., Kirkendall, L.R., 2010. Evidence for a new lineage of primary ambrosia fungi in *Geosmithia* Pitt (Ascomycota: Hypocreales). Fungal Biol. 114, 676–689.

Kuschel, G., 1995. A phylogenetic classification of the Curculionoidea to families and subfamilies. Mem. Entomol. Soc. Wash. 14, 5–33.

Kuschel, G., Leschen, R.A.B., Zimmerman, E.C., 2000. Platypodidae under scrutiny. Invertebr. Taxon. 14, 771–805.

Lanier, G.N., 1970. Biosystematics of North American *Ips* (Coleoptera: Scolytidae): Hopping's group IX. Can. Entom. 102, 1139–1163.

Lanier, G.L., 1981. Cytotaxonomy of *Dendroctonus*. In: Stock, M.W. (Ed.), Application of Genetics and Cytology in Insect Systematics and Evolution. Forest, Wildlife, and Range Experiment Station, University of Idaho, Moscow, ID, pp. 33–66.

Lemmon, A.R., Emme, S., Lemmon, E.M., 2012. Anchored hybrid enrichment for massively high-throughput phylogenomics. Syst. Biol. 61 (5), 727–744.

Leschen, R.A.B., Beutel, R.G., 2014. Arthropoda: Insecta: Coleoptera, Volume 3: Morphology and Systematics (Phytophaga). Walter de Gruyter, Berlin.

Linnaeus, C., 1758. Systema Naturae per Regna Tria Naturae, Secundum Classes, Ordines, Genera, Spacies, cum Characteribus, Differentiis, Synonymis, Locis. L. Salvii, Stockholm.

Lyal, C.H.C., 1995. The ventral structures of the weevil head (Coleoptera: Curculionoidea). Mem. Entomol. Soc. Wash. 14, 35–51.

Maiti, P.K., Saha, N., 2004. Fauna of India and the Adjacent Countries: Scolytidae: Coleoptera (Bark and Ambrosia Beetles) Vol. 1, Part 1, Introduction and tribe Xyleborini. Zoological Survey of India, Kolkata.

Maiti, P.K., Saha, N., 2009. Fauna of India and the Adjacent Countries: Scolytidae: Coleoptera (Bark and Ambrosia Beetles), Vol. 1, Part 2. Zoological Survey of India, Kolkata.

Mandelshtam, M.Y., Beaver, R.A., 2003. *Amphiscolytus*—a new genus, and Amphiscolytini—a new tribe of Scolytidae (Coleoptera) for *Dacryophthorus capensis* Schedl. Zootaxa 298, 1–8.

Marvaldi, A.E., 1997. Higher level phylogeny of Curculionidae (Coleoptera: Curculionoidea) based mainly on larval characters, with special reference to broad-nosed weevils. Cladistics 13, 285–312.

Marvaldi, A.E., Sequeira, A.S., O'Brien, C.W., Farrell, B.D., 2002. Molecular and morphological phylogenetics of weevils (Coleoptera, Curculionoidea): do niche shifts accompany diversification? Syst. Biol. 51, 761–785.

Masood, A., Stoeckle, B.C., Kuehn, R., Saeed, S., 2011. Cross species transfer of microsatellite loci in Scolytidae species mostly associated with mango (*Mangifera indica* L., Anacardiaceae) quick decline disease. Pak. J. Zool. 43, 411–414.

McKenna, D.D., 2011. Temporal lags and overlap in the diversification of weevils and flowering plants: recent advances and prospects for additional resolution. Am. Entomol. 57, 54–55.

McKenna, D.D., Sequeira, A.S., Marvaldi, A.E., Farrell, B.D., 2009. Temporal lags and overlap in the diversification of weevils and flowering plants. Proc. Natl. Acad. Sci. U. S. A. 106, 7083–7088.

Narum, S.R., Buerkle, C.A., Davey, J.W., Miller, M.R., Hohenlohe, P.A., 2013. Genotyping-by-sequencing in ecological and conservation genomics. Mol. Ecol. 22, 2841–2847.

Niehuis, O., Hartig, G., Grath, S., Pohl, H., Lehmann, J., Tafer, H., et al., 2012. Genomic and morphological evidence converge to resolve the enigma of Strepsiptera. Curr. Biol. 22, 1309–1313.

Nobuchi, A., 1988. Feeding habit and proventriculus of Platypodidae and Scolytidae (Coleoptera). Konchu to Shizen (Nature and Insects) 23, 24–30.

Normark, B.B., 1996. Polyploidy of parthenogenetic *Aramigus tessellatus* (Say) (Coleoptera: Curculionidae). Coleopterists Bulletin 50, 73–79.

Normark, B.B., Jordal, B.H., Farrell, B.D., 1999. Origin of a haplodiploid beetle lineage. Proc. R. Soc. London Ser. B 266, 2253–2259.

Nuñez, J., Hernández, E., Giraldo, W., Navarro, L., Gongora, C., Cristancho, M.A., et al., 2012. First draft genome sequence of coffee berry borer: the most invasive insect pest of coffee crops. In: Sixth Annual Arthropod Genomics Symposium, Kansas City, MO. Available online: http://www.k-state.edu/agc/symp2012/images/PosterAbstracts-2012.pdf. Last accessed: April 28, 2014.

Oberprieler, R.G., Marvaldi, A.E., Anderson, R.S., 2007. Weevils, weevils, weevils everywhere. Zootaxa 1668, 491–520.

Peterson, A., 1964. Entomological Techniques: How to Work with Insects. Edwards Brothers Inc., Ann Arbor.

Peterson, B.K., Weber, J.N., Kay, E.H., Fisher, H.S., Hoekstra, H.E., 2012. Double digest RADseq: an inexpensive method for de novo SNP discovery and genotyping in model and non-model species. PLoS One 7, e37135.

Petrov, A.V., 2013. New Ichnotaxon *Megascolytinus zherikhini* (Coleoptera: Curculionidae: Scolytinae) from upper cretaceous deposits of Mongolia. Paleontol. J. 47, 597–600.

Pfeffer, A., 1994. Zentral- und Westpalaarktische Borken- und Kernkafer (Coleoptera: Scolytidae, Platypodidae). Entomologia Basiliensia 17, 310.

Richards, S., Tribolium Genome Sequencing Consortium, 2008. The genome of the model beetle and pest *Tribolium castaneum*. Nature 452, 949–955.

Rios-Reyes, A.V., Valdez-Carrasco, J., Equihua-Martínez, A., Moya-Raygoza, G., 2008. Identification of *Dendroctonus frontalis* (Zimmermann) and *D. mexicanus* (Hopkins) (Coleoptera: Curculionidae: Scolytinae) through structures of female genitalia. Coleopterists Bulletin 62, 99–103.

Sallé, A., Kerdelhué, C., Breton, M., Lieutier, F., 2003. Characterization of microsatellite loci in the spruce bark beetle *Ips typographus* (Coleoptera: Scolytinae). Mol. Ecol. Notes 3, 336–337.

Schawaller, W., 1981. Pseudoskorpione (Cheliferidae) phoretisch auf Kafern (Platypodidae) in Dominikanischem Bernstein (Stuttgarter Bernstein-sammlung: Pseudoscorpionidea und Coleoptera). Stuttgarter Beitr. Naturk. D. Ser. B. (Geol. und Paleaeontol.) 71, 1–107.

Schedl, K.E., 1962. New Platypodidae from Mexican amber. J. Paleontol. 36, 1035–1038.

Schrey, N.M., Schrey, A.W., Heist, E.J., Reeve, J.D., 2007. Microsatellite loci for the southern pine beetle (*Dendroctonus frontalis*) and cross-species amplification in *Dendroctonus*. Mol. Ecol. Notes 7, 857–859.

Sequeira, A.S., Normark, B.B., Farrell, B.D., 2000. Evolutionary assembly of the conifer fauna: distinguishing ancient from recent associations in bark beetles. Proc. R. Soc. London Ser. B 267, 2359–2366.

Sequeira, A.S., Normark, B.B., Farrell, B.D., 2001. Evolutionary origins of Gondwanan interactions: how old are Araucaria beetle herbivores? Biol. J. Linn. Soc. 74, 459–474.

Shi, G., Grimaldi, D.A., Harlow, G.E., Wang, J., Wang, J., Yang, M., et al., 2012. Age constraint on Burmese amber based on U-Pb dating of zircons. Cretaceous Res. 37, 155–163.

Six, D.L., Paine, T.D., 1999. Phylogenetic comparison of Ascomycete mycangial fungi and *Dendroctonus* bark beetles (Coleoptera: Scolytidae). Ann. Entomol. Soc. Am. 92, 159–166.

Six, D.L., 2003. Bark beetle-fungus symbioses. In: Bourtzis, K., Miller, T. A. (Eds.), Insect Symbiosis. CRC Press, Boca Raton, pp. 97–114.

Skalski, A.W., Veggiani, A., 1990. Fossil resin in Sicily and the Northern Apennines: Geology and organic content. Pr. Muz. Ziemi 41, 37–49.

Smith, S.M., Cognato, A.I., 2010. A taxonomic revision of *Camptocerus* Dejean (Coleoptera: Curculionidae: Scolytinae). Insecta Mundi 0148, 1–88.

Song, H., Sheffield, N.C., Cameron, S.L., Miller, K.B., Whiting, M.F., 2010. When the phylogenetic assumptions are violated: the effect of base compositional heterogeneity and among-site rate variation in beetle mitochondrial phylogenomics. Syst. Entomol. 35, 429–448.

Stauffer, C., Lakatos, F., Hewitt, G.M., 1997. The phylogenetic relationships of seven European *Ips* (Scolytidae, Ipinae) species. Insect Mol. Biol. 6, 233–240.

Stoeckle, B., Kuehn, R., 2011. Identification of 18 polymorphic microsatellite loci in the spruce bark beetle *Ips typographus* L. (Coleoptera: Scolytidae) using high-throughput sequence data. Eur. J. Entomol. 108, 169–171.

Sullivan, B.T., Niño, A., Moreno, B., Brownie, C., Macías-Sámano, J., Clarke, S.R., et al., 2012. Biochemical evidence that *Dendroctonus frontalis* consists of two sibling species in Belize and Chiapas. Mexico. Ann. Entomol. Soc. Am. 105, 817–831.

Thompson, R.T., 1992. Observations on the morphology and classification of weevils (Coleoptera, Curculionoidea) with a key to major groups. J. Nat. Hist. 26, 835–891.

Totani, K., Sugimoto, T., 1987. Studies on the interspecies relationships of the genus *Xyleborus* Eichhoff (Coleoptera: Scolytidae) by the proventriculus. Research Bulletin of the Plant Protection Service, Japan 23, 39–43.

Wickham, H.F., 1913. Fossil Coleoptera from the Wilson ranch near Florissant, Colorado. State University of Iowa, Laboratories of Natural History, Bulletin 6, 3–29.

Wolfe, A.P., Tappert, R., Muehlenbachs, K., Boudreau, M., McKellar, R. C., Basinger, J.F., Garrett, A., 2009. A new proposal concerning the botanical origin of Baltic amber. Proc. R. Soc. B 276, 3403–3412.

Wood, S.L., 1963. A revision of the bark beetle genus *Dendroctonus* Erichson (Coleoptera: Scolytidae). Great Basin Nat. 23, 1–117.

Wood, S.L., 1978. A reclassification of the subfamilies and tribes of Scolytidae (Coleoptera). Ann. Soc. Entomol. Fr. 14, 95–122.

Wood, S.L., 1982. The bark and ambrosia beetles of North and Central America (Coleoptera: Scolytidae), a taxonomic monograph. Great Basin Nat. Mem. 6, 1–1359.

Wood, S.L., 1986. A reclassification of the genera of Scolytidae (Coleoptera). Great Basin Nat. Mem. 10, 1–126.

Wood, S.L., 2007. Bark and Ambrosia Beetles of South America (Coleoptera: Scolytidae). Brigham Young University, Provo.

Wood, S.L., Bright Jr., D.E., 1992a. A catalog of Scolytidae and Platypodidae (Coleoptera), Part 2: Taxonomic Index. Volume A. Great Basin Nat. Mem. 13, 1–833.

Wood, S.L., Bright Jr., D.E., 1992b. A catalog of Scolytidae and Platypodidae (Coleoptera), Part 2: Taxonomic Index. Volume B. Great Basin Nat. Mem. 13, 835–1553.

Zachos, J., Pagani, M., Sloan, L., Thomas, E., Billups, K., 2001. Trends, rhythms, and aberrations in global climate 65 Ma to present. Science 292, 686–693.

Chapter 3

Evolution and Diversity of Bark and Ambrosia Beetles

Lawrence R. Kirkendall[1], Peter H.W. Biedermann[2], and Bjarte H. Jordal[3]

[1]Department of Biology, University of Bergen, Bergen, Norway [2]Research Group Insect Symbiosis, Max Planck Institute for Chemical Ecology, Jena, Germany [3]University Museum of Bergen, University of Bergen, Bergen, Norway

1. INTRODUCTION

"No other family of beetles shows such interesting habits as do the members of the family Ipidae."
—Milton W. Blackman, 1928

Most wood-boring insect species only tunnel in wood as larvae. Their adults are free-flying insects that must move about the landscape to encounter mates, find food, and locate oviposition sites; in so doing, they face a myriad of invertebrate and vertebrate predators, and must deal with the vagaries of wind, temperature, and precipitation. A few beetles, however, have evolved to spend nearly their entire adult lives inside woody tissues. Bark beetles (Scolytinae) and pinhole borers (Platypodinae)—which we will refer to collectively as bark and ambrosia beetles—are weevils that have lost their snouts and that spend most of their adult existence ensconced in dead wood (occasionally, in other plant tissues), and by many measures, they are the most successful lineages to do so. In this chapter, we will document and discuss the striking variability in biology of these weevils.

Wood is important to humans in many ways, and bark and ambrosia beetles are abundant in forests and plantations, so it is not surprising that there is a long history of interest in these relatively small and nondescript insects. Carl Linnaeus, the father of modern taxonomy and one of the founders of ecology, described five species of Scolytinae, including four of the most common European species, which he described in 1758 (*Trypodendron domesticum*, *Tomicus piniperda*, *Polygraphus poligraphus*, *Ips typographus*, and *Pityogenes chalcographus*) (Linnaeus, 1758). There are Scolytinae and Platypodinae in the beetle collections of the fathers of evolutionary theory, Charles Darwin and Alfred Russel Wallace, and the latter even published on them (Wallace, 1860). It is only in the first half of the 20th century though that we first began to be aware of the wealth of details of their fascinating but cryptic lives.

1.1 Topics and Taxonomic Coverage

We will focus on the evolution and ecology of feeding and breeding biology, especially mating systems and social behavior. Much of this variation is little known outside of a small circle of specialists, as the vast majority of basic and applied research in Scolytinae and Platypodinae is focused on a handful of serious forest and agricultural pests that have considerable economic and ecological impact. Though well deserving of research, these taxa are not representative of bark and ambrosia beetle biology as a whole. We will not cover population ecology or pheromone biology, as these topics are much more widely known and have been thoroughly addressed in many research and review articles, in this book (Chapters 1, 4, and 5), as well as in other books, e.g., Chararas (1962), Berryman (1982), Mitton and Sturgeon (1982), Speight and Wainhouse (1989), Lieutier *et al.* (2004), and Paine (2006). We will also let others review in detail the growing and fascinating topic of relationships with fungi and other symbionts (but see Section 3; Chapter 6).

There are currently 247 genera of recognized Scolytinae (see Appendix), most of which breed predominantly or entirely in angiosperms (Figure 3.1); 86% of these genera are represented in the tropics or subtropics, and 59% are restricted to these warmer regions (Chapter 2). In terms of numbers of species, 79% (four of five) are found primarily in tropical or subtropical ecosystems. Less than 1% of the ca. 6000 Scolytinae species regularly kill healthy standing trees, and from the existing literature it seems unlikely that more than 5 to 10% occasionally do so (but see Section 3.9).

Books dealing with the biology of Scolytinae (or of Scolytinae plus Platypodinae) often reflect biases towards species breeding in temperate conifer forests or vectoring pathogens with significant impact on urban or ornamental broadleaf trees (e.g., Chararas, 1962; Mitton and Sturgeon, 1982; Lieutier *et al.*, 2004). It is hoped that this

Bark Beetles. http://dx.doi.org/10.1016/B978-0-12-417156-5.00003-4

85

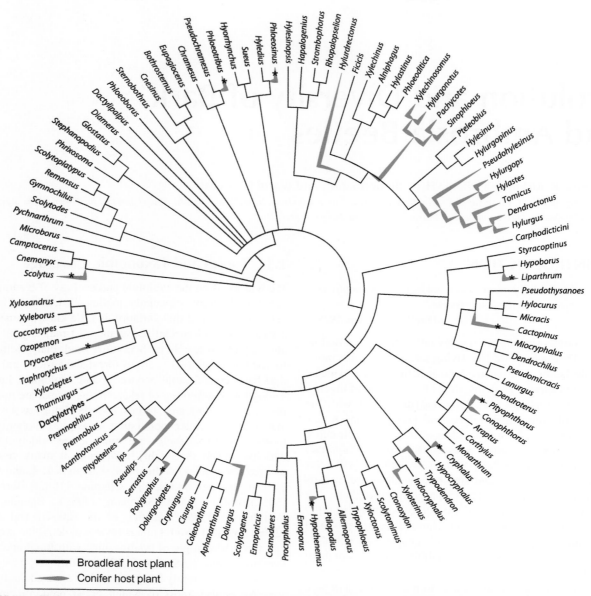

FIGURE 3.1   Phylogenetic tree of Scolytinae indicating associations with broadleaf and conifer host plants. The tree used here and in Figures 3.4 and 3.7 summarizes phylogenetic results based on molecular data with unresolved relationships resolved in part based on morphological evidence (Normark *et al.*, 1999; Farrell *et al.*, 2001; Jordal *et al.*, 2008, 2011; Jordal and Cognato, 2012).

chapter can help to redress this imbalance (see also Chapters 1, 11, and 12).

For less (in)famous bark and ambrosia beetles, information on ecology and behavior can be gleaned from regional or global faunal works by, for example, Thomas Atkinson and colleagues (see references), Roger Beaver (see references), Cyril Beeson (1941), Maulsby Blackman (Blackman, 1922; Blackman and Stage, 1918, 1924), Francis George Browne (1961), Willard Joseph Chamberlin (1939, 1958), Constantin Chararas (1962), L. G. E. Kalshoven (1959, 1960b), Akira Nobuchi (1972), Karl E. Schedl (1961, 1962a, b), James Malcolm

Swaine (1918), and Stephen L. Wood (1982, 2007). Besides the sources of natural history information mentioned above, there are recent, more quantitative treatments of bark and ambrosia beetle ecology, biogeography, and phylogeography: ecological aspects of bark and ambrosia beetle biodiversity (Ødegaard, 2000, 2006; Ødegaard *et al.*, 2000, 2005; Hulcr *et al.*, 2007, 2008a, b; Novotny *et al.*, 2007, 2010); island biogeography (Kirkendall, 1993; Jordal *et al.*, 2001); and phylogeography (Cognato *et al.*, 1999; Jordal *et al.*, 2006; Maroja *et al.*, 2007; Cai *et al.*, 2008; Schrey *et al.*, 2011; Garrick *et al.*, 2013; Jordal and Kambestad, 2014).

Only a few aspects of bark beetle evolutionary biology have been reviewed for Scolytinae (or Scolytinae and Platypodinae) as a group: mating systems (Kirkendall, 1983); inbreeding and other sources of biased sex ratios (Kirkendall, 1993); the evolution of social behavior (Kirkendall *et al.*, 1997); and the evolutionary history of bark beetles and pinholes borers (Jordal *et al.*, 2011; Jordal, 2014a, b, c; Chapter 2).

1.2 Why We include Platypodinae

We have chosen to include Platypodinae ("pinhole borers") in our chapter (as Hulcr *et al.* did in Chapter 2), primarily with respect to mating and social behavior. The extreme morphological similarity of Platypodinae to Scolytinae has bedeviled systematists for decades, and highlights the importance of convergent evolution in wood-tunneling beetles (Figure 3.2). Pinhole borers were long treated by entomologists as a separate family, closely related to "Scolytidae." More recently, phylogenies based on molecular and morphological characters strongly suggest that this group, too, is a highly derived group of weevils, but the Platypodinae may not even be closely related to Scolytinae (reviewed in Chapter 2, but also see McKenna *et al.* (2009), Jordal *et al.* (2011), McKenna (2011), Haran *et al.* (2013), and Gillett *et al.* (2014)).

Virtually all broadly oriented bark beetle specialists have worked with both groups (and usually primarily or exclusively these two), which until recently were considered to be two very closely related but separate families, Platypodidae (ca. 1400 species) and Scolytidae (ca. 6000 species). Platypodine biology seems to only be known to Scolytinae researchers and a few generally oriented forest entomologists: we are not aware of the existence of any specialists who restrict their focus to Platypodinae. It has become common practice to include both Scolytinae and Platypodinae in taxonomic, faunistic and ecological works, and until fairly recently to refer to them jointly as "Scolytoidea" (Hubbard, 1897; Blackman, 1922; Beal and Massey, 1945; Pfeffer, 1955, 1995; Schedl, 1962b, 1974; Chamberlin, 1939, 1958; Kalshoven, 1960a, b; Browne, 1961; Nunberg, 1963; Nobuchi, 1969; Bright and Stark, 1973; Beaver and Browne, 1975; Kirkendall, 1983; Atkinson and Equihua-Martínez, 1985a; Wood and Bright, 1987, 1992 and subsequent supplements; Beaver, 1989; Atkinson and Peck, 1994; Kirkendall *et al.*, 1997). Scolytine and platypodine ambrosia beetles are frequently collected together in dead and dying trees. Platypodinae are strikingly similar to monogynous scolytine ambrosia beetles in gross morphology, tunnel system architecture, use of chemical and acoustic signals, mating behavior, social behavior, and relationships with symbiotic fungi. All but the most basal platypodines are monogynous ambrosia beetles with extensive parental care; one species, *Austroplatypus incompertus* (Schedl), is notable for being eusocial (Kirkendall *et al.*, 1997).

2. WHAT ARE BARK AND AMBROSIA BEETLES?

2.1 Phylogenetics

Why tunnel? Foraging in the green is a dangerous place. Being exposed to parasitoids and predators—and occasionally extreme competition from hyperdiverse insect communities—as well as to wind, rain, and occasional extreme temperatures can generate a selective advantage for a complete life cycle inside dead plant tissues. Although less nutritious, such resources are less hostile in terms of physical and chemical defenses mustered by live plants.

In fact, life under bark has evolved multiple times in weevils (McKenna *et al.*, 2009; Jordal *et al.*, 2011; Haran *et al.*, 2013; Gillett *et al.*, 2014) (Figure 3.2). Most, or

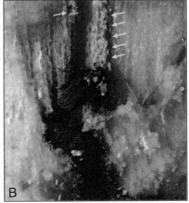

FIGURE 3.2 **Convergence in wood-boring weevils: the genus *Homoeometamelus* (subfamily Baridinae or Conoderinae, tribe Menemachini).** (A) Lateral view; note the lateral socketed teeth on all tibiae (arrows), of the same type as in many Scolytinae. (B) Mating niche with longitudinal egg tunnel; arrows point to eggs laid in niches.

perhaps all, wood-boring groups are old, originating at least some 90–120 millions of years ago (Ma). The oldest scolytine fossil is known from Lebanese amber that dates back to Mid-Cretaceous some 120 Ma (Kirejtshuk et al., 2009). This is about the same age as the oldest known Curculionidae known from the Santana formation in Brazil (116 Ma) (Santos et al., 2011). The Lebanese amber fossil *Cylindrobrotus pectinatus* Kirejtshuk, Azar, Beaver, Mandelshtam and Nel is not closely related to any extant or fossil lineage of Scolytinae, but has all defining morphological characters of a bark beetle. Another fossil (100 Ma) from Burmese amber belongs to the current genus *Microborus*, which may indicate that Scolytinae was well established as a dominant group already at that time (Cognato and Grimaldi, 2009).

Platypodinae has a less documented fossil record, but is represented by tesserocerine and platypodine inclusions in Mexican, Dominican, Sicilian, and Rovno ambers (Schedl, 1972; Bright and Poinar, 1994). This group is much older, however, and two fossils from Burmese amber indicate possible tesserocerine affinity (Cognato and Grimaldi, pers. commun.), in accordance with molecular age estimates (McKenna et al., 2009; Jordal et al., 2011).

The exact phylogenetic position of Scolytinae is uncertain, but it is now well documented that this group originated at the same time as modern phytophagous curculionids (McKenna et al., 2009; Jordal et al., 2011; Haran et al., 2013; Gillett et al., 2014). Soon after the split between the broad-nosed weevils and most other groups of advanced weevils, Scolytinae makes up a consistently monophyletic group closely related to typical long-nosed weevils such as Molytinae, Cryptorrhynchinae, Baridinae, Curculioninae, and Cossoninae. They may not be closely related to Platypodinae, which seem more closely related to Dryophthorinae than to Scolytinae (McKenna et al., 2009; Gillett et al., 2014).

It seems certain that the wood-boring habit evolved from external feeding on green leaves. Herbivorous Scolytinae exist today, but none of these are basal lineages in the phylogenetic tree of bark and ambrosia beetles. The closest match is the Scolytini genus *Camptocerus*, where adults feed on green leaves before tunneling into the bark to breed (Smith and Cognato, 2011). With rare exceptions, Scolytinae are restricted to denser, drier plant tissues such as those in stems and branches of trees and shrubs. Few taxa can deal with the typically soft, very moist tissues associated with herbaceous plants. Even most species categorized as "herbiphagous" breed in the dense supportive tissues of stems or leaf petioles, not in leaves.

2.2 General Morphology

Bark and ambrosia beetles are highly adapted morphologically and ecologically to this unusual lifestyle and to the special challenges of constructing and living nearly their entire adult lives in tunnels. The adaptation to a life in concealed niches in dead lignified plant material apparently followed a distinct selection regime with consequences for morphological change. The change in diet from green leaves to bark, wood or fungi has modified both the external chewing appendages as well as the internal digestive system. Boring in bark and wood also dramatically changed their reproductive biology due to control of valuable resources in the form of durable, protective tunnels. Control of access to the tunnel by the opposite sex has therefore led to a variety of behavioral and morphological changes in the context of optimal mate choice. Maintenance and protection of tunnels has furthermore led to changes in morphology to optimize movement in the tunnel, shoveling of frass, and the blocking of the entrance hole.

Life in tunnels and caves places obvious restrictions on body shapes, since protuberant body parts would limit movement and flexibility. Adult beetles that bore into wood are generally cylindrical, as are bark and ambrosia beetles (Haack and Slansky, 1987). In addition, all bark and ambrosia beetles have large, flattened eyes and short antennae that can be folded into the body. A unique feature involves vertically enlarged eyes, which extend from the vertex to the gula, sometimes slightly or even completely divided where the antennal scape attaches and folds back. It is not known if eye enlargement has evolved due to a life in near darkness, but we note that certain weevil groups, which do not tunnel as adults (such as many conoderines), also have large, flat, contiguous eyes.

Excavation of tunnels requires a considerable biting force, and scolytines and platypodines have larger mandibles than most other weevils. Mandibles are short and thick, and have strong muscles attached (Schedl, 1931). The chewing procedure varies depending on whether the woody tissue is ingested for food or simply chopped up to be removed, as in most ambrosia beetles. These bore new tunnels by cutting with their mandibles during back and forth movements of the head and rotation of the whole body within the tunnel. By contrast, when feeding, they crop the enlarged nutritious spores ("ambrosial growth") of their fungi by horizontal movements of the maxillae (which have comb-like hairs or structures at the end) and swallowing movements of the labrum. Effective chewing of wood bits is enabled by a flexible rotating head with strong muscle attachments.

Wood-boring beetles are generally well equipped with cuticular structures that aid in pushing and scraping, such as various spines and socketed denticles on the tibiae (penultimate leg segment) (Swaine, 1918). While a terminal tibial spine (uncus) is commonly seen across the weevils, many scolytines have additional socketed denticles along the lateral edge of the tibiae. These denticles are typically evolved from ordinary hair-like setae, and their socketed origin is clearly visible (Wood, 1978), although they are sometimes reinforced and overgrown by cuticle (Jordal, 1998). It is unclear how important such denticles are for wood-boring beetles given that several groups are lacking denticles, such as *Scolytus* and close relatives, most wood-boring cossonines, and in Platypodinae, the latter instead have developed sharp ridges and rugae on their protibiae (Strohmeyer, 1918). On the other hand, we do see similarly developed denticles in unrelated wood-boring groups such as Amorphoceriini (Molytinae), Araucariini (Cossoninae), Campyloscelini (Baridinae), and in certain

bostrichid wood borers as well as for digging insects in general (e.g., scarabs) (Figure 3.2).

Some scolytines are cleptoinquilines, and take over ready-made nests of ambrosia beetles, killing or ejecting the original tenants in the process. These species have developed dramatic features such as a sharply prolonged anterior pronotum and various elaborately sculpted sharp elongations of the declivity; the former often takes the form of a pointed hood with or without a terminal hook (Figure 3.3F). Nest parasitism is most common among corthylines in the genera *Tricolus* and *Amphicranus* and

FIGURE 3.3 **Morphology of Scolytinae.** (A, B) Sexual dimorphism, here represented by different shapes of the frons. (A) Male and female *Scolytoplatypus rugosus* Jordal. (B) Male and female *Phrixosoma concavifrons* Jordal. (C–E) Extreme sexual dimorphism in an inbreeding bark beetle. (C) Male (left) and female (right) siblings of *Ozopemon uniseriatus* Eggers. (D) Head features of the male. (E) The male is fully developed and reproductively mature, note the aedeagus. (F, G) Examples of declivity variation. (F) *Amphicranus fastigiatus* Blandford, holotype. (G) *Tomicus piniperda* (L.).

in the xyleborine genus *Sampsonius*. As with declivital teeth and spines in Scolytinae and Platypodinae (Hubbard, 1897; Hamilton, 1979), it is likely that acute developments on the front and back end of the cleptoinquilines are used in fighting and tunnel defense. Other weevil groups also take over ambrosia beetle tunnels, for instance in Brentidae (Kleine, 1931; Beeson, 1941; Roberts, 1969; Sforzi and Bartolozzi, 2004) and in the baridine subtribe Campyloscelina (Schedl, 1972; Thompson, 1996), and these show strikingly similar morphological adaptations.

Life in tunnels has led to multiple origins of fungus farming, including 10 times or more in Scolytinae and once in Platypodinae (Jordal and Cognato, 2012). Shifting from consuming woody to fungal tissues (which are softer and require less chewing) selects for changes in mouthpart and digestive tract morphologies. While phloem-feeding bark beetles have their maxillary laciniae fringed by coarse bristle-like setae, those feeding solely on fungal mycelium and conidia have very fine hair-like setae (Jordal, 2001). We see the same trend in the proventriculus that is situated in the alimentary tract in the prothorax and which functions like the gizzard of birds (Nobuchi, 1969). The normal condition for a bark beetle is to have a strongly sclerotized proventriculus with a large anterior plate containing nodules, teeth or transverse ridges. All ambrosia beetles have their anterior plate strongly reduced or totally absent. Remnants of the anterior plate are most evident in some of the most recently evolved lineages of ambrosia beetles such as Xyleborini and Premnobiina (Ipini), each roughly 20 million years old (Jordal and Cognato, 2012; Cognato, 2013). In each of these groups, the anterior plate is clearly visible, but very short and less sclerotized.

Finally, access to a tunnel for food and reproduction is limited in the sense that the tunnel-initiating individual can control access. This has consequences for mate recognition and mate choice, and for how late arrivals such as nest parasites and predators are rejected. The largest variation in morphological traits is therefore not surprisingly seen in body parts associated with tunnel blocking (discussed further in Section 2.3). Morphological adaptations to blocking the entrance to gallery systems are primarily seen in the declivity. Many taxa have evolved various teeth, knobs, and ridges on the declivity. Though there are few observations and no experiments on the function of these, Hubbard (1897) and Hamilton (1979) have hypothesized that especially the sharp teeth often seen on the borders or apex of the declivity function as weapons of defense against potential rivals and natural enemies. The overall shape of the declivity is likely also an adaptation to burrow blocking, particularly in species with flat or convex declivities, as the back end of the beetle ideally should fit the curvature of the outer bark surface as seamlessly as possible. This hypothesis could be tested by comparing the degree of curvature of the declivity (for the blocking sex) with the surface curvature of preferred host material, where one would expect to find flatter declivities in species regularly breeding in large diameter trunks and more strongly curved declivities in species with strong preferences for twig and small branches or thin stems.

Alternatively, in some taxa, the ventral aspect of the abdomen may be partly or entirely involved in forming the hind end of the beetle. In such cases, the apex rises more or less sharply, involving all or just the last few sternites depending on the group. The venter is only weakly raised in *Xyloctonus* and certain cryphalines, but rises steeply from the second ventrite in *Scolytus* and close relatives. In the latter group, the venter completely takes over the role of the declivity, in forming the hind end, which blocks the entrance. Development of the venter in this manner is extreme in the Platypodinae genera *Doliopygus* and *Mesoplatypus* (Strohmeyer, 1918; Schedl, 1972).

A cryptic lifestyle makes coloration less important for wood-boring beetles compared to those living in the outside world. Very few groups show any coloration beyond shades of brown; the color of mature adults ranges from dark yellow to reddish brown to black. The only significant exception to this pattern is found in three species of *Camptocerus*, a genus closely related to *Scolytus* (Smith and Cognato, 2010, 2011). The metallic green to bronze shine is unique to these species (see Fig. 2.7 in Chapter 2). Although the function of the metallic shine is completely unknown, it is interesting that species in this genus are also unique in spending extended periods aggregating and feeding on green leaves, before moving into wood (Smith and Cognato, 2010, 2011).

Scolytinae beetle bodies are usually 2–3 times as long as wide and fairly parallel sided; they vary in size from ca. 0.5 mm to a little over a centimeter in length, and most species fall in the range 1 to 4 mm long. There is no strong correlation between diameter of breeding material and body size: one finds small species that prefer larger trunks, and medium to large species that breed in branches or even twigs. Platypodines are more slender and on average longer, and they are more frequently confined to trunks and medium to large branches than are scolytines. Browne (1961) has speculated that there may be an evolutionary trend towards small body size, driven by selection for escaping predators that use tunnel entrances to get into gallery systems, especially with respect to ambrosia beetles. This intriguing hypothesis has yet to be tested comparatively. Scolytinae as a group are the smallest of the major groups of wood-boring insects, and platypodines are among the smallest (Haack and Slansky, 1987).

Within species-specific limits, body size of wood-boring beetles such as scolytines and platypodines is generally determined by the quality and amount of food consumed by larvae (Andersen and Nilssen, 1983; Kirkendall, 1983; Haack and Slansky, 1987; Kajimura and Hijii, 1994). Resource quality, in turn, is affected strongly by factors such as how fresh or old the breeding material is, remnants of defensive chemicals, and presence of fungi and microorganisms, while quantity is affected by factors such as inner bark thickness (for phloeophagous species), tunnel length (for ambrosia beetles), and density of competing larvae of the same or different species. Body size is important in natural selection (fecundity, survivorship), and sexual selection (fighting, mate choice), and affects features such as survival in cold temperatures (e.g., *Dendroctonus*; Safranyik, 1976), attractant pheromone production (Anderbrandt *et al.*, 1985), and anti-aggregation pheromone production (Pureswaran and Borden, 2003).

2.3 Sexual Dimorphism

Sexual selection is a powerful evolutionary force (Darwin, 1859; Shuster and Wade, 2003), and has surely been a prime factor in the evolution of sex differences in bark and ambrosia beetles. Dimorphic features are especially common in the frons (Figure 3.3) and declivity, and often in the underside of the abdomen (venter). This is to be expected, since characters involved in mating behavior (primary and secondary sexual characters) often evolve more rapidly than other morphological features (Civetta and Singh, 1999), and the frons, declivity, and venter are directly involved in mating behavior. As is often the case, the features exhibiting sexual dimorphism are frequently the best characters for separating closely related species, and are presumably used by the beetles themselves in species recognition as well as mate recognition.

Courtship in both Scolytinae and Platypodinae involves primarily tactile, chemical, and acoustic stimuli. Typically, the courting sex rubs or bumps the frons against the declivity or venter of the first arriving sex. There is evidence for specific types of setae in these body regions that match between the different sexes of a species, such as in *Scolytus* (Page and Willis, 1983). Species in many outbreeding bark and ambrosia beetle genera are therefore diagnosable mainly based on extravagant sculpturing or ornamentation (such as long setae) seen in only one of the sexes, commonly of the frons (Figure 3.3). Very generally, the frons of the courting sex is frequently flat or concave, while that of the colonizing sex is convex, and frequently the frons of the courting sex has longer or denser setae (S. L. Wood, 1982). In species with a dimorphic frons, individuals of the courted sex from closely related species might be identical in frons features, while frons characters are diagnostic in the courting sex. If there is noticeable sexual dimorphism in features of the declivity, such as degree of concavity or presence and size of teeth or spines, these characters are most developed in the courted (pioneering) sex. However, both frons and

declivity can be monomorphic or nearly so; it is not clear why some species are distinctly sexually dimorphic and others not so.

Characters other than the frons and declivity can be sexually dimorphic as well. Some of these, such as modifications of the antennae, or of the shape or setation of the last ventral abdominal segment, are certainly associated with acquiring mates or with copulation, but others (such as modifications of legs or pronota) may be adaptations to differences in sex roles (including differences in which sex carries symbiotic fungi). The basal antennal segments may differ in shape and setae pattern. For example, individuals of the courting sex (females) in most Micracidini have dense, long setae on the antennal scape, which are not present in the pioneering sex (males); a similar antennal scape dimorphism occurs in *Chramesus* and some *Camptocerus*, but in these genera it is males who court and who bear the long setae on the scape (S. L. Wood, 1982). In *Camptocerus noel* Smith and Cognato, it has been confirmed that the setal brush on the scape is used quite actively in courtship (Smith and Cognato, 2011). In many corthyline ambrosia beetles, the antennal club is enlarged (extremely so in *Corthylus*) and may be different in shape in females (the courting sex). The pronotum is differently shaped in *Trypodendron* (S. L. Wood, 1982), and in some groups (such as *Phloeoborus*, *Scolytoplatypus*, some *Cryphalus*, and some *Scolytodes*) the sexes differ in surface sculpture of the pronotum. The sexes of *Scolytoplatypus* differ dramatically in the protibiae and procoxae (segments of the first pair of legs). The protibiae of females have a rougher surface and more strongly developed teeth (Beaver and Gebhardt, 2006; Jordal, 2013), characters that we speculate might be an adaptation to fungus farming in these ambrosia beetles.

An additional difference between the sexes (occasionally the only one) is body size. Although the pattern has yet to be investigated systematically, it is clear from the average measurements in taxonomic treatments (such as S. L. Wood, 1982, 2007; Jordal, 1998) that, where size differences exist, it is the pioneering sex that is the larger. This is generally associated with mating system, females being the larger sex in monogynous species and males in harem polygynous species (see also Foelker and Hofstetter, 2014). This pattern for size dimorphism in outbreeding species may arise from differences in selection on the two sexes. Females are generally larger in insects, including weevils, probably because of fecundity selection on females being stronger than any selection for large size in males. However, in harem polygynous species, there is likely both intrasexual and intersexual selection for large male size (males being the pioneering sex, and the sex with greater variance in reproductive success), and in these cases this seems to be stronger than fecundity selection on conspecific females.

Sexual selection is presumably weak or absent in extreme inbreeders, where many species frequently or regularly have only one male per brood. Interspecific differences in the frons of females from related species are weak or nonexistent. Declivital differences do exist for females of related inbreeding species, especially in xyleborines, but overall interspecific differences in groups of related inbreeders seem to be much less than those found in groups of related outbreeders.

Sexual dimorphism in extreme inbreeders takes a very different form than that for outbreeding species, and is consistent with patterns found in other regularly inbreeding arthropods (Hamilton, 1967). Males of regularly inbreeding Scolytinae are rare, and are usually smaller (considerably smaller in many species), are less sclerotized, and are differently shaped; they have reduced eyes (Vega *et al.*, 2014) and males cannot fly because the second pair of wings is vestigial. Curiously, there are some striking exceptions. In certain unusually large species of Xyleborini (such as the *Xyleborus princeps* group of species), males are very similar to females in both size and shape. *Cyclorhipidion* males are about the same size as females, but have the pronotum more elongated. *Dendroctonus micans* (Kugelann) and its sister species *D. punctuatus* LeConte are unique among inbreeding Scolytinae in their lack of significant sexual dimorphism; in these species, males are very similar to females in size, and can in fact fly (see Section 4.2). At the other extreme is *Ozopemon*, a genus of haplodiploid, phloeophagous inbreeding scolytines that comprise one of only two examples in Coleoptera of larviform males (Jordal *et al.*, 2002). Sexual dimorphism is so extreme in *Ozopemon* (Figure 3.3C–E) that for about 50 years the rarely collected larva-like males were thought by some leading beetle experts to belong to the family Histeridae (Crowson, 1974).

One rare form of dimorphism in Scolytinae involves the development of horn-like structures on the anterior (rather than posterior) part of the body. Long horns are a particularly striking feature of many *Cactopinus* species, where they originate from the lower part of the frons. Various forms of nodules or carinae are found on the frons of a variety of scolytines, but the large size of these horns is a unique feature for this genus. A few other genera have small spines originating from the mandibles of the courting sex, such as in male *Triotemnus* (Knížek, 2010) and other dryocoetines, female *Styphlosoma* (S. L. Wood, 1982), female *Araptus araguensis* Wood, *Phelloterus* females (Wood, 2007), or female *Diapus* in Platypodinae. At least for *Diapus*, the mandibular teeth are dehiscent and only used during courtship to pull out the pioneering male (Beaver, 2000). Mandibles are greatly enlarged in the courting sex in *Gnatholeptus* females and *Phelloterus* females (Wood, 2007). The role in courtship behavior of these mandibular adaptations is not known.

Dimorphism is also frequently expressed on the declivity of the elytra. The blocking sex can have more strongly developed or a larger number of spines, teeth, or setae, and can have the declivity more flattened or concave than in the other sex. Such differences are so pronounced in, for example, the ambrosia beetle genera *Amphicranus* and *Gnathotrupes* that specialists have occasionally initially assigned males and females to different species or even different genera. However, it should be emphasized that sexual dimorphism of the declivity in many genera is very mild or nonexistent; in our experience, interspecific differences in the declivity are more frequent than intersexual differences, and are a great aid in separating closely related species. Hypothesized functions of features of bark and ambrosia beetle declivities have never been seriously analyzed or studied experimentally, which is unfortunate given the extraordinary variation that can be found within both Scolytinae and Platypodinae.

The various shapes of spines and tubercles on declivities may well serve several purposes, the most obvious possibilities being mate recognition and effective shoveling of frass. Though we can find few mentions of the idea in the literature (as mentioned above, Hubbard, 1897 and Hamilton, 1979), specialists often speculate in conversations that the sharp projections and borders seen in many platypodines and scolytines may be stabbing or cutting weapons useful against conspecific usurpers and natural enemies trying to gain entrance to the gallery system. Hubbard describes finding fragments of "vanquished" males in the tunnel systems of *Euplatypus compositus* (Say), an abundant North American platypodine ambrosia beetle. He writes (p. 14):

The female is frequently accompanied by several males, and as they are savage fighters, fierce sexual contests take place, as a result of which the galleries are often strewn with the fragments of the vanquished. The projecting spines at the end of the wing-cases are very effective weapons in these fights. With their aid a beetle attacked in the rear can make a good defense and frequently by a lucky stroke is able to dislocate the outstretched neck of his enemy.

We mentioned earlier that there are taxa in which the venter takes over part or all of the role of forming the hind end of burrow-blocking bark and ambrosia beetles. Sexual dimorphism in the venter of *Scolytus*, and many platypodine genera, takes the form of differences in spines and setae, exactly as with sexual dimorphism of declivities.

3. EVOLUTIONARY ECOLOGY OF FEEDING

Scolytinae and Platypodinae are components of what are increasingly being termed "saproxylic" beetle communities—species associated with dead wood and associated structures (such as woody fungi) (Ausmus, 1977; Swift, 1977; Ahnlund, 1996; Hammond *et al.*, 2001; Ulyshen *et al.*, 2004; Ødegaard, 2004; Tykarski, 2006; Lachat *et al.*, 2006, 2012; Zanzot *et al.*, 2010). Host trees are usually dead or severely weakened, and their colonization by these beetles, which often carry with them a complex community of fungi, bacteria, yeasts, and mites, initiates the breakdown of plant tissues and recycling of nutrients.

Actually, bark and ambrosia beetles breed in a wide variety of plant tissues. The feeding behavior of Scolytinae and Platypodinae has traditionally been broken down into categories based, first, on whether the larvae feed directly on plant tissues or on cultivated fungus, and second, for the direct plant feeders, on the tissues consumed by developing larvae. Since adults feed within their breeding material, the substances consumed by larvae are normally adult food as well (larvae in some ambrosia beetles feed on fungus-infested wood, whereas adults only feed on fungal tissues, but they here are both regarded as feeding on farmed fungi; see Section 5.3). We adopt the categories that have been standard for over five decades (Table 3.1). However, as Beaver (1986) emphasizes, "[the beetles] do not cooperate very readily in tidy classifications" (quoting Browne, 1961). Though most species can easily be placed in one of these categories, some feeding habits are hard to classify, and our classifications in some cases could be disputed. In this section, we will briefly describe the larval feeding modes of bark and ambrosia beetles, with a focus on more unusual habits, which are less well known than phloem feeding or fungus tending.

As pointed out by many authors, many or most Scolytinae (and all Platypodinae) are associated in one way or

TABLE 3.1 Traditional Classification of Larval Feeding Modes of Scolytinae and Platypodinae (Schedl, 1958; S. L. Wood, 1982, 1986, 2007). The Examples Given are not Exhaustive; for more Details, see Appendix

Larval Feeding Mode	Examples (see Appendix for complete list)	Feeding
Herbiphagy	*Hylastinus obscures* (Marsham) (where invasive), clover roots; *Thamnurgus euphorbiae* (Kuster), stems of *Euphorbia*; *Xylocleptes bispinus* (Duftschmid) in *Clematis*; *Coccotrypes rhizophorae* Eggers, mangrove propagules; petiole-breeding *Scolytodes* species.	Feeding on fresh or dry fleshy plant tissues, including stems of herbaceous plants, leaf petioles, cactus "leaves," grass stems, mangrove viviparous propagules.
Spermatophagy	Most *Coccotrypes*; *Conophthorus*, developing gymnosperm cones; *Araptus*, clade in legume seeds; *Pagiocerus frontalis* (F.), Lauraceae and *Zea* seeds; *Hypothenemus obscures* (F.), macadamia seeds, etc.; *Hypothenemus hampei* (Ferrari), developing *Coffea* fruits; *Dactylotrypes*, palm seeds.	Feeding in large hard seeds and the encompassing fruit tissues.
Mycophagy	*Trischidias* and *Lymantor decipiens* (LeConte), ascomycete fruiting bodies in dry twigs or bark.	Feeding in free-living (not cultivated) fungi (but see Harrington, 2005).
Myelophagy	*Pityophthorus* (some); *Araptus* (some); Bothrosternini (non-xylomycetophagous species); *Cryptocarenus*; *Micracisella*; *Hypothenemus* (a few); *Chramesus* (a few); *Scolytodes* (a few); *Dendrocranulus*, curcubit vines.	Feeding in pith of twigs, small branches or small stems, including small vines (e.g., *Dendrocranulus* in cucurbit vines).
Phloeophagy	Most Scolytinae, no Platypodinae: *Dendroctonus*, *Ips*, *Tomicus*, most *Scolytus*, most *Pityophthorus*, etc.	Feeding in phloem tissues (inner bark), though some larvae engrave outer sapwood; may or may not be regularly associated with fungi which increase nutritional value of the substrate.
Xylomycetophagy (ambrosia beetles)	Platypodinae; Xyleborini; Scolytoplatypodinae; Xyloterini; Hyorrhynchini; Corthylini-Corthylina; *Camptocerus*; *Hypothenemus* (a few); *Premnobius*; *Scolytodes unipunctatus* (Blandford).	Feeding on "farmed" ectosymbiotic fungi growing in wood; larvae of some species also ingest wood. Schedl's (1958) original definition: "larvae...feeding...upon the mycelia of fungi cultivated on the walls of their tunnels."
Xylophagy	*Dactylipalpus*; *Hylocurus, Micracis, Thysanoes*; *Chramesus xylophagus* Wood; *Dendrosinus*; *Phloeoborus*; some *Lymantor, Scolytodes multistriatus* (Marsham).	Feeding in xylem tissues (sapwood, never heartwood) but not cultivating symbiotic fungus.

another with fungi and other microorganisms (Six, 2013). Phloeophagous bark beetle-vectored fungi have long been known to be important in overcoming host defenses of live trees, but their role in nutrition is only now being puzzled out for a few model species. As more is learned about the roles microorganisms play, we will be able to make finer distinctions in feeding categories: one could separate out species of *Ips* and *Dendroctonus* that feed in phloem enriched with symbiotic beetle-borne fungi as "phloeomycophagous," for example (Six, 2012), and distinguish between ambrosia beetles whose larvae feed purely on fungus and those that also consume wood (Roeper, 1995; Hulcr *et al.*, 2007; Chapter 2). These distinctions make sense biologically and reflect different morphological, physiological, and behavioral adaptations, but the usefulness of such fine distinctions will remain limited until we have investigated a broad selection of species. Oversimplified as it is, our categorization of larval feeding habits does have considerable heuristic value and has been essential in documenting and explaining major ecological and evolutionary trends in these two subfamilies (Beaver, 1979a; Kirkendall, 1983, 1993; Atkinson and Equihua-Martínez, 1986a).

Larval feeding habits have consequences for patterns of host usage. Generally, species breeding in live trees tend to be relatively host specific, sometimes very narrowly so (Section 3.9). Phloeophagous and herbiphagous species are more host specific than species breeding in wood, pith, seeds, or as ambrosia beetles (Beaver, 1979a; Atkinson and Equihua-Martínez, 1986b; Hulcr *et al.*, 2007).

Larval feeding habits also have consequences for fecundity, and thus for suites of interrelated life history traits. Plant tissues are generally a poor resource from the point of view of nutritional quality, being much lower in nitrogen than beetle bodies (White, 1993; Ayres *et al.*, 2000). Fresh and particularly living phloem is a better resource than older, dead inner bark (Kirkendall, 1983; Reid and Robb, 1999). Inner bark and seeds are much higher in nitrogen than wood or pith. However, ambrosia fungi and some fungi associated with phloem feeders (Section 3.1) are rich in nitrogen (French and Roeper, 1975; Ayres *et al.*, 2000); ambrosia fungi concentrate nitrogen, and have much higher amounts than the wood itself (French and Roeper, 1975). That pith, wood, and woody leafstalks are unusually poor in nutrition is reflected in the fact that scolytines breeding in these substrates have considerably lower fecundity than those breeding in inner bark or seeds (Kirkendall, 1983, 1984; Jordal and Kirkendall, 1998).

For detailed insight into the ecology of bark and ambrosia beetle feeding see general resources such as the works in our reference list by Beeson, Blackman, Browne, Kalshoven, Schedl, or Wood, review papers by Kirkendall (1983, 1993; Kirkendall *et al.*, 1997) and Beaver (1977, 1979a, b) and the research papers by, for example, Atkinson

and collaborators (Mexico, S. E. US), Blackman (eastern US), Beaver (worldwide), Cognato and collaborators (Hulcr, Smith, and others) (worldwide), and, for fungus farming in particular, by Hulcr, Cognato, Jordal and collaborators Six, and Harrington.

3.1 Phloeophagy (Breeding in Inner Bark)

Of woody tissues, inner bark is the richest, especially in nitrogen (Cowling and Merrill, 1966; Kirkendall, 1983), so it is no surprise that the most primitive Scolytinae breed in dead inner bark of trunks and branches (Figures 3.4 and 3.5), or that phloem feeding is the most widespread larval feeding mode. Roughly half of all Scolytinae genera are wholly or partly phloeophagous, and 20 of 26 tribes have at least some phloeophagous species in them (Table 3.2; Figure 3.4). Only phloeophagous species are known from Hylastini, Phloeotribini, and Polygraphini, and several other tribes are primarily phloeophagous (Appendix).

3.1.1 Phloeophagous with Some Consumption of Wood

In certain phloeophagous species in hardwoods, older larvae (often the final instar) tunnel in the outermost sapwood, and pupate in the wood. Thus, late-stage larvae of *Scolytus muticus* Say, which breeds in *Celtis* (hackberry), burrow "for some distance" in the sapwood, "...and if they are at all numerous soon reduce the outer part of the wood and bark to a mere shell" (Blackman, 1922). *Triotemnus pseudolepineyi* Knížek larvae consume all phloem and sapwood, when breeding in branches of the shrub *Bupleurum spinosum* Gouan (Apiaceae) in Morocco (Knížek, 2010). Other examples include *Chramesus hicoriae* LeConte (Blackman and Stage, 1924); *Phloeosinus sequoia* (Hopkins) (De Leon, 1952); *Strombophorus ericius* (Schaufuss) (Browne, 1963); and species of *Hylurgonotus* and *Xylechinosomus* breeding in *Araucaria* (Rühm, 1981; Jordal and Kirkendall, pers. observ.).

Sapwood is roughly an order of magnitude lower in nitrogen than inner bark and more heavily lignified (Cowling and Merill, 1966; Haack and Slansky, 1987); therefore, phloeophagous larvae should avoid feeding on it, if possible. It is possible that fungi nutritionally improve the wood quality for beetles, but this has not been studied. One possible hypothesis for "late-stage xylophagy" is that in thin-barked hosts, larvae simply are forced to consume wood as they get larger (Browne, 1963); in many species, bark beetle larvae are small enough to be able to feed entirely in inner bark, but in others, the amount of wood consumed will be inversely proportional to the diameter of the breeding material. A second hypothesis is that burrowing into the wood makes it more difficult for parasitoid

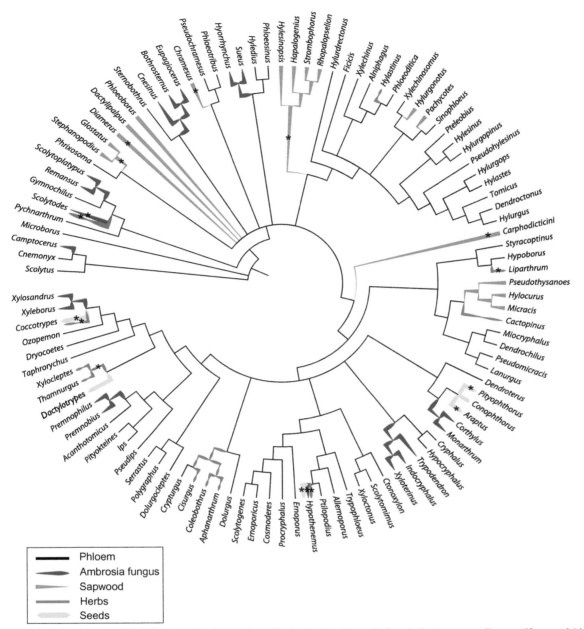

FIGURE 3.4 **Phylogenetic tree of Scolytinae with feeding modes indicated (see inset legend).** Stars indicate genera or lineages (if on a node) in which the feeding mode is rare (one or just a few species).

wasps to locate and parasitize larvae. Additionally, wood might be less strongly infested by potentially harmful microbial pathogens than more nutrient-rich phloem. Many bark beetles pupate in the sapwood, in some cases tunneling directly inwards to do so; this likely is an adaptation to reducing parasitism. Testing the second hypothesis is self-evident; a test for the first hypothesis would be to compare resultant body size of offspring that do not feed on sapwood as larvae (larval tunnels do not engrave the wood) with those that do consume much sapwood as larvae (their tunnels clearly etching the wood).

3.1.2 Feeding on Phloem Nutritionally Improved by Fungi

Insects breeding in dead woody tissues will always have constant interactions with a variety of mites, nematodes, fungi, and bacteria (Hamilton, 1978). Bark and ambrosia beetles are an optimal vehicle for transport of mites, nematodes, fungi, and bacteria from old host material to new, and many hitch rides on them (Stone, 1990; Paine *et al.*, 1997; Six, 2003, 2012; Harrington, 2005; Cardoza *et al.*, 2006a; Hofstetter *et al.*, 2006; Knee *et al.*, 2013;

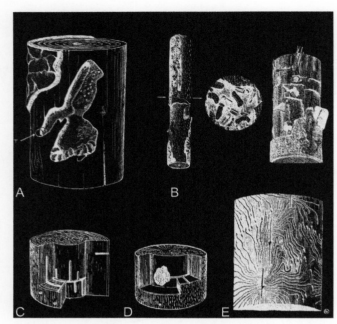

FIGURE 3.5 Variation in gallery systems made by bark and ambrosia beetles. (A, B, E) Engravings in phloem. (A, B) Cave-type galleries of inbreeding polygynous species with communal larval feeding for *Dendroctonus micans* (Kugelann) (from Chararas, 1962) and in *Hypothenemus colae* (Schedl) (from Schedl, 1961b). (C, D) Ambrosia beetle tunnel systems in sapwood for fungus cultivation for inbreeding polygynous *Xyleborus dispar* (F.) and *Xyleborinus saxeseni* (Ratzeburg) (from Balachowsky, 1949). (E) Monogynous egg tunnels of *Kissophagus granulatus* Lepesme in *Ficus* (from Schedl, 1959).

structures are ambrosia beetles (discussed below and in Section 5.1), but some breed in inner bark and feed on phloem they have inoculated with fungi they have introduced. There are also phloeophagous species with consistent associations with fungi but which have no special structures for transporting them, including *D. pseudotsugae* Hopkins, *D. rufipennis* Kirby, *Ips avulsus* (Eichhoff), and *Tomicus minor* (Hartig) (Beaver, 1988).

Many species even have structures for transporting fungi from host to host in more or less sophisticated cuticular invaginations or pits known as mycetangia (Francke-Grosmann, 1956a) or mycangia (Batra, 1963). Larvae of phloeophagous species that are associated with fungi feed (at least in most stages) in woody tissues, not on mats of fungal hyphae. The earliest research into mycetangia revealed their presence in phloeophagous as well as xylomycetophagous species (Francke-Grosmann, 1956a, b, 1963a, b, 1965, 1966; see Kirisits, 2004). Francke-Grosmann (cited above) reported mycetangia in typical phloeophagous species in *Hylastes*, *Hylurgops*, and *Ips*. The potential nutritional benefits of fungi in species that are not ambrosia beetles are now being explored in detail (Six and Paine, 1998; Ayres *et al.*, 2000; Bentz and Six, 2006; Adams and Six, 2007; recent reviews by Six, 2012, 2013; Chapters 6 and 8).

Several phloeophagous *Ips* species have mycangia, including the Eurasian *Ips acuminatus* (Gyllenhal) (Francke-Grosmann, 1963a). This *Ips* transports two mycetangial fungi (Francke-Grosmann, 1963a, 1967; Guérard *et al.*, 2000). Larval mines in phloem that is obviously discolored by fungi are notably shorter than those in phloem with no discoloration. When the fungus is clearly well established, one can see that larvae double-back in their own feeding tunnels and feed on the white fungus growing on the tunnel walls; the first action of eclosed young adults is to completely graze white fungal conidia and hyphae, which have grown on the walls of their pupal chambers (Kirkendall, unpubl.). Several unrelated North American *Ips* species seem to have a similar biology (summarized in Harrington, 2005). *Tomicus minor* is a common Eurasian scolytine breeding in pine trunks and thicker branches; first

Shimizu *et al.*, 2013; Susoy and Herrmann, 2014; Chapter 6). Some small organisms perform useful functions from the beetle's point of view, and many bark and ambrosia beetles have morphological adaptations that increase the likelihood of successful transport of helpful symbionts. In particular, a wide variety of species have developed external crevices, pits, simple pockets, or complex invaginations somewhere on the body, for transporting fungi (and perhaps other microorganisms) (Beaver, 1988; Harrington, 2005; Six, 2012); these structures are often bordered by setae, which help in combing fungal spores into the receptacle. Most species with such

TABLE 3.2 Number of Scolytinae Genera and Tribes with at Least one Species Exhibiting the Given Larval Feeding Mode (247 total genera, 26 total tribes)

Number of Taxa with at Least One Species	Phl	Xym	Spm	Myc	Mye	Xyl	Hbv	?
Genera	121	63	9	2	14	21	17	31
Tribes	20	10	5	1	6	11	9	14

Some genera and tribes are represented in more than one category. Phl=phloeophagous (feeding in inner bark); Xym=xylomycetophagous (ambrosia beetles); Spm=spermatophagous (feeding in seeds, fruits); Myc, mycophagous (feeding on non-symbiotic fungi); Mye=myelophagous, feeding on pith; Xyl=xylophagous, feeding in sapwood; Hbv=herbiphagous (herbivorous), feeding in non-woody plant tissues; "?," unknown larval feeding habits. Data from Appendix.

and second instar larvae feed in inner bark, but later instars move into the xylem where they become strict fungus feeders (Harrington, 2005). Both *I. acuminatus* and *T. minor* would seem to be intermediate between true phloeophages and obligate fungus feeders. They are both associated with *Ambrosiella* fungi (as well as bluestain fungi), which are ambrosia fungi in xylomycetophagous species.

The relationship of symbiotic fungi with certain species of *Dendroctonus* and *Ips* is not an obligate one, but successful establishment of their fungi definitely enhances larval fitness in some species. Southern pine beetle (*Dendroctonus frontalis* Zimmermann) larvae feeding in the absence of their two mycetangial fungi have significantly reduced offspring survivorship (Barras, 1973), and females breeding without these mutualistic fungi lay only half as many eggs as controls (Goldhammer *et al.*, 1990). Similar fitness effects of mutualistic fungi are seen in the mountain pine beetle, *D. ponderosae* Hopkins (Six and Paine, 1998). Southern pine beetle mutualistic fungi raise the nitrogen content of the phloem and increase its digestibility (Ayres *et al.*, 2000). Fox *et al.* (1992) found evidence for enhanced growth of *Ips paraconfusus* Lanier larvae when the associated fungus

was present in the phloem, and Yearian *et al.* (1972) found that reproduction by females of *I. avulsus* (but not for two other *Ips* species) is increased by the establishment of their associated fungus. The terms "phloemycetophagous" and "mycophloeophagous" have been suggested for inner bark-breeding species that regularly feed on phloem plus fungus (Kirisits, 2004; Six, 2012).

We have focused on fungi here, but the nutritional quality of substrates consumed by bark beetles results from a complex interaction between the physical and biochemical attributes of the tissues consumed (Kirkendall, 1983; Haack and Slansky, 1987; Reid and Robb, 1999; Six, 2012) and a complex community of fungi, yeasts, bacteria, and other microbes (Cardoza *et al.*, 2006b; Hofstetter *et al.*, 2006; Six, 2012, 2013; Chapter 6).

3.2 Xylomycetophagy (Ambrosia Beetles)

The larvae and adults of xylomycetophagous species eat cultivated fungi growing on woody tissues (Schedl, 1958; Browne, 1961; S. L. Wood, 1982: see Box 3.1, Table 3.1, and Figure 3.5C, D), and are referred to as ambrosia beetles

---

BOX 3.1 Terminology

Most specialized terms are defined in the text. However, there are a few that are not, or that deserve special comment. We largely follow well-established conventions in bark and ambrosia beetle research (e.g., S. L. Wood, 1982), but have tried to align terms regarding mating systems and social behavior with the vocabulary being used more generally in behavioral ecology (Wilson, 1975; Shuster and Wade, 2003).

Alloparental—Refers to parenting by individuals other than the biological parents of the offspring, such as of ambrosia beetle larvae by siblings or aunts.

Ambrosia beetles—Ambrosia beetles are those Scolytinae (plus all Platypodinae) whose larvae feed primarily on co-evolved symbiotic "ambrosia fungi," which adult females cultivate in tunnel systems in woody tissues. They may consume wood in the process (xylomycetophagy *sensu* Hulcr et al. in Chapter 2) or not (mycophagy *sensu* Hulcr et al. in Chapter 2), but we will not make this distinction (see also "xylomycetophagy," below).

Bark—Shorthand for inner bark, the secondary phloem tissue of woody dicots.

Bark beetles—In the literature, this term is used (confusingly) in two senses, with three different meanings. Taxonomically, "bark beetles" refers to the subfamily Scolytinae; for clarity, we will avoid this usage. The expression is used two ways in an ecological sense: it can mean species breeding in inner bark (live and dead phloem tissues), but many authors also use it in apposition to ambrosia beetles (that is, to include all species that are not xylomycetophagous). To avoid confusion, we will mainly use "phloeophagous" to indicate Scolytinae that

breed in inner bark; occasionally, as in discussions primarily focused on ambrosia beetles, we use bark beetles (or "non-ambrosia beetles") as an umbrella term for all feeding modes other than xylomycetophagy. We will not use it taxonomically.

Bark and ambrosia beetles—This expression is often used as a collective term for Scolytinae. "Bark beetles," in this phrase, refers to all feeding modes other than obligate fungus feeding. We use this compound phrase broadly, to encompass both Scolytinae and Platypodinae, in order to avoid the excessively long "bark and ambrosia beetles and pinhole borers" when referring collectively to these two lineages.

Declivity—The downward-sloping posterior portion of the elytra: the back end of the beetle.

Frass—boring dust; the variegated mixture of feces and wood bits (digested or not) resulting from the tunneling activities of wood-boring insect larvae or adults.

Frons—Front of the head: the area between the eyes, from the vertex (top of the head) to epistoma (upper margin of the mandibles).

Hardwoods—Non-monocot angiosperm trees, as opposed to conifers. We use "broadleaf trees" synonymously, though technically this term also includes monocots.

Harem polygyny—Also known as simultaneous polygyny (as opposed to serial polygyny) in anthropology and behavioral ecology literature; in a harem polygynous scolytine, at least some gallery systems have multiple females. "Polygamy" (see below) is often used incorrectly as a synonym.

Continued

BOX 3.1 Terminology—cont'd

Herbiphagy—Biologists often call feeding on any plant tissue "herbivory." Bark and ambrosia beetle researchers use the related term "herbiphagy" for taxa feeding on fleshy (not woody) plant tissues, such as plant leaves, leaf stalks, or stems and branches of non-woody plants.

Monocots—Monocots are one of the two major groups of flowering plants, the other being dicots. Monocots comprise a monophyletic clade of plants that develop from a single cotyledon; monocot host plants of bark and ambrosia beetles include grasses (especially bamboos), palms, agaves, lilies (*Yucca* trees), and orchids.

Mycophagy—Used by us in a very narrow sense, for feeding on free-living fungi; other authors use this term broadly for any form of feeding on fungal hyphae and conidia (e.g., Harrington, 2005).

Monogyny—In monogynous species, only one female breeds in a gallery system.

Parasitoids—Parasitoids are insects that live on or in their hosts for some time before eventually killing them. Parasitoids of bark and ambrosia beetles are usually wasps, most commonly chalcidoids, pteromaloids, proctotrupoids, or ichneumonoids.

Pinhole borers—Currently, "pinhole borer" is often used to refer to Platypodinae as a group, though in older literature it may refer to any ambrosia beetle. "Shothole borer" has also been used as a generic term for ambrosia beetles, though at some point it seems to have been co-opted by North American entomologists for the phloeophagous bark beetle *Scolytus rugulosus* (Müller), a minor pest of fruit trees.

Polygamy—Also known as communal breeding, colonial breeding, or promiscuous breeding; in Scolytinae, a mating system where several males and several females are involved in constructing egg tunnel systems. In zoology, usually refers to a mating system in which both sexes mate with multiple partners, and have roughly equal variation in mating success.

Spermatophagy—Used (only) by Scolytinae researchers to classify species breeding in seeds and their encasing fruit tissues, and the viviparous propagules of mangrove trees. In the latter two cases, spermatophagy overlaps with herbiphagy, as the beetles are breeding in fleshy tissues. Other biologists call insects breeding in seeds "seed predators" or "seed parasites." Outside of bark beetle research, the term refers to phagocytosis of spermatozoa.

Xylomycetophagy—We use this term to refer collectively to the feeding category for ambrosia beetles: taxa whose larvae and adults feed primarily on cultivated co-evolved fungi. We do not distinguish between fungus farming species that do and do not ingest wood as well as fungus. Tunnel elongation, egg niche enlargement, and construction of pupal chambers (such as by all Platypodinae) may lead to ingesting wood, and in some taxa, species may be consuming wood incidentally while feeding on mycelia. For many ambrosia beetles, wood consumption is an aspect of their feeding ecology that is simply unknown; if "xylomycetophagy" is used narrowly to refer to ambrosia beetles known to feed on wood as well as fungi, and "mycophagy" used for taxa known to ingest fungi exclusively, then there remains no formal term (of the sort "phloeophagy," "xylophagy," etc.) to categorize feeding behavior of all ambrosia beetles, or to refer to ambrosia beetles for which relevant feeding behavior details are not known.

Xylophagy—Scolytinae that breed in tunnels in sapwood, and do not cultivate fungi.

(Schmidberger, 1836; Hubbard, 1897). Ambrosia beetles actively cultivate coevolved mutualistic fungi. The fungus forms layers of nutritious ambrosial growth within a few days (Francke-Grosmann, 1967). This growth is predominantly composed of fruiting structures of a single species of ascomycete fungus, which serves as major food source for adults and larvae. These fungi typically grow as mycelia, but form fruiting structures in the presence of the tending beetles (Batra, 1967; French and Roeper, 1972a; Biedermann, 2012).

Xylomycetophagy (cultivation of fungi growing in wood) is found in 63 genera in 10 tribes of Scolytinae (Table 3.2) and in all but the most basal Platypodinae. Based on the most recent phylogenetic analyses (Jordal and Cognato, 2012), it has evolved 10 or 11 times in Scolytinae, depending on details of the analysis (Figure 3.4), and it has originated once in Platypodinae (Jordal et al., 2011). Two of these origins are recent, being single species in large scolytine genera (*Hypothenemus*, *Scolytodes*). Ambrosia beetles usually tunnel in sapwood or pith, but some can breed in seeds, leafstalks, or the tissues of woody monocots. Several corthyline ambrosia beetle species, for example, have only been collected from the woody petioles of large, fallen *Cecropia* leaves (Wood, 1983, 2007; Jordal and Kirkendall, 1998), which are also utilized by generalist ambrosia beetles such as *Xylosandrus morigerus* (Blandford) (Andersen et al., 2012) and *X. crassiusculus* (Motschulsky) (Kirkendall and Ødegaard, 2007). All Platypodinae are tightly associated with fungi and usually colonize broadleaf trees; all but *Schedlarius* (xylophagous in rotted wood) and *Mecopelmus* (phloeophagous) are ambrosia beetles.

Most xylomycetophagous species transport their fungi in mycetangia or the gut (Schneider-Orelli, 1911; Francke-Grosmann, 1975). Vectoring of fungi within the gut is probably the ancestral mode of spore transmission, but still seems to be the dominant mechanism in some ambrosia beetles, including examples of both Scolytinae and Platypodinae that have no or reduced mycetangia. *Xyleborinus saxesenii*, for example, has very small elytral mycetangia (Francke-Grosmann, 1956a) and transmits its principal ambrosial fungus via the gut (Francke-Grosmann, 1975). In others like *Anisandrus dispar* (F.) with well-developed mycetangia, mycetangia and gut may harbor different fungi (see also *X. saxesenii*: Biedermann *et al.*, 2013); such redundancy may serve as an insurance mechanism in case one of the organs is infected by parasites. However, some others lack mycetangia completely because they rely on the fungal gardening of neighboring beetles of other species.

Fungus stealing was suspected by Kalshoven (1960a) and Beaver (1976), but was first thoroughly documented by Hulcr and Cognato (2010), who termed it "mycocleptism." The latter researchers found mycocleptism to be the main foraging strategy for at least 16 species mainly from the xyleborine genera *Ambrosiophilus* (eight species) and *Diuncus* (five species), but also including *Xylosandrus hulcri* Dole and Cognato, the scolytine *Camptocerus suturalis* (F.), and one Platypodinae, *Crossotarsus imitatrix* (Schedl). The "mycocleptae" tunnel close to the tunnel of an established "provider" species, in some instances breaking into the adjacent gallery system and destroying neighboring brood. The walls of the mycocleptae's tunnels then begin to produce ambrosia fungus, which had been introduced by the provider species. At least the genus *Diuncus* has lost mycetangia all together, and is completely dependent on this parasitic strategy.

3.3   Xylophagy (Breeding in Wood)

Species in which larvae feed wholly in sapwood occur in only 21 genera spread among 11 tribes. The most species-rich xylophagous lineage occurs in the Micracidini, in which three genera of wood feeders, *Hylocurus*, *Micracis*, and *Thysanoes*, include 119 species. Four Hylesinini genera seem to be entirely xylophagous, *Dactylipalpus*, *Hapalogenius*, *Phloeoborus* and *Rhopalopselion*, and *Hylesinopsis* partially so (see Appendix). The remainder of xylophagous examples is single species or small clades. Xylophagy has originated about nine times (Figure 3.4; see the more detailed phylogeny in Jordal and Cognato, 2012). Wood is nutritionally a very poor resource for insects (Cowling and Merrill, 1966; Kramer and Kozlowsky, 1979; Haack and Slansky, 1987). Many organisms feeding on wood are known to be dependent on the contributions of gut microbes. This has long been suspected to be the case for xylophagous bark beetles as well, but there has been relatively little research on this aspect of their biology. Xylophagous species often have low fecundity, relative to phloeophagous species (Kirkendall, 1984). The primary benefit to adopting xylophagy in these beetles would seem to be lower larval mortality from predators and parasites, but it may also be important that the physical environment (temperature,

wood moisture, food quality, persistence of resource quality) is relatively stable, much more so than would be expected for inner bark.

Browne (1961) treats pith and twig breeders as xylophagous; we prefer to separate the two, since pith and sapwood are considerably different in structure, density and hardness, and possibly in nutritional quality, though levels of nitrogen are roughly similar (Cowling and Merrill, 1966; Kramer and Kozlowsky, 1979).

3.3.1   Breeding in Wood Nutritionally Improved by Fungi

Currently, this is a hypothetical group, as no wood-breeding scolytines have been studied in any detail. The xylophagous genera *Dactylipalpus* and *Phloeoborus* have distinctive mycetangia, but do not appear to be true ambrosia beetles. Beaver and Löyttyniemi (1985) report that *Dactylipalpus camerunus* Hagedorn is polyphagous, monogynous, and xylophagous, and attacks moderate to large logs and dying or dead stems. Females have pronotal mycetangia, suggesting that they may be closely associated with fungi. In addition, Browne (1963) reports *Dactylipalpus* as xylophagous. Similarly, as far as is known, *Phloeoborus* are xylophagous, but females have mycetangia (Wood, 1986).

3.4   Herbiphagy

Some genera or single species breed in herbaceous plant tissues, and are classified as herbiphagous (Box 3.1, Table 3.1). It is a rare feeding strategy in Scolytinae, being found in only 17 genera (6%) in nine tribes (Table 3.2), and has evolved only about eight times (Figure 3.4). Half of the genera in which herbiphagy is represented are specialized to this lifestyle (Appendix). One radiation in the Dryocoetini accounts for about two-thirds of all herbiphagous species.

Feeding habits in this category include breeding in herbaceous plants, ivy, *Clematis*, grass stems including bamboos, cacti and succulent euphorbs, leaf petioles, and the viviparous propagules of mangrove trees. We include here two species that breed in roots of herbaceous plants: (1) *Hylastinus obscurus* (Marsham) is a minor pest of clover in North America, where it is an introduced species, though there are no records of it breeding in clover from Europe where it is native (Webster, 1910; Koehler *et al.*, 1961); and (2) the recently discovered *Dryocoetes krivolutzkajae* Mandelshtam, which breeds in roots of *Rhodiola rosea* (Crassulaceae), the only bark beetle of treeless tundra landscapes (Mandelshtam, 2001; Smetanin, 2013). And we include the only galling bark beetle, *Scolytodes ageratinae* Wood, which attacks live plants of a herbaceous montane species of *Ageratina* (Asteraceae) in Costa Rica (Wood, 2007; Kirkendall, unpubl.; see Section 3.9)

Thamnurgus is a typical example of a herbiphagous genus. *Thamnurgus euphorbiae* Küster has been approved for biological control of *Euphorbia esula* L. (leafy spurge), an invasive weed in the USA (Campobasso *et al.*, 2004). Females oviposit in the stem, starting at the top of the plants. Apparently, females have high lifetime fecundity (88 eggs) but lay relatively few eggs per plant. Colonized plants are weakened structurally and break easily, producing fewer seeds. *Thamnurgus pegani* Eggers breeds in stems of *Peganum harmala* L. (Nitrariaceae), a perennial plant toxic to grazing animals (Güclü and Özbek, 2007). One or a few eggs are laid between the stem and a lateral branch junction, and larvae tunnel down the inside of the stem in the pith. The tissue on which larvae are feeding becomes blackish-brown due to presence of *Fusarium oxysporum* Schltdl.; the fungus was also isolated from the bodies of the bark beetles. A couple of weeks after eggs are laid, larval tunnels are still very short (6 mm); this and the presence of white mycelia on the surface of the stained pith tissues suggest that the species may be gaining significant nutrition from the fungus.

An entire scolytine community (29 species, six genera, three tribes) can be found in the cactus-like, shrubby, and tree-like euphorbs of the Canary Islands, Madeira, Cape Verde, and North Africa (Jordal, 2006). Species are narrowly host specific, but up to half a dozen species could be found in one branch. Like the *Thamnurgus* mentioned above, these herbiphagous species are characterized by unusually low (within-plant) fecundity, though they likely oviposit in several plants. The scolytines breed only in dead branches and twigs, but differ ecologically in moisture preferences and host diameter.

The seeds of some mangrove trees (like those of *Rhizophora* or *Bruguiera*) grow while still on the mother plant; these viviparous propagules later drop from the tree and float until they strand on muddy sediments, after which they begin to root. *Coccotrypes* species breeding in the propagules of mangrove trees are sometimes categorized as spermatophagous, but we classify them here as herbiphagous since they are actually breeding in live, non-ligneous (not woody) plant tissues and not in seeds or fruit tissues. Hanging or (usually) newly beached seedlings are attacked by *Coccotrypes rhizophorae* (Hopkins), *C. fallax* (Eggers), and *C. littoralis* (Beeson) (Beeson, 1939, 1941; Kalshoven, 1958; Browne, 1961). These species specialize in mangroves, as opposed to most *Coccotrypes*, which are host generalists and breed in seeds, bark, or leafstalks with some, such as *C. cyperi* (Beeson), breeding on all three. The mangrove *Coccotrypes* are not found in other hosts, or in branches or trunks of mangroves. Interpreting this feeding behavior as herbaceous gets some support from the observation that *C. rhizophorae* also attacks the soft, growing tips of aerial roots of *Rhizophora mangle* L.; it does not, however, breed in the older, woody portions of the roots

(Aktinson and Equihua-Martínez, 1985b). In Neotropical mangroves, only *C. rhizophorae* is found; it occurs in mangrove forests throughout the world, and may have dispersed to the New World on its own, as have the mangrove species in these forests (Atkinson and Peck, 1994). Little has been published on the biology of mangrove *Coccotrypes*, but there have been two ecological studies of the effects of *C. rhizophorae* in the Neotropics, where it seems that the high levels of propagule attacks can have significant effects on the mangrove ecosystem (Rabinowitz, 1977; Sousa *et al.*, 2003).

Herbiphagy is a difficult category to define precisely, especially without detailed knowledge of plant anatomy. *Dendrocranulus*, for example, breeds in stems of cucurbit vines. We choose to classify *Dendrocranulus* as myelophagous (as do Atkinson and Peck, 1994) but it could also have been classified as herbiphagous. Which is more important physiologically, ecologically, and evolutionarily? That it breeds in non-woody plants (herbiphagous), or that it colonizes pithy tissues (myelophagous)? Petioles, too, are problematic. Those of large fallen *Cecropia* leaves are very woody, at one extreme, in contrast to those of *Gunnera*, which although stiff, are quite moist and rather fleshy (Figure 3.6). *Scolytodes*, a large neotropical genus comprised primarily of phloeophagous and myelophagous species, has radiated into both.

Lineages moving from bark to herbaceous tissues probably are moving to food with similar or even higher nutritional quality (with the exception of petioles: Jordal and Kirkendall, 1998), but herbaceous tissues differ from those of trees and woody shrubs tissues in their anatomy, biochemistry, and especially in moisture content. The distribution of herbiphagy in Scolytinae, and what we know of the biology of herbiphagous species, suggest that adopting herbiphagy is not readily accomplished and demands a suite of new adaptations (including major life history adjustments), though perhaps less so in those cases that most resemble woody branches (such as the highly lignified petioles of *Cecropia* leaves).

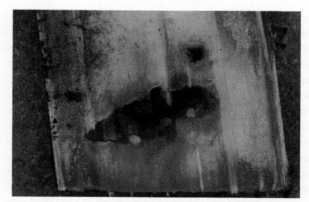

FIGURE 3.6 An example of herbiphagy: cave-type egg gallery of *Scolytodes gunnerae* Wood in live fleshy leafstalk of *Gunnera insignis* in Costa Rica. Eggs are laid loose in the gallery; the leafstalk is ca. 3 cm in diameter.

3.5 Myelophagy (Pith Breeders)

Pith breeding is very uncommon in Scolytinae. Only 14 genera in six tribes have species that regularly breed in pith (Table 3.2). Pith is composed of undifferentiated parenchyma cells, which function in storage of nutrients, and in eudicots is located in the center of the stem. It is mainly present in young growth; in older branches and stems it is often replaced by woodier xylem cells. Pith is poor in nutrients, being about equivalent to young sapwood in terms of nitrogen content (Cowling and Merrill, 1966) or somewhere in between sapwood and inner bark (Haack and Slansky, 1987). It is, however, easy to tunnel through. This combination of features is illustrated by the breeding biology of *Scolytodes atratus* Wood and Bright in *Cecropia* petioles, the centers of which are composed of a relatively large cylinder of soft white pith: tunnels can be several tens of cm in length yet produce only four or five offspring (Wood, 1983; Jordal and Kirkendall, 1998).

Typically, pith breeders construct irregular chambers or meandering egg tunnels, often going both up and down the twig from the entrance. Twig breeders are generally monogynous, even in otherwise harem polygynous genera such as *Pityophthorus*, *Araptus*, or *Scolytodes* (Kirkendall, 1983).

Twigs of many woody plants are largely pith, so twig breeders are classified as myelophagous; often, an entire twig is hollowed out by adult and larval feeding, but most of the tissue consumed is pith. There are a handful of *Pityophthorus* species that breed mainly or only in twigs and that are categorized here as myelophagous. In tropical hardwoods, the tribe Bothrosternini comprises mainly pith borers (some *Cnesinus* are phloeophagous), some of which have evolved fungus farming in pith (Beaver, 1973; S. L. Wood, 1982, 2007; Kolarik and Kirkendall, 2010; Section 3.2, Appendix).

3.6 Spermatophagy (Seed Breeders)

Spermatophagy (or spermophagy) as used by bark beetle researchers denotes species breeding in seeds and the surrounding fruit tissues. This term has been applied very broadly to encompass true seed predators (Janzen, 1971) but also species collected from fleshy fruits, woody seedpods, mangrove propagules (which we treat as herbiphagous), or cones (Schedl, 1958; Browne, 1961; S. L. Wood, 1982, 2007; Kirkendall, 1983; Atkinson and Equihua-Martínez, 1986b). As such, the category is rather heterogeneous with respect to actual feeding adaptations. Normally, exposed seeds from fallen fruits (or defecated seeds) are preferred both by seed specialists and by generalists when they breed in seeds.

Nine genera in five tribes have spermatophagous species, and true seed breeding has originated at least eight times (Table 3.2; Figure 3.4). Two genera of Scolytinae only breed in seeds (*Pagiocerus*, neotropical, five species; *Dactylotrypes*, one species endemic to the Canary Islands), as does possibly *Spermophthorus* (Wood, 2007).

3.6.1 Pagiocerus

Pagiocerus frontalis (F.), found in Central and South America, is often collected from seeds of Lauraceae, including commercial avocado (*Persea americana* Mill.). In Mexico, it bores into partially or completely exposed seeds lying on the ground and does not attack fruits on the tree (Atkinson and Equihua-Martínez, 1985b; Atkinson *et al.*, 1986). In South America, it has been recorded as a pest of maize since at least 1930; the seeds are attacked on the plant and in storage, and it has been collected from coffee berries in Ecuador (Yust, 1957; Okello *et al.*, 1996b; Gianoli *et al.*, 2006). In the laboratory, it can be bred on cassava chips as well as maize (Okello *et al.*, 1996a). The biology of other *Pagiocerus* species is not known, except that *P. punctatus* Eggers has been collected from male strobili of *Araucaria angustifolia* (Bertol.) Kuntze in Brazil (Mecke and Galileo, 2004).

3.6.2 Coccotrypes

Many species of *Coccotrypes* breed in small hard seeds, especially palms. Most *Coccotrypes* that breed in seeds also breed in bark, leafstalks, or other tissues, but some are known to be seed specialists (e.g., *C. carphophagus* (Hornung), *C. dactyliperda* F.), and there are many species that are not often collected but have only been found in seeds (Beeson, 1939, 1941; Browne, 1962). *Coccotrypes* only colonize seeds that have fallen, i.e., seeds that are at least partly exposed or completely bare of fruit tissues. Within seeds, beetles experience similar selective pressures as many ambrosia beetles (e.g., Xyleborini) by inhabiting a "bonanza type" resource that is protectable and may provide ample food for several offspring generations. Hence, this habitat favors the evolution of inbreeding, biased sex ratios, dispersal polymorphism, and advanced social behavior (Hamilton, 1978, 1979), which characterizes *Coccotrypes* (Herfs, 1950; 1959; Gottlieb *et al.*, 2014) and many *Hypothenemus* species (see below) as well as Xyleborini.

3.6.3 Other Seed Breeders

Most *Araptus* species are phloeophagous or myelophagous, but at least 19 species breed in seeds (S. L. Wood, 1982, 2007); half of these are apparently legume seed specialists. Most *Hypothenemus* are highly polyphagous, but a few regularly or most commonly breed in seeds (Beeson, 1941; Browne, 1961; S. L. Wood, 1982, 2007; Atkinson and Equihua-Martínez, 1985c; Chapter 11) and a few species in other genera at least sometimes breed in seeds (see Appendix). In addition, species of *Conophthorus* that breed in developing cones of Pinaceae are also classified as spermatophagous.

3.6.4 Economically Significant Seed Breeders

Only one example of a spermatophagous species attacking fruits still on the plant is known to us: the coffee berry borer, *Hypothenemus hampei* (Ferrari) (Chapter 11). The coffee berry borer is the most serious pest of coffee in most coffee growing countries (LePelley, 1968; Benavides *et al.*, 2005; Jaramillo *et al.*, 2006; Chapter 11). It attacks healthy coffee berries, and breeds in the developing endosperm. This is the only example known to us of scolytines attacking live, attached fruits, and is by far the most economically important spermatophage and the most widely known tropical bark beetle. The congeneric tropical nut borer (*Hypothenemus obscurus* (F.)) is a pest of macadamia in Hawaii and Australia (Jones, 1992; Delate, 1994; Mitchell and Maddox, 2010; Chapter 11). It breeds in both seeds and bark, but primarily breeds in seeds and nuts of a wide variety of plants (S. L. Wood, 1982, 2007).

3.6.5 Cone Breeders

Conophthorus (Chapter 12) have the unique habit of breeding in the developing cones of gymnosperms (Miller, 1915; Lyons, 1956; Chamberlin, 1958; Keen, 1958; Ruckes, 1963; Hedlin *et al.*, 1980; Flores and Bright, 1987; Furniss, 1997). Females bore in from the base of a developing cone, severing the conductive tissues and killing the cone whether or not brood is successfully produced (Ruckes, 1963; Godwin and Odell, 1965; Hedlin *et al.*, 1980). Seed crop loss to *Conophthorus* species can be over 50% (Cognato *et al.*, 2005). Conifer seeds are particularly high in nitrogen, higher than bark (Kramer and Kozlowsky, 1979). *Conophthorus* are relatively host specific; each species breeds in one *Pinus* host, or in a few closely related *Pinus* species (Hedlin *et al.*, 1980; Cognato *et al.*, 2005). *Conophthorus ponderosae*, the one species that is recorded from many pine host species, may be a species complex (Cognato *et al.*, 2005; but see Menard and Cognato, 2007).

Curiously, regular breeding in gymnosperm cones has evolved only once, in North America (Cognato *et al.*, 2005). *Conophthorus* has likely evolved from a *Pityophthorus* ancestor such as the closely related *P. schwerdtfergeri* (Schedl), which breeds in both twigs and cones (Cognato *et al.*, 2005). It should be noted that some *Conophthorus* feed on shoots, e.g., *C. coniperda* Schwarz, especially when all cones are occupied (Morgan and Mailu, 1976), and *C. resinosae* Hopkins both feeds and breeds in shoots as well as cones (McPherson *et al.*, 1970; de Groot and Borden, 1992). Additionally, several *Pityophthorus* species have been collected from cones in North America (Godwin and Odell, 1965). Given these facts, it seems odd that the habit has not also evolved in Eurasian conifer scolytines.

3.7 Mycophagy (Fungus Feeders)

Other than galling (one species), mycophagy is the rarest feeding mode in Scolytinae, known from only two genera in two tribes. At least some species in the rarely collected genus *Trischidias* breed in the fruiting bodies of ascomycete fungus growing in dead twigs or wood (Deyrup, 1987). Similarly, the rare *Lymantor decipiens* (LeConte) (but not other *Lymantor*) is found in dry sapwood with black fungi, upon which they are thought to feed (Swaine, 1918; Blackman, 1922; S. L. Wood, 1982; Kirkendall, unpubl.).

3.8 Breeding in Monocots

Interestingly, there are only a few host-specific phloeophages breeding regularly in the outer tissues of monocots, and there seem to be relatively few records of polyphagous ambrosia beetles breeding in woody monocots. Generally, for bark and ambrosia beetles, the preferred tissues of woody plants are the vascular tissues: cambium plus phloem for phloeophages, and xylem for xylophages and most ambrosia beetles. The vascular tissues taken together constitute a thick cylinder in gymnosperm trees and dicot angiosperms. In monocot angiosperms, xylem and phloem occur together in small bundles scattered in a matrix of nutrient-poor ground tissue. Thus, in monocots, there are no thick rings of relatively favorable tissue for phloeophages as there are in dicots and gymnosperms. It may also be that this radically different distribution of vascular tissues precludes normal phloeophagous gallery construction by bark beetles, and may also hinder normal fungus development in ambrosia beetles. Monocot specialists include few species of *Chramesus* (a genus with phloeophagous and xylophagous species) and of *Corthylus* (ambrosia beetles) that breed in native bamboos in the neotropics (S. L. Wood, 1982, 2007; Atkinson and Equihua-Martínez, 1986b). Otherwise, breeding by non-ambrosia beetle scolytines in monocots is restricted to leaves of yuccas and agaves (species of *Chramesus, Cactopinus, Pseudothysanoes, Hypothenemus*: Atkinson and Equihua-Martínez, 1985a, b, c; Atkinson, 2010) and stems, pseudobulbs, or flowering stalks of bromeliads and orchids (*Chramesus annectans* (Wood), Atkinson *et al.*, 1986; *Tricolus coloreus* Wood, an ambrosia beetle, Wood, 2007; *Xylosandrus* ambrosia beetles, Reitter, 1916, Dekle and Kuitert, 1968, and Dole *et al.*, 2010). In addition, several *Hypothenemus* species and *Chramesus exilis* Wood breed in woody *Smilax* vines (Atkinson and Equihua-Martínez, 1985a, b); *Hypothenemus pubescens* (Hopkins) breeds in the stems of grasses (S. L. Wood, 1982; Atkinson and Peck, 1994). With the exception of the *Hypothenemus* and *Xylosandrus* ambrosia beetles, all of these seem to be monocot specialists, though some are rarely collected, so their true host breadth is not known. Trunks and woody parts of palm leaves are colonized by generalist (polyphagous) ambrosia beetles, but the species richness of ambrosia beetles in palms seems to be much lower than that in dicots in the same forests. Sufficiently large, hard monocot seeds, on the other hand, which have similar structure to those of angiosperms, are readily colonized by both seed specialists and seed generalists.

3.9 Breeding in Live Hosts

Although bark and ambrosia beetles are primarily adapted to colonizing recently dead woody plants, many lineages have evolved to find and breed in living tissues. For species feeding directly on plant tissues (not cultivating fungi), living resources have the advantages of being generally more nutritious than dead tissues, and may have fewer intraspecific and interspecific competitors. Older dead resource units may also have experienced a buildup of predators, parasites, and potentially hostile microbes. The disadvantages of breeding in live resources are that they not only have an array of preformed anatomical and chemical defenses but can also mobilize further physical and chemical weapons. In this section, we present information on Scolytinae and Platypodinae that can tackle living tissues, e.g., wood, seeds, or seedlings. We discuss tree killing, but not the mass attacks on conifers by *Dendroctonus* or *Ips*, which are covered in Chapters 8 and 9, respectively, or in other recent works (Raffa *et al.*, 2008; Kausrud *et al.*, 2011, 2012; Lindgren and Raffa, 2013). We will focus instead on the less well-known instances of bark and ambrosia beetles killing hardwoods or breeding in living plant parts.

Insects breeding in live as opposed to dead plant tissues must adapt to active plant defenses. A clear consequence is that those regularly colonizing living tissues are more host specific than species breeding in the same tissue type but only in dead tissues. *Coccotrypes* and *Hypothenemus*, which breed in seeds, attack seeds of many plant families as long as they are big enough and hard enough (Browne, 1961; Schedl, 1960b, 1961). *Coccotrypes* breeding in mangrove propagules do not breed in any other hosts, or even in branches or trunks of mangrove trees. *Hypothenemus hampei* is the only *Hypothenemus* species that can breed in developing *Coffea* seeds, though many other species have been collected from *Coffea* trees; interestingly, it has been collected from hard seeds and woody pods produced by plants of several different families, but the only live fruits it is known to regularly colonize are those of *Coffea* (Schedl, 1960b, 1961; Vega *et al.*, 2012). A very few ambrosia beetles are known only to attack standing, live trees, and in each case they are unusually host specific. The rare species *Xyleborus vochysiae* Kirkendall has only been collected from one host species (see below), in

contrast to other tropical *Xyleborus*, which usually can be found in dead hosts of several to many different plant families. Three platypodine ambrosia beetles breed exclusively in live trees. The West African *Trachyostus ghanaensis* Schedl breeds only in *Triplochiton scleroxylon* K. Schum. (Sterculiaceae) (Roberts, 1960), while the Malayan *Dendroplatypus impar* (Schedl) breeds only in the certain *Shorea* species (Dipterocarpaceae) (Browne, 1965). The Australian *A. incompertus* is restricted to one genus, *Eucalyptus* (Kent, 2002). As with *Xyleborus*, platypodine ambrosia beetles are usually quite polyphagous. Another West African platypodine, *Doliopygus dubius* (Sampson), is polyphagous when colonizing felled trees and logs, but attacks live (apparently healthy) trees of only one species, *Terminalia superba* Engls. and Diels (Combretaceae) (Browne, 1961). There is one exception to this trend, however. *Corthylus columbianus* Hopkins breeds in live trees, but does not seem to be very host specific (Crozier and Giese, 1967a, b).

3.9.1 Killing Entire Trees

Relatively few bark and ambrosia beetles are able to colonize and kill entire trees, but those that do can have major ecological and economic impacts. Species of *Dendroctonus* (Chapter 8) and *Ips* (Chapter 9), in particular, kill millions of trees each year in North America, Europe, and Asia. Given the worldwide local and regional importance of tree killing by *Dendroctonus* and *Ips*, there is an erroneous but widespread notion that tree killing is by and large restricted to Pinaceae, as reflected in the title of a paper by the Australian forest entomologist Clifford P. Ohmart, who asks "Why are there so few tree-killing bark beetles associated with angiosperms?" (Ohmart, 1989). The article's claim, that the ability to kill trees has only evolved in taxa breeding in Pinaceae, seems to have been accepted uncritically in the few papers citing this work (e.g., Hulcr and Dunn, 2011). Ohmart (1989) argues for a key difference in how angiosperm vs. conifer host trees react physiologically to beetle attack. However, the article is flawed by a bias towards temperate (primarily North American) Scolytinae; this bias is frequently encountered in discussions of bark and ambrosia beetles by forest entomologists (e.g., Stark, 1982). Ohmart's (1989) hypothesis depends on assumptions about differences in temperate vs. tropical scolytine–host tree interactions, but not one article on tropical scolytine biology is cited.

The main tree-killing bark beetle in Europe is *Ips typographus* L., which breeds in spruce (*Picea*), but it is nevertheless not clearly a primary attacker. It mainly kills healthy trees during irregular outbreaks triggered by massive population buildups; otherwise, it kills trees that are highly stressed or attacks recently dead and dying trees

(Berryman, 1982; Kausrud *et al.*, 2012; Chapter 9). *Sphaerotrypes hagedorni* Eggers (Diamerini) can kill its savannah host tree *Anogeissus leiocarpus* (DC.) Guill. and Perr. (Combretaceae), but does so only in the dry season, when trees are water stressed; attacks on living trees in the wet season fail due to active tree defenses, i.e., gum exudation (Roberts, 1969).

A century ago, the hickory bark beetle *Scolytus quadrispinosus* Say was a focus of attention by forest entomologists. It was causing huge losses of hickory timber, particularly trees under moisture stress, in the eastern USA (Schwarz, 1901; Hopkins, 1904, 1908; Blackman, 1924; Blackman and Stage, 1924; Beal and Massey, 1945). During periods of drought, this species kills large tracts of hickory trees in the eastern USA. Normally, it attacks only weakened trees; galleries started in vigorous trees soon fill with sap, and fail (Blackman, 1924; Blackman and Stage, 1924). Felt (1914) and Blackman (1924) used precipitation data to show that significant tree killing only occurred in years with deficiencies of rainfall.

Early in the 20th century, *S. rugulosus* was reported to be regularly killing "large numbers" of scrubby wild plum (*Prunus serotina* Ehrh.), with highest densities on trees injured by ground fires used to clear weeds (Blackman, 1922). Normally, these bark beetles colonize injured branches or trunks, but when numerous they attack healthy hosts (Blackman, 1922; Beal and Massey, 1945). Orchard practices have since changed considerably, and *S. rugulosus* is no longer considered an important pest of *Prunus* fruit trees.

Similarly, the peach bark beetle *Phloeotribus liminaris* (Harris) was studied in the early 1900s because it was damaging and even killing peach, black cherry, wild cherry trees, and mulberry in the northeast USA (Wilson, 1909; Beal and Massey, 1945). Though it was originally collected and described because of its association with "peach yellows" in the 1850s (Harris, 1852), it was not considered an economic problem until the turn of the century, when plantings of peach and cherry had grown (Wilson, 1909). Population buildups due to breeding in slash or windthrown trees can lead to massive attacks on healthy trees during breeding, but normally the main damage is due to gum spotting (gumosis), the result of the tree's reaction to beetles overwintering under the bark in healthy tissues (Beal and Massey, 1945); gum spot defects reduce the veneer value of black cherry by 50–90% (Hanavan *et al.*, 2012). Beetles tunneling in healthy trees usually are either pitched out or killed by the gum reaction (Rexrode, 1982).

These are just a few of many examples of phloeophagous bark beetles locally killing native or ornamental trees, regularly or in occasional outbreaks. A few hardwood examples not yet mentioned include species of *Alniphagus aspericollis* (LeConte) killing alders (Chamberlin, 1958; Borden, 1969); *Dryocoetes betulae* Hopkins killing birches (Hopkins, 1904); four *Phloeotribus* species that can occasionally kill *Prunus* trees (Blackman, 1922; Atkinson and Equihua-Martínez, 1985a; Atkinson *et al.*, 1986); *Scolytodes guyanensis* Schedl killing thousands of mahogany trees "of all sizes" (*Swietenia*) in plantations (Gruner, 1974); *Scolytus ratzeburgi* Jansen killing birches (Tredl, 1915); and *Taphrorhychus villifrons* Dufour killing dwarfed oaks ("nains"; Balachowsky, 1949).

Ambrosia beetles, too, occasionally or regularly attack and kill live hardwood trees. The newly described xyleborine ambrosia beetle *Coptoborus ochromactonus* Smith and Cognato was discovered and named because it was killing large proportions of young trees in commercial balsa plantations in Ecuador (Stilwell *et al.*, 2014). Most mortality occurred in the dry season and to the smallest trees; deaths were attributed to the establishment of the beetles' primary ambrosia fungus, a *Fusarium* (Stilwell *et al.*, 2014). A few ambrosia beetle species such as this one can colonize live trees, though usually hosts are stressed or diseased. If their ambrosia fungus thrives in live trees, when density of attacks is high enough, the fungus's rapid spread in xylem tissues can disable water conduction and effectively throttle the host. In a similar fashion, laurel wilt disease is caused by the symbiotic *Raffaelea* fungus of the Asian ambrosia beetle *Xyleborus glabratus* Eichhoff, which is called the redbay ambrosia beetle in the USA. Laurel wilt disease is killing thousands of mature forest, ornamental, and plantation trees in the family Lauraceae (particularly redbay *Persea borbonia* (L.) Spreng. and sassafras, *Sassafras albidum* (Nutt.) Nees) and is a potential threat to two endangered species and to the southeastern US avocado industry (Fraedrich *et al.*, 2008; Hanula *et al.*, 2008).

Other examples of ambrosia beetles killing hardwoods include *Xylosandrus germanus* (Blandford) (oaks: Heidenreich, 1960); *Xyloterinus politus* Say (birches: Schwartz, 1891); *Euplatypus parallelus* (F.) (Beaver, 2013); *Platypus quercivorus* (Murayama) (oaks: Kamata *et al.*, 2002); *Platypus subgranosus* Schedl (*Nothofagus*: Howard, 1973); and *Euplatypus hintzi* (Schaufuss) (*Eucalyptus* in plantations: Roberts, 1969).

A few examples of gymnosperms being killed by otherwise innocuous species include *Pseudohylesinus grandis* Swaine, which normally breeds in weakened or dying Douglas-fir but occasionally attacks and kills "a considerable quantity of young timber" (Chamberlin, 1918); and *Phloeosinus rubundicollis* Swaine, which has been observed killing thousands of ornamental *Chamaecyparis* (Chamberlin, 1958).

Some species that are considered harmless in their native ecosystems ("secondary") become deadly when introduced to naive forests (Kühnholz *et al.*, 2001; Ploetz *et al.*, 2013). *Dendroctonus valens* LeConte females breed singly or in small numbers at the bases and in the roots of pines, and attacks by this species have no impact on trees in their native forests in North America. Meanwhile, in China, where the species has recently become established, it kills thousands of pines each year (Yan *et al.*, 2005; Sun *et al.*, 2013; Chapter 8). Similarly, the secondary North American bark beetle *Ips grandicollis* (Eichhoff) is a lethal pest of exotic *Pinus resinosa* in plantations in Australia (Morgan, 1967).

Considering these examples, it is important to be cautious in concluding that only certain bark beetle species have evolved to kill trees, or that bark and ambrosia beetles only kill Pinaceae (Ohmart, 1989). While a handful of notorious *Dendroctonus* species are specialists at tree killing, there is a continuum of aggressiveness in Scolytinae and Platypodinae, from species that only breed in live tissues to species that come to a tree months after its death. Many species can and do kill their hosts under the right conditions, even perfectly healthy individuals. In many of the examples cited above, the individual trees that were killed were known or suspected to be stressed. The point is, however, that these trees would likely have survived had the above-mentioned bark or ambrosia beetles not colonized them.

While the greatest ecological and economic impacts of tree killing are by *Dendroctonus* species in low diversity, widespread conifer forests, there is a large and growing number of instances of serious tree pathogens vectored by Scolytinae and Platypodinae in forests around the world, primarily involving angiosperm hosts (Hulcr and Dunn, 2011; Ploetz *et al.*, 2013). There has been considerable research into a few examples, such as Dutch elm disease, vectored in North America by both the native elm bark beetle *Hylurgopinus rufipes* (Eichhoff) and the invasive *Scolytus multistriatus* (Marsham), and in Europe by several native species of *Scolytus*. Other cases, many of which are only recently documented, are just beginning to be investigated (Hulcr and Dunn, 2011; Ploetz *et al.*, 2013). The impacts of attacks by these beetle–fungus partnerships vary from mild economic losses due to wood discoloration to major ecological and economic consequences due to massive tree mortality, mainly mortality of angiosperms, *contra* Ohmart (1989).

3.9.2 Killing Plant Parts, Seedlings, and Seeds

Much less appreciated are the impacts of perhaps hundreds of species, which affect live host plants in more subtle ways, by killing branches or twigs, patches of bark, seedlings, or seeds (Blackman 1922; Beeson, 1941; Chamberlin 1958; Browne, 1961; S. L. Wood, 1982; Postner, 1974). These bark beetles can nonetheless significantly reduce the growth and reproduction of their hosts and repeated branch killing can lead to death of entire trees.

A number of phloeophagous Scolytinae have been described as progressive branch killers. Several ash bark beetle species (*Hylesinus*) kill branches year after year,

eventually moving onto the trunk, perhaps because branch losses have crippled the tree's defenses (Doane, 1923; McKnight and Aarhus, 1973; Postner, 1974; Gast *et al.*, 1989). Progressive branch killing has also been reported for *Hylesinus oleiperda* F. in olive trees and ashes (Postner, 1974; Graf, 1977), *S. ratzeburgi* in birches (Tredl, 1915), and *Pityophthorus costatulus* Wood in *Thevetia* (Apocynaceae) (Atkinson *et al.*, 1986a), to give just a few examples.

A few bark beetles that attack branches have been researched because the damage they cause is of sufficient economic import to warrant attention. The ambrosia beetle known now as the black twig borer, *Xylosandrus compactus* (Eichhoff), is well known as a pest of coffee and cocoa in West Africa, and tea in Asia (Kalshoven, 1925; Brader, 1964; Kaneko *et al.*, 1965; Entwistle, 1972), and where introduced is a pest of a wide variety of ornamental and native trees (Kalshoven, 1958; Browne, 1961, 1968; Beaver, 1988; Chong *et al.*, 2009). The adults bore into healthy young stems, branches or twigs; concentrated attacks can lead to death of the plant (Brader, 1964). Sadly for coffee aficionados, the black twig borer is a major impediment to coffee production in the Kona region of Hawaii (Greco and Wright, 2013.)

A palearctic phloeophagous species reproducing harmlessly in trunks of dead or dying pines is *Tomicus piniperda* (L.) (Chapter 10). Like most other *Tomicus* (Kirkendall *et al.*, 2008), its impact is due not to its breeding habits, but rather to the behavior of recently emerged young adults, which feed in the pith of healthy tree tops and branch tips (maturation feeding), killing them (Chararas, 1962; Långström, 1983; Långström and Hellqvist, 1993; Amezaga, 1997). Shoot pruning by *T. piniperda* in Nordic pine forests has been estimated to reduce forest productivity by up to 45% of the annual volume growth (Eidmann, 1992). Maturation feeding is especially intense in the Chinese *Tomicus yunnanensis* Kirkendall and Faccoli, and trees are so weakened by it that they can later be attacked and killed by this species (Ye, 1997; Ye and Ding, 1999; Lieutier *et al.*, 2003).

A number of phloeophagous and myelophagous species have evolved to breed in small plant parts incapable of defending themselves (such as the *Tomicus* described above). These species either tolerate local host defenses, or can mechanically disable or overwhelm them. Whether or not microbes are an important weapon (as they clearly are in almost all tree killers) is not generally known but is to be expected.

Pityophthorus puberulus (LeConte) offers an example of apparent tolerance. Females breeding in terminal twigs can be seen to be practically swimming in resin, and use a mixture of frass and resin to plug the entrance (males being absent in this parthenogenetic species) (Deyrup and Kirkendall, 1983; Kirkendall, unpubl.). At least several

Pityophthorus species originally described in *Myeloborus* seem to have the same biology, breeding in and killing pitchy twigs of pine trees (Blackman, 1928).

The monogynous ambrosia beetle *Corthylus punctatissimus* (Zimmermann) girdles stems and roots of saplings of a wide variety of angiosperm trees in eastern North America (Merriam, 1883; Schwarz, 1891; Roeper *et al.*, 1987a, b). At high population densities, such girdling could have significant ecological effects: as recounted by Merriam (1883): "...in Lewis county [New York, USA] alone hundreds of thousands of young sugar maples perished from the ravages of this Scolytid during the summer of 1882." However, Schwarz (1891) commented that *C. punctatissimus* pairs destroy the underground stems but not the roots, and that plants later re-sprout. Regardless, the loss of a significant amount of biomass at such a young stage must severely affect plant fitness.

Anisandrus dispar (F.) girdles and kills branches and young trees in fruit tree orchards in the USA, where it is introduced (Hubbard, 1897). It is likely that there are other ambrosia beetles with similar behavior.

Conophthorus females bore in from the base of a developing pinecone and girdle it, cutting the conductive tissues and killing the cone whether or not a brood is successfully produced (Hedlin *et al.*, 1980; Mattson, 1980). After girdling the cone, they tunnel in a more or less straight line along the cone axis.

Curiously, unlike with other wood borers, there seem to be few species, which have been recorded as girdling branches, twigs, or the stems of seedlings or saplings. Girdling not only disables plant defenses, but it also alters physical and nutritional qualities of the resource (Forcella, 1982; Dussourd and Eisner, 1987; Hanks, 1999). Girdling is a widely used strategy in Cerambycidae (Forcella, 1982; Ferro *et al.*, 2009) and in addition to mitigating plant defenses such as sap flow (e.g., *Sthenias grisator*: Duffy, 1968), girdling may alter favorably the nutritional quality of the girdled twig by trapping and concentrating nutrients normally transported from the leaves (Forcella, 1982). Interestingly, Forcella (1982) reports that the cerambycid *Oncideres cingulata* (Say) cuts phloem tissues when girdling, but not xylem, so parts distal to the girdle remain alive. To our knowledge, nobody has investigated this behavior in bark and ambrosia beetles to determine if girdling concentrates nutrients, or simply disarms plant defenses (Dussourd and Eisner, 1987; Hanks, 1999).

On the surface, it would seem that girdling by cerambycids and scolytines are not analogous, in that cerambycids girdle a branch or twig first and oviposit distal to the girdle afterwards, while the girdling of scolytines is primarily during egg gallery construction and goes on over days. Indeed, that scolytines girdle small diameter breeding material in the course of constructing egg tunnels may simply be the optimal behavior for spacing of offspring in the resource medium. Nevertheless, the girdling benefits mentioned above are substantial, and could select for such behavior in scolytines: there are species of *Carphobius* and *Thysanoes* that seem to be specialized to breeding in twigs and branches girdled by cerambycids (S. L. Wood, 1982), suggesting that scolytines reap the same girdling benefits as do longhorn beetles. Depending on the temperature and the size of the beetle with respect to the diameter of the host material, a tunnel that completely severs phloem tissues (the first 360-degree turn) might take only a day or two to complete. It seems clear that girdling is an adaptive strategy in at least *Conophthorus*. If girdling is more than incidental in, for example, twig-breeding species or species breeding in herbaceous stems or vines, we would expect to see that spiraling tunnel construction is always outwards from the initial spot of entry (as described for *Xylocleptes bispinus* (F.) in *Clematis* vines: Lövendal, 1898), while it would be random if girdling was not important.

Herbiphagy is relatively rare in Scolytinae, but many herbiphagous species do attack live plants (see Section 3.4). Attacks on stems can kill the plants.

Spermatophagous species (Section 3.6) usually kill the live seeds in which they breed, and may well have significant impacts on regeneration of certain host trees (Janzen, 1971, 1972; Wood, 2007). Palm seed mortality due to *Coccotrypes* can be up to 100%, though it varies much from place to place and year to year (Janzen, 1972; Kirkendall, unpubl.). Other *Coccotrypes* species breed in and often kill live seedlings (the viviparous propagules) of mangrove trees, affecting mangrove forest communities (Sousa *et al.*, 2003; see Section 3.4).

3.9.3 Breeding in Live Plant Parts without Causing much Damage

In exceptional cases, bark and ambrosia beetles breed in live plants seemingly with little or no damage to the host. Two unique examples can be found in the large neotropical genus *Scolytodes*, both in Costa Rica; both were discovered by the extraordinary young naturalist Kenji Nishida, who was then doing his Master's research at the University of Costa Rica. *Scolytodes ageratinae* Wood galls a small, high elevation herbaceous plant, *Ageratina* cf. *ixiocladon* (Asteraceae) (Nishida, pers. commun.); galled plants seem otherwise healthy, but may have lower fitness than ungalled. No other galling Scolytinae are known anywhere in the world. The congener *Scolytodes gunnerae* Wood breeds in the leaf petioles of two montane *Gunnera* species. The plants, known locally as poor man's umbrella (*la sombrilla de pobre*), have extremely large, rounded leaves 1–2 m in diameter and sprout in a whorl from a very short central stem. The beetles breed in irregular cave-type galleries in the several-cm-thick, fleshy petioles of healthy leaves (Figure 3.6). Old tunnels heal over, and though plant fitness has not been measured, the large leaves seem unaffected by the presence of a few small bark beetle galleries, and plants with colonized leaves seem to flower and fruit normally (Kirkendall, unpubl.). Again, this feeding mode, i.e., breeding in the fleshy petioles of large leaves, was totally unexpected and is unique to *S. gunnerae*.

In addition, a handful of ambrosia beetles tunnel in the wood of healthy live trees. *Xyleborus vochysiae* Kirkendall is a large inbreeding ambrosia beetle that has only been observed to colonize standing live *Vochysia ferruginea* Mart. (Vochysiaceae) in Costa Rica (Kirkendall, 2006). About three-quarters of the standing trees in a 7-year-old plantation were attacked (but multiple felled trees were not), and almost every tree surveyed in a 20-year-old secondary forest had the characteristic entry holes of this species, although it appeared that most attempted colonizations had failed. The interaction between the beetles and their host plants was not studied, but there were no signs of wilting or loss of leaves in the affected trees as might be the case if they were vectoring an aggressive fungus. This rare species has only been collected from this one host species, in contrast to other tropical *Xyleborus*, which usually can be found in hosts of several to many different plant families.

Corthylus columbianus is a common ambrosia beetle species in hardwood forests of eastern North America (S. L. Wood, 1982; Majka *et al.*, 2007), where it breeds in trunks of healthy, vigorous trees. Hosts appear to be unaffected, and old beetle entrance tunnels are gradually covered over by secondary tree growth. Fungal staining from old tunnel systems remains in the wood, making possible the study of historical distributions and population density fluctuations (Crozier and Giese, 1967b; McManus and Giese, 1968; Milne and Giese, 1969). Interestingly, a different *Corthylus* with similar biology does kill its host trees. *Corthylus zulmae* Wood breeds in the trunks of live native alders (*Alnus acuminata* Kunth; Betulaceae) in plantations in Colombia (Gil *et al.*, 2004; Jaramillo *et al.*, 2011). Fungi associated with this species seem to be responsible for tree death. Their biologies being so similar, the lack of harm caused by *C. columbianus* must be due to the low virulence of its ambrosial fungus.

In most of the examples of Scolytinae or Platypodinae breeding in live trees there is little damage to the tree itself, though the value as a timber resource may be reduced. However, the tunneling of *Megaplatypus mutatus* (Chapuis) in the trunks of various hardwoods can weaken the structural integrity of its hosts to result in stem breakage and mortality, and it is considered a pest of plantations (Santoro, 1963; Giménez and Etiennot, 2003; Girardi *et al.*, 2006; Alfaro *et al.*, 2007; Zanuncio *et al.*, 2010).

4. EVOLUTIONARY ECOLOGY OF REPRODUCTIVE BEHAVIOR

Bark and ambrosia beetles do not dazzle the eye as do longhorn and jewel beetles, or please the ear as do crickets and katydids, but few if any insect groups exhibit such an intriguing variety of reproductive behavior as do bark and ambrosia beetles (Kirkendall, 1983, 1993; Kirkendall *et al.*, 1997; Costa, 2010). In most insects, males leave females immediately or soon after copulation; in most bark beetle species, males remain with females in their tunnel systems until most or all eggs have been laid. Only a few examples are known where males do not join females in galleries and remain for at least a week or more. Most insects, and most bark and ambrosia beetles, outbreed, and the dangers of inbreeding are well documented; nonetheless, species reproducing by brother/sister mating are widespread and abundant, and have been mating incestuously for tens of millions of years. Outbreeding taxa vary in how the two sexes meet (mate location), how long males stay with females (male residency), and with how many females individual males are mated simultaneously (mating systems). Among outbreeders we find male/female pairs (which in some species mate for life), males with harems, and numerous instances of bigyny, i.e., species in which males nearly always mate with exactly two females, a mating system virtually unheard of outside of Scolytinae. There are also four forms of parthenogenesis (clonal reproduction) in this group: thelytoky, in which females produce only daughters; pseudogamy (also known as gynogenesis), in which females mate with males but produce only daughters, and only the mother's genes are passed on to offspring; arrhenotoky, in which daughters are formed sexually and are diploid, but sons are produced by the hatching of unfertilized eggs and are haploid; and pseudoarrhenotoky, or paternal genome elimination, in which daughters are formed sexually and are diploid, and males arise from fertilized eggs but express and pass on only genes from their mothers.

4.1 Mating Behavior

4.1.1 Fighting

Newly arriving conspecifics are easily repelled by bark and ambrosia beetles ensconced in tunnel entrances. Physical combat between members of the same sex takes place primarily early in the colonization phase, usually while a member of the pioneering sex is beginning to tunnel or shortly after pairs have formed (Blackman, 1931; Goeden and Norris, 1965; Fockler and Borden, 1972; Salonen, 1973; Beaver, 1976; Petty, 1977; Vernoff and Rudinsky, 1980; Kirkendall, 1983; Swedenborg *et al.*, 1988, 1989; Jordal, 2006; Smith and Cognato, 2011). Wandering males will also try to enter active gallery systems, but are blocked from entering by resident males (McGehey, 1968; Oester and Rudinsky, 1975; Rudinsky and Ryker, 1976; Oester *et al.*, 1978, 1981). Only rarely do intruding males succeed in replacing males already in tunnels (Vernoff and Rudinsky, 1980). Male/male competition is common in female-initiated mating systems (such as in *Tomicus*, *Dendroctonus*, *Pseudohylesinus*, or *Scolytus*) but females have been observed fighting in male-initiated mating systems (Nord, 1972).

4.1.2 Courtship

Courtship in both Scolytinae and Platypodinae takes place with both individuals facing forward, so physical interactions during courtship are between the front of the courting individual and the back end of the courted one. Ancestrally, males court females, as is the general rule in insects and other arthropods (and indeed in animals as a whole). However, females court males in all known cases of harem polygyny and in some monogynous species as well, such as in all Platypodinae, monogynous species of *Scolytodes*, and the monogynous genera of Corthylini; it has been hypothesized that, for most cases, monogynous species with such sex role reversal are likely derived from harem polygynous lineages (Kirkendall, 1983).

Acoustic communication is a key component of intersexual selection during courtship, but may not always be sufficient by itself for species discrimination (Lewis and Cane, 1992). It appears that almost all Scolytinae and Platypodinae stridulate (Barr, 1969; Sasakawa and Yoshiyasu, 1983; Lyal and King, 1996), though stridulation has been secondarily lost in some species (Barr, 1969; Sasakawa and Yoshiyasu, 1983). Stridulation at the entrance to or inside the gallery system is a key component of courtship in Scolytinae (Barr, 1969; Swaby and Rudinsky, 1976; Rudinsky *et al.*, 1978; Rudinsky, 1979; Rudinsky and Vallo, 1979; Oester *et al.*, 1981; Ryker, 1984; Garraway, 1986; Ytsma, 1988; Swedenborg *et al.*, 1989; Lewis and Cane, 1992; Ohya and Kinuura, 2001), and in Platypodinae (Chapman, 1870; Ytsma, 1988; Ohya and Kinuura, 2001; Kobayashi and Ueda, 2002). Stridulation is also used in male/male and female/female competition (Rudinsky and Michael, 1974; Rudinsky, 1976; Swaby and Rudinsky, 1976; Oester and Rudinsky, 1979; Rudinsky and Vallo, 1979; Oester *et al.*, 1981; Swedenborg *et al.*, 1988, 1989) and when predators attempt to enter a gallery system (Roberts, 1960); Wood (2007) reports that *Dendrosinus bourreriae* Schwarz adults working under bark in a branch "buzzed" for several minutes when the branch was disturbed, sounding like a nest of bees had been disturbed.

Courtship involves an interaction between acoustic and chemical communication (Rudinsky *et al.*, 1976; Rudinsky,

1979), and where it has been studied in detail, courtship behavior also may include bumping (frons to declivity), antennal tapping or drumming on the declivity, brushing of antennae or the antennal scape setae against the elytra, and mandibular gnawing (Blackman and Stage, 1924; Petty, 1977; Oester *et al.*, 1981; Swedenborg *et al.*, 1988; Jordal, 2006; Smith and Cognato, 2011). In the platypodine *Doliopygus conradti* Strohmeyer, females and males engage in a "tug-of-war," where females attempt to pull males out of newly started tunnels with their mandibles and males resist; if they ultimately succeed, the female can then enter the gallery, and mating takes place with the male on the surface and the tip of the female's abdomen protruding from the entrance (Browne, 1962). In a similar fashion, courting females tug on male *Platypus quercivorus* Murayama (Ohya and Kinuura, 2001), so this behavior may be common in Platypodinae.

Besides the tactile components of bumping, brushing, stroking, and other rhythmic forms of physical contact between males and females during courtship, there is likely an olfactory or "taste" component as well: though little investigated in bark and ambrosia beetles, interspecific differences in cuticular hydrocarbons are important in species recognition in other insects (Singer, 1998; Howard and Blomquist, 2005) and such differences have been found when looked for in bark and ambrosia beetles (Page *et al.*, 1990a, b, 1997; Sullivan *et al.*, 2012).

Although courtship mostly occurs at or in the entrance or nuptial chamber, for at least some Scolytinae, mating can also occur during pre-dispersal feeding in the previous year's breeding material, hibernating sites, or feeding tunnels in branches or twigs (Kirkendall, 1993; McNee *et al.*, 2000). Although it is likely that courtship patterns, including which sex courts, are similar to those that occur around or in gallery systems of the same species, nothing is known about mating behavior before dispersal and colonization of fresh breeding material.

4.1.3 Copulation

Females of at least outbreeding Scolytinae and Platypodinae copulate more than once, even if with the same individual male. Evidence comes from both watching individuals in nature and observing beetles in semi-natural conditions such as thick sheets of bark between plates of glass. Many authors have reported that bark and ambrosia beetles mate repeatedly during gallery construction (Gossard, 1913; Blackman and Stage, 1924; Doane and Gilliland, 1929; Hadorn, 1933; Hansen, 1956; Reid, 1958; Gouger *et al.*, 1975; Petty, 1977; Garraway, 1986). In some cases, copulation seems to be restricted to the period when females are still on or near the surface or only in the early stages of oviposition (Hadorn, 1933; Gouger *et al.*, 1975; Campobasso *et al.*, 2004).

Copulations themselves are brief, lasting from 10 seconds to a few minutes at most (Blackman and Stage, 1924; Hadorn, 1933; Reid, 1958; Gouger *et al.*, 1975; Garraway, 1986). In the mountain pine beetle (*Dendroctonus ponderosae* Hopkins), coupling lasts 10–60 seconds and is repeated about once per day, and less frequently after egg laying commences (Reid, 1958). For two species of harem polygynous *Ips*, Garraway (1986) reports that copulation takes ca. 10 seconds, and that females beginning oviposition are mated "frequently." In *I. avulsus*, copulation averages 35 seconds and females mated three times at 10-minute intervals, after which the female walled herself off from the nuptial chamber with tightly packed frass in the egg arm (Gouger *et al.*, 1975).

Platypodinae presumably mate only in the earliest stages of tunneling; copulation is probably not possible inside the gallery system, and takes place with the male on the bark surface and the female in the tunnel entrance. Courtship and copulation in Platypodinae is described and illustrated in Jover (1952). There is no nuptial chamber in the tunnel systems of these ambrosia beetles, and copulation is accomplished by the male exiting the tunnel entrance and allowing the courting female to enter, then copulating with the male on the surface and the female in the tunnel entrance. No deviations from this general pattern have been reported for Platypodinae.

4.1.4 Repeated Mating: the Key to Evolution of Prolonged Male Residency?

The fact that females are receptive during part or most of the egg laying period provides an explanation for the evolution of mate guarding, and ultimately of male residence. Lissemore (1997) attributed male residency in *Ips pini* (Say) to the need for males to copulate repeatedly with females in order to fully displace sperm from previous matings. Many *Ips* females joining males already have sperm stored in their spermathecae; in such cases, Lissemore (1997) found that males require about 5–7 days of repeated copulations to attain near-complete paternity. Repeated copulations may function generally as paternity assurance: half of all *T. piniperda* females colonizing breeding material have been inseminated during the previous year's maturation feeding in shoots or while overwintering at the bases of trees (Janin and Lieutier, 1988), and in an Israeli population Mendel (1983) found nearly all females of *Orthotomicus erosus* (Wollaston) had been inseminated after overwintering in dense aggregations. Much lower levels of pre-colonization insemination are probably more usual (reviewed by Kirkendall, 1993; Bleiker *et al.*, 2013), but there is a clear potential for sperm competition in many Scolytinae, and the evolutionary response has been repeated copulation. Mating prior to gallery system construction may, however, be largely confined to species that are not ambrosia beetles, species in which aggregations of young adults occur during fall maturation feeding or in overwintering clusters before young adults emerge and disperse. The importance of this distinction will become apparent in our discussion of the evolution of alloparental care (Section 5.3).

We have discussed repeated copulation from the point of view of males. From a female point of view, repeated copulations (continuous sexual receptivity) may be an adaptation for extending male residency, thus gaining the benefits of male burrow blocking and frass removal. But it also may increase the fitness of her offspring by diluting and eventually replacing sperm from pre-dispersal matings; this becomes an advantage when some proportion of early matings are with relatives, and using early sperm then produces offspring with inbreeding depression.

Thus, males that stay in order to mate repeatedly with the same female gain offspring through increased paternity as well as increased female oviposition rates, while females gain in fecundity (as long as males remove frass) and produce outbred offspring. It is a short step from males staying long enough to ensure maximum paternity to the evolution of paternal care (Section 5.3).

4.2 Mating Systems

Most bark beetles outbreed, as do most insects, but both regular inbreeding and parthenogenesis (clonal reproduction) have evolved in Scolytinae. Outbreeding taxa vary in how the two sexes meet (mate location), how long males stay with females (male residency), and with how many females individual males are mated simultaneously (mating systems).

Mating system diversity and evolution has been reviewed by Kirkendall (1983; see also Kirkendall, 1993). Outbreeding bark and ambrosia mating systems are classified by how many females breed simultaneously with the same male: one, monogyny; regularly two, bigyny; several to many, harem polygyny. For consistency, inbreeding is referred to as inbreeding polygyny, when classifying mating systems based on numbers of females (Kirkendall, 1983). In a handful of species, it appears that both multiple males and multiple females are in contact and mating is indiscriminate: these systems are referred to as colonial mating or polygamy.

Another mating system factor is male residency, how long males remain with females after copulation. Males do stay with females in most species. The species in which males do not stay for an appreciable amount of time are scattered among four unrelated tribes (Hylesinini, Diamerini, Scolytini, and Corthylini (subtribe Pityophthorina) (Kirkendall, 1983). We will treat male residency in Section 5, where we discuss it in the contexts of the evolution of subsocial behavior and paternal care. A detailed overview of variation in how long males remain in gallery systems can be found in Kirkendall (1983), and arguments for the evolution of prolonged male residency are developed in that review and in Kirkendall (1993), and in Section 5.

Generally, the pioneering sex initiates tunneling in fresh breeding material, and is located by the following sex; the second-arriving sex is attracted either to host odors, odors from boring dust, pheromones, or a combination. Members of the pioneer sex are also attracted, which often results in densely colonized host material. In the vast majority of species, males stay for a week or more, guarding the entrance and removing frass; commonly, males depart near or after females have finished ovipositing, and they may even die in the tunnel system (Kirkendall, 1983).

4.2.1 Monogyny

The ancestral mating system for Scolytinae is almost certainly female-initiated monogyny, and it is still the predominant mating system in bark and ambrosia beetles (Kirkendall, 1983; Figure 3.4). Nearly half of all genera have monogynous species, and nearly all tribes (Table 3.3), and most of these (especially in more basal lineages) are female initiated.

Male-initiated monogyny is the rule in Platypodinae, but rare in Scolytinae. In Bothrosternini, it is found at least in pith-breeding species, in *Sternobothrus* and certain *Cnesinus* (Beaver, 1973). The sex initiating mating is not known for most species in the tribe, but it does seem that all species are monogynous (Kirkendall, 1983). Monogynous species of *Scolytodes* (a genus with both monogyny and harem polygyny) are male initiated (Kirkendall, 1983; Kirkendall, pers. observ., *Scolytodes* species in *Cecropia* petioles). The remaining examples are from the Corthylini, a tribe with both monogynous and harem polygynous genera. As far as is known, almost all Corthylini are male initiated, including the monogynous genera, both those that are phloeophagous and those that are xylomycetophagous (Kirkendall, 1983). Exceptions occur in the large phloeophagous genus *Pityophthorus*, where female initiation may have re-evolved in a few twig breeders; cone beetles in the close related genus *Conophthorus* are also monogynous and female initiated (see next subsection).

As far as is known, without exception, Platypodinae are monogynous, and males seek out new host material and initiate tunnel construction (Jover, 1952; Kalshoven, 1960b; Browne, 1961; Kirkendall, 1983). That almost all Platypodinae are male-initiated monogynous species suggests that, once evolved, male initiation is evolutionarily stable (Kirkendall, 1983). Details of mating systems are not known for *Mecopelmus*, which is phloeophagous, and *Schedlarius*, which breeds in fungus-rotted wood of *Bursera* (Wood, 1993). Jover (1952) describes the outcome (apparently with several platypodine species) of introducing a second female to tunnel systems occupied by mated pairs. These females were accepted by the male, but the second female soon abandoned the gallery and left. His observations suggest that monogyny in Platypodinae may be maintained by the decisions of secondary females, rather than by any active resistance by males or primary females, but it would be informative to see if similar experiments confirmed these briefly reported results.

Male-initiated monogyny in Scolytinae tends to occur in species or genera that otherwise are dominated by harem polygyny (Kirkendall, 1983). These species breed in resources where more than one female cannot breed simultaneously without dramatic larval mortality due to intrabrood competition; hence, it is not advantageous for females to join already-mated males.

4.2.2 Reversions to Female-initiated Mating Systems

Kirkendall (1983) argued that colonization by males should be a stable strategy, especially when sex attractant pheromones are involved. Females coming to already established males avoid considerable risks and time investment associated with finding usable host material; when they can enter tunnels immediately, they also reduce their risk of being consumed by surface-active predators such as checkered beetles (Cleridae), ants, and foraging birds. Nonetheless, reversion to female colonization has occurred in cone beetles (*Conophthorus*) and in a few twig-breeding *Pityophthorus* species, both corthylines. In almost all Corthylini, males initiate gallery construction. Cone beetles

TABLE 3.3 Number of Scolyinae Genera and Tribes with at Least One Species having the Given Mating System (247 total genera, 26 total tribes)

Number of Taxa with at Least One Species	MG	BG	HP	Col	Inbr	?
Genera	118	19	27	3	54	45
Tribes	24	8	8	2	9	17

Some genera and tribes are represented in more than one category. MG=monogyny; BG=bigyny; HP=harem polygyny; Col=colonial polygyny (polygamy); Inbr=inbreeding; "?"=mating system unknown. Data from Appendix.

are a monophyletic corthyline group similar to *Pityophthorus* genetically, morphologically, and in pheromone components (pityol, conophthorin) (S. L. Wood, 1982; Dallara *et al.*, 2000; Rappaport *et al.*, 2000; Cognato *et al.*, 2005; *Conophthorus* biology is also discussed in Section 3.6). An example from *Pityophthorus* is *P. pubescens* (Marsham). Most *Pityophthorus* are harem polygynous phloeophages in branches or trunks of hardwoods and conifers, and are distributed around the world; males initiate gallery systems, and where known, produce attractant pheromones. *Pityophthorus pubescens* is one of several twig breeders that have reverted to monogyny, and in this species females initiate gallery construction and also produce a sex pheromone (López *et al.*, 2013).

What these species seem to have in common is that females spread their oviposition among many host resource units, rather than putting a large number of eggs in one resource over a long period of time as is the case in most bark and ambrosia beetles. Perhaps the short female residency time reduces advantages to males of staying with females, which in turn leads to females needing to initiate at least subsequent galleries alone. Once that behavior is in place, it is possible for female initiation of the first egg tunnel to evolve, though it is not clear what balance of selective forces would lead to its evolution.

Females also colonize in parthenogenetic (thelytykous) species (*Pityophthorus puberulus* (LeConte): Deyrup and Kirkendall, 1983) and of course in inbreeders (since males do not disperse), including lineages likely derived from outbreeders with male initiation such as the *Araptus laevigatus* Wood complex. In *Pityophthorus* and *Araptus*, this may have evolved either after female initiation re-evolved, or directly from male initiation (which predominates in these genera and their allies).

4.2.3 Bigyny

Regular bigyny has evolved repeatedly in Scolytinae, from both harem polygynous and monogynous ancestors. Systems in which males regularly have two females are found in 19 different genera, in eight tribes; seven genera are from the Micracidini, in which most genera are bigynous. Several otherwise monogynous genera have one or a few species that are bigynous. In the Phloeosinini, bigynous species are found in two otherwise monogynous and (mainly) phloeophagous genera, *Phloeosinus* and *Chramesus*. *Chramesus* has the bigynous species *C. incomptus* Wood, which makes biramous diagonal galleries in *Clematis* shrub stems (S. L. Wood, 1982).

We can find no references to regular, simultaneous bigyny in other animals. In fish and birds, at least, occasional bigyny in monogynous species seems to occur when male territories are of sufficient size and quality to overlap territories of two females. In such cases, most males are monogynous, and a few (in fish, usually larger males) are bigynous. That regular bigyny is only known from Scolytinae must, then, be related to geometric constraints on egg tunnel construction (situations that force tunnels to diverge at 180°). But this does not explain why the vast majority of bark beetles with transverse or longitudinal biramous tunnel systems (i.e., systems in which the two tunnels do diverge at or nearly at 180°) remain monogynous or are only occasionally, not regularly, bigynous.

That bigynous species rarely have more than two females is easier to understand. When egg tunnels are constrained to run either parallel to the wood grain or perpendicular to it, adding the work of a new female means adding a tunnel parallel to, and close by, the tunnel of another ovipositing female, which should result in extremely high larval mortality in species where larvae must tunnel long distances to acquire enough resources (Kirkendall, 1984; Løyning and Kirkendall, 1999). In most such situations, females should be selected to avoid joining bygynous systems.

By the same reasoning, for species where egg tunnels are all longitudinal or all transverse, polygyny is only possible where resource quality is high and larval tunnels correspondingly short; in such cases, most males have four (maximum of two parallel arms running in one direction) or less females. In cases where females join males with four females, the joining female(s) will suffer large losses of offspring due to competition (Schlyter and Zhang, 1996; Latty *et al.*, 2009; Kirkendall, 1989). This constraint on harem size is weak or nonexistent in harem polygynous species producing star-shaped gallery systems with long egg tunnels, however; egg tunnels diverge more and more, as they progress, steadily reducing intraharem competition for resources. Star-shaped systems are especially common in genera such as *Pityophthorus*, *Scolytodes*, *Pityogenes*, *Pityokteines*, and *Polygraphus*.

4.2.4 Harem Polygyny

Simultaneous polygyny (harem polygyny and bigyny) has evolved only sporadically in more basal taxa (Figure 3.7; reviewed in detail in Kirkendall, 1983). Altogether, 39 genera in 11 tribes have species that are harem polygynous or bigynous (Appendix). Based on Figure 3.7, it appears that polygyny has evolved at least 12 times in Scolytinae; the number of independent origins is certainly higher, given that there are multiple occurrences of polygyny in each of the predominantly monogynous genera *Scolytus* and *Phloeosinus*, and at least some of the polygynous species are not related to other polygynous species in the same genus. Harem polygyny is found in 26 genera in eight tribes. It is the predominant mating system in Ipini, and common in Corthylini and Polygraphini.

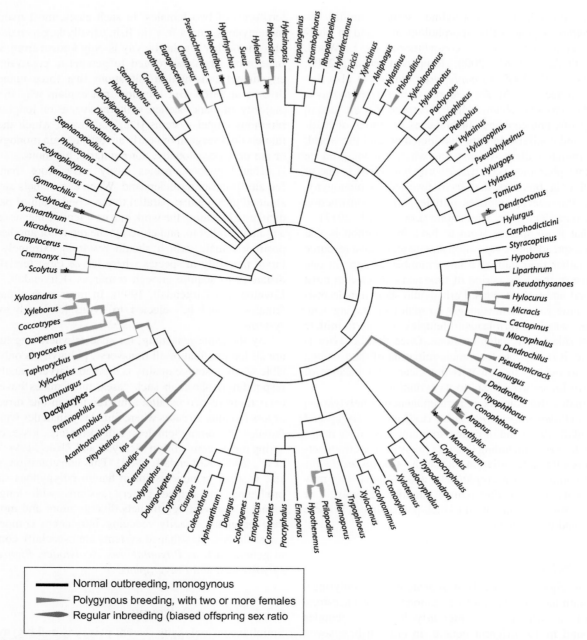

FIGURE 3.7   Phylogenetic tree of Scolytinae with mating systems indicated (see inset legend). Harem polygyny includes bygyny. Stars indicate genera or lineages in which the mating system is rare (one or just a few species).

Harem polygyny is relatively rare in animals. In bark and ambrosia beetles, polygyny takes the form of resource defense polygyny, where males accrue multiple mates because they control critical breeding resources capable of supporting the reproduction of more than one female (Emlen and Oring, 1977; see also Searcy and Yasukawa, 1989). The key question in polygynous mating systems is why females join already mated males, if unmated males are available. Females joining a mated male rather than an unmated (or less mated) one may suffer decreased fecundity in more crowded systems and decreased offspring

survivorship due to within-harem competition (Kirkendall, 1989). This must be outweighed by the costs in time, energy, and predation risk of searching for an unmated (or less mated) male. If mated males control sufficiently high quality breeding resources, the positive effects of resource quality on fitness can outweigh the costs of joining a mated male. This resource-based argument for the evolution of simultaneous polygyny is encapsulated in the polygyny threshold model, which though developed and tested in the context of bird mating systems, would seem to apply well to bark and ambrosia beetles (Kirkendall,

1983). Kirkendall (1983) postulates that the harem polygyny threshold model is most likely to lead to the evolution of polygyny in these beetles when resource quality is highly variable (not uniformly high or low). Variable resource quality leads to some males being in high quality resource patches capable of supporting high fecundity of several females, while other males sit in low quality patches and will be largely ignored by searching females. See Kirkendall (1983) for a more detailed development of the argument and for data supporting it.

For species where egg tunnels are all longitudinal or all transverse and hence run parallel to each other if on the same side of the gallery system, polygyny is only possible where resource quality is sufficiently high and larval tunnels correspondingly short. Support for this hypothesis comes from observations that males in fact refuse entry to additional females after having acquired their normal complement (Borden, 1967; Swaby and Rudinsky, 1976), and that once having achieved large harems, males of several species have been shown to be less attractive or to reduce pheromone emission (Kirkendall et al., 1997).

Kirkendall (1983) suggests that females in large harems do not suffer a fitness cost to joining harems. Available data also suggest that in harems with only three of four females, it is possible for females to avoid within-harem competition if they space their egg galleries optimally, even in systems where egg tunnels run parallel to each other as they do in *Ips* (Kirkendall, 1989; Schlyter and Zhang, 1996; Latty et al., 2009).

In the extreme case of no available solo-male territories, the only option for females is to join mated males (i.e., harems). Mortality of the initiating sex is thought to be quite high in bark and ambrosia beetles, due to the difficulties of locating breeding material before exhausting energy resources, mortality from above-bark predation, and deaths due to residual or active host tree defenses. If males are the pioneer sex, and if mortality is high enough, then one would expect considerable pressure from late-arriving females on blocking males to allow them entry, even when one female is already in the gallery system. Polygyny can then evolve as long as the net change to male fitness is positive and the fitness of joining females greater than zero, and assuming that the first-arriving females cannot prevent entry of further females. Put more simply, polygyny can evolve if it pays males to allow more than one female to enter, and if females joining mated males can successfully produce offspring. Note, however, that in current harem polygynous species, unmated males are relatively frequent (review and original data in Kirkendall, 1983; Schlyter and Zhang, 1996; Latty et al., 2009).

4.2.5 Colonial Polygyny

We have categorized three genera in two tribes as having species with colonial polygyny (Table 3.3): *Aphanarthrum*

and *Crypturgus* (Crypturgini), and *Cyrtogenius* (Dryocoetini). None of the species we call colonial have been studied in detail, but they appear to have multiple males and multiple females in the tunnel systems. It is possible that some of these instances are of multiple male/female pairs sharing a system of tunnels, but it seems more likely that no pair bonds are formed and both sexes mate with several individuals of the opposite sex. It must be difficult for males to maintain exclusive access to females in a many-branched tunnel system. In the phloeophagous *Cyrtogenius brevior* (Eggers) in Fiji, gallery systems are described as having many branches, with several adults in each branch; Roberts (1976) collected 20 males and 32 females from 11 or 12 galleries. Other species in the genus are monogynous, and phloeophagous or xylophagous (Browne, 1961, 1963; Roberts, 1976). Similarly, all *Crypturgus* species are found in networks of interconnected tunnels with many females and males in the same colony (Blackman and Stage, 1918; Chararas, 1962). Jordal (2006) reported on systems of interconnected tunnels with multiple individuals in *Aphanarthrum* species breeding in succulent *Euphorbia* species and suggested that promiscuous systems such as these evolve in lineages of inquilines, i.e., species that regularly use tunnels of other species as a starting point for their own egg galleries. This behavior is the norm, for *Crypturgus* species, and has also been observed in *Aphanarthrum* (Jordal, 2006).

4.2.6 Inbreeding Polygyny

Inbreeding polygyny is not unique to Scolytinae; regular brother/sister mating is found in a wide range of organisms, ranging from eyelash mites to naked mole rats, but it has evolved especially often in bark beetles. Extreme inbreeding has evolved eight times in Scolytinae, and is represented in nine different tribes (Table 3.3). About 27% of all described Scolytinae are thought to breed regularly by brother/sister mating. Of all inbreeding species, 97% come from two major species radiations. The largest inbreeding clade is that of 1336 species from 37 genera of Xyleborini plus three genera of inbreeding Dryocoetini, 22% of all Scolytinae (Tables 3.3 and 3.4, Appendix). This clade has been inbreeding regularly for about 20 million years (Jordal and Cognato, 2012). The second largest clade, the inbreeding Cryphalini, comprises 238 species divided among six genera. Its age is estimated to be ca. 50 million years (Jordal and Cognato, 2012). Despite the evolutionary success of the two major clades and ecological success of many inbreeding species, there is no evidence that inbreeding leads to diversification (Jordal and Cognato, 2012).

While many inbreeding clades are ambrosia beetles, there is no evidence that ambrosia feeding in itself predisposes a lineage to evolving inbreeding. Inbreeding has not

TABLE 3.4 Occurrences of Inbreeding in Scolytinae (after Kirkendall, 1993; see also Phylogeny in Jordal and Cognato, 2012)

Lineage	Tribe	Inbr. spp.	Biology
Bothrosternus	Bothrosternini	11	Ambrosia beetles
Araptus laevigatus complex	Corthylini	9	Seeds, pods, leafstalks, fruits
Cryptocarenus, Hypothenemus, Margadillius, Periocryphalus, Ptilopodius, Trischidias	Cryphalini	238	Highly variable, but few strictly phloeophagous (see text); one ambrosia beetle
Coccotrypes, Ozopemon, Dryocoetiops	Dryocoetini	168	Seeds, fruits; many highly polyphagous; *Ozopemon* is phloeophagous.
+ Xyleborini	Xyleborini	1168	Ambrosia beetles
Dendroctonus micans, D. punctatus	Hylurgini	2	Phloeophagous
Sueus	Hyorrhynchini	5	Ambrosia beetles
Premnobius, Premnophilus	Ipini	25	Ambrosia beetles
Xyloterinus	Xyloterini	1	Ambrosia beetles

evolved in Platypodinae, Corthylina (the ambrosia beetle subtribe of Corthylini), or *Camptocerus*. In six lineages in which ambrosia feeding and inbreeding co-occur, fungus farming preceded inbreeding in only three (Jordal and Cognato, 2012). Actually, the highest transition rates to xylomycetophagy are from lineages with regular inbreeding (Jordal and Cognato, 2012).

What is striking from Table 3.4 is that inbreeders mainly have feeding modes other than the predominant one of phloeophagy. Kirkendall (1993) analyzed the association between inbreeding and larva feeding modes for the bark and ambrosia beetles of North and Central America and for those of Thailand, Malaysia, and Indonesia. For both regions, inbreeders are most commonly ambrosia beetles. For North and Central America, 93% breed either as ambrosia beetles or in pith, seeds, and fruits, or "diverse" tissues. In the Southeast Asia fauna, too, those that are not ambrosia beetles are mainly myelophagous and spermatophagous. Conversely, inbreeding has never evolved in purely herbiphagous lineages (Kirkendall, 1993), though a few inbreeders have evolved herbiphagy and some generalists (both ambrosia beetles and phloeophages) are able to breed in herbaceous tissues. Both *Xylosandrus*, which breed in orchids (Dole *et al.*, 2010), and *Hypothenemus*, which breed in fleshy tissues, are tissue generalists, with the exception of *H*. pubescens, which may breed exclusively in grass stems (Atkinson and Peck, 1994). *Coccotrypes*, which attack mangrove propagules, are also herbiphagous (Sections 3.4 and 3.9). These examples all come from inbreeding clades.

The few phloeophagous inbreeders are atypical for species breeding in inner bark: both *Ozopemon* and

Dendroctonus breed in large chambers with larvae feeding communally, as do phloeophagous *Hypothenemus* and *Coccotrypes* (Kirkendall, 1993). Communal larval feeding is a common theme in inbreeders, and one of the most important factors in the evolution of regular inbreeding. Communal feeding is associated with all inbreeding lineages whether they are ambrosia beetles, pith breeders, seed breeders, or phloem feeders (Kirkendall, 1993; Jordal and Cognato, 2012). In seeds, if colonized only once, a single family develops in close contact within the confines of a single seed. In pith, the larvae feed in close proximity in one long cylinder.

As argued by Kirkendall (1983, 1993), the first step in the evolution of inbreeding must be pre-dispersal mating. However, for pre-dispersal mating to be incestuous, young adults must have developed in close proximity. In most bark beetle systems, larvae tunnel away from the maternal egg gallery, and most bark beetles breed in relatively dense aggregations: any mating before dispersing will usually be between offspring of different broods. Inbreeding can only evolve in an outbreeding species if young adults are in close contact with each other when they mature, as will happen if they develop together in a common nest as larvae of one family.

Inbreeding is characterized by two major ecological patterns (Kirkendall, 1993; Jordal *et al.*, 2001). There is a latitudinal gradient in close inbreeding: the proportion of inbreeders in the Scolytinae fauna increases from just a few species in the far north or far south to being roughly half of the fauna of lowland tropics. It is likely that there is also a corresponding elevational gradient (inbreeding decreases with increasing altitude), though this has not been thoroughly investigated (but see Kirkendall, 1993). Inbreeding is also disproportionately common on small islands, not because outbreeders evolve incestuous mating, but because inbreeders are more successful colonizers (Kirkendall, 1993; Jordal et al., 2001). The species–area relationship differs for the two mating behaviors: numbers of outbreeding species decrease more rapidly with area than do numbers of inbreeding species. Jordal et al. (2001) showed that this pattern was not due to differences between outbreeders and inbreeders in resource utilization (feeding modes) or by sampling biases (undercollecting). Rather, outbreeders are poor colonizers because they are constrained by Allee effects, density-dependent behavioral and ecological factors disproportionately impacting small populations (Gascoigne et al., 2009; Kramer et al., 2009). Jordal et al. (2006) postulate that outbreeders have difficulties successfully establishing new populations because they are more vulnerable to random extinctions of small populations, suffer inbreeding depression, and have difficulties finding mates. Inbreeders, by virtue of investing minimally in males, and by not expending time and energy on mate location, have higher intrinsic rates of increase, and thus are exposed to the dangers of stochastic extinctions for a shorter period than are outbreeders.

Repeated inbreeding rapidly produces homozygotic genomes, which are then passed on intact from one generation to the next. Regular inbreeding, then, can be considered to be quasiclonal reproduction, "quasi-" because outbreeding is always a possibility in inbeeding lineages, while in most cases truly clonal, parthenogenetic organisms cannot suddenly shuffle their genes in a bout of sexual reproduction. Reproduction by extreme inbreeders (species for which interfamilial inbreeding is the norm), then, is "clonal" as long as inbreeding continues, but is "reset" for females mating with unrelated males who manage to get into a foreign gallery system. Outbreeding individuals then produce a genetically variable brood with a burst of heterozygosity.

How often does outbreeding occur in inbreeding lineages, and how does it happen? This is a key question for understanding why inbreeding has been so successful in these beetles. Population structure has been investigated recently for the seed borers Coccotrypes dactyliperda (F.) (Gottlieb et al., 2009; Holzman et al., 2009) and H. hampei (Benavides et al., 2005); all found low rates of genetic variation, and large genetic differences between populations, patterns consistent with high rates of inbreeding. Experiments with X. germanus (Peer and Taborsky, 2005) found outbreeding depression but no inbreeding depression, as expected for regular inbreeders. Gottlieb et al. (2009) estimated inbreeding rates and found that they vary highly between populations but generally reflect high amounts of inbreeding.

Extreme inbreeders, then, potentially reap the benefits of clonal reproduction, i.e., replication of successful genomes from one generation to the next, preserving combinations of genes that work well together and conserving local adaptation. All inbreeders that have been studied in any detail have evolved adaptive, strongly female-biased sex ratios, further increasing advantages to inbreeding; outbreeders invest half their resources in males, while inbreeders invest minimally. However, this nearly two-fold advantage in reproductive rate would be largely mitigated if males significantly increase the reproductive output of their partners. For this reason, the fitness effects of male residency are particularly relevant, in understanding the factors favoring or disfavoring the evolution of regular inbreeding. It should also be more difficult for inbreeding to evolve in species where male presence significantly increases female fecundity or the survivorship of the male's offspring (see Section 5.3). There should then be strong selection on females to bind males to them, which they do by being continuously receptive, even if females have mated previously and have sufficient sperm to fertilize all their eggs. Females breeding alone (and hence using only sperm from a pre-dispersal mating) would have low reproductive success relative to outbreeding females.

4.2.7 Partial Inbreeding

As far as can be determined from the literature on inbreeding Solytinae, almost all instances are examples of extreme inbreeding, and reproduce regularly by brother/sister mating (Kirkendall, 1993). The likely exception is the Palearctic D. micans and its Nearctic sister species D. punctatus LeConte. These two species may be the best examples of scolytine species with populations that regularly experience intermediate levels of inbreeding (but see Holzman et al., 2009). All other inbreeding lineages in Scolytinae exhibit most characteristics of what Hamilton (1967) termed a biofacies of extreme inbreeding, but this Dendroctonus clade does not; their males are only statistically shorter and lighter in weight than females, and can fly (Kirkendall, 1993; Meurisse et al., 2008). Further, D. micans seem to produce more than minimum numbers of males per brood; typical families have 10–30 males (Kirkendall, 1993). Dendroctonus punctuatus have similarly large broods with multiple males, and have an average of about five females per male ($N = 37$ broods: M. Furniss, pers. commun.). Other inbreeding scolytines normally produce broods with just one or very few males, sufficient to fertilize all their sisters (Kirkendall, 1993). However, as with all other inbreeding scolytines, mating in D. micans and D. punctatus occurs before females disperse, and males do not participate in gallery construction. Inbreeding as a breeding strategy has not been studied in D. micans or D. punctatus, but the only genetic study (using protein electrophoresis) supports a hypothesis of intermediate levels of inbreeding, with modest but reduced levels of heterozygosity found in both species (Kegley *et al.*, 1997).

That *D. micans* and *D. punctatus* do not seem to be fully committed to inbreeding could have arisen in two ways. First, it is possible that they are under strong selection to inbreed and do so most of the time, but that inbreeding has evolved too recently for males to fully adapt. In this case, the relatively high numbers of males might be maladaptive, but females might have poor control over the sex of their eggs. Alternatively, these two species may indeed be balancing inbreeding and outbreeding, and the numbers of males produced may be optimal for the levels of outbreeding occurring in natural populations as well as for regular partial brood mortality due to *Rhizophagus grandis* Gyllenhal predation (discussed in Kirkendall, 1993). We lean towards the latter hypothesis. Inbreeding must have evolved before *D. micans* and *D. punctatus* split. This split may have been as recent as the Wisconsin glaciation 85,000–11,000 years ago, as argued by Furniss (1996), but as he pointed out, they differ in 10 discrete morphological characters, and they also differ in karyotype (Zúñiga *et al.*, 2002a, b). Whether or not this level of differentiation can occur in such a relatively short time is an interesting question.

Broods regularly merge in epidemic outbreaks of *D. micans*, and under endemic conditions, males are fully capable of wandering from one brood gallery to another, or even flying to another colonized tree (Meurisse *et al.*, 2008). Whether or not interfamily matings represent genetic outbreeding is not known. When there are multiple broods on a single tree, these may often stem from related females (Grégoire, 1988).

4.2.8 Parthenogenetic Reproduction

Four forms of parthenogenetic reproduction have evolved in Scolytinae. In thelytoky and pseudogamy, females are produced clonally and are genetic copies of their mothers. In the former, no males are involved and populations consist solely of females, while in the latter, fertilization is required (by males of the same or a related species) but male genomes are not used to build the phenotype and are not passed along to offspring. In arrhenotoky and pseudo-arrhenotoky, daughters are sexually produced, but males express and pass on only genes from their mother. Males are thus functionally haploid in both, though in pseudo-arrhenotoky, fertilization takes place but then the paternal genome is eliminated. Since male genomes are produced by meiosis, males are not clonal.

Obligate or facultative thelytoky is relatively frequent in weevils, and occurs at least sporadically in over 80 families of Hexapoda (Normark and Kirkendall, 2009); it has arisen at least once in Scolytinae, in Corthylini, though there are

several lineages in which it may also occur. Deyrup and Kirkendall (1983) examined over 500 *Pityophthorus puberulus* (LeConte) individuals collected from Indiana, Michigan, and Maine; *P. puberulus* is the most common scolytine in dead twigs of native and exotic pines. All were female, and none contained sperm, not even those taken from galleries with eggs and larvae. In no case were two parent adults found in a gallery system. In taxonomic treatments of the genus, Bright (1981) and S. L. Wood (1982) describe *P. puberulus* males simply as being identical to females; but as found by Deyrup and Kirkendall (1983), Bright (1981) reports that galleries contain only one individual. Thus, while it is possible that one or more sexual populations exist in this widespread species, no confirmed males are known, and *P. puberulus* should be considered parthenogenetic. There have been no subsequent investigations into this interesting case of thelytoky.

A possible second instance of all-female lineages comes from *S. rugulosus*. Gurevitz (1975) reported breeding repeated all-female generations in the laboratory, from beetles collected in Israel. *Scolytus rugulosus* has been occasionally studied as a pest of fruit trees in its native Eurasia and as an invasive species in North America, but no deviations from 1:1 sex ratios have been reported by other authors, and it seems to be a normally reproducing monogynous bark beetle everywhere else other than the Middle East (Gossard, 1913; Kemner, 1916; Chodjaï, 1963; Kirkendall, unpubl.). Parthenogenesis in *S. rugulosus* needs to be confirmed.

Finally, there are several groups (*Bothrosternus foveatus* Wood and Bright; *Dryocoetiops*) that have been treated as inbreeding (Kirkendall, 1993) but where males have still not been found; in both, close relatives are inbreeders. Given that males of inbreeders are often tiny and are rarely collected, it is possible that these groups inbreed.

In pseudogamy, a form of sperm-dependent parthenogenesis, eggs must be fertilized to develop, but sperm do not contribute genetically to the offspring and inheritance is strictly mother to daughter (Beukeboom and Vrijenhoek, 1998; Schlupp, 2005). It is a rare reproductive system, occurring among hexapods in just a few other orders; in beetles, it is found in a spider beetle (Ptinidae). Pseudogamy as a reproductive system has evolved at least twice in Scolytinae, in North American spruce-breeding *Ips*, and in a Eurasian pine-breeding *Ips* (Lanier and Kirkendall, 1986).

In North America, pseudogamy occurs in the *tridens* complex of spruce-breeding *Ips* (Hopping, 1964; Lanier and Oliver, 1966; Lanier and Kirkendall, 1986). Three types of individuals, pseudogamous females, sexual females, and males, are found in *Ips borealis* Swaine, *I. tridens* (Mannerheim), *I. pilifrons* Swaine, and *I. perturbatus* (Eichhoff). In all four "species," pseudogamous females are triploid

(Lanier and Kirkendall, 1986). As is usually the case (Schlupp, 2005), pseudogamy has probably originated via interspecific hybridization. These pseudogamous lineages form a monophyletic clade (Cognato and Sperling, 2000), so the phenomenon may have evolved only once in this species group. For the most part, only the taxonomy and systematics of these pseudogamous populations have been studied; virtually nothing is known of the nature of pseudogamy in these lineages. It seems likely that predominantly or entirely sexual populations were studied in the ecological investigations of *I. perturbatus*, which in these papers seems to have typical *Ips* biology and equal sex ratios (Gobeil, 1936; Robertson, 2000).

Interspecific hybridization is much less likely the origin in *I. acuminatus*, the Eurasian example of pseudogamy, though here, too, the parthenogenetic females are triploid (Lanier and Kirkendall, 1986). Though this pine breeder occurs from Western Europe to eastern Siberia, China, and Japan, its reproductive behavior and population dynamics have only been studied in Europe (Bakke, 1968b; Kirkendall, 1989, 1990; Kirkendall and Stenseth, 1990; Løyning and Kirkendall, 1996; Løyning, 2000; Meirmans *et al.*, 2006).

Arrhenotoky has arisen once in Scolytinae, producing the remarkably successful haplodiploid clade comprising Xyleborini (nearly 1200 species in 37 genera) and three inbreeding genera previously placed in Dryocoetini, *Coccotrypes* (129 spp.), *Ozopemon* (21 spp.), and *Dryocoetiops* (18 spp.) (Jordal *et al.*, 2002; Cognato *et al.*, 2011; Jordal and Cognato, 2012). It is well known that bees, wasps, and ants are haplodiploid, but this system is also found in the one species of Micromalthidae, many thrips, a few whiteflies and scale insects, most rotifers, most mites, and some nematodes. The entire scolytine clade is considered haplodiploid, but this is based on the observations of just a few species of *Coccotrypes* and *Xylosandrus* (Kirkendall, 1993). However, there are no data that falsify the hypothesis that the entire clade is haplodiploid, and finding all-male broods in many species supports the hypothesis (these represent reproduction by unfertilized females).

Pseudo-arrhenotoky is known from one inbreeding lineage, *Hypothenemus*, having been demonstrated in *H. hampei* (Brun *et al.*, 1995a, b; Borsa and Kjellberg, 1996a, b; Chapter 11). The phenomenon was discovered serendipitously while studying the evolution of resistance to insecticides, when it was observed that males always had the resistance phenotype of their mother, regardless of the father's phenotype. Worldwide, there are 181 described species in the genus (Chapter 11), but only *H. hampei* has been studied in this context. It is believed that the entire genus inbreeds, since in all cases where broods have been examined, males are rare, and all males known are reduced in size and flightless, which are characteristics of regular inbreeders. Further, the most closely related genera also inbreed. Pseudo-arrhenotoky is a rare breeding system, but is known in some mites and in mealy bugs (Coccidae). Variation in reproductive systems among closely related coccids, though, raises a red flag, and the hypothesis that pseudo-arrhenotoky is characteristic of all inbreeding *Hypothenemus* its relatives should be verified.

4.3 Gallery System Form

The general hiking and camping public, including most entomologists, rarely encounter the insects themselves, but may well be aware of the consequences of their activities: dead and dying trees during bark beetle outbreaks, and the striking engravings seen in older dead wood. Forest entomologists have long classified these etchings based on their general form (Barbey, 1901; Swaine, 1918; Blackman, 1922; Chamberlin, 1939), and they are still used today.

Generally, one can deduce the mating system of a species (especially those breeding in inner bark) from the form of the tunnel system: single egg tunnels result from monogyny, a variable number of egg tunnels per system from polygyny (Figure 3.6). However, when there are two tunnels, this may result either from a single female working in two directions or, in a few lineages, from bigyny. Variously shaped large chambers lacking defined egg tunnels—cave-type systems—are formed by monogynous species.

Females of phloeophagous Scolytinae disperse eggs in a wide variety of ways. Most commonly, all eggs laid by a single female during a given bout of reproduction are deposited in a single long gallery or in two long galleries bored in opposite directions. In a few genera, a single female makes several short tunnels leading away from a central nuptial chamber; an especially interesting example is *Ips latidens* (LeConte), a monogynous species in an otherwise uniformly harem polygynous genus (Reid, 1999). Most Cryphalini, and a few genera or species from other lineages, make elongate to roundish chambers, where the eggs are either spaced around the periphery in egg niches (as in *Procryphalus mucronatus* (LeConte), *Dacryostactus kolbei* Schauffuss, or *Styracoptinus murex* (Blandford)), deposited in egg pockets (*Cryphalus kurilensis* Krivolutskava, *C. exiguus* Blandford), or simply laid in clusters loose in the gallery (*Cryptocarenus*, some *Cryphalus*, many *Hypothenemus*, *Trypophloeus populi* Hopkins). Some specialists, such as Blackman (1922), Browne (1961), and S. L. Wood, (1982), have thought that cave-type galleries were "primitive" in bark beetles, but this seems unlikely given that basal taxa in current phylogenies all make long egg tunnels.

Ambrosia beetle tunnel systems also show variation in egg deposition strategies (Browne, 1961). Again, in most groups, each female constructs a single tunnel or a few long branches. Tunnels constructed in smaller branches may completely encircle the branch, and in very small-diameter breeding material tunnel systems usually spiral. Eggs are placed in niches constructed by the mother beetle in *Camptocerus*, Corthylini, and Xyloterini, reflecting their derivation from phloeophagous ancestors with egg niches; they are laid in batches loose in tunnels or lenticular chambers in Xyleborini, as they are in other members of this clade (spermatophagous or phloeophagous *Coccotrypes*, *Ozopemon*, and relatives), suggesting that the ancestors of this large inbreeding clade lost the practice of placing eggs singly in niches. Interestingly, Platypodinae lays eggs in clusters and larvae feed in the tunnel system, but the last instar larvae form cradles in which they pupate singly.

Long tunnels give females options for optimally dispersing offspring in space, both with respect to resource quality, resource quantity, host plant defenses, and intrabrood competition. At the same time, females themselves must feed continuously in order to produce large, protein-rich eggs. In Scolytinae, eggs are generally one-quarter to one-third or more their mother's body length (Kirkendall, unpublished data), while plant tissues they consume are critically low in nitrogen (White, 1993; Kirkendall, 1983; Haack and Slansky, 1987; Ayres *et al.*, 2000). Tunneling, then, fulfills both needs: spacing of eggs and acquiring nutrients for oviposition. Spatial orientation of tunnels, placement of eggs in the tunnels, and spacing of eggs all are highly variable in Scolytinae and almost certainly are adaptations, but there has been little research in this area.

4.3.1 Spatial Orientation

Broadly considered, phloeophagous tunnel systems are classified by the number of egg tunnels (arms) in a system and by the orientation of the tunnels. Most monogynous systems can have one (uniramous), two (biramous), or (exceptionally) more egg tunnels (polyramous). These tunnels can run with the wood grain (longitudinal or vertical), or perpendicular to the grain of the wood (transverse or horizontal); biramous systems can also be V-shaped. Tunnels of some *Dendroctonus* are long and very irregular in shape. *Chaetophloeus* species make single nearly circular egg tunnels; *Pseudips* construct C- or S-shaped systems (uni- or biramous). Polygynous systems are also classified by egg tunnel orientation, although less often so than are monogynous systems. In many polygynous taxa, egg tunnels are clearly oriented either longitudinally (e.g., most *Ips*) or transversely (*Pityokteines*), but others are simply fairly evenly spaced from each other and form star-shaped patterns (*Polygraphus*, most *Pityophthorus*). Gallery systems of regularly bigynous species usually are biramous, with each arm being the work of one female, and these run directly opposite one another, either longitudinally or transversely, though some species make V-shaped systems (one

female in each arm). Females of *Pseudothyanoes* each make both arms of a "V," the system as a whole resembling an "X" or "H."

The adaptive significance of variation in egg tunnel orientation has not been rigorously analyzed, though hypotheses have been proposed (see Kirkendall, 1983). It is clear that there are associations with host plants: bark beetles in oaks (*Quercus*) and firs (*Abies*) tunnel horizontally, for example, even though congeners tunnel vertically in other hosts (Kirkendall, unpubl.). Since newly hatched larvae tunnel at least initially perpendicularly to the egg tunnel, the best orientation of the egg tunnel may be determined by factors selecting for larval tunneling direction: if it is optimal for larvae to tunnel with the grain of the wood, for example, then the egg tunnel should be transverse. Selection on adult or larval tunneling direction can result from host–plant defenses and physical characteristics of the host. If the inner bark is too thin to completely contain larvae as they feed, or if there are other reasons for larvae to tunnel deeper such that they begin to feed partly in sapwood, then the more fibrous nature of sapwood becomes a significant factor, as pointed out by Trägårdh (1930). Trägårdh (1930) also found that larval mines when engraved into the sapwood run strictly longitudinally and parallel to each other, but if the larval mines are purely in the inner bark, the mines can wander and can be transverse to one another. Both oaks and the woody leafstalks of *Cecropia* are quite fibrous, and especially small larvae probably cannot tunnel transversely; tunnels of bark beetles in oaks, and the smaller species in the cortex of *Cecropia* petioles, are transverse, and larvae feed perpendicularly to the gallery arms (with the wood fibers rather than across them).

It is also conceivable that, in some hosts, it is adaptive for females to oviposit where larvae are forced to partly chew through sapwood. This forces larvae to tunnel in straight lines (and thus they do not accidentally cross paths with neighbors), and allows for females to lay eggs right next to each other, if tighter egg packing is advantageous. A test of this hypothesis would be to compare related species with different egg arm orientations, transverse vs. longitudinal. Egg spacing (eggs per mm gallery) should be "tighter" (i.e., a higher number of eggs/mm) in species with transverse galleries.

4.3.2 Placement of Eggs

Most commonly, including among most basal taxa, eggs are placed in niches (egg-sized pockets) along both sides of an egg tunnel. Though usually these are evenly spaced on either side, some species (such as *Dendroctonus simplex* LeConte: Hopkins, 1909) alternate laying several eggs on one side. In species in which tunnels characteristically curve strongly, eggs are placed exclusively on the outer side of the curve. But also when generally straight or mildly curved tunnels curve more strongly, eggs will be placed only on the outer side. In at least some species, but possibly all or most, phloeophagous females seem to be able to adjust their egg placement adaptively. For example, *Pseudoips mexicanus* (Hopkins) tunnels are usually curved but can be straight; eggs are laid on both sides if straight, but just the outer side if curved (Smith *et al.*, 2009). An advantage of a curved gallery is that larvae emanating from the outer side can fan out, reducing the chance of accidentally coming in contact with each other (though two cases of incidental cannibalism were seen by Smith *et al.* (2009)).

The vast majority of phloeophagous species construct egg-sized niches, and lay just one egg in a niche. Interestingly, a few scattered instances of multiple eggs per niche have evolved: examples include *Chaetophloeus heterodoxus* (Casey) (Swaine, 1918); *Pseudoips* (Trimble, 1924; Chamberlin, 1958; S. L. Wood, 1982; Smith *et al.*, 2009; Zhang *et al.*, 2010); *Orthotomicus caelatus* (Eichhoff) (Swaine, 1918); and *Liparthrum mexicanum* Wood (Atkinson and Equihua-Martínez, 1985b). Such egg pockets are wider and deeper than normal egg niches, and may contain a few eggs (e.g., 1–4 in *Pseudoips*) or many (e.g., 6–12 in *C. heterodoxus*).

Furthermore, some species deposit eggs in widened portions of the egg tunnel rather than in single egg niches or egg pockets. Examples include *Dendroctonus pseudotsugae* Hopkins, *D. piceaperda* Hopkins, *D. rufipennis* (=*D. engelmanni* Hopkins), *D. micans*, *D. punctatus*, *D. valens*, *D. terebrans* (Olivier), *Hylurgops pinifex* (Fitch), *Dryocoetes americanus* Hopkins (=*D. autographus* (Ratzeburg)), and *Orthotomicus laricis* (F.) (Lövendal, 1898; Hopkins, 1909; Swaine, 1918; Balachowsky, 1949). In some cases, these are protected by a layer of frass, just as are normal egg niches. Intermediate between these grooves and the egg pockets is the pattern of *D. simplex*, which places three or four eggs side by side at the bottom of an elongate shallow pocket or groove (Swaine, 1918), and *Xylechinosomus valdivianus* (Eggers), where clusters of up to 30 eggs are placed in shallow troughs along the tunnel wall (Rühm, 1981). *Pseudothysanoes dislocatus* (Blackman) seems to have similar behavior (Blackman, 1922). *Hylurdrectonus piniarius* Schedl lays eggs loose in frass in indefinite tunnels in the cortex of *Araucaria* branches (Brimblecombe, 1953).

The selective advantages of clustering single niches or laying more than one egg in an egg pocket are not obvious. At least with regards to egg pockets, there would seem to be a cost of greater intra-family resource competition when larvae hatch so close to each other, and a risk of accidental siblicide. Perhaps clues can be found in the biology of the inbreeding *D. micans*, where clustered larval feeding significantly increases growth (Figure 3.6A). *Dendroctonus micans* breed solitarily in trunks of live spruces or other conifers. Storer *et al.* (1997) suggest that larval aggregation might be important in dealing with host defenses. This hypothesis suggests that egg clustering in phloeophagous species will be associated with breeding in live (vs. dead) plant tissues, particularly in hosts with strong chemical defenses.

5. SOCIAL EVOLUTION

It is not generally known that bark and ambrosia beetles exhibit an extraordinary diversity of social systems. Higher forms of sociality have evolved repeatedly in these insects, and the only eusocial beetle is the platypodine ambrosia beetle, *A. incompertus* (Kent and Simpson, 1992). Bark and ambrosia beetles are also the only social insects with closely related haplodiploid and diploid social lineages (Normark *et al.*, 1999; Jordal *et al.*, 2000), which would allow comparative studies to contribute to the long-debated role of haplodiploidy for social evolution (Hamilton, 1964; Bourke, 2011). Unfortunately, our knowledge of the detailed behaviors of social species is still superficial, as sociality has rarely been the primary focus of researchers working with these insects. This is primarily because studying insect behavior in tunnel systems under the bark or in the wood of trees is almost impossible. Exciting progress is now being made, however, because evolutionary biologists interested in social behavior have discovered these beetles as an illustrative alternative to classical hymenopteran model systems (Hamilton, 1967, 1978; Kirkendall, 1983, 1993; Kirkendall *et al.*, 1997; Peer and Taborsky, 2007; Biedermann *et al.*, 2009, 2011, 2012; Biedermann and Taborsky, 2011, submitted; Jordal *et al.*, 2011; Boomsma, 2013), and because several ambrosia beetles have been successfully reared in artificial media (Saunders and Knoke, 1967; French and Roeper, 1972b; Roeper *et al.*, 1980b; Mizuno and Kajimura, 2002; Biedermann *et al.*, 2009; Lake Maner *et al.*, 2013), which allows behavioral observations, experimental manipulations, and, due to their often short generation times, even artificial selection experiments (Biedermann and Taborsky, submitted).

5.1 Social Behaviors and Ecology of Bark and Ambrosia Beetles: an Overview

Animal social systems range from simple gregariousness, to family groups with parental care, to complex cooperative breeding or eusocial societies with reproductive altruism (Wilson, 1971; Alexander *et al.*, 1991; Costa, 2006; Boomsma, 2013). In bark and ambrosia beetles, all these forms are present: (1) gregarious feeding is typical for the phloeophagous larval offspring in certain *Dendroctonus* species, many cryphalines, *Ozopemon*, and some phloeophagous and some spermatophagous *Coccotrypes* species;

it is the norm for Xyleborini and Platypodinae. Gregariousness of adults is particularly apparent in cooperative mass attack in some primary *Dendroctonus* and *Ips* species, but also is found during overwintering or maturation feeding of many species. (2) Parental investment in the form of brood care by the mother, the father, or both—also termed "subsociality"—is ancestral for bark and ambrosia beetles and thus typical for the whole group. (3) Adult offspring refrain from dispersal and engage in "alloparental" brood care of young siblings at the natal nest, which is likely confined to some ambrosia beetles (see below) and potentially also *Coccotrypes* species breeding in seeds. Some of these species may form true beetle "societies" with division of labor between adult and immature offspring present in communal tunnel systems. These can be further split into "facultatively eusocial" or "obligately eusocial" societies, depending on whether adult offspring refrain temporarily or permanently from reproduction (Boomsma, 2009). Currently, we know of three facultatively eusocial (*Xyleborinus saxesenii* (Ratzeburg), *X. affinis* Eichhoff, *Trachyostus ghanaensis* Schedl) and one possibly obligately eusocial ambrosia beetle (*A. incompertus*), but there are likely more eusocial species awaiting discovery.

Larvae of phloem-feeding bark beetles construct their own tunnels in the phloem during feeding and gradually move away from their maternal tunnel. As larvae also pack these mines with frass, there is often no physical contact between the parents and their offspring. This is not true for all bark beetles, however, as some *Dendroctonus* species and also many non-phloem feeders like *Hypothenemus* and *Coccotrypes* species live in communal galleries. Communal galleries are also present in many ambrosia beetles, but tunnels may or may not be altered by ambrosial grazing by larvae and adults. Larvae and adults can move and interact freely in such galleries. However, this is not true for all ambrosia beetles; in *Camptocerus*, Xyloterini and Corthylini, larvae are separated from each other because they develop in individual larval niches and do not move freely in the galleries. Nevertheless, they still closely interact with their parents that freely move within the galleries. Consequently, as there are many more interactions between individuals in galleries of many ambrosia beetles and non-phloem feeders than in galleries of true bark beetles, the potential for advanced sociality to evolve is much higher in the first groups (Kirkendall *et al.*, 1997).

5.2   Basic Concepts of Social Evolution Theory

The evolution of behavior is fundamentally based on maximizing the direct fitness of individuals (i.e., individual-level selection; Alexander, 1974; Clutton-Brock, 2009). As Darwin (1859) realized, this cannot explain the evolution of alloparental care and eusociality, however, because the beneficiaries of care are not offspring of the caregivers but rather kin to them with varying degrees of relatedness. This problem was resolved by William Hamilton's theory of inclusive fitness (kin selection theory), which incorporates both the direct and indirect fitness effects of costly behaviors: an altruistic behavior can evolve if it benefits the spread of a cooperative gene, not necessarily by self reproduction (direct fitness), but also through the reproduction of relatives bearing that gene (indirect fitness) (Hamilton, 1964). More precisely, altruism is selected, for if the genetic relatedness (r) between social partners is greater than the ratio of fitness costs (c) to the performer over the fitness benefits (b) to the recipient: $r > c/b$. Accordingly, social behaviors typically arise in kin groups and under ecological conditions that yield higher inclusive fitness gains when remaining in the natal nest.

Ever since the publication of Hamilton's paper (Hamilton, 1964), several ecological conditions have been identified to generally facilitate social evolution across various animal groups, which can be roughly grouped in two categories: environmental constraints on solitary breeding and benefits of philopatry (Korb and Heinze, 2008; Bourke, 2011). Aiding kin becomes a viable alternative to breeding oneself when independent breeding is very costly. Environmental factors that increase the costs of solitary breeding include high mortality during dispersal, breeding sites being limiting, and high population densities (Emlen, 1982). Philopatry (not dispersing before breeding) can be beneficial if there is an opportunity to inherit the nest or a possibility of co-breeding (direct fitness benefits), or by helping related individuals to increase their reproductive output (indirect fitness benefits) (Stacey and Ligon, 1991).

5.3   Subsociality and Parental Care in Bark and Ambrosia Beetles

Subsociality is characterized by reproductive investment of parents beyond egg laying: post-ovipositional care increasing survival, growth, and development of offspring (Wilson, 1971). In insects, it has evolved repeatedly, typically in connection with abundant but ephemeral resources and high competition or predation pressure (Tallamy and Wood, 1986). The bark of dead trees is a prime example of an environment facilitating subsocial life. Wood suitable for insect attack is unpredictably distributed and difficult to locate, but offers an abundant, defendable resource, which may persist for several generations. The physical properties of woody tissues and plant defenses like resin flow and toxic chemical metabolites are likely major obstacles for small larvae, problems more easily overcome with the help of adult individuals (Hamilton, 1978; Chapters 1 and 5). Parents can also assist with food provisioning, in particular by increasing the quality or digestibility of food. By inoculating the wood with microorganisms, they can increase its nitrogen content and can make plant tissues easier to assess and assimilate. Wood-feeding insects can only utilize lignocellulosic resources by engaging in symbioses with bacteria, fungi, or protozoa (Tallamy and Wood, 1986). As parent beetles can significantly reduce physical and nutritional limitations for their offspring (see below), it is not surprising that wood is one of the most favorable habitats for the origin of subsociality in insects as well as of insect–microbe associations (Hamilton, 1978, 1996; Tallamy, 1994; Jordal et al., 2011).

Excavation of tunnels by adults for reproduction is universal in bark and ambrosia beetles. One or both parents typically remain in the tunnel system, providing nest protection and removing frass. This behavior is not common among other weevils that typically lay their eggs singly on the outside of plants or in small pre-bored cavities, where the larvae feed solitarily (Lengerken, 1939). The parental care of bark and ambrosia beetles is no exception in this habitat, as subsociality has evolved repeatedly in other weevil clades that bore in wood, such as Cossoninae and Conoderinae (Kuschel, 1966; Jordal et al., 2011) and Bostrychidae and Ciidae (Hamilton, 1979; Kirkendall, unpubl.). Parental care takes similar forms in these groups, being characterized by one or both sexes boring oviposition tunnels, keeping them free of frass, and protecting them against predators and competitors (Kuschel, 1966; Jordal et al., 2011). This suggests that selective factors specific to wood, like the difficulties faced by immature offspring mentioned above and pressures from competitors and natural enemies, have repeatedly selected for adult beetles which bore oviposition tunnels through the outer bark instead of laying their eggs freely on the plant surface or in simple slits. Following the successful excavation of a tunnel in the phloem, there is no reason for a female to leave this proto-nest after laying the first egg; tunnel excavation is energetically costly and the habitat offers a nutritious, defendable, and abundant food resource, which can support both her own nutritional needs and those of many more offspring. Studies on predation pressure within and outside the gallery are rare, but it is likely that, once under the bark surface, females are also much safer from predation by vertebrates and invertebrates alike. Beetles in tunnels are invisible to foraging vertebrates such as birds or lizards, and invertebrate wood borer predators like ants or checkered beetles (Cleridae) preferentially attack adult beetles on the bark as they have considerable difficulty with extracting them from tunnels (Wichmann, 1967).

Bark and ambrosia beetle females invest relatively heavily in individual offspring, via egg provisioning and maternal care. Eggs are unusually large, ranging from one-sixth the length of the female's body in Tomicus pilifer (Spessivtsev) (Wang, 1981) to one-third the size in X. affinis (Roeper et al., 1980b), T. populi (Petty, 1977),

and Pagiocerus frontalis (F.) (Yust, 1957). Clutch sizes are modest (commonly, 70–90 eggs, but often smaller: Browne, 1961), and some bark beetles (such as those colonizing woody petioles of large leaves) lay fewer than a dozen eggs (Beaver, 1979b; Jordal and Kirkendall, 1998); these are among the insects with the lowest recorded fecundity (Hinton, 1981; Nyland, 1995).

Many males and females commit to one or two breeding sites (Kirkendall, 1983). For holarctic, outbreeding non-xylomycetophagous species, it is often reported that females re-emerge after finishing their first egg tunnel (Kirkendall, 1983); Browne (1961), however, believed that in the humid tropics, females of most species breed in only one bout. Pairs often die in their gallery system in species from a variety of genera: Conophthorus lambertianae Hopkins (Chamberlin, 1958), Scolytus unispinosus LeConte (Chamberlin, 1918), Pseudohylesinus nebulosus (LeConte) (Chamberlin, 1918), Dactylipalpus camerunus Hagedorn (Browne, 1963), T. populi, Procryphalus mucronatus (LeConte) (Petty, 1977), and C. columbianus (Milne and Giese, 1969). Committing strongly to a bout of breeding selects for increased parental investment (Wilson, 1975; Tallamy and Wood, 1986). Where it has been investigated, scolytine beetles as diverse as Dendroctonus, Phloeosinus, Ips, Hypothenemus, and Conophthorus digest (histolyze) their wing muscles once they have begun breeding (Chapman, 1956; Reid, 1958; Lekander, 1963; Borden and Slater, 1969; Morgan and Mailu, 1976; Garraway, 1986; Robertson and Roitberg, 1998; López-Guillén et al., 2011); whether or not Platypodinae do this as well is not known, and how common the phenomenon is within the Scolytinae is similarly unknown. For females, autolysis of wing muscles must free up quantities of protein for egg production; the advantages to males are less clear (Robertson, 1998b). While some females can regenerate their muscles after a post-ovipositional period of feeding, in many species most or all re-emerging females cannot fly, e.g., Dendroctonus (Lawko and Dyer, 1974; Langor, 1987; Grégoire, 1988); Hypothenemus (Ticheler, 1961; López-Guillén et al., 2011); and Phloeosinus (Garraway and Freeman, 1981). Regeneration in Ips males depends on body size and time spent in the tunnel system (Robertson, 1998a). Scolytines that do not regenerate wing muscles can and often do walk to new sites on the same host to start a second egg tunnel, though they cannot disperse to new breeding material (Fuchs, 1907; Reid, 1958; Sauvard, 2004).

In at least some Platypodinae, both sexes lose their tarsal segments after some weeks in the gallery and are thought to be incapable of dispersing anew (reviewed in Kirkendall et al., 1997). Here, commitment to one bout of reproduction seems assured.

This variation in reproductive strategies reflects varying optimal solutions to the problem of balancing the number of eggs laid with investment in offspring being produced, and (especially for males) balancing investment in current vs. future offspring. Over time, breeding material degrades, intraspecific and interspecific competition increase, and pressure from parasites and predators increases. Offspring produced late are smaller and consequently have lower fitness than those produced earlier in the same host (Kajimura and Hiiji, 1994). At some point, these factors shift the balance in favor away from laying more eggs towards either investing more in maternal care, or departing the brood and attempting further reproduction elsewhere.

5.3.1 Removing or Packing Frass

The simplest and most widespread form of parental care common to all bark and ambrosia beetles is clearing frass from the egg tunnel. Females and their offspring produce large amounts of frass during tunneling and feeding, which is pushed back towards the entrance by the mother; it is expelled from the gallery system by the male, if present, or by the female, or in some cases packed tightly into the base of the tunnel. Though modifications exist, frass is typically shuffled out of the nest by sliding it backwards beneath their body with the legs and then using their elytral declivity as a shovel to eject it (Wichmann, 1967). Although the fitness benefits of frass removal have not been studied, it is likely highly advantageous, as it is invariably present in all wood-boring weevils. Apart from enabling free movements within the gallery (females face forward while tunneling but must turn around and back up to lay eggs in newly constructed niches), keeping egg galleries free of frass likely serves two major purposes: ventilation of the nest, and nest hygiene by removing potential substrate for parasites and pathogens (e.g., mites, nematodes, fungi, bacteria). Ventilation of phloeophagous tunnels is important enough that it has been proposed as one possible function for entrance blocking, nuptial chambers, and especially holes bored upwards through the bark from the oviposition tunnels (Swaine, 1918; Blackman, 1922; Morgan, 1998; see below). Both ventilation and hygiene are especially important in ambrosia beetles, as fungus cultures grow only under specific moisture content and are very sensitive to pathogens (Francke-Grosmann, 1967). Likewise *X. saxesenii* females have been shown not only to shuffle frass and sawdust outside of the gallery but may also remove intruding mites, spores of fungal pathogens, and diseased individuals (Biedermann, 2012).

Although females of almost all species begin by removing frass, in at least a few species it has a been observed that, some time after commencing oviposition, females instead begin packing the frass behind them, forming an impenetrable plug between the active part of the egg tunnel and the nuptial chamber or even the tunnel entrance. Oviposition tunnels become plugged with tightly packed frass in *D. ponderosae* (Reid, 1958); some *Ips*

species (Morgan, 1967; Gouger *et al.*, 1975; Garraway, 1986); and some *Pityophthorus* species (Blackman, 1922). When frass blocks off the nuptial chamber, females chew small lateral (sometimes vertical) tunnel extensions in which they can turn around.

Reid and Roitberg (1994) and Robertson (1998b) used male removal experiments to study the effects of male residence on female reproduction in the harem polygynous *I. pini*. Males usually remain with females for several weeks, during most of the time that females are ovipositing. Reid and Roitberg (1994) found that after only 3 or 4 days, females breeding without males present had laid 11% fewer eggs. Robertson (1998b) found that there was considerably more frass in the tunnels of females in systems with no male, and that females with no male present laid fewer eggs and produced fewer emerging offspring. Kirkendall *et al.* (1997) reported similar effects of frass-removing males on female reproduction in a different harem polygynous species, *Pityogenes chalcographus* (L.). They also summarized published field studies on monogynous *Hylesinus*, *Scolytus* (two species), *Trypodendron*, and *Camptocerus*, in which data could be found for both females breeding with a male present and females breeding alone. In all cases, females produced many more eggs when a male was present (Kirkendall *et al.*, 1997). Existing data, then, though covering relatively few genera and species, all support the hypothesis that the most important feature of prolonged male residency is the benefits to offspring production of aiding females with frass removal.

5.3.2 Burrow Blocking or Plugging

Males staying with females is likely ancestral in bark and ambrosia beetles; there are few species (and no entire genera) of outbreeding Scolytinae or Platypodinae in which males do not remain at least some days, and they block the burrow entrance while there (Kirkendall, 1983). Furthermore, male residence and in some cases egg tunnel guarding, seems to have evolved in unrelated insect groups in which females tunnel to oviposit, such as passalids, bostrychids, ciids, subsocial cockroaches, and lower termites (Hamilton, 1979; Tallamy, 1994).

However, males are not always present during periods when it is beneficial to block. In a very few species, males may or may not guard females after surface copulations, but stay with females for at most a few days (reviewed in Kirkendall, 1983); these include species of *Dendroctonus* (Reid, 1958), *Strombophorus* (Schedl, 1960a; Browne, 1963), *Scolytus* (Gossard, 1913; Blackman, 1922; Fisher, 1931; McMullen and Atkins, 1962), an *Alniphagus* (Borden, 1969), a *Pityophthorus* (Hedlin and Ruth, 1970), and a *Conophthorus* (Mattson, 1980). As far as is known, *P. puberulus* is parthenogenetic, and males do not exist (Deyrup and Kirkendall, 1983). Even in species in which males do normally stay, some males may leave early or die, leaving the female alone. When no males are available during some or all oviposition, females may either block the entrance themselves (especially if oviposition is complete), or plug the entrance solidly with frass mixed with resin or possibly oral secretions (Kirkendall, 1984; Kirkendall *et al.*, 1997). Further evidence for the importance of blocking entrances (even late in the breeding cycle) comes from the only ambrosia beetle species in the large inbreeding genus *Hypothenemus*, i.e., *H. curtipennis* (Schedl). If an *H. curtipennis* mother dies or departs, the entrance is blocked by adult offspring (Beaver, 1986). In several platypodines, males block the entrance with long cylinders of wood fibers; if these are removed experimentally, they are rapidly replaced (Jover, 1952; Husson, 1955). Females or males die blocking the entrance in a number of species (Kirkendall, 1984), suggesting a role for blocking even late in the reproductive cycle.

Blocking of the entrance has long been hypothesized to have a protective function. Burrow blocking was discussed at length by Blackman (1922). He hypothesized that it serves to exclude parasites and predators that might otherwise harm eggs and young larvae (see discussion in Kirkendall *et al.*, 1997) and observed that any disturbance of the entrance or even the passing of a shadow over the opening would cause a male deeper in the gallery system to promptly return to his post. The clearest example of parental protection was given by Darling and Roberts (1999), who observed guarding males of the platypodine *Crossotarsus barbatus* Chapuis killing planidia larvae of *Monacon robertsi* Boucek (Hymenoptera: Perilampidae), parasitoids that try to enter the galleries. In *I. pini* male removal experiments, both Robertson (1998b) and Reid and Roitberg (1994) found much higher mortality of females in harems from which males had been removed, suggesting that male presence indeed has an important protective effect. To the extent that males staying with females protect offspring, male residency can be interpreted as paternal care.

In the case of inbreeding species, blocking by mothers or daughters may also hinder the intrusion of unrelated males (Peer and Taborsky, 2005), and in ambrosia beetles with communally feeding offspring they have been shown to protect larvae from accidentally leaving the nest (*X. saxesenii*: Biedermann and Taborsky, 2011).

Blocking could also be important for microclimate. By plugging and unplugging the entrance with their bodies, individuals can possibly regulate the microclimate within the nest, which (as argued above) is especially important in ambrosia beetles. This too, would be a form of paternal care when carried out by males, as is normally the case in outbreeding species. Kalshoven (1959) observed male *Scolytoplatypus eutomoides* Blandford (an outbreeding ambrosia beetle) to perform "...pumping movements, rapidly jerking to and fro..." in the gallery entrance, which he interpreted to serve the ventilation of the nest.

Prolonged male residency (during which they accrue fitness benefits from both blocking and frass removal) could also be favored by intrasexual selection, if males who leave too soon risk being supplanted by new males, or if males of harem polygynous species who leave early forgo opportunities to acquire further mates. However, there is little support for this hypothesis (Kirkendall *et al.*, 1997; Robertson, 1998b), and it likely only applies to the first week or so of gallery construction. Colonization of new breeding material in most species seems to be highly synchronized. When aggregation pheromones are not involved, the attractiveness of colonized breeding material seems to decline rapidly, and for taxa with pheromone systems it is often found that "masking" or "anti-aggregation" pheromones are produced after pairing (Rudinsky, 1969; D. L. Wood, 1982; Birch, 1984; Borden, 1985). Thus, the likelihood of new males entering an open tunnel is low after just a few days, and for harem polygynous species there are few or no new females arriving after a short period. Thus, in a study of harem polygynous *I. pini*, Reid and Roitberg (1994) found just a 4% replacement rate over 6 days, for gallery systems from which males had been removed experimentally.

5.3.3 Ventilation Holes

As mentioned above, females of many species chew special openings to the outside from the egg tunnels, usually referred to as ventilation holes or ventilation tunnels. In species that pack frass rather than expelling it, these must also serve as turning niches. If they do function as ventilation holes then they likely increase the survivorship of young larvae, and hence represent maternal care; however, they also present possible new entry points for natural enemies. Melnikova (1964) demonstrated experimentally that for *Scolytus ratzeburgi* Janson breeding in beech, these holes regulate humidity, after rejecting the hypothesis that they could be used for copulation. Broods with sealed ventilation holes were flooded with sap. The holes were only made by females, and were still being constructed or enlarged after the female was finished ovipositing (=maternal care). The observations of McKnight and Aarhus (1973) support this view: in two *Hylesinus* species breeding in ash, the species breeding in live tissues (*H. californicus* (Swaine)) makes ventilation holes, while the species breeding in dead tissues (*H. criddlei* (Swaine)) does not.

5.3.4 Fungus Tending as Maternal Care

The most elaborate forms of maternal care are found in ambrosia beetles. In these, offspring survival and growth is largely dependent upon female fungus farming. Ambrosia beetle females plant and maintain a fungal food supply and hold pathogens in check. During construction of the egg tunnels, they disseminate fungal spores from their mycetangia to the tunnel walls. Subsequent beetle tending behavior strongly stimulates the growth in unknown ways (Francke-Grosmann, 1966; Happ *et al.*, 1975, 1976). *Crossotarsus japonicus* Blandford ambrosia beetles with oral mycetangia have been observed to spread oral secretions containing fungal spores on other individuals and on tunnel walls, via grooming and tending (Nakashima, 1971). The mother's cleaning and tending activity is essential for keeping fungal garden pathogens in check and to keep the ambrosia fungus from overgrowing immobile eggs and pupae (Hadorn, 1933; L. R. Batra, 1966; Francke-Grosmann, 1967; Biedermann and Taborsky, 2011). Oral applications of secretions to pathogen-infested areas by *D. rufipennis* females (not an ambrosia beetle) have clear antimicrobial effects (Cardoza *et al.*, 2006b). In ambrosia beetles of the tribe Xyleborini, mothers frequently groom their eggs, larvae, and pupae with their mouth parts and relocate brood with behaviors similar to shuffling frass (French and Roeper, 1975; Kingsolver and Norris, 1977a; Roeper *et al.*, 1980a; Biedermann and Taborsky, 2011). A remarkable development of relocation behavior is seen in females of some *Crossotarsus* species (Platypodinae) that have deep hollows in the frons, in which they can carry their eggs (and maybe small larvae) through the tunnel systems (Browne, 1961; Darling and Roberts, 1999). Finally, there are hints of active food provisioning in *Monarthrum fasciatum* (Say) and *Gnathotrichus* species (Scolytinae), in which larvae live in separate niches, where females have been observed to feed them with pieces of fungal mycelium (Hubbard, 1897; Doane and Gilliland, 1929).

5.3.5 Paternal Care

The benefits of prolonged male residency can be attributed to a mixture of sexual and natural selection, as indicated above. Some of the consequences of burrow blocking and frass removal increase the number of offspring, and some increase the survival of those offspring and hence can be considered paternal care. Paternal care is rare in insects and hence is of special interest; in the vast majority of species, males leave females after copulating with them and are not present when eggs are laid, precluding the evolution of males contributing directly to offspring survivorship. Given that mate abandonment is the norm, it is striking that male postcopulatory residency is so common in bark and ambrosia beetles and that male residency seems to significantly increase offspring survivorship as well as male fecundity (Kirkendall 1983; Reid and Roitberg, 1994, 1995; Kirkendall *et al.*, 1997; Lissemore, 1997; Robertson, 1998a, b; Robertson and Roitberg, 1998).

In *I. pini*, Robertson (1998b) found that the longer that males stay with females, the more eggs are laid, the longer the female egg tunnels are, and the less competition there is between larval progeny thus increasing offspring

survivorship. Such paternal care has long-lasting effects, as competition during larval development affects adult size; larger males attract more females (Robertson and Roitberg, 1998) and larger males and females produce larger broods (Foelker and Hofstetter, 2014). Experimental male removal in this species also had dramatic effects on increased predation by tenebrionid and colydiid beetles (Reid and Roitberg, 1994).

Thus, evidence from *I. pini*, *Crossotarsus*, and *Scolytoplatypus* supports interpreting male residence as being a form of paternal care. It is not at all clear yet if this conclusion applies more generally. As emphasized by Kirkendall *et al.* (1997), male residence is a key feature of almost all bark and ambrosia beetle mating systems, and the vast majority of outbreeding species are monogynous. Is male residency selected more strongly by sexual or natural selection? Comparative studies in genera with large variation in male behavior (such as in *Scolytus*, which includes a few species with no male residency at all) could provide key insights into the features selecting for and against postcopulatory residency, and the extent to which paternal care is a significant factor in species in which males stay for all or most of a female's reproduction.

The variability and evolution of male residency is discussed in detail in Kirkendall (1983) and in Kirkendall *et al.* (1997). There is little support in general for the hypothesis that males remain to increase their own mating success via mate guarding or attracting further females. Mate guarding, however, may be important in species in which males leave before oviposition commences; more importantly, mate guarding may have been the initial selective advantage to remaining some time with females after copulating with them. Mate guarding is posited to have preceded evolution of offspring care in other insects (Tallamy, 1994; Costa, 2010), and this is likely the case for bark and ambrosia beetles. Once males are regularly highly related to the offspring they are guarding (as would be assured by strict monogamy or repeated copulations with the same male), male and female reproductive interests are fully aligned, and division of labor between the sexes can evolve.

5.3.6 Males in Inbreeders

Despite the importance of males in outbreeding bark and ambrosia beetles, in inbreeding taxa, the significance of males for productivity appears to be negligible. Although all cooperative behaviors that are shown by adult females other than blocking are also present in male offspring of *X. saxesenii* and *X. affinis*, and all-male colonies (arising from unfertilized females that lay haploid eggs) are almost as productive as normal colonies, there are typically only one to three males per nest. With so few males present, male behaviors can have little impact on nest productivity (Biedermann, 2010, 2012; Biedermann and Taborsky, 2011).

What selects for diminished roles of males in these taxa? Sibling mating within subdivided family groups provides an arena for mate competition between relatives (Charnov, 1982), which favors producing lower numbers of males (i.e., local mate competition *sensu* Hamilton, 1967). In the most extreme cases this may lead to neoteny, as all resources that would have been utilized by males can be invested in dispersing females instead (Jordal *et al.*, 2002). Fighting may be expected among brothers, as in some sib-mating parasitoid wasps (Hamilton, 1978, 1979), but currently there is no evidence that sibling fighting takes place in inbreeding bark or ambrosia beetles; in many species it is extremely unlikely because they regularly produce only one male per brood, unless broods become so large that one male may not be able to fertilize all of his sisters (Kirkendall, 1983, 1993). The pronotal horns in males of several Xyleborini species (see Hubbard, 1897), which have been proposed to have a fighting role (Hamilton, 1979; Costa, 2006), have been observed to function as hooks to attach to the tunnel wall during copulation in *X. affinis* males (Biedermann, 2012). The only exception might be species with less biased sex ratios, in which male dispersal and outbreeding occurs regularly (*D. micans*, *D. punctatus*: Section 4.2). The possibility remains that the unusually large males of some inbreeders, or hooks or horns on inbreeding males, might be important in competition with unrelated, intruding males, but little is known as yet of how often unrelated males successfully enter nests and inseminate non-sisters.

5.4 Delayed Dispersal and Alloparental Care

The evolution of parental care given by siblings (a form of alloparental care) requires that generations overlap, i.e., that immatures are still present when the first offspring reach adulthood, and that caring for related juveniles for some time results in higher inclusive fitness than dispersing immediately and producing own offspring. The first stage in the evolution of alloparental care, given overlap of generations, is delayed dispersal. Once adults are present in the nest because they have delayed leaving, there is a potential for evolving to aid the reproductive efforts of their mother. No new behavior need evolve, beyond not dispersing or delaying dispersal: simply by carrying out the same behaviors they would normally employ while breeding themselves (burrow blocking, frass removal, fungus tending), they can increase the survivorship of their mother's family and perhaps increase their mother's total reproductive output as well by relieving her of some of her duties. Because bark and ambrosia beetles produce large eggs over a period of weeks or even months, overlap of generations is universal, so the potential for the evolution of alloparental care is high. What factors might create delayed dispersal, and can delayed dispersal lead to

significant alloparental care giving? What do we know about the costs and benefits of delayed dispersal for young adults? Can non-dispersing individuals breed further in the same host material?

5.4.1 Delayed Dispersal

Any plant tissues break down once dead, but wood degrades slowly, and woody tissues can potentially support several generations of wood-boring insects. The wood of live trees, however, is well protected, and can potentially support insect colonies as long as the tree lives. Dead wood, even in small branches, is a very large resource unit for tiny beetles, but one that is scattered and unpredictable in the environment. For scolytines and platypodines, locating new breeding material is energetically costly and associated with high levels of mortality. Dead wood degrades slowly but surely, with the rate of deterioration of individual resource units depending on the temperature and what other organisms have colonized the wood. Inner bark degrades much more rapidly than sapwood so more advanced forms of social behavior are more likely to evolve in xylomycetophagous species than phloeophagous species. Unconsumed, usable woody food resources are often still available for further breeding, even while offspring of the current brood are still maturing. Young adult beetles, then, have the options of (1) remaining in the current tunnel system for at least some time; (2) extending the current tunnel system and breeding in it; or (3) leaving and attempting to breed elsewhere. If they do remain, do they do anything that increases the survivorship of current juveniles?

Delayed dispersal of adult offspring is not common in nature, though characteristic of social taxa (Wilson, 1971; Costa, 2006). Adult offspring of bark and ambrosia beetle species are commonly observed to remain in the natal gallery system for days, weeks, or months after maturation (Fuchs, 1907; Kalshoven, 1962; Kirkendall *et al.*, 1997; McNee *et al.*, 2000; Peer and Taborsky, 2007; López-Guillén *et al.*, 2011; Biedermann *et al.*, 2012).

Direct and indirect fitness benefits at the natal nest can select for prolonged delayed dispersal. Delayed dispersal of females in *X. saxesenii* is affected by the quality and amount of ambrosia fungi (Biedermann and Taborsky, submitted). These gains might be either (1) direct through feeding up body reserves for later reproduction, through co-breeding within the natal nest, or through becoming the lone breeder, or (2) indirect through engaging in brood care and fungus tending, thus helping relatives to produce more brood and increasing offspring survivorship. Benefit (2) is especially relevant in species with extended egg laying periods of the mother, where adult and immature offspring stages overlap considerably and brood that is dependent on adult care are still present when the mother dies (e.g., *X. saxesenii*; Biedermann *et al.*, 2012).

The delayed dispersal of adult offspring in bark and ambrosia beetles was recognized by the pioneers of bark beetle research (Ratzeburg, 1839; Eichhoff, 1881; Hopkins, 1909). Typically, it has been attributed to having to build up energy reserves before dispersal, and this period of the life cycle is termed pre-emergence feeding or maturation feeding in these beetles (Eichhoff, 1881; Botterweg, 1982; McNee et al., 2000). Other reasons for at least short delays are adverse environmental conditions (especially cold temperatures or strong winds) that do not allow dispersal or host finding. Typical for poikilothermic animals, bark and ambrosia beetles are only active above certain temperatures and the favorable season for host finding of temperate species is typically in spring or early summer; the beetles are active year round in subtropical and tropical forests, barring prolonged dry seasons. Hence, adults often hibernate within their natal galleries instead of dispersing immediately after reaching adulthood.

Evidence that maturation feeding promotes delayed dispersal comes from phloem-feeding mountain pine beetles, where females were experimentally prevented from feeding after molting to adult. They matured normally, but were less likely to breed successfully, and laid smaller eggs (Elkin and Reid, 2005). Weather has been repeatedly shown to limit dispersal: dispersal is facilitated by sunny weather with little wind, minimum temperatures being in the range of 10–20 °C depending on the species (Bakke, 1968a, 1992; Salom and McLean, 1989, 1991; Faccoli and Rukalski, 2004) and high air pressure (Biedermann, 2012). Adults of some species live through unfavorable seasons or weather periods within their natal nests (other species survive in leaf litter, under bark of live trees, or in twigs of live branches).

Although maturation feeding and waiting for favorable environmental conditions are of primary importance in many taxa, especially in bark beetles in which individuals feed separately in their own cradles, it certainly cannot explain the extraordinarily long philopatric periods of adults in some ambrosia beetle species. Female ambrosia beetles were found to lay eggs only after growing their own fungus garden on which they fed (French and Roeper, 1975; Kingsolver and Norris, 1977b; Roeper et al., 1980a; Beaver, 1989). Hence, it is unlikely that reserves accumulated before emergence will raise the productivity of those beetles sufficiently to outweigh the fitness costs of delayed dispersal (the time lost from potential breeding). In some xyleborine ambrosia beetles, daughters even remain all their lives within their natal nests, e.g., X. affinis (Schneider, 1987) and X. saxesenii (Peer and Taborsky, 2007; Biedermann et al., 2012). Laboratory studies with X. affinis galleries in artificial medium showed that remaining adult females are fully capable of breeding independently when they are experimentally removed from their natal nest (Biedermann et al., 2011), which suggests that maturation feeding is not essential for egg laying. Surprisingly, and in contrast to the maturation feeding hypothesis, delayed dispersal comes at a cost for females. Xyleborus affinis females that disperse after their philopatric period produced fewer eggs than females removed from the gallery before their philopatric period (Biedermann et al., 2011). This cost may result from co-breeding or from engaging in alloparental brood care during the philopatric period at the natal nest.

It is likely that a combination of both direct and indirect benefits select for delayed dispersal in many ambrosia beetle species, as (1) ovary dissections revealed that one-quarter of staying females in X. saxesenii field galleries (Biedermann et al., 2012) and one-half in X. affinis laboratory galleries (Biedermann et al., 2011) lay eggs in the natal nest during their philopatric period, and (2) correlative studies indicate that staying and helping in the nest is triggered by demands of brood dependent on care. The latter was shown by increased social behavior of staying females and later dispersal in relation to both increasing numbers of sibling larvae and pupae (which depend on brood care) and decreasing numbers of adult "helpers" in both species (Biedermann et al., 2011; Biedermann and Taborsky, 2011). Numbers of egg layers correlated with neither the number of staying adult females nor with the number of eggs, which suggests that egg numbers are regulated and adjusted to fungus productivity (Biedermann et al., 2012).

A selection experiment on timing of dispersal in X. saxesenii showed that delayed dispersal and helping in this species and are probably genetically linked (Biedermann and Taborsky, submitted).

Finally, helping in adults can probably evolve relatively easily, as it seems not to strongly curtail a helper's future reproduction because helping is risk free and does not reduce a helper's energy stores. The tradeoff between helping and future reproduction (Queller and Strassmann, 1998; Korb and Heinze, 2008) may thus be weak in such ambrosia beetles. This may also explain why helping is even present in male offspring of the haplodiploid X. saxesenii and X. affinis. Unexpectedly, recent observations indicate that they take part in all cooperative behaviors that are shown by adult females except for blocking (Biedermann, 2010, 2012; Biedermann and Taborsky, 2011), which suggests that relatedness asymmetries caused by haplodiploidy, which would favor female-biased help, are probably offset by inbreeding in these species (Hamilton, 1972). Nevertheless, because of strong local male competition, there are only one or two males and up to 80 females per gallery, and thus their help is of minor importance.

Several factors disfavor the evolution of delayed dispersal of adult offspring, even when food conditions would allow adult offspring to establish a second generation at the natal nest site (Gandon, 1999): (1) a buildup during the breeding period of predators, parasites (e.g., mites, nematodes, parasitoids) and pathogens (e.g., fungal saprobes) (Dahlsten, 1982; Hofstetter *et al.*, 2006; Cardoza *et al.*, 2008; Hofstetter and Moser, 2014); (2) problems in relation to inbreeding, if unrelated mates are not available (Thornhill, 1993; Gandon, 1999); (3) competition among closely related individuals (Kirkendall *et al.*, 1997; West *et al.*, 2002); and (4) the relatively small potential for indirect fitness benefits at the natal nest for beetles that live within their food compared to other social insects that need to forage for their food (Mueller *et al.*, 2005; Biedermann, 2012).

These four factors may all present serious obstacles that might often hinder the evolution of forms of sociality beyond parental care, although the importance of these factors has not been studied in bark and ambrosia beetles. Consequently, bark and ambrosia beetle social systems exceeding subsociality must have evolved mechanisms to overcome or handle these obstacles. Mechanisms increasing social immunity (blocking out of predators and parasites, and gallery hygienic tasks to keep pathogens and diseases in check), and fungiculture techniques that assure a long-term food supply, have likely improved in the course of bark and ambrosia beetle social evolution, as seen in other fungus-farming social insects (Cremer *et al.*, 2007; Hölldobler and Wilson, 2009; Wilson-Rich *et al.*, 2009). Pseudo-arrhenotoky in *H. hampei* (Brun *et al.*, 1995a, b; Borsa and Kjellberg, 1996a, b) and haplodiploidy in the Xyleborini clade (Normark *et al.*, 1999) may mitigate the potential hindrance of inbreeding, by allowing the purging of deleterious mutations through haplodiploid males (Hamilton, 1967; Smith, 2000).

5.5 Larval Cooperation

Some bark and ambrosia beetles not only have adult helpers at the natal nest, but can also have larvae that cooperate and may engage in division of labor with the adults (Biedermann and Taborsky, 2011). Although data on larval behavior in these beetles are mostly anecdotic, it could be a common phenomenon in species with gregariously feeding offspring and in which adults and larvae can move freely within their nests. Larval cooperation has been experimentally proven only in the ambrosia beetle *X. saxesenii* (Biedermann and Taborsky, 2011), but observations suggesting larval cooperation come also from the phloem feeding *D. micans*, *D. valens*, and *D. punctatus* (Grégoire *et al.*, 1981; Deneubourg *et al.*, 1990; Furniss, 1995) and from other ambrosia feeding Xyleborini (*X. affinis*: Biedermann, 2012) and Platypodinae, *Platypus cylindrus* (F.) (Strohmeyer, 1906), *Trachyostus ghanaensis* Schedl, *T. aterrimus* (Schaufuss), *T. schaufussi* Schedl (Roberts, 1968), *Doliopygus conradti* (Strohmeyer), and *D. dubius* (Sampson) (Browne, 1963).

Remarkably, this division of labor between adult and immature stages is almost unique among social insects. Helper or worker castes in insects without metamorphosis (Hemimetabola), like aphids or termites, are always formed by immature individuals, whereas in insects with metamorphosis (Holometabola), such as beetles and Hymenoptera, workers are typically adults, as immature individuals in ant, wasp, and bee societies are largely immobile, helpless, and often dependent on adults to be moved and fed (Wilson, 1971; Choe and Crespi, 1997). There are very few exceptions of cooperatively behaving immatures in Hymenoptera, including nest-building-silk producing weaver ant larvae (Wilson and Hölldobler, 1980) and nutrient and enzyme producing larvae of some wasp and ant species (Ishay and Ikan, 1968; Hunt *et al.*, 1982).

What does larval cooperation in bark and ambrosia beetles look like? In phloem feeders larvae cooperate primarily by feeding side by side, which helps them to overcome plant defenses, and aggregation is effected by pheromones (Grégoire *et al.*, 1981; Deneubourg *et al.*, 1990; Storer *et al.*, 1997). Gregarious feeding is also known from the ambrosia beetle genus *Xyleborinus*, in which larvae feed not only on fungal mycelia (as typical), but also on fungus-infested wood. Aggregation pheromones have not been studied in ambrosia beetle larvae, but it is likely that gregarious feeding may more effectively control fungal saprobes threatening their primary ambrosia food fungus (Biedermann, unpubl.; Biedermann and Taborsky, 2011). Like gregarious feeding on plants, gregarious feeding on fungi has been repeatedly found to be an adaptation of arthropods to overcome the induction of secondary fungal defenses (Rohlfs, 2005; Rohlfs and Churchill, 2011).

Larvae take part in gallery hygiene, by relocating frass and by grooming eggs, pupae, each other, and adults; these behaviors have been widely reported from different bark and ambrosia beetle species. In *X. saxesenii*, larvae ball up frass, which can then be more easily removed by their adult siblings (Biedermann and Taborsky, 2011). In *D. micans*, larvae pack frass at specific locations, allowing free movement within the brood chamber (Grégoire *et al.*, 1981); they also block tunnels to hinder access by *R. grandis* predators (Koch, 1909). Fifth instar larvae in some Platypodinae also relocate frass to unused gallery parts or for plugging artificial nest openings (Hadorn, 1933; Beeson, 1941; Kalshoven, 1959) and expel frass and parasitoid planidia through the nest entrance (Darling and Roberts, 1999). These larvae have a plug-like last abdominal segment, which can be used both as a shovel and as a device to fully plug the gallery entrance against intruders (Strohmeyer, 1906). These larvae have been observed to overtake the role of entrance blocker during times when their parents are deep inside the nest (Strohmeyer, 1906; Roberts, 1968). In both

Platypodinae and many Xyleborini, larvae also engage in excavation of new tunnels or chambers to create more surface for the developing ambrosia fungus (Strohmeyer, 1906; Kent, 2002; Biedermann and Taborsky, 2011). The flat brood chambers that are typically found in the genus *Xyleborinus* are almost exclusively accomplished by the larval habit of feeding on fungus-infested wood (Biedermann and Taborsky, 2011; De Fine Licht and Biedermann, 2012). The same is true for the long transverse tunnels in nests of several Platypodinae that are bored by fifth instar larvae (Roberts, 1962, 1968; Browne, 1972).

The ultimate cause for the larval specialization for tunneling shown by many ambrosia beetles may relate to their repeated molting: mandibles of adults gradually wear down during excavation, and adult females that bore extensively would suffer from substantial long-term costs. In contrast, larval mandibles regenerate at each molt (Biedermann and Taborsky, 2011).

Xyleborinus saxesenii larvae that feed on fungus-infested wood likely fertilize the growing ambrosia fungus with the finely fragmented woody sawdust in their feces, which gets smeared on the gallery walls after defecation (Hubbard 1897; Biedermann and Taborsky, 2011). This larval frass probably also contains enzymes for further wood degradation, as a recent study showed that *X. saxesenii* larvae possess hemicellulases, which are not found in their adult siblings (De Fine Licht and Biedermann, 2012). Furthermore, bark and ambrosia beetle larvae may spread associated bacterial and fungal symbionts within the galleries, which have been shown to have defensive functions against pathogens, detoxify poisonous plant metabolites, degrade lignocellulose plant cell walls, or fix nitrogen from the air (Cardoza *et al.*, 2006b; Adams *et al.*, 2008; Scott *et al.*, 2008; Morales-Jiménez *et al.*, 2013; Chapter 6). This suggests that cooperation, and division of labor among larvae and adults, goes far beyond behavioral interactions, but may also include microbial, biochemical, and enzymatic processes.

Larval contributions to gallery extension and to hygiene reduce the workload for adults. Indeed, and against the common preconception that larvae only compete for resources among each other, positive effects of larval numbers on group productivity have been observed in *X. saxesenii* (Biedermann and Taborsky, 2011), *D. micans* (Storer *et al.*, 1997), and several Platypodinae species, in which females only lay second egg clutches in the presence of fifth instar larval helpers (Roberts, 1968).

In summary, larvae in some bark and many ambrosia beetle species are free to move within the natal nest, and are not confined to small areas or brood cells like those of most hymenopteran social societies (Wilson, 1971; Hölldobler and Wilson, 1990). This, in combination with different capabilities of larvae and adults, predisposes especially ambrosia beetles for division of labor between

larval and adult stages. Importance and specific roles of larvae in the galleries appear to vary between species (Biedermann, 2012).

One aspect that has not been studied at all in bark and ambrosia beetles is the possibility of delayed development of larvae. If larvae play such an important role in the nests of many gregarious bark and ambrosia beetle species and there are possibilities for larvae to gain indirect fitness benefits by cooperating in the natal nest, selection may favor prolonged development (e.g., by additional larval instars). Prolonged development or even permanently immature helper/worker castes are the rule in hemimetabolous social insects like termites, aphids, or thrips, in which individuals only mature to become reproductive queens or kings (Choe and Crespi, 1997; Korb and Heinze, 2008). There are two hints for prolonged development also in larvae of bark and ambrosia beetles. First, the number of larval instars varies between two and five among species in bark and ambrosia beetles; it is unknown what factors select for more or fewer instars. The numbers of instars are sometimes, but not always, related to size of the adult (Lekander, 1962; Lekander, 1968a, b). Second, among species with helping larvae (*Dendroctonus*, Xyleborini, Platypodinae) and for reasons that remain unclear, there appears to be high variability in the developmental periods of larvae (Wichmann, 1927). Koch (1909) observed that from *D. micans* eggs laid the same day, the progeny pupated over a period of 44 days without any obvious reasons. While the first larval instars are typically short and quite fixed in time, the length of the last instar is highly plastic and in some cases two to four times longer than all previous instars together (Koch, 1909; Baker, 1963; Browne, 1963; Biedermann *et al.*, 2009). Generally, the last instar is typically the one that overtakes most helping and has evolved even some morphological adaptations for helping (see above). The maximum of five instars and the longest development of larvae (which can be several years) relative to the oviposition period of adults are both found in Platypodinae (Kirkendall *et al.*, 1997). Unfortunately, researchers have rarely reported larval numbers when dissecting galleries, and experimental studies are lacking, so prolonged development of larvae as an investment in siblings must remain speculative.

5.6 The Evolution of Reproductive Altruism

The frequent occurrence of overlapping generations and cooperative brood care in this group of beetles suggests that reproductive altruism may be more widespread than currently known. In Xyleborini, Corthylini, and Platypodinae, there are several species in which adult females have been observed to delay reproduction. In a single species, *X. affinis*, delayed dispersal and helping at the natal nest have been experimentally shown to involve fitness costs on future independent breeding. Adult daughters remaining longer in their mother's nest produced a significantly smaller brood when given their own choice to breed, than adult females experimentally removed from the nest before their delayed dispersal period (Biedermann *et al.*, 2011). As only some of the females that delayed breeding bred together with the mother, this implies that helping at the natal nest is costly for adult females in ambrosia beetles. Similarly, in *X. saxesenii*, there are hints that some daughters remain, never breed, and die within their mother's nests (Peer and Taborsky, 2007). Sterile adult female worker castes seem to be present in *A. incompertus* (Harris *et al.*, 1976; Kent and Simpson, 1992), although it has not yet been fully proven that sterility is non-reversible in the case when the mother dies (Kirkendall *et al.*, 1997). Furthermore, while many cooperative behaviors of larvae and adults are probably relatively inexpensive in terms of fitness, blocking of the gallery entrance is dangerous and costly (Kirkendall *et al.*, 1997). Feeding and blocking are incompatible and blocking individuals have been observed to be attacked by parasitoids (Beaver, 1986) or killed by predators (Wichmann, 1967). Hence, blocking can be interpreted as self-sacrificing altruism in those Cryphalini, Xyleborini, and Platypodinae in which larvae (*P. cylindrus*: Strohmeyer, 1906) or non-reproducing adult offspring (*H. curtipennis*: Beaver, 1986; *X. saxesenii*: Biedermann and Taborsky, 2011; *X. germanus*: Peer and Taborsky, 2004; *A. incompertus*: Kent, 2002) have been observed to take turns in blocking of the nest. This suggests that facultative (or even obligate) eusociality, defined by overlap of parental and offspring generations, alloparental brood care, and facultative (or permanent) reproductive altruism of some individuals (S. W. T. Batra, 1966; Wilson, 1971) have evolved multiple times in ambrosia beetles.

How is reproductive altruism favored by natural selection? Similar factors that facilitate the evolution of alloparental care also predispose for reproductive altruism. Kin selection is certainly essential, and all current evidence indicates that altruism can only evolve in groups of relatives, in which individuals invest in the reproduction of own genes via related individuals (Hamilton, 1964; Boomsma, 2013). More specifically, studies have shown that permanently sterile castes can only evolve if colony foundation is by a single, monogamously mated female, which assures high relatedness within her offspring group. This way, relatedness between colony females equals relatedness of a female to her own potential offspring; then, any constraint on individual reproduction can favor the evolution of staying, helping, and ultimately (under the right conditions) of sterility of helpers (Boomsma, 2009, 2013). Single gallery foundation and monogamy can be found in some bark beetles and is the rule in ambrosia beetles (Kirkendall, 1993), which suggests that the precondition for altruism to evolve is present in many species.

There are severe constraints on dispersal and individual reproduction in bark and ambrosia beetles. Costs of dispersal depend on the species, but in general it seems difficult for beetles to find suitable host trees and establish galleries (Berryman, 1982). Mortality during dispersal flight is about 50% for bark beetles (Klein *et al.*, 1978; Garraway and Freeman, 1981) and 70–80% for an ambrosia beetle (Milne and Giese, 1970), and survival decreases rapidly after the first day of host search (Pope *et al.*, 1980), typically because individuals are exposed to predation pressure and adverse weather conditions, but also because they exhaust fat reserves necessary for flying. Successful gallery establishment is also difficult as bark and ambrosia beetles have specific requirements for their breeding material, like plant taxon, size of material, moisture content, and the presence or absence of certain fungi or other microorganisms. Although ambrosia beetles are typically less specialized to host taxa (Browne, 1958; Beaver, 1977, 1979a; Atkinson and Equihua-Martínez, 1986b), boring in solid wood, overcoming host tree defenses (e.g., resins), and planting of fungal cultivars are risky tasks. Often, less than half of females successfully manage the last step (Fischer, 1954; Hosking, 1972; Nord, 1972; Weber and McPherson, 1983; Biedermann *et al.*, 2009), typically because either the ambrosia fungus does not grow or fungal pathogens overgrow the initial cultures (Biedermann, 2012; Biedermann *et al.*, 2013). All these factors render pre-dispersal cooperation and altruism more profitable, if longevity of the natal gallery allows adults to gain inclusive fitness benefits.

The longevity of the breeding material is likely the crucial factor that will affect evolution of cooperation and reproductive altruism. This depends on competition with other ambrosia beetles and microorganisms, timing of beetle attack in the dying process of a tree (in cases where breeding is in dead hosts), and size and type of host material that is attacked. Reproductive altruism without sterility can evolve in species attacking dying or dead trees of large diameter as long as they provide resources for several generations of offspring, as seen in *X. affinis*, *X. saxesenii*, and probably other Platypodinae and Xyleborini (see above; Biedermann, 2012). Facultative suppression of oviposition assures that females can disperse and breed independently should the breeding substrate degenerate, and permits further inclusive fitness gains from helping at the natal nest. In *X. saxesenii*, many galleries need to be abandoned after a single generation, despite the fact that other galleries are productive for several offspring cycles. Obligate sterility of adults, however, is expected only to evolve under conditions that consistently provide non-breeding females with indirect fitness gains. This is the case when beetles colonize living trees, which can provide food for many consecutive offspring generations. The only currently known case of obligate eusociality in beetles is found in *A. incompertus*, which attacks living trees and constructs galleries that may last for more than 30 years (Kent and Simpson, 1992). Several more ambrosia beetles breed in living trees, so more cases of obligate eusociality may be discovered in the future (Kirkendall *et al.*, 1997). These systems should have evolved elaborate techniques for maintaining long-term fungiculture and social immunity, such as mechanisms to suppress the spread of fungus-garden pathogens and insect diseases, as have evolved in societies of fungus-farming ants (Currie, 2001). Unexpected discoveries are likely when more researchers have started to work with platypodine ambrosia beetles, especially those in living trees.

6. INTRACELLULAR BACTERIA AND BARK BEETLE EVOLUTION

Because of their potential influences on the evolution of bark and ambrosia beetles, we conclude with a brief discussion of what little we know about intracellular bacteria in Scolytinae (nothing is known for Platypodinae). Intracellular symbionts in the alpha-proteobacterial genera *Wolbachia* and *Rickettsia* are widespread in arthropods and nematodes, with *Wolbachia* present in 70% of all insects (Werren *et al.*, 2008). Bark beetles are no exception and despite the lack of a detailed survey, single screenings have identified *Wolbachia* bacteria in Ipini (*I. typographus*: Stauffer *et al.*, 1997; *P. chalcographus*: Arthofer *et al.*, 2009), Xyleborini (*X. germanus*: Kawasaki *et al.*, 2010), Dryocoetini (*H. hampei*: Vega *et al.*, 2002), and Cryphalini (*Coccotrypes dactiliperda*: Zchori-Fein *et al.*, 2006). In the evolution of insect mating systems, these symbionts are important, as they have repeatedly been shown to be able to manipulate host reproductive biology and evolution (see review by Werren *et al.*, 2008).

Wolbachia, the best studied of these intracellular parasites, is vertically transmitted with the egg from an infected female to her progeny and not via males. *Wolbachia* has a variety of phenotypic effects on its host, including (1) feminization (genetic males develop into females); (2) parthenogenesis; (3) selective male killing; and (4) cytoplasmic incompatibility (prevents infected males from successfully fertilizing eggs of females that lack the same *Wolbachia* types) (Werren *et al.*, 2008). In bark beetles, the role of *Wolbachia* and other intracellular symbionts for host reproduction remains largely unstudied.

It would be interesting to determine if extreme sex ratios in inbreeding Scolytinae are in any way caused by *Wolbachia* infections. This is unlikely, however, given that the extremely female biased sex ratios in regular inbreeders are predicted by local mate competition theory, and in most cases are extremely precise (Hamilton, 1967; Kirkendall, 1983, 1993; Borsa and Kjellberg, 1996a, b; Biedermann, 2010). In the only study on this topic, Zchori-Fein et al. (2006) found no evidence for an influence of *Wolbachia*

on sex ratios in *C. dactyliperda*. Instead, these authors showed that the elimination of both *Wolbachia* and *Rickettsia* by antibiotic treatment led to unfertile females with no sign of oogenesis. Accordingly, also *Xyleborus ferrugineus* (F.) ambrosia beetles cannot reproduce after elimination of their unknown intracellular symbionts (Peleg and Norris, 1973; Norris and Chu, 1980). This may indicate that *Wolbachia* have changed their phenotype from reproductive parasitism to obligate mutualism in these inbreeding scolytids and the hosts are now dependent on the symbionts for oogenesis and/or nutrition, as clearly shown for other arthropods (Dedeine *et al.*, 2001; Hosokawa *et al.*, 2010). However, does *Wolbachia* also affect the evolution of their hosts? Generally, there is strong evidence that infections lead to inbreeding and thus drive speciation (Bordenstein *et al.*, 2001; Brucker and Bordenstein, 2012). Superinfection with up to five different *Wolbachia* strains per female (Kawasaki *et al.*, 2010) is likely responsible for smaller broods produced by females mated with males other than their brothers in the xyleborine ambrosia beetle *X. germanus* (Peer and Taborsky, 2005). This outbreeding depression could be caused by cytoplasmic incompatibility, as egg numbers between outbreeding and inbreeding broods were equal, but hatching rates differed (Peer and Taborsky, 2005). Whether such outbreeding depression is common in other inbreeding bark beetles has not been investigated. Finally, *Wolbachia* have also been hypothesized to play a role in the evolution of haplodiploidy in inbreeding taxa (Normark, 2004). Engelstädter and Hurst (2006) showed that paternal genome exclusion, which can be a predecessor of haplodiploidy, could be caused by cytoplasmic incompatibility-inducing bacteria in eggs of incompatible crosses, rendering the embryo functionally haploid. Paternal genome exclusion as well as *Wolbachia* are present in *H. hampei* (Brun *et al.*, 1995a, b; Vega *et al.*, 2002), which strongly suggests that the genetic system of bark beetles may be influenced by intracellular bacterial symbionts.

The abundance and effect of *Wolbachia* across outbreeding bark and ambrosia beetle is largely unknown. *Wolbachia* are present in *I. typographus* (Stauffer *et al.*, 1997) and *P. chalcographus* (Arthofer *et al.*, 2009) at low titer (35.5% of all sampled individuals infected) and at low density within infected individuals, and no correlation between infection titer and host population or geographic location was found. At least for *P. chalcographus* this suggests either that populations currently evolve towards the loss of *Wolbachia* or unidentified fitness advantages conserve the infection by the symbiont under certain environmental conditions (Arthofer *et al.*, 2009). Hypothetically, bark beetle associated fungi may help beetle hosts to cure themselves from parasitic symbionts (Arthofer *et al.*, 2009), as these fungi are known to produce a rich array of antibiotics (Zrimec *et al.*, 2004). It is possible that *Wolbachia* is repeatedly reacquired by the beetles within their feeding habitat (e.g., Stahlhut *et al.*, 2010).

7. CONCLUSION

Over 100 million years ago, several early lineages of weevils began laying eggs in tunnels under bark rather than in slits cut with their snouts. Two of these, Scolytinae (6000 species) and Platypodinae (1400 species), achieved notable evolutionary and ecological success. Their shift from an herbivorous to a saproxylic lifestyle led rapidly to a series of morphological and behavioral adjustments, adaptations we also see in a variety of other wood-boring beetles. Subsequent key innovations included male residence and monogyny, the development of active fungus cultivation, the evolution of alternative mating systems such as inbreeding and simultaneous polygyny, and haploidiploidy. Central was the adoption of living in tunnels within their food source: tunnels in wood are easily defended, and encourage long residency, which in turn fosters various elements of social behavior.

The variation we have documented in this chapter illustrates the potential for testing a multitude of general hypotheses in behavioral ecology and evolutionary biology. We will soon have the tools to test such hypotheses using the comparative method. Until very recently, most phylogenetic work has been limited in resolution and extent and is therefore of limited value for this purpose. These problems will be resolved in the next few years by several current projects dealing with large-scale weevil and scolytine phylogenetics. The 1000-Curculionidae project is based in part on phylogenomics work using conserved anchored genome regions; it is expected that most weevil relationships will be well resolved, including the position of Scolytinae, Platypodinae, and Cossoninae. The same technology is currently used to develop data matrices for Cryphalini and Xyleborini, and ultimately to develop further a soon-to-be published 20-gene phylogeny of Scolytinae (Pistone and Jordal, in progress).

Hopefully, advances being made by applying the comparative method to a broad selection of taxa will be accompanied by (or will inspire) complementary experimental research. From the perspective of evolutionary biology, four areas discussed in detail in this chapter seem especially promising for such a combined approach: mating system evolution, sexual selection, inbreeding, and social behavior. But in addition, for a topic not covered by us, we would point out that the application of sound phylogenies to existing data on pheromone components will generate important insights into how pheromone systems evolve over time, and into the broad question of how such signaling behavior does or does not constrain the adoption of new hosts (since some components of pheromones are modified plant compounds). Analyses such as these would also point out the major gaps in our knowledge of bark and ambrosia beetle pheromones: almost nothing is known, for example, of the pheromone systems of tropical genera.

7.1 Mating System Evolution

As we have documented, bark and ambrosia beetles provide behavioral ecologists with multiple origins of mating systems otherwise rare in invertebrates (and often rare or nonexistent in vertebrates). Surely, both comparative and experimental studies of selected Scolytinae (but also of conoderine and cossonine weevils with convergent biology) would contribute considerably to our general understanding of mating system evolution and allow testing of hypotheses largely investigated only in birds or fishes. As noted above, there are genera and even species (or species complexes) that vary in their mating systems, and that make tempting targets for such research. There are many abundant and widespread temperate species that are amenable to research into the details of monogyny, harem polygyny, and inbreeding. Phloeophagous and spermatophagous species in particular are easily reared in the laboratory, and both fecundity and egg to adult survivorship easily measured. The fact that most species commonly occur in dense breeding aggregations makes it easy to gather large amounts of data and facilitates thorough replication of experimental treatments (such as removal or addition of males or females).

7.2 Sexual Selection

Although complete sexual role reversal is rare in insects, there are surprisingly many cases of males being selective about which females they mate with. Male mate choice is believed to occur in at least 58 insect species from 37 families and 11 orders, including *I. pini* and *I. acuminatus*, which we discussed earlier (Bonduriansky, 2001). Bonduriansky (2001) finds that male choosiness in Coleoptera is favored, for example, when both sexes occur in dense aggregations and there are low search costs, a common scenario for bark and ambrosia beetles. Also favoring male choosiness is costly male investment in mating, which could be the case with male-initiated tunnels and subsequent helping activities. Male choosiness can evolve if there is large variation in female quality; in bark beetles, this can be reflected in body size variation (strongly correlated with fecundity). Investigating the extent and nature of sex role reversal in Scolytinae and Platypodinae should be a priority for bark beetle behavioral ecologists. This should be done both as a broad comparative study and by the close study of key genera with such variation (such as *Scolytus*, *Phloeosinus*, *Hylesinus*, and *Pityophthorus + Conophthorus*) and species in which role reversal seems to be actively evolving (e.g., *H. varius*). Whether males select females or vice versa is controversial for *I. pini*, a common and widespread North American species deserving further attention in this regard.

We rely heavily on features of the declivity for identifying species of bark and ambrosia beetles, yet we know little of the adaptive significance of the enormous variation we encounter in this key feature. Extreme developments of sharp points and edges combined with deep declivities seems to be associated with taking over the nests of other species, but exactly how such structures are employed is unknown. It is tempting to attribute more modest variation in declivity form and ornamentation to sexual selection in the context of courtship, but there is considerable variation in the declivities of female Xyleborini as well, all of which are inbreeders, which as far as is known have only rudimentary courtship and presumably no intersexual selection. So, the questions arise, how do sexual selection and natural selection interact in sculpting this part of beetle bodies, in outbreeding species, and how does the adoption of inbreeding impact selection on declivities?

7.3   Inbreeding

We have only begun to understand the evolution and ecology of inbreeding in insects, and in these beetles in particular. There are several outstanding questions with regards to Scolytinae that inbreed.

Generally, the distribution of genetic variation (at single loci, but also variation in combinations of alleles over several loci) within individuals, families, populations, and regions has important consequences for the evolutionary fate and ecological impact of species. Extreme inbreeding is expected to generate homozygotic genotypes, and small populations should lose variation among genotypes to genetic drift. Small amounts of outbreeding, however, could have enormous consequences. How often do inbreeders outbreed? How often do males disperse, and how often do they succeed in entering other nests and mating with non-sisters? Are matings between non-siblings "effective" outbreeding: in a local population, what is the degree of relatedness between females and foreign males? How are populations of regular inbreeders structured? Besides these key questions, it is important to investigate the extent of outbreeding depression in regular inbreeders.

A few genetic studies of inbreeders are mentioned in Section 4.2, but these only begin to scratch the surface. We need ecological genetics studies of both indigenous and invasive species, and of lineages with a wide variety of ecological specializations.

We repeat that the highly unusual paternal genome loss system reported for *H. hampei* has only been demonstrated for that one species. Taken together with the related inbreeding genera, this is a lineage of over 200 species. It would be interesting to know if other inbreeders from this clade share this rare breeding system.

7.4   Social Evolution

As with inbreeding, we are only beginning to explore the rich variation in adult and larval social behaviors in these beetles. Only a few of the many potentially social species have been studied behaviorally. The most interesting forms of cooperative behavior seem to be in ambrosia beetles, but these are particularly difficult to observe since they tunnel deep in wood. Observing ambrosia beetle behavior requires establishing them on semi-artificial media in the laboratory, which is quite labor intensive. The last decade has seen major advances in the ability to rear and observe ambrosia beetles, making this group more accessible to researchers interested in social behavior, and should lead to the development of several more potential model systems. Thus far, though, only xyleborine ambrosia beetles have been reared, and a broader understanding of the ecology of social behavior in bark and ambrosia beetles will depend on establishing species from other lineages in the laboratory.

The relative importance of genetic and ecological factors in social evolution is still unclear. Scolytinae and Platypodinae vary in the way they colonize new breeding material (in large aggregations, or single individuals), uni- or biparental care, alloparental care by larvae or adults, and occurrence of division of labor. Further, subsocial species breed in a wide variety of substrates and ecosystems.

Fungus farming seems to provide a variety of opportunities for division of labor, hence the repeated evolution of alloparental care and forms of larval cooperation in ambrosia beetles. Future research, using well-established model systems, should investigate the mechanisms by which these beetles can actively promote the growth of their fungal cultivars and protect them from pathogens, and can induce the specialized "ambrosial" growth forms seen in their tunnels. Careful observations of larvae and adults can elucidate the roles they each play, and look for previously unknown expressions of altruistic behavior.

APPENDIX

Larval feeding modes and adult mating systems in Scolytinae, with the total number of species given for each genus. Rare occurences in a genus (one or a few species) are coded "(x)" and unknown mating behavior or feeding modes are indicated by "?". Abbreviations, larval (and usually adult) feeding: Phl, phloeophagy (feeding in inner bark); Xlm, xylomycetophagy (farming fungus); Spm, spermatophagy (feeding in seeds); Myc, feeding on free-living fungi; Mye, myelophagy (feeding in pith); Xyl, xylophagy (feeding in wood); Hbv, herbiphagy (feeding in non-woody plants); feed?, unknown larval feeding habits. Abbreviations, mating systems: MG, monogyny; HP, harem polygyny; BG, bigyny; Col, colonial polygyny (several males and several females in a gallery system); Inbr, inbreeding; MS?, unknown mating system. The list of tribes and genera and the numbers of species were compiled by T. H. Atkinson (see Chapter 2).

Tribe	Genus	Phl	Xlm	Spm	Myc	Mye	Xly	Hbv	feed?	MG	HP	BG	Col	Inbr	MS?	Spp
Amphiscolytini	Amphiscolytus								?						?	1
Bothrosternini	Akrobothrus								?						?	1
	Bothrosternus					x								(x)		11
	Cnesinus	x	(x)			x				x						95
	Eupagiocerus		(x)			x									?	3
	Pagiocerus			x						x						5
	Sternobothrus					x				x						16
Cactopinini	Cactopinus							x		x						21
Carphodicticini	Carphodicticus	x								x						1
	Craniodicticus						x								?	3
	Dendrodicticus								?						?	1
Corthylini	Amphicranus		x							x	(x)					66
	Araptus			(x)					?		x	(x)		(x)	?	172
	Brachyspartus								?							1
	Conophthorus			x						x						13
	Corthylocurus		x							x						15
	Corthyloxiphus		x							x						21
	Corthylus		x							x	(x)					159
	Dacnophthorus	x								x						5
	Dendroterus	x									x					15
	Glochinocerus									x						2
	Gnatharus								?							1
	Gnatholeptus	x								x						4
	Gnathotrichus		x							x	x					16
	Gnathotrupes		x							x						30
	Metacorthylus		x							x						13
	Microcorthylus		x							x						38
	Mimiocurus	x					x			x						15
	Monarthrum		x								x				x	140

continued

Tribe	Genus	Phl	Xlm	Spm	Myc	Mye	Xly	Hbv	feed?	MG	HP	BG	Col	Inbr	MS?	Spp
	Phelloterus	x									x					3
	Phloeoterus	x									x					1
	Pityoborus		x							x						7
	Pityodendron								?						?	1
	Pityophthorus	x				x				x	x	(x)				385
	Pityotrichus	x								x						3
	Pseudopityophthorus	x								x						27
	Sauroptilius								?						?	1
	Spermophthorus			x											?	2
	Styphlosoma	x								x	x					4
	Tricolus		x						?	x						50
	Urocorthylus								?						?	1
Cryphalini	*Acorthylus*	x								x						6
	Allemoporus	x								x					?	1
	Allothenemus	x							?						?	1
	Coriacephilus	x													?	5
	Cosmoderes					?									?	20
	Cryphalogenes							x		x						2
	Cryphalus	x								x						190
	Cryptocarenus		no?	(x)		x								x		16
	Eidophelus	x										?				5
	Ernocladius	LK								x						2
	Ernoporicus	x								x						15
	Ernoporus	x								x						16
	Hemicryphalus														?	7
	Hypocryphalus	x								x						52
	Hypothenemus	x	(x)	x				(x)	?					x		183
	Margadillius			x					?					x		13
	Neocryphus								?						?	2

Genus	No.	C1	C2	C3	C4	C5	C6	C7	C8	C9	C10	C11	C12	C13	C14
Periocryphalus	2		x								x				
Procryphalus	3														x
Ptilopodius	17		1?				x								x
Scolytogenes	107						x				x				x
Stegomerus	7														x
Stephanopodius	6						x			x					x
Trischidias	7		x									x			
Trypophloeus	17						x								x
Crypturgini															
Aphanarthrum	29			(x)			x		x						
Cisurgus	9														
Crypturgus	15			x			x		x						x
Deropria	1	?								(x)					x
Dolurgus	1						x								x
Diamerini															
Acacicis	11						x								x
Bothrosternoides	1	?						?							
Diamerus	34						x		x						x
Peronophorus	5						x								x
Pseudodiamerus	3						x								x
Sphaerotrypes	47						x								x
Strombophorus	31						x			x					x
Dryocoetini															
Chiloxylon	1	?						?					x		
Coccotrypes	129		x										x		x
Cynanchophagus	1							?					x		
Cyrtogenius	106			x			x			x					x
Dactylotrypes	1						x								
Dendrocranulus	43				(x)				(x)						
Dryocoetes	37					x					x				
Dryocoetiops	18		x?								x			x	
Lymantor	4						x								
Ozopemon	21		x												x

continued

Tribe	Genus	Phl	Xlm	Spm	Myc	Mye	Xly	Hbv	feed?	MG	HP	BG	Col	Inbr	MS?	Spp
	Peridryocoetes	?													?	6
	Pseudothamnurgus								?							5
	Taphronurgus	x						x								1
	Taphrorychus										x					19
	Thamnurgus							x								33
	Tiarophorus	?													?	8
	Triotemnus	x													?	15
	Xylocleptes							x		x						26
Hexacolini	*Gymnochilus*	x								x						9
	Microborus	x								x						8
	Pycnarthrum	x								x						18
	Scolytodes	x	(x)					(x)		x	x					207
Hylastini	*Hylastes*	x								x						32
	Hylurgops	x								x						21
	Scierus	x								x						2
Hylesinini	*Alniphagus*	x								x						3
	Cryptocurus								?						?	1
	Dactylipalpus						x			x						11
	Ficicis	x								x						14
	Hapalogenius						x			x					?	32
	Hylastinus	x						x		x						4
	Hylesinopsis	x					(x)			x						16
	Hylesinus	x					x			x		(x)				37
	Kissophagus	x								x						3
	Longulus	x										x				1
	Neopteleobius	x													?	1
	Phloeoborus				?					x						27
	Pteleobius	x								x						2
	Rhopalopselion						x			x						11

Tribe	Genus	No.								
Hylurgini	Chaetoptelius	7			x					x
	Dendroctonus	20	(x)		x					x
	Dendrotrupes	2	?							x
	Hylurdrectonus	4			x					x
	Hylurgonotus	4			x			x		
	Hylurgopinus	1			x					x
	Hylurgus	3			x					x
	Pachycotes	9			x			x		
	Pseudohylesinus	13			x					x
	Pseudoxylechinus	9			x					x
	Sinophloeus	1	?		x					x
	Tomicus	7			x					x
	Xylechinosomus	11			x					x
	Xylechinus	40		x	x					x
Hyorrhynchini	Hyorrhynchus	11			x				x	
	Pseudohyorrhynchus	3			x				x	
	Sueus	5	x		x				x	
Hypoborini	Chaetophloeus	24			x					x
	Cryphyophthorus	2	?			?				
	Dacryostactus	1			x					x
	Glochiphorus	2	?			?				
	Hypoborus	2			x					x
	Liparthrum	37			x		x			x
	Styracoptinus	4			x					x
	Trypanophellos	1	?			?				
	Zygophloeus	1				?				
Ipini	Acanthotomicus	94		x	(x)			(x)		x
	Dendrochilus	9	?					x		
	Ips	45		x	x					x
	Orthotomicus	20		x	x					x

continued

Tribe	Genus	Phl	Xlm	Spm	Myc	Mye	Xly	Hbv	feed?	MG	HP	BG	Col	Inbr	MS?	Spp
	Pityogenes	x									x					24
	Pityokteines	x									x					10
	Premnobius		x											x		23
	Premnophilus		x											x		2
	Pseudips	x										x				3
Micracidini	*Afromicracis*								?						?	17
	Hylocurus	(x)					x				(x)					78
	Lanurgus	x						x		x		x				22
	Micracis						x					x				26
	Micracisella					x				x						20
	Miocryphalus	x				(x)						?				
	Phloeocleptus	x										x				11
	Phloeocurus	x													?	1
	Pseudomicracis	x								x	x	x				8
	Pseudothysanoes	x					(x)	(x)								92
	Saurotocis	x														2
	Stenoclyptus	x									x	x				2
	Stevewoodia								?						?	1
	Thysanoes						x					x				15
Phloeosinini	*Asiophilus*								?						?	2
	Carphotoreus	x								x						1
	Catenophorus								?						?	1
	Chramesus	x				(x)	(x*)	(x)		x		(x)				92
	Cladoctonus	x							?	x						14
	Cortisinus						x								?	1
	Dendrosinus									x						7
	Hyledius	x								x	(x)					24
	Hyleops	x								x						1
	Microdictica								?						?	1

Tribe	Genus	No.	C1	C2	C3	C4	C5	C6	C7	C8	C9	C10	C11
	Phloeocranus	1					x						x
	Phloeoditica	2			x								x
	Phloeosinopsioides	3						?	?				
	Phloeosinus	66			(x)		x			(x*)			x
	Pseudochramesus	11					x						x
Phloeotribini	*Aricerus*	3					x						x
	Dryotomicus	4						?	?				
	Phloeotribus	103			(x)		x						x
Phrixosomatini	*Phrixosoma*	25					x						x
	Bothinodroctonus	3					x						x
Polygraphini	*Cardroctonus*	3						?	?				
	Carphobius	3					x						x
	Carphoborus	34				x	x						x
	Chortastus	5				x							x
	Dolurgocleptes	2						?	?				
	Halystus	2				x	x						x
	Polygraphus	100				x	(x)						x
	Serrastus	2					x						x
Scolytini	*Camptocerus*	31					x					x	
	Ceratolepis	7					x						x
	Cnemonyx	23					x						x
	Loganius	16					x						x
	Scolytopsis	6					x						x
	Scolytus	126			(x)	(x)	x						x
Scolytoplatypodini	*Remansus*	4					?					x	
	Scolytoplatypus	49					x					x	
Xyleborini	*Amasa*	41	x									x	
	Ambrosiodmus	80	x									x	
	Ambrosiophilus	8	x									x	
	Ancipitis	1	x									x	

continued

Tribe	Genus	Phl	Xlm	Spm	Myc	Mye	Xly	Hbv	feed?	MG	HP	BG	Col	Inbr	MS?	Spp
	Anisandrus		x											x		14
	Arixyleborus		x											x		32
	Beaverium		x											x		7
	Cnestus		x											x		32
	Coptoborus		x											x		23
	Coptodryas		x											x		36
	Cryptoxyleborus		x											x		18
	Cyclorhipidion		x											x		86
	Debus		x											x		16
	Diuncus		x											x		17
	Dryocoetoides		x											x		25
	Dryoxylon		x											x		1
	Eccoptopterus		x											x		6
	Euwallacea		x											x		45
	Fortiborus		x											x		6
	Hadrodemius		x											x		3
	Immanus		x													2
	Leptoxyleborus		x											x		6
	Microperus		x											x		16
	Planiculus		x											x		7
	Pseudowebbia		x											x		6
	Sampsonius		x											x		22
	Schedlia		x											x		6
	Stictodex		x											x		2
	Streptocranus		x											x		11
	Taurodemus		x											x		15
	Theoborus		x											x		11
	Truncaudum		x											x		7
	Wallacellus		x											x		3

Evolution and Diversity of Bark and Ambrosia Beetles **Chapter | 3** 141

Tribe / Genus						Count
Webbia	x				x	38
Xyleborinus	x				x	76
Xyleborus	x				x	404
Xylosandrus	x				x	39
Xyloctonini						
Cryphalomimus			?	?		3
Ctonoxylon		x	x			28
Glostatus		x	x			18
Scolytomimus		x	x			14
Xyloctonus		x	x			15
Xyloterini						
Indocryphalus	x		x			8
Trypodendron	x		x		x	13
Xyloterinus	x				x	1

*Feeding and reproductive behavior for Scolytinae genera of the world. For most genera, larval feeding mode and mating system information are in the review of world genera by (Wood, 1986; see also 1982, 2007), or regional works of Beeson (1941), Schedl (1958, 1959, 1977), Browne (1961, 1963), and Chararas (1962); overlooked or recent information for poorly known genera or exceptional species is in Kalshoven (1958), Roberts (1969, 1976), Beaver and Löyttyniemi (1985), Noguera-Martínez and Atkinson (1990), and Jordal (2006).

ACKNOWLEDGMENTS

The taxonomic distributions of feeding behaviors and reproductive systems would not have been possible without the taxonomic data compilation kindly supplied by T. H. Atkinson. Roger Beaver, Sarah Smith, and Anthony Cognato replied promptly to repeated requests for natural history information, which greatly improved the manuscript. We thank them and all the others who supplied facts or research reprints. P.H.W.B. is funded by an SNSF Postdoctoral Research Grant (P300P3_151134).

REFERENCES

Adams, A.S., Six, D.L., 2007. Temporal variation in mycophagy and prevalence of fungi associated with developmental stages of *Dendroctonus ponderosae* (Coleoptera: Curculionidae). Environ. Entomol. 36, 64–72.

Adams, A.S., Six, D.L., Adams, S.M., Holben, W.E., 2008. *In vitro* interactions between yeasts and bacteria and the fungal symbionts of the mountain pine beetle (*Dendroctonus ponderosae*). Microb. Ecol. 56, 460–466.

Ahnlund, H., 1996. Saproxylic insects on a Swedish dead aspen. Entomologisk Tidskrift 117, 137–144.

Alexander, R.D., 1974. The evolution of social behavior. Annu. Rev. Ecol. Syst. 5, 325–383.

Alexander, R.D., Noonan, K.M., Crespi, B.J., 1991. The evolution of eusociality. In: Sherman, P.W., Jarvis, J.U.M., Alexander, R.D. (Eds.), The Biology of the Naked Mole Rat. Princeton University Press, Princeton, pp. 1–44.

Alfaro, R.I., Humble, L.M., Gonzalez, P., Villaverde, R., Allegro, G., 2007. The threat of the ambrosia beetle *Megaplatypus mutatus* (Chapuis) (=*Platypus mutatus* Chapuis) to world poplar resources. Forestry 80, 471–479.

Amezaga, I., 1997. Forest characteristics affecting the rate of shoot pruning by the pine shoot beetle (*Tomicus piniperda* L.) in *Pinus radiata* D. Don and *P. sylvestris* L. plantations. Forestry 70, 129–137.

Anderbrandt, O., Schlyter, F., Birgersson, G., 1985. Intraspecific competition affecting parents and offspring in the bark beetle *Ips typographus*. Oikos 45, 89–98.

Andersen, J., Nilssen, A.C., 1983. Intrapopulation size variation of free-living and tree-boring Coleoptera. Can. Entomol. 15, 1453–1464.

Andersen, H., Jordal, B., Kambestad, M., Kirkendall, L.R., 2012. Improbable but true: the invasive inbreeding ambrosia beetle *Xylosandrus morigerus* has generalist genotypes. Ecol. Evol. 2, 247–257.

Arthofer, W., Riegler, M., Avtzis, D.N., Stauffer, C., 2009. Evidence for low-titre infections in insect symbiosis: *Wolbachia* in the bark beetle *Pityogenes chalcographus* (Coleoptera, Scolytinae). Environ. Microbiol. 11, 1923–1933.

Atkinson, T.H., 2010. New species and records of *Cactopinus* Schwarz with a key to species (Coleoptera, Curculionidae, Scolytinae). ZooKeys. 17–33.

Atkinson, T.H., Equihua-Martínez, A., 1985a. Lista comentada de los coleópteros Scolytidae y Platypodidae del Valle de México. Folia Entomol. Mexic. 65, 63–108.

Atkinson, T.H., Equihua-Martínez, A., 1985b. Notes on biology and distribution of Mexican and Central American Scolytidae (Coleoptera). I. Hylesininae, Scolytinae except Cryphalini and Corthylini. Coleopterists Bulletin 39, 227–238.

Atkinson, T.H., Equihua-Martínez, A., 1985c. Notes on biology and distribution of Mexican and Central American Scolytidae (Coleoptera). II. Scolytinae: Cryphalini and Corthylini. Coleopterists Bulletin 39, 355–363.

Atkinson, T.H., Equihua-Martínez, A., 1986a. Biology of the Scolytidae and Platypodidae (Coleoptera) in a tropical deciduous forest at Chamela, Jalisco. Mexico. Fla. Entomol. 69, 303–310.

Atkinson, T.H., Equihua-Martínez, A., 1986b. Biology of bark and ambrosia beetles (Coleoptera, Scolytidae and Platypodidae) of a tropical rain forest in southeastern Mexico with an annotated checklist of species. Ann. Entomol. Soc. Am. 79, 414–423.

Atkinson, T.H., Peck, S.B., 1994. Annotated checklist of the bark and ambrosia beetles (Coleoptera: Platypodidae and Scolytidae) of tropical southern Florida. Fla. Entomol. 77, 313–329.

Atkinson, T.H., Saucedo Céspedes, E., Martínez Fernández, E., Burgos Solorio, A., 1986. Coleópteros Scolytidae y Platypodidae asociados con las comunidades vegetales de clima templado y frío en el estado de Morelos, México. Acta Zool. Mex. 17, 1–58 (ns).

Ausmus, B.S., 1977. Regulation of wood decomposition rates by arthropod and annelid populations. Ecol. Bull. 25, 180–192.

Ayres, M.P., Wilkens, R.T., Ruel, J.J., Lombardero, M.J., Vallery, E., 2000. Nitrogen budgets of phloem-feeding bark beetles with and without symbiotic fungi. Ecology 81, 2198–2210.

Baker, J.M., 1963. Ambrosia beetles and their fungi, with particular reference to *Platypus cylindrus* Fab. In: Mosse, B., Nutman, P.S. (Eds.), Symbiotic Associations; Thirteenth Symposium of the Society for General Microbiology. The Syndics of the Cambridge University Press, London, pp. 232–265.

Bakke, A., 1968a. Ecological studies on bark beetles (Coleoptera: Scolytidae) associated with Scots pine (*Pinus sylvestris* L.) in Norway with particular reference to the influence of temperature. Meddelelser Norske Skogforsøksvesen 21, 441–602.

Bakke, A., 1968b. Field and laboratory studies on sex ratio in *Ips acuminatus* (Coleoptera: Scolytidae) in Norway. Can. Entomol. 100, 640–648.

Bakke, A., 1992. Monitoring bark beetle populations: effects of temperature. J. Appl. Entomol. 114, 208–211.

Balachowsky, A., 1949. Coléoptères Scolytidae. Libraire de la Faculte des Sciences, Paris.

Barbey, A., 1901. Les Scolytides de L'Europe Centrale. H. Kündig, Geneva.

Barr, B.A., 1969. Sound production in Scolytidae (Coleoptera) with emphasis on the genus *Ips*. Can. Entomol. 101, 636–672.

Barras, S.J., 1973. Reduction of progeny and development in southern pine beetle following removal of symbiotic fungi. Can. Entomol. 105, 1295–1299.

Batra, L.R., 1963. Ecology of ambrosia fungi and their dissemination by beetles. Trans. Kansas Acad. Sci. 66, 213–236.

Batra, L.R., 1966. Ambrosia fungi: extent of specificity to ambrosia beetles. Science 153, 193–195.

Batra, L.R., 1967. Ambrosia fungi—a taxonomic revision and nutritional studies of some species. Mycologia 59, 976–1017.

Batra, S.W.T., 1966. Nests and social behaviour of halictine bees of India. Indian J. Entomol. 28, 375–393.

Beal, J.A., Massey, C.L., 1945. Bark beetles and ambrosia beetles (Coleoptera: Scolytoidea) with special reference to the species occurring in North Carolina. Duke University School of Forestry Bulletin 10, 1–178.

Beaver, R.A., 1973. Biological studies of Brazilian Scolytidae and Platypodidae (Coleoptera) II. The tribe Bothrosternini. Pap. Avulsos Dep. Zool. 26, 227–236.

Beaver, R.A., 1976. Biological studies of Brazilian Scolytidae and Platy-podidae (Coleoptera). 5. The tribe Xyleborini. Z. angew. Entomol. 80, 15–30.

Beaver, R.A., 1977. Bark and ambrosia beetles in tropical forests. Biotrop Special Publication 2, 133–149.

Beaver, R.A., 1979a. Host specificity in temperate and tropical animals. Nature 281, 139–141.

Beaver, R.A., 1979b. Leafstalks as a habitat for bark beetles (Col.: Scoly-tidae). Z. angew. Entomol. 88, 296–306.

Beaver, R.A., 1986. The taxonomy, mycangia and biology of *Hypothe-nemus curtipennis* (Schedl), the first known cryphaline ambrosia beetle (Coleoptera: Scolytidae). Entomol. Scand. 17, 131–135.

Beaver, R.A., 1988. Biological studies on ambrosia beetles of the Sey-chelles (Col., Scolytidae and Platypodidae). J. Appl. Entomol. 105, 62–73.

Beaver, R.A., 1989. Insect-fungus relationships in the bark and ambrosia beetles. In: Wilding, N., Collins, N.M., Hammond, P.M., Webber, J.F. (Eds.), Insect-Fungus Interactions. Academic Press, London, pp. 121–143.

Beaver, R.A., 2000. Studies on the genus *Diapus* Chapuis (Coleoptera: Pla-typodidae) new species and new synonymy. Serangga 5, 247–260.

Beaver, R.A., 2013. The invasive neotropical ambrosia beetle *Euplatypus parallelus* (Fabricius, 1801) in the oriental region and its pest status. The Entomologist's Monthly Magazine 149, 143–154.

Beaver, R.A., Browne, F.G., 1975. The Scolytidae and Platypodidae (Cole-optera) of Thailand. A checklist with biological and zoogeographical notes. Oriental Insects 9, 283–311.

Beaver, R.A., Gebhardt, H., 2006. A review of the Oriental species of *Sco-lytoplatypus* Schaufuss (Coleoptera, Curculionidae, Scolytinae). Dtsch. Entomol. Z. 53, 155–178.

Beaver, R.A., Löyttyniemi, K., 1985. Bark and ambrosia beetles (Coleoptera: Scolytidae) of Zambia. Revue zoologiques africaine 99, 63–85.

Beeson, C.F.C., 1939. New species and biology of *Coccotrypes* and *Tham-nurgides* (Scolytidae, Col.). *Indian Forest Records (N. S.).* Ento-mology 5, 279–308.

Beeson, C.F.C., 1941. The Ecology and Control of the Forest Insects of India and the Neighbouring Countries. The Vasant Press, India.

Benavides, P., Vega, F.E., Romero-Severson, J., Bustillo, A.E., Stuart, J.J., 2005. Biodiversity and biogeography of an important inbred pest of coffee, coffee berry borer (Coleoptera: Curculionidae: Scolytinae). Ann. Entomol. Soc. Am. 98, 359–366.

Bentz, B.J., Six, D.L., 2006. Ergosterol content of fungi associated with *Dendroctonus ponderosae* and *Dendroctonus rufipennis* (Coleoptera: Curculionidae, Scolytinae). Ann. Entomol. Soc. Am. 99, 189–194.

Berryman, A.A., 1982. Population dynamics of bark beetles. In: Mitton, J.B., Sturgeon, K.B. (Eds.), Bark Beetles in North American Conifers. A System for the Study of Evolutionary Biology. University of Texas Press, Austin.

Beukeboom, L.W., Vrijenhoek, R.C., 1998. Evolutionary genetics and ecology of sperm-dependent parthenogenesis. J. Evol. Biol. 11, 755–782.

Biedermann, P.H.W., 2010. Observations on sex ratio and behavior of males in *Xyleborinus saxesenii* Ratzeburg (Scolytinae, Coleoptera). ZooKeys 56, 253–267.

Biedermann, P.H.W., 2012. The evolution of cooperation in ambrosia beetles. Ph.D. thesis, University of Bern.

Biedermann, P.H.W., Taborsky, M., 2011. Larval helpers and age poly-ethism in ambrosia beetles. Proc. Natl. Acad. Sci. U. S. A. 108, 17064–17069.

Biedermann, P. H. W., and Taborsky, M. (submitted). Experimental social evolution: philopatry and cooperation co-evolve in an ambrosia beetle.

Biedermann, P.H.W., Klepzig, K.D., Taborsky, M., 2009. Fungus culti-vation by ambrosia beetles: behavior and laboratory breeding success in three xyleborine species. Environ. Entomol. 38, 1096–1105.

Biedermann, P.H.W., Klepzig, K.D., Taborsky, M., 2011. Costs of delayed dispersal and alloparental care in the fungus-cultivating ambrosia beetle *Xyleborus affinis* Eichhoff (Scolytinae: Curculionidae). Behav. Ecol. Sociobiol. 65, 1753–1761.

Biedermann, P.H.W., Peer, K., Taborsky, M., 2012. Female dispersal and reproduction in the ambrosia beetle *Xyleborinus saxesenii* Ratzeburg (Coleoptera; Scolytinae). Mitteilungen der Deutschen Gesellschaft für allgemeine und angewandte Entomologie 18, 231–235.

Biedermann, P.H.W., Klepzig, K.D., Taborsky, M., Six, D.L., 2013. Abun-dance and dynamics of filamentous fungi in the complex ambrosia gardens of the primitively eusocial beetle *Xyleborinus saxesenii* Rat-zeburg (Coleoptera: Curculionidae, Scolytinae). FEMS Microbiol. Ecol. 83, 711–723.

Birch, M.C., 1984. Aggregation in bark beetles. In: Bell, W.J., Cardé, R.T. (Eds.), Chemical Ecology of Insects. Chapman and Hall, New York, pp. 331–353.

Blackman, M.W., 1922. Mississippi bark beetles. Miss., Agric. Exp. Stn., Tech. Bull, 11,130 pp., 18 pls.

Blackman, M.W., 1924. The effect of deficiency and excess in rainfall upon the hickory bark beetle (*Eccoptogaster quadrispinosus* Say). J. Econ. Entomol. 17, 460–470.

Blackman, M.W., 1928. The genus *Pityophthorus* Eichh. in North America: a revisional study of the Pityophthori, with descriptions of two new genera and seventy-one new species, 25, New York State College of Forestry at Syracuse, Technical Publication, 1–184.

Blackman, M.W., 1931. The Black Hills beetle (*Dendroctonus ponderosae* Hopk.). Bulletin of the New York State College of Forestry at Syracuse University 4, 1–97.

Blackman, M.W., Stage, H.H., 1918. Notes on insects bred from the bark and wood of the American larch, 10,Technical Publication, New York State College of Forestry at Syracuse University, 1–115.

Blackman, M.W., Stage, H.H., 1924. On the succession of insects living in the bark and wood of dying, dead and decaying hickory. New York State College of Forestry at Syracuse University, Technical Publi-cation 17, 1–268.

Bleiker, K.P., Heron, R.J., Braithwaite, E.C., Smith, G.D., 2013. Preemer-gence mating in the mass-attacking bark beetle, *Dendroctonus pon-derosae* (Coleoptera: Curculionidae). Can. Entomol. 145, 12–19.

Bonduriansky, R., 2001. The evolution of male mate choice in insects: a synthesis of ideas and evidence. Biol. Rev. 76, 305–339.

Boomsma, J.J., 2009. Lifetime monogamy and the evolution of eusociality. Philos. Trans. R. Soc., B 364, 3191–3207.

Boomsma, J.J., 2013. Beyond promiscuity: mate-choice commitments in social breeding. Philos. Trans. R. Soc. B 368,20120050.

Borden, J.H., 1967. Factors influencing the response of *Ips confusus* (Coleoptera: Scolytidae) to male attractant. Can. Entomol. 99, 1164–1193.

Borden, J.H., 1969. Observations on the life history and habits of *Alni-phagus aspericollis* (Coleoptera: Scolytidae) in southwestern British Columbia. Can. Entomol. 101, 870–878.

Borden, J.H., 1985. Aggregation pheromones. In: Kerkut, G.A. (Ed.), Behaviour. Pergamon Press, Oxford, pp. 257–285.

Borden, J.H., Slater, C.E., 1969. Flight muscle volume change in *Ips con-fusus* (Coleoptera: Scolytidae). Can. J. Zool. 47, 29–31.

Bordenstein, S.R., O'Hara, F.P., Werren, J.H., 2001. *Wolbachia*-induced incompatibility precedes other hybrid incompatibilities in *Nasonia*. Nature 409, 707–710.

Borsa, P., Kjellberg, F., 1996a. Experimental evidence for pseudo-arrhenotoky in *Hypothenemus hampei* (Coleoptera: Scolytidae). Heredity 76, 130–135.

Borsa, P., Kjellberg, F., 1996b. Secondary sex ratio adjustment in a pseudo-arrhenotokous insect, *Hypothenemus hampei* (Coleoptera: Scolytidae). Comptes Rendus de l'Academie des Sciences Serie Iii-Sciences de la Vie-Life Sciences 319, 1159–1166.

Botterweg, P.F., 1982. Dispersal and flight behavior of the spruce bark beetle *Ips typographus* in relation to sex, size and fat-content. Z. angew. Entomol. 94, 466–489.

Bourke, A.F.G., 2011. Principles of Social Evolution. Oxford University Press, Oxford.

Brader, L., 1964. Étude de la relation entre le scolyte des rameaux du caféir, *Xyleborus compactus* Eichh. (*X. morstatti* Hag.), et sa plantehôte. Mededelingen van de Landbouwhogeschool te Wageningen 64, 1–109.

Bright, D.E., 1981. Taxonomic monograph of the genus *Pityophthorus* Eichhoff in North and Central America (Coleoptera: Scolytidae). Mem. Entomol. Soc. Can. 118, 1–378.

Bright, D.E., Poinar Jr., G.O., 1994. Scolytidae and Platypodidae (Coleoptera) from Dominican Republic Amber. Ann. Entomol. Soc. Am. 87, 170–194.

Bright, D.E., Stark, R.W., 1973. The bark and ambrosia beetles of California. Coleoptera: Scolytidae and Platypodidae. Bulletin of the California Insect Survey 16, 1–169.

Brimblecombe, A.R., 1953. An annotated list of the Scolytidae occurring in Australia. Queensl. J. Agric. Sci. 10, 167–205.

Browne, F.G., 1958. Some aspects of host selection among ambrosia beetles in the humid tropics of south-east Asia. Malayan Forest Records 21, 164–182.

Browne, F.G., 1961. The Biology of Malayan Scolytidae and Platypodidae. Malayan Forest Records 22, 1–255.

Browne, F.G., 1962. The emergence, flight and mating behaviour of *Doliopygus conradti*. West African Timber Borer Research Unit 5, 21–27.

Browne, F.G., 1963. Notes on the habits and distribution of some Ghanaian bark beetles and ambrosia beetles (Coleoptera: Scolytidae and Platypodidae). Bull. Entomol. Res. 54, 229–266.

Browne, F.G., 1965. Types of ambrosia beetle attack on living trees. Proc. Int. Cong. Entomol. 12, 680.

Browne, F.G., 1968. Pests and Diseases of Forest Plantation Trees. Clarendon, Oxford.

Browne, F.G., 1972. Larvae of principal old world genera of Platypodinae. Trans. R. Entomol. Soc. Lond. 124, 167–190.

Brucker, R.M., Bordenstein, S.R., 2012. Speciation by symbiosis. Trends Ecol. Evol. 27, 443–451.

Brun, L.O., Borsa, P., Gaudichon, V., Stuart, J.J., Aronstein, K., Coustau, C., Ffrench-Constant, R.H., 1995a. 'Functional' haplodiploidy. Nature 374, 506.

Brun, L.O., Stuart, J.J., Gaudichon, V., Aronstein, K., Ffrench-Constant, R.H., 1995b. Functional haplodiploidy—a mechanism for the spread of insecticide resistance in an important international insect pest. Proc. Natl. Acad. Sci. U. S. A. 92, 9861–9865.

Cai, Y., Cheng, X., Xu, R., Duan, D., Kirkendall, L.R., 2008. Genetic diversity and biogeography of red turpentine beetle *Dendroctonus valens* in its native and invasive regions. Insect Sci. 15, 291–301.

Campobasso, G., Terragitti, G., Mann, K., Quimby, P.C., 2004. Field and laboratory biology of the stem-feeding beetle *Thamnurgus euphorbiae* (Kuster) (Col., Scolytidae) in Italy, a potential biological control candidate of leafy spurge in the USA and Canada. J. Appl. Entomol. 128, 1–5.

Cardoza, Y.J., Paskewitz, S., Raffa, K.F., 2006a. Travelling through time and space on wings of beetles: a tripartite insect-fungi-nematode association. Symbiosis 41, 71–79.

Cardoza, Y.J., Klepzig, K.D., Raffa, K.F., 2006b. Bacteria in oral secretions of an endophytic insect inhibit antagonistic fungi. Ecol. Entomol. 31, 636–645.

Cardoza, Y.J., Moser, J.C., Klepzig, K.D., Raffa, K.F., 2008. Multipartite symbioses among fungi, mites, nematodes, and the spruce beetle, *Dendroctonus rufipennis*. Environ. Entomol. 37, 956–963.

Chamberlin, W.J., 1918. Bark beetles infesting the Douglas fir. Bull. Oregon Agric. College Exp. Sta. 147, 1–40.

Chamberlin, W.J., 1939. The Bark and Timber Beetles of North America. North of Mexico, Oregon State College Cooperative Association, Corvallis, Oregon.

Chamberlin, W.J., 1958. The Scolytoidea of the Northwest: Oregon, Washington, Idaho, and British Columbia. Oregon State College, Corvallis.

Chapman, T.A., 1870. On the habits of *Platypus cylindrus*. Entomologists Monthly Magazine 7, 103–107, 132–135.

Chapman, J.A., 1956. Flight muscle changes during adult life in a scolytid beetle, *Trypodendron*. Nature 177, 1183.

Chararas, C., 1962. Étude biologique des scolytidae des conifères. P. Lechevalier, Paris.

Charnov, E.L., 1982. The Theory of Sex Allocation. Princeton University Press, Princeton.

Chodjaï, M., 1963. Étude écologique de *Rugoscolytus mediterraneus* Eggers (Col. Scolytidae) en Iran. Rev. Path. Vég. Entomol. Agr. Fr. 42, 139–160.

Choe, J.C., Crespi, B.J., 1997. The Evolution of Social Behaviour in Insects and Arachnids. Cambridge University Press, Cambridge.

Chong, J.-H., Reid, L., Williamson, M., 2009. Distribution, host plants, and damage of the black twig borer, *Xylosandrus compactus* (Eichhoff), in South Carolina. J. Agric. Urban Entomol. 26, 199–208.

Civetta, A., Singh, R.S., 1999. Broad-sense sexual selection, sex gene pool evolution, and speciation. Genome 42, 1033–1041.

Clutton-Brock, T., 2009. Cooperation between non-kin in animal societies. Nature 462, 51–57.

Cognato, A.I., 2013. Molecular phylogeny and taxonomic review of Premnobiini Browne 1962 (Coleoptera: Curculionidae: Scolytinae). Front. Ecol. Evol. 1http://dx.doi.org/10.3389/fevo.2013.00001.

Cognato, A.I., Grimaldi, D., 2009. 100 million years of morphological conservation in bark beetles (Coleoptera: Curculionidae: Scolytinae). Syst. Entomol. 34, 93–100.

Cognato, A.I., Sperling, F.A.H., 2000. Phylogeny of *Ips* DeGeer species (Coleoptera: Scolytidae) inferred from mitochondrial cytochrome oxidase I DNA sequence. Mol. Phylogenet. Evol. 14, 445–460.

Cognato, A.I., Seybold, S.J., Sperling, F.A.H., 1999. Incomplete barriers to mitochondrial gene flow between pheromone races of the North American pine engraver, *Ips pini* (Say) (Coleoptera: Scolytidae). Proc. R. Soc. London B 266, 1843–1850.

Cognato, A.I., Gillette, N.E., Bolaños, R.C., Sperling, F.A.H., 2005. Mitochondrial phylogeny of pine cone beetles (Scolytinae, *Conophthorus*) and their affiliation with geographic area and host. Mol. Phylogenet. Evol. 36, 494–508.

Cognato, A.I., Hulcr, J., Dole, S.A., Jordal, B.H., 2011. Phylogeny of haplo-diploid, fungus-growing ambrosia beetles (Curculionidae: Scolytinae: Xyleborini) inferred from molecular and morphological data. Zool. Scr. 40, 174–186.

Costa, J.T., 2006. The Other Insect Societies. Belknapp Press of Harvard University. Press, Cambridge.

Costa, J.T., 2010. Social evolution in "other" insects and arachnids. In: Breed, M.D., Moore, J. (Eds.), Encyclopedia of Animal Behavior. Academic Press, Oxford, pp. 231–241.

Cowling, E.B., Merrill, W., 1966. Nitrogen in wood and its role in wood deterioration. Can. J. Bot. 44, 1533–1544.

Cremer, S., Armitage, S.A.O., Schmid-Hempel, P., 2007. Social immunity. Curr. Biol. 17, R693–R702.

Crowson, R.A., 1974. Observations on Histeroidea, with descriptions of an apterous male, and of the internal anatomy of male *Sphaerites*. J. Entomol. Series B, Taxonomy 42, 133–140.

Crozier, R.G., Giese, R.L., 1967a. The Columbian timber beetle, *Corthylus columbianus* (Coleoptera: Scolytidae). III. Definition of epiphytotics. J. Econ. Entomol 60, 55–58.

Crozier, R.G., Giese, R.L., 1967b. The Columbian timber beetle, *Corthylus columbianus* (Coleoptera: Scolytidae). IV. Intrastand population distribution. Can. Entomol. 99, 1203–1214.

Currie, C.R., 2001. A community of ants, fungi, and bacteria: a multilateral approach to studying symbiosis. Annu. Rev. Microbiol. 55, 357–380.

Dahlsten, D.L., 1982. Relationship between bark beetles and their natural enemies. In: Mitton, J.B., Sturgeon, K.B. (Eds.), Bark Beetles in North American Conifers. A System for the Study of Evolutionary Biology. University of Texas Press, Austin, pp. 140–182.

Dallara, P.L., Seybold, S.J., Meyer, H., Tolasch, T., Francke, W., Wood, D.L., 2000. Semiochemicals from three species of *Pityophthorus* (Coleoptera: Scolytidae): identification and field response. Can. Entomol. 132, 889–906.

Darling, D.C., Roberts, H., 1999. Life history and larval morphology of *Monacon* (Hymenoptera: Perilampidae), parasitoids of ambrosia beetles (Coleoptera: Platypodidae). Can. J. Zool. 77, 1768–1782.

Darwin, C., 1859. The Origin of Species. John Murray, London.

De Fine Licht, H.H., Biedermann, P.H.W., 2012. Patterns of functional enzyme activity in fungus farming ambrosia beetles. Front. Zool. 9, 13.

de Groot, P., Borden, J.H., 1992. Host acceptance behaviour of the jack pine tip beetle, *Conophthorus banksianae* and the red pine cone beetle. C. resinosae. Entomol. Exp. Appl. 65, 149–155.

De Leon, D., 1952. Insects associated with *Sequoia sempervirens* and *Sequoia gigantea* in California. Pan-Pac. Entomol. 28, 75–91.

Dedeine, F., Vavre, F., Fleury, F., Loppin, B., Hochberg, M.E., Boulétreau, M., 2001. Removing symbiotic *Wolbachia* bacteria specifically inhibits oogenesis in a parasitic wasp. Proc. Natl. Acad. Sci. U. S. A. 98, 6247–6252.

Dekle, G.W., Kuitert, L.C., 1968. Orchid insects, related pests and control. Florida Department of Agriculture, Bulletin of the Division of Plant Industry 8, 1–28.

Delate, K.M., 1994. Postharvest control treatments for *Hypothenemus obscurus* (F.) in macadamia nuts. J. Econ. Entomol. 87, 120–126.

Deneubourg, J.L., Grégoire, J.-C., LeFort, E., 1990. Kinetics of larval gregarious behaviour in the bark beetle *Dendroctonus micans* (Coleoptera: Scolytidae). J. Insect Behav. 3, 169–182.

Deyrup, M., 1987. *Trischidias exigua* Wood, new to the United States, with notes on the biology of the genus. Coleopterists Bulletin 41, 339–343.

Deyrup, M., Kirkendall, L.R., 1983. Apparent parthenogenesis in *Pityophthorus puberulus* (Coleoptera: Scolytidae). Ann. Entomol. Soc. Am. 76, 400–402.

Doane, R.W., 1923. *Leperisinus californicus* Sw. killing ash trees. Can. Entomol. 55, 217.

Doane, R.W., Gilliland, O.J., 1929. Three California ambrosia beetles. J. Econ. Entomol. 22, 915–921.

Dole, S.A., Jordal, B.H., Cognato, A.I., 2010. Polyphyly of *Xylosandrus* Reitter inferred from nuclear and mitochondrial genes (Coleoptera: Curculionidae: Scolytinae). Mol. Phylogenet. Evol. 54, 773–782.

Duffy, E.A.J., 1968. A monograph of the immature stages of oriental timber beetles (Cerambycidae). British Museum (Natural History), London.

Dussourd, D.E., Eisner, T., 1987. Vein-cutting behavior: insect counterploy to the latex defense of plants. Science 237, 898–901.

Eichhoff, W., 1881. Die Europäischen Borkenkäfer. Julius Springer, Berlin.

Eidmann, H.H., 1992. Impact of bark beetles on forests and forestry in Sweden. J. Appl. Entomol. 114, 193–200.

Elkin, C.M., Reid, M.L., 2005. Low energy reserves and energy allocation decisions affect reproduction by mountain pine beetles, *Dendroctonus ponderosae*. Funct. Ecol. 19, 102–109.

Emlen, S.T., 1982. The evolution of helping. I. An ecological constraints model. Am. Nat. 119, 29–39.

Emlen, S.T., Oring, L.W., 1977. Ecology, sexual selection, and the evolution of mating systems. Science 197, 215–223.

Engelstädter, J., Hurst, G.D.D., 2006. Can maternally transmitted endosymbionts facilitate the evolution of haplodiploidy? J. Evol. Biol. 19, 194–202.

Entwistle, P.F., 1972. Pests of Cocoa. Longman, London.

Faccoli, M., Rukalski, J.P., 2004. Attractiveness of artificially killed red oaks (*Quercus rubra*) to ambrosia beetles (Coleoptera, Scolytidae). In: Ceretti, P., Hardersen, S., Mason, F., Nardi, G., Tisato, M., Zapparoli, M. (Eds.), Invertibrati di una Foresta della Pianura Padana, Bosco della Fontana. Cierre Grafica Editore, Verona, pp. 171–179.

Farrell, B.D., Sequeira, A.S., O'Meara, B.C., Normark, B.B., Chung, J.H., Jordal, B.H., 2001. The evolution of agriculture in beetles (Curculionidae: Scolytinae and Platypodinae). Evolution 55, 2011–2027.

Felt, E.P., 1914. Notes on forest insects. J. Econ. Entomol. 7, 373–375.

Ferro, M.L., Gimmel, M.L., Harms, K.E., Carlton, C.E., 2009. The beetle community of small oak twigs in Louisiana, with a literature review of Coleoptera from fine woody debris. Coleopterists Bulletin 63, 239–263.

Fischer, M., 1954. Untersuchungen über den kleinen Holzbohrer (*Xyleborus saxeseni*). Pflanzenschutzberichte 12, 137–180.

Fisher, R.C., 1931. Notes on the biology of the large elm bark-beetle, *Scolytus destructor* Ol. Forestry 5, 120–131.

Flores, L.J., Bright, D.E., 1987. A new species of *Conophthorus* from Mexico: descriptions and biological notes. Coleopterists Bulletin 41, 181–184.

Fockler, C.E., Borden, J.H., 1972. Sexual behavior and seasonal mating activity of *Trypodendron lineatum* (Coleoptera: Scolytidae). Can. Entomol. 104, 1841–1853.

Foelker, C.J., Hofstetter, R.W., 2014. Heritability, fecundity, and sexual size dimorphism in four species of bark beetles (Coleoptera: Curculionidae: Scolytinae). Ann. Entomol. Soc. Am. 107, 143–151.

Forcella, F., 1982. Twig nitrogen content and larval survival of twig-girdling beetles, *Oncideres cingulata* (Say) (Coleoptera: Cerambycidae). Coleopterists Bulletin 35, 211–212.

Fox, J.W., Wood, D.L., Akers, R.P., Parmeter, J.R., 1992. Survival and development of *Ips paraconfusus* Lanier (Coleoptera, Scolytidae) reared axenically and with tree–pathogenic fungi vectored by cohabiting *Dendroctonus* species. Can. Entomol. 124, 1157–1167.

Fraedrich, S.W., Harrington, T.C., Rabaglia, R.J., Ulyshen, M.D., Mayfield, A.E., Hanula, J.L., et al., 2008. A fungal symbiont of the redbay ambrosia beetle causes a lethal wilt in redbay and other Lauraceae in the southeastern United States. Plant Dis. 92, 215–224.

Francke-Grosmann, H., 1956a. Hautdrüsen als Träger der Pilzsymbiose bei Ambrosiakäfern. Zeitschrift für Ökologie und Morphologie der Tiere 45, 275–308.

Francke-Grosmann, H., 1956b. Zur Übertragung der Nährpilze bei Ambrosiakäfern. Naturwissenschaften 43, 286–287.

Francke-Grosmann, H., 1963a. Die Übertragung der Pilzflora bei dem Borkenkäfer *Ips acuminatus*. Z. angew. Entomol. 52, 355–361.

Francke-Grosmann, H., 1963b. Some new aspects in forest entomology. Annu. Rev. Entomol. 8, 415–438.

Francke-Grosmann, H., 1965. Ein Symbioseorgan bei dem Borkenkäfer *Dendroctonus frontalis* Zimm. (Coleoptera Scolytidae). Naturwissenschaften 52, 143.

Francke-Grosmann, H., 1966. Über Symbiosen von xylo-mycetophagen und phloeophagen Scolitoidea mit holzbewohnenden Pilzen. Material und Organismen 1, 503–522, Beiheft.

Francke-Grosmann, H., 1967. Ectosymbiosis in wood-inhabiting beetles. In: Henry, S.M. (Ed.), Symbiosis. Academic Press, New York, pp. 141–205.

Francke-Grosmann, H., 1975. Zur epizoischen und endozoischen Übertragung der symbiotischen Pilze des Ambrosiakäfers *Xyleborus saxeseni* (Coleoptera: Scolitidae). Entomologica Germanica 1, 279–292.

French, J.R.J., Roeper, R.A., 1972a. Interactions of ambrosia beetle, *Xyleborus dispar* (Coleoptera, Scolytidae), with its symbiotic fungus *Ambrosiella hartigii* (Fungi imperfecti). Can. Entomol. 104, 1635–1641.

French, J.R.J., Roeper, R.A., 1972b. In vitro culture of the ambrosia beetle *Xyleborus dispar* (Coleoptera: Scolytidae) with its symbiotic fungus, *Ambrosiella hartigii*. Ann. Entomol. Soc. Am. 65, 719–721.

French, J.R.J., Roeper, R.A., 1975. Studies on the biology of the ambrosia beetle *Xyleborus dispar* (F.) (Coleoptera: Scolytidae). J. Appl. Entomol. 78, 241–247.

Fuchs, G., 1907. Über die Fortpflanzungsverhältnisse der rindenbrütenden Borkenkäfer verbunden mit einer geschichtlichen und kritischen Darstellung der bisherigen Literatur. Ludwig-Maximilians-University, Munich, Doctoral Dissertation.

Furniss, M.M., 1995. Biology of *Dendroctonus punctatus* (Coleoptera: Scolytidae). Ann. Entomol. Soc. Am. 88, 173–182.

Furniss, M.M., 1996. Taxonomic status of *Dendroctonus punctatus* and *D. micans* (Coleoptera: Scolytidae). Ann. Entomol. Soc. Am. 89, 328–333.

Furniss, M.M., 1997. *Conophthorus ponderosae* (Coleoptera: Scolytidae) infesting lodgepole pine cones in Idaho. Environ. Entomol. 26, 855–858.

Gandon, S., 1999. Kin competition, the cost of inbreeding and the evolution of dispersal. J. Theor. Biol. 200, 345–364.

Garraway, E., 1986. The biology of *Ips calligraphus* and *Ips grandicollis* (Coleoptera: Scolytidae) in Jamaica. Can. Entomol. 118, 113–121.

Garraway, E., Freeman, B.E., 1981. Population-dynamics of the juniper bark beetle *Phloeosinus neotropicus* in Jamaica. Oikos 37, 363–368.

Garrick, R.C., Nason, J.D., Fernandez-Manjarres, J.F., Dyer, R.J., 2013. Ecological co-associations influence species' responses to past climatic change: an example from a Sonoran Desert bark beetle. Mol. Ecol. 22, 3345–3361.

Gascoigne, J., Berec, L., Gregory, S., Courchamp, F., 2009. Dangerously few liaisons: a review of mate-finding Allee effects. Popul. Ecol. 51, 355–372.

Gast, S.J., Furniss, M.M., Johnson, J.B., Ivie, M.A., 1989. List of Montana Scolytidae (Coleoptera) and notes on new records. Great Basin Nat. 49, 381–386.

Gianoli, E., Ramos, I., Alfaro-Tapia, A., Valdéz, Y., Echegaray, E.R., Yábar, E., 2006. Benefits of a maize-bean-weeds mixed cropping system in Urubamba Valley, Peruvian Andes. Int. J. Pest Manage. 4, 283–289.

Gil, Z.N., Bustillo, A.E., Gómez, D.E., Marín, P., 2004. *Corthylus* n. sp. (Coleoptera: Scolytidae), plaga del aiso en la cuenca del Río Blanco en Colombia. Revista Colombiana de Entomología 30, 171–178.

Gillett, C.P.D.T., Crampton-Platt, A., Timmermans, M.J.T.N., Jordal, B., Emerson, B.C., Vogler, A.P., 2014. Bulk *de novo* mitogenome assembly from pooled total DNA elucidates the phylogeny of weevils (Coleoptera: Curculionoidea). Mol. Biol. Evol. http://dx.doi.org/10.1093/molbev/msu154.

Giménez, R.A., Etiennot, A.E., 2003. Host range of *Platypus mutatus* (Chapuis, 1865) (Coleoptera: Platypodidae). Entomotropica 18, 89–94.

Girardi, G.S., Giménez, R.A., Braga, M.R., 2006. Occurrence of *Platypus mutatus* Chapuis (Coleoptera: Platypodidae) in a Brazilwood experimental plantation in southeastern Brazil. Neotrop. Entomol. 35, 864–867.

Gobeil, A.R., 1936. The biology of *Ips perturbatus* Eichhoff. Can. J. Res. 14, 181–204.

Godwin, G., Odell, T.M., 1965. The life history of the white-pine cone beetle, *Conophthorus coniperda*. Ann. Entomol. Soc. Am. 58, 213–219.

Goeden, R.D., Norris, D.M., 1965. Some biological and ecological aspects of ovipositional attack in *Carya* spp. by *Scolytus quadrispinosus* (Coleoptera: Scolytidae). Ann. Entomol. Soc. Am. 58, 771–777.

Goldhammer, D.S., Stephen, F.M., Paine, T.D., 1990. The effect of the fungi *Ceratocystis minor* (Hedgecock) Hunt, *Ceratocystis minor* (Hedgecock) Hunt var. Barrasii Taylor, and SJB 122 on reproduction of the southern pine beetle, *Dendroctonus frontalis* Zimmermann (Coleoptera: Scolytidae). Can. Entomol. 122, 407–418.

Gossard, H.A., 1913. Orchard bark beetles and pin hole borers. Bulletin of the Ohio Agricultural Experiment Station 264, 1–68.

Gottlieb, D., Holzman, J.P., Lubin, Y., Bouskila, A., Kelley, S.T., Harari, A.R., 2009. Mate availability contributes to maintain the mixed-mating system in a scolytid beetle. J. Evol. Biol. 22, 1526–1534.

Gottlieb, D., Lubin, Y., Harari, A.R., 2014. The effect of female mating status on male offspring traits. Behav. Ecol. Sociobiol. 68, 701–710.

Gouger, R.J., Yearian, W.C., Wilkinson, R.C., 1975. Feeding and reproductive behaviour of *Ips avulsus*. Fla. Entomol. 58, 221–229.

Graf, P., 1977. A contribution on the biology and control of *Hylesinus oleiperda* F. (Coleopt., Scolytidae) on olive in the Tadla (Morocco). Z. angew. Entomol. 83, 52–62.

Greco, E.B., Wright, M.G., 2013. Dispersion and sequential sampling plan for *Xylosandrus compactus* (Coleoptera: Curculionidae) infesting Hawaii coffee plantations. Environ. Entomol. 42, 277–282.

Grégoire, J.-C., 1988. The greater European spruce bark beetle. In: Berryman, A.A. (Ed.), Dynamics of Forest Insect Populations. Patterns, Causes, Implications. Plenum, New York, pp. 455–478.

Grégoire, J.-C., Braekman, J.-C., Tondeur, A., 1981. Chemical communication between the larvae of *Dendroctonus micans* Kug (Coleoptera: Scolytidae). Les colloques de l'INRA, 7. Les Médiateurs chimiques, 253–257.

Gruner, L., 1974. Biologie et dégats d'*Hexacolus guyanensis* Schedl, dans les plantations d'acajou rouge (*Swietenia macrophylla* King., Méliacée) en Guadeloupe (Coleoptera: Scolytidae). Ann. Sci. Forest. 31, 111–128.

Güçlü, C., Özbek, H., 2007. Biology and damage of *Thamnurgus pegani* Eggers (Coleoptera: Scolytidae) feeding on *Peganum harmala* L. in Eastern Turkey. Proc. Entomol. Soc. Wash. 109, 350–358.

Guérard, N., Dreyer, E., Lieutier, F., 2000. Interactions between Scots pine, *Ips acuminatus* (Gyll.) and *Ophiostoma brunneo-ciliatum* (Math.): estimation of the critical thresholds of attack and inoculation densities and effects on hydraulic properties in the stem. Ann. For. Sci. 57, 681–690.

Gurevitz, E., 1975. Contribution a l'étude des Scolytidae I. Comportement de différents stades du scolyte méditerranéen, *Scolytus* (*Rugoscolytus*) *mediterraneus* Eggers en Israël. Annales de zoologie—écologie animale 7, 477–489.

Haack, R.A., Slansky, F., 1987. Nutritional ecology of wood-feeding Coleoptera, Lepidoptera, and Hymenoptera. In: Slansky, F., Rodriguez, J.G. (Eds.), The Nutritional Ecology of Insects, Mites, Spiders, and Related Invertebrates. Wiley, New York, pp. 449–486.

Hadorn, C., 1933. Recherches sur la morphologie, les stades evolutifs et l'hivernage du bostryche lisere (*Xyloterus lineatus* Oliv.). Beiheft zu den Zeitschriften des Schweizerischen Forstvereins 11, 1–120.

Hamilton, W.D., 1964. The genetical evolution of social behavior, I and II. J. Theor. Biol. 7 (1–16), 17–52.

Hamilton, W.D., 1967. Extraordinary sex ratios. Science 156, 477–488.

Hamilton, W.D., 1972. Altruism and related phenomena, mainly in social insects. Annu. Rev. Ecol. Syst. 3, 193–232.

Hamilton, W.D., 1978. Evolution and diversity under bark. In: Mound, L.A., Waloff, N. (Eds.), Diversity of Insect Faunas. Blackwell, Oxford, pp. 154–175.

Hamilton, W.D., 1979. Wingless and fighting males in fig wasps and other insects. In: Blum, M.S., Blum, N.A. (Eds.), Reproductive Competition, Mate Choice, and Sexual Selection in Insects. Plenum, New York, pp. 167–220.

Hamilton, W.D., 1996. Funeral feasts: evolution and diversity under bark. In: Narrow Roads of Gene Land: The Collected Papers of W. D. Hamilton: Volume 1: Evolution of Social Behaviour. Spektrum Academic Publishers, Oxford, pp. 387–420.

Hammond, H.E., Langor, D.W., Spence, J.R., 2001. Early colonization of *Populus* wood by saproxylic beetles (Coleoptera). Can. J. For. Res. 31, 1175–1183.

Hanavan, R.P., Adams, K.B., Allen, D.C., 2012. Abundance and distribution of peach bark beetle in northern hardwood stands of New York. N. J. Appl. Forest 29, 128–132.

Hanks, L.M., 1999. Influence of the larval host plant on reproductive strategies of cerambycid beetles. Annu. Rev. Entomol. 44, 483–505.

Hansen, V., 1956. Biller. XVIII Barkbiller. Danmarks Fauna bind 62, G. E. C. Gads Forlag, Copenhagen.

Hanula, J.L., Mayfield, A.E., Fraedrich, S.W., Rabaglia, R.J., 2008. Biology and host associations of redbay ambrosia beetle (Coleoptera: Curculionidae: Scolytinae), exotic vector of laurel wilt killing redbay trees in the southeastern United States. J. Econ. Entomol. 101, 1276–1286.

Happ, G.M., Happ, C.M., Barras, S.J., 1975. Bark beetle fungal symbioses III. Ultrastructure of conidiogenesis in a Sporothrix ectosymbiont of the southern pine beetle. Can. J. Bot, 53, 2702–2711.

Happ, G.M., Happ, C.M., Barras, S.J., 1976. Bark beetle fungal symbioses. II. Fine structure of a basidiomycetous ectosymbiont of the southern pine beetle. Can. J. Bot. 1049–1062.

Haran, J., Timmermans, M.J.T.N., Vogler, A.P., 2013. Mitogenome sequences stabilize the phylogenetics of weevils (Curculionoidea) and establish the monophyly of larval ectophagy. Mol. Phylogenet. Evol. 67, 156–166.

Harrington, T.C., 2005. Ecology and evolution of mycophagous bark beetles and their fungal partners. In: Vega, F.E., Blackwell, M. (Eds.), Insect-Fungal Associations: Ecology and Evolution. Oxford University Press, New York, pp. 257–295.

Harris, T.W., 1852. A Treatise on some of the Insects of New England which are Injurious to Vegetation, Second ed. White & Potter, Boston.

Harris, J.A., Campbell, K.G., Wright, G.M., 1976. Ecological studies on the horizontal borer *Austroplatypus incompertus* (Schedl) (Coleoptera: Platypodidae). J. Entomol. Soc. Aust. (N. S. W.) 9, 11–21.

Hedlin, A.F., Ruth, D.S., 1970. A Douglas-fir twig mining beetle, *Pityophthorus orarius* (Coleoptera: Scolytidae). Can. Entomol. 102, 105–108.

Hedlin, A.F., Yates, H.O., Tovar, D.C., Ebel, B.H., Koerber, T.W., Merkel, E.P., 1980. Cone and seed insects of North American conifers. Environment Canada, Canadian Forest Service, Ottawa.

Heidenreich, E., 1960. Primärbefall durch *Xylosandrus germanus* an Jungeichen. Anz. Schädlingskd. 23, 5–10.

Herfs, A., 1950. Studien an dem Steinnußborkenkäfer, *Coccotrypes tanganus* Eggers. Höfchen-Briefe für Wissenschaft und Praxis 3, 3–31.

Herfs, A., 1959. Über den Steinnußborkenkäfer *Coccotrypes dactyliperda* F. Anz. Schädlingskd. 32, 1–4.

Hinton, H.E., 1981. Biology of Insect Eggs, Vol. 1, Pergamon Press, New York.

Hofstetter, R.W., Moser, J.C., 2014. The role of mites in insect-fungus associations. Annu. Rev. Entomol. 59, 537–557.

Hofstetter, R.W., Cronin, J.T., Klepzig, K.D., Moser, J.C., Ayres, M.P., 2006. Antagonisms, mutualisms and commensalisms affect outbreak dynamics of the southern pine beetle. Oecologia 147, 679–691.

Hölldobler, B., Wilson, E.O., 1990. The Ants. The Belknap Press of Harvard University Press, Cambridge.

Hölldobler, B., Wilson, E.O., 2009. The Superorganism: The Beauty, Elegance, and Strangeness of Insect Societies. W. W. Norton & Company, Inc., New York.

Holzman, J., Bohonak, A., Kirkendall, L., Gottlieb, D., Harari, A., Kelley, S., 2009. Inbreeding variability and population structure in the invasive haplodiploid palm-seed borer (*Coccotrypes dactyliperda*). J. Evol. Biol. 22, 1076–1087.

Hopkins, A.D., 1904. Insects injurious to hardwood forests. U.S. Department of Agriculture, Yearbook for 1903, 313–328.

Hopkins, A.D., 1908. Notable depredations by forest insects. U.S. Department of Agriculture, Yearbook for 1907, 149–164.

Hopkins, A.D., 1909. Contributions toward a monograph of the scolytid beetles. I. The genus *Dendroctonus*, 17 *U.S. Department of Agriculture, Bureau of Entomology, Technical Bulletin* 1–164.

Hopping, G.R., 1964. The breeding evidence indicating two genetic types of females of *Ips tridens*. Can. Entomol. 96, 117–118.

Hosking, G.B., 1972. *Xyleborus saxeseni*, its life-history and flight behaviour in New Zealand. N. Z. J. For. Sci. 3, 37–53.

Hosokawa, T., Koga, R., Kikuchi, Y., Meng, X.-Y., Fukatsu, T., 2010. *Wolbachia* as a bacteriocyte-associated nutritional mutualist. Proc. Natl. Acad. Sci. U. S. A. 107, 769–774.

Howard, T.M., 1973. Accelerated tree death in mature *Nothofagus cunninghamii* Oerst forests in Tasmania. Victorian Natur. 90, 343–345.

Howard, R.W., Blomquist, G.J., 2005. Ecological, behavioral, and biochemical aspects of insect hydrocarbons. Annu. Rev. Entomol. 50, 371–393.

Hubbard, H.G., 1897. The ambrosia beetles of the United States. In: Howard, L.O. (Ed.), Some Miscellaneous Results of the Work of the Division of Entomology, pp. 9–13, U.S. Dept. of Agriculture Bureau of Entomology Bull. No. 7.

Hulcr, J., Cognato, A., 2010. Repeated evolution of theft in fungus farming ambrosia beetles. Evolution 64, 3205–3211.

Hulcr, J., Dunn, R.R., 2011. The sudden emergence of pathogenicity in insect-fungus symbioses threatens naive forest ecosystems. Proc. R. Soc. B 278, 2866–2873.

Hulcr, J., Mogia, M., Isua, B., Novotny, V., 2007. Host specificity of ambrosia and bark beetles (Col., Curculionidae: Scolytinae and Platypodinae) in a New Guinea rainforest. Ecol. Entomol. 32, 762–772.

Hulcr, J., Beaver, R.A., Puranasakul, W., Dole, S.A., Sonthichai, S., 2008a. A comparison of bark and ambrosia beetle communities in two forest types in northern Thailand (Coleoptera: Curculionidae: Scolytinae and Platypodinae). Environ. Entomol. 37, 1461–1470.

Hulcr, J., Novotny, V., Maurer, B.A., Cognato, A.I., 2008b. Low beta diversity of ambrosia beetles (Coleoptera: Curculionidae: Scolytinae and Platypodinae) in lowland rainforests of Papua New Guinea. Oikos 117, 214–222.

Hunt, J.H., Baker, I., Baker, H.G., 1982. Similarity of amino-acids in nectar and larval saliva—the nutritional basis for trophallaxis in social wasps. Evolution 36, 1318–1322.

Husson, R., 1955. Sur la biologie du Coléoptère xylophage *Platypus cylindrus* Fabr. Annales Univeristatis Saraviensis—Scientia 4, 348–356.

Ishay, J., Ikan, R., 1968. Food exchange between adults and larvae in *Vespa orientalis* F. Anim. Behav. 16, 298–303.

Janin, J.L., Lieutier, F., 1988. Early mating in the life-cycle of *Tomicus piniperda* L. (Coleoptera, Scolytidae) in the forest of Orleans (France). Agronomie 8, 169–172.

Janzen, D.H., 1971. Seed predation by animals. Annu. Rev. Ecol. Syst. 2, 465–492.

Janzen, D.H., 1972. Association of a rainforest palm and seed-eating beetles in Puerto Rico. Ecology 53, 258–261.

Jaramillo, J., Borgemeister, C., Baker, P., 2006. Coffee berry borer *Hypothenemus hampei* (Coleoptera: Curculionidae): searching for sustainable control strategies. Bull. Entomol. Res. 96, 223–233.

Jaramillo, J.L., Ospina, C.M., Gil, Z.N., Montoya, E.C., Benavides, P., 2011. Advances on the biology of *Corthylus zulmae* (Coleoptera: Curculionidae) in *Alnus acuminata* (Betulaceae). Revista Colombiana de Entomología 37, 48–55.

Jones, V.P., 1992. Effect of harvest interval and cultivar on damage to macadamia nuts caused by *Hypothenemus obscurus*. J. Econ. Entomol. 85, 1878–1883.

Jordal, B.H., 1998. New species of *Scolytodes* (Coleoptera: Scolytidae) from Costa Rica and Panama. Rev. Biol. Trop. 46, 407–419.

Jordal, B.H., 2001. The origin and radiation of sib-mating haplodiploid beetles (Coleoptera, Curculionidae, Scolytinae), PhD thesis, University of Bergen.

Jordal, B.H., 2006. Community structure and reproductive biology of bark beetles (Coleoptera: Scolytinae) associated with Macaronesian *Euphorbia* shrubs. Eur. J. Entomol. 103, 71–80.

Jordal, B.H., 2013. Deep phylogenetic divergence between *Scolytoplatypus* and *Remansus*, a new genus of Scolytoplatypodini from Madagascar (Coleoptera, Curculionidae, Scolytinae). ZooKeys 9–33.

Jordal, B.H., 2014a. Cossoninae. In: Leschen, R.A.B., Beutel, R. (Eds.), Handbook of Zoology, Band IV Arthropoda: Insecta. Part 38: Coleoptera, Beetles, Vol. 3. de Gruyter, Berlin, pp. 345–349.

Jordal, B.H., 2014b. Platypodinae. In: Leschen, R.A.B., Beutel, R. (Eds.), Handbook of Zoology, Band IV Arthropoda: Insecta. Part 38: Coleoptera, Beetles, Vol. 3. de Gruyter, Berlin, pp. 358–364.

Jordal, B.H., 2014c. Scolytinae. In: Leschen, R.A.B., Beutel, R. (Eds.), Handbook of Zoology, Band IV Arthropoda: Insecta. Part 38: Coleoptera, Beetles, Vol. 3. de Gruyter, Berlin, pp. 349–358.

Jordal, B., Cognato, A., 2012. Molecular phylogeny of bark and ambrosia beetles reveals multiple origins of fungus farming during periods of global warming. BMC Evol. Biol. 12, 133.

Jordal, B.H., Kambestad, M., 2014. DNA barcoding of bark and ambrosia beetles reveals excessive NUMTs and consistent east-west divergence across Palearctic forests. Mol. Ecol. Resour. 14, 7–17.

Jordal, B.H., Kirkendall, L.R., 1998. Ecological relationships of a guild of tropical beetles breeding in *Cecropia* petioles in Costa Rica. J. Trop. Ecol. 14, 153–176.

Jordal, B.H., Normark, B.B., Farrell, B.D., 2000. Evolutionary radiation of an inbreeding haplodiploid beetle lineage (Curculionidae, Scolytinae). Biol. J. Linn. Soc. 71, 483–499.

Jordal, B.H., Beaver, R.A., Kirkendall, L.R., 2001. Breaking taboos in the tropics: incest promotes colonization by wood-boring beetles. Glob. Ecol. Biogeogr. 10, 345–357.

Jordal, B.H., Beaver, R.A., Normark, B.B., Farrell, B.D., 2002. Extraordinary sex ratios and the evolution of male neoteny in sib-mating *Ozopemon* beetles. Biol. J. Linn. Soc. 75, 353–360.

Jordal, B.H., Emerson, B.C., Hewitt, G.M., 2006. Apparent "sympatric" speciation in ecologically similar herbivorous beetles facilitated by multiple colonizations of an island. Mol. Ecol. 15, 2935–2947.

Jordal, B., Gillespie, J.J., Cognato, A.I., 2008. Secondary structure alignment and direct optimization of 28S rDNA sequences provide limited phylogenetic resolution in bark and ambrosia beetles (Curculionidae: Scolytinae). Zool. Scr. 37, 43–56.

Jordal, B.H., Sequeira, A.S., Cognato, A.I., 2011. The age and phylogeny of wood boring weevils and the origin of subsociality. Mol. Phylogenet. Evol. 59, 708–724.

Jover, H., 1952. Note préliminaire su la biologie des Platypodidae de basse Côte d'Ivoire. Revue de pathologie végétale et d'entomologie agricole de France 31, 73–81.

Kajimura, H., Hijii, N., 1994. Reproduction and resource utilization of the ambrosia beetle, *Xylosandrus mutilatus*, in field and experimental populations. Entomol. Exp. Appl. 71, 121–132.

Kalshoven, L.G.E., 1925. Primaire aantasting van houtige gewassen door *Xyleborus*-soorten. *Overdruk uilhet Verslag van de resde Vergadering van de Vereeniging van Proefstation-personeel Gehouden te Djocja*, 7 Oct 1925, 1–14.

Kalshoven, L.G.E., 1958. Studies on the biology of Indonesian Scolytoidea I. Tijdschr. Entomol. 101, 157–184.

Kalshoven, L.G.E., 1959. Studies on the biology of Indonesian Scolytoidea II. Tijdschr. Entomol. 102, 135–173.

Kalshoven, L.G.E., 1960a. A form of commensalism occurring in *Xyleborus* species? Entomologische Berichten 20, 118–120.

Kalshoven, L.G.E., 1960b. Studies on the biology of Indonesian Platypodidae. Tijdschr. Entomol. 103, 31–53.

Kalshoven, L.G.E., 1962. Note on the habits of *Xyleborus destruens* Bldf., the near-primary borer of teak trees on Java. Entomologische Berichten 22, 7–18.

Kamata, N., Esaki, K., Kato, K., Igeta, Y., Wada, K., 2002. Potential impact of global warming on deciduous oak dieback caused by ambrosia fungus *Raffaelea* sp. carried by ambrosia beetle *Platypus quercivorus* (Coleoptera: Platypodidae) in Japan. Bull. Entomol. Res. 92, 119–126.

Kaneko, T., Tamaki, Y., Takagi, K., 1965. Preliminary report on the biology of some Scolytid beetles, the tea root borer, *Xyleborus germanus* Blanford, attacking tea roots, and the tea stem borer, *Xyleborus compactus* Eichhoff, attacking tea twigs. Japanese Journal of Applied Entomology and Zoology 9, 23–28.

Kausrud, K.L., Grégoire, J.-C., Skarpaas, O., Erbilgin, N., Gilbert, M., Økland, B., Stenseth, N.C., 2011. Trees wanted dead or alive! Host selection and population dynamics in tree-killing bark beetles. PLoS One 6, e18274.

Kausrud, K., Okland, B., Skarpaas, O., Grégoire, J.C., Erbilgin, N., Stenseth, N.C., 2012. Population dynamics in changing environments: the case of an eruptive forest pest species. Biol. Rev. 87, 34–51.

Kawasaki, Y., Ito, M., Miura, K., Kajimura, H., 2010. Superinfection of five *Wolbachia* in the alnus ambrosia beetle, *Xylosandrus germanus* (Coleoptera: Curculionidae). Bull. Entomol. Res. 100, 231–239.

Keen, F.P., 1958. Cone and Seed Insects of Western Forest Trees. U.S. Department of Agriculture, Technical Bulletin no. 1169, Washington, DC.

Kegley, S.J., Furniss, M.M., Grégoire, J.-C., 1997. Electrophoretic comparison of *Dendroctonus punctatus* LeConte and *D. micans* (Kugelann) (Coleoptera: Scolytidae). Pan-Pac. Entomol. 73, 40–45.

Kemner, N.A., 1916. Några nya eller mindre kända skadedjur på fruktträd. *Medd. nr. 133 från Centralanstalten för Försöksväsendet på Jordbruksområdet, Entomol. avd. nr. 25*, 21 pp.

Kent, D., 2002. Biology of the ambrosia beetle *Austroplatypus incompertus* Aust. J. Entomol. 41, 378.

Kent, D.S., Simpson, J.A., 1992. Eusociality in the beetle *Austroplatypus incompertus* (Coleoptera: Platypodidae). Naturwissenschaften 79, 86–87.

Kingsolver, J.G., Norris, D.M., 1977a. The interaction of the female ambrosia beetle, *Xyleborus ferrugineus* (Coleoptera: Scolytidae), with her eggs in relation to the morphology of the gallery system. Entomol. Exp. Appl. 21, 9–13.

Kingsolver, J.G., Norris, D.M., 1977b. The interaction of *Xyleborus ferrugineus* (Fabr.) (Coleoptera: Scolytidae) behavior and initial reproduction in relation to its symbiotic fungi. Ann. Entomol. Soc. Am. 70, 1–4.

Kirejtshuk, A.G., Azar, D., Beaver, R.A., Mandelshtam, M.Y., Nel, A., 2009. The most ancient bark beetle known: a new tribe, genus and species from Lebanese amber (Coleoptera, Curculionidae, Scolytinae). Syst. Entomol. 34, 101–112.

Kirisits, T., 2004. Fungal associates of European bark beetles with special emphasis on the ophiostomatoid fungi. In: Lieutier, F., Keith, R.D., Battisti, A., Grégoire, J.-C., Evans, H.F. (Eds.), Bark and Wood Boring Insects in Living Trees in Europe, a Synthesis. Springer, Dordrecht, pp. 181–237.

Kirkendall, L.R., 1983. The evolution of mating systems in bark and ambrosia beetles (Coleoptera: Scolytidae and Platypodidae). Zool. J. Linn. Soc. 77, 293–352.

Kirkendall, L.R., 1984. Notes on the breeding biology of some bigynous and monogynous Mexican bark beetles (Scolytidae: *Scolytus*, *Thysanoes*, *Phloeotribus*) and records for associated Scolytidae (*Hylocurus*, *Hypothenemus*, *Araptus*) and Platypodidae (*Platypus*). Z. angew. Entomol. 97, 234–244.

Kirkendall, L.R., 1989. Within-harem competition among *Ips* females, an overlooked component of density-dependent larval mortality. Holarctic Ecol. 12, 477–487.

Kirkendall, L.R., 1990. Sperm is a limiting resource in the pseudogamous bark beetle *Ips acuminatus* (Scolytidae). Oikos 57, 80–87.

Kirkendall, L.R., 1993. Ecology and evolution of biased sex ratios in bark and ambrosia beetles. In: Wrensch, D.L., Ebbert, M.A. (Eds.), Evolution and Diversity of Sex Ratio in Insects and Mites. Chapman & Hall, New York, pp. 235–345.

Kirkendall, L., 2006. A new host-specific, *Xyleborus vochysiae* (Curculionidae: Scolytinae), from Central America breeding in live trees. Ann. Entomol. Soc. Am. 99, 211–217.

Kirkendall, L., Ødegaard, F., 2007. Ongoing invasions of old-growth tropical forests: establishment of three incestuous beetle species in southern Central America (Curculionidae: Scolytinae). Zootaxa. 53–62.

Kirkendall, L.R., Stenseth, N.C., 1990. Ecological and evolutionary stability of sperm-dependent parthenogenesis: effects of partial niche overlap between sexual and asexual females. Evolution 44, 698–714.

Kirkendall, L.R., Kent, D.S., Raffa, K.F., 1997. Interactions among males, females and offspring in bark and ambrosia beetles: the significance of living in tunnels for the evolution of social behavior. In: Choe, J.C., Crespi, B.J. (Eds.), The Evolution of Social Behavior in Insects and Arachnids. Cambridge University Press, Cambridge, pp. 181–215.

Kirkendall, L.R., Faccoli, M., Ye, H., 2008. Description of the Yunnan shoot borer, *Tomicus yunnanensis* Kirkendall & Faccoli sp n. (Curculionidae, Scolytinae), an unusually aggressive pine shoot beetle from southern China, with a key to the species of *Tomicus*. Zootaxa 1819, 25–39.

Klein, W.H., Parker, D.L., Jensen, C.E., 1978. Attack, emergence, and stand depletion trends of mountain pine beetle in a lodgepole pine stand during an outbreak. Environ. Entomol. 7, 732–737.

Kleine, R., 1931. Die Biologie der Brenthidae. Entomologische Rundschau 48, 149–167, 189–193.

Knee, W., Forbes, M.R., Beaulieu, F., 2013. Diversity and host use of mites (Acari: Mesostigmata, Oribatida) phoretic on bark beetles (Coleoptera: Scolytinae): global generalists, local specialists? Ann. Entomol. Soc. Am. 106, 339–350.

Knížek, M., 2010. Five new species of *Triotemnus* (Coleoptera, Curculionidae, Scolytinae) from Morocco and Yemen. ZooKeys 191–206.

Kobayashi, M., Ueda, A., 2002. Preliminary study of mate choice in *Platypus quercivorus* (Murayama) (Coleoptera: Platypodidae). Appl. Entomol. Zool. 37, 451–457.

Koch, R., 1909. Das Larvenleben des Riesenbastkäfers *Hylesinus (Dendroctonus) micans*. Naturwissenschaftliche Zeitschrift für Land- und Forstwirtschaft 7, 319–340.

Koehler, C.S., Gyrisco, G.G., Newsom, L.D., Schwardt, H.H., 1961. Biology and control of the clover root borer, *Hylastinus obscurus* (Marsham). Memoirs of the Cornell University Agricultural Experiment Station 376, 1–36.

Kolarik, M., Kirkendall, L.R., 2010. Evidence for a new lineage of primary ambrosia fungi in *Geosmithia* Pitt (Ascomycota: Hypocreales). Fungal Biol. 114, 676–689.

Korb, J., Heinze, J., 2008. Ecology of Social Evolution. Springer, Berlin.

Kramer, P., Kozlowsky, T.T., 1979. Physiology of Woody Plants. Academic Press, New York.

Kramer, A.M., Dennis, B., Liebhold, A.M., Drake, J.M., 2009. The evidence for Allee effects. Popul. Ecol. 51, 341–354.

Kühnholz, S., Borden, J.H., Uzunovic, A., 2001. Secondary ambrosia beetles in apparently healthy trees: adaptations, potential causes and suggested research. Integrated Pest. Manag. Rev. 6, 209–219.

Kuschel, G., 1966. A cossonine genus with bark-beetle habits with remarks on relationships and biogeography. N. Z. J. Sci. 9, 3–29.

Lachat, T., Nagel, P., Cakpo, Y., Attignon, S., Goergen, G., Sinsin, B., Peveling, R., 2006. Dead wood and saproxylic beetle assemblages in a semi-deciduous forest in southern Benin. Forest Ecol. Manage. 225, 27–38.

Lachat, T., Wermelinger, B., Gossner, M.M., Bussler, H., Isacsson, G., Müller, J., 2012. Saproxylic beetles as indicator species for dead-wood amount and temperature in European beech forests. Ecol. Indic. 23, 323–331.

Lake Maner, M., Hanula, J.L., Kristine Braman, S., 2013. Rearing redbay ambrosia beetle, *Xyleborus glabratus* (Coleoptera: Curculionidae: Scolytinae), on semi-artificial media. Fla. Entomol. 96, 1042–1051.

Langor, D.W., 1987. Flight muscle changes in the eastern larch beetle, *Dendroctonus simplex* LeConte. Coleopterists Bulletin 41, 351–357.

Långström, B., 1983. Life cycle and shoot feeding of the pine shoot beetles. Stud. For. Suec. 163, 1–29.

Långström, B., Hellqvist, C., 1993. Induced and spontaneous attacks by pine shoot beetles on young Scots pine trees—tree mortality and beetle performance. J. Appl. Entomol. 115, 25–36.

Lanier, G.N., Kirkendall, L.R., 1986. Karyology of pseudogamous *Ips* bark beetles. Hereditas 105, 87–96.

Lanier, G.N., Oliver Jr., J.H., 1966. "Sex-ratio" condition: unusual mechanisms in bark beetles. Science 150, 208–209.

Latty, T.M., Magrath, M.J.L., Symonds, M.R.E., 2009. Harem size and oviposition behaviour in a polygynous bark beetle. Ecol. Entomol. 34, 562–568.

Lawko, C.M., Dyer, E.D.A., 1974. Flight ability of spruce beetles emerging after attacking frontalin-baited trees. Canada Department of the Environment, Canadian Forestry Service, Bi-monthly Research Note 30, 17.

Lekander, B., 1962. Die Borkenkäerlarven, ein vernachlässigter Teil der Forstentomologie. Dtsch. Entomol. Z. 9, 428–432.

Lekander, B., 1963. *Xyleborus cryptographus* Ratzb. (Col. Ipidae), Ein Beitrag zur Kenntnis seiner Verbreitung und Biologie. Entomologisk Tidskrift 84, 96–109.

Lekander, B., 1968a. Scandinavian bark beetle larvae. Department of Forest Zoology, Stockholm.

Lekander, B., 1968b. The number of larval instars in some bark beetle species. Entomologisk Tidskrift 89, 25–34.

Lengerken, H., 1939. Die Brutfürsorge- und Brutpflegeinstinkte der Käfer. Akademische Verlagsgesellschaft m.b.H, Leipzig.

LePelley, R.H., 1968. Pests of Coffee. Longmans, Green and Co., London.

Lewis, E.E., Cane, J.H., 1992. Inefficacy of courtship stridulation as a premating ethological barrier for *Ips* bark beetles (Coleoptera: Scolytidae). Ann. Entomol. Soc. Am. 85, 517–524.

Lieutier, F., Ye, H., Yart, A., 2003. Shoot damage by *Tomicus* sp (Coleoptera: Scolytidae) and effect on *Pinus yunnanensis* resistance to subsequent reproductive attacks in the stem. Agr. Forest Entomol. 5, 227–233.

Lieutier, F., Day, K.R., Battisti, A., Grégoire, J.-C., Evans, H.F. (Eds.), 2004. Bark and Wood Boring Insects in Living Trees in Europe, a Synthesis. Kluwer Academic Publishers, Dordrecht.

Lindgren, B.S., Raffa, K.F., 2013. Evolution of tree killing in bark beetles (Coleoptera: Curculionidae): trade-offs between the maddening crowds and a sticky situation. Can. Entomol. 145, 471–495.

Linnaeus, C., 1758. *Systema Naturae per Regna Tria Naturae: Secundum Classes, Ordines, Genera, Species, Cum Characteribus, Differentiis, Synonymis, Locis* (10th ed.) (in Latin). Laurentius Salvius, Stockholm.

Lissemore, F.M., 1997. Frass clearing by male pine engraver beetles (*Ips pini*; Scolytidae): paternal care or paternity assurance? Behav. Ecol. 8, 318–325.

López, S., Quero, C., Iturrondobeitia, J.C., Guerrero, A., Goldarazena, A., 2013. Electrophysiological and behavioural responses of Pityophthorus pubescens (Coleoptera: Scolytinae) to (E, E)-α-farnesene, (R)-(+)-limonene and (S)-(−)-verbenone in Pinus radiata (Pinaceae) stands in northern Spain. Pest Manage. Sci. 69, 40–47.

López-Guillén, G., Carrasco, J.V., Cruz-López, L., Barrera, J.F., Malo, E.A., Rojas, J.C., 2011. Morphology and structural changes in flight muscles of *Hypothenemus hampei* (Coleoptera: Curculionidae) females. Environ. Entomol. 40, 441–448.

Lövendal, E.A., 1898. De danske barkbiller (Scolytidæ et Platypodidæ danicæ) og deres betydning for skov- og havebruget. Det Schubotheske Forlag, J.L, Lybecker og E.A. Hirschsprung, Copenhagen.

Løyning, M.K., 2000. Reproductive performance of clonal and sexual bark beetles (Coleoptera: Scolytidae) in the field. J. Evol. Biol. 13, 743–748.

Løyning, M., Kirkendall, L., 1996. Mate discrimination in a pseudogamous bark beetle (Coleoptera: Scolytidae): male *Ips acuminatus* prefer sexual to clonal females. Oikos 77, 336–344.

Løyning, M.K., Kirkendall, L.R., 1999. Notes on the mating system of *Hylesinus varius* (F.) (Col., Scolytidae), a putatively bigynous bark beetle. J. Appl. Entomol. 123, 77–82.

Lyal, C.H.C., King, T., 1996. Elytro-tergal stridulation in weevils (Insecta: Coleoptera: Curculionoidea). J. Nat. Hist. 30, 703–773.

Lyons, L.A., 1956. Insects affecting seed production in red pine: Part I *Conophthorus resinosae* Hopk. (Coleoptera: Scolytidae). Can. Entomol. 88, 599–608.

Majka, C.G., Anderson, R.S., McCorquodale, D.B., 2007. The weevils (Coleoptera: Curculionoidea) of the Maritime Provinces of Canada, II: New records from Nova Scotia and Prince Edward Island and regional zoogeography. Can. Entomol. 139, 397–442.

Mandelshtam, M.Y., 2001. New synonymy and new records in Palaearctic Scolytidae (Coleoptera). Zoosystematica Rossica 9, 203–204.

Maroja, L.S., Bogdanowicz, S.M., Wallin, K.F., Raffa, K.F., Harrison, R. G., 2007. Phylogeography of spruce beetles (*Dendroctonus rufipennis* Kirby) (Curculionidae: Scolytinae) in North America. Mol. Ecol. 16, 2560–2573.

Mattson, W.J., 1980. Cone resources and the ecology of the red pine cone beetle, *Conophthorus resinosae* (Coleoptera, Scolytidae). Ann. Entomol. Soc. Am. 73, 390–396.

McGehey, J.H., 1968. Territorial behaviour of bark-beetle males. Can. Entomol. 100, 1153.

McKenna, D.D., 2011. Towards a temporal framework for "inordinate fondness": Reconstructing the macroevolutionary history of beetles (Coleoptera). Entomologica Americana 117, 28–36.

McKenna, D.D., Sequeira, A.S., Marvaldi, A.E., Farrell, B.D., 2009. Temporal lags and overlap in the diversification of weevils and flowering plants. Proc. Natl. Acad. Sci. U. S. A. 106, 7083–7088.

McKnight, M.E., Aarhus, D.G., 1973. Bark beetles, *Leperisinus californicus* and *L. criddlei* (Coleoptera: Scolytidae), attacking green ash in North Dakota. Ann. Entomol. Soc. Am. 66, 955–957.

McManus, M.L., Giese, R.L., 1968. Columbian timber beetle *Corthylus columbianus*. VII. *Effect of climatic integrants on historic density fluctuations*. For. Sci. 14, 242–253.

McMullen, L.H., Atkins, M.D., 1962. The life history and habits of *Scolytus unispinosus* Leconte (Coleoptera: Scolytidae) in the interior of British Columbia. Can. Entomol. 94, 17–25.

McNee, W.R., Wood, D.L., Storer, A.J., 2000. Pre-emergence feeding in bark beetles (Coleoptera: Scolytidae). Environ. Entomol. 29, 495–501.

McPherson, J.E., Stehr, F.W., Wilson, L.F., 1970. A comparison between *Conophthorus* shoot-infesting beetles and *Conophthorus resinosae* (Coleoptera: Scolytidae). 1. Comparative life history studies in Michigan. Can. Entomol. 102, 1008–1015.

Mecke, R., Galileo, M.H.M., 2004. A review of the weevil fauna (Coleoptera, Curculionoidea) of *Araucaria angustifolia* (Bert.) O. Kuntze (Araucariaceae) in South Brazil. Revista Brasileira de Zoologia 21, 505–513.

Meirmans, S., Skorping, A., Løyning, M.K., Kirkendall, L.R., 2006. On the track of the Red Queen: bark beetles, their nematodes, local climate and geographic parthenogenesis. J. Evol. Biol. 19, 1939–1947.

Melnikova, N.I., 1964. Biological significance of the air holes in egg tunnels of *Scolytus ratzeburgi* Jans. (Coleoptera, Ipidae). Entomol. Rev. 43, 16–23.

Menard, K.L., Cognato, A.I., 2007. Mitochondrial haplotypic diversity of pine cone beetles (Scolytinae: *Conophthorus*) collected on food sources. Environ. Entomol. 36, 962–966.

Mendel, Z., 1983. Seasonal history of *Orthotomicus erosus* (Coleoptera: Scolytidae) in Israel. Phytoparasitica 11, 13–24.

Merriam, C.H., 1883. Ravages of a rare scolytid beetle in the sugar maples of northeastern New York. Am. Nat. 17, 84–86.

Meurisse, N., Couillien, D., Grégoire, J.-C., 2008. Kairomone traps: a tool for monitoring the invasive spruce bark beetle *Dendroctonus micans* (Coleoptera: Scolytinae) and its specific predator, *Rhizophagus grandis* (Coleoptera: Monotomidae). J. Appl. Entomol. 45, 537–548.

Miller, J.M., 1915. Cone beetles: Injury to sugar pine and western yellow pine. U.S. Department of Agriculture Bulletin 243, 1–12.

Milne, D.H., Giese, R.L., 1969. Biology of the Columbian timber beetle, *Corthylus columbianus* (Coleoptera: Scolytidae). IX. Population biology and gallery characteristics. Entomol. News 80, 225–237.

Milne, D.H., Giese, R.L., 1970. Biology of the Columbian timber beetle, *Corthylus columbianus* (Coleoptera: Scolytidae). X. Comparison of yearly mortality and dispersal losses with population densities. Entomol. News 81, 12–24.

Mitchell, A., Maddox, C., 2010. Bark beetles (Coleoptera: Curculionidae: Scolytinae) of importance to the Australian macadamia industry: an integrative taxonomic approach to species diagnostics. Aust. J. Entomol. 49, 104–113.

Mitton, J.B., Sturgeon, K.B., 1982. Biotic interactions and evolutionary change. In: Mitton, J.B., Sturgeon, K.B. (Eds.), Bark Beetles in North American Conifers. A System for the Study of Evolutionary Biology. University of Texas Press, Austin, pp. 3–20.

Mizuno, T., Kajimura, H., 2002. Reproduction of the ambrosia beetle, *Xyleborus pfeili* (Ratzeburg) (Col., Scolytidae), on semi-artificial diet. J. Appl. Entomol. 126, 455–462.

Morales-Jiménez, J., Vera-Ponce de León, A., García-Domínguez, A., Martínez-Romero, E., Zúñiga, G., Hernández-Rodríguez, C., 2013. Nitrogen-fixing and uricolytic bacteria associated with the gut of *Dendroctonus rhizophagus* and *Dendroctonus valens* (Curculionidae: Scolytinae). Microb. Ecol. 66, 200–210.

Morgan, F.D., 1967. *Ips grandicollis* in South Australia. Aust. For. 31, 137–155.

Morgan, C., 1998. The assessment of potential attractants to beetle pests: improvements to laboratory pitfall bioassay methods. J. Stored Prod. Res. 34, 59–74.

Morgan, F.D., Mailu, M., 1976. Behavior and generation dynamics of the white pine cone beetle *Conophthorus coniperda* (Schwarz) in central Wisconsin. Ann. Entomol. Soc. Am. 69, 863–871.

Mueller, U.G., Gerardo, N.M., Aanen, D.K., Six, D.L., Schultz, T.R., 2005. The evolution of agriculture in insects. Annu. Rev. Ecol. Evol. Syst. 36, 563–595.

Nakashima, T., 1971. Notes on the associated fungi and the mycetangia of the ambrosia beetle, *Crossotarsus niponicus*. Appl. Entomol. Zool. 6, 131–137.

Nobuchi, A., 1969. A comparative morphological study of the proventriculus in the adult of the superfamily Scolytoidea (Coleoptera). Bull. Gov. For. Exp. Stn. (Jpn.) 224, 39–110.

Nobuchi, A., 1972. The biology of Japanese Scolytidae and Platypodidae (Coleoptera). Rev. Plant Prot. Res. 5, 61–75.

Noguera-Martínez, F.A., Atkinson, T.H., 1990. Biogeography and biology of bark and ambrosia beetles (Coleoptera: Scolytidae and Platypodidae) of a mesic montane forest in Mexico, with an annotated checklist of species. Ann. Entomol. Soc. Am. 83, 453–466.

Nord, J.C., 1972. Biology of the Columbian timber beetle, *Corthylus columbianus* (Coleoptera: Scolytidae). Ann. Entomol. Soc. Am. 65, 350–358.

Normark, B.B., 2004. Haplodiploidy as an outcome of coevolution between male-killing cytoplasmatic elements and their hosts. Evolution 58, 790–798.

Normark, B.B., Kirkendall, L.R., 2009. Parthenogenesis in insects and mites. In: Resh, V.H., Cardé, R.T. (Eds.), Encyclopedia of Insects, Second Edition. Academic Press, Amsterdam, pp. 753–757.

Normark, B.B., Jordal, B.H., Farrell, B.D., 1999. Origin of a haplodiploid beetle lineage. Proc. R. Soc. B 266, 2253–2259.

Norris, D.M., Chu, H.M., 1980. Symbiote-dependent arrhenotokous parthenogenesis in the eukaryote *Xyleborus*. In: Schwemmler, W., Schenk, H.E.A. (Eds.), Endocytobiology: Endosymbiosis and Cell Biology. Walter de Gruyter & Co., Berlin, pp. 453–460.

Novotny, V., Miller, S.E., Hulcr, J., Drew, R.A.I., Basset, Y., Janda, M., et al., 2007. Low beta diversity of herbivorous insects in tropical forests. Nature 448, 692–698.

Novotny, V., Miller, S.E., Baje, L., Balagawi, S., Basset, Y., Cizek, L., et al., 2010. Guild-specific patterns of species richness and host specialization in plant-herbivore food webs from a tropical forest. J. Anim. Ecol. 79, 1193–1203.

Nunberg, M., 1963. Zur Systematik und Synonomie der Scolytoidea (Coleoptera). *Ann. Zool.* Warszawa 20, 357–361.

Nyland, B.C., 1995. Lowest lifetime fecundity. In: Walker, T.J. (Ed.), Insect Book of Records. Available online:http://entnemdept.ifas.ufl. edu/walker/ufbir/chapters/chapter_17.shtml, Last accessed: June 27, 2015.

Ødegaard, F., 2000. The relative importance of trees versus lianas as hosts for phytophagous beetles (Coleoptera) in tropical forests. J. Biogeogr. 27, 283–296.

Ødegaard, F., 2004. Species richness of phytophagous beetles in the tropical tree *Brosimum utile* (Moraceae): the effects of sampling strategy and the problem of tourists. Ecol. Entomol. 29, 76–88.

Ødegaard, F., 2006. Host specificity, alpha- and beta-diversity of phytophagous beetles in two tropical forests in Panama. Biodiv. Conserv. 15, 83–105.

Ødegaard, F., Diserud, O.H., Engen, S., Aagaard, K., 2000. The magnitude of local host specificity for phytophagous insects and its implications for estimates of global species richness. Conserv. Biol. 14, 1182–1186.

Ødegaard, F., Diserud, O.H., Østbye, K., 2005. The importance of plant relatedness for host utilization among phytophagous insects. Ecol. Lett. 8, 612–617.

Oester, P.T., Rudinsky, J.A., 1975. Sound production in Scolytidae: stridulation by "silent" *Ips* bark beetles. Z. angew. Entomol. 79, 421–427.

Oester, P.T., Rudinsky, J.A., 1979. Acoustic behavior of three sympatric species of *Ips* (Coleoptera-Scolytidae) co-inhabiting Sitka spruce. Z. angew. Entomol. 87, 398–412.

Oester, P.T., Rykar, L.C., Rudinsky, J.A., 1978. Complex male premating stridulation of the bark beetle *Hylurgops rugipennis* (Mann). Coleopterists Bulletin 32, 93–98.

Oester, P.T., Rudinsky, J.A., Ryker, L.C., 1981. Olfactory and acoustic behavior of *Pseudohylesinus nebulosus* (Coleoptera, Scolytidae) on Douglas-fir bark. Can. Entomol. 113, 645–650.

Ohmart, C.P., 1989. Why are there so few tree-killing bark beetles associated with angiosperms? Oikos 54, 242–245.

Ohya, E., Kinuura, H., 2001. Close range sound communications of the oak platypodid beetle *Platypus quercivorus* (Murayama) (Coleoptera: Platypodidae). Appl. Entomol. Zool. 36, 317–321.

Okello, S., Reichmuth, C., Schulz, F.A., 1996a. Observations on the biology and host specificity of *Pagiocerus frontalis* (Fabricius) (Coleoptera: Scolytidae) at 20 °C and 25 °C and 75% rh. Zeitschrift für Pflanzenkrankheiten und Pflanzenschutz 103, 377–382.

Okello, S., Reichmuth, C., Schulz, F.A., 1996b. Laboratory investigations on the developmental rate at low relative humidity and the developmental temperature limit of *Pagiocerus frontalis* (Fab) (Col, Scolytidae) at high and low temperatures. Anz. Schädlingskd. 69, 180–182.

Page, R.E., Willis, M.A., 1983. Sexual dimorphism in ventral abdominal setae in *Scolytus multistriatus* (Coleoptera: Scolytidae): possible role in courtship behavior. Ann. Entomol. Soc. Am. 76, 78–82.

Page, M., Nelson, L.J., Haverty, M.I., Blomquist, G.J., 1990a. Cuticular hydrocarbons as chemotaxonomic characters for bark beetles: *Dendroctonus ponderosae*, *D. jeffreyi*, *D. brevicomis*, and *D. frontalis* (Coleoptera: Scolytidae). Ann. Entomol. Soc. Am. 83, 892–902.

Page, M., Nelson, L.J., Haverty, M.I., Blomquist, G.J., 1990b. Cuticular hydrocarbons of eight species of North American cone beetles, *Conophthorus* Hopkins. J. Chem. Ecol. 16, 1173–1198.

Page, M., Nelson, L.J., Blomquist, G.J., Seybold, S.J., 1997. Cuticular hydrocarbons as chemotaxonomic characters of pine engraver beetles (*Ips* spp.) in the grandicollis subgeneric group. J. Chem. Ecol. 23, 1053–1099.

Paine, T.D. (Ed.), 2006. Invasive Forest Insects, Introduced Forest Trees, and Altered Ecosystems. Springer, Dordrecht.

Paine, T.D., Raffa, K.F., Harrington, T.C., 1997. Interactions among scolytid bark beetles, their associated fungi, and live conifers. Annu. Rev. Entomol. 42, 179–206.

Peer, K., Taborsky, M., 2005. Outbreeding depression, but no inbreeding depression in haplodiploid ambrosia beetles with regular sibling mating. Evolution 59, 317–323.

Peer, K., Taborsky, M., 2007. Delayed dispersal as a potential route to cooperative breeding in ambrosia beetles. Behav. Ecol. Sociobiol. 61, 729–739.

Peleg, B., Norris, D.M., 1973. Oocyte activation in *Xyleborus ferrugineus* by bacterial symbionts. J. Insect Physiol. 19, 137–145.

Petty, J.L., 1977. Bionomics of two aspen bark beetles, *Trypophloeus populi* and *Procryphalus mucronatus* (Coleoptera: Scolytidae). Western Great Basin Nat. 37, 105–127.

Pfeffer, A., 1955. Fauna CSR, Kurovci—Scolytoidea. Nakaldatelství CSAV, Praha.

Pfeffer, A., 1995. Zentral- und westpaläarktische Borken- und Kernkäfer (Coleoptera: Scolytidae, Platypodidae). Entomologica Brasiliensia 17, 5–310.

Ploetz, R.C., Hulcr, J., Wingfield, M.J., de Beer, Z.W., 2013. Destructive tree diseases that are associated with ambrosia and bark beetles: black swan events in tree pathology? Plant Dis. 97, 856–872.

Pope, D.N., Coulson, R.N., Fargo, W.S., Gagne, J.A., Kelly, C.W., 1980. The allocation process and between-tree survival probabilities in *Dendroctonus frontalis* infestations. Res. Popul. Ecol. 22, 197–210.

Postner, M., 1974. Scolytidae (=Ipidae), Borkenkäfer. In: Schwenke, W. (Ed.), Die Forstschädlinge Europas. Bind 2, Käfer. Paul Parey, Hamburg, pp. 334–482.

Pureswaran, D.S., Borden, J.H., 2003. Is bigger better? Size and pheromone production in the mountain pine beetle, *Dendroctonus ponderosae* Hopkins (Coleoptera: Scolytidae). J. Insect Behav. 16, 765–782.

Queller, D.C., Strassmann, J.E., 1998. Kin selection and social insects. BioScience 48, 165–175.

Rabinowitz, D., 1977. Effects of a mangrove borer, *Poecilips rhizophorae*, on propagules of *Rhizophora harrisonii* in Panamá. Fla. Entomol. 60, 129–134.

Raffa, K.F., Aukema, B.H., Bentz, B.J., Carroll, A.L., Hicke, J.A., Turner, M.G., Romme, W.H., 2008. Cross-scale drivers of natural disturbances prone to anthropogenic amplification: the dynamics of bark beetle eruptions. BioScience 58, 501–517.

Rappaport, N.G., Stein, J.D., del Rio Mora, A.A., Debarr, G.L., De Groot, P., Mori, S., 2000. Cone beetle (*Conophthorus* spp.) (Coleoptera: Scolytidae) responses to behavioral chemicals in field trials: a transcontinental perspective. Can. Entomol. 132, 925–937.

Ratzeburg, J.T.C., 1839. Die Forst-Insecten. Nicolaische Buchhandlung, Berlin.

Reid, R.W., 1958. Internal changes in the female mountain pine beetle, *Dendroctonus monticolae* Hopk., associated with egg laying and flight. Can. Entomol. 90, 464–468.

Reid, M.L., 1999. Monogamy in the bark beetle *Ips latidens*: ecological correlates of an unusual mating system. Ecol. Entomol. 24, 89–94.

Reid, M.L., Robb, T., 1999. Death of vigorous trees benefits bark beetles. Oecologia 120, 555–562.

Reid, M.L., Roitberg, B.D., 1994. Benefits of prolonged male residence with mates and brood in pine engravers (Coleoptera, Scolytidae). Oikos 70, 140–148.

Reid, M.L., Roitberg, B.D., 1995. Effects of body-size on investment in individual broods by male pine engravers Can. J. Zool. 73, 1396–1401.

Reitter, E., 1916. Fauna Germanica. V Band. Die Käfer des Deutschen Reiches, K. G. Lutz, Stuttgard.

Rexrode, C.O., 1982. Bionomics of the peach bark beetle *Phloeotribus liminaris* (Coleoptera: Scolytidae) in black cherry. J. Georgia Entomol. Soc. 17, 388–398.

Roberts, H., 1960. *Trachyostus ghanaensis* Schedl (Col., Platypodidae) an ambrosia beetle attacking Wawa, *Triplochiton scleroxylon* K. Schum. West Afr. Timber Borer Res. Unit Tech. Bull. No. 3, 1–17.

Roberts, H., 1962. A description of the developmental stages of *Trachyostus aterrimus* a West African Platypodina, and some remarks on its biology. Fifth Report West African Timber Borer Research Unit 1961–62, 29–46.

Roberts, H., 1968. Notes on biology of ambrosia beetles of genus *Trachyostus* Schedl (Coleoptera—Platypodidae) in West Africa. Bull. Entomol. Res. 58, 325–352.

Roberts, H., 1969. Forest insects of Nigeria with notes on their biology and distribution. Commonwealth Forestry Institute Paper 44, 1–206.

Roberts, H., 1976. Observations on the biology of some tropical rain forest Scolytidae (Coleoptera) from Fiji I. Subfamilies—Hylesininae, Ipinae (excluding Xyleborini). Bull. Entomol. Res. 66, 373–388.

Robertson, I.C., 1998a. Flight muscle changes in male pine engraver beetles during reproduction: the effects of body size, mating status and breeding failure. Physiol. Entomol. 23, 75–80.

Robertson, I.C., 1998b. Paternal care enhances male reproductive success in pine engraver beetles. Anim. Behav. 56, 595–602.

Robertson, I.C., 2000. Reproduction and developmental phenology of *Ips perturbatus* (Coleoptera: Scolytidae) inhabiting white spruce (Pinaceae). Can. Entomol. 132, 529–537.

Robertson, I.C., Roitberg, B.D., 1998. Duration of paternal care in pine engraver beetles: why do larger males care less? Behav. Ecol. Sociobiol. 43, 379–386.

Roeper, R.A., 1995. Patterns of mycetophagy in Michigan ambrosia beetles. Michigan Academian 27, 153–161.

Roeper, R.A., Treeful, L.M., Foote, R.A., Bunce, M.A., 1980a. *In vitro* culture of the ambrosia beetle *Xyleborus affinis* (Coleoptera: Scolytidae). Great Lakes Entomologist 13, 33–35.

Roeper, R., Treeful, L.M., O'Brien, K.M., Foote, R.A., Bunce, M.A., 1980b. Life history of the ambrosia beetle *Xyleborus affinis* (Coleoptera: Scolytidae) from *in vitro* culture. Great Lakes Entomologist 13, 141–144.

Roeper, R.A., Zestos, D.V., Palik, B.J., Kirkendall, L.R., 1987a. Distribution and host plants of *Corthylus punctatissimus* (Coleoptera: Scolytidae) in the lower peninsula of Michigan. Great Lakes Entomologist 20, 69–70.

Roeper, R.A., Palik, B.J., Zestos, D.V., Hesch, P.G., Larsen, C.D., 1987b. Observations of the habits of *Corthylus punctatissimus* (Coleoptera, Scolytidae) infesting maple saplings in Central Michigan. Great Lakes Entomologist 20, 173–176.

Rohlfs, M., 2005. Density-dependent insect–mold interactions: effects on fungal growth and spore production. Mycologia 97, 996–1001.

Rohlfs, M., Churchill, A.C.L., 2011. Fungal secondary metabolites as modulators of interactions with insects and other arthropods. Fungal Genet. Biol. 48, 23–34.

Ruckes Jr., H., 1963. Cone beetles of the genus *Conophthorus* in California (Coleoptera: Scolytidae). Pan-Pac. Entomol. 39, 43–50.

Rudinsky, J.A., 1969. Masking of the aggregating pheromone in *Dendroctonus pseudotsugae* Hopk. Science 166, 884–885.

Rudinsky, J.A., 1976. Various host–insect interrelations in host-finding and colonization behavior of bark beetles on coniferous trees. Symp. Biol. Hung. 16, 229–235.

Rudinsky, J.A., 1979. Chemoacoustically induced behavior of *Ips typographus* (Col, Scolytidae). Z. angew. Entomol. 88, 537–541.

Rudinsky, J.A., Michael, R.R., 1974. Sound production in Scolytidae: "rivalry" behaviour of male *Dendroctonus* beetles. J. Insect Physiol. 20, 1219–1230.

Rudinsky, J.A., Ryker, L.C., 1976. Sound production in Scolytidae—rivalry and premating stridulation of male Douglas-fir beetle. J. Insect Physiol. 22, 997–999.

Rudinsky, J.A., Vallo, V., 1979. The ash bark beetles *Leperisinus fraxini* and *Hylesinus oleiperda*: stridulatory organs, acoustic signals, and pheromone production. Z. angew. Entomol. 87, 417–429.

Rudinsky, J.A., Ryker, L.C., Michael, R.R., Libbey, L.M., Morgan, M.E., 1976. Sound production in Scolytidae: female sonic stimulus of male pheromone release in two *Dendroctonus* beetles. J. Insect Physiol. 22, 1675–1681.

Rudinsky, J.A., Vallo, V., Ryker, L.C., 1978. Sound production in Scolytidae: attraction and stridulation of *Scolytus mali* (Col., Scolytidae). Z. angew. Entomol. 86, 381–391.

Rühm, W., 1981. Zur Biologie und Ökologie von *Pteleobius* (*Xylechinus*) *valdivianus* (Eggers, 1942) (Col., Scolytidae), einer vorwiegend an unterständigen Araukarien, *Araucaria araucana* (Mol.) Koch, brütenden Borkenkäferart. Entomologische Mitteilungen aus dem Zoologischen Museum Hamburg 7, 13–20.

Ryker, L.C., 1984. Acoustic and chemical signals in the life cycle of a beetle. Sci. Am. 250, 112–123.

Safranyik, L., 1976. Size- and sex-related emergence, and survival in cold storage, of mountain pine beetle adults. Can. Entomol. 108, 209–212.

Salom, S.M., McLean, J.A., 1989. Influence of wind on the spring flight of *Trypodendron lineatum* (Olivier) (Coleoptera, Scolytidae) in a second-growth coniferous forest. Can. Entomol. 121, 109–119.

Salom, S.M., McLean, J.A., 1991. Environmental influences on dispersal of *Trypodendron lineatum* (Coleoptera, Scolytidae). Environ. Entomol. 20, 565–576.

Salonen, K., 1973. On the life cycle, especially on the reproduction biology of *Blastophagus piniperda* L. (Col., Scolytidae). Acta For. Fenn. 127, 1–72.

Santoro, F.H., 1963. Bioecología de *Platypus sulcatus* Chapuis (Coleoptera, Platypodidae). Revista de Investigaciones Forestales 4, 47–79.

Santos, M.F., de, A., Mermudes, J.R.M., da Fonseca, V.M.M., 2011. A specimen of Curculioninae (Curculionidae, Coleoptera) from the Lower Cretaceous, Araripe Basin, north-eastern Brazil. Palaeontology 54, 807–814.

Sasakawa, M., Yoshiyasu, Y., 1983. Stridulatory organs of the Japanese pine bark beetles (Coleoptera, Scolytidae). Kontyu 51, 493–501.

Saunders, J.L., Knoke, J.K., 1967. Diets for rearing the ambrosia beetle *Xyleborus ferrugineus* (Fabricius) in vitro. Science 157, 460–463.

Sauvard, D., 2004. General biology of bark beetles. In: Lieutier, F., Day, K.R., Battisti, A., Grégoire, J.-C., Evans, H.F. (Eds.), Bark and Wood Boring Insects in Living Trees in Europe, A Synthesis. Kluwer Academic Publishers, Dordrecht, pp. 63–88.

Schedl, K.E., 1931. Morphology of the bark beetles of the genus *Gnathotrichus* Eichhoff. Smithson. Misc. Collect. 82, 1–88.

Schedl, K.E., 1958. Breeding habits of arboricole insects in Central Africa. Tenth International Congress of Entomology, Proceedings, 183–197.

Schedl, K.E., 1959. Scolytidae und Platypodidae Afrikas. Band 1. Familie Scolytidae. Revista de Entomologia de Moçambique 2, 357–422.

Schedl, K.E., 1960a. Scolytidae und Platypodidae Afrikas. Band 1 (forts.). Familie Scolytidae. Revista de Entomologia de Moçambique 3, 75–154.

Schedl, K.E., 1960b. Insectes nuisibles aux fruits et aux graines. Publications de l'Institut National pour l'étude agronomique du Congo Belge, Série Scientifique 82, 1–133.

Schedl, K.E., 1961. Scolytidae und Platypodidae Afrikas. Band 1 (forts.). Familie Scolytidae. Revista de Entomologia de Moçambique 4, 335–742.

Schedl, K.E., 1962a. Scolytidae und Platypodidae Afrikas. Band 2. Familie Scolytidae. Revista de Entomologia de Moçambique 5, 1–594.

Schedl, K.E., 1962b. Scolytidae und Platypodidae Afrikas. Band 3. Familie Platypodidae. Revue Entomologique de Moçambique 5, 595–1352.

Schedl, K.E., 1972. Monographie der Familie Platypodidae Coleoptera. W. Junk, Den Haag.

Schedl, K.E., 1974. Scolytoidea from the Galapagos and Cocos Islands. Studies of the Neotropical Fauna 9, 47–53.

Schedl, K.E., 1977. Die Scolytidae und Platypodidae Madagaskars und einiger naheliegender Inselgruppen. 303 Bietrag. Mitt. Forstl. Bundes-Versuchanstalt, Wien 119, 5–326.

Schlupp, I., 2005. The evolutionary ecology of gynogenesis. Annu. Rev. Ecol. Evol. Syst. 36, 399–417.

Schlyter, F., Zhang, Q.-H., 1996. Testing avian polygyny hypotheses in insects: harem size distribution and female egg gallery spacing in three *Ips* bark beetles. Oikos 76, 57–69.

Schmidberger, J., 1836. Naturgeschichte des Apfelborkenkäfers *Apate dispar*. Beiträge zur Obstbaumzucht und zur Naturgeschichte der den Obstbäumen schädlichen Insekten 4, 213–230.

Schneider, I., 1987. Distribution, fungus-transfer and gallery construction of the ambrosia beetle *Xyleborus affinis* in comparison with *X. mascarensis* (Coleoptera, Scolytidae). Entomologia Generalis 12, 267–275.

Schneider-Orelli, O., 1911. Die Uebertragung und Keimung des Ambrosia-pilzes von *Xyleborus (Anisandrus) dispar* F. Naturwissenschaftliche Zeitschrift für Land- und Forstwirtschaft 8, 186–192.

Schrey, N.M., Schrey, A.W., Heist, E.J., Reeve, J.D., 2011. Genetic heterogeneity in a cyclical forest pest, the southern pine beetle, *Dendroctonus frontalis*, is differentiated into east and west groups in the southeastern United States. J. Insect Sci. 11, 110.

Schwarz, E.A., 1891. Notes on the breeding habits of some scolytids. Proc. Entomol. Soc. Wash. 2, 77–80.

Schwarz, E.A., 1901. Note on *Scolytus quadrispinosus*. Proc. Entomol. Soc. Wash. 4, 344.

Scott, J.J., Oh, D.C., Yuceer, M.C., Klepzig, K.D., Clardy, J., Currie, C.R., 2008. Bacterial protection of beetle-fungus mutualism. Science 322, 63.

Searcy, W.A., Yasukawa, K., 1989. Alternative models of territorial polygyny in birds. Am. Nat. 134, 323–343.

Sforzi, A., Bartolozzi, L., 2004. Brenthidae of the World. Museo Regionale di Scienze Naturali. Torino, Italy.

Shimizu, A., Tanaka, R., Akiba, M., Masuya, H., Iwata, R., Fukuda, K., Kanzaki, N., 2013. Nematodes associated with *Dryocoetes uniseriatus* (Coleoptera: Scolytidae). Environ. Entomol. 42, 79–88.

Shuster, S.M., Wade, M.J., 2003. Mating Systems and Strategies. Princeton University Press, Princeton and Oxford.

Singer, T., 1998. Roles of hydrocarbons in the recognition systems of insects. Am. Zool. 38, 394–405.

Six, D.L., 2003. Bark beetle-fungus symbioses. In: Bourtzis, K., Miller, T.A. (Eds.), Insect Symbiosis. CRC Press, Boca Raton, pp. 97–114.

Six, D.L., 2012. Ecological and evolutionary determinants of bark beetle–fungus symbioses. Insects 3, 339–366.

Six, D., 2013. The bark beetle holobiont: Why microbes matter. J. Chem. Ecol. 1–14.

Six, D.L., Paine, T.D., 1998. Effects of mycangial fungi and host tree species on progeny survival and emergence of *Dendroctonus ponderosae* (Coleoptera: Scolytidae). Environ. Entomol. 27, 1393–1401.

Smetanin, A.N., 2013. A new bark beetle species, *Dryocoetes krivolutzkajae* Mandelshtam, 2001 (Scolytidae), discovered in Kamchatka. Contemp. Probl. Ecol. 6, 43–44.

Smith, N.G.C., 2000. The evolution of haplodiploidy under inbreeding. Heredity 84, 186–192.

Smith, S.M., Cognato, A.I., 2010. A taxonomic revision of *Camptocerus* Dejean (Coleoptera: Curculionidae: Scolytinae). Insecta Mundi 148, 1–88.

Smith, S.M., Cognato, A.I., 2011. Observations of the biology of *Camptocerus* Dejean (Coleoptera: Curculionidae: Scolytinae) in Peru. Coleopterists Bulletin 65, 27–32.

Smith, G.D., Carroll, A.L., Lindgren, B.S., 2009. Life history of a secondary bark beetle, *Pseudips mexicanus* (Coleoptera: Curculionidae: Scolytinae), in lodgepole pine in British Columbia. Can. Entomol. 141, 56–69.

Sousa, W.P., Quek, S.P., Mitchell, B.J., 2003. Regeneration of *Rhizophora mangle* in a Caribbean mangrove forest: interacting effects of canopy disturbance and a stem-boring beetle. Oecologia 137, 436–445.

Speight, M.R., Wainhouse, D., 1989. Ecology and Management of Forest Insects. Clarendon Press, Oxford.

Stacey, P.B., Ligon, J.D., 1991. The benefits-of-philopatry hypothesis for the evolution of cooperative breeding variation in territory quality and group-size effects. Am. Nat. 137, 831–846.

Stahlhut, J.K., Desjardins, C.A., Clark, M.E., Baldo, L., Russell, J.A., Werren, J.H., Jaenike, J., 2010. The mushroom habitat as an ecological arena for global exchange of *Wolbachia*. Mol. Ecol. 19, 1940–1952.

Stark, R.W., 1982. Generalized ecology and life cycle of bark beetles. In: Mitton, J.B., Sturgeon, K.B. (Eds.), Bark Beetles in North American Conifers. A System for the Study of Evolutionary Biology. University of Texas Press, Austin, pp. 21–45.

Stauffer, C., van Meer, M.M.M., Riegler, M., 1997. The presence of the protobacteria *Wolbachia* in European *Ips typographus* (Col, Scolytidae) populations and the consequences for genetic data. Mitteilungen der Deutschen Gesellschaft für allgemeine und angewandte Entomologie 11, 709–711.

Stilwell, A.R., Smith, S.M., Cognato, A.I., Martinez, M., Flowers, R.W., 2014. *Coptoborus ochromactonus*, n. sp. (Coleoptera: Curculionidae: Scolytinae), an emerging pest of cultivated balsa (Malvales: Malvaceae) in Ecuador. J. Econ. Entomol. 107, 675–683.

Stone, C., 1990. Parasitic and phoretic nematodes associated with *Ips grandicollis* (Coleoptera, Scolytidae) in New-South-Wales. Nematologica 36, 478–480.

Storer, A.J., Wainhouse, D., Speight, M.R., 1997. The effect of larval aggregation behaviour on larval growth of the spruce bark beetle *Dendroctonus micans*. Ecol. Entomol. 22, 109–115.

Strohmeyer, H., 1906. Neue Untersuchungen über Biologie, Schädlichkeit und Vorkommen des Eichenkernkäfers, *Platypus cylindrus* var. *cylindriformis*. Naturwissenschaftliche Zeitschrift fur Land- und Forstwirtschaft 4, 329–341, 408–421, 506–511.

Strohmeyer, H., 1918. Die Morphologic des Chitinskeletts der Platypo-diden. Archiv fuer Naturgeschichte Berlin Abt A 84, 1–42.

Sullivan, B.T., Niño, A., Moreno, B., Brownie, C., Macías-Sámano, J., Clarke, S.R., et al., 2012. Biochemical evidence that *Dendroctonus frontalis* consists of two sibling species in Belize and Chiapas. Mexico. Ann. Entomol. Soc. Am. 105, 817–831.

Sun, J.H., Lu, M., Gillette, N.E., Wingfield, M.J., 2013. Red turpentine beetle: innocuous native becomes invasive tree killer in China. Annu. Rev. Entomol. 58, 293–311.

Susoy, V., Herrmann, M., 2014. Preferential host switching and codivergence shaped radiation of bark beetle symbionts, nematodes of *Micoletzkya* (Nematoda: Diplogastridae). J. Evol. Biol. 27, 889–898.

Swaby, J.A., Rudinsky, J.A., 1976. Acoustic and olfactory behaviour of *Ips pini* (Say) (Coleoptera: Scolytidae) during host invasion and colonisation. Z. angew. Entomol. 81, 421–432.

Swaine, J.M., 1918. Canadian bark-beetles, part II. A preliminary classification, with an account of the habits and means of control. Canadian Department of Agriculture, Division of Entomology, Bulletin 14, 1–143.

Swedenborg, P.D., Jones, R.L., Ascerno, M.E., Landwehr, V.R., 1988. *Hylurgopinus rufipes* (Eichhoff) (Coleoptera: Scolytidae): attraction to broodwood, host colonization behavior, and seasonal activity in central Minnesota. Can. Entomol. 120, 1041–1050.

Swedenborg, P.D., Jones, R.L., Ryker, L.C., 1989. Stridulation and associated behavior of the native elm bark beetle *Hylurgopinus rufipes* (Eichhoff) (Coleoptera: Scolytidae). Can. Entomol. 121, 245–252.

Swift, M.J., 1977. The roles of fungi and animals in the immobilization and release of nutrient elements from decomposing branchwood. Ecol. Bull. 25, 193–202.

Tallamy, D.W., 1994. Nourishment and the evolution of paternal investment in subsocial arthropods. In: Hunt, J.H., Nalepa, C.A. (Eds.), Nourishment and Evolution in Insect Societies. Westview Press, Boulder, pp. 21–55.

Tallamy, D.W., Wood, R.F., 1986. Convergence patterns in subsocial insects. Annu. Rev. Entomol. 31, 369–390.

Thompson, R.T., 1996. The species of *Phaenomerus* Schönherr (Coleoptera: Curculionidae: Zygopinae) of the Australian region. Invertebrate Taxonomy 10, 937–993.

Thornhill, N.W., 1993. The Natural History of Inbreeding and Outbreeding: Theoretical and Empirical Perspectives. University of Chicago Press, Chicago.

Ticheler, J.H.G., 1961. An analytical study of the epidemiology of the coffee berry borer in the Ivory Coast. Mededelingen van de Landbouwhogeschool te Wageningen 61 (11), 1–49, 1–13.

Trägårdh, I., 1930. Studies on the galleries of bark beetles. Bull. Entomol. Res. 21, 469–480.

Tredl, R., 1915. Aus dem Leben des Birkensplintkäfers, *Scolytus ratzeburgi*. Entomologische Blätter 11, 146–154.

Trimble, F.M., 1924. Life history and habits of two Pacific coast bark beetles. Ann. Entomol. Soc. Am. 17, 382–391.

Tykarski, P., 2006. Beetles associated with scolytids (Coleoptera, Scolytidae) and the elevational gradient: diversity and dynamics of the community in the Tatra National Park. Poland. Forest Ecol. Manage. 225, 146–159.

Ulyshen, M.D., Hanula, J.L., Horn, S., Kilgo, J.C., Moorman, C.E., 2004. Spatial and temporal patterns of beetles associated with coarse woody debris in managed bottomland hardwood forests. Forest Ecol. Manage. 199, 259–272.

Vega, F.E., Benavides, P., Stuart, J.A., O'Neill, S.L., 2002. *Wolbachia* infection in the coffee berry borer (Coleoptera: Scolytidae). Ann. Entomol. Soc. Am. 95, 374–378.

Vega, F.E., Davis, A.P., Jaramillo, J., 2012. From forest to plantation? Obscure articles reveal alternative host plants for the coffee berry borer, *Hypothenemus hampei* (Coleoptera: Curculionidae). Biol. J. Linn. Soc. 107, 86–94.

Vega, F.E., Simpkins, A., Bauchan, G., Infante, F., Kramer, M., Land, M.F., 2014. On the eyes of male coffee berry borers as rudimentary organs. PLoS One 9 (1), e85860.

Vernoff, S., Rudinsky, J.A., 1980. Sound production and pairing behavior of *Leperisinus californicus* Swaine and *L. oreganus* Blackman (Coleoptera: Scolytidae) attacking Oregon ash. Z. angew. Entomol. 90, 58–74.

Wallace, A.R., 1860. Notes on the habits of Scolytidae and Bostrichidae. Transactions of the Entomological Society of London (n. s.) 5 (Part IV), 218–220.

Wang, C.S., 1981. A study of the Korean pine bark beetle (*Blastophagus pilifer* Spess). Kunchong Zhishi [Insect Knowledge] 18, 165–167 [in Chinese, with English abstract].

Weber, B.C., McPherson, J.E., 1983. Life history of the ambrosia beetle *Xylosandrus germanus* (Coleoptera: Scolytidae). Ann. Entomol. Soc. Am. 76, 455–462.

Webster, F.M., 1910. The clover root-borer (*Hylastinus obscurus* Marsham.). U.S. Department of Agriculture, Bureau of Entomology Circular no 119, 1–5.

Werren, J.H., Baldo, L., Clark, M.E., 2008. *Wolbachia*: master manipulators of invertebrate biology. Nat. Rev. Microbiol. 6, 741–751.

West, S.A., Pen, I., Griffin, A.S., 2002. Cooperation and competition between relatives. Science 296, 72–75.

White, T.C.R., 1993. The Inadequate Environment. Nitrogen and the Abundance of Animals. Springer-Verlag, Berlin.

Wichmann, H.E., 1927. Ipidae. In: Schulze, P. (Ed.), Biologie der Tiere Deutschlands. Gebrüder Bornträger, Berlin, pp. 347–381.

Wichmann, H.E., 1967. Die Wirkungsbreite des Ausstoßreflexes bei Borkenkäfern. J. Pest Sci. 40, 184–187.

Wilson, H.F., 1909. The peach-tree barkbeetle. *U.S.* Department of Agriculture, Bureau of Entomology, Bulletin 68, 91–108.

Wilson, E.O., 1971. The Insect Societies. Belknap Press of Harvard University, Cambridge.

Wilson, E.O., 1975. Sociobiology: The New Synthesis. Harvard University Press, Cambridge.

Wilson, E.O., Hölldobler, B., 1980. Sex-differences in cooperative silk-spinning by weaver ant larvae. Proc. Natl. Acad. Sci. U. S. A. 77, 2343–2347.

Wilson-Rich, N., Spivak, M., Fefferman, N.H., Starks, P.T., 2009. Genetic, individual, and group facilitation of disease resistance in insect societies. Annu. Rev. Entomol. 54, 405–423.

Wood, S.L., 1978. A reclassification of the subfamilies and tribes of Scolytidae (Coleoptera). Annales de la Société entomologique de France (Neauveau Series) 14, 95–122.

Wood, D.L., 1982. The role of pheromones, kairomones, and allomones in the host selection and colonization behavior of bark beetles. Annu. Rev. Entomol. 27, 411–446.

Wood, S.L., 1982. The bark and ambrosia beetles of North and Central America (Coleoptera: Scolytidae), a taxonomic monograph. Great Basin Nat. Mem. 6, 1–1359.

Wood, S.L., 1983. *Scolytodes atratus panamensis* (Escarbajito de guarumo, *Cecropia* petiole borer). In: Janzen, D.H. (Ed.), Costa Rican Natural History. University of Chicago Press, Chicago, pp. 768–769.

Wood, S.L., 1986. A reclassification of the genera of Scolytidae (Coleoptera). Great Basin Nat. Mem. 10, 1–126.

Wood, S.L., 1993. Revision of the genera of Platypodidae (Coleoptera). Great Basin Nat. 53, 259–281.

Wood, S.L., 2007. Bark and Ambrosia Beetles of South America (Coleoptera, Scolytidae). Brigham Young University, Provo.

Wood, S.L., Bright, D.E., 1987. A catalog of Scolytidae and Platypodidae (Coleoptera), Part 1: Bibliography. Great Basin Nat. Mem. 11, 1–685.

Wood, S.L., Bright Jr., D.E., 1992. A catalog of Scolytidae and Platypodidae (Coleoptera), Part 2: Taxonomic Index. Volumes A and B. Great Basin Nat. Mem. 13, 1–1553.

Yan, Z.L., Sun, J.H., Don, O., Zhang, Z.N., 2005. The red turpentine beetle, *Dendroctonus valens* LeConte (Scolytidae): an exotic invasive pest of pine in China. Biodivers. Conserv. 14, 1735–1760.

Ye, H., 1997. Mass attack by *Tomicus piniperda* L. (Col., Scolytidae) on *Pinus yunnanensis* tree in the Kunming region, southwest China. In: Grégoire, J.-C., Liebhold, A.M., Stephen, F.M., Day, K.R., Salom, S.M. (Eds.), Proceedings: Integrating Cultural Tactics into the Management of Bark Beetle and Reforestation Pests, pp. 225–227, USDA Forest Service General Technical Report NE-236.

Ye, H., Ding, X.S., 1999. Impacts of *Tomicus minor* on distribution and reproduction of *Tomicus piniperda* (Col., Scolytidae) on the trunk of the living *Pinus yunnanensis* trees. J. Appl. Entomol. 123, 329–333.

Yearian, W.C., Gouger, R.J., Wilkinson, R.C., 1972. Effects of the blue-stain fungus, *Ceratocystis ips*, on development of Ips bark beetles in pine bolts. Ann. Entomol. Soc. Am. 65, 481–487.

Ytsma, G., 1988. Stridulation in *Platypus apicalis*, *P. caviceps*, and *P. gracilis* (Col., Platypodidae). J. Appl. Entomol. 11, 256–261.

Yust, H.R., 1957. Biology and habits of *Pagiocerus fiorii* in Ecuador. J. Econ. Entomol. 50, 92–96.

Zanuncio, A.J.V., Pastori, P.L., Kirkendall, L.R., Lino-Neto, J., Serrão, J.E., Zanuncio, J.C., 2010. *Megaplatypus mutatus* (Chapuis) (Coleoptera: Curculionidae: Platypodinae) attacking hybrid *Eucalyptus* clones in southern Espirito Santo, Brazil. Coleopterists Bulletin 61, 81–83.

Zanzot, J.W., Matusick, G., Eckhardt, L.G., 2010. Ecology of root-feeding beetles and their associated fungi on longleaf pine in Georgia. Environ. Entomol. 39, 415–423.

Zchori-Fein, E., Borad, C., Harari, A.R., 2006. Oogenesis in the date stone beetle, *Coccotrypes dactyliperda,* depends on symbiotic bacteria. Physiol. Entomol. 31, 164–169.

Zhang, L.L., Chen, H., Ma, C., Tian, Z.G., 2010. Electrophysiological responses of *Dendroctonus armandi* (Coleoptera: Curculionidae: Scolytinae) to volatiles of Chinese white pine as well as to pure enantiomers and racemates of some monoterpenes. Chemoecology 20, 265–275.

Zrimec, M.B., Zrimec, A., Slanc, P., Kac, J., Kreft, S., 2004. Screening for antibacterial activity in 72 species of wood-colonizing fungi by the *Vibrio fisheri* bioluminescence method. J. Basic Microbiol. 44, 407–412.

Zúñiga, G., Cisneros, R., Hayes, J.L., Macias-Samano, J., 2002a. Karyology, geographic distribution, and origin of the genus *Dendroctonus* Erichson (Coleoptera: Scolytidae). Ann. Entomol. Soc. Am. 95, 267–275.

Zúñiga, G., Salinas-Moreno, Y., Hayes, J.L., Grégoire, J.C., Cisneros, R., 2002b. Chromosome number in *Dendroctonus micans* and karyological divergence within the genus *Dendroctonus* (Coleoptera: Scolytidae). Can. Entomol. 134, 503–510.

Chapter 4

Population Dynamics of Bark Beetles

Aaron S. Weed[1], Matthew P. Ayres[2], and Barbara J. Bentz[3]

[1] Department of Biological Sciences, Dartmouth College, Hanover, NH, USA, [2] Department of Biological Sciences, Dartmouth College, Hanover, NH, USA, [3] USDA Forest Service, Rocky Mountain Research Station, Logan, UT, USA

1. INTRODUCTION

"The population trend of subcortical-feeding insects seems to be less predictable than that of any other ecological group of forest insects. In many instances, there does not seem to exist a numerical relation between the offspring and the number of trees subsequently infested, and statistical comparison between the offspring on individual trees, even of similar size, is seldom possible."

—Rudinsky (1962)

Bark beetle population dynamics have been the focus of intense study, fascinating and frustrating many for generations worldwide (Rudinsky, 1962; Berryman, 1974, 1982; Coulson, 1979; Paine *et al.*, 1997; Reeve, 1997; Klepzig *et al.*, 2001; Raffa *et al.*, 2005, 2008; Coulson and Klepzig, 2011; Kausrud *et al.*, 2012). The high interest is because some species exhibit fluctuations in abundance that sometimes have consequences on natural resources and agricultural productivity and because these organisms are central to our knowledge of coevolutionary processes such as interactions with plant secondary metabolism and mutualistic associations (Raffa and Berryman, 1987; Farrell *et al.*, 2001). Moreover, conifer-feeding bark beetles play a critical role in the disturbance ecology and biodiversity of temperate forests with sometimes dramatic effects on entire forest ecosystems (Bentz *et al.*, 2009).

Despite centuries of interest and decades of study, there is a sentiment in the quote above that still resonates today—bark beetle population dynamics are inherently complex and difficult to predict. Over the past decades, however, advances in theoretical and mathematical ecology, experimentation, chemical and molecular ecology, and forest health management have played critical roles in deciphering factors that influence fluctuations in bark beetle population abundance. These developments have not only uncovered a fascinating world of coevolutionary processes but have provided important general contributions to applied ecology with broad relevance for pest management.

In this chapter, we focus our discussion on the demographic processes that affect abundance and stability of bark beetle populations. Where possible we review examples of species colonizing non-woody tissues but most of the discussion is concentrated on species attacking woody plants, particularly conifer-feeding species. The influence that the host defensive system has played on life history trait evolution in bark beetles and its resulting signature on the feedback processes in their population systems is examined and integrated into a formal representation of bark beetle aggression.

2. CONCEPTS IN POPULATION DYNAMICS

Population dynamics is the study of the change in abundance, dispersion, and age structure (Royama, 1992; Berryman, 2003; Turchin, 2003). In this chapter we will focus specifically on topics related to change in abundance, but changes in age structure and gene frequency are also important variables considered in the study of population dynamics (Berryman, 1999). Theory from population dynamics provides a robust basis for understanding why population abundance grows or declines in time and space. Thus, its principles have numerous practical applications when the abundance of a focal organism is consequential to management objectives, such as in species conservation, defining optimal harvesting limits, and pest management (Berryman, 1999; Lande *et al.*, 2003). In this section, we present some fundamental concepts to population ecology that we will later address in relation to processes that affect bark beetle ecology.

2.1 Population Growth

The first fundamental principle in population ecology is related to how populations replace themselves. Biological populations tend to increase or decrease on a per capita

Bark Beetles. http://dx.doi.org/10.1016/B978-0-12-417156-5.00004-6

basis (exponential change), which can be represented mathematically as:

$$N_t = N_0 \cdot e^{Rt} \qquad (4.1)$$

where N_0 is the initial population abundance, N_t the population abundance at time t, and R is the instantaneous per capita growth rate. In practice, per capita growth rate can be estimated as:

$$R = [\ln(N_t) - \ln(N_0)]/t \qquad (4.2)$$

From Eqs. (4.1) and (4.2) it is clear that a population declines over time when $R < 0$, stays the same when $R = 0$, and grows when $R > 0$. Because population growth rate (R) is so important to the behavior of populations over time, the study of population dynamics is deeply invested in estimating R and understanding the processes that affect R.

2.2   Feedbacks and Exogenous Effects

Exponential growth is a fundamental property of all population systems, but populations do not grow indefinitely at a constant rate of change because birth and death rates rarely remain constant (Lande *et al.*, 2003). Birth and death rates change over time due to forces (Figure 4.1) that are commonly divided into density-dependent (or endogenous) and density-independent (or exogenous) processes (Royama, 1992; Ranta *et al.*, 2006). Density-dependent processes are those that change in response to abundance of the focal population, frequently via effects on populations that they consume (bottom-up effects) and specialist enemies that consume them (top-down effects) (Royama, 1992; Berryman and Chen, 1999; Turchin, 1999, 2003). Density-independent processes also have demographic effects that influence population change (R) but independently of population size, e.g., generalist predators and weather (Andrewartha and Birch, 1954; May, 1976; Royama, 1992; Berryman, 1999; Turchin, 2003).

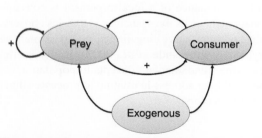

FIGURE 4.1   Diagram showing the influence of a primary consumer and exogenous effects (e.g., generalist predators or weather) on a prey population system (e.g., bark beetles). In this system the prey has a positive effect on its own abundance and the abundance of the consumer. A negative, endogenous feedback loop is created in the prey dynamics due to interactions with the consumer. *Adapted from Berryman (1982, 1999).*

Almost universally, rising abundance of a focal population generates negative feedback on further growth via increased competition for limiting resources and/or increasing abundance of enemies. As populations grow toward an equilibrium abundance that can be supported by the environment (K), this negative feedback slows population growth (Ranta *et al.*, 2006) by causing birth rates to decline, death rates to increase, or both. The logistic growth function Eq. (4.3) provides a simple description of population behavior regulated around an equilibrium:

$$N_{t+1} = N_t \cdot e^{r(1 - N_t/K)}, \qquad (4.3)$$

where r is the maximum per capita population growth rate, K is the equilibrium abundance in the environment, and the realized population growth, $r(1 - N_t/K)$, can be equated to R in Eq. (4.2). A general mathematical approach for studying population dynamics is to model the influence of endogenous feedbacks and exogenous effects on per capita growth rates (R) and couple this with Eq. (4.3) or a variant of Eq. (4.3) (May, 1976; Royama, 1992; Berryman, 1999; Turchin, 2003). For example:

$$R = f(N_{t-i}) + \varepsilon_i, \qquad (4.4)$$

where N_t is the population abundance at some time lag i (density dependence) and ε is a random variable $\approx N(0, \sigma_\varepsilon)$ representing exogenous effects due to density-independent effects and estimation error.

2.3   Stability

The tendency of a population to return to its equilibrium after perturbation is termed stability. Stability is strongly a product of the form and strength of feedback between the population and its environment (Figure 4.1). That is, stability tends to be a function of density dependence rather than density-independent effects. However, interactions between endogenous and exogenous processes can affect stability in population systems (Berryman and Lima, 2006; Friedenberg *et al.*, 2008). In general, negative feedback tends to be stabilizing over time whereas positive feedback is destabilizing and enhances risk of both extinctions and outbreaks (Berryman, 1982). Destabilizing positive feedbacks can arise in species such as bark beetles that demonstrate cooperative resource procurement (Klapwijk *et al.*, 2012). The dynamics of population systems dominated by rapid negative feedback (e.g., due to resource competition) tend to be relatively stable (Berryman, 1999). However, negative feedback that involves delays, e.g., as result of coupled dynamics with specialist enemies, can promote outbreak behavior via cyclical dynamics (Royama, 1992; Berryman and Chen, 1999; Murdoch *et al.*, 2003; Turchin, 2003).

3. HOST USE AND PEST ECOLOGY OF BARK BEETLES

3.1 Seed-feeding Species

The majority of bark beetles developing on seeds are found in the tropics (Wood, 1982b). Knowledge of the biology and ecology of seed-feeding taxa is restricted primarily to a few very important pests. The coffee berry borer, *Hypothenemus hampei* (Ferrari), is the most serious pest of coffee production worldwide (Chapter 11) and *Conophthorus* spp. are important consumers of pine seeds in temperate zones of North America (Cognato *et al.*, 2005). Members of the genus *Pagiocerus* can be pests of maize (Yust, 1957; Eidt-Wendt and Schulz, 1990); *Pagiocerus frontalis* (F.), for example, has a broad geographic distribution extending from the southeastern USA to South America and can be a serious pest of stored maize products in the Andes (Eidt-Wendt and Schulz, 1990). Some ambrosia beetles may attack seeds opportunistically (e.g., *Xylosandrus*), but these species typically feed within the bark of woody plants. Like most scolytids, however, seed feeders may transmit or create entry sites for invading microbes into host plants (Hoover *et al.*, 1995). For example, feeding behavior of *H. hampei* may create sites for bacterial and fungal infection of coffee berries and seeds (Damon, 2000) and *Conophthorus radiatae* Hopkins may vector pitch canker disease to *Pinus radiata* D. Don. (Hoover *et al.*, 1996).

Wood and Bright (1992) recognized 13 species of *Conophthorus* as occurring in North America, all associated with *Pinus*. *Conophthorus* are univoltine and adults emerging from overwintering sites attack new shoots for maturation feeding and oviposit in small, developing cones (Godwin and Odell, 1965; McPherson *et al.*, 1970; Morgan and Mailu, 1976; Mattson, 1980; Furniss, 1997). Mating is coordinated by pheromones (Birgersson *et al.*, 1995; de Groot and DeBarr, 1998). Females typically bore into the base of developing cones and construct egg galleries along a tunnel leading towards the cone apex (Lyons, 1956; Godwin and Odell, 1965; Morgan and Mailu, 1976; Mattson, 1980). This tunneling typically severs the vascular connection ceasing further cone development and prevents resin flow into the cone (Mattson, 1980). Larvae develop through two instars and the adults of most species overwinter in the attacked cones; however, newly emerging adults of *C. resinosae* Hopkins bore into terminal shoots that then fall to the ground where the adults remain until the following spring (Mattson, 1980). In some years seed herbivory by *Conophthorus* may affect well over half of the seed crop (Keen, 1958). Cone beetle damage may affect pine recruitment following forest disturbances from fire, other pests, or logging (Godwin and Odell, 1965; Cognato *et al.*, 2005) and apparently complicates

restoration management of soft pines, which have declined due to the introduction of white pine blister rust into North America (Cognato *et al.*, 2005). Cone beetles compete with and are depredated by a complex of other insects associated with pine cones (e.g., Lepidoptera and Coleoptera; Godwin and Odell, 1965; Mattson, 1980). Early season emergence that is typical of cone beetles may reduce competition and ensure that attack coincides during a period when cones are high in nutritional quality (Mattson, 1980). One study on *C. resinosae* determined that its dynamics were fairly stable over an 11-year period (Berryman, 1999), implying that this cone beetle is regulated primarily by first order feedback due to competition for a limited resource. Population fluctuations appear to be influenced by exogenous processes that affect annual pine cone production because beetle damage does not affect cone production (Berryman, 1999).

Hypothenemus hampei is another well-known seed-feeding bark beetle that is a serious pest of coffee around the world (Damon, 2000; Vega *et al.*, 2009). Left untreated, populations of this beetle can expand quickly in plantations and cause considerable yield losses. Female beetles bore into developing or mature coffee berries and construct egg galleries if the endosperm is sufficiently developed (Baker *et al.*, 1992). In cooler regions, *H. hampei* may complete two to three generations per year but nine generations may be completed in warmer regions (Damon, 2000). Infestations typically begin from infested berries leftover from the previous crop (Irulandi *et al.*, 2008) and start as small aggregations of attacked trees in a field that serve later as foci for future spread. Beetle survival is strongly affected by humidity (Baker *et al.*, 1994) and extended dry seasons between fruiting periods may reduce infestations. *Hypothenemus hampei* has been the subject of biological control investigations for decades and numerous natural enemies have been evaluated and released with varying levels of success throughout its range (Damon, 2000; Vega *et al.*, 2009). A comprehensive treatment of the coffee berry borer is presented in Chapter 11.

3.2 Tree Infesting Species

Bark beetles most commonly exploit the cambial tissues of woody plants (Wood, 1982b; Paine *et al.*, 1997) and much of the remaining discussion on bark beetle population ecology is focused on this group. The majority of these taxa develop on host tissues after attacking dead or recently wounded trees, but as will be discussed in more detail later, some taxa regularly kill heavily defended healthy trees by means of mass attack. Bark beetles are the leading cause of expansive outbreaks within conifer forests worldwide (Raffa *et al.*, 2008; Kausrud *et al.*, 2012), are critically important pathogen vectors (Holmes, 1980;

Hoover *et al.*, 1996; Kolarik *et al.*, 2011), and important pests of agriculture (e.g., *Hypocryphalus mangiferae* Stebbing in mango; Masood *et al.*, 2009). Native bark beetles are the most important disturbance agents in most forested regions around the globe. Over the past two decades, bark beetles alone have affected millions of hectares of conifer forests in western North America (Meddens *et al.*, 2012). Bark beetles are of demonstrable importance to forestry in Mexico (Mendoza and Zuñiga, 2011) and Asia (Chen and Tang, 2007; Wang *et al.*, 2010). The recent introduction of the red turpentine beetle *Dendroctonus valens* LeConte into China from North America has resulted in considerable *Pinus* mortality (Yan *et al.*, 2005; Liu *et al.*, 2013). In Europe, natural forests and plantations of conifers are constantly managed to mitigate bark beetle impacts because of the widespread and persistent threat from *Ips* spp. (Colombari *et al.*, 2012; Kausrud *et al.*, 2012; Marini *et al.*, 2013). In the southeastern USA, *D. frontalis* Zimmermann is a chronic issue in managing pine plantations with economic costs to the timber and pulp industry exceeding $1 billion over a 28-year period (Pye *et al.*, 2011). Although tree mortality is often undesirable and counter to forest management objectives (Greenwood and Weisberg, 2008), forest disturbances by bark beetles are critically important processes in forest dynamics (Edburg *et al.*, 2012; Hicke *et al.*, 2012). Bark beetle-caused tree mortality affects forest structure and composition and thus influences multiple ecosystem services derived from forest resources (Weed *et al.*, 2013).

Phloem-feeding bark beetles are accompanied by a plethora of microbes, nematodes, and mites during attack and colonization of host tissues (Klepzig *et al.*, 1996; Klepzig and Six, 2004; Moser *et al.*, 2005; Hofstetter *et al.*, 2006; Six and Wingfield, 2011). Community interactions with these beetles can range from neutral (Hofstetter *et al.*, 2006) to antagonistic (Reeve, 1997; Kopper *et al.*, 2004) to beneficial (Klepzig *et al.*, 2001). Brood of bark beetle species feed directly on the host phloem, yet they can also augment their diet by feeding upon symbiotic fungi that they vector and inoculate egg galleries with fungi from mycangia (Farrell *et al.*, 2001; Klepzig *et al.*, 2001). Symbiotic fungi often concentrate phloem nutrients (Ayres *et al.*, 2000; Bentz and Six, 2006) and some are pathogenic to trees (Six and Wingfield, 2011; Lahr and Krokene, 2013). Many bark beetles and microbial associates have been transported via global trade into regions containing host plants that do not share a coevolutionary history. These introductions sometimes have widespread and persistent effects on forest ecosystems. The introduction of Dutch elm disease into North America (*Ulmus americana* L.), for example, is a classic case illustrating the importance of bark beetles as vectors of non-native diseases (Holmes, 1980). Similarly, twig beetles (*Pityophthorus*) can be important vectors of disease-causing agents such as

Fusarium (Hoover *et al.*, 1996; Sakamoto *et al.*, 2012) and the transmission of these pathogens onto novel hosts may lead to the demise of susceptible host plants (e.g., *Pityophthorus juglandis* Blackman on *Juglans* in eastern North America; Cranshaw, 2011; Kolarik *et al.*, 2011).

3.3   Other Guilds

A small number of bark beetles develop on herbaceous plants and grasses. Species attacking herbaceous plants belong almost exclusively to the Dryocoetini and Hylesinini, are sometimes important crop pests (Wood, 1982b), and have been considered as biological control agents of noxious weeds (e.g., *Thamnurgus*; Campobasso *et al.*, 2004a; Güclü and Özbek, 2007). Some *Hypothenemus* spp. can apparently develop in grasses and herbaceous plants, while *Hylastinus* spp. develop on legumes, and all *Dendrocranulus* attack weakened plants in the Cucurbitaceae (Wood, 1982b). Members of the genus *Thamnurgus* breed in stems of *Euphorbia*, *Delphinium*, *Consolida*, *Aconitum*, *Salvia*, *Tamarix*, and *Peganum* (Campobasso *et al.*, 2004a; Güclü and Özbek, 2007; Jordal *et al.*, 2013). The European species *Thamnurgus euphorbiae* Kuster, for example, was evaluated as a biological control agent to control leafy spurge (*Euphorbia esula* L.) in North America (Campobasso *et al.*, 2004a) but never released. Bark beetles that attack herbaceous plants do not appear to have associations with symbiotic fungi like some seed and many phloem feeders, but plant pathogens commonly enter beetle attack sites. *Hylastinus obscurus* (Marsham) causes dramatic reductions in crop yields of clover (*Trifolium* spp.) because adult and larval root feeding not only directly impact plant performance, but feeding damage also increases the probability of infection by *Fusarium* or other pathogens (Jin *et al.*, 1992; Alarcón *et al.*, 2010). Similarly, *Fusarium* infection occurs commonly after *Thamnurgus* bores into plant stems near the base of leaf petioles (Campobasso *et al.*, 2004b; Güclü and Özbek, 2007; Mandelshtam *et al.*, 2012).

Population studies of herbaceous feeding bark beetles are primarily restricted to observations of seasonal activity and host selection behavior of the European species *Hylastinus obscurus* (clover root borer). Most of the life cycle of this univoltine species is spent within the root system of its host plant (commonly *Trifolium* and *Medicago*). In the eastern USA, *H. obscurus* adults emerge and disperse from overwintering sites into new clover fields for a short period in the spring. Infestations spread within fields after the initial flight period by crawling, which peaks around June (Culk and Weaver, 1994). Adult attraction to host plant volatiles changes with plant age (Quiroz *et al.*, 2005) and may be moderated by infection by root pathogens (Leath and Byers, 1973). Programs are ongoing in Chile, where *H. obscurus* is the main pest of clover (forage and seed yield), to develop effective techniques for trapping adults. Natural enemies have been noted (Webster, 1899) but to our knowledge their importance on *H. obscurus* population dynamics has not been studied.

4. ECOLOGY OF TREE-INFESTING BARK BEETLES

4.1 Life History Strategies

Bark beetles that attack trees are frequently characterized based on their "aggression," which is generally defined as the tendency to kill healthy trees vs. colonizing dying or recently dead trees. For example, species described as primary or "aggressive" are able to kill live trees via mass attack. Secondary species ("moderately aggressive") are facultative parasites that preferentially colonize recently killed or weakened trees (Rankin and Borden, 1991; Paine *et al.*, 1997). Finally, the non-aggressive taxa or saprophytes colonize dead wood (Paine *et al.*, 1997). Despite these differences, most phloem-feeding bark beetles have very similar life histories. In general, adults disperse within appropriate habitats to find hosts. The initial colonizing adults, which may be female or male depending on the species, bore an entrance hole through the host bark. In some species host-derived pheromones are emitted to initiate mating and attract congeners (i.e., mass attack), and in other species mating occurs prior to emergence from brood host trees, and adult aggregating pheromones appear to be lacking. Females then construct a tunnel near the cambium where she lays eggs. Brood develops in the under bark and pupates in the phloem or outer bark. Beetles may overwinter in a variety of stages in the subcortical region or as adults in leaf litter or soil near the tree base.

Selective pressures from host tree resistance have promoted the diversification of behavioral and physiological traits within this general life history strategy. For example, the mass-attack strategy has evolved in some groups to overwhelm host defense but not in others (Reeve *et al.*, 2012), and host volatile concentrations, which are correlated with host vigor, may elicit opposite behavioral responses (e.g., repellence vs. attraction) in bark beetles (Wallin and Raffa, 2000, 2004 Erbilgin *et al.*, 2003). A scheme of classifying species based on their aggression can be misleading because so-called secondary species can sometimes kill healthy trees and species that frequently kill trees can become less aggressive when rare (Martinson *et al.*, 2013). A classification of beetle aggression is needed that reflects the processes affecting aggression that ultimately predict its outcome. Here we offer a framework for defining bark beetle aggression as a function of population density and the feedback systems that influence population dynamics. The suggestion that density-dependent processes are a part of aggression is not new (Berryman, 1982;

Raffa, 2001; Wallin and Raffa, 2004; Boone *et al.*, 2011), but a theoretical construct that is generally applicable to all phloem-feeding bark beetle systems is lacking.

For the remainder of this chapter we develop a formal representation of bark beetle aggression that considers how life history traits of bark beetles that have evolved in response to the host defensive system result in characteristic feedback processes in their population systems. First, we briefly explore host plant resistance to bark beetles in tree-infesting, phloem-feeding species that primarily attack conifers. Second, we describe variation in the life history strategies that have evolved in response to host plant resistance (e.g., capacity to aggregate, capacity to tolerate host plant defensive compounds, and capacity to overwhelm defense). Finally, we integrate these into a general model of the variable feedback systems existing in bark beetle populations that influence their tendency to attack and kill living trees that is dependent on their capacity to overcome host plant resistance.

4.2 Host Plant Resistance to Bark Beetles

Tree defenses pose a challenge to bark beetle colonization and reproduction (Raffa and Berryman, 1983, 1987), which is presumably why the vast majority of subcortical species attack weakened or dead trees. It is clear that bark beetles have a played a strong role in the evolution of conifer host defensive systems and that these defenses have in turn exerted strong selection pressure on bark beetle behavior, life history, and population ecology (Raffa and Berryman, 1987; Seybold *et al.*, 2006). Conifers resist bark beetle attacks using resin contained with a system of vertical ducts; variation in the number and size of resin ducts are key determinants in the defensive system against bark beetle attacks in pines (Kane and Kolb, 2010; Ferrenberg *et al.*, 2013). Resin is toxic to beetles (Smith, 1961) and their associates (Kopper *et al.*, 2005) and acts as a physical and toxic barrier to beetle attack. Once trees are attacked, they immediately exude resin from constitutive reserves to fend off beetles. If attack continues, trees will actively synthesize and mobilize defensive chemicals to the site of beetle attack. Lignified stone cells are an additional component of the defense system in spruce, but are absent in pine (Wainhouse *et al.*, 1998). Interactions between bark beetles and host tree defenses are further covered in Chapter 5.

The likelihood of a tree surviving beetle attack is related to tree defense and to the abundance of attacking bark beetles (Paine *et al.*, 1997; Boone *et al.*, 2011). To be successful, bark beetles need to attack with sufficient numbers to overwhelm the host response—an attack threshold. This threshold varies among conifers due to variation in the production and chemical composition of resin (constitutive and inducible) and due to behavioral and physiological adaptations of the attacking bark beetles and their accompanying microbes. The attack threshold may vary due to anything that affects tree defenses, including genetic effects, abiotic gradients, and stressors such as lightning, fire, drought, crown loss, and mechanical wounding (Cates and Alexander, 1982; Reeve *et al.*, 1995; Lombardero *et al.*, 2000b). It is not uncommon to observe considerable variation in resin flow among individuals within the same stand (Martinson *et al.*, 2007). Stress factors that compromise tree defenses have been frequently linked to the initiation of epidemics. For example, *Ips* outbreaks can begin following localized windthrow or regional drought events that strongly weaken or diminish tree defenses (Marini *et al.*, 2012, 2013). However, our understanding of how sensitive attack thresholds are to environmental variation is complicated by nonlinearities in environmental effects on tree defense (Reeve *et al.*, 1995) and high intrinsic variation among trees within stands.

4.3 Life History Traits Affecting Aggression

In general, herbivore adaptations to plant defense can be behavioral, physiological, and density dependent; they can involve symbioses and manipulations of gene expression in plants (Karban and Agrawal, 2002). A few bark beetle species have evolved most of these tactics to gain access to subcortical host resources, and thereby avoid the high levels of competition found in non-living tissue (Raffa *et al.*, 2005). The most important strategies for overcoming host plant resistance that show some level of variation among bark beetle taxa are: (1) adult aggregation capacity; (2) capacity to overwhelm tree defense; and (3) ability to tolerate host defenses.

4.3.1 Capacity of Adults to Aggregate

Aggregations of adult bark beetles are extremely important for reproduction of species that attack live trees because they are the primary mechanism for depleting host defenses and mating (found in tribes Ipini, Scolytini, and Tomicini). Bark beetles coordinate mass attacks using combinations of pheromones and host plant volatiles that are released by initial tissue damage (Raffa *et al.*, 2005; Seybold *et al.*, 2006). Once colonizing individuals locate a suitable host, the male or female chews into the bark and begins releasing pheromones to attract conspecifics. The period of a mass attack on an individual tree varies among bark beetle species (Coulson, 1979; Paine *et al.*, 1997; Pureswaran *et al.*, 2006). A variety of mechanisms may be involved in termination of attacks to a tree: the release of inhibitory compounds may disrupt attacks as host defenses are depleted, and attacks may be redirected to neighboring trees that are within the plume of aggregation pheromones (Bentz *et al.*, 1996; Paine *et al.*, 1997). Negative feedback on beetle

population growth from intraspecific competition within trees is thereby limited (Byers, 1989). Flexible, highly coordinated mass-attacking behavior favors beetles partly because the number of attacking adults necessary to overcome defenses can vary among alternative host species and among individual trees in a stand (Raffa, 2001; Raffa *et al.*, 2008). Moreover, beetle survival, flight, and seasonal patterns of adult emergence synchrony are temperature-dependent processes that are prone to vary annually and spatially (Régnière and Bentz, 2007; Trân *et al.*, 2007; Bentz *et al.*, 2014). Hence, highly evolved aggregative behavior can buffer against environmental variation in forest structure (Ylioja *et al.*, 2005) and promote positive population growth in local populations (Ayres *et al.*, 2011).

Relatively few bark beetle species actually mass attack to gain access to defended host tissues (Wood, 1982a); the majority are restricted to attacking dead or severely weakened trees and, to a lesser extent, some are highly tolerant of host defensive compounds (Wood, 1982b). *Dendroctonus micans* (Kugelann), *D. terebrans* (Olivier), and *D. valens*, for example, are highly tolerant of resin and capable of accessing host tree resources and attack trees individually or in small groups, but the host tree is not always killed (Grégoire, 1988; Owen *et al.*, 2012; Reeve *et al.*, 2012). Mass attacking in *Dendroctonus* is apparently ancestral and it has been suggested that the loss of adult aggregation behavior may lessen effects from competitors and natural enemies that are attracted to bark beetle aggregations (Reeve *et al.*, 2012). Mass-attacking behavior then is not a condition for successful host colonization for all bark beetle species, but tends to be prevalent in those species that regularly attack and kill large numbers of healthy trees (Raffa *et al.*, 2008; Kausrud *et al.*, 2012; Reeve *et al.*, 2012).

The aggressiveness of a bark beetle species is in part a product of its capacity to aggregate using pheromones. For example, attack densities of solitary species that do not have an aggregating pheromone but are highly tolerant of host defense (e.g., *D. micans*) should be strongly related to local beetle density (Figure 4.2, blue line). Mass-attacking species, on the other hand, should show a strong ability to aggregate even at low population densities by using highly effective aggregation pheromones (Figure 4.2, red line). The latter prediction was tested in the Bankhead National Forest in Alabama by measuring landing rates of *D. frontalis* per decimeter of bole surface area in all trees within and around a local infestation (Ayres, unpubl.). Daily, local infestation size was estimated by counting the number of adults landing on sticky traps placed on all trees within the forest stand. This demonstrated remarkable efficacy of *D. frontalis* in coordinating high tree-specific attack rates (Figure 4.3). Despite high variation during the summer in the total abundance of adult beetles, tree-specific attack rates remained high and relatively uniform

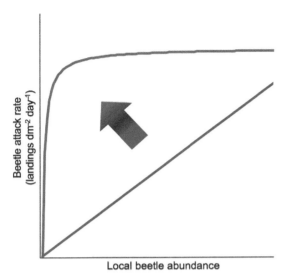

FIGURE 4.2 **Bark beetle aggression is related to their capacity to form aggregations.** Species with a high capacity to aggregate (red line) can coordinate high attacking density at relatively low abundance due to pheromone communication. Beetle attack rate is predicted to be proportional to local density for species with a lower capacity to aggregate (blue line). Arrow indicates direction of increasing aggression.

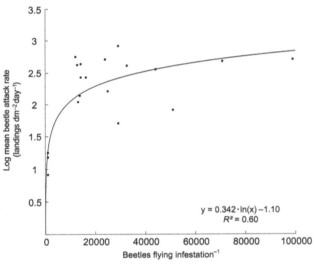

$$y = 0.342 \cdot \ln(x) - 1.10$$
$$R^2 = 0.60$$

FIGURE 4.3 Empirical relationship between stand-level abundance (x-axis) and tree-specific landing rates of *Dendroctonus frontalis* on *Pinus taeda* L. in the Bankhead National Forest, AL.

because beetles adjusted by varying the number of trees that were being attacked at any given time (Figure 4.3). All aggressive taxa of bark beetles seem to have highly evolved pheromone-mediated adult aggregation systems.

4.3.2 Tolerance of Host Defense

Bark beetles utilize behavioral or physiological adaptations to counter tree host defense, and variation among species in the traits that affect tolerance to host defenses has clear

consequences on whether a species can attack and successfully reproduce only in dead vs. living trees. Once bark beetles attack living trees, for example, they and their microbial associates are exposed almost immediately to an array of host defensive chemicals that are toxic to the attacking adults and their developing brood. Adult bark beetles must then penetrate through a viscous sea of crystallizing resin and tolerate its toxic vapors to successfully enter the subcortical tissues of living trees (Paine *et al.*, 1997). When successful at entering the phloem, if not already mated, beetles then mate and excavate brood galleries within this same hostile environment.

Bark beetles show variation in their ability to tolerate the chemical and physical properties of the resin defense system (Hodges *et al.*, 1979; Coulson *et al.*, 1986; Wallin and Raffa, 2000, 2004). Species vary in their tolerance to resin vapors (Smith, 1961; Cook and Hain, 1988; Raffa *et al.*, 2005) and some species are better able to physically maneuver through this flowing, viscous substance (Berryman, 1982). For example, *D. terebrans*, *D. valens*, and *D. micans* are extremely tolerant of resin and capable of developing within living trees. All species that are able to mass attack and kill healthy trees are also capable of tolerating high resin flows that many other species (*Hylurgops*, *Ips*, *Pityophthorus*, and *Scolytus*) could not (Flamm *et al.*, 1987; Safranyik *et al.*, 2000; Ayres *et al.*, 2001). Moreover, microbial associates of bark beetles also vary in their tolerance to host defenses, which is likely to be important for bark beetle survival and fecundity (Paine and Hanlon, 1994; Klepzig *et al.*, 1996; Kopper *et al.*, 2005; Raffa *et al.*, 2005; Adams *et al.*, 2011).

Reproductive success of bark beetles is also related to host tree defenses (Raffa and Berryman, 1983), but only a few tests have examined the effect of resin defense directly on beetle fitness. Reeve *et al.* (1995) demonstrated that fitness (eggs per attack) of the highly resin-tolerant *D. frontalis* decreased with increasing resin flow because beetles spent more time removing resin from their galleries in well-defended trees than constructing new galleries and laying eggs. Laboratory studies have also indicated that fitness generally decreases with increased monoterpene concentration (Wallin and Raffa, 2000, 2004). These experiments also observed different functional relationships between beetle gallery lengths of *D. rufipennis* Kirby and *Ips pini* (Say) in medium amended with increasing concentration of alpha-pinene, suggesting that *D. rufipennis* can tolerate much higher concentrations of this host defense compound compared to *I. pini*.

Collectively, these studies indicate that there is variation among bark beetle species in their tolerance to resin defenses, which should generate variation among species in population response to host defenses. In general, species that are able to sustain attacks on healthy trees are highly tolerant of host defenses whereas intolerant species attack

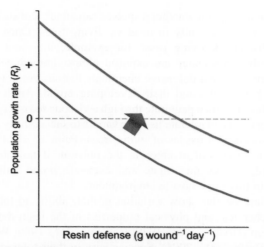

FIGURE 4.4 Theoretical effects of resin flow rate on the fitness of bark beetles demonstrating variance in their tolerance to host defense. Population growth rates of tolerant (red line) and less tolerant (blue line) bark beeetles are predicted to decline at a deccelerating rate with resin flow. Arrow indicates direction of increasing aggression. Note that more aggressive species can sustain positive population growth rates in the presence of stronger resin defenses.

severely weakened or dead trees (Paine et al., 1997). Some of the less tolerant species may be able to mount successful mass attacks and kill healthy trees following events that diminish physiological vigor of host trees (e.g., D. pseudotsugae Hopkins, I. pini, and Scolytus), but their intolerance of resin prevents these populations from sustaining outbreaks once host vigor returns or weakened resources have been exhausted. Thus, population growth rate of less tolerant species is predicted to be more negatively affected by increasing resin production compared to more tolerant species (Figure 4.4).

4.3.3 Capacity to Overwhelm Tree Defense

In addition to tolerating plant secondary metabolites when they are present, herbivores can also suppress or deplete plant defenses. There was a longstanding hypothesis that phytotoxins produced by fungal symbionts of bark beetles were crucial to the tree-killing process by disrupting host physiology and interfering with tree defenses (DeAngelis et al., 1986). Alternative roles for symbionts of many of the most aggressive bark beetle species are emerging and suggest that direct phytopathogenicity is not necessarily the main role of symbionts (Bridges et al., 1985; Six and Wingfield, 2011) and that tree defenses can be overwhelmed even before disruption of tree vascular function (Wullschleger et al., 2004; Martinson et al., 2007). Instead, the most aggressive bark beetles seem to physically deplete host defenses via large numbers of wounds from swarms of simultaneously attacking beetles (Figure 4.2 in Martinson et al., 2007), although some fungi may also aid in depletion of host tree resources used in the defense process (Lahr and

Krokene, 2013). Some Dendroctonus species that do not mass attack as adults may form larval feeding aggregations under the bark (Grégoire, 1988; Reeve et al., 2012) that also serve to overcome host defense as in other herbivores (Karban and Agrawal, 2002; Fordyce, 2003). In D. micans, which does not mass attack its host (Picea abies L. (Karst.)), pheromones promote aggregation of larvae (Grégoire et al., 1981; Deneubourg et al., 1990), which is beneficial to larval performance (Storer et al., 1997). Thus efficient aggregation in either the adult or larval stage seems fundamental to the capacity for overwhelming host defenses in bark beetles (Reeve et al., 2012).

Mass attack is a form of social facilitation that promotes positive feedback in bark beetle population dynamics (Berryman, 1999). However, species vary in their capacity for aggregation, and therefore their ability to overwhelm host defenses. Furthermore, there are costs of cooperative exploitation from intraspecific competition for a finite resource (Berryman and Pienaar, 1973; Raffa, 2001). Studies of reproductive success in mass-attacking bark beetles offer some of the best evidence for the ecological tradeoffs associated with overcoming host defense via cooperation. Due to changes in the relative importance of competition vs. cooperation, maximum fitness is typically achieved at higher attack densities in species that attack live trees vs. colonizing dying or recently dead trees (Raffa, 2001). Density-dependent oviposition behavior of some species can reduce competitive interactions (e.g., overdispersion of galleries within the phloem, Coulson, 1979; Berryman, 1982), but this behavior is not universal and can only reduce competition to an extent.

The interacting effects on beetle fitness of tree defenses, beetle cooperation during mass attack, and beetle competition after attack, suggests a third dimension of aggressiveness in bark beetles (Figure 4.5). Aggressive species that are efficient at killing trees are predicted to have high maximum reproductive success at moderately high attack densities (Figure 4.5, red line). When they attempt to attack living trees, species with less efficient aggregation systems are predicted to have generally lower reproduction that is only weakly related to attack densities (Figure 4.5, blue line).

4.3.4 High Aggression involves Multiple Traits

In sum, bark beetle species vary in their aggressiveness (tendency for tree killing). Their aggressiveness can be characterized in terms of life history traits that have evolved in response to host defenses (Table 4.1): colonization behavior (Figure 4.2); tolerance of host defensive system (Figure 4.4); and the ability to overwhelm host defenses (Figure 4.5). The genus Dendroctonus has many representatives that set the limits for aggressiveness in all three dimensions—especially the tendency and capacity for

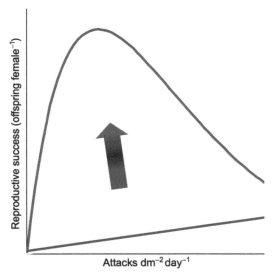

FIGURE 4.5 **Theoretical relationship between attack density and bark beetle fitness for species varying in their ability to overwhelm host defense.** Species with a greater ability to overwhelm host defenses (red line) can attain very high reproductive success at moderate attack densities due to highly synchronous mass attack. Per capita reproductive success declines from intraspecific competition for phloem after this point, but still remains higher than less aggressive species attempting to colonize a healthy tree. Arrow indicates direction of increasing aggression.

simultaneous mass attack of host trees (Reeve *et al.*, 2012). Because successful mass attack is facilitated by high abundance, those species that employ mass attacks tend to have a destabilizing positive feedback in their population

dynamics that promotes outbreaks and exacerbates their impacts on forests. On the other hand, species that preferentially colonize weakened, dead or dying trees may have evolved a portion but not all traits related to aggression and are thus unable to attack healthy trees (Table 4.1). For the most part these species are non-aggressive although some may occasionally attain densities permitting attack of healthy trees for short time periods (*Scolytus ventralis* LeConte and some *Ips* spp.).

5. FIXED, STABLE POINTS, TRANSIENT DYNAMICS, AND BARK BEETLE AGGRESSION

Based on the previous discussion, it is clear that bark beetle populations are under selection from host plant resistance and this has led to the evolution of variable life history strategies. These life history traits permit some species to attack healthy trees when at high density whereas demographic barriers from host plant resistance always limit others to dead or weakened trees, i.e., they affect bark beetle aggression. In this section, we address how varying levels of aggression produce diagnostic effects on bark beetle population dynamics. We present a scheme for defining bark beetle aggression based on a comparison of the feedback processes and equilibrium abundance of species varying in their capacity to overcome host resistance.

TABLE 4.1 Summary of Life History Traits Associated with Aggression of Phloem-feeding Bark Beetles

Aggression Category	Synonyms	Capacity for Adult Aggregation	Tolerance to Host Defenses	Capacity to Overwhelm Host Defense	Kill Healthy Trees	Examples
Aggressive	Primary, near-obligate parasites	High	High	High	Yes	*Dendroctonus frontalis* (Zimmermann), *Dendroctonus rufipennis* (Kirby), *Dendroctonus ponderosae* Hopkins, *Ips typographus* (L.)
Opportunistically aggressive	Secondary, moderately aggressive, facultative parasites	High	Low	Moderate	Yes	*Dendroctonus pseudotsugae* Hopkins, *Ips confusus* (LeConte), *Ips pini* (Say), *Scolytus ventralis* LeConte, *Tomicus piniperda* (L.)
Non-aggressive	Near-obligate parasites	Low	High	High	No	*Dendroctonus micans* (Kugelann)*, *Dendroctonus terebrans* (Olivier), *Dendroctonus valens* LeConte*
	Saprophytic, obligate scavengers	Low	Low	Low	No	*Hylastes* sp., *Hylurgops* sp., *Pityogenes* sp., *Pityopthorus* sp.

*May aggregate as adults and/or kill trees in regions outside of coevolved community.

5.1 Life History Traits, Abundance, and Bark Beetle Aggression

The ability of bark beetles to locate and colonize suitable host trees is a function of their local density, host tree abundance, and host tree vigor. Purely saprophytic species may be an exception in that their local abundance and aggregation behavior primarily favors mate finding (Wood, 1982a; Kirkendall, 1983). For the remaining species, it makes sense that fewer attacking beetles are required to overcome host defense of weakened or severely stressed trees compared to heavily defended trees (Berryman, 1982). This is why the host defense system of healthy trees cannot be surpassed when bark beetles are at low abundance. All phloem-feeding bark beetles at low population abundance require weakened or stressed trees to avoid local extinction. Some species (non-aggressive species) are always maintained at low population density and rely on stochastic processes (e.g., windthrow events, regional droughts, and lightning) or exploit forestry operations (e.g., logging operations) that weaken trees and create an opportunity for colonization and reproduction. Populations of other species, typically *Ips*, *Scolytus*, and some *Dendroctonus*, may increase dramatically in size following these events leading to outbreaks that die out quickly (opportunistically aggressive species) (Furniss, 1962; Santos and Whitham, 2010). For other species, similar stochastic events coupled with the favorable climatic and resource conditions may permit persistent, landscape-scale outbreaks (aggressive species) (Raffa *et al.*, 2008; Bentz *et al.*, 2009) (Table 4.1).

The above are general verbal descriptions of the characteristic dynamics of bark beetle population systems that vary in their aggression. These systems are all subject to multiple sources of feedback as well as stochastic processes that can influence feedback systems. The most aggressive species tend to have strong nonlinearities in their density dependence, including regions of positive feedback (increasing N leads to increasing R leading to increasing N, etc.). Under some circumstances, these nonlinearities can produce dynamics in which populations tend to be regulated at either a low or high abundance, with occasional switches that are a promoted by exogenous effects and resisted by locally stabilizing negative feedbacks (Berryman, 1982). In contrast, species that are less aggressive in terms of Figures 4.2, 4.4, and 4.5 will tend to have population dynamics with greater stability and a single equilibrium abundance (Figure 4.6).

In the following sections, we describe the dynamics and feedback processes characteristic of bark beetle population systems varying in their aggression using mathematical models. Each of these models (or some variant of them) has been used to describe population dynamics of bark beetles by other authors, e.g., Berryman (1982),

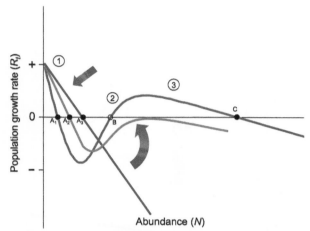

FIGURE 4.6 Three theoretical models of population regulation for bark beetle population systems varying in their aggression. Population growth rate (R_t) is a function of abundance (N). Non-aggressive species (blue line) are regulated at one stable equilibrium (A_3) created by linear, density dependence (1). Populations of more aggressive species (orange and red lines) also have stable equilibria at relatively low abundances (A_1 and A_2), representing non-outbreak conditions, subject to stabilizing, negative feedback (1). Orange and red lines depict two variants of a model with nonlinear density dependence that includes a region of positive density dependence (R increases with increasing N) at intermediate abundances where beetles can begin to escape resource limitations via mass attack of healthy trees (2). In aggressive taxa (red line), but not the blue or orange lines, there is an escape threshold (unstable equilibrium) at B. If B is exceeded, the deterministic tendency of the population is to grow to the upper locally stable equilibrium at C, where it could persist due to negative feedback (3). The orange line indicates a situation where the abundance of opportunistically aggressive species enters the unstable region, but the effects of social facilitation are not strong enough to promote positive population growth. The evolution of increased capacity for synchronous attacks of individual host trees and increased resin tolerance would both tend to move the species represented in blue towards the condition represented in red, which is to say, toward greater tendency for epidemics with sustained abundance at C and shifts in low abundance ($A_3 \rightarrow A_1$). Thus, the arrow indicates the direction of increasing aggression.

Mawby *et al.* (1989), Turchin *et al.* (1991), Økland and Bjørnstad (2006), and Martinson *et al.* (2013). We first introduce population behavior of species that are unable to attain stable, outbreak densities and later contrast these dynamics with population systems exhibiting a high, stable equilibrium (outbreak density) resulting from mass-attacking behavior of healthy trees. Our purpose is to demonstrate that variable feedback systems exist in bark beetle populations that are related to aggression, which is a product of the life history traits presented previously.

5.2 Population Systems Regulated at One Equilibrium

Many biological populations behave in accordance with the model of a single equilibrium abundance from approximately linear, immediate negative feedback in their population systems (May, 1976; Turchin, 2003) (Figure 4.6, blue line). These dynamics can be represented as:

$$R_t = r \cdot (1 - (N_{ti}/K))^q + \varepsilon \qquad (4.5)$$

where R_t is per capita growth rate, r is the maximum growth rate, K is population size at equilibrium, N_t is the abundance at some time lag i with nonlinear effects occurring when $q > 1$, and ε_t represents exogenous effects. Populations that show attraction to one point may also demonstrate less stable behavior due to high intrinsic growth rates (r), non-linearity ($q > 1$) or delays in the feedback ($i > 1$) (Turchin, 2003), or due to high exogenous variance (σ_ε) (May, 1976; Bellows, 1981). Dramatic fluctuations in abundance of bark beetles could be due to the ubiquitous presence of nonlinear negative feedback due to scramble competition (Berryman and Pienaar, 1973; Anderbrant et al., 1985; Zhang et al., 1992; Reeve et al., 1998; Raffa, 2001; Faccoli, 2009; Kausrud et al., 2012). Alternatively, delays in feedback caused by numerical responses from natural enemies (Berryman and Chen, 1999; Dwyer et al., 2004), host plant resistance (Haukioja, 2005), or maternal effects (Ginzburg and Taneyhill, 1994) may create cyclical or oscillatory behavior, but are rarely reported in bark beetles. A delay in feedback attributed to the lagged numerical response of the clerid predator *Thanasimus dubius* (F.) was detected in the time series of the southern pine beetle *D. frontalis* in Texas (Turchin et al., 1991); however, later analyses of a longer time series (Friedenberg et al., 2008) and more time series (Martinson et al., 2013) failed to detect second order feedbacks strong enough to produce cyclical behavior.

5.2.1 Non-aggressive Species

The single equilibrium abundance model seems a reasonable match with the majority of non-aggressive bark beetle species that rely on dead or weakened trees as a resource and are regulated at low equilibrium abundance (Figure 4.6, point A_3). The stabilizing negative feedback that creates the equilibrium (Figure 4.6, region 1) can plausibly arise from interspecific competition and predation (Byers, 1989; Rankin and Borden, 1991; Schlyter and Anderbrant, 1993; Light et al., 2009). Interspecific competition can also play a role because there is frequent overlap among species in host use and cross-attraction of pheromones (Schroeder, 1999; Ayres et al., 2001; Aukema and Raffa, 2002; Schroeder, 2007; Boone et al., 2008; Martinson et al., 2013). Experiments with *Ips* spp. have confirmed the presence of strong interspecific competition (Byers, 1993; Schlyter and Anderbrant, 1993; Boone et al., 2008; Light et al., 2009) and species attacking the same tree tend to avoid areas with galleries or detect pheromones of congeners (Byers, 1989) and effectively partition the resource (Coulson et al., 1976; Byers, 1989; Ayres et al., 2001). Natural enemies, especially predators, are also notable sources of negative feedback on low abundance populations because they respond strongly to bark beetle aggregation pheromones (Reeve, 1997; Erbilgin et al., 2002; Aukema and Raffa, 2005; Costa and Reeve, 2011) and sometimes from multiple prey species (Aukema and Raffa, 2002; Boone et al., 2008; Reeve et al., 2009). Typically, these predators are a complex of clerid, histerid, and trogossitid beetles (Weslien and Regnander, 1992; Schroeder and Weslien, 1994; Reeve, 1997; Aukema and Raffa, 2005; Boone et al., 2008). Some of these predators may demonstrate prey preferences (Reeve et al., 2009), but predation is a potentially strong source of negative feedback due to prey switching (Boone et al., 2008; Martinson et al., 2013). Some predators inflict density-dependent mortality on bark beetles under some conditions (Grégoire et al., 1985; Reeve, 1997; Turchin et al., 1999b). Non-aggressive species are regulated around a low equilibrium abundance and subject to strong negative feedback from community associates (Figure 4.6, point A_3). Unfortunately, empirical validation of the stability and processes affecting it are lacking for non-aggressive species due to a paucity of time series data. Population behavior of non-aggressive species is distinct from that of tree-killing species because they do not mass attack and are typically highly susceptible to host resistance mechanisms (Figures 4.2, 4.4, and 4.5).

5.2.2 Opportunistically Aggressive Species

Bark beetles that undergo punctuated or short-lived outbreaks, rather than sustained epidemics (*D. pseudotsugae, I. confusus* (LeConte), *I. pini*), are characteristic of population systems demonstrating high variation around one fixed point due to high exogenous variance. We consider this to be population behavior of opportunistically aggressive species. In these species, stochastic perturbations, such as drought or windthrow events, create dramatic increases in bark beetle abundance by causing an increase in the availability of weakened trees and a coincident dampening of feedback from resource limitation and shared natural enemies. As will be discussed below in more detail, the presence of a nonlinear region separating two stable points is biologically plausible for these bark beetles because they demonstrate social facilitation (positive density dependence) (Berryman, 1982) (Figure 4.6, region 2). Stochastic perturbations may move populations of opportunistically aggressive species temporarily to a region of abundance where cooperative effects are possible (Figure 4.6, orange line). However, opportunistically aggressive species are unable to establish a stable, high equilibrium because once the perturbation is removed the positive density dependence will weaken and their population will decline from the demographic barriers imposed by host defense (Figure 4.6, orange line). As a result, the population will tend to return to low abundance (Figure 4.6, point A_2). A recent illustration was the dramatic population increases of *I. confusus* in response to regional pinyon pine mortality from drought in the southwest USA, which permitted a short-lived bout of mass attacks on healthy trees (Breshears *et al.*, 2005; Santos and Whitham, 2010). Once drought stress was removed, however, bark beetles were unable to maintain positive growth rates colonizing healthy trees, and their populations returned to their previous low levels. Outbreaks of fir engravers and others (Berryman and Ferrell, 1988; Powers *et al.*, 1999; Gall *et al.*, 2003) that occur in response to resource pulses created from drought, pest or fire disturbances, windthrow, or logging operations and die rapidly afterward are further representatives of this population behavior. We consider these species to be opportunistically aggressive because mechanisms affecting positive density dependence are not strong enough to exceed the escape threshold and permit attraction to the high population abundance once the perturbation is removed.

5.3 Aggressive Species Exist at Low and High Abundance (Alternate Attractors)

It has been hypothesized that population systems of some dramatically outbreaking forest insects are characterized by two stable equilibria (Berryman, 1979, 1982; Raffa and Berryman, 1986; Mawby *et al.*, 1989). According to this model, populations tend to be regulated at low or high abundance, with occasional switches between equilibria due to stochastic effects (Figure 4.6, red line). An escape threshold within the nonlinear region (Figure 4.6, point B) separates two sources of negative feedback that give rise to the low and high stable points (Figure 4.6, points A_1 and C). The nonlinear region represents that as bark beetles increase in abundance they are better able to overwhelm host defense via social facilitation. Following Martinson *et al.* (2013) this region of positive density dependence can be mathematically introduced with a logistic function:

$$F = \frac{1}{1 + e^{\beta(F_{mid}-N)}} \tag{4.6}$$

where F is an index of facilitation that ranges from 0 to 1, F_{mid} is the abundance when $F = 0.5$, and β is the strength of facilitation. Because social facilitation enables bark beetles to attack healthy trees that were inaccessible at low density, each stable state (K_N) is a function of beetle population size (N), expressed as:

$$K_N = (1 - F) \cdot K_{low} + K_{high} \cdot F \tag{4.7}$$

As N approaches the unstable equilibrium, stochastic processes can bump the abundance above the escape threshold or may weaken the density dependence (e.g., loss of tree vigor) causing the lower equilibrium to disappear. Once that happens, the population will grow

deterministically to upper equilibrium, representing an outbreak that is potentially self-sustaining (Figure 4.6, point C). This alternate attractors model is a satisfying theoretical representation of the feedback processes and bimodal patterns of abundance that are known to exist in the most aggressive species of bark beetles—including those producing vast and globally notable epidemics of tree mortality (Økland and Berryman, 2004; Carroll *et al.*, 2006; Meddens *et al.*, 2012).

According to the alternate attractor model, rare populations remain rare due to strong, negative density dependence (Figure 4.6, region 1) and move between attractors by stochastic perturbations that affect either abundance directly, variance in R around the density-dependent function (via ε), or the form of the density-dependent function itself. Regulating negative feedback around the lower attractor (Figure 4.6, point A_1) can arise from predators and interspecific competitors as discussed above (Byers, 1989; Rankin and Borden, 1991; Schlyter and Anderbrant, 1993; Light *et al.*, 2009). In the southern pine beetle system, for example, the clerid predator *T. dubius* and co-occurring complex of *Ips* spp. can produce this effect because they commonly co-occur with endemic populations of *D. frontalis*, and are highly attracted to frontalin, the principal pheromone of *D. frontalis* (Martinson *et al.*, 2013). Similar communities of predators and competitors coexist with other aggressive species such as *D. ponderosae* in western North America and *I. typographus* L. in Europe and show similar community interactions mediated by pheromones, and host plant volatiles (Berryman, 1966; Cole, 1981; Weslien, 1992; Boone *et al.*, 2008). Stochastic effects that create notable spatial and temporal variation in the availability of weakened trees and colonization ability by these community associates across a landscape provide a mechanism for local populations of beetles to exceed the escape threshold, grow to locally high abundance, and spread outward to produce spatially extensive epidemics (Martinson *et al.*, 2013).

An important feature of this model is that deterministic forces alone are incapable of moving populations between attractors (Berryman, 1999). Stochastic perturbations are required to permit transitions between attractors. We define population systems of bark beetles as aggressive when they can exceed the escape threshold and attain a high equilibrium abundance. Environmental stochasticity could plausibly weaken the density dependence (e.g., loss in host vigor) and populations would grow deterministically to the high abundance steady state. However, more typically, as populations increase interactions between endogenous and stochastic processes become critical determinants of whether populations will exceed the escape threshold to the high abundance state or decrease back to low population density. When bark beetle populations increase they are better able to mass attack and overwhelm tree defense

(Rudinsky, 1966; Raffa and Berryman, 1983; Martinson *et al.*, 2007) or may swamp predator functional response (Reeve, 1997; Aukema and Raffa, 2004) giving rise to the region of positive density dependence separating each attractor (Figure 4.6, region 2). Epidemics of aggressive species could theoretically be initiated by lightning (Hodges and Pickard, 1971), disturbances from other pests (Hadley and Veblen, 1993), fire (Furniss, 1965; but see Lombardero and Ayres, 2011; Davis *et al.*, 2012; Powell *et al.*, 2012), and severe storms (Schroeder and Lindelöw, 2002) that increase local resource supplies. Epidemics of the European spruce beetle *I. typographus* begin typically after storm events and the volume of windthrown trees strongly predicts the number of healthy trees that come under attack (Gall *et al.*, 2003; Marini *et al.*, 2013). Unlike most *Ips* spp., *I. typographus* can maintain epidemics for years even when windthrow events cease (Kausrud *et al.*, 2012); in some cases epidemics can be prolonged by drought (Marini *et al.*, 2013). Local infestations of *D. frontalis* are more likely to form, persist, and reinforce each other with increasing abundance in the broader forest landscape (Friedenberg *et al.*, 2007).

The ability of bark beetles to exceed this threshold is governed by factors influencing their ability to mass attack trees. These can be generally grouped into factors affecting the coordination of a mass attack and host tree vigor—each is affected by environmental stochasticity and forest composition (Raffa *et al.*, 2008). Host tree vigor affects the density of attacking beetles needed to successfully colonize trees, and, as discussed above, there can be considerable variation among trees in a stand in their ability to resist bark beetles (Lombardero *et al.*, 2000b; Tisdale *et al.*, 2003; Boone *et al.*, 2011) and regional stochastic events may compromise vigor of entire stands (Raffa *et al.*, 2008). Landscape factors that influence tree susceptibility to bark beetles by dampening attack thresholds have been a long-standing focus in forest health management (Edmonds *et al.*, 2000).

Similar to host tree defense, environmental stochasticity also influences local beetle population abundance and can move populations into the zone of positive feedback (Figure 4.6, region 2). The attack density in a stand is related to the productivity of the previous generation and/or the reemergence of adults of previously attacked trees (Coulson, 1979; Berryman, 1982). Variance in temperature may also influence productivity and beetle-attacking behavior by affecting adult flight, brood survival, and development (Bentz *et al.*, 1991; Lombardero *et al.*, 2000a; Régnière and Bentz, 2007; Trân *et al.*, 2007). Seasonal temperatures that promote univoltine broods in *D. ponderosae* and *D. rufipennis*, for example, often lead to greater synchrony in adult emergence (Hicke *et al.*, 2006; Werner *et al.*, 2006; Powell and Bentz, 2014), which creates positive feedback by enhancing the probability

of coordinating mass attacks (Friedenberg *et al.*, 2007; Powell and Bentz, 2009). Once populations of *D. frontalis*, *D. ponderosae*, and *D. rufipennis* are large enough, host resistance mechanisms are no longer constraining and beetles readily attack and kill healthy trees (Bentz *et al.*, 2009). Higher per capita reproduction is associated with higher attack rates on trees that lead to more rapid depletion of tree defenses (Raffa and Berryman, 1983; Martinson *et al.*, 2007) and volatile concentrations correlated with high host vigor that are repellent to small populations are attractive to growing populations reaching outbreak status (Wallin and Raffa, 2004; Boone *et al.*, 2008, 2011).

Processes that affect forest structure and composition (e.g., succession, silviculture, disturbance) are also regarded as important factors affecting the aggregative behavior and productivity of bark beetles (Fettig *et al.*, 2014). For example, disturbance and management practices that produced extensive tracts of highly susceptible age classes of lodgepole pine were key to the development and persistence of the *D. ponderosae* epidemic in British Columbia (Taylor and Carroll, 2003). Processes that modify age structure influence beetle fitness because larger trees generally support greater reproduction and bark beetles tend to avoid trees below a particular size class (Coulson, 1979; Safranyik and Carroll, 2006). Stand to landscape-scale processes that influence host tree species composition, density and spacing, such as thinning or natural succession to mixed species stands, may disrupt pheromone communication (Turchin *et al.*, 1999a; Fettig *et al.*, 2007) and influence short distance dispersal (Ayres *et al.*, 2011) that are consequential for population growth rates in stands (Ylioja *et al.*, 2005).

The final theoretical requirement for the alternate attractors model is the presence of stabilizing, negative feedback (Figure 4.6, region 3) on populations at the high equilibrium point (Figure 4.6, point C) (May, 1977; Berryman, 1999). The most likely sources of regulation on outbreaking bark beetle populations are from resource depletion and natural enemies. Indeed, resource depletion is constantly invoked as the primary source of endogenous feedback detected during time series analyses of *I. typographus* (Økland and Berryman, 2004; Marini *et al.*, 2013), *D. frontalis* (Martinson *et al.*, 2013) (Figure 4.7), *D. ponderosae* (Berryman, 1999; MacQuarrie and Cooke, 2011), and *D. rufipennis* (Allen *et al.*, 2006). The importance of natural enemies on large bark beetle populations is hardly studied, exceptions being for *I. typographus* and *D. frontalis*. In Norway, *T. formicarius* (L.) has an undetectable effect on *I. typographus* at the forest or landscape scale (Kausrud *et al.*, 2012; Marini *et al.*, 2013) whereas population growth rates of *D. frontalis* are negatively correlated with the abundance of the predator *T. dubius* in Texas (Reeve, 1997; Turchin *et al.*, 1999b) and across the southeastern USA (Figure 4.7).

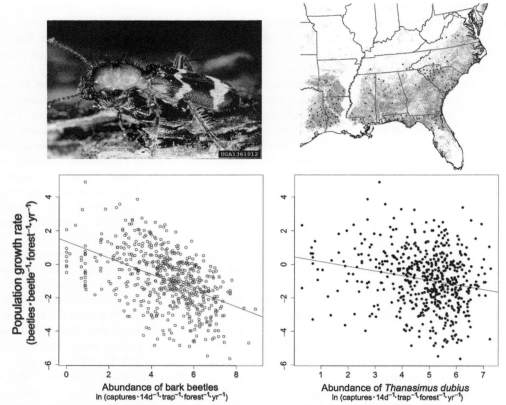

FIGURE 4.7 Per capita change in southern pine beetles, *Dendroctonus frontalis*, as a function of its own abundance (lower left panel) and that of its clerid predator, *Thanasimus dubius* (lower right panel). Model coefficients: Intercept: 1.8 ± 0.3, $N_{SPBt} = -0.50 \pm 0.04$, $N_{SPBt-1} = 0.11 \pm 0.5$, $N_{Tdubt-1} = -0.20 \pm 0.06$; $R^2 = 0.25$; $P < 0.0001$). Data are from 133 pine forests in the southeastern USA (triangles in map in upper right) during 1986 to 2010 (south-wide trapping program maintained by the Texas Forest Service). Map also shows distribution of *Pinus* (green) extracted from forest types of the USA (Ruefenacht *et al.*, 2008). *Photo of* T. dubius *in upper left courtesy of USDA Forest Service Archive, USDA Forest Service, Bugwood.org.*

6. CONCLUSIONS

Bark beetles are important consumers of multiple plant forms and tissues types, and some bark beetles are globally important pests to agriculture and forestry. All bark beetles are embedded within complex communities involving microbes and other associates, encounter a diverse array of host resistance mechanisms, and are often subject to diverse resource and abiotic conditions. These forces have important effects on their abundance in space and time and are best understood for phloem-feeding species, and particularly those that kill healthy trees and especially for populations at high abundance. Factors affecting changes in population abundance of seed-feeding taxa and species attacking herbaceous plants are less understood, but undeniably need further study since some species are notable pests. Although logistically challenging, future work evaluating factors that maintain low abundance populations of phloem-feeding species is a critical but understudied aspect of their ecology.

Host tree resistance mechanisms (primarily resin) have been critical to the life history adaptations of conifer-attacking bark beetle populations. In response, bark beetles have evolved the following behavioral and physiological adaptations to counter defenses: (1) capacity to form aggregations (social facilitation); (2) tolerance to toxic host plant defenses; and (3) capability to overwhelm host defense. Aggressive taxa are highly evolved tree killers utilizing all of these traits; however, not all bark beetles have evolved all of these traits limiting their ability to kill healthy trees and thus their aggression (Table 4.1). Selection towards more aggressive behavior in bark beetles has important considerations for how density-dependent and stochastic processes affect population growth. Under the appropriate resource conditions, tree killing or aggressive taxa are capable of generating self-sustaining ecosystem-wide outbreaks once stochastic perturbations push them to the upper equilibrium. Effects from weather, landscape factors or management are required to push populations back to non-outbreak status. Thus, interactions between density-dependent feedbacks and stochastic processes are key in determining population responses—a virtue that often makes predicting outbreaks difficult because populations can grow so quickly to epidemic sizes once favorable conditions prevail after being at a state where they were virtually undetectable for many years. Populations of opportunistically aggressive species may undergo dramatic population increases that permit occasional mass attack of healthy trees for short time periods, but are regulated at a low abundance equilibrium. The majority of bark beetle species, however, are non-aggressive and unable to attain or sustain outbreak status because strong demographic constraints imposed by host defense restrict them to dying or very weak trees. These species are able to maintain low, relatively stable population sizes across forested landscapes.

REFERENCES

Adams, A.S., Boone, C.K., Bohlmann, J., Raffa, K.F., 2011. Responses of bark beetle-associated bacteria to host monoterpenes and their relationship to insect life histories. J. Chem. Ecol. 37, 808–817.

Alarcón, D., Ortega, F., Perich, F., Fernando, P., Parra, L., Quiroz, A., 2010. Relationship between radical infestation of *Hylastinus obscurus* (Marsham) and the yield of cultivars and experimental lines of red clover (*Trifolium pratense* L.). Revista de la Ciencia del Suelo y Nutrición Vegetal 10, 115–125.

Allen, J., Wesser, S., Markon, C., Winterberger, K., 2006. Stand and landscape level effects of a major outbreak of spruce beetles on forest vegetation in the Copper River Basin Alaska. For. Ecol. Manag. 227, 257–266.

Anderbrant, O., Schlyter, F., Birgersson, G., 1985. Intraspecific competition affecting parents and offspring in the bark beetle *Ips typographus*. Oikos 45, 89–98.

Andrewartha, H.G., Birch, L.C., 1954. The Distribution and Abundance of Animals. University of Chicago Press, Chicago.

Aukema, B.H., Raffa, K.F., 2002. Relative effects of exophytic predation, endophytic predation, and intraspecific competition on a subcortical herbivore: consequences to the reproduction of *Ips pini* and *Thanasimus dubius*. Oecologia 133, 483–491.

Aukema, B.H., Raffa, K.F., 2004. Does aggregation benefit bark beetles by diluting predation? Links between a group-colonisation strategy and the absence of emergent multiple predator effects. Ecol. Entomol. 29, 129–138.

Aukema, B.H., Raffa, K.F., 2005. Selective manipulation of predators using pheromones: responses to frontalin and ipsdienol pheromone components of bark beetles in the Great Lakes region. Agr. Forest Entomol. 7, 193–200.

Ayres, M.P., Wilkens, R.T., Ruel, J.J., Lombardero, M.J., Vallery, E., 2000. Nitrogen budgets of phloem-feeding bark beetles with and without symbiotic fungi. Ecology 81, 2198–2210.

Ayres, B.D., Ayres, M.P., Abrahamson, M.D., Teale, S.A., 2001. Resource partitioning and overlap in three sympatric species of *Ips* bark beetles (Coleoptera: Scolytidae). Oecologia 128, 443–453.

Ayres, M.P., Martinson, S.J., Friedenberg, N.A., 2011. Southern pine beetle ecology: populations within stands. Gen. Tech. Rep. SRS-140. In: Coulson, J.C., Klepzig, K. (Eds.), Southern Pine Beetle II. USDA Forest Service, Southern Research Station, Asheville, NC, pp. 75–89.

Baker, P.S., Barrera, J.F., Rivas, A., 1992. Life-history studies of the coffee berry borer (*Hypothenemus hampei*, Scolytidae) on coffee trees in southern Mexico. J. Appl. Ecol. 29, 656–662.

Baker, P.S., Rivas, A., Balbuena, R., Ley, C., Barrera, J.F., 1994. Abiotic mortality factors of the coffee berry borer (*Hypothenemus hampei*). Entomol. Exp. Appl. 71, 201–209.

Bellows, T.S., 1981. The descriptive properties of some models for density dependence. J. Anim. Ecol. 50, 139–156.

Bentz, B.J., Six, D.L., 2006. Ergosterol content of fungi associated with *Dendroctonus ponderosae* and *Dendroctonus rufipennis* (Coleoptera: Curculionidae, Scolytinae). Ann. Entomol. Soc. Am. 99, 189–194.

Bentz, B.J., Logan, J.A., Amman, G.D., 1991. Temperature-dependent development of the mountain pine beetle (Coleoptera: Scolytidae) and simulation of its phenology. Can. Entomol. 123, 1083–1094.

Bentz, B. J., Powell, J. A., and Logan, J. A., 1996. Localized spatial and temporal attack dynamics of the mountain pine beetle. USDA Forest Service Research Paper, INT-494.

Bentz, B., Logan, J., MacMahon, J., Allen, C.D., Ayres, M.P., Berg, E., et al., 2009. Bark beetle outbreaks in western North America: causes and consequences. Bark Beetle Symposium, Snowbird, UtahUniversity of Utah Press, Salt Lake City.

Bentz, B.J., Vandygriff, J.C., Jensen, C., Coleman, T., Maloney, P., Smith, S., et al., 2014. Mountain pine beetle voltinism and life history characteristics across latitudinal and elevational gradients in the western United States. Forest Sci. 60, 434–449.

Berryman, A., Lima, M., 2006. Deciphering the effects of climate on animal populations: Diagnostic analysis provides new interpretation of Soay sheep dynamics. Am. Nat. 168, 784–795.

Berryman, A.A., 1966. Studies on the behavior and development of *Enoclerus lecontei* (Wolcott), a predator of the western pine beetle. Can. Entomol. 98, 519–526.

Berryman, A.A., 1974. Dynamics of bark beetle populations—towards a general productivity model. Environ. Entomol. 3, 579–585.

Berryman, A.A., 1979. Dynamics of bark beetle populations: analysis of dispersal and redistribution. Bulletin de la Societe Entomologique Suisse 52, 227–234.

Berryman, A.A., 1982. Population dynamics of bark beetles. In: Mitton, J. B., Sturgeon, K.B. (Eds.), Bark Beetles in North American Conifers. University of Texas Press, Austin, pp. 264–314.

Berryman, A.A., 1999. Principles of of Population Dynamics and their Application. Stanley Thornes, Cheltenham.

Berryman, A.A., 2003. On principles, laws and theory in population ecology. Oikos 103, 695–701.

Berryman, A.A., Pienaar, L.V., 1973. Simulation of intraspecific competition and survival of *Scolytus ventralis* broods (Coleoptera: Scolytidae). Environ. Entomol. 2, 447–460.

Berryman, A.A., Ferrell, G.T., 1988. The fir engraver beetle in western states. In: Berryman, A.A. (Ed.), Population Dynamics of Forest Insects. Plenum, New York, pp. 556–577.

Berryman, A.A., Chen, X., 1999. Population cycles: the relationship between cycle period and reproductive rate depends on the relative dominance of bottom-up or top-down control. Oikos 87, 589–593.

Birgersson, G., Debarr, G.L., Groot, P., Dalusky, M.J., Pierce, J., Harold, D., et al., 1995. Pheromones in white pine cone beetle, *Conophthorus coniperda* (Schwarz) (Coleoptera: Scolytidae). J. Chem. Ecol. 21, 143–167.

Boone, C.K., Six, D.L., Raffa, K.F., 2008. The enemy of my enemy is still my enemy: competitors add to predator load of a tree-killing bark beetle. Agr. Forest Entomol. 10, 411–421.

Boone, C.K., Aukema, B.H., Bohlmann, J., Carroll, A.L., Raffa, K.F., 2011. Efficacy of tree defense physiology varies with bark beetle population density: a basis for positive feedback in eruptive species. Can. J. Forest Res. 41, 1174–1188.

Breshears, D.D., Cobb, N.S., Rich, P.M., Price, K.P., Allen, C.D., Balice, R.G., et al., 2005. Regional vegetation die-off in response to global-change-type drought. Proc. Natl. Acad. Sci. U. S. A. 102, 15144–15148.

Bridges, J.R., Nettleton, W.A., Connor, M.D., 1985. Southern pine beetle (Coleoptera, Scolytidae) infestations without the bluestain fungus, *Ceratocystis minor*. J. Econ. Entomol. 78, 325–327.

Byers, J.A., 1989. Behavioral mechanisms involved in reducing competition in bark beetles. Holarct. Ecol. 12, 466–476.

Byers, J.A., 1993. Avoidance of competition by spruce bark beetles, *Ips typographus* and *Pityogenes chalcographus*. Experientia 49, 272–275.

Campobasso, G., Terragitti, G., Colonnelli, E., Spencer, N.R., 2004a. Host specificity of *Thamnurgus euphorbiae* Kuster (Coleoptera: Scolytidae): a potential biological control agent of leafy spurge *Euphorbia esula* L. (Euphorbiaceae) in the United States. Environ. Entomol. 33, 1673–1680.

Campobasso, G., Terragitti, G., Mann, K., Quimby, P.C., 2004b. Field and laboratory biology of the stem-feeding beetle *Thamnurgus euphorbiae* (Kuster) (Col., Scolytidae) in Italy, a potential biological control candidate of leafy spurge in the USA and Canada. J. Appl. Entomol. 128, 1–5.

Carroll, A. L., Aukema, B. H., Raffa, K. F., Linton, D. A., Smith, G. D., and Lindgren, B. S. (2006). Mountain pine beetle outbreak development: the endemic–incipient epidemic transition. *Mountain Pine Beetle Initiative* Project # 1.03, 1–22.

Cates, R., Alexander, H., 1982. Host resistance and susceptibility. In: Mitton, J.B., Sturgeon, K.B. (Eds.), Bark Beetles in North American Conifers. University of Texas Press, Austin, pp. 212–263.

Chen, H., Tang, M., 2007. Spatial and temporal dynamics of bark beetles in Chinese white pine in Qinling Mountains of Shaanxi Province. China. Environ. Entomol. 36, 1124–1130.

Cognato, A.I., Gillette, N.E., Bolaños, R.C., Sperling, F.A., 2005. Mitochondrial phylogeny of pine cone beetles (Scolytinae, *Conophthorus*) and their affiliation with geographic area and host. Mol. Phylogenet. Evol. 36, 494–508.

Cole, W.E., 1981. Some risks and causes of mortality in mountain pine-beetle populations—a long-term analysis. Res. Popul. Ecol. 23, 116–144.

Colombari, F., Schroeder, M.L., Battisti, A., Faccoli, M., 2012. Spatio-temporal dynamics of an *Ips acuminatus* outbreak and implications for management. Agr. Forest Entomol. 15, 34–42.

Cook, S.P., Hain, F.P., 1988. Toxicity of host monoterpenes to *Dendroctonus frontalis* and *Ips calligraphus* (Coleoptera, Scolytidae). J. Entomol. Sci. 23, 287–292.

Costa, A., Reeve, J.D., 2011. Olfactory experience modifies semiochemical responses in a bark beetle predator. J. Chem. Ecol. 37, 1166–1176.

Coulson, R.N., 1979. Population dynamics of bark beetles. Annu. Rev. Entomol. 24, 417–447.

Coulson, R.N., Klepzig, K.D., 2011. Southern Pine Beetle II. USDA Forest Service, Southern Research Station, Asheville, NC. Gen. Tech. Rep. SRS-140.

Coulson, R.N., Mayyasi, A.M., Foltz, J.L., Hain, F.P., Martin, W.C., 1976. Resource utilization by the southern pine beetle, *Dendroctonus frontalis* (Coleoptera: Scolytidae). Can. Entomol. 108, 353–362.

Coulson, R.N., Flamm, R.O., Pulley, P.E., Payne, T.L., Rykiel, E.J., Wagner, T.L., 1986. Response of the southern pine bark beetle guild (Coleoptera, Scolytidae) to host disturbance. Environ. Entomol. 15, 850–858.

Cranshaw, W., 2011. Recently recognized range extensions of the walnut twig beetle, *Pityophthorus juglandis* Blackman (Coleoptera: Curculionidae: Scolytinae), in the western United States. Coleopt. Bull. 65, 48–49.

Culk, M.P., Weaver, J.E., 1994. Seasonal crawling activity of adult root-feeding insect pests (Coleoptera, Curculionidae, Scolytidae) of red-clover. Environ. Entomol. 23, 68–75.

Damon, A., 2000. A review of the biology and control of the coffee berry borer, *Hypothenemus hampei* (Coleoptera: Scolytidae). Bull. Entomol. Res. 90, 453–465.

Davis, R.S., Hood, S., Bentz, B.J., 2012. Fire-injured ponderosa pine provide a pulsed resource for bark beetles. Can. J. Forest Res. 42, 2022–2036.

de Groot, P., DeBarr, G.L., 1998. Factors affecting capture of the white pine cone beetle, *Conophthorus coniperda* (Schwarz) (Col., Scolytidae) in pheromone traps. J. Appl. Entomol. 122, 281–286.

DeAngelis, J.D., Hodges, J.D., Nebeker, T.E., 1986. Phenolic metabolites of *Ceratocystis minor* from laboratory cultures and their effects on transpiration in loblolly pine seedlings. Can. J. Bot. 64, 151–155.

Deneubourg, J.L., Grégoire, J.-C., LeFort, E., 1990. Kinetics of larval gregarious behavior in the bark beetle *Dendroctonus micans* (Coleoptera, Scolytidae). J. Insect Behav. 3, 169–182.

Dwyer, G., Dushoff, J., Yee, S.H., 2004. The combined effects of pathogens and predators on insect outbreaks. Nature 430, 341–345.

Edburg, S.L., Hicke, J.A., Brooks, P.D., 2012. Cascading impacts of bark beetle-caused tree mortality on coupled biogeophysical and biogeochemical processes. Front. Ecol. Environ. 10, 416–424.

Edmonds, R.L., Agee, J.K., Gara, R.I., 2000. Forest Health and Protection. McGraw Hill, Boston.

Eidt-Wendt, J., Schulz, F.A., 1990. Studies on the biology and ecology of *Pagiocerus frontalis* (Fabricius) (Coleoptera: Scolytidae) infesting stored maize in Ecuador. In: Fleurat-Lessard, F., Ducom, P. (Eds.), Proceedings of the Fifth International Working Conference on Stored-Product Protection. Bordeaux, France, pp. 61–69.

Erbilgin, N., Nordheim, E.V., Aukema, B.H., Raffa, K.F., 2002. Population dynamics of *Ips pini* and *Ips grandicollis* in red pine plantations in Wisconsin: within- and between-year associations with predators, competitors, and habitat quality. Environ. Entomol. 31, 1043–1051.

Erbilgin, N., Powell, J.S., Raffa, K.F., 2003. Effect of varying monoterpene concentrations on the response of *Ips pini* (Coleoptera: Scolytidae) to its aggregation pheromone: implications for pest management and ecology of bark beetles. Agr. Forest Entomol. 5, 269–274.

Faccoli, M., 2009. Breeding performance of *Tomicus destruens* at different densities: the effect of intraspecific competition. Entomol. Exp. Appl. 132, 191–199.

Farrell, B.D., Sequeira, A.S., O'Meara, B.C., Normark, B.B., Chung, J.H., Jordal, B.H., 2001. The evolution of agriculture in beetles (Curculionidae: Scolytinae and Platypodinae). Evolution 55, 2011–2027.

Ferrenberg, S., Kane, J.M., Mitton, J.B., 2013. Resin duct characteristics associated with tree resistance to bark beetles across lodgepole and limber pines. Oecologia 174, 1283–1292.

Fettig, C.J., Klepzig, K.D., Billings, R.F., Munson, A.S., Nebeker, T.E., Negrón, J.F., Nowak, J.T., 2007. The effectiveness of vegetation management practices for prevention and control of bark beetle infestations in coniferous forests of the western and southern United States. For. Ecol. Manag. 238, 24–53.

Fettig, C.J., Gibson, K.E., Munson, A.S., Negrón, J.F., 2014. Cultural practices for prevention and mitigation of mountain pine beetle infestations. Forest Sci. 60, 450–463.

Flamm, R.O., Wagner, T.L., Cook, S.P., Pulley, P.E., Coulson, R.N., McArdle, T.M., 1987. Host colonization by cohabiting *Dendroctonus frontalis*, *Ips avulsus*, and *I. calligraphus* (Coleoptera: Scolytidae). Environ. Entomol. 16, 390–399.

Fordyce, J., 2003. Aggregative feeding of pipevine swallowtail larvae enhances host plant suitability. Oecologia 135, 250–257.

Friedenberg, N.A., Powell, J.A., Ayres, M.P., 2007. Synchrony's double edge: transient dynamics and the Allee effect in stage structured populations. Ecol. Lett. 10, 564–573.

Friedenberg, N.A., Sarkar, S., Kouchoukos, N., Billings, R.F., Ayres, M.P., 2008. Temperature extremes, density dependence, and southern pine beetle (Coleoptera: Curculionidae) population dynamics in east Texas. Environ. Entomol. 37, 650–659.

Furniss, M.M., 1962. Infestation patterns of Douglas-fir beetle in standing and windthrown trees in southern Idaho. J. Econ. Entomol. 55, 486–491.

Furniss, M.M., 1965. Susceptibility of fire-injured Douglas-fir to bark beetle attack in southern Idaho. J. For. 63, 8–11.

Furniss, M.M., 1997. *Conophthorus ponderosae* (Coleoptera: Scolytidae) infesting lodgepole pine cones in Idaho. Environ. Entomol. 26, 855–858.

Gall, R., Meier, F., Meier, A.L., Forster, B., 2003. Regional distribution of standing tree infestation by bark beetle (*Ips typographus*) following storm damage in Canton Berne, Switzerland. Schweizerische Zeitschrift fur Forstwesen 154, 442–448.

Ginzburg, L.R., Taneyhill, D.E., 1994. Population-cycles of forest Lepidoptera—a maternal effect hypothesis. J. Anim. Ecol. 63, 79–92.

Godwin, P.A., Odell, T.M., 1965. The life history of the white-pine cone beetle, *Conophthorus coniperda*. Ann. Entomol. Soc. Am. 58, 213–219.

Greenwood, D.L., Weisberg, P.J., 2008. Density-dependent tree mortality in pinyon-juniper woodlands. For. Ecol. Manag. 255, 2129–2137.

Grégoire, J.-C., 1988. The greater European spruce beetle. In: Berryman, A.A. (Ed.), Population Dynamics of Forest Insects. Plenum, New York, pp. 455–478.

Grégoire, J.-C., Braekman, J.C., Tondeur, A., 1981. Chemical communication between the larvae of *Dendroctonus micans* Kug (Coleoptera: Scolytidae). Les Colloques de l'INRA 7, 253–257.

Grégoire, J.-C., Merlin, J., Pasteels, J.M., Jaffuel, R., Vouland, G., Schvester, D., 1985. Biocontrol of *Dendroctonus micans* by *Rhizophagus grandis* Gyll. (Col., Rhizophagidae) in the Massif Central (France). J. Appl. Entomol. 99, 182–190.

Güclü, C., Özbek, H., 2007. Biology and damage of *Thamnurgus pegani* Eggers (Coleoptera: Scolytidae) feeding on *Peganum harmala* L. in eastern Turkey. Proc. Entomol. Soc. Wash. 109, 350–358.

Hadley, K.S., Veblen, T.T., 1993. Stand response to western spruce budworm and Douglas-fir bark beetle outbreaks, Colorado Front Range. Can. J. Forest Res. 23, 479–491.

Haukioja, E., 2005. Plant defenses and population fluctuations of forest defoliators: mechanism-based scenarios. Ann. Zool. Fenn. 42, 313–325.

Hicke, J.A., Logan, J.A., Powell, J., Ojima, D.S., 2006. Changing temperatures influence suitability for modeled mountain pine beetle (*Dendroctonus ponderosae*) outbreaks in the western United States. J. Geophys. Res. 111, G02019.

Hicke, J.A., Allen, C.D., Desai, A.R., Dietze, M.C., Hall, R.J., Hogg, E.H., et al., 2012. Effects of biotic disturbances on forest carbon cycling in the United States and Canada. Glob. Chang. Biol. 18, 7–34.

Hodges, J.D., Pickard, L.S., 1971. Lightning in the ecology of the southern pine beetle, *Dendroctonus frontalis* (Coleoptera: Scolytidae). Can. Entomol. 103, 44–51.

Hodges, J.D., Elam, W.W., Watson, W.F., Nebeker, T.E., 1979. Oleoresin characteristics and susceptibility of four southern pines to southern pine beetle (Coleoptera: Scolytidae) attacks. Can. Entomol. 111, 889–896.

Hofstetter, R., Cronin, J., Klepzig, K., Moser, J., Ayres, M.P., 2006. Antagonisms, mutualisms and commensalisms affect outbreak dynamics of the southern pine beetle. Oecologia 147, 679–691.

Holmes, F.W., 1980. Bark beetles, *Ceratocystis ulmi* and Dutch elm disease. In: Harris, K.F., Maramorosch, K. (Eds.), Vectors of Plant Pathogens. Academic Press, New York, pp. 133–147.

Hoover, K., Wood, D.L., Fox, J.W., Bros, W.E., 1995. Quantitative and seasonal association of the pitch canker fungus, *Fusarium subglutinans* f. sp. *pini* with *Conophthorus radiatae* (Coleoptera: Scolytidae) and *Ernobius punctulatus* (Coleoptera: Anobiidae) which infest *Pinus radiata*. Can. Entomol. 127, 79–91.

Hoover, K., Wood, D.L., Storer, A.J., Fox, J.W., Bros, W.E., 1996. Transmission of the pitch canker fungus, *Fusarium subglutinans* f. sp. *pini*, to Monterey pine, *Pinus radiata*, by cone- and twig-infesting beetles. Can. Entomol. 128, 981–994.

Irulandi, S., Kumar, P., Samuel, S.D., Rajendran, R., 2008. Population dynamics of the coffee berry borer and its pathogen in the Pulney hills of Tamil Nadu. J. Coffee Res. 36, 30–37.

Jin, X., Morton, J., Butler, L., 1992. Interactions between *Fusarium avenaceum* and *Hylastinus obscurus* (Coleoptera: Scolytidae) and their influence on root decline in red clover. J. Econ. Entomol. 85, 1340–1346.

Jordal, B.H., Gebhardt, H., Mandelshtam, M.Y., 2013. The red-listed species *Thamnurgus rossicus* in East Europe is a synonym of the rare Central European species, *T. petzi* (Curculionidae: Scolytinae). Zootaxa 3750, 83–88.

Kane, J.M., Kolb, T.E., 2010. Importance of resin ducts in reducing ponderosa pine mortality from bark beetle attack. Oecologia 164, 601–609.

Karban, R., Agrawal, A.A., 2002. Herbivore offense. Annu. Rev. Ecol. Syst. 33, 641–664.

Kausrud, K., Økland, B., Skarpaas, O., Grégoire, J.-C., Erbilgin, N., Stenseth, N.C., 2012. Population dynamics in changing environments: the case of an eruptive forest pest species. Biol. Rev. Camb. Philos. Soc. 87, 34–51.

Keen, F.P., 1958. Cone and seed insects of western forest trees. US Deparment of Agriculture, Washington, DC, Technical Bulletin 1169.

Kirkendall, L.R., 1983. The evolution of mating systems in bark and ambrosia beetles (Coleoptera: Scolytidae and Platypodidae). Zool. J. Linn. Soc. 77, 293–352.

Klapwijk, M.J., Ayres, M.P., Battisti, A., Larsson, S., 2012. Assessing the impact of climate change on outbreak potential. In: Barbosa, P., Letourneau, D., Agrawal, A.A. (Eds.), Insect Outbreaks Revisited. Academic Press, Inc., New York, pp. 429–450

Klepzig, K.D., Six, D.L., 2004. Bark beetle-fungal symbiosis: context dependency in complex associations. Symbiosis 37, 189–205.

Klepzig, K.D., Smalley, E.B., Raffa, K.F., 1996. Interactions of ecologically similar saprogenic fungi with healthy and abiotically stressed conifers. For. Ecol. Manag. 86, 163–169.

Klepzig, K.D., Moser, J.C., Lombardero, M.J., Ayres, M.P., Hofstetter, R. W., Walkinshaw, C.H., 2001. Mutualism and antagonism: ecological interactions among bark beetles, mites, and fungi. In: Jeger, M.J., Spence, N.J. (Eds.), Biotic Interactions in Plant–Pathogen Associations. CABI Publishing, New York, pp. 237–269.

Kolarik, M., Freeland, E., Utley, C., Tisserat, N., 2011. *Geosmithia morbida* sp. nov., a new phytopathogenic species living in symbiosis with the walnut twig beetle (*Pityophthorus juglandis*) on *Juglans* in USA. Mycologia 103, 325–332.

Kopper, B.J., Klepzig, K.D., Raffa, K.F., 2004. Components of antagonism and mutualism in *Ips pini*-fungal interactions: relationship to a life history of colonizing highly stressed and dead trees. Environ. Entomol. 33, 28–34.

Kopper, B.J., Illman, B.L., Kersten, P.J., Klepzig, K.D., Raffa, K.F., 2005. Effects of diterpene acids on components of a conifer bark beetle-fungal interaction: tolerance by *Ips pini* and sensitivity by its associate *Ophiostoma ips*. Environ. Entomol. 34, 486–493.

Lahr, E.C., Krokene, P., 2013. Conifer stored resources and resistance to a fungus associated with the spruce bark beetle *Ips typographus*. PLoS One 8, e72405.

Lande, R., Engen, S., Sæther, B.-E., 2003. Stochastic Population Dynamics in Ecology and Conservation. Oxford University Press, Oxford.

Leath, K.T., Byers, R.A., 1973. Attractiveness of diseased red clover roots to the clover root borer. Phytopathology 63, 428–431.

Light, D.M., Birch, M.C., Paine, T.D., 2009. Laboratory study of intraspecific and interspecific competition within and between two sympatric bark beetle species. Ips pini and I. paraconfusus. J. Appl. Entomol. 96, 233–241.

Liu, Z., Xu, B., Miao, Z., Sun, J., 2013. The pheromone frontalin and its dual function in the invasive bark beetle *Dendroctonus valens*. Chem. Senses 38, 485–495.

Lombardero, M.J., Ayres, M.P., 2011. Factors influencing bark beetle outbreaks after forest fires on the Iberian Peninsula. Environ. Entomol. 40, 1007–1018.

Lombardero, M.J., Ayres, M.P., Ayres, B.D., Reeve, J.D., 2000a. Cold tolerance of four species of bark beetle (Coleoptera: Scolytidae) in North America. Environ. Entomol. 29, 421–432.

Lombardero, M.J., Ayres, M.P., Lorio, P.L., Ruel, J.J., 2000b. Environmental effects on constitutive and inducible resin defences of *Pinus taeda*. Ecol. Lett. 3, 329–339.

Lyons, L.A., 1956. Insects affecting seed production in red pine: Part I *Conophthorus resinosae* Hopk. (Coleoptera: Scolytidae). Can. Entomol. 88, 599–608.

MacQuarrie, C.J.K., Cooke, B.J., 2011. Density-dependent population dynamics of mountain pine beetle in thinned and unthinned stands. Can. J. Forest Res. 41, 1031–1046.

Mandelshtam, M.Y., Petrov, A.V., Korotyaev, B.A., 2012. To the knowledge of the herbivorous scolytid genus *Thamnurgus* Eichhoff (Coleoptera, Scolytidae). Entomol. Rev. 92, 329–349.

Marini, L., Ayres, M.P., Battisti, A., Faccoli, M., 2012. Climate affects severity and altitudinal distribution of outbreaks in an eruptive bark beetle. Clim. Change 115, 327–341.

Marini, L., Lindelöw, Å., Jönsson, A.M., Wulff, S., Schroeder, L.M., 2013. Population dynamics of the spruce bark beetle: a long-term study. Oikos 122, 1768–1776.

Martinson, S., Hofstetter, R.W., Ayres, M.P., 2007. Why does longleaf pine have low susceptibility to southern pine beetle? Can. J. Forest Res. 37, 1966–1977.

Martinson, S.J., Ylioja, T., Sullivan, B.T., Billings, R.F., Ayres, M.P., 2013. Alternate attractors in the population dynamics of a tree-killing bark beetle. Popul. Ecol. 55, 95–106.

Masood, A., Saeed, S., Sajjad, A., Ali, M., 2009. Life cycle and biology of mango bark beetle, *Hypocryphalus mangiferae* (Stebbing), a possible vector of mango sudden death disease in Pakistan. Pak. J. Zool. 41, 281–288.

Mattson, W.J., 1980. Cone resources and the ecology of the red pine cone beetle, *Conophthorus resinosae* (Coleoptera: Scolytidae). Ann. Entomol. Soc. Am. 73, 390–393.

Mawby, W.D., Hain, F.P., Doggett, C.A., 1989. Endemic and epidemic populations of southern pine beetle: implications of the two-phase model for forest managers. Forest Sci. 35, 1075–1087.

May, R.M., 1976. Simple mathematical-models with very complicated dynamics. Nature 261, 459–467.

May, R.M., 1977. Thresholds and breakpoints in ecosystems with a multiplicity of stable states. Nature 269, 471–477.

McPherson, J.E., Wilson, L.F., Stehr, F.W., 1970. A comparison between *Conophthorus* shoot-infesting beetles and *Conphthorus resinosae* (Coleoptera: Scolytidae). I. Comparative life history studies in Michigan. Can. Entomol. 102, 1008–1015.

Meddens, A.J.H., Hicke, J.A., Ferguson, C.A., 2012. Spatiotemporal patterns of observed bark beetle-caused tree mortality in British Columbia and the western United States. Ecol. Appl. 22, 1876–1891.

Mendoza, M.G., Zuñiga, G., 2011. Factors influencing the geographical distribution of *Dendroctonus rhizophagus* (Coleoptera: Curculionidae: Scolytinae) in the Sierra Madre Occidental, México. Environ. Entomol. 40, 549–559.

Morgan, F.D., Mailu, M., 1976. Behavior and generation dynamics of the white pine cone beetle *Conophthorus coniperda* (Schwarz) in central Wisconsin. Ann. Entomol. Soc. Am. 69, 863–871.

Moser, J.C., Konrad, H., Kirisits, T., Carta, L.K., 2005. Phoretic mites and nematode associates of *Scolytus multistriatus* and *Scolytus pygmaeus* (Coleoptera: Scolytidae) in Austria. Agr. Forest Entomol. 7, 169–177.

Murdoch, W.W., Briggs, C.J., Nisbet, R.M., 2003. Consumer-Resource Dynamics. Princeton University Press, Princeton.

Økland, B., Berryman, A.A., 2004. Resource dynamic plays a key role in regional fluctuations of the spruce bark beetles *Ips typographus*. Agr. Forest Entomol. 6, 141–146.

Økland, B., Bjørnstad, O.N., 2006. A resource-depletion model of forest insect outbreaks. Ecology 87, 283–290.

Owen, D.R., Wood, D.L., Parmeter, J.R., 2012. Association between *Dendroctonus valens* and black stain root disease on ponderosa pine in the Sierra Nevada of California. Can. Entomol. 137, 367–375.

Paine, T.D., Hanlon, C.C., 1994. Influence of oleoresin constituents from *Pinus ponderosa* and *Pinus jeffreyi* on growth of mycangial fungi from *Dendroctonus ponderosae* and *Dendroctonus jeffreyi*. J. Chem. Ecol. 20, 2551–2563.

Paine, T.D., Raffa, K.F., Harrington, T.C., 1997. Interactions among scolytid bark beetles, their associated fungi, and live host conifers. Annu. Rev. Entomol. 42, 179–206.

Powell, E.N., Townsend, P.A., Raffa, K.F., 2012. Wildfire provides refuge from local extinction but is an unlikely driver of outbreaks by mountain pine beetle. Ecol. Monogr. 82, 69–84.

Powell, J.A., Bentz, B.J., 2009. Connecting phenological predictions with population growth rates for mountain pine beetle, an outbreak insect. Landsc. Ecol. 24, 657–672.

Powell, J.A., Bentz, B.J., 2014. Phenology and density-dependent dispersal predict patterns of mountain pine beetle (*Dendroctonus ponderosae*) impact. Ecol. Model. 273, 173–185.

Powers, J.S., Sollins, P., Harmon, M.E., Jones, J.A., 1999. Plant-pest interactions in time and space: a Douglas-fir bark beetle outbreak as a case study. Landsc. Ecol. 14, 105–120.

Pureswaran, D., Sullivan, B., Ayres, M.P., 2006. Fitness consequences of pheromone production and host selection strategies in a tree-killing bark beetle (Coleoptera: Curculionidae: Scolytinae). Oecologia 148, 720–728.

Pye, J.M., Holmes, T.P., Prestemon, J.P., Wear, D.N., 2011. Economic impacts of the southern pine beetle. In: Coulson, R.N., Klepzig, K.D. (Eds.), Southern Pine Beetle II. USDA Forest Service, Southern Research Station, Asheville, NC, pp. 213–222, Gen. Tech. Rep. SRS-140.

Quiroz, A., Ortega, F., Ramírez, C.C., Wadhams, L.J., Pinilla, K., 2005. Response of the beetle *Hylastinus obscurus* Marsham (Coleoptera: Scolytidae) to red clover (*Trifolium pratense* L.) volatiles in a laboratory olfactometer. Environ. Entomol. 34, 690–695.

Raffa, K.F., 2001. Mixed messages across multiple trophic levels: the ecology of bark beetle chemical communication systems. Chemoecology 11, 49–65.

Raffa, K.F., Berryman, A.A., 1983. The role of host plant-resistance in the colonization behavior and ecology of bark beetles (Coleoptera, Scolytidae). Ecol. Monogr. 53, 27–49.

Raffa, K.F., Berryman, A.A., 1986. A mechanistic computer model of mountain pine beetle populations interacting with lodgepole pine stands and its implications for forest managers. Forest Sci. 32, 789–805.

Raffa, K.F., Berryman, A.A., 1987. Interacting selective pressures in conifer-bark beetle systems—a basis for reciprocal adaptations. Am. Nat. 129, 234–262.

Raffa, K.F., Aukema, B.H., Erbilgin, N., Klepzig, K.D., Wallin, K.F., 2005. Interactions among conifer terpenoids and bark beetles across multiple levels of scale: an attempt to understand links between population patterns and physiological processes. In: Romeo, J.T. (Ed.), Recent Advances in Phytochemistry: Chemical Ecology and Phytochemistry of Forest Ecosystems. Elsevier, London, pp. 79–118.

Raffa, K.F., Aukema, B.H., Bentz, B.J., Carroll, A.L., Hicke, J.A., Turner, M.G., Romme, W.H., 2008. Cross-scale drivers of natural disturbances prone to anthropogenic amplification: the dynamics of bark beetle eruptions. Bioscience 58, 501–517.

Rankin, L.J., Borden, J.H., 1991. Competitive interactions between the mountain pine beetle and the pine engraver in lodgepole pine. Can. J. Forest Res. 21, 1029–1036.

Ranta, E., Lundberg, P., Kaitala, V., 2006. Ecology of Populations. Cambridge University Press, Cambridge.

Reeve, J.D., 1997. Predation and bark beetle dynamics. Oecologia 112, 48–54.

Reeve, J.D., Ayres, M.P., Lorio Jr., P.L., 1995. Host suitability, predation, and bark beetle population dynamics. In: Cappuccino, N., Price, P.W. (Eds.), Population Dynamics: New Approaches and Synthesis. Academic Press, San Diego, pp. 339–357.

Reeve, J.D., Rhodes, D.J., Turchin, P., 1998. Scramble competition in the southern pine beetle, *Dendroctonus frontalis*. Ecol. Entomol. 23, 433–443.

Reeve, J.D., Strom, B.L., Rieske, L.K., Ayres, B.D., Costa, A., 2009. Geographic variation in prey preference in bark beetle predators. Ecol. Entomol. 34, 183–192.

Reeve, J.D., Anderson, F.E., Kelley, S.T., 2012. Ancestral state reconstruction for *Dendroctonus* bark beetles: evolution of a tree killer. Environ. Entomol. 41, 723–730.

Régnière, J., Bentz, B., 2007. Modeling cold tolerance in the mountain pine beetle. Dendroctonus ponderosae. J. Insect Physiol. 53, 559–572.

Royama, T., 1992. Analytical Population Dynamics. Chapman & Hall, New York.

Rudinsky, J.A., 1962. Ecology of Scolytidae. Annu. Rev. Entomol. 7, 327–348.

Rudinsky, J.A., 1966. Host selection and invasion by the Douglas-fir beetle, *Dendroctonus pseudotsugae* Hopkins, in coastal Douglas-fir forests. Can. Entomol. 98, 98–111.

Ruefenacht, B., Finco, M.V., Nelson, M.D., Czaplewsk, R., Helmer, E.H., Blackard, J.A., et al., 2008. Conterminous U.S. and Alaska forest type mapping using forest inventory and analysis dataUSDA Forest Service, Forest Inventory and Analysis, Arlington, VA.

Safranyik, L., Carroll, A.L., 2006. The biology and epidemiology of the mountain pine beetle in lodgepole pine forests. In: Safranyik, L., Wilson, W.R. (Eds.), The Mountain Pine Beetle: A Synthesis of Biology, Management, and Impacts on Lodgepole Pine. Natural Resources Canada, Canadian Forest Service, Pacific Forestry Centre, Victoria, British Columbia, pp. 3–66.

Safranyik, L., Linton, D.A., Shore, T.L., 2000. Temporal and vertical distribution of bark beetles (Coleoptera: Scolytidae) captured in barrier traps at baited and unbaited lodgepole pines the year following attack by the mountain pine beetle. Can. Entomol. 132, 799–810.

Sakamoto, J.M., Gordon, T.R., Storer, A.J., Wood, D.L., 2012. The role of *Pityophthorus* spp. as vectors of pitch canker affecting *Pinus radiata*. Can. Entomol. 139, 864–871.

Santos, M.J., Whitham, T.G., 2010. Predictors of *Ips confusus* outbreaks during a record drought in southwestern USA: implications for monitoring and management. Environ. Manag. 45, 239–249.

Schlyter, F., Anderbrant, O., 1993. Competition and niche separation between two bark beetles: existence and mechanisms. Oikos 68, 437–447.

Schroeder, L.M., 1999. Population levels and flight phenology of bark beetle predators in stands with and without previous infestations of the bark beetle *Tomicus piniperda*. For. Ecol. Manag. 123, 31–40.

Schroeder, L.M., 2007. Escape in space from enemies: a comparison between stands with and without enhanced densities of the spruce bark beetle. Agr. Forest Entomol. 9, 85–91.

Schroeder, L.M., Weslien, J., 1994. Interactions between the phloem-feeding species *Tomicus piniperda* (Col.: Scolytidae) and *Acanthocinus aedilis* (Col.: Cerambycidae), and the predator *Thanasimus formicarius* (Col.: Cleridae) with special reference to brood production. BioControl 39, 149–157.

Schroeder, L.M., Lindelöw, Å., 2002. Attacks on living spruce trees by the bark beetle *Ips typographus* (Col. Scolytidae) following a storm-felling: a comparison between stands with and without removal of wind-felled trees. Agr. Forest Entomol. 4, 47–56.

Seybold, S.J., Huber, D.P.W., Lee, J.C., Graves, A.D., Bohlmann, J., 2006. Pine monoterpenes and pine bark beetles: a marriage of convenience for defense and chemical communication. Phytochem. Rev. 5, 143–178.

Six, D.L., Wingfield, M.J., 2011. The role of phytopathogenicity in bark beetle-fungus symbioses: a challenge to the classic paradigm. Annu. Rev. Entomol. 56, 255–272.

Smith, R.H., 1961. The fumigant toxicity of three pine resins to *Dendroctonus brevicomis* and *D. jeffreyi*. J. Econ. Entomol. 54, 359–365.

Storer, A., Wainhouse, D., Speight, M., 1997. The effect of larval aggregation behaviour on larval growth of the spruce bark beetle *Dendroctonus micans*. Ecol. Entomol. 22, 109–115.

Taylor, S.W., Carroll, A.L., 2003. Disturbance, forest age, and mountain pine beetle outbreak dynamics in BC: a historical perspective. In: Shore, T.L., Brooks, J.E., Stone, J.E. (Eds.), Proceedings of the Mountain Pine Beetle Symposium: Challenges and Solutions. Canadian Forest Service, Victoria, Canada, pp. 67–94.

Tisdale, R.A., Nebeker, T.E., Hodges, J.D., 2003. Role of oleoresin flow in initial colonization of loblolly pine by southern pine beetle (Coleoptera: Scolytidae). J. Entomol. Sci. 38, 576–582.

Trân, J., Ylioja, T., Billings, R., Régnière, J., Ayres, M.P., 2007. Impact of minimum winter temperatures on the population dynamics of *Dendroctonus frontalis*. Ecol. Appl. 17, 882–899.

Turchin, P., 1999. Population regulation: a synthetic view. Oikos 84, 153–159.

Turchin, P., 2003. Complex Population Dynamics: A Theoretical/Empirical Synthesis. Princeton University Press, Princeton.

Turchin, P., Lorio, P.L., Taylor, A.D., Billings, R.F., 1991. Why do populations of southern pine beetles (Coleoptera, Scolytidae) fluctuate? Environ. Entomol. 20, 401–409.

Turchin, P., Davidson, J., Hayes, J.L., 1999a. Effects of thinning on development of southern pine beetle infestations in old growth stands. South. J. Appl. For. 23, 193–196.

Turchin, P., Taylor, A., Reeve, J.D., 1999b. Dynamical role of predators in population cycles of a forest insect: an experimental test. Science 285, 1068–1071.

Vega, F.E., Infante, F., Castillo, A., Jaramillo, J., 2009. The coffee berry borer, *Hypothenemus hampei* (Ferrari) (Coleoptera: Curculionidae): a short review, with recent findings and future research directions. Terrestrial Arthropod Reviews 2, 129–147.

Wainhouse, D., Ashburner, R., Ward, E., Boswell, R., 1998. The effect of lignin and bark wounding on susceptibility of spruce trees to *Dendroctonus micans*. J. Chem. Ecol. 24, 1551–1561.

Wallin, K.F., Raffa, K.F., 2000. Influences of host chemicals and internal physiology on the multiple steps of postlanding host acceptance behavior of *Ips pini* (Coleoptera: Scolytidae). Environ. Entomol. 29, 442–453.

Wallin, K.F., Raffa, K.F., 2004. Feedback between individual host selection behavior and population dynamics in an eruptive herbivore. Ecol. Monogr. 74, 101–116.

Wang, X., Chen, H., Ma, C., Li, Z., 2010. Chinese white pine beetle, *Dendroctonus armandi* (Coleoptera: Scolytinae), population density and dispersal estimated by mark-release-recapture in Qinling Mountains, Shaanxi China. Appl. Entomol. Zool. 45, 557–567.

Webster, F., 1899. The clover root-borer. Ohio Agriculural Experiment Station, Bulletin 112Columbus, OH.

Weed, A.S., Ayres, M.P., Hicke, J.A., 2013. Consequences of climate change for biotic disturbances in North American forests. Ecol. Monogr. 83, 441–470.

Werner, R.A., Holsten, E.H., Matsuoka, S.M., Burnside, R.E., 2006. Spruce beetles and forest ecosystems in south-central Alaska: a review of 30 years of research. For. Ecol. Manag. 227, 195–206.

Weslien, J., 1992. The arthropod complex associated with *Ips typographus* (L.) (Coleoptera, Scolytidae): species composition, phenology, and impact on bark beetle productivity. Entomol. Fenn. 3, 205–213.

Weslien, J., Regnander, J., 1992. The influence of natural enemies on brood production in *Ips typographus* (Col.: Scolytidae) with special reference to egg-laying and predation by *Thanasimus formicarius* (Col.: Cleridae). BioControl 37, 333–342.

Wood, D.L., 1982a. The role of pheromones, kairomones, and allomones in the host selection and colonization behavior of bark beetles. Annu. Rev. Entomol. 27, 411–446.

Wood, S.L., 1982b. The bark and ambrosia beetles of North and Central America (Coleoptera: Scolytidae), a taxonomic monograph. Great Basin Nat. Mem. 6, 1–1359.

Wood, S.L., Bright Jr., D.E., 1992. A catalog of Scolytidae and Platypodidae (Coleoptera), Part 2: Taxonomic Index, Volumes A and B. Great Basin Nat. Mem. 13 (1–833), 835–1553.

Wullschleger, S.D., McLaughlin, S.B., Ayres, M.P., 2004. High-resolution analysis of stem increment and sap flow for loblolly pine trees attacked by southern pine beetle. Can. J. Forest Res. 34, 2387–2393.

Yan, Z.L., Sun, J.H., Don, O., Zhang, Z.N., 2005. The red turpentine beetle, *Dendroctonus valens* LeConte (Scolytidae): an exotic invasive pest of pine in China. Biodivers. Conserv. 14, 1735–1760.

Ylioja, T., Slone, D., Ayres, M.P., 2005. Mismatch between herbivore behavior and demographics contributes to scale-dependence of host susceptibility in two pine species. Forest Sci. 51, 522–531.

Yust, H.R., 1957. Biology and habits of *Pagiocerus fiorii* in Ecuador. J. Econ. Entomol. 50, 92–96.

Zhang, Q.H., Byers, J.A., Schlyter, F., 1992. Optimal attack density in the larch bark beetle, *Ips cembrae* (Coleoptera, Scolytidae). J. Appl. Ecol. 29, 672–678.

Chapter 5

Conifer Defense and Resistance to Bark Beetles

Paal Krokene

Norwegian Forest and Landscape Institute, Ås, Norway

1. INTRODUCTION

1.1 Bark Beetles, Symbionts, and Tree Defenses

Bark beetles include some of the world's most devastating tree killers and some species may kill enormous numbers of trees over huge areas during intermittent outbreaks (Raffa *et al.*, 2008). However, most of the time even notorious tree-killing species remain at low population levels and can only attack weakened and dying trees. One important factor regulating bark beetle populations at low, endemic levels is the trees' elaborate defense systems (Lieutier, 2004; Franceschi *et al.*, 2005; Raffa *et al.*, 2008) (Figure 5.1). In this chapter, I describe the multitude of tree defense mechanisms that bark beetles, or rather the bark beetle–symbiont complex, have to deal with when they attack healthy trees. All bark beetles are associated with a large number of symbiotic organisms, including microorganisms such as fungi and bacteria (Paine *et al.*, 1997; Six and Wingfield, 2011), and consequently, attacked trees are always facing a beetle–symbiont complex. The focus in this chapter is on conifer trees. This is because (1) widespread mass attacks and tree killing only occurs in conifer forests, (2) most bark beetle pests of economic importance attack conifers, and (3) most of the truly aggressive, tree-killing bark beetles in the world colonize conifers (Ohmart, 1989). The focus is further restricted to stem defenses, i.e., defenses residing in the bark and sapwood of the stem, because this is the part of the tree that is attacked by tree-killing bark beetles.

The defenses of a healthy tree are a formidable barrier and the vast majority of the world's tree-colonizing bark beetle species avoid most tree defenses by breeding in dead or dying trees (Raffa *et al.*, 1993; Krokene *et al.*, 2013). Only a very small minority, consisting of less than a dozen so-called aggressive species, are able to breach the potent defenses of healthy trees (Figure 5.2). A couple of dozen additional species are able to kill trees that are more or less

severely stressed by drought or disease (semi-aggressive or facultative parasitic species; Raffa *et al.*, 1993; Krokene *et al.*, 2013). However, most of the time, even the intermittent tree killers among the aggressive and semi-aggressive species are regulated at low population densities and colonize dead or dying trees (Figure 5.2). For example, when its populations are low, the notorious mountain pine beetle *Dendroctonus ponderosae* Hopkins colonizes weakened lodgepole pine trees, following in the wake of non-aggressive bark beetles such as *Pseudips mexicanus* (Hopkins) (Smith *et al.*, 2011).

Bark beetles are associated with a wide diversity of other organisms, including fungi, bacteria, mites, and others (Whitney, 1982; Six, 2013; Chapters 6 and 7). These associations range from mutualistic, such as the mutually beneficial associations between ambrosia beetles and their nutritional fungi (Chapter 6), to detrimental, including interactions with natural enemies (Chapter 7) and competitors (Chapter 4). Some of the associations are symbiotic, where the beetles live in close proximity with other organisms that are carried on or inside the beetle body. Some symbiotic associates of tree-killing bark beetles have been suggested to contribute to overwhelming tree defenses, and thus act as mutualistic associates in tree killing (Craighead, 1928; Franceschi *et al.*, 2005; Lieutier *et al.*, 2009). Historically, research on microorganisms involved in tree killing has focused on ascomycete bluestain fungi (Harrington, 1993; Wingfield *et al.*, 1993; Krokene and Solheim, 1996; Seifert *et al.*, 2013), but several recent papers have demonstrated that bacteria also may play a role in neutralizing tree defenses (Adams *et al.*, 2013; Boone *et al.*, 2013).

The fundamental challenge shared by tree-killing bark beetles and their symbionts is to cope with the powerful defenses of a living conifer tree (Franceschi *et al.*, 2005). During outbreaks, both partners are early successional colonizers of living trees and face potent tree defenses. When aggressive bark beetles attack living trees during epidemics

Bark Beetles. http://dx.doi.org/10.1016/B978-0-12-417156-5.00005-8

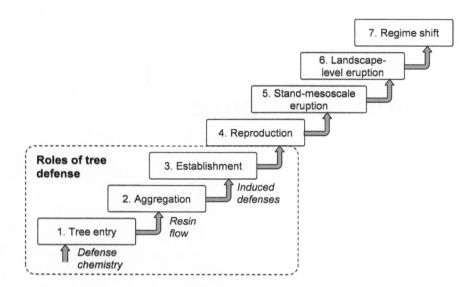

FIGURE 5.1   Tree-killing bark beetles must cross a sequence of thresholds to produce landscape-level eruptions. Tree defenses influence the thresholds at the bottom of this hierarchical chain of threshold processes and are thus an important regulator that can keep beetle populations at low, endemic levels. Managing forests in ways that increase tree resistance may help prevent population buildup and outbreaks of bark beetles. *Adapted from Raffa et al. (2008).*

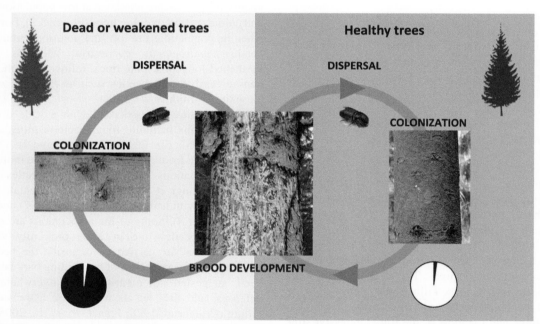

FIGURE 5.2   The epidemic and endemic variant of the bark beetle life cycle. An overwhelming majority of bark beetle species only colonize dead or severely stressed trees (left). Tree-killing species may colonize healthy trees during intermittent epidemics (right), but most of the time they are endemic and utilize dead and weakened trees (left). All bark beetle species depend on dead or dying tissues for brood development. Pie charts show (in black) the proportion of species capable of attacking healthy trees (right) and those restricted to dead/dying trees (left) among bark beetles that colonize trees in northern hemisphere forests. *From Krokene et al. (2013).*

they must be able to neutralize tree defenses in the bark, since the larvae of all bark beetles need dead or dying bark tissues for successful development (Raffa *et al.*, 1993). The evolution of partnerships between tree-killing bark beetles and microbial symbionts may have started when beetles and microorganisms met on the same substrate (Chapter 3). This may have led to the development of a symbiotic relationship and to the evolution of mechanisms that increased the stability of the associations. No matter what the evolutionary origin of the association was, both partners have adapted the ability to deal with tree defenses, after millions of years of coevolutionary interaction with conifer trees. Extant species of conifers and bark beetles have likely coexisted for at least three million years (Seybold *et al.*, 2000), providing ample time for coevolution of subtle species interactions. Many of the adaptations that enable bark beetles and symbionts to colonize living trees seem to complement each other, and each partner may thus be able to reduce the difficulties encountered by the other in tree colonization. Thus, the two partners can be hypothesized to constitute an additive or synergistic tree-killing complex (Franceschi *et al.*, 2005; Lieutier *et al.*, 2009; Krokene *et al.*, 2013).

1.2 The Bark Beetle Life Cycle, Tree Colonization, and Mass Attack

Bark beetles spend most of their life inside their host plant or in hibernation sites, such as the forest litter. The bark beetle life cycle can be divided into three main phases: the dispersal phase, the colonization phase, and the development phase (Wood, 1972; Borden, 1982). During the dispersal phase, the new generation emerges from the brood tree or hibernation site and flies to a new tree (Figure 5.2). The colonization phase includes host selection and tunneling into the bark. Tree colonization begins when a beetle lands on and enters the bark of an unattacked tree. These so-called pioneer beetles (Alcock, 1982; Latty and Reid, 2009) must assess the tree's defensive capability and nutritional quality, but we know very little about the precise mechanisms involved in bark beetle host recognition and acceptance. Inside the bark, the pioneers emit pheromones that attract additional attackers, a process known as secondary attraction (Wood, 1982; Vanderwel and Oehlschlager, 1987). The development phase occurs within the bark and includes mating, gallery construction, oviposition, and brood development. For bark beetles that attack live trees, the defenses in the bark must be neutralized when brood development begins. Most of the vascular cambium will then be destroyed, and the tree is usually irreversibly stressed (Lieutier *et al.*, 2009). However, the sapwood may still be functional, the canopy alive with green foliage, and it may take weeks or months until the canopy fades and the whole tree dies (Paine *et al.*, 1997).

Bark beetles are minuscule in size compared to their often very large host trees, and are dwarfed by a factor of 1 to 10^8 or more in terms of body volume. Thus, a single beetle cannot inflict much damage to a tree. The key to tree killing by bark beetles is that the beetles engage in coordinated mass attacks where hundreds of them attack a single tree within a short time (Chapters 1 and 4). Such coordinated mass attacks may overwhelm the defense capacity of even vigorous trees (Berryman, 1982; Mulock and Christiansen, 1986). Bark beetle mass attacks are coordinated by aggregation pheromones that may be made from precursors in the tree or produced *de novo* by the beetles (Blomquist *et al.*, 2010). The pheromone plume emitted by the initial attackers recruits additional non-pioneer attackers among the flying beetles in the surroundings. In many species, attraction to pheromones is synergized by monoterpenes emitted from the attacked trees, at least up to moderately high monoterpene levels (Rudinsky *et al.*, 1971; Miller and Borden, 2000; Erbilgin *et al.*, 2007). The adaptive explanation for this synergism may be that the combination of pheromones and host defense compounds signals a successful ongoing attack on a vigorous, high quality host tree. Many bark beetle species that do not colonize live trees also produce aggregation pheromones and engage in mass attacks (discussed in Franceschi *et al.*, 2005). Mass attack in such species is probably a result of breeding in a resource that can support many individuals once it is located, but that cannot easily be monopolized by one or a few individuals (Kirkendall *et al.*, 1997).

A tree can successfully defend itself against a certain number of simultaneous attacks, but if the number of attackers exceeds this threshold, tree defenses at each attack site are insufficient to fend off the attackers (Christiansen, 1985; Christiansen *et al.*, 1987). This simple relationship between tree resistance and the number of attacking beetles is the basis for the conceptual model of the threshold of successful attack, which states that every tree has a critical threshold density of beetle attacks it can withstand (Thalenhorst, 1958; Berryman, 1982). The model can be extended to the forest stand and beetle population level: for any given combination of bark beetle population density and average tree resistance in the stand the beetle–host tree system will be either in the epidemic phase (where the beetles can successfully overcome tree resistance) or the endemic phase (where the trees resist beetle attack and the beetles must breed in dead or severely weakened trees). Because the beetles overwhelm tree defenses by engaging in pheromone-mediated mass attacks they have an advantage in numbers, and the higher the beetle population density is, the easier it is for beetles to successfully colonize a tree (Boone *et al.*, 2011). The capacity of mass attacks to exhaust tree defenses is probably reinforced by the ability of the beetles' microbial symbionts to engage tree defenses (Paine *et al.*, 1997; Franceschi *et al.*, 2005; Lieutier *et al.*, 2009, but see Six and Wingfield (2011) for an opposite view). Plant pathogenic microbial symbionts probably increase the impact of each beetle attack on exhausting tree resistance, thereby lowering the threshold of successful attack. Pheromone-coordinated mass attacks and phytopathogenic symbionts may thus act synergistically to overwhelm tree defenses.

1.3 Conifers and their Defenses

The conifer species that exist today are relicts of a once more ecologically dominant group. The total number of extant conifer species is only about 630, and several genera and families contain only a handful of species (Farjon, 2001). Taxonomically, all conifer species reside in the division Pinophyta, which contains a single class (Pinopsida) with a single order (Pinales or Coniferales) containing seven to eight families. Of these, the pine family (Pinaceae) is the most species rich and geographically widespread. The pine family contains about 225 species in 11 genera, including pine (*Pinus*), spruce (*Picea*), larch (*Larix*), and fir (*Abies*). These genera include many of the world's most important commercial timber species. Other species-rich and ecologically important conifer families are the Cupressaceae (with around 135 species), the Podocarpaceae ($\approx$185 species), and the Araucariaceae (41 species), all of which include many tropical or southern hemisphere species (Farjon, 2001). Although the global ecological dominance of conifers is much reduced from its peak, many extant species are very abundant. They dominate many boreal, temperate, and mountainous ecosystems and play important ecological and economic roles. Part of the reason for the continued ecological success of many conifer species appears to be their ability to defend themselves effectively against natural enemies.

The conifers include some of the largest and oldest organisms on Earth (Krokene *et al.*, 2008b), and such large and long-lived organisms are highly apparent and bound to be located by most natural enemies (Feeny, 1976). To survive they must be masters of defense and conifers integrate multiple preformed and inducible defenses into a coordinated, multi-purpose defense strategy (Phillips and Croteau, 1999; Franceschi *et al.*, 2005; Keeling and Bohlmann, 2006). These defenses impede fungal establishment and inflict considerable beetle mortality during tree colonization. Conifer defenses generally seem to be of a quantitative and general nature, and may be considered relatively unsophisticated compared to the elaborate gene-for-gene interactions in angiosperm systems. However, the multi-purpose defenses of conifers are effective against a wide range of potential attackers and have stood the test of time through 200 million years of exposure to natural enemies.

"Plant resistance" includes all plant traits that decrease the performance or host preference of an aggressor, whereas "plant defense" may be defined as the subset of plant resistance traits that actually benefits the plant and increases its fitness (Karban and Myers, 1989). The definition of plant resistance/defense is thus context dependent, since it is focusing on the negative effects a putative resistance or defense trait has on an attacking organism (Karban and Baldwin, 1997; Larsson, 2002). Plant resistance or defense mechanisms are manifested as the interaction between a plant trait (the resistance or defense trait) and the aggressor's response to this trait (Larsson, 2002). Strictly speaking, plant resistance and defense can therefore be precisely characterized only if both the plant trait and the aggressor are studied. However, it is convenient to classify all traits that negatively affect at least some aggressors (or that on logical grounds can be supposed to do so) as resistance or defense traits (Larsson, 2002). For example, the defensive role of anatomically based defenses in conifer bark is often inferred from their spatial distribution within the bark and sometimes from results of inoculation/pathogenicity trials (Franceschi *et al.*, 2000, Hudgins *et al.*, 2003b).

The ultimate function of conifer stem defenses is to maintain the tree's integrity by protecting the nutrient-rich bark, the vascular cambium, and the transpiration stream in the sapwood. The first line of conifer defenses consists of preformed (constitutive) defenses with a mechanical or chemical mode of action. Mechanical defenses are structural elements that deter invaders by providing toughness or thickness to tissues. This may involve impregnation of tissues with lignin and suberin polymers that enhance resistance to penetration, degradation, and ingestion (Franceschi *et al.*, 2005). Chemical defenses include substances with toxic or inhibitory effects, such as specialized plant metabolites, proteins, and enzymes (Franceschi *et al.*, 2005; Keeling and Bohlmann, 2006). Since defense is costly and drains resources away from other vital plant functions, such as growth and reproduction (Herms and Mattson, 1992), not all defenses are expressed constitutively. Trees have therefore evolved the capacity to up-regulate additional, inducible defenses when they are attacked. The integration of preformed and inducible defenses provides a cost-effective and flexible response to attack (Steppuhn and Baldwin, 2008; Cipollini and Heil, 2010). Preformed defenses inhibit initial attacks, and inducible defenses ensure that possible invasions are recognized and dealt with effectively.

A successful tree defense reaction, integrating preformed and inducible defenses, typically goes through one to four successive stages (Franceschi *et al.*, 2005). The first stage is to repel or inhibit attacks by means of effective preformed defenses. In many cases, this may be sufficient, but if the attackers are not deterred by the preformed defenses, the next stage is to kill or compartmentalize the attackers by initiating inducible defenses. Usually, the first two stages operate in parallel, since inducible defenses are initiated immediately after an attacking organism penetrates into the living bark and comes in contact with preformed defenses there. The third defense stage is to seal and repair the damaged area to ensure that the tree can continue to function normally and to prevent secondary infections by opportunistic organisms. Finally, acquired resistance may be induced locally and systemically so that future attacks can be dealt with more effectively (Krokene *et al.*, 2008b; Eyles *et al.*, 2010).

2. ANATOMICAL AND CHEMICAL COMPONENTS OF CONIFER DEFENSES

2.1 Anatomical Layout of Conifer Defenses

Many of the defense mechanisms in conifer stems have a clear anatomical basis and reside in characteristic anatomical structures. Here I will describe the basic anatomy and development of the cell and tissue types involved in conifer stem defenses, beginning at the stem surface and moving towards the heartwood (Figure 5.3). The defense mechanisms residing in these cells and organs and their effect on insects and microorganisms will be treated in more detail in Section 3. A comprehensive general overview of the anatomy of conifer bark defenses can be found in Franceschi *et al.* (2005). More restricted reviews on the

anatomy of specific defense structures, such as resin ducts and polyphenolic parenchyma cells, can be found in Krokene *et al.* (2008a) and Krokene and Nagy (2012).

2.1.1 General Anatomy of Conifer Stems

Conifer stem tissues are highly organized, with different cell types laid out in concentric layers around the stem circumference (Franceschi *et al.*, 2005). Many of these cell types play important roles in conifer defense. Before we go into the different cell types involved in defense, I will briefly summarize the main anatomical structure of conifer stems. Conifer stems are made up of three main tissue regions: the phloem or bark, the water-conducting sapwood, and an inner core of supporting, non-conducting heartwood. The bark includes all tissues outside the vascular cambium

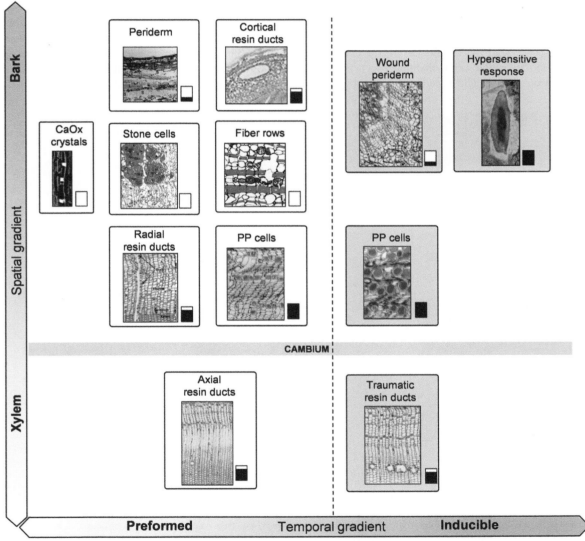

FIGURE 5.3 The major anatomically based defenses in conifer stems ordered along a spatial (bark surface to sapwood) and temporal gradient (preformed to inducible). The small black and white boxes indicate to what extent a defense structure's mode of action is mechanical (white) or chemical (black).

and is anatomically more complex than the rather simple wood. The bark includes several different cell types and can be subdivided into the outer periderm, the cortex (representing primary phloem growth), and the secondary phloem. The secondary phloem is the largest section of the bark and is a major site of bark defenses (Franceschi *et al.*, 2005). It also contains the sieve cells, the tubing system that distributes photosynthates from the needles to the roots and other parts of the tree. The thin vascular cambium at the interface of the sapwood and bark is only a few cell layers thick, but is the main source of new cells in the stem. It produces bark cells towards the outside of the stem and wood cells (tracheids) towards the inside. If the cambium is destroyed by tunneling bark beetles or invading fungi, the tree is no longer able to produce new cells in the damaged area. If the cambium is destroyed around the full stem circumference, the tree will eventually die, although it may stay alive for months or years until the roots die, water flow ceases, and the foliage fades. Beneath the cambium is the sapwood, which is structurally relatively simple and largely consists of dead, water-conducting tracheid cells. These elongated tube-like cells channel water from the roots to the tree crown. The innermost tissue region of

the stem, the heartwood, is structurally similarly to the sapwood but consists of only dead cells and the tracheids are no longer functional and do not transport water. Most of the cell types in the bark and wood are laid out in concentric rings around the stem circumference. In addition to these concentrically arranged cell and tissue layers, there is one tissue type, the radial rays, that runs radially and spans from the secondary phloem into the sapwood.

2.1.2 Periderm—the Outer Defense Layer

The first defense line in conifer stems is the outer surface of the bark, the periderm. This is a multi-purpose complex made up of several cell layers that provides resistance to penetration by insects and fungi (Franceschi *et al.*, 2005). It also protects the tree from abiotic disturbances, such as desiccation and fire. Inside the periderm is the cork cambium (or phellogen), a secondary meristem that produces cork tissue (phellem) outwards and secondary cortex (phelloderm) inwards. Cork is the brownish dry bark layer that can be several centimeters thick especially in old pine trees, but may be comparably thin in many other conifers (Figure 5.4G). It consists of multiple layers of mostly dead

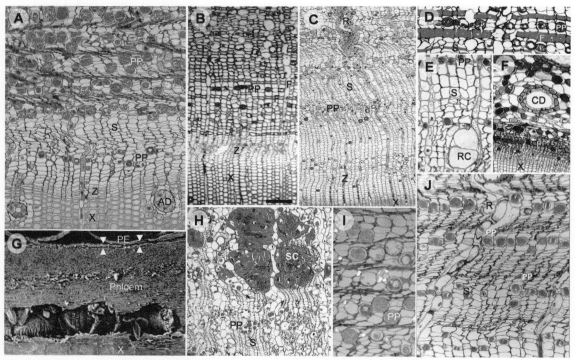

FIGURE 5.4 **Anatomy of preformed defenses in the bark and sapwood of conifer stems (cross-sections).** (A) Bark and sapwood of Scots pine (*Pinus sylvestris*) with scattered axial resin ducts and PP cells. (B) Bark and sapwood of giant sequoia (*Sequoiadendron giganteum*) with closely spaced layers of PP cells, fiber cells, and sieve cells. (C) Bark and sapwood of lodgepole pine (*Pinus contorta*) with PP cells and a large radial ray containing a radial resin duct. (D) Bark of western red cedar (*Thuja plicata*) with layers of PP cells, sieve cells, and fibers. (E) Bark of grand fir (*Abies grandis*) with PP cells and large resin cell. (F) Bark and sapwood of young balsam fir (*Abies balsamea*) with large cortical resin duct in the outer bark (cortex). (G) Bark of Norway spruce (*Picea abies*) with tunneling bark beetles (radial section). Arrowheads mark the very thin periderm layer with cork bark on the outside. (H) Bark of blue spruce (*Picea pungens*) with large stone cell aggregates. (I) Scots pine bark with PP cells and calcium oxalate crystals (white). (J) Norway spruce bark with PP cells and a radial ray. Abbreviations: AD=axial resin duct, CD=cortical resin duct, F=fiber row, PE=periderm with cork bark, PP=polyphenolic parenchyma cells, R=radial ray, RC=resin cell, S=sieve cells, SC=stone cells, X=xylem, Z=vascular cambium.

cells with lignified or suberized walls, usually encrusted with calcium oxalate crystals, and often containing large amounts of phenolic materials (Franceschi *et al.*, 2005) (Figure 5.3). The periderm is not a continuous armor covering the entire external surface of the bark, but is penetrated by loose cell aggregates, called lenticels, that allow gas exchange over the bark surface. The lenticels are not wide open but represent a weak point in the outer defense and may be used as entry points by small bark beetles, such as the six-toothed bark beetle *Pityogenes chalcographus* L. colonizing Norway spruce (Rosner and Führer 2002).

2.1.3 Cortex—Defense in Young Tissues

The cortex is situated just inside the phelloderm, the innermost part of the periderm, and is an important defensive barrier in young tissues. It is produced during the primary development of the stem and is usually maintained for several years of secondary growth. However, as the secondary phloem underneath it is gradually crushing and pushing the cortex outwards, the cortex's function is gradually taken over by the secondary phloem. Anatomically, the cortex consists of large, undifferentiated cells with characteristic phenolic inclusions, and is sometimes interspersed with sclerenchyma and calcium oxalate crystals. In addition, many species in the pine family have very large axial resin ducts in the cortex (Alfaro *et al.*, 1997; Franceschi *et al.*, 2005; Krokene *et al.*, 2008b) (Figures 5.3 and 5.4F). These axial resin ducts are formed in a circumferential ring as the cortex is produced during primary stem development and are the primary source of resin in young stems (Abbot *et al.*, 2010). They can remain functional for at least 25 years (Christiansen *et al.*, 1999a).

2.1.4 Secondary Phloem—the Primary Site of Stem Defenses

Most of the living bark of mature trees is made up by the secondary phloem, which is the main site of preformed and inducible defenses in conifer stems. Three cell types in the secondary phloem play important roles in defense in most conifer species: (1) lignified sclerenchyma cells, (2) cells with calcium oxalate crystals, and (3) polyphenolic parenchyma (PP) cells with characteristic polyphenolic inclusions (Franceschi *et al.*, 2005) (Figure 5.3). In addition to these concentrically arranged cell types, the secondary phloem contains the radial rays with associated resin ducts that extend radially into the sapwood (Franceschi *et al.*, 1998, Krekling *et al.*, 2000) (Figure 5.3). Furthermore, resin-producing structures in the form of resin ducts, blisters, or resin cells (Figure 5.4A, E) are a prominent feature in many species in the pine family but are generally absent in non-pine conifers. The anatomical structure and development of all these different cell types are described in more detail below.

The secondary phloem in the pine family on the one hand and all other conifer families on the other hand have a highly organized concentric arrangement of cell types, but there are important anatomical differences between the two groups. In the pine family, concentric rings of polyphenolic parenchyma (PP) cells, which play major roles in preformed and inducible defenses, are separated by multiple (9–12) rows of sieve cells and associated albuminous cells interspersed with a few scattered stone cells (Franceschi *et al.*, 1998; Krekling *et al.*, 2000) (Figure 5.4J). In the non-pines, the block of sieve cells that separate the PP cells is only a few cell rows thick and is intersected by a tangential row of fiber cells (Figure 5.4B, D) (Hudgins *et al.*, 2003a, 2004).

2.1.4.1 Polyphenolic Parenchyma (PP) Cells

The most abundant type of living cells in the secondary phloem is the PP cells, and the PP cells' ubiquitous occurrence and multiple defensive roles make them the single most important cell type in conifer defense (Franceschi *et al.*, 2005). They seem to be equally important in both pine and non-pine conifers (Franceschi *et al.*, 1998, 2000; Hudgins *et al.*, 2003a, 2004). Polyphenolic parenchyma cells are specialized for synthesis and storage of phenolic compounds and contain very large vacuoles with characteristic polyphenolic content (Figure 5.4A, I, J). Considering their abundance and importance in conifer defense, surprisingly little was known about their structure and function before 1998, when Franceschi and co-workers published the first in a series of papers on PP cell anatomy, development, and roles in preformed and inducible defenses (Franceschi *et al.* 1998, 2000; 2005; Krekling *et al.*, 2000, 2004; Krokene *et al.*, 2003; Hudgins *et al.*, 2003a, 2004; Nagy *et al.*, 2004). Although Norway spruce has served as a model species for most of the research on PP cells, studies on non-pine conifers have demonstrated the general importance of PP cells in conifer defense (Hudgins *et al.*, 2003a, 2004).

PP cells are produced during normal phloem development, but undergo extensive changes in response to attack, and are thus a major site of both preformed and induced defenses (Franceschi *et al.*, 1998, 2000) (Figure 5.3). At the start of each growth season, a new concentric layer of PP cells is differentiated from cells produced at the end of the previous growth season (Krekling *et al.*, 2000). Thus, at least in temperate conifers the PP cells form annual rings in the phloem (Alfieri and Evert, 1973; Krekling *et al.*, 2000). Each annual ring forms an almost complete cylinder or curtain that extends around the circumference of the tree (Figure 5.4J). Norway spruce trees begin to produce PP cells at the earliest stages of primary growth, and by the second year of growth, trees produce regular annual PP cell layers (Krekling *et al.*, 2000). Interestingly, a new PP cell layer is also produced quickly after major disruptions of cambial activity in Norway spruce, such as insect or pathogen attack (Krekling *et al.*, 2000, 2004; Krokene *et al.*, 2003).

Maturation of PP cells proceeds slowly and may take 5 years or more in Norway spruce (Krekling *et al.*, 2000). As the PP cells mature they increase in size and gradually become more rounded in cross-section, the cell walls thicken, and the phenolic bodies become more extensive. Additional PP cells may develop from undiffer-entiated axial parenchyma cells in between the regular annual layers during the first 5–8 years after formation (Krekling *et al.*, 2000). As the PP cells gradually increase in size during maturation, the surrounding sieve cells are crushed and compacted and eventually become non-functional (Figure 5.4A). The PP cells themselves, however, stay alive for many years. For example, in 100-year-old Norway spruce trees PP cell rows formed more than 70 years ago contained living cells (Krekling *et al.*, 2000). Polyphenolic parenchyma cells in older phloem layers in the outer parts of the secondary phloem may undergo cell division, and this is necessary to maintain complete layers of PP cells as the stem diameter increases during growth (Krekling *et al.*, 2000). New PP cells may also be recruited in the older layers of the sec-ondary phloem by differentiation of existing residual parenchyma cells or ray cell derivatives (Fahn, 1990; Krekling *et al.*, 2000).

2.1.4.2 Lignified Sclerenchyma Cells and Cells with Calcium Oxalate Crystals

In addition to the very dynamic PP cells, the secondary phloem contains some cell types with inert mechanical defenses. One such cell type, the sclerenchyma cells, has lignified secondary wall thickenings. Sclerenchyma cells occur as large stone cells in members of the pine family, or as rows of fiber cells in non-pine conifers (Franceschi *et al.*, 2005). Stone cells are massive, irregularly shaped cells with very thick lignified cell walls and occur as single cells or in small clusters in all species in the pine family and in the genus Araucaria among the non-pines (Figure 5.4H) (Franceschi *et al.*, 2005). The spatial distri-bution of stone cells in the secondary phloem suggests that they are derived from PP cells (Franceschi *et al.*, 2005). Sclerenchyma cells in non-pines occur as densely spaced concentric rows of fiber cells (Figure 5.4B, D). These cells are not derived from PP cells but develop from a layer of precursor cells. Fiber cells need 2–3 years to become fully lignified (Hudgins *et al.*, 2004; Franceschi *et al.*, 2005).

In addition to the lignified sclerenchyma cells, the sec-ondary phloem of all conifers contains cells with calcium oxalate crystals inside or outside their cell walls. Although such crystals are present in all conifers, they are much more prominent in non-pine species such as the

Taxodiaceae, where many cell walls in the secondary phloem are covered by small extracellular crystals. In the pine family large calcium oxalate crystals are found inside modified PP cells (Hudgins *et al.*, 2003b) (Figure 5.3). Unlike normal PP cells, these modified cells are dead at maturity and have suberized cell walls. Calcium oxalate crystals may also occur in the vacuole of regular PP cells along with the polyphenolic material (Franceschi *et al.*, 1998).

2.1.4.3 Resin-producing Structures

The secondary phloem of all members of the pine family contains preformed resin structures, in the form of resin cells or radially oriented resin ducts. Resin ducts are long intercellular spaces lined with plastid-enriched epithelial cells that produce and secrete resin into the duct lumen, where it is stored under pressure (Charon *et al.*, 1987; Gershenzon and Croteau, 1990; Nagy *et al.*, 2000). Resin ducts in the secondary phloem are always oriented radially, and are located within the multiseriate radial rays (Fahn *et al.*, 1979). Resin ducts form schizogenously as the epi-thelial cells pull apart during resin duct formation (Nagy *et al.*, 2000). The simpler resin cells accumulate resin inter-nally under pressure and may expand into quite large struc-tures (Figure 5.4E). The epithelial cells lining the resin ducts are usually thin-walled and long-lived, in contrast to the epithelial cells of resin cavities, which are short-lived and gradually become lignified during development (Bannan, 1936; Fahn, 1979).

2.1.4.4 Radial Rays

The radial rays consist of parenchyma cells forming radial plates and are the only cell type in conifer stems that are radially oriented (Figure 5.4C, J) (Franceschi *et al.*, 2005). Radial rays may be either multiseriate or uniseriate. Multiseriate rays are spindle-shaped in tangential sections, being several cell layers thick in the middle and a single cell layer thick at the ends (Nagy *et al.*, 2006). In the middle, they contain a single radial resin duct. The mu-ltiseriate rays are uniformly distributed in the secondary phloem and sapwood. Their density in tangential sapwood sections ranges from 0.15–0.7 ducts mm^{-2} in *Larix* to 0.5–2.0 ducts mm^{-2} in *Pinus* (Wu and Hu, 1997). The much more abundant uniseriate rays typically consist of a single row of 5–8 parenchyma cells without any associated resin duct. At their outermost end in the secondary phloem, the resin duct in the multiseriate rays is enlarged into a cyst-like vesicle. The inner end of the subset of radial resin ducts that extend into the sapwood is connected to an axial resin duct. The lumen of the radial resin ducts appears to be closed in the cambial region (Fahn *et al.*, 1979).

2.1.5 The Vascular Cambium—a Defenseless Cell Factory

The vascular cambium is the main meristem in the stem, producing undifferentiated wood cells inwards and bark cells outwards. The thickness of the vascular cambium varies from around six cells during dormant periods to around 14 during the most active periods of growth (Figure 5.4A–C). Being a meristem the cambium consists of flattened, undifferentiated cells. These undifferentiated cells possess no defense capabilities, although the cambium quickly can be reprogrammed to produce cells that are differentiated into PP cells or traumatic resin ducts. Since the cambium itself is defenseless, but crucial for maintaining stem growth and tree integrity, it must be protected by the different defense structures in the secondary phloem, cortex, and periderm.

2.1.6 Sapwood—Tracheids, Rays, and Resin Ducts

The sapwood is mostly made up of dead cells and is much less active in defense than the bark. The two only defense structures that are found constitutively in the sapwood are axial resin ducts and radial rays (Krokene et al., 2013). Axial resin ducts are usually found in the outer region of the earlywood and in the first-formed latewood in each annual ring of sapwood growth (Fahn et al., 1979; Wu and Hu, 1997). Under normal conditions there are only a few, scattered axial resin ducts in the sapwood (Figure 5.4A), but abundant so-called traumatic resin ducts may be produced in response to various stresses, such as wounding, infection, insect attack (Figure 5.5), or drought (Kane and Kolb, 2010). Traumatic resin ducts are described in more detail under induced defenses, below.

Pines have more abundant axial resin ducts in the sapwood than most other conifers, with a density of 4–5 ducts mm^{-2} (compared with 2.4–3.5 ducts mm^{-2} in Picea and 0.6–4.1 in *Pseudotsuga*) (Wu and Hu, 1997). Individual ducts can be more than 40 cm long in old pine trees (Bannan, 1936; Reid and Watson, 1966). Different axial ducts may be connected via the radial resin ducts through connection of their duct lumen. These connections only occur in the same radial plane; therefore, strictly speaking, the network of resin ducts found in conifer sapwood is only two dimensional (Fahn et al., 1979).

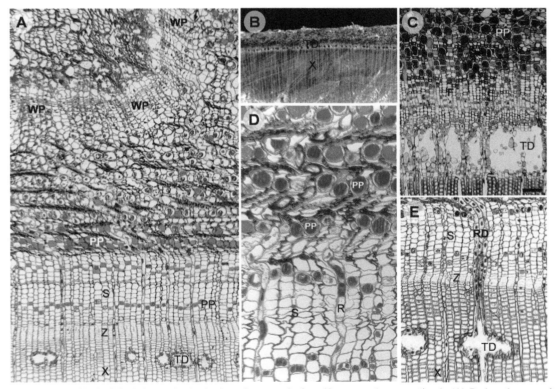

FIGURE 5.5   Anatomy of inducible defenses in the bark and sapwood of conifer stems (cross sections). (A) Bark and sapwood of Norway spruce (*Picea abies*) with traumatic resin ducts, activated PP cells, and wound periderm formation. (B) Low magnification of a Norway spruce stem with a continuous ring of traumatic resin ducts in the young sapwood after treatment with methyl jasmonate. (C) Bark and sapwood of giant sequoia (*Sequoiadendron giganteum*) with very large traumatic resin ducts and activated PP cells after treatment with methyl jasmonate. (D) Bark of Norway spruce with activated PP cells. (E) Bark and sapwood of Norway spruce with interconnection between a radial and traumatic resin duct. Abbreviations: PP = polyphenolic parenchyma cells, R = radial ray, RD = radial resin duct, S = sieve cells, TD = traumatic resin duct, WP = wound periderm, X = xylem, Z = vascular cambium.

2.1.7 Pines vs. Non-pines—Two Fundamentally Different Defense Strategies

As described above, there are several important differences in the stem defenses of members of the pine family on the one hand and all other conifer families (the non-pines) on the other hand. The division between the species-rich pine family (ca. 230 species) and the other conifer families (ca. 400 species) appears to represent a major split in the evolution of conifers, separating fundamentally different defense strategies. The two groups differ in their reliance on resin-based defenses, the nature of their sclerenchyma cells, and the nature and layout of calcium oxalate crystals (Hudgins *et al.*, 2003b; Franceschi *et al.*, 2005). The division between pines and non-pines is also a division between conifers that regularly experience outbreaks of tree-killing bark beetles (the pines) and those that are generally spared from such attacks (the non-pines).

Most strikingly, the non-pine conifers in the families Podocarpaceae, Araucariaceae, Taxaceae, Taxodiaceae, and Cupressaceae have little or no resin-based stem defenses. Most species in these families altogether lack preformed resin structures in the bark and sapwood (Bannan, 1936; Fahn, 1979; Wu and Hu, 1997), and inducible resin structures are only present in some genera (Hudgins *et al.*, 2004; Franceschi *et al.*, 2005) (Figure 5.6). The non-pine conifers seem instead to rely on a combination of massive preformed bark defenses such as PP cells, abundant sclerenchymatic fiber cells, and extracellular calcium oxalate crystals (Franceschi *et al.*, 2005; Hudgins *et al.*, 2003a,b) (Figure 5.6). Species in the pine family on the other hand have a combination of preformed and/or inducible resin producing structures in the bark and/or the sapwood, stone cell sclerenchyma in the bark, and scattered intracellular calcium oxalate crystals in the bark (Franceschi *et al.*, 2005) (Figure 5.6).

It may seem paradoxical that the pines, having the most elaborate resin-based defenses, are also most vulnerable to destructive bark beetle outbreaks, whereas the non-pines, relying more heavily on mechanical defenses, rarely suffer from bark beetle attacks (Hudgins *et al.*, 2004; Franceschi *et al.*, 2005). The answer to this paradox is probably coevolution (Franceschi *et al.*, 2005). When selecting trees for attack, tree-killing bark beetles are attracted to the volatile terpenoids in the tree's resin, in combination with beetle-produced pheromones that are often derived from the same host volatiles. The strong reliance on resin-based defenses has thus been both a blessing and a curse for the members of the pine family. The pines are obviously ecologically very successful, being the most species-rich conifer family with many very abundant species, and their resin-based defenses have probably contributed to their ecological success. However, resin-based defenses have also made the trees vulnerable to coevolved insect pests that have turned the trees' primary chemical defense to their own advantage, illustrating how coevolution can turn a defense into a weakness. Still, since most conifers, both pines and non-pines, are able to resist bark beetles most of the time, these contrasting strategies may also serve as an example of how very different defense strategies can be effective against the same pest (Franceschi *et al.*, 2005).

2.2 Chemical Traits of Conifer Defenses

Terpenes and other specialized metabolites stored in resin ducts and PP cells play important roles in conifer resistance against bark beetles and fungi (Raffa and Berryman, 1983; Franceschi *et al.*, 2005; Keeling and Bohlmann, 2006). Terpenes are considered to be important in tree defense since they occur in large quantities, are metabolically costly to produce and maintain, and are toxic or inhibitory to both bark beetles and their symbionts (Reid *et al.*, 1967; Raffa *et al.*, 1985; Gijzen *et al.*, 1993; Gershenzon, 1994; Keeling and Bohlmann, 2006). The highly viscous and sticky resin is stored under pressure and can physically flush out or trap invading organisms and be involved in wound sealing (Bordasch and Berryman, 1977; Gijzen *et al.*, 1993). The two classes of specialized metabolites that have been most studied with respect to conifer resistance against the bark beetle–symbiont complex are terpenoids and phenolics.

Conifers in the pine family produce a structurally diverse mixture of terpenes stored in resin ducts or cavities in the bark and sapwood (Phillips and Croteau, 1999; Keeling and Bohlmann, 2006). Although terpenes are important in conifer resistance due to their physical and chemical properties (Keeling and Bohlmann, 2006), some terpenes are also essential in bark beetle tree colonization since they are used as primary chemical host-finding cues and/or precursors for the beetle's aggregation pheromones (Wood, 1982; Blomquist *et al.*, 2010). Next to terpenes, phenolics are the group of specialized metabolites that have been most extensively investigated for their role in conifer resistance. Phenolic compounds are abundant in the secondary phloem and may be toxic or repellent to bark beetles and microorganisms.

2.2.1 Terpenes

Terpenoid resin is a complex mixture of monoterpenes (C_{10}), sesquiterpenes (C_{15}), and diterpenes (C_{20}), with small amounts of other compounds (Phillips and Croteau, 1999; Keeling and Bohlmann, 2006). The basic building blocks of terpenes are C_5 units that are fused to produce monoterpenes, sesquiterpenes, diterpenes, and higher-order terpenes. Conifer resin is composed of large and roughly equal proportions of mono- and diterpenes, with a much smaller fraction of sesquiterpenes. When a tunneling bark beetle ruptures the resin ducts in the bark, it releases resin, which is stored under pressure. This resin flow may

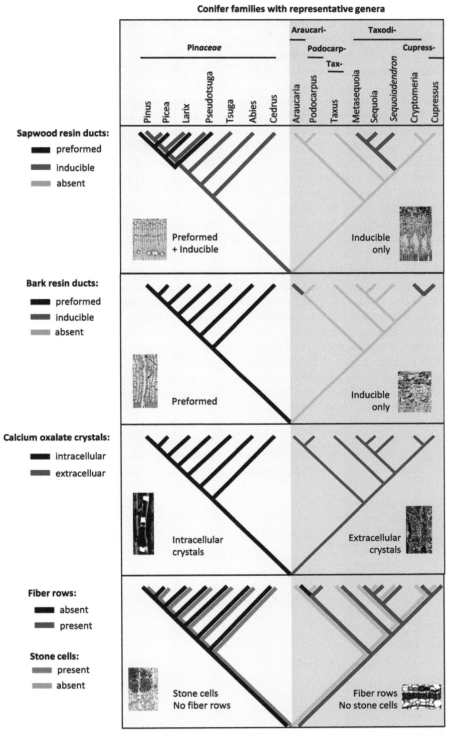

FIGURE 5.6 **Conifers form two phylogenetic groups with fundamentally different defense strategies.** Species in the pine family (left) generally have elaborate resin-based defenses in the bark and sapwood, large intracellular calcium oxalate crystals, and large stone cells but no fiber rows in the bark. Species in all the other conifer families (right) have no preformed (but sometimes inducible) resin ducts in the bark and sapwood, and small abundant extracellular calcium oxalate crystals and closely spaced fiber rows in the bark. *Adapted from Franceschi* et al. *(2005).*

physically force the beetle out of the bark. Upon exposure to air, the volatile monoterpenes in the resin evaporate, while the diterpenes polymerize and contribute to wound sealing. Terpenoid resin thus combines chemical toxicity and repellency with important mechanical defense properties.

Terpenes constitute the largest group of natural plant products, with about 30,000 known compounds (Keeling and Bohlmann, 2006). Most of this enormous chemical diversity is derived from three simple building blocks, called prenyl diphosphates, that are subsequently modified into many different mono-, sesqui- and diterpenes. The C_{10} monoterpene building block is geranyl diphosphate (GPP); the C_{15} sesquiterpene building block is farnesyl diphosphate (FPP); and the C_{20} diterpene building block is geranyl-geranyl diphosphate (GGPP) (Keeling and Bohlmann, 2006) (Figure 5.7). These C_{10} to C_{20} prenyl diphosphate building blocks are formed by the fusion of two types of C_5 units; one unit of dimethylallyl diphosphate (DMAPP) and one, two or three units of isopentenyl diphosphate (IPP). Specific enzymes known as prenyltransferases (or isoprenyl diphosphate synthases (IDS)) catalyze the formation of the different prenyl diphosphate building blocks (GPP, FPP, and GGPP synthase; Figure 5.7). These linear molecules are then used as substrates by different terpene synthases (mono-, sesqui-, and diterpene synthases) and cytochrome P450 monooxygenases (Cyp P450) to form the huge diversity of terpenes present in conifers (Keeling and Bohlmann, 2006).

The composition of the terpenoid resin of individual trees is very diverse, with dozens of individual mono-, sesqui- and diterpenes (Schiebe *et al.*, 2012) (Figure 5.8). The reasons for this chemical diversity are that trees have multiple genes coding for terpene synthases and Cyp P450s and that those individual enzymes often produce multiple products. In grand fir (*Abies grandis* Lindl.), for example, a single terpene synthase (γ-humulene synthase) may produce 52 different sesquiterpene products (Steele *et al.*, 1998). Such multifunctional enzymes contribute to the high biochemical diversity of conifer resin, and maintaining this diversity seems to be an important element in the chemical defense strategy of conifers (Ro *et al.*, 2005; Keeling *et al.*, 2008).

2.2.2 Phenolics

Phenolics are involved in conifer resistance against bark beetles and their associated fungi (Lieutier, 2004; Ralph *et al.*, 2006; Faccoli and Schlyter, 2007). Plant phenolics include around 4000 structurally diverse compounds that are biosynthesized by different pathways. The major pathway in conifers appears to be the shikimic acid pathway, linking carbohydrate metabolisms to the biosynthesis of aromatic amino acids (Ralph *et al.*, 2006). The aromatic amino acid phenylalanine is a precursor for the formation of most secondary phenolic compounds, including flavonoids, stilbenes, condensed tannins, and other polyphenolics, as well as the structural polymer lignin (Dixon *et al.*, 2001). The soluble phenolics, which include stilbenes and flavonoids (Figure 5.8), have been most studied in relation to conifer defense. These relatively simple ring structures can be identified and quantified by HPLC analyses (e.g., Schiebe *et al.*, 2012). The

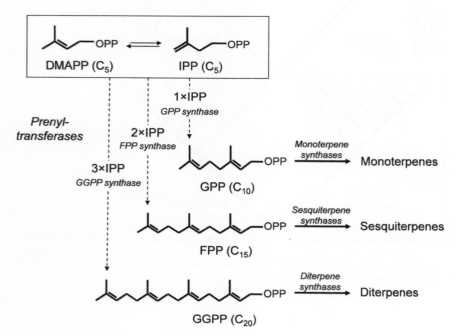

FIGURE 5.7 Biosynthesis of mono-, sesqui-, and diterpenes in conifers. The basic building blocks for the different terpene classes are formed by the fusion of two C_5 units; one unit of dimethylallyl diphosphate (DMAPP) and one, two or three units of isopentenyl diphosphate (IPP). The resulting prenyl diphosphates (geranyl diphosphate (GPP), farnesyl diphosphate (FPP), and geranyl-geranyl diphosphate (GGPP)) are subsequently modified into a large number of different mono-, sesqui-, and diterpenes. Specific enzymes (the prenyltransferases GPP, FPP, and GGPP synthase) catalyze the formation of the different prenyl diphosphate building blocks. These linear precursors are then modified by different terpene synthases (mono-, sesqui-, and diterpene synthases) and cytochrome P450 mono-oxyenases (Cyp P450) to form the huge diversity of terpenes present in conifers.

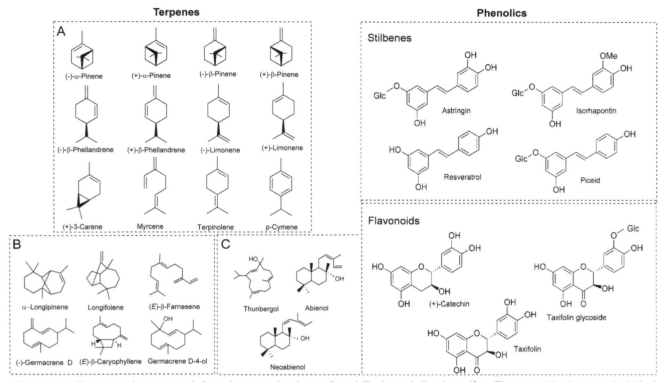

FIGURE 5.8 **Representative compounds from the two major classes of specialized metabolites in conifers.** The terpenoid resin present in the bark and sapwood of most species in the pine family consists mainly of C_{10} monoterpenes (A) and C_{20} diterpenes (C), with smaller amounts of C_{15} sesquiterpenes (B). Phenolics, including soluble phenolics such as stilbenes and flavonoids, are abundant in the bark of all conifers.

identification of more complex phenolic compounds is more challenging and studies to determine their roles in conifer defense have only just started (Li *et al.*, 2012; Hammerbacher *et al.*, 2013, 2014).

Important enzymes involved in phenol synthesis, such as chalcon and stilbene synthases, have been found to increase in whole bark samples after fungal infection, suggesting that phenolics play a role in bark defenses (Brignolas *et al.*, 1995a; Richard *et al.*, 2000; Nagy *et al.*, 2004). Further evidence for a role of phenolics in conifer defense is that whole conifer bark often contains high concentrations of flavonoids, stilbenes, and other soluble phenolics (Lieutier *et al.*, 1991, 1996, 2003; Lindberg *et al.*, 1992; Brignolas *et al.*, 1995a; 1995b; 1998; Viiri *et al.*, 2001; Zeneli *et al.*, 2006; Schiebe *et al.*, 2012). However, these compounds do not tend to display strong induced responses and are often found to be only weakly inducible by fungal inoculation, mechanical wounding, or bark beetle attack (Brignolas *et al.*, 1995a, 1998; Evensen *et al.*, 2000; Lieutier *et al.*, 2003; Erbilgin *et al.*, 2006; Zeneli *et al.*, 2006; Schiebe *et al.*, 2012). The weak induction of soluble phenolics could indicate that these compounds are not crucial for defense, but a recent study of the fate of soluble phenolics in Norway spruce following fungal infection suggests another explanation. The lack of induction of stilbenes following infection with a bark beetle-associated

bluestain fungus was a result of fungal metabolism that converted the stilbenes to other compounds (Hammerbacher *et al.*, 2013). Thus, although stilbene biosynthesis was up-regulated in the tree, fungal metabolism prevented stilbenes from accumulating in the bark.

3. PREFORMED AND INDUCED CONIFER DEFENSES

As described above, conifer defenses come in many different shapes and sizes and this diversity of defense mechanisms and their mode of action are often categorized using various dichotomous classification schemes. Defenses may be classified as preformed or inducible, mechanical or chemical, local or systemic, and specific or general (Lieutier, 2004; Franceschi *et al.*, 2005). While these subdivisions are convenient simplifications to help us make sense of the wide array of defense responses found in plants, we should keep in mind that they are indeed simplifications and thus come with some limitations. There may, for example, be unclear boundaries and overlap between different categories, and one and the same defense structure may contribute to several defense categories. Resin ducts in the sapwood can be either preformed or inducible and the mode of action of their resin content is both chemical and mechanical. An important chemical effect of resin is its toxicity to bark beetles and fungi (Smith, 1961; Raffa et al., 1985; Everaerts et al., 1988). Mechanical effects of resin include that it is being stored under pressure and therefore may physically expel beetles that attempt to tunnel into the bark, and that resin is sticky and may trap beetles. The ubiquitous PP cells in the bark also combine preformed and inducible defense roles, being formed constitutively in the bark at the start of each growth season (Krekling et al., 2000), but also undergoing extensive changes in response to attack (Franceschi et al., 2005). The following presentation of defense mechanisms in conifers focuses on the dichotomy of preformed and induced defenses, but highlights the mixed nature of many of the defenses. The same cell or tissue type may be addressed under different headings if it contributes to both preformed and induced defenses.

3.1 Preformed Defenses

Conifers integrate multiple defense mechanisms into a multitier defense strategy that includes both preformed and inducible defense mechanisms (Phillips and Croteau, 1999; Franceschi et al., 2005; Keeling and Bohlmann, 2006). Preformed defenses form the first line of defense and are produced in the absence of any attack. Usually, these defenses do not need to undergo modifications and are thus active in the same state as they were before an attack (Lieutier, 2002). They represent a baseline, broad-scale resistance that the tree must have to be able to deal with the many different threats it faces throughout its often long life. Preformed defenses may be either mechanical or chemical, or they may combine mechanical and chemical properties. Preformed chemical defenses in conifers often include pools of stored terpenoids and phenolics that can be released or mobilized in response to an attack.

It has been argued that very few plants in nature will have purely preformed defenses, since every plant probably has experienced multiple attacks in its lifetime and hence is induced to a larger or smaller extent (Walters, 2009). However, it also seems to be true that the defenses of every plant can be induced to a much higher level than this background induction if it is experimentally treated with defense elicitors such as methyl jasmonate (e.g., Zeneli et al., 2006). Thus, although it might be difficult to find entirely non-induced plants in nature, it is useful to distinguish between preformed and inducible defenses.

3.1.1 Mechanical Defenses

The first line of preformed defense against attackers, the outer periderm, combines mechanical elements, chemical defenses, and suberization, which provides a hydrophobic

barrier that reduces water loss (Franceschi et al., 2005). The periderm does not seem to be an important barrier for bark beetles, although attacking beetles tend to avoid areas with thick cork and preferentially enter through bark cracks where the periderm is thinner. For beetle-associated bluestain fungi and other symbionts the periderm probably is an impenetrable barrier (Franceschi et al., 2000), and most symbionts are therefore very dependent on the beetles for transport into the bark.

Inside the periderm, lignified sclerenchyma cells and calcium oxalate crystals in the bark provide another mechanical defense line. Due to their physical toughness, stone cells can be a deterrent to tunneling insects like bark beetles (Wainhouse et al., 1990, 1997; Hudgins et al., 2004). However, the abundant fiber cells of the non-pine conifers appear to be a particularly effective mechanical defense barrier (Hudgins et al., 2003a, 2004). The very orderly arrangement of densely spaced fiber rows separated by layers of PP cells found in these species probably constitutes a formidable physical barrier to organisms that attempt to penetrate the bark (Hudgins et al., 2004). This may help explain the relative immunity of non-pine conifers to attack by tree-killing bark beetles. Calcium oxalate crystals in the bark can probably also deter tunneling bark beetles (Hudgins et al., 2003b), but because they are chemically inert the crystals are unlikely to be effective against microorganisms. The large intracellular calcium oxalate crystals in the bark of members of the pine family seem to be less important in defense than the much more abundant extra-cellular crystals of the non-pine conifers (Hudgins et al., 2003b).

3.1.2 Preformed Resin

Preformed resin ducts are present in the bark, sapwood, and needles of many conifers in the pine family (Wu and Hu, 1997; Franceschi et al., 2005). In stem tissues, preformed resin ducts reach their highest complexity and abundance in the pines (*Pinus*), which tend to have more axial and radial resin ducts than other conifers (Wu and Hu, 1997).

The resin ducts in the outermost layers of the living bark, the cortex, seem to be purely preformed. Their number appears to remain constant over time, in contrast to axial resin ducts in the sapwood that may be induced by biotic and abiotic disturbances (Nagy et al., 2000, 2006). The very large cortical resin ducts form an independent resin duct system that is not connected to the radial resin ducts or the rest of the resin canal system. Cortical resin ducts are important in the defense against insects that attack young stem tissues. One well-studied example is the white pine weevil *Pissodes strobi* Peck (Coleoptera: Curculionidae) that attacks more than 20 conifer species in North America (Furniss and Carolin, 1977). Resistant trees have been shown to have significantly more cortical resin ducts in the leader shoots than susceptible trees, particularly in the outer part of the cortex (Tomlin and Borden, 1994; Alfaro *et al.*, 1997). Like the cortical resin ducts, the radial resin ducts in the secondary phloem seem to be predominantly preformed, and do not increase in number after fungal infection (Nagy *et al.*, 2006). However, there are very few studies of the function and structure of the radial rays and their associated radial resin ducts and these tissues are an understudied component of conifer defenses. Since the rays are the only radially oriented tissue type in the conifer stem, they probably have important roles in transport of signals and metabolites (Franceschi *et al.*, 2000).

The sapwood is the main site of preformed resin ducts in the four genera (pine, spruce, larch, and Douglas fir) that have preformed axial resin ducts (Wu and Hu, 1997) (Figure 5.6). In these genera, the axial resin ducts are the largest resin reservoir and play an important role in preformed resin-based defenses. For instance, Kane and Kolb (2010) found that trees containing a greater proportion of xylem resin ducts were more likely to survive during a severe drought and subsequent bark beetle attack. The axial and radial resin ducts are connected through numerous plasmodesmata in cell wall regions where epithelial cells of the two resin duct systems connect (Benayoun and Fahn, 1979). This allows transport of terpene precursors and other metabolites from the bark through the radial rays and further into the stem (Fahn *et al.*, 1979). When a tree is attacked by bark beetles, preformed resin from the axial ducts in the sapwood may flow to the attack sites in the bark via the radial resin ducts. In pines, the epithelial cells lining the axial resin ducts have thin and non-lignified walls and remain alive for several years (Bannan, 1936). In other conifers with preformed axial resin ducts in the sapwood, the epithelial cells are reported to become lignified (and hence non-functional) with time (Bannan, 1936; Fahn *et al.*, 1979). Characteristics of the preformed resin, such as its quantity, chemical composition, exudation pressure, and crystallization rate, have been shown to correlate with tree resistance in some pine species (Raffa *et al.*, 1993; Lieutier, 2004). However, in the case of rapid bark beetle mass attacks, the preformed resin reserves may quickly become depleted and the tree may succumb to the attack (Raffa and Berryman, 1983).

In addition to the preformed resin ducts found in the stem, all conifers in the pine family have resin ducts in the needle mesophyll. These ducts run along the longitudinal needle axis and there can be as many as 12 ducts per needle in some pine species (Wu and Hu, 1997). The resin composition in the needles may differ qualitatively from that in the resin duct systems in the cortex, secondary phloem and sapwood (Werker and Fahn, 1969). Needle resin is important in defense against defoliating insects and other herbivores (Speight and Wainhouse, 1989; Björkman *et al.*, 1997).

3.1.3 Polyphenolic Parenchyma (PP) Cells

The PP cells in the secondary phloem synthesize and store phenolic compounds. The characteristic polyphenolic content of the PP cells is stored inside a large vacuole (the phenolic body) that may fill most of the PP cell volume. The polyphenolic content is highly dynamic and changes in appearance over the year, as well as after pathogen attack or mechanical wounding (Figure 5.5A, D). Attempts have been made to classify PP cells based on the appearance of the vacuolar content (Franceschi *et al.*, 1998), but any relationship between vacuolar morphology and tree resistance to infection remains unclear. Polyphenolic parenchyma cells are produced constitutively, but also have important roles in induced defenses (Franceschi *et al.*, 1998, 2000). Polyphenolic parenchyma cells are produced in the earliest phases of both primary and secondary growth and the tree spends considerable resources on their development and maintenance, indicating that these cells are of crucial importance to the trees.

The effectiveness of the PP cells as a preformed defense against microorganisms has been demonstrated by inoculation experiments with bark beetle-associated bluestain fungi. If the fungus is inoculated into wounds extending all the way to the vascular cambium, the preformed PP cell barrier is circumvented, and the fungus is able to colonize the bark and outer sapwood (Franceschi *et al.*, 1998). However, if the fungus is inoculated into shallow bark wounds that end within the PP cell layers, fungal growth is completely restricted (Franceschi *et al.*, 2000). This demonstrates the power of the defenses residing in the preformed PP cell layers (Franceschi *et al.*, 2000). It also suggests that bark beetle symbionts need the beetles to carry them past the powerful PP cell barriers in the bark (Paine *et al.*, 1997; Franceschi *et al.*, 2000, 2005).

The polyphenolic nature of the PP cells suggests that they are important in resistance, although the exact composition of the phenolic material inside the PP cells is not known. The conclusion that the vacuolar material is polyphenolic in nature is based largely on microscopy evidence, such as structural similarity to polyphenolics in other plant cell types, bright yellow fluorescence of aldehyde fixated sections when exposed to blue light (450–490 nm), strong staining (and quenching of fluorescence) by osmium tetroxide and the periodic acid-Schiff procedure, and localization of phenylalanine ammonia lyase (PAL), a key enzyme in phenolic synthesis, to the plasma membrane and cytoplasm of PP cells by immunolocalization (Parham and Kaustinen, 1976; Franceschi *et al.*, 1998). Additional evidence that the PP cells contain phenolic material comes from chemical and molecular analyses of whole conifer bark. Although whole bark consists of a mosaic of different cell types, most of the living cells in the bark are PP cells (the remainder consisting of more or less empty sieve cells and a few radial rays). Conifer bark contains high concentrations of soluble phenolics, such as flavonoids and stilbenes (Lieutier et al., 1991, 1996, 2003; Lindberg et al., 1992; Brignolas et al., 1995a, b, 1998; Viiri et al., 2001; Zeneli et al., 2006; Schiebe et al., 2012) (Figure 5.8), as well as important enzymes in phenol synthesis, such as chalcon and stilbene synthases (Brignolas et al., 1995a; Richard et al., 2000; Nagy et al., 2004). Recent chemical analyses of laser micro-dissected PP cells in Norway spruce revealed increased levels of different phenolics following fungal infection and confirmed that the PP cells is a principal site of phenolic accumulation in spruce bark, although the sensitivity of the chemical analyses was limited due to the minute amounts of tissue that could be harvested (Li et al., 2012).

In addition to being a major storage site for phenolic compounds, PP cells also accumulate and store carbohydrates in the form of starch grains and lipids (Alfieri and Evert, 1973; Krekling et al., 2000). Starch reserves build up through the summer months in Norway spruce, are completely absent in the winter, and begin to accumulate again in the spring (Krekling et al., 2000). In addition to forming important energy reserves, the starch and lipid of the PP cells can probably be used for rapid synthesis of defensive chemicals in response to attack.

3.2   Induced Defenses

Overlaid with their preformed defenses, conifers have the capacity to up-regulate additional, inducible defenses in response to an attack. Inducible defenses represent a variable resistance that is turned on when it is needed and is thus a cost-effective and flexible response to attack (Steppuhn and Baldwin, 2008). The response time of inducible defenses from attack to full induction can vary from minutes to weeks, depending on the type of defense that is activated. Examples of rapid responses are changes in gene transcription and up-regulation of pathogenesis-related proteins that may take place within minutes to hours of an attack (Karlsson et al., 2007). Slower processes are the formation of functional traumatic resin ducts in the sapwood, which may require 2–3 weeks in Norway spruce (Nagy et al., 2000), and formation of a wound periderm, which may require several weeks to be completed (Franceschi et al., 2000). The study of induced defenses in conifers dates back to the pioneering work by Reid and co-workers (Reid et al., 1967) and Berryman's seminal 1972 synthesis paper (Berryman, 1972).

Conifers have two basic types of inducible defenses that differ in their level of organizational complexity (Franceschi et al., 2005). At the simplest level are the defense responses resulting from changed metabolism of pre-existing cells, such as cell wall lignification, induction of resin production in existing resin ducts, and activation of the hypersensitive response (Krokene et al., 2008b). At a more complex organizational level are induced defense responses that involve changes in cell division and differentiation, leading to formation of traumatic resin ducts and a wound periderm (Franceschi et al., 2005). Some cell types can be involved in both types of inducible defense; pre-existing PP cells are mobilized following infection and alter their phenolic content, but PP cells are also involved in more complex responses such as production of additional PP cells and formation of a wound periderm (Krokene et al., 2008b). Axial resin ducts in the sapwood are another example; biotic or abiotic challenges induce resin production in pre-existing axial ducts (Ruel et al., 1998), but may also induce the formation of new axial resin ducts (traumatic resin ducts; Bannan, 1936; Nagy et al., 2000).

3.2.1   Induced Resinosis

In addition to the preformed resin structures found in all conifers in the pine family, many species have the ability to form new resin ducts following abiotic and biotic challenges (Bannan, 1936; Krokene et al., 2008b). All conifers in the pine family can form traumatic resin ducts in the sapwood following wounding or infection (Krokene et al., 2008b). Interestingly, such induced resin ducts may also form in several genera of non-pine conifers that do not have any preformed resin ducts (Figure 5.6). Axial resin ducts can be induced following stem treatment with methyl jasmonate, either in the bark (Araucaria, Cupressus, and Cryptomeria) or in the sapwood (Metasequoia, Sequoia, Sequoiadendron) (Hudgins et al., 2004). Induced resin ducts are constrained by the nature of the vascular cambium to be axially and not radially oriented. The cambium can quickly be reprogrammed to produce functional axial resin ducts several meters long (Christiansen et al., 1999a, b; Nagy et al., 2000), but it would require many years to produce radial ducts that were more than a few millimeters long.

3.2.1.1   Traumatic Resin Ducts

Traumatic resin ducts are formed in tangential bands in the sapwood, in contrast to normal axial sapwood ducts that have a more scattered distribution (Bannan, 1936; Nagy et al., 2000). The tangential bands of traumatic resin ducts are often surrounded by small tracheids with thickened cell walls and may be visible to the naked eye as a punctuated line within the annual ring (false annual rings). An almost complete tangential ring of traumatic resin ducts can be formed around the stem circumference in spruce if the inducing stimulus is strong, whereas in *Abies*, *Tsuga*, *Cedrus*, *Pseudolarix*, and to some extent *Pinus*, traumatic resin ducts mainly form in the vicinity of wounds or infection sites (Bannan, 1936; Wu and Hu, 1997; Nagy *et al.*, 2006). The traumatic ducts represent a considerably larger resin volume than the normal axial resin ducts, and contribute significantly to the tree's overall resin production. Each traumatic resin duct is much larger than a preformed duct, with a volume that is about four times greater per unit duct length in Norway spruce (Krokene *et al.*, 2008b). Very wide traumatic ducts are sometimes formed through coalescence of neighboring ducts (Hudgins *et al.*, 2003a). There are numerous connections between the traumatic resin ducts and the radial resin ducts, and this enables resin to flow both axially along the trunk, as well as radially towards the bark surface (Nagy *et al.*, 2000; Krokene *et al.*, 2003).

Traumatic resin duct formation is induced through the octadecanoid pathway, involving jasmonate and ethylene signaling (Hudgins and Franceschi 2004; Ralph *et al.*, 2007; Schmidt *et al.*, 2011). In response to the inducing signal, xylem mother cells in the cambium that would normally develop into tracheids are reprogrammed to form resin duct epithelial cells (Werker and Fahn, 1969; Nagy *et al.*, 2000; Krekling *et al.*, 2004). The fact that traumatic resin ducts are formed in strict tangential bands implies a spatially coordinated induction and differentiation of the xylem mother cells. The incipient traumatic resin ducts develop in the same way as normal axial resin ducts, by schizogenesis between incipient epithelial cells when these are still close to the cambial zone. The formation of functional traumatic resin ducts with secretory activity may require 2–4 weeks in Norway spruce *Picea abies* (L.) Karst. in Norway (Nagy *et al.*, 2000), but could be faster in warmer climates. As you move up or down along the stem away from the point of induction, the traumatic resin ducts are gradually reduced in size and number, and are situated closer and closer to the cambium (Krekling *et al.*, 2004). This suggests that there is a time lag in the production and differentiation of traumatic resin ducts away from the inducing stimulus. Careful anatomical studies in Norway spruce indicate that a wave of traumatic resin duct development moves along the stem axis at a speed of about 2.5 cm per day (Krekling *et al.*, 2004).

Traumatic resin duct formation is probably an important component of induced defenses in conifers, since it greatly increases resin production. This induced resin production may protect the trees against opportunistic wound colonizers following bark injury and may kill developing bark beetle egg or larvae in the bark (Franceschi *et al.*, 2005). However, since trees may need a few weeks to form functional traumatic resin ducts, these ducts will usually appear too slowly to prevent bark beetle mass attacks, which may

be completed in a few days (Paine *et al.*, 1997). Still, traumatic resin duct formation may be important when mass attacks are slowed down by unfavorable weather conditions (Franceschi *et al.*, 2000).

3.2.1.2 Activation of Existing Axial Resin Ducts

Even the preformed axial resin ducts in the sapwood are not exclusively involved in preformed resin production. At least in some species the epithelial cells lining these ducts can be induced to produce more resin following wounding and presumably insect attack. In loblolly pine (*Pinus taeda* L.), depletion of resin from axial ducts following wounding induces refilling of the ducts to about twice their previous capacity (Ruel *et al.*, 1998; Lombardero *et al.*, 2000). Immediately after wounding there is a decrease in resin flow as the preformed resin stores are drained, but after 2–3 days resin flow increases again (Ruel *et al.*, 1998). Most likely, this increased resin flow is a result of *de novo* synthesis of terpenoids by the epithelial cells around the preformed ducts. Lombardero *et al.* (2000) hypothesize that when an axial duct is ruptured this signals an increase in the level to which the duct is filled before resin biosynthesis in the epithelial cells is turned off. The fact that pines, in contrast to most other conifers (Wu and Hu, 1997), usually have thin-walled and unlignified epithelial cells in their axial resin ducts suggests that activation of resin production in these ducts may be specific to pines. Even in pines, the response has not been found in all species investigated. Gaylord *et al.* (2011) found no evidence of refilling of preformed axial resin ducts in ponderosa pine (*Pinus ponderosa* Dougl. ex. Lawson) using similar methodology to Ruel *et al.* (1998). Still, since resin flow was only monitored for the first 7 days after wounding it is possible that ponderosa pine, growing in a much cooler climate, simply had a slower response than loblolly pine (Gaylord *et al.*, 2011).

It certainly makes adaptive sense that the epithelial cell machinery of the preformed axial resin ducts also are used for induced resin production. The elaborate resin duct system of conifers is a costly structure to build and maintain and it seems rational to utilize this investment not only constitutively. Induction of resin production in the existing epithelial cells can probably happen very quickly, and the speed of the response may be rapid enough to be effective against ongoing bark beetle attacks. At least in pines, activation of preformed resin ducts is thus an additional defense line against invasion by bark beetles, in addition to the other preformed and inducible defenses in the bark and sapwood.

3.2.1.3 Induction of Radial Stem Ducts

The radial rays and their associated radial resin ducts, which run from the bark to the sapwood, are preformed but may also undergo induced defense reactions. Induced responses in radial rays have been little studied, but work on Scots pine suggests that pathogen infection induces swelling of ray parenchyma cells in the sapwood (Nagy *et al.*, 2006). Three months after infection the total area of ray cells in tangential sapwood sections was 68% greater in fungus-infected sapwood than in uninfected control samples. The increased cell area was caused by swelling of existing ray cells and not by the formation of new rays. Swelling of ray parenchyma cells is probably related to increased cellular activity and resin production. Swelling and activation of the ray cells was particularly pronounced in the latewood, which also has more abundant axial resin ducts than the earlywood (Nagy *et al.*, 2006). Induced responses in radial rays have also been observed in Douglas fir (*Pseudotsuga menziesii* (Mirb.) Franco), where the lumen of the radial resin ducts increased in size after treatment with methyl jasmonate, suggesting that resin production was induced (Hudgins and Franceschi, 2004). Considering their anatomical layout, it is logical that changes in the radial rays and their associated radial resin ducts accompany the induced resinosis occurring in the preformed and induced resin ducts in the sapwood. The radial resin ducts are the main pathway for resin flow from the sapwood to the bark surface, as each radial duct is connected to many preformed and traumatic axial resin ducts along its length.

The radial rays do not only provide the tree with a useful transport route, but could also serve as an easy and nutrient-rich infection route into the stem for invading pathogens. However, as studies of pathogen infection in conifers demonstrate, the radial rays do not appear to be an easy infection route for pathogens (Franceschi *et al.*, 2000), probably because the rays are very well defended. Immunolabeling studies of Norway spruce have demonstrated that radial rays have very high levels of phenylalanine ammonia lyase (PAL) activity, a key enzyme in phenolic synthesis (Franceschi *et al.*, 1998). In fact, PAL activity in the rays was much stronger than that in PP cells, suggesting that ray cells may be among the best defended tissues in conifer stems. However, as already noted there are very few studies on the structure and function of ray cells in conifers, so this is an area were more research is needed.

3.2.1.4 Cortical Resin Ducts

The cortical resin ducts in the outer bark are preformed and do not seem to change anatomically after an attack, but they still appear to be capable of some induced responses. Up-regulation of genes involved in terpene biosynthesis was observed in cortical resin ducts in white spruce (*Picea glauca* (Moench) Voss) 8 days after stem treatment with methyl jasmonate (Abbott *et al.*, 2010). This shows that

the epithelial cells of the cortical ducts rapidly can be induced to produce more terpenes following wounding or insect attack, possibly also involving qualitative changes in terpene composition (Abbott *et al.*, 2010; Zulak *et al.*, 2010). The cortical ducts' capacity to modify their resin production is compatible with the observation that the epithelial and sheath cells surrounding these ducts may remain alive and thin walled for many years (Wu and Hu, 1997) and thus remain metabolically active.

3.2.2 Polyphenolic Parenchyma (PP) Cells

As noted earlier, the PP cells are the most abundant living cell type in conifer bark and are considered the single most important cell type in conifer defense (Franceschi *et al.*, 2005). The PP cells are involved in numerous inducible defense responses, including activation of preformed PP cells, production of extra PP cells, and wound periderm formation.

Activation or swelling of preformed PP cells begins a few days after an attack and results in a four-fold increase in cell volume and a change in the appearance (and presumably chemical composition) of the phenolic content (Franceschi *et al.*, 1998, 2000, 2005; Krokene *et al.*, 2003) (Figure 5.5A, C, D). As the PP cells expand in size, the surrounding sieve cells are gradually compressed, and the phloem is transformed into dense blocks of cell walls separated by layers of swollen PP cells. This combined cell wall/PP cell barrier appears to be reinforced by phenolics released from the induced PP cells and deposited in the surrounding sieve cell walls, adding to the tissues' role as a physical and chemical barrier to penetration by invading organisms (Franceschi *et al.*, 2000).

As noted above, the PP cells appear to be the major reservoir of phenolics in conifer stems (Franceschi *et al.*, 2005). Phenylalanine ammonia lyase has, for example, been localized to the PP cells by immunolocalization (Franceschi *et al.*, 1998). Polyphenolic parenchyma cells undergo dramatic anatomical responses in response to infection but the chemical nature of these anatomical changes remains to be determined. The changes do not appear to be related to induction of soluble phenolics, since these are not very responsive to fungal infection, as discussed above. The anatomical changes in the PP cells may be related to modifications of more complex phenolics, such as high molecular weight condensed tannins or substances that are bound to the cell walls (Schmidt *et al.*, 2005; Li *et al.*, 2012), but this remains to be shown.

Following insect attack or pathogen infection new or extra PP cells may develop from undifferentiated parenchyma cells in between the regular annual PP cell layers. This is not part of the normal developmental program of the secondary phloem and is only seen after severe challenges such as massive fungal infection

(Krekling *et al.*, 2000, 2004; Krokene *et al.*, 2003). Production of extra PP cells may also be induced by treatment with methyl jasmonate (Krokene *et al.*, 2008b) (Franceschi *et al.*, 2002; Martin *et al.*, 2002; Hudgins *et al.*, 2004; Krokene *et al.*, 2008b). With time, extra PP cells may be formed several centimeters away from the point of induction and can be quite abundant, covering 50% of the phloem circumference in Norway spruce 15 weeks after a massive fungal infection (Krokene *et al.*, 2003). Because differentiation of extra PP cells is a relatively slow process compared to activation of existing PP cells, it probably enhances long-term resistance rather than resistance to an ongoing infection.

Activation of existing PP cells and production of extra PP cells seem to be involved in the phenomenon of systemic induced resistance in Norway spruce (Krokene *et al.*, 2003). Trees that were pretreated with a sublethal fungal infection 3–9 weeks before they were subjected to a massive challenge infection showed strong PP cell responses, and were much more resistant to the subsequent challenge inoculation than untreated trees (Krokene *et al.*, 1999, 2003). Trees pretreated 1 week before mass infection showed no activation of PP cells and were equally susceptible as untreated control trees (Krokene *et al.*, 2003). Thus, tree resistance developed over a similar time frame as that required to elicit induced responses in the PP cells, suggesting that these responses may be involved in enhancing tree resistance (Krokene *et al.*, 2003).

3.2.3 *Wound Periderm Formation — Repair Mechanisms*

Yet another inducible defense response involving the PP cells is the process of wound periderm formation (Franceschi *et al.*, 2000). Wound periderms form around any damaged tissue, whether it is caused by fungal infection, unsuccessful bark beetle attacks, or mechanical injury. Formation of a wound periderm is an essential final stage in a successful defense reaction, as it walls off the damaged tissue region and re-establishes a continuous surface barrier. Wound periderms are produced by activation of existing PP cells, which begin to divide to form a new cork cambium (Figure 5.5A). Just like the normal cork cambium this meristem produces cork tissue (phellem) outwards and phelloderm inwards and thus regenerates the stem's continuous surface barrier (Franceschi *et al.*, 2000, 2005). Damaged tissues outside this new periderm are isolated from the rest of the bark and will eventually be pushed outwards and shed as new phloem is produced underneath it by the vascular cambium. In addition to its wound healing role, the wound periderm may stop the spread of invading organisms, if it is established rapidly enough and at some distance from the attack site.

3.2.4 *Acquired Resistance — Long-term Defense Induction*

Acquired or systemic induced resistance (Bonello *et al.*, 2006; Eyles *et al.*, 2010) in conifer stems increases tree resistance to future attacks and may be a longer-term consequence of inducible defenses. Acquired resistance has been most extensively studied in Norway spruce (Christiansen and Krokene, 1999; Christiansen *et al.*, 1999b; Krokene *et al.*, 1999, 2001, 2003; Zeneli *et al.*, 2006), but the mechanism has also been characterized in a number of pine species, including Scots pine, loblolly pine, ponderosa pine, Monterey pine (*Pinus radiata* D. Don), and Austrian pine (*Pinus nigra* J. F. Arnold) (Reglinski *et al.*, 1998; Enebak and Carey, 2000; Krokene *et al.*, 2000; Bonello *et al.*, 2001; Bonello and Blodgett 2003; McNee *et al.*, 2003; Luchi *et al.*, 2005). Acquired resistance in different forms is probably present in most species in the pine family, and perhaps in all conifers (Eyles *et al.*, 2010).

Acquired resistance in conifers was first described in 1999 (Christiansen *et al.*, 1999b). Norway spruce trees that had been mechanically wounded and inoculated with fungi 1–2 weeks beforehand were found to be much more resistant to a massive fungal inoculation than intact control trees (Christiansen *et al.*, 1999b). Subsequently, it has been shown that resistance to fungal infection in Norway spruce can be boosted by prior mechanical wounding (Krokene *et al.*, 1999), sublethal fungal inoculations (Christiansen *et al.*, 1999b; Krokene *et al.*, 1999), or external application of methyl jasmonate (Franceschi *et al.*, 2002; Martin *et al.*, 2002; Zeneli *et al.*, 2006; Krokene *et al.*, 2008a) prior to a challenge inoculation. Acquired resistance seems to involve inducible defense reactions, such as traumatic resin duct formation and PP cell activation (Bonello and Blodgett, 2003; Krokene *et al.*, 2003).

Boosting conifer resistance by pretreatment with sublethal fungal infection or methyl jasmonate treatment also increases resistance to attack by bark beetles (Christiansen and Krokene, 1999; Erbilgin *et al.*, 2006). Spruce bark beetles (*Ips typographus* L.) colonizing methyl jasmonate-treated trees make shorter galleries, lay fewer eggs, and produce smaller offspring than beetles colonizing untreated control bark (Erbilgin *et al.*, 2006). Interestingly, beetles tunneling in methyl jasmonate-treated Norway spruce bark also release negligible amounts of methyl butenol and cis-verbenol, the two main components in the beetle's aggregation pheromone, compared to beetles tunneling in untreated bark (Zhao *et al.*, 2011a). The poor beetle performance in methyl jasmonate-treated bark is probably due to dramatically increased terpene levels and other induced defenses. By itself, methyl jasmonate treatment of Norway spruce only led to a moderate (ca. three-fold) increase in diterpene levels compared to untreated bark, but when methyl jasmonate-treated bark was mechanically wounded diterpene levels increase more than 26-fold within just 24 hours (Zhao *et al.*, 2011a). It thus appears that bark beetles entering methyl jasmonate-treated bark will quickly face induced defenses that will place the beetles under severe physiological stress and interfere with their ability to produce effective aggregation pheromones.

3.2.5 Defense Priming—Defend Now, Pay Later?

As discussed above, treatment with methyl jasmonate may boost tree resistance to subsequent fungal infection or bark beetle attack (Erbilgin *et al.*, 2006; Zhao *et al.*, 2011a). The increased tree resistance observed in these studies can to some extent be explained by the fact that methyl jasmonate triggers induced defense responses, such as traumatic resin duct formation and terpene biosynthesis (Martin *et al.*, 2002; Zeneli *et al.*, 2006). However, although it consistently boosts tree resistance, methyl jasmonate treatment induces very modest increases in terpenoid production in many trees (Erbilgin *et al.*, 2006; Zhao *et al.*, 2011a). Thus, methyl jasmonate treatment seems to have the dual roles of inducer of tree defenses and trigger of defense priming in conifers. Plant defense priming is a physiological process by which a plant is prepared to respond more quickly or aggressively to future biotic or abiotic stress (Conrath *et al.*, 2006; Frost *et al.*, 2008). Priming in itself does not lead to increased resistance, but rather initiates a state of readiness that allows for accelerated induced resistance if an attack occurs. One presumed benefit of priming is that it saves the tree from the costs associated with full implementation of an induced defense response should the attack not come (Frost *et al.*, 2008). Although the phenomenon of defense priming has been known for decades (Ross, 1961; Kuć, 1982) most progress in understanding the underlying mechanisms has been made in recent years from studies of angiosperm model plants such as *Arabidopsis* (Conrath *et al.*, 2006; Conrath, 2011; Luna *et al.*, 2012). In conifers, the underlying molecular mechanisms of defense priming are completely unknown, but primed trees seem to be in a state where they are able to respond very efficiently to subsequent attacks. For example, Norway spruce trees primed with methyl jasmonate mobilized a much stronger increase in terpenoids in the bark within 24 hours after mechanical wounding than untreated trees (Zhao *et al.*, 2011a; Krokene and Zhao, unpubl.).

When methyl jasmonate was first discovered to enhance conifer resistance to bark beetles and fungi its effects were supposed to closely mimic biotic attack, since it induced broadly similar symptoms of induced resinosis, traumatic resin duct formation, and activation of PP cells (Franceschi *et al.*, 2002; Martin *et al.*, 2002; Zeneli *et al.*, 2006; Erbilgin

et al., 2006; Krokene *et al.*, 2008a; Schmidt *et al.*, 2011). However, when examined at a more detailed level, methyl jasmonate has been found to induce a qualitatively different terpene response than inoculation with bluestain fungus in Norway spruce (Zhao *et al.*, 2010). The extent to which methyl jasmonate is acting as an inducer or a primer of tree defenses seems to be genetically controlled, as the relative strength of the two responses varies between spruce genotypes (Erbilgin *et al.*, 2006; Zeneli *et al.*, 2006). A re-examination of the results from previous studies reveals that behind the average tree response to methyl jasmonate treatment there is much individual variation between trees. In some trees, methyl jasmonate treatment is clearly inducing defenses, as seen by extensive formation of traumatic resin ducts, up-regulation of defense-related gene transcripts, and highly elevated terpene levels in the bark and sapwood (Erbilgin *et al.*, 2006; Zeneli *et al.*, 2006; Schmidt *et al.*, 2011; Krokene *et al.*, unpubl.). In other trees, however, there are no obvious anatomical responses to methyl jasmonate treatment, no induction of terpene levels, and very little up-regulation of various transcript markers of resistance (Krokene, unpubl.). Still, trees with no obvious anatomical or chemical induction were also resistant to bark beetle attack and fungal infection, suggesting that tree resistance had been primed by methyl jasmonate.

3.3 Relative Importance of Preformed vs. Induced Defenses

Both preformed and induced defenses are important in conifer defense, and function together as an integrated whole in the trees' multitiered defense response (Franceschi *et al.*, 2005). However, molecular, chemical, and anatomical studies suggest that a hallmark of tree resistance is the rate of defense induction, i.e., trees that are able to mount a high defense level are more resistant than trees with a more moderate or slower response (Hietala *et al.*, 2004; Fossdal *et al.*, 2006; Krokene *et al.*, 2008a; Zhao *et al.*, 2011a, b). Similarly, recent field studies have shown that a tree's ability to launch effective induced defenses correlates better with resistance to bark beetle attack than do characteristics of the preformed defenses (Boone *et al.*, 2011; Schiebe *et al.*, 2012). Thus, a tree's ability to mount effective induced defenses seems to be a better marker of tree resistance than the quantity or quality of its preformed defenses. This insight may be utilized to select resistant trees for inclusion in breeding programs (Krokene *et al.*, 2008a). Resistant trees may be identified by treating the stem bark locally with methyl jasmonate, wait a few weeks for defense priming to develop, wound the treated stem section to unleash the primed defenses, sample the wounded bark after 24 hours, and quantify induced terpene levels or other tree defenses. Trees with higher induced terpene levels in the bark would be relatively more resistant than trees with lower terpene levels.

The relative investment in preformed and induced defenses appears to vary between different conifer groups. Within the pine family, *Pinus* is generally considered to have weaker inducible resin-based defenses than other species and to rely more on preformed resin-based defenses (Christiansen *et al.*, 1987; Raffa and Berryman, 1987; Raffa, 1991). Induced resinosis in *Pinus* tend to be more quantitative (i.e., producing more of the same compounds), compared to genera such as *Abies*, which tend to have more pronounced qualitative responses, adding new compounds to their resin after a challenge or greatly increasing the production of compounds that are only present in trace amounts constitutively (Raffa, 1991). Consistent with this pattern, pines seem to have a weaker and more localized traumatic resin duct response that is more quickly attenuated away from the infection point. In Scots pine, for example, there was very little traumatic resin duct formation following wounding or fungal infection, and extensive traumatic ducts were only found very close to the inoculation sites (Nagy *et al.*, 2006). The localized nature of the response in Scots pine is very different from that in Norway spruce, where traumatic resin ducts may form an almost complete ring around the stem circumference (Franceschi *et al.*, 2002; Krokene *et al.*, 2003) and gradually extend several meters from the point of induction (Christiansen *et al.*, 1999b).

There is, however, much variation to these broad patterns, both between and within different genera. Some pines, like whitebark pine (*Pinus albicaulis* Engelm.) growing at high elevations in western North America, show the typical pattern of weak induction of resin-based defenses (Raffa *et al.*, 2013). Ponderosa pine also seems to have limited inducible defenses to wounding or pathogen infection (Wallin *et al.*, 2003; Gaylord *et al.*, 2011). Other pine species, however, may be highly inducible. For example, induction of monoterpene levels in Scots pine started one day after inoculation and increased up to 800-fold 4 weeks after inoculation with the bluestain fungus *Leptographium wingfieldii* M. Morelet (Fäldt *et al.*, 2006). The large differences in inducible resin-based defenses occurring between different pine species illustrate the difficulties in making broad generalizations about the defense strategies of different conifer genera.

3.4 Conifer Defenses as a Regulator of Bark Beetle Outbreaks

The ability of bark beetles and their symbionts to colonize living trees is determined by tree resistance, including induced defense responses and defense priming (Franceschi *et al.*, 2005; Eyles *et al.*, 2010). Most of the time, tree defenses regulate bark beetle populations at endemic levels and prevent the beetles from reaching outbreak population densities (Raffa and Berryman, 1983, 1987). Because healthy, vigorous trees are more resistant to bark beetle attack, managing forests in ways that increase tree resistance may be a good strategy to prevent population buildup and outbreaks of bark beetles (Raffa *et al.*, 2008).

Bark beetle epidemics may develop only if the beetles are able to cross a series of hurdles or threshold processes as they interact with their environment (Raffa *et al.*, 2008) (Figure 5.1). Tree resistance strongly influences the thresholds at the bottom of this hierarchical chain of threshold processes. These thresholds regulate the early phases of beetle attack on individual trees and include processes such as tree entry, aggregation of local beetle populations, and establishment within the tree (Raffa *et al.*, 2008). The outcome of these three initial threshold processes depends on the relative strength of the beetle–symbiont complex and the tree's defense reactions. Preformed defenses, such as PP cells and preformed resin ducts, may, for example, impede fungal establishment and inflict considerable beetle mortality during tree entry and establishment (Franceschi *et al.*, 2000; Raffa *et al.*, 2005). If the tree gains the upper hand, the attack fails and the beetles abandon the attack or die, but if the beetle–symbiont complex gains the upper hand, the attack can proceed and the tree might die (Raffa and Berryman, 1987).

Tree colonization always starts when one or a few initial attackers, the so-called pioneer beetles (Alcock, 1982; Latty and Reid, 2009), enter the bark of a previously unattacked tree and establish breeding sites. Pioneering can be very risky since the first attackers must face the full force of the tree's defenses if they are not able to elicit aggregation and initiate a mass attack (Latty and Reid, 2009). Before they establish in the bark, the pioneers must assess the tree's defense capacity, including its capacity to launch effective induced defenses (Schiebe *et al.*, 2012). The beetles' need to be discriminative and select the least vigorous trees may, however, vary with population phase. Pioneering mountain pine beetles were shown to preferentially attack low-defense hosts under endemic or incipient population levels, but as the beetle population increased, the beetles became less cautious and preferentially attacked high defense/high reward hosts (Boone *et al.*, 2011). Secondary attraction of bark beetles to ongoing attacks may also be influenced by tree defenses. More spruce bark beetles were attracted to pheromone-baited Norway spruce trees that had been primed by methyl jasmonate treatment (Krokene, unpubl.), probably because such trees emit high levels of monoterpenes that synergize attraction to the beetles' aggregation pheromones (Erbligin *et al.*, 2007). However, very few of the beetles that landed on the primed trees tunneled into the bark. The few that entered only made very short tunnels before they aborted the attack, probably because they were deterred by vigorous induced defenses (Erbilgin *et al.*, 2006; Zhao *et al.*, 2011a; Krokene, unpubl.).

Tree defenses may also affect the beetles' ability to elicit aggregation and mass attack. Observations in the field have shown that beetle entry into the bark does not necessarily lead to pheromone release and mass attack. Mountain pine beetles tunneling in trees with copious resin flow were unable to elicit a mass attack, even though the beetle population in the surrounding area was epidemic (Raffa and Berryman, 1983). Spruce bark beetles tunneling in bark that had been primed with methyl jasmonate attracted significantly fewer conspecifics than beetles tunneling in untreated control bark (Erbilgin *et al.*, 2006). The reason for this reduced attraction is probably that beetles tunneling in primed bark produce very little aggregation pheromones (Zhao *et al.*, 2011a). Induced or primed tree defenses may thus interfere directly with pheromone production and disrupt beetle aggregation and mass attack (Erbilgin *et al.*, 2006; Zhao *et al.*, 2011a).

After the beetles have successfully entered a tree, elicited aggregation, and commenced breeding, the tree's induced defenses may still prevent beetle establishment and reproduction. To ensure successful reproduction, the beetles must block the tree's ability to launch induced defenses that may kill the developing beetle brood (Franceschi *et al.*, 2005). Beetle symbionts that engage tree defenses away from the immediate vicinity of beetle galleries probably play important roles in this process, as discussed in the following section.

4. FAILURE OF CONIFER DEFENSES

Despite their powerful multifaceted defenses, conifers periodically face extensive outbreaks of tree-killing bark beetles. During outbreaks, beetles overwhelm the trees' total defense capacity through pheromone-coordinated mass attacks (Paine *et al.*, 1997; Franceschi *et al.*, 2005; Raffa *et al.*, 2008). The effect of such mass attacks is probably reinforced by the beetles' microbial symbionts that, like the beetles themselves, engage tree defenses and thus enhance the effect of each attack. Because beetles and symbionts break down tree resistance in different ways, together they seem to present a more formidable challenge to the tree than either partner does alone. Bark beetles obviously play a central role in tree killing, since they are the ones who actively select suitable host trees and short-circuit the powerful defenses in the bark by boring straight into the defenseless cambial area. In addition, the beetles' tunneling activity inflicts considerable mechanical damage in the bark. However, beetle symbionts such as bluestain fungi seem to be obvious contributors to tree death. Nearly all tree-killing bark beetles are associated with bluestain fungi, these fungi are able to colonize and kill healthy bark and

sapwood beyond the beetle tunnels, and several species can kill healthy trees on their own if they are inoculated in sufficient numbers (Paine *et al.*, 1997; Franceschi *et al.*, 2005).

4.1 Contributions of Bark Beetles to Tree Death

The bark beetles are of course central in tree killing, as they are the active part in selecting and colonizing healthy trees. The beetle is the "bus" that arrives at the tree and delivers its many external and internal symbionts into the phloem. Furthermore, beetles carefully select trees of the right species and susceptibility to attack. They also bore through the bark and make tunnels deep inside the secondary phloem, thus bringing their symbionts close to the defenseless cambium zone and effectively short-circuiting the concentric layers of defense structures in the bark (Franceschi *et al.*, 2000). The beetles also produce the all-important aggregation pheromones that probably are essential to tree killing. Pheromone-coordinated mass attacks involving hundreds or thousands of beetles on a tree can rapidly overwhelm the total defense capacity of the tree and put it on a trajectory that eventually leads to tree death (Wood, 1982).

In addition to being essential for host selection, entry, and mass attack, the beetles damage the bark mechanically as they excavate their breeding galleries. However, during the critical first few weeks of an attack, before the eggs hatch and the larvae begin to develop, the extent of this mechanical damage might not be sufficient to severely stress the trees. Because the beetles' egg tunnels are only a few millimeters in diameter and usually run parallel to the stem axis, the amount of bark circumference that is destroyed by the tunnels is quite moderate. Horizontally oriented larval tunnels are not produced until the eggs hatch some weeks later, when the decisive phase in the interaction between tree defenses and the beetle–symbiont complex already may be over (Paine *et al.*, 1997; Lieutier *et al.*, 2009). The combined mechanical damage inflicted by the beetles' egg tunnels in a mass-attacked tree can be illustrated using the spruce bark beetle as an example. If we assume a bark thickness of 5 mm, a typical breeding density of 500 egg tunnels m^{-2} bark surface (Weslien, 1989), and that each egg tunnel is 10 cm long and 2.5 mm in diameter, this corresponds to the destruction of about 5% of the bark volume. Studies on Norway spruce suggest that this level of mechanical damage has little effect on tree vigor. Trees that were mechanically wounded by removing 5.5% of the bark volume 1–2 weeks prior to massive fungal inoculation did not become more susceptible to infection (Christiansen *et al.*, 1999b). On the contrary, trees that were inoculated a few times with a bluestain fungus and then wounded extensively became much more resistant to the subsequent mass inoculation (Christiansen *et al.*, 1999b). Thus, Norway spruce and probably other conifers appear to tolerate mechanical damage quite well and remain vigorous even in the face of extensive mechanical damage.

Indeed, a classic argument against a decisive role of beetle tunneling in exhausting tree defenses is the fact that even if trees are girdled by removing the bark around 100% of the bark circumference, they may continue to live (and even grow) for up to 2 years. Girdling only kills the trees when the roots starve and die from lack of photosynthate (Craighead, 1928). This illustrates that even massive destruction of bark tissue by tunneling bark beetles is unlikely to kill trees within the timeframe commonly seen in beetle-attacked trees in nature. Moderate amounts of mechanical damage due to beetle tunneling are in fact more likely to increase tree resistance than to decrease it, because wounding may trigger defense priming that boosts tree resistance (Christiansen *et al.*, 1999b; Zhao *et al.*, 2011a).

4.2 Contributions of Symbionts to Tree Death and the Case for the Beetle–Symbiont Complex

Some bark beetle symbionts are adapted to colonizing living bark and sapwood and may thus contribute to overwhelming tree defenses and conquering trees. Most importantly, many symbionts of tree-killing bark beetles may help deplete the trees' preformed and induced defenses (reviewed in Paine *et al.*, 1997; Lieutier *et al.*, 2009). Bluestain fungi associated with tree-killing bark beetles are able to colonize and kill tree tissues far beyond the beetle galleries, and some species can kill healthy trees in experimental mass inoculations (Horntvedt *et al.*, 1983; Christiansen, 1985; Krokene *et al.*, 2003). Because symbionts can colonize bark and sapwood beyond the reach of the beetles themselves, they probably increase the impact of each beetle attack on breaking down tree defenses. In addition, the fact that different symbionts interact with tree defenses in different ways may increase the overall challenge to tree resistance. Recent molecular and biochemical studies have, for example, shown that beetle-associated fungi and bacteria can metabolize specialized tree metabolites such as terpenes and phenolics (DiGuistini *et al.*, 2011; Adams *et al.*, 2013; Boone *et al.*, 2013; Hammerbacher *et al.*, 2013; Wang *et al.*, 2013). These exciting new studies provide direct molecular and biochemical evidence for a role of beetle symbionts in neutralizing tree defenses.

Some tree-killing bark beetles are associated with symbiotic fungi that may mobilize nutrients from the sapwood to the bark, where they can be utilized by the developing beetle larvae (Ayres *et al.*, 2000; Bleiker and Six, 2007). In these systems, the beetle symbionts clearly may provide important nutritional benefits to their beetle vectors

(Six and Wingfield, 2011), but this function is not in conflict with a role in neutralizing tree defenses (Lahr and Krokene, 2013). In order to gain access to tree resources, the nutritional symbionts must be able to break down tree defenses, alone or in combination with bark beetles and/or other symbionts. Phytopathogenicity in bluestain fungi may have evolved to mediate competitive interactions with other fungi by supporting survival and efficient resource capture in living, defensive trees (Harrington, 1993; Six and Wingfield, 2011). Under this scenario, fungal pathogenicity surely evolved because it increased fungal fitness, but the fungi's ability to metabolize tree defenses would still be beneficial to the beetles and could provide the basis for a mutualistic symbiosis between beetles and fungi (Chapter 3).

In principle, the relative contribution of bark beetle symbionts to tree defense exhaustion could range from zero to a 100%. It could be the case that symbionts do not contribute at all to breaking down tree resistance, as argued by Six and Wingfield (2011), or it could be that they are fully responsible for tree killing once their beetle vectors have introduced them into the bark. Most probably, however, the contribution of beetle symbionts to defense exhaustion is somewhere in between these two extremes, suggesting that both bark beetles and symbionts contribute to defense depletion. Because they contribute in different ways and thus seem to complement each other, beetles and symbionts probably constitute an additive or synergistic tree-killing complex (Franceschi *et al.*, 2005; Lieutier *et al.*, 2009; Krokene *et al.*, 2013). However, the relative contribution of beetles and symbionts to breaking down tree resistance probably varies much between different bark beetle–symbiont systems, and the symbionts' contribution may turn out to be quite low in some systems (Six and Wingfield, 2011).

4.3 Tree Death—Must the Trees be Killed?

Even though interactions between conifers, tree-killing bark beetles, and symbionts have been studied for more than a century, we have little precise knowledge about how beetle-attacked trees are dying. The fact that the actual tree-killing process is unclear has contributed to the confusion about the relative contribution of bark beetles and symbionts in tree death. What seems clear is that the decisive stages in the interaction between tree defenses and the beetle–symbiont complex take place long before the tree dies or develops clear symptoms (Paine *et al.*, 1997; Lieutier *et al.*, 2009). In their review on bark beetle–fungus interactions, Paine *et al.* (1997) stressed that when discussing tree killing by bark beetles it is critical to distinguish between the early phase of overcoming host defenses in the bark and later symptoms such as sapwood colonization and occlusion, fading of foliage, and death of the tree. Death of the whole tree, as manifested by completely bluestained sapwood, may occur long after the critical interactions between tree defenses and aggressors have taken place.

Death of the whole tree does not therefore seem to be the critical process to look at when studying bark beetle host colonization, and tree killing may in fact not be of primary importance to the bark beetles. What is critical to the reproductive success of the bark beetles seems to be that tree defenses are immobilized so the tree's induced defense responses do not become activated and kill the developing beetle brood (Franceschi et al., 2005). However, the only way to immobilize tree defenses is to kill the cambium and other living cells that are capable of mounting induced defenses. Disruption of water transport, which may be the proximate cause of death in a beetle-attacked tree, may be considered a later consequence of the immobilization of the induced defenses. When tree resistance in the bark has been overwhelmed, bluestain fungi and other symbionts may begin to colonize the sapwood. Thus, sapwood blue-staining is a very late manifestation of the important interactions between tree defenses and the bark beetle–symbiont complex that have taken place much earlier in the colonization process.

4.4 Naïve Host Trees

Increasing international trade and travel are accelerating the rate of new species introductions and giving rise to new interactions between plants and pests (Aukema et al., 2010). When species without a coevolutionary history meet, their interaction may be unstable, since there will often be an unbalance between the infection pressure of the pest and the resistance of the plant. Plants that are not coevolved with new pests have been termed naïve, since they tend to lack effective defenses to resist attack (Raffa et al., 2008). In conifers, many of the defense mechanisms described in this chapter may not work against new pests, which may be able to invade host tissues almost unrestrictedly (Hudgins et al., 2005). In forestry, such heightened host tree vulnerability often results in devastating losses from invasive insects (Liebhold et al., 1995). However, interactions between naïve plants and new pest species will not necessarily be disastrous for the plant. Exotic pests can also be considered coevolutionary naïve, since non-adapted pests may lack the ability to colonize native host plants. However, instances where native plants successfully resist attack and prevent exotic pests from establishing will usually go unnoticed, and the literature is likely to be enriched with cases where invasive pests get the upper hand and cause widespread plant mortality.

A prominent example of the potential importance of naïve hosts involving bark beetles is the spread of the mountain pine beetle across the Canadian Rocky Mountains

into Alberta, where the beetle has made contact with pine species that have had little exposure to the beetle previously (Logan and Powell, 2001; de la Giroday et al., 2012). Mountain pine beetles have been found to perform better in jack pine (*Pinus banksiana* Lamb.), a novel host tree, than in the historical host lodgepole pine *P. contorta* var. latifolia (Erbilgin et al., 2014). Beetles tunneling in jack pine bark attracted more conspecifics than beetles tunneling in lodgepole pine and produced larger offspring. Similarly, the mountain pine beetle has been found to have higher reproductive success in naïve lodgepole pine populations in areas that have not previously experienced frequent outbreaks (Cudmore et al., 2010). Further south in its range, the mountain pine beetle has expanded into high-elevation whitebark pine stands, where it is causing widespread tree mortality in a very vulnerable ecosystem (Logan and Powell, 2001; Logan et al., 2010). Compared to the historical host lodgepole pine, whitebark pine has very weak induction of terpenoid defenses and the terpene composition in its bark appears less able to inhibit the beetle's pheromone communication (Raffa et al., 2013).

4.5 Effects of Biotic and Abiotic Disturbances on Tree Defense

When trees are healthy and vigorous their preformed and induced defenses usually contribute to regulate bark beetle populations at low, endemic levels. However, various biotic and abiotic disturbance factors may stress trees and increase their susceptibility to attack. Disturbances that may induce tree susceptibility and trigger bark beetle outbreaks include severe drought, storm damage, flooding, lightning strikes, defoliation, and diseases (Paine and Baker, 1993). Human interventions, including thinning and other silvicultural measures, may promote tree vigor and increase forest resilience to disturbances (Fettig et al., 2007).

4.5.1 Biotic Disturbances: Defoliation and Disease

Pathogen and insect attack can reduce tree resistance and predispose trees to bark beetle colonization (Larsson, 1989). Bark beetle performance may be better in stressed trees with reduced resistance, since fewer beetles are required to overwhelm tree defenses, resulting in lower attack densities and higher offspring production. Root diseases are well known to increase the risk of bark beetle attack in several North American bark beetle–host tree systems (Paine and Baker, 1993). Similarly, defoliation by insects or pathogens may stress trees directly by reducing photosynthesis and carbon stores, and indirectly by reducing fine root growth and water uptake (Paine and Baker, 1993). Tree decline and mortality may also be caused by combined damage to both roots and foliage by interacting complexes of pathogens and insects, as exemplified by red pine decline in Illinois, Michigan, and Wisconsin (Klepzig *et al.*, 1991). Red pine *Pinus resinosa* Aiton is colonized by a closely associated complex of pathogens and insects that increase the trees' susceptibility to attack by the pine engraver *Ips pini* Say.

The effects of biotic disturbances on tree resistance do not always follow simple dose–response relationships but may be nonlinear (Bonello *et al.*, 2006). Trees in early, presymptomatic stages of infection may show increased resistance compared to healthy trees, due to activation of systemic induced resistance. However, if the disease organism becomes established and the tree turns symptomatic, tree resistance usually declines and trees become predisposed to attack by bark beetles or other mortality agents (Bonello *et al.*, 2006).

4.5.2 Abiotic Disturbances: Drought and High Temperatures

Drought is another well-documented factor triggering outbreaks of bark beetles and other insects (Mattson and Haack, 1987; Fettig *et al.*, 2007). Climate change is likely to accelerate drought-induced susceptibility to insect attack in many tree species (McDowell *et al.*, 2008; Allen *et al.*, 2010). Drought and accompanying high temperatures may benefit insects both directly, by increasing their growth and development rates, and indirectly, by increasing host stress and tree susceptibility (Mattson and Haack, 1987; Larsson, 1989). As with biotic disturbance factors the relationship between drought stress and tree susceptibility to bark beetles is not necessarily linear. Moderate drought stress has been observed to increase tree resistance to infection by bark beetle-associated bluestain fungi (Christiansen and Glosli, 1996), consistent with the growth-differentiation balance hypothesis (Herms and Mattson, 1992).

Severe drought exacerbated by high temperatures triggered a broad-scale eruption of the pinyon ips beetle *Ips confusus* (LeConte) in pinyon pine-juniper woodlands of southwestern North America in 2002 and 2003 (Raffa *et al.*, 2008). The outbreak quickly subsided when precipitation returned to normal levels and the drought stress was relaxed (Raffa *et al.*, 2008). This inability to sustain outbreaks in healthy, non-stressed stands is typical for semi-aggressive bark beetles like the pinyon ips. It contrasts with aggressive species, such as the mountain pine beetle, that can have self-amplifying outbreaks that are maintained even when the initial eliciting factors are relaxed (Raffa *et al.*, 2008).

4.5.3 Tree Phenology

Tree resistance to insect and pathogen attack is strongly influenced by phenology, with trees being less resistant during periods of active shoot growth. For example, from early spring to mid-summer, Norway spruce becomes progressively less resistant to an important fungal associate of the spruce bark beetle (Horntvedt, 1988). Similarly, young Norway spruce trees are less resistant to fungal infection during shoot elongation in the early summer, compared with at the dormant bud stage in the early spring or when shoot elongation has been completed in mid-summer (Krokene *et al.*, 2012). In the youngest trees (2 years old) changes in tree resistance occurred very rapidly, with only 2 weeks separating the growth stages with minimum and maximum resistance. These rapid changes in tree resistance may be related to trade-offs between allocation of starch and other stem carbohydrates to shoot growth versus defense, consistent with the growth-differentiation balance hypothesis (Herms and Mattson, 1992; Krokene *et al.*, 2012).

Variation in tree resistance with phenology is relevant in the context of climate change. Increasing temperatures will almost certainly alter the voltinism of bark beetle species by increasing the number of generations that are completed per year and shifting the first attack period earlier in the spring (Ayres and Lombardero, 2000; Berg *et al.*, 2006; Bentz *et al.*, 2010; Chapter 13). Depending on how tree phenology and resistance are affected by climate change this may either increase or reduce tree susceptibility to bark beetle attack. For the spruce bark beetle in Scandinavia, climate change is projected to alter the voltinism from one to two generations per year (Jönsson *et al.*, 2009). A second beetle generation that attacks trees in late summer, when tree resistance may be much lower than in the spring (Horntvedt, 1988), could increase tree killing by the spruce bark beetle in the future.

4.5.4 Effects of Thinning and other Silvicultural Measures on Tree Resistance

A variety of management practices may reduce the risk of bark beetle outbreaks and prevent outbreaks from expanding. Many of these practices are aimed at reducing stress levels in the trees by reducing competition among trees for water, nutrients, and growing space (Fettig *et al.*, 2007). Thinning stands by reducing tree density is a well-documented measure to increase tree vigor and mitigate negative impacts of bark beetles (reviewed in Fettig *et al.*, 2007). Because even-aged forest monocultures are more susceptible to bark beetle outbreaks, efforts to increase forest heterogeneity with respect to tree age and species composition are also likely to reduce the risk and impact of bark beetle outbreaks (Fettig *et al.*, 2007; Raffa *et al.*, 2008).

Thinning and other management activities (e.g., road construction, logging) may also increase the risk of bark beetle attacks if they are not done properly. Thinning materials that are left in the forest can provide beetles with necessary breeding material and mechanical damage to standing trees may promote infections that predispose trees to subsequent beetle attacks (Paine and Baker, 1993). Furthermore, although thinning increases the long-term resilience of forest stands, initial tree responses to thinning are not always positive and include reduced growth and symptoms associated with sudden exposure of shade-adapted foliage. These transient negative responses are referred to as "thinning shock" (Sharma et al., 2006).

Stand age may be a contributing factor to bark beetle outbreaks, as older trees often provide attractive and nutritious breeding substrates with thick phloem and relatively low defense capacity (Boone et al., 2011). Increasing areas of aging stands with high susceptibility to beetle attack was an important contributing factor to the enormous outbreak of the mountain pine beetle in Canada in 2000–2010 (Kurz et al., 2008; Raffa et al., 2008). Thus, if the focus is solely on reducing the risk of bark beetle outbreaks, forest stands should be rejuvenated either by logging, prescribed burning, or forest fires before trees reach an age when they are highly susceptible to attack.

5. CONCLUSIONS

Tree-killing bark beetles are a major mortality factor in many conifer species and are sometimes claimed to kill more trees than any other natural factor. Keys to the beetles' ability to kill relatively healthy trees are (1) their effective population aggregation pheromones that coordinate rapid mass attacks on individual host trees, and (2) their association with symbiotic organisms that may engage tree defenses and thus increase the impact of each attack. Although the exact contribution of bark beetles and symbionts in overwhelming tree defenses is not known, trees are always facing a combined attack from beetles and symbionts that probably constitute an additive or synergistic tree-killing complex. In the long periods that separate the relatively sporadic bark beetle outbreaks, the beetle populations are regulated at low, endemic levels and colonize weakened and dying trees. Conifer defense is considered a key regulator of bark beetle populations, and because well-functioning tree defenses may prevent bark beetle outbreaks, we need to understand tree defense mechanisms and how they interact with the bark beetle–symbiont complex. Important new insights into the three-way interactions between beetles, symbionts, and host trees are likely to come from new technological developments, such as the ongoing DNA sequencing and bioinformatics revolution. The rapid development of these technologies allows for massive genome and transcriptome sequencing of non-model organisms such as tree-killing bark beetles and their symbionts. Detailed characterization of the interactions between genomes of bark beetles, symbionts, and host trees may yield important new knowledge and contribute to improved management tools, such as silvicultural practices to maintain healthy, vigorous stands, and development of more resistant trees through breeding. Anthropogenic disturbances like climate change and associated insect range expansion reduce tree vigor and create new pest–conifer associations at an increasing rate. The continuing range expansion of the notorious mountain pine beetle and other tree-killing bark beetles is one example accentuating the pressing need for a more precise understanding of conifer resistance mechanisms and their interaction with bark beetles and symbionts.

REFERENCES

Abbott, E., Hall, D., Hamberger, B., Bohlmann, J., 2010. Laser microdissection of conifer stem tissues: isolation and analysis of high quality RNA, terpene synthase enzyme activity and terpenoid metabolites from resin ducts and cambial zone tissue of white spruce (Picea glauca). BMC Plant Biol. 10, 106.

Adams, A.S., Aylward, F.O., Adams, S.M., Erbilgin, N., Aukema, B.H., Currie, C.R., et al., 2013. Mountain pine beetles colonizing historical and naive host trees are associated with a bacterial community highly enriched in genes contributing to terpene metabolism. Appl. Environ. Microbiol. 79, 3468–3475.

Alcock, J., 1982. Natural selection and communication among bark beetles. Fla. Entomol. 65, 17–32.

Alfaro, R.I., He, F.L., Tomlin, E., Kiss, G., 1997. White spruce resistance to white pine weevil related to bark resin canal density. Can. J. Bot. 75, 568–573.

Alfieri, F.J., Evert, R.F., 1973. Structure and seasonal development of secondary phloem in Pinaceae. Bot. Gaz. 134, 17–25.

Allen, C.D., Macalady, A.K., Chenchouni, H., Bachelet, D., McDowell, N., Vennetier, M., et al., 2010. A global overview of drought and heat-induced tree mortality reveals emerging climate change risks for forests. For. Ecol. Manage. 259, 660–684.

Aukema, J.E., McCullough, D.G., Von Holle, B., Liebhold, A.M., Britton, K., Frankel, S.J., 2010. Historical accumulation of nonindigenous forest pests in the continental United States. Bioscience 60, 886–897.

Ayres, M.P., Lombardero, M.J., 2000. Assessing the consequences of global change for forest disturbance from herbivores and pathogens. Sci. Total Environ. 262, 263–286.

Ayres, M.P., Wilkens, R.T., Ruel, J.J., Lombardero, M.J., Vallery, E., 2000. Nitrogen budgets of phloem-feeding bark beetles with and without symbiotic fungi. Ecology 81, 2198–2210.

Bannan, M.W., 1936. Vertical resin ducts in the secondary wood of the Abietineae. New Phytol. 35, 11–46.

Benayoun, J., Fahn, A., 1979. Intracellular transport and elimination of resin from epithelial duct-cells of Pinus halepensis. Ann. Bot. 43, 179–181.

Bentz, B.J., Régnière, J., Fettig, C.J., Hansen, E.M., Hayes, J.L., Hicke, J.A., et al., 2010. Climate change and bark beetles of the Western United States and Canada: direct and indirect effects. Bioscience 60, 602–613.

Berg, E.E., Henry, J.D., Fastie, C.L., De Volder, A.D., Matsuoka, S.M., 2006. Spruce beetle outbreaks on the Kenai Peninsula, Alaska, and Kluane National Park and Reserve, Yukon Territory: relationship to summer temperatures and regional differences in disturbance regimes. For. Ecol. Manage. 227, 219–232.

Berryman, A.A., 1972. Resistance of conifers to invasion by bark beetle-fungus associations. Bioscience 22, 598–602.

Berryman, A.A., 1982. Biological control, thresholds, and pest outbreaks. Environ. Entomol. 11, 544–549.

Björkman, C., Larsson, S., Bommarco, R., 1997. Oviposition preferences in pine sawflies: a trade-off between larval growth and defence against natural enemies. Oikos 79, 45–52.

Bleiker, K.P., Six, D.L., 2007. Dietary benefits of fungal associates to an eruptive herbivore: potential implication of multiple associates on host population dynamics. Environ. Entomol. 36, 1384–1396.

Blomquist, G.J., Figueroa-Teran, R., Aw, M., Song, M., Gorzalski, A., Abbott, N.L., et al., 2010. Pheromone production in bark beetles. Insect Biochem. Mol. Biol. 40, 699–712.

Bonello, P., Blodgett, J.T., 2003. *Pinus nigra-Sphaeropsis sapinea* as a model pathosystem to investigate local and systemic effects of fungal infection of pines. Physiol. Mol. Plant Pathol. 63, 249–261.

Bonello, P., Gordon, T.R., Storer, A.J., 2001. Systemic induced resistance in Monterey pine. Forest Pathol. 31, 99–106.

Bonello, P., Gordon, T.R., Herms, D.A., Wood, D.L., Erbilgin, N., 2006. Nature and ecological implications of pathogen-induced systemic resistance in conifers: a novel hypothesis. Physiol. Mol. Plant Pathol. 68, 95–104.

Boone, C.K., Aukema, B.H., Bohlmann, J., Carroll, A.L., Raffa, K.F., 2011. Efficacy of tree defense physiology varies with bark beetle population density: a basis for positive feedback in eruptive species. Can. J. Forest Res. 41, 1174–1188.

Boone, C.K., Keefover-Ring, K., Mapes, A.C., Adams, A.S., Bohlmann, J., Raffa, K.F., 2013. Bacteria associated with a tree-killing insect reduce concentrations of plant defense compounds. J. Chem. Ecol. 39, 1003–1006.

Bordasch, R.P., Berryman, A.A., 1977. Host resistance to the fir engraver beetle, *Scolytus ventralis* (Coleoptera: Scolytidae). 2. Repellency of *Abies grandis* resins and some monoterpenes. Can. Entomol. 109, 95–100.

Borden, J.H., 1982. Aggregating pheromones. In: Mitton, J.B., Sturgeon, K.B. (Eds.), Bark Beetles in North American Conifers. A System for the Study of Evolutionary Biology. University of Texas Press, Austin, pp. 74–139.

Brignolas, F., Lacroix, B., Lieutier, F., Sauvard, D., Drouet, A., Claudot, A.-C., et al., 1995a. Induced responses in phenolic metabolism in two Norway spruce clones after wounding and inoculations with *Ophiostoma polonicum*, a bark beetle-associated fungus. Plant Physiol. 109, 821–827.

Brignolas, F., Lieutier, F., Sauvard, D., Yart, A., Drouet, A., Claudot, A.-C., 1995b. Changes in soluble-phenol content of Norway-spruce (*Picea abies*) phloem in response to wounding and inoculation with *Ophiostoma polonicum*. Eur. J. Forest Pathol. 25, 253–265.

Brignolas, F., Lieutier, F., Sauvard, D., Christiansen, E., Berryman, A.A., 1998. Phenolic predictors for Norway spruce resistance to the bark beetle *Ips typographus* (Coleoptera: Scolytidae) and an associated fungus, *Ceratocystis polonica*. Can. J. Forest Res. 28, 720–728.

Charon, J., Launay, J., Carde, J.-P., 1987. Spatial organization and volume density of leucoplasts in pine secretory cells. Protoplasma 138, 45–53.

Christiansen, E., 1985. *Ips/Ceratocystis*-infection of Norway spruce: what is a deadly dosage? Zeitschrift für Angewandte Entomologie 99, 6–11.

Christiansen, E., Glosli, A.M., 1996. Mild drought enhances the resistance of Norway spruce to a bark beetle-transmitted blue-stain fungus. USDA For. Serv. Gen. Tech. Rep. NC 192–199.

Christiansen, E., Krokene, P., 1999. Can Norway spruce trees be "vaccinated" against attack by *Ips typographus*? Agr. Forest Entomol. 1, 185–187.

Christiansen, E., Krokene, P., Berryman, A.A., Franceschi, V.R., Krekling, T., Lieutier, F., et al., 1999b. Mechanical injury and fungal infection induce acquired resistance in Norway spruce. Tree Physiol. 19, 399–403.

Christiansen, E., Franceschi, V.R., Nagy, N.E., Krekling, T., Berryman, A. A., Krokene, P., Solheim, H., 1999a. Traumatic resin duct formation in Norway spruce after wounding or infection with a bark beetle-associated blue-stain fungus, *Ceratocystis polonica*. In: Lieutier, F., Mattson, W.J., Wagner, M.R. (Eds.), Physiology and Genetics of Tree-Phytophage Interactions. INRA Editions, Versailles, pp. 79–89.

Christiansen, E., Waring, R.H., Berryman, A.A., 1987. Resistance of conifers to bark beetle attack: searching for general relationships. For. Ecol. Manage. 22, 89–106.

Cipollini, D., Heil, M., 2010. Costs and benefits of induced resistance to herbivores and pathogens in plants. CAB Rev.—Perspectives in Agriculture, Veterinary Science Nutrition and Natural Resources 5, 1–25.

Conrath, U., 2011. Molecular aspects of defence priming. Trends Plant Sci. 16, 524–531.

Conrath, U., Beckers, G.J.M., Flors, V., García-Agustín, P., Jakab, G., Mauch, F., et al., 2006. Priming: getting ready for battle. Mol. Plant-Microbe Interact. 19, 1062–1071.

Craighead, F.C., 1928. Interrelation of tree-killing bark beetles (*Dendroctonus*) and blue stains. J. For. 26, 886–887.

Cudmore, T.J., Björklund, N., Carroll, A.L., Lindgren, B.S., 2010. Climate change and range expansion of an aggressive bark beetle: evidence of higher beetle reproduction in naïve host tree populations. J. Appl. Ecol. 47, 1036–1043.

de la Giroday, H.-M.C., Carroll, A.L., Aukema, B.H., 2012. Breach of the northern Rocky Mountain geoclimatic barrier: initiation of range expansion by the mountain pine beetle. J. Biogeogr. 39, 1112–1123.

DiGuistini, S., Wang, Y., Liao, N.Y., Taylor, G., Tanguay, P., Feau, N., et al., 2011. Genome and transcriptome analyses of the mountain pine beetle-fungal symbiont *Grosmannia clavigera*, a lodgepole pine pathogen. Proc. Natl. Acad. Sci. U. S. A. 108, 2504–2509.

Dixon, R.A., Chen, F., Guo, D., Parvathi, K., 2001. The biosynthesis of monolignols: a "metabolic grid", or independent pathways to guaiacyl and syringyl units? Phytochemistry 57, 1069–1084.

Enebak, S.A., Carey, W.A., 2000. Evidence for induced systemic protection to fusiform rust in loblolly pine by plant growth-promoting rhizobacteria. Plant Dis. 84, 306–308.

Erbilgin, N., Krokene, P., Christiansen, E., Zeneli, G., Gershenzon, J., 2006. Exogenous application of methyl jasmonate elicits defenses in Norway spruce (*Picea abies*) and reduces host colonization by the bark beetle *Ips typographus*. Oecologia 148, 426–436.

Erbilgin, N., Krokene, P., Kvamme, T., Christiansen, E., 2007. A host monoterpene influences *Ips typographus* (Coleoptera: Curculionidae, Scolytinae) responses to its aggregation pheromone. Agr. Forest Entomol. 9, 135–140.

Erbilgin, N., Ma, C., Whitehouse, C., Shan, B., Najar, A., Evenden, M., 2014. Chemical similarity between historical and novel host plants promotes range and host expansion of the mountain pine beetle in a naïve host ecosystem. New Phytol. 201, 940–950.

Evensen, P.C., Solheim, H., Høiland, K., Stenersen, J., 2000. Induced resistance of Norway spruce, variation of phenolic compounds and their effects on fungal pathogens. For. Pathol. 30, 97–108.

Everaerts, C., Grégoire, J.-C., Merlin, J., 1988. The toxicity of Norway spruce monoterpenes to two bark beetle species and their associates.

In: Mattson, W.J., Levieux, J., Bernard-Dagan, C. (Eds.), Mechanisms of Woody Plant Defenses Against Insects—Search for Pattern. Springer-Verlag, New York, pp. 333–344.

Eyles, A., Bonello, P., Ganley, R., Mohammed, C., 2010. Induced resistance to pests and pathogens in trees. New Phytol. 185, 893–908.

Faccoli, M., Schlyter, F., 2007. Conifer phenolic resistance markers are bark beetle antifeedant semiochemicals. Agr. Forest Entomol. 9, 237–245.

Fahn, A., 1979. Secretory Tissues in Plants. Academic Press, London.

Fahn, A., 1990. Plant Anatomy. Pergamon Press, Oxford.

Fahn, A., Werker, E., Ben-Tzur, P., 1979. Seasonal effects of wounding and growth substances on development of traumatic resin ducts in *Cedrus libani*. New Phytol. 82, 537–544.

Fäldt, J., Solheim, H., Långström, B., Borg-Karlson, A.-K., 2006. Influence of fungal infection and wounding on contents and enantiomeric compositions of monoterpenes in phloem of *Pinus sylvestris*. J. Chem. Ecol. 32, 1779–1795.

Farjon, A., 2001. World Checklist and Bibliography of Conifers, Second ed. Royal Botanical Gardens, Kew, Richmond, UK.

Feeny, P., 1976. Plant apparency and chemical defense. In: Wallace, J.W., Mansell, R.T. (Eds.), Biochemichal Interaction between Plants and Insects. Plenum Press, New York, pp. 1–40.

Fettig, C.J., Klepzig, K.D., Billings, R.F., Munson, A.S., Nebeker, T.E., Negrón, J.F., Nowak, J.T., 2007. The effectiveness of vegetation management practices for prevention and control of bark beetle infestations in coniferous forests of the western and southern United States. For. Ecol. Manage. 238, 24–53.

Fossdal, C.G., Hietala, A.M., Kvaalen, H., Solheim, H., 2006. Changes in host chitinase isoforms in relation to wounding and colonization by *Heterobasidion annosum*: early and strong defense response in 33-year-old resistant Norway spruce clone. Tree Physiol. 26, 169–177.

Franceschi, V.R., Krekling, T., Berryman, A.A., Christiansen, E., 1998. Specialized phloem parenchyma cells in Norway spruce (Pinaceae) bark are an important site of defense reactions. Am. J. Bot. 85, 601–615.

Franceschi, V.R., Krokene, P., Krekling, T., Christiansen, E., 2000. Phloem parenchyma cells are involved in local and distant defense responses to fungal inoculation or bark-beetle attack in Norway spruce (Pinaceae). Am. J. Bot. 87, 314–326.

Franceschi, V.R., Krekling, T., Christiansen, E., 2002. Application of methyl jasmonate on *Picea abies* (Pinaceae) stems induces defense-related responses in phloem and xylem. Am. J. Bot. 89, 602–610.

Franceschi, V.R., Krokene, P., Christiansen, E., Krekling, T., 2005. Anatomical and chemical defenses of conifer bark against bark beetles and other pests. New Phytol. 167, 353–375.

Frost, C.J., Mescher, M.C., Carlson, J.E., De Moraes, C.M., 2008. Plant defense priming against herbivores: getting ready for a different battle. Plant Physiol. 146, 818–824.

Furniss, R.L., Carolin, V.M., 1977. Western forest insects. Miscellaneous publication no. 1339. US Department of Agriculture, Forest Service, Washington DC.

Gaylord, M.L., Hofstetter, R.W., Kolb, T.E., Wagner, M.R., 2011. Limited response of ponderosa pine bole defenses to wounding and fungi. Tree Physiol. 31, 428–437.

Gershenzon, J., 1994. Metabolic costs of terpenoid accumulation in higher plants. J. Chem. Ecol. 20, 1281–1328.

Gershenzon, J., Croteau, R., 1990. Regulation of monoterpene biosynthesis in higher plants. In: Towers, G.H.N., Stafford, H.A. (Eds.), Biochemistry of the Mevalonic Acid Pathway to Terpenoids.In: Recent Advances in Phytochemistry 24, pp. 99–160.

Gijzen, M., Lewinsohn, E., Savage, T.J., Croteau, R.B., 1993. Conifer monoterpenes—biochemistry and bark beetle chemical ecology. ACS Symp. Ser. 525, 8–22.

Hammerbacher, A., Schmidt, A., Wadke, N., Wright, L.P., Schneider, B., Bohlmann, J., et al., 2013. A common fungal associate of the spruce bark beetle metabolizes the stilbene defenses of Norway spruce. Plant Physiol. 162, 1324–1336.

Hammerbacher, A., Paetz, C., Wright, L.P., Fischer, T.C., Bohlmann, J., Davis, A.J., et al., 2014. Flavan-3-ols in Norway spruce: biosynthesis, accumulation, and function in response to attack by the bark beetle-associated fungus *Ceratocystis polonica*. Plant Physiol. 164, 2107–2122.

Harrington, T.C., 1993. Diseases of conifers caused by species of *Ophiostoma* and *Leptographium*. In: Wingfield, M.J., Seifert, K.A., Webber, J.F. (Eds.), *Ceratocystis* and *Ophiostoma*. Taxonomy, Ecology, and Pathogenicity. APS Press, St. Paul, pp. 161–172.

Herms, D.A., Mattson, W.J., 1992. The dilemma of plants: to grow or defend. Q. Rev. Biol. 67, 283–335.

Hietala, A.M., Kvaalen, H., Schmidt, A., Jøhnk, N., Solheim, H., Fossdal, C.G., 2004. Temporal and spatial profiles of chitinase expression by Norway spruce in response to bark colonization by *Heterobasidion annosum*. Appl. Environ. Microbiol. 70, 3948–3953.

Horntvedt, R., 1988. Resistance of *Picea abies* to *Ips typographus*: tree response to monthly inoculations with *Ophiostoma polonicum*, a beetle transmitted blue-stain fungus. Scand. J. Forest Res. 3, 107–114.

Horntvedt, R., Christiansen, E., Solheim, H., Wang, S., 1983. Artificial inoculation with *Ips typographus*-associated blue-stain fungi can kill healthy Norway spruce trees. Meddelelser fra Norsk Institutt for Skogforskning 38, 1–20.

Hudgins, J.W., Franceschi, V.R., 2004. Methyl jasmonate-induced ethylene production is responsible for conifer phloem defense responses and reprogramming of stem cambial zone for traumatic resin duct formation. Plant Physiol. 13, 2134–2149.

Hudgins, J.W., Christiansen, E., Franceschi, V.R., 2003a. Methyl jasmonate induces changes mimicking anatomical and chemical defenses in diverse members of the Pinaceae. Tree Physiol. 23, 361–371.

Hudgins, J.W., Krekling, T., Franceschi, V.R., 2003b. Distribution of calcium oxalate crystals in the secondary phloem of conifers: a constitutive defense mechanism? New Phytol. 159, 677–690.

Hudgins, J.W., Christiansen, E., Franceschi, V.R., 2004. Induction of anatomically based defense responses in stems of diverse conifers by methyl jasmonate: a phylogenetic perspective. Tree Physiol. 24, 251–264.

Hudgins, J.W., McDonald, G.I., Zambino, P.J., Klopfenstein, N.B., Franceschi, V.R., 2005. Anatomical and cellular responses of *Pinus monticola* stem tissues to invasion by *Cronartium ribicola*. Forest Pathology 35, 423–443.

Jönsson, A.M., Appelberg, G., Harding, S., Bärring, L., 2009. Spatio-temporal impact of climate change on the activity and voltinism of the spruce bark beetle, *Ips typographus*. Glob. Chang. Biol. 15, 486–499.

Kane, J.M., Kolb, T.E., 2010. Importance of resin ducts in reducing ponderosa pine mortality from bark beetle attack. Oecologia 164, 601–609.

Karban, R., Baldwin, I.T., 1997. Induced Responses to Herbivory. University of Chicago Press, Chicago.

Karban, R., Myers, J.H., 1989. Induced plant responses to herbivory. Annu. Rev. Ecol. Syst. 20, 331–348.

Karlsson, M., Hietala, A.M., Kvaalen, H., Solheim, H., Olson, Å., Stenlid, J., Fossdal, C.G., 2007. Quantification of host and pathogen DNA and RNA transcripts in the interaction of Norway spruce with *Heterobasidion parviporum*. Physiol. Mol. Plant Pathol. 70, 99–109.

Keeling, C.I., Bohlmann, J., 2006. Genes, enzymes and chemicals of terpenoid diversity in the constitutive and induced defence of conifers against insects and pathogens. New Phytol. 170, 657–675.

Keeling, C.I., Weisshaar, S., Lin, R.P.C., Bohlmann, J., 2008. Functional plasticity of paralogous diterpene synthases involved in conifer defense. Proc. Natl. Acad. Sci. U. S. A. 105, 1085–1090.

Kirkendall, L.R., Kent, D.S., Raffa, K.F., 1997. Interaction among males, females and offspring in bark and ambrosia beetles: the significance of living in tunnels for the evolution of social behavior. In: Choe, J.C., Crespi, B.J. (Eds.), The Evolution of Social Behavior in Insects and Arachnids. Cambridge University Press, Cambridge, pp. 181–215.

Klepzig, K.D., Raffa, K.F., Smalley, E.B., 1991. Association of an insect-fungal complex with red pine decline in Wisconsin. For. Sci. 37, 1119–1139.

Krekling, T., Franceschi, V.R., Berryman, A.A., Christiansen, E., 2000. The structure and development of polyphenolic parenchyma cells in Norway spruce (*Picea abies*) bark. Flora 195, 354–369.

Krekling, T., Franceschi, V.R., Krokene, P., Solheim, H., 2004. Differential anatomical response of Norway spruce stem tissues to sterile and fungus infected inoculations. Trees 18, 1–9.

Krokene, P., Nagy, N.E., 2012. Anatomical aspects of resin-based defences in pine. In: Fett-Neto, A.G., Rodrigues-Corrêa, K.C.S. (Eds.), Pine Resin: Biology, Chemistry and Applications. Research Signpost, Kerala, pp. 67–86.

Krokene, P., Solheim, H., 1996. Fungal associates of five bark beetle species colonizing Norway spruce. Can. J. Forest Res. 26, 2115–2122.

Krokene, P., Christiansen, E., Solheim, H., Franceschi, V., Berryman, A., 1999. Induced resistance to pathogenic fungi in Norway spruce. Plant Physiol. 121, 565–570.

Krokene, P., Solheim, H., Långström, B., 2000. Fungal infection and mechanical wounding induce disease resistance in Scots pine. Eur. J. Plant Pathol. 106, 537–541.

Krokene, P., Solheim, H., Christiansen, E., 2001. Induction of disease resistance in Norway spruce (*Picea abies*) by necrotizing fungi. Plant Pathol. 50, 230–233.

Krokene, P., Solheim, H., Krekling, T., Christiansen, E., 2003. Inducible anatomical defense responses in Norway spruce stems and their possible role in induced resistance. Tree Physiol. 23, 191–197.

Krokene, P., Nagy, N.E., Krekling, T., 2008a. Traumatic resin ducts and polyphenolic parenchyma cells in conifers. In: Schaller, A. (Ed.), Induced Plant Resistance to Herbivory. Springer, Berlin, pp. 147–169.

Krokene, P., Nagy, N.E., Solheim, H., 2008b. Methyl jasmonate and oxalic acid treatment of Norway spruce: anatomically based defense responses and increased resistance against fungal infection. Tree Physiol. 28, 29–35.

Krokene, P., Lahr, E., Dalen, L.S., Skrøppa, T., Solheim, H., 2012. Effect of phenology on susceptibility of Norway spruce (*Picea abies*) to fungal pathogens. Plant Pathol. 61, 57–62.

Krokene, P., Fossdal, C.G., Nagy, N.E., Solheim, H., 2013. Conifer defense against insects and fungi. In: Seifert, K.A., de Beer, Z.W., Wingfield, M.J. (Eds.), Ophiostomatoid Fungi: Expanding Frontiers. CBS Biodiversity Series 12. CBS Press, Utrecht, pp. 141–154.

Kuć, J., 1982. Induced immunity to plant disease. Bioscience 32, 854–860.

Kurz, W.A., Dymond, C.C., Stinson, G., Rampley, G.J., Neilson, E.T., Carroll, A.L., et al., 2008. Mountain pine beetle and forest carbon feedback to climate change. Nature 452, 987–990.

Lahr, E.C., Krokene, P., 2013. Conifer stored resources and resistance to a fungus associated with the spruce bark beetle *Ips typographus*. PLoS One 8 (8), e72405.

Larsson, S., 1989. Stressful times for the plant stress: insect performance hypothesis. Oikos 56, 277–283.

Larsson, S., 2002. Resistance in trees to insects—an overview of mechanisms and interactions. In: Wagner, M.R., Clancy, K.M., Lieutier, F., Paine, T.D. (Eds.), Mechanisms and Deployment of Resistance in Trees to Insects. Kluwer Academic Publishers, Dordrecht, pp. 1–29.

Latty, T.M., Reid, M.L., 2009. Who goes first? Condition and danger dependent pioneering in a group-living bark beetle (*Dendroctonus ponderosae*). Behav. Ecol. Sociobiol. 64, 639–646.

Li, S.-H., Nagy, N.E., Hammerbacher, A., Krokene, P., Niu, X.-M., Gershenzon, J., Schneider, B., 2012. Localization of phenolics in phloem parenchyma cells of Norway spruce (*Picea abies*). Chembiochem 13, 2707–2713.

Liebhold, A.M., MacDonald, W.L., Bergdahl, D., Mastro, V.C., 1995. Invasion by exotic forest pests: a threat to forest ecosystems. Forest Science Monograph 30, 1–49.

Lieutier, F., 2002. Mechanisms of resistance in conifers and bark beetle attack strategies. In: Wagner, M.R., Clancy, K.M., Lieutier, F., Paine, T.D. (Eds.), Mechanisms and Deployment of Resistance in Trees to Insects. Springer, Netherlands, pp. 31–77.

Lieutier, F., 2004. Host resistance to bark beetles and its variations. In: Lieutier, F., Day, K.R., Battisti, A., Grégoire, J.-C., Evans, H.F. (Eds.), Bark and Wood Boring Insects in Living Trees in Europe, a Synthesis. Kluwer Academic Publishers, Dordrecht, pp. 135–180.

Lieutier, F., Yart, A., Jay-Allemand, C., Delorme, L., 1991. Preliminary investigations on phenolics as a response of Scots pine phloem to attacks by bark beetles and associated fungi. Eur. J. Forest Pathol. 21, 354–364.

Lieutier, F., Sauvard, D., Brignolas, F., Picron, V., Yart, A., Bastien, C., Jay-Allemand, C., 1996. Changes in phenolic metabolites of Scots-pine phloem induced by *Ophiostoma brunneo-ciliatum*, a bark-beetle-associated fungus. Eur. J. Forest Pathol. 26, 145–158.

Lieutier, F., Brignolas, F., Sauvard, D., Yart, A., Galet, C., Burnet, M., Van de Sype, H., 2003. Intra- and inter-provenance variability in phloem phenols of *Picea abies* and relationship to a bark beetle-associated fungus. Tree Physiol. 23, 247–256.

Lieutier, F., Yart, A., Salle, A., 2009. Stimulation of tree defenses by Ophiostomatoid fungi can explain attack success of bark beetles on conifers. Ann. Forest Sci. 66, 801.

Lindberg, M., Lundgren, L., Gref, R., Johansson, M., 1992. Stilbenes and resin acids in relation to the penetration of *Heterobasidion annosum* through the bark of *Picea abies*. Eur. J. Forest Pathol. 22, 95–106.

Logan, J.A., Powell, J.A., 2001. Ghost forests, global warming, and the mountain pine beetle (Coleoptera: Scolytidae). American Entomologist 47, 160–172.

Logan, J.A., Macfarlane, W.W., Willcox, L., 2010. Whitebark pine vulnerability to climate-driven mountain pine beetle disturbance in the Greater Yellowstone Ecosystem. Ecol. Appl. 20, 895–902.

Lombardero, M.J., Ayres, M.P., Lorio, P.L.J., Ruel, J.J., 2000. Environmental effects on constitutive and inducible resin defences of *Pinus taeda*. Ecol. Lett. 3, 329–339.

Luchi, N., Ma, R., Capretti, P., Bonello, P., 2005. Systemic induction of traumatic resin ducts and resin flow in Austrian pine by wounding and inoculation with *Sphaeropsis sapinea* and *Diplodia scrobiculata*. Planta 221, 75–84.

Luna, E., Bruce, T.J.A., Roberts, M.R., Flors, V., Ton, J., 2012. Next-generation systemic acquired resistance. Plant Physiol. 158, 844–853.

Martin, D., Tholl, D., Gershenzon, J., Bohlmann, J., 2002. Methyl jasmonate induces traumatic resin ducts, terpenoid resin biosynthesis, and terpenoid accumulation in developing xylem of Norway spruce stems. Plant Physiol. 129, 1003–1018.

Mattson, W.J., Haack, R.A., 1987. The role of drought in outbreaks of plant-eating insects. Bioscience 37, 110–118.

McDowell, N., Pockman, W.T., Allen, C.D., Breshears, D.D., Cobb, N., Kolb, T., et al., 2008. Mechanisms of plant survival and mortality during drought: why do some plants survive while others succumb to drought? New Phytol. 178, 719–739.

McNee, W.R., Bonello, P., Storer, A.J., Wood, D.L., Gordon, T.R., 2003. Feeding response of *Ips paraconfusus* to phloem and phloem metabolites of *Heterobasidion annosum*-inoculated ponderosa pine, *Pinus ponderosa*. J. Chem. Ecol. 29, 1183–1202.

Miller, D.R., Borden, J.H., 2000. Dose-dependent and species-specific responses of pine bark beetles (Coleoptera: Scolytidae) to monoterpenes in association with pheromones. Can. Entomol. 132, 183–195.

Mulock, P., Christiansen, E., 1986. The threshold of successful attack by *Ips typographus* on *Picea abies*: a field experiment. For. Ecol. Manage. 14, 125–132.

Nagy, N.E., Franceschi, V.R., Solheim, H., Krekling, T., Christiansen, E., 2000. Wound-induced traumatic resin duct formation in stems of Norway spruce (Pinaceae): anatomy and cytochemical traits. Am. J. Bot. 87, 302–313.

Nagy, N.E., Fossdal, C.G., Krokene, P., Krekling, T., Lönneborg, A., Solheim, H., 2004. Induced responses to pathogen infection in Norway spruce phloem: changes in polyphenolic parenchyma cells, chalcone synthase transcript levels and peroxidase activity. Tree Physiol. 24, 505–515.

Nagy, N.E., Krokene, P., Solheim, H., 2006. Anatomical-based defense responses of Scots pine (*Pinus sylvestris*) stems to two fungal pathogens. Tree Physiol. 26, 159–167.

Ohmart, C.P., 1989. Why are there so few tree-killing bark beetles associated with angiosperms? Oikos 54, 242–245.

Paine, T.D., Baker, F.A., 1993. Abiotic and biotic predisposition. In: Schowalter, T.D., Filip, G.M. (Eds.), Beetle-Pathogen Interactions in Conifer Forests. Academic Press, London, pp. 61–79.

Paine, T.D., Raffa, K.F., Harrington, T.C., 1997. Interactions among scolytid bark beetles, their associated fungi, and live host conifers. Annu. Rev. Entomol. 42, 179–206.

Parham, R.A., Kaustinen, H.M., 1976. Differential staining of tannin in sections of epoxy-embedded plant cells. Stain Technol. 51, 237–240.

Phillips, M.A., Croteau, R.B., 1999. Resin-based defense in conifers. Trends in Plant Sci. 4, 184–190.

Raffa, K.F., 1991. Induced defensive reactions in conifer-bark beetle systems. In: Tallamy, D.W., Raupp, M.J. (Eds.), Phytochemical Induction by Herbivores. John Wiley and Sons, New York, pp. 245–276.

Raffa, K.F., Berryman, A.A., 1983. The role of host plant resistance in the colonization behaviour and ecology of bark beetles (Coleoptera: Scolytidae). Ecol. Monogr. 53, 27–49.

Raffa, K.F., Berryman, A.A., 1987. Interacting selective pressures in conifer-bark beetle systems: a basis for reciprocal adaptations? Am. Nat. 129, 234–262.

Raffa, K.F., Berryman, A.A., Simasko, J., Teal, W., Wong, B.L., 1985. Effects of grand fir monoterpenes on the fir engraver, Scolytus ventralis (Coleoptera: Scolytidae), and its symbiotic fungus. Environ. Entomol. 14, 552–556.

Raffa, K.F., Phillips, T.W., Salom, S.M., 1993. Strategies and mechanisms of host colonization by bark beetles. In: Schowalter, T.D., Filip, G.M.

(Eds.), Beetle–Pathogen Interactions in Conifer Forests. Academic Press, London, pp. 103–128.

Raffa, K.F., Aukema, B.H., Erbilgin, N., Klepzig, K.D., Wallin, K.F., 2005. Interactions among conifer terpenoids and bark beetles across multiple levels of scale: an attempt to understand links between population patterns and physiological processes. In: Romeo, J. (Ed.), Chemical Ecology and Phytochemistry of Forest Ecosystems. In: Recent Adv. Phytochem. 39, 79–118.

Raffa, K.F., Aukema, B.H., Bentz, B.J., Carroll, A.L., Hicke, J.A., Turner, M.G., Romme, W.H., 2008. Cross-scale drivers of natural disturbances prone to anthropogenic amplification: the dynamics of bark beetle eruptions. Bioscience 58, 501–517.

Raffa, K.F., Powell, E.N., Townsend, P.A., 2013. Temperature-driven range expansion of an irruptive insect heightened by weakly coevolved plant defenses. Proc. Natl. Acad. Sci. U. S. A. 110, 2193–2198.

Ralph, S.G., Yueh, H., Friedmann, M., Aeschliman, D., Zeznik, J.A., Nelson, C.C., et al., 2006. Conifer defence against insects: microarray gene expression profiling of Sitka spruce (*Picea sitchensis*) induced by mechanical wounding or feeding by spruce budworms (*Choristoneura occidentalis*) or white pine weevils (*Pissodes strobi*) reveals large-scale. Plant Cell Environ. 29, 1545–1570.

Ralph, S.G., Hudgins, J.W., Jancsik, S., Franceschi, V.R., Bohlmann, J., 2007. Aminocyclopropane carboxylic acid synthase is a regulated step in ethylene-dependent induced conifer defense. Full-length cDNA cloning of a multigene family, differential constitutive, and wound- and insect-induced expression, and cellular and subcellular localization in spruce and Douglas fir. Plant Physiol. 143, 410–424.

Reglinski, T., Stavely, F.J.L., Taylor, J.T., 1998. Induction of phenylalanine ammonia lyase activity and control of *Sphaeropsis sapinea* infection in *Pinus radiata* by 5-chlorosalicylic acid. Eur. J. Forest Pathol. 28, 153–158.

Reid, R.W., Watson, J.A., 1966. Sizes, distributions, and numbers of vertical resin ducts in lodgepole pine. Can. J. Bot. 44, 519–525.

Reid, R.W., Whitney, H.S., Watson, J.A., 1967. Reactions of lodgepole pine to attack by *Dendroctonus ponderosae* Hopkins and blue stain fungi. Can. J. Bot. 45, 1115–1126.

Richard, S., Lapointe, G., Rutledge, G., Seguin, A., 2000. Induction of chalcone synthase expression in white spruce by wounding and jasmonate. Plant Cell Physiol. 41, 982–987.

Ro, D.-K., Arimura, G.-I., Lau, S.Y.W., Piers, E., Bohlmann, J., 2005. Loblolly pine abietadienol/abietadienal oxidase PtAO (CYP720B1) is a multifunctional, multisubstrate cytochrome P450 monooxygenase. Proc. Natl. Acad. Sci. U. S. A. 102, 8060–8065.

Rosner, S., Führer, E., 2002. The significance of lenticels for successful *Pityogenes chalcographus* (Coleoptera: Scolytidae) invasion of Norway spruce trees [*Picea abies* (Pinaceae)]. Trees 16, 497–503.

Ross, A.F., 1961. Systemic acquired resistance induced by localized virus infections in plants. Virology 14, 340–358.

Rudinsky, J.A., Novák, V., Švihra, P., 1971. Pheromone and terpene attraction in the bark beetle *Ips typographus* L. Experientia 27, 161–162.

Ruel, J.J., Ayres, M.P., Lorio Jr., P.L., 1998. Loblolly pine responds to mechanical wounding with increased resin flow. Can. J. Forest Res. 28, 596–602.

Schiebe, C., Hammerbacher, A., Birgersson, G., Witzell, J., Brodelius, P. E., Gershenzon, J., et al., 2012. Inducibility of chemical defenses in Norway spruce bark is correlated with unsuccessful mass attacks by the spruce bark beetle. Oecologia 170, 183–198.

Schmidt, A., Zeneli, G., Hietala, A.M., Fossdal, C.G., Krokene, P., Christiansen, E., Gershenzon, J., 2005. Induced chemical defenses in conifers: biochemical and molecular approaches to studying their function. In: Romeo, J. (Ed.), Chemical Ecology and Phytochemistry of Forest Ecosytems.In: Recent Adv. Phytochem, 39, pp. 1–28.

Schmidt, A., Nagel, R., Krekling, T., Christiansen, E., Gershenzon, J., Krokene, P., 2011. Induction of isoprenyl diphosphate synthases, plant hormones and defense signalling genes correlates with traumatic resin duct formation in Norway spruce (Picea abies). Plant Mol. Biol. 77, 577–590.

Seifert, K.A., de Beer, Z.W., Wingfield, M.J., 2013. Ophiostomatoid Fungi: Expanding Frontiers. CBS Biodiversity Series 12, CBS Press, Utrecht.

Seybold, S.J., Bohlmann, J., Raffa, K.F., 2000. Biosynthesis of coniferophagous bark beetle pheromones and conifer isoprenoids: evolutionary perspective and synthesis. Can. Entomol. 132, 697–753.

Sharma, M., Smith, M., Burkhart, H.E., Amateis, R.L., 2006. Modeling the impact of thinning on height development of dominant and codominant loblolly pine trees. Ann. Forest Sci. 63, 349–354.

Six, D.L., 2013. The bark beetle holobiont: why microbes matter. J. Chem. Ecol. 39, 989–1002.

Six, D.L., Wingfield, M.J., 2011. The role of phytopathogenicity in bark beetle-fungus symbioses: a challenge to the classic paradigm. Annu. Rev. Entomol. 56, 255–272.

Smith, G.D., Carroll, A.L., Lindgren, B.S., 2011. Facilitation in bark beetles: endemic mountain pine beetle gets a helping hand. Agr. Forest Entomol. 13, 37–43.

Smith, R.H., 1961. The fumigant toxicity of three pine resins to Dendroctonus brevicomis and D. jeffreyi. J. Econ. Entomol. 54, 365–369.

Speight, M.R., Wainhouse, D., 1989. Ecology and Management of Forest Insects. Clarendon Press, Oxford.

Steele, C.L., Crock, J., Bohlmann, J., Croteau, R., 1998. Sesquiterpene synthases from grand fir (Abies grandis)—comparison of constitutive and wound-induced activities, and cDNA isolation, characterization and bacterial expression of delta-selinene synthase and gamma-humulene synthase. J. Biol. Chem. 273, 2078–2089.

Steppuhn, A., Baldwin, I.T., 2008. Induced defences and the cost-benefit paradigm. In: Schaller, A. (Ed.), Induced Plant Resistance to Herbivory. Springer, Berlin, pp. 61–83.

Thalenhorst, W., 1958. Grundzuge der Populationdynamik des grossen Fichtenborkenkafers Ips typographus L. Schriften aus der Forstlichen Fakultät der Universität Göttingen 21, 1–126.

Tomlin, E.S., Borden, J.H., 1994. Relationship between leader morphology and resistance or susceptibility of Sitka spruce to the white pine weevil. Can. J. Forest Res. 24, 810–816.

Vanderwel, D., Oehlschlager, A.C., 1987. Biosynthesis of pheromones and endocrine regulation of pheromone production in Coleoptera. In: Prestwich, G.D., Blomquist, G.J. (Eds.), Pheromone Biochemistry. Academic Press, New York, pp. 175–215.

Viiri, H., Annila, E., Kitunen, V., Niemelä, P., 2001. Induced responses in stilbenes and terpenes in fertilized Norway spruce after inoculation with blue-stain fungus, Ceratocystis polonica. Trees 15, 112–122.

Wainhouse, D., Cross, D.J., Howell, R.S., 1990. The role of lignin as a defense against the spruce bark beetle Dendroctonus micans—effect on larvae and adults. Oecologia 85, 257–265.

Wainhouse, D., Rose, D.R., Peace, A.J., 1997. The influence of preformed defences on the dynamic wound response in spruce bark. Funct. Ecol. 11, 564–572.

Wallin, K.F., Kolb, T.E., Skov, K.R., Wagner, M.R., 2003. Effects of crown scorch on ponderosa pine resistance to bark beetles in northern Arizona. Environ. Entomol. 32, 652–661.

Walters, D.R., 2009. Are plants in the field already induced? Implications for practical disease control. Crop Protect. 28, 459–465.

Wang, Y., Lim, L., DiGuistini, S., Robertson, G., Bohlmann, J., Breuil, C., 2013. A specialized ABC efflux transporter GcABC-G1 confers monoterpene resistance to Grosmannia clavigera, a bark beetle-associated fungal pathogen of pine trees. New Phytol. 197, 886–898.

Werker, E., Fahn, A., 1969. Resin ducts of Pinus halepensis Mill.—their structure, development and pattern of arrangement. Bot. J. Linn. Soc. 62, 379–411.

Weslien, J., 1989. The role of trapping in population management of the spruce bark beetle Ips typographus (L.). PhD thesis, Division of Forest Entomology, Swedish University of Agricultural Sciences, Uppsala.

Whitney, H.S., 1982. Relationships between bark beetles and symbiotic organisms. In: Mitton, J.B., Sturgeon, K.B. (Eds.), Bark Beetles in North American Conifers. A System for the Study of Evolutionary Biology. University of Texas Press, Austin, pp. 183–211.

Wingfield, M.J., Seifert, K.A., Webber, J.F., 1993. Ceratocystis and Ophiostoma. Taxonomy, Ecology, and Pathogenicity, APS Press, St. Paul.

Wood, D.L., 1972. Selection and colonization of ponderosa pine by bark beetles. In: van Emden, H.F. (Ed.), Insect/Plant Relationships. Blackwell Scientific, London, pp. 101–117.

Wood, D.L., 1982. The role of pheromones, kairomones, and allomones in the host selection and colonization behaviour of bark beetles. Annu. Rev. Entomol. 27, 411–446.

Wu, H., Hu, Z., 1997. Comparative anatomy of resin ducts of the Pinaceae. Trees 11, 135–143.

Zeneli, G., Krokene, P., Christiansen, E., Krekling, T., Gershenzon, J., 2006. Methyl jasmonate treatment of mature Norway spruce (Picea abies) trees increases the accumulation of terpenoid resin components and protects against infection by Ceratocystis polonica, a bark beetle-associated fungus. Tree Physiol. 26, 977–988.

Zhao, T., Krokene, P., Björklund, N., Långström, B., Solheim, H., Christiansen, E., Borg-Karlson, A.-K., 2010. The influence of Ceratocystis polonica inoculation and methyl jasmonate application on terpene chemistry of Norway spruce, Picea abies. Phytochemistry 71, 1332–1341.

Zhao, T., Borg-Karlson, A.-K., Erbilgin, N., Krokene, P., 2011a. Host resistance elicited by methyl jasmonate reduces emission of aggregation pheromones by the spruce bark beetle, Ips typographus. Oecologia 167, 691–699.

Zhao, T., Krokene, P., Hu, J., Christiansen, E., Björklund, N., Långström, B., et al., 2011b. Induced terpene accumulation in Norway spruce inhibits bark beetle colonization in a dose-dependent manner. PLoS One 6 (10), e26649.

Zulak, K.G., Dullat, H.K., Keeling, C.I., Lippert, D., Bohlmann, J., 2010. Immunofluorescence localization of levopimaradiene/abietadiene synthase in methyl jasmonate treated stems of Sitka spruce (Picea sitchensis) shows activation of diterpenoid biosynthesis in cortical and developing traumatic resin ducts. Phytochemistry 71, 1695–1699.

Chapter 6

Symbiotic Associations of Bark Beetles

Richard W. Hofstetter[1], Jamie Dinkins-Bookwalter[2], Thomas S. Davis[3], and Kier D. Klepzig[2]

[1] *School of Forestry, Northern Arizona University, Flagstaff, AZ, USA,* [2] *United States Department of Agriculture Forest Service Station, Southern Research Station, Asheville, NC, USA,* [3] *Plant, Soil and Entomological Sciences, University of Idaho, Moscow, ID, USA*

1. INTRODUCTION

1.1 Why Symbioses are Important to Understand

Symbiotic interactions are prevalent in all bark beetle communities. For many species, the ability to associate with multiple partners enables species to persist through fluctuations in climate, resources, predation, and partner availability. Symbionts, particularly mutualistic species associated with bark beetles, can increase bark beetle fitness by providing nutrition (Francke-Grosmann, 1952; Barras, 1970; Ayres *et al.*, 2000) or protection (Klepzig *et al.*, 2001a; Six, 2003), exhaust or detoxify tree defenses (DiGuistini *et al.*, 2007, 2011), enhance communication (Brand *et al.*, 1976; Whitney, 1982; Leufven *et al.*, 1984), and promote or discourage other organisms (Cardoza *et al.*, 2008; Hofstetter *et al.*, 2013). Alternatively, symbiotic species that are antagonistic to bark beetles can negatively affect bark beetle fitness directly (e.g., pathogens of bark beetles; Chapter 7) or indirectly (e.g., competing with mutualistic microbes within trees; Hofstetter *et al.*, 2006b; Adams *et al.*, 2009; Cardoza *et al.*, 2012). Symbionts associated with bark beetles have also been used to better understand the basic field of science such as mutualism theory (Klepzig and Six, 2004; Klepzig *et al.*, 2009), evolutionary and ecology adaption (e.g., horizontal gene transfer, Acuña et al., 2012), and drivers of population dynamics (Martinson *et al.*, 2013).

In general, bark beetle symbionts are known to affect mechanisms of evolution, coadaptation and speciation (Aanen *et al.*, 2009; Chapter 3), tree defenses (Six and Paine, 1998; Davis *et al.*, 2011), chemical communication (Raffa, 2001), population dynamics (Yearian *et al.*, 1966, 1972; Kirisits, 2004; Hofstetter *et al.*, 2006a, b), range expansion (Safranyik *et al.*, 2010; Hulcr and Dunn, 2011), and pest management (Wegensteiner, 2004). Symbionts can have multiple roles, and interactions can change as species and environments change. Thus, simply categorizing symbionts as mutualistic, antagonistic, commensal (Table 6.1), etc. can be misleading (Adams *et al.*, 2009, 2011). The combinations of genomic, behavioral, and ecological research approaches that incorporate symbionts will help us better understand how symbionts affect bark beetles (Cardoza *et al.*, 2012; Six, 2012). These interactions and effects will be discussed in more detail in this chapter.

1.2 General Description of Major Symbionts

All bark beetles harbor gut microbes and many bark beetle species are obligate associates with external microbes. These microbes include species of fungi, bacteria, viruses, and algae. Microbes can also be transported on beetle body surfaces or within highly specialized structures of the exoskeleton called mycangia (Figures 6.1 and 6.2) (Francke-Grosmann, 1967; Whitney, 1982; Six, 2003; Kirisits, 2004; Wegensteiner, 2004; Six, 2012). Economic impacts of bark beetles can be partly attributed to fungal associates, as some fungi are tree pathogens, causing vascular wilt or vascular stain diseases (Wingfield *et al.*, 1984; Harrington, 1993). Fungal associates affect the wood industry by causing discoloration called bluestain of the sapwood (Seifert, 1993; Solheim, 1995). Many fungal species are pathogenic to bark beetles and might be useful as biological control agents (Popa *et al.*, 2012; Chapter 7). Non-fungal microbial species, such as bacteria, can play a role in overcoming plant defenses (Adams *et al.*, 2013; Boone *et al.*, 2013) or are tree disease agents (Sinclair and Lyon, 2005), but have been less researched.

Mites, protozoa, nematodes, and other motile organisms are important symbiotic associates of bark beetles. Over a hundred species of mites (Lindquist, 1969a; Hofstetter *et al.*, 2013) and nematodes (Massey, 1974) have been documented as phoretic on adult beetles and other invertebrates in trees infested by bark beetles. Here, over 250 mite species are listed as phoretic on bark beetles (Table 6.2). Many animals are dependent on bark beetles or other insects for transportation between or within trees (Moser, 1985;

Bark Beetles. http://dx.doi.org/10.1016/B978-0-12-417156-5.00006-X

TABLE 6.1 Terms and Definitions Relating to Symbioses

Term	Definition
Symbiome	The complete set of symbionts with which an organism is associated
Symbiosis (plural: symbioses)	The living together of two organisms from different species, regardless of the impact that each organism has on each another
Mutualism	Two associated organisms of different species that benefit each other
Commensalism	Two associated organisms where one organism benefits without affecting the other
Antagonism/competition	Interaction where both organisms inflict harm on one another
Parasitism/ predation	Interaction where one organism benefits at the expense of another (e.g., predator or parasitoid)
Obligate	When one organism is fully dependent on the other for survival, reproduction, or nutrients, etc.
Facultative	When one organism is not fully dependent on the other for survival, reproduction, or nutrients, etc.
Fungus (plural: fungi)	Eukaryotic organisms with cell walls that contain chitin. This kingdom of organisms also contains the yeasts and molds
Yeast	Fungi that are unicellular, although strings of cells can be connected via budding cells
Bacterium (plural: bacteria)	Prokaryotic microorganisms, typically a few micrometers in length. Bacteria have a number of shapes, ranging from spheres to rods and spirals
Virus (plural: viruses)	Small infectious agents that replicate inside the living cells of other organisms. Viruses consist of either DNA or RNA, a protein coat that protects these genes; and in some cases an envelope of lipids
Nematode	Small and slender worms, typically less than 2.5 mm long and 5 to 100 μm thick. Also called "roundworms"; in the phylum nematode
Mycangium (plural: mycangia)	A structure adapted for the transport of fungi (usually in spore form or yeast form). Mycangia can take various forms (e.g., pits, punctures, setae, sacs) and can be complex cuticular invaginations with glands that secrete substances to support and nourish the fungi during transport
Sporotheca (plural: sporothecae)	Envelope or pocket on the exoskeleton of mites that carries (transport) fungal spores

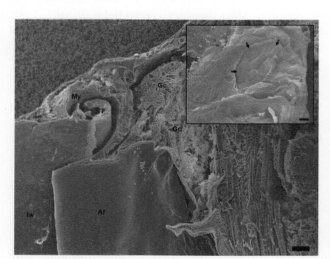

FIGURE 6.1 **Section of the prothoracic mycangium of *Dendroctonus frontalis* adult female.** Mycangial cavity (my); fungal spores in tube-like mass within mycangium (F, and see inset image); glandular cells (Gc); inner wall of mycangium (Iw); and anterior fold (Af) of mycangium.

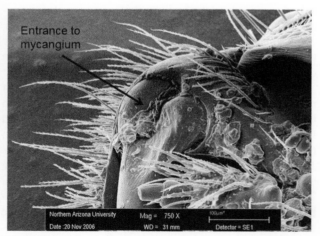

FIGURE 6.2 Mandibular mycangium of *Dryocoetes confusus* collected near Flagstaff, Arizona. *Photo by R. W. Hofstetter.*

Moser *et al.*, 2010). Mites can also carry and transport protozoa and nematodes (Ciancio and Mukerji, 2010; Perotti and Braig, 2011). Bark beetle excavation through the bark and phloem provides a means for other animals to move freely and feed in a somewhat protected habitat.

1.3 Definition of Terms

Symbiosis was defined by de Bary in 1879 to mean the "living together of two differently named organisms" (Sapp, 1994). A more recent definition was posed by Klepzig *et al.* (2001a) modified only slightly from Zook (1998): the acquisition and maintenance of one or more organisms by another that results in novel structures and (or) metabolism (Klepzig *et al.*, 2001a). This comes closer to describing the mutualistic and commensal types of relationships we focus on here. Antagonistic, parasitic (such as microsporidia), and predatory interactions are covered in Chapter 7. Definitions of the terms used in this chapter can be found in Table 6.1.

2. FUNGI

2.1 Biodiversity of Fungi

Given the niche available, that is, the highly nutritious, moist, and uncolonized phloem, it is not surprising that many fungi have evolved adaptations for insect dispersal (Klepzig *et al.*, 2009) such as spores produced in sticky droplets at the tip of fungal fruiting structures, long perithecial necks, spores that easily adhere to the bodies of the insects and/or possess well-developed sheaths, which may protect the spore from digestion in the gut of the beetles (Kirisits, 2004; Harrington, 2005; Seifert *et al.*, 2013). Based on observations of beetle galleries and isolations from adult beetles, the most successful genera of filamentous fungi associated with conifer bark beetles are *Ophiostomatales* (Harrington, 1993; Seifert *et al.*, 2013). These fungi may have a late Cretaceous origin (Harrington, 2005). Bark beetles are also believed to have their origins in the late Cretaceous (estimated at 67–93 million years before present), feeding on the coniferous genus *Araucaria*, and there may have been a diversification of bark beetles on the Pinaceae sometime near the Cretaceous/Paleocene border (Harrington, 2005).

Most bark beetles are associated with a few to many filamentous fungi (Kirisits, 2004). This assemblage can vary depending on tree host and environment (especially temperature; Evans *et al.*, 2011; Klepzig and Hofstetter, 2011). While there is general fidelity in beetle–fungal associations, some hold that notions of specific associations between individual beetle species and certain fungi are "overstated" (Kirisits, 2004). DNA-based analyses

and recent research organizes some of the better known fungi associated with bark beetles into two phylum: Ascomycota (genera *Ophiostoma, Ceratocystiopsis, Grosmannia, Ceratocystis* as well as associated anamorphs such as *Leptographium*) and Basidiomycota (genera *Phlebiopsis* and *Entomocorticium*) (Jacobs and Wingfield, 2001; Harrington, 2005; Zipfel *et al.*, 2006; Rehner, 2009). Within Ascomycota, recent phyletic maps using small subunit RNA or other DNA sequence data divides *Ophiostoma, Grosmannia*, and *Ceratocystiopsis* into the Ophiostomatales and *Ceratocystis* into *Microascales* (Spatafora and Blackwell, 1994), with *Ceratocystiopsis* and *Grosmannia* distinct from *Ophiostoma* (Zipfel *et al.*, 2006).

2.2 Ecology of Fungi

There exists a relatively rich literature reviewing the many associations between bark beetles and fungi. For example, Harrington (2005) lists the many species of bark beetles thought to be mycophagous, their hosts, mycangial types, and the principal ascomycetous and basidiomycetous fungi upon which they feed or carry. Most of these fungal associates are saprophytes on wood and inner bark, often in association with coniferous bark beetles (Harrington, 2005). Due to their melanistic hyphae and habit of staining infected wood a blue-gray/black color, these fungi are often referred to as bluestain fungi (Harrington, 1993; Seifert, 1993) (Figures 6.3 and 6.5). Besides giving these fungi their common name, melanin has also been implicated as a factor in their ability to competitively exclude other fungi (Klepzig, 2006).

Basidiomycete fungi associated with bark beetles are much less studied than the Ophiostamatales, and not as well known. The main basidiomycete genus associated with bark beetles is *Entomocorticium* (Hsiau and Harrington, 2003). Species in this genus are most often nutritional mutualists (Figure 6.4) carried in mycangia (Figures 6.1 and 6.2), specialized cuticular structures on the beetle's integument, on the surface of the exoskeleton, or within the gut (Paine *et al.*, 1997; Hsiau and Harrington, 2003; Klepzig and Hofstetter, 2011). Mycangia range from simple pits to complex, glandularized invaginations (Six, 2003), which may house antibiotic producing bacteria (Scott *et al.*, 2008). A particularly well-described example is found in the southern pine beetle *Dendroctonus frontalis* Zimmermann (Yuceer *et al.*, 2011).

Consistent fungal associations may largely be driven by the host tissue within which the beetles develop. While bark beetles are able to gain some of their nutrients directly from the host, feeding on fungus (mycophagy) is necessary for many bark beetle species to meet their nutritional requirements (Bleiker and Six, 2007; Six, 2012).

FIGURE 6.4 Hyphal growth of mutualistic fungi in a larval feeding chamber of *Dendroctonus frontalis* with close-up image (inset image) of hyphal growth of fungus in a larval feeding chamber of *D. frontalis*. Larva is also present in the image.

FIGURE 6.3 (A) Larval feeding galleries of *Dendroctonus frontalis*. Note the limited amount of phloem consumed, short larval feeding galleries terminating in chambers with mutualistic fungi for feeding, and lack of bluestain fungi in areas of larval development. (B) Larval feeding galleries of *Ips pini* (pine engraver). Note the long feeding galleries and the presence of bluestain fungi throughout the phloem. *Photos by (A) K. D. Klepzig and (B) R. W. Hofstetter.*

2.3 Impacts of Fungi on Bark Beetle Biology and Population Dynamics

Bark beetles, their fungal symbionts, and their tree hosts are complex organisms. Bark beetle–fungal associations can be more complicated than other fungus-dependent beetles (e.g., ambrosia beetles). Most bark beetle species colonize dead or dying trees, some can reproduce in healthy trees without killing them, and a few are able to overwhelm and kill healthy trees (Six, 2012). These diverse interactions range from positive, to neutral, or to negative; however,

FIGURE 6.5 *Ophiostoma minus* **in loblolly pine.** (A) Artificial mass inoculations with hyphae of *O. minus* sampled after 30 days' growth, and (B) patches of bluestain (=fungus *Ophiostoma minus* growth) in tree naturally attached by *Dendroctonus frontalis*. *Photos by: (A) K. D. Klepzig and (B) M. P. Ayres.*

most interactions studied are commensal (Table 6.1) to the host beetle (Paine *et al.*, 1997; Six, 2003, 2012; Harrington, 2005; Klepzig and Hofstetter, 2011; Six and Wingfield, 2011). Adding to the challenge of theoretical approaches, the effects of these fungal relationships can be thought of in terms of temporal (and thus physiological and population) overlapping phases within the beetle life cycle: beetles attacking and killing trees, larval development within successfully attacked trees, and population dynamics of bark beetles over the course of multiple generations (Bleiker and Six, 2008). The physiology, ecology, dynamics, and genetics of any one participant in these multiple partner relationships resist generalization. Nevertheless, numerous attempts have been made to construct unifying theories about the interactions between fungi and bark beetles.

Fungi vary widely in their degree of virulence toward the host tree. It is worth noting that the problems of using a term like virulence to describe fungi, which can apparently only succeed (in terms of causing extensive symptoms including tree death) when introduced by beetles, can lead to misinterpretations. Complicating the issue is the fact that these fungi are notoriously difficult to artificially inoculate into trees. Nevertheless, there has been considerable debate on the degree to which these interactions are mutualistic to beetles (Six and Wingfield, 2011). Early thinking focused on the role of bark beetles inoculating trees with pathogenic fungi, which must kill the tree for the beetles to succeed. This thinking likely reflected patterns observed in Dutch elm disease. More recently (Lieutier, 2002), researchers have de-emphasized the role of fungi as tree killers, not referring to the fungal associates as being phytopathogenic, but rather on their role in exhausting host defenses in a way that allows the beetles to succeed. In this way, fungi that are too aggressive/virulent would cause beetles to fail through too rapid a stimulation of host defense. This line of inquiry has become more prevalent of late (see Six and Wingfield (2011) for a contrary view). A well-understood system illustrating these points is *Dendroctonus frontalis* and its associates. One of the most destructive bark beetles in the world, the southern pine beetle carries all or one of the following: *Ophiostoma minus* (Hedge.) H. and P. Sydow, carried phoretically on the exoskeleton, and two mutualists *Ceratocystiopsis ranaculosus* Bridges and Perry and *Entomocorticium* sp. A. carried in its mycangium. While *O. minus* may play a role in tree killing, in artificial inoculations none of these three fungi are capable of killing trees on their own. All three fungi do grow within the phloem, and sporulate heavily in beetle tunnels within which the beetle larvae feed. The initial inoculation of *O. minus* could contribute to reduced resin flow, helping to compromise the tree's defense system (Klepzig and Hofstetter, 2011). The relationship with this fungus becomes antagonistic once *D. frontalis* larvae begin to develop and the host tree dies,

as the larval mycangial fungi (*C. ranaculosus* and *Entomocorticium* sp. A.) and *O. minus* are competitors for uncolonized phloem. *Dendroctonus frontalis* larvae require mycangial fungi for successful development and emergence (Klepzig and Wilkens, 1997) and it is suggested that mycangial fungi concentrate dietary nitrogen for the feeding larvae (Klepzig *et al.*, 2001b; Klepzig and Hofstetter, 2011). However, abiotic factors such as water potential, phloem chemistry, and temperature can also govern the nature of these relationships (Hofstetter *et al.*, 2007; Klepzig and Hofstetter, 2011). It may be that the role of some bark beetle-associated fungi, e.g., *O. minus*, may be that of cofactors (Lieutier, 2002), i.e., biotic agents that are not pathogenic in and of themselves but do function in compromising host defenses (Kopper *et al.*, 2004; Klepzig and Hofstetter, 2011) (Figure 6.6).

Ultimately, we may be best served by a "more comprehensive paradigm" (Paine *et al.*, 1997) in which a tree is killed "as a result of simultaneous actions and interactions of both components rather than successive actions of vector and pathogen." In terms of sublethal effects, there seems little disagreement that bark beetle-associated fungi are capable of detoxifying host chemicals, which would have deleterious effects on beetle larvae (Paine *et al.*, 1997; Lindgren and Raffa, 2013). For example, microorganisms associated with *Dendroctonus ponderosae* Hopkins and *D. frontalis* convert verbenols to verbenone, a chemical signal used by the beetles. However, these interactions are not entirely beneficial to the fungi or beetles. Natural enemies can use microbe-based odors to locate their insect hosts (Paine *et al.*, 1997), and this secondary chemistry appears to mediate the growth of some bark beetle microbial mutualists (Adams *et al.*, 2011; Davis and Hofstetter, 2012).

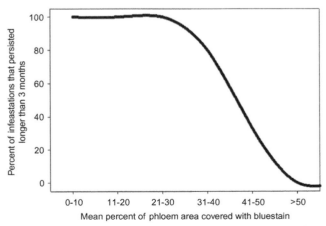

FIGURE 6.6 Relationship between bluestain (=*Ophiostoma minus* fungus) levels within loblolly pine trees and *Dendroctonus frontalis* local population persistence in Alabama in 2000 and 2001. Mean levels of bluestain per tree per infestation were sampled in June and monitored until October. Twenty-five infestations were monitored. *Adapted from Hofstetter (2004).*

3. YEASTS

3.1 Biodiversity of Yeasts

Yeasts (Kingdom Fungi) are well-known endosymbionts of insects, including bark beetles (Vega and Dowd, 2005), and appear to be present throughout most bark beetle life stages (Six, 2003). Yeasts are frequently isolated from the subcortical gallery environment or within the beetle midgut epithelium or hemolymph of *Ips* and *Dendroctonus*. In a study designed to investigate the mechanisms and structures by which the Douglas-fir beetle (*Dendroctonus pseudotsugae* Hopkins) acquires and transports fungal associates, Lewinsohn *et al.* (1994) found that yeasts were isolated from 100% of the substrates they tested at 100% frequency, including adults and teneral adults of both sexes, eggs, larvae, pupal chambers, frass, and phloem tissues adjacent to oviposition galleries. Even after rinsing beetles in ethanol, yeasts were still isolated from samples. Bacteria, by contrast, were only recovered from about 13% of the adult beetles they sampled on average, and fungal symbionts such as *Ophiostoma pseudotsugae* (Rumb.) and *Leptographium* spp. were recovered from approximately 43% and 23% of adult beetles, respectively. Unfortunately, Lewinsohn *et al.* (1994) did not identify the yeast species they isolated. In a more recent study, Davis *et al.* (2011) isolated the yeast *Ogataea* (=*Pichia*) *pini* Holst from thoracic mycangia of female western pine beetles (*Dendroctonus brevicomis* LeConte) at a frequency of 56% after noting that isolations of filamentous fungi transported in the mycangia of *D. brevicomis* (e.g., *Entomocorticium* spp.) were frequently accompanied by yeast growth (Figure 6.7). Similarly, in a study comparing fungal communities associated with the maxillary mycangia and integuments of mountain pine beetles (*D. ponderosae*), Six (2003) recorded a high frequency of yeast isolation from

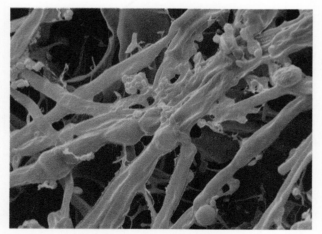

FIGURE 6.7 Scanning electron image of yeast and mycelial fungi from *Dendroctonus brevicomis* (western pine beetle) growing on malt-extract agar media.

both mycangia (80%) and exoskeletons (91%). However, the author was unable to determine whether particular yeast communities were specific to mycangia or integuments. Previous studies suggest that the yeasts most commonly associated with *D. ponderosae* are *O. pini*, *Kuraishia capsulata* (=*Hansenula capsulata*) Wick., and *Pichia* (=*Hansenula*) *holstii* Wick. (Whitney, 1982). Studies of the yeasts associated with *D. frontalis* by Bridges *et al.* (1984) yielded a similar pattern: yeasts were associated with all larval stages, and the most common yeasts (of 23 species total) were *O. pini*, *Candida tenuis* Diddens and Lodder, and *K. capsulata*. The authors also noted that yeast abundances, in terms of yeast cells/mg phloem or frass, greatly exceeded the number of bacterial cells present. In Europe, similar yeast genera (*Pichia holstii*, *K. capsulata*, *Candida diddensii* (Phaff, Mrak and Williams), *C. nitratophila* (Shifrine and Phaff), *Pichia pinus* (Holst) Phaff, *Crytococcus* spp., and *Metschnikowia* spp.) have been found associated with several *Ips* species and *Dendroctonus micans* (Kugelann), and several species appear to have both Palearctic and Nearctic distributions (Leufvén and Nehls, 1986; Wegensteiner and Weiser, 1998; Händel *et al.*, 2003; Weiser *et al.*, 2003; Unal *et al.*, 2009; Lukášová *et al.*, 2013).

3.2 Ecology of Yeasts

Yeasts are prolific metabolizers, and byproducts of yeast metabolism often include a suite of chemicals that are highly bioactive to arthropods (Davis *et al.*, 2013). For example, gaseous emissions by yeast cultures can act to attract or repel bark beetles, as well as associated predators and parasites (Brand *et al.*, 1976; Hunt and Borden, 1990; Boone *et al.*, 2008). Alternatively, some yeast species may actually utilize toxic tree chemicals such as terpenoids as carbon sources for proliferation, or even produce these chemicals as byproducts, which may coincide with beetle performance (Adams *et al.*, 2008; Davis and Hofstetter, 2011). In some contexts, a single yeast species or strain has been demonstrated to fulfill several of these potential functions simultaneously, suggesting that the ecological roles of bark beetle yeasts are not mutually exclusive of one another. In these capacities, yeasts are likely to be important in the chemical ecology of bark beetles, and until recently, the majority of ecological studies on bark beetle yeasts were concerned only with yeast volatile emissions.

Hunt and Borden (1990) demonstrated that two yeasts isolated from *D. ponderosae*, *K. capsulata* and *O. pini*, were able to metabolically convert cis- and trans-verbenol, which are aggregation pheromones of *D. ponderosae*, into verbenone, which acts as an anti-aggregation pheromone. The authors surmised that this pattern suggests that high levels of colonization by yeasts in host trees signals to beetles that the substrate may no longer be suitable for colonization or reproduction. Leufven *et al.* (1984) found that yeasts associated with *Ips typographus* (L.) could covert cis-verbenol to verbenone. In effect, the volatile signals released by yeasts in this context may act as infochemicals that tell beetles that a tree has "no vacancy." Yeast volatiles may also be important to the behavior of predators and parasites of bark beetles. A study by Boone *et al.* (2008) revealed that logs inoculated with *Pichia scolyti* (Phaff and Yoney) Kreger isolated from *Ips pini* (Say) attracted many more predatory flies (Diptera: Dolichopodidae) than uninoculated logs. However, an effect of inoculation with *P. scolyti* on captures of parasitic wasps (Hymenoptera: Pteromalidae) was not evident.

Yeast volatiles appear to play a role in mediating the growth of fungi that are present in the tree environment. Davis and Hofstetter (2011) found that *O. pini* isolated from *D. brevicomis* produced ethanol, carbon disulfide, and Δ-3-carene in headspace, and that these volatiles, when produced by yeast cultures, had significant effects on the growth of symbiotic fungi associated with *D. brevicomis*. For instance, they demonstrated that the radial growth rates of the mutualistic fungus *Entomocorticium* sp. B were enhanced when yeast volatiles were present. Conversely, they also found that growth of the entomopathogenic fungus *Beauveria bassiana* (Bals.-Criv.) Vuill. was inhibited in the presence of yeast volatiles; however, there was no clear evidence from their experiments that *O. pini* volatiles influenced the growth of the antagonistic fungus *O. minus* or growth of an opportunistic *Aspergillus* sp. The authors proposed that some yeast volatiles might have generally positive effects on beetle performance in the gallery environment by promoting mutualistic fungi and inhibiting or delaying pathogen establishment. Adams *et al.* (2008) showed that volatiles produced by three yeasts (*Pichia scolyti*, *Candida* sp., and an unidentified basidiomycetous yeast) had divergent effects on the growth of fungal symbionts isolated from the mycangia of *D. ponderosae*: yeast volatiles promoted the growth of one symbiont, *Ophiostoma montium* (Rumbold) Arx, while inhibiting the growth of another fungal symbiont, *Grosmannia clavigera* Zipfel, de Beer and Wingfield. This suggests that the effects of yeast volatile on the microbial symbiome of bark beetles could be context dependent, and yeast volatiles may have substantially different influences on bark beetle population performance depending on the species composition of fungal communities.

Some yeasts are able to metabolize terpenoids (Sutherland, 2004), the primary defensive chemicals constitutively present in the phloem resins of conifers, which may be important to beetle tolerance of phytotoxins. Leufvén *et al.* (1988) found that yeasts, particularly *K. capsulata* and *C. nitratophila*, produced oxygenated monoterpenes such as α-terpineol, borneol, terpinene-4-ol, myrtenol, and *trans*-pinocarveol. Interestingly, most of the yeasts

associated with *I. typographus* were negatively affected or sensitive to the presence of α-pinene. Davis and Hofstetter (2011) tested whether the yeast *O. pini* could tolerate monoterpenes in artificial media, as well as whether the yeast could alter the monoterpene content of tree tissues over time. They found that several monoterpenes (α-pinene, β-pinene, Δ-3-carene, and limonene) modestly inhibited yeast growth; however, *O. pini* biomass growth was substantially enhanced by two monoterpenes (terpinolene and myrcene) that entirely inhibited the establishment of mycangial fungi (Davis, 2011). They hypothesized that *O. pini* might thus be able to persist in chemical environments that would otherwise be unsuitable for beetle survival. They also found evidence that *O. pini* altered the monoterpene content of tree tissues during time periods consistent with the latent stages of bark beetle colonization: 7 days after inoculating phloem with *O. pini*, concentrations of the monoterpene Δ-3-carene were reduced by approximately 57% relative to uninoculated phloem.

3.3 Impacts on Bark Beetle Biology and Population Dynamics

Yeasts as nutritional symbionts have remained an attractive hypothesis regarding beetle–yeast interactions (Callaham and Shifrine, 1960; Francke-Grosmann, 1967); however, this concept has not yet been demonstrated in the Scolytinae. Presently, there have been no direct investigations of yeasts as a food source for beetles, though Grosmann (1930) conjectured that beetles likely did not digest yeast cells, as viable cells were often isolated from beetle digestive tracts.

4. BACTERIA

4.1 Biodiversity of Bacteria

The bacterial gut community of bark beetles is relatively low (e.g., six to 17 bacterial species in *Dendroctonus valens* LeConte; Morales-Jiménez *et al.*, 2009) compared to other invertebrates. The gut bacterial species have been surveyed for several bark beetle species, including *D. frontalis* (Moore, 1971, 1972; Vasanthakumar *et al.*, 2006), *D. ponderosae* (Adams *et al.*, 2013), *D. rhizophagus* Thomas and Bright (Morales-Jiménez *et al.*, 2012), *D. valens* (Adams *et al.*, 2010), *Ips grandicollis* Eichhoff (Bungey, 1966), *I. typographus* (Muratoglu *et al.*, 2011), and *I. sexdentatus* Boern (Sevim *et al.*, 2012). Bacteria commonly found in bark beetle guts include species in the genera *Rahnella*, *Bacillus*, *Chryseobacterium*, *Acinetobacter*, *Enterobacter*, *Klebsiella*, *Pantoae*, *Pseudomonas*, and *Serratia* (Moore, 1971, 1972; Bridges *et al.*, 1984; Vasanthakumar *et al.*, 2006; Adams *et al.*, 2009; Muratoglu *et al.*, 2011; Sevim *et al.*, 2012). Many of the bacteria found in bark beetles are common in other insects and are recognized insect pathogens, such as *Bacillus* spp. (Moore, 1971, 1972). Bacteria have also been found in the ovarian epithelial tissue of *Pityogenes calcographus* L. (Arthofer *et al.*, 2010), on the exoskeleton of over 20 bark beetle species (*Streptomyces*; Hulcr *et al.*, 2011), and on the gallery walls (Hulcr *et al.*, 2011).

Wolbachia species, responsible for skewed sex ratios favoring females in many insects as well as other organisms, have been reported in *Coccotrypes dactyliperda* F. (Zchori-Fein *et al.*, 2006), *Hypothenemus hampei* (Ferrari) (Vega *et al.*, 2002), *Ips typographus* (Stauffer *et al.*, 1997), *Pityogenes chalcographus* L. (Arthofer *et al.*, 2009), and *Xylosandrus germanus* Blandford (Peer and Taborsky, 2005). *Wolbachia* is described in more detail in Chapter 3 as these bacteria likely influence bark beetle mating systems and evolution.

4.2 Ecology of Bacteria

Bacteria range from obligate mutualist to commensal to parasitic (Kikuchi, 2009) with their bark beetle hosts. Bacteria play important roles in bark beetle development and colonization of trees, as some of the bacteria species have cellulolytic activity (Delalibera *et al.*, 2007; Hu *et al.*, 2014), can fix nitrogen (Bridges, 1981; Morales-Jiménez *et al.*, 2013), provide vital nutrients (Gibson and Hunter 2010), are pathogenic (Moore, 1972), produce pheromones (Brand *et al.*, 1975; Hunt and Borden, 1990), or influence the growth of associated fungi (Adams *et al.*, 2009). Interestingly, no *obligate* anaerobic bacteria occur in bark beetles, but are frequently found in the gut of other insects (Broderick *et al.*, 2004). However, *facultative* anaerobic bacteria such as *Serratia* and *Erwinia* species occur within bark beetles and could contribute to nitrogen fixation and carbohydrate fermentation (Morales-Jiménez *et al.*, 2009). Several bacterial species, such as *Rahnella aquatilis* (Gavini), *Enterobacter*, and *Aerogenes* appear to be common in bark beetle guts, suggesting that these bacteria might have important symbiotic roles with bark beetles (Moore, 1972; Bridges *et al.*, 1984; Vasanthakumar *et al.*, 2006). Uricolytic bacteria have also been found in the gut of bark beetles (Morales-Jiménez *et al.*, 2013), and these bacteria have the capability to use uric acid as a sole source of nitrogen or carbon. Nitrogen fixation and uric acid recycling by gut bacteria are likely important to the nitrogen budgets of bark beetles.

Actinobacteria species have been isolated from the guts (Delalibera *et al.*, 2007; Morales-Jiménez *et al.*, 2009; Hulcr *et al.*, 2011) and mycangia (Scott *et al.*, 2008; Hulcr *et al.*, 2011) of bark beetles. Scott *et al.* (2008) found that a *Streptomyces* actinobacteria found in the mycangium of *D. frontalis* inhibited the growth of *O. minus*, an antagonistic fungal associated with *D. frontalis*. In addition to the production of antibiotics, Hulcr *et al.* (2011) suggest that

Streptomyces likely degrade cellulose and are common in the midgut of beetles.

The gut microbiota of bark beetle adults and larvae can differ substantially in composition (Morales-Jiménez *et al.*, 2012; Hu *et al.*, 2013). For instance, Vasanthakumar *et al.* (2006) found that adult *D. frontalis* typically contain γ-Proteobacteria while larvae contain α- and γ-Proteobacteria, and Firmicutes bacteria in their guts. The presence of Gram-positive bacteria in larval but not adult guts suggests that these bacteria could be important for larval growth and development (Vasanthakumar *et al.*, 2006). Hu *et al.* (2013) found that bacteria likely play important roles at different developmental stages of the Chinese white pine beetle *D. armandi* Tsai and Li. They propose that gut-associated bacteria could have potential as a vector for a biocontrol agent by interfering with beetle development.

Some bacteria affect fungal symbionts and interact with tree chemistry to affect fungal growth and sporulation (Adams *et al.*, 2009). For example, Cardoza *et al.* (2006) found that bacteria species isolated from oral secretions of *Dendroctonus rufipennis* (Kirby) inhibited growth of several fungi that are detrimental to beetle development, and Adams *et al.* (2008) showed that volatiles from bacteria inhibited the growth of *Grosmannia clavigera* a common mutualist of *D. ponderosae*. Interestingly, the presence of particular host tree terpenes can alter the interactions between bacteria and fungi (Adams *et al.*, 2009, 2011) and high concentrations of terpenes are toxic to bacteria (Adams *et al.*, 2013). For bark beetles that colonize living trees, bacteria help degrade or detoxify tree defensive chemicals. For example, gut bacteria of *D. ponderosae* can significantly reduce levels of monoterpenes *in vitro* (Boone *et al.*, 2013).

Horizontal gene transfer, the process whereby genes move across species boundaries by non-sexual means, has been shown to occur in the coffee berry borer beetle (Acuña *et al.*, 2012). This gene encodes for the mannanase enzyme that hydrolyzes the polysaccharide galactomannan. Galactomannan is the most abundant polysaccharide in coffee beans, where it acts as a carbohydrate storage reserve. The transfer of this gene allows the beetle to occupy a unique ecological niche and feed exclusively on coffee beans.

Thus, there are many important effects of bacteria on bark beetles, but many other ecological traits, such as adaptation to various non-biotic environments, facilitation of the beetle invasiveness, nutritional role for beetles, horizontal gene transfers, and insect population-level regulation and insect parasite manipulation, remain to be discovered.

5. MITES

5.1 Biodiversity of Mites

Mites (Phylum Arthropoda: Class Arachnida: Subclass: Acari) are associated with all species of bark beetles. More than 100 species of mites are known to associate with bark beetles or are found within trees (Hofstetter *et al.*, 2013) or crops (such as coffee; Vega *et al.*, 2007) colonized by bark beetles. Here, we list 270 mite species and their associated bark beetle hosts in Table 6.2. Mite communities associated with bark beetles are composed of not only species considered as regular associates of bark beetles, but species that are typically recorded from trees, mushrooms, the forest, litter, and other insects such as ants (Kaliszewski, 1993; Walter and Proctor, 1999). Extensive literature exists on phoretic mite composition associated with several bark beetle species such as *Dendroctonus frontalis* (Kinn and Witcosky, 1978; Hofstetter, 2011; Hofstetter *et al.*, 2013), *D. rufipennis* (Cardoza *et al.*, 2008), *Ips typographus* (Takov *et al.*, 2009), *Pityokteines* species (Pernek *et al.*, 2008, 2012), and *Scolytus* species (Moser *et al.*, 2010), but the mites species associated with many bark and ambrosia beetle species remain to be studied.

Mites are characterized by pincer-like mouthparts called chelicerae, and the absence of antennae, mandibles, and maxillae, which are common in other arthropods such as scorpions, insects, and spiders. Mites also differ from insects because adults have four pairs of legs and lack wings. Mites are subdivided into two superorders: the Parasitiformes and Acariformes (Krantz and Walter, 2009). The Parasitiformes contain about 12,500 species within four orders: Holothyrida and Opilioacaridae (not found on bark beetles), Ixodida, and the Mesostigmata. The Ixodid, commonly called ticks, and are rarely found on bark beetles. Species of Mesostigmata, including many of the genera found in decaying fungi, are phoretic on beetles (Hofstetter *et al.*, 2013; McGraw and Farrier, 1969; Kinn, 1971), and prominent predators of nematodes and mites or mycetophagous on ophiostomatoid fungi (Hofstetter *et al.*, 2013; Hofstetter and Moser, 2014; Chapter 7). Superorder Acariformes are divided into the orders Trombidiformes and Sacroptiformes, and are the most diverse and abundant of the two superorders, with over 30,000 described species. Acariformes mites occur in most terrestrial habitats. Mites in the Trombidiformes are the most frequent residents of bark beetle habitats (Lindquist, 1969b; Kinn, 1971; Moser and Roton, 1971; Bridges and Moser, 1983; Moser, 1985; Moser *et al.*, 1989a, 1989b; Lombardero *et al.*, 2000). Mite genera commonly found on bark beetles are shown in Figure 6.8.

5.2 Ecology of Mites

Mites have the potential to alter interactions between bark beetles and microbes, influence tree diseases, affect beetle fitness, and thus influence the structure, diversity, and robustness of bark beetle communities. Antagonistic, commensalistic, and mutualistic behaviors of mites have evolved as adaptations of mites within bark beetle

communities. Mites exploit or compete with conspecifics and heterospecifics for resources, leading to a complex array of interactions and associations (Hofstetter and Moser, 2014). Individual species evolve in the context of a community, often resulting in coevolution (although this is difficult to prove; Kim, 1985). The evolution of mite–insect–fungal communities is dynamic in that fungi and bark beetles can have different mechanisms that drive evolution (Hofstetter and Moser, 2014). Additional complexity arises due to the significant differences in developmental rates and generation times between mites and bark beetles (Lombardero *et al.*, 2000), and potential host switching by mites to other insect species or tree species (Hofstetter and Moser, 2014).

In order for mites to reach bark beetle habitats, they must either be blown by wind or hitchhike on insects (Athias-Binche, 1991). A mite carried by an animal is called a phoretic mite (termed "phoresy" by Lesne in 1896) (Macchioni, 2007) (Figure 6.9). Phoresy is also known as phoresis, phorecy, or phoresia (Perotti and Braig, 2009). Phoresy by mites typically occurs via transmission on adult bark beetles or associated insects (Athias-Binche, 1993; Binns, 1984; Hofstetter *et al.*, 2013). For instance, Hofstetter (2004) found that 49% of *D. frontalis* across multiple infestations and 25% of arthropods other than *D. frontalis* (such as *Thanasmus dubius* F., *Aulonium*, *Cossonus*, *Platysoma*, and *Hylastes* species) emerging from trees had phoretic mites. However, the relative number of beetles with phoretic mites can vary greater across beetle populations.

Phoretic mites can play an important role in the transmission of microbes and interactions among microbes and beetles (Pernek *et al.*, 2008; Hofstetter and Moser, 2014) or microbes and trees (Moser *et al.*, 2010). For instance, several mite species possess specialized structures, called sporothecae, to transmit fungi (Lindquist, 1985; Moser, 1985). Sporothecae have been recorded in species within the heterostigmatic mite families Siteroptiidae, Trochometridiidae, Tarsonemidae, and Scutacaridae (Lindquist, 1985; Moser, 1985; Magowski and Moser, 2003; Ebermann and Hall, 2004). Different types of sporothecae (e.g., a pouch on the hysterosoma just behind the fourth pair of legs in Siteroptes mites or "C-flaps" on tergite 1C on lateral sides of *Tarsonemus* mites) demonstrate their analogous character, as they have evolved multiple times (Hofstetter and Moser, 2014). Why particular mite species develop sporothecae whereas other mite species do not may relate to the specific preference for one or a few fungal species as a food source. For a mite species that specialized on a fungus, the colonization of trees involves the risk of absence of the specific fungus. Such situations would lead to mortality or failure of reproduction but can be avoided if mites carry fungal spores. In addition to a sporotheca, mites frequently carry fungal spores on their body surface (Moser

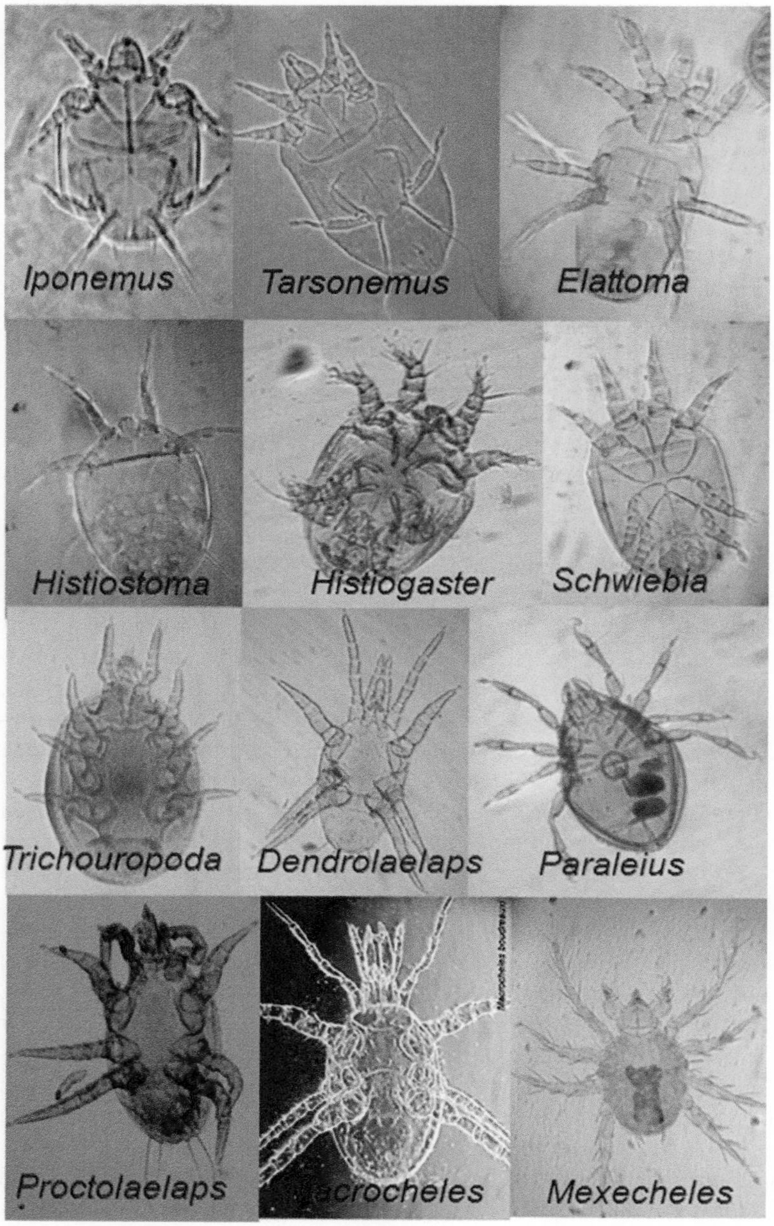

FIGURE 6.8 Common mite genera associated with bark beetles; arranged from smallest (upper left) to largest (lower right) in body size. *Photos by R. W. Hofstetter.*

et al., 1989b). The extremely adhesive conidiospores of many fungi can adhere to the mite's body and are thus dispersed by mites both within and between trees.

Fungi commonly associated with bark beetles and bark beetle-infested trees are dispersed or fed upon by some mite species (Bridges and Moser, 1983; Levieux *et al.*, 1989; Klepzig *et al.*, 2001a; Lombardero *et al.*, 2003; Moser *et al.*, 2010). These genera include *Ceratocystis, Ceratocystiopsis, Cornuvesica, Gondwanamyces, Grosmannia*, and *Ophiostoma* (Kirisits, 2004; Alamouti *et al.*, 2011), and related anamorph genera such as *Geosmithia* (Jankowiak

and Kot, 2011), *Leptographium* (Jacobs and Wingfield, 2001), and *Sporothrix* (Musvuugwa, 2014). It is unknown whether yeasts such as *Candida, Pichia, Saccharomyces*, and *Cryptococcus* are fed upon by mites. Basidiomycetes also occur with bark beetles (Weber and McPherson, 1983) belonging to the genera *Entomocorticium* (Klepzig *et al.*, 2001b), *Gloeocystidium* (Solheim, 1992), and *Heterobasidion* (Kirschner, 2001) and are fed upon by mite species in the genera *Elattoma, Dendrolaelaps*, and *Histiogaster*. For example, *Elattoma bennetti* Cross & Moser has been observed to feed on *Entomocorticium* associated with

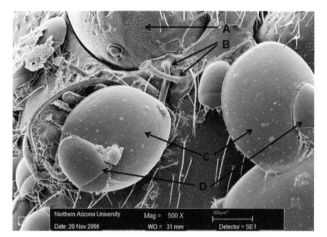

FIGURE 6.9 Scanning electron image of phoretic mites on body of *Dryocoetes confusus* (western balsam bark beetle). (A) A coxa of *D. confusus*; (B) anal stalks from *Trichouropoda* mites; (C) *Trichouropoda* mites; and (D) *Histiogaster* mites. *Photo by R. W. Hofstetter.*

Ips avulsus Eichhoff. Female *E. bennetti* become physogastric (swollen) as they feed, and nourish developing mite larvae within their abdomens (Klepzig *et al.*, 2001a). The developing mites mate while still inside the parent female mite who eventually ruptures to give birth to reproductively mature adult mites (Klepzig *et al.*, 2001b). Newly hatched mites may continue to feed on *Entomocorticium* and transmit it to the next tree.

5.3 Impacts on Bark Beetle Biology and Population Dynamics

Although most mites are passive inhabitants of bark beetle communities, mite species can impact bark beetle population dynamics via predation or parasitism (see Chapter 7) or by influencing microbial interactions (e.g., Moser, 1985; Klepzig *et al.*, 2001a, b; Hofstetter *et al.*, 2006b; Hofstetter and Moser, 2014). Some phoretic mites

TABLE 6.2 Mite Species and their Phoretic Bark Beetle Hosts

Species	Order, Family	Feeding Behavior[1]	Beetle hosts
Acrochelya implita Smiley and Moser	Tromibidiformes, Cheyletidae	Predacious	Dendroctonus frontalis, D. terebrans, Ips avulsus, I. grandicollis, I. calligraphus
Acrochelya virginiensis (Baker)	Tromibidiformes, Cheyletidae	Predacious	D. frontalis, I. avulsus, I. calligraphus, I. cribricollis, I. grandicollis
Aethiophenax (Paracarophenax) ipidarius (Redikortsev)	Heterostigmata, Acarophenacidae	Predacious	I. typographus
Amblyseiulus clausae Muma	Mesostigmata, Phytoseiidae	Predacious	D. frontalis, I. avulsus
Amblyseius guatemalensis (Chant)	Mesostigmata, Phytoseiidae	Predacious	D. frontalis, I. avulsus, I. grandicollis
Ameroseius cetratus Sellnick	Mesostigmata, Ameroseiidae	?	I. bonanseai
Ameroseius longitrichus Hirschmann	Mesostigmata, Ameroseiidae	Mycetophagous	I. avlusus, I. calligraphus, Hylotrupes bajalus, Hylurgops palliatus
Ameroseius sculptilis Berlese	Mesostigmata, Ameroseiidae	Mycetophagous	I. typographus
Ameroseius semicissus Berlese	Mesostigmata, Ameroseiidae	?	I. cribicolis
Androlaelaps casalis (Berlese)	Mesostigmata, Laelapidae	Predacious	D. frontalis. D. rhizophagus, I. avulsus, I. calligraphus, I. grandicollis
Arctoseius centratus Hirschmann	Mesostigmata, Ascidae	?	I. bonanseai
Arctoseius longispinosus Hirschmann	Mesostigmata, Ascidae	?	I. typographus
Arctoseius semicissus (Berlese)	Mesostigmata, Ascidae	?	I. cribicolis

Continued

TABLE 6.2 Mite Species and their Phoretic Bark Beetle Hosts—cont'd

Species	Order, Family	Feeding Behavior	Beetle hosts
Asca pini (Hurlbutt)	Mesostigmata, Ascidae	?	I. bonanseai, I. cribicollis
Bakerdania hylophila (Cooreman)	Trombidiformes, Pygmephoridae	Mycetophagous	I. typographus
Bakerdania sellnicki (Krczal)	Trombidiformes, Pygmephoridae	Mycetophagous	D. frontalis, I. avulsus, I. calligraphus, Gnathotrichus materiarius
Blattisocius dentriticus (Berlese)	Mesostigmata, Ascidae	Predacious	D. frontalis, I. avulsus, I. grandicollis
Blattisocius keegani Fox	Mesostigmata, Ascidae	Predacious	D. frontalis, I. avulsus, I. grandicollis
Blattisocius tarsalis (Berlese)	Mesostigmata, Ascidae	Predacious	I. grandicollis
Bonomoia pini Scheucher	Sarcoptiformes, Histiostomatidae	Omnivorous	I. typographus
Boletoglyphus boletophagi (Turk)	Sarcoptiformes, Acaridae	Mycetophagous	I. typographus
Carabodes labyrinthicus (Michael)	Sarcoptiformes, Carabodidae	Mycetophagous	I. typographus
Cepheus latus C.L. Koch	Sarcoptiformes, Cepheidae	Omnivorous	I. typographus
Cercoleipus coelonotus Kinn	Mesostigmata, Cercomegistidae	Predacious	D. frontalis, I. avulsus, I. calligraphus, I. confusus, I. emarginatus, I. grandicollis, I. monantus, I. sexdentatus, T. destruens
Ceratoppia bipilis (Hermann)	Mesostigmata, Peloppiidae	Predacious	I. typographus
Chamobates borealis (Tragardh)	Mesostigmata, Oribatulidae	Mycetophagous	I. typographus
Chelacheles michalskii Samsinak	Tromibidiformes, Cheyletidae	Predacious	I. typographus
Conchogneta traegardhi (Forsslund)	Sarcoptiformes, Autognetidae	Omnivorous	I. typographus
Crytograthus barrasi Smiley and Moser	Trombidiformes, Cryptognathidae	?	D. frontalis
Crytograthus capreolus (Berlese)	Trombidiformes, Cryptognathidae	?	D. frontalis
Crytograthus taurus (Kramer)	Trombidiformes, Cryptognathidae	?	D. frontalis
Cunaxoides andrei Baker and Hoffmann	Trombidiformes, Cunaxidae	?	I. avulsus, I. calligraphus, I. grandicollis
Cydnodromus mckenziei (Schuster and Pritchard)	Mesostigmata, Phytoseiidae	?	I. grandicollis
Dendrolaelaps armatus Hirschmann	Mesostigmata, Digamasellidae	Predacious	D. frontalis, I. avulsus, I. grandicollis
Dendrolaelaps apopthyseus Hirschmann	Mesostigmata, Digamasellidae	Predacious	I. sexdentatus
Dendrolaelaps brachypoda (Hurlbutt)	Mesostigmata, Digamasellidae	Predacious	D. frontalis

TABLE 6.2 Mite Species and their Phoretic Bark Beetle Hosts—cont'd

Species	Order, Family	Feeding Behavior	Beetle hosts
Dendrolaelaps carolinensis McGraw & Robert	Mesostigmata, Digamasellidae	Predacious	D. terebrans
Dendrolaelaps concinna (adelaideae) Womersley	Mesostigmata, Digamasellidae	Predacious	unknown beetle
Dendrolaelaps cornutulus Hirschmann	Mesostigmata, Digamasellidae	Predacious	unknown beetle
Dendrolaelaps cornutus (Kramer)	Mesostigmata, Digamasellidae	Predacious	D. frontalis, I. typographus
Dendrolaelaps cuniculus (Chant)	Mesostigmata, Digamasellidae	Predacious	D. frontalis, D. approximatus, I. avulsus, I. calligraphus, Hylastes spp.,Temnochila chlorodia, woodborers
Dendrolaelaps disetosimilis Hirschmann	Mesostigmata, Digamasellidae	Predacious	Hylurgops palliatus
Dendrolaelaps disetus Hirschmann	Mesostigmata, Digamasellidae	Predacious	I. typographus
Dendrolaelaps hexaspinosus Hirschmann	Mesostigmata, Digamasellidae	Predacious	D. valens, I. grandicollis, Gnathotrichus materiarius, Hylastes spp.
Dendrolaelaps isodentatus (Hurlbutt)	Mesostigmata, Digamasellidae	Predacious	D. frontalis, I. avulsus, I. calligraphus, I. grandicollis, G. materiarius, T. scabricollis
Dendrolaelaps louisianae Hirschmann & Wisniewski	Mesostigmata, Digamasellidae	Predacious	I. grandicollis
Dendrolaelaps moseri (Hurlbutt)	Mesostigmata, Digamasellidae	Predacious	Scolytus multristriatus
Dendrolaelaps neocornutus (Hurlbutt)	Mesostigmata, Digamasellidae	Predacious	D. frontalis, D. brevicomis, D. rhizophagus, D. terebrans, D. valens, I. avulsus, I. bonanseai, I. calligraphus, I. grandicollis, Pityopthorus sp.
Dendrolaelaps neodisetosimilis McGraw & Robert	Mesostigmata, Digamasellidae	Predacious	D. frontalis
Dendrolaelaps neodisetus (Hurlbutt)	Mesostigmata, Digamasellidae	Predacious	Acanthocinus obsoletus, D. adjunctus, D. frontalis, D. brevicomis, D. simplex, D. terebrans, D. valens, I. avulsus, I. bonanseai, I. calligraphus, I. cribricollis, I. grandicollis, I. pini, Temnchila chlorodia, Monochamus titillator, Xylotrechus sagittatus
Dendrolaelaps nostricornumtus Hirschmann & Wisniewski	Mesostigmata, Digamasellidae	Predacious	I. typographus
Dendrolaelaps pilospatulatus	Mesostigmata, Digamasellidae	Predacious	D. frontalis
Dendrolaelaps pini Hirschmann	Mesostigmata, Digamasellidae	Predacious	I. typographus

Continued

TABLE 6.2 Mite Species and their Phoretic Bark Beetle Hosts—cont'd

Species	Order, Family	Feeding Behavior	Beetle hosts
Dendrolaelaps quadrisetus Berlese	Mesostigmata, Digamasellidae	Predacious	D. frontalis, D. valens, D. adjunctus, Dry. Confusus, I. avulsus, I. bonanseai, I. calligraphus, I. grandicollis, I. pini, I. sexdentatus, I. typographus, Pityokteines spp., Pseudips mexicnaus
Dendrolaelaps quadrisetosimilis (Hirschmann)	Mesostigmata, Digamasellidae	Predacious	D. frontalis, I. avulsus, I. calligraphus
Dendrolaelaps quadritorus Robillard	Mesostigmata, Digamasellidae	Predacious	D. frontalis, D. valens, Hylastes spp., I. avulsus, I. grandicolis
Dendrolaelaps rotoni (Hurlbutt)	Mesostigmata, Digamasellidae	Predacious	D. frontalis, D. pseudotsugae, I. avulsus, I. calligraphus
Dendrolaelaps terebrans (Hurlbutt)	Mesostigmata, Digamasellidae	Predacious	D. valens
Dendrolaelaps tetraspinosus Hirschmann	Mesostigmata, Digamasellidae	Predacious	D. frontalis, I. avulsus, I. grandicollis
Dendrolaelaps uncinatus Hirschmann	Mesostigmata, Digamasellidae	Predacious	Hylurgops palliatus
Dendrolaelaps varipunctatus (Hurlbutt)	Mesostigmata, Digamasellidae	Predacious	D. frontalis, I. avulsus, I. calligraphus, T. scabricollis, G. materiarius, Platysoma sp.
Diapterobates humerlis (Hermann)	Sarcoptiformes, Humerobatidae	?	I. typographus
Digamasellus rotoni (Hurlbutt)	Mesostigmata, Digamasellidae	?	D. frontalis
Dolicheremaeus dorni (Balogh)	Sarcoptiformes, Tetracondylidae	?	Pityokteines spp.
Elattoma bennetti (Cross and Moser)	Trombidiformes, Pyemotidae	Mycetophagous	D. frontalis, I. avulsus, I. calligraphus, I. grandicollis
Elattoma karafiati (Krczal)	Trombidiformes, Pyemotidae	Mycetophagous	Xyloterus lineatus
Eporibatula gessneri Willmann	Sarcoptiformes, Oribatulidae	?	I. typographus
Ereynetes propescutulis Hunter & Rosario	Trombidiformes, Ereynetidae	Predacious	I. pini, I. typographus
Ereynetes scutulis Hunter	Trombidiformes, Ereynetidae	Predacious	D. frontalis, D. terebrans, I. avulsus, I. bonanseai, I. calligraphus, I. confusus, I. grandicollis, I. pini, I. sexdentatus, I. typographus, Scolytus ventralis, Pityokteines spp.
Eugamasus lyriformis McGraw and Farrier	Mesostigmata, Parasitidae	Predacious	D. frontalis, D. simplex, D. pseudotsugae, D. terebrans, I. avulsus, I. bonanseai, I. borealis, I. calligraphus, I. confusus, I. grandicollis, I. pilifrons, I. pini
Eutogenes vicinus Summers and Price	Trombidiformes, Cheyletidae	Predacious	D. frontalis, I. calligraphus
Gaeolaelaps ninabregus McGraw & Farrier	Mesostigmata, Laelapidae	?	D. frontalis, D. terebrans, I. grandicollis
Gamasellodes rectiventris Lindquist	Mestigmata, Ascidae	?	D. frontalis, I. avulsus, I. grandicollis

TABLE 6.2 Mite Species and their Phoretic Bark Beetle Hosts—cont'd

Species	Order, Family	Feeding Behavior	Beetle hosts
Gamasolaelaps subcorticalis McGraw & Farrier	Mesostigmata, Veigaiidae	Predacious	D. frontalis, I. avulsus, I. calligraphus, I. grandicollis, I. lecontei
Garmania fiseri (Samsinak)	Mesostigmata, Ascidae	?	Xyleborus eurygraphus
Garmania (Garmaniella) eccoptogasteris Vitzthum	Mesostigmata, Ascidae	?	unknown beetle
Gilselia arizonica Magowski, Lindquist & Moser	Trombidiformes, Tarsonemidae	?	Pseudopityophthorus sp.
Haemolaelaps fenilis (Megnin)	Mesostigmata, Laelapidae	?	D. fronalis, D. ponderosae, I. avulsus, I. grandicollis
Haemolaelaps megaventralis (Strandtmann)	Mesostigmata, Laelapidae	?	D. frontalis, D. terebrans, I. avulsus, I. grandicollis
Heterotarsonemus bicornis Schaarschmidt	Trombidiformes, Tarsonemidae	Mycetophagous	D. ponderosae, D. jefferyi, Ips spp.
Heterotarsonemus coleopterorum Schaarschmidt	Trombidiformes, Tarsonemidae	Mycetophagous	Pityogenes bidentatus
Heterotarsonemus lindquisti Smiley	Trombidiformes, Tarsonemidae	Mycetophagous	D. frontalis, Corticeus rosei, I. avulsus
Histiogaster anops Woodring	Sarcoptiformes, Acaridae	Omnivorous	D. frontalis, D. brevicomis, D. valens, I. pini, Enoclerus sp.
Histiogaster arborsignis Woodring	Sarcoptiformes, Acaridae	Mycetophagous	D. frontalis, D. ponderosae, D. terebrans, D. valens, Enoclerus sp., I. avulsus, I. calligraphus, I. grandicollis, I. pini, Scolytus multistriatus, Temnochila chlorodia, woodborers
Histiogaster rotundus Woodring	Sarcoptiformes, Acaridae	Omnivorous	D. brevicomis, D. ponderosae, I. avulsus, I. calligraphus, I. grandicollis, G. materiarius, T. scabricollis
Histiostoma (=Anoetus) abietus (Scheucher)	Sarcoptiformes, Histiostomatidae	?	I. curvidens
Histiostoma conjuncta Woodring and Moser	Sarcoptiformes, Histiostomatidae	?	D. frontalis, I. avulsus, I. calligraphus, I. crebricollis, I. grandicollis
Histiostoma crassipes (Oudemans)	Sarcoptiformes, Histiostomatidae	?	I. stebbingi
Histiostoma crypturgi (Scheucher)	Sarcoptiformes, Histiostomatidae	?	Rhagium sp.
Histiostoma dryoceoti (Scheucher)	Sarcoptiformes, Histiostomatidae	?	Dry. autographus
Histiostoma gordius (Vitzthum)	Sarcoptiformes, Histiostomatidae	?	I. laricis
Histiostoma himalayae (Vitzthum)	Sarcoptiformes, Histiostomatidae	?	Polygraphus minor
Histiostoma media Woodring and Moser	Sarcoptiformes, Histiostomatidae	?	D. frontalis, D. brevicomis, D. terebrans, D. valens, I. avulsus, I. calligraphus, Cortecius sp., Platysoma sp.

Continued

TABLE 6.2 Mite Species and their Phoretic Bark Beetle Hosts—cont'd

Species	Order, Family	Feeding Behavior	Beetle hosts
Histiostoma ovalis (Gerv)	Sarcoptiformes, Histiostomatidae	?	I. sexdentatus
Histiostoma pannonicus (Mahunka)	Sarcoptiformes, Histiostomatidae	?	Dry. villosus
Histiostoma piceae (Scheucher)	Sarcoptiformes, Histiostomatidae	?	I. typographus, Pityokteines spp.
Histiostoma pini (Scheucher)	Sarcoptiformes, Histiostomatidae	?	Hylastes ater, H. cunicularius
Histiostoma sordida Woodring and Moser	Sarcoptiformes, Histiostomatidae	?	D. frontalis, I. avulsus, I. calligraphus, I. grandicollis, Platysoma sp., Temnochila chlorodia
Histiostoma ulmi (Scheucher)	Sarcoptiformes, Histiostomatidae	?	Hypophloeus bicolor, Scolytus sp.
Histiostoma varia Woodring	Sarcoptiformes, Histiostomatidae	filter feeder	D. frontalis, D. terebrans, Dry. Confusus, I. avulsus, I. calligraphus, I. grandicollis, I. pini, Cortecius sp., Platysoma sp.
Histiostoma vitzthumi (=*serrata*) Scheucher	Sarcoptiformes, Histiostomatidae	?	I. typographus, D. micans, Dry. autographus
Hoplocheylus pickardi Smiley and Moser	Trombidiformes, Tarsocheylidae	?	D. frontalis
Hypoaspis disjuncta Hunter and Yeh	Mestigmata, Ascidae	Mycetophagous	D. frontalis
Hypoaspis krantzi Hunter	Mestigmata, Ascidae	Mycetophagous	D. frontalis, I. calligraphus
Hypoaspis lubricoides Karg	Mestigmata, Ascidae	?	Hylurgops palliatus
Hypoaspis ninabregus McGraw & Robert	Mestigmata, Ascidae	Predacious	D. frontalis, I. avulsus, I. grandicollis
Hypoaspis sp. nr. praesternalis Willman	Mestigmata, Ascidae	Predacious	D. frontalis
Hypoaspis ca vacua Michael	Mestigmata, Ascidae	?	D. valens, I. bonanseai, I. calligraphus, I. integer, I. lecontei, Pseudips mexicanus
Hypoaspis vitzthumi (Womersley)	Mestigmata, Ascidae	Predacious	D. frontalis
Insectolaelaps armatus (Hirschmann)	Mesostigmata, Digamasellidae	?	D. pseudotsugae, Hylurgops palliatus, I. pini, Buprestis lineata
Insectolaelaps quadrisetus (Berlese)	Mesostigmata, Digamasellidae	?	D. brevicomis, D. pseudotsugae, I. confusus, I. typographus, Monchamus titillator, Orthotomicus latidens
Ipiduropoda australis Hirschmann	Mesostigmata, Trematuridae	?	D. frontalis, I. bonanseai, Cerambycidae, Pissodes nemorensis
Ipiduropoda polytrichasimilis Hirschmann	Mesostigmata, Trematuridae	?	D. frontalis
Iponemus calligraphi Lindquist	Trombidiformes, Tarsonemidae	Predacious	D. frontalis, I. grandicollis, I. calligraphus, I. knausi

TABLE 6.2 Mite Species and their Phoretic Bark Beetle Hosts—cont'd

Species	Order, Family	Feeding Behavior	Beetle hosts
Iponemus confusus (Lindquist)	Trombidiformes, Tarsonemidae	Predacious	D. frontalis, I. avulsus, I. calligraphus, I. grandicollis, I. pini
Iponemus gaebleri (Schaarschmidt)	Trombidiformes, Tarsonemidae	Predacious	I. borealis, I. pilifrons, I. typographus
Iponemus leinotum Lindquist	Trombidiformes, Tarsonemidae	Predacious	I. sexdentatus, Orthotomicus longicollis
*Iponemus nahua*Lindquist	Trombidiformes, Tarsonemidae	Predacious	I. grandicollis
*Iponemus plastographus*Lindquist	Trombidiformes, Tarsonemidae	Predacious	D. frontalis
Iponemus punctatus Lindquist	Trombidiformes, Tarsonemidae	Predacious	Orthotomicus erosus, Pityogenes bistridentatus, Pityogenes calcaratus
Iponemus truncatus Lindquist	Trombidiformes, Tarsonemidae	Predacious	D. frontalis, I. pini
Iponemus truncatum (Ewing)	Trombidiformes, Tarsonemidae	Predacious	D. frontalis
Lasioseius arboreus Chant	Mestigmata, Ascidae	Predacious	D. frontalis, I. avulsus
Lasioseius corticeus Lindquist	Mestigmata, Ascidae	Predacious	D. frontalis, I. avulsus, I. calligraphus, I. grandicollis, Scolytus mundus
Lasioseius cuniculus Chant	Mestigmata, Ascidae	Predacious	D. frontalis, I. avulsus
Lasioseius dentatus Fox	Mestigmata, Ascidae	Predacious	D. frontalis
Lasioseius epicriodopsis DeLeon	Mestigmata, Ascidae	Predacious	D. frontalis
Lasioseius furcisetus Athias-Henriot	Mestigmata, Ascidae	Predacious	beetles of *Betula pendula* trees
Lasioseius hystrix Vitzthum	Mestigmata, Ascidae	Predacious	D. frontalis
Lasioseius imitans Berlese	Mestigmata, Ascidae	Predacious	D. rhizophagus, D. valens
Lasioseius neometes McGraw and Farrier	Mestigmata, Ascidae	Predacious	D. frontalis, D. terebrans, I. avulsus
Lasioseius ometes (Oudemans)	Mestigmata, Ascidae	Predacious	D. terebrans, Hylurgops palliatus, I. avulsus, I. typographus
Lasioseius safroi Ewing	Mestigmata, Ascidae	Predacious	D. adjunctus, D. frontalis, D. brevicomis, D. simplex, D. terebrans, D. valens, Dry. Confusus, I. avulsus, I. bonanseai, I. calligraphus, I. grandicollis, I. pini, Temnchila chlorodia, woodborers
Lasioseius tubiculiger (Berlese)	Mestigmata, Ascidae	Predacious	D. frontalis
Ledermulleria segnis Koch	Trombidiformes,Stigmaeidae	?	D. frontalis
Leptogamasus suecicus Tragardh	Mesostigmata, Parasitidae	?	I. typographus
Licnocepheus reticulatus	Trombidiformes, Eremellidae	?	D. frontalis
Longoseius brachypoda Hurlbutt	Mesostigmata, Digamasellidae	?	D. frontalis, I. avulsus

Continued

TABLE 6.2 Mite Species and their Phoretic Bark Beetle Hosts—cont'd

Species	Order, Family	Feeding Behavior	Beetle hosts
Longoseius ciniculus Chant	Mesostigmata, Digamasellidae	?	D. fronalis
Longoseius cuniculus Chant	Mesostigmata, Digamasellidae	?	D. frontalis, I. avulsus, Cerambycidae spp.
Lucoppia burrowsi (Michael)	Sarcoptifomes. Oribatulidae	?	I. sexdentatus
Macrocheles boudreauxi Krantz	Mesostigmata, Macrochelidae	Predacious	D. frontalis, D. rhizophagus, D. terebrans, G. materiarius, I. avulsus, I. calligraphus, I. grandicollis, T. scabricollis
Macrocheles glaber (Muller)	Mesostigmata, Macrochelidae	Predacious	I. typographus
Macrocheles mammifer Berlese	Mesostigmata, Macrochelidae	Predacious	D. frontalis
Macrocheles monochami (Lindquist)	Mesostigmata, Macrochelidae	Predacious	D. frontalis
Macrocheles shaeferi Walter	Mesostigmata, Macrochelidae	Predacious	D. ponderosae
Melichares monochami (Lindquist)	Mesostigmata, Melicharidae	?	D. frontalis, I. avlusus, I. grandicollis
Mexecheles virginiensis (Baker)	Trombidiformes, Cheyletidae	Predacious	D. frontalis, D. approximatus, D. brevicomis, D. mexicanus, Dry. Confusus, I. pini, I. typographus
Micreremus brevipes (Michael)	Mesostigmata, Micreremidae	?	I. typographus
Microgynium rectangulatum Tragardh	Mesostigmata, Microgyniidae		Hylurgops palliatus
Mucroseius monochami Lindquist	Mesostigmata, Ascidae	?	Cerambycidae
Multidendrolaelaps hexaspinosus Hirschmann	Mesostigmata, Digamasellidae	?	Hylurgops palliatus
Multidendrolaelaps isodentatus Hirschmann	Mesostigmata, Digamasellidae	?	D. frontalis, I. avlusus, I. grandicollis, Scolytus multistriatus, Cerambycidae
Multidendrolaelaps tetraspinosus Hirschmann	Mesostigmata, Digamasellidae	?	Corticeus glaber, D. frontalis, I. avulsus, I. calligraphus
Nenteria moseri Hirschmann	Mesostigmata, Nenteriidae	?	D. frontalis
Nenteria orri Moser & Roton	Mesostigmata, Nenteriidae		D. frontalis
Neojordensia tennesseensis De Leon	Mesostigmata, Blattisociidae	?	D. frontalis, D. terebrans, I. avulsus, I. grandicollis
Neophyllobius lorioi Smiley and Moser	Trombidiformes, Camerobiidae	?	D. frontalis
Neoraphignathus howei Smiley and Moser	Trombidiformes, Raphignathidae	Predacious	D. frontalis
Oodinychus hirsuta Hirschmann	Mesostigmata, Trematuridae	?	Cerambycidae

TABLE 6.2 Mite Species and their Phoretic Bark Beetle Hosts—cont'd

Species	Order, Family	Feeding Behavior	Beetle hosts
Oppia splendens (C.L. Koch)	Mesostigmata, Oppiidae	Predacious	I. typographus
Oribatella calcarata (C.L. Koch)	Mesostigmata, Oribatellidae	?	I. typographus
Paracarophaenax spp.	Trombidiformes, Pygmephoridae	Predacious	Dry. confusus, I. pini
Paracheyletia wellsi (Baker)	Trombidiformes, Cheyletidae	Predacious	D. frontalis, I. avulsus, I. calligraphus
Paraleius leontonychus Berlese	Sarcoptiformes, Oribatulidae	?	D. frontalis, D. valens, Dry. Confusus, Gnathogricus materiarius, Hylastes porculus, I. grandicollis, I. sexdentatus, I. typographus, Pityokteines spp.
Paraeupalopsellus hodgesi Smiley and Moser	Trombidiformes, Eupalopsellidae	?	D. frontalis, I. avulsus, I. calligraphus
Parawinterschmidtia furnissi Khaustov	Sarcoptiformes, Winterschmidtiidae	?	D. frontalis, D. brevicomis, D. valens, D. mexicanus, D. adjunctus
Paraswinterschmidtia michiganeosis Khaustov	Sarcoptiformes, Winterschmidtiidae	?	D. valens
Pergamasus crassipes (L.)	Parasitiformes, Parasitidae	?	Hylurgopinus rufipes
Pergamasus vagabundus Karg	Parasitiformes, Parasitidae		Hylurgops palliatus
Phauloppia lucorum	Mesostigmata, Oribatulidae	?	Pityokteines spp.
Phauloppia rauschenensis (Sellnick)	Mesostigmata, Oribatulidae	?	I. typographus
Phthiracarus nitens (Nicolet)	Mesostigmata, Mesoplophoridae	?	I. typographus
Pleuronectocelaeno austriaca Vitzthum	Mesostigmata, Neotenogyniidae	Predacious	I. typographus
Pleuronectocelaeno barbara Kinn (= Pleuronectocelaeno drymoecetes)	Mesostigmata, Neotenogyniidae	Predacious	D. frontalis I. avulsus, I. calligraphus, I. confusus, I. grandicollis, I. crebricollis, I. typographus, Orthotomicus sabinianae, Pissodes nemorensis
Pleuronectocelaeno drymoecetes Kinn	Mesostigmata, Neotenogyniidae	Predacious	I. typographus, Pityokteines curvidens
Pleuronectocelaeno japonica Kinn	Mesostigmata, Neotenogyniidae	Predacious	I. typographus, Pityokteines spp.
Podocinum pacificum Berlese	Mesostigmata, Podocinidae	Predacious	D. frontalis, I. avulsus, I. grandicollis
Poecilohirus carabi G. & R. Canestrini		?	I. typographus
Proctogastrolaelaps libris McGraw and Farrier	Mesostigmata, Melicharidae	Predacious	D. frontalis, I. avulsus, I. grandicollis
Proctolaelaps bickleyi Bram	Mesostigmata, Melicharidae	Predacious	D. frontalis, D. terebrans, Hylurgops palliatus, I. avulsus

Continued

TABLE 6.2 Mite Species and their Phoretic Bark Beetle Hosts—cont'd

Species	Order, Family	Feeding Behavior	Beetle hosts
Proctolaelaps dendroctoni Lindquist and Hunter	Mesostigmata, Melicharidae	Predacious	D. fronatlis, D. mexicanus, D. terebrans, D. ponderosae, D. valens, I. avulsus, I. calligraphus, I. grandicollis, Cortecius sp., Hylastes spp.
Proctolaelaps eccoptogasteris (Vitzthum)	Mesostigmata, Melicharidae	Predacious	I. typographus, S. multistratus
Proctolaelaps fiseri Samsinak	Mesostigmata, Melicharidae	Predacious	D. frontalis, D. terebrans, D. valens, Hylurgops palliatus, I. avulsus, I. calligraphus, I. grandicollis, I. typographus, Hylastes sp., Temnocila chlorodia
Proctonlaelaps hystricoides Lindquist & Hunter	Mesostigmata, Melicharidae	Omnivorous	D. frontalis, D. mexicanus, D. terebrans, D. ponderosae, D. valens, I. avulsus, I. calligraphus, I. grandicollis, Pityokteines spp.
Proctolaelaps hystrix (Vitzthum)	Mesostigmata, Melicharidae	Predacious	D. adjunctus D. frontalis, D. terebrans, D. micans, D. valens, D. rhizophagus, D. mexicanus, Dryocoetes spp., I. avulsus, I. pini, Hylastes sp., Enoclerus sphegus
Proctolaelaps kielczewskii Wisniewski	Mesostigmata, Melicharidae	Predacious	Hylurgops palliatus, Hylastes ater, Orthotomicus laricis, Xyloterus signatus
Proctolaelaps libris McGraw and Farrier	Mesostigmata, Melicharidae	Predacious	D. frontalis, S. multistriatus
Proctolaelaps moseri Wisniewski	Mesostigmata, Melicharidae	Predacious	Dry. autographus, Xyloterus lineatus
Proctolaelaps pini Hirschmann	Mesostigmata, Melicharidae	Predacious	Trypodendron sp.
Proctolaelaps pygmaeus (Muller)	Mesostigmata, Melicharidae	Omnivorous	D. terebrans
Proctolaelaps scolyti Evans	Mesostigmata, Melicharidae	?	I. typographus, Scolytus spp.
Proctolaelaps subcorticalis Lindquist	Mesostigmata, Melicharidae	Predacious	Cortecius sp., D. adjunctus, D. brevicomis, D. frontalis, D. mexicanus, D. ponderosae, D. rhizophagus, D. valens, I. cribicolis, I. bonanseai, I. mexicanus, I. pini, I. integer, I. lecontei, Hylurgops spp., Scolytus mundos, Temnochila chlorodia
Proctolaelaps xyloteri Samsinak	Mesostigmata, Melicharidae	Predacious	Gnathotrichus materiarius, Hylurgops palliatus, Hylastes ater, Orthotomicus laricis, Xyloterus signatus
Prosocheyla acanthus Smiley and Moser	Trombidiformes, Cheyletidae	?	D. frontalis
Prozercon kochi Sellnick	Mesostigmata, Zerconidae	?	Hylurgops palliatus
Pleuronectocelaeno barbara Athias-Henriot	Mesostigmata, Celaenopsidae	?	I. typographus
Pleuronectocelaeno japonica Kinn	Mesostigmata, Celaenopsidae	?	I. typographus japonicus
Pseudoparasitus thatcheri Hunter and Moser	Mesostigmata, Pachylaelapidae	Predacious	D. frontalis, I. avulsus, I. calligraphus
Pseudoparasitus (Hypoaspis) vitzthumi (Womersley)	Mesostigmata, Pachylaelapidae	?	I. typographus

TABLE 6.2 Mite Species and their Phoretic Bark Beetle Hosts—cont'd

Species	Order, Family	Feeding Behavior	Beetle hosts
Pseudopygmephorus bogenschutzi Mahunka and Moser	Trombidiformes, Pygmephoridae	?	I. typographus, X. lineatus
Pseudotarsonemoides eccoptogasteri Vitzthum	Trombidiformes, Tarsonemidae	?	Scolytus spp.
Pyemotes dryas (Vitzthum)	Trombidiformes, Pyemotidae	?	I. typographus
Pyemotes herfsi (Oudemans)	Trombidiformes, Pyemotidae	Predacious	Scolytus spp.
Pyemotes parviscolyti Cross & Moser	Trombidiformes, Pyemotidae	Predacious	D. frontalis, I. avulsus, I. calligraphus, I. grandicollis, Pityophthorus annectans, P. bisulcatus
Pyemotes scolyti (Oudemans)	Trombidiformes, Pyemotidae	Predacious	Scolytus spp.
Scapheremaeus palustris (Sellnick)	Sarcoptiformes, Cymbaeremaeidae	?	D. frontalis, D. terebrans, I. avulsus, I. calligraphus, I. grandicollis, I. pini, Scolytus multistriatus
Schweibea spp.	Sarcoptiformes, Acaridae	?	D. frontalis, D. valens, D. mexicanus, D. ponderosae, I. pini, I. typographus, Pityokteines curvidens
Schizosthetus (Schizosthetus) lyriformis (McGraw & Farrier)	Mesostigmata, Parasitidae	?	D. frontalis, D. mexicanus, D. simplex, D. terebrans, D. valens, Dry. confusus, Cortecius sp., I. avulsus, I. grandicollis, I. bonanseai, I. calligraphus, I. confusus, I. lecontei, I. mexicanus, I. sexdentatus, I. typographus, P. mexicanus, Try. lineatum
Schizosthetus simulatrix (Athias-Henriot)	Mesostigmata, Parasitidae	?	Hylurgops palliatus, I. typographus, Pityokteines spp.
Scutacarus scolyti Mahunka and Moser	Trombidiformes, Scutacaridae	?	I. typographus
Sejus boliviensis Hirschmann & Kaczmarek	Sejidae	?	D. valens
Siculobata leontonycha (Berlese)	Sarcoptiformes, Hemileiidae	?	I. typographus
Spinibdella depressa (Ewing)	Tromibidiformes, Bdellidae	?	D. ponderosae, I. grandicollis
Tarsonemus curiosus Livshitz, Mitrofanov et Sharonov	Trombidiformes, Tarsonemidae	Mycetophagous	Scolytus mali, Scolytus rugulosus, Taphrorychus villifrons
Tarsonemus egregius Livshitz, Mitrofanov et Sharonov	Trombidiformes, Tarsonemidae	Mycetophagous	Xylocleptes bispinus
Tarsonemus endophloeus Lindquist	Trombidiformes, Tarsonemidae	Mycetophagous	D. ponderosae
Tarsonemus fuserii Cooreman	Trombidiformes, Tarsonemidae	Mycetophagous	D. frontalis, Scolytus mali, Orthotomicus longicollis
Tarsonemus ips Lindquist	Trombidiformes, Tarsonemidae	Mycetophagous	D. frontalis, D. brevicomis, D. mexicanus, D. adjunctus, D. ponderosae, I. acuminatus, I. avulsus, I. calligraphus, I. confusus, I. grandicollis, I. pini, P. annectans, Enoclerus sphegus, woodborers

Continued

TABLE 6.2 Mite Species and their Phoretic Bark Beetle Hosts—cont'd

Species	Order, Family	Feeding Behavior	Beetle hosts
Tarsonemus krantzii Smiley and Moser	Trombidiformes, Tarsonemidae	Mycetophagous	D. frontalis, D. approximatus, D. brevicomis, D. mexicanus, D. adjuctus, Temnochila chlorodia
Tarsonmeus metacinops Kaliszewski	Trombidiformes, Tarsonemidae	Mycetophagous	Orthotomicus longicollis, Tomicus minor, Rhagium inquisitor
Tarsonemus minimax (Vitzthum)	Trombidiformes, Tarsonemidae	Mycetophagous	Pityokteines spp.
Tarsonemus primus Suski	Trombidiformes, Tarsonemidae	Mycetophagous	Hypoborus ficus, Scolytus pygmaeus, Kylesinus oleipterda
Tarsonemus pseudolacustris Kaliszewski	Trombidiformes, Tarsonemidae	Mycetophagous	Scolytus rugulosus
Tarsonmeus spathulaphorus Magowski & Khaustov	Trombidiformes, Tarsonemidae	Mycetophagous	Hypoborus ficus
Tarsonmeus subcorticallis Lindquist	Trombidiformes, Tarsonemidae	Mycetophagous	D. frontalis, D. approximatus, D. terebrans, Dry. confusus, I. avulsus, I. bonanseai, I. calligraphus, I. grandicollis, I. typographus, Pityokteines curvidens
Tarsonmeus triarcus Lindquist	Trombidiformes, Tarsonemidae	Mycetophagous	I. mexicanus, I. bonanseai, I. concinnus
Thyreophagus corticalis (Michael)	Sarcoptiformes, Acaridae	Mycetophagous	I. typographus
Trichouropoda adjuncti Wisniewski & Hirschmann	Mesostigmata, Trematuridae	?	D. adjunctus, D. valens
Trichouropoda alascae Hirschmann & Wisniewski	Mesostigmata, Trematuridae	?	D. obesus, D. rufipennis
Trichouropoda australis Hirschmann	Mesostigmata, Trematuridae	Omnivorous	D. frontalis, D. brevicomis, D. ponderosae, D. simplex, D. terebrans, D. valens, I. avulsus, I. bonanseai, I. calligraphus, I. confusus, I. mexicanus, I. lecontei, I. pini, I. grandicollis
Trichouropoda bipilis (Vitzthum)	Mesostigmata, Trematuridae	?	Hylesinus aculeatus, H. varius, Scolytus pygmaeus
Trichouropoda californica (Wisniewski & Hirschmann)	Mesostigmata, Trematuridae	?	I. confusus
Trichouropoda dalarenaensis (Hirschmann & Zirngiebl-Nicol)	Mesostigmata, Trematuridae	?	D. ponderosae
Trichouropoda denticulata Hirschmann	Mesostigmata, Trematuridae	?	I. avulsus, I. calligraphus, I. cribricollis, I. grandicollis, Pissodes nemorensis, woodborers
Trichouropoda fallax Vitzthum	Mesostigmata, Trematuridae	?	D. adjunctus, D. mexicanus, Hylurgops ater, H. cunicularius, H. interstitialis, Hylurgops pinifex
Trichouropoda guatemalensis Hirschmann	Mesostigmata, Trematuridae	?	D. frontalis

TABLE 6.2 Mite Species and their Phoretic Bark Beetle Hosts—cont'd

Species	Order, Family	Feeding Behavior	Beetle hosts
Trichouropoda hirsuta Hirschmann	Mesostigmata, Trematuridae	Omnivorous	D. frontalis, D. approximatus, D. brevicomis, D. valens, I. avulsus, I. calligraphus, I. grandicollis, I. pini, I. typographus, Pityopthorus sp., Temnochila chloridia, T. scabricollis, Monochamus spp.
Trichouropoda hondurasae Hirschmann & Wisniewski	Mesostigmata, Trematuridae	?	D. frontalis
Trichouropoda idahoensis (Hirschmann & Wisniewski)	Mesostigmata, Trematuridae	?	I. pini
Trichouropoda lamellosa Hirschmann	Mesostigmata, Trematuridae	Omnivorous	D. frontalis, D. ponderosae, D. pseudotsugae, D. valens, Dry. confusus, I. avulsus, I. calligraphus, I. cribricollis, I. grandicollis, Hylastes sp., Pityokteines spp., woodborers
Trichouropoda moseri Hirschmann	Mesostigmata, Trematuridae	?	D. simplex
Trichouropoda ovalis (C. L. Koch)	Mesostigmata, Trematuridae	?	D. adjunctus, D. rhizophagus, D. valens, I. sexdentatus, I. typographus
Trichouropoda parisiana (Hirschmann & Wisniewski)	Mesostigmata, Trematuridae	?	Gnathotrichus materiarius, I. sexdentatus, I. typographus
Trichouropoda polygraphi (Vitzthum)	Mesostigmata, Trematuridae	?	D. mexicanus, Polygraphus minor, I. bonanseai, I. sexdentatus, I. typographus japonicus
Trichouropoda polytricha (Vitzthum)	Mesostigmata, Trematuridae	?	D. frontalis, D. mexicanus, D. pseudotsugae, D. rhyzophagus, D. valens, Dry. Autogrpahus, Hylurgops palliatus, I. amitinus, I. typographus, Pityogenes chalcographus
Trichouropoda shcherbakae Hirschmann	Mesostigmata, Trematuridae	?	D. rhyzophagus
Trichouropoda tegucigalpae Hirschmann and Wisniewski	Mesostigmata, Trematuridae	?	D. frontalis, I. bonansi, I. cribricollis
Typhlodromus guatemalensis Chant	Sarcoptiformes, Acaridae	?	I. bonanseai
Tyrophagus putrescentiae (Schrank)	Sarcoptiformes, Acaridae	Mycetop./ Predatory	D. frontalis, G. materiarius, I. avulsus, I. calligraphus, I. grandicollis, I. pini, T. scabricollis
Ununguitarsonemus peacocki Smiley & Moser	Trombidiformes, Tarsonemidae	?	unknown beetle
Ununguitarsonemus rarus Magowski, DiPalma & Khaustov	Trombidiformes, Tarsonemidae	?	Dryocoetes villosus
Ununguitarsonemus tremulae Magowski & Khaustov	Trombidiformes, Tarsonemidae	?	Cerambycidae
Uroobovella americana Hirschmann	Mesostigmata, Urodinychidae	Predacious	D. frontalis, D. terebrans, D. valens, Hylastes porculus, G. materiaius, I. avulsus, I. calligraphus, I. grandicollis

Continued

TABLE 6.2 Mite Species and their Phoretic Bark Beetle Hosts—cont'd

Species	Order, Family	Feeding Behavior	Beetle hosts
Uroobovella dryocoetes Hirschmann & Zirngiebl-Nicol	Mesostigmata, Urodinychidae	?	Dry. autographus, Hylastes porculus, D. valens, I. grancollis, I. sexdentatus
Uroobovella ipidis (Vitzthum)	Mesostigmata, Urodinychidae	?	I. sexdentatus, I. typographus, Pityokteines spp.
Uroobovella laciniata Berlese	Mesostigmata, Urodinychidae	?	D. frontalis, I. avulsus, I. calligraphus
Uroobovella moseri Hirschmann	Mesostigmata, Urodinychidae	?	D. frontalis
Uroobovella neoamericana Hirschmann	Mesostigmata, Urodinychidae	?	D. valens, Temnocila cholorodia
Uroobovella orri Hirschmann	Mesostigmata, Urodinychidae	?	D. brevicomis, D. frontalis, D. obesus, D. pseudotsugae, D. valens, Dry. affaber, Dry. confusus, G. materiaius, I. avulsus, I. calligraphus, I. cribricollis, I. grandicollis, I. pini, Cylistix cylindrica, Temnochila cholorodia, Platysoma sp., Cortecius sp., woodborers
Uroobovella pulchella (Berlese)	Mesostigmata, Urodinychidae	?	I. typographus
Uroobovella varians Hirschmann	Mesostigmata, Urodinychidae	?	I. sexdentatus
Uroobovella vinicolora (Vilzthum)	Mesostigmata, Urodinychidae	?	I. typographus, D. valens
Veigaia kochi Tragardh	Mesostigmata, Veigaiidae	?	Hylurgops palliatus
Vulgarogamasus lyriformis Oudemans	Parasitiformes, Parasitidae	?	D. frontalis, D. valens, I. bonanseai, I. confusus, I. lecontei, I. mexicanus, I. pini, I. sexdentatus
Vulgarogamasus oudemansi (Berlese)	Parasitiformes, Parasitidae	?	I. typographus
Winterschmidtia spp.	Sarcoptiformes, Winterschmidtiidae	?	D. brevicomis, D. frontalis, D. ponderosae, D. valens, I. pini
Zygoribatula exilis (Nicolet)	Sarcoptifomes, Oribatulidae	?	I. typographus

[1]*Mite feed behavior for some species is based on knowledge of that genera, rather than that specific species*
Other insect hosts such as woodborers and predators are sometimes included. Feeding behaviors are listed, if known. Compiled from Lindquist (1969 a,b), McGraw and Farrier (1969), Moser and Roton (1971), Khaustov et al. (2003), Knee et al. (2012), Grijalva (2013), Hofstetter et al. (2013), Knee et al. (2013), Pfammatter et al. (2013), Mercado et al. (2014), and references within these publications, and Ohio State University Acarology Collection (http://www.biosci. ohio-state.edu/~acarolog/collection/index.html).

are predators and parasitoids of the bark beetles' immature stages, especially egg and early larval stages (Lindquist, 1969a; Kinn, 1983) and are covered in Chapter 7. Several mite species interact with antagonistic fungi associated with bark beetles (Kinn, 1967, 1971; Bridges and Moser, 1983; Moser, 1985; Hofstetter *et al.*, 2007). For instance, the presence of a bluestain fungus, *O. minus*, in phloem of trees colonized by *D. frontalis* is known to negatively affect the growth and survival of *D. frontalis* larvae (Figure 6.3)

(Bridges and Perry, 1985; Goldhammer *et al.*, 1990; Lombardero *et al.*, 2000). The abundance of *Tarsonemus* mites positively influences *O. minus* abundance within trees (Lombardero *et al.*, 2003; Hofstetter *et al.*, 2006a) thus affecting *D. frontalis* survival and reproductive rates (Lombardero *et al.*, 2003; Hofstetter *et al.*, 2006a, b). The mobility of mites and their association with microbes likely contribute to the transmission of microbes both among bark beetles within a population and across beetle species. Thus, mites likely facilitate the movement of microbes across trees and tree species (Hofstetter and Moser, 2014), and these fungi could aid bark beetles by overcoming the resistance of the host tree or by transmitting plant pathogens. For example, the mite *Thyreophagus corticalis* (Michael) feeds on and transmits the chestnut blight fungus *Cryphonectria parasitica* (Murr.) Barr (Simoni *et al.*, 2014) resulting in tree death (Levieux *et al.*, 1989; Moser *et al.*, 2010).

The association between beetles and mites may indicate a long coevolutionary relationship, or reveal a history of host switching and ecological tracking on the part of the mite (Knee *et al.*, 2012). For instance, uropodoid mites show little evidence of coevolution with their beetle hosts or tracking of ecologically similar beetle species even with overlapping geographic ranges (Knee *et al.*, 2012). The majority of uropodoid mite species have a narrow host range (≈60% of mites are on one or two beetle hosts) but at a global scale, bark beetle mites have a broad host range while at a local scale many species are host specific (Knee *et al.*, 2013). In general, we know very little about the phylogeny of most mite taxa, and future investigations require extensive sampling and an improved understanding of mite ecology and taxonomy.

6. NEMATODES

6.1 Biodiversity of Nematodes

Nematodes, commonly known as roundworms, are an incredibly diverse and numerous Metozoan group with an estimated one million species within the phylum Nematoda. These organisms can be parasitic or free living and inhabit virtually every environment in which there is sufficient moisture. Nematodes are structurally simple with long narrow bodies and a worm-like appearance. Most are microscopic, but species as long as 8 m in length occur (Gubanov, 1951). In general, the phylum is underdescribed and understudied. An exception is the species *Caenorhabditis elegans* Maupas. With its simple body plan and fully functioning nervous system, this nematode has emerged as a model organism used extensively in the fields of genetics, developmental biology, toxicology, and neurobiology (Wood, 1982, 1988).

Nematodes can be difficult to identify, as many different species tend to be morphologically similar (Perera *et al.*, 2005). In recent decades, phylogenetic maps based on morphological and ecological characteristics (Chitwood, 1937) have given way to classification systems based on DNA-derived data. As with the fungal and mite phylum, recent studies disagree on the phylogenetic framework for Nematoda. For example, studies using small subunit ribosomal RNA presented different phylogenetic trees than studies based on mitochondrial DNA, even at the level of suborder and class (Blaxter *et al.*, 1998; De Ley and Blaxter, 2004; Holterman *et al.*, 2006; Liu *et al.*, 2013). Current species identification requires both morphological and molecular data. Taxonomists currently use a variety of molecular approaches: phylogenetic analyses of the ITS and D2D3 expansion segments of rDNA, mitochondrial cytochrome c oxidase subunit I, and cytochrome b (Ferri *et al.*, 2009; Ma *et al.*, 2012; Valadas *et al.*, 2013).

An estimated 40,000 to 500,000 species of insect-associated nematodes exist (Giblin-Davis *et al.*, 2013). Many of these are in the family Diplogastridae (Order: Rhabditida). The genera *Fuchsnema* and *Micoletzkya* are known to be symbionts with bark beetles, and several *Micoletzkya* inhabit bark beetle galleries (Fuchs, 1915; Massey, 1960, 1974; Poinar, 1975; Ruhm, 1956, 1965; Susoy *et al.*, 2013). *Micoletzkya* species form a monophyletic group within the Diplogastridae. It is debatable, however whether *Micoletzkya* and their bark beetle hosts codiverged together and are host specific (Susoy *et al.*, 2013).

Many species of bark beetles are commonly found infected with nematodes. Massey (1974) reported counting approximately 2500 nematodes of one species from an individual bark beetle. Interestingly, he reported only two of the bark beetle species examined in large numbers were devoid of nematodes: *Dryocoetes confusus* L. and *Ips integer* Eichhoff. The percentage of nematode infestation within a bark beetle species and individual populations can vary from 0 to 100% and can fluctuate from year to year and generation to generation. Population fluxes are difficult to predict and are probably influenced by environmental factors such as moisture (Choo *et al.*, 1987; Massey, 1974; Shimizu *et al.*, 2013).

6.2 Ecology of Nematodes

Commensal, mutualistic, and parasitic nematode–bark beetle relationships occur. Many of the documented relationships are complicated and understudied, and involve multipartite symbioses. Multipartite symbioses are the beneficial, harmful, and neutral relationships that can change over time among multiple organisms (adapted from Relman, 2008). The nematodes associated with bark beetles can be endoparasites (transported internally) or ectoparasites (transported externally on the body surface or transported in nematangia, specialized pocket-like structures on the jugal wing folds of the bark beetles (Cardoza *et al.*, 2008).

6.2.1 Commensals

Many of the commensal nematode/bark beetle associates are examples of phoresy: a commensal relationship where the nematode is transported by the bark beetle from an area of declining resources to a more favorable area, and the bark beetle is not harmed. For example, two *D. rufipennis* ectoparasitic nematodes, *Micoletzkya ruminis* Massey and *Parasitorhabditis* sp., were successfully maintained on a microbial culture that most likely originated from fungal spores transferred with the nematodes. This indicated that these microbes, rather than the beetles themselves, could be the primary nematode nutrition source. *Dendroctonus rufipennis* may act as a transport vehicle for both the nematodes and the nematode's microbial nutritional source (Cardoza *et al.*, 2008). Other studies dealing with unknown *Micoletzkya* species have shown similar findings. For example, a *Micoletzkya* sp. isolated from the elytra of the bark beetle *Dryocoetes uniseriatus* Eggers was cultured on bacteria, a probable example of simple phoresy (Shimizu *et al.*, 2013).

Other bark beetle–nematode relationships seem to be more complex, and describing a relationship as "phoretic" could be misleading. Kanzaki *et al.* (2007) isolated *Bursaphelenchus clavicauda* Kanzaki, Maehara and Masuya from the elytra of *Cryphalus* sp. and successfully reared the nematode on the gray mold fungus *Botrytis cinerea* De Bary. *Bursaphelenchus clavicauda* is thought to feed on its bark beetle's mutualistic *Fusarium* fungi, thereby possibly competing with the beetle host for sustenance and invalidating a simple "phoretic" relationship label (Kanzaki *et al.*, 2007). Another complicated relationship is that between the nematodes *Bursaphelenchus* sp. and *Aphelenchoides* sp. and *D. rufipennis*. These two nematode species were successfully maintained on *Leptographium abietinum* (Peck) M. J. Wingfield, the ophiostomatoid bluestain fungus most frequently associated with *D. rufipennis* (Cardoza *et al.*, 2008). Bark beetle and bluestain fungi symbioses are context and system dependent and mainly multipartite (Paine *et al.*, 1997; Klepzig and Six, 2004; Six, 2012). While it may appear that *Bursaphelenchus* sp. and *Aphelenchoides* sp. only benefit from transportation on the host, the association would be more complicated if the nematodes feed upon *L. abietinum*. In that case, the effect of *Bursaphelenchus* sp. and *Aphelenchoides* sp. cohabiting with *D. rufipennis* could depend on the relationship between *L. abietinum* and *D. rufipennis*. It has been suggested that *L. abietinum* and *D. rufipennis* compete for resources during gallery construction, but provide benefit during beetle colonization (Cardoza *et al.*, 2006). The antagonistic or beneficial effects of *Bursaphelenchus* sp. and *Aphelenchoides* sp. feeding on *L. abietinum* could be time and space dependent, and are parts of a multipartite system. Thus, labeling the relationship between other mycophagous nematodes and their bark beetle hosts, such as the nematode *Bursaphelenchus rainulfi* Brassch and Burgermeister and its host *D. uniseriatus* as phoretic (i.e., commensal), could be misleading (Shimizu *et al.*, 2013).

Most phoretic relationships, including those mentioned above, involve ectoparasitic nematodes or nematodes transported in nematangia. While phoretic endoparasitic nematodes of bark beetles are uncommon, at least one example is known. *Rhabditis obtusa* (Fuchs) is transported to bark beetle galleries in the gut of various bark beetles, including *D. rufipennis*, *D. ponderosae*, *D. monticule* Hopkins, *D. pseudotsugae*, *D. borealis* Hopkins, *Ips pilifrons* Swaine, and *I. borealis* Swaine, none of which seem to be affected by the nematode (Massey, 1956).

6.2.2 Mutualism

Pine wilt disease, caused by the nematode *Bursaphelenchus xylophilus* (Steiner and Buhrer) Nickle, is a devastating disease of pines especially in areas where it has been introduced, such as Japan (Mamiya, 1983, 1988). Pine wilt disease is closely associated with pine sawyer beetles in the genus *Monochamus* (Cerambycidae) (Himelick, 1982; Linit, 1988; Ryss *et al.*, 2005). Both pine sawyer beetles and bark beetles oviposit on weakened trees at around the same time, and bark beetles transport a polyphyletic group of ophiostomatoid fungi commonly known as bluestain fungi (Wingfield *et al.*, 1984; Zhou *et al.*, 2006). *Bursaphelenchus xylophilus* can feed on both the fungus and the epithelial cells of the resin ducts (Wingfield *et al.*, 1984) and reproduces more successfully on bluestain fungi than all other fungi tested (Fukushige, 1991; Maehara and Futai, 1997). A mutualistic relationship between the pine sawyer beetles, bark beetles, and the nematode *B. xylophilus* could possibly occur as the bluestain fungi and the pine wood nematode help in overcoming the host tree's natural defense system thereby creating a more suitable breeding site for both the bark beetle and the pine sawyer beetles (Mamiya, 1983; Christiansen *et al.*, 1987; Paine *et al.*, 1997).

6.2.3 Parasitic

Most nematode parasites of bark beetles are obligate parasites (Chapter 7), incapable of completing an individual generation in a free-living state (Massey, 1974). While some nematode parasites of bark beetles can kill their host, most do not, instead causing a variety of non-lethal conditions. These include reduced fertility (Massey, 1956; Tomalak *et al.*, 1984), fecundity (Macguidwin *et al.*, 1980b; Castillo *et al.*, 2002; Poinar et al., 2004), modified gallery construction and flight pattern behavior (Atkins, 1961; Nickle, 1963, 1971; Ashraf *et al.*, 1971); Macguidwin *et al.*, 1980b), delayed emergence (Nickle, 1963; Ashraf and Berryman, 1970), and other behavioral modifications.

Some nematode parasites of bark beetles can kill their host. One of the best known examples of this interaction is between the fir engraver beetle *Scolytus ventralis* LeConte and the nematode *Sulphuretylenchus* (=*Parasitylenchus*) *elongatus* (Massey). *Scolytus ventralis* most often infests *Abies* (true firs), including *A. concolor* (Gord. and Glend.) Lindl. ex Hildebr (white fir), *A. magnifica* Murray (California red fir), and *A. grandis* Lindl. (grand fir) and has caused heavy losses of firs in western USA (Berryman and Ferrell, 1988). *Sulphuretylenchus elongatus* has been shown to delay emergence, limit flight, cause aberrant host attack behavior and gallery construction, as well as host death and sterilization (Massey, 1964; Ashraf and Berryman, 1970). The life cycle is generally synchronized with the host. Free-living stage nematodes exit though intersegmental membranes, and oral, genital, and anal openings of the beetle host. The nematodes then mate in beetle larval galleries, and infective-stage inseminated females move through moist, dead wood searching for new *S. ventralis* beetle larvae. The nematodes then use a well-developed stylet to penetrate the beetle larvae, usually through the cuticle. The *S. ventralis* larvae have been shown to defend against nematode parasitism through encapsulation of nematode egg and young larvae (Ashraf and Berryman, 1970).

Another well-studied parasitic nematode–bark beetle association that can possibly lead to bark beetle population declines involves the three major beetle vectors of Dutch elm disease: *Scolytus scolytus* F., *S. multistriatus* (Marsham), and *Hylurgopinus rufipes* (Eichhoff) (Mazzone and Peacock, 1985). Dutch elm disease, one of the most destructive vascular wilt diseases of 20th century, has devastated elms across North America and Europe, particularly American elm (*Ulmus americana* L.). Dutch elm disease is caused by *O. ulmi* and subspecies and hybrids of *Ophiostoma novo-ulmi* Brasier (Brasier and Kirk, 2001; Konrad *et al.*, 2002), which produce several toxins that together cause shoot wilting and necrosis (Claydon *et al.*, 1980). In laboratory studies, species of the nematode genera *Neoplectana* and *Heterorhabditis* were lethal to larvae and adults of *S. scolytus* (Finney and Mordue, 1976; Poinar and Deschamps, 1981). In another study, the nematode *Parasitaphelenchus oldhami* Rühm and a Gram-negative rod-shaped bacterium concurrently infected the bark beetle *H. rufipes*, depleting fat bodies, changing structure of individual cells, and possibly contributing to a population decline (Tomalak *et al.*, 1988). However, replicating laboratory results in the field have been less successful, possibly because survival and spread of nematodes depend on sufficient moisture and nematodes cannot reach larvae through dry galleries (Tomalak and Welch, 1982). Another approach to controlling Dutch elm disease involves inducing sterility of *S. multistriatus* and *S. scolytus* by *Neoparasitylenchyus* sp. nematodes (Oldham, 1930). However, other studies have indicated conflicting results (Moser *et al.*, 2005). Further biocontrol studies of Dutch elm disease are needed as the

disease continues to confound tree breeders and scientists (Schelfer *et al.*, 2008).

Interactions between nematode species in the field have yet to be broadly documented. Bark beetle galleries are excellent habitats for nematode development, and more than one species or genus of nematodes are often found attached to or within the bark beetle body or in the galleries (Massey, 1974; Poinar, 2011). While nematode–nematode associations are not well documented, there are suggestions of interactions among nematode species by exclusion interaction (two species unable to occupy the same geographic area because one species will eliminate or outcompete the other) and sympatry (two species occupying overlapping geographic areas without interbreeding). For example, while individuals of the bark beetle *D. rufipennis* did not have hemocoel infections of the nematode *Sphaerulariopsis dendroctoni* Lindquist and Hunter when concurrent with *C. reversus* infections (possible exclusive interaction) (Thong and Webster, 1983). In collections of nematodes isolated from the bark beetle *D. uniseriatus*, no significant evidence for interactions among the nematode species was found and most nematode species "seemed to use the insect body independently depending on their preference among body organs (space)" (Shimizu *et al.*, 2013). However, researchers did find three instances of exclusive interactions (the nematodes *Devibursaphelenchus* cf. *eproctatus* and *Contortylenchus* sp. rarely occurred together on an individual beetle, etc.) and one instance of sympatric interaction (two nematode species exist in the *D. uniseriatus* nematangia and did not compete for food) (Shimizu *et al.*, 2013).

Some species of nematode parasites of bark beetles can be nematophagous when food is scarce or even cannibalistic when food is absent. For example, in laboratory studies, the bacteriophagous bark beetle-associated nematode *Micoletzkya masseyi* Susoy, Kanzaki and Herrmann will prey upon *Parasitorhabditis obtusa* Fuchs and *C. elegans* when food resources are low, and even become predatory among its own species when no food is available (Susoy *et al.*, 2013).

6.3 General Life History/Cycle of Nematode Parasites

There are usually four juvenile stages and one adult stage in a nematode life cycle. An embryotic nematode differentiates into a larva (first stage) and molts within the egg (second stage). Many species emerge from the egg as a second-stage larva and molt twice more into an adult (National Research Council Committee on Plant and Animal Pests, 1968). Parasitic nematodes generally synchronize their life cycle with their host to minimize time spent in more vulnerable free-living stages (Wharton, 2004), and most nematode parasites of bark beetle are no different; female nematodes and their bark beetle hosts usually reach sexual maturity concurrently. The life cycle of *Contortylenchus elongatus* (=*Aphelenchulus elongatus*), an endoparasite of *Ips confusus* and *I. lecontei*, and *C. brevicomi* (Massey), a parasite of *D. frontalis*, are typical examples. Female *C. elongatus* and *C. brevicomi* reach sexual maturity while living inside the body cavity of their host beetle, depositing their eggs into the host body cavity. The eggs hatch and the larval nematodes migrate to the gut of the host and are depositing out of the host and into the egg gallery with the host's fecal matter. Free-living male nematodes develop into sexually viable adults, and mate with immature free-living females. These impregnated females enter the host body cavity, most likely by penetrating the cuticle or anus of a host bark beetle larva. Under normal conditions, female *C. elongatus* and *C. brevicomi* and their bark beetle hosts attain sexual maturity simultaneously and release eggs into the body cavity, thereby completing the life cycle (Nickle, 1963; Massey, 1974; Macguidwin *et al.*, 1980a). This life cycle can vary. For example, males of *S. dendroctoni* reach sexual maturity within the body cavity of the host, and species of the genus *Contortylenchus* is oviparous (egg-laying with little to no embryo development inside the mother's body) while *Parasitylenchus* is ovoviviparous (eggs are retained within the mother's body and either hatch in or immediately after expulsion) (Massey, 1974).

For most parasitic nematode–bark beetle interactions under normal conditions, one nematode generation per bark beetle host generation is observed (Massey, 1974). In fact, the relatively longer-living bark beetle *D. rufipennis* often takes 2 years to complete a generation (Werner and Holsten, 1985) and its nematode parasite *C. reversus* can complete a 2-year generation time as well. However, multiple generations per host generation can occur (Massey, 1974). For example, in the field, *C. elongatus* has three to four generations when infecting *Ips confusus*, a bark beetle with three to four generations per year (Chansler, 1964; Massey, 1974). Under controlled laboratory conditions, *C. elongatus* can have as many as 12 generations per year (Massey, 1974).

Nematodes change body form in different stages of their life cycle, especially once inside the host beetle body cavity. After female *P. elongatus* enter the host beetle, "the lip region degenerates rapidly and the stylet becomes nonfunctional and displaced, and food apparently is absorbed through the body wall of the parasite" (Massey, 1974). The body of female *C. elongatus* elongates and broadens after entering the host body cavity (Massey, 1974). Once the females of *S. dendroctoni* (a nematode parasite of *D. rufipennis*) are sexually mature, "the uterus, ovary, and oviduct are extruded through the vulva and enlarge to more than 100 times their original size, dwarfing the body of the female nematode to which they are still attached" (Thong and Webster, 1983).

Another important body transformation of parasitic nematodes of bark beetles is the dauer juvenile. This specialized form is morphologically and physiologically distinct from the other juvenile stages, and is more resistant to food and water limitations. The dauer juvenile is adapted to withstand travel with the beetle host, allowing for dispersal, invasion, and access to a new food source (Poinar, 1975, 2011; Penas *et al.*, 2006). For some genera of parasitic nematodes of bark beetles (particularly those often transported phoretically such as *Bursaphelenchus* sp.), this is an especially important step of their life cycle (Ryss *et al.*, 2005; Penas *et al.*, 2006; Kanzaki *et al.*, 2008; Shimizu *et al.*, 2013).

6.4 Impacts on Bark Beetle Biology and Population Dynamics

There are very few reports of success using nematodes as biological control agents against bark beetles. The nematodes associated with *Scolytus* seem to have the most promise, or at least comprise some of the more studied interactions. As mentioned earlier, the nematode *S. elongatus* is reported to kill the fir engraver beetle *S. ventralis*; however, *S. ventralis* continues to kill fir trees in western USA (Owen, 2003; Negrón *et al.*, 2008).

Successful reductions of other forest insect pests using parasitic nematodes have been achieved for wood-boring insects such as the European woodwasp *Sirex noctilio* in Australia and South America (Bedding and Akhurst, 1974; Bedding, 2009), the red palm weevil (*Rhynchophorus ferrugineus* Olivier) and the plum curculio (*Conotrachelus nenuphar* Herbst), and are available commercially (Olthof and Hagley, 1993; Llácer *et al.*, 2009; Dembilio *et al.*, 2011; Manachini *et al.*, 2013) but no work has been done for managing bark beetles.

7. VIRUSES

Most research on viruses associated with the bark beetles has occurred in Europe (Wegensteiner, 2004; Chapter 7) and has focused on entomopathogenic viruses. For instance, *Ips typographus* often harbors entomopox virus *ItEPV* (Wegensteiner, 2004; Burjanadze and Goginashvili, 2009; Yaman and Baki, 2011) in the cells of the midgut epithelium (Wegensteiner *et al.*, 1996). The *ItEPV* virus is found in some beetles, such as *I. amitinus* (Händel *et al.*, 2003) but does not occur in all populations and has not been found in many bark beetle species (Lukášová and Holuša, 2012). In North America, viruses in all life stages of *D. frontalis* have been identified (Sikorowski *et al.*, 1996).

8. OTHER ARTHROPOD SYMBIONTS

Other arthropods, not reported above, are associated with bark beetles or within bark beetle-infested trees. These include Collembola (Collembola: Hypogastruridae, Entomobryidae), which have been found in the brood galleries of bark beetles (Dahlsten, 1970) and likely feed on fungi or are saprophagous (Stone and Simpson, 1990); Symphyla, which are small clear myriapods without eyes (Stone and Simpson, 1990) that likely eat detritus; Pseudoscorpions (Arachnida: Chernetidae), which are predators, feeding on many types of small animals within beetle galleries, including beetle larvae, flies, nematodes, and mites, and are common phoretics of many bark beetles and associated insects (Berryman, 1970; DeMars *et al.*, 1970; Haack and Wilkinson, 1987); bark lice (Psocoptera) that may prey on bark beetle eggs (Ashraf and Berryman, 1970); and duff millipedes (Diplopoda) that occur under bark of beetle-infested trees and are believed to feed on wood decay and fungi (R. W. Hofstetter, unpubl.).

9. CONCLUSIONS

Symbionts, particularly fungi and bacteria, are critical for the development of many bark beetles (Norris *et al.*, 1969; Bridges, 1981; Six and Paine, 1997, 1998; Ayres *et al.*, 2000). However, the symbiome varies with temperature (Klepzig *et al.*, 2001a; Hofstetter *et al.*, 2007; Six and Bentz, 2007; Evans *et al.*, 2011), host tree (Hofstetter *et al.*, 2005; Lindgren and Raffa, 2013), geographic location (Hofstetter *et al.*, 2006a), and beetle population phase (Aukema *et al.*, 2005). These associations may also exert costs (Klepzig *et al.*, 2004; Kopper *et al.*, 2004), and many symbionts are antagonistic to beetles (Ayres *et al.*, 2000; Klepzig and Six, 2004). Some microbial symbionts play a role in overcoming host defenses (Plattner *et al.*, 2008; Lieutier *et al.*, 2009; Adams *et al.*, 2013; DiGuistini *et al.*, 2011), but can later negatively impact beetle colonization or development (Lombardero *et al.*, 2000; Hofstetter *et al.*, 2006b; Moser *et al.*, 2010; Adams *et al.*, 2011).

Although hundreds of symbiotic species associated with bark beetles have been identified and described, we understand little about how and why these species benefit or harm each other. Some of these gaps in knowledge include the total net effects of these organisms on one another; the degree of reliance or host specificity of bark beetles and symbionts; the processes that promote or discourage particular symbioses; the energy costs and investments needed to promote particular symbioses; changes in symbiont assemblages over time and space; relationships between beetle and symbiont phylogenies, and how evolutionary history of a symbioses affects the associations (e.g., mutualism vs. antagonism) (Aanen *et al.*, 2009). The context within which the interactions are considered is key to describing the nature of each relationship (Klepzig and Six, 2004). Bark beetle communities serve as a particularly useful model for exploring cross-scale interactions (Adams *et al.*, 2013) because of their widespread and diverse associations at multiple spatial scales (e.g., interactions among microbial symbionts to landscape level outbreaks), temporal scales (e.g., interactions within beetle generations to evolutionary timescales), and significant impacts to forests ecosystems (e.g., species invasions and habitat alteration) (Lindgren and Raffa, 2013).

9.1 Interesting Questions and Challenges

Bark beetles exert strong ecological impacts on forests and ecosystem processes such as forest succession, forest structure and composition, nutrient cycling, fire, hydrology, microclimate, and biodiversity, and as a result present many environmental and socioeconomic challenges (Bentz *et al.*, 2010; Six, 2012). Advances in technology (e.g., newly developed mating type markers for fungi; Duong *et al.*, 2014) have allowed for the better identification of symbiotic species, but the role of many symbionts in beetle ecology and evolution are unknown. More studies are needed to investigate the structure and composition of the bark beetle microbial communities. In addition, we need to better study multi-species interactions among symbionts and how they influence beetle nutrition, development, reproduction, and survival. For instance, what is the role of microbial volatiles in mediating beetle behavior and symbiotic interactions?; what role do viruses play in host tree colonization by beetles?; how do mites locate phoretic hosts?; how will climate change alter the roles of bark beetles and their symbionts in forest ecosystems?; do microbial symbioses shape the path of diversification by facilitating the invasion into novel ecological zones?; do particular clades of symbionts track particular clades of bark beetles?; how common is host switching by symbiotic microbes, bark beetles, and mites?

Most studies of bark beetle symbionts have involved species identifications, mortality impacts, or investigations of tree defensive responses to individual fungi or bacteria species. Multi-partite symbiotic interactions and how such interactions affect bark beetle systems are complex and understudied. For instance, trees contain an impressive diversity of endophytic fungi (Redman *et al.*, 2002; Arnold and Lutzoni, 2007; Vega *et al.*, 2010) and bacteria (Adams *et al.*, 2008, 2009), in addition to their well-known mycorrhizal (Smith and Read, 1997) and nitrogen-fixing bacterial mutualists (Gresshoff, 1990) that can influence host tree defenses or directly interact with microbes introduced by bark beetles. In addition, more studies need to focus on interactions between non-aggressive bark beetles and their symbionts from other regions of the world, particularly the tropics, to develop a more general and comprehensive understanding of the role of symbionts.

REFERENCES

Aanen, D.K., Slippers, B., Wingfield, M.J., 2009. Biological pest control in beetle agriculture. Trends Microbiol. 17, 179–182.

Acuña, R., Padilla, B.E., Flórez-Ramos, C.P., Rubio, J.D., Herrera, J.C., Benavides, P., et al., 2012. Adaptive horizontal transfer of a bacterial gene to an invasive insect pest of coffee. Proc. Natl. Acad. Sci. U. S. A. 109, 4197–4202.

Adams, A.S., Adams, S.M., Currie, C.R., Gillette, N.E., Raffa, K.F., 2010. Geographic variation in bacterial communities associated with the red turpentine beetle (Coleoptera: Curculionidae). Environ. Entomol. 39, 406–414.

Adams, A.S., Aylward, F.O., Adams, S.M., Erbilgin, N., Aukema, B.H., Currie, C.R., Raffa, K.F., 2013. Mountain pine beetles colonizing historical and naïve host trees are associated with a bacterial community highly enriched in genes contributing to terpene metabolism. Appl. Environ. Microbiol. 79, 3468–3475.

Adams, A.S., Boone, C.K., Bohlmann, J., Raffa, K.F., 2011. Responses of bark beetle-associated bacteria to host monoterpenes and their relationship to insect life histories. J. Chem. Ecol. 37, 808–817.

Adams, A.S., Currie, C.R., Cardoza, Y.J., Klepzig, K.D., Raffa, K.F., 2009. Effects of symbiotic bacteria and tree chemistry on the growth and reproduction of bark beetle fungal symbionts. Can. J. Forest Res. 39, 1133–1147.

Adams, A.S., Six, D.L., Adams, S.M., Holben, W.E., 2008. In vitro interactions between yeasts and bacteria and the fungal symbionts of the mountain pine beetle (*Dendroctonus ponderosae*). Microb. Ecol. 56, 460–466.

Alamouti, S.M., Wang, V., DiGuistini, S., Six, D.L., Bohlmann, J., Hamelin, R.C., et al., 2011. Gene genealogies reveal cryptic species and host preferences for the pine fungal pathogen *Grosmannia clavigera*. Mol. Ecol. 20, 2581–2602.

Arnold, A.E., Lutzoni, F., 2007. Diversity and host range of foliar fungal endophytes: are tropical leaves biodiversity hotspots? Ecology 88, 541–549.

Arthofer, W., Avtzis, D.N., Riegler, M., Stauffer, C., 2009. Low titer *Wolbachia* infection of mitochondrial defined European *Pityogenes chalcographus* (Coleoptera, Scolytinae) populations. Environ. Microbiol. 11, 1923–1933.

Arthofer, W., Avtzis, D.N., Riegler, M., Stauffer, C., 2010. Mitochondrial phylogenies in the light of pseudogenes and *Wolbachia*: re-assessment of a bark beetle dataset. Zookeys 56, 269–280.

Ashraf, M., Berryman, A.A., 1970. Biology of *Sulphuretylenchus elongatus* (Nematoda Sphaerulariidae), and its effect on its host, *Scolytus ventralis* (Coleoptera-Scolytidae). Can. Entomol. 102, 197–213.

Ashraf, M., Mayr, W., Sybers, H.D., 1971. Ultrastructural pathology of flight muscles of *Scolytus ventralis* (Coleoptera-Scoltidae) infected by a nematode parasite. J. Invertebr. Pathol. 18, 363–372.

Athias-Binche, F., 1991. Ecology and evolution of phoresy in mites. In: Dusbabek, F., Bukva, V. (Eds.), Modern Acarology: Proceedings of the VIII International Congress of Acarology, České Budějovice, Czechoslovakia, 1990. SPB Academic Publishing, The Hague and Czechoslovak Academy of Sciences, Prague, pp. 27–41.

Athias-Binche, F., 1993. Dispersal in varying environments: the case of phoretic uropodid mites. Can. J. Zool. 71, 1793–1798.

Atkins, M.D., 1961. A study of the flight of the Douglas-fir beetle *Dendroctonus pseudotsugae* Hopk. (Coleoptera: Scolytidae). III. Flight capacity. Can. Entomol. 93, 467–474.

Aukema, B.H., Werner, R.A., Haberkern, K.E., Illman, B.L., Clayton, M. K., Raffa, K.F., 2005. Quantifying sources of variation in the frequency of fungi associated with spruce beetles: implications for hypothesis testing and sampling methodology in bark beetle–symbiont relationships. Forest Ecol. Manag. 217, 187–202.

Ayres, M.P., Wilkens, R.T., Ruel, J.J., Lombardero, M.J., Vallery, E., 2000. Nitrogen budgets of phloem-feeding bark beetles with and without symbiotic fungi. Ecology 81, 2198–2210.

Barras, S.J., 1970. Antagonism between *Dendroctonus frontalis* and fungus *Ceratocystis minor*. Ann. Entomol. Soc. Amer. 63, 1187–1190.

Bedding, R., 2009. Controlling the pine-killing woodwasp, *Sirex noctilio*, with nematodes. In: Hajek, A., Glare, T., O'Callaghan, M. (Eds.), Use of Microbes for Control and Eradication of Invasive Arthropods. Springer, Netherlands, pp. 213–235.

Bedding, R.A., Akhurst, R.J., 1974. Use of the nematode *Deladenus siricidicola* in the biological control of *Sirex noctilio* in Australia. J. Aust. Entomol. Soc. 13, 129–135.

Bentz, B.J., Régnière, J., Fettig, C.J., Hansen, E.M., Hayes, J.L., Hick, J.A., et al., 2010. Climate change and bark beetles in the western United States and Canada: direct and indirect effects. Bioscience 60, 602–613.

Berryman, A.A., 1970. Evaluation of insect predators of the western pine beetle. In: Stark, R.W., Dahlsten, D.L. (Eds.), Studies on the Population Dynamics of the Western Pine Beetle, *Dendroctonus brevicomis* LeConte (Coleoptera: Scolytidae). University of California, Berkeley, pp. 102–112.

Berryman, A.A., Ferrell, G., 1988. The fir engraver beetle in western states. In: Berryman, A.A. (Ed.), Dynamics of Forest Insect Populations: Patterns, Causes, Implications. Plenum Press, New York, pp. 555–577.

Binns, E.S., 1984. Phoresy as migration: some functional aspects of phoresy in mites. Biol. Rev. 57, 571–620.

Blaxter, M.L., De Ley, P., Garey, J.R., Liu, L.X., Scheldeman, P., Vierstraete, A., et al., 1998. A molecular evolutionary framework for the phylum Nematoda. Nature 392, 71–75.

Bleiker, K., Six, D.L., 2007. Dietary benefits of fungal associates to an eruptive herbivore: potential implications of multiple associates on host population dynamics. Environ. Entomol. 36, 1384–1396.

Bleiker, K., Six, D.L., 2008. Competition and coexistence in a multi-partner mutualism: interactions between two fungal symbionts of the mountain pine beetle in beetle-attacked trees. Microb. Ecol. 57, 191–202.

Boone, C.K., Six, D.L., Zheng, Y., Raffa, K.F., 2008. Parasitoids and dipteran predators exploit volatiles from microbial symbionts to locate bark beetles. Environ. Entomol. 37, 150–161.

Boone, C., Keefover-Ring, K., Mapes, A.C., Adams, A.S., Bohlmann, J., Raffa, K.F., 2013. Bacteria associated with a tree-killing insect reduce concentrations of plant defense compounds. J. Chem. Ecol. 39, 1003–1006.

Brand, J.M., Bracke, J.W., Britton, L.N., Markovetz, A.J., Barras, S.J., 1976. Bark beetle pheromones: production of verbenone by a mycangial fungus of *Dendroctonus frontalis*. J. Chem. Ecol. 2, 195–199.

Brand, J.M., Bracke, J.W., Markovetz, A.J., Wood, D.L., Browne, L.E., 1975. Production of verbenol pheromone by a bacterium isolated from bark beetles. Nature 254, 136–137.

Brasier, C.M., Kirk, S.A., 2001. Designation of the EAN and WAN races of *Ophiostoma novo-ulmi* as subspecies. Mycol. Res. 105, 547–554.

Bridges, J.R., 1981. Nitrogen-fixing bacteria associated with bark beetles. Microb. Ecol. 7, 131–171.

Bridges, J.R., Moser, J.C., 1983. Role of two phoretic mites in transmission of bluestain fungus, *Ceratocystis minor*. Ecol. Entomol. 8, 9–12.

Bridges, J.R., Perry, T.J., 1985. Effects of mycangial fungi on gallery construction and distribution of bluestain in southern pine beetle-infested pine bolts. J. Entomol. Sci. 20, 271–275.

Bridges, J.R., Marler, J.E., McSparrin, B.H., 1984. A quantitative study of the yeast and bacteria associated with laboratory-reared *Dendroctonus frontalis* Zimm. (Coleopt., Scolytidae). Z. Angew. Entomol. 97, 261–267.

Broderick, N.A., Raffa, K.F., Goodman, R.M., Handelsman, J., 2004. Census of the bacterial community of the gypsy moth larval midgut using culturing and culture-independent methods. Appl. Environ. Microbiol. 70, 293–300.

Bungey, R.S., 1966. The biology, behavior and chemical control of *Ips grandicollis* Eichhoff in pine slash. PhD thesis, University of Adelaide.

Burjanadze, M., Goginashvili, N., 2009. Occurrence of pathogens and nematodes in the spruce bark beetles, *Ips typographus* (Col., Scolytidae) in Borjomi Gorge. Bulletin of the Georgian National Academy of Sciences 3, 145–150.

Callaham, R.Z., Shifrine, M., 1960. The yeasts associated with bark beetles. For. Sci. 6, 146–154.

Cardoza, Y.J., Hofstetter, R.W., Vega, F.E., 2012. Insect-associated microorganisms and their possible role in Outbreaks. In: Barbosa, P., Letourneau, D.K., Agrawal, A.A. (Eds.), Insect Outbreaks Revisited. John Wiley & Sons, pp. 155–174.

Cardoza, Y.J., Klepzig, K.D., Raffa, K.F., 2006. Bacteria in oral secretions of an endophytic insect inhibit antagonistic fungi. Ecol. Entomol. 31, 636–645.

Cardoza, Y.J., Moser, J.C., Klepzig, K.D., Raffa, K.F., 2008. Multipartite symbioses among fungi, mites, nematodes, and the spruce beetle, *Dendroctonus rufipennis*. Environ. Entomol. 37, 956–963.

Castillo, A., Infante, F., Barrera, J.F., Carta, L., Vega, F.E., 2002. First field report of a nematode (Tylenchida: Sphaerularioidea) attacking the coffee berry borer, *Hypothenemus hampei* (Ferrari) (Coleoptera: Scolytidae) in the Americas. J. Invertebr. Pathol. 79, 199–202.

Chansler, J.F., 1964. Overwintering habits of *Ips lecontei* Sw. and *Ips confusus* (Lec.) in Arizona and New Mexico, U.S. Department of Agriculture, Forest Service, Rocky Mountain Forest and Range Experiment Station.

Chitwood, B.G., 1937. A revised classification of the Nematoda. Papers on Helminthology: Published in Commemoration of the 30 Year Jubileum of the Scientific, Educational and Social Activities of the Honoured Worker of Science K. J. Skrjabin. All-union Lenin Academy of Agricultural Sciences, Moscow, pp. 69–80.

Choo, H., Kaya, H.K., Shea, P., Noffsinger, E.M., 1987. Ecological study of nematode parasitism in *Ips* beetles from California and Idaho. J. Nematol. 19, 495–502.

Christiansen, E., Waring, R.H., Berryman, A.A., 1987. Resistance of conifers to bark beetle attack—searching for general relationships. For. Ecol. Manage. 22, 89–106.

Ciancio, A., Mukerji, K.G. (Eds.), 2010. Integrated Management of Arthropod Pests and Insect Borne Diseases, Volume 5. Springer, New York.

Claydon, N., Elgersma, D.M., Grove, J.F., 1980. The phytotoxicity of some phenolic metabolic productsof *Ophiostoma Ulmi* to *Ulmus* sp. Neth. J. Plant Pathol. 86, 229–237.

Dahlsten, D.L., 1970. Parasites, predators, and associated organisms reared from western pine beetle infested bark samples. In: Stark, R.W., Dahlsten, D.L. (Eds.), Studies on the Population Dynamics of the Western Pine Beetle, *Dendroctonus brelicomis* LeConte (Coleoptera: Scolytidae). University of California, Berkeley, pp. 75–79.

Davis, T.S., 2011. Response of a beetle-microbe complex to variation in host tree phytochemistry. PhD thesis, Northern Arizona University.

Davis, T.S., Hofstetter, R.W., 2011. Reciprocal interactions between the bark beetle-associated yeast *Ogataea pini* and host plant chemistry. Mycologia 103, 1201–1207.

Davis, T.S., Hofstetter, R.W., 2012. Plant secondary chemistry mediates the performance of a nutritional symbiont associated with a tree-killing herbivore. Ecology 93, 421–429.

Davis, T.S., Crippen, T.L., Hofstetter, R.W., Tomberlin, J.K., 2013. Microbial volatile emissions as insect semiochemicals. J. Chem. Ecol. 39, 840–859.

Davis, T.S., Hofstetter, R.W., Foster, J.T., Foote, N.E., Keim, P., 2011. Interactions between the yeast *Ogataea pini* and filamentous fungi associated with the western pine beetle. Microb. Ecol. 61, 626–634.

De Ley, P., Blaxter, M.L., 2004. A new system for Nematoda: combining morphological characters with molecular trees, and translating clades into ranks and taxa. In: Cook, R.C., Hunt, D.J. (Eds.), Proceeding of the Fourth International Congress of Nematology. Brill Academic Publishers, Leiden, pp. 633–653.

Delalibera Jr., I., Vasanthakumar, A., Burwitz, B.J., Schloss, P.D., Klepzig, K.D., Handelsman, J., Raffa, K.F., 2007. Composition of the bacterial community in the gut of the pine engraver, *Ips pini* (Say) (Coleoptera), colonizing red pine. Symbiosis 43, 97–104.

DeMars, C.J., Dahlsten, D.L., Stark, R.W., 1970. Survivorship curves for eight generations of the western pine beetle in California, 1962–1965, and a preliminary life table. In: Stark, R.W., Dahlsten, D.L. (Eds.), Studies on the Population Dynamics of the Western Pine Beetle, *Dendroctonus brelicomis* LeConte (Coleoptera: Scolytidae). University of California, Berkeley, pp. 134–146.

Dembilio, O., Karamaouna, F., Kontodimas, D.C., Nomikou, M., Jacas, J.A., 2011. Short communication. Susceptibility of *Phoenix theophrasti* (Palmae: Coryphoideae) to *Rhynchophorus ferrugineus* (Coleoptera: Curculionidae) and its control using *Steinernema carpocapsae* in a chitosan formulation. Spanish J. Agr. Res. 9, 623–626.

DiGuistini, S., Ralph, S.G., Lim, Y.W., Holt, R., Jones, S., Bohlmann, J., Breuil, C., 2007. Generation and annotation of lodgepole pine and oleoresin-induced expressed sequences from the blue-stain fungus *Ophiostoma clavigerum*, a mountain pine beetle-associated pathogen. FEMS Microbiol. Lett. 267, 151–158.

DiGuistini, S., Wang, Y., Liao, N.Y., Taylor, G., Tanguay, P., Feau, N., et al., 2011. Genome and transcriptome analyses of the mountain pine beetle–fungal symbiont *Grosmannia clavigera*, a lodgepole pine pathogen. Proc. Natl. Acad. Sci. U. S. A. 108, 2504–2509.

Duong, T.A., Beer, Z.W., Wingfield, B.D., Eckhardt, L.G., Wingfield, M.J., 2014. Microsatellite and mating type markers reveal unexpected patterns of genetic diversity in the pine root-infecting fungus *Grosmannia alacris*. Plant Path. http://dx.doi.org/10.1111/ppa.12231 (in press).

Ebermann, E., Hall, M., 2004. A new species of scutacarid mites transferring fungal spores (Acari, Tarsonemina). Rev. Suisse Zool. 111, 941–950.

Evans, L.M., Hofstetter, R.W., Ayres, M.P., Klepzig, K.D., 2011. Temperature alters the relative abundance and population growth rates of species within the *Dendroctonus frontalis* (Coleoptera: Curculionidae) community. Environ. Entomol. 40, 824–834.

Ferri, E., Barbuto, M., Bain, O., Galimberti, A., Uni, S., Guerrero, R., et al., 2009. Integrated taxonomy: traditional approach and DNA barcoding for the identification of filarioid worms and related parasites (Nematoda). Front. Zool. 6, 12.

Finney, J.R., Mordue, W., 1976. Susceptibility of elm bark beetle *Scolytus scolytus* to Dd-136 strain of *Neoaplectana* sp. Ann. Appl. Biol. 83, 311–312.

Francke-Grosmann, H., 1952. Uber die Ambrosiazucht der beiden Kiefern-borkenkafer *Myelophilus minor* Htg. und *Ips acuminatus* Gyll. Medde-landen fran Statens Skogforskningsinstitutut 41, 1–52.

Francke-Grosmann, H., 1967. Ectosymbiosis in wood-inhabiting insects. In: Henry, S.M. (Ed.), Symbiosis, Volume 2: Associations of Inverte-brates, Birds, Ruminants, and Other Biota. Academic Press, New York, pp. 141–206.

Fuchs, G., 1915. Die Naturgeschichte der Nematodon und einiger anderer Parasiten. 1, des *Ips typographus* L. des *Hylobius abietis* L. Zoolo-gische Jahrbuecher Abteilungen f Systematik 38, 109–222.

Fukushige, H., 1991. Propagation of *Bursaphelenchus xylophilus* (Nematoda, Aphelenchoididae) on fungi growing in pine-shoot seg-ments. Appl. Entomol. Zoolog. 26, 371–376.

Giblin-Davis, R.M., Kanzaki, N., Davies, K.A., 2013. Nematodes that ride insects: unforeseen consequences of arriving species. Fla. Entomol. 96, 770–780.

Gibson, C.M., Hunter, M.S., 2010. Extraordinarily widespread and fantas-tically complex: comparative biology of endosymbiotic bacterial and fungal mutualists of insects. Ecol. Lett. 13, 223–234.

Goldhammer, D.S., Stephen, F.M., Paine, T.D., 1990. The effect of the fungi *Ceratocystis minor*, *Ceratocystis minor* var. *barussii* and SJB 122 on reproduction of the southern pine beetle, *Dendroctonus frontalis*. Can. Entomol. 122, 407–418.

Gresshoff, P.M., 1990. The Molecular Biology of Symbiotic Nitrogen Fixation. CRC Press, Boca Raton, FL.

Grijalva, M.P.C., 2013. Mesotigmados (Acari: Mesostigmata) asociados a Scolytinae (Coleoptera: Curculiondiae) de importancia forestall en Mexico, PhD thesis, Montecillo, Texcoco, Edo. de Mexico.

Grosmann, H., 1930. Beitrage zur Kenntnis der Lebensgemeinsc-haft zwischen Borkenkafern und Pilzen. Z. f. Parasitenkunde 3, 56–102.

Gubanov, N.M., 1951. Gigantic nematode from the placenta of cetacean *Placentonema gigantissima* nov. gen., nov. sp. Doklady Akademii Nayuk SSSR 77, 1123–1125.

Haack, R.A., Wilkinson, R.C., 1987. Phoresy by *Dendrochernes* Pseudo-scorpions on Cerambycidae (Coleoptera) and Aulacidae (Hyme-noptera) in Florida. Am. Midl. Nat. 117, 369–373.

Händel, U., Wegensteiner, R., Weiser, J., Zizka, Z., 2003. Occurrence of pathogens in associated living bark beetles (Col., Scolytidae) from dif-ferent spruce stands in Austria. J. Pest Sci 76, 22–32.

Harrington, T.C., 1993. Diseases of conifers caused by species of *Ophiostoma* and *Leptographium*. In: Wingfield, M.J., Seifert, K.A., Webber, J.F. (Eds.), *Ceratocystis* and *Ophiostoma*: Taxonomy, Ecology and Pathogenicity. American Phytopathological Society, St. Paul, pp. 161–172.

Harrington, T.C., 2005. Ecology and Evolution of Mycophagous Bark Beetles and Their Fungal Partners. Oxford University Press.

Himelick, E.B., 1982. Pine blue-stain associated with the pine wilt syn-drome. Journal of Arboriculture 8, 212–216.

Hofstetter, R.W., 2004. Interspecific Interactions and Population Dynamics of the Southern Pine Beetle, PhD dissertation. Dartmouth College, Hanover, New Hampshire, USA.

Hofstetter, R.W., 2011. Mutualists and phoronts of the southern pine beetle. In: Klepzig, K.D., Coulson, R. (Eds.), Southern Pine Beetle II. United States Dept. of Agriculture Forest Service, pp. 161–181, Southern Research Station General Technical Report SRS-140.

Hofstetter, R.W., Moser, J.C., 2014. Role of mites in insect-fungus associ-ations. Ann. Rev. Entomol. 59, 537–557.

Hofstetter, R.W., Cronin, J.J., Klepzig, K.D., Moser, J.C., Ayres, M.P., 2006a. Antagonisms, mutualisms and commensalisms affect outbreak dynamics of the southern pine beetle. Oecologia 147, 679–691.

Hofstetter, R.W., Dempsey, T.D., Klepzig, K.D., Ayres, M.P., 2007. Tem-perature-dependent effects on mutualistic and phoretic associations. Comm. Ecol. 8, 47–56.

Hofstetter, R.W., Klepzig, K.D., Moser, J.C., Ayres, M.P., 2006b. Seasonal dynamics of mites and fungi and their interaction with southern pine beetle. Environ. Entomol. 35, 22–30.

Hofstetter, R.W., Mahfous, J., Klepzig, K.D., Ayres, M.P., 2005. Effects of tree phytochemistry on the interactions between endophloedic fungi associated with the southern pine beetle. J. Chem. Ecol. 31, 551–572.

Hofstetter, R.W., Moser, J.C., Blomquist, S., 2013. Mites associated with bark beetles and their hypophoretic Ophiostomatoid fungi. In: Wingfield, Seifert (Eds.), The Ophiostomatoid Fungi: Expanding Frontiers. CBS-KNAW Fungal Biodiversity Centre, Utrecht, The Netherlands, pp. 165–176.

Holterman, M., van der Wurff, A., van den Elsen, S., van Megen, H., Bongers, T., Holovachov, O., et al., 2006. Phylum-wide analysis of SSU rDNA reveals deep phylogenetic relationships among nematodes and accelerated evolution toward crown clades. Mol. Biol. Evol. 23, 1792–1800.

Hsiau, P.-T.W., Harrington, T.C., 2003. Phylogenetics and adaptations of basidiomycetous fungi fed upon by bark beetles (Coleoptera: Scoly-tidae). Symbiosis 34, 111–131.

Hu, X., Wang, C., Chen, H., Ma, J., 2013. Differences in the structure of the gut bacteria communities in development stages of the Chinese white pine beetle (*Dendroctonus armandi*). Int. J. Mol. Sci. 14, 21006–21020.

Hu, X., Yu, J., Wang, C., Chen, H., 2014. Cellulolytic bacteria associated with the gut of *Dendroctonus armandi* larvae (Coleoptera: Curculio-nidae: Scolytinae). Forests 5, 455–465.

Hulcr, J., Adams, A.S., Raffa, K.F., Hofstetter, R.W., Klepzig, K.D., Currie, C.R., 2011. Presence and diversity of *Streptomyces* in *Den-droctonus* and sympatric beetle galleries across North America. Mol. Ecol. 61, 759–768.

Hulcr, J., Dunn, R.R., 2011. The sudden emergence of pathogenicity in insect-fungus symbiosis threatens naïve forest ecosystems. P. Roy. Soc. Lond. B Bio. 278, 2866–2873.

Hunt, D.W.A., Borden, J.H., 1990. Conversion of verbenols to verbenone by yeasts isolated from *Dendroctonus ponderosae* (Coleoptera: Scoly-tidae). J. Chem. Ecol. 16, 1385–1397.

Jacobs, K., Wingfield, M.J., 2001. *Leptographium* Species: Tree Path-ogens, Insect Associates and Agents of Blue Stain. APS Press, St. Paul, MN.

Jankowiak, R., Kot, M., 2011. Ophiostomatoid fungi associated with bark beetles (Coleoptera: Scolytidae) colonizing branches of *Pinus syl-vestris* in southern Poland. Pol. Bot. J. 56, 287–293.

Kaliszewski, M., 1993. Key to palearctic (sic) species of the genus *Tarso-nemus*. Zoologia, Wydawnictwo Naukowe UAM, Poznan, Poland, Seria, 14. 204 pp.

Kanzaki, N.K., Maehara, N., Masuya, H., 2007. *Bursaphelenchus clavicauda* n. sp (Nematoda: Parasitaphelenchidae) isolated from *Cryphalus* sp emerged from a dead *Castanopsis cuspidata* (Thunb.) Schottky var. *sieboldii* (Makino) Nakai in Ishigaki Island, Okinawa, Japan. Nematology 9, 759–769.

Kanzaki, N., Giblin-Davis, R.M., Cardoza, Y.J., Ye, W.M., Raffa, K.F., Center, B.J., 2008. *Bursaphelenchus rufipennis* n. sp (Nematoda: Parasitaphelenchinae) and redescription of *Ektaphelenchus obtusus* (Nematoda: Ektaphelenchinae), associates from nematangia on the hind wings of *Dendroctonus rufipennis* (Coleoptera: Scolytidae). Nematology 10, 925–955.

Khaustov, A.A., Magowski, W.L., Xayctob, A.A., Marobcknn, B.N., 2003. New data on Tarsonemid mites (Acari: Tarsonemidae) associated with subcortical beetles (Coleoptera) in Ukraine and Russia. Acarina 11, 241–245.

Kikuchi, Y., 2009. Endosymbiotic bacteria in insects: their diversity and culturability. Environ. Microbiol. 24, 195–204.

Kim, K.C., 1985. Coevolution and Parasitic Arthropods and Mammals. Wiley-Interscience, New York.

Kinn, D.N., 1967. Notes on the life cycle and habits of *Digamaselus quadrisetus* (Mesostigmata: Digamasellidae). Ann. Entomol. Soc. Am. 60, 862–865.

Kinn, D.N., 1971. The life cycle and behavior of *Cercoleipus coelonotus* (Acarina: Mesostigmata). U. Calif. Pub. Entomol. 65, 1–63.

Kinn, D.N., 1983. The life cycle of *Proctolaelaps dendroctoni* Lindquist and Hunter (Acari: Ascidae): a mite associated with pine bark beetles. Intern. J. Acarol. 9, 205–210.

Kinn, D.N., Witcosky, J.J., 1978. Variation in southern pine beetle attack height associated with phoretic uropodid mites. Can. Entomol. 110, 249–251.

Kirisits, T., 2004. Fungal associates of European bark beetles with special emphasis on the ophiostomatoid fungi. In: Lieutier, F., Day, K.R., Battisti, A., Gregoire, J.-C., Evans, H.F. (Eds.), Bark and Wood Boring Insects in Living Trees in Europe, a Synthesis. Kluwer Academic Publishers, Dordrecht, The Netherlands, pp. 181–236.

Kirschner, R., 2001. Diversity of filamentous fungi in bark beetle galleries in central Europe. In: Misra, J.K., Horn, B.W. (Eds.), Trichomycetes and Other Fungal Groups: Robert W. Lichtwardt Commemoration Volume. Science, Enfield, NJ, pp. 175–196.

Klepzig, K.D., 2006. Melanin and the southern pine beetle-fungus symbiosis. Symbiosis 40, 137–140.

Klepzig, K.D., Hofstetter, R.W., 2011. From attack to emergence: interactions between southern pine beetle, mites, microbes and trees. In: Klepzig, K.D., Coulson, R. (Eds.), Southern Pine Beetle II. United States Dept. of Agriculture Forest Service, pp. 141–152, Southern Research Station General Technical Report SRS-140.

Klepzig, K.D., Six, D.L., 2004. Bark-beetle-fungal symbiosis: context dependency in complex associations. Symbiosis 37, 189–205.

Klepzig, K.D., Adams, A.S., Handelsman, J., Raffa, K.F., 2009. Symbioses: a key driver of insect physiological processes, ecological interactions, evolutionary diversification, and impacts on humans. Environ. Entomol. 38, 67–77.

Klepzig, K.D., Wilkens, R.T., 1997. Competitive interactions among symbiotic fungi of the southern pine beetle. Appl. Env. Microbiol. 63, 621–627.

Klepzig, K.D., Flores-Otero, J., Hofstetter, R.W., Ayres, M.P., 2004. Effects of available water on growth and competition of southern pine beetle associated fungi. Mycol. Res. 108, 183–188.

Klepzig, K.D., Moser, J.C., Lombardero, F.J., Ayres, M.P., Hofstetter, R.W., Walkinshaw, C.J., 2001a. Mutualism and antagonism: ecological interactions among bark beetles, mites and fungi. In: (Jeger, M.J., Spence, N.J. (Eds.), Biotic Interactions in Plant-Pathogen Associations. CAB International, pp. 237–267.

Klepzig, K.D., Moser, J.C., Lombardero, F.J., Hofstetter, R.W., Ayres, M.P., 2001b. Symbiosis & competition: complex interactions among beetles, fungi, and mites. Symbiosis 30, 83–96.

Knee, W., Beaulieu, F., Skevington, J.H., Kelso, S., Cognato, A.I., Forbes, M.R., 2012. Species boundaries and host range of tortoise mites (Uropodoidea) phoretic on bark beetles (Scolytinae), using morphometric and molecular markers. PLoS One 7, e47243.

Knee, W., Forbes, M.R., Beaulieu, F., 2013. Diversity and host use of mites (Acari: Mesostigmata, Oribatida) phoretic on bark beetles (Coleoptera: Scolytinae): global generalists, local specialists? Ann. Entomol. Soc. Amer. 106, 339–350.

Konrad, H., Kirisits, T., Riegler, M., Halmschlager, E., Stauffer, C., 2002. Genetic evidence for natural hybridization between the Dutch elm disease pathogens *Ophiostoma novo-ulmi* ssp *novo-ulmi* and *O-novo-ulmi* ssp *americana*. Plant Pathol. 51, 78–84.

Kopper, B.J., Klepzig, K.D., Raffa, K.F., 2004. Components of antagonism and mutualism in *Ips pini*–fungal interactions: relationship to a life history of colonizing highly stressed and dead trees. Environ. Entomol. 33, 28–34.

Krantz, G.W., Walter, D.E., 2009. A Manual of Acarology. Tech University Press, Texas.

Leufvén, A., Nehls, L., 1986. Quantification of different yeasts associated with the bark beetle, *Ips typographus*, during its attack on a spruce tree. Microbial Ecol. 12, 237–243.

Leufvén, A., Bergstrom, G., Falsen, E., 1988. Oxygenated monoterpenes produced by yeasts, isolated from *Ips typographus* (Coleoptera: Scolytidae) and grown in phloem medium. J. Chem. Ecol. 14, 353–362.

Leufven, A., Bergstrom, G., Falsen, E., 1984. Interconversion of verbenols and verbenone by identified yeasts associated from the spruce bark beetle *Ips typographus*. J. Chem. Ecol. 10, 1349–1361.

Levieux, J., Lieutier, F., Moser, J.C., Roques, A., 1989. Transportation of phytopathogenic fungi by the bark beetle *Ips sexdentatus* Boerner and associated mites. Z. Angew. Entomol. 108, 1–11.

Lewinsohn, D., Lewinsohn, E., Bertagnolli, C.L., Partridge, A.D., 1994. Blue-stain fungi and their transport structures on the Douglas-fir beetle. Can. J. Forest Res. 24, 2275–2283.

Lieutier, F., 2002. Mechanisms of resistance in conifers and bark beetle attack strategies. In: Wagner, M.R., Clancy, K.M., Lieutier, F., Paine, T.D. (Eds.), Mechanisms and Deployment of Resistance in Trees to Insects. Kluwer, Dordrecht, pp. 31–75.

Lieutier, F., Yart, A., Salle, A., 2009. Stimulation of tree defenses by Ophiostomatoid fungi can explain attack success of bark beetles on conifers. Ann. For. Sci. 66, 801.

Lindgren, B.S., Raffa, K.F., 2013. Evolution of tree killing in bark beetles (Coleoptera: Curculionidae): trade-offs between the maddening crowds and a sticky situation. Can. Entomol. 145, 471–495.

Lindquist, E.E., 1969a. Mites and the regulation of bark beetle populations. In: Evans, G.O. (Ed.), Proceedings of the 2nd International Congress of Acarology, Sutton Bonington, UK, 19th–25th July, 1967. Section VIII, Biological Control. Akademie Kiado, Budapest, pp. 389–399.

Lindquist, E.E., 1969b. New species of *Tarsonemus* (Acarina: Tarsonemidae) associated with bark beetles. Can. Entomol. 101, 1291–1314.

Lindquist, E.E., 1985. Discovery of sporothecae in adult female Trochometridium Cross, with notes on analogous structures in *Siteroptes* Amerling (Acari: Heterostigmata). Exp. Appl. Acarol. 1, 73–85.

Linit, M.J., 1988. Nematode vector relationships in the pine wilt disease system. J. Nematol. 20, 227–235.

Liu, G.H., Shao, R.F., Li, J.Y., Zhou, D.H., Li, H., Zhu, X.Q., 2013. The complete mitochondrial genomes of three parasitic nematodes of birds: a unique gene order and insights into nematode phylogeny. BMC Genomics 14, 13.

Llácer, E., Martinez de Altube, M.M.M., Jacas, J.A., 2009. Evaluation of the efficacy of *Steinernema carpocapsae* in a chitosan formulation against the red palm weevil, *Rhynchophorus ferrugineus*, in *Phoenix canariensis*. Biocontrol 54, 559–565.

Lombardero, M.J., Hofstetter, R.W., Ayres, M.P., Klepzig, K.D., Moser, J.C., 2003. Strong indirect interactions among *Tarsonemus* mites (Acarina: Tarsonemidae) and *Dendroctonus frontalis* (Coleoptera: Scolytidae). Oikos 102, 342–352.

Lombardero, M.J., Klepzig, K.D., Moser, J.C., Ayres, M.P., 2000. Biology, demography and community interactions of *Tarsonemus* (Acarina: Tarsonemidae) mites phoretic on *Dendroctonus frontalis* (Coleoptera:Scolytidae). Agr. Forest Entomol. 2, 193–202.

Lukášová, K., Holuša, J., 2012. Patogeny lýkožroutů rodu *Ips* (Coleoptera: Curculionidae: Scolytinae): review. Zprávy Lesnického Výzkumu 57, 230–240.

Lukášová, K., Holuša, J., Turčáni, M., 2013. Pathogens of *Ips amitinus*: new species and comparison with *Ips typographus*. J. Appl. Entomol. 137, 188–196.

Ma, J., Chen, S., De Clercq, P., Waeyenberge, L., Han, R., Moens, M., 2012. A new entomopathogenic nematode, *Steinernema xinbinense* n. sp (Nematoda: Steinernematidae), from north China. Nematology 14, 723–739.

Macchioni, F., 2007. Importance of phoresy in the transmission of Acari. Parasitologia 49, 17–22.

Macguidwin, A.E., Smart, G.C., Allen, G.E., 1980a. Redescription and life-history of *Contortylenchus brevicomi*, a parasite of the southern pine beetle *Dendroctonus frontalis*. J. Nematol. 12, 207–212.

Macguidwin, A.E., Smart, G.C., Wilkinson, R.C., Allen, G.E., 1980b. Effect of the nematode *Contortylenchus brevicomi* on gallery construction and fecundity of the southern pine beetle. J. Nematol. 12, 278–282.

Maehara, N., Futai, K., 1997. Effect of fungal interactions on the numbers of the pinewood nematode, *Bursaphelenchus xylophilus* (Nematoda: Aphelenchoididae), carried by the Japanese pine sawyer, *Monochamus alternatus* (Coleoptera: Cerambycidae). Fundam. Appl. Nematol. 20, 611–617.

Magowski, W.L., Moser, J.C., 2003. Redescription of *Tarsonemus minimax* and definition of its species-group in the genus *Tarsonemus* (Acari: Tarsonemidae) with descriptions of two new species. Ann. Entomol. Soc. Am. 96, 345–368.

Mamiya, Y., 1983. Pathology of the pine wilt disease caused by *Bursaphelenchus xylophilus*. Annu. Rev. Phytopathol. 21, 201–220.

Mamiya, Y., 1988. History of pine wilt disease in Japan. J. Nematol. 20, 219–226.

Manachini, B., Schillaci, D., Arizza, V., 2013. Biological responses of *Rhynchophorus ferrugineus* (Coleoptera: Curculionidae) to *Steinernema carpocapsae* (Nematoda: Steinernematidae). J. Econ. Entomol. 106, 1582–1589.

Martinson, S.J., Ylioja, T., Sullivan, B.T., Billings, R.F., Ayres, M.P., 2013. Alternate attractors in the population dynamics of a tree-killing bark beetle. Popul. Ecol. 55, 95–106.

Massey, C.L., 1956. Nematode parasites and associates of the Engelmann spruce beetle (*Dendroctonus engelmanni* Hopk.). P. Helm. Soc. Wash. 23, 14–24.

Massey, C.L., 1960. Nematode parasites and associates of the California five-spined engraver, *Ips confusus* (Lee). P. Helm. Soc. Wash. 27, 14–22.

Massey, C.L., 1964. Nematode parasites + associates of the fir engraver beetle *Scolytus ventralis* Leconte in New Mexico. Journal of Insect Pathology 6, 133–155.

Massey, C.L., 1974. Biology and taxonomy of nematode parasites and associates of bark beetles in the United States. Forest Service, US Dept. of Agriculture, Washington, DC, Washington, DC.

Mazzone, H.M., Peacock, J.W., 1985. Prospects for control of Dutch elm disease- biological considerations. Journal of Arboriculture 11, 285–292.

McGraw, J.R., Farrier, M.H., 1969. Of the Superfamily Parasitoidea (Acarina: Mesostigmata) Associated with *Dendroctonus* and *Ips* (Coleoptera: Scolytidae). USDA, North Carolina Agricultural Experiment Station. Tech. Bull, 192.

Mercado, J.E., Hofstetter, R.W., Reboletti, D.M., Negrón, J.R., 2014. Phoretic symbionts of the mountain pine beetle (*Dendroctonus ponderosae* Hopkins). For. Sci. 60 (in press).

Moore, G.E., 1971. Mortality factors caused by pathogenic bacteria and fungi of the southern pine beetle in North Carolina. J. Invert. Path. 17, 28–37.

Moore, G.E., 1972. Pathogenicity of ten strains of bacteria to larvae of the southern pine beetle. J. Invertebr. Pathol. 20, 41–45.

Morales-Jiménez, J., Zúñiga, G., Ramirez-Saad, H.C., Hernández-Rodríguez, C., 2012. Gut-associated bacteria throughout the life cycle of the bark beetle *Dendroctonus rhizophagus* Thomas and Bright (Curculionidae: Scolytinae) and their cellulolytic activities. Microb. Ecol. 64, 268–278.

Morales-Jiménez, J., de León, A.V.P., García-Domínguez, A., Martínez-Romero, E., Zúñiga, G., Hernández-Rodríguez, C., 2013. Nitrogen-fixing and uricolytic bacteria associated with the gut of *Dendroctonus rhizophagus* and *Dendroctonus valens* (Curculionidae: Scolytinae). Microb. Ecol. 66, 200–210.

Morales-Jiménez, J., Zúñiga, G., Villa-Tanaca, L., Hernández-Rodríguez, C., 2009. Bacterial community and nitrogen fixation in the red turpentine beetle, *Dendroctonus valens* LeConte (Coleoptera: Curculionidae: Scolytinae). Microb. Ecol. 58, 879–891.

Moser, J.C., 1985. Use of sporothecae by phoretic Tarsonemus mites to transport ascospores of coniferous bluestain fungi. Trans. Br. Mycol. Soc. 84, 750–753.

Moser, J.C., Roton, L.M., 1971. Mites associated with southern pine bark beetles in Allen Parish. Louisiana. Can. Entomol. 103, 1775–1798.

Moser, J.C., Eidmann, H.H., Regnander, J.R., 1989a. The mites associated with *Ips typographus* in Sweden. Ann. Entomol. Fennici 55, 23–27.

Moser, J.C., Konrad, H., Blomquist, S.R., Kirisits, T., 2010. Do mites phoretic on elm bark beetles contribute to the transmission of Dutch elm disease? Naturwissenschaften 97, 219–227.

Moser, J.C., Perry, T.J., Solheim, H., 1989b. Ascospores hyperphoretic on mites associated with *Ips typographus*. Myc. Res. 93, 513–517.

Moser, J.C., Konrad, H., Kirisits, T., Carta, L.K., 2005. Phoretic mites and nematode associates of *Scolytus multistriatus* and *Scolytus pygmaeus* (Coleoptera: Scolytidae) in Austria. Agr. Forest. Entomol. 7, 169–177.

Muratoglu, H., Sezen, K., Demirbag, Z., 2011. Determination and pathogenicity of the bacterial flora associated with the spruce bark beetle, *Ips typographus* (L.) (Coleoptera: Curculionidae: Scolytinae). Turk. J. Biol. 35, 9–20.

Musvuugwa, T., 2014. Biodiversity and ecology of Ophiostomatoid fungi associated with trees in the Cape Florist Region of South Africa. PhD thesis. Stellenbosch University, South Africa.

National Research Council Committee on Plant and Animal Pests, 1968. Control of Plant-Parasitic Nematodes. National Academy of Sciences, Washington, DC, p. 172.

Negrón, J.F., Bentz, B.J., Fettig, C.J., Gillette, N., Hansen, E.M., Hayes, J. L., et al., 2008. US Forest Service bark beetle research in the western United States: Looking toward the future. J. For. 106, 325–331.

Nickle, W.R., 1963. Observations on the effect of nematodes on *Ips confusus* (Leconte) and other bark beetles. Journal of Insect Pathology 5, 386–389.

Nickle, W.R., 1971. Behavior of the shothole borer, *Scolytus rugulosus*, altered by the nematode parasite *Neoparasitylenchus rugulosi*. Ann. Entomol. Soc. Am. 64, 751.

Norris, D.M., Baker, J.M., Chu, H.M., 1969. Symbiotic interrelationships between microbes and ambrosia beetles III: Ergosterol as the source of sterol to the insect. Ann. Entomol. Soc. Am. 62, 413–414.

Oldham, J.N., 1930. On the infestation of elm bark-beetles (Scolytidae) by a nematode. Parasitylenchus scolyti n. sp. J. Helminthol. 8, 239–248.

Olthof, T.H.A., Hagley, E.A.C., 1993. Laboratory studies of the efficacy of *Steinernematid* nematodes against the plum curculio (Coleoptera, Curculionidae). J. Econ. Entomol. 86, 1078–1082.

Owen, D., 2003. The Fir Engraver Beetle. Produced by California Department of Forestry and Fire Protection. Tree Notes 10, 1–4.

Paine, T.D., Raffa, K.F., Harrington, T.C., 1997. Interactions among scolytid bark beetles, their associated fungi, and live host conifers. Annu. Rev. Entomol. 42, 179–206.

Peer, K., Taborsky, M., 2005. Outbreeding depression, but no inbreeding depression in haplodiploid ambrosia beetles with regular sibling mating. Evolution 59, 317–323.

Penas, A.C., Bravo, M.A., Naves, P., Bonifácio, L., Sousa, E., Mota, M., 2006. Species of *Bursaphelenchus* Fuchs, 1937 (Nematoda: Parasitaphelenchidae) and other nematode genera associated with insects from *Pinus pinaster* in Portugal. Ann. Appl. Biol. 148, 121–131.

Perera, M.R., Vanstone, V.A., Jones, M.G., 2005. A novel approach to identify plant parasitic nematodes using matrix-assisted laser desorption/ionization time-of-flight mass spectrometry. Rapid Commun. Mass Spectrom. 19, 1454–1460.

Pernek, M., Hrasovec, B., Matosevic, D., Pilas, I., Kirisits, T., Moser, J.C., 2008. Phoretic mites of three bark beetles (*Pityokteines* spp.) on Silver fir. J. Pest Sci. 81, 35–42.

Pernek, M., Wirth, S., Blomquist, S.R., Avtzis, D.N., Moser, J.C., 2012. New associations of phoretic mites on *Pityokteines curvidens* (Coleoptera, Curculionidae, Scolytinae). Cent. Eur. J. Biol. 7, 63–68.

Perotti, M.A., Braig, H.R., 2009. Phoretic mites associated with animal and human decomposition. Exp. Appl. Acarol. 49, 85–124.

Perotti, M.A., Braig, H.R., 2011. Eukaryotic ectosymbionts of Acari. J. Appl. Entomol. 135, 514–523.

Pfammatter, J.A., Moser, J.C., Raffa, K.F., 2013. Mites phoretic on *Ips pini* (Coleoptera: Curculionidae: Scolytinae) in Wisconsin red pine stands. Ann. Entomol. Soc. Amer. 106, 204–213.

Plattner, A., Kim, J.-J., Diguistini, S., Breuil, C., 2008. Variation in pathogenicity of a mountain pine beetle-associated blue-stain fungus, *Grosmannia clavigera*, on young lodgepole pine in British Columbia. Canadian Journal of Plant Pathology 30, 1–10.

Poinar Jr., G.O., 1975. Entomogenous Nematodes. A Manual and Host List of Insect-Nematode Associations. Koninklijke Brill NV, Leiden, The Netherlands.

Poinar Jr., G.O., 2011. The Evolutionary History of Nematodes: As Revealed in Stone, Amber and Mummies. Koninklijke Brill NV, Leiden, The Netherlands.

Poinar Jr., G.O., Deschamps, N., 1981. Susceptibility of *Scolytus multistriatus* to neoaplectanid and heterorhabditid nematodes. Environ. Entomol. 10, 85–87.

Poinar Jr., G., Vega, F.E., Castillo, A., Chavez, I.E., Infante, F., 2004. *Metaparasitylenchus hypothenemi* n. sp. (Nematoda: Allantonematidae), a parasite of the coffee berry borer, *Hypothenemus hampei* (Curculionidae: Scolytinae). J. Parasitol. 90, 1106–1110.

Popa, V., Deziel, E., Lavallee, R., Bauce, E., Guertin, C., 2012. The complex symbiotic relationships of bark beetles with microorganisms: a potential practical approach for biological control in forestry. Pest Manag. Sci. 68, 963–975.

Raffa, K.F., 2001. Mixed messages across multiple trophic levels: the ecology of bark beetle chemical communication systems. Chemoecology 11, 49–65.

Redman, R.S., Sheehan, K.B., Stout, R.G., Rodriguez, R.J., Henson, J.M., 2002. Thermotolerance conferred to plant host and fungal endophyte during mutualistic symbiosis. Science 298, 1581.

Rehner, S.A., 2009. Molecular systematics of entomopathogenic fungi. In: Stock, S.P., Vandenberg, J., Glazer, I., Boemare, N. (Eds.), Insect Pathogens: Molecular Approaches and Techniques. CAB International, Oxfordshire (UK), pp. 145–158.

Relman, D.A., 2008. "Til death do us part": coming to terms with symbiotic relationships—Foreword. Nat. Rev. Microbiol. 6, 721–724.

Ruhm, W., 1956. Die Nematoden der Ipiden. Parasitologische Schriftenreihe 6, 1–437.

Ruhm, W., 1965. Widening of the number of possible hosts of a nematode species related to bark beetles (Scolytoidea, Coleoptera). Z. Parasitenk. 26, 230–253.

Ryss, A., Vieira, P., Mota, M., Kulinich, O., 2005. A synopsis of the genus *Bursaphelenchus* Fuchs, 1937 (Aphelenchida: Parasitaphelenchidae) with keys to species. Nematology 7, 393–458.

Safranyik, L., Carroll, A.L., Regniere, J., Langor, D.W., Riel, W.G., Shore, T.L., et al., 2010. Potential for range expansion of mountain pine beetle into the northern boreal forest of North America. Can. Entomol. 142, 415–442.

Sapp, J., 1994. Evolution by Association: a History of Symbiosis. Oxford University Press.

Schelfer, R.J., Voeten, J., Guries, R.P., 2008. Biological control of Dutch elm disease. Plant Dis. 92, 192–200.

Scott, J.J., Dong-Chan, O., Yuceer, M.C., Klepzig, K.D., Clardy, J., Currie, C.R., 2008. Bacterial protection of beetle-fungus mutualism. Science 322, 63.

Seifert, K.A., 1993. Sapstain of commercial lumber by species of *Ophiostoma* and *Ceratocystis*. In: Wingfield, M.J., Seifert, K.A., Webber, J.F. (Eds.), *Ceratocystis* and *Ophiostoma*: Taxonomy, Ecology and Pathogenicity. American Phytopathological Society, St. Paul, MN, USA, pp. 141–151.

Seifert, K.A., De Beer, Z.W., Wingfield, M.J., 2013. Ophiostomatoid Fungi: Expanding Frontiers. CBS Biodiversity Series 12. The Netherlands: CBS Fungal Diversity Centre, Utrecht, The Netherlands.

Sevim, A., Gokce, C., Erbas, Z., Ozkan, F., 2012. Bacteria from *Ips sexdentatus* (Coleoptera: Curculionidae) and their biocontrol potential. J. Basic Microb. 52, 695–704.

Shimizu, A., Tanaka, R., Akiba, M., Masuya, H., Iwata, R., Fukuda, K., Kanzaki, N., 2013. Nematodes associated with *Dryocoetes uniseriatus* (Coleoptera: Scolytidae). Environ. Entomol. 42, 79–88.

Sikorowski, P.P., Lawrence, A.M., Nebeker, T.E., Price, T.S., 1996. Virus and virus-like particles found in southern pine beetle adults in Mississippi and Georgia. Technical Bulletin, 212, 1–9, Mississippi Agricultural & Forestry Experiment Station.

Simoni, S., Nannelli, R., Roversi, P.F., Turchetti, T., Bouneb, M., 2014. *Thyreophagus corticalis* as a vector of hypovirulence in *Cryphonectria parasitica* in chestnut stands. Exp. Appl. Acarol. 62, 363–375.

Sinclair, W.A., Lyon, H.H., 2005. Diseases of Trees and Shrubs, second ed. Comstock Publishing Associates, Cornell University Press.

Six, D.L., 2003. Bark beetle-fungus symbioses. In: Bourtzis, K., Miller, T.A. (Eds.), Insect Symbiosis. Contemporary Tropics in Entomology Services. CRC Press, Boca Raton, London, New York, Washington, DC, pp. 97–114.

Six, D.L., 2012. Ecological and evolutionary determinants of bark beetle–fungus symbioses. Insects 3, 339–366.

Six, D.L., Bentz, B., 2007. Temperature determines symbiont abundance in a multipartite bark beetle-fungus ectosymbiosis. Microb. Ecol. 54, 112–118.

Six, D.L., Paine, T.D., 1997. *Ophiostoma clavigerum* is the mycangial fungus of the Jeffrey pine beetle, *Dendroctonus jeffreyi* (Coleoptera: Scolytidae). Mycologia 89, 858–866.

Six, D.L., Paine, T.D., 1998. Effects of mycangial fungi and host tree species on progeny survival and emergence of *Dendroctonus ponderosae* (Coleoptera: Scolytidae). Environ. Entomol. 27, 1393–1401.

Six, D.L., Wingfield, M.J., 2011. The role of phytopathogenicity in bark beetle-fungus symbioses: a challenge to the classic paradigm. Annu. Rev. Entomol. 56, 255–272.

Smith, S.E., Read, D.J., 1997. Mycorrhizal Symbiosis. Academic Press, San Diego.

Solheim, H., 1992. Fungal succession in sapwood of Norway spruce infested by the bark beetle *Ips typographus*. Eur. J. For. Pathol. 22, 136–148.

Solheim, H., 1995. Early stages of blue-stain fungus invasion of lodgepole pine sapwood following mountain pine beetle attack. Can. J. Bot. 73, 70–74.

Spatafora, J.W., Blackwell, M., 1994. The polyphyletic origins of ophiostomatoid fungi. Mycol. Res. 98, 1–9.

Stauffer, C., van Meer, M.M.M., Riegler, M., 1997. The presence of the proteobacteria *Wolbachia* in European *Ips typographus* (Col., Scolytidae) populations and the consequences for genetic data. Proceedings of the German Society for General and Applied Entomology 11, 709–711.

Stone, C., Simpson, J.A., 1990. Species associations in *Ips grandicollis* galleries in *Pinus taeda*. New Zeal. J. Forest. Sci. 20, 75–96.

Susoy, V., Kanzaki, N., Herrmann, M., 2013. Description of the bark beetle associated nematodes *Micoletzkya masseyi* n. sp and *M. japonica* n. sp (Nematoda: Diplogastridae). Nematology 15, 213–231.

Sutherland, J.B., 2004. Degradation of hydrocarbons by yeast and filamentous fungi. In: Arora, D.K., Bridge, P.D., Bhatnagar, D. (Eds.), Fungal Biotechnology in Agricultural, Food and Environmental Applications. CRC Press, Marcel Dekker, New York, pp. 443–445.

Takov, D., Pilarska, D., Moser, J., 2009. Phoretic mites associated with spruce bark beetle *Ips typographus* L. (Curculionidae: Scolytinae) from Bulgaria. Acta Zool. Bulgaria 61, 293–296.

Thong, C.H.S., Webster, J.M., 1983. Nematode parasites and associates of *dendroctonus spp* and *Trypodendron-Lineatum* (Coleoptera, Scolytidae), with a description of *Bursaphelenchus-Varicauda* n-sp. J. Nematol. 15, 312–318.

Tomalak, M., Welch, H.E., 1982. *Neoplectana carpocapsae* DD-136 as a potential biological agent for control of *Hylurgopinus rufipes* (Eichhoff). In: Kondo, E.S., Hirateuka, Y., Denyer, W.B.G. (Eds.), Proceedings of the Dutch elm disease workshop and symposium. Winnipeg, Manitoba, Canada, pp. 14–23.

Tomalak, M., Michalski, J., Grocholski, J., 1984. The influence of nematodes on the structure of genitalia of *Tomicus piniperda* (Coleoptera, Scolytidae). J. Invertebr. Pathol. 43, 358–362.

Tomalak, M., Welch, H.E., Galloway, T.D., 1988. Interaction of parasitic nematode *Parasitaphelenchus oldhami* (Nematoda, Aphelenchoididae) and a bacterium in dutch elm disease vector, *Hylurgopinus rufipes* (Coleoptera, Scolytidae). J. Invertebr. Pathol. 52, 301–308.

Unal, S., Yaman, M., Tosun, O., Aydin, C., 2009. Occurrence of *Gregarina typographi* (Apicomplexa, Gregarinidae) and *Metschnikowia typographi* (Ascomycota, Metschnikowiaceae) in *Ips sexdentatus* (Coleoptera: Curculionidae, Scolytinae) populations in Kastamonu (Turkey). J. Anim. Vet. Adv. 8, 2687–2691.

Valadas, V., Laranjo, M., Mota, M., Oliveira, S., 2013. A survey of entomopathogenic nematode species in continental Portugal. J. Helm, 1–15.

Vasanthakumar, A., Delalibera, I., Handelsman, J., Klepzig, K.D., Schloss, P.D., Raffa, K.F., 2006. Characterization of gut-associated bacteria in larvae and adults of the southern pine beetle, *Dendroctonus frontalis* Zimmermann. Environ. Entomol. 35, 1710–1717.

Vega, F.E., Dowd, P.F., 2005. The role of yeasts as insect endosymbionts. In: Vega, F.E., Blackwell, M. (Eds.), Insect-Fungal Associations: Ecology and Evolution. Oxford University Press, New York, pp. 211–243.

Vega, F.E., Benavides, P., Stuart, J.A., O'Neill, S., 2002. *Wolbachia* infection in the Coffee berry borer (Coleoptera: Scolytidae). Ann. Entomol. Soc. Am. 95, 374–378.

Vega, F.E., Ochoa, R., Astorga, C., Walter, D.E., 2007. Mites (Arachnida: Acari) inhabiting coffee domatia: A short review and recent findings from Costa Rica. Internat. J. Acarol. 33, 291–295.

Vega, F.E., Simpkins, A., Aime, M.C., Posada, F., Peterson, S.W., Rehner, S.A., et al., 2010. Fungal endophyte diversity in coffee plants from Colombia, Hawai'i, Mexico, and Puerto Rico. Fungal Ecology 3, 122–138.

Walter, D.E., Proctor, H.C., 1999. Mites: Ecology. Evolution and Behavior, CAB International, Wallingford, UK.

Weber, B.C., McPherson, J.E., 1983. Life history of the ambrosia beetle *Xylosandrus germanus* (Coleoptera: Scolytidae). Ann. Entomol. Soc. Am. 76, 455–462.

Wegensteiner, R., 2004. Pathogens in bark beetles. In: Lieutier, F., Day, K.R., Battisti, A., Gregoire, J.-C., Evans, H.F. (Eds.), Bark and Wood Boring Insects in Living Trees in Europe, a Synthesis. Kluwer Academic Publishers. Dordrecht, The Netherlands, pp. 291–313.

Wegensteiner, R., Weiser, J., 1998. Infection of *Ips typographus* from Finland with the Ascomycete *Metschnikowia* cf. bicuspidata, 6th European Congress of Entomology, Ceske Budejovice Proceedings, p. 667.

Wegensteiner, R., Eiser, J.W., Ührer, E.F., 1996. Observations on the occurrence of pathogens in the bark beetle *Ips typographus* L. (Coleoptera. Scolytidae). J. Appl. Entomol. 120, 199–204.

Weiser, J., Wegensteiner, R., Händel, U., Žižka, Z., 2003. Infections with the Ascomycete *Metschnikowia typographi* n.sp. in the bark beetle *Ips typographus* and *Ips amitinus* (Col., Scolytidae). Folia Microbiol. 48, 611–618.

Werner, R.A., Holsten, E.H., 1985. Factors influencing generation times of spruce beetles in Alaska. Can. J. Forest Res. 15, 438–443.

Wharton, D.A., 2004. Survival strategies. In: Gaugler, R., Bilgram, A.L. (Eds.), Nematode Behaviour. CABBI, Cambridge, MA, pp. 371–399.

Whitney, H.S., 1982. Relationships between bark beetles and symbiotic organisms. In: Mitton, J.B., Sturgeon, K.B. (Eds.), Bark Beetles in North American Conifers. Univ. Texas Press, Austin, pp. 183–211.

Wingfield, M.J., Blanchette, R.A., Nicholls, T.H., 1984. Is the pine wood nematode an important pathogen in the United States? J. For. 82, 232–235.

Wood, S.L., 1982. The bark and ambrosia beetles of North and Central America (Coleoptera: Scolytidae), a taxonomic monograph. Great Basin Nat. Mem. 6, 1–1359.

Wood, W.B., 1988. The nematode *Caenorhabditis elegans*. Cold Spring Harbor Monograph Series 17, 215–241.

Yaman, M., Baki, H., 2011. First record of entomopoxvirus of *Ips typographus* (Linnaeus) (Coleoptera: Curculionidae, Scolytinae) for Turkey. Acta Zoolog. Bulg. 63, 199–202.

Yearian, W.C., 1966. Relations of the blue stain fungus, *Ceratocystis ips* (Rumbold) C. Moreau, to *Ips* bark beetles (Coleoptera: Scolytidae) occurring in Florida. PhD thesis, University of Florida.

Yearian, W.C., Gouger, R.J., Wilkinson, R.C., 1972. Effects of the blue-stain fungus *Ceratocystis ips* on development of *Ips* bark beetles in pine bolts. Ann. Entomol. Soc. Amer. 65, 481–487.

Yuceer, C.M., Chuan-Yu, H., Erbilgin, N., Klepzig, K.D., 2011. Ultrastructure of the mycangium of the southern pine beetle, *Dendroctonus frontalis* (Coleoptera: Curculionidae, Scolytinae): complex morphology for complex interactions. Acta Zoologica 92, 216–224.

Zchori-Fein, E., Borad, C., Harari, A.R., 2006. Oogenesis in the date stone beetle, *Coccotrypes dactyliperda*, depends on symbiotic bacteria. Physiol. Entomol. 31, 164–169.

Zhou, X., de Beer, Z.W., Wingfield, M.J., 2006. DNA sequence comparisons of *Ophiostoma* spp., including *Ophiostoma aurorae* sp. nov., associated with pine bark beetles in South Africa. Stud. Mycol. 55, 269–277.

Zipfel, R.D., de Beer, Z.W., Jacobs, K., Wingfield, B.D., Wingfield, M.J., 2006. Multi-gene phylogenies define *Ceratocystiopsis* and *Grosmannia* distinct from *Ophiostoma*. Stud. Mycol. 55, 75–97.

Zook., D., 1998. A new symbiosis language. Symbiosis News 1, 1–3.

Chapter 7

Natural Enemies of Bark Beetles: Predators, Parasitoids, Pathogens, and Nematodes

Rudolf Wegensteiner[1], Beat Wermelinger[2], and Matthias Herrmann[3]

[1] *University of Natural Resources and Life Sciences, BOKU–Vienna, Department of Forest and Soil Sciences, Institute of Forest Entomology, Forest Pathology and Forest Protection, Vienna, Austria,* [2] *Swiss Federal Institute for Forest, Snow and Landscape Research WSL, Forest Dynamics, Birmensdorf, Switzerland,* [3] *Max Planck Institute for Developmental Biology, Department of Evolutionary Biology, Tuebingen, Germany*

1. INTRODUCTION

Natural enemies, such as predators, parasitoids, and pathogens, play an important role in the population dynamics and ecology of bark beetles. Until now, relatively few publications have comprehensively considered all groups of bark beetle natural enemies (Bałazy, 1968; Stark and Dahlsten, 1970; Mills, 1983; Stephen et al., 1993; Fuxa et al., 1998; Gokturk et al., 2010). In this chapter, we review information on the biology and ecology of natural enemies and their significance in bark beetle population dynamics with a particular focus on European and North American species from coniferous forests. The natural enemies of bark beetles in other systems, such as coffee where the coffee berry borer (*Hypothenemus hampei* (Ferrari)) is an important bark beetle pest, are covered in Chapter 11.

2. PREDATORS AND PARASITOIDS OF BARK BEETLES

Even as early as the late 19th century, woodpeckers were recognized as important antagonists of bark beetles in Europe (Sperling, 1878). However, systematic scientific research on both avian and insect natural enemies only really began in the second half of the last century, in response to outbreaks of bark beetles in European and North American coniferous forests (De Leon, 1934; Sachtleben, 1952; Nuorteva, 1957; Hedqvist, 1963; Bushing, 1965; Dahlsten and Stephen, 1974). Most literature on this topic still originates from these regions and is, understandably, focused on those bark beetle species that pose the greatest economic and ecological threat to managed forests; with a few exceptions, such as *Scolytus intricatus* (Ratzeburg) on oak, the most damaging bark beetles are associated with conifers. The most recent comprehensive synopsis of the species assemblages and significance of predators and parasitoids of European bark beetles by Kenis et al. (2004) and the reviews of Dahlsten (1982) and Mills (1983) for North America and Europe, respectively, form the foundation for this chapter and are complemented by newer literature.

2.1 Woodpeckers and other Avian Predators

A number of different bird groups are known to include bark beetles in their diet, particularly catching and consuming them in-flight when beetles are swarming (Otvos and Stark, 1985). However, unlike other birds, saproxylic woodpecker species (Picidae) also feed on adult beetles at the bark surface and can access immature stages concealed beneath the bark or in the sap wood. Woodpeckers are a diverse group and, as such, have a broad diet range; some species are not insectivorous at all, some feed only occasionally on saproxylic insects and prefer to consume ground-dwelling ants, while others only prey on insects on the surface of bark and leaves (Otvos, 1965). However, a significant group of woodpeckers specialize on subcortical insects such as the larvae of woodwasps (Siricidae), longhorned beetles (Cerambycidae), jewel beetles (Buprestidae), deathwatch beetles (Anobiidae), weevils (Curculionidae), and bark beetles (Curculionidae: Scolytinae). These are the species we focus on in this chapter (Table 7.1). How woodpeckers locate their prey beneath the bark is still not fully understood, but acoustics may play a major role. Often they seem to prefer prey species that are larger than bark beetle larvae and adults (Nuorteva and Saari, 1980), and, among bark beetle larvae, they prefer the larger instars (Kroll and Fleet, 1979).

While woodpeckers are widely acknowledged as efficient predators of bark beetles, their effect on population regulation relative to that of insect predators and parasitoids

Bark Beetles. http://dx.doi.org/10.1016/B978-0-12-417156-5.00007-1

TABLE 7.1 Woodpeckers (Picidae) Reported as Predators of Bark Beetle Species

Species	Distribution[1]	Dendroctonus	Ips	Orthotomicus	Phloeosinus	Pityogenes	Scolytus	Tomicus	Trypodendron	Other
Red-shafted flicker Colaptes cafer (Gmelin)	ne	x								
Great spotted woodpecker Dendrocopos major (L.)	pa		x				x			
Syrian woodpecker Dendrocopos syriacus [Hemprich&Ehrenberg])	pa			x	x	x	x	x		x
Black-backed woodpecker Dryocopus martius (L.)	pa		x							
Pileated woodpecker Dryocopus pileatus (L.)	ne									
White-headed woodpecker Picoides albolarvatus (Cassin)	ne	x	x				x		x	
American three-toed woodpecker Picoides dorsalis Baird[2]	ne	x	x							x
Downy woodpecker Picoides pubescens (L.)	ne	x	x				x		x	x
Eurasion three-toed woodpecker Picoides tridactylus (L.)	pa		x							
Hairy woodpecker Picoides villosus (L.)	ne	x	x	x						x
References		1-4, 6, 8, 10	1, 3, 5, 7, 10-12	2, 3, 9	9	9	2, 3, 9	9	2, 3	7, 9

[1]pa, palaearctic; ne, nearctic
[2]until recently synonymized with P. tridactylus.
References: (1) Otvos, 1965; (2) Otvos, 1979; (3) Otvos and Stark, 1985; (4) Otvos, 1970; (5) Schimitschek, 1931a; (6) Knight, 1958; (7) Koplin, 1969; (8) Koplin and Baldwin, 1970; (9) Mendel, 1985; (10) Shook and Baldwin, 1970; (11) Pechacek and Kristin, 2004; (12) Pechacek, 1994

is considered limited for several reasons (Fayt *et al.*, 2005): (1) there is a lack of synchronization between bark beetle and woodpecker generations; (2) increases in woodpecker populations in response to bark beetle outbreaks are delayed due to their lower reproductive rate relative to that of their prey; (3) limited nesting and roosting sites in conjunction with territoriality prevent woodpecker populations from exceeding a critical density necessary for bark beetle regulation; and (4) woodpeckers may switch to other diets depending on the nesting season and availability of alternative food sources (Pechacek and Kristin, 2004; Fayt *et al.*, 2005). Nevertheless, woodpeckers can show both functional and numerical responses to bark beetle density. The proportion of bark beetles in their diet increases with greater availability of this prey (Koplin and Baldwin, 1970; Koplin, 1972; Fayt *et al.*, 2005). Woodpeckers may

rapidly invade infested stands from adjacent forests and they are known to aggregate towards local outbreaks during winter (Otvos, 1979). For example, during an outbreak of the spruce beetle, *Dendroctonus rufipennis* Kirby, a 50-fold increase in woodpecker density was observed (Koplin, 1969). However, the reproductive capacity of woodpeckers does not increase with increasing prey availability (Edworthy *et al.*, 2011). Furthermore, they are believed to have their greatest impact when beetles are at endemic levels, and when they can accelerate beetle population collapse, delay the onset of outbreaks, and expand the periods between outbreaks (Otvos, 1979).

Bark beetle mortality inflicted by woodpeckers is twofold, i.e., by direct consumption of larvae, pupae and adults, but also through indirect effects (Otvos, 1965, 1979; McCambridge and Knight, 1972). Indirect effects can result from desiccation of the bark after intensive woodpecker activity, the change in temperatures of the remaining thinned bark, or the loss of brood with detached bark flakes. Brood in detached bark flakes can desiccate or be subject to predation by other avian or ground predators. Additionally, woodpeckering (puncturing, loosening, and removal of bark) enables other birds to feed on exposed larvae. Foraging birds may also transmit pathogens of bark beetles between infested trees (Otvos, 1979). The density of predatory insects is generally lower in woodpeckered areas; this could be because they are also consumed by the woodpeckers or be due to associated indirect effects (Otvos, 1979). However, there are also positive interactions between woodpeckers and parasitoids; when pecking in search of pine bark beetles, woodpeckers remove bark material to access their prey and reduce bark thickness to approximately 5 mm. This in turn allows parasitoids with short ovipositors to access hosts that they would otherwise be unable to reach, facilitating a 10-fold increase in parasitism (Otvos, 1979).

There is some evidence that overwintering generations of bark beetles suffer greater mortality due to woodpeckers than summer generations. In winter, alternative prey is scarce and woodpecker diets can consist of up to 99% bark beetles (Otvos, 1965; Baldwin, 1968). Furthermore, the energy requirement to maintain a constant body temperature in winter is greater than in summer. Baldwin (1968) calculated that at −12°C, some 3200 larvae per day are required to meet the caloric needs of woodpeckers. In addition, overwintering bark beetle generations are exposed to woodpecker predation for a longer time than summer generations.

Woodpeckers feed more or less uniformly along an infested bole, but once the host trees have reached a certain infestation density, the woodpeckers continue at the section with the highest prey density, usually in the middle of the bole (Otvos, 1965; DeMars et al., 1970; Kroll and Fleet, 1979).

In North American conifer forests, studies have been conducted on woodpeckers predating four bark beetle species from the genus Dendroctonus, mostly in epidemic population phases (Otvos, 1965, 1979). The most important woodpecker species in these forests are the hairy, downy, white-headed, and American three-toed woodpeckers (Table 7.1). The hairy woodpecker is the key species and the first to find newly infested trees (Otvos, 1965, 1970). While the hairy woodpecker is the most generalist species foraging in dead wood of varying dimensions and at different stages of decay, the downy woodpecker prefers relatively small branches of freshly-killed spruce and fir (Koplin, 1969), and the three-toed woodpecker is almost entirely restricted to spruce forests (Baldwin, 1968). Until recently, the American three-toed woodpecker was considered the same species as the three-toed woodpecker in Europe, but the former is now regarded a distinct species,

i.e., *Picoides dorsalis* Baird (Table 7.1). In Europe, the most important species foraging on bark beetles are the European three-toed, the great spotted and, to a lesser extent (with regional differences), the black-backed woodpecker (Schimitschek, 1931a; Fayt et al., 2005). The three-toed woodpecker is a particularly important, though often rare, forager on *Ips typographus* (L.).

Correlations between woodpeckering intensity and resulting bark beetle mortality have shown that an average bark beetle mortality of 72% is achieved in bark areas that have been completely worked by woodpeckers (Otvos, 1970). Open stands seem to promote the impact of woodpeckers on *Dendroctonus* beetles compared to dense stands (Shook and Baldwin, 1970). Even isolated trees infested with beetles can be detected by woodpeckers (DeMars et al., 1970). In the case of the western pine beetle, *Dendroctonus brevicomis* LeConte, woodpecker-mediated mortality was estimated at 32% at the beginning of an outbreak. In spruce forests, the reduction of bark beetles by woodpecker foraging ranged between 19 and 98% (Knight, 1958; Fayt et al., 2005). Woodpecker-inflicted mortality of bark beetle populations was dependent on the phase of the outbreak (Baldwin, 1968). During a *D. rufipennis* outbreak, mortality rates were low at low beetle densities, increased with higher bark beetle populations, but then decreased again with still higher beetle densities.

Although the most important species feeding on bark beetles in both North American and European spruce forests are the three-toed woodpeckers (*Picoides tridactylus* (L.), *P. dorsalis*), their populations are limited by bark beetle availability. In a German study, the droppings of *P. tridactylus* consisted of 89% *I. typographus* remains, representing a consumption rate of 1200 bark beetles per day (Pechacek, 1994). The population density of *P. dorsalis* increased more than 40-fold after a wildfire associated with an increase in bark beetle attack of the resulting dead wood (Baldwin, 1968). A conceptual framework for the ecological relationship between woodpeckers and their prey has been developed by Fayt et al. (2005) and bioenergetic models have been built calculating the consumption rate of bark beetles by different woodpecker species based on food requirement, population density, home range and temperature (Koplin, 1972; Bütler et al., 2004). Such models improve our understanding of the relationships between three-toed woodpeckers and bark beetles and highlight the significance of woodpeckers in the control of bark beetles.

Although woodpeckers are the key avian predators of bark beetles, there are other avian predators that forage on bark beetles in-flight or on the bark surface of standing or fallen trees, and they merit a brief mention here. These include, among others, tyrant flycatchers (Tyrannidae), tree creepers (Certhiidae), jays (Corvidae), nuthatches (Sittidae), chickadees (Paridae), bluebirds (Turdidae), and juncos (Emberizidae). These passerine birds are estimated to cause up to 30% mortality in bark beetle populations (Baldwin, 1968; Otvos, 1979). The mountain chickadee, *Poecile gambeli* (Ridgway), showed a clear numerical response to an outbreak of *Dendroctonus ponderosae* Hopkins (Norris *et al.*, 2013).

Forest management can take advantage of the regulatory power of woodpeckers by retaining or providing the necessary habitats for these birds such as old trees with cavities for nesting, and trees or dead wood to provide subcortical insects, including bark beetles, as a food resource.

2.2 Arthropod Predators and Parasitoids

Predators and parasitoids emerging from bark beetle-infested logs are often assumed to be the natural enemies of the most common or most aggressive bark beetle species present in the logs. However, they may have been feeding or parasitizing other scolytine species, other pest species, inquilines in the galleries, or even on associated fungi. In this chapter we only consider the species that have been reliably confirmed as, at least, facultative predators and parasitoids of bark beetles (Tables 7.2–7.7). This is not an exhaustive inventory of the natural enemies of bark beetles and is focused on European and North American forest ecosystems where the vast majority of studies have occurred. Many species have been renamed or synonymized over the years, and so to achieve consistency in taxonomic attribution we have used the following online databases: Catalogue of Life (2013); ZipcodeZoo (2013); BioLib (2014); EOL (2014); EUNIS (2014); ION (2014); and ITIS (2014).

Each tree species harbors a characteristic composition of predatory and parasitic guilds that is distinct from other tree species infested with the same bark beetle. Arthropod natural enemies locate their prey using semiochemicals, i.e., they can use bark beetle pheromones as host-specific kairomones or the volatiles emitted by the tree in response to bark beetle attack. The volatile profile produced by one tree species in response to bark beetle attack is different to the volatile profile produced by another tree species, and therefore attracts a different guild of natural enemies (Aukema *et al.*, 2004). When bark beetles attack a tree some invertebrate antagonists arrive almost immediately and others much later, following a distinct sequence; predators typically arrive before parasitoids and lay their eggs alongside the bark beetles, ensuring synchrony between predatory larvae and the bark beetle eggs and larvae they feed on (Stephen and Dahlsten, 1976; Dahlsten, 1982; Ohmart and Voigt, 1982; Linit and Stephen, 1983; Aukema *et al.*, 2004). Most parasitoids require later instars of bark beetle larvae to oviposit in and thus appear later when suitable hosts are present.

2.2.1 Predators

Predators are carnivorous animals that consume more than one, often many, prey during their developmental and adult life stages. For instance, checkered beetles (Cleridae) forage on adult bark beetles entering or leaving their brood trees, while the larvae devour immature bark beetles developing in the phloem (Reeve *et al.*, 1995). Other predatory beetles, dipteran flies, and snakeflies are only predatory on bark beetles during their larval stages (Wichmann, 1957). Most predators are generalists while parasitoids are usually more specific with regard to host species and host size. Predators can be very efficient because they are often more mobile than their prey and can remain active during winter on warm days. Although predators are generally the first to arrive at an infested tree, some important predatory taxa do emerge later than their prey (Moore, 1972a); a good example of this is the spring emergence of dolichopodid flies preying on *I. typographus* (Wermelinger *et al.*, 2012).

The taxonomic range of predators attacking bark beetles is very broad and includes, among others, beetles (Coleoptera), flies (Diptera), true bugs (Heteroptera), snake flies (Raphidioptera), and mites (Acari) (Tables 7.2–7.4). Some of the most important taxa are discussed below. Spiders and earwigs feed occasionally or coincidentally on bark beetles (Kenis *et al.*, 2004) but they are not included in this compilation because they are unlikely to have a significant impact on bark beetle population dynamics.

2.2.1.1 Coleoptera (Beetles)

The order Coleoptera includes some very diverse and well-known predatory groups, some of which can have substantial impacts on bark beetle population dynamics. Coleopteran predators are quite mobile, and can feed on prey at the surface and beneath the bark; depending on their size they can consume all life stages of the bark beetles from eggs to adults. Since the larvae of coleopteran predators are usually quite large they are frequently found in the lower sections of the tree bole where the bark is thicker (Otvos, 1965; Stephen and Dahlsten, 1976; Gargiullo and Berisford, 1981; Wermelinger, 2002).

Key components of attraction for host location in coleopteran predators are host tree-emitted compounds such as ethanol and alpha-pinene, and bark beetle-synthesized monoterpenes that form the aggregation pheromone, for example frontalin, cis-verbenol, and ipsdienol (Schroeder and Weslien, 1994; Aukema *et al.*, 2000b; Aukema and Raffa, 2005). While known to locate infested trees via a combination of these semiochemicals (e.g., Erbilgin and Raffa, 2001), coleopteran predators also respond to anti-aggregation pheromones that are produced by bark beetles when their colonization density has reached an optimum level (Lindgren and Miller, 2002). There is also evidence for visual attraction to particular sites within a tree (Shepherd and Goyer, 2003). In an infestation of *D. brevicomis*, coleopteran predators only deposited their eggs on sections of the tree where bark beetle attack was greatest demonstrating accurate host location (DeMars *et al.*, 1970).

In populations of bivoltine bark beetle species, their coleopteran predators (except for nitidulid species) overwinter in the tree alongside their prey, emerging in spring concomitantly with their prey (Wermelinger *et al.*, 2012). In histerid species, while emergence is concomitant with their prey, it also continues for up to a further 4 weeks (Shepherd and Goyer, 2003). In high altitudes where the climate is cool, the bivoltine bark beetle *I. typographus* can only achieve one generation in a year and becomes effectively univoltine; when this happens more of its predators emerge in autumn than in the following spring (Wermelinger *et al.*, 2012).

The most conspicuous coleopteran predators of bark beetles are the checkered beetles (Cleridae), a small family that includes some very important predatory species that are active very early in the season (Stephen and Dahlsten, 1976; Hérard and Mercadier, 1996; Lawson *et al.*, 1997). Both *Thanasimus formicarius* (L.) (Figure 7.1) from Europe and *Thanasimus dubius* (F.) from North America are abundant generalist species (Table 7.2). Gauss (1954) lists more than 20 species of bark beetles as prey for *T. formicarius*. Several *Enoclerus* species are also present in America. Three well-studied clerid species associated with bark beetles are present both in Europe and in North America: *Allonyx quadrimaculatus* (Schaller), *Clerus mutillarius* F., and *Enoclerus lecontei* (Wolcott) (Table 7.2).

Since both adult and larval stages of clerid beetles predate bark beetles, their biology has attracted much attention (Person, 1940; Gauss, 1954; Reeve *et al.*, 1995; Ye and Bakke, 1997). Most clerid beetles start flying early in the season and forage on bark beetles throughout the summer, attacking a wide range of different species. They respond to prey pheromones and host-tree volatiles and are able to discriminate between different prey species (Bakke and Kvamme, 1981; Tømmerås, 1985). They feed

on alighting bark beetles and oviposit in bark crevices near bark beetle entrance holes (Aukema and Raffa, 2002). In Central Europe, oviposition by *Thanasimus* species begins in early April and continues until late August (Gauss, 1954). Clerid species are highly fecund, with reports of 100–300 eggs being produced per female for *Thanasimus* species (Thatcher and Pickard, 1966; Dippel *et al.*, 1997) and up to 1000 eggs per female for *E. lecontei* (Berryman, 1966). Following hatching, larvae enter bark beetle galleries and forage on immature stages. The larval period of *T. formicarius* lasts for between 30 and 60 days (Person, 1940) while the duration from egg to adult emergence for *T. dubius* is 70 – 100 days (Lawson and Morgan, 1992; Aukema and Raffa, 2002). Once mature, the majority of larvae bore out of the bark, crawl down to the base of the tree, and pupate in bark near the ground or in the soil, although a proportion of larvae pupate in the outer bark near their development site. The majority of the population emerges in the summer of the same year while the remaining individuals (5–20% for *T. formicarius*) overwinter as prepupae in the bark and emerge the following spring, or even later (Person, 1940; Reeve, 1997). Adult *T. formicarius* live for 4–10 months and the entire life cycle takes a year (Gauss, 1954), or 2 years in Scandinavia (Schroeder, 1999).

As adults the consumption of adult bark beetles by clerids becomes very important; consumption rates of adult bark beetles achieved by adult clerids can greatly exceed consumption rates of larval bark beetles by larval clerids (Reeve *et al.*, 1995; Aukema and Raffa, 2002). *Thanasimus formicarius* consumes approximately 50 prey larvae during its larval development (Mills, 1985; Dippel *et al.*, 1997) and *E. lecontei* consumes up to 150 prey items during its entire life (Berryman, 1966). Larvae of *T. formicarius* were estimated to kill between 57 and 627 prey larvae per 1000 cm^2 (Kenis *et al.*, 2004). This makes clerids effective predators; they are generally considered to be among the most efficient predators, particularly in infestations of *Dendroctonus* species in North America (Moore, 1972a). Experimental studies showed that *T. formicarius* could reduce the size of *Tomicus piniperda* (L.) broods by 81% (Schroeder, 1997)

FIGURE 7.1 The clerid beetle *Thanasimus formicarius* is predatory both as a larva and as an adult. (A) Larva consuming a bark beetle larva; (B) adult feeding on adult bark beetle of *Ips typographus*. Copyright: Beat Wermelinger.

TABLE 7.2 Beetles (Coleoptera) Reported as Predators of Bark Beetle Species

Species	Family	Distribution[1]	Dendroctonus	Hylurgops	Ips	Orthotomicus	Pityogenes	Pityokteines	Polygraphus	Scolytus	Taphrorychus	Tomicus	Trypodendron	Other
Apristus subsulcatus Dejean	Carabidae	ne	x		x							x		
Calodromius spilotus (Illiger) (= Dromius quadrinotatus)	Carabidae	pa												
Cymindis platicollis (Say) (= Pinacodera platicollis)	Carabidae	ne												
Dromius piceus Dejean	Carabidae	ne	x											
Dromius quadrimaculatus (L.)	Carabidae	pa			x							x		
Allonyx quadrimaculatus (Schaller)	Cleridae	pa,ne										x		
Clerus mutillarius F.	Cleridae	pa,ne			x	x						x		
Cymatodera ovipennis LeConte	Cleridae	ne	x											
Enoclerus barri Knull	Cleridae	ne			x									
Enoclerus lecontei (Wolcott) (= Thanasimus lecontei, = T. nigriventris)	Cleridae	pa,ne	x		x		x							
Enoclerus muttkowskii (Wolcott)	Cleridae	ne			x									
Enoclerus nigrifrons (Say)	Cleridae	ne			x				x					x
Enoclerus nigripes (Say)	Cleridae	ne			x									
Enoclerus pinus (Schaeffer)	Cleridae	ne	x		x									
Enoclerus sphegeus (F.)	Cleridae	ne	x		x		x							
Thanasimus dubius (F.)	Cleridae	ne	x		x									
Thanasimus femoralis (Zett.) (= T. rufipes)	Cleridae	pa			x									
Thanasimus formicarius (L.)	Cleridae	pa	x		x	x	x			x		x		x
Thanasimus undulatus Say	Cleridae	ne							x					x
Cucujus clavipes F.	Cucujidae	ne	x											
Narthecius simulator Casey	Cucujidae	ne	x											
Eblisia minor (Rossi) (= Platysoma frontale)	Histeridae	pa			x							x		

Species	Family	Region											
Paromalus bistriatus Erichson	Histeridae	ne	X										
Paromalus mancus Casey (= *Isomalus mancus*)	Histeridae	ne	X										
Paromalus parallelepipedus (Herbst)	Histeridae	pa	X			X						X	
Platysoma attenuatum LeConte (= *Cylistix attenuata*)	Histeridae	ne	X			X						X	
Platysoma cornix Marseul (= *Cylister cornix*)	Histeridae	pa	X			X						X	
Platysoma cylindricum (Paykull) (= *Cylistix cylindrica*, = *Hister cylindricus*)	Histeridae	ne	X			X							
Platysoma elongatum (Thunberg) (= *Cylister elongatus*)	Histeridae	pa				X						X	
Platysoma gracile (LeConte) (= *Cylistix gracilis*)	Histeridae	ne							X				
Platysoma parallellum (Say)	Histeridae	ne	X		X	X							
Platysoma punctigerum (LeConte)	Histeridae	ne	X		X	X	X					X	
Plegaderus discisus Erichson	Histeridae	pa			X	X	X						
Plegaderus nitidus Horn	Histeridae	ne	X		X	X	X						
Plegaderus transversus (Say)	Histeridae	ne	X	X	X	X							
Plegaderus vulneratus (Panzer)	Histeridae	pa		X	X	X		X			X		X
Cryptolestes fractipennis (Motschulsky)	Laemophloeidae	pa			X	X	X					X	
Cryptolestes spartii (Curtis)	Laemophloeidae	pa										X	
Placonotus testaceus (F.) (= *Laemophloeus testaceus*)	Laemophloeidae	pa	X							X			
Rhizophagus bipustulatus (F.)	Monotomidae	pa		X	X	X	X		X			X	X
Rhizophagus cribratus Gyll.	Monotomidae	pa		X	X							X	X
Rhizophagus depressus (F.)	Monotomidae	pa	X	X	X	X	X		X		X	X	
Rhizophagus dimidiatus Mannerheim	Monotomidae	ne	X			X							
Rhizophagus dispar (Paykull)	Monotomidae	pa	X	X	X						X	X	
Rhizophagus ferrugineus (Paykull)	Monotomidae	pa	X	X	X						X		X
Rhizophagus grandis Gyll.	Monotomidae	pa	X										
Rhizophagus nitidulus (F.)	Monotomidae	pa			X						X	X	X
Rhizophagus parvulus Paykull	Monotomidae	pa							X			X	
Rhizophagus perforatus Erichson	Monotomidae	pa			X						X		X

Continued

TABLE 7.2 Beetles (Coleoptera) Reported as Predators of Bark Beetle Species—cont'd

Species	Family	Distribution	Dendroctonus	Hylurgops	Ips	Orthotomicus	Pityogenes	Pityokteines	Polygraphus	Scolytus	Taphrorychus	Tomicus	Trypodendron	Other
Rhizophagus procerus Casey	Monotomidae	ne	x											x
Rhizophagus puncticollis Sahlberg	Monotomidae	pa												
Rhizophagus sculpturatus Mannerheim	Monotomidae	ne	x											x
Epuraea angustula Sturm	Nitidulidae	pa		x	x									
Epuraea marseuli Reitter (= *E. pusilla*)	Nitidulidae	pa		x	x		x					x	x	
Epuraea pygmaea (Gyll.)	Nitidulidae	pa			x									x
Epuraea rufomarginata (Stephens)	Nitidulidae	pa			x							x		
Epuraea silacea (Herbst)	Nitidulidae	pa			x							x		
Epuraea unicolor (Olivier)	Nitidulidae	ne									x			
Glischrochilus lecontei Brown	Nitidulidae	ne	x											
Glischrochilus vittatus (Say)	Nitidulidae	ne	x											
Ipidia binotata Reitter (= *Silpha quadrimaculata*)	Nitidulidae	pa										x		
Pityophagus ferrugineus (L.)	Nitidulidae	pa			x						x	x		
Pytho depressus L.	Pythidae	pa										x		
Pytho planus (Olivier)	Pythidae	ne	x											
Rabdocerus foveolatus (Ljungh)	Salpingidae	pa			x									
Rabdocerus gabrieli (Gerhardt)	Salpingidae	pa			x									
Salpingus planirostris (F.) (= *Rhinosimus planirostris*)	Salpingidae	pa			x		x						x	
Salpingus ruficollis (L.) (= *Rhinosimus ruficollis*)	Salpingidae	pa			x		x						x	
Sphaeriestes castaneus (Panzer) (= *Salpingus castaneus*)	Salpingidae	pa			x		x						x	
Aleochara sparsa Heer	Staphylinidae	pa			x		x						x	

Species	Family													
Atrecus macrocephalus (Nordmann)	Staphylinidae	ne	x											
Nudobius cephalus (Say)	Staphylinidae	ne	x											
Nudobius corticalis Casey	Staphylinidae	ne	x											
Nudobius lentus (Gravenhorst)	Staphylinidae	pa		x	x		x		x		x			x
Nudobius luridipennis Casey	Staphylinidae	ne			x									
Phacophallus parumpunctatus (Gyll.) (= Leptacinus parumpunctatus)	Staphylinidae	ne	x		x									
Phloeonomus punctipennis Thomson	Staphylinidae	pa			x									
Phloeonomus pusillus (Gravenhorst)	Staphylinidae	pa	x		x									
Phloeopora corticalis (Gravenhorst)	Staphylinidae	pa			x									
Phloeopora testacea (Mannerheim)	Staphylinidae	pa		x	x						x			
Placusa adscita Erichson	Staphylinidae	pa			x						x			
Placusa atrata (Mannerheim)	Staphylinidae	pa		x	x					x				x
Placusa depressa (Maeklin)	Staphylinidae	pa		x	x		x				x	x		x
Placusa tachyporoides (Waltl)	Staphylinidae	pa			x									
Quedius laevigatus (Gyll.)	Staphylinidae	ne	x											
Quedius plagiatus Mannerheim (= Q. laevigatus, = Q. longipennis)	Staphylinidae	pa	x	x	x									
Zeteotomus brevicornis (Erichson) (= Metoponcus brevicornis)	Staphylinidae	pa						x						x
Corticeus fraxini (Kugelann) (= Hypophloeus fraxini)	Tenebrionidae	pa		x	x	x	x				x			
Corticeus glaber(LeConte) (= Hypophloeus glaber)	Tenebrionidae	ne	x		x									
Corticeus linearis (F. (= Hypophloeus linearis)	Tenebrionidae	pa		x	x	x	x				x	x		
Corticeus parallelus (Melsheimer) (= Hypophloeus parallelus)	Tenebrionidae	ne	x		x									
Corticeus praetermissus (Fall)	Tenebrionidae	ne								x				
Corticeus suturalis (Paykull) (= Hypophloeus praetermissus)	Tenebrionidae	pa			x									
Nemozoma cornutum Sturm	Trogossitidae	pa				x								
Nemozoma elongatum (L.)	Trogossitidae	pa			x		x						x	
Nemozoma pliginskyi Reitter	Trogossitidae	pa			x									x

Continued

TABLE 7.2 Beetles (Coleoptera) Reported as Predators of Bark Beetle Species—cont'd

Species	Family	Distribution	Dendroctonus	Hylurgops	Ips	Orthotomicus	Pityogenes	Pityokteines	Polygraphus	Scolytus	Taphrorychus	Tomicus	Trypodendron	Other
Temnochila caerulea (Olivier)	Trogossitidae	pa			x	x		x				x		
Temnochila chlorodia (Mannerheim)	Trogossitidae	ne	x		x		x							
Temnochila virescens (F.)	Trogossitidae	ne	x		x									
Tenebroides collaris (Sturm)	Trogossitidae	ne	x		x									x
Tenebroides marginatus (Palisot de Beauvois)	Trogossitidae	ne	x		x									
Aulonium ferrugineum Zimmermann	Zopheridae	ne	x		x									
Aulonium longum LeConte	Zopheridae	ne	x											
Aulonium ruficorne (Olivier)	Zopheridae	pa,ne			x	x	x	x				x		x
Aulonium trisulcum (Geoffroy)	Zopheridae	pa								x				
Aulonium tuberculatum LeConte	Zopheridae	ne	x		x									
Bitoma crenata (F.)	Zopheridae	pa			x							x		
Colydium elongatum (F.)	Zopheridae	pa				x		x				x		
Lasconotus complex LeConte	Zopheridae	ne	x											
Lasconotus pusillus LeConte	Zopheridae	ne	x		x									
Lasconotus referendarius Zimmermann	Zopheridae	ne	x		x									
Lasconotus subcostulatus Kraus	Zopheridae	ne	x		x									
Lasconotus tuberculatus Kraus	Zopheridae	ne	x											
References			1, 2, 4, 8-10, 15, 20-25, 28, 30, 31, 33, 38-40	4, 27	4-10, 14 16, 17-19, 21, 24, 26, 29, 32-36	3, 4, 13, 17	4, 9, 13, 27	17, 41	11	4, 37	4	4, 12, 17	4, 27	4, 11, 29

pa, palaearctic; ne, nearctic;

References: (1) Otvos, 1965; (2) DeMars et al., 1970; (3) Mendel, 1985; (4) Kenis et al., 2004; (5) Aukema et al., 2000a; (6) Aukema et al., 2000b; (7) Aukema and Raffa, 2004; (8) Boone et al., 2008a; (9) Dahlsten et al., 2003; (10) Erbilgin and Raffa, 2001; (11) Haberkern and Raffa, 2003; (12) Pishchik, 1980; (13) Podoler et al., 1990; (14) Raffa and Dahlsten, 1995; (15) Reeve, 1997; (16) Rohlfs and Hyche, 1984; (17) Sarikaya and Avci, 2009; (18) Shepherd and Goyer, 2003; (19) Raffa, 1991; (20) McCambridge and Knight, 1972; (21) Miller, 1986; (22) Stephen and Dahlsten, 1976; (23) Dahlsten and Stephen, 1974; (24) Moore, 1972a-c; (25) Chansler, 1967; (26) Nuorteva, 1971; (27) Wermelinger et al., 2012; (28) Overgaard, 1968; (29) Wermelinger et al., 2013; (30) Stein and Coster, 1977; (31) Linit and Stephen, 1983; (32) Riley and Goyer, 1986; (33) Gokturk et al., 2010; (34) Hedgren and Schroeder, 2004; (35) Hilszczanski et al., 2007; (36) Martin et al., 2013; (37) Bright, 1996; (38) Langor, 1991; (39) De Leon, 1934; (40) Dahlsten, 1970; (41) Capek, 1957 and those of *I. typographus* by 18% (Mills, 1985). Adult *T. dubius* exhibit both functional and numerical responses to their prey (Frazier *et al.*, 1981).

Most bark-gnawing beetles (Trogossitidae) spend their entire life beneath the bark. However, only a few species are reported to predate bark beetles. In North America, the trogossitids *Temnochila virescens* (F.) and *Temnochila chlorodia* (Mannerheim) are commonly found in pine forests infested with *Dendroctonus* and *Ips* species. While clerid beetles are typically generalist predators, trogossitids are considered more specialized (Kohnle and Vite, 1984; Lawson and Morgan, 1992); *Temnochila virescens* responds almost exclusively to attractants produced by *Ips* species (Billings and Cameron, 1984). Females of *Temnochila* species can produce up to 200 eggs each (Mignot *et al.*, 1970) and they are usually univoltine, overwintering both as larvae and adults. Adults of *Temnochila* species feed on bark beetle adults while their larvae forage on bark beetle larvae and pupae in the tree phloem. Typical of other predatory beetles, both larvae and adults can be cannibalistic (Mignot *et al.*, 1970). *Temnochila virescens* has an extended development time and adult longevity of up to 2 years, which is approximately four times that of the clerid *T. dubius* (Lawson and Morgan, 1992). *Temnochila chlorodia* typically occurs in pine stands infested with either *Dendroctonus* or *Ips* species, although it is more attracted to the pheromone components of *Ips* species than *Dendroctonus* species; the tree volatile alpha-pinene significantly increased captures of *T. chlorodia* in traps (Hofstetter *et al.*, 2012). Interestingly, Boone *et al.* (2008a) found *T. chlorodia* to be strongly attracted to *D. ponderosae* pheromones. *Temnochila chlorodia* shows a close synchrony in arrival time with its prey (Stephen and Dahlsten, 1976). *Temnochila caerulea* is another species in this group and is one of the predators of *Ips sexdentatus* (Börner) in southern Europe (Martín *et al.*, 2013).

In contrast to North America, species in the genus *Nemozoma* are much more important than *Temnochila* species in Europe, with *Nemozoma elongatum* (L.) being the most widespread predator with a broader diet range than *Temnochila* species. *Nemozoma elongatum* predates bark beetles infesting both conifers and broadleaf trees, but is considered a particularly important predator of *Pityogenes chalcographus* (L.) on spruce (Dippel *et al.*, 1997). As such the biology and ecology of *N. elongatum* have been investigated quite extensively (Dippel, 1995, 1996). Like most other beetle predators, *N. elongatum* responds to prey pheromones and the wood dust produced by boring bark beetles; it can represent 20% of the catches in traps baited with *P. chalcographus* pheromones (Wigger, 1993). Long-term abundance is closely connected with that of *P. chalcographus* while its seasonal phenology shows much variation (Baier, 1991). However, in spring, oviposition of both prey and predator is simultaneous.

The root-eating beetles (Monotomidae = Rhizophagidae) are well represented in Europe by a number of species in the genus *Rhizophagus* (Table 7.2). Many are facultative predators and respond to both the pheromones of their prey and to ethanol (Byers, 1992; Grégoire *et al.*, 1992), but are still able to detect and exploit a high proportion of available bark beetle broods (van Averbeke and Grégoire, 1995). In particular, *Rhizophagus depressus* (F.) and *Rhizophagus dispar* (Paykull) feed on various prey species. In the laboratory, Hanson (1937) observed a single adult *Rhizophagus ferrugineus* (Paykull) consuming 79 eggs of a *Hylastes* species. The larvae of *R. depressus* are facultative predators on, among other species, *T. piniperda*. They consumed 14 prey larvae during their 10-week larval development period (Hérard and Mercadier, 1996) with all stages also feeding on the eggs of bark beetles. In an exclusion experiment, *R. depressus* was reported to reduce *T. piniperda* broods by 41% (Schroeder, 1996).

A rare example of a specialized predator is *Rhizophagus grandis* Gyll. This species has received much interest and is exploited as a biological control agent against *Dendroctonus micans* (Kugelann) in France (Grégoire *et al.*, 1985; van Averbeke and Grégoire, 1995). Its oviposition is regulated both by chemical stimuli and inhibitors (Grégoire *et al.*, 1991). Adults and larvae feed on the eggs, larvae, pupae, and teneral adults of *D. micans*. During its entire larval life, each individual *R. grandis* consumes the equivalent of one fully grown *D. micans* larva. Prepupae of *R. grandis* become photopositive, leave the brood chamber, and pupate in the ground or on the bark at the base of trees (Kenis *et al.*, 2004). There is at least one generation per year (King *et al.*, 1991).

The taxonomy of hister beetles (Histeridae) has been updated in recent years (Table 7.2). Most species associated with bark beetles now belong to the holarctic genera *Platysoma* and *Plegaderus* (Table 7.2), which are attracted to bark beetle pheromones and plant volatiles confirming them as bark beetle predators (Schroeder and Weslien, 1994; Shepherd, 2004). There are more potentially predatory hister species associated with bark beetles (Kenis *et al.*, 2004); however, their feeding behavior was considered too uncertain to confirm them as bark beetle predators and include them in this chapter.

In North American pine forests, *Platysoma* species are the most common hister beetles and have been investigated in detail (Struble, 1930; Shepherd, 2004). *Platysoma* species arrive shortly after the start of tree colonization by *Ips* species and enter through the entrance holes of the bark beetles (Struble, 1930). The short-range location of infested trees is, in addition to chemical cues, affected by visual orientation; *Platysoma parallelum* (Say) prefers horizontal logs and *Platysoma attenuatum* LeConte prefers vertical trees. *Platysoma parallelum* is a large species and causes greater mortality in populations of *Ips* species than smaller species such as *Plegaderus transversus* (Say). *Platysoma cylindricum* (Paykull) is another important species in this genus; adults feed in the galleries on adult *I. pini*, and, in contrast to other histerids, also on bark beetle eggs. Larvae of *P. cylindricum* predate bark beetle larvae (Aukema and Raffa, 2004). The foraging behavior of the European species, *Eblisia minor* (Rossi), has been studied in some detail (Hérard and Mercadier, 1996); during its three larval stages, this species consumes an average of 44 bark beetle larvae and the adults are also predatory. Some *Platysoma* species only predate bark beetles when they are adults (Shepherd, 2004).

Among the many species of darkling beetles (Tenebrionidae), only a few are active predators. The genus *Corticeus* (=*Hypophloeus*) includes some well-known European and North American species of bark beetle predators, all of which are associated with conifers. *Corticeus glaber* (LeConte) and *Corticeus parallelus* (Melsheimer) are facultative predators of eggs and young larvae of *Dendroctonus* species, although they mainly feed on beetle frass and mycelium of the bluestain fungus, *Grosmannia clavigera* (Robinson-Jeffrey and Davidson), which infects the trees (Goyer and Smith, 1981; Smith and Goyer, 1982). The biology of *C. fraxini* (Hérard and Mercadier, 1996) and *C. glaber* (Smith and Goyer, 1982) has been investigated in some detail. Adult *C. glaber* enter the galleries through the bark beetle entrance or through ventilation holes, and mating occurs in the galleries. Their life cycle is completed within approximately 5 weeks (Smith and Goyer, 1982).

A considerable number of rove beetle species (Staphylinidae) are thought to predate bark beetles and some species have been caught in traps baited with bark beetle pheromone. However, they feed on a wide range of prey species and the details of their ecology are unclear. It is striking that relatively few species have been reported from North America. They are difficult to identify and thus often only reported as Staphylinidae spp. The main genera, both in Europe and North America, are *Nudobius*, *Phloeonomus*, *Phloeopara*, and *Placusa*. The species found most frequently in Europe are *Nudobius lentus* and *Placusa depressa*. They forage facultatively on adults and larvae of a wide range of bark beetle species, and *Nudobius lentus* (Gravenhorst) is frequently found in traps baited with the pheromones of spruce bark beetles (Zumr, 1983).

Of the sap beetles (Nitidulidae) six European species in the genus *Epuraea* and two North American *Glischrochilus* species have been recorded as bark beetle predators (Table 7.2). Other bark beetle-associated *Epuraea* species are not considered as confirmed predators because they are thought more likely to be detritivorous or fungivorous. A number of the predatory species are attracted to the pheromones of bark beetles (Zumr, 1983; Faccoli, 2001) or to plant volatiles (Schroeder and Weslien, 1994). While *G. lecontei* is thought to be a specialist predator of

Dendroctonus species, *Epuraea* species feed on a broader range of prey species including bark beetles. Both adults and larvae feed on the eggs of bark beetles (Nuorteva, 1956).

The family Zopheridae (formerly Colydiidae) are the ironclad beetles. In this family, species associated with bark beetles are more numerous in North America than in Europe. The two most important genera are *Aulonium*, which are predatory on many different prey species, and *Lasconotus*, known to predate mainly on *Dendroctonus* and *Ips* species of bark beetles. The most noticeable predator of *Orthotomicus erosus* (Wollaston) and *Pityogenes calcaratus* (Eichhoff) is *Aulonium ruficorne* (Olivier), which has been studied in detail by Podoler *et al.* (1990). Both adults and larvae of this species feed on all the immature stages of bark beetles and also on young adults; bark beetle mortality rates of up to 90% have been inferred for *A. ruficorne*.

Very little information is available on the other coleopteran families containing bark beetle predators (Table 7.2) and they are unlikely to have a significant impact on bark beetle dynamics. Within the Carabidae, a family of numerous predatory species, only *Dromius* and *Calodromius* species are frequently associated with bark beetles. Some other predators such as *Salpingus planirostris* (F.) (Rauhut *et al.*, 1993) and *Pytho depressus* (L.) (Schroeder and Weslien, 1994) are attracted to the pheromones of bark beetles, but there is no information on their impact on bark beetle population dynamics.

2.2.1.2 Diptera (Flies)

With the exception of the robber flies (Asilidae), all dipteran species that predate bark beetles only do so in their larval stages (Table 7.3). *Medetera* species are predatory as adults but they only feed on small insects with soft integuments and not bark beetles (Nuorteva, 1956). Dipteran predators can be very numerous and they often outnumber other subcortical predatory taxa (Morge, 1961) (Table 7.3). They do not feed exclusively on bark beetles, but also on larvae of Hymenoptera, Cerambycidae, Curculionidae, and Diptera. The two most important families are the Dolichopodidae and the Lonchaeidae, which will be described in more detail below.

The long legged flies (Dolichopodidae) are certainly important predators; as early as 1934 dolichopodid flies were recognized as the "most valuable predators [of bark beetles] in lodgepole, western white pine and probably western yellow pine" (De Leon, 1934) and, while some reports question their impact (Mills, 1985, 1986), this largely remains the case. The most important genus is *Medetera* (Table 7.3) and *Medetera* species are the most common insect predators in both conifer and broadleaf forests (Morge, 1961; Beaver, 1966a; Lawson *et al.*,

TABLE 7.3 Flies (Diptera) Reported as Predators of Bark Beetle Species

Species	Family	Distribution[1]	Dendroctonus	Hylurgops	Ips	Orthotomicus	Pityogenes	Polygraphus	Scolytus	Taphrorychus	Tomicus	Trypodendron	Other
Choerades gilva (L.) (= Laphria gilva)	Asilidae	pa			×								×
Holcocephala fusca Bromley	Asilidae	ne											×
Laphria flava (L.)	Asilidae	pa			×								
Tolmerus atricapillus (Fallén) (= Machimus atricapillus)	Asilidae	pa			×								
Gymnopternus politus Loew	Dolichopodidae	ne			×								
Medetera adjaniae (Gosseries) (= M. breviseta)	Dolichopodidae	pa		×	×		×	×					×
Medetera aldrichii Wheeler	Dolichopodidae	ne							×				×
Medetera ambigua (Zett.)	Dolichopodidae	pa	×		×								
Medetera bistriata Parent	Dolichopodidae	ne	×		×								
Medetera dendrobaena Kowarz	Dolichopodidae	pa			×								
Medetera dichrocera Kowarz	Dolichopodidae	pa		×	×		×			×	×	×	
Medetera excellens Frey	Dolichopodidae	pa			×								
Medetera fumida Negrobov	Dolichopodidae	pa			×								
Medetera gaspensis Bickel	Dolichopodidae	ne	×										
Medetera impigra Collin	Dolichopodidae	pa			×				×	×			
Medetera infumata Loew	Dolichopodidae	pa			×						×		
Medetera maura Wheeler	Dolichopodidae	ne	×		×								
Medetera melancholica Lundbeck	Dolichopodidae	pa			×		×						
Medetera nitida (Macquart)	Dolichopodidae	pa			×				×				×
Medetera pinicola Kowarz (= M. piceae, = M. nuortevai)	Dolichopodidae	pa		×	×				×		×		
Medetera prjachinae Negrobov	Dolichopodidae	pa			×								
Medetera setiventris Thuneberg	Dolichopodidae	pa		×	×		×						×
Medetera signaticornis Loew	Dolichopodidae	pa			×				×				

Continued

TABLE 7.3 Flies (Diptera) Reported as Predators of Bark Beetle Species—cont'd

Species	Family	Distribution	Dendroctonus	Hylurgops	Ips	Orthotomicus	Pityogenes	Polygraphus	Scolytus	Taphrorychus	Tomicus	Trypodendron	Other
Medetera striata Parent	Dolichopodidae	pa				x	x						
Medetera thunebergi Negrobov	Dolichopodidae	pa			x								
Medetera zinovjevi Negrobov	Dolichopodidae	pa			x								
Earomyia viridana (Meigen) (= Lonchaea viridana)	Lonchaeidae	ne	x										
Lonchaea auranticornis Mcalpine	Lonchaeidae	ne	x										
Lonchaea bruggeri Morge	Lonchaeidae	pa			x								
Lonchaea collini Hackman	Lonchaeidae	pa			x						x		
Lonchaea coloradensis Malloch	Lonchaeidae	ne	x										
Lonchaea corticis Taylor	Lonchaeidae	ne			x								
Lonchaea fugax Becker	Lonchaeidae	pa			x								
Lonchaea helvetica MacGowan	Lonchaeidae	pa			x								
Lonchaea polita Say	Lonchaeidae	ne	x										
Lonchaea scutellaris Rondani	Lonchaeidae	pa			x								
Lonchaea seitneri Hendel	Lonchaeidae	pa			x								
Palloptera ustulata Fallén	Pallopteridae	pa			x				x				
Toxoneura usta (Meigen) (= Palloptera usta)	Pallopteridae	pa			x								
Zabrachia polita Coquillett	Stratiomyidae	ne			x								
Zabrachia tenella (Jaennicke)	Stratiomyidae	pa			x								
Xylophagus cinctus (De Geer) (= Erinna abdominalis; = X. abdominalis)	Xylophagidae	ne	x										
References			4, 6, 7, 10, 11, 17, 1,8, 20	1, 9	1-3, 5, 8, 9, 12-14	1	1, 9, 14	1, 9	1, 15, 16	1, 19	1, 9	1	1, 9, 15

[1]pa, palaearctic; ne, nearctic;
References: (1) Kenis et al., 2004; (2) Morge, 1967; (3) Aukema and Raffa, 2004; (4) McCambridge and Knight, 1972; (5) Miller, 1986; (6) Moore, 1972a–c; (7) Chansler, 1967; (8) Wermelinger et al., 2012; (9) Nuorteva, 1971; (10) Overgaard, 1968; (11) Linit and Stephen, 1983; (12) Feicht, 2004; (13) Hedgren and Schroeder, 2004; (14) Hulcr et al., 2005; (15) Nagel and Fitzgerald, 1975; (16) Bright, 1996; (17) Langor, 1991; (18) Massey and Wygant, 1973; (19) Chandler, 1991; (20) Dahlsten, 1970

1996; Wermelinger, 2002; Aukema *et al.*, 2004). Bark beetle consumption rates by *Medetera* species show a functional response, i.e., consumption increases with increasing bark beetle density (Beaver, 1966a; Nicolai, 1995). In an *I. typographus* infestation, *Medetera* species were the only predators whose populations could be correlated with the mortality of bark beetle larvae (Lawson *et al.*, 1996). In a rare study in broadleaves (elm), the majority of *Scolytus multistriatus* (Marsham) mortality (6–25%) could be attributed to *Medetera nitida* (Macquart) (Schröder, 1974). Furthermore, considerably higher mortalities of up to 90% were achieved by *Medetera* species predating *D. pseudotsugae* (Hopping, 1947).

In North America, *Medetera aldrichii* Wheeler is commonly found in populations of *Dendroctonus* species; *Medetera bistriata* Parent is found in populations of both *Dendroctonus* and *Ips* species. It is likely that both these species use the pheromones of bark beetles and the volatiles of their associated microbial symbionts as kairomones since they find attacked trees very soon after bark beetle infestation has begun (Stephen and Dahlsten, 1976; Ohmart and Voigt, 1982; Aukema *et al.*, 2004; Boone *et al.*, 2008b). Mating occurs on the trunks of infested trees (Hopping, 1947) and subsequently, females deposit their eggs in crevices and under scales near the bark beetle entrance holes. Following hatching, the larvae enter the galleries and begin feeding on the eggs and young larvae of bark beetles (Nagel and Fitzgerald, 1975). *Medetera* species can also predate larger larval stages and pupae; they rupture the prey's integument with their mandibular hooks and suck out the fluid within (Aukema and Raffa, 2004). Interestingly, Aukema and Raffa (2004) suggest that, before feeding, larvae of *Medetera* species immobilize their prey with a toxin. With the exception of Acari (Gerson *et al.*, 2003), this is the only indication of toxins being involved in bark beetle predation. At low densities of the bark beetle *Dendroctonus pseudotsugae* Hopkins, when the phloem of the tree still remains intact, predatory *M. aldrichii* larvae are unable to access prey larvae (Nagel and Fitzgerald, 1975; but see Hopping, 1947). Only after additional feeding by the bark beetle larvae are the physical barriers of the intact phloem removed, and *M. aldrichii* can move freely and successfully forage among the prey larvae.

In Europe, there are more species of *Medetera* foraging on bark beetle species than there are in North America. Most *Medetera* species studied are associated with *I. typographus* and they are attracted to their prey by a mixture of tree volatiles and prey pheromones (Hulcr *et al.*, 2005). In a study with *I. typographus*, not only did the bark beetles reproduce more successfully on standing trees than on felled trees, but the *Medetera* species attacking them on standing trees were also 10 times more abundant (Hedgren and Schroeder, 2004). Once the defensive mechanisms of an infested tree are overcome, the phloem quality

of a living tree becomes more nutritious to bark beetles with subsequent positive effects on their predators. Some *Medetera* species have very precise ecological requirements. In a Swedish study, *Medetera zinovjewi* Negrobov was the most abundant species from this genus infesting standing trees, but was completely absent on felled trees (Hedgren and Schroeder, 2004). A similar pattern has also been reported for *Medetera signaticornis* Loew (Hedgren and Schroeder, 2004).

Medetera species typically oviposit in the lower bole of a tree (Wermelinger, 2002). Females have a relatively high fecundity; *Medetera dendrobaena* Kowarz, for example, produces up to 120 eggs per female (Dippel *et al.*, 1997). Larvae feed on eggs, larvae, pupae and even teneral adult bark beetles; depending on prey size they can consume between five and 20 individuals during their larval developmental stage (Hopping, 1947; Kenis *et al.*, 2004). At low prey densities they can be cannibalistic (Beaver, 1966a). *Medetera* species can be uni- or bivoltine, and it is not always clear how they overwinter. In an infestation of bivoltine *I. typographus* populations, dolichopodid larvae overwintered with their prey in the trees (Beaver, 1966a; Lieutier, 1979; Wermelinger *et al.*, 2012). However, in a univoltine population of *I. typographus*, adult flies emerged and left the trees in autumn to overwinter in an unknown location (Wermelinger *et al.*, 2012). Winter mortality in dolichopodid larvae can be substantial (Hopping, 1947; Nuorteva, 1959; Beaver, 1966a).

Among the lance flies (Lonchaeidae), only species in the genus *Lonchaea* live subcortically. The feeding behavior of these species remains controversial, with at least some species reported as saprophagous or coprophagous (Morge, 1961; Lieutier, 1979). It is believed that most predatory species in this genus have evolved from saprophagous ancestors. The biology of some species has been well investigated (Morge, 1961, 1963; Hérard and Mercadier, 1996). Like the Dolichopodidae, they are very voracious, feeding on eggs, larvae, and adults of bark beetles (Morge, 1967) and often kill more prey individuals than they can eat; at low prey densities cannibalism occurs. *Lonchaea* species are considered to be specialists, as each species only attacks a single bark beetle genus or species (Table 7.3). However, they are closely associated with particular tree species and so the host tree, rather than the bark beetle species, may be responsible for this apparent specialism. Surprisingly, more species of *Lonchaea* are reported in broadleaf forests than in conifer forests (Morge, 1961). However, while there is less species diversity in conifer forests, as a genus, the *Lonchaea* are more abundant in conifer forests and some species are known to be obligate predators and can reach very high numbers. In univoltine populations of *I. typographus*, a proportion of the lance fly population leave the bark beetle galleries and overwinter elsewhere (Wermelinger *et al.*, 2012).

The two species in the family Pallopteridae (Martinek, 1977; Chandler, 1991) predate bark beetle eggs, larvae, pupae, and even adult bark beetles, also killing many more prey than they can actually eat (Table 7.3). The Asilidae are the only dipteran group with adults that predate bark beetles; however, they are typically generalists feeding on a wide range of insect species including other subcortical species (Wichmann, 1956; Dennis, 1979). Adult asilid flies catch beetles in-flight, paralyze them with saliva, and suck up the liquefied body contents.

2.2.1.3 Hemiptera and Heteroptera (True Bugs)

The minute pirate bugs (Anthocoridae) are the only hemipteran family that includes bark beetle predators. Predatory anthocorids are primarily found in the genera *Lyctocoris* (particularly in North America), *Scoloposcelis*, and *Xylocoris* (Heidger, 1994; Hérard and Mercadier, 1996; Dippel *et al.*, 1997) (Table 7.4). *Scoloposcelis pulchella* (Zett.) is bivoltine and attracted to the pheromones of bark beetles (Heidger 1994). Both larval and adult stages are very voracious, killing more prey than they can consume, but little information is available on their impact on bark beetle populations. In North America, *Scoloposcelis flavicornis* Reuter (=*Scoloposcelis mississippensis*) is one of the most abundant predators of bark beetles (Riley and Goyer, 1986). It arrives quickly to new infestations and is reported to significantly contribute to mortality in *Dendroctonus frontalis* Zimmermann populations (Moore, 1972a).

While *Lyctocoris campestris* (F.) is a generalist predator (even of stored-products insects), the other *Lyctocoris* species primarily predate eggs, larvae, and pupae of *Dendroctonus* species. *Lyctocoris elongates* (Reuter) has a higher prey consumption rate than *S. flavicornis*, but is less abundant (Schmitt and Goyer, 1983). In the laboratory, adult *L. elongatus* consumed approximately 20 *Ips grandicollis* Eichoff eggs per day.

2.2.1.4 Other Insect Predators

The remaining predatory groups of insects have less impact on bark beetle population dynamics and have attracted less research attention (Table 7.4). The larvae of snakeflies (Raphidioptera) are generally epicortical and feed on, among other things, eggs of bark beetle predators (Wichmann, 1957). A few species forage non-specifically on cerambycids, bark beetles, and other subcortical species (Schimitschek, 1931b). Snakeflies can only access bark beetle galleries when the bark has been loosened, e.g., by woodpeckering or by the feeding activity of mature bark beetles (Wichmann, 1957). A few other predatory insect groups such as Dermaptera, Formicidae, and Chrysopidae are briefly mentioned in Kenis *et al.* (2004).

2.2.1.5 Acari (Mites)

Bark beetle galleries have a large and diverse acarine fauna that is described in detail in Chapter 6; for example, there are 38 species associated with *I. typographus* in Sweden (Moser *et al.*, 1989a) and 96 species associated with *D. frontalis* in Louisiana (Moser and Roton, 1971). Most of these mites are phoretic and not all are likely to be predatory or parasitic on bark beetles (Lindquist, 1964; Moser and Bogenschütz, 1984; Hofstetter *et al.*, 2013; Pfammatter *et al.*, 2013). The ecological role of those species that are thought to be predatory is poorly understood and it is likely that only a small fraction are truly antagonistic to bark beetles. Most also forage on fungi, nematodes, other mites, or on the eggs of other insects, and so may even be beneficial to bark beetles (Hirschmann and Wisniewski, 1983; Hofstetter *et al.*, 2013; see Chapter 6). The species with confirmed predatory behavior (Table 7.4) forage on bark beetle eggs, larvae, and pupae, but not on adults (Moser, 1975).

The most important genera of mites predating bark beetles are *Iponemus*, *Pyemotes*, *Proctolaelaps*, and *Dendrolaelaps* (Moser and Roton, 1971; Moser, 1975; Moser *et al.*, 1978, 2005; Hofstetter *et al.*, 2009). *Iponemus* and *Pyemotes* species are technically parasitoids because they need only one host to complete their development and are fatal to that host (Gerson *et al.*, 2003). These mites parasitize eggs and are host specific although this specificity is in response to habitat rather than host species directly (Lindquist, 1969). They are relatively abundant; for example, 41% of *I. pini* carried the egg parasitoid *Iponemus confusus* (Lindquist) (Pfammatter *et al.*, 2013). Females of *Iponemus* species produce 40–80 eggs each (Gerson *et al.*, 2003). *Pyemotes* species appear only as free-living adults as the juveniles develop within their mother's body (Gerson *et al.*, 2003). The adults feed on larvae and pupae of bark beetles. While *Iponemus* species depend on only a few host species, species in other genera feed on almost any species of bark beetle. *Proctolaelaps dendroctoni* Lindquist and Hunter, formerly believed to feed on bark beetles, is now thought to exclusively predate nematodes (Kinn, 1983).

Little is known of the impact of mites on bark beetle population dynamics but it is generally considered to be substantial. They can cause egg mortalities of up to 90% (Gäbler, 1947; Moser *et al.*, 1978, 1989a; Kielczewski *et al.*, 1983; Gerson *et al.*, 2003).

2.2.2 Parasitoids

2.2.2.1 General Ecology

The role of parasitic wasps in bark beetle population regulation has been the focus of much research, and comprehensive species lists and notes on their biology have been published (Nuorteva, 1957; Hedqvist, 1963, 1998; Bushing, 1965; Berisford *et al.*, 1970; Ball and Dahlsten, 1973; Pettersen, 1976). Parasitoids (almost exclusively

TABLE 7.4 Other Arthropods Reported as Predators of Bark Beetle Species

Species	Group	Distribution[1]	Cryphalus	Dendroctonus	Hylurgops	Ips	Leperisinus	Orthotomicus	Pityogenes	Pityokteines	Pityophtorus	Polygraphus	Scolytus	Taphrorychus	Tomicus	Trypodendron	Other
Dichrostigma flavipes (Stein) (= Raphidia flavipes)	Raphidioptera	pa				x											
Phaeostigma notata (F.) (= Raphidia notata)	Raphidioptera	pa				x											
Puncha ratzeburgi (Brauer)	Raphidioptera	pa				x											
Raphidia ophiopsis L.	Raphidioptera	pa				x		x							x		
Lyctocoris campestris (F.)	Heteroptera	pa		x													
Lyctocoris doris Van Duzee	Heteroptera	ne		x													
Lyctocoris elongatus (Reuter)	Heteroptera	ne		x		x											
Lyctocoris okanaganus Kelten & Anderson	Heteroptera	ne		x													
Lyctocoris stalii (Reuter)	Heteroptera	ne		x													
Scoloposcelis flavicornis Reuter (= S. mississipensis)	Heteroptera	ne		x		x											
Scoloposcelis obscurella (Zett.)	Heteroptera	pa				x											
Scoloposcelis pulchella (Zett.)	Heteroptera	pa			x	x		x	x			x			x		
Xylocoris cursitans (Fallén)	Heteroptera	pa				x									x		
Aethiophenax ipidarius (Redikortsev) (= Paracarophenax ipidarius)	Acari	pa				x											
Androlaelaps casalis (Berlese)	Acari	ne		x													
Chelacheles michalskii Samsinak	Acari	pa													x		
Dendrolaelaps apophyseosimilis Hirschmann	Acari	pa														x	
Dendrolaelaps cornutus (Kramer)	Acari	ne		x													
Dendrolaelaps isodentatus (Hurlbutt)	Acari	ne		x													

Continued

TABLE 7.4 Other Arthropods Reported as Predators of Bark Beetle Species—cont'd

Species	Group	Distribution	Cryphalus	Dendroctonus	Hylurgops	Ips	Leperisinus	Orthotomicus	Pityogenes	Pityokteines	Pityophtorus	Polygraphus	Scolytus	Taphrorychus	Tomicus	Trypodendron	Other
Dendrolaelaps neocornutus (Hurlbutt)	Acari	ne		x													
Dendrolaelaps neodisetus (Hurlbutt)	Acari	ne		x													
Dendrolaelaps quadrisetus Berlese	Acari	pa				x											
Dendrolaelaps rotoni (Hurlbutt)	Acari	ne		x													
Dendrolaelaps varipunctatus (Hurlbutt)	Acari	ne		x													
Gamasolaelaps subcorticalis McGraw & Farrier	Acari	ne		x													
Histiogaster rotundus Woodring	Acari	ne		x													
Iponemus calligraphi Lindquist	Acari	ne		x		x											
Iponemus confusus (Lindquist)	Acari	ne		x		x					x						
Iponemus gaebleri (Schaarschmidt)	Acari	pa				x											x
Iponemus truncatum (Ewing)	Acari	ne		x		x											
Lasioseius dentatus Fox	Acari	ne		x													
Lasioseius epicriodopsis DeLeon	Acari	ne		x													
Lasioseius ometes Oudemans	Acari	pa														x	
Lasioseius tubiculiger (Berlese)	Acari	ne		x													
Macrocheles boudreauxi Krantz	Acari	ne		x													
Macrocheles mammifer Berlese	Acari	ne		x													

Species	Order	Distribution	C1	C2	C3	C4	C5	C6	C7	C8	C9	C10	C11	C12	C13	C14
Mexecheles virginiensis (Baker)	Acari	ne	x													x
Pleuronectocelaeno barbara Kinn (= *Pleuronectocelaeno drymocetes*)	Acari	ne	x													
Proctolaelaps bickleyi Bram	Acari	ne	x													
Proctolaelaps eccoptogasteris (Vitzthum)	Acari	pa	x			x									x	x
Proctolaelaps fiseri Samsinak	Acari	pa	x			x									x	
Proctolaelaps hystrix (Vitzthum)	Acari	ne	x													
Proctolaelaps pini Hirschmann	Acari	pa												x		
Proctolaelaps xyloteri Samsinak	Acari	pa, ne													x	
Pseudoparasitus vitzthumi (Womersley) (= *Hypoaspis viththumi*)	Acari	ne	x													
Pseudotarsonemoides eccoptogasteri Vitzthum	Acari	pa												x		
Pyemotes dryas (Vitzthum)	Acari	pa	x	x		x		x	x	x	x					
Pyemotes herfsi (Oudemans)	Acari	pa	x			x					x	x				
Pyemotes parviscolyti Cross & Moser	Acari	ne	x			x								x		
Pyemotes scolyti (Oudemans)	Acari	pa	x									x		x		
Schizosthetus lyriformis (McGraw & Farrier (= *Eugasmus lyriformis*)	Acari	ne	x			x										
Trichouropoda lamellosa Hirschmann	Acari	ne	x													
Uroobovella americana Hirschmann	Acari	ne	x													
References			8	6, 7, 9-11, 13, 14, 16, 19	1	1, 2, 4, 5, 7, 12, 15, 17, 18	8	1, 2	1	8	7, 8	1, 3	1	1, 2, 3	1, 7	1

*pa, palaearctic; ne, nearctic;
References: (1) Kenis et al., 2004; (2) Sarikaya and Avci, 2009; (3) Moser et al., 2005; (4) Burjanadze et al., 2008; (5) Miller, 1986; (6) Stephen and Dahlsten, 1976; (7) Moser and Roton, 1971; (8) Moser et al., 1978; (9) Moore, 1972a–c; (10) Chansler, 1967; (11) Moser et al., 1971; (12) Wermelinger et al., 2013; (13) Stein and Coster, 1977; (14) Linit and Stephen, 1983; (15) Riley and Goyer, 1986; (16) Langor, 1991; (17) Fernández et al., 2013; (18) Hofstetter et al., 2009; (19) Moser, 1975

hymenopterans; Tables 7.5–7.7) are organisms that are only parasitic during their juvenile stages and, during their development, kill their host. All stages of bark beetles, i.e., eggs, larvae, pupae, and adults, are subject to parasitoid attack. Eggs are the least commonly parasitized stage, perhaps due to their small size. Beside some acarine parasitoids, *Trichogramma* species (see Table 7.7) are the only true egg parasitoids of bark beetles such as *Hylesinus (=Leperisinus) fraxini* (Panzer) and *Hylesinus crenatus* (F.) (Michalski and Seniczak, 1974). The eulophid *Entedon ergias* also parasitizes eggs but continues its development in the host larva (Beaver, 1966b; Yates, 1984).

Most species of Braconidae, Pteromalidae, Eupelmidae, Eurytomidae, and other families live ectoparasitically on bark beetle larvae or pupae. To access their host, the females of most parasitoids penetrate the bark with their ovipositor. Thus, their oviposition sites are determined by bark thickness and ovipositor length. Alternatively, females may use crevices in the bark to get closer to potential hosts. After they have successfully located a host, they first paralyze the host larva or pupa by injecting venom and, subsequently, deposit one egg on the host surface. A few species enter galleries through the bark beetle entrance holes in search for hosts. This is the case, for example, in the two pteromalids *Roptrocerus xylophagorum* (Ratz.) (Samson, 1984) and *Cerocephala eccoptogastri* Masi (Beaver, 1967). Following hatching, the parasitoid larva eats the interior of the host's body leaving just the cuticle and head capsule.

A third category of parasitoids attacks adult bark beetles when they alight and prepare to bore into the host tree, or when they emerge from it. Parasitoids of adult beetles are recorded from the Braconidae (*Cosmophorus* or *Ropalophorus* spp.; Table 7.5) and the Pteromalidae (*Tomicobia* spp.; Table 7.6). The parasitoid female oviposits directly through the thorax or elytra of its host; the parasitized bark beetle continues to bore into the bark and begins depositing eggs. The parasitoid larva developing within the beetle feeds on the host's body tissue eventually killing it. The adult wasp gnaws its way out of the bark beetle remains and leaves the gallery.

Some parasitoids that attack adult and larval beetles are obligate or facultative hyperparasitoids (Mills, 1991). The pteromalid *Mesopolobus typographi* (Ruschka) parasitizes the adult parasitoid *Tomicobia seitneri* (Ruschka) (Figure 7.2D) on *I. typographus* (Seitner, 1924), and *Dinotiscus eupterus* (Walker) is a facultative hyperparasitoid of the primary parasitoid *Dendrosoter middendorffii* (Ratz.) (Sachtleben, 1952). Other obligate or facultative hyperparasitoids are the eupelmids *Calosota aestivalis* Curtis and *Eupelmus urozonus* Dalman (Kenis and Mills, 1994), and some *Eurytoma* species (Dahlsten, 1982; Kenis *et al.*, 2004). Cleptoparasitism is thought to be quite common and has been observed in Pteromalidae and Eurytomidae (Mills, 1991; Hougardy and Grégoire, 2003; Kenis *et al.*,

2004). Cleptoparasitoids usually have poor host location abilities and steal hosts from other species by displacing the ovipositing females. However, super- and multiparasitism by parasitoids of bark beetles are rare (Dahlsten, 1982).

In general, parasitoids are more host specific than predatory insects. Parasitoids of beetle eggs and adults in particular have quite a narrow host range, often comprising only one or a few species (Kenis *et al.*, 2004). *Tomicobia seitneri*, for example, is strongly associated with *I. typographus* (Sachtleben, 1952; Krüger and Mills, 1990). Parasitoid mites that attack bark beetle eggs are also host specific. Very few parasitoid species have host ranges broad enough to span both coniferous and broadleaf bark beetle species; among the most abundant of these are the two pteromalids *Heydenia praetiosa* Forster and *D. eupterus* and some species associated with ambrosia beetles (Table 7.6). Most parasitoids are oligophagous. In coniferous forests, many parasitoids are known to attack bark beetles from various genera. One of the most abundant parasitoids, *R. xylophagorum*, parasitizes at least 10 bark beetle genera, including some important pest species: *D. frontalis*, *D. ponderosae*, *D. rufipennis*, *D. brevicomis*, and *Ips pini* (Say) in North America and *Ips acuminatus* (Gyllenhal), *I. typographus*, *Ips duplicatus* (Sahlberg), and *T. piniperda* in Europe. Interestingly, when monophagous bark beetle species switch tree species, a large proportion of their associated parasitoid complex also make this switch (Kenis *et al.*, 2004).

Adult feeding behavior, dispersal capacity, and longevity of parasitoids are not well studied. To meet their energy needs for egg production, parasitoids feed on pollen, nectar, and honeydew; longevity and egg load of *Coeloides bostrichorum* Giraud were significantly increased when allowed access to wildflowers that provided them with pollen and nectar (Hougardy and Grégoire, 2000). Apparently, enough of these resources are present in forests, although nearby cultivated land with melliferous plants does increase parasitism rates (Manojlovic *et al.*, 2000). Host feeding by parasitoids of adult beetles is unlikely (Mendel, 1988).

2.2.2.2 Host Location

Relatively little information is available on how parasitoids locate trees infested with suitable hosts (long-range host location). Unlike coleopteran predators, most parasitoids do not rely on bark beetle pheromones to find hosts. Females of most species utilize late larval instars for oviposition. At this point in time tree colonization is complete and pheromone production by bark beetles is likely to have ceased. Parasitoids of adult bark beetles do exploit beetle pheromones in the same way as the predatory guild; they depend on newly arriving host beetles and thus rapidly respond to their highly specific aggregation infochemicals

TABLE 7.5 Parasitoid Wasps in the Family Braconidae that have been Reported to Attack Bark Beetles

Species	Distribution[1]	Cryphalus	Dendroctonus	Hylesinus	Hylurgops	Ips	Leperisinus	Orthotomicus	Phloeosinus	Phloeotribus	Pityogenes	Pityokteines	Polygraphus	Scolytus	Tomicus	Trypodendron	Other
Atanycolus comosifrons Shenefelt	ne		X														
Blacus humilis (Nees)	pa														X		
Blacus koenigi Fischer	pa														X		
Bracon hylobii Ratz.	pa		X														
Bracon obscurator Nees	pa				X	X		X			X		X		X	X	X
Bracon palpebrator Ratz.	pa																
Bracon stabilis Wesmael	pa			X			X										
Bracon tenuicornis Wesmael	pa									X							
Caenopachys caenopachoides (Ruschka) (= Dendrosoter caenopachoides)	pa							X			X						
Caenopachys hartigii (Ratz.)	pa																
Cenocoelius nigrisoma (Rohwer)	ne		X			X					X						
Centistes cuspidatus (Haliday)[2]	pa						X										
Coeloides abdominalis (Zett.)	pa	X				X					X				X		
Coeloides bostrichorum Giraud	pa					X					X	X			X		
Coeloides filiformis Ratz.	pa			X			X			X							
Coeloides melanotus Wesmael	pa			X			X			X							
Coeloides pissodis (Ashmead)	ne		X			X								X			
Coeloides rufovariegatus (Provancher) (= C. dendroctoni)	ne		X			X		X									
Coeloides scolyticida Wesmael	pa					X	X							X			
Coeloides sordidator (Ratz.)	pa					X	X				.			X	X		
Coeloides subconcolor (Russo)	pa									X				X			
Coeloides ungularis Thomson	pa													X			

Continued

TABLE 7.5 Parasitoid Wasps in the Family Braconidae that have been Reported to Attack Bark Beetles—cont'd

Species	Distribution	Cryphalus	Dendroctonus	Hylesinus	Hylurgops	Ips	Leperisinus	Orthotomicus	Phloeosinus	Phloeotribus	Pityogenes	Pityoktenes	Polygraphus	Scolytus	Tomicus	Trypodendron	Other
Coeloides vancouverensis (Dalle Torre) (= C. brunneri)	ne		×														
Cosmophorus cembrae Ruschka[2]	pa	×									×						
Cosmophorus klugii Ratz.[2]	pa					×						×	×				
Cosmophorus regius Niezabitowski[2]	pa				×	×		×			×		×			×	
Cryptoxilos convergens Muesebeck[2]	ne									×							
Cryptoxilos cracoviensis (Capek and Capecki)[2]	pa	×															
Caenopachys hartigii (Ratz.) (= Dendrosoter hartigi, = Dendrosoter flaviventris, = D. caenopachoides)	pa	×				×		×			×	×	×		×		×
Dendrosoter middendorffii (Ratz.)	pa		×			×		×		×	×		×		×		
Dendrosoter protuberans (Nees)	pa, ne			×			×		×					×	×		
Dendrosoter scolytivorus (Viereck & Rohwer)	ne		×			×											
Dendrosoter sulcatus Muesebeck	ne		×			×											
Dendrosotinus ferrugineus (Marshall) (= Dendrosoter ferrugineus)	pa									×							
Dendrosotinus similes Boucek	pa	×															
Doryctes pomarius Reinhard	pa													×			
Ecphylus caudatus Ruschka	pa	×															
Ecphylus eccoptogastri (Ratz.)	pa									×				×			
Ecphylus hylesini (Ratz.)	pa					×					×	×	×		×		

Species[1]		15	2, 4–10, 13, 15, 23–25	4	19	4	3, 4, 19, 27	4	28	4	3, 4, 12, 19	4	4, 19	4, 17, 21, 22	1, 4	4, 19	4, 19, 26
Ecphylus silesiacus (Ratz.)	pa	x															
Hecabolus sulcatus Curtis	pa													x			
Heterospilus ater Fischer	pa					x											
Heterospilus incompletus (Ratz.)	pa													x			
Heterospilus sicanus (Marshall)	pa	x				x											
Lysitermus pallidus Foerster	pa												x				
Meteorus consimilis (Nees)	pa		x														
Meteorus hypophloei Cushman	ne												x				
Meteorus obfuscatus (Nees)	pa													x			
Monolexis fuscicornis Foerster (= *Hecabolus dederoi*)	pa								x								
Ontsira antica (Wollaston)	pa		x		x		x				x		x				
Perilitus rutilus (Nees)[2]	pa							x					x				
Rhoptrocentrus piceus Marshall	pa								x								
Ropalophorus clavicornis (Wesmael)[2]	pa			x													
Spathius brevicaudis Ratz.	pa													x			
Spathius canadensis Ashmead	ne		x														
Spathius curvicaudis Ratz.	pa								x					x			
Spathius pallidus Ashmead	ne			x						x							
Spathius rubidus (Rossi)	pa													x			
Spathius sequoiae Ashmead (= *S. californicus*)	ne														x		

[1] pa, palaearctic; ne, nearctic

[2] obligate adult parasitoid

References: (1) Schimitschek, 1931a; (2) Knight, 1958; (3) Mendel, 1985; (4) Kenis et al., 2004; (5) Gargiullo and Berisford, 1981; (6) Dahlsten, 1982; (7) McCambridge and Knight, 1972; (8) Miller, 1986; (9) Moore, 1972a–c; (10) Chansler, 1967; (11) Wermelinger et al., 2012; (12) Nuorteva, 1971; (13) Overgaard, 1968; (14) Ball and Dahlsten, 1973; (15) Linit and Stephen, 1983; (16) Riley and Goyer, 1986; (17) Schröder, 1974; (18) Feicht, 2004; (19) Gibb et al., 2008; (20) Hedgren and Schroeder, 2004; (21) Markovic and Stojanovic, 2003; (22) Bright, 1996; (23) Langor, 1991; (24) Massey and Wygant, 1973; (25) De Leon, 1934; (26) Mendel, 1986; (27) De Leon, 1935; (28) Bushing, 1965; (29) Berisford et al., 1970

TABLE 7.6 Parasitoid Wasps in the Family Pteromalidae that have been Reported to Attack Bark Beetles

Species	Distribution[1]	Cryphalus	Dendroctonus	Hylesinus	Ips	Leperisinus	Orthotomicus	Phloeosinus	Phloeotribus	Pityogenes	Pityoktenes	Pityophtorus	Polygraphus	Scolytus	Tomicus	Trypodendron	Other
Acrocormus semifaciatus Thomson	pa													X			
Agrilocida ferrierei Steffan	pa													X			
Cerocephala cornigera Westwood	pa					X			X					X			
Cerocephala eccoptogastri Masi	pa, ne					X		X	X					X			X
Cheiropachus quadrum (F.)	pa, ne				X	X			X					X			
Cleonymus brevis Boucek	pa													X			
Cleonymus obscurus Walker	pa													X			
Dibrachys cavus (Walker)	ne				X												
Dinotiscus aponius (Walker)	pa			X		X								X			
Dinotiscus colon (L.) (= Cheiropachus colon)	pa				X				X					X	X		
Dinotiscus dendroctoni Asmead (= Cecidostiba dendroctoni, = C. burkei)	ne		X		X									X			
Dinotiscus eupterus (Walker)[2] (= Cecidostiba acutus, = C. polygraphi)	pa, ne	X	X		X					X	X	X	X				
Habritys brevicornis (Ratz.)	pa															X	
Heydenia praetiosa Forster	pa				X	X	X	X	X		X			X	X		
Heydenia unica Cook & Davis	ne		X		X												
Macromesus americanus Hedqvist	ne		X		X												
Macromesus amphiterus Walker	pa				X					X		X		X			
Mesopolobus typographi (Ruschka)[3] (= Amblymerus typographi)	pa				X					X			X				
Metacolus azureus (Ratz.)	pa				X		X			X					X		
Metacolus fasciatus Girault	ne		X		X												
Metacolus unifasciatus Forster	pa				X		X	X							X		

Species	Type[1]
Pemiphora robusta Ruschka	pa
Platygerrhus affinis (Walker)	pa
Platygerrhus dolosus (Walker)	pa
Platygerrhus ductilis (Walker)	pa
Platygerrhus maculatus Erdos	pa
Pteromalus abieticola Ratz.	pa
Pteromalus brunnicans Ratz.	pa
Rhaphitelus ladenbergii (Ratz.)	pa
Rhaphitelus maculatus Walker	pa, ne
Rhopalicus guttatus (Ratz.)	pa
Rhopalicus pulchripennis (Crawford)	ne
Rhopalicus quadratus (Ratz.)	pa
Rhopalicus tutela (Walker)	pa, ne
Roptrocerus brevicornis Thomson	pa
Roptrocerus mirus (Walker)	pa
Roptrocerus xylophagorum (Ratz.) (= R. eccoptogastri, = Pachyceras xylophagorum, = P. eccoptogasteri)	pa, ne
Tomicobia acuminati Hedqvist[4]	pa
Tomicobia pityophthori (Boucek)[4] (= Karpinskiella pityophthori)	pa
Tomicobia seitneri (Ruschka)[4]	pa
Tomicobia tibialis Ashmead[4]	ne
Trigonoderus princeps Westwood	pa

References (by column): 2; 1, 2, 6–9, 12, 14, 16, 23–25, 27; 2; 2–4, 5, 7, 10, 13, 15, 18–20, 28; 2; 2; 2; 2; 2, 17, 29; 2; 2, 7, 17; 2; 2, 21, 22; 2; 2; 11, 26

[1] pa, palaearctic; ne, nearctic
[2] facultative hyperparasitoid
[3] obligate hyperparasitoid
[4] obligate adult parasitoid

References: (1) Otvos, 1970; (2) Kenis et al., 2004; (3) Aukema and Raffa, 2004; (4) Raffa and Dahlsten, 1995; (5) Boone et al., 2009; (6) Gargiullo and Berisford, 1981; (7) Dahlsten and Stephen, 1974; (8) Moore, 1972a–c; (9) Chansler, 1967; (10) Wermelinger et al., 2012; (11) Nuorteva, 1971; (12) Overgaard, 1968; (13) Wermelinger et al., 2013; (14) Stein and Coster, 1977; (15) Ball and Dahlsten, 1973; (16) Linit and Stephen, 1983; (17) Lobinger and Feicht, 1999; (18) Feicht, 2004; (19) Hedgren and Schroeder, 2004; (20) Hilszczanski et al., 2007; (21) Markovic and Stojanovic, 2003; (22) Bright, 1996; (23) Langor, 1991; (24) Massey and Wygant, 1973; (25) De Leon, 1934; (26) Mendel, 1986; (27) Dahlsten, 1970; (28) Berisford et al., 1970; (29) Capek, 1957

FIGURE 7.2   Pteromalid parasitoids develop endo- and ectoparasitically in bark beetles. (A) Adult *Rhopalicus tutela* parasitoid; (B) ectoparasitic larva of *R. tutela* with the remaining head capsule of an *Ips typographus* host larva; (C) *Roptrocerus xylophagorum*, a frequent parasitoid on both conifer and broadleaf bark beetles; (D) adult of *Tomicobia seitneri* and the empty body of its host, *I. typographus*, from which it has emerged. Copyright: WSL/Beat Fecker.

(Kenis *et al.*, 2004). This has been demonstrated for the parasitoid *Tomicobia pityophthori* (Boucek) and its host *P. chalcographus* (Lobinger and Feicht, 1999); *T. seitneri* and its host *I. typographus* (Faccoli, 2000); and *Tomicobia tibialis* Ashmead and its host *I. pini* (Raffa *et al.*, 2007). There is also evidence that parasitoids are capable of differentiating between parasitized and unparasitized hosts (Rice, 1968).

Parasitoids of larval beetles do not respond to aggregation pheromones (Boone *et al.*, 2008b) but are particularly attracted to allelochemicals that are produced by fungi and microorganisms in the host galleries. The pteromalids (*Roptrocerus* and *Rhopalicus* species) and the braconid *C. bostrichorum*, which all prefer late instar host larvae for oviposition, are attracted to infested bark by the volatiles (oxygenated monoterpenes) of microbes that are emitted from infested conifers (Pettersson *et al.*, 2000; Sullivan *et al.*, 2000; Pettersson, 2001b; Boone *et al.*, 2008b). The odor of the host larvae themselves is not attractive, in contrast to their frass, which is highly attractive (Sullivan *et al.*, 2000).

Our understanding of short-range host finding in these parasitoids is ambiguous. For example, the question of how exactly the females know where to insert their ovipositors to successfully parasitize concealed larvae is unclear. The first evidence for acoustic/vibrational cues (De Leon, 1935; cf. Meyhöfer and Casas, 1999) were placed in doubt by the fact that *Coeloides vancouverensis* (Dalle Torre)

could also identify the presence of motionless dead larvae of *D. pseudotsugae* (Richerson, 1972). Therefore, this parasitoid must perceive and respond to other stimuli. Females were then observed to change their searching behavior on the bark surface at the end of larval galleries and it was discovered that this was in response to changes in temperature associated with the presence or absence of bark beetle larvae. *Coeloides vancouverensis* females identified and probed in "hot spots" where host larvae were present, and could be induced to probe and oviposit in artificially heated areas in the absence of host larvae (Richerson, 1972). In studies with other braconids and the pteromalid *Rhopalicus tutela* (Walker) parasitizing *I. typographus*, however, both vibration and heat were ruled out as oviposition stimulants and volatiles suggested instead (Mills *et al.*, 1991). Parasitoids are likely to exploit the same volatiles for short-range host location as previously described for long-range host location (Pettersson, 2001a). Indeed, the anntennal structures previously thought to be infrared detectors have chemoreceptive functions (Pettersson *et al.*, 2001). However, the situation is complex; the role of volatiles from bark beetle and bluestain stain fungi, presented together or separately, on long- and short-range host location by *Roptrocerus xylophagorum* (Ratz.) provided ambiguous results in bioassays and Y-tube olfactometer experiments (Sullivan and Berisford, 2004).

In their search for suitable hosts, parasitoids prefer the upper parts of tree boles where they can penetrate the thinner bark with their ovipositors (Otvos, 1965; Ball and Dahlsten, 1973; Stephen and Dahlsten, 1976; Dahlsten, 1982; VanLaerhoven and Stephen, 2002; Wermelinger, 2002). This has been demonstrated for the braconids *Ropalophorus clavicornis* (Wesmael) and *Spathius pallidus* Ashmead (Gargiullo and Berisford, 1981). However, in *D. brevicomis* infestations, parasitoids were most numerous at mid-tree height, although their density was still negatively correlated with bark thickness (DeMars *et al.*, 1970). Interestingly, the pteromalid *R. xylophagorum*, which accesses its host larvae through bark beetle entrance holes, is still more abundant in the upper parts of the tree where the bark is thinner (Gargiullo and Berisford, 1981; VanLaerhoven and Stephen, 2002; Wermelinger, 2002). It is likely, therefore, that this species also parasitizes hosts through thin bark (Dahlsten, 1982).

As previously discussed, parasitoids (e.g., *Tomicobia* species) of adult bark beetles arrive concomitantly with their hosts at newly-infested trees (Aukema *et al.*, 2004) in response to host pheromones. Parasitoids of larval beetles arrive later at infestations than predators or parasitoids of adult beetles (Stephen and Dahlsten, 1976; Ohmart and Voigt, 1982), and the abundant parasitoid of late-instar beetle larvae, *R. xylophagorum*, is usually the last to arrive (Aukema *et al.*, 2004). Little information is available on the overwintering of parasitoids. In an infestation of *I. typographus*, the majority of the parasitoids (except *C. bostrichorum* and *T. seitneri*) left infested trees in fall, both in univoltine and bivoltine host generations (Wermelinger *et al.*, 2012). There is no information about where they overwinter, but it is suggested that they may be more tolerant to cold temperatures than their hosts (Dahlsten, 1982).

Most parasitoids of bark beetles belong to the Braconidae and Pteromalidae. Here we will describe these two families in more detail and direct the reader to further reading for some of the most important species (Schimitschek, 1931a; Sachtleben, 1952; Reid, 1957; Berisford *et al.*, 1970; Samson, 1984; Eck, 1990; Hougardy and Grégoire, 2004).

2.2.2.3 Braconidae

Nearly 60 species of braconid parasitoids have been reported associated with forest bark beetles (Table 7.5). The most frequently occurring genus is *Coeloides*, which is present both in Europe and North America. Many species such as *Dendrosoter middendorffii* (Ratz.) have a very broad host range, but remain restricted to bark beetle species in either conifer or broadleaf forests. The braconids include some important parasitoids of adult beetles that develop endoparasitically in the body of their host. This feeding behavior is common in the genera *Cosmophorus* and *Cryptoxilos* and in some particular species such as *R. clavicornis*, which can be very abundant (Wermelinger, 2002). However, the majority of species in the Braconidae

are ectoparasitoids on late-instar host larvae. Females of *Coeloides* species possess a particularly long ovipositor, which enables them to access hosts beneath thick bark (Krüger and Mills, 1990). *Coeloides* species are therefore evenly distributed along the tree trunk (Dahlsten, 1982). On occasion, braconids are parasitized by hyperparasitoids from other hymenopteran families (Boone *et al.*, 2009 and references therein).

One of the most abundant species in Europe is *C. bostrichorum* (Eck, 1990; Feicht, 2004). This species is stenotopic, preferring warmer climates and occurring only sporadically (Thalenhorst, 1958); it was absent at higher elevations (1300 m above sea level) (Wermelinger, 2002) and some other particular sites (Feicht, 2004), and was abundant in standing trees but not in felled or fallen trees (Thalenhorst, 1958; Hedgren and Schroeder, 2004). Like the similarly abundant *D. middendorffii*, *C. bostrichorum* oviposits in third instar larvae (Krüger and Mills, 1990). In North America, *Coeloides pissodes* (Ashmead) and *Coeloides rufovariegatus* (Provancher) are the most abundant species. The latter species is the most important parasitoid of *D. ponderosae* and, accordingly, its biology has been extensively studied (De Leon, 1935).

2.2.2.4 Pteromalidae

The most abundant and widespread bark beetle parasitoids belong to the family Pteromalidae (Wermelinger, 2002; Hedgren and Schroeder, 2004), and the majority of the species are ectoparasitic parasitoids of larval beetles (Figure 7.2). Two species, *Mesopolobus typographi* (Ruschka) and *D. eupterus* (Table 7.6), are obligate or facultative hyperparasitoids. Some of the few parasitoids occurring both in coniferous and broadleaf forests belong to the Pteromalidae, for example *H. praetiosa*, *Dinotiscus colon* (L.), and the holarctic species *Raphitelus maculatus* Walker, *Rhopalicus tutela* (Walker) (Figure 7.2A, B), *R. xylophagorum* (Figure 7.2C), and *Cheiropachus quadrum* (F.). *Tomicobia* species are adult endoparasitoids (Faccoli, 2000) (Figure 7.2D). Senger and Roitberg (1992) evaluated the effect of *T. tibialis* on *I. pini*; parasitized females produced 50% fewer offspring than unparasitized females.

One of the most important pteromalid species, both in Europe and North America, is *R. xylophagorum*, which has an exceedingly large host range. *Roptrocerus* species are among the few parasitoids known to enter bark beetle galleries for oviposition, but they can also penetrate bark that is thin enough in the upper regions of the tree, or that has been worked by woodpeckers (Otvos, 1979; Dahlsten, 1982). Since its target hosts are late-instar larvae of bark beetles, *R. xylophagorum* is also one of the last to arrive on infested trees (Aukema *et al.*, 2004); each female parasitizes an average of 33 larvae or pupae during its lifetime (Samson, 1984).

In North America, the pteromalids *Heydenia unica* Cook & Davis and *Rhopalicus pulchripennis* (Crawford) are abundant in infestations of *Dendroctonus* and *Ips* species. In a study with *I. pini*-infested logs, *H. unica* represented 29% of all pteromalids (Boone *et al.*, 2009). *Rhopalicus pulchripennis* prefers trees and tree sections with smooth bark surfaces, i.e., in the crown region (Ball and Dahlsten, 1973). Not all species are restricted to bark beetles. For example, *Dibrachys cavus* (Walker) is a generalist parasitoid that also attacks other beetles (Boone *et al.*, 2009 and references therein).

2.2.2.5 Other Parasitoids

Other parasitic Hymenoptera, mainly in the families Bethylidae, Eulophidae, Eupelmidae, and Eurytomidae, and predominantly from Europe, are known to attack bark beetles (Table 7.7). There is only one ichneumonid, *Dolichomitus terebrans* (Ratz.), which parasitizes the large bark beetle *D. micans*. Two egg parasitoids are fairly abundant, the holarctic eulophid wasp *Entedon ergias* Walker and *Trichogramma semblidis* (Aurivillius). They enter the galleries for oviposition (Beaver, 1966b). While *E. ergias* parasitizes eggs it completes its development in the host larva. In contrast, *T. semblidis* is a true egg parasitoid completing its entire development in the host egg. Parasitism rates of up to 98% have been reported for *T. semblidis* attacking *H. fraxini* eggs and superparasitism is common (Michalski and Seniczak, 1974). Some species are facultative hyperparasitoids; *E. urozonus* and *Eurytoma morio* Bohman, for example, attack bark beetle larvae in both coniferous and broadleaf forests, and at the same time parasitize *Rhopalicus* and *Coeloides* species, respectively (Sachtleben, 1952).

2.2.3 Impact of Arthropod Natural Enemies

There is a wealth of information and data concerning the mortality caused by antagonistic taxa and guilds based on laboratory experiments, exclusion experiments, and field observations. Some of the taxa-specific data have already been discussed; arthropod predators can cause significant mortality in bark beetle populations. Many predatory species are generalists and do not rely solely on bark beetle prey but may include other insects in their diet when a specific food source is scarce. This ensures that there is always a population of predatory beetles present that can respond quickly to an emerging food resource, such as bark beetles.

The density of predatory insects usually exceeds that of parasitic species (Riley and Goyer, 1986; Wermelinger *et al.*, 2012). During their development, and usually also as adults, every individual predator can kill several prey items. Many predators exhibit functional responses to prey density. For example, in exclusion experiments with logs colonized by *D. frontalis*, the consumption rate by predators

linearly increased with increasing density of *D. frontalis* (Linit and Stephen, 1983).

A range of parasitism rates in bark beetle populations has been reported in the literature. In one study parasitoids were responsible for 9–20% mortality in *S. intricatus* populations on oak (Markovic and Stojanovic, 2003). The braconid *D. rufovariegatus* caused up to 90% mortality in *D. ponderosae* populations (De Leon, 1935) and *T. seitneri* caused 40–70% (in some cases 100%) mortality in adult *I. typographus* populations (Sachtleben, 1952; Faccoli, 2000). Despite these impressive reports, parasitoids are still considered to have only a minor impact on bark beetle populations in general (Dahlsten, 1982), because each parasitoid developing in or on a host results in only one bark beetle killed. Moreover, adults do not feed on bark beetles. Therefore, even high parasitoid densities do not result in the same level of bark beetle mortality as comparable densities of predators. This is particularly evident for parasitoids of adult bark beetles, since the parasitized hosts continue to produce progeny for some time prior to death.

Interestingly, parasitoids are capable of responding quickly to changes in host abundance. For example, *I. typographus* populations exploiting ephemeral windthrows within stands of Norway spruce were rapidly found by the parasitoid fauna resulting in highly synchronous populations (Wermelinger *et al.*, 2013). In *D. brevicomis* infested trees, however, both positive and negative correlations have been found between parasitoid and bark beetle densities (DeMars *et al.*, 1970). Mortality rates due to parasitism also differed between generations, being lower in winter generations than in summer generations of *D. brevicomis* (Otvos, 1965; DeMars *et al.*, 1970); parasitoid efficacy is strongly dependent on the time within an outbreak (Wermelinger, 2002).

The overall impact of the combined actions of woodpeckers, predatory insects, mites, and parasitoids has been debated for a long time and is obviously variable. It is the result of the associations and interactions between many different taxa and functional groups, and interspecific competition can influence host/prey interactions in complex ways (Schroeder, 1996; Boone *et al.*, 2008a). For example, predators such as woodpeckers or large predatory insects not only consume bark beetles but also predatory and parasitic insects. However, woodpecker activity also provides parasitoids with better access to concealed host larvae (Otvos, 1979). Furthermore, the complex interactions between abiotic and biotic factors vary between regions and change with time and bark beetle density (Beaver, 1967; McCambridge and Knight, 1972); the dynamics and impact of natural enemies are likely to be affected by forest management strategies (Weslien and Schroeder, 1999; Hilsczcanski *et al.*, 2007).

There are numerous examples of the wide range in bark beetle mortalities that can be attributed to natural enemies.

TABLE 7.7 Hymenopteran Parasitoids from other Families that have been Reported to Attack Bark Beetles

Species	Family	Distribution[1]	Dendroctonus	Hylesinus	Ips	Leperisinus	Orthotomicus	Phloeosinus	Phloeotribus	Pityogenes	Pityokteines	Polygraphus	Scolytus	Tomicus	Trypodendron
Cephalonomia cursor Westwood	Bethylidae	pa											×		
Cephalonomia hypobori Kiefer	Bethylidae	pa						×	×						
Laelius elisae Russo	Bethylidae	pa				×			×						
Plastanoxus westwoodi (Kiefer)	Bethylidae	pa							×						
Sclerodermus brevicornis Kiefer	Bethylidae	pa							×						
Sclerodermus domesticus Klug	Bethylidae	pa													
Aprostocetus hedqvisti Graham	Eulophidae	pa											×	×	
Aulogymnus bivestigatus (Ratz.)	Eulophidae	pa				×									
Baryscapus hylesini Graham	Eulophidae	pa				×		×							
Entedon ergias Walker[2] (= E. leucogramma)	Eulophidae	pa, ne	×										×		
Entedon methion Walker	Eulophidae	pa			×										
Entedon pinetorum Ratz.	Eulophidae	pa			×										
Entedon tibialis (Nees) (= E. euphorion)	Eulophidae	pa					×	×							
Calosota aestivalis Curtis[3]	Eupelmidae	pa											×		
Calymmochilus russoi Gibson	Eupelmidae	pa					×		×						
Eupelmus annulatus Nees	Eupelmidae	pa											×		
Eupelmus cyaniceps Ashmead	Eupelmidae	ne	×												
Eupelmus pini Taylor (= E. sculpturatus, = E. aloysii)	Eupelmidae	pa							×				×		
Eupelmus urozonus Dalman[3]	Eupelmidae	pa			×	×			×		×	×	×	×	
Eupelmus vesicularis (Retzius)	Eupelmidae	pa				×			×				×		
Eusandalum merceti (Bolivar & Pieltain)	Eupelmidae	pa											×		
Bruchophagus maurus (Boheman)	Eurytomidae	pa											×		

Continued

276 Bark Beetles

TABLE 7.7 Hymenopteran Parasitoids from other Families that have been Reported to Attack Bark Beetles—cont'd

Species	Family	Distribution	Dendroctonus	Hylesinus	Ips	Leperisinus	Orthotomicus	Phloeosinus	Phloeotribus	Pityogenes	Pityokteines	Polygraphus	Scolytus	Tomicus	Trypodendron
Eurytoma aloisifilippoi (Russo)	Eurytomidae	pa							×						
Eurytoma arctica Thomson[3] (= E. auricoma)	Eurytomidae	pa		×	×	×				×			×	×	
Eurytoma blastophagi Hedqvist	Eurytomidae	pa			×										
Eurytoma cleri Ashmead	Eurytomidae	ne	×		×										
Eurytoma conica Provancher[3]	Eurytomidae	ne	×		×										
Eurytoma flavoscapularis Ratz.	Eurytomidae	pa				×									
Eurytoma morio Boheman[3]	Eurytomidae	pa		×	×	×	×	×	×	×			×	×	
Eurytoma phloeotribi Asmead	Eurytomidae	ne	×												
Eurytoma polygraphi (Asmead) (= E. spessivtsevi)	Eurytomidae	pa			×										×
Eurytoma tomici Ashmead	Eurytomidae	ne	×												
Dolichomitus terebrans (Ratz.)	Ichneumonidae	pa, ne	×												
Torymus arundinis (Walker)	Torymidae	pa				×									
Torymus hylesini Graham	Torymidae	pa				×									
Trichogramma semblidis (Aurivillius)[2]	Trichogrammatidae	pa		×		×									
References			1, 2, 3, 5, 7-9, 11, 15	3	3, 10, 12, 13, 16	3	3	3	3	3	3	3	3, 4, 6, 14	3	3

[1]pa, palaearctic; ne, nearctic
[2]egg parasitoid
[3]facultative hyperparasitoid.
References: (1) DeMars et al., 1970; (2) Otvos, 1970; (3) Kenis et al., 2004; (4) Yates, 1984; (5) Gargiullo and Berisford, 1981; (6) Dahlsten, 1982; (7) Dahlsten and Stephen, 1974; (8) Moore, 1972a-c; (9) Chansler, 1967; (10) Wermelinger et al., 2012; (11) Overgaard, 1968; (12) Ball and Dahlsten, 1973; (13) Hedgren and Schroeder, 2004; (14) Markovic and Stojanovic, 2003; (15) Dahlsten, 1970; (16) Berisford et al., 1970

The combined activity of predators and parasitoids achieved 23–28% mortality in *D. frontalis* populations on pine (Moore, 1972a; Linit and Stephen, 1983); 31% in populations of *Ips* species on pines (Riley and Goyer, 1986); in excess of 80% in populations of *I. typographus* on Norway spruce (Wermelinger, 2002); 90% in populations of *Ips* spp. on loblolly pines (Miller, 1984); and 80% in *I. typographus* populations in Sweden (Weslien, 1992). These mortality rates have been assessed at the level of bark beetle galleries, individual trees, and in some cases, entire stands. Nonetheless, it is still very difficult to determine how these individual examples translate into the temporal dynamics of bark beetle populations in a forest, or even a landscape, over time.

What these examples demonstrate is that natural enemies play an essential role in the regulation of bark beetle populations. It is evident that it is not individual taxa but the combination of many taxa that are the top-down drivers of bark beetle population dynamics. However, key taxa are woodpeckers, clerid beetles, dolichopodid flies (*Medetera* species), and, among the parasitoids, *Coeloides* species and *R. xylophagorum*. The importance of predatory mites is difficult to assess, but is probably underestimated. The most effective arthropods arrive at the very beginning of bark beetle colonization and therefore cause higher mortality than the late arriving species (Linit and Stephen, 1983; Nebeker *et al.*, 1984).

3. PATHOGENS OF BARK BEETLES

Insect pathogens play diverse roles in managed ecosystems such as forests; they can infect pest species (e.g. bark beetles), biological control agents (predators and parasitoids), and other beneficial species (e.g., bees) (Vega and Kaya, 2012). The basic biology, ecology, taxonomy, morphology, and mode of action of different entomopathogens have been widely reviewed (Adams and Bonami, 1991; Hajek, 2004; Vega and Kaya, 2012), and in some cases shown to play a major role in the population dynamics of their hosts (Vincent *et al.*, 2007; Vega and Kaya, 2012).

Despite their potential as microbial control agents, there are relatively few papers on pathogens of bark beetles and a limited understanding of their role in bark beetle population dynamics under natural conditions; most studies have been on the pathogens of *I. typographus*. However, pathogens of bark beetles have been reported widely from both Europe and North America (Postner, 1974; Mills, 1983; Bathon, 1991; Nierhaus-Wunderwald, 1993; Stephen *et al.*, 1993; Fuxa *et al.*, 1998; Wegensteiner, 2004; Takov *et al.*, 2007, 2010, 2011; Wegensteiner and Weiser, 2009; Lukášová and Holuša, 2012). Burjanadze and Kereselidze (2003) compared the prevalence of infection in field-collected beetles compared with laboratory-reared beetles,

while others have examined pathogen diversity in different bark beetle hosts including species occurring in the same biotope (Händel *et al.*, 2003; Händel and Wegensteiner, 2005; Holuša *et al.*, 2009, 2013; Lukášová *et al.*, 2013). The epizootiology of bark beetle pathogens in relation to host density, time since establishment of bark beetles within an area, climatic conditions, geographical location, and elevation are, however, understudied.

3.1 Pathogen Groups and their Modes of Action

Insect pathogens include viruses, bacteria, fungi, and protists. The method of entry of pathogens varies with the type of pathogen. Fungi generally invade their host directly through the cuticle, whereas other pathogen groups need to be ingested. Some pathogens are intracellular, developing within cells of the midgut epithelium or other tissues, while others are extracellular and destroy the midgut epithelium via toxins such as parasporal crystals (e.g., *Bacillus thuringiensis* [Berliner]). Fungi are extracellular and develop in the hemocoel. In all these examples, infection causes death of the host. Some pathogens can be transmitted horizontally (between conspecifics of the same generation) or vertically from parents to their offspring (via transovarial transmission) and, particularly in the latter, do not always cause mortality. Some pathogens are also thought to modify the behavior of infected hosts to encourage transmission (MacLeod, 1963; Evans, 1989; Roy *et al.*, 2006). The virulence of a pathogen and the progression of disease can also have premortality effects on the host, such as reduced fecundity (Roy *et al.*, 2006). Both positive and negative effects due to the progression of disease have been reported in bark beetles. For example, infection with *Gregarina typographi* (Fuchs) increased swarming of *I. typographus* compared with uninfected individuals (Wegensteiner *et al.*, 2010).

The main challenge for microbial control of bark beetles is determining the method by which the beetles can be inoculated with the pathogen. The most promising method would be to infect adult beetles during host colonization. Successfully infected adult beetles may then distribute the pathogen among conspecifics and to their offspring. Furthermore, predators, parasitoids or inquilines within the galleries may also contribute to the distribution of pathogens within and between bark beetle galleries (Karpinski, 1935; Doane, 1959; Weiser, 1961b; Moser *et al.*, 1989a; Dowd and Vega, 2003; Dromph, 2003; Vega and Blackwell, 2005).

3.1.1 Viruses

A virus has only one type of nucleic acid, either DNA or RNA. In some groups (e.g., Baculoviridae, Entomopoxviridae), the virions (nucleocapsids) are occluded in a crystalline protein matrix (occlusion body). This occlusion body contributes to the stability and persistence of the virions when they are released from the host into the environment. When a host consumes the occlusion body, the protein matrix dissolves in the midgut, releasing the virions, which attach to, and cross, the membranes of the midgut epithelium. The virions replicate in the cytoplasm, or in the nucleus. Insect viruses are obligate pathogens (they need a living cell and kill this cell during replication) and most of them are relatively host specific. Pathogen transfer occurs mainly by ingestion of virions during feeding (Harrison and Hoover, 2012).

In the first study with viruses and bark beetles, *S. scolytus* larvae were fed the *Oryctes* baculovirus (from *Oryctes rhinoceros* L.) but the results were inconclusive; the virus particles present in the bark beetles were not the same as the original particles that had been inoculated, and no effect of virus on beetle mortality was observed (Arnold and Barson, 1977). Since then viruses have been reported within living bark beetles but the first records of virus-killed cadavers being found in field-collected bark beetle adults and larvae were for an entomopoxvirus in *Dendroctonus armandi* Tsai and Li from northern China (Fan *et al.*, 1987; Tang and Sun, 1990). In North America, Sikorowski *et al.* (1996) found five different types of virus-like particles in the midgut epithelium of *D. frontalis* and, though similar to other types of insect virus, they were not associated with overt disease.

The *Ips typographus* entomopoxvirus (Weiser and Wegensteiner) (*It*EPV) was the first record of a viral disease in the most economically important bark beetle species in Europe, *I. typographus*, and it was found in the cells of the midgut epithelium of living beetles (Weiser and Wegensteiner, 1994; Wegensteiner and Weiser, 1995). *It*EPV has subsequently been detected in *I. typographus* from a number of European countries: Germany (Wegensteiner and Weiser, 1996a, 2009), Austria (Wegensteiner *et al.*, 1996; Haidler *et al.*, 1998, 2003; Gasperl, 2002; Wegensteiner and Weiser, 2009; Gasperl and Wegensteiner, 2012), Bulgaria (Takov *et al.*, 2006, 2007, 2012; Takov and Pilarska, 2009; Wegensteiner and Weiser, 2009), Georgia (Burjanadze and Goginashvili, 2009), Czech Republic (Wegensteiner and Weiser, 2009; Lukášová *et al.*, 2012, 2013), Denmark, Finland, Greece, Italy, Romania, Slovakia and Switzerland (Wegensteiner and Weiser, 2009), and Turkey (Gokturk *et al.*, 2010; Yaman and Baki, 2011).

Since the discovery of *It*EPV, a number of similar viruses have been reported. Novotný and Turčani (2000) reported a new virus infection in *I. typographus* larvae. A subsequent detailed investigation using transmission electron microscopy confirmed that the internal structure of the *It*EPV was different to all formerly described entomopoxvirus type A species (Žižka *et al.*, 2001). Händel

et al. (2001, 2003) reported a virus very similar to the *It*EPV in *Ips amitinus* (Eichhoff) from Austria, which has subsequently also been found in the Czech Republic (Lukášová *et al.*, 2013). Entomopoxvirus-like particles, unlike those known from *I. typographus*, have been found in the cytoplasm of midgut epithelium cells from *Polygraphus poligraphus* (L.) (Händel *et al.*, 2003), and baculovirus-like particles in the nuclei of midgut epithelium cells from *P. poligraphus* Händel (2001).

*It*EPV is the most extensively studied virus of bark beetles and, in Bulgaria, has been found in populations of *I. typographus* from different locations and altitudes (1450, 1510, and ≥1100 mm.a.s.l.), although its incidence varied widely (1–41%) between two consecutive years and between different sampling dates in the same year (Takov *et al.*, 2006, 2007). In another study, *It*EPV infection rates in *I. typographus* before and after overwintering were less variable (Händel *et al.*, 2001). In a wilderness area in Austria, *It*EPV was found in *I. typographus* populations at all sampling plots, but always at very low levels (0.4–4.0%) compared with other reports; this may be because, in undisturbed forests that are more diverse than managed forests, the pathogen complex and the key species changes (Wegensteiner *et al.*, 2014).

In a study by Tonka *et al.* (2007) artificial infection of *I. typographus* by *It*EPV was never achieved in the laboratory, and virus occlusion bodies were never found in larvae or any other preadult stage during their experiments. To date, there has only been one report presenting laboratory data on successful infection of *I. typographus* with *It*EPV (52% of 1142 insects died due to infection), and this was achieved by feeding a suspension of spheroid inclusion bodies to the beetles on bark chips; however, there was large variability in the time to kill (9–64 days) with mortality occurring mainly from days 4–16 post inoculation (Tonka and Weiser, 2009). Furthermore, there has been only one preliminary field experiment showing successful introduction of *It*EPV to field populations of *I. typographus*; up to 40% mortality in *I. typographus* populations was achieved after spraying trap logs with a suspension of *It*EPV (Pultar and Weiser, 1999). Although *It*EPV is the most extensively studied virus of bark beetles, these studies demonstrate that more fundamental studies are required to understand the relationships between pathogen virulence, host infection, and mortality.

3.1.2 Bacteria

Bacteria are unicellular organisms that lack a defined nucleus (Prokaryotes). They are common microorganisms associated with insects, and *B. thuringiensis* has received considerable attention as a microbial control agent against several insect groups (e.g., Hajek, 2004; Jurat-Fuentes and Jackson, 2012). Bacteria infect insects mostly through the mouth and digestive tract, following ingestion with food. Most bacteria are extracellular with the exception of pathogenic species in the genus *Rickettsia* that are obligate intracellular parasites (Jurat-Fuentes and Jackson, 2012). Several bacteria are associated with bark beetles, but whether they are pathogens or symbionts is often unclear (Bridges *et al.*, 1984; French *et al.*, 1984; Tomalak *et al.*, 1988; Canganella *et al.*, 1994; Vasanthakumar *et al.*, 2006.

Pesson *et al.* (1955) were the first to describe bacteria in bark beetles in Europe, and reported that some bacteria caused up to 100% mortality in *S. multistriatus* within 72 h. Subsequent reports of bacteria in European bark beetles include: Lysenko (1959) in *Pityokteines curvidens* (Germar) and *Trypodendron lineatum* (Olivier); Novak (1960) in *T. lineatum*; Bałazy (1967) and Imnadze (1978) in *I. typographus*; Imnadze (1978, 1984) and Yilmaz *et al.* (2006) in *D. micans*; and Canganella *et al.* (1994) and Natali *et al.* (1994) in *Anisandrus dispar* (F.). Bacteria have also been reported in North American bark beetles: Doane (1960) and Wood (1961) in *S. multistriatus*; Moore (1971, 1972b), Bridges *et al.* (1984), and Vasanthakumar *et al.* (2006) in *D. frontalis*; and Tomalak *et al.* (1988) in *Hylurgopinus rufipes* (Eichhoff). In Australia, French *et al.* (1984) reported the presence of bacteria in *S. multistriatus*.

Infection studies with different bacteria have had highly variable results confirming that some may be symbionts rather than pathogens, or that some may be secialists with restricted host ranges. However, in comparisons between several different bacteria against *S. scolytus* larvae, up to 91% mortality was achieved for some species and some doses; 80% mortality was achieved against *S. multistriatus* larvae (Jassim *et al.*, 1990). Bacteria isolated from dead adults of *D. micans* from Turkey were also pathogenic in subsequent bioassays against *D. micans* adults and larvae (Yaman *et al.*, 2010a).

One species, *Serratia marcescens* (Bizio), was responsible for high mortality in *S. multistriatus* larvae within a 4-day period, and levels of mortality were correlated with larval density; this was interpreted as a crowding effect (Doane, 1960). In one population of adult *Ips calligraphus* (Germar) from Florida, the incidence of *S. marcescens* was 46% following ingestion of bacteria-contaminated phloem from old logging debris or from contact between crowded adults during the predispersal and maturation feeding period. The incidence of *S. marcescens* was greater in young, inexperienced adults than parental beetles of *Ips calligraphus* (Germar), and higher in adults than in their associated larvae (Jouvenaz and Wilkinson, 1970). In bioassays against *D. frontalis* larvae, six isolates of *S. marcescens*, originally isolated from diseased *D. frontalis* beetles, were classified as pathogenic (Moore, 1972c). Eight different species of bacteria have been identified in *I. typographus* using DNA analysis. However, only one species, *Serratia*

liquifaciens (Grimes and Hennerty), caused significant mortality (53%) in *I. typographus* adults in subsequent laboratory bioassays (Muratoğlu *et al.*, 2011).

A bacterial preparation of *B. thuringiensis* that included bitoxibacillin showed activity against *D. micans* and *I. typographus*, but the effect was attributed to the β-exotoxin (Imnadze, 1978). Infection experiments with *B. thuringiensis* var. *insectus* (Guk.) were partially successful against *Ips subelongatus* (Motschulsky), but this was probably also as a consequence of exotoxins (Gusteleva, 1980, 1982a, b). In the laboratory, *B. thuringiensis* was not effective against *I. calligraphus* or *D. frontalis* (perhaps due to contamination), but two bacterial metabolites, avermectin B1 and *Bt* metabolite R003, were effective (Cane *et al.*, 1995) when applied to field collected larvae. *Bacillus thuringiensis* var. *tenebrionis* (Krieg, Huger, Langenbruch and Schnetter) caused no significant mortality in adult *I. typographus* in field experiments where the bark surface of spruce logs had been sprayed with bacteria prior to beetle colonization (Novotný and Turčani, 2000). Multiple failures testing field-collected *B. thuringiensis* isolates in the past motivated Vlasek *et al.* (2012) to mutate the δ-endotoxin of the *B. thuringiensis* Cry3A crystal protein, resulting in increased toxicity to *I. typographus* larvae.

3.1.3 Algae

Non-photosynthetic algae from the genus *Helicosporidium* occur in insects as pathogens (Lange and Lord, 2012). *Helicosporidium* is classified within the class Trebouxiophyceae as a non-photosynthetic green alga (Phylum: Chlorophyta). *Helicosporidium* species can be easily identified from the very typical cysts that they form. Cysts are ingested *per os* and develop in the midgut epithelium and in the hemolymph of their hosts. They have been described from at least 23 insect species worldwide and are present in bark beetles. However, other than their descriptions that have been made using light and electron microscopy techniques, and some molecular studies, little is known about their pathogenic effect on hosts or whether they contribute to bark beetle population regulation (Lange and Lord, 2012; Tartar, 2013).

Purrini (1980) reported *Helicosporidium parasiticum* (Keilin) in *Hylurgops palliatus* (Gyllenhal) from Germany, but did not include a description. *Helicosporidium* sp. has also been isolated from the hemolymph of *D. micans* from Turkey (Yaman and Radek, 2005, 2008a). Yaman (2008) surveyed *D. micans* adults from seven plots in four areas in Turkey over a number of years and, in at least one plot, the infection rate was high (72%), while in another plot it was very low (1.3%); the authors concluded that climatic conditions may have played an important role in infection success. Furthermore, slightly more *D. micans* male beetles were infected than females, and the pathogen was never found in larvae (Yaman, 2008).

280 Bark Beetles

3.1.4 Fungi

3.1.4.1 Ascomycota: Hypocreales

Entomopathogenic fungi attacking bark beetles belong to the phylum Ascomycota. The infective units are the conidia; sporulation and germination require high humidity and adequate temperatures. Fungi gain access to the insect directly through the insect's integument. After germination of the conidium on the insect's cuticle, the fungus penetrates the integument and proliferates throughout the host, ultimately resulting in death of the host. Hyphae then emerge from the cadaver and produce conidia for subsequent transmission. Host specificity of entomopathogenic fungi varies considerably; some species have a broad host range and others are more restricted (Inglis et al., 2001; Vega et al., 2012).

There are numerous reports of entomopathogenic fungi infecting bark beetles and in all cases they are able to infect and kill all life stages (Müller-Kögler, 1965; Jurc, 2004; Wegensteiner, 2004). Many species of fungi have been found in the galleries of bark beetles from different tree species, even when beetles are no longer present; several of these fungi are entomopathogenic species (Bałazy, 1965; Kirschner, 2001). The first records of fungal pathogens of bark beetles were from Scolytus spp. in France and the Netherlands at the beginning of the 20th century (Gösswald 1938). In the 1930s, Beauveria bassiana (Bals.-Criv.) Vuill. was reported from Hylastes ater (Paykull) in Great Britain (Petch, 1932), and from I. typographus and Ips duplicatus (Sahlberg) in Poland (Karpinski, 1935; Siemaszko, 1939). Since then the presence of entomopathogenic fungi from different bark beetle species has been widely reported (Table 7.8).

In contrast to all other pathogen groups associated with bark beetles, entomopathogenic fungi have been well studied and evaluated as bark beetle control agents (Popa et al., 2012). Some molecular studies of field-collected isolates have also been made (Landa et al., 2001b). The requirement of entomopathogenic fungi for high humidity has been noted by several authors (Schvester, 1957; Bałazy, 1963); when S. multistriatus larvae were collected in a damp location, Doane (1959) found larval mortality due to B. bassiana was greater in beetles from shaded trees (93–99%) than in beetles from trees with a sunny aspect (1–8%).

Numerous bioassays have been done with different fungal species (and isolates) and against different species of bark beetle, all demonstrating the potential to achieve high mortalities in bark beetle populations in the laboratory. The most widely studied species is B. bassiana, which has been bioassayed against a number of different bark beetle species. Doane (1959) reported B. bassiana-induced mortality in larvae (92%), pupae (87%), and adults (61%) of Scolytus multistriatus (Marsham). Beauveria bassiana also caused 100% mortality in T. lineatum larvae within 6 to 8 days (Novak and Samsinakova, 1962) and was highly

pathogenic to I. typographus and Scolytus ratzeburgi (Janson) (Bałazy, 1966). High B. bassiana-induced mortality was observed in S. scolytus; the LD_{50} was reported as 1×10^6 conidia/ml after 5 days (at 23°C and 100% relative humidity) (Barson, 1976, 1977). Beauveria bassiana caused high mortality rates in bioassays against D. micans ($\geq$90%) (Tanyeli et al., 2010). Bioassays with a field-collected isolate of B. bassiana were effective against Dryocoetes confusus (Swaine) (85–96% mortality), although there were also high levels of infection in the untreated control group (Whitney et al., 1984). An isolate of B. bassiana from I. duplicatus sprayed on to log sections infested with I. typographus caused ca. 60% mortality (Andrei et al., 2013). An isolate of B. bassiana from I. duplicatus caused 100% mortality in the same species within 4 days in the laboratory (Dinu et al., 2012). Evaluation of B. bassiana against Dendroctonus valens (LeConte) resulted in up to 100% mortality within 4.6 days (Zhang et al., 2011). Inoculation of I. sexdentatus with dry conidial powders or conidial suspensions of B. bassiana caused more than 90% mortality within a few days in young, immature adults and also old adults (Steinwender et al., 2010). Spraying B. bassiana conidial suspensions onto the bark of I. typographus-infested logs caused 29% mortality (Jakuš and Blaženec, 2011).

Commercial isolates of B. bassiana have also been evaluated. Gusteleva (1980, 1984) evaluated the commercial product Boverin against adult Ips subelongatus (Motschulsky), and achieved 89% mortality. Bioassays with a number of commercial products of B. bassiana and one of Metarhizium brunneum Petch against Xylosandrus germanus (Blandford) also achieved high levels of mortality, especially with one of the B. bassiana isolates and the M. brunneum isolate (Castrillo et al., 2011, 2013).

As in the studies of Castrillo et al. (2011, 2012) that were done with commercial isolates, virulence of B. bassiana has been widely compared with other fungal species against a range of bark beetle species. Doberski (1981a, b) found that isolates of B. bassiana were more virulent against larvae and adults of S. scolytus than Isaria farinosa (Holmsk.) and Metarhizium anisopliae (Metschn.) Sorokin. In contrast, B. bassiana, M. anisopliae, and Isaria fumosorosea (Wize) (formerly Paecilomyces fumosoroseus (Wize) Brown and Smith) were equivalent in virulence against S. multistriatus, all causing 100% mortality at high doses (Houle et al., 1987). Markova (2000) compared the virulence of B. bassiana, M. anisopliae, I. farinosa, and C. confragosa against I. typographus and found that the most aggressive species was M. anisopliae (100% mortality in 4 days) and the least aggressive was C. confragosa (still causing 90% mortality after 5 days). In another study with I. typographus it was highly susceptible to both B. bassiana and M. anisopliae; 90–99% mortality was achieved in inoculated individuals (Mudrončeková et al., 2013). In direct

TABLE 7.8 Entomopathogenic Fungi Isolated from Field-collected Bark Beetles

	Crypturgus	Dendroctonus	Dryocoetes	Gnathotrichus	Hylastes	Hylesinus	Hylurgops	Hylurgus	Ips	Leperesinus	Orthotomicus	Pityogenes	Pityokteines	Polygraphus	Scolytus	Tomicus	Trypodendron
Beauveria bassiana (Bals.-Criv) Vuill	×	×	×		×		×		×	×	×		×		×	×	×
Beauveria globulifera (Speg.) Picard									×								
Beauveria densa (Link) Picard									×								
Beauveria caledonica Bissett and Widden			×	×	×		×	×	×		×	×		×			
Beauveria brongniartii (Sacc.) Petch							×		×								
Metarhizium anisopliae (Metschn.) Sorokin		×							×								
Metarhizium flavoviride var. pemphigi Driver and Milner					×												
Isaria farinosa (Holmsk.)		×				×			×								×
Isaria fumosorosea (Wize)									×								
Lecanicillium lecanii complex	×	×							×					×	×		
Hirsutella guignardii (Mahau) Samson, Rombach and Seifert					×												
Tolypocladium cylindrosporum W. Gams												×					
References	3, 11	7, 16	9, 11, 15	11	6, 20, 22	24	11, 17, 19	20	1, 4, 11, 12, 14, 15, 19, 21, 24	11	11, 19	11	13	11, 23	2, 3, 24, 25	5, 19	8, 11

References: (1) Neužilová, 1956; (2) Doane, 1959; (3) Bałazy, 1962; (4) Bałazy et al., 1967; (5) Nuorteva and Salonen, 1968; (6) Leatherdale, 1970; (7) Moore, 1971; (8) Magema et al., 1981; (9) Whitney et al., 1984; (10) Bałazy et al., 1987; (11) Kirschner, 2001; (12) Landa et al., 2001a; (13) Pernek, 2007; (14) Takov et al., 2007; (15) Draganova et al., 2010; (16) Tanyeli et al., 2010; (17) Takov et al., 2011; (18) Dinu et al., 2012; (19) Takov et al., 2012; (20) Glare et al., 2008; (21) Keller et al., 2004; (22) Brownbridge et al., 2010; (23) Bałazy, 1963; (24) Bałazy, 1973; (25) Barson, 1976.

inoculation bioassays of *I. typographus*, high mortality was achieved with *B. bassiana* (84–98%) and *M. anisopliae* (88–92%), but was slightly lower with *I. fumosorosea* (74–82%) (Herrmann and Wegensteiner, 2010). Neither *B. bassiana* nor *I. farinosa* isolates were very aggressive against *Ips acuminatus* (Gyllenhal) (5.5%) but caused high mortality in *I. sexdentatus*, especially the *B. bassiana* isolates (≥90% mortality c.f. 45–66% for *I. farinosa*) (Draganova *et al.*, 2007). *Beauveria bassiana* was slightly more pathogenic (72–100%) than *Beauveria brongniartii* (Sacc.) Petch (70–81%) against *P. poligraphus* when they were exposed to inoculum on treated bark (Wegensteiner, 2000). In bioassays against *D. micans* one isolate of *B. bassiana* was more virulent than *M. anisopliae* isolates, *I. fumosorosea*, *Evlachovaea* sp., and other isolates of *B. bassiana*, particularly against larvae (Sevim *et al.*, 2010).

Other species of fungi have also been evaluated. In the only study with an isolate from the *Lecanicillium lecanii* complex it was found to be infective to *S. scolytus* larvae, achieving 100% mortality within 5 days at a conidial concentration of 4.5×10^6 conidia/ml (Bałazy, 1963; Barson, 1976). In New Zealand *Beauveria caledonica* Bissett and Widden was virulent against *Hylurgus ligniperda* (F.) and *H. ater* (Glare *et al.*, 2008; Reay *et al.*, 2008).

Apart from the well-known effects of humidity on infection (Doberski, 1981a, b) there are other abiotic and biotic factors that influence infectivity, the progression of disease, and the outcome of bioassays and field trials. *Beauveria bassiana* was most effective against adult *D. frontalis* at 15–20°C; in this experiment ambient humidity was suspected to be less important because contact moisture on the bark beetle cuticle was sufficient for germination of conidia (Moore, 1973). Dose, temperature, and bark humidity had the greatest effects on susceptibility of larvae, pupae, and adults of *P. chalcographus* (Wulf, 1979, 1983), *T. lineatum* (Prazak, 1988, 1991, 1997), and *S. scolytus* (Barson, 1976, 1977) to *B. bassiana*. Hunt *et al.* (1984) and Hunt (1986) found that germination of *B. bassiana* conidia was inhibited on the cuticle of *D. ponderosae* and hypothesized that this was due to insufficient nutrients at the cuticle surface.

Infection can have premortality effects on host behavior that influence transmission. For example, females of *P. chalcographus* and *T. lineatum* infected with *B. bassiana* lay significantly fewer eggs than uninfected females (Wulf, 1979, 1983; Prazak, 1988, 1991, 1997). Infection of *X. germanus* with either *B. bassiana* or *M. brunneum* also reduced reproductive rate, and mortality was dose dependent (Castrillo *et al.*, 2011, 2013). Females of *D. confusus* that had been killed by *B. bassiana* were walled off in their galleries with frass plugs, which resulted in reduced dissemination of the fungus (Whitney *et al.*, 1984).

Biotic and abiotic variables, particularly temperature and humidity, can fluctuate widely in the field and have strong influences on the outcomes of field trials evaluating entomopathogenic fungi as biological control agents of bark beetles. Up to 62% of *Tomicus piniperda* (L.) were killed by *B. bassiana* sprayed onto log sections as conidial suspensions; the highest mortality occurred when beetles were placed onto the treated bark surface immediately after spraying and decreasing to 45% when insects were placed on the treated bark surface a week after spraying (Nuorteva and Salonen, 1968). *Beauveria bassiana* was also responsible for 81–100% mortality in other bioassays against *T. piniperda* (Burjanadze, 2010). Bychawska and Swiezynska (1979) sprayed *B. bassiana* conidia onto bait logs in the field but were unable to infect *T. piniperda*; abiotic conditions (rain, snow, and solar radiation) were identified as the major problems reducing fungal viability. However, when bait logs were sprayed with *B. bassiana* conidia and then covered with polyethylene foil to improve abiotic conditions, 71–100% of the larvae in the logs became infected (Lutyk and Swiezynska, 1984). Nevertheless, in a Spanish field trial, spraying pine bait logs with either *C. confragosa* or *B. bassiana* suspensions led to 55% and 85% mortality in *Tomicus minor* (Hartig), respectively (Ruiz-Portero *et al.*, 2002) and Batta (2007) achieved 80% mortality with *B. bassiana* sprayed onto peach trees against *Scolytus amygdali* (Guerin-Meneville). These results suggest great potential and also opportunities for improvement.

The life-stage of the beetles targeted can be influencial in improving the results of field applications of entomopathogenic fungi. For example, as bark beetles leave their cryptic locations within trees to hibernate in the soil, fungal treatment of soil where beetles are hibernating can achieve good control; 88–100% mortality in hibernating *T. lineatum* was achieved when *B. bassiana* conidia were applied to the soil of their overwintering sites (Prazak, 1988). In laboratory experiments that simulated hibernation conditions, up to 70% mortality in *I. typographus* was achieved when they were hibernating in *B. bassiana*-treated litter at 20, 17, and 12°C after 28, 35, and 41 days, respectively (Hallet *et al.*, 1994). Spraying the soil surface with *B. bassiana* conidial suspensions (10^{10} conidia/m^2) next to the stem base of a spruce tree infested with *I. typographus* also caused 41% mortality in beetles emerging from the soil in spring, although there were also high levels of fungus-induced mortality in the control group (24%) suggesting natural levels of infection are high in overwintering beetles (Hein, 1995). High levels of *B. bassiana* infection have been found in *I. typographus* emerging from fungus-treated plots; again there were high levels of infection in control plots, indicating the natural presence of *B. bassiana* in soil, and that targeting populations in the soil could be an effective control strategy (Arbesleitner, 2002).

Inoculation methods can also be improved. While application of inoculum directly to beetles causes the highest levels of infection, e.g., 78–100% in *I. typographus* inoculated with different *Beauveria* spp. and incubated at 13, 23, and 33°C, infection levels can also be high when the beetles receive inoculum indirectly. For example, 60% of *I. typographus* walking over bark treated with *B. bassiana* conidia died due to infection and 99% of healthy beetles contacting sporulating cadavers of *I. typographus* died (Wegensteiner, 1992, 1996). Direct inoculation with *M. anisopliae* resulted in 95% mortality in adult *P. chalcographus* but indirect inoculation via treated bark still resulted in 27% mortality (Pehl and Kehr, 1994). Matha and Weiser (1985) were also able to successfully infect 100% of *I. typographus* with a commercial *B. bassiana* preparation (Boverol) via indirect inoculation when beetles walked on conidia-treated filter paper or contacted sporulating cadavers. This indirect contamination has potential for effective delivery of inoculum, especially when combined with semiochemicals. *Ips typographus* attracted to autoinoculation devices containing *B. bassiana* conidia became infected and 71–97% died; inoculated beetles had significantly shorter lifespans than uninoculated beetles, the length of maternal galleries was reduced, as were the number of offspring (≥70%) (Kreutz, 1995; Vaupel and Zimmermann, 1996; Kreutz et al., 2000a, b, 2004a, b; Kreutz, 2001). Brownbridge et al. (2012) successfully established *B. bassiana* as an endophyte in *Pinus radiata* (D. Don) seeds and seedlings with the aim of providing inoculum systemically within the plant that could infect bark beetles feeding internally within young trees; these beetles would otherwise be impossible to target given their cryptic feeding site.

3.1.4.2 Yeasts (Ascomycota: Saccharomycotina)

Little is known about the occurrence of pathogenic ascomycete yeasts (Saccharomycotina) in bark beetles. Many yeasts associated with bark beetle galleries are involved in symbioses with bark beetles, including increasing the attractiveness of trees to bark beetles (Callaham and Shifrine, 1960); these aspects are covered in Chapter 6.

Moore (1972b) identified a *Candida* sp. in adults and larvae of *D. frontalis* and a *Pichia* sp. has been detected in *I. typographus* and *Ips duplicatus* (Eichhoff) from Poland and the Czech Republic (Holusa et al., 2007). In the USA, 28 different yeast species were found in laboratory reared *D. frontalis* larvae and in their galleries (Bridges et al., 1984), Leufvén and Nehls (1986) described yeast associates of *I. typographus* in Denmark, and Moser et al. (1989b) reported that ascospores of several species were transported on mites associated with *I. typographus*; whether these species have any negative effects on bark beetles is unclear.

In Finland, a new *Metschnikowia* species (*Metschnikowia* cf. *bicuspidata* (Metschn.) Kamienski) was isolated from *I. typographus* midgut epithelium and hemolymph (Wegensteiner and Weiser, 1998). Following

further sampling this species was described as *Metschnikowia typographi* (Weiser, Wegensteiner, Händel and Žižka) and, from a number of studies, has been reported from *I. typographus* (0.3–4.7%), *I. amitinus* (5–13%), *I. sexdentatus* (0.9–51%), and *D. micans* (Händel et al., 2003; Weiser et al., 2003; Wegensteiner et al., 2007a; Yaman and Radek, 2008a, b; Unal et al., 2009). Subsequent field sampling has shown that it is not common, and not always found in all available host species. For example, from extensive samples of different bark beetle species made in several European countries, it was only found in *I. typographus* from Finland (Wegensteiner and Weiser, 2009). However, in some sites in the Czech Republic, it was found only in *I. amitinus* and not in *I. typographus* (Lukášová et al., 2013). Furthermore, a *Metschnikowia* sp. was found in *I. typographus* from two sites in Georgia (Burjanadze et al., 2011).

3.1.4.3 Microsporidia

Previously, the phylum Microsporidia was considered closely related to the Protozoa, because of the parallels in their life histories (see section 3.1.6) but is now known to belong to the Fungi (Solter et al., 2012). They are obligate, spore-forming pathogens that only multiply in living cells. Spores can occur in two forms: thick-walled forms termed environmental spores and thin-walled forms termed primary spores. Microsporidian spores contain an uninucleate or binucleate sporoplasm, and an extrusion apparatus that includes a polar filament and a polar cap; they do not have mitochondria. Microsporidia are ingested with the food and infection is initiated in the midgut epithelium (Solter et al., 2012). A number of microsporidia have been reported from bark beetles (Table 7.9).

Weiser (1954a, b, 1955) was the first to detect *Haplosporidium typographi* (Weiser), syn. *Chytridiopsis typographi* (Weiser), in the gut epithelium of *I. typographus* from the Czech Republic and Slovakia. Since then a number of microsporidian genera (*Chytridiopsis, Nosema, Unikaryon, Canningia, Larsoniella*, etc.) have been described from different bark beetle species. In the USA high prevalences (up to 87.5%) of *Unikaryon minutum* (Knell and Allen) have been reported in populations of pheromone trap-caught *D. frontalis* (Knell and Allen, 1978) and also in laboratory colonies (up to 79%) (Atkinson and Wilkinson, 1979). *Unikaryon minutum* was later found in other *D. frontalis* populations in the USA, occasionally at high prevalence levels (4–55% and 65% infection), although it was suspected to have limited deleterious effects on host populations (Bridges, 1987; MacGuidwin et al., 1980).

Ips typographus appears highly susceptible to infection by *C. typographi*. In laboratory bioassays 100% mortality was achieved within 60 days (Tonka et al., 2007) and it

TABLE 7.9 Microsporidia Isolated from Field-collected Bark Beetles

Cryphalus	*Dendroctonus*	*Dryocoetes*	*Hylastes*	*Hylurgops*	*Hylurgus*	*Ips*	*Orthotomicus*	*Pityogenes*	*Pityokteines*	*Polygraphus*	*Scolytus*	*Tomicus*	*Taphrorychus*	*Trypodendron*
36	8	11, 29	13, 29	10, 13, 28, 29	36, 48	1, 2, 3, 14, 15, 17, 18, 19, 20, 21, 22, 23, 24, 25, 27, 28, 29, 31, 32, 33, 34, 35, 36, 37, 38, 39, 41, 42, 43, 44, 45, 46, 47, 48, 49, 50, 51, 52, 53, 54	36, 48, 51	12, 22, 28, 29, 54	4, 16, 40	26	5, 6, 7, 9	30, 47	48	11

References: (1) Weiser, 1954a; (2) Weiser, 1954b; (3) Weiser, 1955; (4) Weiser 1961b; (5) Weiser, 1966; (6) Lipa, 1968; (7) Weiser, 1968; (8) Weiser, 1970; (9) Purrini, 1975; (10) Purrini, 1978b; (11) Purrini and Ormieres, 1981; (12) Purrini and Halperin, 1982; (13) Purrini and Weiser, 1984; (14) Purrini and Weiser, 1985; (15) Wegensteiner, 1994; (16) Weiser et al., 1995; (17) Wegensteiner and Weiser, 1996a; (18) Weiser et al., 1997; (19) Wegensteiner and Weiser, 1996b; (20) Wegensteiner et al., 1996; (21) Laucius and Zolubas, 1997; (22) Haidler et al., 1998; (23) Weiser et al., 1998; (24) Gasperl, 2002; (25) Skuhravy, 2002; (26) Weiser et al., 2002; (27) Zitterer, 2002; (28) Händel et al., 2003; (29) Haidler et al., 2003; (30) Kohlmayr et al., 2003; (31) Wegensteiner and Weiser, 2004; (32) Händel and Wegensteiner, 2005; (33) Takov et al., 2006; (34) Weiser et al., 2006; (35) Holuša et al., 2007; (36) Takov et al., 2007; (37) Wegensteiner et al., 2007b; (38) Burjanadze, 2009; (39) Burjanadze and Goginashvili, 2009; (40) Pernek et al., 2009; (41) Wegensteiner and Weiser, 2009; (42) Takov and Pilarska, 2009; (43) Holuša et al., 2009; (44) Gokturk et al., 2010; (45) Kereselidze et al., 2010; (46) Tonka et al., 2010; (47) Burjanadze et al., 2011; (48) Takov et al., 2011; (49) Lukášová et al., 2012; (50) Michalková et al., 2012; (51) Takov et al., 2012; (52) Holuša et al., 2013; (53) Lukášová et al., 2013; (54) Wegensteiner et al., 2014.

was possible to maintain consistent levels of infection over several generations (22–67.5%) (Wegensteiner and Weiser, 1996b). However, while it can be found in *I. typographus* populations from both managed and unmanaged forests (Kereselidze *et al.*, 2010), infection rates in the field varied considerably between years, between sampling dates within a year, and with elevation (1–80% infection) (Wegensteiner and Weiser, 1996b; Gasperl and Wegensteiner, 2012). In a study by Händel *et al.* (2001) *C. typographi* infections in *I. typographus* were more frequent in the spring, although they did also occur in the fall, and Kereselidze *et al.* (2010) reported that *C. typographi* infections in *I. typographus* only occurred at elevations ≥1100 m.a.s.l. *Chytridiopsis typographi* has also been reported from populations of other *Ips* species such as *I. duplicatus* (Holuša *et al.*, 2009) and *I. amitinus* (Händel and Wegensteiner, 2005); 3–26% of *I. amitinus* were infected on *Pinus cembra* (L.) from three elevations in the Austrian Alps (3–26%) (Händel and Wegensteiner, 2005). From comparisons of *C. typographi* infection rates in *I. typographus* collected in pheromone traps over a 10-year period, it has been suggested that infection hinders flight ability and may interfere with pheromone perception (Wegensteiner *et al.*, 2010).

In contrast to *C. typographi*, other microsporidian species seem more limited in their host range and distribution. *Unikaryon montanum* (Weiser, Wegensteiner and Žižka) infections of *I. typographus* were only found in samples taken in the fall (Händel *et al.*, 2001) and in some studies have only been reported from females, and always at very low levels (Wegensteiner and Weiser, 2004).

Comparisons of the microsporidian species complex of *I. duplicatus* and *I. typographus* showed that *Larssoniella*

duplicati (Weiser, Holuša and Žižka) only infected *I. duplicatus* (Holuša *et al.*, 2009). In an extensive multi-site study by Gasperl and Wegensteiner (2012) *Nosema typographi* (Weiser) and *U. montanum* were each only found in beetles from a single site (perhaps related to elevation) and only in a few beetles. In the Czech Republic, *N. typographi* was only reported from *I. typographus* and not *I. amitinus* (Lukášová *et al.*, 2013).

3.1.6 Protista

In the past, the Kingdom Protista included the phylum Microsporidia but molecular studies have revealed that microsporidia are more closely related to the Kingdom Fungi, even though this placement continues to be debated (Corradi and Keeling, 2009; Solter *et al.*, 2012). For this reason early reviews of protists grouped microsporidia and protozoa together; Fuchs (1915) was the first to describe diseases caused by protists in *I. typographus* and Weiser (1954a, b, 1955, 1961a, 1966, 1977) the first to review them in bark beetles. Sprague (1977a, b) and Levine (1988) included all protist species described from bark beetles in their systematic compendia, and Führer and Purrini (1981) emphasized the importance of more detailed investigations because of their possible, though incalculable, effects on bark beetle regulation.

Entomopathogenic protists are unicellular organisms. They can cause infections in insects, but are generally inapparent or chronic infections that, nonetheless, may play a role in regulating insect populations. The majority of protists enter the insect through the mouth and digestive tract. The infective stage is generally a spore or a cyst. Those that

TABLE 7.10 Protistan Pathogens Isolated from Field-collected Bark Beetles

	Cryphalus	*Dendroctonus*	*Dryocoetes*	*Hylastes*	*Hylurgops*	*Ips*	*Leperesinus*	*Orthotomicus*	*Pityogenes*	*Pityokteines*	*Pityophthorus*	*Polygraphus*	*Tomicus*
Eugregarinida	x		x	x	x	x		x	x	x	x	x	x
Neogregarinida		x	x	x	x	x	x		x	x	x		
References	19	1, 48, 49	2, 11, 12, 44, 45	8, 11, 12	2, 11, 12, 45	2, 5, 6, 7, 8, 9, 10, 11, 12, 13, 14, 15, 16, 17, 18, 19, 20, 21, 23, 24, 25, 26, 27, 30, 31, 32, 33, 34, 36, 37, 38, 39, 40, 41, 42, 43, 46, 47	3	19, 27, 30	4, 8, 11, 12, 14, 19, 27, 33	22	12, 19	19	35

References: (1) Weiser, 1970; (2) Purrini, 1978b; (3) Purrini and Ormieres, 1981; (4) Purrini and Halperin, 1982; (5) Wegensteiner, 1994; (6) Wegensteiner and Weiser, 1996a; (7) Wegensteiner et al., 1996; (8) Haidler et al., 1998; (9) Gasperl, 2002; (10) Zitterer, 2002; (11) Händel et al., 2003; (12) Haidler et al., 2003; (13) Wegensteiner and Weiser, 2004; (14) Händel and Wegensteiner, 2005; (15) Takov et al., 2006; (16) Weiser et al., 2006; (17) Holuša et al., 2007 (18) Holuša et al., 2009; (19) Takov et al., 2007; (20) Burjanadze, 2009; (21) Burjanadze and Goginashvili, 2009; (22) Pernek et al., 2009; (23) Wegensteiner and Weiser, 2009; (24) Gokturk et al., 2010; (25) Kereselidze et al., 2010; (26) Burjanadze et al., 2011; (27) Takov et al., 2011; (28) Lukášová et al., 2012; (29) Michalková et al., 2012; (30) Takov et al., 2012; (31) Holuša et al., 2013; (32) Lukášová et al., 2013; (33) Wegensteiner et al., 2014; (34) Händel et al., 2001; (35) Kohlmayr, 2001; (36) Wegensteiner et al., 2007a; (37) Yaman, 2007; (38) Tonka et al., 2007; (39) Kereselidze and Wegensteiner, 2007; (40) Takov and Pilarska, 2008; (41) Unal et al., 2009; (42) Yaman and Baki, 2010; (43) Gasperl and Wegensteiner, 2012; (44) Purrini, 1977; (45) Purrini, 1980; (46) Žižka et al., 1997; (47) Žižka et al., 1998; (48) Yaman and Radek, 2008a; (49) Yaman and Radek, 2012.

remain in the gut lumen are attached to the midgut epithelium or enter appendages associated with the digestive tract (Lange and Lord, 2012). A number of protists from different phyla have been reported from bark beetles (see section 3.1.6.1 and Table 7.10).

3.1.6.1 Rhizopoda

The first rhizopodan species was identified from a bark beetle (*D. autographus*) in Germany and thought to be a variant of *Malamoeba locustae* (King and Taylor) (Purrini, 1978a, b). It was subsequently described as *Malamoeba scolyti* (Purrini) from the cells of the midgut epithelium and the Malpighian tubules of both *D. autographus* and *H. palliatus* (Purrini, 1980; Kirchhoff and Führer, 1985) and has been widely reported from *I. typographus* in Austria (Wegensteiner, 1994; Wegensteiner et al., 1996; Händel et al., 2001, 2003; Haidler et al., 2003; Wegensteiner and Weiser, 2009).

Purrini and Žižka (1983) described the life cycle of *M. scolyti* based on light and electron microscopic studies. Significant reductions in the longevity of *D. autographus* have been reported in response to infection with *M. scolyti*; at 20°C, time to death was 5–7 weeks post inoculation, compared with 3–4 months in control beetles (Kirchhoff and Führer, 1990). The host range of *M. scolyti* includes a number of different bark beetles that could be experimentally infected in the laboratory (Purrini and Führer, 1979;

Kirchhoff and Führer, 1990; Kirchhoff et al., 2005) and it has been reported from a number of different bark beetle species in the field. For example, in Austria it has been reported from *D. autographus*, *Hylastes cunicularius* (Erichson), *H. glabratus*, *H. palliatus*, *I. acuminatus*, *I. amitinus*, and *I. typographus* (Zitterer 2002; Haidler et al., 2003; Händel et al., 2003; Händel and Wegensteiner, 2005; Wegensteiner and Weiser, 2009); from *Pityophthorus pityographus* (Ratzeburg) in Bulgaria (Takov et al., 2007); and from *I. typographus* in Slovakia (Michalková et al., 2012). However, in a comprehensive survey of *M. scolyti* in *I. typographus* from 22 European countries it was always at a low prevalence and only in beetles from Austria (Wegensteiner and Weiser, 2009). Interestingly, some inhibition in the manifestation of disease can occur if nematodes are co-infesting the gut of *D. autographus* (Kirchhoff and Führer, 1990). With respect to other rhizopodan species, there has only been one passing mention of an *Endamoeba* sp. in *Pityogenes calcaratus* Eichhoff (Purrini and Halperin, 1982).

3.1.6.2 Apicomplexa: Eugregarinida

Fuchs (1915) was the first to describe *Gregarina typographi* (Fuchs) from the midgut lumen of *I. typographus* collected in Germany and it has subsequently been reported in *I. typographus*, *I. sexdentatus*, and *I. acuminatus*, from a number of European countries (Théodoridès, 1960; Wegensteiner *et al.*, 2007a; Yaman, 2007; Takov and Pilarska, 2008; Unal *et al.*, 2009; Gokturk *et al.*, 2010). Geus (1969) has listed several gregarine species from different bark beetle species: *Gregarina pityokteinidis* (Rauchalles) in *Pityokteines curvidens* (Germ.); *Gregarina hylastidis* (Rauchalles) in *H. ater*, *H. cunicularius*, and *H. opacus*; *Gregarina pityogenidis* (Rauchalles) in *Pityogenes bidentatus* (Herbst); *Acanthospora crypturgi* (Geus) in *Crypturgus pusillus* (Gyllenhal); and *G. typographi* in *D. autographus*.

The levels of infection achieved by *G. typographi* in the field vary depending on host species (*I. typographus*, *I. sexdentatus* or *I. acuminatus*) and also host sex; the mechanisms driving this variability are not fully understood. For example, in two Austrian studies where samples were taken from 10 different locations, levels of infection were high in both *I. typographus* and *I. sexdentatus*, though more variable in *I. typographus* (reaching a maximum of 89%); in *I. sexdentatus* infections were predominantly in females (Wegensteiner *et al.*, 2007a, b). In a Bulgarian study higher *G. typographi* infection rates were found in *I. sexdentatus* than in *I. acuminatus*, and in contrast to the previous studies, infections were predominantly in male beetles of both species (Takov and Pilarska, 2008). In Turkey, *G. typographi* infection rates were higher in *I. typographus* (41–49.5%) than in *I. sexdentatus* (17%) (Gokturk *et al.*, 2010).

During a 10-year study in Austria, Wegensteiner *et al.* (2010) found explicit evidence that *G. typographi* infection levels in adult *I. typographus* sampled using pheromone traps were higher than infection levels in the surrounding population sampled conventionally; the authors suggested that infection increased the beetles' motivation to fly and acted as a stimulus for migration. These behavioral changes could be important with respect to the migration capacity of beetles and the spread of disease to distant forests. Lukášová and Holuša (2011) investigated *G. typographi* transmission in *I. typographus* and concluded that transmission of *G. typographi* occurred between beetles in the nuptial chambers.

3.1.6.3 Apicomplexa: Neogregarinida

Fuchs (1915) was the first to report the neogregarine protist *Telosporidium typographi* (Fuchs) in *I. typographus* in Germany, and argued that this protist may be an important mortality factor. Weiser (1955) described the occurrence of another neogregarine, *Menzbieria chalcographi* (Weiser), in *P. chalcographus*. Protozoan entomopathogens belonging to the Apicomplexa (Neogregarinida) have been described from several different bark beetle species and mainly occur in the Malpighian tubules or in the adipose tissue (Table 7.10).

Pityogenes chalcographus from different localities varied in their susceptibility to *M. chalcographi* (0–12% infection); furthermore, they were not susceptible to

infection if nematodes were present in their guts (Purrini and Führer, 1979). Yaman and Radek (2012) reported *M. chalcographi* in *D. micans* from Turkey (0.4–27% infection). A *Mattesia* species has been found in only a few *I. typographus* (4.6%) from Austria (Gasperl and Wegensteiner, 2012). Infection rates for *Mattesia schwenkei* (Purrini) in *I. typographus* were 34% in the Czech Republic (Lukášová *et al.*, 2012).

3.2 Issues Relating to Our Current Understanding of Insect Pathogens in Bark Beetles

Questions concerning the possible side effects on other insects, of augmenting bark beetle pathogens, have been raised and there are specific concerns around negative impacts on arthropod natural enemies and bees. Unfortuntely, there is limited knowledge of the host range of bark beetle pathogens and their infectivity to non-target insects such as predators and parasitoids (Mills, 1983; Furlong and Pell, 2005). Bałazy (1962) found that, in the field, predators and parasitoids exhibited minimal fungal infection compared to bark beetles in the same tree. However, in the laboratory fungi caused high mortality in arthropod natural enemies; the predator *R. grandis* was highly sensitive to *B. bassiana* and *M. anisopliae* (Kulhavy and Miller, 1989). In Turkey, the occurrence of *Helicosporidium* sp. and *Mattesia* sp. in both the bark beetle, *D. micans*, and its predator *R. grandis* demonstrate the potential for cross-infection (Yaman and Radek, 2007; Yaman *et al.*, 2009, 2010b, 2011, 2012). In contrast, a *B. bassiana* isolate virulent against *I. sexdentatus* had limited virulence to the predator *T. formicarius* (Steinwender *et al.*, 2010).

Most research on bark beetle pathogens has been confined to their morphological descriptions using light or electron microscopy, and there are relatively few studies on pathogen epizootiology or the physiological results of infection on hosts, particularly for those pathogens that are less virulent and exist at chronic levels in their hosts' populations. Unlike other insect species, horizontal transmission of pathogens is only likely to be possible between female and male beetles of the parental generation, especially in polygamous species, simply because of the bark beetles' life history. Vertical pathogen transmission from beetles of the parental generation to beetles of the filial generation is conceivable (trans ovum or in the course of contact during maturation feeding). Infection during feeding is unlikely to occur in beetle larvae because most species feed in uncontaminated regions of the tree (phloem), and individuals avoid contact with each other.

Pathogen incidence varies widely between studies and this is likely to be due to sampling period (time of the year), host generation (parental or filial generation), and method of sampling (collecting beetles by hand from logs, collecting beetles from pheromone traps or collecting independently emerging beetles). There are no precise records of the numbers of dead infected beetles in the field to effectively quantify natural mortality levels in relation to local population density; a shortened life expectancy in infected beetles can be assumed, although there is almost no knowledge concerning where the beetles die. Without standardized and complementary sampling methods it is likely that pathogen occurrence in the field is underestimated.

For microbial control to be effective, it is necessary to dose (by feeding or via cuticle contact) target insects with sufficient inoculum to guarantee successful infection and subsequent mortality. The time window for this process, especially for artificial infection (through biological control), is normally very short for bark beetles due to their relatively short swarming periods. Once in the bark, beetles are "cryptic" and protected from inoculation, with the exception of horizontal transmission between breeding partners, vertical transmission, or transmission by vectors. One promising possibility could be the use of pheromone traps as delivery systems; beetles attracted to the traps would be inoculating with inoculum held within the trap and then released to carry the inoculum to the rest of the population. Alternatively, bait logs could be contaminated with inoculum targeted to infect beetles attacking those logs. Fungal pathogens are the most promising candidates for microbial control of bark beetles. A large number of species have been described from bark beetles and several cause rapid mortality in laboratory experiments, demonstrating their potential as efficient control agents.

Evaluating the efficacy of single pathogens is relatively easy in the laboratory, but it is more difficult to estimate the action of pathogens under field conditions; more extensive field trials are necessary. Furthermore, survival of resistant pathogen stages in the environment should not be neglected. Even if the number of pathogens isolated from bark beetles is relatively high, few studies have dealt with their practical exploitation as control agents, or with understanding their underlying biology and ecology, which is necessary to underpin their effective exploitation.

4. NEMATODES

Nematodes are a widespread (Bongers, 1988) and species-rich taxon (Lambshead, 1993). There are many species that can be found in association with bark beetles and their life histories range from purely phoretic to parasitic, with a number of intermediates (see Chapter 6); despite this, our in-depth knowledge of these nematodes is limited. Since the development of DNA sequencing technology the taxonomy and phylogeny of nematodes (which previously relied on morphological characteristics, which are limited) has become clearer and will continue to illuminate their ecology and evolution. It is clear from the phylogeny of nematodes that associations with bark beetles evolved several times independently (Blaxter *et al.*, 1998). Observation of the dynamics of nematodes parasitizing their hosts can seldom be made in nature and depend greatly on the use of laboratory cultures of both bark beetles and nematodes.

4.1 Historical Background

The first observations of nematodes associated with bark beetles were made in the late 18th century when von Linstow (1890) described *Allantonema diplogaster* (current name: *Contortylenchus diplogaster* (von Linstow)) from *I. typographus*. Rühm (1956) described 77 new taxa, examining 59 species of bark beetles from Europe, and explored methods for culture of these nematodes in the laboratory. He also proposed the potential for close co-evolution of beetles and nematodes based on his observation that closely related beetles seem to host closely related nematodes (Rühm, 1965). Massey's (1974) monograph provides an impressive list of 32 parasitic and 112 phoretic nematodes from 51 genera that were isolated from 45 different bark beetle species in North America. Kaya (1984), for the first time, illustrated life cycles of many important taxa and, most recently, Grucmanová and Holuša (2013) published a summary of the nematodes associated with the genus *Ips* in Europe.

Generally, the best-studied nematode systems are from the genera of bark beetles that cause the greatest damage, namely *Ips* and *Dendroctonus* (Thorne, 1935; Massey, 1956, 1957, 1960, 1964a, b, 1966a, b, 1969; Nickle, 1963, 1970; Slankis, 1969; Thong and Webster, 1975, 1983; Hoffard and Coster, 1976; Lieutier and Laumond, 1978; Kinn and Stephen, 1981; Lieutier, 1982, 1984a, b; Takov *et al.*, 2006; Choo *et al.*, 1987; Cardoza *et al.*, 2008; Kereselidze *et al.*, 2010; Michalková *et al.*, 2012). While Africa, Australia, Asia, Central and South America are largely *terra incognita* for bark beetle nematode studies, many studies have been published in Europe and North America. The most important comprehensive contributions, by country, are as follows: Germany: Fuchs (1914a, b, 1915, 1928, 1930, 1938) and Rühm (1956, 1965); France: Lieutier (1982, 1984a,b); Russia: Yatsenkovsky (1924) and Filipjev and Schuurmans Stekhoven (1941); Republic of Georgia: Burjanadze (2009a), Burjanadze and Goginashvili (2009b), and Burjanadze *et al.* (2011, 2012); United States: Massey (1956, 1957, 1960, 1964a, b, 1966a, b, 1969, 1974), Kaya (1984), and Nickle (1963, 1970, 1971).

4.2 General Biology

There are many more phoretic (non-parasitic) and mutualistic nematode species found in association with bark beetles than parasitic taxa; indeed, they often appear together in the same galleries (Rühm, 1956, Massey, 1974, Poinar, 1975). However, only parasitic nematodes can be considered as natural enemies of bark beetles. Phoretic nematodes depend on bark beetles (and other organisms in the galleries) purely for transportation to new habitats where they then feed on tree tissues (Penas et al., 2006; Susoy and Herrmann, 2014) and mutualistic nematodes benefit bark beetles (see Chapter 6). The most important parasitic nematodes belong to the orders Tylenchida (Siddiqui, 2000) and Aphelenchida (Poinar, 1975) (Table 7.11), and can be recognized by their syringe-like stylets. This structure is used to pierce and suck nutrients from the host. Prior to host entry, parasitic nematodes can be found on the host cuticle but, having entered the host, they reside in the Malpighian tubes (e.g., *Cryptaphelenchus* species) or internally (e.g., *Parasitaphelenchus* species).

In the genus *Contortylenchus* (Tylenchidae), adult, mated females are found in bark beetle galleries and actively seek out beetle larvae and pupae. These females can penetrate the beetle directly through the integument, although entry through the oral or anal openings followed by penetration through the gut into the hemocoel is also possible. Once in the hemocoel, female nematodes are predominantly found in the thorax or abdomen. Infected beetle larvae continue their development, pupate, and become adults. The female nematodes also continue to develop. Female nematodes with fully developed eggs in their uterus have been observed during dissections of beetle larvae, although they do not usually begin depositing eggs until the beetle has reached the adult stage. Nematode eggs hatch and develop through to fourth stage juveniles within the adult beetle, migrating to the alimentary canal, where they will eventually be passed into the galleries in frass. There they molt into adults and mate. Males quickly die and mated females seek out beetle larvae and pupae to begin the cycle again (Kaya, 1984). Interestingly, in the Tylenchidae, endoparasitic females within the host have a very different morphology to the females of the same species that are found external to the host (Kaya, 1984). In the genus *Parasitaphelenchus*, third and fourth stage juveniles are the only parasitic stages; the third stage juvenile infects the beetles and develops into the fourth stage juvenile in the hemocoel of larvae, pupae, and adults before exiting the beetle in the frass (Kaya, 1984).

Parasitic nematodes can infest a significant proportion of any bark beetle population and can have negative impacts on fitness. For *Dendroctonus adjunctus* Blandford, 1–91% of the beetles sampled were nematode infested (Takov and Pilarska, 2008); 30 species of nematodes are associated with this bark beetle species (Massey, 1974). The number of individual nematodes inside a single beetle can easily be in the hundreds (Nickle, 1973). It is therefore not surprising that infested beetles show signs of reduced fitness. For example, nematodes will affect the brood size of beetles by directly feeding on the fat bodies or gonads and consequently delaying development. The number of bark beetle generations per year can be reduced (Massey, 1974) and, depending on the host–nematode pairing and the density of nematodes present, can even result in death of the beetle (Yatsenkowsky, 1924; Kaya, 1984). While

TABLE 7.11 Nematode Taxa Commonly Associated with Bark Beetles

Tylenchida		
Nematode family	**Nematode genus**	**Hosts**
Contortylenchidae	*Contortylenchus*	*Ips, Dendroctonus, Cryphalus, Hylastes, Hylurgops, Orthotomicus* (Slankis, 1967)
	Bovienema	*Cryphalus, Ips, Pityogenes, Scolytus, Xyleborus*
Parasitylenchidae	*Sulphuretylenchus* (syn. *Parasitylenchus*)	*Dendroctonus, Dryocoetes, Hylastes, Ips, Pityogenes, Poligraphus, Pseudohylesinus, Scolytus* (Ashraf, 1968, Ashraf and Berryman, 1970, Tomalak et al., 1989)
Aphelenchida		
Aphelenchoididae	*Cryptaphelenchus*	*Cryphalus, Dendroctonus, Dryocoetes, Hylastes, Hylurgops*
Ektaphelenchidae	*Ektaphelenchus*	*Alniphagus, Dendroctonus, Hylastes Dryocoetes, Ips*
Parasitaphelenchidae	*Parasitaphelenchus*	*Ips, Dendroctonus, Scolytus, Polygraphus, Pityogenes, Hylastes, Alniphagus, Tomicus*
	Bursaphelenchus	*Dendroctonus, Ips, Hylesinus, Hylurgops, Scolytus, Pityogenes, Leperesinus, Pityokteines, Orthotomicus, Tomicus*

Parasitylenchus, *Sulphuretylenchus*, and *Neoparasitylenchus* species are capable of killing bark beetles under particular circumstances, studies on *Contortylenchus* species have not demonstrated any lethal effects (Kaya, 1984). The fatal effects of nematodes on bark beetles are often overlooked in nature as dead beetles disappear rapidly, while living, but heavily-infested beetles are easy to spot. The classical entomopathogenic nematodes of the families Steinernematidae and Heterorhabditidae (Nguyen and Hunt, 2007) have been used in experiments to control bark beetles (Finney and Walker, 1977; Gaugler and Kaya, 1990) although they are not found attacking bark beetles in nature.

To screen for parasitic nematodes, beetles should be studied under dissecting microscopes, starting with the cuticle, where invading nematodes can be found. The beetles should then be submerged in physiological saline solution for dissection with fine forceps and, if present, manual extraction of endoparasitic nematodes using fine needles. For microscopic examination, nematodes can be fixed in formalin-acetic acid (FAA). Morphological identification of nematodes can be very difficult; for the non-specialist, identification to genus level might be feasible using the excellent online key by Nguyen (2010). Molecular methods for species identification are improving at a rapid pace and will be the gold standard in the future (e.g., Kanzaki *et al.*, 2012; Ragsdale *et al.*, 2013). By sequencing certain molecular markers even larval stages and eggs could be identified and cryptic species detected (Kanzaki *et al.*, 2012).

5. CONCLUSION

When bark beetles attack trees, a significant number of natural enemies (predators and parasitoids) are attracted to the breeding beetles and developing broods. Undoubtedly, active and directed searching for prey or hosts must be considered an important advantage for predators and parasitoids. Despite considerable knowledge of the action of several of these natural enemies, mass rearing and release of predators or parasites to successfully suppress bark beetle populations has been done in only a very few cases. Under favorable conditions nematodes are also able to move some distance in search of prey, but unfortunately, almost nothing is known of their potential regulatory effects on bark beetles at the population level. Pathogens cannot actively search for their hosts but have other strategies to aid transmission and can also be manipulated within biological control.

The significant number of natural enemy species and the vast abundance in which they are sometimes encountered in infested trees supports the hypothesis that they play a major role in the regulation of bark beetle populations. This has relevance to pest management strategies in forests. For example, depending on timing, the removal of infested trees and logs could actually remove large numbers of natural enemies from the system which, if left *in situ*, could disperse to other bark beetle populations and prohibit their establishment.

In addition to increasing our knowledge of the biology and epidemiology of bark beetle natural enemies, there remains an urgent need to identify the key factors that influence their efficiency in pest population regulation. We also need to understand the interactions between different natural enemies and between the guilds of natural enemies and bark beetles at different population densities. The biology of the pest species, and the responses of the natural enemies, must be taken into account to optimize the beneficial effect of natural enemies within biological control.

ACKNOWLEDGMENTS

We acknowledge our colleagues who provided papers, particularly those that would otherwise have been very difficult to obtain. Furthermore, we thank the reviewers for their valuable comments and suggestions and especially J. K. Pell Consulting for a thorough linguistic revision of the manuscript.

REFERENCES

Adams, J.R., Bonami, J.R., 1991. Atlas of Invertebrate Viruses. CRC Press, Inc., Boca Raton.

Andrei, A.-M., Lupăştean, D., Ciornei, C., Fătu, A.-C., Dinu, M.M., 2013. Laboratory *Beauveria bassiana* (Bals.) Vuill. bioassays on spruce bark beetle (*Ips typographus* L.). IOBC/wprs Bull 90, 93–96.

Arbesleitner, G., 2002. Der insektenpathogene Pilz *Beauveria bassiana*—seine Wirkung auf den Borkenkäfer *Ips typographus* bei der Überwinterung im Boden und Überprüfung der keimungs- und wachstumshemmenden Wirkungen unterschiedlicher Substanzen. Diploma thesis, University of Natural Resources, Vienna.

Arnold, M.K., Barson, G., 1977. Occurrence of virus like particles in midgut epithelial cells of the large elm bark beetle, *Scolytus scolytus*. J. Invertebr. Pathol. 29, 373–381.

Ashraf, M., 1968. Biological studies of *Scolytus ventralis* Le Conte (Coleoptera: Scolytidae) with particular reference to nematode parasite *Sulphuretylenchus elongatus* Massey. Ph.D. thesis, Washington State University, Pullman.

Ashraf, M., Berryman, A.A., 1970. Biology of *Sulphuretylenchus elongatus* (Nematoda: Spaerulariidae) and its effect on its host *Scolytus ventralis* (Coleoptera: Scolytidae). Can. Entomol. 102, 197–213.

Atkinson, T.H., Wilkinson, R.C., 1979. Microsporidian and nematode incidence in live-trapped and reared southern pine beetle adults. Fla. Entomol. 62, 169–175.

Aukema, B.H., Raffa, K.F., 2002. Relative effects of exophytic predation, endophytic predation, and intraspecific competition on a subcortical herbivore: consequences to the reproduction of *Ips pini* and *Thanasimus dubius*. Oecologia 133, 483–491.

Aukema, B.H., Raffa, K.F., 2004. Behavior of adult and larval *Platysoma cylindrica* (Coleoptera: Histeridae) and larval *Medetera bistriata*

(Diptera: Dolichopodidae) during subcortical predation of *Ips pini* (Coleoptera: Scolytidae). J. Insect Behav. 17, 115–128.

Aukema, B.H., Raffa, K.F., 2005. Selective manipulation of predators using pheromones: responses to frontalin and ipsdienol pheromone components of bark beetles in the Great Lakes region. Agric. For. Entomol. 7, 193–200.

Aukema, B.H., Dahlsten, D.L., Raffa, K.F., 2000a. Exploiting behavioral disparities among predators and prey to selectively remove pests: maximizing the ratio of bark beetles to predators removed during semiochemically based trap-out. Environ. Entomol. 29, 651–660.

Aukema, B.H., Dahlsten, D.L., Raffa, K.F., 2000b. Improved population monitoring of bark beetles and predators by incorporating disparate behavioral responses to semiochemicals. Environ. Entomol. 29, 618–629.

Aukema, B.H., Richards, G.R., Krauth, S.J., Raffa, K.F., 2004. Species assemblage arriving at and emerging from trees colonized by *Ips pini* in the Great Lakes region: partitioning by time since colonization, season, and host species. Ann. Entomol. Soc. Am. 97, 117–129.

van Averbeke, A., Grégoire, J.-C., 1995. Establishment and spread of *Rhizophagus grandis* Gyll (Coleoptera: Rhizophagidae) 6 years after release in Forêt domaniale Mézenc (France). Ann. Sci. For. 52, 243–250.

Baier, P., 1991. Zur Biologie des Borkenkäferräubers *Nemosoma elongatum* (L.) (Col.: Ostomidae). Z. Ang. Zool. 78, 421–431.

Bakke, A., Kvamme, T., 1981. Kairomone response in *Thanasimus* predators to pheromone components of *Ips typographus*. J. Chem. Ecol. 7, 305–312.

Bałazy, S., 1962. Observations on appearing of some entomogenous fungi of fungi imperfecti group of forest insects. Polskie Pismo Entomologiczne, 149–164, Ser. B, 3–4.

Bałazy, S., 1963. The fungus *Cephalosporium (Acrostalagmus) lecanii* Zimm., a pathogen of beetle larvae. Acta Societatis Botanicorum Poloniae 32, 69–80.

Bałazy, S., 1965. Entomopathogenous fungi from the order Hyphomycetes damaging forest insects in Poland. Roczniki Wyzszej Szkoly Rolniczej w Poznaniu 27, 21–30.

Bałazy, S., 1966. Living organisms regulating population density of bark beetles in spruce forests, with special reference to entomopathogenic fungi I. Poznanskie Towarzystwo Przyjaciol Nauk Wydzial Nauk Rolniczych I Lesnych 21, 3–50.

Bałazy, S., 1967. Mortality of mature beetles *Ips typographus* (L.) (Col., Scolytidae) in the galleries and its causes. Polskie Pismo Entomologiczne 37, 201–205.

Bałazy, S., 1968. Analysis of bark beetle mortality in spruce forests in Poland. Ekologia Polska-Ser. A 16, 657–687.

Bałazy, S., 1973. A review of entomopathogenic species of the genus *Cephalosporium* Corda (Mycota, Hyphomycetales). Bulletin de la societe des Amis des Sciences et des Lettres de Poznan, Ser. D 14, 101–137.

Bałazy, S., Bargielski, J., Ziolkowski, G., Czerwinska, C., 1967. Mortality of mature beetles *Ips typographus* (L.) (Col., Scolytidae) in the galleries and its causes. Polskie Pismo Entomologiczne 37, 201–205.

Bałazy, S., Michalski, J., Ratajczak, E., 1987. Materials to the knowledge of natural enemies of *Ips acuminatus* Gyll. (Coleoptera; Scolytidae). Polskie Pismo Entomologiczne 57, 735–745.

Baldwin, P.H., 1968. Predator-prey relationships of birds and spruce beetles. Proc. North Central Branch E.S.A. 23, 90–99.

Ball, J.C., Dahlsten, D.L., 1973. Hymenopterous parasites of *Ips paraconfusus* (Coleoptera: Scolytidae) larvae and their contribution to mortality I. Influence of host tree and tree diameter on parasitization. Can. Entomol. 105, 1453–1464.

Barson, G., 1976. Laboratory studies on the fungus *Verticillium lecanii*, a larval pathogen of the large elm bark beetle (*Scolytus scolytus*). Ann. Appl. Biol. 83, 207–214.

Barson, G., 1977. Laboratory evaluation of *Beauveria bassiana* as a pathogen of the larval stage of the large elm bark beetle, *Scolytus scolytus*. J. Invertebr. Pathol. 29, 361–366.

Bathon, H., 1991. Möglichkeiten der biologischen Bekämpfung von Borkenkäfern. Mitteilungen aus der Biologischen Bundesanstalt für Land- und Forstwirtschaft. Berlin-Dahlem 267, 111–117.

Batta, Y.A., 2007. Biocontrol of almond bark beetle (*Scolytus amygdali* Geurin-Meneville, Coleoptera: Scolytidae) using *Beauveria bassiana* (Bals.) Vuill. (Deuteromycotina: Hyphomycetes). J. Appl. Microbiol. 103, 1406–1414.

Beaver, R.A., 1966a. The biology and immature stages of two species of *Medetera* (Diptera: Dolichopodidae) associated with the bark beetle *Scolytus scolytus* (F.). Proc. Roy. Entomol. Soc. London (A) 41, 145–154.

Beaver, R.A., 1966b. The biology and immature stages of *Entedon leucogramma* (Ratzeburg) (Hymenoptera: Eulophidae), a parasite of bark beetles. Proc. Roy. Entomol. Soc. London (A) 41, 37–41.

Beaver, R.A., 1967. The regulation of population density in the bark beetle *Scolytus scolytus* (F.). J. Anim. Ecol. 36, 435–451.

Berisford, C.W., Kulman, H.M., Pienkowski, R.L., 1970. Notes on the biologies of hymenopterous parasites of *Ips* spp. bark beetles in Virginia. Can. Entomol. 102, 484–490.

Berryman, A.A., 1966. Studies on the behavior and development of *Enoclerus lecontei* (Wolcott), a predator of the western pine beetle. Can. Entomol. 98, 519–526.

Billings, R.F., Cameron, R.S., 1984. Kairomonal responses of Coleoptera, *Monochamus titillator* (Cerambycidae), *Thanasimus dubius* (Cleridae), and *Temnochila virescens* (Trogositidae), to behavioral chemicals of southern pine bark beetles (Coleoptera: Scolytidae). Environ. Entomol. 13, 1542–1548.

BioLib, 2014. Biological library. Available online: http://www.biolib.cz, Last accessed: April 14, 2014.

Blaxter, M.L., De Ley, P., Garey, J.R., Liu, L.X., Scheldeman, P., Vierstraete, A., et al., 1998. A molecular evolutionary framework for the phylum Nematoda. Nature 392, 71–75.

Bongers, T., 1988. De Nematoden Van Nederland. Koninklijke Nederlandse Naturhistorische Vereiniging, Utrecht, The Netherlands.

Boone, C.K., Six, D.L., Raffa, K.F., 2008a. The enemy of my enemy is still my enemy: competitors add to predator load of a tree-killing bark beetle. Agric. For. Entomol. 10, 411–421.

Boone, C.K., Six, D.L., Zheng, Y.B., Raffa, K.F., 2008b. Parasitoids and dipteran predators exploit volatiles from microbial symbionts to locate bark beetles. Environ. Entomol. 37, 150–161.

Boone, C.K., Six, D.L., Krauth, S.J., Raffa, K.F., 2009. Assemblage of Hymenoptera arriving at logs colonized by *Ips pini* (Coleoptera: Curculionidae: Scolytinae) and its microbial symbionts in western Montana. Can. Entomol. 141, 172–199.

Bridges, J.R., 1987. Prevalence of *Unikaryon minutum* (Microsporidia: Nosematidae) infection in outbreak populations of the southern pine beetle (Coleoptera: Scolytidae). J. Invertebr. Pathol. 49, 334–335.

Bridges, J.R., Marler, J.E., McSparrin, B.H., 1984. A quantitative study of the yeasts and bacteria associated with laboratory-reared *Dendroctonus frontalis* Zimm. (Coleopt., Scolytidae). Z. Ang. Entomol. 97, 261–267.

Bright, D.E., 1996. Notes on native parasitoids and predators of the larger pine shoot beetle, *Tomicus piniperda* (Linnaeus) in the Niagara region of Canada (Coleoptera: Scolytidae). Proc. Entomol. Soc. Ontario 127, 57–62.

Brownbridge, M., Reay, S.D., Cummings, N.J., 2010. Association of entomopathogenic fungi with exotic bark beetles in New Zealand pine plantations. Mycopathologia 169, 75–80.

Brownbridge, M., Reay, S.D., Nelson, T.L., Glare, T.R., 2012. Persistence of *Beauveria bassiana* (Ascomycota: Hypocreales) as an endophyte following inoculation of radiata pine seed and seedlings. Biol. Control 61, 194–200.

Burjanadze, M., 2009. Pathogen and nematode occurrence in the spruce bark beetle *Ips typographus* (Col., Scolytidae) in two different region of Georgia. IOBC/wprs Bull 45, 505–508.

Burjanadze, M., 2010. Efficacy of *Beauveria bassiana* isolate against pine shoot beetle *Tomicus piniperda* L. (Coleoptera, Scolytidae) in laboratory. Bull. Georgian Natl. Acad. Sci. 4, 119–122.

Burjanadze, M., Goginashvili, N., 2009. Occurrence of pathogens and nematodes in the spruce bark beetle, *Ips typographus* (Col., Scolytidae) in Borjomi Gorge. Bull. Georgian Natl. Acad. Sci. 3, 145–150.

Burjanadze, M., Kereselidze, M., 2003. The role of entomopathogene microorganisms in number regulation of *Ips typographus* L. V. Gulisashvili Institute of Mountain Forestry Proceedings 39, 238–241.

Burjanadze, M., Moser, J.C., Zimmermann, G., Kleespies, R.G., 2008. Antagonists of the spruce bark beetle *Ips typographus* L. (Coleoptera: Scolytidae) of German and Georgian populations. IOBC/wprs Bull 31, 245–250.

Burjanadze, M., Lortkipanidze, M., Supatashvili, A., Gorgadze, O., 2011. Occurrence of pathogens and nematodes of bark beetles (Coleoptera, Scolytidae) from coniferous forests in different region of Georgia. IOBC/wprs Bull. 66, 351–354.

Burjanadze, M., Lortkipanidze, M., Gorgadze, O., 2012. Influence of ecological factors on the formation of nematode fauna of bark beetles (Coleoptera Scolytidae). Bull. Georgian Natl. Acad. Sci 6, 133–136.

Bushing, R.W., 1965. A synoptic list of the parasites of Scolytidae (Coleoptera) in North America north of Mexico. Can. Entomol. 97, 449–492.

Bütler, R., Angelstam, P., Ekelund, P., Schlaepfer, R., 2004. Dead wood threshold values for the three-toed woodpecker presence in boreal and sub-Alpine forest. Biol. Conserv. 119, 305–318.

Bychawska, S., Świeżyńska, H., 1979. Attempts of *Myelophilus piniperda* L. control with the use of entomopathogenous fungus *Beauveria bassiana* (Bals.). Vuill. Sylwan 12, 59–64.

Byers, J.A., 1992. Attraction of bark beetles, *Tomicus piniperda*, *Hylurgops palliatus*, and *Trypodendron domesticum* and other insects to short-chain alcohols and monoterpenes. J. Chem. Ecol. 18, 2385–2402.

Callaham, R.Z., Shifrine, M., 1960. The yeasts associated with bark beetles. Forest Sci. 6, 146–154.

Cane, J.H., Cox, H.E., Moar, W.J., 1995. Susceptibility of *Ips calligraphus* (Germar) and *Dendroctonus frontalis* (Zimmermann) (Coleoptera: Scolytidae) to coleopteran-active *Bacillus thuringiensis*, a *Bacillus* metabolite, and Avermectin B1. Can. Entomol. 127, 831–837.

Canganella, F., Paparatti, B., Natali, V., 1994. Microbial species isolated from the bark beetle *Anisandrus dispar* F. Microbiol. Res. 149, 123–128.

Capek, M., 1957. Beitrag zur Kenntnis der Entomophagen von *Pityokteines vorontzovi* Jac. und anderen Tannenborkenkäfern. Z. Ang. Entomol. 41, 277–284.

Cardoza, Y.J., Moser, J.C., Klepzig, K.D., Raffa, K.F., 2008. Multipartite symbioses among fungi, mite, nematodes and the spruce beetle, *Dendroctonus rufipennis*. Environ. Entomol. 37, 956–963.

Castrillo, L.A., Griggs, M.H., Ranger, C.M., Reding, M.E., Vandenberg, J.D., 2011. Virulence of commercial strains of *Beauveria bassiana* and *Metarhizium brunneum* (Ascomycota: Hypocreales) against adult *Xylosandrus germanus* (Coleoptera: Curculionidae) and impact on brood. Biol. Control 58, 121–126.

Castrillo, L.A., Griggs, M.H., Vandenberg, J.D., 2013. Granulate ambrosia beetle, *Xylosandrus crassiusculus* (Coleoptera: Curculionidae), survival and brood production following exposure to entomopathogenic and mycoparasitic fungi. Biol. Control 67, 220–226.

Catalogue of Life, 2013. Available online: http://www.catalogueoflife.org, Last accessed: April 14, 2014.

Chandler, P.J., 1991. Attraction of *Palloptera usta* Meigen (Diptera: Pallopteridae) to recently cut conifer wood and other notes on Pallopteridae. Br. J. Entomol. Nat. Hist. 4, 85–86.

Chansler, J.F., 1967. Biology and life history of *Dendroctonus adjunctus* (Coleoptera: Scolytidae). Ann. Entomol. Soc. Am. 60, 760–767.

Choo, H.Y., Kaya, H.K., Shea, P., Noffsinger, E.M., 1987. Ecological study of nematode parasitism in *Ips* beetles from California and Idaho. J. Nematol. 19, 495–502.

Corradi, N., Keeling, P.J., 2009. Microsporidia: a journey through radical taxonomical revisions. Fungal Biology Reviews 23, 1–8.

Dahlsten, D.L., 1970. Parasites, predators, and associated organisms reared from western pine beetle infested bark samples. In: Stark, R.W., Dahlsten, D.L. (Eds.), Studies on the Population Dynamics of the Western Pine Beetle, *Dendroctonus brevicomis* LeConte (Coleoptera: Scolytidae). Univ. California Press, Berkeley, pp. 75–79.

Dahlsten, D.L., 1982. Relationships between bark beetles and their natural enemies. In: Mitton, J.B., Sturgeon, K.B. (Eds.), Bark beetles in North American Conifers. University of Texas Press, Austin, pp. 140–182.

Dahlsten, D.L., Stephen, F.M., 1974. Natural enemies and insect associates of the mountain pine beetle, *Dendroctonus ponderosae* (Coleoptera: Scolytidae), in sugar pine. Can. Entomol. 106, 1211–1217.

Dahlsten, D.L., Six, D.L., Erbilgin, N., Raffa, K.F., Lawson, A.B., Rowney, D.L., 2003. Attraction of *Ips pini* (Coleoptera: Scolytidae) and its predators to various enantiomeric ratios of ipsdienol and lanierone in California: implications for the augmentation and conservation of natural enemies. Environ. Entomol. 32, 1115–1122.

De Leon, D., 1934. An annotated list of the parasites, predators, and other associated fauna of the mountain pine beetle in western white pine and lodgepole pine. Can. Entomol. 66, 51–61.

De Leon, D., 1935. The biology of *Coeloides dendroctoni* Cushman (Hymenoptera-Braconidae) an important parasite of the mountain pine beetle (*Dendroctonus monticolae* Hopk.). Ann. Entomol. Soc. Am. 28, 411–424.

DeMars, C.J., Berryman, A.A., Dahlsten, D.L., Otvos, I.S., Stark, R.W., 1970. Mortality factors and their interactions. In: Stark, R.W., Dahlsten, D.L. (Eds.), Studies on the Population Dynamics of the Western Pine Beetle, *Dendroctonus brevicomis* LeConte (Coleoptera: Scolytidae). Univ. California Press, Berkeley, pp. 80–101.

Dennis, D.S., 1979. Ethology of *Holcocephala fusca* in Virginia (Diptera: Asilidae). Proc. Entomol. Soc. Wash. 81, 366–378.

Dinu, M.M., Lupăştean, D., Cardaş, G., Andrei, A.-M., 2012. New *Beauveria bassiana* (Bals.) Vuill. isolate from *Ips duplicatus* (Sahlberg). Romanian Journal of Plant Protection 5, 12–15.

Dippel, C., 1995. Zur Bionomie des Borkenkäferantagonisten *Nemosoma elongatum* L. (Col., Ostomidae). Mitt. Dtsch. Ges. Allg. Angew. Entomol. 10, 67–70.

Dippel, C., 1996. Investigations on the life history of *Nemosoma elongatum* L. (Col., Ostomidae), a bark beetle predator. J. Appl. Entomol. 120, 391–395.

Dippel, C., Heidger, C., Nicolai, V., Simon, M., 1997. The influence of four different predators on bark beetles in European forest ecosystems (Coleoptera: Scolytidae). Entomol. Gener. 21, 161–175.

Doane, C.C., 1959. *Beauveria bassiana* as a pathogen of *Scolytus multistriatus*. Ann. Entomol. Soc. Am. 52, 109–111.

Doane, C.C., 1960. Bacterial pathogens of *Scolytus multistriatus* Marsham as related to crowding. J. Insect Pathol. 2, 24–29.

Doberski, J.W., 1981a. Comparative laboratory studies on three fungal pathogens of the elm bark beetle, *Scolytus scolytus*: Pathogenicity of *Beauveria bassiana*, *Metarhizium anisopliae*, and *Paecilomyces farinosus* to larvae and adults of *S. scolytus*. J. Invertebr. Pathol. 37, 188–194.

Doberski, J.W., 1981b. Comparative laboratory studies on three fungal pathogens of the elm bark beetle, *Scolytus scolytus*: effects of temperature and humidity on infection by *Beauveria bassiana*, *Metarhizium anisopliae*, and *Paecilomyces farinosus*. J. Invertebr. Pathol. 37, 195–200.

Dowd, P.F., Vega, F., 2003. Autodissemination of *Beauveria bassiana* by sap beetles (Coleoptera: Nitidulidae) to overwintering sites. Biocontrol Sci. Tech. 3, 65–75.

Draganova, S., Takov, D., Doychev, D., 2007. Bioassays with isolates of *Beauveria bassiana* (Bals.) Vuill. and *Paecilomyces farinosus* (Holm.) Brown & Smith against *Ips sexdentatus* Boerner and *Ips acuminatus* Gyll. (Coleoptera: Scolytidae). Plant Sci. 44, 24–28.

Draganova, S.A., Takov, D.I., Doychev, D.D., 2010. Naturally occurring entomopathogenic fungi on three bark beetle species (Coleoptera: Curculionidae) in Bulgaria. Pestic. Phytomed. (Belgrade) 25, 59–63.

Dromph, K.M., 2003. Collembolans as vectors of entomopathogenic fungi. Pedobiologia 47, 245–256.

Eck, R., 1990. Die parasitischen Hymenopteren des *Ips typographus* in der Phase der Progradation; Artenspektrum und Parasitierungsraten in einigen Waldgebieten der ehemaligen DDR. Entomol. Abh. Mus. Tierkd. Dresden 53, 151–178.

Edworthy, A.B., Drever, M.C., Martin, K., 2011. Woodpeckers increase in abundance but maintain fecundity in response to an outbreak of mountain pine bark beetles. For. Ecol. Manage. 261, 203–210.

EOL, 2014. Encyclopedia of life (EOL). Available online: http://eol.org, Last accessed: April 14, 2014.

Erbilgin, N., Raffa, K.F., 2001. Modulation of predator attraction to pheromones of two prey species by stereochemistry of plant volatiles. Oecologia 127, 444–453.

EUNIS, 2014. European Nature Information System (EUNIS). Available online: http://eunis.eea.europa.eu, Last accessed: April 14, 2014.

Evans, H.C., 1989. Mycopathogens of insects of epigeal and aerial habitats. In: Wilding, N., Collins, N.M., Hammond, P.M., Webber, J.F. (Eds.), Insect-Fungus Interactions. Academic Press, London, pp. 205–238.

Faccoli, M., 2000. Osservazioni bio-ecologiche relative a *Tomicobia seitneri* (Ruschka) (Hymenoptera Pteromalidae), un parassitoide di *Ips typographus* (L.) (Coleoptera Scolytidae). Frustula Entomol 23, 47–55.

Faccoli, M., 2001. Catture di coleotteri "non-target" mediante alberi esca allestiti contro *Ips typographus* (L.) (Coleoptera Scolytidae). Redia 84, 105–118.

Fan, M., Kuo, C., Lu, X., 1987. Tentative study on entomopoxvirus of *Dendroctonus armandi* Tsai et Li. Journal Disinsectional Microorganism 1, 140–141.

Fayt, P., Machmer, M.M., Steeger, C., 2005. Regulation of spruce bark beetles by woodpeckers—a literature review. For. Ecol. Manage. 206, 1–14.

Feicht, E., 2004. Parasitoids of *Ips typographus* (Col., Scolytidae), their frequency and composition in uncontrolled and controlled infested spruce forest in Bavaria. J. Pest Sci. 77, 165–172.

Fernández, M., Diez, J., Moraza, M.L., 2013. Acarofauna associated with *Ips sexdentatus* in northwest Spain. Scand. J. For. Res. 28, 358–362.

Filipjev, I.N., Schuurmans Stekhoven, J.H., 1941. A Manual of Agricultural Helminthology. E. J. Brill, Leiden.

Finney, J.R., Walker, C., 1977. The DD-136 strain of *Neoaplectana* sp. as a potential biological control agent for the European elm bark beetle, *Scolytus scolytus*. J. Invertebr. Pathol. 29, 7–9.

Frazier, J.L., Nebeker, T.E., Mizell, R.F., Calvert, W.H., 1981. Predatory behavior of the clerid beetle *Thanasimus dubius* (Coleoptera: Cleridae) on the southern pine beetle (Coleoptera: Scolytidae). Can. Entomol. 113, 35–43.

French, J.R.J., Robinson, P.J., Minko, G., Pahl, P.J., 1984. Response of the European elm bark beetle, *Scolytus multistriatus*, to host bacterial isolates. J. Chem. Ecol. 10, 1133–1149.

Fuchs, G., 1914a. *Tylenchus dispar curvidentis* m. und *Tylenchus dispar cryphali* m. Zool. Anz. 45, 195–207.

Fuchs, G., 1914b. Über Parasiten und andere biologisch an die Borkenkäfer gebundene Nematoden. *85*. Verhandlung der Gesellschaft Deutscher Naturforscher und Ärzte 2, 688–692.

Fuchs, G., 1915. Die Naturgeschichte der Nematoden und einiger anderer Parasiten 1. des *Ips typographus* L. 2. des *Hylobius abietis* L. *Zool. Jahrb.* Abt. Syst. 38, 109–222.

Fuchs, G., 1930. Neue an Borken- und Rüsselkäfer gebundene Nematoden, halbparasitische und Wohnungseinmieter. Zoologische Jahrbücher, Abteilung fur Systematik 59, 505–646.

Fuchs, G., 1938. Neue Parasiten und Halbparasiten bei Borkenkäfern und einige andere Nematoden. Zoologische Jahrbücher, Abteilung fur Systematik 71, 123–190.

Führer, E., Purrini, K., 1981. Protozoan parasites of bark beetles—a vacancy in research work as to population dynamics. Proceedings, XVII IUFRO World Congress, Ibaraki, Japan, pp. 501–511.

Furlong, M.J., Pell, J.K., 2005. Interactions between entomopathogenic fungi and arthropod natural enemies. In: Vega, F.E., Blackwell, M. (Eds.), Insect-Fungal Associations. Ecology and Evolution. Oxford University Press, New York, pp. 51–73.

Fuxa, J.R., Ayyappath, R., Goyer, R.A., 1998. Pathogens and Microbial Control of North American Forest Insect Pests. USDA, Forest Health Technology Enterprise Team, Morgantown.

Gäbler, H., 1947. Milbe als Eiparasit des Buchdruckers. Nachrichtenbl. Deut. Pflanzenschutzd. 1, 113–115.

Gargiullo, P.M., Berisford, C.W., 1981. Effects of host density and bark thickness on the densities of parasites of the southern pine beetle. Environ. Entomol. 10, 392–399.

Gasperl, H., 2002. Grundlegende Erhebungen zum Borkenkäferauftreten an Fichten-Fangbäumen und zum Pathogenauftreten in Borkenkäfern aus dem geplanten Nationalpark Gesäuse. Diploma thesis, University of Natural Resources, Vienna.

Gasperl, H., Wegensteiner, R., 2012. Untersuchungen zum höhenabhängigen Auftreten von Borkenkäfern und von Pathogenen in *Ips typographus* (L. 1758) (Coleoptera, Curculionidae) im Bereich des Nationalparks Gesäuse (Steiermark). Mitt. Dtsch. Ges. Allg. Angew. Entomol. 18, 413–417.

Gaugler, R., Kaya, H.K., 1990. Entomopathogenic Nematodes in Biological Control. CRC Press, Boca Raton.

Gauss, R., 1954. Der Ameisenbuntkäfer *Thanasimus (Clerus) formicarius* Latr. als Borkenkäferfeind. In: Wellenstein, G. (Ed.), Die grosse Borkenkäferkalamität in Südwestdeutschland 1944–1951. Forstschutzstelle Südwest, Ringingen, pp. 417–429.

Gerson, U., Smiley, R.L., Ochoa, R., 2003. Mites (Acari) for Pest Control. Blackwell Science, Oxford.

Geus, A., 1969. Sporentierchen, Sporozoa. Die Gregarinida der land- und süßwasserbewohnenden Arthropoden Mitteleuropas. Die Tierwelt Deutschlands und der angrenzenden Meeresteile nach ihrer Lebensweise 57, 1–608.

Gibb, H., Hilszczanski, J., Hjältén, J., Danell, K., Ball, J.P., Pettersson, R.B., Alinvi, O., 2008. Responses of parasitoids to saproxylic hosts and habitat: a multi-scale study using experimental logs. Oecologia 155, 63–74.

Glare, T.R., Reay, S.D., Nelson, T.L., Moore, R., 2008. *Beauveria caledonica* is a naturally occurring pathogen of forest beetles. Mycol. Res. 112, 352–360.

Gokturk, T., Burjanadze, M., Aksu, Y., Supatashvili, A., 2010. Nature enemies—predators, pathogens and parasitic nematodes of bark beetles in Hatila Valley National Park of Turkey. *Proc. Georgian Acad. Sci., Biol. Ser.* Bull. Am. Meteorol. Soc. 8, 59–71.

Gösswald, K., 1938. Über den insektentötenden Pilz *Beauveria bassiana* (Bals.) Vuill. bisher Bekanntes und eigene Versuche. Arbeiten aus der Biologischen Reichsanstalt für Land- und Forstwirtschaft 22, 399–452.

Goyer, R.A., Smith, M.T., 1981. The feeding potential of *Corticeus glaber* and *Corticeus parallelus* (Coleoptera: Tenebrionidae), facultative predators of the southern pine beetle, *Dendroctonus frontalis* (Coleoptera: Scolytidae). Can. Entomol. 113, 807–811.

Grégoire, J.-C., Merlin, J., Pasteels, J.M., Jaffuel, R., Vouland, G., Schvester, D., 1985. Biocontrol of *Dendroctonus micans* by *Rhizophagus grandis* Gyll. (Col., Rhizophagidae) in the Massif Central (France). Z. Ang. Entomol. 99, 182–190.

Grégoire, J.-C., Baisier, M., Drumont, A., Dahlsten, D.L., Meyer, H., Francke, W., 1991. Volatile compounds in the larval frass of *Dendroctonus valens* and *Dendroctonus micans* (Coleoptera: Scolytidae) in relation to oviposition by the predator, *Rhizophagus grandis* (Coleoptera: Rhizophagidae). J. Chem. Ecol. 17, 2003–2019.

Grégoire, J.-C., Couillien, D., Drumont, A., Meyer, H., Francke, W., 1992. Semiochemicals and the management of *Rhizophagus grandis* Gyll. (Col., Rhizophagidae) for the biocontrol of *Dendroctonus micans* Kug. (Col., Scolytidae). J. Appl. Entomol. 114, 110–112.

Grucmanová, S., Holuša, J., 2013. Nematodes associated with bark beetles, with focus on the genus *Ips* (Coleoptera: Scolytinae) in Central Europe. Acta Zool. Bulg. 65, 547–556.

Gusteleva, L.A., 1980. *Beauveria bassiana* (Bals) Vuill preparation effect on *Ips subelongatus* Motsch. Leatsch results of the test. Izvestiya Sibirskogo Otdeleniya Akademii Nauk Sssr Seriya Biologicheskikh Nauk 1980, 49–54.

Gusteleva, L.A., 1982a. Prospects for using microbial preparations against *Ips subelongatus*. Lesnoe Khozyaistvo 9, 67.

Gusteleva, L.A., 1982b. The interaction of wood decomposing insects with microorganisms. In: Isaev, A.S. (Ed.), Konsortivnye svyazi dereva i dendrofil nykh nasekomykh. Novosibirsk, USSR, pp. 56–67.

Gusteleva, L.A., 1984. Virulence of *Beauveria bassiana* (Bals.) Vuill to the larvae of the bark beetle *Ips subelongatus* Motsch. (Coleoptera, Scolytidae). Revue d'Entomologie de l'URSS 63, 40–42.

Haberkern, K.E., Raffa, K.F., 2003. Phloeophagous and predaceous insects responding to synthetic pheromones of bark beetles inhabiting white spruce stands in the great lakes region. J. Chem. Ecol. 29, 1651–1663.

Haidler, B., Wegensteiner, R., Sänger, K., Weiser, J., 1998. Pathogen occurrence in bark beetles (Col., Scolytidae) living associated on Norway spruce (*Picea abies*). IOBC/wprs Bull 21 (4), 263–264.

Haidler, B., Wegensteiner, R., Weiser, J., 2003. Occurrence of microsporidia and other pathogens in associated living spruce bark beetles (Coleoptera: Scolytidae) in an Austrian forest. IOBC/wprs Bull. 26, 257–260.

Hajek, A.E., 2004. Natural Enemies. An Introduction to Biological Control. Cambridge University Press, New York.

Hallet, S., Grégoire, J.-C., Coremans-Pelseneer, J., 1994. Prospects in the use of the entomopathogenous fungus *Beauveria bassiana* (Bals.) Vuill. (Deuteromycetes: Hyphomycetes) to control the spruce bark beetle *Ips typographus* L. (Coleoptera: Scolytidae). Mededelingen Faculteit Landbouwkundige en Toegepaste Biologische Wetenschappen Universiteit Gent 59, 379–383.

Händel, U., 2001. Untersuchungen zum Gegenspielerkomplex assoziiert lebender Fichtenborkenkäfer (Col., Scolytidae) aus naturnahen und sekundären Fichtenbeständen unter besonderer Berücksichtigung der Pathogene. PhD thesis, University of Natural Resources, Vienna.

Händel, U., Wegensteiner, R., 2005. Occurrence of pathogens in bark beetles (Coleoptera, Scolytidae) from Alpine pine (*Pinus cembra* L.). IOBC/wprs Bull 28, 155–158.

Händel, U., Kenis, M., Wegensteiner, R., 2001. Untersuchungen zum Vorkommen von Pathogenen und Parasiten in Populationen überwinternder Fichtenborkenkäfer (Col., Scolytidae). Mitt. Dtsch. Ges. Allg. Angew. Entomol. 13, 423–428.

Händel, U., Wegensteiner, R., Weiser, J., Žižka, Z., 2003. Occurrence of pathogens in associated living bark beetles (Col., Scolytidae) from different spruce stands in Austria. J. Pest Sci. 76, 22–32.

Hanson, H.S., 1937. Notes on the ecology and control of pine beetles in Great Britain. Bull. Entomol. Res. 28, 185–236.

Harrison, R., Hoover, K., 2012. Baculovirus and other occluded insect viruses. In: Vega, F.E., Kaya, H.K. (Eds.), Insect Pathology, Second Edition. Academic Press, San Diego, pp. 73–131.

Hedgren, P.O., Schroeder, L.M., 2004. Reproductive success of the spruce bark beetle *Ips typographus* (L.) and occurrence of associated species: a comparison between standing beetle-killed trees and cut trees. For. Ecol. Manage. 203, 241–250.

Hedqvist, K.J., 1963. Die Feinde der Borkenkäfer in Schweden I. Erzwespen (Chalcidoidea). Stud. For. Suec. 11, 1–176.

Hedqvist, K.J., 1998. Bark beetle enemies in Sweden II. Braconidae (Hymenoptera). Entomol. Scand Suppl. 52, 3–87.

Heidger, C.-M., 1994. Die Ökologie und Bionomie der Borkenkäfer-Antagonisten *Thanasimus formicarius* L. (Cleridae) und *Scoloposcelis pulchella* Zett. (Anthocoridae): Daten zur Beurteilung ihrer prädatorischen Kapazität und der Effekte beim Fang mit Pheromonfallen. Dissertation, Philipps-Universität Marburg, Germany.

Hein, C., 1995. Versuche zu biologischen Bekämpfung des Buchdruckers (*Ips typographus*) mit entomophagen Pilzen. Diploma thesis,

Fachhochschule Hildesheim/Holzminden, Fachbereich Forstwirtschaft, Göttingen.

Hérard, F., Mercadier, G., 1996. Natural enemies of *Tomicus piniperda* and *Ips acuminatus* (Col., Scolytidae) on *Pinus sylvestris* near Orléans, France: temporal occurrence and relative abundance, and notes on eight predatory species. Entomophaga 41, 183–210.

Herrmann, F., Wegensteiner, R., 2010. Infecting *Ips typographus* (Coleoptera, Curculionidae) with *Beauveria bassiana*, *Metarhizium anisopliae* or *Isaria fumosorosea* (Ascomycota). IOBC/wprs Bull. 66, 209–212.

Hilszczanski, J., Gibb, H., Bystrowski, C., 2007. Insect natural enemies of *Ips typographus* (L.) (Coleoptera, Scolytinae) in managed and unmanaged stands of mixed lowland forest in Poland. J. Pest Sci. 80, 99–107.

Hirschmann, W., Wisniewski, J., 1983. Gangsystematik der Parasitiformes. Teil 30. Lebensräume der *Dendrolaelaps*- und *Longoseius*-Arten. Acarologie 30, 21–33.

Hoffard, W.H., Coster, J.E., 1976. Endoparasitic nematodes of *Ips* Bark beetles in eastern Texas. Environ. Entomol. 5, 128–132.

Hofstetter, R.W., Moser, J.C., McGuire, R., 2009. Observations on the mite *Schizosthetus lyriformis* (Acari: Parasitidae) preying on bark beetle eggs and larvae. Entomol. News 120, 397–400.

Hofstetter, R.W., Gaylord, M.L., Martinson, S., Wagner, M.R., 2012. Attraction to monoterpenes and beetle-produced compounds by syntopic *Ips* and *Dendroctonus* bark beetles and their predators. Agric. For. Entomol. 14, 207–215.

Hofstetter, R.W., Moser, J.C., Blomquist, S.R., 2013. Mites associated with bark beetles and their hyperphoretic ophiostomatoid fungi. In: Seifert, K.A., Wilhelm de Beer, Z., Wingfield, M.J. (Eds.), The Ophiostomatoid Fungi: Expanding Frontiers. CBS-KNAW Fungal Biodiversity Centre, Utrecht, pp. 165–176.

Holuša, J., Weiser, J., Drapela, K., 2007. Pathogens of *Ips duplicatus* (Coleoptera: Scolytidae) in three areas in Central Europe. Acta Protozool. 46, 157–167.

Holuša, J., Weiser, J., Žižka, Z., 2009. Pathogens of the spruce bark beetles *Ips typographus* and *Ips duplicatus*. Central Eur. J. Biol. 4, 567–573.

Holuša, J., Lukášová, K., Wegensteiner, R., Grodzki, W., Pernek, M., Weiser, J., 2013. Pathogens of the bark beetle *Ips cembrae*: microsporidia and gregarines also known from other *Ips* species. J. Appl. Entomol. 137, 181–187.

Hopping, G.R., 1947. Notes on the seasonal development of *Medetera aldrichii* Wheeler (Diptera, Dolichopodidae) as a predator of the Douglas fir bark- beetle, *Dendroctonus pseudotsugae* Hopkins. Can. Entomol. 79, 150–153.

Hougardy, E., Grégoire, J.-C., 2000. Spruce stands provide natural food sources to adult hymenopteran parasitoids of bark beetles. Entomol. Exp. Appl. 96, 253–263.

Hougardy, E., Grégoire, J.-C., 2003. Cleptoparasitism increases the host finding ability of a polyphagous parasitoid species, *Rhopalicus tutela* (Hymenoptera: Pteromalidae). Behav. Ecol. Sociobiol. 55, 184–189.

Hougardy, E., Grégoire, J.-C., 2004. Biological differences reflect host preference in two parasitoids attacking the bark beetle *Ips typographus* (Coleoptera: Scolytidae) in Belgium. Bull. Entomol. Res. 94, 341–347.

Houle, C., Hartmann, G.C., Wasti, S.S., 1987. Infectivity of eight species of entomogenous fungi to the larvae of the elm bark beetle, *Scolytus multistriatus* (Marsham). J. N. Y. Entomol. Soc. 95, 14–18.

Hulcr, J., Pollet, M., Ubik, K., Vrkoc, J., 2005. Exploitation of kairomones and synomones by *Medetera* spp. (Diptera: Dolichopodidae), predators of spruce bark beetles. Eur. J. Entomol. 102, 655–662.

Hunt, D.W.A., 1986. Absence of fatty acid germination inhibitors for conidia of *Beauveria bassiana* on the integument of the bark beetle *Dendroctonus ponderosae* (Coleoptera: Scolytidae). Can. Entomol. 118, 837–838.

Hunt, D.W.A., Borden, J.H., Rahe, J.E., Whitney, H.S., 1984. Nutrient-mediated germination of *Beauveria bassiana* conidia on the integument of the bark beetle *Dendroctonus ponderosae* (Coleoptera: Scolytidae). J. Invertebr. Pathol. 44, 304–314.

Imnadze, T.S., 1978. Characterization of strains of *Bacillus thuringiensis* serotype I isolated from some bark beetles in Georgia. Bulletin of the Academy of the Georgian SSR 92, 457–460.

Imnadze, T.S., 1984. Role of entomopathogenous microflora in number regulation of *Dendroctonus micans* Kugel. in Georgian SSR. Abstract. International Congress of Entomology, Proceedings, Hamburg, p. 603.

Inglis, G.D., Goettel, M.S., Butt, T.M., Strasser, H., 2001. Use of hyphomycetous fungi for managing insect pests. In: "Fungi as Biocontrol Agents. Progress, Problems and Potential" (T. M. Butt, C. W. Jackson, and N. Magan). CABI Publishing, Wallingford, pp. 23–69.

ION, 2014. Index to organism names (ION). Available online: http://www.organismnames.com, Last accessed: April 14, 2014.

ITIS, 2014. Integrated Taxonomic Information System (ITIS). Available online: http://www.itis.gov, Last accessed: April 14, 2014.

Jakuš, R., Blaženec, M., 2011. Treatment of bark beetle attacked trees with entomopathogenic fungus *Beauveria bassiana* (Balsamo) Vuillemin. Folia Forestalia Polonica, Series A 53, 150–155.

Jassim, H.K., Foster, H.A., Fairhurst, C.P., 1990. Biological control of Dutch elm disease: *Bacillus thuringiensis* as a potential control agent for *Scolytus scolytus* and *S. multistriatus*. J. Appl. Bacteriol. 69, 563–568.

Jouvenaz, D.P., Wilkinson, R.C., 1970. Incidence of *Serratia marcescens* in wild *Ips calligraphus* populations in Florida. J. Invertebr. Pathol. 16, 295–296.

Jurat-Fuentes, J.L., Jackson, T.A., 2012. Bacterial entomopathogens. In: Vega, F.E., Kaya, H.K. (Eds.), Insect Pathology, Second Edition. Academic Press, San Diego, pp. 265–349.

Jurc, M., 2004. Insect pathogens with special reference to pathogens of bark beetles (Col., Scolytidae, *Ips typographus* L.). Preliminary results of isolation of entomopathogenic fungi from two spruce bark beetles in Slovenia. Zbornik gozdarstva in lesarstva 74, 97–124.

Kanzaki, N., Ragsdale, E.J., Herrmann, M., Mayer, W.E., Sommer, R.J., 2012. Description of three *Pristionchus* species (Nematoda: Diplogastridae) from Japan that form a cryptic species complex with the model organism *P. pacificus*. Zool Sci 29, 403–417.

Karpinski, J. J. (1935). Przyczyny ograniczające rozmnażanie się korników drukarzy (*Ips typographus* L. i *Ips duplicatus* Sahlb.) w lesie pierwotnym. *Instytut Badawczy Lasów Państwowych, Seria A—Rozprawy I sprawozdania* 15, 1–86.

Kaya, H.K., 1984. Nematode parasites of bark beetles. In: Nickle, W.R. (Ed.), Plant and Insect Nematodes. Marcel Dekker, New York, pp. 727–754.

Keller, S., Epper, C., Wermelinger, B., 2004. *Metarhizium anisopliae* as a new pathogen of the spruce bark beetle *Ips typographus*. Mitt. Schweiz. Entomol. Ges. 77, 121–123.

Kenis, M., Mills, N.J., 1994. Parasitoids of European species of the genus *Pissodes* (Col: Curculionidae) and their potential for the biological control of *Pissodes strobi* (Peck) in Canada. Biol. Control 4, 14–21.

Kenis, M., Wermelinger, B., Grégoire, J.-C., 2004. Research on parasitoids and predators of Scolytidae—a review. In: Lieutier, F., Day, K.R., Battisti, A., Grégoire, J.-C., Evans, H.F. (Eds.), Bark and Wood Boring

Insects in Living Trees in Europe—A Synthesis. Kluwer Academic Publishers, Dordrecht, pp. 237–290.

Kereselidze, M., Wegensteiner, R., 2007. Occurrence of pathogens and parasites in *Ips typographus* L. from spruce stands (*Picea orientalis* L.) in Georgia. IOBC/wprs Bull 30, 207–210.

Kereselidze, M., Wegensteiner, R., Goginashvili, N., Tvaradze, M., Pilarska, D., 2010. Further studies on the occurrence of natural enemies of *Ips typographus* (Coleoptera: Curculionidae: Scolytinae) in Georgia. Acta Zool. Bulg. 62, 131–139.

Kielczewski, B., Moser, J.C., Wisniewski, J., 1983. Surveying the acarofauna associated with Polish Scolytidae. *Bull. Soc. Amis Sci. Lettres Poznan.* Bull. Soc. Amis Sci. Lettres Poznan Serie D 22, 151–159.

King, C.J., Fielding, N.J., O'Keefe, T., 1991. Observations on the life cycle and behaviour of the predatory beetle, *Rhizophagus grandis* Gyll. (Col., Rhizophagidae) in Britain. J. Appl. Entomol. 111, 286–296.

Kinn, D.N., 1983. The life cycle of *Proctolaelaps dendroctoni* Lindquist and Hunter (Acari: Ascidae): a mite associated with pine bark beetles. Intern. J. Acarol. 9, 205–210.

Kinn, D.N., Stephen, F.M., 1981. The incidence of endoparasitism of *Dendroctonus frontalis* Zimm. (Coleoptera: Scolytidae) by *Contortylenchus brevicomi* (Massey) Rühm (Nematoda: Sphaerulidae). Z. Angew. Entomol. 91, 452–458.

Kirchhoff, J.-F., Führer, E., 1985. Häufigkeit und Verbreitung von *Malamoeba scolyti* Purrini bei *Dryocoetes autographus* in einigen Gebieten Nord- und Nordwestdeutschlands. Forstwiss. Cent. bl. 104, 373–380.

Kirchhoff, J.-F., Führer, E., 1990. Experimentelle Analyse der Infektion und des Entwicklungszyklus von *Malamoeba scolyti* in *Dryocoetes autographus* (Coleoptera: Scolytidae). Entomophaga 35, 537–544.

Kirchhoff, J.-F., Wegensteiner, R., Weiser, J., Führer, E., 2005. Laboratory evaluation of *Malamoeba scolyti* (Rhizopoda, Amoebidae) in different bark beetle hosts (Coleoptera, Scolytidae). IOBC/wprs Bull. 28, 159–162.

Kirschner, R., 2001. Diversity of filamentous fungi in bark beetle galleries in Central Europe. In: Misra, J.K., Horn, B.W. (Eds.), Trichomycetes and Other Fungal Groups: Professor Robert W. Lichtwardt Commemoration Volume. B. W. Science Publishers, Inc., Enfield, pp. 175–196.

Knell, J.D., Allen, G.E., 1978. Morphology and ultrastructure of *Unikaryon minutum* sp.n. (Microsporidia: Protozoa), a parasite of the Southern Pine Beetle, *Dendroctonus frontalis*. Acta Protozool. 17, 271–278.

Knight, F.B., 1958. The effects of woodpeckers on populations of the Engelmann spruce beetle. J. Econ. Entomol. 51, 603–607.

Kohlmayr, B., 2001. Zum Auftreten von Krankheitserregern in den Europäischen Waldgärtnerarten *Tomicus piniperda*, *Tomicus minor* und *Tomicus destruens* (Coleoptera; Scolytidae). PhD thesis, University of Natural Resources, Vienna.

Kohlmayr, B., Weiser, J., Wegensteiner, R., Händel, U., Žižka, Z., 2003. Infection of *Tomicus piniperda* (Col., Scolytidae) with *Canningia tomici* sp.n. (Microsporidia, Unikaryonidae). J. Pest Sci. 76, 65–73.

Kohnle, U., Vité, J.P., 1984. Bark beetle predators: strategies in the olfactory perception of prey species by clerid and trogositid beetles. Z. Ang. Entomol. 98, 504–508.

Koplin, J.R., 1969. The numerical response of woodpeckers to insect prey in a subalpine forest in Colorado. Condor 71, 436–438.

Koplin, J.R., 1972. Measuring predator impact of woodpeckers on spruce beetles. J. Wildl. Manage. 36, 308–320.

Koplin, J.R., Baldwin, P.H., 1970. Woodpecker predation on an endemic population of Engelmann spruce beetles. Am. Midl. Nat. 83, 510–515.

Kreutz, J., 1995. Untersuchungen zur Überwinterung des Buchdruckers, *Ips typographus* (L.) (Scolytidae) im Boden. Diploma thesis, Universität des Saarlandes, Germany.

Kreutz, J., 2001. Möglichkeiten zur biologischen Bekämpfung des Buchdruckers, *Ips typographus* L. (Col., Scolytidae), mit insektenpathogenen Pilzen in Kombination mit Pheromonfallen. PhD thesis, Universität Saarbrücken, Germany.

Kreutz, J., Zimmermann, G., Marohn, H., Vaupel, O., Mosbacher, G., 2000a. Möglichkeiten des Einsatzes von *Beauveria bassiana* (Bals.) Vuill. und anderen Kontrollmethoden zur biologischen Bekämpfung des Buchdruckers *Ips typographus* L. (Col., Scolytidae) im Freiland. Mitt. Dtsch. Ges. Allg. Angew. Entomol. 12, 119–125.

Kreutz, J., Zimmermann, G., Marohn, H., Vaupel, O., Mosbacher, G., 2000b. Preliminary investigations on the use of *Beauveria bassiana* (Bals.) Vuill. and other control methods against the bark beetle *Ips typographus* (Col., Scolytidae) in the field. IOBC/wprs Bull 23, 167–173.

Kreutz, J., Vaupel, O., Zimmermann, G., 2004a. Efficacy of *Beauveria bassiana* (Bals.) Vuill. against the spruce bark beetle, *Ips typographus* L., in the laboratory under various conditions. J. Appl. Entomol. 128, 384–389.

Kreutz, J., Zimmermann, G., Vaupel, O., 2004b. Horizontal transmission of the entomopathogenic fungus *Beauveria bassiana* among the spruce bark beetle, *Ips typographus* (Col., Scolytidae) in the laboratory and under field conditions. Biocont. Sci. Technol. 14, 837–848.

Kroll, J.C., Fleet, R.R., 1979. Impact of woodpecker predation on overwintering within-tree populations of the southern pine beetle (*Dendroctonus frontalis*). In: Dickson, J.G., Connor, R.N., Fleet, R. R., Kroll, J.C., Jackson, J.A. (Eds.), The Role of Insectivorous Birds in Forest Ecosystems. Academic Press, Inc., London, pp. 269–281.

Krüger, K., Mills, N.J., 1990. Observations on the biology of three parasitoids of the spruce bark beetle, *Ips typographus* (Col., Scolytidae): *Coeloides bostrychorum, Dendrosoter middendorffii* (Hym., Braconidae) and *Rhopalicus tutela* (Hym., Pteromalidae). J. Appl. Entomol. 110, 281–291.

Kulhavy, D.L., Miller, M.C., 1989. Potential for Biological Control of *Dendroctonus* and *Ips* Bark Beetles. Center for Applied Studies, School of Forestry, Stephen S, Austin State University, Texas.

Lambshead, P.J.D., 1993. Recent developments in marinebenthic biodiversity research. Oceanis 19, 5–24.

Landa, Z., Horňák, P., Osborne, L.S., Nováková, A., Bursová, E., 2001a. Entomogenous fungi associated with spruce bark beetle *Ips typographus* L. (Coleoptera, Scolytidae) in the Bohemian Forest. Silva Gabreta 6, 250–272.

Landa, Z., Bieliková, L., Osborne, L.S., Bobková, E., 2001b. Assesment of isolates of entomogenous fungi collected from spruce bark beetle *Ips typographus* L. (Coleoptera, Scolytidae) in the Bohemian Forest. Silva Gabreta 6, 273–286.

Lange, C.E., Lord, J.C., 2012. Protistan entomopathogens. In: Vega, F.E., Kaya, H.K. (Eds.), Insect Pathology, Second Edition. Academic Press, San Diego, pp. 367–394.

Langor, D.W., 1991. Arthropods and nematodes co-occurring with the eastern larch beetle, *Dendroctonus* simplex [Col: Scolytidae], in Newfoundland. Entomophaga 36, 303–313.

Laucius, S., Zolubas, P., 1997. Spruce bark beetle (*Ips typographus* L.) dynamics in forest reserve in 1995–1996. Miskininkyste 1, 84–92.

Lawson, S.A., Morgan, F.D., 1992. Rearing of two predators, *Thanasimus dubius* and *Temnochila virescens*, for the biological control of *Ips grandicollis* in Australia. Entomol. Exp. Appl. 65, 225–233.

Lawson, S.A., Furuta, K., Katagiri, K., 1996. The effect of host tree on the natural enemy complex of *Ips typographus japonicus* Niijima (Col., Scolytidae) in Hokkaido, Japan. J. Appl. Entomol. 120, 77–86.

Lawson, S.A., Furuta, K., Katagiri, K., 1997. Effect of natural enemy exclusion on mortality of *Ips typographus japonicus* Niijima (Col, Scolytidae) in Hokkaido, Japan. J. Appl. Entomol. 121, 89–98.

Leatherdale, D., 1970. The arthropod hosts of entomogenous fungi in Britain. Entomophaga 15, 419–435.

Leufvén, A., Nehls, L., 1986. Quantification of different yeasts associated with the bark beetle, *Ips typographus*, during its attack on a spruce tree. Microbial Ecol. 12, 237–243.

Levine, N.D., 1988. Protozoan phylum Apicomplexa, two volumes. CRC Press, Boca Raton.

Lieutier, F., 1979. Les diptères associés à *Ips typographus* et *Ips sexdentatus* (Coleoptera: Scolytidae) en région parisienne, et les variations de leurs populations au cours du cycle annuel. Bull. Ecol. 10, 1–13.

Lieutier, F., 1982. Weight variations of adipose tissue and ovaries and variations in the length of ovocytes in *Ips sexdentatus* Boern. (Coleoptera: Scolytidae); relation to parasitism by nematodes. Ann. Parasitol. Hum. Comp. 57, 407–418.

Lieutier, F., 1984a. Disturbances in the digestive tract of *Ips sexdentatus* (Insecta: Scolytidae) induced by *Parasitorhabditis ipsophila* (Nematoda: Rhabditidae). Ann. Parasitol. Hum. Comp. 59, 597–605.

Lieutier, F., 1984b. Parasitism of *Ips sexdentatus* (Insecta: Scolytidae) by *Parasitorhabditis ipsophila* (Nematoda: Rhabditidae). Ann. Parasitol. Hum. Comp. 59, 507–520.

Lieutier, F., Laumond, C., 1978. Nematode parasites and associates of the *Ips sexdentatus* and *Ips typographus* (Coleoptera, Scolytidae). En Région Parisienne Nematologica 24, 184–200.

Lindgren, B.S., Miller, D.R., 2002. Effect of verbenone on attraction of predatory and woodboring beetles (Coleoptera) to kairomones in lodgepole pine forests. Environ. Entomol. 31, 766–773.

Lindquist, E.E., 1964. Mites parasitizing eggs of bark beetles of the genus *Ips*. Can. Entomol. 96, 125–126.

Lindquist, E.E., 1969. Mites and the regulation of bark beetle populations. Proc. 2nd Internat. Congr. Acarol 1967, 389–399.

Linit, M.J., Stephen, F.M., 1983. Parasite and predator component of within-tree southern pine beetle (Coleoptera: Scolytidae) mortality. Can. Entomol. 115, 679–688.

von Linstow, O.F.B., 1890. Über *Allantonema* und *Diplogaster*. Centralblatt für Bakteriologie 8, 489–493.

Lipa, J.J., 1968. *Stempellia scolyti* Weiser com.nov. and *Nosema scolyti* n. sp. microsporidian parasites of four species of *Scolytus* (Coleoptera). Acta Protozool. 6, 69–78.

Lobinger, G., Feicht, E., 1999. Schwärmverhalten und Abundanzdynamik der Erzwespe *Karpinskiella pityophthori* (Bouček) (Hym., Pteromalidae), eines Parasitoiden des Kupferstechers (*Pityogenes chalcographus* L., Col., Scolytidae). J. Pest Sci. 72, 65–71.

Lukášová, K., Holuša, J., 2011. *Gregarina typographi* (Eugregarinorida: Gregarinidae) in the bark beetle *Ips typographus* (Coleoptera: Curculionidae): changes in infection level in the breeding system. Acta Protozool. 50, 311–318.

Lukášová, K., Holuša, J., 2012. Patogeny lýkožroutů rodu *Ips* (Coleoptera: Curculionidae: Scolytinae): review. Pathogens of bark beetles of the genus *Ips* (Coleoptera: Curculionidae: Scolytinae): review. Zprávy Lesnického Výzkumu 57, 230–240.

Lukášová, K., Holuša, J., Grucmanová, Š., 2012. Reproductive performance and natural antagonists of univoltine population of *Ips typographus* (Coleoptera, Curculionidae, Scolytinae) at epidemic level: a study from Šumava Mountains, Central Europe. Beskydy 5, 153–162.

Lukášová, K., Holuša, J., Turčáni, M., 2013. Pathogens of *Ips amitinus*: new species and comparison with *Ips typographus*. J. Appl. Entomol. 137, 188–196.

Lutyk, P., Świeżyńska, H., 1984. Trials to control the larger pine-shoot beetle (*Tomicus piniperda* L.) with the use of the fungus *Beauveria bassiana* (Bals.) Vuill. on piled wood. Sylwan R 128, 41–45.

Lysenko, O., 1959. Report on diagnosis of bacteria isolated from insects (1954–1958). Entomophaga 4, 15–22.

MacGuidwin, A.E., Smart, G.C., Wilkinson, R.C., Allen, G.E., 1980. Effect of the nematode *Contortylenchus brevicomi* on gallery construction and fecundity of the southern pine beetle. J. Nematol. 12, 278–282.

MacLeod, D.M., 1963. Entomophthorales Infections. In: Steinhaus, E.A. (Ed.), Insect Pathology. An Advanced Treatise. Volume 2. Academic Press, New York, pp. 189–231.

Magema, N., Verstraeten, C., Gaspar, C., 1981. Les ennemis naturels du scolyte *Trypodendron lineatum* (Olivier, 1795) (Coleopetra, Scolytidae) dans la forest de Hazeilles et des Epioux. Annales de la Societe Royale Zoologique de Belgique 111, 89–95.

Manojlovic, B., Zabel, A., Kostic, M., Stankovic, S., 2000. Effect of nutrition of parasites with nectar of melliferous plants on parasitism of the elm bark beetles (Col., Scolytidae). J. Appl. Entomol. 124, 155–161.

Markova, G., 2000. Pathogenicity of several entomogenous fungi to some of the most serious forest insect pests in Europe. IOBC/wprs Bull. 23, 231–239.

Markovic, C., Stojanovic, A., 2003. Significance of parasitoids in the reduction of oak bark beetle *Scolytus intricatus* Ratzeburg (Col., Scolytidae) in Serbia. J. Appl. Entomol. 127, 23–28.

Martín, A., Etxebeste, I., Pérez, G., Álvarez, G., Sánchez, E., Pajares, J., 2013. Modified pheromone traps help reduce bycatch of bark-beetle natural enemies. Agric. For. Entomol. 15, 86–97.

Martinek, V., 1977. Species of genus *Palloptera* Fallen, 1820 (Dipt., Pallopteridae) in Czechoslovakia. Stud. Entomol. For. 2, 203–220.

Massey, C.L., 1956. Nematode parasites and associates of the Engelmann spruce beetle (*Dendroctonus engelmanni* Hopk.). Proc. Helminthol. Soc. Wash. 23, 14–24.

Massey, C.L., 1957. Four new species of *Aphelenchus* (Nematoda) parasitic in bark beetles in the United States. Proc. Helminthol. Soc. Wash. 24, 29–34.

Massey, C.L., 1960. Nematode parasites and associates of the California five-spined engraver, *Ips confusus* (Lee). Proc. Helminthol. Soc. Wash. 27, 14–22.

Massey, C.L., 1964a. The nematode parasites and associates of the fir engraver beetle, *Scolytus ventralis* Le Conte, in New Mexico. J. Insect Pathol. 6, 133–155.

Massey, C.L., 1964b. Two new species of the nematode genus *Ektaphelenchus* (Nematoda: Aphelenchoididae) parasites of bark beetles in the south-western United States. Proc. Helminthol. Soc. Wash. 31, 37–40.

Massey, C.L., 1966a. The nematode parasites and associates of *Dendroctonus adjunctus* (Coleoptera: Scolytidae) in New Mexico. Ann. Entomol. Soc. Am. 59, 424–440.

Massey, C.L., 1966b. Genus *Mikoletzkya* (Nematoda) in the United States. Proc. Helminthol. Soc. Wash. 33, 13–19.

Massey, C.L., 1969. New species of tylenchs associated with bark beetles in New Mexico and Colorado. Proc. Helminthol. Soc. Wash. 64, 43–52.

Massey, C.L., 1974. Biology and Taxonomy of Nematode Parasites and Associates of Bark Beetles in the United States. U.S. Department of Agriculture, Forest Service, Agriculture Handbook No. 446.

Massey, C.L., Wygant, N.D., 1973. Biology and control of the Engelmann spruce beetle in Colorado. U.S. Dept. Agriculture Circular 944, 35.

Matha, V., Weiser, J., 1985. Effect of the fungus *Beauveria bassiana* on adult bark beetles *Ips typographus*. Presentation at the Conference Biological and Biotechnical Control of Forest Pests. September 10–12, 1985, Tabor (CSFR).

McCambridge, W.F., Knight, F.B., 1972. Factors affecting spruce beetles during a small outbreak. Ecology 53, 830–839.

Mendel, Z., 1985. Predation of *Orthotomicus erosus* (Col., Scolytidae) by the Syrian woodpecker (*Picoides syriacus*, Aves, Picidae). Z. Ang. Entomol. 100, 355–360.

Mendel, Z., 1986. Hymenopterous parasitoids of bark beetles [Scolytidae] in Israel: relationships between host and parasitoid size and sex ratio. Entomophaga 31, 127–137.

Mendel, Z., 1988. Effect of food, temperature and breeding conditions on the life span of adults of three cohabiting bark beetle (Scolytidae) parasitoids (Hymenoptera). Environ. Entomol. 17, 293–295.

Meyhöfer, R., Casas, J., 1999. Vibratory stimuli in host location by parasitic wasps. J. Insect Physiol. 45, 967–971.

Michalková, V., Krascsenitsová, E., Kozánek, M., 2012. On the pathogens of the spruce bark beetle *Ips typographus* (Coleoptera: Scolytinae) in the Western Carpethians. Biologia 67, 217–221.

Michalski, J., Seniczak, S., 1974. *Trichogramma semblidis* [Chalcidoidea: Trichogrammatidae] as a parasite of the bark beetle eggs [Coleoptera: Scolytidae]. Entomophaga 19, 237–242.

Mignot, E.C., Anderson, R.F., Struble, G.R., 1970. Bionomics of the bark beetle predator *Temnochila virescens* Mann. (Coleoptera: Ostomidae). Entomol. News 81, 85–91.

Miller, M.C., 1984. Mortality contribution of insect natural enemies to successive generations of *Ips calligraphus* (Germar) (Coleoptera, Scolytidae) in loblolly pine. Z. Ang. Entomol. 98, 495–500.

Miller, M.C., 1986. Survival of within-tree *Ips calligraphus* (Col.: Scolytidae): effect of insect associates. Entomophaga 31, 39–48.

Mills, N.J., 1983. The natural enemies of scolytids infesting conifer bark in Europe in relation to the biological control of *Dendroctonus* spp. in Canada. Biocontrol News and Information 4, 305–328.

Mills, N.J., 1985. Some observations on the role of predation in the natural regulation of *Ips typographus* populations. Z. Ang. Entomol. 99, 209–215.

Mills, N.J., 1986. A preliminary analysis of the dynamics of within tree populations of *Ips typographus* (L.) (Coleoptera: Scolytidae). Z. Ang. Entomol. 102, 402–416.

Mills, N.J., 1991. Searching strategies and attack rates of parasitoids of the ash bark beetle (*Leperisinus varius*) and its relevance to biological control. Ecol. Entomol. 16, 461–470.

Mills, N.J., Krüger, K., Schlup, J., 1991. Short-range host location mechanisms of bark beetle parasitoids. J. Appl. Entomol. 111, 33–43.

Moore, G.E., 1971. Mortality factors caused by pathogenic bacteria and fungi of the southern pine beetle in North Carolina. J. Invertebr. Pathol. 17, 28–37.

Moore, G.E., 1972a. Southern pine beetle mortality in North Carolina caused by parasites and predators. Environ. Entomol. 1, 58–65.

Moore, G.E., 1972b. Microflora from the alimentary tract of healthy southern pine beetles, *Dendroctonus frontalis* (Scolytidae), and their possible relationship to pathogenicity. J. Invertebr. Pathol. 19, 72–75.

Moore, G.E., 1972c. Pathogenicity of ten strains of bacteria to larvae of the southern pine beetle. J. Invertebr. Pathol. 20, 41–45.

Moore, G.E., 1973. Pathogenicity of three entomogenous fungi to the southern pine beetle at various temperatures and humidities. Environ. Entomol. 2, 54–57.

Morge, G., 1961. Die Bedeutung der Dipteren im Kampf gegen die Borkenkäfer. Arch. Forstwes. 10, 505–511.

Morge, G., 1963. Die Lonchaeidae und Pallopteridae Österreichs und der angrenzenden Gebiete 1.Teil: Die Lonchaeidae. Naturkundl. Jahrb. d. Stadt Linz 9, 123–312.

Morge, G., 1967. Die Lonchaeidae und Pallopteridae Österreichs und der angrenzenden Gebiete 2.Teil: Die Pallopteridae. Naturkundl. Jahrb. d. Stadt Linz 13, 141–188.

Moser, J.C., 1975. Mite predators of the southern pine beetle. Ann. Entomol. Soc. Am. 68, 1113–1116.

Moser, J.C., Bogenschütz, H., 1984. A key to the mites associated with flying *Ips typographus* in South Germany. Z. Ang. Entomol. 97, 437–450.

Moser, J.C., Roton, L.M., 1971. Mites associated with southern pine bark beetles in Allen Parish. Louisiana. Can. Entomol. 103, 1775–1798.

Moser, J.C., Thatcher, R.C., Pickard, L.S., 1971. Relative abundance of southern pine beetle associates in east Texas. Ann. Entomol. Soc. Am. 64, 72–77.

Moser, J.C., Kiełczewski, B., Wiśniewski, J., Bałazy, S., 1978. Evaluating *Pyemotes dryas* (Vitzthum 1923) (Acari: Pyemotidae) as a parasite of the southern pine beetle. Int. J. Acarol. 4, 67–70.

Moser, J.C., Eidmann, H.H., Regnander, J.R., 1989a. The mites associated with *Ips typographus* in Sweden. Ann. Entomol. Fenn. 55, 23–27.

Moser, J.C., Perry, T.J., Solheim, H., 1989b. Ascospores hyperphoretic on mites with *Ips typographus*. Mycol. Res. 93, 513–517.

Moser, J.C., Konrad, H., Kirisits, T., Carta, L.K., 2005. Phoretic mites and nematode associates of *Scolytus multistriatus* and *Scolytus pygmaeus* (Coleoptera: Scolytidae) in Austria. Agric. For. Entomol. 7, 169–177.

Mudrončeková, S., Mazáň, M., Nemčovič, M., Šalomon, I., 2013. Entomopathogenic fungus species *Beauveria bassiana* (Bals.) and *Metarhizium anisopliae* (Metsch.) used as mycoinsecticide effective in biological control of *Ips typographus* (L.). J. Microbiol., Biotech. Food Sci. 2, 2469–2472.

Müller-Kögler, E., 1965. Pilzkrankheiten bei Insekten. Anwendung zur biologischen Schädlingsbekämpfung und Grundlagen der Insektenmykologie. Parey, Berlin.

Muratoğlu, H., Sezen, K., Demirbağ, Z., 2011. Determination and pathogenicity of the bacterial flora associated with the spruce bark beetle, *Ips typographus* (L.) (Coleoptera: Curculionidae: Scolytinae). Turk. J. Biol. 35, 9–20.

Nagel, W.P., Fitzgerald, T.D., 1975. *Medetera aldrichii* larval feeding behavior and prey consumption (Dipt.: Dolichopodidae). Entomophaga 20, 121–127.

Natali, V., Paparatti, B., Canganella, F., 1994. Microorganisms carried by *Xyleborus dispar* (F.) (Coleoptera Scolytidae) females, collected on European Hazel trees in the area surrounding the lake of Vico (Viterbo, Central Italy). Redia 77, 285–295.

Nebeker, T.E., Mizell, R.F.I., Bedwell, N.J., Garner, W.Y., Harvey, J.J., 1984. Management of bark beetle populations. Impact of manipulating predator cues and other control tactics. Chemical and biological controls in forestry, Seattle, pp. 25–33.

Neužilová, A., 1956. Ein Beitrag zur Kenntnis der parasitischen Pilze bei *Ips typographus* L. Preslia 28, 273–275.

Nguyen, K.B., 2010. Insect nematodes. Available at: http://entnemdept. ifas.ufl.edu/nguyen/insectnema/insect-nematodes.html, Last accessed: April 14, 2014.

Nguyen, K.B., Hunt, D.J., 2007. Entomopathogenic Nematodes: Systematics, Phylogeny and Bacterial Symbionts. Nematology Monographs and Perspectives 5, Brill, Leiden, Netherlands.

Nickle, W.R., 1963. Observations on the effect of nematodes on *Ips confusus* (LeConte) and other bark beetles. J. Insect Pathol. 5, 386–389.

Nickle, W.R., 1970. A taxonomic review of the genera of the Aphelenchoidea (Fuchs, 1937) Thorne, 1949 (Nematoda: Tylenchida). J. Nematol. 2, 375–392.

Nickle, W.R., 1971. Behavior of the shothole borer, *Scolytus rugulosus*, altered by the nematode parasite *Neoparasitylenchus rugulosi*. Ann. Entomol. Soc. Am. 64, 751.

Nicolai, V., 1995. The impact of *Medetera dendrobaena* Kowarz (Dipt., Dolichopodidae) on bark beetles. J. Appl. Entomol. 119, 161–166.

Nierhaus-Wunderwald, D., 1993. Die natürlichen Gegenspieler der Borkenkäfer. Wald und Holz 1 (93), 8–14.

Norris, A.R., Drever, M.C., Martin, K., 2013. Insect outbreaks increase populations and facilitate reproduction in a cavity-dependent songbird, the Mountain Chickadee *Poecile gambeli*. Ibis 155, 165–176.

Novák, V., 1960. Die natürlichen Feinde und Krankheiten des gemeinen Nutzholzborkenkäfers *Trypodendron lineatum* Oliv. Zoologicke Listy, Folia Zoologica 9, 309–322.

Novák, V., Samšiňaková, A., 1962. Les essais d'aplication du champignon parasite *Beauveria bassiana* (Bals.) Vuill. dans la lutte contre les parasites en agriculture et sylviculture en CSSR. Colloques Internationaux de Pathologique des Insectes. Paris, 133–135.

Novotný, J., Turčani, M., 2000. The results of experimental treatment of biopreparation on the basis of *Bacillus thuringiensis* var. *tenebrionis* Berl. for control of *Ips typographus* L. (Col., Scolytidae). Lesnicky casopis 46, 303–310.

Nuorteva, M., 1956. Über den Fichtenstamm-Bastkäfer, *Hylurgops palliatus* Gyll., und seine Insektenfeinde. Acta Entomol. Fenn. 13, 1–116.

Nuorteva, M., 1957. Zur Kenntnis der parasitischen Hymenopteren der Borkenkäfer Finnlands. Suomen Hyönteistieteellinen Aikakauskirja 23, 47–71.

Nuorteva, M., 1959. Untersuchungen über einige in den Frassbildern der Borkenkäfer lebende *Medetera*-Arten (Dipt., Dolichopodidae). Suomen Hyöönteistieteellinen Aikakauskirja 25, 192–210.

Nuorteva, M., 1971. Die Borkenkäfer (Col., Scolytidae) und deren Insektenfeinde im Kirchspiel Kuusamo, Nordfinnland. Ann. Entomol. Fenn. 37, 65–72.

Nuorteva, M., Saari, L., 1980. Larvae of *Acanthocinus, Pissodes* and *Tomicus* (Coleoptera) and the foraging behaviour of woodpeckers (Picidae). Ann. Entomol. Fenn. 46, 107–110.

Nuorteva, M., Salonen, M., 1968. Versuche mit *Beauveria bassiana* (Bals.) Vuill. gegen *Blastophagus piniperda* L. (Col., Scolytidae). Ann. Ent. Fenn. 34, 49–55.

Ohmart, C.P., Voigt, W.G., 1982. Temporal and spatial arrival of *Ips plastographus maritimus* (Coleoptera: Scolytidae) and its insect associates on freshly felled *Pinus radiata* in California. Can. Entomol. 114, 337–348.

Otvos, I.S., 1965. Studies on avian predators of *Dendroctonus brevicomis* LeConte (Coleoptera: Scolytidae) with special reference to Picidae. Can. Entomol. 97, 1184–1199.

Otvos, I.S., 1970. Avian predation of the western pine beetle. In: Stark, R.W., Dahlsten, D.L. (Eds.), Studies on the Population Dynamics of the Western Pine Beetle, *Dendroctonus brevicomis* LeConte (Coleoptera: Scolytidae). Univ. California Press, Berkeley, pp. 119–127.

Otvos, I.S., 1979. The effects of insectivorous bird activities in forest ecosystems: an evaluation. In: Dickson, J.G., Connor, R.N., Fleet, R.R., Kroll, J.C., Jackson, J.A. (Eds.), The Role of Insectivorous Birds in Forest Ecosystems. Academic Press, Inc., London, pp. 341–374.

Otvos, I.S., Stark, R.W., 1985. Arthropod food of some forest-inhabiting birds. Can. Entomol. 117, 971–990.

Overgaard, N.A., 1968. Insects associated with the southern pine beetle in Texas, Louisiana, and Mississippi. J. Econ. Entomol. 61, 1197–1201.

Pechacek, P., 1994. Reaktion des Dreizehenspechts auf eine Borkenkäfergradation. Allg. Forst Z. 49, 661.

Pechacek, P., Kristin, A., 2004. Comparative diets of adult and young three-toed woodpeckers in a European alpine forest community. J. Wildl. Manage. 68, 683–693.

Pehl, L., Kehr, R., 1994. Biologische Bekämpfung von Borkenkäfern. Allgemeine Forst Zeitschrift 19, 1065–1067.

Penas, A.C., Bravo, M.A., Naves, P., Bonifácio, L., Sousa, E., Mota, M., 2006. Species of *Bursaphelenchus* Fuchs, 1937 (Nematoda: Parasitaphelenchidae) and other nematode genera associated with insects from *Pinus pinaster* in Portugal. Ann. Appl. Biol. 148, 121–131.

Pernek, M., 2007. Influence of the entomopathogenic fungi *Beauveria bassiana* on the mortality of fir bark beetles *Pityokteines spinidens* and *Pityokteines curvidens*. Radovi-Sumarskog Instituta Jastrebarsko 42, 143–153.

Pernek, M., Matošević, D., Hrašovec, B., Kučinić, M., Wegensteiner, R., 2009. Occurrence of pathogens in outbreak populations of *Pityokteines* spp. (Coleoptera, Curculionidae, Scolytinae) in silver fir forests. J. Pest Sci. 82, 343–349.

Person, H.L., 1940. The clerid *Thanasimus lecontei* (Wolc.) as a factor in the control of the western pine beetle. J. For. 38, 390–396.

Pesson, P., Toumonoff, C., Chararas, C., 1955. Étude des epizooties bactériennes observées dans les élevages d'insectes xylophages. Annales Épiphyties 6, 315–328.

Petch, T., 1932. A list of the entomogenous fungi of Great Britain. T. Brit. Mycol. Soc. 17, 170–178.

Pettersen, H., 1976. Parasites (Hym., Chalcidoidea) associated with bark beetles in Norway. Norw. J. Entomol. 23, 75–78.

Pettersson, E.M., 2001a. Volatiles from potential hosts of *Rhopalicus tutela* a bark beetle parasitoid. J. Chem. Ecol. 27, 2219–2231.

Pettersson, E.M., 2001b. Volatile attractants for three Pteromalid parasitoids attacking concealed spruce bark beetles. Chemoecology 11, 89–95.

Pettersson, E.M., Sullivan, B.T., Anderson, P., Berisford, C.W., Birgersson, G., 2000. Odor perception in bark beetle parasitoid *Roptrocerus xylophagorum* (Ratzeburg) (Hymenoptera: Pteromalidae) exposed to host associated volatiles. J. Chem. Ecol. 26, 2507–2525.

Pettersson, E.M., Hallberg, E., Birgersson, G., 2001. Evidence for the importance of odour-perception in the parasitoid *Rhopalicus tutela* (Walker) (Hym., Pteromalidae). J. Appl. Entomol. 125, 293–301.

Pfammatter, J.A., Moser, J.C., Raffa, K.F., 2013. Mites phoretic on *Ips pini* (Coleoptera: Curculionidae: Scolytinae) in Wisconsin red pine stands. Ann. Entomol. Soc. Am. 106, 204–213.

Pishchik, A.A., 1980. An insect predator of *Blastophagus [Tomicus] piniperda* and *B. [T.] minor*. Lesnoe Khozyaistvo 11, 55–57.

Podoler, H., Mendel, Z., Livne, H., 1990. Studies on the biology of a bark beetle predator, *Aulonium ruficorne* (Coleoptera: Colydiidae). Environ. Entomol. 19, 1010–1016.

Poinar Jr., G.O., 1975. Entomogenous Nematodes: a Manual and Host List of Insect-Nematode Associations. E. J. Brill, Leiden.

Popa, V., Déziel, E., Lavallée, Bauce, E., Guertin, C., 2012. The complex symbiotic relationships of bark beetles with microorganisms: a potential practical approach for biological control in forestry. Pest Manag. Sci. 68, 963–975.

Postner, M., 1974. Scolytidae (=Ipidae), Borkenkäfer. In: Schwenke, W. (Ed.), Die Forstschädlinge Europas, Vol. 2. P. Parey, Hamburg, pp. 334–482.

Prazak, R., 1988. Die Wirkung des insektenpathogenen Pilzes *Beauveria bassiana* (Bals.) Vuill. auf den Gestreiften Nutzholzborkenkäfer *Trypodendron lineatum* Oliv. (Coleoptera: Scolytidae). PhD thesis, University of Natural Resources, Vienna.

Prazak, R., 1991. Studies on indirect infection of *Trypodendron lineatum* Oliv. with *Beauveria bassiana* (Bals.) Vuill. J. Appl. Entomol. 111, 431–441.

Prazak, R., 1997. Laboratory evaluation of *Beauveria bassiana* (Bals.) Vuill. (Deuteromycotina: Hyphomycetes) against *Trypodendron lineatum* Oliv. (Coleopetra: Scolytidae). *J. Plant Dis*. Proc. Natl. Acad. Sci. U. S. A. 104, 459–465.

Pultar, O., Weiser, J., 1999. Infection of *Ips typographus* with entomopoxvirus in the forest. Poster at 7th IOBC/wprs meeting in Vienna.

Purrini, K., 1975. Zur Kenntnis der Krankheiten des Großen Ulmensplintkäfers, *Scolytus scolytus* F. im Gebiet von Kosovo, Jugoslawien. J. Pest Sci. 48, 154–156.

Purrini, K., 1977. Über eine neue Schizogregarinen-Krankheit der Gattung *Mattesia* Naville (Sporozoa, Dischizae) des Zottigen Fichtenborkenkäfers, *Dryocoetes autographus* Ratz (Coleoptera, Scolytidae). J. Pest Sci. 50, 132–135.

Purrini, K., 1978a. Über *Malamoeba locustae* King and Taylor (Protozoa, Rhizopoda, Amoebidae) beim Zottigen Fichtenborkenkäfer, *Dryocoetes autographus* Ratz (Col.: Scolytidae). J. Pest Sci. 51, 139–141.

Purrini, K., 1978b. Protozoen als Krankheitserreger bei einigen Borkenkäferarten (Col., Scolytidae) im Königsee-Gebiet, Oberbayern. J. Pest Sci. 51, 171–175.

Purrini, K., 1980. *Malamoeba scolyti* sp.n. (Amoebidae, Rhizopoda, Protozoa) parasitizing the bark beetles *Dryocoetes autographus* Ratz. and *Hylurgops palliatus* Gyll. (Scolytidae, Col.). Arch. Protistenkd. 123, 358–366.

Purrini, K., Führer, E., 1979. Experimentelle Infektion von *Pityogenes chalcographus* L. (Col.: Scolytidae) durch *Malamoeba scolyti* Purrini (Amoebina, Amoebidae) und *Menzbieria chalcographi* Weiser (Neogregarina, Ophryocystidae). J. Pest Sci. 52, 167–173.

Purrini, K., Halperin, J., 1982. *Nosema calcarati* n.sp. (Microsporidia), a new parasite of *Pityogenes calcaratus* Eichhoff (Col., Scolytidae). Z. Ang. Entomol. 94, 87–92.

Purrini, K., Ormieres, R., 1981. On three new sporozoan parasites of bark beetles (Scolytidae, Coleoptera). Z. Ang. Entomol. 91, 67–74.

Purrini, K., Weiser, J., 1984. Light- and electron microscopic studies of *Chytridiopsis typographi* (Weiser 1954) Weiser 1970 (Microspora), parasitizing the bark beetle *Hylastes cunicularius* Er. Zool. Anz. 212, 369–376.

Purrini, K., Weiser, J., 1985. Ultrastructural study of the microsporidian *Chytridiopsis typographi* (Chytridiopsida: Microspora) infecting the bark beetle *Ips typographus* (Scolytidae: Coleoptera), with new data on spore dimorphism. J. Invertebr. Pathol. 45, 66–74.

Purrini, K., Žižka, Z., 1983. More on the life cycle of *Malamoeba scolyti* (Amoebidae: Sarcomastigophora) parasitizing the bark beetle *Dryocoetes autographus* (Scolytidae, Coleoptera). J. Invertebr. Pathol. 42, 96–105.

Raffa, K.F., 1991. Temporal and spatial disparities among bark beetles, predators, and associates responding to synthetic bark beetle pheromones: *Ips pini* (Coleoptera: Scolytidae) in Wisconsin. Environ. Entomol. 20, 1665–1679.

Raffa, K.F., Dahlsten, D.L., 1995. Differential responses among natural enemies and prey to bark beetle pheromones. Oecologia 102, 17–23.

Raffa, K.F., Hobson, K.R., LaFontaine, S., Aukema, B.H., 2007. Can chemical communication be cryptic? Adaptations by herbivores to natural enemies exploiting prey semiochemistry. Oecologia 153, 1009–1019.

Ragsdale, E.J., Kanzaki, N., Röseler, W., Herrmann, M., Sommer, R.J., 2013. Three new species of *Pristionchus* (Nematoda: Diplogastridae) show morphological divergence through evolutionary intermediates of a novel feeding-structure polymorphism. Zool. J. Linn. Soc. 168, 671–698.

Rauhut, B., Schmidt, G.H., Schmidt, L., 1993. Das Coleopteren-Spektrum in Borkenkäfer-Pheromonfallen eines heterogenen Waldgebietes im Landkreis Hannover. Braunschw. naturkdl. Schr. 4, 247–278.

Reay, S.D., Brownbridge, M., Cummings, N.J., Nelson, T.L., Souffre, B., Lignon, C., Glare, T.R., 2008. Isolation and characterization of *Beauveria* spp. associated with exotic bark beetles in New Zealand *Pinus radiata* plantation forest. Biol. Control 46, 484–494.

Reeve, J.D., 1997. Predation and bark beetle dynamics. Oecologia 112, 48–54.

Reeve, J.D., Ayres, M.P., Lorio, P.L.J., 1995. Host suitability, predation, and bark beetle population dynamics. In: Cappuccino, N., Price, P. W. (Eds.), Population Dynamics: New Approaches and Synthesis. Academic Press, London, pp. 339–357.

Reid, R.W., 1957. The bark beetle complex associated with lodgepole pine slash in Alberta. Part II—Notes on the biologies of several hymenopterous parasites. Can. Entomol. 89, 5–8.

Rice, R.E., 1968. Observations on host selection by *Tomicobia tibialis* Ashmead (Hymenoptera: Pteromalidae). Contrib. Boyce Thompson Inst. 24, 53–56.

Richerson, J.V., 1972. Host finding mechanisms of *Coeloides brunneri* Viereck (Hymenoptera: Braconidae). PhD thesis, Simon Fraser University, Canada.

Riley, M.A., Goyer, R.A., 1986. Impact of beneficial insects on *Ips* spp. (Coleoptera: Scolytidae) bark beetles in felled loblolly and slash pines in Louisiana. Environ. Entomol. 15, 1220–1224.

Rohlfs, W.M.I., Hyche, L.L., 1984. Observations on activity and development of *Lasconotus pusillus* and *L. referendarius* (Coleoptera: Colydiidae) following arrival at *Ips* spp. infested southern pines. J. Georgia Entomol. Soc. 19, 114–119.

Roy, H.E., Steinkraus, D.C., Eilenberg, J., Hajek, A.E., Pell, J.K., 2006. Bizarre interactions and endgames: entomopathogenic fungi and their arthropod hosts. Annu. Rev. Entomol. 51, 331–357.

Rühm, W., 1956. Die Nematoden der Ipiden. Parasitologische Schriftenreihe 6, 1–437.

Rühm, W., 1965. Zur "Wirtskreiserweiterung" einer mit Borkenkäfern (Scolytoidea, Col.) vergesellschafteten Nematodenart. Z. Parasitenkd. 26, 230–253.

Ruiz-Portero, C., Barranco, P., de la Peña, J., Cabello, T., 2002. Bioensayo con entomopatógenos para el control de escolídos plagas forestales (Col.: Scolytidae). Boletin de Sanidad Vegetal, Plagas 28, 367–373.

Sachtleben, H., 1952. Die parasitischen Hymenopteren des Fichtenborkenkäfers Ips typographus L. Beitr. Entomol. 2, 137–189.

Samson, P.R., 1984. The biology of Roptrocerus xylophagorum [Hym.: Torymidae], with a note on its taxonomic status. Entomophaga 29, 287–298.

Sarikaya, O., Avci, M., 2009. Predators of Scolytinae (Coleoptera: Curculionidae) species of the coniferous forests in the Western Mediterranean Region, Turkey. Turkiye Entomoloji Dergisi—Turkish J. Entomol. 33, 253–264.

Schimitschek, E., 1931a. Der achtzähnige Lärchenborkenkäfer Ips cembrae Heer. Zur Kenntnis seiner Biologie und Ökologie sowie seines Lebensvereines. Z. Ang. Entomol. 17, 253–344.

Schimitschek, E., 1931b. Forstentomologische Untersuchungen aus dem Gebiete von Lunz. I. Standortsklima und Kleinklima in ihren Beziehungen zum Entwicklungsablauf und zur Mortalität von Insekten. Z. Ang. Entomol. 18, 460–491.

Schmitt, J.J., Goyer, R.A., 1983. Consumption rates and predatory habits of Scoloposcelis mississippensis and Lyctocoris elongatus (Hemiptera: Anthocoridae) on pine bark beetles. Environ. Entomol. 12, 363–367.

Schröder, D., 1974. Untersuchungen über die Aussichten einer biologischen Bekämpfung von Scolytiden an Ulmen als Mittel zur Einschränkung des "Ulmensterbens." Z. Ang. Entomol. 76, 150–159.

Schroeder, L.M., 1996. Interactions between the predators Thanasimus formicarius (Col.: Cleridae) and Rhizophagus depressus (Col.: Rhizophagidae), and the bark beetle Tomicus piniperda (Col.: Scolytidae). Entomophaga 41, 63–75.

Schroeder, L.M., 1997. Impact of natural enemies on Tomicus piniperda offspring production. In: Grégoire, J.-C., Liebhold, A.M., Stephen, F.M., Day, K.R., Salom, S.M. (Eds.), Integrating Cultural Tactics into the Management of Bark Beetle and Reforestation Pests. USDA For. Serv. Gen. Tech. Rep, Radnor, PA, pp. 204–214.

Schroeder, L.M., 1999. Prolonged development time of the bark beetle predator Thanasimus formicarius (Col.: Cleridae) in relation to its prey species Tomicus piniperda (L.) and Ips typographus (L.) (Col.: Scolytidae). Agric. For. Entomol. 1, 127–135.

Schroeder, L.M., Weslien, J., 1994. Reduced offspring production in bark beetle Tomicus piniperda in pine bolts baited with ethanol and α-pinene, which attract antagonistic insects. J. Chem. Ecol. 20, 1429–1444.

Schvester, D., 1957. Contribution a l'etude ecologique des Coleopteres Scolytides. Essai d'analyse des facteurs de fluctuation des populatins chez Ruguluscolytus rugulosus Müller 1818. Annales Epiphyties, Paris, 8 np. Hors ser.

Seitner, M., 1924. Beobachtungen und Erfahrungen aus dem Auftreten des achtzähnigen Fichtenborkenkäfers Ips typographus L. in Oberösterreich und Steiermark in den Jahren 1921 bis einschl. 1923. 5. Parasiten und Räuber. Cent.bl. Gesamte Forstwes 50, 2–23.

Senger, S.E., Roitberg, B.D., 1992. Effects of parasitism by Tomicobia tibialis Ashmead (Hymenoptera: Pteromalidae) on reproductive parameters of female pine engravers, Ips pini (Say). Can. Entomol. 124, 509–513.

Sevim, A., Demir, I., Tanyeli, E., Demirbag, Z., 2010. Screening of entomopathogenic fungi against the European spruce bark beetle,

Dendroctonus micans (Coleoptera: Scolytidae). Biocontrol Sci. Tech. 20, 3–11.

Shepherd, W.P., 2004. Biology and host finding of predaceous hister beetles (Coleoptera: Histeridae) associated with Ips spp. (Coleoptera: Scolytidae) in loblolly pine (Pinus taeda L.). PhD thesis Louisiana State University, Baton Rouge.

Shepherd, W.P., Goyer, R.A., 2003. Seasonal abundance, arrival and emergence patterns of predaceous hister beetles (Coleoptera: Histeridae) associated with Ips engraver beetles (Coleoptera: Scolytidae) in Louisiana. J. Entomol. Sci. 38, 612–620.

Shook, R.S., Baldwin, P.H., 1970. Woodpecker predation on bark beetles in Engelmann spruce logs as related to stand density. Can. Entomol. 102, 1345–1354.

Siddiqui, M.R., 2000. Tylenchida. Parasites of Plants and Insects. CABI, Wallingford.

Siemaszko, W., 1939. Fungi associated with bark beetles in Poland. Planta Polonica 7, 1–54.

Sikorowski, P.P., Lawrence, A.M., Nebeker, T.E., Price, T.S., 1996. Virus and virus-like particles found in southern pine beetle adults in Mississippi and Georgia. Mississippi Agricultural & Forestry Experiment Station, Technical Bulletin 212, 1–9.

Skuhravy, V., 2002. Lykozrout smrkovy a jeho kalamity (The Norway spruce bark beetle and its outbreaks). Agrospoj, Praha, 196 pp.

Slankis, A., 1967. Contortylenchus cylindricus sp. n. and Contortylenchus rarus sp. n. (Tylenchida: Contortylenchidae). Parasites of bark beetles and taxonomic notes on the genus Contortylenchus Rühm 1956. Trudy Gel'mintologischeskoi Laboratorii 18, 111–118.

Slankis, A. (1969). A new species of endoparasitic nematode, Contortylenchus pseudodiplogaster n.sp. from the bark beetle, Ips sexdentatus.—Materialy nauchn Konf vsesoiuz Obshchestva Gel'mint 2, 302–305.

Smith, M.T., Goyer, R.A., 1982. The life cycle of Corticeus glaber (Coleoptera: Tenebrionidae), a facultative predator of the southern pine beetle, Dendroctonus frontalis (Coleoptera: Scolytidae). Can. Entomol. 114, 535–537.

Solter, L.F., Becnel, J.J., Oi, D.H., 2012. Microsporidian entomopathogens. In: Vega, F.E., Kaya, H.K. (Eds.), Insect Pathology, Second Edition. Academic Press, San Diego, pp. 221–263.

Sperling, P., 1878. Unsere Spechte und ihre forstliche Bedeutung. Julius Springer, Berlin.

Sprague, V., 1977a. Annotated list of species of microsporidia. In: Bulla Jr., L.A., Cheng, T.C. (Eds.), Comparative Pathobiology, Volume 2: Systematics of the Microsporidia. Plenum Press, New York, pp. 31–334.

Sprague, V., 1977b. The zoological distribution of microsporidia. In: Bulla, L.A., Cheng, T.C. (Eds.), Comparative Pathobiology, Volume 2: Systematics of the Microsporidia. Plenum Press, New York, pp. 335–385.

Stark, R.W., Dahlsten, D.L., 1970. Studies on the population dynamics of the western pine beetle, Dendroctonus brevicomis Le Conte (Coleoptera: Scolytidae). University of California, Division of Agricultural Sciences, 174 pp.

Stein, C.R., Coster, J.E., 1977. Distribution of some predators and parasites of the southern pine beetle in two species of pine. Environ. Entomol. 6, 689–694.

Steinwender, B., Krenn, H., Wegensteiner, R., 2010. Different effects of the insect pathogenic fungus Beauveria bassiana (Deuteromycota) on the bark beetle Ips sexdentatus (Coleoptera: Curculionidae) and on its predator *Thanasimus formicarius* (Coleoptera: Cleridae). J. Plant Dis. Protect. 117, 33–38.

Stephen, F.M., Dahlsten, D.L., 1976. The arrival sequence of the arthropod complex following attack by *Dendroctonus brevicomis* (Coleoptera: Scolytidae) in ponderosa pine. Can. Entomol. 108, 283–304.

Stephen, F.M., Berisford, C.W., Dahlsten, D.L., Fenn, P., Moser, J.C., 1993. Invertebrate and microbial associates. In: Schowalter, T.D., Filip, G.M. (Eds.), Beetle-Pathogen Interactions in Conifer Forests. Academic Press, San Diego, pp. 129–153.

Struble, G.R., 1930. The biology of certain Coleoptera associated with bark beetles in western yellow pine. Univ. California Publ. Entomol. 5, 105–134.

Sullivan, B.T., Berisford, C.W., 2004. Semiochemicals from fungal associates of bark beetles may mediate host location behavior of parasitoids. J. Chem. Ecol. 30, 703–717.

Sullivan, B.T., Pettersson, E.M., Seltmann, K.C., Berisford, C.W., 2000. Attraction of the bark beetle parasitoid *Roptrocerus xylophagorum* (Hymenoptera: Pteromalidae) to host-associated olfactory cues. Environ. Entomol. 29, 1138–1151.

Susoy, V., Herrmann, M., 2014. Preferential host switching and codivergence shaped radiation of bark beetle symbionts, nematodes of *Micoletzkya* (Nematoda: Diplogastridae). J. Evol. Biol.

Takov, D., Pilarska, D., 2008. Prevalence of *Gregarina typographi* Fuchs (Apicomplexa: Gregarinidae) and Nematodes (Nematoda) in bark beetles (Coleoptera: Scolytinae) from Bulgaria depending on the host gender. Acta Zool. Bulg. 60, 227–232.

Takov, D., Pilarska, D., 2009. Single and mixed infections in *Ips typographus* (Coleoptera: Scolytinae) caused by the entomopathogens *Entomopoxvirus typographi* (Virales), *Gregarina typographi* (Sporozoa) and *Chytridiopsis typographi* (Microsporidia). Acta Zool. Bulg. 61, 45–48.

Takov, D., Pilarska, D., Wegensteiner, R., 2006. Entomopathogens in *Ips typographus* (Coleoptera: Scolytidae) from several spruce stands in Bulgaria. Acta Zool. Bulg. 58, 409–420.

Takov, D., Doychev, D., Wegensteiner, R., Pilarska, D., 2007. Study of bark beetle (Coleoptera, Scolytidae) pathogens from coniferous stands in Bulgaria. Acta Zool. Bulg. 59, 87–96.

Takov, D., Pilarska, D., Wegensteiner, R., 2010. List of protozoan and microsporidian pathogens of economically important bark beetle species (Coleoptera: Curculionidae: Scolytinae) in Europe. Acta Zool. Bulg. 62, 201–209.

Takov, D., Doychev, D., Linde, A., Draganova, S., Pilarska, D., 2011. Pathogens of bark beetles (Coleoptera: Curculionidae) in Bulgarian forests. Phytoparasitica 39, 343–352.

Takov, D.I., Dimitrov Doychev, D., Linde, A., Atanasova Draganova, S., Kirilova Pilarska, D., 2012. Pathogens of bark beetles (Curculionidae: Scolytinae) and other beetles in Bulgaria. Biologia 67, 1–7.

Tang, X.C., Sun, F.L., 1990. Study on DNA and polypeptides of a smallpox virus in *Dendroctonus armandi* Tsai et Li. Microbiology (Beijing) 17, 258–261.

Tanyeli, E., Sevim, A., Demirbag, Z., Eroglu, M., Demir, I., 2010. Isolation and virulence of entomopathogenic fungi against the great spruce bark beetle, *Dendroctonus micans* (Kugelann) (Coleoptera: Scolytidae). Biocontrol Sci. Tech. 20, 695–701.

Tartar, A., 2013. The non-photosynthetic algae *Helicosporidium* spp.: emergence of a novel group of insect pathogens. Insects 4, 375–391.

Thalenhorst, W., 1958. Grundzüge der Populationsdynamik des grossen Fichtenborkenkäfers *Ips typographus* L. Schriftenreihe der Forstlichen Fakultät der Universität Göttingen und Mitteilungen der Niedersächsischen Forstlichen Versuchsanstalt 21, 1–126.

Thatcher, R.C., Pickard, L.S., 1966. The clerid beetle, *Thanasimus dubius*, as a predator of the southern pine beetle. J. Econ. Entomol. 59, 955–957.

Théodoridès, J., 1960. Parasites et phoretiques de coleopteres et de myriapodes de Richelieu (Indre-et-Loire). Annales de Parasitologie 35, 488–581.

Thong, C.H.S., Webster, J.M., 1975. Effects of the bark beetles nematode *Contortylenchus reversus* on gallery construction, fecundity and egg viability of the Douglas fir beetle *Dendroctonus pseudotsugae* (Coleoptera: Scolytidae). J. Invertebr. Pathol. 26, 235–238.

Thong, C.H.S., Webster, J.M., 1983. Nematode parasites and associates of *Dendroctonus* spp. and *Trypodendron lineatum* (Coleoptera: Scolytidae) with description of *Bursaphelenchus varicauda* n. sp. J. Nematol. 15, 312–318.

Thorne, G., 1935. Nemic parasites and associates of the mountain pine beetle (*Dendroctonus monticolae*) in Utah. J. Agric. Res. 51, 131–144.

Tomalak, M., Welch, H.E., Galloway, T.D., 1988. Interaction of parasitic nematode *Parasitaphelenchus oldhami* (Nematoda: Aphelenchoididae) and a bacterium in Dutch elm disease vector, *Hylurgopinus rufipes* (Coleoptera: Scolytidae). J. Invertebr. Pathol. 52, 301–308.

Tomalak, M., Welch, H.E., Galloway, T.D., 1989. Nematode parasites of bark beetles (Scolytidae) in southern Manitoba, with descriptions of three new species of *Sulphuretylenchus* Ruhm (Nematoda: Allanotonematidae). Can. J. Zool. 67, 2497–2505.

Tømmerås, B.Å., 1985. Specialization of the olfactory receptor cells in the bark beetle *Ips typographus* and its predator *Thanasimus formicarius* to bark beetle pheromones and host tree volatiles. J. Comp. Physiol. A 157, 335–341.

Tonka, T., Weiser, J., 2009. Entomopoxvirus in the spruce bark beetle, *Ips typographus* and its laboratory management, Proceedings of the IUFRO WP 7.03.10 meeting, 92–95, in ŠtrbskéPleso, Slovakia, Sept. 15–19, 2008.

Tonka, T., Pultar, O., Weiser, J., 2007. Survival of the spruce bark beetle, *Ips typographus*, infected with pathogens or parasites. IOBC/wprs Bull. 30, 211–215.

Tonka, T., Weiser Jr., J., Weiser, J., 2010. Budding: a new stage in the development of *Chytridiopsis typographi* (Zygomycetes: Microsporidia). J. Invertebr. Pathol. 104, 17–22.

Unal, S., Yaman, M., Tosun, O., Aydin, C., 2009. Occurrence of *Gregarina typographi* (Apicomplexa, Gregarinidae) and *Metschnikowia typographi* (Ascomycota, Metschnikowiaceae) in *Ips sexdentatus* (Coleoptera: Curculionidae, Scolytinae) populations in Kastamonu (Turkey). J. Anim. Vet. Adv. 8, 2687–2691.

VanLaerhoven, S.L., Stephen, F.M., 2002. Height distribution of adult parasitoids of the southern pine beetle complex. Environ. Entomol. 31, 982–987.

Vasanthakumar, A., Delalibera Jr., I., Handelsman, J., Klepzig, K.D., Schloss, P.D., Raffa, K.F., 2006. Characterization of gut-associated bacteria in larvae and adults of the southern pine beetle, *Dendroctonus frontalis* Zimmermann. Environ. Entomol. 35, 1710–1717.

Vaupel, O., Zimmermann, G., 1996. Orientierende Versuche zur Kombination von Pheromonfallen mit dem insektenpathogenen Pilz *Beauveria bassiana* (Bals.) Vuill. gegen die Borkenkäferart *Ips typographus* L. (Col., Scolytidae). J. Pest Sci. 69, 175–179.

Vega, F.E., Blackwell, M. (Eds.), 2005. Insect-Fungal Associations. Ecology and Evolution. Oxford University Press, New York.

Vega, F.E., Kaya, H.K. (Eds.), 2012. Insect Pathology. Second Edition. Academic Press, San Diego, 490 pp.

Vega, F.E., Meyling, N.V., Luangsaard, J.J., Blackwell, M., 2012. Fungal entomopathogens. In: Vega, F.E., Kaya, H.K. (Eds.), Insect Pathology, Second Edition. Academic Press, San Diego, pp. 171–220.

Vincent, C., Goettel, M.S., Lazarovits, G., 2007. Biological Control: A Global Perspective. CABI, Wallingford.

Vlasák, J., Bříza, J., Pavingerová, D., Modlinger, R., Knížek, M., Malá, J., 2012. Cry3A δ-endotoxin gene mutagenized for enhanced toxicity to spruce bark beetle in a receptor binding loop. Afr. J. Biotechnol. 11, 15236–15240.

Wegensteiner, R., 1992. Untersuchungen zur Wirkung von Beauveria-Arten auf Ips typographus (Coleoptera, Scolytidae). Mitt. Dtsch. Ges. Allg. Angew. Entomol. 8, 104–106.

Wegensteiner, R., 1994. Chytridiopsis typographi (Protozoa, Microsporidia) and other pathogens in Ips typographus (Coleoptera, Scolytidae). IOBC/wprs Bull. 17, 39–42.

Wegensteiner, R., 1996. Laboratory evaluation of Beauveria bassiana (Bals.) Vuill. against Ips typographus (Coleoptera, Scolytidae). IOBC/wprs Bull 19, 186–189.

Wegensteiner, R., 2000. Laboratory evaluation of Beauveria bassiana (Bals.) Vuill. and Beauveria brongniartii (Sacc.) Petch against the four eyed spruce bark beetle, Polygraphus poligraphus (L.) (Coleopetra, Scolytidae). IOBC/wprs Bull 23, 161–166.

Wegensteiner, R., 2004. Pathogens in bark beetles. In: Lieutier, F., Day, K. R., Battisti, A., Grégoire, J.-C., Evans, H.F. (Eds.), Bark and Wood Boring Insects in Living Trees in Europe, A Synthesis. Kluwer Academic Publishers, Dordrecht, pp. 291–313.

Wegensteiner, R., Weiser, J., 1995. A new entomopoxvirus in the bark beetle Ips typographus (Coleoptera: Scolytidae). J. Invertebr. Pathol. 65, 203–205.

Wegensteiner, R., Weiser, J., 1996a. Untersuchungen zum Auftreten von Pathogenen bei Ips typographus (Coleoptera, Scolytidae) aus einem Naturschutzgebiet im Schwarzwald (Baden-Württemberg). J. Pest Sci. 69, 162–167.

Wegensteiner, R., Weiser, J., 1996b. Occurrence of Chytridiopsis typographi (Microspora, Chytridiopsida) in Ips typographus L. (Coleoptera, Scolytidae) field populations and in a laboratory stock. J. Appl. Entomol. 120, 595–602.

Wegensteiner, R., Weiser, J., 1998. Infection of Ips typographus from Finland with the Ascomycete Metschnikowia cf. bicuspidata, Proceedings, 6th European Congress of Entomology, Ceske Budejovice, p. 667.

Wegensteiner, R., Weiser, J., 2004. Annual variation of pathogen occurrence and pathogen prevalence in Ips typographus (Coleoptera, Scolytidae) from the BOKU University Forest Demonstration Centre. J. Pest Sci. 77, 221–228.

Wegensteiner, R., Weiser, J., 2009. Geographische Verbreitung und Häufigkeit von Pathogenen im Fichtenborkenkäfer Ips typographus L. (Coleoptera, Curculionidae) in Europa. Mitt. Dtsch. Ges. Allg. Angew. Entomol. 17, 159–162.

Wegensteiner, R., Weiser, J., Führer, E., 1996. Observations on the occurrence of pathogens in the bark beetle Ips typographus L. (Coleoptera, Scolytidae). J. Appl. Entomol. 120, 199–204.

Wegensteiner, R., Pernek, M., Weiser, J., 2007a. Occurrence of Gregarina typographi (Sporozoa, Gregarinidae) and of Metschnikowia typographi (Ascomycota, Metschnikowiaceae) in Ips sexdentatus (Coleoptera, Scolytidae) from Austria. IOBC/wprs Bull. 30, 217–220.

Wegensteiner, R., Epper, C., Wermelinger, B., 2007B. Untersuchungen über das Auftreten und die Dynamik von Pathogenen bei Ips typographus (Coleoptera, Scolytidae) in Befallsherden unter besonderer Berücksichtigung der Protozoen. Mitt. Schweiz. Entomol. Ges. 80, 79–90.

Wegensteiner, R., Dedryver, C.-A., Pierre, J.-S., 2010. The comparative prevalence and demographic impact of two pathogens in swarming Ips typographus adults: a quantitative analysis of long term trapping data. Agric. For. Entomol. 12, 49–57.

Wegensteiner, R., Stradner, A., Händel, U., 2014. Occurrence of pathogens in Ips typographus (Coleoptera, Curculionidae) and in other spruce bark beetles from the wilderness reserve Dürrenstein (Lower Austria). Biologia 69, 92–100.

Weiser, J., 1954a. Beitrag zur Kenntnis der Parasiten des Borkenkäfers Ips typographus. I. Vestnik Ceskoslovenske Zoologicke Spolecnosti. Acta Societatis Zoologicae Bohemoslovenicae 18, 217–227.

Weiser, J., 1954b. Prispevek k systematizaci Schizogregarin. Ceskoslovenske Parasitologie 1, 179–212.

Weiser, J., 1955. Beitrag zur Kenntnis der Parasiten des Borkenkäfers Ips typographus. II. Vestnik Ceskoslovenske Zoologicke Spolecnosti Acta Societatis Zoologicae Bohemoslovenicae 19, 374–380.

Weiser, J., 1961a. A new microsporidian from the bark beetle Pityokteines curvidens Germar (Col. Scolytidae) in Czechoslovakia. J. Insect Pathol. 3, 324–329.

Weiser, J., 1961b. Die Mikrosporidien als Parasiten der Insekten. 17, Monographien zur angewandten Entomologie. P. Parey, Hamburg und Berlin.

Weiser, J., 1966. Nemoci hmyzu. Academia, Prague.

Weiser, J., 1968. Plistophora scolyti sp.n. (Protozoa, Microsporidia) a new parasite of Scolytus scolytus F. (Col., Scolytidae). Folia Parasitol. 15, 11–14.

Weiser, J., 1970. Three new pathogens of the Douglas fir beetle, Dendroctonus pseudotsugae: Nosema dendroctoni n.sp., Ophryocystis dendroctoni n.sp. and Chytridiopsis typographi n.comb. J. Invertebr. Pathol. 16, 436–441.

Weiser, J., 1977. An Atlas of Insect Diseases, 2nd ed. Academia, Prague.

Weiser, J., Wegensteiner, R., 1994. A new entomopoxvirus in the bark beetle Ips typographus (Coleoptera, Scolytidae) in Czechoslovakia. Z. Angew. Zool. 80, 425–434.

Weiser, J., Wegensteiner, R., Žižka, Z., 1995. Canningia spinidentis gen. et sp.n. (Protista: Microspora), a new pathogen of the fir bark beetle Pityokteines spinidens. Folia Parasit 42, 1–10.

Weiser, J., Wegensteiner, R., Žižka, Z., 1997. Ultrastructures of Nosema typographi Weiser 1955 (Microspora: Nosematidae) of the bark beetle Ips typographus (Coleoptera: Scolytidae). J. Invertebr. Pathol. 70, 156–160.

Weiser, J., Wegensteiner, R., Žižka, Z., 1998. Unikaryon montanum sp. n. (Protista, Microspora), a new pathogen of the spruce bark beetle, Ips typographus (Coleoptera: Scolytidae). Folia Parasit 45, 191–195.

Weiser, J., Händel, U., Wegensteiner, R., Žižka, Z., 2002. Unikaryon polygraphi sp.n. (Protista: Microspora), a new pathogen of the four-eyed spruce bark beetle, Polygraphus poligraphus (Col., Scolytidae). J. Appl. Entomol. 126, 148–154.

Weiser, J., Wegensteiner, R., Händel, U., Žižka, Z., 2003. Infections with the Ascomycete Metschnikowia typographi n.sp. in the bark beetle Ips *typographus* and *Ips amitinus* (Col., Scolytidae). Folia Microbiol. 48, 611–618.

Weiser, J., Holuša, J., Žižka, Z., 2006. *Larssoniella duplicati* n.sp. (Microsporidia, Unikaryonidae), a newly described pathogen infecting the double-spined spruce bark beetle, *Ips duplicatus* (Coleoptera, Scolytidae) in the Czech Republic. J. Pest Sci. 79, 127–153.

Wermelinger, B., 2002. Development and distribution of predators and parasitoids during two consecutive years of an *Ips typographus* (Col., Scolytidae) infestation. J. Appl. Entomol. 126, 521–527.

Wermelinger, B., Epper, C., Kenis, M., Ghosh, S., Holdenrieder, O., 2012. Emergence patterns of univoltine and bivoltine *Ips typographus* (L.) populations and associated natural enemies. J. Appl. Entomol. 136, 212–224.

Wermelinger, B., Obrist, M.K., Baur, H., Jakoby, O., Duelli, P., 2013. Synchronous rise and fall of bark beetle and parasitoid populations in windthrow areas. Agric. For. Entomol. 15.

Weslien, J., 1992. The arthropod complex associated with *Ips typographus* (L.) (Coleoptera, Scolytidae): species composition, phenology, and impact on bark beetle productivity. Entomol. Fenn. 3, 205–213.

Weslien, J., Schroeder, L.M., 1999. Population levels of bark beetles and associated insects in managed and unmanaged spruce stands. For. Ecol. Manage. 115, 267–275.

Whitney, H.S., Ritchie, D.C., Borden, J.H., Stock, A.J., 1984. The fungus *Beauveria bassiana* (Deuteromycotina: Hyphomycetaceae) in the western balsam beetle, *Dryocoetes confusus* (Coleoptera: Scolytidae). Can. Entomol. 116, 1419–1424.

Wichmann, H.E., 1956. Untersuchungen über *Ips typographus* L. und seine Umwelt—Asilidae. Raubfliegen. Z. Ang. Entomol. 39, 58–62.

Wichmann, H.E., 1957. Untersuchungen an *Ips typographus* L. und seiner Umwelt—Die Kamelhalsfliegen. Z. Ang. Entomol. 40, 433–440.

Wigger, H., 1993. Ökologische Bewertung von Räuber-Beifängen in Borkenkäfer-Lockstofffallen. Anz. Schädl.kd. Pflanzenschutz Umweltschutz 66, 68–72.

Wood, D.L., 1961. The occurrence of *Serratia marcescens* Bizio in laboratory populations of *Ips confusus* (LeConte) (Coleoptera, Scolytidae). J. Insect Pathol. 3, 330–331.

Wulf, A., 1979. Der insektenpathogene Pilz *Beauveria bassiana* (Bals.) Vuill. als Krankheitserreger des Kupferstechers *Pityogenes chalcographus* L. (Col., Scolytidae). PhD thesis, Georg-August-Universität Göttingen, Germany.

Wulf, A., 1983. Untersuchungen über den insektenpathogenen Pilz *Beauveria bassiana* (Bals.) Vuill. als Parasit des Borkenkäfers *Pityogenes chalcographus* L. (Col., Scolytidae). Z. Ang. Entomol. 95, 34–46.

Yaman, M., 2007. *Gregarina typographi* Fuchs, a gregarine pathogen of the six-toothed pine bark beetle, *Ips sexdentatus* (Boerner) (Coleoptera: Curculionidae, Scolytinae) in Turkey. Turk. J. Zool. 31, 359–363.

Yaman, M., 2008. First results on distribution and occurrence of the insect pathogenic alga *Helicosporidium* sp. (Chlorophyta: Trebouxiophyceae) in the populations of the great spruce bark beetle, *Dendroctonus micans* (Kugelann) (Coleoptera: Curculionidae, Scolytinae). N. West. J. Zool. 4, 99–107.

Yaman, M., Baki, H., 2010. The first record of *Gregarina typographi* Fuchs (Protista: Apicomplexa: Gregarinidae) from the European spruce bark

beetle, *Ips typographus* (Linnaeus) (Coleoptera: Curculionidae, Scolytinae) in Turkey. Türkiye Parazitoloji Dergisi 34, 179–182.

Yaman, M., Baki, H., 2011. First record of entomopoxvirus of *Ips typographus* Linnaeus (Coleoptera: Curculionidae, Scolytinae) for Turkey. Acta Zool. Bulg. 63, 199–202.

Yaman, M., Radek, R., 2005. *Helicosporidium* infection of the European spruce bark beetle, *Dendroctonus micans* (Coleoptera: Scolytidae). Eur. J. Protistol. 41, 203–207.

Yaman, M., Radek, R., 2007. Infection of the predator beetle *Rhizophagus grandis* Gyll. (Coleoptera, Rhizophagidae) with the insect pathogenic alga *Helicosporidium* sp. (Chlorophyta: Trebouxiophyceae). Biol. Control 41, 384–388.

Yaman, M., Radek, R., 2008a. Pathogens and parasites of adults of the great spruce bark beetle, *Dendroctonus micans* (Kugelann) (Coleoptera: Curculionidae, Scolytinae) from Turkey. J. Pest Sci. 81, 91–97.

Yaman, M., Radek, R., 2008b. Identification, distribution and occurrence of the Ascomycete *Metschnikowia typographi* in the great spruce bark beetle, *Dendroctonus micans*. Folia Microbiol. 53, 427–432.

Yaman, M., Radek, R., 2012. *Menzbieria chalcographi*, a new neogregarina pathogen of the great spruce bark beetle, *Dendroctonus micans* (Kugelann) (Curculionidae, Scolytinae). Acta Parasitol. 57, 216–220.

Yaman, M., Radek, R., Aydin, C., Tosun, O., Ertürk, Ö., 2009. First record of the insect pathogenic alga *Helicosporidium* sp. (Chlorophyta: Trebouxiophyceae) infection in larvae and pupae of *Rhizophagus grandis* Gyll. (Coleoptera, Rhizophaginae) from Turkey. J. Invertebr. Pathol. 102, 182–184.

Yaman, M., Ertürk, Ö., Aslan, I., 2010a. Isolation of some pathogenic bacteria from the Great Spruce Bark Beetle, *Dendroctonus micans* and its specific predator, *Rhizophagus grandis*. Folia Microbiol. 55, 35–38.

Yaman, M., Radek, R., Weiser, J., Aydin, C., 2010b. A microsporidian pathogen of the predatory beetle *Rhizophagus grandis* (Coleoptera: Rhizophagidae). Folia Parasit. 57, 233–236.

Yaman, M., Tosun, O., Aydin, C., Ertürk, Ö., 2011. Distribution and occurrence of the insect pathogenic alga *Helicosporidium* sp. (Chlorophyta: Trebouxiophyceae) in the predator beetle *Rhizophagus grandis* Gyll. (Coleoptera: Rhizophagidae)-rearing laboratories. Folia Microbiol. 56, 44–48.

Yaman, M., Radek, R., Linde, A., 2012. A new neogregarine pathogen of *Rhizophagus grandis* (Coleoptera: Monotomidae). North-Western J. Zool. 8, 353–357.

Yates, M.G., 1984. The biology of the oak bark beetle, *Scolytus intricatus* (Ratzeburg) (Coleoptera: Scolytidae), in southern England. Bull. Entomol. Res. 74, 569–579.

Yatsenkowsky, A.W., 1924. The castration of *Blastophagus* of pines by roundworms and their effect on the activity and life phenomena of the Ipidae. Publ. Agric. Inst. 3, 1–19, *Western White Russian*.

Ye, H., Bakke, A., 1997. Development and reproduction of *Thanasimus formicarius* (L.) (Coleoptera, Cleridae) at three constant temperatures. Can. Entomol. 129, 579–583.

Yilmaz, H., Sezen, K., Kati, H., Demirbağ, Z., 2006. The first study on the bacterial flora of the European spruce bark beetle, *Dendroctonus micans* (Coleoptera: Scolytidae). Biologia 61, 679–686.

Zhang, L.-W., Liu, Y.-J., Yao, J., Wang, B., Huang, B., Li, Z.-Z., et al., 2011. Evaluation of *Beauveria bassiana* (Hyphomycestes) isolates as potential agants for control of *Dendroctonus valens*. Insect Sci. 18, 209–216.

ZipcodeZoo, 2014. Available online: http://zipcodezoo.com, Last accessed: April 14, 2014.

Zitterer, P.M., 2002. Antagonists of *Ips acuminatus* (Gyllenhall) with special consideration of pathogens. Diploma thesis, University of Natural Resources, Vienna.

Žižka, Z., Weiser, J., Wegensteiner, R., 1997. Ultrastructures of oocysts of *Mattesia* sp. in *Ips typographus*. J. Eukaryotic Microbiol. 44, 25A, no. 98.

Žižka, Z., Weiser, J., Wegensteiner, R., 1998. Ultrastructures of syzygies and gametocysts of *Mattesia* sp. in the bark beetle *Ips typographus*. J. Eukaryotic Microbiol. 45, 8A, no. 42.

Žižka, Z., Weiser, J., Wegensteiner, R., 2001. Ultrastructures of the Entomopoxvirus Ab in the bark beetle *Ips typographus*. In: Berger, J. (Ed.), Cells III. Kopp Publ, Ceske Budejovice, pp. 214–215.

Zumr, V., 1983. Effect of synthetic pheromones Pheroprax on the coleopterous predators of the spruce bark beetle *Ips typographus* (L.). Z. Ang. Entomol. 95, 47–50.

Chapter 8

Dendroctonus

Diana L. Six and Ryan Bracewell

Department of Ecosystem and Conservation Sciences, University of Montana, Missoula, MT, USA

1. INTRODUCTION

Dendroctonus Erichson is a relatively small genus (19 described species), but one with a disproportionately large impact on coniferous forests. Its name is particularly apt (*Dendro-* tree, *-tonus* destroyer) for a group that contains most of the major conifer-killing bark beetles in the world. Several members of the genus develop outbreaks that make them important economically as well as ecologically. The importance of this group of beetles has generated considerable interest that has resulted in a number of taxonomic revisions (Hopkins, 1909; Chamberlin, 1939; Wood, 1963, 1982) and in-depth syntheses of the state of knowledge on particular species (Miller and Keen, 1960; Thatcher *et al.*, 1980; Grégoire, 1988; Raffa, 1988; Safranyik and Carroll, 2006; Coulson and Klepzig, 2011). A number of key characteristics of *Dendroctonus* are summarized in Table 8.1.

1.1 Life History

All *Dendroctonus* species share a number of basic life history attributes. The general life cycle begins with a female colonizing a tree and constructing a nuptial chamber. Except for two or three species that sib mate in the natal host, a male beetle is attracted to the female's location by various cues that differ by species and joins her under the bark. The pair mate and then construct a tunnel (gallery) in the phloem in which the female deposits her eggs. Most *Dendroctonus* are considered monogamous although some re-emergence and re-mating occurs. The female is responsible for most tunneling while the male packs boring dust and frass into older portions of the gallery or pushes it out of the gallery. Gallery shape and oviposition pattern along with host tree species can be used to confidently identify many species (Wood, 1982).

Egg hatch usually occurs soon after oviposition and first instar larvae feed perpendicular to the parental gallery (also called oviposition gallery). For many, the larvae feed and develop on a mixture of phloem and symbiotic fungi, although several species spend a portion of their development feeding in the outer bark, probably on fungi (Wood, 1982; Six, 2013). The fourth instar constructs a small chamber, stops feeding, voids its gut, and transforms into a pupa. Upon eclosion, the teneral adult maturation feeds on phloem, or for those with close symbiotic associations with fungi, on fungal spores, prior to exiting the tree (Six, 2003). All species reproduce sexually, although the degree of outcrossing varies among species (Grégoire, 1988; Bleiker *et al.*, 2013). The time required to complete a generation is influenced by local thermal conditions and the specific physiological properties of each species.

Dendroctonus have developed two distinct strategies for colonizing trees. One involves a pheromone-mediated mass attack of the bole that results in the death of the tree, larvae that feed individually, and for a few, the potential to develop massive outbreaks (Raffa *et al.*, 2008). The second strategy involves attacking singly or in pairs on the lower portion of the bole including the root collar and sometimes the roots themselves, gregarious larval feeding, and usually development in a living tree (Reeve *et al.*, 2012). Members of the latter group are true parasites that use the host for nutrition, but do not kill the host except under unusual conditions. Host location is based on tree chemical volatiles and mate location is effected with sex pheromones rather than aggregation pheromones. Members of the former group can be divided into early successional saprophages and facultative predators (Lindgren and Raffa, 2013) (Table 8.1). Early successional saprophages attack recently killed or weakened trees. These species are not capable of overwhelming and killing healthy trees and do not develop outbreaks. Facultative predators, on the other hand, are efficient tree killers that use weak trees in a manner similar to early successional saprophages during non-outbreak periods, but are capable of killing healthy trees during outbreaks (Lindgren and Raffa, 2013). Only a few predators can sustain outbreaks in healthy trees once their numbers surpass a particular threshold (Boone *et al.*, 2011).

1.2 Morphology, Body Size, and Sexual Size Dimorphism

Species in the genus *Dendroctonus* are morphologically very similar (Wood, 1982) (Figure 8.1). The body is oblong

Bark Beetles. http://dx.doi.org/10.1016/B978-0-12-417156-5.00008-3

TABLE 8.1 Distribution of *Dendroctonus* species by Phylogenetic Relatedness, Karyotype, Presence or Absence of Sexual Size Dimorphism, Host Tree Genus, Primary Location of Tree Colonized, Type of Larval Feeding, and Life History Strategy

Phylogenetic Group	Beetle Species	Karyotype	Host Genus	Location	Larval Feeding	Host Use Strategy
Clade 1	*D. armandi*	unknown	*Pinus*	Mid-lower bole	Individual	Facultative predator
Clade 2	**D. rufipennis*	14AA + Xyp	*Picea*	Mid-bole	Individual	Facultative predator
	D. micans	10AA + Xyp	*Picea*	Lower bole	Gregarious	Parasite
	D. punctatus	14AA + Xyp	*Picea*	Lower bole	Gregarious	Parasite
	D. murrayanae	14AA + Xyp	*Pinus*	Lower bole	Gregarious	Parasite
Clade 3	*D. simplex*	14AA + Xyp	*Larix*	Mid-bole	Individual	Facultative predator
	*D. pseudotsugae	14AA + Xyp	*Pseudotsuga*	Mid-bole	Individual	Facultative predator
Clade 4	*D. terebrans*	12AA + Xyp	*Pinus*	Lower bole-roots	Gregarious	Parasite
	*D. valens	13AA + Xyp	*Pinus*	Lower bole-roots	Gregarious	Parasite
	D. parallelocollis	13AA + Xyp	*Pinus*	Lower bole-roots	Individual	Parasite
	D. rhizophagus	13AA + Xyp	*Pinus*	Lower bole-roots	Gregarious	Parasite/Predator
Clade 5	**D. ponderosae*	11AA + Neo XY	*Pinus*	Mid-bole	Individual	Facultative predator
	D. jeffreyi	11AA + Neo XY	*Pinus*	Mid-bole	Individual	Facultative predator
Clade 6	*D. vitei*	unk.	*Pinus*	Mid-bole	Individual	Early successional saprophage
	* D. frontalis	7AA + Xyp (5AA + Xyp)	*Pinus*	Mid-bole	Individual	Facultative predator
	*D. mexicanus	5AA + Xyp	*Pinus*	Mid-bole	Individual	Early successional saprophage
	D. adjunctus	6AA + Xyp	*Pinus*	Mid-bole	Individual	Facultative predator
	*D. approximatus	5AA + Neo XY[+]	*Pinus*	Mid-bole	Individual	Early successional saprophage
	*D. brevicomis	5AA + Neo XY	*Pinus*	Mid-bole	Individual	Facultative predator

*Species suspected or shown to include cryptic species. [+] One of two cryptic groups within this currently described species.

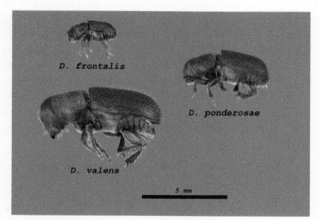

FIGURE 8.1 Three *Dendroctonus* species showing the broad range of adult size in the genus. *Photo credit: Erich Vallery, USDA Forest Service.*

with a steep convex declivity. Adult color ranges from dark brown to black with some species being reddish brown or possessing reddish-brown elytra. The antennae end in strongly flattened round clubs (Wood, 1982) (see Chapter 2).

One conspicuous trait that varies significantly among species is their average body size (Figure 8.1). For example, *Dendroctonus frontalis* Zimmermann, the smallest species, averages 2.8 mm long, in comparison with the largest species, *D. valens* LeConte, which at nearly three times that length, averages 7.3 mm (Wood, 1982). Large species generally restrict their feeding to tree regions with thicker phloem (i.e., the root collar, stump, and lower bole), while species of small-to-medium size tend to colonize higher portions of the tree stem (i.e., the lower to mid-bole).

Although this, in part, may be due to niche partitioning (Paine *et al.*, 1981), these colonization and feeding patterns are also driven by physical constraints as large beetles cannot fit into thin-phloemed portions of the tree (Pureswaran and Borden, 2003). Additionally, small species may have difficulty overcoming the strong resin defenses present at the base of a tree. Body size is also linked to life history as many of the larger species (e.g., *D. ponderosae* Hopkins, *D. rufipennis* Kirby) are univoltine or semivoltine, whereas smaller species (e.g., *D. brevicomis* LeConte, *D. frontalis*) are bivoltine or multivoltine. This may indicate some level of constraint on size due to developmental rate.

Within a particular species, body size is thought to be under strong selection as this trait can significantly affect fitness. Studies indicate that larger individuals can fly further (Thompson and Bennett, 1971; Hedden and Billings, 1977; Williams and Robertson, 2008; Chen *et al.*, 2011), produce more pheromones (Pureswaran and Borden, 2003), have greater overwintering success (Safranyik, 1976), and produce more offspring (Reid, 1962; Amman and Cole, 1983; Elkin and Reid, 2005; Manning and Reid, 2013). However, there is substantial phenotypic plasticity for size, and expression of this trait is ultimately determined by a mixture of both environmental and genetic effects (Bentz *et al.*, 2011; Foelker and Hofstetter, 2014). There is clear evidence of population-level genetic differentiation in body size in *D. ponderosae*, the only species where it has rigorously been examined (Bentz *et al.*, 2001, 2011; Bracewell *et al.*, 2013), suggesting local evolutionary processes are important in shaping variation in this trait.

Several species of *Dendroctonus* exhibit female-biased sexual size dimorphism (Hedden and Billings, 1977; Pureswaran and Borden, 2003) while others show no differences between the sexes (Furniss and Kegley, 2008; Safranyik, 2011). Unfortunately, the presence (or absence) of sexual size dimorphism and the magnitude of the difference between sexes has only been characterized in a small number of species (Foelker and Hofstetter, 2014). There does not appear to be a strong phylogenetic pattern to the presence/absence of this trait. For species that do show dimorphism, there is a remarkably consistent difference among the sexes even when comparing different populations (Bentz *et al.*, 2001, 2011; Bracewell *et al.*, 2013) or different cohorts emerging at different times of the year within the same population (Hedden and Billings, 1977).

The forces that drive differences in size between the sexes are not well understood (Pureswaran and Borden, 2003; Foelker and Hofstetter, 2014). Selection directly on female size seems an obvious candidate, although it has repeatedly been found that there is no size assortative mating in the sexually dimorphic species *D. ponderosae*

(Pureswaran and Borden, 2003; Reid and Baruch, 2010). Fecundity selection on females is expected (Honek, 1993), but cannot necessarily explain why males are smaller (Fairbairn, 2007). It has been hypothesized that there is strong selection for greater female size as females are the "pioneering" sex in *Dendroctonus* and are, therefore, the first to encounter tree defenses (Foelker and Hofstetter, 2014). However, in some species, pioneering females are the same size as males (Safranyik, 2011). Clearly, any explanation for what drives sexual size dimorphism in *Dendroctonus* needs to account for why sex-specific differences exist in one species but not another, and occurs differentially among species that have similar mating systems and life histories.

1.3 Host Range and Specialization

All members of the genus are restricted to Pinaceae as hosts (Wood, 1982). Their distribution is primarily Nearctic with 17 species in North and Central America, one species in China, and one species in Eurasia (Wood, 1982). The geographic origin of the genus is clearly Nearctic although where in North America the genus arose remains uncertain. Wood (1982) suggested a Mexican origin due to the high diversity of *Dendroctonus* in that region. In contrast, Zúñiga *et al.* (2002a, b) postulated that the genus originated north of Mexico with beetles tracking the dispersal of the main host genus *Pinus* from the boreal region southward into Mexico and across Beringia into Asia.

In general, the genus is associated with montane systems rich in *Pinus*. Most species have relatively broad ecological amplitudes that may account for the high degree of sympatry that is observed. However, factors such as temperature, moisture, and elevation also clearly exert strong constraints resulting in smaller realized geographic ranges relative to the distribution of host trees (Salinas-Moreno *et al.*, 2004).

Within *Dendroctonus*, species specialize at least at the generic level. All but five species colonize *Pinus*. The remainder colonizes *Picea*, *Pseudotsuga*, or *Larix*. The number of host species used varies from one (e.g., *Dendroctonus jeffreyi* Hopkins) to more than 20 (e.g., *D. valens*) (Wood, 1982; Salinas-Moreno *et al.*, 2004). However, even highly polyphagous species typically show a preference for one or a few host species. In addition, while *Pinus* is the primary host for most *Dendroctonus*, not all *Pinus* within the range of *Dendroctonus* are used. For example, in Mexico, where the highest diversity of *Dendroctonus* occurs as well as that of *Pinus*, only 24 of 47 species of *Pinus* are hosts (Salinas-Moreno *et al.*, 2004). Preferred species of *Pinus* are predominantly found in subsections *Leiophyllae*, *Ponderosae*, and *Oocarpae*. A few species of *Dendroctonus* will attack, and occasionally successfully colonize, trees in non-host genera, but typically only during severe outbreaks (Reed *et al.*, 1986; Huber *et al.*, 2009). Overall, *Dendroctonus* have remained conservative in their use of Pinaceae and *Pinus*. There have been relatively few shifts to genera other than *Pinus* and then only to *Larix*, *Picea*, and *Psuedotsuga*. Only one species, *D. murrayanae* Hopkins, appears to have reverted to using *Pinus* (Kelley and Farrell, 1998).

As with all phytophagous insects, the degree of specialization and generalization in host plant use must be placed within the context of the actual host range to that of available hosts. Clear constraints in host use occur for *Dendroctonus* given no species exploit hosts outside of Pinaceae. However, when considering host breadth within Pinaceae, *Dendroctonus* species can be categorized as specialist or generalist by using the ratio of potential hosts they encounter to the actual number of hosts used. Approximately 70% of *Dendroctonus* can be considered generalists under this criterion (Kelley and Farrell, 1998). Kelley and Farrell (1998) explicitly investigated the evolution of specialization in *Dendroctonus* within this context and found that the degree of host specialization in the genus was bimodally distributed; generalist species used greater than 60% of congeneric hosts within their geographic range while specialists used less than 40%.

1.4 Communication and Host Location

Dendroctonus must communicate in order to locate and recognize mates and, for some, to initiate and coordinate a mass attack on a host tree. For *Dendroctonus* that do not kill trees and attack singly or in pairs, volatile host tree compounds are commonly used to locate a suitable tree (Erbilgin *et al.*, 2007). These compounds vary greatly in type and concentration allowing the insects to distinguish between different tree species as well as to assess the "condition" of trees of the same species (Pureswaran and Borden, 2003; Sheperd *et al.*, 2008, and references therein).

For species that kill trees, tree chemistry is important in the initial location and choice of a tree as well as in stimulating the production of pheromones involved in mass attack, either *de novo* or from tree defensive compounds (Byers, 1995; Raffa, 2001). In *Dendroctonus,* the female selects the host tree. For mass attacking species, the female will also initiate a mass attack through the release of an aggregation pheromone. These pheromones can attract thousands of beetles to the tree within just a few days. Once tree defenses are depleted, the beetles cease producing the aggregation pheromone and begin production of anti-aggregation pheromones that act to reduce the potential for intra-specific competition (Byers, 1995; Raffa, 2001).

Communication is also required for mate location and recognition at the tree except for the few species that mate as sibs in the natal host and colonize trees singly as mated females. For *Dendroctonus*, mate location can involve pheromones, acoustic signals, or both. In some cases, females and males release similar or different pheromones that initiate courtship and mating (Phillips *et al.*, 1989; Byers, 1995). In *Dendroctonus*, both males and females stridulate using opposing structures on the elytra and abdomen (Flemming *et al.*, 2013). The process produces simple and interrupted chirps and vibrations in the phloem. Males produce chirps when stressed or interacting with females or other males. Females stridulate less frequently and typically in response to the presence of males or when they sense other females boring nearby (Rudinsky and Michael, 1973). Once a male lands at the opening of a new gallery, he begins to stridulate. These chirps may indicate to the female that a potential mate is present and can cause females to cease producing the aggregation pheromone (Fleming *et al.*, 2013).

While pheromones have long been thought to have evolved through gradual changes in proportions and structures of the chemicals involved (Roelofs and Brown, 1982), recent evidence suggests that evolution of pheromones in bark beetles, including *Dendroctonus*, has involved saltational shifts, with sibling species often possessing pheromones that are more different from one another than to those of more distantly related species (Symonds and Elger, 2004). Because many *Dendroctonus* share the same host trees and geographical range, these differences are unlikely to be linked to differences in diet or environment (Symonds and Elger, 2004).

1.5 Phylogeny and Taxonomy

The genus *Dendroctonus* is placed within the subfamily Scolytinae of the Curculionidae (weevils). There are currently 19 recognized species and there have been a number of significant revisions within this group over time (Hopkins, 1909; Chamberlin, 1939; Wood, 1963, 1982). Only two phylogenetic analyses of *Dendroctonus* have been completed to date. Both have, by and large, been conducted on the same mtDNA COI sequences (Kelley and Farrell, 1998; Reeve *et al.*, 2012). The phylogenies developed in these studies were inferred using maximum parsimony (Kelley and Farrell, 1998) and maximum likelihood and Bayesian methods (Reeve *et al.*, 2012) (Figure 8.2). The newer phylogeny developed by Reeve *et al.* (2012) is similar to that of Kelley and Farrell (1998) except for shifts in the positions of three species. Some of the deeper nodes in the tree remain ambiguous. However, many major nodes are well supported by morphological and chromosomal data (Wood, 1982; Zúñiga *et al.*, 2002b).

Fossil evidence of larval feeding consistent with *Dendroctonus* galleries suggests that the genus originated more than 45 million years ago (Labandeira *et al.*, 2001). Molecular data appear to generally corroborate this date although these estimates tend to support a more recent

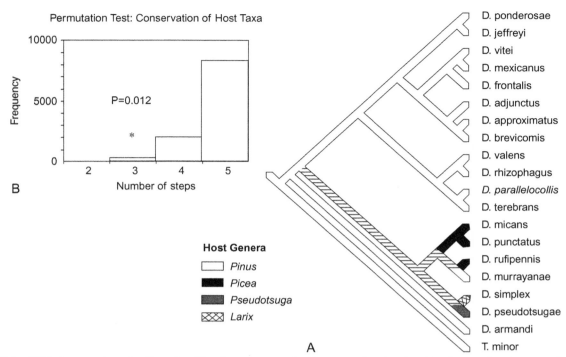

FIGURE 8.2   (A) Phylogeny estimate for 19 species of *Dendroctonus* mapped with most-parsimonious transitions between host tree genera in Pinaceae. (B) Histogram showing results of a permutation tail probability test randomizing the character "host genus" across beetle species holding topology constant.

(Continued)

origin (Sequeira *et al.*, 2000; McKenna *et al.*, 2009). Macro-evolutionary patterns in the genus suggest some level of host use conservatism since all species only use tree genera within the Pinaceae, and most only use *Pinus* (Kelley and Farrell, 1998). However, there is no evidence of strict co-speciation between beetle and host tree and there have been shifts in host use indicating that transitions to alternate genera (i.e., *Larix*, *Pseudotsugae*, and *Picea*) can occur (Kelley and Farrell, 1998). It is also important to note that specialist lineages of beetles have evolved repeatedly from generalist ancestors and are often in derived positions in the tree (Kelley and Farrell, 1998). Ancestral state reconstruction of the genus suggests that ancestral *Dendroctonus* likely mass attacked the bole of *Pinus* and were capable of developing outbreaks (Reeve *et al.*, 2012) (Figure 8.2).

1.6   Population Genetics and Cryptic Species

The importance of many *Dendroctonus,* both ecologically and economically, has spurred a number of population genetic analyses aimed at describing current genetic structure, demographic history, and phylogeography (reviewed in Avtzis *et al.*, 2012). From these studies, a number of general patterns have emerged. Nearly all have found evidence of significant population structuring indicating Pleistocene glaciations and refugia strongly

impacting the distribution of modern-day genetic variation. Additionally, most studies have found evidence of substantial post-Pleistocene range expansion and population expansion that mirrors the expansion pattern of their host trees. Many also find evidence of an isolation-by-distance gene flow pattern whereby more geographically distant populations tend to be less closely related than more geographically proximal populations. These gene flow patterns have been attributed to the fragmented distribution of host trees, the somewhat limited dispersal distance of these insects, and post-Pleistocene migration routes of host trees. Because of the strong effect of geography and host tree distribution, isolated populations in small fragmented forests at the periphery of the distribution tend to be genetically differentiated from populations in the core of the distribution.

One of the more surprising results from the many population analyses of *Dendroctonus* is the number of cryptic species that have been revealed. In nearly every species that has been investigated, there is some evidence of genetically isolated populations and reproductive isolation. For example, cryptic species have been suggested to occur in *D. ponderosae* (Mock *et al.*, 2007; Bracewell *et al.*, 2011), *D. rufipennis* (Maroja *et al.*, 2007), *D. pseudotsugae* Hopkins (Ruíz *et al.*, 2009), *D. valens* (Cai *et al.*, 2008), *D. brevicomis* (Kelley *et al.*, 1999), *D. frontalis* (Armendáriz-Toledano *et al.*, 2014), and *D. mexicanus*

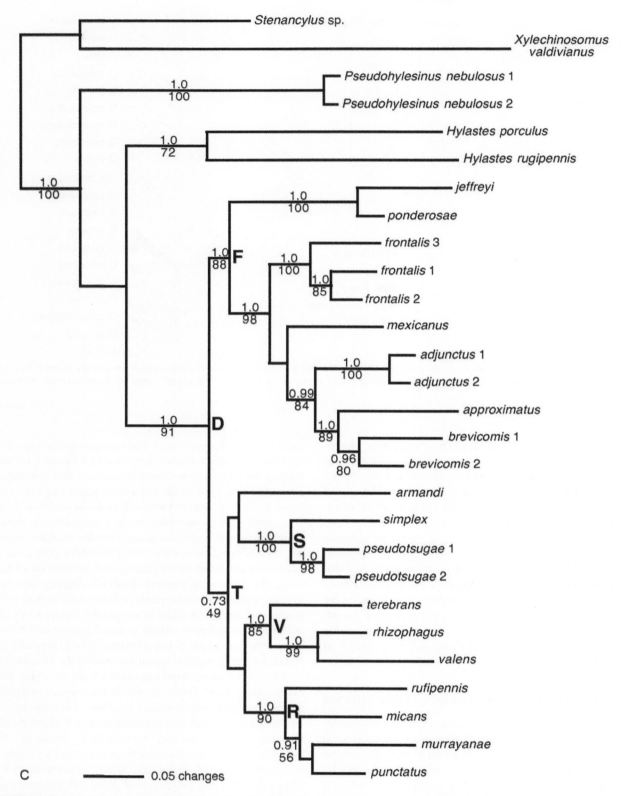

FIGURE 8.2, Cont'd (C) Phylogenetic (BKL—best known likelihood) tree estimating relationships among *Dendroctonus*. Values at nodes are posterior probabilities (top) and maximum likelihood bootstrap support values (lower). *(B) Taken from Kelley and Farrell (1998) with permission. (C) Taken from Reeve et al. (2012) with permission.*

Hopkins (Zúñiga *et al.*, 2006). These remained undetected until recent molecular analyses because of the overall morphological similarity between many species and because many field-based studies have often been geographically restricted. It is quite likely that additional population genetic analyses of several poorly studied *Dendroctonus* species will uncover more cryptic species.

The drivers of speciation within *Dendroctonus* are not well understood. There is evidence that geographic distance and isolation and host tree distribution play roles in facilitating genetic differentiation and possibly allow for allopatric speciation (Kelley *et al.*, 1999; Maroja *et al.*, 2007; Ruíz *et al.*, 2009). Host use and in particular host race formation via host shifts are thought to be important in speciation in phytophagous insects (Dres and Mallet, 2002), and some have proposed that changes in host use in *Dendroctonus* may facilitate speciation (Sturgeon and Mitton, 1982; Kelley and Farrell, 1998). However, evidence of host use alone causing differentiation and promoting reproductive isolation in the genus is mixed (Sturgeon and Mitton, 1986; Langor and Spence, 1990, 1991; Mock *et al.*, 2007).

One difficulty in determining the relative roles of host use and geography in *Dendroctonus* speciation is that the two factors are often confounded (Sturgeon and Mitton, 1986). Interestingly, the earliest stage of postzygotic reproductive isolation has been observed between populations with low levels of genetic divergence and from the same host tree species suggesting that reproductive isolation may evolve quickly and be unrelated to current host use (Bracewell *et al.*, 2011). Sexual selection is also considered by many to play a large role in speciation (Panhuis *et al.*, 2012), although this has largely been unexplored as a mechanism driving species formation in *Dendroctonus*.

1.7 Karyotypic Diversity

A noteworthy feature of *Dendroctonus* is the large range in chromosome number among species and numerous changes in sex chromosome configuration (Lanier *et al.*, 1981) (Table 8.1). Chromosome numbers range from $2n = 30$ in species like *D. rufipennis*, to $2n = 12$ in species like *D. brevicomis*. This diversity highlights significant chromosomal evolution (Zúñiga *et al.*, 2002b). Phylogenetic analyses suggest the ancestral condition in *Dendroctonus* is $2n = 30$ and many of the so-called "primitive" species retain this karyotype (Table 8.1). Many of the more derived lineages feature fewer chromosomes, indicating multiple chromosomal fusions and a general trend towards chromosome number reduction (Zúñiga *et al.*, 2002b). All species of *Dendroctonus* have heteromorphic sex chromosomes with the male being the heterogametic sex. The sex chromosomes show significant turnover through time with many species being described as a sex chromosome

bivalent Xyp yet with multiple species having neo-XY chromosomes (Lanier, 1981). Neo-XY chromosomes result when a fusion between a sex chromosome and an autosome occurs (Kaiser and Bachtrog, 2010). The evolutionary forces promoting these transitions in karyotype are unknown. Although there is considerable diversity among species, there is also very little evidence of intra-species chromosomal polymorphisms (Zúñiga *et al.*, 2002a, but see Lanier, 1981). Because of this karyotypic stability, differences in chromosome size and number have repeatedly been used to distinguish among species (Lanier and Wood, 1968; Lanier, 1981; Lanier *et al.*, 1988; Armendáriz-Toledano *et al.*, 2014).

2. EFFECTS OF THE ENVIRONMENT

2.1 The Abiotic Environment

Abiotic factors exert substantial effects on the population dynamics and distribution of *Dendroctonus*. These include temperature, moisture, and physiographic site conditions such as soils, elevation, and aspect that may affect tree vigor and susceptibility.

Like all insects, *Dendroctonus* are poikilothermic and thus all aspects of their lives are highly influenced by temperature. Temperature controls their development rate within the tree, generation length, initiation of and maintenance of dispersal flights, mating activity, and oviposition rates (Bentz *et al.*, 1991; Gaylord *et al.*, 2008). In general, as temperature increases, all of these rates increase, while cooling results in the reverse effect. However, the response is not linear. For example, development of *D. ponderosae* larvae increases with increasing temperature but rates are instar specific which aids the beetle in synchronizing emergence and mass attack, and entry into winter at the proper time (Bentz *et al.*, 1991). In *D. rufipennis*, cold induces diapause in overwintering larvae preventing progression to the pupal stage that is highly susceptible to freezing (Hansen *et al.*, 2011).

Temperature plays an important role in determining the number of generations a species produces per year. Some species, particularly those living in warmer regions, are highly labile, producing additional generations when and where temperatures are warmer. For these species, generations are often not synchronized and exhibit considerable overlap (Cibrían Tovar *et al.*, 1995). Multiple life stages often overwinter and in some cases, development continues year round. However, for species that live in areas with cold winters, adaptive seasonality is of the utmost importance. These beetles must survive very cold winters and it is critical that they enter winter in a stage capable of cold hardening and surviving subfreezing temperatures (Logan and Bentz, 1999; Logan and Powell, 2001). Temperatures that are too warm or too cool to support the appropriate length and timing of a life cycle can result in many beetles entering winter in stages susceptible to freezing (Logan and Powell, 2001).

For many *Dendroctonus*, temperature plays an important role delimiting geographic distribution (Mendoza *et al.*, 2011). A lack of cold tolerance in *D. frontalis* likely limits its range in the USA to primarily the southern region although suitable hosts exist further north (Tran *et al.*, 2007). High temperatures also reduce survival, further restricting its geographic range (Friedenberg *et al.*, 2008). Likewise, cold temperatures limit the extent of *D. ponderosae* in Canada despite the existence of suitable hosts further north and east (but see Section 4).

Moisture influences the dynamics of *Dendroctonus* by affecting the susceptibility of host trees to attack and their suitability for brood development post-attack. Drought is increasingly being implicated as a major initiation factor for outbreaks of facultative predators (Powers *et al.*, 1999; Hebertson and Jenkins, 2008; Sherrif *et al.*, 2011; Derose and Long, 2012; Creeden *et al.*, 2014). For many of these species, a relationship between beetle attack and the slow growth of trees, either related to genetics, dense stand conditions or to drought, has been found (Fischer *et al.*, 2010; DeRose and Long, 2012; Millar *et al.*, 2012; Knapp *et al.*, 2013). Some parasitic *Dendroctonus* may also respond to drought-affected trees, but this attraction does not appear to be universal. For example, *D. micans* Kug. has been reported to build up in drought-affected stands (Grégoire, 1988) while *D. murrayanae* prefers wetter sites and trees (Six *et al.*, 2011a).

The amount of moisture within a tree influences oviposition behavior of females as well as development (Webb and Franklin, 1978; Wagner *et al.*, 1979), and desiccation of brood can be a major cause of mortality (McCambridge and Knight, 1972). Studies have shown that moisture levels in trees killed by facultative predators shift considerably over time although not always in a predictable manner. Moisture may decrease continually in a tree over the development period of the beetle (Bleiker and Six, 2008a), or it may initially decline, but increase again near eclosion and dispersal of brood adults (Wagner *et al.*, 1979). Such variable outcomes may be due to differences in humidity and the incidence and timing of rainfall, which affects the rate and timing of moisture loss or rehydration of wood.

Moisture also influences the growth rates and survival of symbiotic fungi within the sapwood and phloem affecting nutrient availability to developing larvae and teneral adults and vectoring of the fungi by brood adults to new trees (Bleiker and Six, 2008a, b). For example, Bleiker and Six (2008a) found that high phloem moisture at the beginning of colonization just after beetle attack inhibited the growth of both symbionts of *D. ponderosae*, although one fungus was inhibited more than the other.

Both fungi grew well during the mid-stage of colonization, but by the end of the beetle's development period, growth was reduced or halted and the fungi began to die in some areas within the tree. In areas where fungi die or survive but cannot grow and sporulate, teneral beetles cannot maturation feed on spores and cannot acquire spores to carry the next tree and their brood. For beetles that are obligately dependent upon their fungal associates, this can potentially have strong negative effects on the survival of the next generation of beetles and on overall population dynamics.

The physical characteristics of sites can vary in a number of ways that affect their suitability for *Dendroctonus*. Garrison-Johnston *et al.* (2003) found that bedrock soil type highly influenced endemic and epidemic populations of *D. pseudotsugae* probably due to effects on nutrient availability to host trees. A number of studies have found that elevation, topography, and microsite variability influences the activity of facultative predators (Amman, 1973; Powers *et al.*, 1999; Williams *et al.*, 2008). For example, locations on the landscape that receive more solar radiation and have greater moisture stress can be more likely to support increases in tree mortality due to *D. pseudotsugae* (Powers *et al.*, 1999).

Fire is an additional abiotic factor that can influence populations of *Dendroctonus*. While fire can directly reduce populations of *Dendroctonus* when infested stands burn, it can also indirectly increase abundance by predisposing trees to infestation. Responses by parasites to fire-affected trees appear to be variable. *Dendroctonus valens* is clearly attracted to fire-damaged trees (Parker *et al.*, 2006; Six and Skov, 2009; Youngblood *et al.*, 2009; Fettig *et al.*, 2008, 2010; Hood *et al.*, 2010), however, its sister species, *D. terebrans*, also a parasite, does not show a response and is often lower in numbers in burned than in unburned sites (Hanula *et al.*, 2002; Sullivan, *et al.*, 2003; Campbell *et al.*, 2008).

Responses of facultative predators to fire-damaged trees also vary considerably. *Dendroctonus brevicomis*, *D. pseudotsugae*, and *D. adjunctus* Blanford respond positively to fire-damaged stands (McHugh *et al.*, 2003; Parker *et al.*, 2006; Six and Skov, 2009; Davis *et al.*, 2012). However, responses of *D. ponderosae*, *D. jeffreyi*, and *D. rufipennis* are variable with positive, neutral, and negative responses to fire all being reported (Ryan and Amman, 1996; Bradley and Tueller, 2001; Parker *et al.*, 2006; Six and Skov, 2009; Fettig *et al.*, 2010; Davis *et al.*, 2012). Part of the variability of responses within species may be due to whether flight periods occurred prior to or after burns, the degree and type of damage to the tree, and the size of the beetle population at the site (Fettig *et al.*, 2008; Six and Skov, 2009). Spread into healthy trees after fire-damaged trees have been depleted is not typical (Hanula *et al.*, 2002; Six and Skov, 2009; Davis *et al.*, 2012), but can occur if drought or some other stressing agent is present.

2.2   The Biotic Environment

Each *Dendroctonus* species interacts with a large number of other organisms that influence their fitness and dynamics to a greater or lesser degree. These include the host tree, other bark beetles including conspecifics, microbial symbionts, natural enemies, and phoronts including mites and nematodes. Here we review how these factors affect *Dendroctonus*. For the host tree, natural enemies including pathogens, and mite and nematode phoronts, we provide only a brief overview and refer the reader to more comprehensive treatments of these topics in Chapters 5, 6, and 7. Intra- and interspecific interactions of *Dendroctonus* species and other bark beetles and interactions with microbial symbionts are treated here in more detail.

2.2.1   The Host Tree

The host tree is one of the most important biotic factors affecting the life history of *Dendroctonus*. A key component of the interactions of the tree with these insects involves the tree's physical and chemical defenses. These take the form of preformed (primarily constitutive resin) and induced (resin with elevated concentrations of toxins) defenses that influence many aspects of host location and acceptance, survival and fitness of parental adults and offspring, pheromone production, and the proliferation of symbionts.

Dendroctonus use host tree volatiles to locate appropriate hosts, although the degree to which these are used varies by life history strategy. Parasites, which do not use aggregation pheromones, are typically strongly attracted to host volatiles, often those emanating from damaged trees, or in some cases from trees that have been previously attacked by previous generations of the parasite (Grégoire, 1988; Gilbert *et al.*, 2001; Erbilgin *et al.*, 2007).

For facultative predators, pioneering females locate hosts and initiate mass attacks. How pioneers choose hosts is still poorly understood and likely varies considerably among species. For less aggressive predators such as *D. pseudotsugae* and *D. rufipennis* that prefer weakened hosts, host tree volatiles may act as kairomones and synergize the effects of aggregation pheromones (Pureswaran and Borden, 2005). For highly aggressive species such as *D. ponderosae* that can maintain outbreaks in healthy trees once particular population thresholds are exceeded, it appears that random landing and close range acceptance or rejection of hosts may be more common (Pureswaran and Borden, 2005; Boone *et al.*, 2011). Host acceptance can also be mediated by female condition and population density where females in high-density populations and in poor condition exhibit lower selectivity in relation to tree defenses (Wallin and Raffa, 2004; Latty and Reid, 2010).

Entry into the tree involves exposure to various levels and types of tree defenses. For parasites that develop in living trees and that must contend with a continually defensive host, several behaviors are used to mediate the negative effects of defenses including laying eggs in frass mats, strategic angling of oviposition galleries, and gregarious feeding by larvae (Grégoire, 1988; Storer *et al.*, 1997). While these activities reduce exposure to resin, these beetles still require high tolerances to toxic tree resin compounds, which may be achieved through the use of bacterial symbionts in the gut (see Section 2.8.2).

Facultative predators, on the other hand, must contend with tree defenses only relatively early on in the colonization of the tree. Mass attack results in a depletion of preformed resin and shuts down the production of induced defensive compounds (Paine *et al.*, 1997). However, at least initially, defenses pose a serious challenge and, in at least some hosts, chemical defenses may remain high for a substantial period of time after attack. For example, in lodgepole pine, levels of monoterpenes may increase in the period after attack and remain at elevated levels for a month or more (Clark *et al.*, 2012). Host tree defenses have strong effects on beetle fitness though direct effects on survival of attacking adults as well as on production of brood (Raffa *et al.*, 2006; Seybold *et al.*, 2006; Manning and Reid, 2013). They also influence the proliferation of filamentous fungal associates within the host, potentially influencing the outcomes of symbioses over time (see Section 2.8.2).

Early successional saprophytes face lower levels of tree defenses because they colonize weak or recently killed trees. However, this strategy likely comes with trade-offs where reduced exposure to defenses is offset by lower food quality and higher levels of competition (Lindgren and Raffa, 2013).

The quality of the host tree as food is an important factor influencing beetle fitness. Thicker phloem tends to support greater productivity and size of offspring regardless of beetle life history strategy (Amman, 1982; Cerezke, 1995). Higher nitrogen availability in soils can also affect beetle productivity, either directly by increasing phloem nitrogen content or indirectly through the concentration of greater pools of nitrogen from sapwood by symbiotic fungi (Bleiker and Six, 2007; Cook *et al.*, 2010; Goodsman *et al.*, 2012).

For more comprehensive reviews on how *Dendroctonus* and other bark beetles interact with the host tree, see Chapters 1 and 5.

2.2.2   Interactions Within and Among *Dendroctonus* and other Bark Beetles

Interactions within and among bark beetle species can have substantial effects on fitness. The outcome and degree of impact of such interactions vary greatly in relation to host colonization strategy, host tree quality, the diversity of bark beetle species involved, geographic location, environmental conditions, timing of colonization, and population density. Overall, there has been a tendency to perceive all interactions among bark beetles as competition. However, evidence is accruing that substantial variability in outcomes occurs and that some interactions involve facilitation (Chen and Tang, 2007; Smith et al., 2011).

Co-occurring bark beetle species tend to segregate to a degree within a tree. This segregation is linked to both the size of the beetle and phloem thickness. Smaller species tend to colonize small diameter trees or tops of larger trees where phloem is thin, while larger beetles tend to colonize larger trees or low on the bole where phloem is thickest (Paine et al., 1981; Wood, 1982). However, phloem thickness does not completely determine the distribution of species within a tree. For example, distributions often shift depending on whether the tree is attacked by one species alone, attacked concurrently by multiple species, or colonized by a number of species over time (Paine et al., 1981; Cibrían Tovar et al., 1995). When species attack alone they may exploit their full niche. However, when multiple species co-occur, realized niches may be much smaller (Paine et al., 1981). Such contractions of niche may, in part, account for the failure of some otherwise aggressive species to develop outbreaks in areas of their range where bark beetle species diversity is particularly high.

Beetles use a number of chemical cues, including pheromones of conspecific and heterospecific beetles, to make decisions on where to locate on a tree, as well as to orient to trees that have been made more suitable by previous arriving species or to avoid areas within the tree that contain efficient competitors (Paine et al., 1981; Byers et al., 1984). However, competition is still often the outcome when bark beetle species co-occur. For example, Paine et al. (1981) found that when D. frontalis and I. avulsus inhabited the same portion of the tree, both species colonized less area. Often co-habitation results in asymmetrical effects on the species involved due to differences in their timing of entry into the tree, life history, and development rate (i.e., rate of resource consumption). Davis and Hofstetter (2009) found that when D. frontalis and D. brevicomis co-occurred, only D. frontalis suffered a reduction in fitness. In some cases, co-habitation may have a negative effect on one species, while having a positive effect on the other. For example, when Ips pini Say colonizes trees soon after they are killed by D. ponderosae, they benefit from reduced tree defenses and often exhibit high levels of productivity. However, their presence has a negative effect on D. ponderosae brood because of their more rapid consumption of resources (Safranyik et al., 1999; Boone et al. 2008).

Whether a species benefits or suffers because of the presence of other species of bark beetles appears to be species specific and may vary within a single community.

For example, D. armandi Tsai and Li is the most aggressive of a complex of bark beetles that colonize Pinus armandi Franchet. It is usually the first to attack a host tree (Chen and Tang, 2007). This facilitates the entry of a number of other bark beetle species. In the tree, niche partitioning appears to reduce competition among some of the species. Dendroctonus armandi and Hylurgops longipilis Reitter primarily concentrate attacks in the lower bole and root collar while the other species tend to attack from the mid-bole to the tree top. While spatial segregation may reduce competition, it does not completely alleviate it. The presence of three co-occurring bark beetles, Polygraphus sinensi Eggers, Cryphalus lipingensis Tsai and Li, and C. chinlingensis Tsai and Li, reduce survival of D. armandi. However, the presence of Ips accuminatus Gyll. positively influences D. armandi productivity, although the mechanism by which this occurs is not known (Chen and Tang, 2007).

Facilitation can also occur when non-aggressive facultative predators affect trees in such a way that they support more aggressive beetles during endemic phases. Smith et al. (2011) and Carroll et al. (2006) found that endemic populations of D. ponderosae in Canada often attack Pinus contorta Dougl. that have been previously infested by weak facultative predators and early successional saprophytes. This relationship was particularly strong with Pseudips mexicanus Hopkins (Smith et al., 2011). Pseudips mexicanus appears to condition trees in a way that aids in population maintenance of D. ponderosae when their numbers are too low to attack vigorous trees. In fact, in trees previously infested by P. mexicanus, larvae of D. ponderosae required less phloem resource to complete development while suffering no reduction in size (Smith et al., 2011).

The presence of multiple species within a tree can also have strong effects on beetle fitness through effects of natural enemies. For example, Boone et al. (2008) found that in trees killed by D. ponderosae, subsequent colonization by I. pini added to the predator load experienced by D. ponderosae. Chemical signals released by I. pini were highly attractive to its primary predator, Enoclerus lecontei Wolcott, which increased its predation on D. ponderosae in co-colonized trees. The later arrival of I. pini to the tree also extended the period of time that predators were attracted, further increasing predation on D. ponderosae. Overall, the presence of I. pini decreased D. ponderosae productivity by 35% due to direct and indirect effects on predation load.

Intra-specific interactions of facultative predators, not surprisingly, are typically competitive; there is complete niche overlap and reproductive success is dependent upon the attraction of large numbers of conspecifics to a tree for successful mass attack. Intra-specific competition has been observed to be a major factor affecting fitness of several Dendroctonus species (Coulson, 1979; Raffa and

Berryman, 1983; Amman, 1984; Davis and Hofstetter, 2009). However, the effect of intra-specific competition on individual beetles within a tree and on the beetle population as a whole is more complex than merely the additive negative effects of sharing a limited resource. Outcomes vary considerably over the course of an individual attack, through a generation, and by population phase. For example, Reeve *et al.* (1998) found that for *D. frontalis*, contest competition prevails during gallery construction and oviposition, while larvae exhibit scramble competition during development.

For facultative predators, arrival time at the tree can influence survival and the degree of intra-specific competition that occurs. Pioneer beetles that initiate aggregation must contend with strong tree defenses and are at much higher risk of mortality than later arrivals. However, pioneers have been predicted to have a greater competitive edge in accessing resources for their young relative to late arriving beetles. Latty and Reid (2010) investigated fitness effects of the order of arrival of *D. ponderosae* at individual attacked trees as well as of date of colonization. They found that surviving pioneers did not have a competitive advantage, but rather had significantly smaller broods than later arriving females. In fact, the later a beetle arrived in the sequence of attack on a tree, the higher its success. However, beetles colonizing trees earlier in the season had an advantage over those colonizing trees later in the season. The benefit of an earlier colonization date may be due to a longer period for the female to lay eggs prior to winter and greater adaptive seasonality allowing more of the brood to enter winter as cold tolerant larvae. Likewise, Pureswaran *et al.* (2006) found that timing affected reproductive productivity of *D. frontalis*. However, with this species, beetles that arrived midpoint in the colonization of a tree had the highest reproductive output.

Habitat selection by pioneers varies by beetle population density and condition of female beetles that can further influence the degree to which competition occurs. Female *D. ponderosae* pioneers in high-density populations and those with poorer condition are less selective in choosing trees, perhaps as a response to high levels of intra-specific competition (Elkin and Reid, 2010). *Dendroctonus rufipennis* are repelled by high concentrations of host monoterpenes such as those found in healthy vigorous *Picea* hosts; however, with an increase in numbers of attacking beetles, this avoidance behavior is reduced (Wallin and Raffa, 2004). Such density-dependent behaviors may lead to differential responses of beetles to hosts of varying quality and may alter the potential for intra-specific competition within tree hosts by population phase.

For parasites that tend to attack singly or in pairs, intra-specific competition is unlikely except during rare times when populations undergo large increases. In fact, for some species such as *D. micans*, prior attacks by conspecifics may actually increase host suitability either through effects on phloem quality or tree defenses, and facilitate subsequent colonization (Lieutier *et al.*, 1992; Storer and Speight, 1996; Wainhouse *et al.*, 1998). *Dendroctonus rhizophagus* Thomas and Bright is a strong exception. While this beetle has the typical behavior of a parasite, it kills its host due to the low resources available in its seedling and sapling hosts. In this species, the size of the brood appears to be the direct outcome of strong intra-specific completion among members of the same brood; the larger the diameter of the host, the greater the number of surviving brood. This effect is not due to inter-brood competition because only one brood is typically present in a tree (Sánchez-Martínez *et al.*, 2009).

2.2.3 Microbial Symbioses

Microbial symbioses are an important aspect of *Dendroctonus* biology (Six, 2013). *Dendroctonus* exploit an array of microbes that aid them in basic life functions. In turn, the beetles are exploited by a broad spectrum of commensal microbes as well as antagonists. The two main groups of microbial symbionts associated with *Dendroctonus* are fungi (filamentous and yeast) and bacteria. All are ectosymbiotic and are vectored on the exoskeleton or in the gut. Some of these symbionts appear to live and propagate only in the gut, while others grow on gallery walls or within host tree phloem, bark, and sapwood.

2.2.4 The Filamentous Fungi

The filamentous fungal partners are the best studied of the symbionts of *Dendroctonus*. However, even after more than a century of research, the associates of most *Dendroctonus* remain nearly or completely undescribed. Most of our knowledge is based on the symbioses involving a few aggressive tree-killing species. This narrow focus limits our ability to develop a broad theoretical framework for understanding how these symbioses have evolved and function within *Dendroctonus* as a whole.

There are four means by which fungi are transported by beetles: sac mycangia, pit mycangia, incidentally on the exoskeleton, and in the gut (Table 8.2). Sac mycangia are complex structures consisting of invaginations of the exoskeleton. They are associated with glands and are highly selective carrying only one or a few mutualistic filamentous associates as well as yeasts and bacteria (Six, 2003). In *Dendroctonus*, two distinct types of sac mycangia have evolved: maxillary mycangia located on the cardines of the mouthparts and present and functional in both sexes, and pronotal mycangia found on the anterior portion of the pronotum of females only (Whitney and Farris, 1970; Barras and Perry, 1971, 1972). Pit mycangia are distinct depressions on the exoskeleton of adults that may or may

TABLE 8.2 Symbiotic Fungal Communities of *Dendroctonus*

Phylogenetic Clade	Beetle Species	Mycangia	Fungal Symbionts
Clade 1	D. armandi*	Uninvestigated	*Leptographium qinlingensis*
Clade 2	D. simplex	Uninvestigated	**Grosmannia americana**
	D. pseudotsugae*	Pit (female biased)	*L. abietinum*[b] **Ophiostoma pseudotsugae**[b]
Clade 3	D. rufipennis	Uninvestigated	**L. abietinum**
	D. micans*	Uninvestigated	*O. canum*
	D. punctatus*	Uninvestigated	Unknown
	D. murrayanae	Uninvestigated	**G. aurea**
Clade 4	D. terebrans*	Uninvestigated	*L. terebrantis, L. serpens, L. procerum* *G. huntii*
	D. valens	Uninvestigated	*L. procerum, L. sinoprocerum, L. serpens,* *L. terebrantis, O. minus*
	D. rhizophagus*	Uninvestigated	Unknown
	D. parallelocollis*	Uninvestigated	Unknown
Clade 5	D. ponderosae	Maxillary (male, female)	**G. clavigera**[ab] *Entomocorticium dendroctoni* **O. montium**[ab] **L. longiclavatum**[ab] *O. minutum* *Ceratocystiopsis sp.*
	D. jeffreyi	Maxillary (male, female)	**G. clavigera**[ab]
Clade 6	D. vitei*	Unknown (pronotal likely)	Unknown
	D. frontalis	Pronotal (female)	**Ceratocystiopsis ranaculosus**[ab] **E. sp. A**[ab] *O. minus*[c]
	D. mexicanus*	Pronotal (female)	Unknown
	D. adjunctus	Pronotal (female)	**L. pyrinum**[b]
	D. approximatus*	Pronotal (female)	Unknown
	D. brevicomis	Pronotal (female)	**C. brevicomi**[ab] **E. sp. B**[ab] *O. minus*[c]

Species listed are those that have been found to be most commonly associated with each Dendroctonus species, and thus, most likely to be symbiotic. Incidental species are not included. Species in bold are reported to be highly consistent associates. For species with asterisks, communities are uninvestigated or records are extremely sparse, limiting our ability to infer whether the fungi are true symbionts. The roles of most, including many mycangial associates, remain uninvestigated. [a] nutritional mutualist, [b] mycangial associate, [c] antagonist.

not be associated with setae (Lewinsohn, *et al.*, 1994). These pits are often associated with glands and waxy substances that may facilitate spore acquisition and maintenance (Lewinsohn *et al.*, 1994). While many pits occur on the surface of beetles, not all act as mycangia (Lewinsohn *et al.*, 1994; Bleiker *et al.*, 2009). Spores of some fungi adhere more or less randomly on the exoskeleton. This is likely the main mode of dispersal for commensals.

Many filamentous fungi can be isolated from *Dendroctonus*. However, only those associated with a beetle at a relatively high frequency can be considered symbiotic. The symbiotic fungi associated with *Dendroctonus* are in the Ascomycota and the Basidiomycota (Six and Klepzig, 2004). While some of the ascomycetes are phytopathogenic, none are biotrophs that require a living tree as a host. Instead, they are bionectrotrophs, capable of invading living tree tissues, but best suited to rapidly exploiting tree tissues once they die (Six, 2013). The non-phytopathogenic ascomycetes and the basidomycetes grow as saprobes. The basidomycetes differ from the ascomycetes in exhibiting substantial cellulytic activity and are able to grow in the outer bark as well as in phloem and sapwood.

The majority of ascomycetes associated with *Dendroctonus* are in the Ophiostomatales, a large group of arthropod-associated fungi found worldwide. Many of these fungi have specialized structures for acquisition and transport by arthropods including sticky spores. The Ophiostomatales associated with *Dendroctonus* include members of *Grosmannia* (*Leptographium* anamorph), *Ceratocystiopsis* (*Sporothrix* anamorph), and *Ophiostoma* (*Hyalorhinocladiella*, *Pesotum* anamorphs) (Zipfel *et al.*, 2006). Other fungi associated with some *Dendroctonus* include members of the ascomycete genus *Ceratocystis* (Microascales) and the basidiomycete genus *Entomocorticium*.

For decades, the general perception was that the symbioses between *Dendroctonus* and filamentous fungi functioned similarly across species. The main paradigm was that the association between fungi and beetles is a mutualism; that tree-killing bark beetles require phytopathogenic fungi to overcome their tree hosts and the fungi benefit in return by being transported reliably to an ephemeral resource (Paine *et al.*, 1997; Lieutier *et al*, 2009). This hypothesis was based entirely on aggressive tree-killing beetle systems. It did not account for the lack of pathogenicity of many of the most consistent fungal associates or the incidental nature of the most virulent associates. It also did not consider fungal associations with parasites and early successional saprophages. The hypothesis that *Dendroctonus* require virulent phytopathogenic fungi to overwhelm trees is now being questioned (Six and Wingfield, 2011).

An alternate hypothesis is that the beetles gain nutritional supplementation from their filamentous associates. This hypothesis has been supported by several studies on mycangium-possessing *Dendroctonus* (Goldhammer *et al.*, 1990; Six and Paine, 1998; Ayres *et al.*, 2000, Bleiker and Six, 2007). For mycangial beetles, fungus feeding appears to be obligate and results in reduced feeding on tree tissues, and for some species, the ability to develop in nutrient devoid outer bark. Not surprisingly, given the diversity of life histories in *Dendroctonus*, interactions with filamentous fungi are turning out to be quite varied and include mutualisms, commensalisms, and antagonisms (Six, 2003).

Table 8.2 lists fungal associates of *Dendroctonus* for which data support a symbiotic relationship. Except for a few *Dendroctonus* species, comprehensive sampling has not been conducted. Here, we review what is known about these symbioses using a phylogenetic framework based on the host beetles (Reeve *et al.*, 2012) (Figure 8.2, Table 8.1).

2.3 Host Beetle Clade 1: *Dendroctonus armandi*

Dendroctonus armandi vectors *Leptographium qinlingensis* on its exoskeleton (Xiao-Jun *et al.*, 2012). This fungus produces a number of metabolites, three of which exhibit phytotoxicity resulting in depressions in chlorophyll content of needles. Whether this fungus affects beetle nutrition has not been tested; however, it has been shown to exhibit some cellulytic activity and to alter the carbohydrate and fat profiles of infected trees (Pu and Chen, 2007; Ruixia *et al.*, 2009).

2.4 Host Beetle Clade 2: *Dendroctonus simplex, D. pseudotsugae*

Two studies have reported *D. simplex* to be associated at high rates with *Grosmannia americana* Zipfel (Jacobs and Wingfield, 2001; D. L. Six, unpubl.). However, more isolations from beetles from across the extensive range of this insect are needed. It is notable that *G. americana* has not been isolated from any other source and may be specific to this insect. The effect of the fungus on the host beetle has not been investigated.

Dendroctonus pseudotsugae is associated with *Ophiostoma pseudotsugae* Rumbold and *Leptographium abietinum* Peck (Harrington, 1988; Lewinsohn *et al.*, 1994; Ross and Solheim, 1997). This beetle possesses pit mycangia (Lewinsohn *et al.*, 1994), which appear to be more functional in females than in males (Lewinsohn *et al.*, 1994). Whether the fungi influence beetle nutrition is not known. Inoculation of seedlings or saplings with *L. abietinum* does not result in disease symptoms or death (Harrington and Cobb, 1983). However, *O. pseudotsugae* produces greater lesion lengths and sapwood occlusion than *L. abietinum* suggesting it is better suited to invading freshly colonized tree tissues (Ross and Solheim, 1997).

2.5 Host Beetle Clade 3. *Dendroctonus rufipennis, D. micans, D. punctatus, D. murrayanae*

The fungi associated with *D. rufipennis* have been well sampled from across the beetle's geographic range. The dominant associate is *L. abietinum* that has been isolated from 80–100% of beetles from sites in Alaska, Colorado, Utah, and Minnesota (Reynolds, 1992; Six and Bentz, 2003; Aukema *et al.*, 2004). Several other fungi were isolated, but at a much lower prevalence and were highly variable in their association with the beetle among sites and years.

Although *L. abietinum* is a highly consistent associate of *D. rufipennis*, how it interacts with its host beetle is not well understood. Only one study has been conducted investigating its effect on the beetle (Cardoza *et al.*, 2008). Adult beetles fed on malt extract agar or phloem with or without *L. abietinum* exhibited no difference in weight or in survival. However, in phloem containing the fungus, the beetles excavated fewer and shorter galleries and laid fewer eggs. This was interpreted as antagonism of the fungus to the beetle. However, entering phloem already colonized by fungi is unnatural for adult females and the presence of fungi may provide a signal that a tree is already colonized by other beetles. Studies are needed to determine how the fungus affects larval and teneral adult nutrition as those are the stages most likely to be influenced by nutritional supplementation by the fungus. *Leptographium abietinum* has been shown to produce high concentrations of ergosterol, a compound that may be important in hormone production by the beetle (Bentz and Six, 2006); however, its potential role in supporting nutrition of the beetle has not been explicitly tested.

Another fungus that has been considered an important symbiont of *D. rufipennis* is *Ceratocystis rufipenni* Wingfield. This fungus has only been found in association with this beetle (Wingfield *et al.*, 1997). It is one of the few bark beetle-associated fungi that are highly virulent and capable of killing trees (Solheim and Safranyik, 1997). Due to its virulence, it has been considered a critical partner enabling the beetle to overcome tree defenses. However, surveys conducted in Alaska (Haberkern *et al.*, 2002; Six and Bentz, 2003; Aukema *et al.*, 2004), Colorado, Utah (Six and Bentz, 2003), and Minnesota (Haberkern *et al.*, 2002; Six and Bentz, 2003) have failed to detect this fungus indicating it may be only a sporadic associate and, as such, may have little impact on the beetle's fitness or dynamics (Six and Bentz, 2003).

Only one study has investigated the fungal associates of *D. micans* (Lieutier *et al.*, 1992). The location of beetle collection was not provided. *Ophiostoma canum* Münch was isolated at fairly high but variable rates, while three other ophiostomatoid fungi were isolated only rarely. Rates of isolation of *O. canum* ranged from 32 to 92% from flying beetles, 0.5 to 90% from beetles artificially inserted into trees, and 52 to 56% from beetles taken from natural attacks. The overall conclusion of the study was that the beetle does not have symbiotic fungi. However, further investigations including more locations are needed.

Dendroctonus murrayanae consistently (74–100%) carries *Grosmannia aurea* Zipfel, although a substantial percentage of beetles (29–75%) also carry *Ophiostoma abietinum* Marm. and Butin (a common commensal found with many bark beetles) (Six *et al.*, 2011a). This is the only parasitic beetle known to have a consistent filamentous fungus associate. If and how the fungus affects the beetle is not known.

Dendroctonus punctatus remains uninvestigated for fungal associates.

2.6 Host Beetle Clade 4: *Dendroctonus terebrans, D. valens, D. parallelocollis, D. rhizophagus*

The fungal associates of *D. terebrans* have only been the focus of a few studies, primarily in relation to its association with root disease in the southern USA (Eckhardt *et al.*, 2007). Overall, evidence indicates that this beetle has no consistent symbiotic partners, but rather carries a number of fungi that vary greatly in their distribution and frequency. The beetle has been reported to carry *Grosmannia huntii* (Rob.-Jeffr.) Zipfel, Z. W. de Beer and M. J. Wingfield, *Leptographium terebrantis* S. J. Barras and T. J. Perry, *L. serpens* (Goid.) M. J. Wingfield, *L. procerum* (Kendrick) M. J. Wingfield, and fungi that resemble *Ophiostoma ips* (Rumbold) Nannfeldt and *G. aurea* but that are likely to be undescribed species (Eckhardt *et al.*, 2007; Zanzot *et al.*, 2010). Most, if not all, of these species are also carried by a number of bark beetles and other weevils, particularly those that attack the root collar. Rates of isolation are also variable with a high proportion of beetles not yielding fungi (Zanzot *et al.*, 2010).

The fungal communities associated with *D. valens* are better described than those of *D. terebrans*, mainly due to its relationship with root disease in North America (Klepzig *et al.*, 1995) and its recent introduction into China where it has become a serious pest (see Section 4). Early studies in California and Wisconsin reported that the beetle carried *L. terebrantis* and *L. procerum* at relatively high rates, as well as a number of less consistent associates (Harrington and Cobb, 1983; Owen *et al.*, 1987; Klepzig *et al.*, 1995). Subsequent work surveying fungi from *D. valens* in 11 states detected a broad array of fungi, but not *L. terebrantis* (Taerum *et al.*, 2013). Taerum *et al.* (2013) also found distinct differences in the fungal communities associated with the eastern and western populations of the beetle in North America. Of 24 fungal species detected, only four were shared indicating that little if any migration occurs between the two populations. Most species were only isolated in low frequencies. In the western USA, only an undescribed *Leptographium* and an undescribed *Ophiostoma* were present on more than 30% of the beetles. In the eastern portion of the USA, only *L. procerum* was present on more than 30% of the beetles. Results of isolations from beetles in China are discussed in Section 4.

The effects of fungi on *D. valens* have been studied in three controlled laboratory experiments. In one, beetles gained less weight when grown on *L. procerum* (the most common associate in eastern USA), *L. sinoprocerum*

Q. Lu, Decock and Maraite (the main fungus associated with the beetle in China), and *Ophiostoma minus* (Hedgc.) Syd. and P. Syd (a fungus occasionally isolated from the beetle in both locations), than on no fungus controls (Wang *et al.*, 2013). All three fungi depleted glucose and fructose in the medium upon which the fungi were grown. Supplementing the medium with these sugars reversed the negative effects of the fungi on beetle growth, suggesting that these fungi may compete with the beetle for nutrients with the tree. However, in other studies, weight gain of larvae was similar among controls and *L. terebrantis*, *L. procerum* (China), *L. procerum* (US), and *L. sinoprocerum* treatments, and larvae lost weight only when fed *O. minus* (Shi *et al.*, 2012; Wang *et al.*, 2012). Larvae feeding on *L. sinoprocerum* and *O. minus* exhibited elevated immune responses relative to controls (Shi *et al.*, 2012). Because immune responses are energetically expensive, it was suggested that this type of response might negatively affect the growth rate of larvae that feed on the fungi. However, lower growth rates only occurred with larvae that fed on *O. minus*, although elevated immune responses occurred in response to both fungi (Shi *et al.*, 2012). Together, these studies suggest that the fungi most commonly found with the beetle have neutral or only slightly antagonistic effects. *Ophiostoma minus* is only found in low frequencies and may have little impact, although it has the greatest potential to be an antagonist.

Fungal associates remain uninvestigated for *D. rhizophagus* and *D. parallelocollis*.

2.7 Host Beetle Clade 5: *Dendroctonus ponderosae, D. jeffreyi*

Dendroctonus ponderosae and *D. jeffreyi* are sister species and share the same type of sac mycangium (Whitney and Farris, 1970; Six and Paine, 1998). The paired mycangia are located on the maxillary cardines (mouthparts) of both sexes and consist of deep U-shaped oblong invaginations and carry the specific filamentous associates of the beetles as well as yeasts and bacteria (Bleiker *et al.*, 2009). The mycangial fungi occur within the mycangia as conidia or compact conidiophores that can extrude from the openings. The growth outward of fungi from the mycangia may result in a continuous supply of inoculum as females excavate egg galleries (Bleiker *et al.*, 2009). The mycangia are the main means of transport of fungi; while fungi can be isolated from both mycangia and the exoskeleton, the prevalence of spores in pits on the body is low, particularly in comparison with species that possess pit mycangia (Lewinsohn *et al.*, 1994; Bleiker *et al.*, 2009).

Both *D. ponderosae* and *D. jeffreyi* carry *Grosmannia clavigera* (Rob.-Jeffr. and R. W. Davidson) Zipfel, Z. W. de Beer and M. J. Wingfield in their mycangia. Genetic

evidence indicates that *G. clavigera* is actually comprised of two cryptic species. One, *G. clavigera* (including the type), is found with *D. jeffreyi* and *D. ponderosae* colonizing *P. ponderosa* (Lee *et al.*, 2007, Massoumi Alamouti *et al.*, 2011). The other, *G. clavigera*, currently designated as Gs, is found with *D. ponderosae* in *P. contorta* and other *Pinus* species (Massoumi Alamouti *et al.*, 2011). Gs and *G. clavigera* (species type) exhibit distinct responses to host tree defensive monoterpenes, with each less tolerant of non-host than host compounds (Paine and Hanlon, 1994).

Dendroctonus ponderosae is also associated with two other mycangial fungi, *Leptographium longiclavatum* Lee, Kim, and Breuil, a species very closely related to *G. clavigera sensu lato*, and *Ophiostoma montium* (Rumbold) von Arx (Whitney and Farris, 1970; Lee *et al.*, 2005). While *G. clavigera sensu lato* and *O. montium* are found across the entire range of the beetle, *L. longiclavatum* appears to be mostly restricted to northern populations (Rice *et al.*, 2008). Other more incidental associates that are not carried in mycangia also occur, including several species of *Entomocorticium* (Whitney *et al.*, 1987; Hsiau and Harrington, 2003; Lee *et al.*, 2006), an *O. minutum*-like sp. (Lee *et al.*, 2006), and a *Ceratocystiopsis* sp. (Khadempour *et al.*, 2010, 2012).

The mycangial fungi associated with *D. ponderosae* are obligate mutualists providing nutritional supplementation to the host beetle (Six and Paine, 1998). The fungi tap into sapwood nitrogen and transport it to the phloem where the larvae feed. Increases in nitrogen due to the fungi can exceed 40% (Bleiker and Six, 2007). *Grosmannia clavigera* is superior at nitrogen translocation, which may account for its greater positive effects on beetle survival and reproduction relative to *O. montium* (Six and Paine, 1998; Bleiker and Six, 2007; Cook *et al.*, 2010; Goodsman *et al.*, 2012). The two mycangial fungi also produce high concentrations of ergosterol, a compound that may be important for beetle growth and hormone production (Bentz and Six, 2006).

The mycangial fungi shift in prevalence within a tree and a population over time (Adams and Six, 2006; Six and Bentz, 2007; but see Khadempour *et al.*, 2012). This is likely due to several factors, including changing nutrient and moisture profiles within the tree, competition among the fungi, and temperature (Bleiker and Six, 2007, 2008b; Six and Bentz, 2007). All three fungi are bionecrotrophs, but each exhibits a different rate of growth and lesion production within the tree as a result of differential tolerances to host defensive chemistry and moisture and oxygen conditions (Solheim and Krokene, 1998; Lee *et al.*, 2006). Although the fungi exhibit scramble competition (Bleiker and Six, 2008a, b), temperature appears to be an equally, if not more important, driver of their relative prevalence within a tree and a population over time (Six and Bentz, 2007;

Addison *et al.*, 2013). *Grosmannia clavigera sensu lato* grows best when conditions are relatively cool, while *O. montium* grows best when conditions are warm (Six and Paine, 1998; Rice *et al.*, 2008). This results in differential rates of capture of space within the tree as well as differential rates of dispersal (Adams and Six, 2006; Six and Bentz, 2007; Bleiker and Six, 2008a, b) and supports the stability of the symbiosis by not allowing any one to dominate over time (Addison *et al.*, 2013). This redundancy of symbionts that operate under different environmental conditions and that provide similar benefits may provide a "back-up" system allowing the beetle to live in highly variable habitats across a broad geographic range (Six and Bentz, 2007).

The interaction between *G. clavigera* and *D. jeffreyi* is expected to function similarly to the symbiosis of *G. clavigera sensu lato* with *D. ponderosae*. However, the D. *jeffreyi*-fungus symbiosis lacks redundancy in symbionts that might be a reflection of the narrower ecological amplitude of this species.

As with their host beetles, the fungi associated with *D. ponderosae* and *D. jeffreyi* also exhibit considerable genetic variability. *Grosmannia clavigera sensu stricto* associated with *D. jeffreyi* exhibits low genetic diversity (Massoumi Alamouti *et al.*, 2011), but still shows typical isolation-by-distance differentiation from the northern to southern reaches of its range (Six and Paine, 1999). The population genetic structure of the fungi associated with *D. ponderosae* is more complex, but in general all three mycangial associates exhibit population differentiation into northern and southern groups (Roe *et al.*, 2011a). Not unexpectedly, genetic variability is lower in areas of range expansion during the recent outbreak (Roe *et al.*, 2011a, b) (see Section 4). Gs associated with *D. ponderosae* further displays four genetic clusters that correspond to four geographical regions (Tsui *et al.*, 2012). These clusters exhibit an isolation-by-distance pattern between, but not within, clusters. The fungus associated with *D. ponderosae* in the area of northern expansion of the current outbreak has experienced an extreme genetic bottleneck as might be expected for a newly established population (Tsui *et al.*, 2012).

2.8 Host Beetle Clade 6: *Dendroctonus vitei, D. frontalis, D. mexicanus, D. adjunctus, D. approximatus, D. brevicomis*

The species in this clade possess pronotal mycangia (their presence is assumed for *D. vitei*, even though it has not been investigated for the structures). These mycangia consist of paired pouches located at the anterior of the pronotum of female beetles that resemble a folded shirt collar (Barras and Perry, 1971; Paine and Birch, 1983). Males have a reduced homologous non-functional structure.

The fungal communities of *D. mexicanus*, *D. vitei* Wood, and *D. approximatus* Dietz have not been described. The fungal associates of *D. adjunctus* have been the focus of only two studies. Davidson (1978) isolated two species of ophiostomatoid fungi from *Pinus* killed by the beetle in New Mexico that he described as *Leptographium pyrinum* R. W. Davidson and *Ceratocystis adjuncti* R. W. Davidson (now *Ophiostoma adjuncti* (R. W. Davidson) T. C. Harr). Six and Paine (1996) conducted isolations from mycangia of beetles collected from multiple locations in Arizona and found mycangia of most beetles contained *L. pyrinum*, although one mycangium yielded *O. adjuncti*. An ophiostomatoid-selective medium was used to make these isolations, which does not support the growth of *Ceratocystiopsis* or *Entomocorticium*, genera that include species associated with some *Dendroctonus* with pronotal mycangia. Further isolations need to be conducted using non-selective media to determine whether species of fungi from these genera are also present. No studies have been conducted to investigate how fungi may affect this insect.

Dendroctonus frontalis carries *Entomocorticium* sp. A and *Ceratocystiopsis ranaculosus* J. R. Bridges and T. J. Perry in its mycangia (Barras and Perry, 1972; Hsiau and Harrington, 1997). These fungi supplement the diet of the beetles by concentrating nitrogen (Ayres *et al.*, 2000). Nitrogen supplementation may be especially important for this and other beetles in this clade that spend part of their development in outer bark that is particularly low in nutrients.

As with *D. ponderosae*, the two mycangial fungi associated with *D. frontalis* differentially affect host fitness. Beetles that develop on *Entomocorticium* sp. A are larger and have greater rates of reproduction than those developing on *C. ranaculosus* (Bridges, 1983; Coppedge *et al.*, 1995). Like the fungi of *D. ponderosae*, these two fungi also fluctuate in their prevalence over time, in part driven by changes in temperature (Hofstetter *et al.*, 2006a). Although the two fungi differentially affect several fitness parameters of the host, changes in their relative prevalence do not appear to greatly influence the population dynamics of the host (Hofstetter *et al.*, 2007). However, the presence of a third, non-mycangial fungus, *O. minus*, has substantial negative effects on brood survival and can lead to the collapse of outbreaks (Lombardero *et al.*, 2003). *Ophiostoma minus* is a mutualist of mites phoretic on the beetle, which carry its spores in pouches on their bodies termed sporothecae (Moser, 1985). In periods of low prevalence of mites, and consequently *O. minus*, beetle population growth rates are high. However, when *O. minus* colonizes large areas of the tree, beetle populations decline (Hofstetter *et al.*, 2006b). The prevalence of the mite vectors, and thus *O. minus*, is apparently regulated by temperature (Hofstetter *et al.*, 2006a). The exposure of the beetle to the antagonist

O. minus appears also to be influenced by the presence of the mycangial fungi. While the two mycangial fungi of *D. frontalis* are roughly equally competitive with one another, *Entomocorticium* sp. A is able to better maintain space (and protect beetle brood) in the presence of *O. minus* than is *C. ranaculosus* (Klepzig and Wilkens, 1997).

While the symbiosis among *D. frontalis* and its fungi is similar to those of *D. ponderosae* and *D. jeffreyi* in that it possesses two symbiotic fungi, there are also substantial differences. In the case of *D. ponderosae* and *D. jeffreyi*, the beetles develop completely within the phloem and both partners are ascomycetes well suited to grabbing up easy assimilable sugars and amino acids present in phloem and sapwood parenchyma. However, the partners of *D. frontalis* are an ascomycete and a white rot basidiomycete. This difference may reflect the different feeding habit of *D. frontalis* within trees. The larvae of *D. frontalis* move into the outer bark during the fourth instar where all subsequent feeding and development occurs. The ascomycete *C. ranaculosus* may best support feeding in the phloem. However, only *Entomocorticium* sp. A., the basidiomycete, is likely to grow well in the outer bark and support beetle nutrition in that substrate.

The symbioses between *D. brevicomis* and filamentous fungi are similar to that occurring among *D. frontalis* and its associates. Its mycangial fungi include *Ceratocystiopsis brevicomi* Hsiau and T. C. Harr. and *Entomocorticium* sp. B, species very closely related to the mycangial fungi carried by *D. frontalis* (Hsiau and Harrington, 1997, 2003). As with *D. frontalis*, *O. minus* is also found with *D. brevicomis* and is expected to act as an antagonist.

2.8.1 Yeasts

Yeasts are ubiquitous with *Dendroctonus*. Yeasts are found in the gut, on the exoskeleton and gallery walls, and in the mycangia of every *Dendroctonus* species thus far investigated (Shifrine and Phaff, 1956; Lu *et al.*, 1957; Whitney and Farris, 1970; Bridges *et al.*, 1984; Lewinsohn *et al.*, 1994; Adams *et al.*, 2008; Rivera *et al.*, 2009; Davis, *et al.*, 2011). In some cases, the combined isolation rates for yeasts from individual beetles exceed those of the obligate symbiotic fungal associates. However, none so far has shown high host specificity or incidence with any one species of *Dendroctonus* with the exception of *Ogataea pini* (Holst) that can be found at about 60% prevalence in the mycangia of *D. brevicomis*, and can be isolated from the beetle's exoskeleton (Davis *et al.*, 2011). Yeasts in the gut do not appear to be specialized to any particular section (Rivera *et al.*, 2009). However, a number of species do appear to be specialists on bark beetles in general, if not with particular species (Rivera *et al.*, 2009). While some are found with a broad array of bark beetle species, some appear limited to species of beetles that infest particular

conifer genera, suggesting they may be specialized to, or inhibited by, particular host tree chemistries (Rivera *et al.*, 2009).

The effects yeasts have on their insect hosts are not well understood. Several yeasts have been shown to be able to convert aggregation pheromones to anti-aggregation pheromones potentially influencing communication among beetles (Leufvén *et al.*, 1984; Hunt and Borden, 1990). Some yeasts may directly affect the growth of the filamentous associates of the beetles. Adams *et al.* (2008) found that an unidentified basiodiomycete yeast isolated from the galleries of *D. ponderosae* enhanced the growth of one mycangial fungus, *O. montium*, while suppressing the growth of the other, *G. clavigera*.

Yeasts may also indirectly affect growth of the filamentous fungus associates of *Dendroctonus* through the production of volatiles. Davis *et al.* (2011) found that *O. pini* isolated from *D. brevicomis* significantly enhanced the growth of a mutualistic fungus, *Entomocorticium* sp. B, while inhibiting the growth of a fungal entomopathogen, *Beauveria bassiana* (Bals.-Criv.) Vuillemin. The presence of *O. pini* also altered the rate of loss of several host tree monoterpenes in phloem over time. While some volatiles declined at a greater rate in the presence of the yeast, others actually increased in concentration, indicating a potential antagonist interaction with the beetle. Yeasts may also elicit attraction to natural enemies of the beetles with a negative impact on the host (Adams and Six, 2008; Boone *et al.*, 2008). Because yeasts can be highly localized in galleries it is not known how the effects observed in experiments may translate to effects on the beetle. For the most part, yeast communities and their interactions with *Dendroctonus* remain poorly described and are in need of further investigation.

2.8.2 Bacteria

Advances in molecular technology are allowing unprecedented insights into the bacterial communities associated with *Dendroctonus* and other bark beetles. Like yeasts, bacteria are ubiquitous and can be found in the gut, mycangia, on the exoskeleton, and growing in galleries. Bacterial communities of *Dendroctonus* appear to be relatively species poor compared to those found on many other insects. This may be due to the toxic nature of the phloem substrate or a reflection of the early successional nature of the colonization of trees by their host beetles as well as the protected nature of the beetle habitat that reduces exposure to environmental species (Delalibera *et al.*, 2005, 2007; Morales-Jiménez *et al.*, 2009; Adams *et al.*, 2010).

The structure and composition of the bacterial communities associated with *Dendroctonus* can be similar or highly variable among individuals or populations of a given species. For example, the bacterial community found in the gut of *D. valens* varies in composition and diversity across the beetle's geographic range (Adams *et al.*, 2010), while the community associated with the gut of *D. ponderosae* is relatively constant across populations (Adams *et al.*, 2013). The reason for these differences is not known.

Bacteria may influence host beetle fitness in a number of ways. One is through effects on nutrition, particularly access to complex carbohydrates and nitrogen. Morales-Jiménez *et al.* (2012) investigated cellulytic activity in the gut bacteria of *D. rhizophagus*. While some of the bacteria exhibited cellulytic activity, none were consistently associated with the insect. Delalibera *et al.* (2005) found no evidence of cellulytic activity in the bacterial gut community of *D. frontalis*. While evidence is lacking that bacteria play a role in accessing sugars from cellulose, it is possible that some will contribute to host nutrition through nitrogen fixation. *Rhanella aquatilis* is a bacterium capable of nitrogen fixation that has been isolated from the guts of many bark beetles including *D. rhizophagus*, *D. frontalis*, and *D. ponderosae* (Vasanthakumar *et al.*, 2006; Morales-Jiménez *et al.*, 2009; Boone *et al.*, 2013). Further study is needed, however, to determine whether gut conditions support nitrogen fixation by this bacterium and, if so, whether it confers a benefit to the host.

Bacteria associated with *Dendroctonus* may also aid beetle hosts in detoxification of host tree defensive compounds. The gut bacterial community of *D. ponderosae* contains a number of species that possess genes involved in terpene degradation (Adams *et al.*, 2013) and several of these bacteria have been shown to reduce levels of monoterpenes *in vitro* (Boone *et al.*, 2013). While some overlap occurs in the ability of the bacteria to degrade particular terpenes, the capabilities of the bacteria can also be complementary (Boone *et al.*, 2013). It may be that *Dendroctonus* are able to exploit a flexible community of bacteria to meet their needs for toxin degradation.

Another way that bacteria may interact with *Dendroctonus* hosts is through protection. Cardoza *et al.* (2006) suggested that *Dendroctonus* might use bacteria as a defense against antagonistic fungi that reduce beetle productivity and survival. They observed beetles smearing oral secretions on gallery walls particularly in the presence of antagonistic fungi. Bacteria isolated from these secretions were found to inhibit two antagonistic molds, *Aspergillus* and *Trichoderma*, as well as the symbiotic filamentous fungus of the beetle *L. abietinum*. Scott *et al.* (2008) observed that a *Streptomyces* sp. isolated from the mycangia of *D. frontalis* inhibited an antagonistic fungus, *O. minus*, more than it did one of the mycangial fungi, *Entomocorticium* sp. A. It was suggested that the bacterium may be involved in the selectivity of mycangia for the mycangial associate (*O. minus* is not acquired or carried in the structure) and protection of the beneficial fungus. Further work is needed

to determine if this bacterium is required for this role. A subsequent study found that a broad diversity of *Streptomyces* occur with *Dendroctonus* (Hulcr *et al.*, 2011).

Bacteria may also affect the beetle host indirectly through their interactions with mutualistic filamentous fungal associates. Adams *et al.* (2008) isolated bacteria from galleries of *D. ponderosae* as well as from uninfested pine sapwood and grew them with the beetle's mycangial fungi *O. montium* and *G. clavigera*. A bacterium from gallery walls facilitated the growth of *O. montium* while inhibiting *G. clavigera*. However, the putative endophytic bacterium from sapwood, *Bacillus subtilis*, inhibited growth of both fungi. Adams *et al.* (2009) tested for effects of bacteria from galleries of *D. valens* and *D. ponderosae* on these beetle's associated fungi and found that while some bacteria facilitated growth, others inhibited it. By adding a host tree compound, alpha-pinene, these effects could be variously amplified, reduced or reversed.

As with yeasts, the bacteria found in *Dendroctonus* galleries are typically localized and may have highly variable effects. Additional research on these symbionts is needed to reveal how bacteria affect *Dendroctonus* and how their roles may vary with life history strategy.

2.8.3 Nematodes

Nematodes have been found with every species of *Dendroctonus* thus far investigated (Massey, 1956, 1966; Khan, 1957; Thong and Webster, 1975; Langor, 1991; Carta *et al.*, 2010). A single beetle species may be associated with a large number of nematodes species from multiple genera, such has been found for *D. rufipennis* (Massey, 1956), or potentially as few as a single species as with *D. simplex* (Langor, 1991). For some, like *D. rufipennis*, the nematode community appears to be relatively constant across the range of the insect (Massey, 1956). Nematodes have a broad array of relationships with the beetle host, ranging from phoretic to parasitic. Accordingly, their effects on the host vary considerably. Most research has involved surveys and descriptions of nematode communities, with only a few describing their life histories and fewer still quantifying their effects on host fitness and behavior.

Phoretic nematodes use the beetle for dissemination from tree to tree. Transport may occur in many locations on the beetle although packing in clusters under the elytra appears common (Langor, 1991; Cardoza *et al.*, 2006). Once in the tree, the nematodes rapidly leave the host and spend most of their life cycle foraging and reproducing within the galleries. This accounts for why they can often be collected from beetles emerging from natal trees or entering new trees, but not from adults in galleries or from re-emerging beetles (Langor, 1991).

Phoretic species have various feeding habits of which mycophagy appears to be common. One of the earliest papers on bark beetle-associated nematodes describes a nematode phoretic on *D. frontalis* and *Ips* spp. that was associated with, and fed upon, beetle-associated bluestain fungi (Steiner and Buhrer, 1934). Hunt and Poinar (1971) similarly reported a nematode that feeds on *O. minus*, a fungus common with some *Dendroctonus* and a number of other bark beetles. Carta *et al.* (2010) found a nematode associated with the prothoracic region and heads of *D. frontalis* and the galleries of this beetle. This nematode was readily reared on fungi but not on bacteria.

Parasitic nematodes can occur within the hemocoel or the gut or use various tissues within the body at different stages of their life cycle (Massey, 1956). Some spend a period of time "free-living" in the galleries and only a portion of their life cycle within the insect (Massey, 1956). Infection rates vary greatly within and among species of nematodes as well as beetle species and stage of beetle development. For example, the parasitic nematode *Sphaerularia dendroctoni* Massey was found infecting 1–76% of *D. rufipennis*, but only the adult stage. In contrast, *Aphelenchulus reversus* Thorne was found in 12–36% of the beetle and in all stages except the egg (Massey, 1956). Different nematode species can occur together in the same host although rates of co-infection tend to be low (Massey, 1956). The numbers of parasitic nematodes in an individual host can range from one to hundreds.

Not all parasites are transported to new trees within the host's body. For example, *Ektaphelenchus obtusus* Massey, a parasite of *D. rufipennis*, may be carried within adult beetles or on the exoskeleton (Massey, 1956). Females can also be carried in pocket-like structures on the undersides of elytra (Cardoza *et al.*, 2006). These structures form in response to the presence of the nematodes. Daurs of a phoretic fungus-feeding species *Bursaphelenchus rufipennis* Kanzaki, Giblin-Davis, Cardoza, Ye and Raffa, and fungal spores, can also be found within the structure (Kanzaki *et al.*, 2008).

The parasitic nematodes of bark beetles have various effects on their beetle hosts. Flight capacity has been found to be impaired by nematode parasitism in *D. pseudotsugae* and *D. brevicomis* (Atkins, 1961; Nickle, 1963). Infections in *D. ponderosae* can also result in slowed "escape" responses, lethargy, and tremors (Reid, 1945). Reductions in oviposition gallery length and egg production have been observed in *D. rufipennis*, *D. frontalis*, and *D. pseudotsugae* (Massey, 1956; Thong and Webster, 1975; MacGuidwin *et al.*, 1980).

2.8.4   Mites

Phoretic mites are common associates of *Dendroctonus*. While they remain poorly studied for most *Dendroctonus* species, it is clear that the diversity of mites associated with this group of beetles is immense. Moser and Roton (1971) found 96 species of mites associated with *D. frontalis* just in Louisiana and 57 species with the beetle in southern Mexico, Honduras, and Guatemala (Moser *et al.*, 1974). The mites have a wide range of feeding habits. Some are predators on bark beetles, primarily on the eggs. Others are predators of nematodes and other organisms that live within *Dendroctonus* galleries (Kinn, 1970). Still others are mycophagous and feed generally on fungi found in galleries, or for some, specific fungi that they carry in specialized structures of the exoskeleton termed sporothecae (Lombardero *et al.*, 2003; Hofstetter *et al.*, 2006a, b; Cardoza *et al.*, 2008). The latter can highly influence the population dynamics of the host beetle through effects on the beetle's symbiotic fungi and larval survival.

2.8.5   Natural enemies

Dendroctonus are associated with large numbers of natural enemies. For example, *D. brevicomis* and *D. ponderosae* have over 60 species (Dahlsten, 1970; Dahlsten and Stephen, 1974) and *D. frontalis* at least 29 species of insect natural enemies (Moser *et al.*, 1971). The main insect natural enemies include predaceous beetles in the families Cleridae (*Enoclerus*, *Thanasimus*), Trogossitidae (*Temnochila*), Colydiidae (*Lasconotus*, *Aulonium*), and Histeridae (*Platysoma*, *Plegaderus*), predacious flies in Dolichopodidae (*Medetera*), and parasitoid wasps (several families) (Dahlsten, 1982).

The insect natural enemies of *Dendroctonus* exhibit distinct arrival sequences to beetle-infested trees (Stephen and Dahlsten, 1976). Host location is achieved with chemical cues associated with the stage of beetle used. Clerids and trogossitids use host beetle aggregation pheromones to locate trees under attack, as does *Tomicobia tibialis* Ashmead, a wasp that parasitizes adult *Dendroctonus* (Reeve, 1997; Raffa *et al.*, 2007). Most other predacious beetles including *Platysoma* and *Lasconotus*, other parasitoids, and the predacious flies *Medetera* prey on *Dendroctonus* larvae, and arrive after pheromone production by the beetle has ceased. The cues used by the beetles are not known. However, at least some parasitoids use volatiles produced by the symbiotic fungi of *Dendroctonus* in host location (Sullivan and Berisford, 2004; Adams and Six, 2008). *Medetera* appears to locate hosts using a combination of host tree and fungal odors (Boone *et al.*, 2008).

Measuring the impact of natural enemies in *Dendroctonus* is difficult and only a few studies have been conducted, mainly on *D. frontalis*, *D. ponderosae*, and *D. brevicomis* (Linit and Stephen, 1983). Linit and Stephen (1983) reported that parasitoids and predators accounted for 23–28% of within-tree mortality of *D. frontalis* in Georgia and Arkansas, respectively. However, with *D. brevicomis*, mortality due to insect natural enemies is highly variable and often low (Berryman, 1970; Dahlsten, 1982). Parasitism rates, in general, range from as low as 1% to as high as 98% in individual trees (Dahlsten, 1982). In general, predators appear to cause greater mortality than do parasitoids (Dahlsten, 1982).

Birds, particularly woodpeckers, are also important predators of *Dendroctonus*. Woodpeckers often consume 20–30% of the beetles in a tree and have been seen to remove up to 98% of beetle brood in standing trees during outbreaks (Otvos, 1970; Koplin and Baldwin, 1970). Natural enemies of bark beetles including *Dendroctonus* are covered in detail in Chapter 7.

3. ECOLOGICAL IMPORTANCE

Dendroctonus are among the most important insects affecting conifer forests. Tree-killing species that develop outbreaks have major economic effects through mortality to wood stocks, and effects on aesthetics and property and recreational incomes. Management efforts are also expensive adding substantially to the massive economic impacts of these insects. While we are beginning to better understand the drivers of outbreaks at many scales (Raffa *et al.*, 2008), we still have only rudimentary knowledge of the ecological effects *Dendroctonus* outbreaks have on forests.

Ecological impacts are highly variable, and depend upon life history strategy, and for tree killers, population phase. Parasites likely play only small roles in forest processes. However, secondary species are important in naturally thinning forests and nutrient cycling, affecting primarily weakened, damaged or drought-affected trees. These species do not develop outbreaks except when severe stress in host trees exists. In these cases, ecological impacts are likely short term except when outbreaks are abnormally severe (see Section 4).

Tree-killing *Dendroctonus* have the greatest influences on forest ecosystems. These beetles are often present in low densities for decades and only occasionally develop outbreaks. During non-outbreak periods, their numbers are low due to climatic conditions and forest structures and compositions that are not conducive to population increases (Bentz *et al.*, 2009). At these times, they are limited to colonizing weak trees with few defenses and have relatively low productivity. Outbreaks develop in response to a combination of appropriate stand conditions (mature hosts, often even aged stands) and a climatic trigger (often hot and dry conditions) that coincidentally support greater beetle survival and productivity while creating stress in trees and lowering their defenses (Bentz *et al.*, 2009). Once triggered, outbreaks can have significant effects on forest succession (Klutsch *et al.*, 2009), fuels (Simard *et al.*, 2011), carbon and nitrogen cycling (Kurz *et al.*, 2008; Griffin *et al.*, 2011; Hansen, 2014), hydrology and snow pack retention (Boon, 2007; Bewley *et al.*, 2010).

Much as wildfire was viewed in the past, outbreaks of *Dendroctonus* have often been perceived as damaging to forests. However, as natural disturbances, like fire, they can also be important in maintaining ecosystem structure and function. However, unlike fire, beetles affect these processes without immediately altering the physical nature of forests and occur over a longer time frame (several years) resulting in different effects on biogeophysical and biogeochemical processes (Edburg *et al.*, 2013).

Tree-killing *Dendroctonus* can act as natural thinning agents and seldom remove all mature trees during outbreaks. These effects are an important part of the beetles' ecological roles (Hansen, 2014). Beetle outbreaks enhance diversity and complexity of forest communities and are important in the reallocation of resources among different components of the ecosystem (Romme *et al.*, 1986). Forest ecosystems show considerable resilience to beetle outbreaks and effects on primary productivity are often not as severe as commonly reported (Romme *et al.*, 1986; Coates *et al.*, 2013). Residual trees, seedlings, and advance regeneration often return to previous levels of primary productivity fairly rapidly, and outbreaks contribute greatly to forest structural and landscape heterogeneity that aids in buffering the severity and extent of future disturbance including fire and insect outbreaks (Collins *et al.*, 2011; Hawkins *et al.*, 2013; Harvey *et al.*, 2013).

Overall, there is a considerable need to improve our knowledge of the ecological roles of *Dendroctonus*, primarily tree-killing species, and how they (and human management in response to them) affect the function and trajectories of affected coniferous forest ecosystems.

4. ANTHROPOGENIC EFFECTS

The activities of humans are creating massive changes in Earth's forests directly through extraction, development, and management, as well as indirectly through effects on climate. Effects of anthropogenic change on *Dendroctonus* species can be placed into three main categories, although considerable overlap can occur: (1) alterations of forest structure and composition, (2) movement into novel habitats, and (3) climate change.

4.1   Alteration of Forests

Deforestation, fragmentation, removal of preferred species, selective growth of preferred species, and suppression of natural disturbances have all contributed to massive alterations of Earth's forests. These alterations have resulted in substantial changes to the population dynamics of many insects that has further affected forest structure and composition. Numerous examples exist where human-induced changes in forests have affected the dynamics of *Dendroctonus* species. For example, in the southern USA, extensive plantings and more rapid regeneration of loblolly pine (a species susceptible to *D. frontalis*) in areas once dominated by longleaf pine (a less susceptible species) is thought to be a major contributing factor in the development of extensive outbreaks of *D. frontalis* (Friedenberg *et al.*, 2007). The spread of *D. micans* into many regions of Europe is thought to have been simultaneously supported by the movement of infested material as well as extensive plantings of the beetle's host tree outside its native range (Grégoire, 1988). In western North America, fire suppression and various logging and management practices have disrupted the natural forest mosaic structure historically maintained by fire, beetles, and other disturbances leading to a greater potential for widespread mortality when conditions occur that support beetle population amplification (Bentz *et al.*, 2009).

4.2 Movement of Species into Novel Environments

Humans are contributing to the movement of *Dendroctonus* into novel environments in several ways. As mentioned previously regarding the spread of *D. micans* in *Picea* in Europe, planting host trees outside their native range can allow beetles to likewise expand their ranges. Another major mode of movement of *Dendroctonus* into new environments is through the transportation of infested wood. For *Dendroctonus*, species that are parasites or early successional saprophages are more likely to be successfully introduced than aggressive predators because founder populations need not be large as would be required for aggressive predators that exploit living trees through mass attacks (Dodds *et al.*, 2010). Likewise, specialists may have more difficulty establishing than generalists due to fewer suitable hosts. For example, *D. pseudotsugae*, a specialist on *Pseudotsuga menziesii* (Mirb.) Franco (but that can use freshly killed *Larix* to develop), was introduced into Minnesota in the early 2000s (Dodds *et al.*, 2010). The beetle was captured in pheromone traps for *D. simplex*, a *Larix*-infesting species, but not in *Larix* logs. *Pseudotsuga menziesii* does not occur in Minnesota and the beetle was apparently not able to establish in local *Larix*. However, *D. valens*, a parasite native to North America and a broad generalist on *Pinus*, is now well established in China where it has become a serious tree killer on native Chinese *Pinus* (Sun *et al.*, 2013). The example of *D. valens* in China also highlights how these insects may act very differently in new environments and host tree species than in their native range.

Changes in the geographic ranges of *Dendroctonus* are also occurring through shifts in suitable habitat due to a changing climate. Two dramatic examples include recent range expansions by *D. ponderosae* and *D. frontalis*.

Dendroctonus ponderosae has in recent years developed the largest bark beetle outbreak recorded to date. Due to global warming, the insect has extended its range several hundred kilometers further north in British Columbia, has breached the historic biogeographic barrier of the northern Rockies, and is now spreading across Alberta (Safranyik *et al.*, 2010; Cullingham *et al.*, 2011; de la Giroday *et al.*, 2012). While climatic suitability remains low for this beetle in many portions of its new range, some areas are capable of supporting endemic populations (Bleiker *et al.*, 2011). The beetle is expected to continue to spread eastward through the boreal forest and eventually into eastern pine forests as warming increases over time. Warming has also allowed the beetle to extend its range higher in elevation where it is threatening the continued existence of *Pinus albicaulis* Engelm. in subalpine exosystems (Logan *et al.*, 2010).

In both the northern and eastern expansion of its range, *D. ponderosae* is infesting *P. contorta*, a common host for the beetle. However, in naïve populations, the tree produces higher concentrations of a chemical precursor the insect uses for production of pheromones while producing lower concentrations of toxic defensive chemicals (Clark *et al.*, 2014). The beetle also exhibits greater brood productivity in naïve *P. contorta* (Cudmore *et al.*, 2010). These results indicate that as global warming continues, these trees will be highly suitable hosts for the beetle.

In the eastern expansion, the beetle has moved into a novel host tree species, *P. banksiana* Lamb., which has also proven to be highly suitable (Cullingham *et al.*, 2011). Furthermore, genetic assessments have found that there has been no apparent loss of genetic variability in the beetle during this expansion, indicating that there has been no reduction in the insect's potential for adaptation due to founder effects (Samarasekera *et al.*, 2012).

For the last several decades, *D. frontalis* has been moving northward and in recent years has moved into the heart of the New Jersey pine barrens (Weed *et al.*, 2013). Record-breaking warm temperatures coupled with below average precipitation has supported rapid growth of outbreaks in this region. Interestingly, in some areas in the southern reaches of the beetle's range, and where the beetle previously developed severe outbreaks, the beetle is now undetectable. Whether a warming climate has resulted in a range contraction in this area is not known (Friedenberg *et al.*, 2008; Weed *et al.*, 2013).

It is highly likely that other *Dendroctonus* are responding to changing climatic conditions through range expansions and contractions, but that shifts thus far have been either subtle or not detected because of a lack of monitoring. In some cases, it may be difficult to determine if new records are due to expansions or a lack of detection in the past. For example, *D. mexicanus* was first recorded in southern Arizona near the Mexican border in 2000 (Moser *et al.*, 2005). Prior to this, the beetle was only known from Mexico. The beetle may have expanded its range northward; however, it is also possible that the beetle has been present in Arizona but remained undetected or was misidentified until recently.

Anthropogenic climate change has also supported the development of outbreaks of unprecedented size by several *Dendroctonus* species. *Dendroctonus rufipennis* developed massive outbreaks extending from Alaska and the Yukon Territory and parts of the western USA in the 1990s and a new outbreak is currently building in Colorado and other parts of the central and northern Rocky Mountains (Raffa *et al.*, 2008; Hart *et al.*, 2014). From the 1990s to the present, *D. ponderosae* has developed a massive outbreak extending from the southern edge of the Yukon Territory southward across much of the Rocky Mountains and in the Black Hills of South Dakota (Raffa *et al.*, 2008; Samarasekera *et al.*, 2012). The regional synchronicity of these outbreaks has been supported by unusually warm temperatures, and in some cases concurrent drought. These shifts supported decreases in life cycle length for *D. rufipennis*, increased adaptive seasonality for the *D. ponderosae* in many areas, increased overwintering survival, and increased tree stress in areas experiencing drought-reducing tree defenses. These factors combined to result in massive tree mortality across extensive regions (Berg *et al.*, 2006; Sherriff *et al.*, 2011; DeRose *et al.*, 2013; Creeden *et al.*, 2014). Mexico is currently experiencing one of the worst droughts in their recorded history with a subsequent rapid development of outbreaks by several bark beetle species including *D. mexicanus* and *D. brevicomis*. These outbreaks are generally occurring in areas of high pine diversity that have been impacted by logging and ecosystem change and as predicted in Salinas-Moreno *et al.* (2010).

While climate change is currently enhancing the success of many *Dendroctonus*, as warming continues, it may act to suppress their fitness as well. As some areas continue to warm, they are expected to cease to be suitable resulting in range contractions (Weed *et al.*, 2013). Changing conditions may also disrupt their symbioses with mutualistic fungi resulting in lowered fitness or inability to develop and reproduce, which may also lead to range contraction or lower habitat suitability (Six *et al.*, 2011b; Addison *et al.*, 2013). The effects of climate change on bark beetles are covered in more detail in Chapter 13.

5. THE BASIC BIOLOGY OF *DENDROCTONUS* SPECIES

In this section, the biology of each *Dendroctonus* species is reviewed using the phylogenetic framework developed by Reeve *et al.* (2012) and Kelley and Farrell (1998) (Figure 8.2). A list of the currently described *Dendroctonus*

by major clades, host tree use, and ecological strategy is presented in Table 8.1. Morphological descriptions and detailed geographic distributions are not included as these are available elsewhere (Wood, 1982).

5.1 Clade 1: *Dendroctonus armandi*

5.1.1 *Dendroctonus armandi* Tsai and Li (Chinese White Pine Beetle)

Dendroctonus armandi is the only species found in eastern Asia, where it infests *P. armandi* and occasionally *P. tabuliformis* Carr. in the Qinling Mountains of China (Chen and Tang, 2007).

Dendroctonus armandi favors the lower bole of the tree and seldom attacks above one and one-half meters (Chen and Tang, 2007). Trees approximately 30 years or older are preferred (Chen and Tang, 2007). While this beetle periodically develops outbreaks during which thousands of trees can be killed, little is known about its population dynamics, the factors that support outbreak development and amplification, or the factors that act to regulate its populations.

Dendroctonus armandi is a facultative predator that mass attacks trees, indicating that it produces aggregation pheromones. Electroantennograms have shown that the beetle is responsive to a number of host tree-produced volatiles, although how these chemicals may be used by the beetle to orient to, or to select, hosts is not known (Zhang *et al.*, 2010). As with all *Dendroctonus*, attacks are initiated by females. Females are soon followed by males. The beetles then mate and construct vertical galleries with eggs deposited into niches on the walls. After hatching, the larvae feed individually in galleries in the phloem. The number of generations produced annually varies by elevation in response to temperature. Three generations per year are typically produced at lower elevations while only one occurs at the highest (Chen and Tang, 2007).

5.2 Clade 2: *Dendroctonus simplex,* *D. pseudotsugae*

5.2.1 *Dendroctonus simplex* LeConte (Eastern Larch Beetle)

This beetle can be found throughout the range of its host from Alaska, across Canada and into the north central and northeastern USA (Wood, 1982). It colonizes the main bole and exposed roots of standing *Larix laricina* (Du Roi) K. Koch or, less commonly, windthrow and logging slash (Wood, 1982; Seybold *et al.*, 1992). This species has not been reported infesting other native *Larix*, but is known to infest a number of exotic *Larix* (Seybold *et al.*, 1992).

Dendroctonus simplex is a facultative predator that is typically present in low numbers but is capable of developing outbreaks under particular conditions. Outbreaks are usually associated with defoliation by sawflies or moths, although drought and flooding injury also predispose stands to the beetle (Langor and Raske, 1989).

Dendroctonus simplex mass attacks trees using a combination of aggregation pheromones and host tree odors (Baker *et al.*, 1977). Like other *Dendroctonus*, females initiate the attack. However, unlike other species, several pairs may share the same entrance hole (Seybold *et al.*, 1992). In addition, up to 90% of parents re-emerge to initiate a second brood in a new tree. In warmer areas, parents may even re-emerge a second time to produce yet a third brood. This pattern of staggered oviposition results in a lack of synchronicity of broods and the beetle can often be found in various stages of development over the course of a year (Seybold *et al.*, 1992). Despite the overlapping nature of broods, one generation a year appears to be typical across the beetle's range.

Oviposition galleries are vertical and linear in standing trees with eggs laid in grooves along the sides. While larvae initiate feeding in individual galleries, by the third instar, feeding within the phloem can be so extensive that galleries may combine into large areas packed with frass. Broods resulting from late attacks overwinter as larvae, while those from early attacks overwinter as newly formed adults. In many areas, new adults remain in brood galleries over the winter (Werner, 1986; Seybold *et al.*, 1992). However, in Alaska, many adults emerge to re-enter the phloem at the root collar where they overwinter below the snowline (Werner, 1986).

5.2.2 *Dendroctonus pseudotsugae* Hopkins (Douglas-fir Beetle)

Dendroctonus pseudotsugae consists of two subspecies, *D. pseudotsugae pseudotsugae* Hopkins and *D. pseudotsugae barragani* Furniss. *Dendroctonus pseudotsugae barrangani* occurs in isolated populations in northern Mexico while *D. pseudotsugae pseudotsugae* occurs in all other areas of the species range including the Pacific Northwest and the Rocky Mountains (Ruíz et al., 2010). The two subspecies differ in morphology and oviposition behavior (Furniss, 2001), and have reduced numbers of progeny when cross-mated (Furniss and Cibrían Tovar, 1980). The two groups also exhibit clear genetic differentiation forming two distinct lineages (Ruíz et al., 2010). This differentiation is not consistent with host tree differentiation (the host also consists of two subspecies, *Ps. menziesii* var. *menziesii* and *Ps. menziesii* var. *glauca*) but rather with long-term geographic isolation of the two groups (Ruíz et al., 2010). The degree of genetic divergence is sufficient to rank each subspecies as a distinct species (Ruíz et al., 2009).

While *D. pseudotsugae* specializes on *Ps. menziesii*, it occasionally attacks *Larix occidentalis* Nutt. when its numbers are very high. In *L. occidentalis*, brood production is seldom successful in standing trees, although some offspring may be produced in downed or highly stressed trees. The beetle prefers large diameter trees. While it is a facultative predator, it is mainly limited to weakened or stressed trees, or trees with reduced rates of growth that have poor defenses (Negrón, 1998). It also colonizes recently killed trees, logs, and fresh stumps (Wood, 1982).

Dendroctonus pseudotsugae is a swift responder to trees damaged by storms (Garrison-Johnston *et al.*, 2003), logging, and fire (Hood and Bentz, 2007; Six and Skov, 2009). However, once damaged hosts are depleted, the insect typically returns to pre-disturbance levels within 1 to 3 years (Schmitz and Gibson, 1996; Six and Skov, 2009). An exception is when other stressors co-occur that also affect host tree susceptibility. For example, after the fires of 1988 in Yellowstone National Park, the beetle remained active for over 4 years due to ongoing drought in the region (Ryan and Amman, 1996).

Build-ups of *D. pseudotsugae* that occur in response to fire or other localized disturbances tend to remain limited to the affected areas (Lejeune *et al.*, 1961; Six and Skov, 2009). However, outbreaks driven by drought can be extensive because hosts are stressed at a regional scale (Furniss *et al.*, 1979). Even during drought, mortality is often patchy, located at lower elevations, and associated with forest edges (along clearcuts and forest fragments), and southern, southwestern, and eastern aspects (Powers *et al.*, 1999). This pattern is likely due to greater temperatures, solar radiation, and higher evapotranspiration rates at such locations (Powers *et al.*, 1999). In addition, trees growing at such sites often have lower growth rates and are more quickly affected by moisture deficits (Powers *et al.*, 1999). The beetles also tend to prefer denser stands and often remove high-density clumps of larger diameter trees further contributing to the patchy nature of mortality during outbreaks (Negrón *et al.*, 2001).

Defoliation by other insects, particularly Lepidoptera, can also result in expansions of Douglas-fir beetle populations (Wright *et al.*, 1984; Lessard and Schmid, 1990; Negrón *et al.*, 2011). Fredricks and Jenkins (1988) observed that *D. pseudotsugae* attacked *Ps. menziesii* that had been very heavily defoliated by *Choristoneura occidentalis* Freeman, but not those that had been only lightly or moderately defoliated. While defoliators may support short-term increases in the beetle, its numbers typically drop sharply once the predisposing defoliator outbreak declines (Fredricks and Jenkins, 1988).

Like other facultative predators in *Dendroctonus*, *D. pseudotsugae* uses a pheromone-mediated mass attack to overwhelm host trees. Oviposition galleries are vertical and linear and eggs are laid in niches in groups along the sides. Larvae develop individually within the phloem (Wood, 1982); however, like *D. simplex*, feeding galleries sometimes anastomose into mass feeding areas prior to adult emergence. This type of mass feeding is not equivalent to gregarious feeding by parasites. The larvae do not exhibit attraction to one another, nor do they feed as a cohesive group. Rather, in this case, group feeding is the result of exploiting the phloem of the tree to such a degree that the integrity of individual larval galleries is lost.

Most *D. pseudotsugae* overwinter as adults. In the spring, brood adults spend a period of time maturation feeding under the bark, which further destroys the integrity of individual feeding galleries. Emergence typically occurs in late spring or early summer depending on temperature (Lessard and Schmid, 1990; Negrón *et al.*, 2011). The life cycle takes 1 year.

5.3 Clade 3: *Dendroctonus rufipennis, D. micans, D. punctatus, D. murrayanae*

5.3.1 *Dendroctonus rufipennis* (Kirby) (Spruce Beetle)

Dendroctonus rufipennis is a facultative predator that occurs throughout the range of its *Picea* hosts in North America from Alaska across Canada and the USA (Wood, 1982). It is the only beetle in this clade to use a non-*Pinus* host. The beetle is differentiated into three genetically distinct groups that have likely been isolated since the early-to-mid Pleistocene (Marjoa *et al.*, 2007). The two northern groups colonize *Picea glauca* (Moench) Voss and extend from Alaska to Newfoundland. The third group mainly infests *Pi. engelmannii* Parry ex Engelm. in the Rocky Mountains. The third group is further subdivided into two groups with populations in Utah and Arizona distinct from populations in British Columbia, Colorado, Montana, and Washington (Marjoa *et al.*, 2007). The northern groups and the southern lineage exhibit about 3–4% differentiation, suggesting they may represent distinct but cryptic species.

Dendroctonus rufipennis is believed to persist during the non-outbreak phase predominantly in weak and damaged trees (Holsten *et al.*, 1999) including fresh slash and stumps from logging operations (Safranyik and Linton, 1999). All sizes of mature trees can be killed, although there tends to be a preference for those of larger diameter and with slow growth rates (Hard *et al.*, 1983; Reynolds and Holsten, 1996; DeRose and Long, 2012).

While this species often kills the host tree during a mass attack, strip attacks are also common where only one side of the bole is killed. In strip attacks, the tree lives but is often attacked by subsequent beetle generations, with some trees hosting more than one generation at the same time (Holsten *et al.*, 1999). Repeatedly attacked trees are eventually killed.

Outbreaks of *D. rufipennis* are not cyclic, but sporadic, and occur in response to a complex suite of environmental

factors. In the past, outbreaks were believed to be triggered by stochastic events such as windthrow resulting from extreme storms, avalanches, and logging (Hebertson and Jenkins, 2008; Schmid, 1981). However, increasingly, evidence suggests that outbreak initiation is more involved (Jenkins *et al.*, 2014). Some disturbance events, including some windthrow events, do not result in outbreaks indicating that a number of factors need to co-occur for an outbreak to initiate (Kulakowski *et al.*, 2003; Heberston and Jenkins, 2008; Jenkins *et al.*, 2014). Recent evidence indicates that warm temperatures and drought play major roles in the development and maintenance of outbreaks (Berg *et al.*, 2006; Hebertson and Jenkins, 2008; Hart *et al.*, 2014) and that outbreaks are now increasingly severe due to changes in these factors (Sherriff *et al.*, 2011) (see Section 4). Environmental drivers such as drought explain the region-wide synchronicity of large outbreaks that cannot result from highly localized events such as windthrow (Dymerski *et al.*, 2001; DeRose and Long, 2012).

Oviposition galleries are linear and vertical within a standing tree with eggs laid in niches in groups along the sides (Wood, 1982). Unlike other predators in *Dendroctonus*, the larvae of *D. rufipennis* often mine in small groups away from the oviposition gallery. However, by the third instar they begin to excavate individual feeding galleries, although these frequently cross one another (Wood, 1982).

In a typical 2-year life cycle, late stage larvae pass the first winter along with surviving parent adults. The next spring or early summer, surviving parents re-emerge to initiate new attacks. The larvae commence development and eventually construct pupal chambers in the phloem or partially within the outer bark. The second winter is usually passed as adults although the location of overwintering can vary. In windthrow, new adults typically remain in their pupal chambers. However, in standing trees, they often emerge and move to the base of the tree where they bore into the bark at the litter line (Holsten *et al.*, 1999). These behaviors increase the likelihood of overwintering beneath the snow, providing protection from extreme cold and woodpeckers (Holsten *et al.*, 1999).

Re-emerging parent *D. rufipennis* are capable of flight and oviposition, and therefore can substantially influence the beetle's population dynamics by contributing brood in multiple years. Parents also fly a week or more earlier than brood adults and may play a primary role in host selection (Hansen and Bentz, 2003).

The length of the life cycle varies from 1 to 2 years depending on the thermal environment. A 2-year life cycle is common across most of its range (however, see Section 4). The length of the life cycle does not appear to affect beetle productivity of individual females (Hansen and Bentz, 2003). Although univoltine females are smaller than semivoltine adults (likely due to shorter feeding times), they are no less fecund than semivoltine females. In fact, fecundity and brood survival is equivalent for univoltine, semivoltine, and re-emerged parents (Hansen and Bentz, 2003). However, a univoltine life cycle greatly increases the potential for outbreak development due to increased rates of reproduction in the population (Hansen *et al.*, 2011).

Dendroctonus rufipennis is the only *Dendroctonus* species that has been clearly demonstrated to exhibit diapause (Hansen *et al.*, 2011, but see Lester and Erwin, 2012). Temperature plays a primary role in spruce beetle diapause induction. The low temperature cue required for induction must occur in the early portion of the fourth larval instar or pupation will proceed regardless of temperature. *Dendroctonus rufipennis* lives in harsh subalpine ecosystems that experience very cold winters. A larval diapause is adaptive in that it prevents the insect from entering winter as pupae, a stage highly susceptible to freezing. It also supports adaptive seasonality, allowing the beetle the opportunity to develop on a univoltine cycle when conditions are favorable, but also supporting a semivoltine life cycle when cooler conditions exist (Hansen *et al.*, 2011).

5.3.2 *Dendroctonus micans* (Kugelann) (Greater European Spruce Beetle)

Dendroctonus micans is the best studied of the parasites in this clade. It primarily attacks *Picea*, although it sometimes also attacks some species of *Pinus*, *Abies*, *Pseudotsuga*, and *Larix* (Grégoire, 1988). It is closely related to *D. punctatus*, and is thought to have arisen from a common ancestor that migrated into Asia from Alaska via Beringia (Zúñiga *et al.*, 2002b). The beetle is believed to be native to Siberia (Schedl, 1932) but has been expanding its range into western Asia and Europe for several decades (Meurisse *et al.*, 2008). This spread has been mainly supported by human transport of infested logs and the extensive planting of *Picea* hosts outside their native range (Gilbert *et al.*, 2003; Rolland and Lemperiere, 2004).

Mating of *D. micans* occurs in the natal tree among siblings (Grégoire, 1988). Sex ratios are highly skewed toward females, ranging from 5:1 to 48:1 (Grégoire, 1988). While inbreeding is the norm, outbreeding does occur, although its occurrence appears to be rare. Outbreeding may occur when the density of attacks on an individual tree is high and brood chambers coalesce allowing mixing of brood. While many males do not leave the natal tree, they are capable of flight and some may disperse and mate with pre-emergent females from other broods. No mating in egg galleries has been observed (Grégoire, 1988).

Lone, mated females initiate attacks and lay their eggs in the phloem layer in a mass of chewed bark and frass.

This mass is thought to absorb toxic resin, protecting eggs and first instar larvae (Grégoire, 1988). The gallery is initially constructed as an oblique tunnel along the sides of which batches of eggs are laid (Grégoire, 1988). Once eggs hatch, larvae immediately aggregate and begin to feed side by side and gradually expand a communal feeding chamber. Gregarious feeding is highly coordinated and mediated by the production of larval aggregation pheromones (Grégoire, 1988; Deneubourg *et al.*, 1990). Group feeding by larvae is thought to have evolved to protect the brood from the continual threat of resin (Storer *et al.*, 1997). Gregarious feeding also appears to confer benefits beyond protection. Storer *et al.* (1997) found an overall positive relationship between the number of larvae feeding in a group and larval growth, even when larvae fed in tree tissues without active resin responses. The mechanism whereby growth and survival, and thus fitness, is enhanced as group size increases is not known, but may be related to heightened feeding rates or higher food use efficiency (Storer *et al.*, 1997).

Once feeding is complete, the beetles pupate in individual niches (Grégoire, 1988). After eclosion, new adults re-aggregate, mate, and maturation feed within the phloem until their flight muscles are fully developed (Vouland *et al.*, 1984; Grégoire, 1988). The life cycle lasts 1–3 years depending on temperature (Grégoire, 1988). The wing muscles of parent females degenerate once they initiate an oviposition gallery and females do not re-emerge to produce subsequent broods in new trees (Reid, 1958; Vouland *et al.*, 1984).

Dendroctonus micans appear to be highly selective in choosing hosts. Larger diameter trees tend to be preferred (Grégoire, 1988) as are trees with higher moisture and lower astringin and stilbene contents in the phloem (Storer and Speight, 1996). Moisture and nitrogen content are positively correlated in *Picea*, which may account for why egg production and larval weights are greater in phloem containing higher levels of moisture (Storer and Speight, 1996).

Previous attack history also influences host tree choice. Initially, within a stand, attacks appear to occur more or less at random. However, over time there is a clear pattern for some trees to be repeatedly colonized while others remain unattacked (Gilbert *et al.*, 2001). Because *D. micans* does not use aggregation pheromones, each dispersing female beetle must individually exert host choice. This choice is heavily influenced by host tree odors. Trees that have been previously attacked may emit chemical volatile cues that indicate a suitable host. Re-attacks may also merely be a response to the release of volatiles from resin tubes and wounding created by prior attacks. *Dendroctonus micans* are known to be highly attracted to felled and damaged trees (Storer and Speight, 1996), and in the lab, the beetle has been shown to be strongly attracted to spruce resin and various monoterpenes (Vasechko, 1978; Grégoire, 1988). A strong attraction to host volatiles associated with wounding may also explain why new attacks on previously attacked trees often occur adjacent to older attacks. It has also been suggested that trees previously attacked by *D. micans* exhibit induced susceptibility making them easier to colonize than trees that have never been attacked (Gilbert *et al.*, 2001). In support of this hypothesis, beetles attacking previously colonized trees and areas adjacent to previous attacks or next to wounds have been observed to have greater rates of survival and higher larval weights (Lieutier *et al.*, 1992; Storer and Speight, 1996; Wainhouse *et al.*, 1998).

In much of its current range, *D. micans* causes little damage. During non-outbreak periods, trees are usually attacked singly and not killed. Even heavily attacked trees often fully recover. However, outbreaks do occur in which numerous attacks can result in girdling and the death of the tree (Grégoire, 1988). Outbreaks are correlated with drought and waterlogged soils (Grégoire, 1988) indicating that host stress, particularly related to water balance, is important in the development of outbreaks. The relationship between *D. micans* activity and stand density remains is unclear. Some studies have found that denser stands experience greater numbers of attacks while others have found the opposite (Grégoire, 1988).

A poorly understood aspect of *D. micans* outbreaks is their common occurrence at the expanding front of the beetle's geographic range. These outbreaks can last several years, but then typically subside and stabilize at low densities with few resurgences, even during periods of drought (Grégoire, 1988). This stable dampening effect is attributed to regulation by a highly specific predator, *Rhizophagus grandis* Gyllenhal, that has tracked the spread of *D. micans* across Europe.

5.3.3 *Dendroctonus punctatus* LeConte (Allgeheny Spruce Beetle)

Dendroctonus punctatus is morphologically nearly identical to *D. micans* (Furniss, 1995); however, there are clear chromosomal (Zúñiga *et al.*, 2002b) and molecular (Kelley and Farrell, 1998) differences supporting their designation as distinct species. This beetle occurs in Alaska, Canada, and eastward across the USA to West Virginia (Furniss and Carolin, 1977; Wood, 1982).

While little is known about *D. punctatus*, much of its life history appears to be similar to *D. micans*. Like *D. micans*, *D. punctatus* is a parasite attacking *Picea*. However, it is restricted to suppressed or weakened trees and outbreaks are unknown. Like *D. micans*, females mate before emerging from their natal trees and attacks on new trees are made by single mated females rather than female–male

pairs. Attacks occur near ground level at low densities (Furniss and Johnson, 1989). Oviposition galleries initially extend upward from the entrance but may turn down after several centimeters. The larvae feed gregariously in a common chamber in the lower bole or root collar. Mature larvae pupate in separate cells, but after transforming to adults, they tunnel actively throughout the brood chamber, enlarging it while maturation feeding. New adults have been described to emerge through ruptures in the weakened bark (Furniss and Carolin, 1977). While the life cycle has been suspected to take 2 years (Furniss and Carolin, 1977), observations in the field indicate that one generation per year is more likely, although broods that originate in late summer may possibly overwinter twice (Furniss and Johnson, 1989).

5.3.4 *Dendroctonus murrayanae* Hopkins (Lodgepole Pine Beetle)

This species occurs in the central and northern Rocky Mountains where it attacks the lower bole and root collar of weakened and damaged *P. contorta* and *P. banksiana* (Furniss and Kegley, 2008). *Dendroctonus murrayanae* also colonize trees infested by other bark beetles, particularly *Pseudips mexicanus* Hopkins, *I. pini*, and *Hylurgops porosus* LeConte (Furniss and Kegley, 2008; Six *et al.*, 2011a). The beetle sometimes increases briefly in abundance in residual *P. contorta* after logging (Safranyik *et al.*, 2004). Attack densities are typically extremely low and trees seldom die (Wood, 1982).

The beetle appears to prefer wetter locations (Safranyik *et al.*, 2004; Six *et al.*, 2011a). Likewise, successful attacks on trees typically occur in areas adjacent to, or within, deep crevices near the soil line that contain wet phloem and sapwood (Six *et al.*, 2011a). Many attacks are unsuccessful (Furniss and Kegly, 2008; Six *et al.*, 2011a). Unsuccessful attacks are often located on the bole more than a meter above the soil line and where the sapwood and phloem is relatively dry. These galleries often terminate with the exit of the beetle without oviposition (Six *et al.*, 2011a).

It is unclear whether these beetles mate as sibs in the natal tree as do *D. micans* and *D. punctatus* or if they colonize the tree as female–male pairs and mate under the bark of the new host. Galleries are initiated near the soil line, are irregular in shape, and tend to angle upward with eggs laid in one or more groups along the side. As with *D. micans* and *D. punctatus*, the larvae of *D. murrayanae* feed gregariously until they complete development at which time they create separate cells for pupation (Furniss and Kegley, 2008). In Utah and Montana, the beetles complete one generation annually, while in British Columbia a generation can take 2 years (Safranyik *et al.*, 2004).

5.4 Clade 4: *Dendroctonus terebrans, D. valens, D. parallelocollis, D. rhizophagus*

5.4.1 *Dendroctonus terebrans* (Olivier) (Black Turpentine Beetle)

Dendroctonus terebrans occurs from Maine to Texas and throughout the southern USA. It colonizes all native *Pinus* species within its range, although *P. rigida* Mill., *P. taeda* L., *P. echinata* Mill., and *P. elliottii* Englm. are preferred (Staeben *et al.*, 2010). It also commonly attacks two species of exotic pines, *P. thunbergii* Parl. and *P. sylvestris* L. Most attacks occur within 1 meter of the soil line, although attacks can extend much further up the bole when populations are high (Wood, 1982; Staeben *et al.*, 2010).

As with other parasites, *D. terebrans* is typically present in forests in low numbers. However, it can increase in response to factors that stress trees such as drought or widespread wounding such as may occur during storms (Staeben *et al.*, 2010). The beetle is also highly attracted to, and readily infests, trees affected by logging and the stumps of freshly cut pines (Staeben *et al.*, 2010). Fire-affected trees have been thought to be attractive to *D. terebrans* and mortality post-fire has often been attributed to its accumulation in scorched trees. However, several controlled studies found no attraction to trees in burned vs. unburned plots and the beetle did not account for greater levels of mortality in fire-affected vs. unaffected trees (Hanula *et al.*, 2002; Sullivan *et al.*, 2003; Campbell *et al.*, 2008).

Outbreaks of *D. terebrans* are seldom severe and attacks on trees are usually limited to scattered groups with few trees killed outright. Outbreaks are typically short (1–2 years) although they may extend up to 5 years if environmental conditions (e.g., drought) are driving their dynamics (Staeben *et al.*, 2010).

Female *D. terebrans*, like other parasites, use tree volatiles to locate suitable hosts. This accounts for the beetle's strong attraction to wounded hosts and logging residues (Staeben *et al.*, 2010). Previously attacked trees are also much more likely to be attacked, either due to attraction to volatiles released from prior attack sites and wounds or because such hosts are more suitable (Smith, 1963; Phillips *et al.*, 1989). *Dendroctonus terebrans* is also a very rapid responder to trees attacked by other bark beetles, particularly *D. frontalis*. This behavior may be in response to pitch tubes and wounding caused by other bark beetles or because these trees have reduced defenses.

Once a female has located a host and has begun to tunnel, she releases frontalin that acts as a sex attractant to male conspecifics. Once a male has located a female, he releases *endo*- and *exo*-brevicomin, which may act to repel other males (Phillips *et al.*, 1989). These pheromones, however, do not exert significant responses to either sex in the absence of host tree odors (Phillips *et al.*, 1989).

Most attacks occur at the soil line. Oviposition galleries often extend upwards for one or more centimeters before angling downward. Galleries can be linear or with multiple branches. Eggs are deposited in groups in a frass-free section of the gallery, which the larvae widen into a large chamber as they feed (Wood, 1982).

The length of the life cycle varies considerably. Under very warm conditions, the beetle can complete its life cycle in as few as 3 to 5 months (Staeben *et al.*, 2010). In the northernmost portion of its range, the beetle has one generation a year, while in the southern portions it may have up to three (Staeben *et al.*, 2010). Generations are not synchronized and exhibit considerable overlap. Adults emerge and initiate new attacks all year in warmer areas. All stages may overwinter. Regardless of the staggered nature of brood initiation and development, the beetle does exhibit peak flight periods, although their periodicity can vary greatly. For example, one study conducted in South Carolina reported peak flight from June through October (Sullivan *et al.*, 2003) while another conducted in Georgia reported peak flight in winter (Zanzot *et al.*, 2010).

5.4.2 *Dendroctonus valens* LeConte (Red Turpentine Beetle)

Dendroctonus valens is quite similar to *D. terebrans* in size, morphology, behavior, and breadth of host range. This species has the broadest host range of all *Dendroctonus*, in part due to its large geographic range, which extends across Canada, the western USA and through Mexico into Guatemala and Honduras (Wood, 1982). The beetle exists as three isolated populations with two separated by the Great Plains and the boreal forest (Taerum *et al.*, 2013) and a third in Central America. The Central American population exhibits strong genetic divergence suggesting the presence of cryptic species (Cai *et al.*, 2008). This population was formerly described as *D. barberi*. Whether it is appropriate to reassign this name, however, will require additional sampling and analysis (Cai *et al.*, 2008). *Dendroctonus valens* is also now well established in China where, unlike in its native range, it has become a serious pest (see Section 4).

Dendroctonus valens is a parasite that typically attacks and develops in living trees. However, it is highly attracted to injured, weakened, and dying trees including those attacked by more aggressive bark beetles (Cibrían -Tovar *et al.*, 1995; Fettig *et al.*, 2004). Host odors are extremely important in host selection. Monoterpenes vary in their release rates from trees in response to damage and drought (Lorio *et al.*, 1995). Higher emission rates of these compounds after thinning, drought, and fire may account for increased attraction and attack rates by *D. valens* in these situations. Preferential attraction of this beetle to (+)-3-carene suggests it is a key stimulus involved in host choice and may explain why some *Pinus* species are preferred, such as *P. ponderosa*, which possesses high levels of this compound (Erbilgin *et al.*, 2007).

Unlike *D. terebrans*, this beetle responds strongly to trees in stands recently affected by fire where it can continue to attack trees for many years (Parker *et al.*, 2006; Fettig *et al.*, 2008; Six and Skov, 2009; Youngblood *et al.*, 2009; Hood *et al.*, 2010). Kelsey and Joseph (2003) found that fire-damaged trees produce greater amounts of ethanol, which was associated with increased landing rates of *D. valens*. Fire-damaged ponderosa pines also produce greater levels of resin in the one to two years after fire, which may increase attraction through increased volatile emissions (Six and Skov, 2009). Trees are very seldom killed unless densities are significantly elevated due to high host availability (extensive fire damage or logging operations that leave many large diameter stumps) (Cibrían Tovar *et al.*, 1995; Youngblood *et al.*, 2009).

The beetle also uses chemical cues in mate location. Females initiate attacks singly and there is no aggregation pheromone. Rather, males and females produce *cis*- and *trans*-verbenol, myrtenol, myrtenal, and verbenone, which influence attraction and courtship behavior (Shi and Sun, 2010). Males also appear to use chemical cues in mate choice. Chen *et al.* (2012) found that male *D. valens* assess and choose females based on odor cues and that pairings between males and preferred females have enhanced reproductive fitness over pairings with non-preferred females.

Oviposition galleries are extremely variable and can be linear, cave-like, or branched. The initial section is often linear and extends upward in the bole. The gallery may continue upward or after a few centimeters may angle downward and extend into the roots (Wood, 1982). Eggs are deposited in groups in clumps of frass. Larval feeding is gregarious and results in a broad communal chamber. Pupation occurs *en masse* or individually in the frass-filled chamber or in individual chambers on the edge of the chamber. Development typically takes 1 year although in the northern parts of its range it may be longer and in the southern portions of its range it may have one and a partial second or two generations a year (Cibrían Tovar *et al.*, 1995).

5.4.3 *Dendroctonus parallelocollis* Chapuis (No Common Name)

Dendroctonus parallelocollis occurs from Chihuahua to Sinaloa, Mexico, to Honduras (Wood, 1982). Very little information exists on this species. Only two hosts, *Pinus leiophylla* Schiede and Deppe and *P. oocarpa* Schiede, are known. The beetle is a parasite that initiates single

attacks at the base of trees from November to February (Cibrían Tovar *et al.*, 1995). It can also sometimes be found colonizing dying trees and fresh stumps (Cibrían Tovar *et al.*, 1995) and occasionally co-occurs in trees with *D. frontalis* and *D. mexicanus* (Cibrían Tovar *et al.*, 1995). While the beetle can cause the death of individual roots and sometimes trees, tree killing is not common. Outbreaks do not occur.

Oviposition galleries are usually constructed in the larger roots that connect directly to the bole, although some may extend for a distance up the bole and sometimes form a criss-cross pattern similar to the galleries of *D. approximatus*. The galleries can extend as deep as 70 cm on bigger roots (Cibrían Tovar *et al.*, 1995). Eggs are laid alternately along the gallery. Larvae construct individual galleries in large diameter roots. The length of a generation is not known.

5.4.5 *Dendroctonus rhizophagus* Thomas and Bright (No Common Name)

This beetle occurs in the Sierra Madre Occidental of Mexico. While other *Dendroctonus* attack mature trees, this species exclusively infests seedlings and young saplings. It colonizes primarily *P. engelmannii* Carr., *P. durangensis* Martínez, *P. leiophylla* Schiede ex Schltdl. and Cham., and *P. arizonica* Engelm. (Thomas and Bright, 1970). The distribution of the beetle does not appear to be limited by the distribution of its hosts. Rather, it appears to possess a relatively narrow niche bounded by highly specific precipitation and temperature requirements (Mendoza *et al.*, 2011).

The attack dynamics and larval feeding patterns of *D. rhizophagus* are similar to those of parasitic *Dendroctonus*; however, unlike the parasites, it kills its host (Cibrían Tovar *et al.*, 1995). Tree-killing in this species, however, is not linked to the mass attack behavior exhibited by facultative predators, but rather is the result of the small diameter of the trees it uses, which are girdled as the larvae feed on the roots.

Unlike most other *Dendroctonus*, this species prefers healthy fast-growing trees (Sánchez-Martínez and Wagner, 2009). In natural forests, this insect is seldom a problem. However, in new plantations or where forests are regenerating after large stand replacement events such as fire or clearcutting, this insect can build up rapidly and cause significant amounts of mortality (Cibrían Tovar *et al.*, 1995).

New attacks on a tree are accomplished by a single female–male pair. The female attacks the tree near soil level and the male soon follows. After mating, an oviposition gallery is constructed, which often spirals around the tree. The eggs are laid in groups in masses of frass. The larvae hatch and begin feeding in groups and eventually form a large common feeding chamber. As they feed, the larvae move upwards into the bole of the tree, sometimes to the tops of seedlings. Once they reach the last instar the larvae turn and move downward until they reach the roots. At the roots, the group may split up, with individual larvae seeking out roots within which to over-winter (Cibrían Tovar *et al.*, 1995). At the beginning of spring, the larvae regroup in the main stem or taproot where they chew individual pupal chambers. Most adults from a single brood emerge out of just a few holes near the root collar. Emergence and attacks occur in mid-summer and there is one generation a year (Cibrían Tovar *et al.*, 1995).

5.5 Clade 5: *Dendroctonus ponderosae, D. jeffreyi*

5.5.1 *Dendroctonus ponderosae* Hopkins (Mountain Pine Beetle)

Dendroctonus ponderosae has historically occurred from central British Columbia, throughout the western USA, the Black Hills of South Dakota, and into Baja California Norte (Wood, 1982). Due to a warming climate, the beetle has recently extended its range northward through northern British Columbia to the southern boundary of the Yukon, and across portions of Alberta (see Section 4).

Dendroctonus ponderosae exhibits considerable genetic variability across its extensive geographic range. Early studies suggested there might be some differentiation by host tree species, although results were variable (Sturgeon and Mitton, 1986; Langor and Spence, 1991). Subsequent studies using more sensitive markers have not found evidence supporting this hypothesis (Mock *et al.*, 2007). However, several studies have found substantial population level subdivision suggestive of gene flow occurring in an isolation-by-distance pattern following the horseshoe shape distribution of the beetle around the Great Basin Desert. Populations in the southern ends of the horseshoe in Arizona and southern California exhibit the greatest differentiation (Mock *et al.*, 2007; Bracewell *et al.*, 2013). Experimental crosses of beetles from populations around the horseshoe revealed the existence of cryptic postzygotic isolation involving hybrid male sterility (Bracewell *et al.*, 2011). This effect occurred abruptly in crosses made between beetles from California and Idaho rather than incrementally with increasingly distant populations. Further, a reproductive barrier appears to exist between populations in Oregon and those in Idaho. Interestingly, neutral genetic markers failed to detect the abrupt decrease in gene flow among populations exhibiting postzygotic isolation (Mock *et al.*, 2007).

Genetic differences in development and size have also been noted for northern vs. southern populations (Bentz *et al.*, 2001; Bracewell *et al.*, 2013). Beetles in northern populations develop at a faster rate than do beetles in southern populations, likely due to selection for different thermal conditions. Genetic differences in body size also exist between different populations. Together, the results of these studies are consistent with observations that substantial differences in life history traits exist among populations of this beetle (Amman, 1982; Stugeon and Mitton, 1986; Langor, 1989; Bentz and Mullins, 1999; Bentz *et al.*, 2001).

Dendroctonus ponderosae is the most aggressive of all the facultative predators in *Dendroctonus*. While, like many *Dendroctonus*, this species is limited to colonizing weak hosts when its numbers are low, once populations have expanded, this beetle can move into and maintain extensive outbreaks in healthy trees (Raffa *et al.*, 2008; Boone *et al.*, 2011). Outbreaks are typically triggered by warm dry periods that simultaneously support greater beetle productivity and survival while compromising host tree defenses (Creeden *et al.*, 2014). Although outbreaks may be triggered fairly rapidly by favorable abiotic conditions, there can also be a substantial lag period between when the abiotic factors that initiate the outbreak occur and when populations actually expand (Thompson and Shrimpton, 1984; Preisler *et al.*, 2012). Once beetle numbers surpass a threshold over which they are able to kill healthy trees, tree defenses become inconsequential and the beetle switches to preferring healthy trees, likely due to their superior qualities for supporting brood (Boone *et al.*, 2011). At this point an outbreak becomes self-propagating. Once initiated, outbreaks can continue even with an alleviation of drought, and collapse often occurs more in response to host depletion than to changes in climate and weather (Creeden *et al.*, 2014). Cold weather events can act to slow outbreaks, but may not completely stop them as surviving beetles may rebuild populations as long as favorable conditions exist (Safranyik, 1978; Creeden *et al.*, 2014).

The dynamics of *D. ponderosae* varies considerably across its range. The development of massive outbreaks is primarily limited to areas north of Arizona and Southern California, although more restricted outbreaks have occurred in Arizona (Parker and Stevens, 1979). Mortality in Mexico tends to be restricted to small groups of trees (Cibrían Tovar *et al.*, 1995).

Dendroctonus ponderosae colonizes all native pines within its range except *P. jeffreyi* (Wood, 1982). It also colonizes a number of exotic pines. The beetle exhibits clear preferences for one species over others where multiple pine species co-occur (Baker *et al.* 1971; Six and Adams, 2007; Raffa *et al.*, 2013). This difference in preference is likely related to varying levels of defense and different monoterpene compositions and concentrations (Raffa *et al.*, 2013). Switching occurs mainly during outbreaks when the availability of preferred hosts of suitable diameter is depleted (Wood, 1963). The most commonly infested host is *P. contorta*. It is in this species that the beetle develops the most extensive outbreaks (Safranyik and Carroll, 2006). Outbreaks also occur in *P. ponderosa*, *P. flexilis* James, and *P. albicaulis* Engelm., although they tend to be more limited in distribution, in part due to the more limited geographic ranges of these host species. The recent extensive outbreak in *P. albicaulis* is unprecedented in size and has been primarily driven by anthropogenic climate change (see Section 4).

Dendroctonus ponderosae prefers standing trees over 15 cm in diameter and only rarely can be found in wind-throw or felled trees. A number of studies have found that the beetle preferentially attacks slow growing trees with more rapidly growing trees having a greater potential to survive an outbreak (Millar *et al.*, 2012; Knapp *et al.*, 2013). These differences in growth rates are most likely linked to tree genotype (Millar *et al.*, 2012). The beetle's response to fire-affected trees is variable. In some studies the beetle has shown no increase in attraction to damaged trees after fire (Geizler *et al.*, 1984; Six and Skov, 2009), while in others, positive responses have been observed (Fettig *et al.*, 2010; Powell *et al.*, 2012; Davis *et al.*, 2012).

Females initiate mass attacks in mid-summer (although there is considerable variability in flight periods at different elevations) (Amman, 1973). In some areas, two peak flights occur. A smaller early flight in late spring or early summer is typically comprised of re-emerged parents while the summer flight is primarily made up of brood adults.

Galleries are J-shaped and vertical within the bole of the tree with eggs laid in niches alternating along the sides. All development occurs in the phloem (Wood, 1982). While this species is expected to be outcrossing due to pupation in individual chambers under bark and the aggregation of males and females on new host trees, substantial inbreeding can occur among sibs under bark of the natal tree (Bleiker *et al.*, 2013).

The life cycle typically takes 1 year (univoltine) although in cooler parts of its range it can be longer (semivoltine). A univoltine life cycle is seasonally adaptive and a critical component in the development of outbreaks (Logan and Bentz, 1999; Logan and Powell, 2001). A univoltine life cycle allows the insect to synchronize emergence of brood adults, which supports the ability of the insect to mass attack trees as well as ensuring offspring will enter winter as larvae, the stage most suited to surviving cold winter temperatures. A semivoltine life cycle is maladaptive because adaptive seasonality is disrupted increasing the number of brood that enters winter in stages vulnerable to freezing. It also requires that brood feed for an extended period of time in a host declining in nutrients and moisture and increases length exposure to natural enemies (Kim *et al.*, 2005; Bleiker and Six, 2008b).

The development rate of D. *ponderosae*, like all bark beetles, is temperature dependent. In this beetle, development rate is stage specific, with different life stages having different thermal thresholds and development rates that act to aid in synchronizing emergence and mass attack (Bentz *et al.*, 1991). Development rate is also genetically determined with northern populations developing faster than southern populations when held at a constant temperature (Bentz *et al.*, 2001, 2011). This difference in development rate is likely linked to maintenance of adaptive seasonality allowing northern beetles to develop rapidly enough, and southern beetles slowly enough, to maintain univoltine life cycles (Bentz *et al.*, 2011).

Overwintering by *D. ponderosae* involves a process called cold-hardening. The beetle overwinters primarily as larvae, which are the stages most suited to surviving winter (Bentz *et al.*, 1991). Eggs and pupae are the most susceptible to freezing although eggs sometimes survive during mild winters (Lester and Irwin, 2012). Larvae progressively gain cold tolerance in response to decreasing temperatures in fall and early winter (Bentz and Mullins, 1999). The overall thermal history experienced by the larvae, as well as daily changes in temperature, is an important factor influencing the cold-hardening process and the degree of supercooling ability that is achieved. The temperature-dependent nature of developing and maintaining cold-hardiness results in considerable variation in the ability of beetles to withstand cold because of variability in weather (Bentz and Mullins, 1999).

Evidence for diapause in *D. ponderosae* remains equivocal. Early experiments rearing *D. ponderosae* in the lab failed to find evidence of diapause (Logan and Amman, 1986). Likewise, winter-collected larvae or those held in cold storage were able to resume development immediately upon warming and individuals collected from the field could be reared through the adult stage without chilling (Safranyik and Whitney, 1985). However, in a recent study, Lester and Irwin (2012) reported a strong mid-winter suppression of metabolic activity in adult beetles correlated with improved supercooling ability. They also found a lack of response to changes in temperature indicating the possibility of a facultative diapause. The overwintering population studied by Lester and Irwin (2012) consisted of adults and eggs. This situation is in stark contrast with most populations of *D. ponderosae* that overwinter as cold-tolerant larvae with some parent adults surviving to re-emerge and initiate new broods the following year (Bentz *et al.*, 1991). The study by Lester and Irwin (2012) indicates that diapause has evolved in adults that have already produced brood, which seems counterintuitive in terms of natural selection and the prevailing manner in which most populations of this beetle overwinter. However, their results indicate a clear need to investigate how different populations of the beetle under different selective pressures may vary in their response to winter.

5.5.2 *Dendroctonus jeffreyi* Hopkins (Jeffrey Pine Beetle)

This beetle is very difficult to distinguish morphologically from its sister species *D. ponderosae*, and both beetles produce similar galleries (Wood, 1982). *Dendroctonus jeffreyi* is sympatric with *D. ponderosae*; however, there is no overlap in host tree use by the two beetles. *Dendroctonus jeffreyi* is highly specialized and colonizes only *P. jeffreyi*, while *D. ponderosae* is highly polyphagous, colonizing all species of *Pinus* within its range including several exotic *Pinus*, but never *P. jeffreyi* (Wood, 1982). The geographic distribution of *D. jeffreyi* follows that of its host extending from northern California through Baja California Norte. Populations in southern California are geographically isolated from northern Sierran populations and exhibit genetic differentiation consistent with long-term isolation (Six *et al.*, 1999). Mexican populations have not been investigated from a population genetics perspective.

Dendroctonus jeffreyi is an aggressive facultative predator and kills trees using an aggregation pheromone-mediated mass attack and is capable of developing widespread outbreaks. As with other aggressive *Dendroctonus*, outbreaks are not cyclic, but rather are sporadic and driven by extended droughts (Smith *et al.*, 2009). The beetle is not attracted to stumps and felled trees and does not appear to respond to fire-affected trees (Fettig *et al.*, 2010).

Galleries are J-shaped, linear, and vertical within standing trees with eggs laid in niches alternating along the sides (Wood, 1982). Larvae feed individually in the phloem where they also pupate. Because these beetles develop individually under bark and form mating pairs on new hosts after they have dispersed, this species is expected to be strongly outcrossing. However, Six *et al.* (1999) using allozymes detected a strong departure from random mating indicating that considerable inbreeding occurs in this species. Dispersal and new attacks typically occur early to mid-summer. The beetle has one generation a year.

5.6 Clade 6: *Dendroctonus vitei, D. frontalis, D. mexicanus, D. adjunctus, D. approximates, D. brevicomis*

5.6.1 *Dendroctonus vitei* Wood (No Common Name)

This is the least known of all *Dendroctonus*. It is found in Guatemala where it infests *Pinus pseudostrobus* Lindley and *P. tenuifolia* Benth., mostly at higher elevations. While its habits are thought to be similar to *D. mexicanus*, the pheromones of the two species are distinct (Wood, 1982). Outbreaks are unknown. It is most likely an early successional saprophage.

5.6.2 *Dendroctonus frontalis* Zimmermann (Southern Pine Beetle)

This beetle has historically occurred across the southern USA from Oklahoma and Pennsylvania south through Florida (Wood, 1982) and has recently expanded its range further north into the New Jersey pine barrens (see Section 4). The beetle also occurs in Arizona and New Mexico, has an extensive distribution in Mexico (Cibrían Tovar *et al.*, 1995), and populations in Honduras and Belize (Sullivan *et al.*, 2012). Two morphotypes occur in Chiapas, Mexico, and in Belize, one of which also occurs in the southern USA. These morphotypes possess distinct cuticular hydrocarbons and pheromones indicating they are likely cryptic species (Sullivan *et al.*, 2012). However, in the southeastern USA, the beetle appears to occur as a large stable metapopulation (Schrey *et al.*, 2011). These studies, as well as the broad distribution of the beetle and the relative isolation of some populations, suggest it is a species complex. Recent evidence indicates that at least two cryptic species exist. Two sympatric morphological variants occurring in Central America and southern Mexico exhibit different chromosomal formulas (Armendáriz-Toledano *et al.*, 2014) and pairings of the beetles from the two morphotypes result in lower frequencies of progeny production. Phylogenetic analysis of COI shows the two groups are distinct with 98% nodal support (Armendáriz-Toledano *et al.*, 2014).

Dendroctonus frontalis is an aggressive facultative predator that colonizes a large number of *Pinus* species across its geographic range, although in the southern USA it prefers *P. taeda* and *P. echinata* (Wood, 1982; Cibrían Tovar *et al.*, 1995). The beetle mass attacks trees and periodically develops outbreaks. It often kills trees in "spots." Spots are comprised of groups of dead trees that occur as a result of infestation of individual trees followed by movement into adjacent trees in subsequent generations. Spots often start in lightning-struck or weakened trees (Clarke and Nowak, 2009). Long-distance dispersal also occurs and can result in the formation of new spots (Clarke and Nowak, 2009). During outbreaks, spots coalesce into large contiguous areas of mortality.

The conditions that initiate outbreaks of *D. frontalis* are not well understood. Bark beetle outbreaks are typically eruptive, driven by numerous interacting factors, but with exogenous factors, particularly climate, often playing a primary role. However, two early studies reported evidence that *D. frontalis* outbreaks were cyclic, involving endogenous delayed density-dependent feedback dynamics that were hypothesized to be primarily driven by natural enemies (Reeve, 1997; Turchin *et al.*, 1999). A later study that incorporated a larger geographic area and a more complex suite of models found that the beetle's dynamics are more likely to be driven by a combination of density-dependent feedback and annual variation in temperature extremes (Friedenberg *et al.*, 2008). Duehl *et al.* (2011) found that while local baseline populations influence outbreak expansions, the actual development of outbreaks is most often related to climatic variables, particularly warmer average annual minimum winter temperatures and high fall precipitation. These studies indicate that, like other outbreaking bark beetles, outbreaks are not cyclic, but rather are driven primarily by stochastic climatic variables.

Of the 16 species of *Dendroctonus* that have been tested for temperature limits, *D. frontalis* is the least cold tolerant and extreme cold temperatures likely play an important role in regulating its populations and geographic distribution (Evans *et al.*, 2011). In addition, the complex community of microbes and mites found with this beetle has also been found to influence outbreak dynamics (Lombardero *et al.*, 2003; Hofstetter *et al.*, 2006a, 2007; Evans *et al.*, 2011). The beetle's mutualistic fungal partners, and its phoretic mites and their symbiotic fungi, are regulated by temperature whose relative prevalence feeds back to affect beetle dynamics (Hofstetter *et al.*, 2006b) and may even lead to the collapse of some outbreaks (Lombardero *et al.*, 2003).

Dendroctonus frontalis has three to seven generations a year depending on temperature (Wood, 1982; Cibrían Tovar *et al.*, 1995). Generations are often overlapping and many different stages can be found in trees at a single location over time. All stages can be found overwintering. However, prepupae are more cold tolerant, and at least in the northern parts of its range, are the dominant stage found in winter (Tran *et al.*, 2007). The dominance of prepupae in winter is not due to differential mortality of other life stages, but rather appears to be a case of adaptive seasonality (Tran *et al.*, 2007). In contrast, in the southernmost portions of its range, *D. frontalis* may remain active and even attack trees throughout the year. Some parents remerge to produce brood in new trees (Clarke and Nowak, 2009).

The beetle typically attacks standing pines larger than 15 cm in diameter. Most attacks are concentrated on the upper half or mid-bole of the tree. Galleries are sinuous and angular, and sometimes are branched. Eggs are laid in niches in an alternate pattern along the sides. The larvae feed individually in the phloem until they reach the third or fourth instar when they move into the outer bark where they complete development. Pupation occurs in the outer bark (Wood, 1982).

5.6.3 *Dendroctonus mexicanus* Hopkins (Mexican Pine Beetle)

This species is morphologically similar to, and constructs nearly identical galleries as, *D. frontalis*. There is considerable genetic variability and it is likely a complex of cryptic species (Zúñiga *et al.*, 2006).

The beetle colonizes a very broad range of *Pinus* hosts (Wood, 1982; Cibrían Tovar *et al.*, 1995). Its geographic range has historically been described as occurring from Chihuahua (Mexico) southward through Honduras (Wood, 1982). However, in 2001, the beetle was found infesting *P. leiophylla* var. *chihuahuana* (Engelm.) Shaw in southeastern Arizona, where both *D. mexicanus* and *D. frontalis* were found co-infesting pines at multiple sites, but with *D. mexicanus* predominating (Moser *et al.*, 2005). While the new record was reported as a recent range expansion for this species, it is possible that *D. mexicanus* had been overlooked or misidentified as *D. frontalis* in this area in the past (Moser *et al.*, 2005).

Dendroctonus mexicanus is a facultative predator. It attacks trees of all diameters except small diameter regeneration. It periodically develops outbreaks where thousands of trees are killed (Cibrían Tovar *et al.*, 1995). Most outbreaks are associated with drought.

Dendroctonus mexicanus produces three to six generations a year depending on thermal conditions. Populations often overlap and flight periods can extend over several months. Females typically initiate mass attacks on weakened trees that do not produce strong defenses. Some females remerge and produce an additional brood in another tree (Cibrían Tovar *et al.*, 1995).

The oviposition galleries are sinuous and eggs are laid individually along the sides. Larvae tunnel and feed in the phloem until the third or fourth instar when they move into the outer bark. Pupation occurs in the outer bark (Cibrían Tovar *et al.*, 1995).

5.6.4 *Dendroctonus adjunctus* Blandford (Roundheaded Pine Beetle)

Dendroctonus adjunctus is a facultative predator found from southern Utah to Colorado to Guatemala (Wood, 1982). It colonizes a wide range of *Pinus* hosts (Wood, 1982; Cibrían Tovar *et al.*, 1995). It is not highly aggressive, although it can develop outbreaks when drought weakens large numbers of hosts (Negrón, 1997). Fischer *et al.* (2010) found there appears to be a genetic component to host choice during outbreaks. Trees that have slower growth rates, particularly in response to drought, are significantly more likely to be killed than co-occurring trees with more rapid growth rates. Likewise, Vargas *et al.* (2002) found a correlation between the frequencies of particular alleles of pines and their likelihood of attack further indicating that attacks by the beetle are non-random and that host choice is affected by tree genotype.

In endemic periods, the beetle is most often found in suppressed trees or trees affected by fire or infested by other bark beetles (Cibrían Tovar *et al.*, 1995). Trees over 25 cm in diameter are preferred, although stumps may also be colonized. It is not attracted to felled trees or logs

(Wood, 1982). The beetle is sometimes restricted to small areas of a tree due to competition with other bark beetles, but in the absence of other species, it can colonize the majority of the bole (Wood, 1982; Cibrían Tovar *et al.*, 1995).

Females initiate mass attacks although in some cases only a few attacks occur when trees are already colonized by other bark beetle species. Galleries are sinuous and sometimes branched with eggs laid in alternating niches along the sides. The galleries and beetle can easily be confused with co-occurring *D. frontalis* and *D. brevicomis*.

The larvae feed individually. Some larvae complete development and pupate in the phloem while others move into the outer bark for pupation (Wood, 1982). Dispersal can occur over a relatively long period and typically occurs in spring or late summer through fall (Wood, 1982). This beetle usually has one generation a year, although some variability may occur.

5.6.5 *Dendroctonus approximatus* Dietz (Larger Mexican Pine Beetle)

Dendroctonus approximatus is an early successional saprophage that occurs from Colorado through Honduras and colonizes a wide range of *Pinus* including *P. ayacahuite* Ehrenb. ex Schltdl., *P. engelmannii*. *P. hartwegii* Lindl., *P. leiophylla*, *P. montezumae* Lamb., *P. ponderosa*, and *P. teocote* Schied. ex Schltdl. and Cham. (Wood, 1982). At least two genetically divergent groups exist in Mexico (Sánchez-Sánchez *et al.*, 2012).

Dendroctonus approximatus usually relies on other more aggressive bark beetles to overwhelm the tree before it enters the host (Wood, 1982). Galleries are produced in standing trees or the undersides of fallen trees. Emergence from trees can be extended over many months. In the northern part of its range, the beetle flies from June to October while in the southern part it likely flies all year (Wood, 1982). The pheromone system is unknown.

Attacks in the northern part of the range are seldom higher than two meters above the soil line while in the southern portion they can reach as high as four meters (Wood, 1982; Cibrían Tovar *et al.*, 1995). This difference in distribution within the bole may be due to competition with *D. adjunctus* and other beetles in the north, which tend to colonize at mid-bole. Trees over 30 cm in diameter appear to be preferred (Cibrían Tovar *et al.*, 1995). When injured trees are used as hosts, small numbers of attacks may recur over time until the tree is finally girdled and killed (Wood, 1982).

Oviposition galleries of *D. approximatus* are unique among *Dendroctonus*. They can be linear but most often form a matrix of criss-crossing tunnels with considerable branching and anastomosis (Wood, 1982). The egg niches are also distinctive. They are large and cup-shaped and extend into the outer bark (Cibrían Tovar *et al.*, 1995). Usually, niches contain one egg but may contain up to four

eggs (Wood, 1982; Cibrían Tovar *et al.*, 1995). The larvae mine mostly in the outer bark producing very short feeding galleries. Pupation also occurs in the outer bark (Wood, 1982). One generation appears to be common across the beetle's range. Two generations a year may occur in very warm locations.

5.6.6 *Dendroctonus brevicomis* LeConte (Western Pine Beetle)

This beetle occurs from British Columbia, throughout the western USA, and northern Mexico (Wood, 1982; Cibrían Tovar *et al.*, 1995). It is highly specialized using only *P. ponderosa* and *P. coulteri* D. Don. Large genetic differences exist between beetles in western (California, Idaho, Oregon, and British Columbia) and eastern (Arizona, Colorado, New Mexico, and Utah) regions indicating these groups are distinct species (Kelley *et al.*, 1999). These groups correspond to the distribution of *P. ponderosa* var. *scopulorum* and *P. coulteri* (western population) and *P. ponderosa* var. *ponderosa* (eastern population).

Dendroctonus brevicomis is a facultative predator, and like *D. ponderosae* can kill trees across large areas during outbreaks. Outbreaks are usually associated with drought (Page, 1981). The beetle also responds to fire-damaged trees, but not those killed by fire. It does not maintain elevated numbers once damaged trees are killed unless underlying stressors such as drought also exist (Miller and Keen, 1960; Six and Skov, 2009; Davis *et al.*, 2012).

These beetles prefer large (over 30 cm in diameter) standing trees although they often colonize smaller diameter trees during non-outbreak periods. Trees that are slower growing are preferred over those that are growing rapidly (Miller and Keen, 1960). History of exposure of beetles to host tree defensive chemistry also affects oviposition gallery construction and fecundity with beetles exhibiting superior performance in host trees with similar chemical profiles as their natal host. However, rates of acceptance of hosts do not appear to be affected by host tree genotype (Davis and Hofstetter, 2011b).

The flight period can extend from May through November (Miller and Keen, 1960; Furniss and Carolin, 1977). Mass attacks on individual trees are often spread out over many weeks. The beetle has one to four generations a year depending on thermal conditions. There is considerable overlap of generations in warmer regions. In Mexico, the beetle mostly overwinters as adults (Cibrían Tovar *et al.*, 1995). In North America north of Mexico, most overwintering occurs as larvae (Miller and Keen, 1960).

Oviposition galleries are sinuous, vertical or horizontal and often branching. Eggs are laid in niches at irregular spacings along the sides. Only first instars feed in phloem. Second instars move into the outer bark where all subsequent development occurs (Wood 1982).

6 CONCLUSION

Despite the importance of the genus *Dendroctonus*, many species remain poorly studied, and for those that are well investigated, serious gaps still exist in our understanding of even their basic life histories. One of the biggest gaps is in our understanding of their ecological roles. Considerable work has focused on managing their populations, but very little on understanding how they influence ecosystem function and process. Except for *D. valens* in China, and *D. ponderosae* and *D. frontalis* in areas of recent range expansions in North America, all species occupy native ranges where several act as natural disturbance agents. It is well recognized in many forests that other natural disturbances such as fire are important in regulating productivity and structure, composition, and function. Similar roles may be expected for tree-killing bark beetles. However, few *Dendroctonus* have been investigated within this context (Romme *et al.*, 1986).

For many species, little to no information exists on their ecology, and in many cases these species are assumed to act similarly to closely related beetles. However, as more information has begun to accrue over time, we find that even sister species with similar overall life strategies can exhibit very different responses to their environment indicating that extrapolation should be done with caution. Even for well-studied species, such as *D. ponderosae* and *D. frontalis*, there remains much to be learned, particularly on how the insects respond to environmental change. Recent approaches that have moved beyond case studies and short-term plot-scale experiments to incorporate remote sensing and large-scale long-term retrospective studies are starting to reveal patterns that have not been detectable at smaller scales and sometimes yield results in conflict with conventional wisdom.

A common theme that is emerging is that abiotic environmental conditions (i.e., temperature and precipitation) play major roles in the population dynamics of most, if not all, *Dendroctonus* species. This should come as no surprise. *Dendroctonus*, like all insects, are poikilothermic and temperature is one of the most important factors influencing their development and population dynamics. Likewise, host tree defenses have strong effects on bark beetle population dynamics and are strongly influenced by moisture availability.

While it has long been recognized that stress in host trees, particularly due to drought, can increase susceptibility to beetle attack, our understanding of what constitutes stress in trees, as well as our ability to detect it, is still very poor. There are numerous examples in the literature where *Dendroctonus* are described to have developed outbreaks in apparently healthy trees. However, bark beetle outbreaks do not develop spontaneously without cause, but require a trigger (Bentz *et al.*, 2009). A more likely scenario is that the "healthy" trees in these studies were stressed in some manner, but that it was unrecognized by the investigators. Additionally, a tree's response to stressors are influenced by its genetics, a subject even less well understood than the role of tree physiology in regard to host tree susceptibility. Research on the roles of tree genetics, physiology, and tree stress will be paramount to developing a better understanding of *Dendroctonus* dynamics, particularly as forests are increasingly affected by anthropogenic change.

Many new powerful tools are providing us with an unprecedented ability to study *Dendroctonus* at all levels. For example, beetle and symbiotic fungus DNA sequences have been used to aid in determining where in North America *D. valens* populations in China may have originated (Cognato *et al.*, 2005; Taerum *et al.*, 2013) as well as to track the origins of *D. ponderosae* expansions in Canada (Massoumi Alamouti *et al.*, 2011; Roe *et al.*, 2011a; Tsui *et al.*, 2012). Tools such as RAD-seq are now being used to conduct intensive assessments of population structure using many thousands of markers rather than a mere handful (Bracewell and Six, unpubl.). The delimitation of the many cryptic species of *Dendroctonus* that have recently been detected would not be possible without multi-gene sequencing.

Genomics is a particularly exciting field of study that is beginning to provide remarkable insights into *Dendroctonus* on many levels. Genomics is already being used for gene and gene family discovery in *D. ponderosae* and is contributing to our understanding of chromosomal evolution, sex determination, metabolism, chemical detoxification, and pheromone production (Aw *et al.*, 2010; DiGuistini *et al.*, 2011; Keeling *et al.*, 2012). Likewise, genomic analyses of *Dendroctonus*-associated symbiotic fungi have allowed detection of sexuality in species previously thought to be asexual (Tsui *et al.*, 2013), and the discovery of genes associated with phytopathogenicity and metabolism (Lah *et al.*, 2013). Comparative genomics of pathogenic and saprophytic symbionts are providing insights into how these fungi differ in their interactions with host trees (Haridas *et al.*, 2013). As we accumulate genomes of more *Dendroctonus* (currently only two have been sequenced, *D. ponderosae* (Keeling *et al.*, 2013) and *D. brevicomis* (R. Bracewell, J. Good, D. Six, unpubl.)), we will also be able to resolve the currently ambiguous internal nodes in the phylogeny of these beetles leading to a better understanding of the evolutionary relationships among species.

Transcriptomics can be used to detect how gene expression varies with external environmental conditions and can provide a picture of how, and how strongly, an organism responds to factors such as temperature and exposure to tree defenses and carbon and nitrogen sources. Transcriptomic approaches have been used to investigate how fungi symbiotic with *Dendroctonus* respond to host tree chemicals and to the presence of various carbon compounds within the tree (DiGuistini *et al.*, 2011; Cano-Ramírez *et al.*, 2013; Haridas *et al.*, 2013; Lah *et al.*, 2013) and for defense gene discovery in lodgepole and jack pine, a co-evolved and novel host, respectively, for *D. ponderosae*, to better understand differences in their responses to the beetle (Hall *et al.*, 2013).

Proteomics can take our understanding of gene expression an additional step by allowing us to measure actual protein synthesis. Because relatively small changes in gene expression can produce large changes in the amount of the corresponding protein present in a cell, proteomics can provide a more accurate estimate of expression and can be used to compare protein profiles as insects encounter and respond to different conditions. Thus far, proteomics has been used to profile the responses of overwintering and developing larvae of *D. ponderosae* to better understand the insect's ability to cold-harden (Bonnett *et al.*, 2012).

Mathematical modeling is increasingly being applied in aiding our understanding of many aspects of *Dendroctonus* biology and ecology including outbreak dynamics and spread, response to climate change, and symbiosis stability (Aukema *et al.*, 2006, 2008; Samarasekera *et al.*, 2012; Coops *et al.*, 2012; Preisler *et al.*, 2012; Addison *et al.*, 2013; Powell and Bentz, 2014). Increases in computing power are simultaneously improving our ability to run increasingly complex models and to deal with massive data sets associated with remote sensing, genomics, and other approaches.

The near future promises to be an exciting and illuminating time in the study of *Dendroctonus*. Applying traditional and new tools with scientific rigor and hypothesis testing will allow us to build a more complete picture of how these fascinating and complex insects function and are responding to a rapidly changing world.

REFERENCES

Adams, A.S., Six, D.L., 2006. Temporal variation in mycophagy and prevalence of fungi associated with developmental stages of the mountain pine beetle, *Dendroctonus ponderosae* (Coleoptera: Curculionidae, Scolytinae). Environ. Entomol. 36, 64–72.

Adams, A.S., Six, D.L., 2008. Detection of host habitat by parasitoids using cues associated with mycangial fungi of the mountain pine beetle, *Dendroctonus ponderosae*. Can. Entomolo. 140, 124–127.

Adams, A.S., Six, D.L., Adams, S., Holben, W., 2008. *In vitro* interactions among yeasts, bacteria and the fungal symbionts of the mountain pine beetle, *Dendroctonus ponderosae*. Microb. Ecol. 56, 460–466.

Adams, A.S., Currie, C.R., Cardoza, Y., Klepzig, K., Raffa, K.F., 2009. Effects of symbiotic bacteria and tree chemistry on the growth and reproduction of bark beetle fungal symbionts. Can. J. Forest Res. 39, 1133–1147.

Adams, A.S., Adams, S.M., Curie, C.R., Gillette, N.E., Raffa, K.F., 2010. Geographic variation in bacteria commonly associated with the red

turpentine beetle (Coleoptera: Curculionidae). Environ. Entomol. 39, 406–414.

Adams, A.S., Aylward, F.O., Adams, S.M., Erbilgin, N., Aukema, B.H., Currie, C.R., et al., 2013. Mountain pine beetles colonizing historical and naïve host trees are associated with a bacterial community highly enriched in genes contributing to terpene metabolism. Appl. Environ. Microbiol. 79, 3468–3475.

Addison, A., Powell, J.A., Six, D.L., Moore, M., Bentz, B.J., 2013. The role of temperature variability in stabilizing the mountain pine beetle-fungus mutualism. J. Theor. Biol. 335, 40–50.

Amman, G.D., 1973. Population changes of the mountain pine beetle in relation to elevation. Environ. Entomol. 2, 541–547.

Amman, G.D., 1982. Characteristics of mountain pine beetle reared in four pine hosts. Environ. Entomol. 2, 389–405.

Amman, G.D., 1984. Mountain pine beetle (Coleoptera: Scolytidae) mortality in three types of infestations. Environ. Entomol. 13, 184–191.

Amman, G.D., Cole, W.E., 1983. Mountain pine beetle dynamics in lodgepole pine forests. Part II: Population dynamics, USDA Forest Service General Technical Report INT-145.

Armendáriz-Toledano, F., Niño, A., Sullivan, B.T., Macías-Sámano, J., Víctor, J., Clarke, S.B., Zúñiga, G., 2014. Two species within *Dendroctonus frontalis* (Coleoptera: Curculionidae): evidence from morphological, karyological, molecular, and crossing studies. Ann. Entomol. Soc. Am. 107, 11–27.

Atkins, M.D., 1961. A study of the flight of the Douglas-fir beetle, *Dendroctonus pseudotsugae* Hopk. 3. Flight capacity. Can. Entomol. 93, 467–474.

Aukema, B.H., Werner, R.A., Haberkearn, K.E., Illman, B.L., Clayton, M.K., Raffa, K.F., 2004. Quantifying sources of variation and frequency of fungi associated with spruce beetles: Implications for hypothesis testing and sampling methodology in bark beetle-symbiont relationships. Forest Ecol. Manag. 217, 187–202.

Aukema, B.H., Carroll, A.L., Zhu, J., Raffa, K.F., Sickley, T.A., Taylor, S.W., 2006. Landscape level analysis of mountain pine beetle in British Columbia, Canada: spatiotemporal development and spatial synchrony within the present outbreak. Ecography 29, 427–441.

Aukema, B.H., Carroll, A.L., Zheng, Y., Zhu, J., Raffa, K.F., Moore, R.D., et al., 2008. Movement of outbreak populations of mountain pine beetle: Influences of spatiotemporal patterns and climate. Ecography 31, 348–358.

Avtzis, D.N., Bertheau, C., Stauffer, C., 2012. What is next in bark beetle phylogeography? Insects 3, 453–472.

Ayres, M.P., Wilkens, R.T., Ruel, J.J., 2000. Nitrogen budgets of phloem-feeding bark beetles with and without symbiotic fungi. Ecology 8, 2198–2210.

Aw, T., Schlauch, K., Keeling, C.I., Young, S., Bearfield, J.C., Blomquist, G.J., Tittiger, C., 2010. Functional genomics of mountain pine beetle (*Dendroctonus ponderosae*) midguts and fat bodies. BMC Genomics 11, 215.

Baker, B.H., Amman, G.D., Trostle, G.C., 1971. Does the mountain pine beetle change hosts in lodgepole and whitebark pine stands? USDA For. Serv. Res. Note INT-151.

Baker, B.H., Hostettler, B.B., Furniss, M.M., 1977. Response of eastern larch beetle (Coleoptera: Scolytidae) in Alaska to its natural attractant and to Douglas-fir beetle pheromones. Can. Entomol. 109, 289–299.

Barras, S.J., Perry, T., 1971. Gland cells and fungi associated with the prothoracic mycangium of *D. adjunctus* (Coleoptera: Scolytidae). Ann. Entomol. Soc. Am. 64, 123–126.

Barras, S.J., Perry, T., 1972. Fungal symbionts in the prothoracic mycangium of *Dendroctonus frontalis* (Coleoptera: Scolytidae). Z. Ang. Entomol. 71, 95–104.

Bentz, B.J., Mullins, D.E., 1999. Ecology of mountain pine beetle cold hardening in the Intermountain West. Environ. Entomol. 28, 577–587.

Bentz, B.J., Six, D.L., 2006. Ergosterol content of four fungal symbionts associated with *Dendroctonus ponderosae* and *D. rufipennis* (Coleoptera: Curculionidae, Scolytinae). Ann. Entomol. Soc. Am. 99, 189–194.

Bentz, B.J., Logan, J.A., Amman, G.D., 1991. Temperature dependent development of the mountain pine beetle and simulation of its phenology. Can. Ent. 123, 1083–1094.

Bentz, B.J., Logan, J.A., Vandygriff, J.C., 2001. Latitudinal variation in *Dendroctonus ponderosae* (Coleoptera: Scolytidae) development time and adult size. Can. Ent. 133, 375–387.

Bentz, B., Logan, J., MacMahon, J., Allen, C.D., Ayres, M., Berg, E., et al., 2009. Bark Beetle Outbreaks in Western North America: Causes and Consequences. Bark Beetle Symposium, Snowbird, Utah, University of Utah Press, Salt Lake City.

Bentz, B.J., Bracewell, R.R., Mock, K.E., Pfrender, M.E., 2011. Genetic architecture and phenotypic plasticity of thermally regulated traits in an eruptive species, *Dendroctonus ponderosae*. Evol. Ecol. 25, 1269–1288.

Berg, E.E., Henry, J.D., Fastie, C.L., De Volder, A.D., Matsuoka, S.M., 2006. Spruce beetle outbreaks on the Kenai Peninsula, Alaska, and Kluane National Park and Reserve, Yukon Territory: relationship to summer temperatures and regional differences in disturbance regimes. Forest Ecol. Manag. 227, 219–232.

Berryman, A.A., 1970. Insect parasites of western pine beetle. In: Stark, R. W., Dahlsten, D.L. (Eds.), Studies on the Population Dynamics of the Western Pine Beetle, *Dendroctonus brevicomis*, LeConte (Coleoptera: Scolytidae). University of California, Berkeley, pp. 102–112.

Bewley, D., Alila, Y., Varhola, A., 2010. Variability of snow water equivalent and snow energetics across a large catchment subject to Mountain Pine Beetle infestation and rapid salvage logging. J. Hydrol. 388, 464–479.

Bleiker, K., Six, D.L., 2007. Dietary benefits of fungal associates to an eruptive herbivore: potential implications of multiple associates on host population dynamics. Environ. Entomol. 36, 1384–1396.

Bleiker, K., Six, D.L., 2008a. Effects of water potential and solute on growth and interactions of two fungal symbionts of the mountain pine beetle. Mycol. Res. 113, 3–15.

Bleiker, K., Six, D.L., 2008b. Competition and coexistence in a multi-partner mutualism: interactions between two fungal symbionts of the mountain pine beetle in beetle-attacked trees. Microb. Ecol. 57, 191–202.

Bleiker, K.P., Potter, S.E., Lauzon, C.R., Six, D.L., 2009. Transport of fungal symbionts by mountain pine beetle. Can. Entomol. 141, 503–514.

Bleiker, K.P., Carroll, A.L., Smith, G.D., 2011. Mountain pine beetle expansion: assessing the threat to Canada's boreal forest by evaluating the endemic niche, Canadian Forest Service Mountain Pine Beetle Working Paper 2010-2.

Bleiker, K.P., Heron, R.J., Braithwaite, E.C., Smith, G.D., 2013. Preemergence mating in the mass attacking bark beetle, *Dendroctonus ponderosae* (Coleoptera: Scolytidae). Can. Entomol. 145, 12–19.

Bonnet, T.R., Robert, J.A., Pitt, C., Fraser, J.D., Keeling, C.I., Bohlmann, J., Huber, D.P.W., 2012. Global and comparative proteomic profiling of overwintering and developing mountain pine beetle, *Dendroctonus ponderosae* (Coleoptera: Curculionidae), larvae. Insect Biochem. Mol. Biol. 42, 890–901.

Boon, S., 2007. Snow accumulation and ablation in a beetle-killed pine stand in Northern Interior British Columbia. BC Journal of Ecosystems and Management 8, 1–13.

Boone, C.K., Six, D.L., Raffa, K.F., 2008. The enemy of my enemy is still my enemy: competitors add to predator load of primary bark beetles. Agric. Forest Manag. 10, 411–421.

Boone, C.A., Aukema, B.H., Bolmann, J., Carroll, A.L., Raffa, K.F., 2011. Efficacy of tree defense physiology varies with bark beetle population density: a basis for positive feedback in eruptive species. Can. J. Forest Res. 41, 1174–1188.

Boone, C.K., Keefover-Ring, K., Mapes, A.C., Adams, A.S., Bohlmann, J., Raffa, K.F., 2013. Bacteria associated with a tree-killing insect reduce concentrations of plant defensive compounds. J. Chem. Ecol. 39, 1003–1006.

Bracewell, R.R., Pfrender, M.E., Mock, K.E., Bentz, B.J., 2011. Cryptic postzygotic isolation in an eruptive species of bark beetle (*Dendroctonus ponderosae*). Evolution 65, 961–975.

Bracewell, R.R., Pfrender, M.E., Mock, K.E., Bentz, B.J., 2013. Contrasting geographic patterns of genetic differentiation in body size and development time with reproductive isolation in *Dendroctonus ponderosae* (Coleoptera: Curculionidae, Scolytinae). Ann. Entomol. Soc. Am. 106, 385–391.

Bradley, T., Tueller, P., 2001. Effects of fire on bark beetle presence on Jeffrey pine in the Lake Tahoe Basin. Forest Manag. Ecol. 142, 205–214.

Bridges, J.R., 1983. Mycangial fungi of *Dendroctonus frontalis* (Coleoptera: Scolytidae) and their relationship to beetle population trends. Environ. Entomol. 12, 858–861.

Bridges, J.R., Marler, J.E., McSparrin, B.H., 1984. A quantitative study of the yeasts associated with laboratory-reared *Dendroctonus frontalis* Zimm. (Coleopt., Scolytidae). Z. Angew. Entomol. 97, 261–267.

Byers, J.A., 1995. Host tree chemistry affecting colonization in bark beetles. In: Cardé, R.T., Bell, W.J. (Eds.), Chemical Ecology of Insects 2. Chapman & Hall, New York, pp. 154–213.

Byers, J.A., Wood, D.L., Craig, J., Hendry, L.B., 1984. Attractive and inhibitory pheromones produced in the bark beetle, *Dendroctonus brevicomis*, during host colonization: Regulation of inter- and intraspecific competition. J. Chem. Ecol. 10, 861–877.

Cai, Y.-W., Cheng, X.-Y., Xu, R.-M., Duan, D.-H., Kirkendall, L.R., 2008. Genetic diversity and biogeography of red turpentine beetle *Dendroctonus valens* in its native and invasive regions. Insect Sci. 15, 291–301.

Campbell, J.W., Hanula, J.L., Outcalt, K.W., 2008. Effects of prescribed fire and other community restoration treatments on tree mortality, bark beetles, and other saproxylic Coleoptera of longleaf pine, *Pinus palustris* Mill., on the coastal plain of Alabama. Forest Ecol. Manag. 254, 134–144.

Cano-Ramírez, C., López, M.F., Cesar-Ayala, A.K., Pineda-Martínez, V., Sullivan, B.T., Zúñiga, G., 2013. Isolation and expression of cytochrome P450 genes in the antennae and gut of pine beetle *Dendroctonus rhizophagus* (Curculionidae: Scolytinae) following exposure to host monoterpenes. Gene 520, 47–63.

Cardoza, Y.J., Klepzig, K.D., Raffa, K.F., 2006. Bacteria in oral secretions of an endophytic insect inhibit antagonistic fungi. Ecol. Entomol. 31, 636–645.

Cardoza, Y.J., Moser, J.C., Klepzig, K.D., Raffa, K.F., 2008. Multi-partite symbioses among fungi, mites, nematodes, and the spruce beetle, *Dendroctonus rufipennis*. Environ. Entomol. 37, 956–963.

Carroll, A.L., Régnière, J., Logan, J.A., Taylor, S.W., Bentz, B.J., Powell, J.A., 2006. Impacts of climate change on range expansion of mountain pine beetle, Canadian Forest Service Mountain Pine Beetle Initiative Working Paper 2006-14.

Carta, L.K., Bauchan, G., Hsu, C.-Y., Yuceer, C., 2010. Description of *Parasitorhabditis frontali* n. sp. (Nemata: Rhabditida) from *Dendroctonus frontalis* Zimmermann (Coleoptera: Scolytidae). J. Nematol. 42, 46–54.

Cerezke, H.F., 1995. Egg gallery, brood production, and adult characteristics of mountain pine beetle, *Dendroctonus ponderosae* Hopkins (Coleoptera: Scolytidae) in three pine hosts. Can. Ent. 127, 955–965.

Chamberlin, W.J., 1939. The bark and timber beetles of North America north of Mexico. Oregon State College Association, Corvallis.

Chen, H., Tang, M., 2007. Spatial and temporal dynamics of bark beetles in Chinese white pine in Qinling Mountains of Shaanxi Province, China. Environ. Entomol. 36, 1124–1130.

Chen, H., Li, Z., Bu, S.H., Tian, Z.Q., 2011. Flight of the Chinese white pine beetle (Coleoptera: Scolytidae) in relation to sex, body weight, and energy reserves. Bull. Entomol. Res. 101, 53–62.

Chen, H.-F., Salcedo, C., Sun, J.-H., 2012. Male mate choice by chemical cues leads to higher reproductive success in a bark beetle. Anim. Behav. 83, 421–427.

Cibrían Tovar, D., Méndez Montiel, J.T., Campos Bolaños, R., Yates III, H.O., Flores Lara, J.E., 1995. Insectos Forestales de México Publicación No. 6. Universidad Autónoma Chapingo, México.

Clark, E.L., Huber, D.P.W., Carroll, A.L., 2012. The legacy of attack: implications of high phloem resin monoterpene levels in lodgepole pines following mass attack by mountain pine beetle, *Dendroctonus ponderosae* Hopkins. Environ. Entomol. 41, 392–398.

Clark, E.L., Pitt, C., Carroll, A.L., Lindgren, B.S., Huber, D.P.W., 2014. Comparison of lodgepole and jack pine resin chemistry: implications for range expansion by the mountain pine beetle, *Dendroctonus ponderosae* (Coleoptera: Curculionidae). PeerJ. 2, e240.

Clarke, S.R., Nowak, J.T., 2009. Southern Pine Beetle. USDA Forest Service Forest Insect and Disease Leaflet 49, 8 p.

Cognato, A.I., Sun, J.-H., Anducho-Reyes, M.A., Owen, D.R., 2005. Genetic variation and origin of red turpentine beetle (*Dendroctonus valens* LeConte) introduced to the People's Republic of China. Agr. Forest Entomol. 7, 87–94.

Collins, B.J., Rhoades, C.C., Hubbard, R.M., Battaglia, M.A., 2011. Tree regeneration and future stand development after bark beetle infestation and harvesting in Colorado lodgepole pine stands. Forest Ecol. Manag. 261, 2168–2175.

Cook, S.S., Shirley, B.M., Zambino, P., 2010. Nitrogen concentration in mountain pine beetle larvae reflects nitrogen status of tree host and two fungal associates. Environ. Entomol. 39, 821–826.

Coops, N.C., Wulder, M.A., Waring, R.A., 2012. Modeling lodgepole and jack pine vulnerability to mountain pine beetle expansion in to the western Canadian boreal forest. Forest Ecol. Manag. 274, 161–171.

Coppedge, B.R., Stephen, F.M., Felton, G.W., 1995. Variation in southern pine beetle size and lipid content in relation to fungal associates. Can. Ent. 127, 145–154.

Coulson, R.N., 1979. Population dynamics of bark beetles. Annu. Rev. Entomol. 24, 417–447.

Coulson, R.N., Klepzig, K.D., 2011. Southern Pine Beetle II, USDA Forest Service, General Technical Report SRS-140.

Creeden, E.P., Hicke, J.A., Buotte, P.C., 2014. Climate, weather, and recent mountain pine beetle outbreaks in the western United States. Forest Ecol. Manag. 312, 239–251.

Cudmore, T.J., Bjorklund, N.B., Carroll, A.L., Lindgren, B.S., 2010. Climate change and range expansion of an aggressive bark beetle: Evidence of higher beetle reproduction in naïve host tree populations. J. Appl. Ecol. 47, 1036–1043.

Cullingham, C.I., Cooke, J.E.K., Dang, S., Davis, C.S., Cooke, B.J., Coltman, D.W., 2011. Mountain pine beetle host range expansion threatens the boreal forest. Mol. Ecol. 20, 2157–2171.

Dahlsten, D.L., 1970. Parasitoids, predators, and associated organisms reared from western pine beetle infested bark samples. In: Stark, R.W., Dahlsten, D.L. (Eds.), Studies on the Population Dynamics of the Western Pine Beetle, *Dendroctonus brevicomis*, LeConte (Coleoptera: Scolytidae). University of California, Berkeley, pp. 75–79.

Dahlsten, D.L., 1982. Relationships between bark beetles and their natural enemies. In: Mitton, J.B., Sturgeon, K.B. (Eds.), Bark Beetles in North American Conifers: A System for the Study of Evolutionary Biology. University of Texas Press, Austin, pp. 140–182.

Dahlsten, D.L., Stephen, F.M., 1974. Natural enemies and insect associates of the mountain pine beetle, *Dendroctonus ponderosae* (Coleoptera: Scolytidae) in sugar pine. Can. Entomol. 106, 1211–1217.

Davidson, R.W., 1978. Staining fungi associated with *Dendroctonus adjunctus* in pines. Mycologia 70, 35–40.

Davis, T.S., Hofstetter, R.W., 2009. Effects of gallery density and species ratio on the fitness and fecundity of two sympatric bark beetles (Coleoptera: Curculioniodae). Environ. Entomol. 38, 639–650.

Davis, T.S., Hofstetter, R.W., 2011b. Oleoresin chemistry mediates oviposition behavior and fecundity of a tree-killing bark beetle. J. Chem. Ecol. 37, 1177–1183.

Davis, T.S., Hofstetter, R.W., Foster, J.T., Foote, N.E., Keim, P., 2011. Interactions between the yeast *Ogataea pini* and filamentous fungi associated with the western pine beetle. Microb. Ecol. 61, 626–634.

Davis, R.S., Hood, S., Bentz, B.J., 2012. Fire-injured ponderosa pine provide pulsed resource for bark beetles. Can. J. Forest Res. 42, 2022–2036.

de la Giroday, H.-M., Carroll, A.L., Aukema, B.H., 2012. Breach of the northern Rocky Mountain geoclimatic barrier: initiation of range expansion by the mountain pine beetle. J. Biogeog. 39, 1112–1123.

Delalibera Jr., I., Handelsman, J., Raffa, K.F., 2005. Contrasts in cellulytic activities of gut microorganisms between the wood borer, *Saperda vestita* (Coleptera: Cerambycidae), and the bark beetles, *Ips pini* and *Dendroctonus frontalis* (Coleoptera: Curculionidae). Environ. Entomol. 34, 541–547.

Delalibera Jr., I., Vasanthakumar, A., Burwitz, B.J., Schloss, P.D., Klepzig, K.D., Handelsman, J., Raffa, K.F., 2007. Composition of the bacterial community in the gut off the pine engraver, *Ips pini* (Say) (Coleoptera), colonizing red pine. Symbiosis 43, 97–104.

Deneubourg, J.-L., Grégoire, J.-C., Le Fort, E., 1990. Kinetics of larval gregarious behavior in the bark beetle *Dendroctonus micans* (Coleoptera: Scolytidae). J. Insect Behavior 3, 169–182.

DeRose, R.J., Long, J.N., 2012. Factors influencing the spatial and temporal dynamics of Engelman spruce mortality during a spruce beetle outbreak on the Markagunt Plateau Utah. For. Sci. 51, 1–13.

DeRose, R.J., Bentz, B.J., Long, J.N., Shaw, J.D., 2013. Effect of increasing temperatures on the distribution of spruce beetle in Engelmann spruce forests of the interior west, USA. Forest Ecol. Manag. 308, 198–206.

DiGuistini, S., Wang, Y., Liao, N., Taylor, G., Tanguay, P., Feau, N., Henrissat, B., et al., 2011. Genome and transcriptome analyses of the mountain pine beetle—fungal symbiont *Grosmannia clavigera*, a lodgepole pine pathogen. Proc. Natl. Acad. Sci. U. S. A. 108, 2504–2509.

Dodds, K.J., Gilmore, D.W., Seybold, S.J., 2010. The threat posed by indigenous exotics: A case study of two North American bark beetle species. Ann. Entomol. Soc. Am. 103, 39–49.

Drès, M., Mallet, J., 2002. Host races in plant-feeding insects and their importance in sympatric speciation. Trans. Royal Soc. B 357, 471–492.

Duehl, A.J., Koch, F.H., Hain, F.P., 2011. Southern pine beetle regional outbreaks modeled on landscape, climate, and infestation history. Forest Ecol. Manag. 261, 473–479.

Dymerski, A.D., Anhold, J.A., Munson, A.S., 2001. Spruce beetle (*Dendroctonus rufipennis*) outbreak in Engelmann spruce (*Picea engelmannii*) in central Utah, 1986–1998. West. N. Am. Nat. 61, 19–24.

Eckhardt, L.E., Weber, A.M., Menard, R.D., Jones, J.P., Hess, N.J., 2007. Insect-fungal complex associated with loblolly pine decline in central Alabama. For. Sci. 53, 84–92.

Edburg, S.L., Hicke, J.A., Brooks, P.D., Pendall, E.G., Ewars, B.E., Norton, U., et al., 2013. Cascading impacts of bark beetle-caused tree mortality on coupled biogeophysical and biogeochemical processes. Front. Ecol. Environ. 10, 416–424.

Elkin, C.M., Reid, M.L., 2005. Low energy reserves and energy allocation decisions affect reproduction by mountain pine beetles, *Dendroctonus ponderosae*. Funct. Ecol. 19, 102–109.

Elkin, C.M., Reid, M.L., 2010. Sub-lethal effects of monoterpenes on reproduction by mountain pine beetles. Agric. Forest Entomol. 15, 262–271.

Erbilgin, N., Morei, S.R., Sun, J.H., Stein, J.D., Owen, D.R., Merrill, L.D., et al., 2007. Response to host volatiles by native and introduced populations of *Dendroctonus valens* (Coleoptera: Curculionidae, Scolytinae) in North America and China. J. Chem. Ecol. 33, 131–146.

Evans, L.M., Hofstetter, R.W., Ayres, M.P., Klepzig, K.D., 2011. Temperature alters the relative abundance and population growth rates of species within the *Dendroctonus frontalis* (Coleoptera: Curculionidae) community. Environ. Entomol. 40, 824–834.

Fairbairn, D.J., 2007. Introduction: the enigma of sexual size dimorphism. In: Fairbairn, D.J., Blanckenhorn, W.U., Székely, T. (Eds.), Sex, Size & Gender Roles: Evolutionary Studies of Sexual Size Dimorphism. Oxford University Press, Oxford, pp. 1–15.

Fettig, C.J., Shea, P.J., Borys, R.R., 2004. Seasonal flight patterns of four bark beetle species (Coleoptera: Scolytidae) along a latitudinal gradient in California. Pan-Pacific Entomol. 80, 4–17.

Fettig, C.J., Borys, R.R., McKelevy, S.R., Dabney, C., 2008. Blacks mountain experimental forest: bark beetle responses to differences in forest structure and the application of prescribed fire in interior ponderosa pine. Can. J. Forest Res. 38, 924–935.

Fettig, C.J., McKelvey, S.R., Cluck, D.R., Smith, S.L., Otrosina, W.J., 2010. Effects of prescribed fire and season of burn on direct and indirect levels of mortality in ponderosa and Jeffrey pine forests in California, USA. Forest Ecol. Manag. 60, 207–218.

Fischer, M.J., Waring, K.M., Hofstetter, R.W., Kolb, T., 2010. Ponderosa pine characteristics associated with attack by roundheaded pine beetle. Forest Sci. 56, 473–483.

Fleming, A.J., Lindeman, A.A., Carroll, A.L., Yack, J.E., 2013. Acoustics of the mountain pine beetle (*Dendroctonus ponderosae*) (Curculionidae, Scolytinae): Sonic, ultrasonic, and vibration characteristics. Can. J. Zool. 91, 235–244.

Foelker, C.J., Hofstetter, R.W., 2014. Heritability, fecundity, and sexual ssize dimorphism in four species of bark beetle (Coleoptera: Curculionidae: Scolytinae). Ann. Entomol. Soc. Am. 107, 143–151.

Fredricks, S.E., Jenkins, M.J., 1988. Douglas-fir beetle (*Dendroctonus pseudotsugae* Hopkins, Coleoptera: Scolytidae) brood production on Douglas-fir defoliated by western spruce budworm (*Choristoneura occidentalis* Freeman, Lepidoptera: Tortricidae) in Logan Canyon Utah. Great Basin Nat. 48, 348–351.

Friedenberg, N.A., Whited, B.M., Slone, D.H., Martinson, S.H., Ayres, M. P., 2007. Differential impacts of the southern pine beetle, *Dendroctonus frontalis*, on *Pinus palustris* and *Pinus taeda*. Can. J. Forest Res. 37, 1427–1437.

Friedenberg, N.A., Sarkar, S., Kouchoukos, N., Billings, R.F., Ayres, M.P., 2008. Temperature extremes, density dependence, and southern pine beetle (Coleoptera: Curculionidae) population dynamics in east Texas. Environ. Entomol. 37, 650–659.

Furniss, M.M., 1995. Biology of *Dendroctonus punctatus* (Coleoptera: Scoytidae). Ann. Entomol. Soc. Am. 88, 173–182.

Furniss, M.M., 2001. A new subspecies of *Dendroctonus* (Coleoptera: Scolytidae) from Mexico. Ann. Entomol. Soc. Am. 94, 21–25.

Furniss, M.M., Cibrían Tovar, D., 1980. Compatibilidad reproductiva e insectos asociados a *Dendroctonus pseudotsugae* (Coleoptera: Scolytidae) de Chihuahua, Mexico e Idaho, E. U. A. Folia Entomológica Mexicana 44, 129–142.

Furniss, M.M., Johnson, J.B., 1989. Description of the gallery and larva of *Dendroctonus punctatus* LeConte (Coleoptera: Scolytidae). Can. Entomol. 121, 757–762.

Furniss, M.M., Kegely, S.J., 2008. Biology of *Dendroctonus murrayanae* (Curculionidae: Scolytinae) in Idaho and Montana and comparative taxonomic notes. Ann. Entomol. Soc. Am. 101, 1010–1016.

Furniss, M.M., McGregor, M.D., Foiles, M.W., 1979. Chronology and characteristics of a Douglas-fir beetle outbreak in northern Idaho, (1979). USDA Forest Service General Technical Report INT-59.

Furniss, R.L., Carolin, V.M., 1977. Western Forest Insects, USDA Forest Service Miscelaneous Publication 1339.

Garrison-Johnston, M.T., Moore, J.A., Cook, S.P., Niehoff, G.J., 2003. Douglas-fir beetle infestations are associated with certain rock and stand types in the inland northwestern United States. Environ. Entomol. 32, 1354–1363.

Gaylord, M., Williams, K.K., Hofstetter, R.W., McMillin, J.D., deGómez, T.E., Wagner, M.R., 2008. Influence of temperature on spring flight initiation for southwestern ponderosa pine bark beetles (Coleoptera: Curculionidae, Scolytinae). Environ. Entomol. 37, 57–69.

Geizler, D.R., Gara, R.I., Littke, W.R., 1984. Bark beetle infestations of lodgepole pine following a fire in South Central Oregon. J. Appl. Entomol. 98, 389–394.

Gilbert, M., Vouland, G., Grégoire, J.-C., 2001. Past attacks influence host selection by the solitary bark beetle *Dendroctonus micans*. Ecol. Entomol. 26, 133–142.

Gilbert, M., Fielding, N., Evans, H.F., Grégoire, J.-C., 2003. Spatial pattern of invading *Dendroctonus micans* (Coleoptera: Scolytidae) populations in the United Kingdom. Can. J. Forest Res. 33, 712–725.

Goldhammer, D.S., Stephen, F.M., Paine, T.D., 1990. The effects of the fungi *Ceratocystis minor* (Hedgcock) Hunt, *Ceratocystis minor* (Hedgcock) Hunt var. *barrassii*, and SJB 122 on reproduction of the southern pine beetle, *Dendroctonus frontalis* Zimmermann (Coleoptera: Scolytidae). Can. Ent. 122, 407–418.

Goodsman, D.W., Erbilgin, N., Lieffers, V.T., 2012. The impact of phloem nutrients on overwintering mountain pine beetle and their fungal symbionts. Environ. Entomol. 42, 478–486.

Grégoire, J.-C., 1988. The greater European spruce beetle. In: Berryman, A.A. (Ed.), Population Dynamics of Forest Insects. Plenum Press, New York, pp. 455–478.

Griffin, J.M., Turner, M.G., Simard, M., 2011. Nitrogen cycling following mountain pine beetle disturbance in lodgepole pine forests of Greater Yellowstone. Forest Ecol. Manag. 261, 1077–1089.

Haberkern, K.E., Illman, B.L., Raffa, K.F., 2002. Bark beetles and fungal associates colonizing white spruce in the Great Lakes region. Can. J. Forest Res. 32, 1137–1150.

Hall, D.E., Yuen, M.M.S., Jancsik, S., Quesada, A.L., Dullat, H.K., Li, M., et al., 2013. Transcriptome resources and functional characterization of monoterpene synthases for two host species of the mountain pine beetle, lodgepole pine (*Pinus contorta*) and jack pine (*Pinus banksiana*). BMC Plant Biol. 13, 80.

Hansen, E.M., 2014. Forest development and carbon dynamics after mountain pine beetle outbreaks. Forest Sci. in press.

Hansen, E.M., Bentz, B.J., 2003. Comparison of reproductive capacity among univoltine, semivoltine, and re-emerged parent spruce beetles (Coleoptera: Scolytidae). Can. Entomol. 135, 697–712.

Hansen, E.M., Bentz, B.J., Powell, J.A., Gray, D.R., Vandygriff, J.C., 2011. Prepupal diapause and instar IV development rates of the spruce beetle *Dendroctonus rufipennis* (Coleoptera: Curculionidae, Scolytinae). J. Insect Physiol. 57, 1347–1357.

Hanula, J.L., Meeker, J.R., Miller, D.R., Barnard, E.L., 2002. Association of wildfire with tree health and numbers of pine bark beetles, reproduction weevils and their associates in Florida. Forest Ecol. Manag. 170, 233–247.

Hard, J.S., Werner, R.A., Holsten, E.H., 1983. Susceptibility of white spruce to attack by spruce beetles during the early years of an outbreak in Alaska. Can. J. Forest Res. 13, 678–684.

Haridas, S., Wang, Y., Lim, L., Massoumi Alamouti, S., Jackman, S., Docking, R., et al., 2013. The genome and transcriptome of the pine saprophyte Ophiostoma picea, and a comparison with the bark beetle-associated pine pathogen *Grosmannia clavigera*. BMC Genomics 14, 373.

Harrington, T.C., 1988. *Leptographium* species, their distributions, hosts and insect vectors. In: Harrington, T.C., Cobb Jr., F.W. (Eds.), *Leptographium* Root Diseases on Conifers. APS Press, St. Paul, pp. 1–40.

Harrington, T.C., Cobb Jr., F.W., 1983. Pathogenicity of *Leptographium* and *Verticicladiella* spp. isolated from roots of western North American conifers. Phytopathology 73, 596–599.

Hart, S.J., Veblen, T.T., Eisenhart, K.S., Jarvis, D., Kulakowski, D., 2014. Drought induces spruce beetle (*Dendroctonus rufipennis*) outbreaks across northwestern Colorado. Ecology 95, 930–939.

Harvey, B.J., Donato, D.C., Romme, W.H., Turner, M.G., 2013. Influence of recent bark beetle outbreak on fire severity and postfire tree regeneration in montane Douglas-fir forests. Ecology 94, 2475–2486.

Hawkins, C.D.B., Dhar, A., Balliet, N.A., 2013. Radial growth of residual overstory trees and understory saplings after mountain pine beetle attack in central British Columbia. Forest Ecol. Manag. 310, 348–356.

Hebertson, E.G., Jenkins, M.J., 2008. Climate factors associated with historic spruce beetle (Coleoptera: Curculionidae) outbreaks in Utah and Colorado. Environ. Entomol. 37, 281–292.

Hedden, R.L., Billings, R.F., 1977. Seasonal variations in fat content and size of the southern pine beetle in east Texas. Ann. Entomol. Soc. Am. 70, 876–880.

Hofstetter, R.W., Klepzig, K.D., Moser, J.C., Ayres, M.P., 2006a. Seasonal dynamics of mites and fungi and their interaction with southern pine beetle. Environ. Entomol. 35, 22–30.

Hofstetter, R.W., Cronin, J.J., Klepzig, K.D., Moser, J.C., Ayres, M.P., 2006b. Antagonisms, mutualisms and commensalisms affect outbreak dynamics of the southern pine beetle. Oecologia 147, 679–691.

Hofstetter, R.W., Dempsey, T.D., Klepzig, K.D., Ayres, M.P., 2007. Temperature-dependent effects on mutualistic, antagonistic and commensalistic interactions among insects, fungi and mites. Community Ecol. 8, 47–56.

Holsten, E.H., Their, R.W., Munson, A.S., Gibson, K.E., 1999. The spruce beetle. USDA Forest Service Forest Insect and Disease Leaflet No.27.

Honek, A., 1993. Intraspecific variation in body size and fecundity in insects: a general relationship. Oikos 66, 483–492.

Hood, S.H., Bentz, B.J., 2007. Predicting postfire Douglas-fir beetle attacks and tree mortality in the northern Rocky Mountains. Can. J. Forest Res. 37, 1058–1069.

Hood, S.H., Smith, S.L., Cluck, D.R., 2010. Predicting mortality for five California conifers following wildfire. Forest Ecol. Manag. 260, 750–762.

Hopkins, A.D., 1909. Practical information on the scolytid betles of North American forests. I. Barkbeetles of the genus Dendroctonus. USDA Bureau of Entomology Bulletin No 83. .

Hsiau, P.T.W., Harrington, T.C., 1997. *Ceratocystiopsis brevicomi* sp. nov., a mycangial fungus from *Dendroctonus brevicomis* (Coleoptera: Scolytidae). Mycologia 89, 661–669.

Hsiau, P.T.W., Harrington, T.C., 2003. Phylogenetics and adaptations of basidiomycetous fungi fed upon by bark beetles (Coleoptera: Scolytidae). Symbiosis 34, 111–131.

Huber, D.P.W., Aukema, B.H., Hodgkinson, R.S., Lindgren, B.S., 2009. Successful colonization, reproduction, and new generation emergence in live interior hybrid spruce, *Picea engelmannii* x *glauca*, by mountain pine beetle, *Dendroctonus ponderosae*. Agric. Forest Entomol. 11, 83–89.

Hulcr, J., Adams, A.S., Raffa, K.F., Hofstetter, R.W., Klepzig, K.D., Currie, C.R., 2011. Presence and diversity of *Streptomyces* in *Dendroctonus* and sympatric beetle galleries across North America. Mol. Ecol. 61, 759–768.

Hunt, R.S., Poinar Jr., G.O., 1971. Culture of *Parasitorhabditis* sp. (Rhabditida: Protorhabditinae) on a fungus. Nematologica 17, 321–322.

Hunt, D.W.A., Borden, J.H., 1990. Conversion of verbenols to verbenone by yeasts isolated from *Dendroctonus ponderosae* (Coleoptera: Scolytidae). J. Chem. Ecol. 16, 1385–1397.

Jacobs, K., Wingfield, M.J., 2001. *Leptographium* Species: Tree Pathogens, Insect Associates and Agents of Blue-Stain. APS Press, St. Paul.

Jenkins, M.J., Hebertson, E.G., Munson, A.S., 2014. Spruce beetle biology, ecology, and management in the Rocky Mountains: an addendum to the spruce beetle in the Rockies. Forests 5, 21–71.

Kaiser, V.B., Bachtrog, D., 2010. Evolution of sex chromosomes in insects. Annu. Rev. Genetics 44, 91–112.

Kanzaki, N., Giblin-Davis, R.M., Cardoza, Y.J., Ye, W., Raffa, K.F., Center, B.J., 2008. *Bursaphelenchus rufipennis* n. sp. (Nematoda: Parasitaphelenchinae) and redescription of *Ektaphelenchus obtusus* (Nematoda: Ektaphelenchinae), associates from nematangia on the hind wings of *Dendroctonus rufipennis* (Coleoptera: Scolytidae). Nematology 10, 925–955.

Keeling, C.I., Henderson, H., Li, M., Yuen, M., Clarck, E.L., Fraser, J.D., et al., 2012. Transcriptome and full-length cDNA resources for the mountain pine beetle, *Dendroctonus ponderosae* Hopkins, a major insect pest of pine forests. Insect Biochem. Mol. Biol. 42, 525–536.

Keeling, C.I., Yuen, M.M.S., Liau, N.Y., Docking, T.R., Chan, S.K., Taylor, G.A., et al., 2013. Draft genome of the mountain pine beetle, *Dendroctonus ponderosae* Hopkins, a major forest pest. Genome Biology 14, R27.

Kelley, S.T., Farrell, B.D., 1998. Is specialization a dead end? The phylogeny of host use in *Dendroctonus* bark beetles (Scolytidae). Evolution 52, 1731–1743.

Kelley, S.T., Mitton, J.B., Paine, T.D., 1999. Strong differentiation in mitochondrial DNA of *Dendroctonus brevicomis* (Coleoptera: Scolytidae) on different subspecies of ponderosa pine. Ann. Entomol. Soc. Am. 92, 193–197.

Kelsey, R.G., Joseph, G., 2003. Ethanol in ponderosa pine as an indicator of physical injury from fire and its relationship to secondary beetles. Can. J. Forest Res. 33, 870–884.

Khadempour, L., Massoumi Alamouti, S., Hamelin, R.C., Bohlmann, J., Breuil, C., 2010. Target-specific PCR primers can detet and differentiate ophiosstomatoid fungi from microbial communities associated with the mountain pine beetle *Dendroctonus ponderosae*. Fungal Biol. 114, 825–833.

Khadempour, L., LeMay, V., Jack, D., Bohlmann, J., Breuil, C., 2012. The relative abundance of mountain pine beetle fungal assoocuates through the beetle life cycle on pine trees. Microb. Ecol. 64, 909–917.

Kahn, M.A., 1957. Sphaerularia ungulacauda sp. nov. (nematode: Allantonematidae) from the Douglas-fir beetle, Dendroctonus pseudotsugae Hopk, with key to Sphaerularia species. Can. J. Zool. 35, 635–639.

Kim, J.-J., Allen, E.A., Humble, L.M., Breuil, C., 2005. Ophiostomatoid and basidiomycetous fungi associated with green, red, and grey lodgepole pines after mountain pine beetle (*Dendroctonus ponderosae*) infestation. Can. J. Forest Ecol. 35, 274–284.

Kinn, D.N., 1970. Acarine parasites and predators of the western pine beetle. In: Stark, R.W., Dahlsten, D.L. (Eds.), Studies on the Population Dynamics of the Western Pine Beetle, *Dendroctonus brevicomis*, LeConte (Coleoptera: Scolytidae). University of California, Berkeley, pp. 128–131.

Klepzig, K.D., Raffa, K.F., Smalley, E.B., 1995. *Dendroctonus valens* and *Hylastes porculus* (Coleoptera: Scolytidae) vectors of pathogenic fungi (Ophiostomatales) associated with red pine decline disease. Great Lakes Entomol. 28, 81–87.

Klepzig, K.D., Wilkens, R.T., 1997. Competitive interactions among symbiotic fungi of the southern pine beetle. Appl. Environ. Microbiol. 63, 621–627.

Klutsch, J., Negrón, J., Costello, S., Rhoades, C., West, D., Popp, J., Caissie, R., 2009. Stand characteristics and downed woody debris accumulations associated with a mountain pine beetle (*Dendroctonus ponderosae* Hopkins) outbreak in Colorado. Forest Ecol. Manag. 258, 641–649.

Knapp, P.A., Soulé, P.T., Maxwell, J.T., 2013. Mountain pine beetle selectivity in old-growth ponderosa pine forests, Montana, USA. Ecol. Evol. 3, 1141–1148.

Koplin, J.R., Baldwin, P.H., 1970. Woodpecker predation on an endemic population of Engelmann spruce beetles. Am. Midland Nat. 83, 510–515.

Kulakowski, D., Veblen, T.T., Bebi, P., 2003. Effects of fire and spruce beetle outbreak legacies on the disturbance regime of a subalpine forest in Colorado. J. Biogeogr. 30, 1445–1456.

Kurz, W.A., Dymond, C.C., Stinson, G., Rampley, G.J., Neilson, E.T., Carroll, A.L., Safryanyik, L., 2008. Mountain pine beetle and forest carbon feedback to climate change. Nature 452, 987–990.

Labandeira, C.C., LePage, B.A., Johnson, A.H., 2001. A *Dendroctonus* engraving (Coleoptera: Scolitidae) from a middle Eocene *Larix* (Conferales: Pinaceae): early or delayed colonization? Am. J. Bot. 88, 2016–2039.

Lah, L., Haridas, S., Bohlmann, J., Breuil, C., 2013. The cytochrome P450 of *Grosmannia clavigera*. Genome organization, phylogeny, and expression in response to pine host chemicals. *Fungal Gen.* Biol. 50, 72–81.

Langor, D.W., 1989. Host effects on the phenology, development, and mortality of field populations of the mountain pine beetle, *Dendroctonus ponderosae* (Coleoptera: Scolytidae). Can. Entomol. 121, 149–157.

Langor, D.W., 1991. Arthropods and nematodes co-occurring with the eastern larch beetle, *Dendroctonus simplex Coleoptera: Scolytidae)*, in Newfoundland. Entomophaga 36, 303–313.

Langor, D.W., Raske, A.G., 1989. A history of the eastern larch beetle, *Dendroctonus simplex* (Coleoptera: Scolytidae), in North America. Great Lakes Entomol. 22, 139–154.

Langor, D.W., Spence, J.R., 1991. Host effects on allozymes and morphological variation of the mountain pine beetle, *Dendroctonus ponderosae* Hopkins (Coleoptera: Scolytidae). Can. Entomol. 123, 395–410.

Langor, D.W., Spence, J.R., Pohl, G., 1990. Host effects on fertility and reproductive success of *Dendroctonus ponderosae* Hopkins (Coleoptera: Scolytidae). Evolution 44, 609–618.

Lanier, G.N., 1981. Cytotaxonomy of *Dendroctonus*. In: Stock, M.W. (Ed.), Applications of Genetics and Cytology in Insect Systematics and Evolution. University of Idaho, Wildlife and Range Experimental Station, Moscow, Idaho, pp. 33–66.

Lanier, G.N., Wood, D.L., 1968. Controlled mating, karyology, morphology and sex ratio in the *Dendroctonus ponderosae* complex. Ann. Entomol. Soc. Am. 61, 517–526.

Lanier, G.N., Hendrichs, J.P., Flores, J.E., 1988. Biosystematics of the *Dendroctonus frontalis* (Coleoptera: Scolytidae) complex. Ann. Entomol. Soc. Am. 81, 403–418.

Latty, T.M., Reid, M.L., 2010. Who goes first? Condition and danger dependent pioneering in a group-living beetle (*Dendroctonus ponderosae*). Behav. Ecol. Sociobiol. 64, 639–646.

Lee, S., Kim, J.-J., Breuil, C., 2005. *Leptographium longiclavatum* sp. nov., a new species associated with the mountain pine beetle, *Dendroctonus ponderosae*. Mycol. Res. 109, 1162–1170.

Lee, S., Kim, J.-J., Breuil, C., 2006. Diversity of fungi associated with the mountain pine beetle, Dendroctonus ponderosae and infested lodgepole pines in British Columbia. Fungal Divers. 22, 91–105.

Lee, S., Hamelin, R.C., Six, D.L., Breuil, C., 2007. Genetic diversity and the presence of two distinct groups in *Ophiostom clavigerum* associated with *Dendroctonus ponderosae* in BC and the northern Rocky Mountains. Phytopathol. 97, 1177–1185.

Lejeune, R.R., McMullen, L.H., Atkins, M.D., 1961. The influence of logging on Douglas-fir beetle populations. Forest. Chron. 37, 308–314.

Lessard, E.D., Schmid, J.M., 1990. Emergence, attack densities, and host relationships for the Douglas-fir beetle (*Dendroctonus pseudotsugae* Hopkins) in northern Colorado. West. N. Am. Nat. 50, 333–338.

Lester, J.D., Irwin, J.T., 2012. Metabolism and cold tolerance of overwintering adult mountain pine beetles (*Dendroctonus ponderosae*): Evidence of facultative diapause? J. Insect Physiol. 58, 808–815.

Leufvén, A., Bergström, G., Falsen, E., 1984. Interconversion of verbenols and verbenone by identified yeasts isolated from the spruce bark beetle, *Ips typographus*. J. Chem. Ecol. 10, 1349–1361.

Lewinsohn, D., Lewinsohn, E., Bertagolli, C.L., Partidge, A.D., 1994. Blue stain fungi and their transport structures on the Douglas-fir beetle. Can. J. Forest Res. 24, 2275–2283.

Lieutier, F., Vouland, G., Pettinetti, M., Garcia, J., Romary, P., Yart, A., 1992. Defence reactions of Norway spruce (*Picea abies* Karst.) to artificial insertion of *Dendroctonus micans* Kug. (Col. Scolytidae). J. Appl. Entomol. 114, 174–186.

Lieutier, F., Yart, A., Salle, A., 2009. Stimulation of tree defenses by ophiostomatoid fungi can explain attack success of bark beetles on conifers. Ann. Forest Sci. 66, 801–823.

Lindgren, B.S., Raffa, K.F., 2013. Evolution of tree-killing in bark beetles (Coleoptera, Curculionidae): trade-offs between the maddening crowds and a sticky situation. Can. Ent. 145, 471–495.

Linit, M.J., Stephen, F.M., 1983. Parasite and predator component of within-tree southern pine beetle (Coleoptera: Scolytidae) mortality. Can. Ent. 115, 679–688.

Logan, J.A., Amman, G.D., 1986. A distribution model for egg development in mountain pine beetle. Can. Ent. 118, 361–372.

Logan, J.A., Bentz, B.J., 1999. Model analysis of mountain pine beetle (Coleoptera: Scolytidae_ Seasonality. Environ. Entomol. 28, 924–934.

Logan, J.A., Powell, J.A., 2001. Ghost forests, global warming, and the mountain pine beetle (Coleoptera: Scolytidae). Am. Entomol. 47, 160–173.

Logan, J.A., Macfarlane, W.W., Wilcox, L., 2010. Whitebark pine vulnerability to climate-driven mountain pine beetle disturbance in the Greater Yellowstone ecosystem. Ecol. Appl. 20, 895–902.

Lombardero, M.J., Ayres, M.P., Hofstetter, R.W., Moser, J.C., Klepzig, K.D., 2003. Strong indirect interactions of *Tarsonemus* mites (Acarina: Tarsonemidae) and *Dendroctonus frontalis* (Coleoptera: Scolytidae). Oikos 102, 243–252.

Lorio, P., Stephen, F.M., Paine, T.D., 1995. Environment and ontogeny modify loblolly pine response to induced water deficits and bark beetle attack. Forest Ecol. Manag. 73, 97–110.

Lu, K.C., Allen, D.G., Bollen, W.B., 1957. Association of yeasts with the Douglas-fir beetle. Forest Sci. 3, 336–342.

MacGuidwin, A.E., Smart, G.C., Wilkinson, R.C., 1980. Effect of the nematode *Contortylenchus brevicomi* on gallery construction and fecundity of the southern pine beetle. J. Nematol. 12, 278–282.

Manning, C.G., Reid, M.L., 2013. Sub-lethal effects of monoterpenes on reproduction by mountain pine beetles. Agr. For. Entomol. 15, 262–271.

Maroja, L.S., Bogdanowicz, S.M., Wallin, K.F., Raffa, K.F., Harrison, R.G., 2007. Phylogeography of spruce beetles (*Dendroctonus rufipennis* Kirby) (Curulionidae: Scolytinae) in North America. Mol. Ecol. 16, 2560–2573.

Massey, C.L., 1956. Nematode parasites and associates of the Engelmann spruce beetle (*Dendroctonus engelmanni* Hopk.). Proc. Helminthol. Soc. Wash. 23, 14–24.

Massey, C.L., 1966. The influence of nematode parasites and associates on bark beetles in the United States. Bull. Entomol. Soc. Am. 12, 384–386.

Massoumi Alamouti, S., Six, D.L., Wang, V., DiGuistini, S., Bohlman, J., Hamelin, R.C., Feau, N., Breuil, C., 2011. Gene genealogies reveal cryptic speciation and host-specificity for the pine fungal pathogen, *Grosmannia clavigera*. Microb. Ecol. 20, 2581–2602.

McCambridge, W.F., Knight, F.B., 1972. Factors affecting spruce beetle during a small outbreak. Ecology 53, 830–839.

McHugh, C.W., Kolb, T.E., Wilson, J.L., 2003. Bark beetle attacks on ponderosa pine following fire in northern Arizona. Environ. Entomol. 32, 510–522.

McKenna, D.D., Sequeira, A.S., Marvaldi, A.E., Farrell, B.D., 2009. Temporal lags and overlap in the diversification of weevils and flowering plants. Proc. Natl. Acad. Sci. U. S. A. 106, 7083–7088.

Mendoza, M.G., Salinas-Moreno, Y., Olivo-Martínez, A., Zúñiga, G., 2011. Factors influencing the geographical distribution of *Dendroctonus rhizophagus* (Coleoptera: Curculionidae: Scolytinae) in the Sierra Madre Occidental, Mexico. Environ. Entomol. 40, 549–559.

Meurisse, N., Couillien, D., Grégoire, J.-C., 2008. Kairomone traps: a tool for monitoring the invasive bark beetle *Dendroctonus micans* (Coleoptera: Scolytinae) and its specific predator, *Rhizophagus grandis* (Coleoptera: Monotomidae). J. Appl. Ecol. 45, 537–548.

Millar, C.I., Stephenson, N.L., Stephens, S.L., 2012. Climate change and forests of the future: managing in the face of uncertainty. Ecol. Appl. 17, 2145–2151.

Miller, J.M., Keen, F.P., 1960. Biology and Control of the Western Pine Beetle. A Summary of the First Fifty Years of Research. U.S. Department of Agriculture Forest Service, Miscellaneous Publication 800.

Mock, K.E., Bentz, B.J., O'Neil, E.M., Chong, J.P., Orwin, J., Pfrender, M.E., 2007. Landscape scale genetic variation in a forest outbreak species, the mountain pine beetle (*Dendroctonus ponderosae*). Mol. Ecol. 16, 553–568.

Morales-Jiménez, J., Zúñiga, G., Villa-Tanaca, L., Hernández-Rodríguez, C., 2009. Bacterial community and nitrogen fixation in the red turpentine beetle, *Dendroctonus valens* LeConte (Coleoptera: Curculionidae: Scolytinae). Microb. Ecol. 58, 879–891.

Morales-Jiménez, J., Zúñiga, G., Ramírez-Saad, H.C., Hernández-Rodríguez, C., 2012. Gut-associated bacteria throughout the life cycle of the bark beetle *Dendroctonus rhizphagus* Thomas and Bright (Curculionidae: Scolytinae) and their cellulolytic activities. Microb. Ecol. 64, 268–278.

Moser, J.C., 1985. Use of sporothecae by phoretic *Tarsonemous* mitesto transport ascospores of coniferous blue stain fungi. Trans. Brit. Mycol. Soc. 84, 750–753.

Moser, J.C., Roton, L.R., 1971. Mites associated with southern pine beetles in Allan Parish. Louisiana. Can. Ent. 103, 1775–1798.

Moser, J.C., Thatcher, R.C., Pickard, L.S., 1971. Relative abundance of southern pine beetle associates in east Texas. Ann. Entomol. Soc. Am. 64, 72–77.

Moser, J.C., Wilkinson, R.C., Clark, E.W., 1974. Mites associated with *Dendroctonus frontalis* Zimmerman (Scolytidae: Coleoptera) in Central America and Mexico. Turrialba 24, 37–381.

Moser, J.C., Fitzgibbon, B.A., Klepzig, K.D., 2005. The Mexican pine beetle, *Dendroctonus mexicanus*: first record in the United States and co-occurrence with the southern pine beetle—*Dendroctonus frontalis* (Coleoptera: Scolytidae or Curculionidae: Scolytinae). Ent. News 116, 235–243.

Negrón, J., 1997. Estimating probabilities of infestation and extent of damage by the roundheaded beetle in ponderosa pine in the Sacramento Mountains, New Mexico. Can. J. Forest Res. 27, 1936–1945.

Negrón, J.F., 1998. Probability of infestation and extent of mortality associated with the Douglas-fir beetle in the Colorado Front Range. Forest Ecol. Manag. 107, 71–85.

Negrón, J.F., Anhold, J.A., Munson, A.S., 2001. Within-stand distributions of tree mortality caused by the Douglas-fir beetle (Coleoptera: Scolytidae). Environ. Entomol. 30, 215–224.

Negrón, J.F., Schaupp, W.C., Pederson, L., 2011. Flight periodicity of the Douglas-fir beetle, *Dendroctonus pseudostugae* Hopkins (Coleoptera: Curculionidae: Scolytinae) in Colorado, U.S.A. Coleopterists Bull 65, 182–184.

Nickle, W.R., 1963. Observations on the effect of nematodes on *Ips confusus* LeConte, and other bark beetles. J. Insect Pathol. 5, 386–389.

Otvos, I.S., 1970. Avian predation of the western pine beetle. In: Stark, R.W., Dahlsten, D.L. (Eds.), Studies on the Population Dynamics of the Western Pine Beetle, *Dendroctonus brevicomis*, LeConte (Coleoptera: Scolytidae). University of California, Berkeley, pp. 119–127.

Owen, D.R., Lindahl Jr., K.Q., Wood, D.L., Parmeter Jr., J.R., 1987. Pathogenicity of fungi isolated from *Dendroctonus valens, D. brevicomis*, and *D. ponderosae* to ponderosa pine seedlings. Phytopathology 77, 631–636.

Paine, T.D., Birch, M.C., 1983. Acquisition and maintenance of mycangial fungi of *Dendroctonus brevicomis* LeConte (Coleoptera: Scolytidae). Environ. Entomol. 12, 1384–1386.

Paine, T.D., Hanlon, C.C., 1994. Influence of oleoresin constituents from *Pinus ponderosa* and *P. jeffreyi* on growth of the mycangial fungi of *Dendroctonus ponderosae* and *D. jeffreyi*. J. Chem. Ecol. 20, 2551–2563.

Paine, T.D., Birch, M.C., Sivhra, P., 1981. Niche breadth and resource partitioning by four sympatric species of bark beetles (Coleoptera: Scolytidae). Oecologia 48, 1–6.

Paine, T.D., Raffa, K.F., Harrington, T., 1997. Interactions among scolytid bark beetles, their associated fungi, and live host conifers. Ann Rev. Entomol. 42, 179–206.

Panhuis, T.M., Butlin, R., Zuk, M., Tregenza, T., 2012. Sexual selection and speciation. Trends in Ecology and Evolution 16, 364–371.

Parker, D.L., Stevens, R.E., 1979. Mountain pine beetle infestation characteristics in ponderosa pine, Kaibab Plateau, AZ 1975–1977. Rocky Mtn. Res. Station USDA For. Serv. Res Note-RM 367.

Parker, T.J., Clancy, K.M., Mathiasen, R.L., 2006. Interactions among fire, insects and pathogens in coniferous forests of the interior western United States and Canada. Agric. Forest Entomol. 8, 167–189.

Phillips, T.W., Nation, J.L., Wilkinson, R.C., Foltz, J.L., 1989. Secondaary attraction and field activity of beetle-produced volatiles in *Dendroctonus terebrans*. J. Chem. Ecol. 15, 1513–1533.

Powell, E.N., Townsend, P.A., Raffa, K.F., 2012. Wildfire provides refuge from local extinction but is an unlikely driver of outbreaks by mountain pine beetle. Ecol. Monogr. 82, 69–84.

Powell, J.A., Bentz, B.J., 2014. Phenology and density dependent dispersal predicts patterns of mountain pine beetle (*Dendroctonus ponderosae*) impact. Ecol. Modell. 273, 173–185.

Powers, J.S., Sollins, P., Harmon, M.E., Jones, J.A., 1999. Plant-pest interactions in time and space: a Douglas-fir bark beetle outbreak as a case study. Landscape Ecol. 14, 105–120.

Preisler, H.K., Hicke, J.A., Ager, A.A., Hayes, J.L., 2012. Climate and weather influences on spatial temporal patterns of mountain pine beetle populations in Washington and Oregon. Ecology 93, 2421–2434.

Pu, X.J., Chen, H., 2007. Relations between attacking of *Dendroctonus armandi* and nutrition and resistance material of host trees (*Pinus armandi*). Journal of the Northwest Agriculture and Forest University 35, 106–110. (in Chinese).

Pureswaran, D., Borden, J., 2003. Is bigger better? Size and pheromone production in the mountain pine beetle, *Dendroctonus ponderosae* Hopkins (Coleoptera: Scolytidae). J Insect Behavior 16, 765–782.

Pureswaran, D.S., Borden, J.H., 2005. Primary attraction and kairomonal host discrimination in three species of *Dendroctonus* (Coleoptera: Scolytidae). Agric. Forest Entomol. 7, 219–230.

Pureswaran, D.S., Sullivan, B.T., Ayres, M.P., 2006. Fitness consequences of pheromone production and host selection strategies in a tree-killing bark beetle (Coleoptera: Curculionidae: Scolytinae). Oecologia 148, 720–728.

Raffa, K.F., 1988. The mountain pine beetle in western North America. In: Berryman, A.A. (Ed.), Dynamics of Forest Insect Populations: Patterns, Causes, Implications. Plenum Press, New York, pp. 506–550.

Raffa, K.F., 2001. Mixed messages across multiple trophic levels: the ecology of bark beetle chemical communication systems. Chemoecology 11, 49–65.

Raffa, K.F., Berryman, A.A., 1983. The role of host plant resistance in the colonization behavior and ecology of bark beetles (Coleoptera: Scolytidae). Ecol. Monogr. 53, 27–49.

Raffa, K.F., Aukema, B.H., Erbilgin, N., Klepzig, K.D., Wallin, K.F., 2006. Interactions among conifer terpenoids and bark beetles across multiple levels of scale: an attempt to understand links between population patterns and physiological processes. Rec. Adv. Phytochem. 39, 79–118.

Raffa, K.F., Hobson, K.R., LaFontaine, S., Aukema, B.H., 2007. Can chemical communication be cryptic? Adaptations by herbivores to natural enemies exploiting prey semiochemistry. Oecologia 153, 1009–1019.

Raffa, K.F., Aukema, B.H., Bentz, B.J., Carroll, A.L., Hicke, J.A., Turner, M.G., Romme, W.H., 2008. Cross-scale drivers of natural disturbances prone to anthropogenic amplification: the dynamics of bark beetle eruptions. Bioscience 58, 501–517.

Raffa, K.F., Powell, E.N., Townsend, P.A., 2013. Temperature-driven range expansion of an irruptive insect heightened by weakly co-evolved plant defense. Proc. Natl. Acad. Sci. U. S. A. 110, 2193–2198.

Reed, A.N., Hanover, J.W., Furniss, M.M., 1986. Douglas-fir and western larch: chemical and physiological properties in relation to Douglas-fir bark beetle attack. Tree Physiol. 1, 277–287.

Reeve, J.D., 1997. Predation and bark beetle dynamics. Oecologia 112, 48–54.

Reeve, J.D., Rhodes, D.J., Turchin, P., 1998. Scramble competition in the southern pine beetle. Dendroctonus frontalis. Ecol. Entomol. 23, 433–443.

Reeve, J.D., Anderson, F.E., Kelley, S., 2012. Ancestral state reconstruction for *Dendroctonus* bark beetles: Evolution of a tree killer. Environ. Entomol. 41, 723–730.

Reid, R.W., 1945. Nematodes associated with the mountain pine beetle. Bi-monthly Progress Report, Division of Forest Biology, Department of Agriculture, Canada 14, 1–3.

Reid, R.W., 1958. Internal changes in the female mountain pine beetle, *Dendroctonus monticolae* Hopk., associated with egg laying and flight. Can. Entomol. 90, 464–468.

Reid, R.W., 1962. Biology of the mountain pine beetle, *Dendroctonus monticolae*, in the east Kootenay region of British Columbia. I. Life cycle, brood development and flight periods. Can. Ent. 94, 531–538.

Reid, M.L., Baruch, O., 2010. Mutual mate choice by mountain pine beetle: size dependence but not ssize asssotive mating. Ecol. Entomol. 35, 69–76.

Reynolds, K.M., 1992. Relations between activity of *Dendroctonus rufipennis* (Kirby) on Lutz spruce and blue stain associated with *Leptographium abietinum* (Peck) Wingfield. Forest Ecol. Manag. 47, 71–86.

Reynolds, K.M., Holsten, E.H., 1996. Classification of spruce beetle hazard in Lutz and Sitka spruce stands on the Kenai peninsula, Alaska. Forest Ecol. Manag. 84, 251–262.

Rice, A.V., Thormann, M.N., Langor, D.W., 2008. Mountain pine beetle-associated blue-stain fungi are differentially adapted to boreal temperatures. For. Pathol. 38, 113–123.

Rivera, F.N., González, E., Gómez, Z., López, N., Hernández-Rodríguez, C., Berkov, A., Zúñiga, G., 2009. Gut-associated yeast in bark beetles of the genus *Dendroctonus* Erichson (Coleoptera: Curculionidae: Scolytinae). Biol. J. Linn. Soc. 98, 325–342.

Roe, A.D., James, P.M.A., Rice, A.V., Cooke, J.E.K., Sperling, F.A., 2011a. Spatial community structure of mountain pine beetle fungal symbionts across a latitudinal gradient. Microb. Ecol. 62, 347–360.

Roe, A.D., Rice, A.V., Coltman, D.W., Cooke, J.E.K., Sperling, F.A.H., 2011b. Comparative phylogeography, genetic differentiation and contrasting reproductive modes in three fungal symbionts of a multipartite bark beetle symbiosis. Mol. Ecol. 20, 584–600.

Roelofs, W.L., Brown, R.L., 1982. Pheromones and evolutionary relationships of Tortricidae. Annu. Rev. Ecol. Syst. 13, 395–422.

Rolland, C., Lemperiere, G., 2004. Effects of climate on radial growth of Norway spruce and interactions with attacks by the bark beetle *Dendroctonus micans* (Kug., Coleoptera: Scolytidae): a dendroecological study in the French Massif Central. Forest Ecol. Manag. 201, 89–104.

Romme, W.H., Knight, D.H., Yavitt, J.B., 1986. Mountain pine beetle outbreaks in the Rocky Mountains: regulators of primary productivity? Am. Nat. 27, 484–494.

Ross, D.W., Solheim, H., 1997. Pathogenicity to Douglas-fir of *Ophiostoma pseudotsugae* and *Leptographium abietinum*, fungi associated with the Douglas-fir beetle. Can. J. Forest Res. 27, 39–43.

Rudinsky, J.A., Michael, K.R., 1973. Sound production in Scolytidae: stridulation by female *Dendroctonus* beetles. J. Insect Physiol. 19, 689–705.

Ruixia, G., Chen, H., Tang, M., Hui, X., Li, S., Chen, G., 2009. Cellulose produced by the symbiotic fungus *Leptographium qinlingensis* of *Dendroctonus armandi*. Scientia Silvae Sinica 45, 108–112 (in Chinese).

Ruíz, E.A., Victor, J., Hayes, J.L., Zúñiga, G., 2009. Molecular and morphological analysis of *Dendroctonus pseudotsugae* (Coleoptera: Curculionidae: Scolytinae) and assessment of the taxonomic status of subspecies. Ann. Entomol. Soc. Am. 10, 982–997.

Ruíz, E.A., Rinehart, J.E., Hayes, J.L., Zúñiga, G., 2010. Historical demography and phylogeography of a specialist bark beetle, *Dendroctonus pseudotsugae* Hopkins (Curculionidae, Scolytinae). Environ. Entomol. 39, 1685–1697.

Ryan, K., Amman, G.D., 1996. Bark beetle activity and delayed tree mortality in the Greater Yellowstone area following the 1988 fires. In: Keane, R.E., Ryan, K.C., Running, S.W. (Eds.), Ecological implications of fire in Greater Yellowstone Proceedings. International Association of Wildland Fire, Fairland, Washington, pp. 151–158.

Safranyik, L., 1976. Size- and sex-related emergence, and survival in cold storage of mountain pine beetle adults. Can. Entomol. 108, 209–212.

Safranyik, L., 1978. Effects of climate and weather on mountain pine beetle populations. In: Berryman, A.A., Amman, G.D., Stark, R.W. (Eds.), Proceedings, Symposium on Theory and Practice of Mountain Pine Beetle Management in Lodgepole Pine Forests. University of Idaho, Moscow, pp. 77–84.

Safranyik, L., 2011. Development and survival of the spruce beetle, *Dendroctonus rufipennis*, in stumps and windthrow. Canadian Forest Service Information Report BC-X-430.

Safranyik, L., Whitney, H.S., 1985. Development and survival of axenically reared mountain pine beetles, *Dendroctonus ponderosae* (Coleoptera: Scolytidae), at constant temperatures. Can. Ent. 117, 185–192.

Safranyik, L., Linton, D.A., 1999. Survival and development of mountain pine beetle broods in Jack pine bolts from Ontario. Canadian Forest Service Research Note 2, Pacific Forestry Centre, Victoria, BC.

Safranyik, L., Carroll, A.L., 2006. The biology and epidemiology of the mountain pine beetle in lodgepole pine forests. In: Safranyik, L., Wilson, W.R. (Eds.), The Mountain Pine Beetle: A Synthesis of Biology, Management, and Impacts on Lodgepole Pine. Canadian Forest Service, Pacific Forestry Centre, Victoria, British Columbia, pp. 3–66.

Safranyik, L., Barclay, H., Thomson, A., Riel, W.G., 1999. A population dynamics model for the mountain pine beetle, *Dendroctonus ponderosae* Hopk (Coleoptera: Scolytidae). Natural Resources Canada, Can. For. Serv., Pac. For. Centre, Victoria, BC, Canada.

Safranyik, L., Shore, T.L., Carroll, A.L., Linton, D.A., 2004. Bark beetle (Coleoptera: Scolytidae) diversity in spaced and unmanaged mature lodgepole pine (Pinaceae) in southeastern British Columbia. Forest Ecol. Manag. 200, 23–38.

Safranyik, L., Carroll, A.L., Regniere, J., Langor, D.W., Riel, W.G., Shore, T.L., et al., 2010. Potential for range expansion of mountain pine beetle into the northern boreal forest of North America. Can. Ent. 142, 415–442.

Salinas-Moreno, Y., Mendoza, M.G., Barrios, M.A., Cisneros, R., Macías-Sámano, J., Zúñiga, G., 2004. Areography of the genus *Dendroctonus* (Coleoptera: Curculionidae: Scolytinae) in Mexico. J. Biogeogr. 31, 1163–1177.

Salinas-Moreno, Y., Ager, A., Vargas, C.F., Hayes, J.L., Zúñiga, G., 2010. Determining the vulnerability of Mexican pine forests to bark beetles of the genus *Dendroctonus* Erichson (Coleoptera: Curculionidae: Scolytinae). Forest Ecol. Manag. 260, 52–61.

Samarasekera, G.D.N.G., Bartell, N.V., Lindgren, B.S., Cooke, J.E.K., Davis, C.S., James, P.M.A., et al., 2012. Spatial genetic structure of the mountain pine beetle (*Dendroctonus ponderosae*) outbreak in western Canada: historical patterns and contemporary dispersal. Mol. Ecol. 21, 2931–2948.

Sánchez-Martínez, G., Wagner, M.R., 2009. Host preference and attack pattern of *Dendroctonus rhizophagus* (Coleoptera: Curculionidae: Scolytinae): a bark beetle specialist on pine regeneration. Environ. Entomol. 38, 1197–1204.

Sánchez-Sánchez, H., López-Barrera, G., Peñaloza-Ramírez, J.M., Rocha-Ramírez, V., Oyama, K., 2012. Phylogeography reveals routes of colonization of the bark beetle *Dendroctonus approximatus* Dietz in Mexico. J. Hered. 103, 638–650.

Schedl, K., 1932. Scolytoidae. In: Winkler, A. (Ed.), Catalogus Coleopterorum regionis palaearcticae. Winkler & Wagner, Vienna, pp. 1632–1647.

Schmid, J., 1981. Spruce beetles in blowdown. USDA Forest Service, Rocky Mountain Forest and Range Experiment Station, Research Note RM-411.

Schmitz, R., Gibson, K., 1996. Douglas-fir beetle. USDA Forest Service Forest Insect and Disease Leaflet 5, 7pp.

Schrey, N.M., Schrey, A.W., Heist, E.J., Reeve, J.D., 2011. Genetic heterogeneity in a cyclical forest pest, the southern pine beetle, *Dendroctonus frontalis*, is differentiated into east and west groups in the southeastern United States. J. Insect Sci. 11, 110.

Scott, J.J., Dong-Chan, O., Yuceer, M.C., Klepzig, K.D., Clardy, J., Currie, C.R., 2008. Bacterial protection of beetle-fungus mutualism. Science 322, 63.

Sequeira, A.S., Normark, B.B., Farrell, B.D., 2000. Evolutionary assembly of the conifer fauna: distinguishing ancient from recent association in bark beetles. Proc. R. Soc. London B 267, 2359–2366.

Seybold, S.J., Albers, M.A., Katovich, S.A., 1992. Eastern larch beetle. USDA Forest Service Forest Insect and Disease Leaflet 175.

Seybold, S.J., Huber, D.P.W., Lee, J.C., Graves, A.D., Bohlmann, J., 2006. Pine monoterpenes and pine bark beetles: a marriage of convenience for defense amd chemical communication. Phytochem. Rev. 5, 143–178.

Sherriff, R.L., Berg, E.E., Miller, A.E., 2011. Climate variability and spruce beetle (*Dendroctonus rufipennis*) outbreaks in south-central and southwest Alaska. Ecology 92, 1459–1470.

Shepherd, W.P., Huber, D.W.P., Seybold, S.J., Fettig, C.J., 2008. Antennal responses of the western pine beetle, *Dendroctonus brevicomis* (Coleoptera: Curculionidae), to stem volatiles of its primary host, *Pinus ponderosa*, and nine sympatric non-host angiosperms and conifers. Chemoecology 17, 209–221.

Shi, Z.-H., Sun, J.-H., 2010. Immunocompetence of the red turpentine beetle, *Dendroctonus valens* LeConte (Coleoptera: Curculionidae, Scolytinae): variation between developmental stages and sexes in populations in China. J. Insect Physiol. 56, 1696–1701.

Shi, Z.-H., Wang, B., Clarke, S.R., Sun, J.H., 2012. Effect of associated fungi on the immunocompetence of red turpentine beetle larvae, *Dendroctonus valens* (Coleoptera: Curculionidae: Scolytinae). Insect Sci. 9, 579–584.

Shifrine, M., Phaff, H.J., 1956. The association of yeasts with certain bark beetles. Mycologia 48, 41–55.

Simard, M., Romme, W.H., Griffin, J.M., Turner, M.G., 2011. Do mountain pine beetle outbreaks change the probability of active crown fire in lodgepole pine forests? Ecolog. Monogr. 81, 3–24.

Six, D.L., 2003. Bark beetle-fungus symbioses. In: Bourtzis, K., Miller, T. (Eds.), Insect Symbiosis. CRC Press, Boca Raton, pp. 97–114.

Six, D.L., 2013. The bark beetle holobiont: why microbes matter. J. Chem. Ecol. 39, 989–1002.

Six, D.L., Paine, T.D., 1996. *Leptographium pyrinum* is a mycangial fungus of *Dendroctonus adjunctus*. Mycologia 88, 739–744.

Six, D.L., Paine, T.D., 1998. Effects of mycangial fungi and host tree species on progeny survival and emergence of *Dendroctonus ponderosae* (Coleoptera: Scolytidae). Environ. Entomol. 27, 1393–1401.

Six, D.L., Paine, T.D., 1999. Allozyme diversity and gene flow in *Ophiostoma clavigerum* (Ophiostomatales: Ophiostomataceae), the mycangial fungus of the Jeffrey pine beetle, *Dendroctonus jeffreyi* (Coleoptera: Scolytidae). Can. J. Forest Res. 29, 324–331.

Six, D.L., Bentz, B.J., 2003. Fungi associated with the North American spruce beetle, *Dendroctonus rufipennis*. Can. J. Forest Res. 33, 1815–1820.

Six, D.L., Klepzig, K.D., 2004. *Dendroctonus* bark beetles as model systems for studies on symbiosis. Symbiosis 37, 207–232.

Six, D.L., Adams, J.C., 2007. Relationships between white pine blister rust and the selection of individual whitebark pine by the mountain pine beetle. J. Entomol. Sci. 42, 345–353.

Six, D.L., Bentz, B.J., 2007. Temperature determines the relative abundance of symbionts in a multipartite bark beetle-fungus symbiosis. Microb. Ecol. 54, 112–118.

Six, D.L., Skov, K., 2009. Response of bark beetles and their natural enemies to fire and fire surrogate treatments in mixed-conifer forests in western Montana. Forest Ecol. Manag. 258, 761–772.

Six, D.L., Wingfield, M.J., 2011. The role of phytopathogenicity in bark beetle-fungus symbioses: a challenge to the classic paradigm. Annu. Rev. Entomol. 56, 255–272.

Six, D.L., Paine, T.D., Hare, J.D., 1999. Allozyme diversity and gene flow in the bark beetle, *Dendroctonus jeffreyi* (Coleoptera: Scolytidae). Can. J. Forest Res. 29, 315–323.

Six, D.L., de Beer, Z.W., Duong, T., Carroll, A.L., Wingfield, M.J., 2011a. Fungal associates of the lodgepole pine beetle, *Dendroctonus murrayanae*. Antonie Van Leeuwenhoek 100, 231–244.

Six, D.L., Poulsen, M., Hansen, A.K., Wingfield, M.J., Roux, J., Eggleton, P., et al., 2011b. Anthropogenic effects on insect-microbial symbioses in forest and savanna ecosystems. Symbiosis 53, 101–121.

Smith, R.H., 1963. Preferential attack by *Dendroctonus terebrans* on *Pinus elliotii*. J. Econ. Entomol. 56, 817–819.

Smith, S.L., Borys, R.R., Shea, P.J., 2009. Jeffrey pine beetle. USDA Forest Service Insect and Disease Leaflet 11.

Smith, G.D., Carroll, A.L., Lindgren, B.S., 2011. Facilitation in bark beetles: endemic mountain pine beetle gets a helping hand. Agric. Forest. Entomol. 13, 37–43.

Solheim, H., Safranyik, L., 1997. Pathogenicity to Sitka spruce of *Ceratocystis rufipenni* and *Leptographium abietinum* blue-stain fungi associated with the spruce beetle. Can. J. Forest Res. 27, 1336–1341.

Solheim, H., Krokene, P., 1998. Growth and virulence of *Ceratocystis rufipenni* and three blue-stain fungi isolated from the Douglas-fir beetle. Can. J. Bot. 76, 1763–1769.

Staeben, J.C., Clarke, S., Gandhi, K.J.K., 2010. Black turpentine beetle. USDA Forest Service Forest Insect and Disease Leaflet 12.

Steiner, G., Buhrer, E.M., 1934. *Aphelenchoides xylophilus*, n. sp. A nematode associated with blue-stain and other fungi in timber. J. Agric. Res. 48, 948–950.

Stephen, F.M., Dahlsten, D.L., 1976. The arrival sequence of the arthropod complex following attack by Dendroctonus brevicomis (Coleoptera: Scolytidae) in ponderosa pine. Can. Entomol. 108, 283–304.

Storer, A.J., Speight, M.R., 1996. Relationships between *Dendroctonus micans* Kug. (Coleoptera: Scolytidae) survival and development and biochemical changes in Norway spruce, *Picea abies* (L.) Karst., phloem caused by mechanical wounding. J. Chem. Ecol. 22, 559–573.

Storer, A.J., Wainhouse, D., Speight, M.R., 1997. The effect of larval aggregation behavior on larval growth of the spruce bark beetle *Dendroctonus micans*. Ecol. Entomol. 22, 109–115.

Sturgeon, K.B., Mitton, J.B., 1982. Evolution of bark beetle communities. In: Mitton, J.B., Sturgeon, K.B. (Eds.), Bark Beetles in North American Conifers: A System for the Study of Evolutionary Biology. University of Texas Press, Austin, pp. 350–384.

Sturgeon, K.B., Mitton, J.B., 1986. Allozyme and morphological differentiation of mountain pine beetles, *Dendrotconus ponderosae* Hopkins (Coleoptera: Scolytidae), associated with host tree. Evolution 40, 290–302.

Sullivan, B.T., Berisford, C.W., 2004. Semiochemicals from fungal associates of bark beetles may mediate host location behavior of parasitoids. J. Chem. Ecol. 30, 703–717.

Sullivan, B.T., Fettig, C.J., Otrasina, W.J., Dalusky, M.J., Berisford, C.W., 2003. Association between severity of prescribed burns and subsequent activity of conifer-infesting beetles in stands of longleaf pine. Forest Ecol. Manag. 185, 327–340.

Sullivan, B.T., Niño, A., Moreno, B., Brownie, C., Macías-Sámano, J., Clarke, S.R., et al., 2012. Biochemical evidence that *Dendroctonus frontalis* consists of two sibling species in Belize and Chiapas Mexico. Ann. Entomol. Soc. Am. 105, 817–831.

Sun, J., Lu, M., Gillette, N.E., Wingfield, M.J., 2013. Red turpentine beetle: innocuous native becomes invasive tree killer in China. Annu. Rev. Entomol. 58, 293–311.

Symonds, M.R.E., Elger, M.A., 2004. The mode of pheromone evolution: evidence from bark beetles. Proc. R. Soc. Lond. B 271, 839–846.

Taerum, S.J., Duong, T.A., de beer, Z.W., Gillette, N., Sun, J.-H., Owen, D.R., Wingfield, M.J., 2013. Large shift in symbiont assemblage in the invasive red turpentine beetle. PLoS One 8, e78126.

Thatcher, R.C., Searcy, J.L., Coster, J.E., Hertel, G.D. (Eds.), 1980. The Southern Pine Beetle.USDA Forest Service Technical Bulletin 1634.

Thompson, S.N., Bennett, R.B., 1971. Oxidation of fat during flight of male Douglas-fir beetles, Dendroctonus pseudotsugae. J. Insect Physiol. 17, 1555–1563.

Thompson, A.J., Shrimpton, D.M., 1984. Weather associated with the start of mountain pine beetle outbreaks. Can. J. Forest Res. 14, 255–258.

Thomas, J.B., Bright Jr., D.E., 1970. A new species of *Dendroctonus* (Coleoptera: Scolytidae) from Mexico. Can. Entomol. 102, 479–483.

Thong, C.H.S., Webster, J.M., 1975. Effects of the bark beetle nematode, *Contortylenchus reversus*, on gallery construction, fecundity, and egg viability of the Douglas-fir beetle, *Dendroctonus pseudotsugae* (Coleoptera: Scolytidae). J. Invertebr. Pathol. 26, 235–238.

Tran, J.K., Ylioja, T., Billings, R.F., Regniere, J., Ayres, M.P., 2007. Impact of minimum winter temperatures on the population dynamics of *Dendroctonus frontalis*. Ecol. Appl. 17, 882–899.

Tsui, C.K.M., Roe, A.D., El-Kassaby, Y.A., Rice, A.V., Alamouti, S.M., Sperling, F.H.A., et al., 2012. Population structure and migration pattern of a conifer pathogen, *Grosmannia clavigera*, as influenced by its symbiont, the mountain pine beetle. Mol. Ecol. 21, 71–86.

Tsui, C.K.-M., DiGuistini, S., Wang, S., Feau, N., Dhillon, B., Bohlmann, J., Hamelin, R.C., 2013. Unequal recombination and evolution of the mating-type (*MAT*) loci in the pathogenic fungus *Grosmannia clavigera* and relatives. G3 465–480.

Turchin, P., Lorio, P.L., Taylor, A.D., Billings, R.F., 1999. Why do populations of southern pine beetles (Coleoptera: Scolytidae) fluctuate? Environ. Entomol. 20, 401–409.

Vasanthakumar, A., Delalibera, I., Handelsman, J., Klepzig, K., Schloss, P., Raffa, K., 2006. Characterization of gut-associated bacteria in larvae and adults of the southern pine beetle, *Dendroctonus frontalis* Zimmermann. Environ. Entomol. 35, 1710–1717.

Vargas, C.F., López, A., Sánchez, H., Rodríguez, B., 2002. Allozyme analysis of host selection by bark beetles in central Mexico. Can. J. Forest Res. 32, 24–30.

Vasechko, G.I., 1978. Host selection by some bark-beetles (Col. Scolytidae). I. Study of primary attraction to chemical stimuli. Z. Angew. Entomol. 85, 66–76.

Vouland, G., Giraud, M., Schvester, D., 1984. The teneral period and the flight-taking in *Dendroctonus micans* Kug. (Coleoptera:

Scolytidae). In: Pasteels, J.-M., Grégoire, J.-C. (Eds.), Proceedings of the EEC Seminar, Biological Control of Bark Beetles, Brussels. Commission of the European Communities, Brussels, Belgium, pp. 68–79.

Wagner, T.L., Gagne, J.A., Doraiswamy, P., Coulson, R.N., Brown, K.W., 1979. Development time and mortality of *Dendroctonus frontalis* in relation to changes in tree moisture and xylem water potential. Environ. Entomol. 8, 1129–1138.

Wainhouse, D., Ashburner, R., Ward, E., Boswell, R., 1998. The effect of lignin and bark wounding on susceptibility of Spruce trees to *Dendroctonus micans*. J. Chem. Ecol. 24, 1551–1561.

Wallin, K.F., Raffa, K.F., 2004. Feedback between individual host selection behavior and population dynamics in an eruptive herbivore. Ecol. Monogr. 74, 101–116.

Wang, B., Salcedo, C., Lu, M., Sun, J., 2012. Mutual interactions between an invasive bark beetle and its associated fungi. Bull. Entomol. Res. 102, 71–77.

Wang, B., Lu, M., Cheng, C., Salcedo, C., Sun, J., 2013. Saccharide-mediated antagonistic effects of bark beetle fungal associates on larvae. Biol. Lett. 9, 20120787.

Webb, J.W., Franklin, R.T., 1978. Influence of phloem moisture on brood development of the southern pine beetle (Coleoptera: Scolytidae). Environ. Entomol. 7, 405–410.

Weed, A.S., Ayres, M.P., Hicke, J.A., 2013. Consequences of climate change for biotic disturbances in North American forests. Ecol. Monogr. 83, 441–470.

Werner, R.A., 1986. The eastern larch beetle in Alaska. USDA Forest Service Research Paper PNW-357.

Whitney, H.S., Farris, S.H., 1970. Maxillary mycangium in mountain pine beetle. Science 167, 54–55.

Whitney, H.S., Bandoni, R.J., Oberwinkler, F., 1987. *Entomocorticum dendroctoni* gen. et sp. nov. (Basidomycotina), a possible nutritional symbiote of the mountain pine beetle in British Columbia. Can. J. Bot. 65, 95–102.

Williams, W.I., Roberston, I.C., 2008. Using automated flight mills to manipulate fats reserves in Douglas-fir beetles (Coleoptera: Curculionidae). Environ. Entomol. 37, 850–856.

Williams, K.K., McMillin, J.D., Clancy, K.M., Miller, A., 2008. Influence of elevation on bark beetle (Coleoptera: Curculionidae, Scolytinae) community structure and flight periodicity in ponderosa pine forest of Arizona. Environ. Entomol. 37, 94–109.

Wingfield, M.J., Harrington, T.C., Solheim, H., 1997. Two species in the *Ceratocystis coerulescens* complex from conifers in western North America. Can. J. Bot. 75, 827–834.

Wood, S.L., 1963. A revision of the bark beetle genus *Dendroctonus* Erichson (Coleoptera: Scolytidae). Great Basin Nat. 6, 1–117.

Wood, S.L., 1982. The bark and ambrosia beetles of North and Central America (Coleoptera: Scolytidae), a taxonomic monograph. Great Basin Nat. Mem. 6, 1–1359.

Wright, L.C., Berryman, A.A., Wickman, B.E., 1984. Abundance of fir engraver, *Scolytus ventralis*, and the Douglas-fir beetle, *Dendroctonus pseudotsugae*, following tree defoliation by the Douglas-fir tussock moth, *Orgyia pseudotsugata*. Can. Entomol. 116, 293–305.

Xiao-Jun, L., Gao, J.-M., Chen, H., Zhang, A.-L., Tang, M., 2012. Toxins from a symbiotic fungus. Leptographium qinlingensis associated with Dendroctonus armandi and their in vitro toxicities to Pinus armandi seedlings. Eur. J. Plant Pathol. 134, 239–247.

Youngblood, A., Grace, J.B., McIver, J.D., 2009. Delayed conifer mortality after fuel reduction treatments: interactive effects of fuel, fire intensity, and bark beetles. Ecol. Appl. 19, 321–337.

Zanzot, J.W., Matusick, G., Eckhardt, L.G., 2010. Ecology of root-feeding beetles and their associated fungi on longleaf pine in Georgia. Environ. Entomol. 39, 415–423.

Zhang, L., Chen, H., Ma, C., Tian, Z., 2010. Electrophysiological response of *Dendroctonus armandi* (Coleoptera: Curculionidae: Scolytinae) to volatiles of Chinese white pine as well as to pure enantiomers and racemates of some monoterpenes. Chemoecology 20, 265–275.

Zipfel, R.D., de Beer, Z.W., Jacobs, K., Wingfield, B.D., Wingfield, M.J., 2006. Multigene phylogenies define *Ceratocystiopsis* and *Grosmannia* distinct from *Ophiostoma*. Stud. Mycol. 55, 75–97.

Zúñiga, G., Cisneros, R., Hayes, J.L., Macias-Samano, J., 2002a. Karyology, geographic distribution, and the origin of the genus *Dendroctonus* Erichson (Coleoptera: Scolytidae). Ann. Entomol. Soc. Am. 95, 267–275.

Zúñiga, G., Salinas-Moreno, Y., Hayes, J.L., Grégoire, J.-C., Cisneros, R., 2002b. Chromosome number in *Dendroctonus micans* and karyological divergence within the genus *Dendroctonus* (Coleoptera: Scolytidae). Can. Entomol. 134, 503–510.

Zúñiga, G., Cisneros, R., Salinas-Moreno, Y., Hayes, J.L., Rinehart, J.E., 2006. Genetic structure of *Dendroctonus mexicanus* (Coleoptera: Curculionidae: Scolytinae) in the trans-Mexican volcanic belt. Ann. Entomol. Soc. Am. 99, 945–958.

Chapter 9

Biology, Systematics, and Evolution of *Ips*

Anthony I. Cognato

Department of Entomology, Michigan State University, East Lansing, MI, USA

1. INTRODUCTION

1.1 Biology

Ips is one of the better-known bark beetle genera because of several species that are destructive to conifer forests and plantations (Chararas, 1962; Furniss and Carolin, 1992). Species utilize *Pinus*, *Picea*, and uncommonly *Larix*, *Abies*, and *Cedrus* as a food resource for larvae and adults (Wood and Bright, 1992). *Ips* species tend to feed on a specific tree genus or species but host species infidelity and successful brood development in non-hosts have been observed (Cognato, personal observ.). Dead and moribund trees are mostly used, but beetles may colonize environmentally stressed and healthy trees. Epidemic *Ips* populations can destroy thousands of hectares of forest (Furniss and Carolin, 1992). For example, in North America, a large outbreak of *Ips confusus* LeConte occurred during a prolong drought (2001–2004) in southwestern USA (Figure 9.1) causing extensive ecological damage and economical loss. During the peak outbreak, approximately 15–30% of pinyon pine trees were killed throughout 1.6 million hectares (Anonymous, 2005; Breshears *et al.*, 2005; Williams *et al.*, 2010). In Europe, *I. typographus* periodically outbreaks and kills thousands of hectares of healthy spruce trees (Schroeder and Lindelöw, 2002, Stadelmann *et al.*, 2013). In addition, some species, like *Ips pini* (Say) and *I. paraconfusus* Lanier, kill the tops of pine trees, which increase their susceptibility to attack by *Dendroctonus* spp. (Furniss and Carolin, 1992). Management of *Ips* populations is mostly limited to a local scale through improving tree and stand health and removing slash and beetle infested wood (e.g., Wermelinger, 2004; Stadelmann *et al.*, 2013).

Most of the *Ips* life cycle occurs underneath the tree bark within the cambium and phloem (Furniss and Carolin, 1977). Males typically colonize the host tree and creation of a nuptial chamber (Figure 9.2), although females occasionally initiate an entrance into the bark (All and Anderson, 1972). As the males feed, they produce semiochemicals that attract conspecific males and females to the tree (D. L. Wood, 1982). The main semiochemical

components, ipsenol, ipsdienol, and *cis*-verbenol, are either oxidation by-products of host tree terpene compounds or synthesized *de novo* within the beetle's gut (see below for further discussion). *Ips* species are polygamous and depending on the species, two to six females join a male in his nuptial chamber. In some species, females are only admitted entrance upon stridulation via a file on the vertex of the head and a plectrum on the dorsal anterior edge of the pronotum (Barr, 1969). However, the stridulation is not species specific (Lewis and Cane, 1992). After mating, females excavate egg tunnels in phloem and lay 20–30 eggs in niches along the tunnel walls (Figure 9.2) (Chararas, 1962). Depending on the ambient temperature, the eggs hatch after ≈ 7 days and the three larval instars feed under the bark for 3–6 weeks until pupation (Chararas, 1962; Lekander, 1968). Eclosion of teneral adults occurs in 1–3 weeks and the beetles feed for 1–2 weeks before dispersing (Chararas, 1962). Adult males often remain with the females to care for the brood, although some males re-emerge to produce a second brood concurrent with the first brood (Reid and Roitberg, 1994). Beetles complete development in 6–8 weeks given summer temperatures and thus one to five generations per year can occur depending on the climate (Furniss and Carolin, 1992). In colder climates, 2 years are needed for adults to mature (Furniss and Carolin, 1992). Adults typically overwinter under the bark or within forest duff and some species bore under the bark of live trees or within the xylem of braches (Chansler, 1964; Lanier, 1967).

1.2 Pheromones

Since the discovery that *Ips* species were attracted to frass produced by conspecifics (Wood *et al.*, 1967), much research attention has been given to the study of their pheromones. Aggregation pheromones, which are comprised of a bouquet of semiochemicals, are produced as males feed on host phloem (D. L. Wood, 1982). Several chemicals have been identified and three, ipsenol, ipsdienol, and *cis*-verbenol, are greatly attractive but their production varies among

FIGURE 9.1 Outbreak of *Ips confusus* killing pinyon pine (*Pinus edulis*) in southern Colorado 2003.

FIGURE 9.2 Egg galleries of *Ips nitidus*, QingHai, China.

species (Byers, 1989; Birgersson *et al.*, 2012). However, additional compounds such as lanierone, *E*-myrcenol or 2-methyl-3-buten-2-ol are synergistic with other pheromone components of the aggregation pheromone blends and can drastically increase trap capture (Birgersson *et al.*, 2012). These semiochemicals are, in part, produced as the detoxification by-product of the tree's secondary chemistry (i.e., monoterpenes) (Hughes, 1974). The beetles or their microbes may be responsible for the oxidation of the monoterpenes into ipsdienol and *cis*-verbenol (Brand *et al.*, 1975; Byers and Wood, 1981). However, these semiochemicals have been produced when the beetles were fed antibiotics and

when the beetles fed in non-host trees (Elkinton and Wood, 1980; Elkinton *et al.*, 1980; Seybold, 1992; Byers and Birgersson, 1990; Hunt and Borden, 1989). Potentially *de novo* production of these semiochemicals constituted an important part of pheromone production (Byers, 1989). Seybold *et al.* (1995) and Tillman *et al.* (1998) demonstrated feeding *I. paraconfusus* and *I. pini* individuals the *de novo* production of ipsenol and ipsdienol from acetate. The molecular basis for *de novo* synthesis involves the interaction of juvenile hormone III and HMG-CoA reductase in the mevalonate biosynthetic pathway to produce monoterpenoids, the precursors to ipsenol and ipsdienol (Martin *et al.* 2003; Seybold and Tittiger, 2003). Cytochromes P450 hydroxylate the monoterpene myrcene to produce ipsdienol (Sandstrom *et al.*, 2006). An enzyme produced by male *I. pini* likely converts ipsdienone into the final enantiomeric pheromone blend (Figueroa-Teran *et al.*, 2012). However, the molecular details of pheromone production are in need of additional study.

1.3 Taxonomic History

It is surprising that a genus with distinct diagnostic characters such as elytral declivity spines would have its name historically mired in confusion. DeGeer (1775) described *Ips* for seven scolytine species with *Dermestes typographus* (L.) as the first species listed. Fabricius (1776) then applied the name *Ips* to several beetle families excluding scolytines and also considered *Scolytus* (Geoffroy, 1762) a synonym of *Bostrichus* (Geoffroy, 1762). Marsham (1802) recognized *Ips* but did not designate a type species. Latreille (1802) described *Tomicus* (Latreille) for *Dermestes piniperda* (L.). In a later publication, without reason, Latreille included *Dermestes typographus* in *Tomicus* and did not recognize *Ips* (Latreille, 1806). Hence, the nomenclatural actions of Fabricius (1776) and Latreille (1802, 1806) apparently contributed to the dismissal of the name *Ips* from scolytine taxonomy and promoted the use of *Bostrichus* and *Tomicus* to describe scolytine species with elytral declivity spines throughout the 19th century (Table 9.1). Crotch (1870) implicitly and Bergroth (1884) explicitly designated *D. typographus* as the type for *Ips*. The correct application of *Ips* resumed near the end of the 19th century (e.g., Reitter, 1895). Swaine (1909) clearly articulated the priority of *Ips* over the use of *Bostrichus* and *Tomicus*, which helped to firmly establish the current use of the name.

Diagnosis of *Ips* from other Ipini genera was difficult prior to phylogenetic reconstruction. Few morphological characters vary among Ipini species and mostly quantitative differences occurred, thus making attempting to circumscribe "natural groups" (=monophyletic) difficult (Hopping, 1963a, b). Genera were diagnosed with a suite of variable characters; the combination and the importance of characters were the opinion of the

TABLE 9.1 *Ips* Species, Author, Year, and Original Genus Given in Chronological Order of Descriptions

Species	Author	Year	Original Genus
typographus	Linnaeus	1758	Dermestes
sexdentatus	Boerner	1767	Dermestes
calligraphus	Germar	1824	Tomicus
pini	Say	1826	Bostrichus
acuminatus	Gyllenhal	1827	Bostrichus
cembrae	Heer	1836	Bostrichus
duplicatus	Sahlberg	1836	Bostrichus
tridens	Mannerheim	1852	Bostrichus
subelongatus	Motschulsky	1860	Tomicus
avulsus	Eichhoff	1868	Tomicus
grandicollis	Eichhoff	1868	Tomicus
plastographus	LeConte	1868	Tomicus
cribricollis	Eichhoff	1869	Tomicus
integer	Eichhoff	1869	Tomicus
perturbatus	Eichhoff	1869	Tomicus
amitinus	Eichhoff	1872	Tomicus
confusus	LeConte	1876	Tomicus
emarginatus	LeConte	1876	Tomicus
montanus	Eichhoff	1881	Tomicus
hauseri	Reitter	1894	Ips
bonanseai	Hopkins	1905	Tomicus
stebbingi	Strohmeyer	1908	Ips
longifolia	Stebbing	1909	Tomicus
borealis	Swaine	1911	Ips
pilifrons	Swaine	1912	Ips
knausi	Swaine	1915	Ips
perroti	Swaine	1915	Ips
hunteri	Swaine	1917	Ips
lecontei	Swaine	1924	Ips
nitidus	Eggers	1933	Ips
woodi	Thatcher	1965	Ips
chinensis	Kurenzov and Kononov	1966	Ips
hoppingi	Lanier	1970	Ips
paraconfusus	Lanier	1970	Ips
schmutzenhoferi	Holzschuh	1988	Ips
apache	Lanier	1991	Ips
shangrila	Cognato and Sun	2007	Ips

taxonomic expert. Hence, the boundaries among genera were blurred and taxonomic confusion for some species persisted for nearly a hundred years. Two examples illustrate this issue. First, the placement of *Orthotomicus latidens* (LeConte) waivered several times between *Ips* and *Orthotomicus* (e.g., Swaine, 1909; Hopping, 1963a; S. L. Wood, 1982; Cognato and Vogler, 2001). Second, *Pseudips* Cognato was described for *Ips concinnus* (Mannerheim) and *I. mexicanus* (Hopkins) based on monophyly of the two species and its sister relationship to *Pityokteines* Fuchs (Cognato, 2000). Wood (2007) synonymized *Pseudips* with *Orthotomicus* Ferrari citing only a few morphological, behavioral, and cytoplasmic similarities. *Pseudips* was soon removed from synonymy (Alonso-Zarazaga and Lyal, 2009; Knížek 2011; Bright, 2014). Phylogenetic analyses using morphological and DNA sequence data helped to resolve these issues by explicitly testing the monophyly and phylogenetic placement of the genera (Cognato and Sperling, 2000; Cognato, 2000, 2013; Cognato and Vogler, 2001; Jordal and Cognato, 2012).

In light of a multiple gene phylogeny (Figure 9.3) (Cognato, 2013), *Ips* is delimited with a combination of antennal, declivital, and male genitalic characters (Swaine, 1918; Hopping, 1963a, b). The antennae consist of five funicular segments and an oval club. The antennal club is compressed and three sutures divide the anterior face. The sutures vary in shape from straight, bisinuate, to acutely angulate (Cognato, 2000). The elytral declivity originally defined the genus (DeGeer, 1775; Swaine, 1918) and is concave with spines along the lateral margins and ending in an expanded marginal apex. Female spines are reduced in size and spines never occur on the face of the declivity. The parts of the male genitalia include the median lobe, median struts, internal sac, tegmen, seminal trough, and the speculum gastrale. The seminal trough is variable in shape but consistently represented as two rods either held parallel or crossing each other and ending in an acute or broad tip (Cognato, 2000).

1.4 Subgenera

Hopping (1963a) divided the North American *Ips* into 10 species groups and included three Eurasian species. These "natural" (=monophyletic) groups based on antennal, elytral, and male genitalic characters were intended to include closely related species. Postner (1974) placed seven Eurasian species into two groups (Table 9.2). Lanier (1966) tested the validity of several of Hopping's groups through mating experiments and karyology. He also revised the relationships between the North America and Eurasian species (Lanier, 1972). S. L. Wood (1982) revised Hopping's species groups without explicitly citing diagnostic characters for each group. He recognized fewer species,

combined groups, and placed eight Eurasian species in relation to the North American species (Table 9.2). S. L. Wood's (1982) species groups were used as a proxy for a phylogeny of *Ips* and evolutionary inferences were made for all the species within the group based on biological data of one species (Seybold *et al.*, 1995). The obvious issue was that monophyly of these groups was never tested with modern phylogenetic analysis, thus casting doubt on the broad biological conclusions made for the species group.

Several consecutive phylogenetic studies tested the monophyly of these species groups (Cognato and Sperling, 2000; Cognato 2000, 2013; Cognato and Vogler, 2001; Cognato and Sun, 2007). These phylogenies consistently demonstrated that some groups (1) belonged to different genera or groups, (2) were monophyletic, and (3) were paraphyletic (Table 9.2). Consequently, Cognato and Vogler (2001) erected four subgenera for the larger and well-supported clades and placed the subclassification of *Ips* species in the context of monophyly (Figure 9.4). Thus, biological inferences made for a species can potentially be applied for the remaining species within the subgenus. This classification has been recognized (Alonso-Zarazaga and Lyal, 2009), except for Knížek (2011) who did not give a reason for synonymizing these subgenera within *Ips*. Given the evidence of monophyly and the need to unite taxonomic groups with evolutionary lineages, this chapter will recognize these subgenera.

2. HISTORICAL PERSPECTIVE OF SPECIES TAXONOMY

Delimitation of *Ips* species was practiced without an explicit species concept until the late 1960s. Prior to this date, the majority of *Ips* species were described based on unique combination of morphological characters such as setae on the elytral interstriae and the granulation of the frons (Swaine, 1909; Hopping, 1963a). Lanier (1966), in the context of the biological species concept (Mayr, 1963), explicitly tested for post-mating barriers among perceived closely related species. He repeated these interspecific mating experiments for the species of several of Hopping's species groups (Table 9.2) and as a result new species were described and previous synonymies were supported (Lanier, 1970a, b, 1972, 1987; Lanier *et al.*, 1991). Cognato and colleagues applied an evolutionary lineage species concept (Hey *et al.*, 2003) to delimit *Ips* species (Cognato and Sperling, 2000; Cognato and Sun, 2007). Given the criteria of monophyly and diagnostic characters, they explicitly described a new species and implicitly tested the validity of 23 species (Cognato and Sun, 2007). The validity of nearly all 23 species was supported except for the North American spruce-feeding *Ips* species and revision

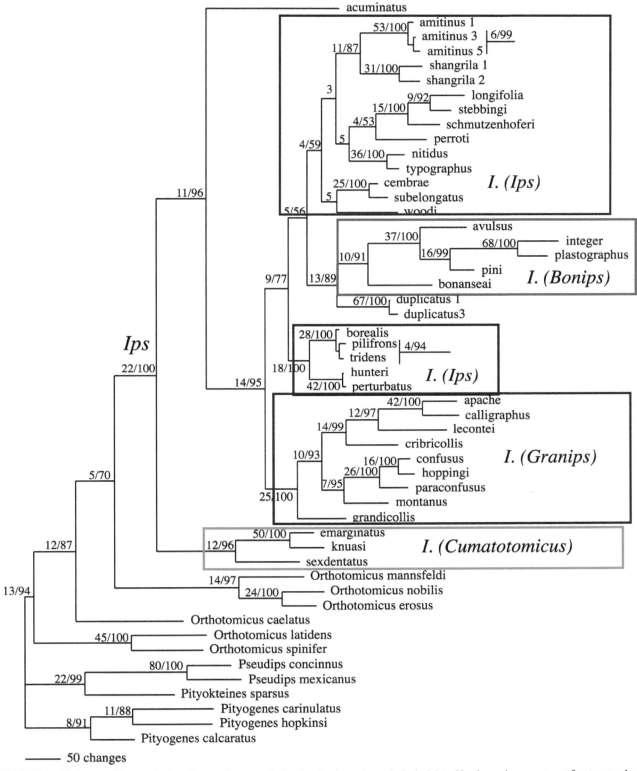

FIGURE 9.3 Phylogeny of *Ips* species based on parsimony analysis of molecular and morphological data. Numbers = bremer support/bootstrap values *(after Cognato and Sun, 2007).*

TABLE 9.2 Various *Ips* Species Groups. Some Species were Synonymized by Subsequent Authors

Natural Groups	Natural Groups	Species Groups	Species Groups	Monophyletic groups	Monophyletic Subgenera
Hopping, 1963a, b, c, 1964, 1965a, b, c, d, e	Lanier, 1970a, b, 1972, 1987; Lanier *et al.*, 1991	Postner, 1974	S. L. Wood, 1982	Cognato and Sperling, 2000, Cognato, 2000	Cognato and Vogler, 2001, Cognato and Sun, 2007
Group I—*I. concinnus* (Mannerheim), I. mexicanus (Hopkins).	Group I—*I. concinnus, I. mexicanus.*	not addressed	*concinnus* group—*I. concinnus, I. mexicanus*	*concinnus* group— moved to *Pseudips*	*Pseudips concinnus, P. mexicanus*
Group II—*I. emarginatus, I. knausi.*	Group II—*I. emarginatus, I. knausi.*	*acuminatus* group— *I. acuminatus, I. duplicatus, I. mannsfeldi*	*emarginatus* group—*I. emarginatus, I. knausi, I. acuminatus*	*emarginatus* group—*I. emarginatus, I. knausi.* Monophyletic.	**Ips (Cumatotomicus) Ferrari** *I. emarginatus, I. sexdentatus I. knausi*
Group III—*I. plastographus.*	Group III—*I. plastographus, I. integer* (Eichhoff).	*typographus* group—*I. typographus*, I. sexdentatus, I. amitinus, I. cembrae*	*plastographus* group—*I. plastographus, I. integer, I. typographus*	*plastographus* group—*I. plastographus, I. integer.* Monophyletic with *pini* group.	**Ips (Bonips) Cognato** *I. avulsus, I. bonanseai, I. pini I. integer, I. plastographus*
Group IV—*I. pini, I. avulsus, I. bonanseai, I. duplicatus.*	Group IV—*I. pini, I. avulsus, I. bonanseai.*	*Ips mannsfeldi* in *acuminatus* group, other species not addressed	*pini* group—*I. pini, I. avulsus, I. bonanseai, I. mannsfeldi* Wachl.	*pini* group—*I. pini, I. avulsus, I. bonanseai.* Paraphyletic.	**Ips (Ips) DeGeer** *I. amitinus, I. borealis, I. cembrae, I. duplicatus I. hunteri, I. longifolia, I. nitidus I. perroti, I. perturbatus, I. pilifrons, I. schmutzenhoferi, I. shangrila, I. stebbingi, I. subelongatus, I. tridens, I. typographus, I. woodi*
Group V—*I. perroti, I. amitinus.*	not addressed	not addressed	Combined with the *perturbatus* group		
Group VI—*I. perturbatus, I. hunteri, I. utahensis* Wood, *I. woodi, I. typographus, I. cembrae.*	not addressed	*Ips cembrae* and *I. amitinus* in *typographus* group, and *I. duplicatus* in *acuminatus* group, other species not addressed	*perturbatus* group—*I. perturbatus, I. hunteri, I. woodi, I. perroti, I. cembrae, I. amitinus, I. duplicatus*	*perturbatus* group— *I. perturbatus, I. hunteri.* Monophyletic with *tridens* group.	**Ips (Granips) Cognato** *I. apache, I. calligraphus, I. confusus, I. cribricollis, I. grandicollis, I. hoppingi, I. lecontei, I. montanus, I. paraconfusus*
Group VII—*I. borealis, I. swainei* R. Hopping, *I. thomasi* G. Hopping.	not addressed	not addressed	Combined with the *tridens* group		

TABLE 9.2 Various *Ips* Species Groups. Some Species were Synonymized by Subsequent Authors—cont'd

Natural Groups	Natural Groups	Species Groups	Species Groups	Monophyletic groups	Monophyletic Subgenera
Group VIII—*I. tridens*, *I. yohoensis* Swaine, *I. pilifrons*, *I. interruptus* (Eichhoff), *I. sulcifrons* Wood, *I. semirostris* G. Hopping, *I. amiskwiensis* G. Hopping, *I. engelmanni* Swaine.	not addressed	not addressed	*tridens* group—*I. tridens*, *I. pilifrons*, *I. borealis*	*tridens* group—*I. tridens*, *I. pilifrons*, *I. borealis*. Monophyletic with *perturbatus* group.	**Incertae sedis** *I. acuminatus*, *I. chinensis*, *I. duplicatus*, *I. hauseri*
Group IX—*I. confusus*, *I. lecontei*, *I. montanus*, *I. cribricollis*, *I. grandicollis*.	Group IX—*I. paraconfusus*, *I. confusus*, *I. hoppingi*, *I. lecontei*, *I. montanus*, *I. cribricollis*, *I. grandicollis*.	not addressed	*grandicollis* group—*I. paraconfusus*, *I. confusus*, *I. hoppingi*, *I. lecontei*, *I. montanus*, *I. grandicollis*	*grandicollis* group—*I. paraconfusus*, *I. confusus*, *I. hoppingi*, *I. lecontei*, *I. montanus*, *I. grandicollis*, *I. cribricollis*. Paraphyletic.	
Group X—*I. calligraphus*, *I. interstitialis* (Eichhoff), *I. ponderosae* Swaine, *I. sexdentatus*.	Group X—*I. calligraphus*, *I. apache*.	*Ips sexdentatus* in *typographus* group, other species not addressed.	*calligraphus* group—*I. calligraphus*, *I. sexdentatus*	*calligraphus* group—*I. calligraphus*, *I. apache*. Monophyletic with *grandicollis* group.	
latidens group in *Orthotomicus*	*latidens* group in *Orthotomicus*	*latidens* group not addressed	*latidens* group—*I. latidens* (LeConte), *I. spinifer* (Eichhoff)	*latidens* group—*I. latidens*, *I. spinifer*. Monophyletic	*I. latidens*, *I. spinifer*, *I. mannsifeldi*, *I. nobilis* moved to *Orthotomicus*.

of the group was suggested. Also, Cognato and Sun (2007) identified two species, *Ips cribricollis* (Eichhoff) and *I. shangrila* Cognato and Sun, with >6% intraspecific mitochondrial cytochrome oxidase I (COI) nucleotide difference that warranted further taxonomic investigation.

3. PHYLOGENETICS AND POPULATION GENETICS

As mentioned throughout this chapter, phylogenetic analysis of *Ips* species provided a quantitative means for assessing generic and specific relationships, which helped to solve taxonomic problems. Early phylogenetic analyses were mostly based on molecular data for a limited sample of species (Cane *et al.*, 1990b; Stauffer *et al.*, 1997; Cognato *et al.*, 1995). Still, these phylogenies were useful in elucidating the evolution of *Ips* biology (Cognato *et al.*, 1997) and many of the species relationships proposed were supported by more comprehensive datasets (e.g., Cognato and Sperling, 2000). Multiple gene and morphology derived

phylogenies including most *Ips* species supported the monophyly of (1) *Ips*, *Orthotomicus*, *Pityogenes*, and *Pityokteines*, (2) *Ips* and *Orthotomicus*, (3) *Ips*, (4) subgenera, and (5) most *Ips* species (Figures 9.3 and 9.4) (Cognato and Vogler, 2001; Cognato and Sun, 2007). Also, the sister relationships between *Ips* (*Cumatotomicus*) and *I. acuminatus* and the group containing *I. acuminatus* and the remaining *Ips* species were well supported (Cognato, 2013). However, the phylogenetic relationships among the remaining subgenera and the species relationships within *I.* (*Ips*) were unresolved or poorly supported (Figure 9.1) (Cognato and Vogler, 2001; Cognato and Sun, 2007). These results suggest that the *I.* (*Ips*) species experienced a relatively rapid radiation that associated with few characters, which would allow for the resolution of these interspecific relationships. The augmentation of the existing datasets with additional DNA sequence data, especially from the nuclear genome, will hopefully result in a fully resolved and well-supported *Ips* phylogeny.

Population genetics of only a few *Ips* species have been studied. Allozymes were first used to characterize the

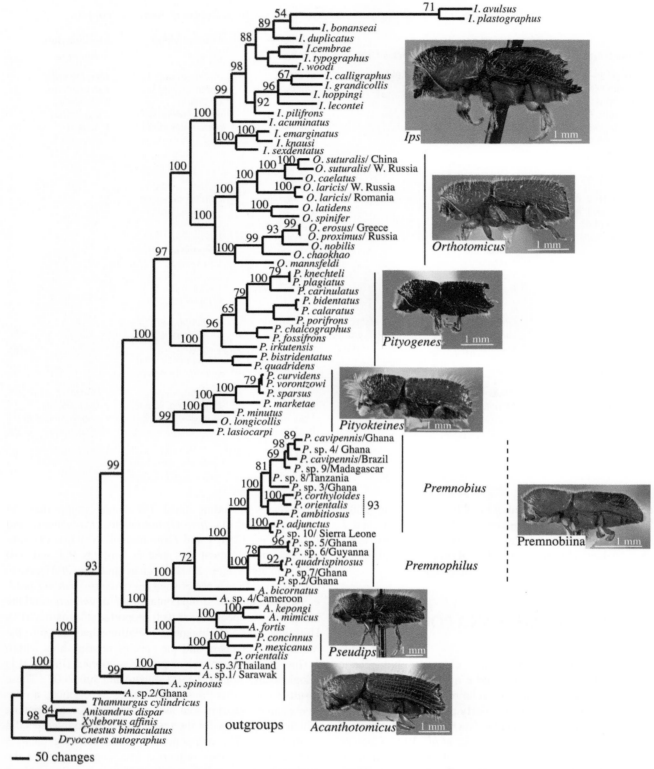

FIGURE 9.4 Phylogeny of Ipini genera based on a Bayesian analysis of nucleotides from five genes. Numbers are posterior probabilities *(after Cognato, 2013)*.

genetic structure of *I. typographus* populations in Central Europe (Stauffer *et al.*, 1992). These data suggested much gene flow among six proximal populations. More detailed studies, which incorporated allozyme and mitochondrial DNA (mtDNA) sequence data for populations distributed throughout Europe, demonstrated that *I. typographus* dispersed from southern and northeastern Europe from Pleistocene glacial refugia to repopulate northern and western Europe (Stauffer *et al.*, 1999; Krascsenitsová *et al.*, 2013). Additional studies concluded there was high gene flow among the European populations (Sallé *et al.*, 2007; Bertheau *et al.*, 2013).

Population genetic structures based on mtDNA for two North American *Ips* species have been conducted. Populations of transcontinental species, *I. pini*, were phylogenetically structured and associated with pheromone races that occurred in eastern, Rocky Mountain, and western regions of North America (Cognato *et al.*, 1999). Individuals of the Eastern lineage occurred in western North America but individuals of the Rocky Mountain and Western lineages were absent from eastern North America. These results reflected the incomplete pre-mating barriers known for *I. pini* pheromone races and suggested that female-controlled assortative mating mediated directional gene flow from east to west. A similar east/west structure was observed with mtDNA phylogeny of *I. confusus*, a species that occurs in southwestern USA and feeds on two species of pinyon pine tree (Cognato *et al.*, 2003). There was little association between monophyletic groups and host tree. Isolation by distance due to Pleistocene habitat fragmentation better explained the pattern of haplotype distribution among localities (Cognato *et al.*, 2003).

4. EVOLUTION

Although the evolution of *Ips* species diversification has yet to be specifically addressed (Cognato and Jordal, in prep.), limited observations are possible given the published phylogenies (Jordal and Cognato, 2012; Cognato, 2013). First, the clade of Pineceae-feeding Ipini containing *Pityokteines*, *Pityogenes*, *Ips*, and *Orthotomicus* derived about ≈45 million years ago (mya) from angiosperm feeding Ipini (Jordal and Cognato, 2012). During this period of time, the early Eocene, the world was mostly tropical (Sloan and Rea, 1995). Angiosperms dominated most habitats and conifers, including pines, were dispersed in fragmented populations (Keeley, 2012). This patchiness of habitats likely placed tropical angiosperms and Pinaceae in close proximity and presented early Ipini species opportunities for host switches. Earth cooled throughout the Oligocene and pines returned to the mid-latitudes and increased in abundance even during the relatively brief "Antarctic thawing" period, which lasted ca. 10 million years (Millar, 1993). During this time (20 mya), conifers expanded into the high latitudes, which may have contributed to the cladogenesis of *Ips*. Diversification of *Ips* (*Ips*) species likely occurred ≈10 mya during a period of the diversification of pine in drier, fire-prone habitats (Keeley, 2012).

Most evolutionary investigations have focused on pheromones given their importance in pest management and as a model of the evolution of communication systems. Ecological studies showed that bark beetle communities segregate based on the production and response to different pheromone blends (Lewis and Cane, 1990). For example, in western US pine forests, *I. paraconfusus*, *I. pini*, *Orthotomicus latidens*, *Dendroctonus brevicomis* LeConte, and *D. valens* LeConte often infest different parts of the same tree. Differential production of and the response to ipsenol and ipsdienol reduce the interspecific competition among these species (D. L. Wood, 1982; Byers, 1989). Also, predator beetles exhibit preferences to certain semiochemicals and track variation in the production of pheromone through space and time (Miller *et al.*, 1997, Raffa and Klepzig, 1989; Costa and Reeve, 2011). Thus, interspecific competition and predator attraction in part contributes to pheromone diversity (D. L. Wood, 1982).

Phylogenetic inertia also influences pheromone evolution. Indirect evidence from ecological studies demonstrated that allopatric sibling species (e.g., *I. paraconfusus*, *I. confusus*, *I. hoppingi* Lanier) produced and were attracted to similar pheromones (Cane *et al.* 1990a, c). Cognato *et al.* (1997) first addressed pheromone evolution concerning these species in a phylogenetic context. Their results demonstrated that these sibling species produced similar pheromone bouquets, although some of the semiochemicals varied.

Pheromone evolution likely results from changes in the production and/or response in a few semiochemicals. Cognato (1998) demonstrated that two important pheromones, ipsenol and ipsdienol, changed only a few times within an *Ips* phylogeny that represented 90% of the species. Both of these semiochemicals were produced by the sister taxa (*Pityogenes*, *Pityokteines*, *Pseudips*, *Orthotomicus*) and most of these changes were due to losses in the production/response of one of the semiochemicals. These changes occurred throughout the phylogeny, thus the production/response to these semiochemicals was not historically constrained. Phylogenetic constraints and ecological selection pressures influence pheromone evolution among *Ips* species, although these studies were limited in scope (Cognato *et al.* 1997; Cognato, 1998).

In a subsequent study, Symonds and Egar (2004a) concluded that pheromone evolution was caused by saltational shifts in semiochemicals, which lead to substantial pheromonal differences among sibling species (in actuality, the comparisons were of putative sister species). However, their analyses and conclusions were made based on a taxonomically culled *Ips* phylogeny representing only half of the species. All of their comparisons were not based on known sister-species relationships (Cognato and Sperling, 2000; Cognato and Vogler, 2001) thus substantial semiochemical differences were expected. Given this methodological issue, the conclusions of the Symonds and Egar (2004a, b) publications require reevaluation.

Other evolutionary studies using *Ips* phylogenies addressed the co-evolution of beetle associated mites and nematodes. Uropodoid mites, which are external phoretic parasites, showed little evidence of co-phylogenetics with *Ips* species (Knee *et al.*, 2012). However, a clade of mites restricted to ambrosia-feeding scolytines suggested that beetle-feeding habit might influence the diversification of these mites. On the contrary, *Micoletzkya* nematode species exhibited a pattern of co-phylogenetics and host switching with sympatric *Ips* species (Susoy and Herrmann, 2014). Potentially, the internal habit of this parasitic nematode promoted co-phylogenetics at deeper phylogenetic levels, which is a hypothesis to be tested with a larger sample of *Ips* species and their nematodes.

5. ANNOTATED LIST OF *IPS* SPECIES

Ips species are difficult to identify. A combination of frontal fovea, granules, punctures and tubercles, punctation and setae of the elytral disk and sculpture of the elytral declivity are diagnostic for many species. However, intraspecific variation often occurs within these characters. Some closely related species may only be identified by cryptic morphology such as the par stridens (Lanier, 1970a). S. L. Wood's (1982) key to North American *Ips* is effective when used with a reference collection but its species coverage is incomplete. Pfeffer (1995) and Yin *et al.* (1984) provide keys in German and Chinese to the Eurasian species but their taxonomy is dated. Buhroo and Lakatos (2011) include a morphological key to three Himalayan species; however, the cited diagnostic characters are inconsistent among the species. At present, DNA-based identification provides an important means for verifying species determinations and should be used to verify all questionable identities (Cognato and Sun, 2007). A morphological key to the world of *Ips* species is being developed (Douglas and Cognato, in prep.). The intention of this section is to summarize from the original descriptions and other publications (i.e., S. L. Wood, 1982; Pfeffer, 1995) the diagnostic characters between sister species and/or morphologically similar species. Synonymized species are list under valid species, which are in bold. Distribution data was gleaned from Wood and Bright (1992), Bright and Skidmore (1997, 2002), Knížek (2011), and Bright (2014).

FIGURE 9.5 Lateral habitus of *Ips* species. (A) *I. acuminatus*, (B) *I. amitinus*, (C) *I. apache*, and (D) *I. pini*.

Ips acuminatus (Gyllenhal), 1827 (Figure 9.5A)
Bostrichus geminatus Zetterstedt
Tomicus heydeni Eichhoff

Distribution: Austria. Belgium. Bosnia-Herzegovina. ulgaria. Belarus. Croatia. China: Fujian, Gansu, Hebei, Heilongjiang, Henan, Hubei, Inner Mongolia, Jilin, Liaoning, Qinghai, Sichuan, Shaanxi, Shandong, Shanxi, Xinjiang. Czech Republic. Denmark. Estonia. Finland. France. Germany. Great Britain. Greece. Hungary. Ireland. Italy. Japan. Kazakhstan. Latvia. Lithuania. Luxembourg. Macedonia. Moldavia. Mongolia. Montenegro. The Netherlands. Norway. North Korea. Poland. Romania.

Russia: ubiquitous. Serbia. South Korea. Slovakia. Slovenia. Spain. Sweden. Switzerland. Syria. Turkey. Taiwan. Ukraine.

Principal hosts: *Pinus* spp.

Diagnosis: This species is distinguished from all other European *Ips* spp. by the elytral declivity that has three spines, of which the third spine is flattened and acuminate in the male. It differs from the North American species *I. emarginatus* and *I. knausi* by its smaller size, 2.2–3.9 mm. This species differs from its sister species *I. chinensis* by the separation of the bases of spines 2 and 3.

Ips amitinus (Eichhoff), 1872 (Figure 9.5B)
Ips montana Fuchs

Distribution: Austria. Belgium. Bosnia-Herzegovina. Bulgaria. Croatia. Czech Republic. Denmark. Estonia. Finland. France. Germany. Greece. Hungary. Italy. Latvia. Lithuania. Macedonia. Montenegro. The Netherlands. Poland. Romania. Russia: western. Serbia. Slovenia. Switzerland. Turkey.

Principal hosts: *Picea* spp. and *Pinus* spp.

Diagnosis: There are four spines on elytral declivity. This species is distinguished from all other Eurasian *Ips* spp. by the straight antennal club sutures. It differs from the morphologically similar North American species *I. perroti* by its larger size, 3.5–4.8 mm.

Ips apache Lanier, 1991 (Figure 9.5C)

Distribution: Guatemala. Honduras. Mexico: Chiapas, Hidalgo, Jalisco, Mexico, Michoacán, Nuevo León, Sinaloa. Nicaragua. Panama (introduced). USA: southeastern Arizona.

Principal hosts: *Pinus* spp.

Diagnosis: There are six spines on the elytral declivity. Distinguished from other *Ips* spp. by the presence of three spines after the third declivity spine. Potentially allopatric in Arizona and northern Mexico with related and morphologically similar species *I. calligraphus*. Differs by the distance between the ridges of the par stridens and the smaller size and pronotal width 1.615 ± 0.021 mm (Lanier *et al.*, 1991).

Ips avulsus (Eichhoff), 1869

Distribution: USA: Alabama, Arkansas, Florida, Georgia, Louisiana, Maryland, Mississippi, New Jersey, North Carolina, Oklahoma, Pennsylvania, South Carolina, Texas, Virginia, West Virginia.

Principal hosts: *Pinus* spp.

Diagnosis: There are four spines on the elytral declivity and its general appearance is similar to *I. pini* (Figure 9.5D). Potentially allopatric in the northern extent of its distribution with related and morphologically similar species *I. pini*. Differs by the noncapitate spine 3 of the male declivity, the short expansion of the declivital apex, and the smaller size, 2.1–2.8 mm (S. L. Wood, 1982).

Ips bonanseai (Hopkins), 1905

Distribution: Guatemala. Honduras. Mexico: Chiapas, Chihuahua, Distrito Federal, Durango, Hidalgo, Jalisco, Mexico, Michoacán, Morelos, Nuevo León, Oaxaca, Puebla, Tamaulipas, Tlaxcala, Veracruz, Zacatecas. USA: southeastern Arizona.

Principal hosts: *Pinus spp.*

Diagnosis: There are four spines on the elytral declivity and its general appearance is similar to *I. pini* (Figure 9.5D). Potentially allopatric in Arizona and northern Mexico with related and morphologically similar species *I. pini*. Differs by the median frontal tubercle connected to the epistomal tubercle, and the smaller size, 2.9–3.4 mm (S. L. Wood, 1982).

Ips borealis Swaine, 1911 (Figure 9.6A)

Distribution: Canada: Alberta, British Columbia, Manitoba, New Brunswick, Newfoundland, Northwest Territories, Nova Scotia, Ontario, Prince Edward Island, Quebec, Saskatchewan. USA: Colorado, Maine, Minnesota, Montana, South Dakota.

Principal hosts: *Picea spp.*

Diagnosis: There are four spines on the elytral declivity. Potentially allopatric with related species *I. tridens*, *I. pilifrons*, *I. perturbatus* and morphologically similar species *I. pini*. It differs from the related species by the even minute punctures on the upper female frons, and smaller size, 2.6–3.8 mm and from *I. pini* by the uniseriately punctured discal interstriae.

Ips calligraphus (Germar), 1828
Bostrichus exesus Say
Tomicus praemorsus Eichhoff
Tomicus interstitialis Eichhoff
Ips ponderosae Swaine

Distribution: Bahama Islands. Canada: Ontario, Quebec. Dominican Republic. Mexico: Hidalgo, Jalisco, Nuevo León. Philippines (introduced). Jamaica. USA: Alabama, Arizona, Arkansas, California, Colorado, Delaware, Florida, Georgia, Indiana, Louisiana, Maryland, Massachusetts, Michigan, Minnesota, Mississippi, Missouri, Montana, Nebraska, New Hampshire, New Jersey, New Mexico, New York, North Carolina, Ohio, Oklahoma, Pennsylvania, South Carolina, South Dakota, Tennessee, Texas, Utah, Virginia, West Virginia, Wisconsin, Wyoming.

Principal hosts: *Pinus spp.*

Diagnosis: There are six spines on the elytral declivity and its general appearance is similar to *I. apache* (Figure 9.5C). Distinguished from other *Ips* spp. by the presence of three spines after the third declivity spine. Potentially allopatric in Arizona and northern Mexico with related and

FIGURE 9.6 **Lateral habitus of *Ips* species.** (A) *I. borealis*, (B) *I. typographus*, (C) *I. confusus*, and (D) *I. duplicatus*.

morphologically similar species *I. apache*. It differs by the distance between the ridges of the par stridens and the larger size, and pronotal width 2.033 ± 0.018 mm (Lanier *et al.*, 1991).

Ips cembrae (Heer), 1836
  Ips engadinensis Fuchs
  Ips fallax Eggers
  Ips shinanoensis Yano

Distribution: Austria. China: Heilongjiang, Jilin. Czech Republic. Denmark. France. Germany. Great Britain.

Greece. Hungary. Italy. Kazakhstan. Liechtenstein. Mongolia. The Netherlands. Poland. Russia: western and eastern. Slovakia. Slovenia. South Korea. Switzerland.
Principal hosts: *Larix* spp.
Diagnosis: There are four spines on the elytral declivity and its general appearance is similar to *I. typographus* (Figure 9.6B). This species is distinguished from *I. typographus* by the shiny elytral declivity and interstrial punctures of the elytral disk. It differs from the morphologically similar North American *Picea*-feeding species and *I. woodi*, by the space between the first and second spines, which is less than the length of the first spine. It differs from its sister-species *I. subelongatus* by the less setose elytral declivity but these species are best diagnosed with DNA data.

Ips chinensis Kurenzov and Kononov, 1966

Distribution: China: Yunnan. Laos. Thailand.
Principal hosts: *Pinus* spp.
Diagnosis: This species is distinguished by the allopatric species *I. sexdentatus* by its three elytral declivital spines. This species differs from its sister species *I. acuminatus* (Figure 9.5A) by the tumescent that joins spines 2 and 3.

Ips cribricollis (Eichhoff), 1869
  Ips cloudcrofti Swaine

Distribution: Dominican Republic. Guatemala. Honduras. Mexico: Chiapas, Hidalgo, Jalisco, Mexico, Michoacán, Nuevo León, Sinaloa. Nicaragua. Jamaica. USA: southern New Mexico.
Principal hosts: *Pinus spp*.
Diagnosis: There are five spines on the elytral declivity and its general appearance is similar to *I. confusus* (Figure 9.6C). Potentially allopatric in Central America and Mexico with related and morphologically similar species *I. grandicollis*. Differs by the width of the par stridens and the absence of a fovea on the male frons (Lanier, 1987).

Ips confusus (LeConte), 1876 (Figure 9.6C)

Distribution: Mexico: Baja California Norte. USA: Arizona, California, Colorado, Idaho, Nevada, New Mexico, Utah, Wyoming.
Principal hosts: *Pinus edulis* Engelm. and *P. monophylla* Torr. and Frém.
Diagnosis: There are five spines on the elytral declivity. Sibling species to *I. hoppingi*, which is diagnosable by the distance between the ridges of the par stridens (Lanier, 1970b) and molecular phylogenetics (Cognato and Sun, 2007). These species are potentially allopatric in southeastern Arizona. Allopatric with *I. paraconfusus* in eastern and southern California. *Ips confusus* differs by the frons devoid of punctures and distance between the ridges of the par stridens.

Ips duplicatus (Sahlberg), 1836 (Figure 9.6D)

Tomicus rectangulus Eichhoff
Bostrichus judeichii Kirsch
Tomicus infucatus Eichhoff
Tomicus infucatus Eichhoff

Distribution: Austria. Belarus. China: Heilongjiang, Inner Mongolia, Jilin, Liaoning, Sichuan, Xinjiang. Czech Republic. Estonia. Finland. France. Germany. Hungary. Japan. Kazakhstan. Latvia. Lithuania. Mongolia. The Netherlands. North Korea. Poland. Russia: ubiquitous. South Korea. Slovakia. Slovenia. Syria. Turkey. Taiwan. Ukraine.
Principal hosts: *Picea* spp.
Diagnosis: There are four spines on the elytral declivity. This species is distinguished from all other European *Ips* spp. by the position of the first spine of the elytral declivity, which is closer to the elytral suture. It differs from the morphologically similar Himalayan species, North American *Picea*-feeding species and *I. woodi*, by the sparsely granulate frons. This species differs from its sister species *I. hauseri* by the close proximity of the bases of spines 2 and 3 (less than the distance between the first and second spines).

Ips emarginatus (LeConte), 1876 (Figure 9.7A)

Distribution: Canada: British Columbia. Mexico: Baja California. USA: California, Idaho, Montana, Oregon, Washington.
Principal hosts: *Pinus jeffreyi* Balf., *P. monticola* Douglas ex D. Don, *P. ponderosa* P. Lawson and C. Lawson.

FIGURE 9.7 Lateral habitus of *Ips* species. (A) *I. emarginatus* and (B) *I. sexdentatus*.

Diagnosis: There are three spines on the elytral declivity. This species distinguished from other *Ips* by its large size (5.5–7.0 mm) and large flatten and emarginated third spine. Sibling species to *I. knausi*, which is diagnosable by the absence of interstrial punctures and the fourth declivital spine.

Ips grandicollis (Eichhoff), 1868

Tomicus cacographus LeConte
Ips chagnoni Swain

Distribution: Australia (introduced). Canada: Manitoba, Nova Scotia, Ontario, Quebec. Honduras. Mexico: Colima, Guerrero, Hidalgo, Jalisco, Nuevo León, Veracruz. USA: Alabama, Arkansas, Connecticut, Delaware, Florida, Georgia, Illinois, Indiana, Louisiana, Maine, Maryland, Massachusetts, Michigan, Minnesota, Mississippi, Missouri, Nebraska, New Hampshire, New Jersey, New York, North Carolina, Ohio, Oklahoma, Pennsylvania, South Carolina, Tennessee, Texas, Utah, Virginia, West Virginia, Wisconsin.
Principal hosts: *Pinus* spp.
Diagnosis: There are five spines on the elytral declivity and its general appearance is similar to *I. confusus* (Figure 9.6C). Potentially allopatric in Central America and Mexico with related and morphologically similar species *I. cribricollis*. Differs by the width of the par stridens and the presence of a fovea on the male frons (Lanier, 1987).

Ips hauseri (Reitter), 1894

Ips ussuriensis Reitter, 1913

Distribution: China: Jilin, Xinjiang. Kyrgyzstan. Kazakhstan. Russia: eastern. Tajikistan. Turkey.
Principal hosts: *Picea* spp.
Diagnosis: There are four spines on the elytral declivity and its general appearance is similar to *I. duplicatus* (Figure 9.6D). This species is distinguished from all other European *Ips* spp. by the position of first spine of the elytral declivity, which is closer to the elytral suture. It differs from the morphologically similar Himalayan species, North American *Picea*-feeding species and *I. woodi* by the sparsely granulate frons. This species differs from its sister species *I. duplicatus* by the separation of the bases of spines 2 and 3 (nearly equal to the distance between the first and second spines).

Ips hoppingi Lanier, 1970b

Distribution: Mexico: Chihuahua, Hidalgo. USA: southeastern Arizona, Western Texas.
Principal hosts: Pinyon pines including *Pinus cembroides* Zucc. and *P. discolor* D. K. Bailey and Hawksw.
Diagnosis: There are five spines on the elytral declivity. Sibling species to *I. confusus* (Figure 9.6C) from which it is diagnosed by the distance between the ridges of the par

stridens (Lanier, 1970b) or molecular phylogenetics (Cognato and Sun, 2007). These species are potentially allopatric in southeastern Arizona.

Ips hunteri Swaine, 1917

Distribution: USA: Arizona, Colorado, Utah.
Principal hosts: *Picea pungens* Engelm.
Diagnosis: There are four spines on the elytral declivity and its general appearance is similar to *I. borealis* (Figure 9.6A). Potentially allopatric with related species *I. pilifrons*, *I. borealis* and morphologically similar species *I. pini*. It differs from the related species by lack of frontal sexual dimorphism and from *I. pini* by uniseriately punctured discal interstriae.

Ips integer (Eichhoff), 1869

Distribution: Canada: British Columbia. Guatemala. Mexico: Chiapas, Chihuahua, Colima, Distrito Federal, Durango, Guerrero, Hidalgo, Jalisco, Mexico, Michoacán, Morelos, Querétaro, Tamaulipas, Veracruz, Zacatecas. USA: Arizona, California, Idaho, Montana, New Mexico, Oregon, South Dakota, Utah, Washington.
Principal hosts: *Pinus* spp.
Diagnosis: There are four spines on the elytral declivity and its general appearance is similar to *I. pini* (Figure 9.5D). Sibling species to *I. plastographus*, which is diagnosable by the connection of the median epistomal and frontal tubercles by a carinate elevation or molecular phylogenetics (Cognato and Sun 2007). These species are potentially allopatric in the Pacific Northwest of North America. However, *I. plastographus* is mostly restricted to two hosts, *P. contorta* and *P. muricata*.

Ips knausi Swaine, 1915

Distribution: USA: Arizona, Colorado, Nevada, New Mexico, South Dakota, Utah.
Principal hosts: *Pinus ponderosa*
Diagnosis: There are four spines on the elytral declivity. This species is distinguished from other *Ips* by its large size (5.0–6.5 mm) and large flatten and emarginated third spine. Sibling species to *I. emarginatus* (Figure 9.7A), which is diagnosable by the presence of interstrial punctures and the fourth declivital spine.

Ips lecontei Swaine, 1924

Distribution: Honduras. Mexico: Chiapas, Chihuahua, Colima, Jalisco, Oaxaca, Sonora. USA: Arizona, New Mexico.
Principal hosts: *Pinus ponderosa*, *P. pseudostrobus* Lindl.
Diagnosis: There are five spines on the elytral declivity and its general appearance is similar to *I. confusus* (Figure 9.6C). Potentially allopatric with morphologically similar species *I. confusus*, *I. cribricollis*, and *I. hoppingi*. This species is distinguished from all other species with five declivital spines by the presence of a pair of medial frontal tubercles on the epistoma.

Ips longifolia (Stebbing), 1909

Distribution: Bhutan. China: Xinjiang. India: Himachal Pradesh, Uttar Pradesh. Nepal. Pakistan.
Principal hosts: *Pinus roxburghii* Sarg.
Diagnosis: There are four spines on the elytral declivity and its general appearance is similar to *I. borealis* (Figure 9.6E). Morphological characteristics are nearly indistinguishable from sister species *I. stebbingi*, the North American *Picea* feeding *Ips* and *I. woodi*. DNA-based diagnosis is recommended.

Ips montanus (Eichhoff), 1881
Ips vancouveri Swaine

Distribution: Canada: Southern British Columbia. USA: northern California, Idaho, Montana, Oregon, Washington.
Principal host: *Pinus monticola*
Diagnosis: There are five spines on the elytral declivity and its general appearance is similar to *I. confusus* (Figure 9.6C). Potentially allopatric with *I. paraconfusus* in northern California and Oregon. Differs by the absence of the frontal fovea; the male major medial frontal tubercle displaced from the epistoma; and larger, 4.6–5.4 mm (S. L. Wood, 1982).

Ips nitidus Eggers, 1923

Distribution: China: Gansu, Qinghai, Sichuan, Xingjiang, Yunnan.
Principal hosts: *Picea* spp.
Diagnosis: There are four spines on the elytral declivity and its general appearance is similar to *I. typographus* (Figure 9.6B). This species is distinguished from allopatric *I. shangrila* by the capitate third declivital spine. Morphologically it differs from its sister species *I. typographus* by the deeper frontal fovea, shiny elytral declivity. However, these characters may vary and DNA-based diagnosis is necessary to confirm identity.

Ips paraconfusus Lanier, 1970b

Distribution: USA: California, western Nevada, western Oregon.
Principal hosts: *Pinus attenuate* Lemmon, *P. coulteri* Lamb. ex D. Don, *P. jeffreyi*, *P. lambertiana* Douglas, and *P. ponderosa*.
Diagnosis: There are five spines on the elytral declivity and its general appearance is similar to *I. confusus* (Figure 9.6C). Central frons with a few small punctures and additional difference among related species *I. confusus*, *I. hoppingi*, and *I. montanus* are given above.

Ips perroti Swaine, 1915

Distribution: Canada: Alberta, Manitoba, New Brunswick, Ontario, Quebec. USA: Michigan, Minnesota.
Principal hosts: *Pinus banksiana* Lamb. and *P. resinosa* Aiton.
Diagnosis: There are four spines on the elytral declivity and its general appearance is similar to *I. amitinus*

(Figure 9.5B). Potentially allopatric with morphologically similar species *I. borealis*, *I. perturbatus*, and *I. pini*. It is distinguished by the nearly straight antennal club sutures, the uniseriately punctured discal interstriae, and lack of frontal sexual dimorphism.

Ips perturbatus (Eichhoff), 1869

Distribution: Canada: Alberta, British Columbia, Manitoba, New Brunswick, Northwest Territories, Ontario, Quebec, Saskatchewan, Yukon Territory. USA: Alaska, Colorado, Maine, Michigan, Minnesota, Montana, Washington.
Principal hosts: *Picea* spp.
Diagnosis: There are four spines on the elytral declivity and its general appearance is similar to *I. borealis* (Figure 9.6A). Potentially allopatric with related species *I. tridens*, *I. pilifrons*, *I. borealis*, and morphologically similar species *I. pini*. It differs from the related species by the nearly impunctate discal interstriae and from *I. pini* by a pair of transverse tubercles on the frons.

Ips pilifrons Swaine, 1912

Distribution: USA: Arizona, Colorado, Idaho, Nevada, New Mexico, Utah, Wyoming.
Principal hosts: *Picea engelmannii* Carrière, *P. pungens*
Diagnosis: There are four spines on the elytral declivity and its general appearance is similar to *I. borealis* (Figure 9.6A). Potentially allopatric with related species *I. borealis*, *I. tridens*, *I. hunter*, and morphologically similar species *I. pini*. It differs from the related species by deep, coarse strial punctures, large size, 4.4–5.0 mm and from *I. pini* by uniseriately punctured discal interstriae.

Ips pini (Say), 1826 (Figure 9.5D)
Bostrichus dentatus Strum
Bostrichus pallipes Strum
Tomicus praefrictus Eichhoff
Tomicus oregonis Eichhoff
Tomicus rectus LeConte
Ips laticollis Swaine

Distribution: Throughout Canada and USA except the southeast. Northern Mexico.
Principal hosts: *Pinus* spp.
Diagnosis: There are four spines on the elytral declivity. This species lacks a major median tubercle on its frons and additional differences among related species *I. avulsus* and *I. bonanseai* are given above.

Ips plastographus (LeConte), 1868

Distribution: Canada: British Columbia. USA: California, Idaho, Montana, Oregon, Washington, Wyoming.
Principal hosts: *Pinus contorta* Douglas ex Loudonand, *P. muricata* D. Don.
Diagnosis: There are four spines on the elytral declivity and its general appearance is similar to *I. pini* (Figure 9.5D).

It lacks a frontal carinate elevation and additional differences compared to related species *I. integer* are given above.

Ips schmutzenhoferi Holzschuh, 1988

Distribution: Bhutan.
Principal hosts: *Larix* sp., *Picea* sp., and *Pinus* sp.
Diagnosis: There are four spines on the elytral declivity and its general appearance is similar to *I. borealis* (Figure 9.6A). This species is generally larger (5.1–6.5 mm) and has a more densely granulated frons than the potentially allopatric species *I. longifolia* and *I. stebbingi*, and also the North American *Picea*-feeding *Ips* and *I. woodi*. DNA-based diagnosis is recommended.

Ips sexdentatus (Boerner), 1767 (Figure 9.7B)
Bostrichus stenographus Duftschmid
Bostrichus pinastri Bechstein
Ips junnanicus Sokanovskiy

Distribution: Austria. Belgium. Bosnia-Herzegovina. Bulgaria. Belarus. Croatia. China: Gansu, Hebei, Heilongjiang, Henan, Hubei, Inner Mongolia, Jilin, Liaoning, Qinghai, Sichuan, Shaanxi, Shanxi, Yunnan. Czech Republic. Denmark. Estonia. Finland. France. Corsica. Germany. Great Britain. Greece. Hungary. Italy. Japan. Kazakhstan. Latvia. Lithuania. Luxembourg. Macedonia. Moldavia. Mongolia. Montenegro. Myanmar. The Netherlands. Norway. North Korea. Poland. Portugal. Romania. Russia: ubiquitous. Serbia. South Korea. Slovakia. Slovenia. Spain. Sweden. Switzerland. Thailand. Turkey. Ukraine.
Principal hosts: *Pinus* spp.
Diagnosis: There are six spines on the elytral declivity. This species is distinguished from all other *Ips* spp. by the largest spine in the fourth position. This species is not related to nor should be confused with the North American six-spine species *I. calligraphus* and *I. apache*, which have the largest spine in the third position.

Ips shangrila Cognato and Sun, 2007

Distribution: China: Gansu, Qinghai, Sichuan, Shaanxi, Xizang, Yunnan.
Principal host: *Picea crassifolia* Kom.
Diagnosis: There are four spines on the elytral declivity. This species is distinguished from related *I. amitinus* (Figure 9.5B) and allopatric *I. nitidus* by the similar sized second and third declivital spines, which are connected by a tumescent base.

Ips stebbingi Strohmeyer, 1908
Tomicus blandfordi Stebbing
Tomicus ribbentropi Stebbing

Distribution: Afghanistan. Bhutan. China: Xinjiang. India: Himachal Pradesh, Uttar Pradesh, Kashmir. Nepal. Pakistan.
Principal hosts: *Picea* spp., *Pinus* spp.

Diagnosis: There are four spines on the elytral declivity and its general appearance is similar to *I. borealis* (Figure 9.6A). See *I. longifolia* and *I. schmutzenhoferi*. DNA-based diagnosis is recommended.

Ips subelongatus (Motschulsky), 1860

Distribution: China: Heilongjiang, Henan, Hubei, Inner Mongolia, Jilin, Liaoning, Shaanxi, Shandong, Shanxi, Xinjiang. Japan. Mongolia. North Korea. Russia: eastern and western (introduced). South Korea. Taiwan.
Principal hosts: *Larix* spp.
Diagnosis: There are four spines on the elytral declivity. This species is distinguished from *I. typographus* (Figure 9.6B) by the shiny elytral declivity and interstrial punctures of the elytral disk, from *I. cembrae* by larger body size and more densely setose elytral declivity. It differs from the morphologically similar North American *Picea*-feeding species and *I. woodi* by the space between the first and second spines, which is less than the length of the first spine.

Ips tridens (Mannerheim), 1852

Distribution: Canada: Alberta, British Columbia, Northwest Territories, Yukon Territory. USA: Alaska, California, Idaho, Montana, Oregon, Washington, Wyoming.
Principal hosts: *Picea* spp.
Diagnosis: There are four spines on the elytral declivity and its general appearance is similar to *I. borealis* (Figure 9.6A). Potentially allopatric with related species *I. pilifrons*, *I. borealis*, *I. perturbatus* and morphologically similar species *I. pini*. It differs from the related species by the coarse irregular punctures on the upper frons and small, shallow strial punctures, and from *I. perturbatus* and *I. pini* by uniseriately punctured discal interstriae.

Ips typographus (Linnaeus), 1758 (Figure 9.6B)
Bostrichus octodentatus Paykull
Ips japonicus Niisima

Distribution: Algeria. Austria. Belgium. Bosnia-Herzegovina. Bulgaria. Belarus. Croatia. China: Heilongjiang, Henan, Inner Mongolia, Jilin, Qinghai, Sichuan, Shaanxi, Xinjiang. Czech Republic. Denmark. Estonia. Finland. France. Germany. Great Britain. Greece. Hungary. Iceland (introduced). Ireland. Italy. Japan. Kazakhstan. Latvia. Liechtenstein. Lithuania. Luxembourg. Moldavia. Mongolia. Montenegro. North Korea. Poland. Portugal. Romania. Russia: ubiquitous. Serbia. South Korea. Slovakia. Slovenia. Spain. Sweden. Switzerland. Turkey. Ukraine.
Principal hosts: *Picea* spp.
Diagnosis: There are four spines on the elytral declivity. This species is distinguished from Eurasian species by its dull elytral declivity and impunctate interstriae on the basal half of the elytral disk. It differs from the morphologically similar Himalayan species, North American spruce-feeding species and *I. woodi* by the presence of a major medial frontal tubercle.

Ips woodi Thatcher, 1965

Distribution: Canada: Alberta. USA: Arizona, California, Colorado, Idaho, Montana, Nevada, New Mexico, Utah, Washington, Wyoming.
Principal hosts: *Pinus flexilis* E. James and *P. strobiformis* Engelm.
Diagnosis: There are four spines on the elytral declivity and its general appearance is similar to *I. borealis* (Figure 9.6A). Potentially allopatric with morphologically similar species *I. borealis*, *I. perturbatus*, and *I. pilifrons*. It is distinguished by the coarse, deep strial punctures, lack of frontal sexual dimorphism and its larger size (3.5–4.7 mm).

6. CONCLUSION

Relative to other scolytine genera, much is known in terms of *Ips* general biology, ecology, and phylogeny. However, details of life histories are only known for a few species (e.g., *I. pini* and *I. typographus*), thus attention of the lesser-studied species would help to identify atypical behaviors or ecologies. Also, additional study of associated organisms such as bacteria, fungi, nematodes, and mites would increase the understanding of the complexity of the subcortical community associated with *Ips* gallery systems. Placement of this new biological data in the context of *Ips* phylogenies will help to elucidate the patterns and processes of the evolution of *Ips* and its community of associated organisms. At present, the basic framework of *Ips* phylogeny is well founded; however, the internal relationships among the subgenera *I. (Ips)*, *I. (Bonips)*, and *I. (Granips)* and among the species within *I. (Ips)* are not well resolved (Figure 9.3). Potentially, the lack of resolution of species relationships is due to a relativity fast radiation (perhaps <1 million years) of *Ips* species 10 million years ago. Given the hasty radiation, likely few DNA sequence changes were associated with lineage diversification and thus limiting phylogenetic reconstruction with the current genetic loci. The burgeoning field of next-generation sequencing may provide a solution through the sequencing of hundreds of genetic loci for all *Ips* species (McCormack et al., 2013). Given good fortune, these loci will provide an abundance of phylogenetically informative data that will completely resolve *Ips* phylogeny and allow credible evolutionary inferences of their biology.

ACKNOWLEDGMENTS

I thank Sarah M. Smith and Milos Knížek for reviewing early versions of this chapter. Sarah Smith imaged and edited specimens for the figures, which helped hasten the completion of this chapter.

REFERENCES

All, J.N., Anderson, R.F., 1972. Initial attack and brood production by females of *Ips grandicollis* (Coleoptera: Scolytidae). Ann. Entomol. Soc. Am. 65, 1293–1296.

Alonso-Zarazaga, M., Lyal, C.H.C., 2009. A catalogue of family and genus group names in scolytinae and platypodinae with nomenclatural remarks (Coleoptera: Curculionidae). Zootaxa 2258, 1–134.

Anonymous, 2005. Forest insect and disease conditions in the United States 2004. United States Department of Agriculture, Forest Service, Forest Health Protection. Report 54, 1–142.

Barr, B., 1969. Sound production in Scolytidae with an emphasis on the genus *Ips*. Can. Entomol. 101, 636–672.

Bergroth, E., 1884. Bemerkungen zur dritten Auflage des Catalogus Coleopterorum Europae auctoribus L. v. Heyden, E. Reitter et J. Weise. Berliner entomologische Zeitschrift 28, 225–230.

Bertheau, C., Schuler, H., Arthofer, W., Avtzis, D., Mayer, F., Krumböck, S., et al., 2013. Divergent evolutionary histories of two sympatric spruce bark beetle species. Mol. Ecol. 22, 3318–3332.

Birgersson, G., Dalusky, M.J., Espelie, K.E., Berisford, C.W., 2012. Pheromone production, attraction, and interspecific inhibition among four species of *Ips* bark beetles in the Southeastern USA, Psyche ID532652.

Boerner, I.C.H., 1767. Beschreibung eines neuen Insects, des *Dermestes sexdentatus*. Okonomische Nachrichten der Patriotrischen Gesellschaft in Schlesien 4, 78–80.

Brand, J.M., Bracke, J.W., Markovetz, A.J., Wood, D.L., Browne, L.E., 1975. Production of verbenol pheromone by a bacterium isolated from bark beetles. Nature 254, 136–137.

Breshears, D.D., Cobb, N.S., Price, P.M., Allen, C.D., Balice, R.G., Romme, W. H., et al., 2005. Regional vegetation die-off in response to global change-type-drought. Proc. Natl. Acad. Sci. U. S. A. 102, 15144–15148.

Bright, D.E., 2014. A catalog of Scolytidae and Platypodidae (Coleoptera), Supplement 3 (2000–2010), with notes on subfamily and tribal reclassifications. Insecta Mundi 356, 1–336.

Bright, D.E., Skidmore, R.E., 1997. A Catalog of Scolytidae and Platypodidae (Coleoptera), Supplement 2 (1990–1994). NRC Research Press, Ottawa, Ontario, Canada.

Bright, D.E., Skidmore, R.E., 2002. A Catalog of Scolytidae and Platypodidae (Coleoptera), Supplement 2 (1995–1999). NRC Research Press, Ottawa, Ontario, Canada.

Buhroo, A.A., Lakatos, F., 2011. Molecular and morphological diagnostic markers for the Himalayan *Ips* DeGeer species (Coleoptera: Curculionidae: Scolytinae). Zootaxa 3128, 47–57.

Byers, J.A., 1989. Chemical ecology of bark beetles. Experientia 45, 271–283.

Byers, J.A., Birgersson, G., 1990. Pheromone production in a bark beetle independent of myrcene precursor in host pine species. Naturwissenschaften 77, 385–387.

Byers, J.A., Wood, D.L., 1981. Antibiotic-induced inhibition of pheromone synthesis in a bark beetle. Science 213, 763–764.

Cane, J., Merrill, L., Wood, D.L., 1990a. Attraction of pinyon pine bark beetle, *Ips hoppingi*, to conspecific and *I. confusus* pheromones (Coleoptera: Scolytidae). J. Chem. Ecol. 16, 2791–2798.

Cane, J.H., Stock, M.W., Wood, D.L., Gast, S.J., 1990b. Phylogenetic relationships of *Ips* bark beetles (Coleoptera: Scolytidae): Electrophoretic and morphometric analysis of the grandicollis group. Biochem. Syst. Ecol. 19, 359–368.

Cane, J.H., Wood, D.L., Fox, J.W., 1990c. Ancestral semiochemical attraction persists for adjoining populations of sibling *Ips* bark beetles. J. Chem. Ecol. 16, 993–1013.

Chansler, J.F., 1964. Overwintering habits of *Ips lecontei* Sw. and *Ips confusus* (Lec.) in Arizona and New Mexico: United States Department of Agriculture Forest Service Research Note RM–27, Rocky Mountain Forest and Range Experiment Station, Fort Collins, CO.

Chararas, C., 1962. Etude biologique des scolytides des conifers. Encyclopedie Entomologique, Ser. A, Nr. 38, Paul Lechevalier, Paris.

Cognato, A.I., 1998. Molecular Phylogenetics and Evolutionary Ecology of *Ips* Bark Beetles (Scolytidae). Ph.D. dissertation, Univ. of California, Berkeley.

Cognato, A.I., 2000. Phylogenetic analysis reveals new genus of Ipini bark beetle (Scolytidae). Ann. Entomol. Soc. Am. 93, 362–366.

Cognato, A.I., 2013. Molecular phylogeny and taxonomic review of Premnobiini Browne 1962 (Coleoptera: Curculionidae: Scolytinae). Frontiers in Ecology and Evolution 1, 1–12.

Cognato, A.I., Sperling, F.A.H., 2000. Phylogeny of *Ips* DeGeer (Coleoptera: Scolytidae) inferred from mitochondrial cytochrome oxidase I sequence. Mol. Phylogenet. Evol. 14, 445–460.

Cognato, A.I., Sun, J.H., 2007. DNA based cladograms augment the discovery of a new *Ips* species from China (Coleoptera: Curculionidae: Scolytinae). Cladistics 23, 539–551.

Cognato, A.I., Vogler, A.P., 2001. Exploring data interaction and nucleotide alignment in a multiple gene analysis of *Ips* (Coleoptera: Scolytinae). Syst. Biol. 50, 758–780.

Cognato, A.I., Rogers, S.O., Teale, S.A., 1995. Species diagnosis and phylogeny of the *Ips grandicollis* group (Coleoptera: Scolytidae) using random amplified polymorphic DNA. Ann. Entomol. Soc. Am. 88, 397–405.

Cognato, A.I., Seybold, S.J., Wood, D.L., Teale, S.A., 1997. A cladistic analysis pheromone evolution in *Ips* bark beetles (Coleoptera: Scolytidae). Evolution 51, 313–318.

Cognato, A.I., Seybold, S.J., Sperling, F.A.H., 1999. Incomplete barriers to mitochondrial gene flow between pheromone races of the North American pine engraver, *Ips pini* (Say). Proc. R. Soc. of London Ser. B. 266, 1843–1850.

Cognato, A.I., Harlin, A.D., Fisher, M.L., 2003. Genetic structure among pinyon pine beetle populations (Scolytinae: *Ips confusus*). Environ. Entomol. 30, 1262–1270.

Costa, A., Reeve, J.D., 2011. Olfactory experience modifies semiochemical responses in a bark beetle predator. J. Chem. Ecol. 37, 1166–1176.

Crotch, G.R., 1870. The genera of Coleoptera studied chronologically (1735–1801). Transactions of the Entomological Society of London 1, 41–52.

DeGeer, C., 1775. Mémoires pour servir à l'histoire des insectes. Vol. 5. L. L. Grefing, Stockholm.

Eggers, H., 1933. Borkenkäfer (Ipidae, Col) aus China. Entomologisches Nachrichtenblatt 7, 97–102.

Eichhoff, W.J., 1868. Neue amerikanische Borkenkäfer-Gattung und Arten. Berliner Entomologische Zeitschrift 11, 399–402.

Eichhoff, W.J., 1869. Neue Borkenkäfer. Berliner Entomologische Zeitschrift 12, 273–280.

Eichhoff, W.J., 1872. Zwei neue deutsche Tomicus-Arten. Berliner Entomologische Zeitschrift 15, 138–139.

Eichhoff, W.J., 1881. Die Europaischen Borkenkäfer Fur Forstleute, Baumzuchter und Entomologen. Julius Springer, Berlin.

Elkinton, J.S., Wood, D.L., 1980. Feeding and boring behavior of the bark beetle *Ips paraconfusus* (Coleoptera: Scolytidae) on the bark of host and non-host tree species. Can. Entomol. 112, 797–809.

Elkinton, J.S., Wood, D.L., Hendry, L.B., 1980. Pheromone production by the bark beetle *Ips paraconfusus*, in the nonhost, white fir. J. Chem. Ecol. 6, 979–987.

Fabricius, J.C., 1776. Genera insectorum eorumque characteres naturales, secundum numerum, figuram, situm et proportionem omnium partium oris adjecta mantissa specierum nuper detectarum. Litteris Mich. Friedr. Bartschii, Chilonii.

Figueroa-Teran, R., Welch, W.H., Blomquist, G.J., Tittiger, C., 2012. Ipsdienol dehydrogenase (IDOLDH): A novel oxidoreductase important for *Ips pini* pheromone production. Insect Biochem. Mol. Biol. 42, 81–90.

Furniss, M.L., Carolin, V.M., 1992. Western forest insects: United States Department of Agriculture Forest Service Miscellaneous Publication, No. 1339, Washington, DC.

Geoffroy, E.L., 1762. Histoire abregée des insectes qui se trouvent aux environs de Paris; dans laquelle ces animaux sont rangés suivant un ordre méthodique. Vol. 1. Durand, Paris.

Germar, E.F., 1824. Insectorum species novae aut minus cognitae, descriptionibus illustratae, Volumen Primum, Coleoptera. Impensis J. C. Hendelli et Filii, Halae.

Gyllenhal, L., 1827. Insecta Svecica descripta a Leonardo Gyllenhal. Classis I. Coleoptera sive Eleuterata. Tomi I, Pars IVFriedericum Fleischer, Lipsiae.

Heer, O., 1836. Observationes entomologicae, continentes metamorphosin Coleopterorum nonnullorum adhuc incognitorum. (VIII. Bostrichus cembrae, pp. 28–31, Taf. V). Zurich.

Hey, J., Waples, R.S., Arnold, M.L., Butlin, R.K., Harrison, R.G., 2003. Understanding and confronting species uncertainty in biology and conservation. Trends in Ecology and Evolution 18, 598–603.

Holzschuh, C., 1988. Eine neue Art der Gattung *Ips* aus Bhutan (Coleoptera: Scolytidae). Entomologica Basiliensia 12, 481–485.

Hopkins, A.D., 1905. Notes on some Mexican Scolytidae, with descriptions of some new species. Proc. Entomol. Soc. Wash. 7, 71–81.

Hopping, G.R., 1963a. Generic characters in the tribe Ipini (Coleoptera: Scolytidae), with a new species, a new combination, and new synonymy. Can. Entomol. 95, 61–68.

Hopping, G.R., 1963b. The North American species in group I of *Ips* DeGeer (Coleoptera: Scolytidae). Can. Entomol. 95, 1091–1096.

Hopping, G.R., 1963c. The North American species in groups II and III of *Ips* DeGeer (Coleoptera: Scolytidae). Can. Entomol. 95, 1202–1210.

Hopping, G.R., 1964. The North American species in group IV and V of *Ips* DeGeer (Coleoptera: Scolytidae). Can. Entomol. 96, 970–978.

Hopping, G.R., 1965a. The North American species in group VI of *Ips* DeGeer (Coleoptera: Scolytidae). Can. Entomol. 97, 533–541.

Hopping, G.R., 1965b. The North American species in group VII of *Ips* DeGeer (Coleoptera: Scolytidae). Can. Entomol. 97, 193–198.

Hopping, G.R., 1965c. The North American species in group VIII of *Ips* DeGeer (Coleoptera: Scolytidae). Can. Entomol. 97, 159–172.

Hopping, G.R., 1965d. The North American species in group IX of *Ips* DeGeer (Coleoptera: Scolytidae). Can. Entomol. 97, 422–434.

Hopping, G.R., 1965e. The North American species in group X of *Ips* DeGeer (Coleoptera: Scolytidae). Can. Entomol. 97, 803–809.

Hughes, P.R., 1974. Myrcene: a precursor of pheromones in *Ips* beetles. J. Insect Physiol. 20, 1271–1275.

Hunt, D.W.A., Borden, J.H., 1989. Conversion of verbenols to verbenone by yeasts isolated from *Dendroctonus ponderosae* (Coleoptera: Scolytidae). J. Chem. Ecol. 16, 1385–1397.

Jordal, B.H., Cognato, A.I., 2012. Molecular phylogeny of bark beetles reveals multiple origins of fungus farming during periods of global warming. BMC Evol. Biol. 12, 133.

Keeley, J.E., 2012. Ecology and evolution of pine life histories. Ann. For. Sci. 69, 445–453.

Knee, W., Beaulieu, F., Skevington, J.H., Kelso, S., Cognato, A.I., Forbes, M.R., 2012. Species boundaries and host range of tortoise mites (Uropodoidea) phoretic on bark beetles (Scolytinae), using morphometric and molecular markers. PLoS One 7, e47243.

Knížek, M., 2011. Scolytinae. In: Löbl, I., Smetana, A. (Eds.), Catalogue of Palaearctic Coleoptera, Curculionoidea I, Vol. 7. Apollo Books, Stenstrup, pp. 204–251.

Krascsenitsová, E., Kozánek, M., Ferenčík, J., Roller, L., Stauffer, C., Bertheau, C., 2013. Impact of the Carpathians on the genetic structure of the spruce bark beetle *Ips typographus*. J. Pest Sci. 86, 669–676.

Kurenzov, A.I., Kononov, D.G., 1966. A new species of bark beetles (Ipidae Coleoptera). In: Cherepanow, A.I. (Ed.), New Species of Fauna of Siberia and Adjoining Regions". Institute of Biology, Academy of Sciences of the USSR, Siberian Branch, Novosibirsk, pp. 29–33 [In Russian].

Lanier, G.N., 1966. Interspecific matings and cytological studies of closely related species of *Ips* and *Orthotomicus* Ferrari (Coleoptera: Scolytidae). Can. Entomol. 98, 175–188.

Lanier, G.N., 1967. *Ips plastographus* (Coleoptera: Scolytidae) tunneling in sapwood in lodgepole pine in California. Can. Entomol. 99, 1334–1335.

Lanier, G.N., 1970a. Biosystematics of the genus *Ips* (Coleoptera: Scolytidae) in North America: Hopping's group III. Can. Entomol. 102, 1404–1423.

Lanier, G.N., 1970b. Biosystematics of North American *Ips* (Coleoptera: Scolytidae): Hopping's group IX. Can. Entomol. 102, 1139–1163.

Lanier, G.N., 1972. Biosystematics of the genus *Ips* (Coleoptera: Scolytidae) in North American: Hopping's groups IV and X. Can. Entomol. 104, 361–388.

Lanier, G.N., 1987. The validity of *Ips cribricollis* (Eichhoff) as distinct from *I. grandicollis* (Eichhoff) and occurrence of both species in Central America. Can. Entomol. 119, 179–187.

Lanier, G.N., Teale, S.A., Parjares, J.A., 1991. Biosystematics of the genus *Ips* (Coleoptera: Scolytidae) in North American: review of *Ips calligraphus* group. Can. Entomol. 23, 1103–1124.

Latreille, P.A., 1802. Histoire Naturelle, Générale et Particulière, des Crustacés et des Insectes. Ouvrage faisant suite aux Oeuvres de Leclercq de Buffon, et partie du Cours complet d'Histoire naturelle rédigé par C. S. Sonnini, membre de plusieurs Sociétés Savantes. Vol. 3, Dufart, Paris.

Latreille, P.A., 1806. Genera Crustaceorum et Insectorum, secundum ordinem naturalem in familias disposita, iconibus exemplisque plurimis explicataVol. 2. Parisiis et Argentorati, Koenig.

LeConte, J.L., 1868. Appendix. *In* "Synopsis of the Scolytidae of America North of Mexico by C. Zimmermann, M. D. with notes and an Appendix by John L. LeConte, M.D.). Transaction of the American Entomological Society 2, 150–178.

LeConte, J.L., 1876. Family IX. Scolytidae. In: LeConte, J.L., Horn, G.H. (Eds.), The Rhynchophora of America North of Mexico. pp. 341–391, Proc. Am. Philos. Soc. 15.

Lekander, B., 1968. Scandinavian bark beetle larvae: descriptions and classification. Department of Forest Zoology, Royal College of Forestry, Stockholm. Res. Notes 4, 1–186.

Lewis, E.E., Cane, J.H., 1990. Pheromonal specificity of southeastern *Ips* pine bark beetles reflects phylogenetic divergence (Coleoptera: Scolytidae). Can. Entomol. 122, 1235–1238.

Lewis, E.E., Cane, J.H., 1992. Inefficacy of courtship stridulation as a premating ethological barrier for *Ips* bark beetles. Ann. Entomol. Soc. Am. 85, 517–524.

Linnaeus, C., 1758. Systema naturae per regna tria naturae secundum classes, ordines, genera, species, cum characteribus, differentiis, synonymis, locis. Edition 10. Holmiae.

Mannerheim, C.G.V., 1852. Zweiter Nachtrag zur Kafer-fauna der Nord-Amerikanischen Laender des Russischen Reiches. Societe Imperiale des Naturalistes de Moscou, Bulletin 25, 283–387.

Marsham, T., 1802. Entomologia Britannica, sistens insecta Britanniae indegena, secundum methodum Linnaeanum disposita. J. White, London.

Martin, D., Bohlmann, J., Gershenzon, J., Francke, W., Seybold, S.J., 2003. A novel sex-specific and inducible monoterpene synthase activity associated with a pine bark beetle, the pine engraver, *Ips pini*. Naturwissenschaften 90, 173–179.

Mayr, E., 1963. Animal Species and Evolution. Harvard University Press, Cambridge.

McCormack, J.E., Hird, S.M., Zellmer, A.J., Carstens, B.C., Brumfield, R. T., 2013. Applications of next-generation sequencing to phylogeography and phylogenetics. Mol. Phylogenet. Evol. 66, 526–538.

Millar, C.I., 1993. Impact of the Eocene on the evolution of Pinus L. Ann. Mo. Bot. Gard. 80, 471–498.

Miller, D., Gibson, K., Raffa, K., Seybold, S., Teale, S., Wood, D., 1997. Geographic variation in response of pine engraver, *Ips pini*, and associated species to pheromone, lanierone. J. Chem. Ecol. 23, 2013–2031.

Motschulsky, V.V., 1860. Coleopteres de la Siberie orientale et en particulier des rivers de l'Amour. In: Leopold von Schrenck, Reisen und Forschungen im Amur-lande in den Jahren 1854–1856, 2, pp. 77–257.

Pfeffer, A., 1995. Zentral- und Westpalärktische Borken- und Kernkäfer. Pro Entomologia, c/o Naturhistorisches Museum, Basel.

Postner, M., 1974. Scolytidae (Ipidae), Borkenkäfer. In: Schwenke, W. (Ed.), Die Forstschadlinge Europas, Vol. 2. Hamburgund, Berlin, pp. 334–487.

Raffa, K.F., Klepzig, K.D., 1989. Chiral escape of bark beetles from predators responding to a bark beetle pheromone. Oecologia 80, 566–569.

Reid, M.L., Roitberg, B.D., 1994. Benefits of prolonged male residence with mates and brood in pine engravers (Coleoptera: Scolytidae). Okios 70, 140–148.

Reitter, E., 1895. Bestimmungs-Tabelle der Borkenkäfer (Scolytidae) aus Europa und den angrenzenden Ländern. Verhandlungen des naturforschenden Vereines in Brünn, [1894] 33, Abhandlungen, pp. 36–97.

Sahlberg, C.R., 1836. Dissertatio entomologica insecta Fennica enumerans. Vol. 2, pars 9 and 10, 144–145.

Sallé, A., Arthofer, W., Lieutier, F., Stauffer, C., Kerdelhué, C., 2007. Phylogeography of a host-specific insect: genetic structure of *Ips typographus* in Europe does not reflect past fragmentation of its host. Biol. J. Linn. Soc. 90, 239–246.

Sandstrom, P., Welch, W.H., Blomquist, G.J., Tittiger, C., 2006. Functional expression of a bark beetle cytochrome P450 that hydroxylates myrcene to ipsdienol. Insect Biochem. Mol. Biol. 36, 835–845.

Say, T., 1826. Descriptions of new species of coleopterous insects inhabiting the United States. J. Acad. Nat. Sci. Phila. 5 (237–284), 293–304.

Schroeder, M.L., Lindelöw, Å., 2002. Attacks on living spruce trees by Ips typographus (Col. Scolytidae) following a storm-felling: comparison between stands with and without removal of wind-felled trees. Agr. Forest Entomol. 4, 47–56.

Seybold, S.J., 1992. The role in the olfactory-directed aggregation behavior of pine engraver beetles in the genus *Ips* (Coleoptera: Scolytidae). Ph. D. dissertation, University of California, Berkeley.

Seybold, S.J., Tittiger, C., 2003. Biochemistry and molecular biology of *de novo* isoprenoid pheromone production in the Scolytidae. Annu. Rev. Entomol. 48, 425–453.

Seybold, S.J., Quilici, D.R., Tillman, J.A., Vanderwel, D., Wood, D.L., Blomquist, G.L., 1995. De novo biosynthesis of the aggregation pheromone components ipsenol and ipsdienol by the pine bark beetle, *Ips paraconfusus* Lanier and *Ips pini* (Say) (Coleoptera: Scolytidae). Proc. Natl. Acad. Sci. U. S. A. 92, 8393–8397.

Sloan, L.C., Rea, D.K., 1995. Atmospheric carbon dioxide and early Eocene climate: a general circulation modeling sensitivity study. Palaeogeography, Palaeoclimatology, and Palaeoecology 119, 275–292.

Stadelmann, G., Bugmann, H., Meier, F., Wermalinger, B., Bigler, C., 2013. Effects of salvage logging and sanitation felling on bark beetle (*Ips typographus* L.) infestations. For. Ecol. Manage. 305, 273–281.

Stauffer, C., Leitinger, R., Simsek, Z., Schreiber, J.D., Führer, E., 1992. Allozyme variation among nine Austrian *Ips typographus* L. (Col., Scolytidae) populations. J. Appl. Ent. 114, 17–25.

Stauffer, C., Lakatos, F., Hewitt, G.M., 1997. The phylogenetic relationships of seven European *Ips* (Scolytidae, Ipinae) species. Insect Mol. Biol. 6, 1–8.

Stauffer, C., Lakatos, F., Hewitt, G.M., 1999. Phylogeography and postglacial colonization routes of *Ips typographus* L. (Coleoptera, Scolytidae). Mol. Ecol. 8, 763–773.

Stebbing, E.P., 1909. On some undescribed Scolytidae of economic importance from the Indian Region, II. Indian Forest Memoirs, Forest Zoology Series 1, 13–32.

Strohmeyer, H., 1908. Neue Borkenkäfer (Ipidae) aus dem westlichen Himalaja, Japan und Sumatra. Entomologisches Wochenblatt (Insekten-Börse) 25 (69–70), 72–73.

Susoy, V., Herrmann, M., 2014. Preferential host switching and codivergence shaped radiation of bark beetle symbionts, nematodes of *Micoletzkya* (Nematoda: Diplogastridae). J. Evol. Biol. 27, 889–898.

Swaine, J.M., 1909. Catalogue of the described Scolytidae of America, north of Mexico. 24th Report of the State Entomologist on Injurious and other Insects of the State of New York 1908, Appendix B. Education Department Bulletin no. 455, Museum Bulletin, New York State Museum 134, 76–159.

Swaine, J.M., 1911. A few new Ipidae. Can. Entomol. 43, 213–224.

Swaine, J.M., 1912. New species of the family Ipidae. Can. Entomol. 44, 349–353.

Swaine, J.M., 1915. Descriptions of new species of Ipidae (Coleoptera). Can. Entomol. 47, 355–369.

Swaine, J.M., 1917. Canadian bark-beetles, Part 1. Descriptions of new species. Dominion of Canadian Department of Agriculture, Entomological Branch, Technical Bulletin, 14, 1–32.

Swaine, J.M., 1918. Canadian bark-beetles, Part 2. A preliminary classification with an account of the habits and means of control. Dominion of Canadian Department of Agriculture, Entomological Branch. Technical Bulletin 14, 1–143.

Swaine, J.M., 1924. The allies of *Ips confusus* LeC in western America (Family Ipidae, Coleoptera). Can. Entomol. 56, 69–72.

Symonds, M.E., Egar, M.A., 2004a. The mode of pheromone evolution: evidence from bark beetles. Proc. R. Soc. London Ser. B. 271, 839–846.

Symonds, M.E., Egar, M.A., 2004b. Species overlap, speciation and the evolution of aggregation pheromones in bark beetles. Ecol. Lett. 7, 202–212.

Thatcher, T.O., 1965. A new species of *Ips* from Utah, Montana, and Alberta. Can. Entomol. 97, 493–496.

Tillman, J.A., Holbrook, G.L., Dallara, P.L., Schal, C., Wood, D.L., Blomquist, G.J., Seybold, S.J., 1998. Endocrine regulation of *de novo* aggregation pheromone biosynthesis in the pine engraver, *Ips pini* (Say) (Coleoptera: Scolytidae). Insect Biochem. Mol. Biol. 28, 705–715.

Wermelinger, B., 2004. Ecology and management of the spruce bark beetle *Ips typographus*—a review of recent research. For. Ecol. Manage. 202, 67–82.

Williams, A.P., Allen, C.D., Millar, C.I., Swetnam, T.W., Michaelsen, J., Stilla, C.J., Leavitt, S.W., 2010. Forest responses to increasing aridity and warmth in the southwestern United States. Proc. Natl. Acad. Sci. U. S. A. 107, 21289–21294.

Wood, D.L., 1982. The role of pheromones, kairomones, and allomones in the host selection and colonization behavior of bark beetles. Annu. Rev. Entomol. 27, 411–446.

Wood, S.L., 1982. The bark and ambrosia beetles of North America (Coleoptera: Scolytidae), A Taxonomic Monograph. Great Basin Nat. Mem. 6, 1–1359.

Wood, S.L., 2007. Bark and Ambrosia Beetles of South America (Coleoptera: Scolytidae). Brigham Young University, Provo.

Wood, S.L., Bright, D.E., 1992. A catalog of Scolytidae and Platypodidae (Coleoptera), Part 2: Taxonomic Index. Great Basin Nat. Mem. 13, 1–1553.

Wood, D.L., Stark, R.W., Silverstein, R.M., Rodin, J.O., 1967. Unique synergistic effects produced by the principal sex attractant compounds of *Ips confusus* (LeConte) (Coleoptera: Scolytidae). Nature 215, 206.

Yin, H.-F., Huang, F.-S., Li, Z.-L., 1984. Economic Insect Fauna of China, Fasc. 29, Coleoptera: Scolytidae, Science Press, Beijing, China.

Chapter 10

The Genus *Tomicus*

François Lieutier[1], Bo Långström[2], and Massimo Faccoli[3]

[1]*Laboratoire de Biologie des Ligneux et des Grandes Cultures, Université d'Orléans, Orléans, France,* [2]*Swedish University of Agricultural Sciences, Department of Ecology, Uppsala, Sweden,* [3]*Department of Agronomy, Food, Natural Resources, Animals and Environment (DAFNAE), Agripolis, Legnaro (PD), Italy*

1. INTRODUCTION

The genus *Tomicus* was established by Latreille in 1802 (Latreille, 1802), but Linnaeus described its oldest species as *Dermestes piniperda* in 1758 (Linnaeus, 1758). In his *magnum opus, Systema naturae,* he writes "*habitat in Europae ramulis inferioribus pini quos perforat, exsiccat, unde naturae hortulanus in hac arbore.*" Translated to English, this Latin description means "lives in younger pine shoots in Europe, which he hollows out, dries out, hence acting as nature's gardener in this tree," and shows that Linnaeus was well aware of the shoot feeding habit of the species and its effect on the appearance of the damaged pine tree.

The genus *Tomicus* (Latreille, 1802) has also been reported as *Dendroctonus* (Erichson, 1836), *Blastophagus* (Eichhoff, 1864), and *Myelophilus* (Eichhoff, 1878). *Tomicus* species develop on Pinaceae (mainly *Pinus*) and present several life traits that give them an unusual biology when compared with those of other bark beetle genera. Larvae develop in the phloem and adults mature in the shoots, leading the insect to occupy at least two habitats during its life cycle, to which a third one can be added in localities where the beetles overwinter at the base of the trunks or in the litter. Moreover, the shoots used for maturation feeding are not always on the same tree used for larval development, and the callow adult feeds successively upon several shoots and trees during its maturation. Contrary to most bark and ambrosia beetle species where the whole life cycle takes place in the same tree, *Tomicus* can thus use several host trees during its life cycle. Another unique life trait of the genus *Tomicus* is the precocity of its reproductive attacks. For example, in the northern part of France, it is not uncommon to observe attacks by *T. piniperda* (L.) on the bole of pines as early as mid-February. In the Mediterranean Basin, attacks can occur in autumn with winter development of the larvae for *T. destruens* (Wollaston). In all localities where the genus *Tomicus* occurs, its species are the most precocious among bark

beetles. All *Tomicus* species have one generation per year but sister broods can occur. As with many other bark beetle genera, each species has a special location along the trunk or branches, depending on adult size and bark thickness. Beetle preferences may, however, differ between regions. For example, *T. minor* (Hartig) prefers the base of the trunk in southern China, while in Europe it prefers the top of the bole and the base of big branches (Bakke, 1968; Ye and Ding, 1999).

Tomicus lives in temperate (including Mediterranean) and boreal coniferous forests, where their host plant is present. Attacks on trunks or branches for reproduction occur mainly on weakened trees, which results in the genus being considered as a "secondary" bark beetle species. However, shoot feeding for maturation occurs mainly in healthy and vigorous pines. Shoot feeding generally causes little damage to trees, without economic consequences, except when significant shoot attacks occur in young plantations or when a large proportion of shoots is destroyed, resulting in loss of growth and pine decline, making the trees suitable for trunk attack by the following beetle generation or by secondary beetles. The economic importance of *Tomicus* species varies depending on localities. Only *T. destruens* (Wollaston) and *T. yunnannensis* Kirkendall and Faccoli have been reported to kill healthy trees during bole attacks (Ye, 1992; Monleón *et al.*, 1996), although in the latter species, bole attacks followed heavy shoot attacks, which had drastically weakened the trees (Lieutier *et al.*, 2003). Figure 10.1 shows adults of *T. minor, T. piniperda,* and *T. yunnanensis.*

2. TAXONOMIC AND PHYLOGENETIC ASPECTS

2.1 Taxonomy

The genus *Tomicus* belongs to the tribe Hylurgini within the subfamily Scolytinae (Coleoptera, Curculionidae) (Alonso-Zarazaga and Lyal, 2009; Knížek, 2011). Eight

Bark Beetles. http://dx.doi.org/10.1016/B978-0-12-417156-5.00010-1

FIGURE 10.1 Adults of three *Tomicus* species. (A) *T. minor*; (B) *T. piniperda*; (C) *T. yunnanensis*. Photos: (A) by Maja Jurc; (B) from Pest and Diseases Image Library, Australia; (C) by Massimo Faccoli from Kirkendall et al. (2008) with the permission of the publisher.

Tomicus species have been described: *T. armandii* Li and Zhang, *T. brevipilosus* (Eggers), *T. destruens*, *T. minor*, *T. pilifer* (Spessivtsev), *T. piniperda*, *T. puellus* (Reitter), and *T. yunnanensis*. A list of synonymies for each species is given in Table 10.1.

Many *Tomicus* species are morphologically similar and difficult to differentiate. Moreover, because of the large distribution of some species (e.g., *T. piniperda* and *T. minor* covering almost the whole Palaearctic region) and the occurrence of small geographic variations of some morphological traits, many local species, subspecies, and varieties have been described for almost every *Tomicus* species, except for those of most recent description (*T. yunnanensis* and *T. armandii*). More detailed morphological and molecular analyses have resulted in subspecies and varieties now considered as synonyms (Table 10.1). The taxonomy of the genus *Tomicus* was characterized by a large uncertainty often resulting in a delay in the validation of new species living in sympatry and even on the same host with other species. For instance, *T. piniperda* and *T. destruens* were considered as synonymous (Schedl, 1932, 1946), rejecting the conclusions

of Wollaston (1865) who described *Hylurgus destruens* as a separate species based on specimens collected in Madeira. Wollaston (1865) reported *H. destruens* as "different from *T. piniperda* in being on the average a little larger and thicker, and its elytra, which are more coarsely rugulose, being always more or less ferruginous. Its antennae are totally pale (brown in *T. piniperda*) with their clubs somewhat longer and more acute." Reitter (1913), apparently unaware of Wollaston's paper and finding differences in the elytra color, described *T. piniperda* var. *rubripennis* (elytra reddish) as a Mediterranean variety of *T. piniperda* (elytra brown). The same character was also used by Krausse (1920), who described *T. piniperda* var. *rubescens* as having reddish elytra. However, these varieties have no taxonomical value as they were described only on the basis of different colors of the elytra, and are not supported by morphological or genetic differences. Lekander (1971), studying the morphological characters of the larvae, accepted *T. destruens* (as *Blastophagus destruens*) as a different species having larvae with three pairs of epipharyngeal setae instead of the four pairs found in *T. piniperda*. Lekander (1971) briefly commented

TABLE 10.1 Species of *Tomicus* Latreille, their Synonymies and Host Species (from Schedl, 1946; Browne, 1968; Pfeffer, 1995; Kirkendall *et al.*, 2008; Li *et al.*, 2010)

Species	Synonymies	Host Species
Tomicus armandii Li and Zhang	None	*Pinus armandii*
T. brevipilosus (Eggers)	*Blastophagus brevipilosus* Eggers *Blastophagus fukiensis* Eggers (nomen nudum) *Blastophagus khasianus* Beeson *Blastophagus multisetosus* Murayama	*Pinus koraiensis, P. insularis, P. sylvestrys parvifolia, P. yunnanensis, P. kesiya*
T. destruens (Wollaston)	*Hylurgus destruens* Wollaston *Blastophagus piniperda* var. *rubripennis* Reitter *Blastophagus piniperda* var. *rubescens* Krausse *Blastophagus piniperda* Schedl *Blastophagus destruens* Lekander	*Pinus halepensis, P. pinaster, P. pinea, P. brutia, P. canariensis, P. radiata,* occasionally reported on *P. nigra*
T. minor (Hartig)	*Hylesinus minor* Hartig *Hylurgus minor* Doebner *Blastophagus minor* Eichhoff *Myelophilus minor* Eichhoff *Myelophilus corsicus* Eggers *Blastophagus minor* var. *flavipennis* Krausse *Blastophagus minor* var. *flavus* Krausse *Blastophagus minor* var. *fuscipennis* Krausse *Blastophagus minor* var. *nigripennis* Mader	All pine species in its range with preference for Scots pine and black pines (*Pinus sylvestris, P. nigra austriaca, P. nigra balcanica, P. nigra cevennensis, P. nigra laricio, P. nigra nigra, P. nigra pallasiana, P. mugo, P. rotundata, P. densiflora, P. halepensis, P. pinaster, P. pinea, P. brutia, P. koraiensis, P. thunbergiana, P. pythusa, P. strobus, P. leucodermis, P. cembra, P. cembra sibirica, P. tabliformis, P. densiflora, P. yunnanensis*)
T. pilifer (Spessivtsev)	*Myelophilus pilifer* Spessivtsew	*Pinus koraiensis, P. armandii, P. tabulaeformis, P. yunnanensis*
T. piniperda (L.)	*Dermestes piniperda* L. *Bostrichus testaceus* L. *Bostrichus abietinus* F. *Hylesnus testaceus* F. *Bostrichus piniperda* Bechstein *Hylurgus piniperda* Latreille *Hylesinus piniperda* Gyllenhal *Dendroctonus piniperda* Erichson *Blastophagus piniperda* Eichhoff *Hylurgus analogus* LeConte *Myelophilus piniperda* Eichhoff *Blastophagus major* Eggers	Continental pine species and maritime pine (*Pinus sylvestris, P. nigra cevennensis, P. nigra nigra, P. nigra austriaca, P. nigra pallasiana, P. mugo, P. pinaster, P. koraiensis, P. thunbergiana, P. pythusa, P. strobus, P. leucodermis, P. cembra, P. densiflora, P. tabulaeformis, P. pentaphylla, P. funebris, P. peuce*)
T. puellus (Reitter)	*Myelophilus puellus* Reitter *Blastophagus starki* Eggers *Blastophagus puellus orientalis* Krivolutskaya	*Picea jezoensis, P. ajanensis, Abies holophylla, A. nephrolepis,* occasionally on *Pinus koraiensis*
T. yunnanensis Kirkendall and Faccoli	None	*Pinus yunnanensis, P. armandii* (in shoots), *P. kesiya* var. *langvianensis* (in shoots), *P. densata* (in shoots)

also on the size and proportions of the two species, but did not provide measurements or other numerical data. The descriptions of Wollaston (1865) and Lekander (1971) were supported recently by genetic and morphological investigations, which confirm *T. destruens* and *T. piniperda* as two species

(Gallego and Galian, 2001; Kerdelhué *et al.*, 2002; Kohlmayr *et al.*, 2002; Faccoli *et al.*, 2005a; Faccoli, 2006).

Populations of *T. piniperda* from Yunnan in southwestern China were recorded as unusually aggressive. The beetles aggregate densely for maturation feeding in

Pinus yunnanensis Franch. trees and subsequently attack the trunks of these trees, leading to their rapid decline and death (Ye and Lieutier, 1997; Lieutier *et al.*, 2003), although tree killing is considered to be atypical for pine shoot beetles (Chararas, 1962; Långström and Hellqvist, 1993b). In the last 30 years outbreaks have decimated over 200,000 ha of pine forest in Yunnan (Ye and Ding, 1999), resulting in considerable research into the ecology and control of these populations (Ye, 1991, 1994; Ye and Li, 1995; Ye and Zhao, 1995; Ye and Lieutier, 1997; Ye and Ding, 1999; Långström *et al.*, 2002; Lieutier *et al.*, 2003; Sun *et al.*, 2005). DNA sequencing revealed that populations from Yunnan, long assumed to be the widespread Eurasian *T. piniperda*, were strongly differentiated from *T. piniperda* from northeast China, whose populations are only weakly differentiated from and clearly conspecific with European *T. piniperda* (Duan *et al.*, 2004). Genetic distances between southern China specimens and those from northeastern China or France were an order of magnitude larger than those within regions or between France and northeastern China. Duan *et al.* (2004) concluded that Yunnan populations comprised an unrecognized, undescribed *Tomicus* species, for years confused with *T. piniperda* due to morphological similarities. The new species was described as *T. yunnanensis* (Kirkendall *et al.*, 2008), and all *T. piniperda* specimens collected up to now in Yunnan were transferred to *T. yunnanensis*. The morphological characteristics of *T. armandii*, described from Yunnan by Li *et al.* (2010), are similar to both the sympatric *T. yunnanensis* and *T. brevipilosus*. The genetic analysis now clearly supports the validation of *T. armandii* as a distinct species (Li *et al.*, 2010). The increasing use of molecular tools in taxonomy is of crucial importance in solving systematic problems due to the occurrence of sibling species.

2.2 Phylogenetic Relationships between Species

Although *Tomicus* species have high morphological affinities, they are genetically well characterized. Li *et al.* (2010) presented a phylogenetic tree including six of the eight known *Tomicus* species. The tree shows three clearly separated clades including *T. armandii* and *T. yunnanensis*, *T. brevipilosus* and *T. piniperda*, and *T. minor* and *T. destruens*, respectively (Figure 10.2). The genetic distance between *T. armandii* and *T. yunnanensis* was lower than that between *T. armandii* and the other *Tomicus* species; moreover, these two species were the only species found on *P. armandii* Franch; finally *T. armandii* was morphologically more similar to *T. yunnanensis* than to other *Tomicus* species. In light of these results, Li *et al.* (2010) suggested that *T. armandii* might be a sibling species of *T. yunnanensis*. The genetic affinity between *T. brevipilosus* and *T. piniperda* and between *T. minor* and *T. destruens* was also reported by Duan *et al.* (2004), who after comparing European and Asian populations of *Tomicus*, obtained the first evidence of a new *Tomicus* species from Yunnan, China (i.e., *T. yunnanensis*). Investigating the phylogenetic relationships of the European *Tomicus* species, Kohlmayr *et al.* (2002) found that *T. destruens* was closer to *T. minor* (same clade) than to *T. piniperda*.

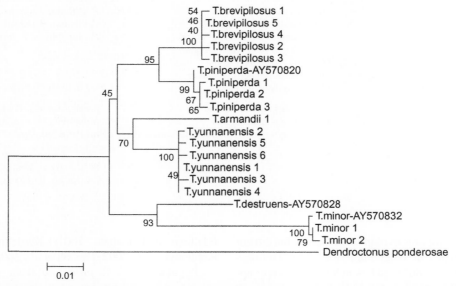

FIGURE 10.2 **Phylogenetic tree of six *Tomicus* species obtained by neighbor joining.** Bootstrap values calculated from 500 replicates are given on the nodes (Li *et al.*, 2010).

Concerning the phylogeography of *T. destruens* in the Mediterranean Basin, preliminary molecular analyses carried out in Italy suggested an east/west genetic differentiation of the species (Faccoli *et al.*, 2005a). The *T. destruens* haplotypes were found to be geographically structured in two clades distributed on the eastern and western part of the Mediterranean Basin. A contact zone was identified along the Adriatic coast of Italy (Horn *et al.*, 2006). The eastern clade was characterized by a significant phylogeographic pattern and low levels of gene flow, whereas the western clade barely showed a spatial structure in haplotype distribution. Moreover, the main pine hosts were different between groups, with the Aleppo–Brutia complex (*Pinus halepensis* Miller and *P. brutia* Tenore) in the east and the maritime pine in the west clade with a significant effect of the host tree. The finding was supported by results from Vasconcelos *et al.* (2006), which suggest the existence of two glacial refugia, from which *T. destruens* recolonized its current range. One refugium was located in Portugal where the beetle probably evolved on *P. pinaster* Aiton. The other refugium was probably in the eastern area of the Mediterranean Basin, where the beetles evolved on *P. halepensis* and *P. pinea* L. (Vasconcelos *et al.*, 2006).

The phylogeography of *T. piniperda* in Eurasia show a different pattern. Ritzerow *et al.* (2004) analyzed European, Asian, and American *T. piniperda* populations by sequencing a region of the mitochondrial cytochrome oxidase subunit I gene (*COI*). Although only a few significant relationships were found in the nested clade analysis, the results suggested that *T. piniperda* is a polymorphic species with numerous haplotypes distributed throughout Europe. It is likely that during different ice ages there were refugial areas in southern Europe and western Russia, although after the last ice age the Pyrenees formed a barrier to migration. Lastly, Chinese and European populations have been separated for at least 0.6 million years. Later, Horn *et al.* (2009) sequenced and analyzed 34 populations of *T. piniperda* sampled throughout the European range. Again, a high-genetic variability was detected in the Iberian Peninsula, with numerous endemic haplotypes. In contrast, the other European populations were less diverse with a single haplotype predominating from the Pyrenees to Scandinavia. Although *T. piniperda* had multiple fragmented refugia in the Iberian Peninsula, these only partly contributed to postglacial recolonizations of northern Europe during interglacials. Nevertheless, few long-range migration events up to northern Europe were detected, mostly originating from the Pyrenees. In the rest of Europe, the phylogeographical patterns were instead unclear, because of repeated cycles of contraction and expansion and the large distribution of Scots pine (*P. sylvestris* L.), the main *T. piniperda* host.

Specimens of *T. minor* from Europe and Asia were also sequenced, and a divergence of 6.4% between the European and the Chinese populations was detected (Ritzerow *et al.*, 2004). Thus, the evolutionary histories of *T. piniperda* and *T. minor* in Europe and Asia are different. According to the molecular clock of arthropod species (DeSalle *et al.*, 1987), the Asian *T. minor* diverged from the European populations about three million years ago whereas the *T. piniperda* populations diverged about 0.6 million years ago. Although these two species occur sympatrically it appears that they have had different colonization histories (Ritzerow *et al.*, 2004).

2.3 General Morphology and Species Separation

Tomicus species have similar morphological features. In general, males and females are not readily distinguishable, but at least in *T. piniperda* and *T. minor*, the sexes can be separated by the shape of the last abdominal tergites, and live beetles can be sexed by the sound produced by stridulating males (Bakke, 1968; Salonen, 1973). The length of adults in different species ranges from 2.9 to 5.4 mm. The frons is shiny, weakly to moderately impressed, with a fine median carina from the middle to just above the epistoma. The lower frons is bordered laterally by narrow carinae from the epistoma to the middle of the eye. The frons cuticle is usually smooth in the central part, with scattered deep setiferous punctures. The antennal funicle has six segments; antenna brown or yellow uniformly colored, although in some species the club and funicle may have different colors; club ovate, narrowly rounded, tip with abundant vestiture of short pale hair-like setae. The pronotum in mature individuals is almost black or reddish-brown, always darker than elytra, stout, usually wider than longer, strongly constricted in anterior third. There is cuticle shining along the dorsal part of the pronotum, finely, fairly densely punctured. There is vestiture of moderately abundant, fine, pale setae, denser along lateral areas of pronotum. The legs are from brown to almost black, usually with the same color as pronotum. The elytral color of mature adults is from reddish to brown or almost black, usually paler than pronotum and body. The disc is shiny, with weak traverse crenulations, these most dense and most pronounced basally. Striae with punctures are spaced by approximately their diameter. Interstriae are wider than striae, with uniseriate rows of widely spaced granules, these more closely spaced on declivity than on disc. Interstriae have erect hair-like setae, scattered on disc but denser on declivity, about as long as distance between striae or shorter. Declivity is broadly rounded, beginning at distal fourth of elytra; interstriae similarly sculptured, all but second interstria with or without a row of small conical setiferous tubercles. Second interstria is usually impressed but broadly convex to flat, punctures evenly spaced, uni-, bi- or even tri-seriate and irregularly distributed along the declivity of second interstria. If second interstria is without tubercles then first and third interstriae are often weakly elevated.

The main traits useful to separate the different species mostly concern adult size, characteristic of the second interstria along the declivity, elytral vestiture, elytral and antennal color (Schedl, 1946; Pfeffer 1995; Faccoli, 2006; Kirkendall *et al.*, 2008). The morphological differences that distinguish *Tomicus* species are, moreover, fully supported by molecular results (Kerdelhué *et al.*, 2002; Kohlmayr *et al.*, 2002; Duan *et al.*, 2004; Faccoli *et al.*, 2005a; Li *et al.*, 2010). Many morphological characters may vary intraspecifically and the extreme character states may overlap interspecifically; however, combinations of these characters reliably distinguish otherwise similar species. The eight *Tomicus* species can be separated by the following key (modified after Kirkendall *et al.*, 2008):

1. Second interstria on declivity with rows of small granules, not impressed   2
   – Second interstria devoid of granules, clearly impressed   4
2. (1) Elytral vestiture consisting of longer, erect interstrial hairs (arising from granules) in uniseriate rows and shorter decumbent hairs (ground vestiture), erect hairs longer on declivity. Elytral declivity with conspicuous interstrial tubercles in regular uniseriate rows. Larger species, length 3.1–5.2 mm. Normal hosts: *Pinus* spp.   3
   – Elytral interstrial hairs and ground vestiture equally short, dense, confused, decumbent or nearly so, not longer on declivity. Elytral interstriae strongly crenulate; interstrial tubercles large, transversely confluent, confused. Interstrial punctures confused on declivity, only slightly larger than strial punctures; interstrial tubercles inconspicuous on declivity. Smallest species, length 2.9–3.5 mm. Maternal gallery monoramous, longitudinal. Distribution: Asia. Normal hosts: *Picea* spp.   *T. puellus* (Reitter)
3. (2) Interstrial punctures on disc and declivity fine points, difficult to see with normal lighting, not dense. Declivital ground vestiture absent or sparse and difficult to see, inconspicuous. Pronotal punctures sparse most separated by much more than their diameter; most specimens with a distinct central impunctate longitudinal median strip. Antennal club pale to medium brown, at most little darker than funicle. Larger, length 3.2–5.2 mm. Maternal gallery biramous, transverse. Distribution: Eurasia. Host: all *Pinus* spp. in its range   *T. minor* (Hartig)
   – Interstrial punctures on disc and declivity conspicuous, uniformly dense, on declivity only slightly smaller than, or equal to, strial punctures. Declivity densely hairy due to abundant conspicuous decumbent ground vestiture (but hairs can be

completely worn away in older specimens). Pronotal punctures dense, separated on average by about their diameter, no impunctate median strip. Antennal club brown to dark brown, distinctly darker than funicle. Smaller, length 3.0–4.3 mm. Maternal gallery monoramous, longitudinal. Distribution: Asia. Hosts: *Pinus koraiensis* Siebold and Zucc., *P. armandii*, *P. tabulaeformis* Carrière   *T. pilifer* (Spessivtsev)
4. (1) Erect elytral hairs on disc longer, about as long as distance between striae; erect hairs on declivity distinctly longer than those on disc   5
   – Erect elytral hairs on disc shorter, about 0.5 × as long as distance between striae; erect hairs on declivity as long as those on disc   6
5. (4) Fine punctures of interstria 2 on declivity uniseriate, very fine, uniseriate, sparse, often widely spaced or even absent on much of declivity. Antennal club brown to dark brown, usually noticeably darker than funicle. Stouter, elytra 1.6 × longer than wide; smaller, length 3.2–4.4 mm. Elytra shorter than twice width of pronotum. Distribution: Asian species. Hosts: *Pinus koraiensis*, *P. insularis* Endl., *P. sylvestris parvifolia* Heer, *P. yunnanensis*   *T. brevipilosus* (Eggers)
   – Fine punctures of interstria 2 on declivity biseriate or triseriate, less than half the size of strial punctures. Antenna uniformly colored, dark brown. Slender elytra 1.9 × longer than wide; larger, length 3.9–4.6 mm. Elytra 1.3 × width of pronotum. Distribution: Yunnan (China). Hosts: *P. armandii*   *T. armandii* Li and Zhang
6. (4) Fine punctures of interstria 2 on declivity confused or appearing biseriate, punctures of striae 1 to 3 on declivity less than or equal to twice as large as fine interstrial punctures. Antenna uniformly colored, yellow to yellow-brown. Distribution: Mediterranean regions or Yunnan (China) Hosts: Mediterranean pine species or *P. yunnanensis*   7
   – Interstria 2 on declivity with uniseriate, regularly spaced fine punctures; punctures of striae 1 to 3 on declivity more than twice as large as fine interstrial punctures. Antenna uniformly colored, brown. Maternal gallery monoramous, longitudinal. Distribution: Eurasia including Japan. Hosts: continental *Pinus* spp. and *P. pinaster*   *T. piniperda* (L.)
7. (6) Mediterranean species. Second interstria on declivity weakly impressed, punctures dense, confused. Granules of second and third interstriae on disc closely spaced, at most by a distance equal to 1.5–2.5 punctures of adjacent striae; granules of first and third interstriae on declivity widely spaced, the distance between adjacent granules within a row equal to ca. 2/3 the distance between rows of granules on first and third interstriae. Only base of elytra dark. Protibia with five or six teeth, usually evenly spaced in one cluster.

Length 4.1–4.9 mm. Maternal gallery monoramous, longitudinal. Distribution: Mediterranean basin and Atlantic coastal regions of Spain, Portugal, and North Africa. Hosts: Mediterranean *Pinus* spp.   *T. destruens* (Wollaston)

- Asian species. Second interstria on declivity strongly impressed but broadly convex to flat, punctures more evenly spaced, appearing biseriate or irregularly uniseriate (appearing to zig-zag down declivity). Granules of first and third interstriae on disc widely spaced, many by about the distance between three to five punctures of adjacent striae; granules of first and third interstriae near base of declivity closely spaced, distance between pairs of granules ≤1/2 distance separating rows of granules of first interstria from those of third interstria. Basal 1/6th to 1/5th of elytral disc darker, often black (easily seen in paler specimens). Protibiae usually with six (five to seven) marginal teeth, the first (closest to body) often separated from the remaining teeth (>2/3 of individuals). Length 4.3–5.3 mm. Maternal gallery monoramous, longitudinal. Distribution: Yunnan (China). Hosts: *Pinus yunnanensis*   *T. yunnanensis* Kirkendall and Faccoli

3. GEOGRAPHIC DISTRIBUTION AND HOST RANGE

Tomicus is a Palaearctic genus with species distribution mainly in Europe and Asia. Five species occur only in Asia (*T. armandii*, *T. brevipilosus*, *T. pilifer*, *T. puellus*, and *T. yunnanensis*), while two species (*T. minor* and *T. piniperda*) are largely spread over both Asia and Europe, suggesting Asia as the continent of origin for the whole genus (Ritzerow *et al.*, 2004). *Tomicus destruens* is strictly distributed in the coastal areas of the circum-Mediterranean countries and Portugal. Lastly, *T. yunnanensis* and *T. armandii* are so far known only from Yunnan, a region of southern China, although their geographic distribution is probably wider. Additional collections and sampling are needed to better understand the natural range of distribution of these new species.

Tomicus piniperda and *T. destruens* have been reported to occur in northwestern Africa, as a probable natural post-glacial colonization from Europe by Spain. Being a thermophilous circum-Mediterranean species spread along the northern and southern rim of the Mediterranean Sea and in Asia Minor, the presence of *T. destruens* in North Africa is not surprising. The records in these regions for *T. piniperda*—a species of cold climates and continental hosts—are certainly due to past identification mistakes between *T. piniperda* and *T. destruens*, which has now been validated as a distinct species (Kerdelhué *et al.*,

2002; Kohlmayr *et al.*, 2002). Although in some cases *T. piniperda* and *T. destruens* may occur in the same country (e.g., Italy, France, Spain, Balkan countries), they usually develop in different environments and on continental and Mediterranean pine species, respectively. The only known case of true sympatry of these two species, i.e., populations of *T. destruens* and *T. piniperda* living in the same area and even on the same tree, refers to populations from northern Portugal, northwestern Spain (Lombardero 2005, pers. comm.), and southwestern France (Kerdelhué *et al.*, 2002) in stands of maritime pine growing at a middle elevation of 400–600 m above sea level (m.a.s.l.). A recent climatic model proposed by Horn *et al.* (2012) supports these findings. In 1992, *T. piniperda* was detected in the United States, close to Cleveland, Ohio (Haack and Kucera, 1993), from where it has spread to several states and to the provinces of Ontario and Quebec in Canada (Haack and Poland, 2001; Haack, 2006; Humble and Allen, 2006).

All *Tomicus* species normally breed on *Pinus* spp. and only occasionally on other conifers (Schedl, 1946), except for *T. puellus*, which usually develops on Asian spruce species (*Picea* spp.), although it was reported occasionally also from pine (Schedl, 1946). A detailed map of the distribution (modified from Knížek, 2011) is presented for each species (Figure 10.3), taking into account the necessary corrections regarding misidentifications of *T. piniperda* in Yunnan and in the Mediterranean basin. A list of the main hosts (modified from Schedl, 1946; Pfeffer, 1995; Kirkendall *et al.*, 2008; Li *et al.*, 2010) is also presented for each *Tomicus* species (Table 10.1).

4. BASIC BIOLOGY AND ECOLOGY

4.1 General Patterns, Gallery Systems, and Localization in Trees

The *Tomicus* life cycle includes phases of dispersal, reproduction, maturation, and hibernation but it is noteworthy that dispersal occurs twice in their life cycle, before and after the reproductive phase. After flight to the breeding material is completed, which depending on species and region takes place from November to May, the females excavate their typical egg galleries under the preferred part of the host tree, which again varies with the species and region.

All *Tomicus* species are monogamous and all but *T. minor* have a typical longitudinal gallery, where the female excavates niches in which the eggs are laid. The larvae feed on phloem and make typical larval galleries ending in a pupal chamber from which the new adults make an exit hole and emerge. Figure 10.4A shows an egg gallery of *T. piniperda*, but could as well describe *T. destruens*, *T. yunnanensis*, *T. brevipilosus*, *T. pilifer* or *T. puellus* (Schedl, 1946). According to Schedl (1946), the egg and

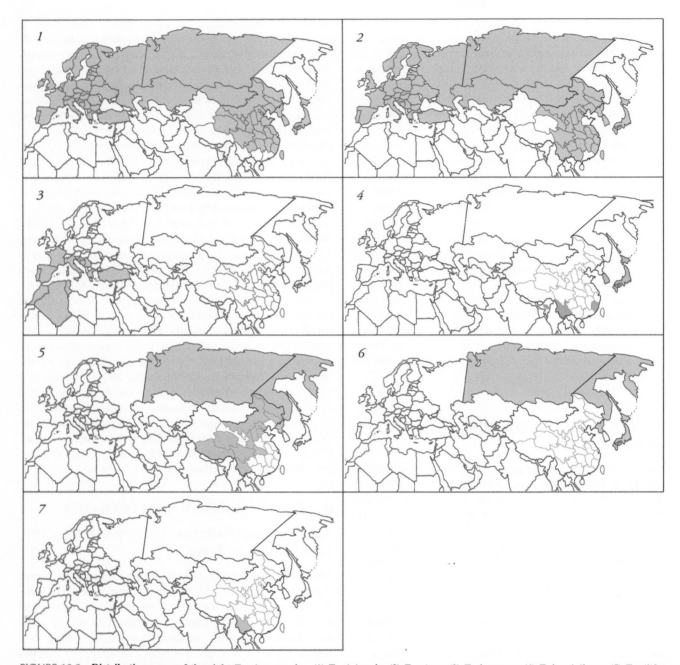

FIGURE 10.3 Distribution maps of the eight *Tomicus* species. (1) *T. piniperda*; (2) *T. minor*; (3) *T. destruens*; (4) *T. brevipilosus*; (5) *T. pilifer*; (6) *T. puellus*; (7) *T. yunnanensis* and *T. armandii*.

larval galleries of *T. pilifer* are deeper in sapwood than those of the other species with vertical galleries. The larval galleries of *T. pilifer* are also clearly shorter than those of *T. piniperda* (Stark 1952). In *T. minor* (Figure 10.4B), the egg gallery is transverse and double extending on both sides of the entrance hole and the larval galleries are short, ending deeper into the xylem than in any other *Tomicus* species.

All *Tomicus* species are considered to be univoltine, but there is some confusion about the voltinism in *T. destruens*

and *T. yunnanensis*, which will be discussed below. The shoot feeding behavior is also similar in all species although the preferred shoot size and crown level may vary between species. Schedl (1946) gives a summary of the biology and damage of all species known at the time. Browne (1968) summarizes the biology of *T. piniperda*, *T. minor*, and *T. brevipilosus* (using the name *T. khasianus*). No details are so far known about *T. armandii* except that it occurs in Yunnan and attacks *P. armandii* (Li *et al.*, 2010).

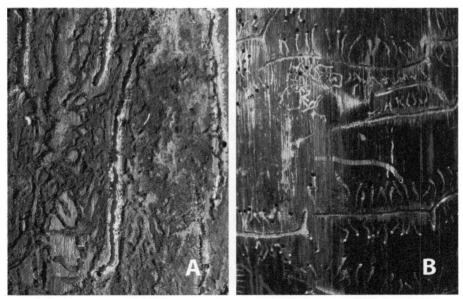

FIGURE 10.4   Egg galleries of *T. piniperda* (A) *and T. minor* (B) on Scots pine. *Photos by Claes Hellqvist.*

The maturation feeding in the shoots is revealed by typical entrance holes, often surrounded by crystallized resin, and the feeding tunnel is always oriented towards the shoot tip (Escherich, 1923). Figure 10.5 presents typical shoot damage by *T. piniperda*. In contrast to many other shoot borers, boring debris are never present in the shoot gallery. The gallery length is determined by the time the beetle spends in the shoot and the distance to the shoot apex, but it is normally a few cm long. In early season, attacks often take place in last year's shoot axis, a few cm below the node, and consequently all the expanding shoots may be affected by one attacking beetle. On the other hand, more than one attack may occur in the same shoot if the population level is high. These early-attacked shoots often wither and turn brown before they break off and fall to ground. Thick shoots may survive and remain in the crown. The new beetle generation mainly attacks the newer green shoots which often fall down. The feeding galleries are often visible in the fallen shoots either fully or partially. In the crown, the broken shoot bases or "pegs" often contain part of the entrance hole or the feeding gallery, and may be recognized as *Tomicus*-attacked shoots even years afterwards.

4.2   *Tomicus piniperda*

Studies on the biology and ecology of *T. piniperda* were initiated by Ratzeburg (1839), and subsequent German work was summarized by Escherich (1923). Since then, hundreds of studies have been conducted, mainly in the northern parts of Europe, including Russia. Our current understanding of the life cycle and general biology of *T. piniperda* in Europe

has evolved via basic studies in the former Czechoslovakia (Srot, 1966, 1968), England (Hanson, 1937, 1940), Finland (Kangas *et al.*, 1971; Salonen, 1973; Saarenmaa, 1985; Annila *et al.*, 1999), France (Chararas, 1962; Sauvard *et al.*, 1988; Hérard and Mercadier, 1996; Lieutier, 2002), Germany (Führer and Kerck, 1978; Vité *et al.*, 1986), Norway (Bakke, 1968), and Sweden (Långström, 1983a; Lanne *et al.*, 1987; Schroeder, 1988). The biology and damage caused by *T. piniperda* has also been studied in many countries in Eastern Europe, e.g., Estonia (Voolma and Luik, 2001), Poland (Gidaszewski, 1974; Borkowski, 2001), Romania (Drugescu, 1980), and Russia (Agafonov and Kuklin, 1979; Bogdanova, 1998; Kolomiets and Bogdanova, 1998; Gninenko and Vetrova, 2002). Most studies on *T. piniperda* in southern Europe deal in fact with *T. destruens*, and hence the role of *T. piniperda* in the Mediterranean area needs to be reconsidered and re-described (Faccoli *et al.*, 2005a). Regarding East Asia, there area few studies on *T. piniperda* from Korea (Park and Lee, 1972; Park and Byun, 1988) and Japan (Masuya *et al.*, 1998). Since its detection in 1992, dozens of North American publications on *T. piniperda* have been published (Haack and Poland, 2001; Kennedy and McCullough, 2002; and references therein).

As described in Section 3, *T. piniperda* is the most widespread of the *Tomicus* species, attacking its principal host, Scots pine, throughout its entire range. It is called the common pine shoot beetle in the UK (Bevan, 1962), but it is sometimes also referred to as the European pine shoot beetle in the USA and Canada. The host range has also been described in Section 3, but several studies in the USA and Canada have shown that its performance is better on pine

FIGURE 10.5   Shoot damage of *T. piniperda* and *T. minor* in Scots pine. (A) Resin tube at entrance hole to feeding tunnel; (B) *T. piniperda* in feeding tunnel; (C) wilting current shoots after attack in last year's shoot in early season; (D) stunted pine crowns after severe shoot damage by mainly *T. piniperda* during several years in the vicinity of a sawmill. *Photos: (A) and (C) by Claes Hellqvist; (B) and (D) by Bo Långström.*

species closely related to Scots pine, such as "red or soft pines" like *P. ponderosa* Laws. or *P. banksiana* Lamb. as compared to "white or hard pines" like *P. strobus* L. (Sadof *et al.*, 1994; Långström *et al.*, 1995; Lawrence and Haack, 1995). In Sweden, lodgepole pine (*P. contorta* Loud.) was clearly less attractive than Scots pine (Långström and Hellqvist, 1985). Occasionally other conifers like spruce and fir can also be attacked (Schedl, 1946). During the dispersal flight, *T. piniperda* responds to host odors, mainly α-pinene, which guide the beetles to suitable host material (Kangas *et al.*, 1971; Vité *et al.*, 1986; Lanne *et al.*, 1987;

Schroeder, 1988; Lindgren, 1997). Until now, there has been a consensus that *T. piniperda* is lacking aggregation pheromones (Schönherr, 1972; Lanne *et al.*, 1987; Löyttyniemi *et al.*, 1988), but Poland *et al.* (2003) concluded that trans-verbenol might play the role of an aggregation pheromone.

In northern Europe and North America, *T. piniperda* adults hibernate in short galleries in the bark at the base of standing pine trees (Escherich, 1923). *T. piniperda* is an early flyer, and in northern Europe, it leaves the hibernation sites in spring when temperature rises to 5°C

(Salonen, 1973), and the flight starts when air temperature exceeds 10–12°C (Bakke, 1968; Salonen, 1973; Långström, 1983a). Thus, flight may start in late March in years when spring arrives early and a month later when spring is late (Lekander, 1984). In central France, beetles fly in February–March (Sauvard, 1989), and in Portugal in March (Vasconcelos *et al.*, 2005). In the USA, spring flight begins in early or late March depending on the latitude (Poland *et al.*, 2002). Depending on the weather conditions, spring flight may be over in a few days or last for several weeks (Långström, 1983a). During the flight period, the females find and colonize suitable host material, i.e., fresh timber or weakened standing trees. In Europe, *T. piniperda* mainly attacks the lower stem covered with thick corky bark (Bakke, 1968; Långström, 1984). The species is monogamous and the female always excavates the egg gallery, which always runs along the wood grain. On standing trees all galleries are oriented upwards, but on fallen stems they may go towards the base or the top. The gallery starts with a short part without eggs, followed by egg niches on both sides at regular intervals, and finished by gallery without eggs. The full gallery length varies with attack density from 4–5 cm at high attack density, up to more than 10 cm at low attack density, and the egg numbers can be derived from gallery lengths and densities (Saarenmaa, 1983).

The larvae feed on the phloem and construct winding larval galleries perpendicular to the egg galleries, ending in a pupal chamber in the bark, or partly in the outer xylem. There are four larval stages (Lekander, 1968). The pupal period is short, and the callow adults emerge via individual exit holes through the bark. In northern Europe, emergence mainly takes place during July, but varies with the weather conditions. The development is faster than in *T. minor*, and in Sweden, the time span from mean flight to mean emergence was 92 ± 12 days with a thermal sum of day-degrees exceeding 0°C of 1016 ± 79 (Långström, 1983a). Salonen (1973) gives a more detailed description of the thermal sum required for each developmental stage, and Saarenmaa (1985) modeled the different phases in the life cycle. The immature stages do not survive the winter in Scandinavia (Bakke, 1968). In more southerly areas in Europe, the emergence takes place earlier, but the more or less regular occurrence of sister broods blurs the picture, as described below. In North America, the development and emergence times are similar to those of central and northern Europe. In Ontario, the corresponding thermal sum for the development period was 1250 ± 73 day-degrees (Ryall and Smith, 2000b).

There has been ample discussion concerning the number of pine shoot beetle generations and sister broods per year. Escherich's (1923) statement that there is only one generation per year but that sister broods (i.e., new brood(s) by the same parent beetles in the same year) are common still seems to hold for as different regions as Sweden

(Långström, 1983a), France (Sauvard, 1989), and Ontario (Ryall and Smith, 2000b). The occurrence of sister broods increases from the north to the south. In Sweden, sister broods are rare although the thermal sum in most years would allow a sister brood in the southern part of the country (Långström, 1983a). Ryall and Smith (2000b) found that two broods occurred in Canada, and that the thermal sum required was lower (856 day-degrees) for the second than for the first one (see above). In France, Sauvard (1993) found five waves of oviposition, i.e., the initial and four sister broods under semi-natural conditions. Also, Srot (1966, 1968) has observed one or more sister broods in central Europe resulting in an extended period of beetle emergence lasting into September.

During the oviposition period, the male stays in the gallery and removes the frass, but towards the end of that period leaves the gallery and flies to the pine crowns to feed in the shoots in order to regain sexual maturity. A few weeks later, the females have finished oviposition and in early summer they are also found feeding in the shoots. This pattern has been observed in Sweden (Långström, 1983a) and Canada (Ryall and Smith, 2000b). At least in Fennoscandia, some of the adult beetles may hibernate a second time after a period of regeneration feeding in the shoots and produce another brood the following year (Långström, 1983a; Schroeder and Risberg, 1989). Whether this phenomenon also occurs in a warmer climate is not known, but considering the univoltine life cycle and the occurrence of post-reproductive adults in the shoots indicates that this could well be the case.

The main shoot feeding period takes place when the callow adults emerge and fly to nearby pine crowns, where they tunnel mainly current shoots at the outer parts of the branches. In Sweden, this takes place from July to October, when the first severe frosts cause the beetles to leave the shoots and move to the duff on the ground where they hibernate (Långström, 1983a). The same pattern has been observed in North America (Kauffman *et al.*, 1998; Ryall and Smith, 2000b; Poland *et al.*, 2002). As stated above, the parent beetles also move to the shoots for a regeneration-feeding period that starts earlier and lasts longer than that of the callow beetles. In Sweden, this regeneration feeding takes place in last year's shoots as the current ones (i.e., those shoots that develop in the same season as the attacks takes place) are seldom attacked while expanding (Långström, 1980). A similar pattern regarding the maturation feeding of re-emerging parent beetles was reported in Canada (Ryall and Smith, 2000b). Also, Långström (1983a) found that some beetles entered the shoots soon after the flight period and stayed there for the whole summer. As these beetles were sexually mature, he concluded that they had turned to the shoots after exhausting their fat reserves during the search for host material. There are no clear data demonstrating the presence of regeneration feeding after flight or after oviposition among parent beetles in the southern parts of the distribution area of *T. piniperda*. It probably exists, but to what extent is unknown. In *T. piniperda*, the shoot feeding preferentially takes place in the upper whorls of the pine crown (Führer and Kerck, 1978; Långström, 1983a; Kauffman *et al.*, 1998). Långström (1980) concluded that the beetles colonize the crown from above, and that the outermost shoots are taken first. In Sweden, the preferred shoot diameter was ca. 3–4 mm, and on average each beetle tunneled one shoot, although multiple attacks (same beetle in several shoots or more than one beetle in the same shoot) were not uncommon (Långström, 1980, 1983a). In general, the previous year's shoots were tunneled in early season, and current shoots in late season. This pattern from Sweden was largely confirmed for North America (Kauffman *et al.*, 1998; Haack *et al.*, 2000, 2001), although Ryall and Smith (2000b) found that the species could consume as many as five to six shoots per beetle.

4.3 *Tomicus minor*

There are few studies devoted to the biology and ecology of the lesser pine shoot beetle *T. minor*. Major research contributions have been made in Scotland (Ritchie, 1917), UK (Bevan, 1962), Ukraine (Greese, 1926; Iljinsky, 1928), Norway (Bakke, 1968), Finland (Kangas *et al.*, 1971; Annila *et al.*, 1999), Sweden (Långström, 1983b; Lanne *et al.*, 1987), and Spain (Fernández Fernández *et al.*, 1999a–c). In the 1990s, the species was recorded as a pest in China (Ye and Ding, 1999; Långström *et al.*, 2002).

Spring flight is the main period of dispersal, which takes place in early spring when temperatures exceed 12°C in the shade. Dependent on the latitude this may occur in November–December in southern China, February–March in southern Europe (Masutti, 1959; Roversi *et al.*, 2004), March–April in central Europe, and April–May in Scandinavia, where the beetles normally fly 1–2 weeks after *T. piniperda* (Långström, 1983b). Flying beetles respond to host odors, including α-terpineol, cis-carveol, and trans-carveol, which are believed to guide beetles to the host material (Kangas *et al.*, 1970, 1971). Lanne *et al.* (1987) concluded that the host and mate finding of *T. minor* is enhanced by a primitive pheromone based on trans-verbenol synergized from host terpenes. Martikainen (2001) found *T. minor* also in traps baited with lineatin. In Europe, *T. minor* mainly attacks the upper part of the pine stem covered with smooth bark, whereas in China it seems to predominate on the lower stem with thick and corky bark (Ye and Ding, 1999). However, in 30-year-old pines, egg galleries occurred mainly on the first 2 m from the ground, but in some cases up to 6 m (Masutti, 1959).

The species is monogamous and the female always excavates the egg gallery under the thin bark on the upper stem or on thick branches, and there is a strong preference for the underside of fallen logs or trees (Långström, 1984). The egg gallery is oriented across the wood grain and is normally two-armed. The female alternates between the arms extending them gradually as oviposition proceeds. The mean egg gallery length is dependent on tree size and competition, but a full size gallery may exceed 10 cm in length (4–20 cm in the Alps; Masutti, 1959) and may contain ca. 100 eggs (Långström, 1983a, 1984). The male stays in the gallery and removes the frass, but towards the end of the oviposition period leaves the gallery and flies to the pine crowns to feed in the shoots in order to regain sexual maturity. A few weeks later the females have finished oviposition and in early summer they are also found feeding in the shoots (Långström, 1983a).

The larvae first feed in the phloem, constructing short larval tunnels perpendicular to the egg galleries. After a few cm they enter the sapwood and the larvae become sessile, and commence feeding on the fungus growing in the galleries (Francke-Grosmann, 1952; for more details see Section 5). *Tomicus minor* has four larval instars (Lekander, 1968). When fully grown, they pupate at the end of their feeding tunnel and emerge through an exit hole as callow adults after a short pupal period. The breeding success is low on the upper side of a fallen tree as compared to that on the lower side, where thousands (in a few cases more than 10,000) beetles emerge per square meter of bark area (Långström, 1983b, 1984). Survivorship curves indicate that 26% of the eggs make it to callow adults (Långström, 1983b). In Sweden, emergence takes place in late summer and the average timespan between median flight and emergence dates was 105 days with an average sum of day-degrees (above 0°C) of ca. 1300 (Långström, 1983a). The immature stages do not survive the winter in Scandinavia (Bakke, 1968). In Spain, the average life cycle duration from egg to adult was 135 days with the first eggs seen in late March and last adults emerging in early September (Fernández Fernández *et al.*, 1999c). In Italy, development lasts from March to the end June both in the Alps (Masutti, 1959) and in central Italy (Roversi *et al.*, 2004). In China, it took 125 days from mean flight to mean emergence in a laboratory study (Chen, 2003). Ritchie's (1917) statement that *T. minor* has only one generation per year but that sister broods may occur seems to hold for different regions such as Sweden (Långström, 1983a), Spain (Fernández Fernández *et al.*, 1999a), and southern China (Långström *et al.*, 2002).

The shoot feeding of *T. minor* is very similar to that of *T. piniperda*, but the callow adults start the main shoot feeding period later, as they emerge later (Långström, 1983a). Both species prefer ca. 3- to 4-mm-thick newer shoots, but *T. minor* was more abundant in the lower parts of the crown, whereas *T. piniperda* preferred the upper whorls (Långström, 1983a). The regeneration feeding described above for *T. piniperda* also occurs in *T. minor*, starting after oviposition and continuing until late autumn (Långström, 1983a). Thus, shoot feeding occurs from early summer to late season until hibernation, and part of the *T. minor* population may survive and breed again the following spring (Långström, 1983a). Whether this phenomenon also occurs in a warmer climate is not known, but considering the univoltine life cycle and occurrence of post-reproductive adults in the shoots even in southern China (Långström *et al.*, 2002), this could well be the case. In any case, shoot feeding of *T. minor* occurred in all months except April–May (Lu *et al.*, 2014). *T. minor* preferred thinner shoots than did *T. yunnanensis* (Chen, 2003; Zhao, 2003), but there was no difference between the species in the vertical attack pattern within the pine crown (Chen, 2003).

In China and southern Europe, *T. minor* hibernates in the shoots (Masutti, 1959; Fernández Fernández *et al.*, 1999b; Lu *et al.*, 2014), but in northern Europe it leaves the shoots after the first frost nights, normally during October, and hibernates in the duff below the canopy (Långström, 1983a). Some beetles fall down with broken shoots, but most of them probably walk down the stem like *T. piniperda* does.

4.4 *Tomicus destruens*

The first detailed studies about biology and ecology of the Mediterranean pine shoot beetle *T. destruens* were carried out in Italy more than 130 years ago (Targioni-Tozzetti, 1886). These studies continued during the first half of the last century (Razzauti, 1921; Russo, 1940, 1946), followed by several studies in the 1960s and 1970s in many Mediterranean countries (Chararas, 1964; Masutti, 1969; Astiaso and Leyva, 1970; Romanyk, 1972; Carle, 1974a, b; 1975; Halperin, 1978). Nevertheless, following the uncertain identification of *T. destruens* and *T. piniperda*, the largest part of the scientific papers published in the Mediterranean countries reports *T. piniperda* as the investigated species although they actually deal with *T. destruens*.

Like all *Tomicus* species, *T. destruens* has a univoltine life cycle. In most of its distribution range *T. destruens* breeds mainly during winter. In October and November mature adults fly in search of suitable declining host material in which to reproduce. Beetles are attracted by a blend of volatiles including ethanol, α-pinene, β-myrcene, and α-terpinolene (Faccoli *et al.*, 2008), which are emitted from fresh undebarked timber, dying, or stressed pine trees (Santos *et al.*, 2005; Branco *et al.*, 2010). *Tomicus destruens* infests mainly Mediterranean pine species, such as *P. halepensis*, *P. pinaster*, *P. pinea*, *P. brutia*, and *P. canariensis* C. Smith, although infestations were recorded also on exotic pine species (*P. radiata*) and occasionally on *P. nigra* in Turkey (Sarikaya and Avci, 2010), Spain

(Gallego *et al.*, 2004), and Italy (Faccoli *et al.*, 2005b). *Tomicus destruens* is a monogamous species, and females penetrate the bark first, which is subsequently reached by males, followed by mating in the nuptial chamber. Adults of *T. destruens* mainly infest the medium-upper part of stem of mature pine trees (Larroche, 1971; Carle, 1974a; Stergulc, 2002). After mating, the female lays eggs along a single longitudinal gallery that runs along the wood grain. Like other *Tomicus* species, at its beginning and at its end the egg gallery exhibits two short lengths without eggs. Female fecundity and the length of the maternal tunnel are strongly affected by colonization density and intraspecific competition, and they may range between 50–90 eggs per female and 4–10 cm, respectively (Faccoli, 2009; Durand-Gillmann, 2014). In an Aleppo pine stand in Algeria, Chakali (2005) reported a mean fecundity of 58.6 eggs per female, ranging from 26 to 139, and a mean tunnel length of 10 cm, with a range of 5.3–17.2 cm. Length of the egg tunnels and female fecundity are affected also by the host species, with Aleppo pine showing values higher than maritime and stone pines (Faccoli, 2007). In Turkey, the length of egg galleries bored in *P. brutia* varied between 6 and 12.4 cm (9.8 cm average), while in *P. nigra* the longest gallery was 7.5 cm (Sarikaya and Avci, 2010). At ca. 17°C, egg hatching takes 12–14 days (Chakali, 2005). The larvae feed in the phloem and bore galleries that develop perpendicular to the egg galleries, ending in a pupal chamber excavated in the bark. There are four larval instars (Sabbatini Peverieri and Faggi, 2005). The pupal period is usually short, and at ca. 17°C, takes 12–22 days (Chakali, 2005), although more than 60% of the population needs only 14–18 days. Callow adults emerge from the bark through individual exit holes. *Tomicus destruens* needs about 80 days to develop and emerge, with small, nonsignificant variations among pine species and insect sex (Chararas, 1964; Faccoli *et al.*, 2005b; Faccoli, 2007).

In many regions of its distribution, *T. destruens* exhibits a less clear phenology, with adults flying and laying eggs both in autumn (October–November) and early spring (February–March). This phenomenon is due to different local climatic conditions and to female re-emergence producing sister generations (Monleón *et al.*, 1996). In central and south Italy, Nanni and Tiberi (1997) and Russo (1940, 1946) reported two breeding periods per year (erroneously called generations), the first beginning in autumn and ending in February and the second starting in February and ending in June. Similar phenology was found also in southern France (Durand-Gillmann, 2014), Spain (Monleón *et al.*, 1996), Algeria (although 70% of the adults infested the bark during the second half of October; Chakali, 2005), and Portugal (Vasconcelos *et al.*, 2005). In Turkey, the adult flight period (autumn or spring) varied with elevation (Sarikaya and Avci, 2010). In Israel (Halperin, 1966; Mendel *et al.*, 1985), Tunisia (Ben Jamâa *et al.*, 2000), and Morocco (Graf and Mzibri, 1994) the insect develops mainly in winter (October–April), although with a common second and even a third sister generation in early spring. In all these cases, the winter is spent under the bark at the same time by eggs, larvae, and adults.

Populations of *T. destruens* living in cold sites from northern latitudes or higher elevations may show a clearly different phenology, with bark colonization and reproduction starting in early spring (Masutti, 1969; Titovšek, 1988; Faccoli *et al.*, 2005b; Sarikaya and Avci, 2010). In these regions, the life cycle of *T. destruens* is very similar to that previously described for *T. piniperda* in central and northern Europe (Escherich, 1923; Chararas, 1962; Bakke, 1968; Långström, 1983a), exhibiting behavioral analogies in tree colonization, reproduction, and adult feeding. In these climatic conditions, the whole life cycle (egg–adult) takes about 12 weeks (from March to June; Faccoli *et al.*, 2005b), although variations in developmental time or flight period could be observed in relation to seasonal temperature. This reproductive behavior was reported for populations living along the northern Italy (Masutti, 1969; Faccoli *et al.*, 2005b) and Slovenia (Titovšek, 1988) coasts, in northern Corsica, and some sites of southern France (Kerdelhué *et al.*, 2002). *Tomicus destruens* usually develops at low altitudes (lower than 500 m) (Carle, 1974a; Abgrall and Soutrenon, 1991; Vasconcelos *et al.*, 2005), but in semi-arid stands of *P. halepensis* in Algeria the species is very active at altitudes between 1200 and 1400 m a.s.l. (Chakali, 2005). Elevation may also affect insect phenology. In Turkey, populations living up to 300 m a.s.l. flew and started ovipositing at the beginning of November and young adults were observed in the middle of April. At middle elevations (300–600 m a.s.l.), the first eggs were observed in the second half of December and the callow beetles emerged at the beginning of May. At higher elevations (600 m a.s.l. or higher), females started to lay their eggs at the beginning of February and young adults appeared the second half of June (Sarikaya and Avci, 2010).

Callow adults of all *Tomicus* species need a period of maturation feeding to reach sexual maturation. Maturation feeding occurs in shoots or twigs of healthy trees usually belonging to the same host species where the beetles developed. After emergence, callow adults of *T. destruens* fly towards the canopy of vigorous pines where they tunnel mainly in the previous year's shoots in early season, and newer shoots in late season. In Algeria, Chakali (2005) reported the initiation of migration from the bark to the pine shoots in April, with mean temperatures of ca. 15°C. A preference is observed for vigorous young trees (Masutti, 1969; Chakali, 2005; Branco *et al.*, 2010; Faccoli, pers. observ.) having shoots releasing an attractive blend of α-pinene and β-myrcene (Faccoli *et al.*, 2008). In tests with paired plants, the number of holes and tunnels excavated by beetles in well-watered pines, together with beetle survival

and fat contents, were significantly higher than in drought stressed maritime pines, whereas in no-choice tests no differences occurred (Branco *et al.*, 2010). When more pine species are available, shoots of maritime pine are preferred (Masutti, 1969; Stergulc, 2002). In field and laboratory tests carried out in central Italy, adults apparently preferred to feed on shoots of the pine species where they developed, although the results were not consistent with other experiments carried out by the same authors (Tiberi *et al.*, 2009). Each shoot usually hosts one or two adults, rarely more. Of 176 shoots of Aleppo pine infested by callow adults of *T. destruens* in Algeria, 60.2% and 22.7% had one or two penetration holes, respectively. Shoots with three or four holes were relatively few, not exceeding 17% overall (Chakali, 2005). However, although the same shoot may host many adults, neither mating nor reproduction occurs within the shoots. Adults stay within shoots for about 7 months (April–October). During maturation and estivation then, each adult consumes a variable number of shoots, which may dry or fall on the litter following strong winds, with a strong pruning effect, although no precise data are available.

Adults of populations reproducing in spring (March–June) show a similar maturation feeding behavior, although shifted in time. In June, the new adults fly toward healthy pine trees and tunnel into shoots to become sexually mature. During shoot tunneling the adults face two different phases: maturation feeding in summer and hibernation in winter. In these areas *T. destruens* overwinters as adults in pine shoots (Masutti, 1969; Faccoli et al., 2005b), and infested pine shoots may be easily found on the litter after strong winter winds or storms (Stergulc, 2002). Adults of *T. destruens* can survive in pine shoots for more than 10 months, from June to March of the following year (Faccoli *et al.*, 2005b). Masutti (1969) reported that a very small part of these populations overwinters as adults in isolated niches excavated in the bark of healthy trees and in short newly formed mother galleries (without eggs).

In general, the breeding performance of *T. destruens*, reported as percentage of eggs producing emerging adults, progressively decreased from 17 to 4% with increasing colonization density going from 0.25 to 1.50 females per squared decimeter of bark (Faccoli, 2009), although differences were statistically significant only at a density of 0.75 females (i.e., 1.3 dm^2 of bark per female) or higher. Within-tree mortality due to intraspecific competition has been reported to be one of the main factors affecting breeding success in *T. destruens* (Chararas, 1964; Faccoli, 2007). Significant differences in the mean number of larvae and emerging adults occur also among populations reared on different host trees. In particular, *P. pinea* and *P. pinaster* had mean number of larvae, but not eggs, lower than *P. halepensis*, suggesting egg mortality occurring in different pine species as the crucial factor affecting the breeding performance of *T. destruens* (Faccoli, 2007).

4.5 *Tomicus yunnannensis*

Tomicus yunnanensis has only been reported in the Yunnan province in southern China (Kirkendall *et al.*, 2008), and therefore we propose to use the Yunnan pine shoot beetle as the common name in English. Due to the earlier confusion with *T. piniperda*, the status of this species in this region was not quite clear, but all earlier studies concerning pine shoot beetle attacks on Yunnan pine (*P. yunnannensis*) are here considered to be caused by *T. yunnanensis*, even though they were reported as *T. piniperda*. The possible presence in northern Yunnan of the true *T. piniperda* needs to be clarified, although so far no scientific findings support this hypothesis (Duan *et al.*, 2004; Kirkendall *et al.*, 2008). The host range of *T. yunnanensis* is given in Table 10.1, but with the recent discovery of *T. armandii* (Li *et al.*, 2010), the host status of *P. armandii* Franchlet for the Yunnan pine shoot beetle needs to be clarified. Breeding experiments have shown that the Yunnan pine shoot beetle can breed successfully in *P. yunnanensis* and *P. armandii*, but smaller beetles were produced in the latter (Zhao and Långström, 2012). Until now, *T. armandii* has only been found in shoots in the field (Li *et al.*, 2010).

The life cycle of *T. yunnanensis* is closer to that of *T. destruens* than to *T. piniperda*, as it has an extended flight period and several sister broods. Pine shoot beetles have been studied for a long time in Yunnan, and the accumulated knowledge has been compiled in a special volume containing 16 contributions on the biology, damage, and control of *T. yunnanensis* (although reported as *T. piniperda*) and *T. minor* (Anonymous, 1997). Like in the Mediterranean area, the flight period of *T. yunnanensis* in Yunnan may start in November, but the main flight occurs in February–March and then there is a sister brood flight in April–May (Ye, 1991; Li *et al.*, 1993). Hence, flying beetles can be found during half of the year (Li *et al.*, 1993). During flight, the beetles are guided to their host material by chemical cues, mainly terpenoid compounds (Zhou *et al.*, 1997). Liu *et al.* (2010) found that infested pine bolts attracted more beetles than uninfested pines, which indicates that attracting compounds were released, but the nature of these is not explained, i.e., whether they were host or pheromonal compounds. As peeled logs of host material attracted more beetles than unpeeled ones (Lu *et al.*, 2012a), it seems obvious that host odors are causing the attraction during the spring flight to the host material.

After finding a host tree, oviposition often takes place in the middle or upper part of the stem of weakened standing trees (Lu *et al.*, 2012b). As *T. yunnanensis* and *T. minor* often coexist on the same trees (Ye and Ding, 1999; Lu *et al.*, 2012b), there may be competition between the two species although the latter species mainly is found on the lower part of the stem (in contrast to the usual situation in Europe, but see Masutti, 1959). The female constructs the longitudinal

egg gallery and oviposits an average of 76 eggs (Ye, 1991). The developmental time decreased with increasing temperature and the egg, larval, and pupal stages required 96, 358, and 116 day-degrees, respectively (Ye, 1994). The development period lasts 1.5–2 months (Ye, 1991), and the time span from mean flight to mean emergence ranged from 86 to 94 days in two laboratory studies (Chen, 2003; Zhao, 2003). The body size of the emerging beetles decreases with time (Zhao and Långström, 2012), probably as an effect of intraspecific competition for food. There is one generation per year, but the parent beetles re-emerge after oviposition and after a shoot feeding period, establish a sister brood with fewer eggs in April–May (Ye, 1991). Upon emergence, the new generation flies to the pine crowns for their maturation feeding in the pine shoots.

Like *T. destruens* in southern Europe, the pine shoot beetles in Yunnan can be found in the shoots from spring to late autumn (Ye, 1991). In early season, the previous year shoots are attacked and when the newer shoots have developed they are preferred by *T. yunnanensis*. All crown levels are affected by the attacks (Ye and Li, 1994; Långström *et al.*, 2002), and each beetle may tunnel several shoots during the extended shoot-feeding phase (Ye, 1991, 1996). In Yunnan pine, the mean diameter of tunneled shoots was 7–8 mm (Ye, 1996), but shoots up to 10 mm are attacked (Ye, 1991). Gao *et al.* (2012) reported that chemical cues guide the beetles to the shoots. It has also been found that *T. yunnanensis* may share the same tree not only with *T. minor* but also with *T. brevipilosus*, and that the species prefer to attack shoots differently in time and space, thus possibly reducing competition (Lu *et al.*, 2014). *Tomicus yunnanensis* attacked earlier than *T. brevipilosus* and preferred the base of the shoot, while *T. minor* attacked close to the apical bud. Shoot damage may be very intense and affect more than 80% of all shoots, and there has been some evidence for an aggregation of beetles to certain trees (Ye and Li, 1994; Ye and Lieutier, 1997). This could explain why some trees were attacked more than others in the same stand (Långström *et al*, 2002). The beetles move directly from the shoots to the trunk and intensive shoot feeding may predispose trees for subsequent stem attacks; thus, the beetles are capable of creating their own breeding materials (Lieutier *et al.*, 2003). Lieutier *et al.* (2003) demonstrated that the level of shoot damage and stem attacks were strongly related and concluded that this phenomenon was the main reason for the aggressiveness of *T. yunnanensis*. It has also been suggested that chemical compounds like trans-verbenol and myrtenol may play a role in the aggregation process (Ye, 1993).

4.6 Other *Tomicus* Species

Tomicus brevipilosus was described by Schedl (1946) as a species very similar to *T. piniperda* and although he did not list any host plants, he stated that it occurs in China and Japan. Browne (1968) summarized the biology of *T. brevipilosus* (using the name *T. khasianus*) as follows. It occurs in northeastern India and attacks *Pinus kesiya* Royle ex Gord. It has a typical *Tomicus* biology with longitudinal egg galleries and it breeds under the bark of weak trees. The shoot feeding in the crowns of its host tree is the principal damage done by this species, resulting in deformed crowns and sometimes in tree death. Recently, a microsatellite library of *T. brevipilosus* was constructed (Cao *et al.*, 2012), and the antennal morphology and sensilla ultrastructure of this species was compared to that of *T. yunnanensis* and *T. minor* (Wang *et al.*, 2012). Very little is known about the shoot-feeding habits of *T. brevipilosus*, although it has been shown that the feeding period of this species lasts longer (9 months) and starts later (June) than for *T. yunnanensis* and *T. minor* (Lu *et al.*, 2014).

Tomicus pilifer is called the Korean pine shoot beetle as it mainly lives on Korean pine (*P. koraiensis* Seib. Ex Succ.) (Wang, 1981), but it is also recorded from other Asian pines (Table 10.1). It occurs in the Far East in the Amur and Ussuri regions in Russia, close to the Chinese border (Schedl, 1946), and also around Vladivostok (Stark, 1952). The biology of *T. pilifer* was described from the Heilongjiang Province in northern China by Wang (1981). It shows some similarities with that of the pine shoot beetles in northern Europe. The adult beetles hibernate in the soil below the tree canopy (like *T. minor*). They emerge in spring (early May) and move to the shoots to complete their maturation feeding in the previous year shoots. The spring flight takes place in early June (occasionally in mid-May), when breeding material is located and the females start excavating the typical longitudinal egg galleries. Pine material of 10 cm diameter is preferred but a wide range of fresh pine (fallen trees, logging waste, or standing weakened trees) can serve as breeding material. After mating in the gallery, the female lays 32–74 eggs. The larvae feed until late June and the pupation stage lasts 5–17 days. Most callow adults emerge in mid-July but emergence continues until October. The shoot damage mainly occurs in the current shoots and the beetles enter at the shoot base. Each beetle may attack several shoots and the beetles fall to the ground with the broken shoots in September, followed by the beetles moving to the soil for hibernation. The beetles mainly attack stressed trees and the shoot damage is more severe in the upper crown and on the sunny side of the crown. The study reports shoot damage ranging from 1 to 14% on mature Korean pines.

Tomicus puellus mainly lives on spruce (*P. jezoensis* (Siebold and Zucc.) Carrière), but occasionally also on Korean pine (Schedl, 1946), and according to Stark (1952) even on two Asian firs (*Abies holophylla* Maxim and *A. nephrolepis* (Trautv. ex Maxim.)). The galleries resemble those of *T. piniperda* but they go deeper into the sapwood. The distribution area follows that of *P. jezoensis* in the Far East (Schedl, 1946). According to Maslov *et al.* (1988), *T. puellus* flies and colonizes fallen or standing host trees in April–May, and the new generation emerges in July. In the mountains and in northerly areas the development takes place later. There is no description on its shoot-feeding habits.

5. BIOTIC ASSOCIATIONS

As for all bark beetle genera, a large diversity of organisms is associated with the genus *Tomicus*. The kind of relationships refers to competition, commensalism, symbiosis, parasitism, parasitoidism, or predation, and involves many different taxonomic groups, such as protozoa, fungi, nematodes, insects, and mites. The biotic associations involving *Tomicus* have been described and studied mainly in *T. piniperda* and *T. minor*, and in much less detail in *T. destruens*. A very limited amount of data is available for *T. yunnanensis* and none for the other Asian *Tomicus* species. The discussion is thus based on the three European species while information on *T. yunnanensis* is given on occasion.

5.1 Viruses, Bacteria, and Unicellular Eukaryotes

According to the synthesis by Wegensteiner (2004) on bark beetle pathogens, no information has been published on the occurrence of viruses and bacteria in the genus *Tomicus*. However, the few papers reporting on bark beetle-associated bacteria reveal a rather high diversity of bacteria in bark beetle species belonging to *Ips*, *Scolytus*, *Dendroctonus*, *Xyloterus*, *Trypodendron*, and *Anisandrus* (Wegensteiner, 2004; Chapter 7). Bacteria are thus very common microorganisms associated with bark beetles, and probably also with *Tomicus*.

Unicellular eukaryotes reported as associated with bark beetles are microsporidia, Apicomplexa, and Rhizopoda (Wegensteiner, 2004; Chapter 7). In microsporidia, *Canningia tomici* (Kohlmayr *et al.*) occurs in cells of indigenous *T. piniperda* from several European localities, as well as in insects introduced in North America (Kohlmayr *et al.*, 2003). This pathogen can infest the midgut epithelium, the fat body, and the gonads. *Chytridiopsis typographi* (Weiser) is a non-specific microsporidium that has been reported from 4.7% of the individuals in *T. piniperda* (Burjanadze *et al.*, 2011). In Apicomplexa, several species of *Gregarina* have been detected in the midgut lumen of various bark beetle species, including *T. minor* (Kohlmayr, 2001). *Malamoebia scolyti* (Purrini) is an intracellular Rhizopoda that has been experimentally and successfully transferred from other bark beetles to *T. piniperda* (Kirchhoff and Führer, 1990; Wegensteiner, 2004). However, it has not been reported from natural *Tomicus* populations.

5.2 Fungi

5.2.1 Non-pathogenic Fungi

Yeasts are common associates of bark beetles and the association can be unspecific (Six, 2003). Some yeasts are known to be nutritionally important for some bark beetles (Whitney, 1982; Six, 2003) and some unidentified yeasts have been reported as associates of *T. minor* and *T. piniperda* (Mathiesen, 1950; Mathiesen-Käarik, 1953). The vast majority of the non-pathogenic fungi associated with bark beetles are ophiostomatoid ascomycetes (Table 10.2).

There are minor differences between the fungal assemblages of the three European species (*T. piniperda*, *T. minor*, and *T. destruens*). Indeed, Table 10.2 shows that 62% of the European fungal species are common to at least two European beetle species and 14% to three, whereas in Europe, *T. piniperda* shares 65% of its fungal species, *T. minor* 100%, and *T. destruens* 83%, with at least one other *Tomicus* species. Only the ubiquitous *Ophiostoma minus* (Hedgc.) Syd. and P. Syd. is shared between the European species and the Asian *T. yunnanensis* (Table 10.2). This sharing of fungal species among European *Tomicus* species is likely due to sharing the same host as well as geographic areas. *Tomicus piniperda* and *T. minor* are largely sympatric while the natural distributions of *T. destruens* and *T. piniperda* overlap at some places in the Mediterranean (Horn *et al.*, 2009, 2012).

Although the situation for Asian species is not known, fungal specificity could be more related to host tree species than to beetle species. Such a possibility has already been reported by Harrington (1993) and Kirisits (2004) for bark beetles in general. It is possible that fungal specificity relates also to a region or a beetle genus. There are some fungal species associated with *Tomicus* that have an extremely wide spectrum of bark beetle hosts and tree hosts. This is the case for *Ceratocystiopsis minuta* (Siemaszko) Upadhyay and W. B. Kendr., *Grosmannia piceiperda* (Rumbold) Gold., *Ophiostoma floccosum* Math.-Käarik, and *Ophiostoma piliferum* (Fr.) Syd. and P. Syd., which can develop on spruce as well as on pines, and especially for the largely unspecific *Ophiostoma piceae* (Münch) Syd. and P. Syd., which has been reported from all European genera of conifers as well as from broadleaved trees (Kirisits, 2004).

Some fungal species are clearly more frequently found with a particular *Tomicus* species (Table 10.2). *Ophiostoma tingens* (Lagerb. and Melin) de Beer and M. J. Wingfield is very often cited as associated with *T. minor* but it can be found occasionally in the galleries of *T. piniperda*. Its frequency of association with *T. minor* averages 45% but it can rise up to 86% in certain regions (Mathiesen, 1950; Rennerfelt, 1950; Mathiesen-Käarik, 1953).

Ophiostoma canum (Münch) Syd. and P. Syd. is very frequently associated with *T. minor* but less frequently

TABLE 10.2 Non Insect-pathogenic Fungal Species Associated with Four *Tomicus* Species

Fungal Species	T. piniperda	T. minor	T. destruens	T. yunnanensis
Ceratocystiopsis minuta (Siemaszko) Upadhyay and W. B. Kendr.	+11,15,23	+23,24,25		
C. autographa Bakshi	+17			
Graphium fragrans Math.-Käarik				+27
G. pseudormiticum M. Mouton and M. J. Wingfield	+11	+9		
G. pycnocephalum Grosmann	+11	+12		
Grosmannia galeiformis (Bakshi) Zipfel, de Beer and M. J. Wingfield	+38			
G. huntii (Rob.-Jeffr.) Zipfel, de Beer and M. J. Wingfield	+3,6			
G. koreana (Masuya, Kim and M. J. Wingfield) Lu, Decock and Maraite	+13			

Continued

TABLE 10.2 Non Insect-pathogenic Fungal Species Associated with Four *Tomicus* Species—cont'd

Fungal Species	T. piniperda	T. minor	T. destruens	T. yunnanensis
G. piceiperda (Rumbold) Gold.	+11,15,33			
G. yunnanensis Yamaoka, Masuya and M. J. Wingfield				+++20,36,37
Leptographium euphyes K. Jacobs and M. J. Wingfield	+6, 7			
L. guttulatum M. J. Wingfield and K. Jacobs	+6, 8	+6,8,15	+31	
L. lundbergii Lagerb. and Melin	+3,(5),6,11,23,25	+12,23,25	+31	
L. procerum (W. B. Kendr.) M. J. Wingfield	+3,(5),6,11	+12		
L. serpens (Goid.) M. J. Wingfield.			+31	
L. wingfieldii M. Morelet	+++3,(5),6, (10),11,15,18,26, 28,33,34		+++1,19,31	
Ophiostoma canum (Münch) Syd. and P. Syd.	+ 11,16,23,25,29	+++ 2,12,14,17,23,24, 25,29		
O. floccosum Math.-Käärik.	+21			
O. ips (Rumbold) Nannf.	+(5),25		+1	
O. minus (Hedgc.) Syd. and P. Syd.	+++4,(5),11,15, 17,18,22,23,25,28, 29,32,33	+4,12,23,25, 29, 35	+1	+35
O. piceae (Münch) Syd. and P. Syd.	+4,11,15,16,22, 23,25,32,33	+2,12,23,25		
O. piliferum (Fr.) Syd. and P. Syd.	+3,11,14,17,23, 25,29,32,33	+2,4,23,25,29,32		
O. quercus (Georgev.) Nannf.				+27
O. tingens (Lagerb. and Melin) de Beer and M. J. Wingfield	+23,25,29	+++2,12,14,23, 2,29,30		

The nomenclature of ophiostomatoid fungi proposed by de Beer et al. (2013) is used. Extremely rare fungal species and those with doubtful taxonomic status or determined at the genus level only (Kirisits, 2004) are not mentioned. No data are available on fungi associated with the Asian endemic Tomicus species others than T. yunnanensis. +: Fungal species present in at least one beetle population; +++: Fungal species very commonly associated with a beetle species (present in all studied populations); ( ): species isolated in North America. Numbers indicate references.
References: *(1) Ben Jamâa et al., 2007[a]; (2) Francke-Grosmann, 1952; (3) Gibbs and Inman, 1991; (4) Grosmann, 1931; (5) Hausner et al., 2005; (6) Jacobs and Wingfield, 2001; (7) Jacobs et al., 2001b; (8) Jacobs et al., 2001a; (9) Jacobs et al., 2003; (10) Jacobs et al., 2004; (11) Jankoviak, 2006; (12) Jankoviak, 2008; (13) Kim et al., 2005; (14) Kirisits, 2004; (15) Kirisits et al., 2000; (16) Kirschner, 1998 in Kirisits, 2004; (17) Kotýnková-Sytchrová, 1966; (18) Lieutier et al., 1989b; (19) Lieutier et al., 2002[a]; (20) Lieutier and Ye, unpublished; (21) Lin, 2003 in Kirisits, 2004; (22) MacCallum, 1922; (23) Mathiesen, 1950; (24) Mathiesen, 1951; (25) Mathiesen-Käärik, 1953; (26) Morelet, 1988; (27) Paciura et al., 2010; (28) Piou and Lieutier, 1989; (29) Rennerfelt, 1950; (30) Rollins et al., 2001; (31) Sabbatini Peverieri et al., 2006; (32) Siemaszko, 1939; (33) Solheim and Långström, 1991; (34) Wingfield and Gibbs, 1991; (35) Ye et al., 2000; (36) Ye et al., 2004; (37) Zhou et al., 2000; (38) Zhou et al., 2004.*
[a]Reference to T. piniperda in the natural area of T. destruens according to Horn et al. (2012) and thus considered to refer to T. destruens.
(modified from Kirisits, 2004)

associated with *T. piniperda* (Table 10.2). Its frequency of association with *T. minor* averages 52% with little variations between regions, whereas it is lower than 4% with *T. piniperda* (Mathiesen, 1950; Rennerfelt, 1950; Mathiesen-Käarik, 1953).

Ophiostoma minus has been reported from all investigated *Tomicus* species (Table 10.2). Its frequency of association with *Tomicus* varies considerably depending on beetle species and regions, and ranges from 0 to 46% in *T. piniperda* (Mathiesen, 1950; Rennerfelt, 1950; Mathiesen-Käarik, 1953; Lieutier *et al.*, 1989b; Solheim and Långström, 1991; Jankoviak, 2006), and from 0 to 9% in *T. destruens* (Ben Jamâa *et al.*, 2007).

Leptographium wingfieldii M. Morelet has been reported only from *T. piniperda* and *T. destruens* (Table 10.2). Its frequency of association with *T. piniperda* varies between localities and years, from 0 to 71% (Lieutier *et al.*, 1989b; Gibbs and Inman, 1991; Solheim and Långström, 1991) and from 0 to 86% with *T. destruens* (Lieutier *et al.*, 2002; Sabbatini-Peverieri *et al.*, 2006; Ben Jamâa *et al.*, 2007). Comparisons of isolates from Sweden and France have shown that *L. wingfieldii* was more adapted to low temperatures than *O. minus* (Lieutier and Yart, 1989; Solheim *et al.*, 2001). In laboratory cultures, it was able to grow rapidly even at 10°C (15°C was necessary for a similar growth of *O. minus*), whereas its optimal temperature was 25°C (30°C for *O. minus*).

Grosmannia yunnanensis Yamaoka, Masuya and M. J. Wingfield is a usual associate of *T. yunnanensis* in Yunnan and can be found at a high frequency of association in all localities where the beetle is present (Ye *et al.*, 2004; Lieutier and Ye, unpubl.). The fungal flora of the other Chinese *Tomicus* has not yet been investigated.

The non-insect-pathogenic fungal flora associated with each bark beetle species (Table 10.2) form a complex of several fungal species that could be interfering with each other, with the beetle, and with other organisms present in the galleries (Klepzig *et al.*, 2001a, b; Klepzig and Six, 2004; Six and Klepzig, 2004). The relative roles of the different species belonging to these fungal complexes are not clear and are subject to controversy (Berryman, 1972; Christiansen *et al.*, 1987; Paine *et al.*, 1997; Lieutier *et al.*, 2009; Six and Wingfield, 2011; Chapters 6 and 8).

The benefits of fungi being associated with *Tomicus* species include transportation and introduction into new host trees. The spores possess special adaptations for transport, such as sticky and thick walls allowing attachment to the insect, resistance to desiccation, and protection from digestive enzymes (Mathiesen-Käarik, 1960; Francke-Grosmann, 1966). However, *O. tingens* is an exception since its spores are thin-walled and delicate, suggesting transportation in a mycangium, but no such structure has been observed on *T. minor* (Kirisits, 2004).

Fungi as a source of nutrients have not been investigated for *Tomicus*. According to Francke-Grosmann (1966),

Tomicus species are either phloeophagous (exclusively feed on phloem) or phloeomycetophagous (feed on both phloem and fungi). *Tomicus minor* belongs to the second category since the old larvae and young adults feed on conidia and mycelium of *O. tingens*, a behavior similar to that of the ambrosia beetles (Francke-Grosmann, 1952, 1966). However, mycangia seem absent in this beetle, although their presence is generally associated with such a feeding behavior (Six, 2003). *Tomicus piniperda*, *T. destruens*, and *T. yunnanensis* are truly phloeophagous as their associated fungi have never been reported as a food source. Nevertheless, some fungi might be used as an additional source of food for larvae and teneral adults, as suggested in other phloeophagous species (Francke-Grosmann, 1967; Whitney, 1982; Six, 2003; Klepzig and Six, 2004; Harrington, 2005).

The frequency of association between *Tomicus* species and *O. minus* is very variable, even in the same forest (Lieutier *et al.*, 1989b). Moreover, colonization of trees by *T. piniperda* can be impeded in the zones of the trunk where *O. minus* is present suggesting that *O. minus* is detrimental for the beetle (Piou and Lieutier, 1989). These observations raise questions about the real role of this fungus as an associate of *Tomicus*. *Ophiostoma minus* resembles a species that uses the beetle as a vector but which becomes a competitor once it has arrived on its substrate. It has also been considered as a bark beetle competitor in trees colonized by *Dendroctonus frontalis* Zimmermann in North America, where mutualistic mycangial fungi act as antagonists of *O. minus* (Klepzig *et al.*, 2001a, b).

When mass inoculated above a certain threshold of inoculation density, *O. minus*, *L. wingfieldii*, and *G. yunnanensis* can invade the sapwood and eventually cause tree death (Solheim *et al.*, 1993; Croisé *et al.*, 1998a; Lieutier *et al.*, 2004; Ben Jamâa *et al.*, 2007; Sallé *et al.*, 2008). The corresponding inoculation density threshold (which measures fungal virulence) has been determined at 800 inoculations points per m^2 for *O. minus* on Scots pine (low virulence), 300 to 400 inoculations points per m^2 for *L. wingfieldii* on Scots pine (high virulence), and 400 inoculations points per m^2 for *G. yunnanensis* on Yunnan pine (Solheim *et al.*, 1993; Croisé *et al.*, 1998a; Sallé *et al.*, 2008). Sapwood occlusion strongly correlates with loss of hydraulic conductivity (Croisé *et al.*, 2001; Sallé *et al.*, 2008). These experiments demonstrate that these fungi can be pathogenic under certain conditions but they do not prove that they are responsible for killing the tree at the moment of beetle attacks. Indeed, there is no relation between fungal virulence and beetle aggressiveness (Paine *et al.*, 1997; Lieutier, 2002; Harrington, 2005; Lieutier *et al.*, 2009). It has even been suggested that blue-stain fungi invade the sapwood only after beetle attack has succeeded (Lieutier *et al.*, 2009).

When isolated inoculations are artificially performed on trees, *O. minus*, *L. wingfieldii*, and *G. yunnanensis* induce the development of a hypersensitive reaction in their host (Lieutier *et al.*, 1989b, 2004; Solheim *et al.*, 2001; Ben Jamâa *et al.*, 2007; Sallé *et al.*, 2008). In most beetle–fungi associations attacking living trees, such an induction would be a benefit brought by fungi to their beetle hosts through contributing in weakening the tree and thus facilitating beetle establishment (Lieutier, 2004; Lieutier *et al.*, 2009). *Leptographium wingfieldii* would seem a valuable candidate to play this role for *T. piniperda* because of its ability to induce very violent tree reactions (Lieutier *et al.*, 1989b). However, for such an induction, the inoculum must contain at least 15,000 spores, a number certainly higher than that carried by an insect (Lieutier *et al.*, 1989a). It is very likely that *T. piniperda* establishes on trees without the help of a fungus (Lieutier *et al.*, 1989a, 1995; Lieutier, 2002, 2004).

5.2.2 Insect-pathogenic Fungi

It is not rare to observe mycosed dead bark beetles under the bark of trees. Beetle galleries are indeed a very favorable medium for the development of fungal pathogens because their humidity is frequently high. Many records can be found in the literature dealing with bark beetle pathogenic fungi, especially *Beauveria bassiana* (Bals.-Criv.) Vuillemin (Ascomycota: Hypocreales), but surprisingly, only two involve *Tomicus* species: *T. piniperda* and *T. destruens* (Kirschner, 2001; Wegensteiner, 2004; Chapter 7). *Beauveria bassiana* has been reported to naturally occur in the galleries and its presence can be quite common (Triggiani, 1984; Jankoviak and Bilanski, 2007; Burjanadze, 2010; Takov *et al.*, 2012). Papers dealing with *Tomicus* and *B. bassiana* are discussed in Section 9.2.

5.3 Nematodes

Bark beetle galleries provide favorable microclimatic conditions as well as abundant and much diversified food resources for nematodes, resulting in a high diversity of saprophytic or parasitic species (Rühm, 1956). Saprophytic species are those whose life cycle takes place entirely inside the galleries (Rühm, 1956; Poinar, 1972). The resistant phoretic third instar (L3) larvae (dauerlarvae) of some species can attach themselves on various places of the insect cuticle and be transported from one tree to another by the adult beetles. The life cycle of parasitic species alternates between the internal tissues of the insect and the gallery (Rühm, 1956; Poinar, 1972). Among them, certain species infect the insect with larval forms (L3) only and are thus generally not considered as truly parasitic, whereas the true parasites infect the insect with adult, eggs, and larval forms. The species that are not truly parasitic are found in the gut or

the body cavity, whereas the true parasites can infest the body cavity, the Malpighian tubules, the fat body, or the ovaries.

Table 10.3 summarizes most *Tomicus*–nematodes associations in their natural areas, together with their location in the insects or their galleries. Certain specificity is recognizable at the nematode species level. This is particularly clear when comparing the nematodes of *T. piniperda* with those of *T. minor*. Only *Micoletzkia* sp. seems common to these two beetle species, although they live in the same forests and can attack the same trees. However, one third of the nematode species associated with *T. piniperda* are associated with *T. destruens*, although the natural areas of the beetles largely differ. Nematode genera exhibit no specificity, even for a beetle genus, as most nematode genera in Table 10.3 are also known from bark beetle genera other than *Tomicus* (Rühm, 1956) and even other insect families.

All nematode species have at least part of their life cycle in the beetle galleries and most have stages in or on the beetle host (Table 10.3). In most cases, only the L3 larvae are found in or on the insect. The true parasitic species (*Allantonema* sp., *Allantonema morosum* (Fuchs), *Neoparasitylenchus* sp., *Parasitorhabditis fuchsi* Blinova and Gurando, *Parasitorhabditis piniperdae* (Fuchs), *Parasitylenchus* sp., *Parasitylenchus macrobursatus* Blinova and Gurando, and *Prothallonema tomici* Nedelchev, Takov and Pilarska) are limited in number (8/33) and have been found in the insects' body cavity or ovaries. A particularity of the *Tomicus* genus is that its gut does not appear to host parasitic nematode species. Even the genus *Parasitorhabditis*, of which the L3 larvae are classical gut parasites of many bark beetle genera, is found in the body cavity in *Tomicus*. In France, the L3 larvae of *P. piniperdae* have been found in 95 to 100% of the emerging beetles and in 35% of the ovipositing beetles, in both *T. piniperda* and *T. destruens* (Laumond and Carle, 1971; Lieutier and Vallet, 1982). In Germany, Rühm (1956) observed 10–15% of parasitized *T. piniperda* adults. *Parasitaphelenchus papillatus* Fuchs is also a classical parasite with L3 larvae in the body cavity of larval and adults stages of the insect. Rühm (1956) reported that 28 to 33% of *T. piniperda* could be infected with nematodes, while Slobodyanyuk (1973) reported 37% parasitism in pupae and 84% in emerging beetles. Meanwhile, Lieutier and Vallet (1982) found L3 larvae in 62% of the adults in shoots and in 40% of the ovipositing beetles. In *T. destruens*, *P. papillatus* L3 larvae have been observed in 20 to 91% of the adults (Laumond and Carle, 1971). *Parasitaphelenchus piniperdae* and *P. papillatus* frequently parasitize the same individual beetles, to their detriment (Lieutier and Vallet, 1982). *Bursaphelenchus piniperdae* (Fuchs) Rühm is also common in *T. piniperda* populations. Adult *B. piniperdae* have been observed in the galleries in all insect stages (Lieutier and Vallet, 1982). The L3 larvae are phoretic under the elytra of 4% of the adults (Rühm, 1956). An undetermined

TABLE 10.3 Main Nematode Species Associated with Three *Tomicus* species and Their Location in the Beetle

Nematode Species	*T. piniperda*	*T. minor*	*T. destruens*	Nematode Location in Beetles
Allantonema sp.		+6,8		Body cavity; ovaries
A. morosum (Fuchs)	+19			Body cavity
Bursaphelenchus sp.			+11	Metathoracic intercoxas
B. hellenicus Skarmoutsos, Braasch and Michalopoulou	+14		+16	Third instar larvae on beetles
B. leoni Baujard			+16	
B. piniperdae (Fuchs) Rühm	+++4,12,15			Under elytra
B. sexdentati Rühm	+2			Third instar larvae on beetles
B. teratospicularis Kakulia and Devdariani		+10		Third instar larvae on beetles
B. xylophilus (Steiner and Buhrer) Nickle	+21			Third instar larvae on beetles
Cryptaphelenchus aedili (Laz.)		+6		Galleries
C. viktoris (Fuchs) Goodey	+4,15			Third instar larvae in body cavity
Ektaphelenchus sp.		+6		
Macrolaimus canadensis Sanwal	+12		+11	Metathoracic intercoxas
Micoletzkia sp.	+12	+6	+11	Galleries
M. cordovector Kakuliya		+9		Under elytra
Neoditylenchus eremus (Rühm)		+6,7		Under elytra and on cuticle of adults; gut of larvae
Neoparasitylenchus sp.	+12			Adults in body cavity
Panagrodontus breviureus Kakuliya		+6		Galleries
Panagrolaimus detritophagus Fuchs	+3			
P. ruehmi (Ivanova)		+6		
P. tigrodon Fuchs	+3,19		+11	Metathoracic spiracles; under elytra
Parasitaphelenchus sp.		+++6,8		Third instar larvae in body cavity; ovaries
P. ateri (Fuchs) Rühm	+19			Third instar larvae in body cavity
P. papillatus Fuchs	+++11,12,15,17,19		+11,20	Third instar larvae in body cavity; seminal vesicle
Parasitorhabditis fuchsi Blinova and Gurando		+1,6		Larvae and adults in body cavity

Continued

TABLE 10.3 Main Nematode Species Associated with Three *Tomicus* species and Their Location in the Beetle—cont'd

Nematode Species	*T. piniperda*	*T. minor*	*T. destruens*	Nematode Location in Beetles
P. piniperdae (Fuchs)	+++11,12,15,18,19		+11,20	Third instar larvae in body cavity; larvae and adults in body cavity
Parasitylenchus sp.		+8		Ovaries
P. macrobursatus Blinova and Gurando		+5,6		Adults in body cavity
Prothallonema tomici Nedelchev, Takov and Pilarska	+13			Larvae and adults in body cavity
Rhabdontolaimus carinthiacus Fuchs		+6		Galleries
Ruidosaphelenchus janasii Launond and Carle			+11	Metathoracic spiracles

Extremely rare species and those with doubtful taxonomic status are not included. No data are available on nematodes associated with the Asian Tomicus *species. +: species present in at least one beetle population; +++: species very commonly associated with a beetle species (present in all studied populations). Numbers indicate references.*

References: *(1) Blinova and Gurando, 1974; (2) Braasch et al., 1999; (3) Fuchs, 1930; (4) Fuchs, 1937; (5) Gurando, 1974; (6) Gurando, 1979; (7) Gurando, 1990; (8) Gurando and Tsarichkova, 1974; (9) Kakuliya, 1966; (10) Kakuliya and Devdariani, 1966; (11) Laumond and Carle, 1971; (12) Lieutier and Vallet, 1982; (13) Nedelchev et al., 2011; (14) Penas et al., 2006; (15) Rühm, 1956; (16) Skarmoutsos et al., 1998[a]; (17) Slobodyanyuk, 1973; (18) Slobodyanyuk, 1974; (19) Tomalak et al., 1984; (20) Triggiani, 1984[a]; (21) Xu et al., 1993.*
[a]Reference to T. piniperda *in the natural area of* T. destruens *according to Horn et al. (2012) and thus considered to refer to* T. destruens.

Parasitaphelenchus frequently found in *T. minor* populations (Table 10.3) has been reported as a true parasite of *T. minor*'s ovaries (Gurando and Tsarichkova, 1974).

The parasitism by L3 dauerlarvae has no or a very limited effect on the host. Tomalak *et al.* (1984) report some damage in seminal vesicles of *T. piniperda* parasitized by larvae of *P. papillatus*. Alternatively, the true parasitism can have more drastic effects. Gurando and Tsarichkova (1974) indicate that species of *Allantonema*, *Parasitaphelenchus*, and *Parasitylenchus* can enter the ovaries of *T. minor*, resulting in partial or complete destruction of these organs. Very few investigations have been conducted on the possibility of using nematodes for biological control of bark beetles. Triggiani (1983) has tested the susceptibility of *T. destruens* to *Heterorhabditis bacteriophora* (Poinar), *Heterorhabditis heliothidis* (Khan, Brooks and Hirschmann) Poinar and *Steinernema feltiae* (Filipjev).

5.4 Mites (Acarina)

Many mite species can be found in the galleries of bark beetles, some feeding on fungi or nematodes, while others are predators or parasites of bark beetles eggs and larvae (Chapter 6). Most mite species can attach themselves to the cuticle of the emerging beetles, which they use as vehicles to find new host trees. Predaceous mites tend to be more specific to a habitat than a host (Lindquist, 1970). Consequently, galleries of *Tomicus* species share a large

number of mite species with other European bark beetles species attacking pines, such as *Ips sexdentatus* (Boerner), *Ips acuminatus* (Gyllenhal) or *Orthotomicus erosus* (Wollaston). However, few mite species have been reported. In Poland, *Pyemotes herfsi* (Oudemans) (Pyemotidae), *Dendrolaelaps krantzi* (Wisniewski) (Digamasellidae), and *Trichouropoda obscura* (Koch) (Trematuridae) have been found in the galleries of *T. piniperda* in Scots pine, with *T. obscura* representing sometimes 80% of all mite individuals (Wisniewski, 1980; Kielcziewski *et al.*, 1983; Kaczmarek *et al.*, 1992). *Proctolaelaps fiseri* (Vitzthum) and *Proctolaelaps xyloteri* (Samsinak) (Ascidae), the first being the most abundant, have been found also in *T. piniperda* galleries in Scots pine in Russia (Andreev, 1988). *Proctolaelaps* species have been reported to be predators of early stages of bark beetles (Lindquist, 1970). *Cercoleipus coelonotus* Kinn (Cercomegistidae) was found in the galleries of *T. destruens* in *P. pinea* in Italy, feeding on nematodes or other acari (Sabbatini-Peverieri and Francardi, 2010).

5.5 Insects

A large number of studies have dealt with insects associated with *Tomicus*, representing a large variety of feeding behaviors. It is often difficult to determine their exact relationships with bark beetles, as a large diversity of feeding resources exist in the beetles' galleries (e.g., frass, decaying wood, dead organisms, fungi, prey such as nematodes or insects, etc.). Owing to their possible use in biological control, parasitoids and predators have received widespread attention. Below, we mainly consider parasitoid and predatory species that have been clearly recognized as natural enemies of bark beetles.

Table 10.4 lists the main parasitoid species associated with *T. piniperda*, *T. minor*, and *T. destruens*. No data are available for the Asian *Tomicus* species. Most parasitoids are common to *T. piniperda* and *T. minor*. Although less parasitoid species have been reported for *T. destruens*, most of them also parasitize the other two beetle species. This supports the hypothesis that parasitoids tend to be more host-tree specific than host specific (Kenis *et al.*, 2004). However, a large diversity of situations exists. *Dendrosoter middendorfi* Ratzeburg, *Metacolus azureus* (Ratzeburg), and *Roptrocerus xylophagorum* (Ratzeburg), are found only in the galleries of bark beetles attacking trees of the Pinaceae family, whereas *Dendrosoter protuberans* (Nees), *Heydenia pretiosa* (Förster), and *Eurytoma morio* Boheman are found in a variety of tree families, including broad leaved trees (Mendel, 1986). Among Braconidae, *Coeloides abdominalis* (Zetterstedt) and *D. middendorfi* are largely represented, but *C. abdominalis* is not found with *T. destruens*, whereas *Dendrosoter flaviventris* (Förster) appears particular to *T. destruens*. Among Pteromalidae, *Metacolus unifasciatus* (Förster), *Rhopalicus tutela* (Walker), and *Roptrocerus brevicornis* Thomson are the most frequent species in *T. piniperda* and *T. minor*, to which *R. xylophagorum* must be added for *T. piniperda*. *Metacolus unifasciatus* has been considered the most important parasitoid of *T. destruens* (Halperin, 1978; Mendel *et al.*, 1986). Most are larval ectoparasitoids and lay their eggs through the bark (Kenis *et al.*, 2004). Because of this oviposition behavior, they can parasitize a very high proportion of beetle larvae in the thin bark zones, sometimes up to 100% (Mendel, 1986). Few species of Eupelmidae and Eurytomidae have been reported as parasitoids of *Tomicus*. *Eurytoma morio* and *E. arctica* Thomson could also be hyperparasitoids of Braconidae or Pteromalidae (Nuorteva, 1957).

Table 10.5 lists the main predators of the European *Tomicus* species. All species found with *T. minor* are associated with *T. piniperda*. Many species associated with *T. piniperda* are, however, not associated with *T. minor*. This is likely a result of research efforts focused mainly on *T. piniperda* due to its economic importance. Similarly, most species associated with *T. destruens* are also associated with *T. piniperda*, with some exceptions. The situation of *Thanasimus dubius* (F.) is of particular interest because it occurs in Asia and in North America, but not in Europe. It has been observed preying on *T. piniperda* in North America (Kennedy and McCullough, 2002). A similar situation exists for *Enoclerus nigripes* (Say), *Cylistix gracilis* (LeConte), *Platysoma cylindrica* (Paykull), *Corticeus parallelus* (Melsheimer), and *C. praetermissus* (Fallén) (Bright, 1996;

Kennedy and McCullough, 2002), clearly demonstrating the general adaptability of predators.

Thanasimus species are very voracious predators, which often play an important role in regulating *Tomicus* populations. *Thanasimus formicarius* (L.) is the most important predator. Their eggs are laid in bark crevices close to the entrance hole of the beetle galleries and the larvae hunt their prey inside galleries. In laboratory rearing, each of its larvae could kill an average of 1.4 beetle larvae per day, whereas an adult could kill 2.9 adult bark beetles (Hérard and Mercadier, 1996). *Thanasimus dubius* appears to play a similar role for the North American bark beetles (Kennedy and McCullough, 2002) and for *T. yunnanensis* (Ye and Liu, 2006). In fact, in a *Tomicus* gallery, *T. formicarius* larvae seem to be able to feed on all kinds of prey, including scavengers or other bark beetle predators such as *Rhizophagus* adults and *Medetera* larvae (Hanson, 1937).

Rhizophagus (Rhizophagidae), *Medetera* (Diptera: Dolichopodidae), and *Scoloposcelis* (Hemiptera: Anthocoridae) are often cited as important and frequent predators of the European *Tomicus*, as well as *Aulonium ruficorne* (Olivier) (Colydiidae) and *Raphidia ophiopsis* (L.) (Raphidioptera) for *T. destruens* (Hanson, 1937; Mendel *et al.*, 1990; Hérard and Mercadier, 1996; Durand-Gillmann, 2014). *Rhizophagus* species have been considered as very efficient predators of eggs and larvae of *T. piniperda* (Hanson, 1937). However, studies of the feeding behavior of *R. dispar* have concluded that it appears to be a scavenger (Merlin *et al.*, 1986). *Aulonium ruficorne* is a common predator of larvae and adults of Mediterranean pine bark beetles (Mendel *et al.*, 1990), whereas *R. ophiopsis* feeds mainly on eggs and adults (Pishchik, 1979). In *T. piniperda* galleries, *Medetera* larvae attack larvae, pupae, and teneral adults (Hérard and Mercadier, 1996) as well as other predators (Mendel *et al.*, 1990). *Scoloposcelis pulchella* (Zetterstedt) is highly voracious, killing large quantities of larvae and adult beetles in laboratory rearing (Hérard and Mercadier, 1996) and is able to feed on other predators in bark beetles gallery (Mendel *et al.*, 1990). Staphylinidae are diverse and abundant (Table 10.5) but most are not predators of bark beetles, except *Nudobius lentus* (Grav.) and some *Placusa* species.

There is good phenological coincidence between bark beetle development in trees and arrivals and emergence of the various predators and parasitoids (Hérard and Mercadier, 1996; Schroeder, 1999; Kennedy and McCullough, 2002). The presence of many predators and parasitoids in the same system results in complex interactions including competition and predation. A particular example is that of *A. ruficorne* in *T. destruens* galleries. Mendel *et al.* (1990) have reported that it could compete with parasitoids such as *D. flaviventris* and *M. unifasciatus* in smooth bark tree sections and with the predators *Nemosoma elongatum* (L.), *Rhizophagus bipustulatus* (F.), and *Corticeus* and *Platysoma* species. They also observed *A. ruficorne* larvae to be preyed upon by *Medetera*

TABLE 10.4 Main Parasitoids of Three *Tomicus* Species

Parasitoids	*T. piniperda*	*T. minor*	*T. destruens*
Braconidae			
Blacus humilis (Nees)	+8		
Bracon palpebrator Ratzeburg	+3,10,13,19	+10,13	
Coeloides abdominalis (Zetterstedt)	++1,3,4,8,9,10,13,15,16,17,19	++8,10,13,16,19	
C. bostrichorum Giraud	+1,10,13	+10,13	
C. melanostigma Strand	++9,17		
C. pissodis (Ashmead)	+(2)		
C. sordidator (Ratzeburg)	++8,10,13	++8,10,13	
Dendrosoter flaviventris Förster			++6,11,12
D. hartigi (Ratzeburg)		++8,10,13,15	
D. middendorfi Ratzeburg	++3,4,8,9,10,13,15,19	++8,10,13,15,19	+11,12,20
D. protuberans (Nees)	+3,4,8,10,13,19	+8,13,19	
Ecphylus hylesini (Ratzeburg)		+13,15,19	
Eubazus atricornis (Ashmead)	+8,13,19		
Eupelmidae			
Calosota vernalis Curtis	+7,13	+7,13	
Eurytomidae			
Eurytoma arctica Thomson	++7,10,13,15,16	++7,10,13,15,16	
E. blastophagi Hedqvist		+7	
E. morio Boheman		+13,18	+6,12
E. rufipes Walker	+10,13		
Pteromalidae			
Dinotiscus colon (L.)	+7,13	++7,13,18	
D. dendroctoni (Ashmead)	+(2)		
Heydenia pretiosa Förster		++7,10,13,18	+11,12
Metacolus azureus (Ratzeburg)	++3,14,15,16	+3,14,15,16	
M. unifasciatus Förster	++7,10,13,15	++7,10,13,18	++5,6,11,12
Pteromalus abieticola Ratzeburg	+3		
Rhaphitelus maculatus Walker			+6
Rhopalicus quadratus (Ratzeburg)	++7,10,13,19	++7,10,13	
R. tutela (Walker)	++(2),3,4,7,9,10,13,14,15,16,19	++7,10,13	
Roptrocerus brevicornis Thomson	++1,3,4,7,9,10,13,15,16	++10,13,15,16,17	
R. mirus (Walker)	+10,		
R. xylophagorum (Ratzeburg)	++(2),3,4,7,9,10,13,14,15,19	+7,14	+6,11,12

No data are available on parasitoids associated with Asian endemic Tomicus species. ++: Particularly reliable association according to Kenis et al. (2004); +: other record. Dubious associations (Kenis et al., 2004) have not been taken into account. ( ): observed in North America. Numbers indicate references.
References: (1) Bogdanova, 1982; (2) Bright, 1996; (3) Chararas, 1962; (4) Hanson, 1937; (5) Halperin, 1978; (6) Halperin and Holzchuh, 1984; (7) Hedqvist, 1963; (8) Hedqvist, 1998; (9) Hérard and Mercadier, 1996; (10) Herting, 1973; (11) Mendel, 1986; (12) Mendel and Halperin, 1981; (13) Mills, 1983; (14) Nuorteva, 1956; (15) Nuorteva, 1957; (16) Nuorteva, 1964; (17) Nuorteva, 1971; (18) Pettersen, 1976; (19) Thompson, 1943; (20) Triggiani, 1984[a].
[a]Reference to T. piniperda in the natural area of T. destruens according to Horn et al. (2012) and thus considered to refer to T. destruens.
(modified from Kenis et al., 2004)

TABLE 10.5 Main Predators of Four *Tomicus* Species

Predators	T. piniperda	T. minor	T. destruens	T. yunnanensis	Feeding habits
COLEOPTERA					
Carabidae					
Calodromius spilotus (Illiger)	+4				a
Dromius quadrimaculatus (L.)	+4				a
Pterostichus oblongopunctatus (F.)	+6,8				a
Cleridae					
Allonyx quadrimaculatus (Schaller)	+4				a
Enoclerus nigripes (Say)	+(5)				p
Thanasimus dubius (F.)	++(5)			++23	p
T. formicarius (L.)	+++3,4,6,8,11,12, 13,18,19,21	++11,17	+++2,17,22		p,s
T. femoralis (Zett.)	+13,20				p,s
T. rufipes (Brahm)	+11,13	++11			p
Colydiidae					
Bitoma crenata (F.)	+4				a
Aulonium ruficorne (Olivier)	+4		+++2,10,17		p
Cucujidae					
Histeridae					
Cylistix gracilis (LeConte)	+(1)				a
Eblisia minor (Rossi)	+++4				p
Paromalus parallelipipedus (Herbst)	++6,7,8,9	+11	+17		a,s
Platysoma angustatum (Thunberg)	++4,6		+10		a,s
P. cornix Marseul			+17		a
P. cylindrica (Paykull)	++(5)				p
P. elongatum (Thunberg)	+4		++17,22		p
P. lineare Erichson	+21				a,s
P. parallelum (Say)	+(5)				p
Plegaderus otti Marseul			+22		p
P. vulneratus (Panzer)	++7,8,9,11,21	+11			p,s
Laemophloeidae					
Cryptolestes fractipennis (Motschulsky)	+4				a
Cryptolestes spartii (Curtis)			+10		a
Mycetophagidae					
Litargus connexus (Geoffroy)	+4				a
Nitidulidae					
Epuraea boreella (Zetterstedt)	+11				a

Continued

TABLE 10.5 Main Predators of Four *Tomicus* Species—cont'd

Predators	T. piniperda	T. minor	T. destruens	T. yunnanensis	Feeding habits
E. marseuli Reitter	++4,6,8,11,19,21	+11			a,p,s
E. pygmaea (Gyllenhal)	+11				p,s
E. rufomarginata (Stephens)	+4				a,s
E. silacea (Herbst)	+4				a
E. thoracica Tournier	+11	+11			a
Glischrochilus quadripunctatus (L.)	++6,8,11,18,21				p,s
Ipidia binotata (Reitter)	+4				a
Pityophagus ferrugineus (L.)	++6,8,11,16	+11			p,s
Ostomidae					
Nemosoma elongatum (L.)			+10		p,s
Temnochila caerulea (Olivier)			++2,17		p
Rhizophagidae					
Rhizophagus bipustulatus (F.)	++3,6,11	+11	+10		p,s
R. depressus (F.)	+++ 4,6,8,9,11,13,19,21	+11	+++2,17		a,p,s
R. dispar (Paykull)	++3,6,11,13				p,s
R. ferrugineus (Paykull)	++3,4,11,13,18,21	+11			p,s
R. nitidulus (F.)	+6				a,s
Silvanidae					
Silvanus bidentatus (F.)	+4				a
S. unidentatus (F.)	+4				a
Staphylinidae					
Nudobius lentus (Grav.)	+6,7,11	+11			p,s
Pholeonomus lapponicus (Zetterstedt)	++6,8,9,11,13	++11			a
P. pusillus (Gravenhorst)	+++6,8,9,11,13	++11			a
Placusa depressa Maekl.	+6,8,9,11	+11			p,s
Quedius laevigatus (Gyllenhal)	+11				p,s
Other Staphilinidae (13 species)	+(5),6,8,9,11				
Tenebrionidae					
Corticeus fraxini Kug.	++4	++14,17	++17,22		p
C. linearis (F.)	+4		+17		p
C. longulus (Gyllenhal)	+16	+16			p
C. pini (Panzer)			+17		p
C. parallelus (Melsheimer)	++(5)				p
C. praetermissus (Fallén)	+(1)				p
Corticeus sp.			+++2		p

TABLE 10.5 Main Predators of Four *Tomicus* Species—cont'd

Predators	T. piniperda	T. minor	T. destruens	T. yunnanensis	Feeding habits
DIPTERA					
Dolichopodidae					
Medetera dichrocera Kowarz	+12				p
M. obscura (Zetterstedt)	+++3,11				p
M. pinicola Kowarz	+(1),11,12				p
M. setiventris Thuneberg	+12	+12			p
M. signaticornis Loew	+(1)				p
M. stackelbergi Parent	+11				p
M. striata Parent	+11	+11	+10		p
Lonchaeidae					
Lonchaea collini Hackman	+4				p
HETEROPTERA (Anthocoridae)					
Lyctocoris campestris (L.)	+4				a
Scoloposcelis obscurella (Zetterstedt)	++4				p
S. pulchella (Zetterstedt)	+4		++10,17		p
Xylocoris cursitans (Fallén)	+4				p
RAPHIDIOPTERA (Raphidiidae)					
Raphidia notata (F.)	+3				a
R. ophiopsis (L.)	++15	++17	+++2,17,22		p

No data are available on predators of the Asian Tomicus species others than T. yunnanensis. +++ Very frequent seen; ++ moderately seen; + rarely seen. ( ): observed in North America. a: present in galleries; p: predators; s: attracted to semiochemicals. Numbers indicate references.
References: (1) Bright, 1996; (2) Durand-Gillmann, 2014; (3) Hanson, 1937; (4) Hérard and Mercadier, 1996; (5) Kennedy and McCullough, 2002; (6) Mazur, 1973; (7) Mazur, 1975; (8) Mazur, 1979; (9) Mazur, 1985; (10) Mendel et al., 1990; (11) Nuorteva, 1956; (12) Nuorteva, 1959; (13) Nuorteva, 1964; (14) Nuorteva, 1971; (15) Pischchik, 1979; (16) Pischchik, 1980; (17) Sarikaya and Avci, 2009; (18) Schroeder, 1988; (19) Schroeder, 1996; (20) Schroeder, 2003; (21) Schroeder and Werslien, 1994a; (22) Triggiani, 1984[(a)]; (23) Ye and Liu, 2006.
[(a)]Reference to T. piniperda in the natural area of T. destruens according to Horn et al. (2012) and thus considered to refer to T. destruens. (modified from Kenis et al., 2004)

striata (Parent) and *S. pulchella*. Other interactions among predators are presented later with their consequences for the population dynamics of *Tomicus*.

5.6   Vertebrates

Woodpeckers can sometimes have a strong impact on *Tomicus* populations, although their impact on brood productivity has rarely been assessed. In a Swedish field experiment, woodpeckers attacked 3 to 44% of experimental logs infested with *T. piniperda* (Långström, 1986). In Israel, *Picoides syriacus* Emprich and Ehrenberg preys on

T. destruens (Mendel, 1985). Impacts on bark beetle survival result from direct consumption by the woodpeckers and from bark removal, which exposes broods to desiccation or to predation (Chapter 7).

6.   POPULATION DYNAMICS

6.1   Principles of Population Dynamics and Strategies for Establishment on Trees

The key factor in the population dynamics of most bark beetle species attacking living trees is tree resistance level, defined as the attack density threshold above which trees are killed (Thalenhorst, 1958; Berryman, 1976; Raffa and Berryman, 1983; Christiansen et al., 1987; Paine et al., 1997; Lieutier, 2004; Chapters 4 and 5). Below the attack density threshold, the trees resist all attacks and survive, while the beetle population cannot establish and stays at endemic status, surviving on very weak or broken trees with very low or no resistance. If the population size becomes sufficiently high to exceed the attack density threshold of multiple trees, then attacks can succeed on those trees, providing the beetle population with an augmented food source and subsequently resulting in an increase in the beetle population. By being able to exceed the attack density threshold, the population status has gone from an endemic to an epidemic state (Berryman, 1982). The passage to the epidemic state can result from large quantities of substrate with no or very low resistance level being suddenly available, such as after storms, or from a decrease in the attack density threshold due to some type of stress on the trees.

The critical threshold of attack density is also a measure of beetle aggressiveness, the level of aggressiveness being inversely related to the threshold level (Berryman, 1976; Raffa and Berryman, 1983; Christiansen et al., 1987; Paine et al., 1997; Lieutier, 2004). This threshold is influenced by genetic and environmental factors affecting the trees' level of resistance. It should thus be defined only for well-specified trees and beetle species in well-defined environmental conditions. However, it is often used to compare the aggressiveness of different beetle species towards their usual hosts. For T. piniperda on Scots pine, the threshold has been evaluated at above 300 attacks/m^2 for 30-year-old severely pruned trees (Långström et al., 1992; Långström and Hellqvist, 1993a), which means it is much higher for healthy trees. This a very low level of beetle aggressiveness compared to the 200–400 attacks/m^2 of *Ips typographus* (L.) on healthy *Picea excelsa* Karlsten (Mulock and Christiansen, 1986) and the 50–120 attacks/m^2 of *Dendroctonus ponderosae* Hopkins on *Pinus contorta* (Waring and Pitman, 1983, 1985; Raffa and Berryman, 1983). In southern China, the critical attack density threshold of T. yunnanense on P. yunnanensis is around 80 attacks/m^2 (Lieutier et al., 2003). In Algeria, P. halepensis killed by T. destruens exhibited averaged attack densities of 178 attacks/m^2 on northern slopes and 137 attacks/m^2 on southern slopes (Chakali, 2007), suggesting a critical attack density threshold slightly lower. Such comparisons may explain why epidemic and damage are rare for T. piniperda but frequent for T. yunnanensis and T. destruens.

The *Tomicus* strategy for establishment on host trees is similar to that of most bark beetle species attacking living trees and refers to exhausting tree defenses (Lieutier, 2002, 2004 for details). Everything contributing to rapidly stimulating tree energy expenditure at the moment of attack

lowers the attack density threshold: mass aggregation, longitudinal maternal galleries, early spring attacks when the tree is budding, and associations with fungi able to stimulate tree defenses (Lieutier, 2002; Lieutier et al., 2009). However, T. piniperda is particular regarding the strategy (Lieutier et al., 1989a; 1995; Lieutier, 2002, 2004) because it does not have an aggregation pheromone (Schroeder, 1987; Löyttyniemi et al., 1988), it attacks trees well before their activity begins, and its associated fungi do not stimulate host's defense reactions (Section 5.2). Stimulation of tree energy expenditure is thus based on aggregation through terpenes and on stimulation of tree defenses by the sole mechanical stress due to beetle's boring activity (Lieutier et al., 1995). Its low ability to stimulate tree energy expenditure is very likely the reason for the very low aggressiveness of T. piniperda and its status as a "secondary" bark beetle species. The attack strategy of T. yunnannensis is quite similar to that of T. piniperda. However, shoot attacks seem to aggregate on the same trees (Ye and Lieutier, 1997), leading to very high levels of defoliation. The consequence is a considerable lowering of the attack density threshold in the stem, which makes subsequent beetle attacks on the trunk able to succeed at very low densities by using the exhausting tree defense strategy (Lieutier et al., 2003). The other *Tomicus* species also use this strategy since they need mass attacks to establish on trees but nothing is known regarding how they proceed.

6.2 Factors involved in Population Dynamics

6.2.1 Fecundity and Brood Productivity

Fecundity plays a basic role in bark beetle population dynamics, as it is the only factor to positively affect population increase. Brood productivity of a generation is the number of ovipositing daughter females per parent female. Because of the existence of shoot maturation, brood productivity of *Tomicus* is most often given as the number of emerging callow adults per m^2 or, better, per parent female, a value overestimated because of mortality during shoot feeding and overwintering. Moreover, all *Tomicus* species are univoltine but often have several sister broods. The annual brood productivity should thus be calculated by summing the emerging callow adults of all sister broods before comparing with the number of female parents during the first oviposition period. Measuring fecundity and calculating brood productivity in optimal laboratory conditions inform on the theoretical rate of population increase, which, compared with data in natural conditions, informs on the impact of regulating factors.

Rearing *Tomicus* under optimal conditions in the laboratory and in absence of limiting factors, five sister broods have been observed in *P. sylvestris* logs, and the number of eggs was estimated at 40 to 50 per fertile gallery (Sauvard, 1993). Due to the observed re-emergence rate, it is possible to estimate the average potential fecundity at 137 to 172 eggs per female. Considering the larval survival data and a 1:1 sex ratio among callow adults, the annual productivity would be 44 to 70 callow females per parent female, with that of the first sister brood alone being 14 to 17.5. Similar values were obtained by Långström and Hellqvist (1985), with 27 to 32 offspring (equivalent to 13.5 to 16 callow females) per parent female for the first sister brood. Based on field data from Sweden on brood production (Långström, 1984), a maximal brood productivity of seven callow females per parent female can be calculated. In other situations where attack densities and brood production/m^2 varied considerably (Långström, 1986), calculations of brood productivities give values ranging from 0 to 9.5 callow females per parent female. In Poland, productivity levels of 0.7 to 1.7 callow females per parent female were observed (Gidaszewski, 1974). In North America, brood productivity varied from 0.5 to 9.5 callow females per parent female depending on years and pine species (Ryall and Smith, 2000a). Although sister broods were not considered, these examples demonstrate the powerfulness of the regulating factors.

Rearing of *T. destruens* under optimal conditions without limiting factors gave an average fecundity of 90.6 eggs per female and a brood productivity of 30 callow females per parent female for the first sister brood (Faccoli, 2007). A field endemic population of *T. destruens* in a *P. halepensis* stand had an average fecundity of 60.2 eggs per female (Durand-Gillmann, 2014) and based on the provided data, a brood productivity of 19 callow females per parent female. No evaluation of the theoretical productivity is available for the other *Tomicus* species. In a pine forest in China, *T. yunnanensis* can lay 50 to 110 eggs per female during the first sister brood and 10 to 30 during the second (Ye, 1991). In a subsequent study in the same forest, field average fecundity of the first sister brood was 61.9 eggs per female and after larval maturation in the laboratory, brood productivity was 13.8 callow female per parent female (Ye and Zhao, 1995). However, in logs placed outdoors and protected from predators and parasites, the productivity of *T. yunnanensis* first sister brood, in numbers of callow females per parent female, was only 8.6 in *P. yunnanensis* and 9.7 in *P. armandii*, whereas that of *T. minor* was 4.6 and 6.6, respectively (calculated from Zhao and Långström, 2012). After natural attacks on Scots pine trees in Poland, field productivity of *T. minor* was 0.1 to 0.8 callow females per parent female, depending on localities (Gidaszewski, 1974). Långström (1983b) found a similarly low brood productivity of *T. minor* on the upper side of fallen pine trees, but a highly variable productivity (range 0–20 daughters per female) on the underside.

6.2.2 Population Regulating Factors

The factors involved in the bark beetle population dynamics can be separated into two categories: (1) those related to the beetles themselves and directly affecting their populations and (2) those related to the tree and affecting its resistance level. In *Tomicus*, they interfere at four essential phases of the life cycle: attack and establishment on host tree, larval development, shoot maturation, and overwintering. They can affect both beetle productivity and offspring quality. Brood productivity was discussed previously. Brood quality is a parameter indicating the ability of the new generation to survive and reproduce (Sauvard, 2004).

6.3 Density-dependent Factors

Intraspecific competition, a strong population-regulating factor in most bark beetle species, has an effect mainly during larval development as a direct consequence of mass attacks. Models have explained how a compromise between increasing attack density to overcome tree defense and minimizing subsequent larval competition defines a maximum brood productivity (Raffa and Berryman, 1983). Regarding this, the situation is particularly dramatic for *T. piniperda*, due to its very high attack density threshold. In some cases, competition occurs during shoot maturation, but its impact on *Tomicus* has been studied mainly after stem attacks. Many papers report its negative effects on brood production through dramatic reductions of larval survival and/or brood quality in *T. piniperda* (Nuorteva, 1954; Saarenmaa, 1983; Sauvard, 1989; Långström and Hellqvist, 1993b; Ryall and Smith, 1997; Amezaga and Garbisu, 2000), *T. destruens* (Chakali, 2007; Faccoli, 2009), and *T. yunnanensis* (Ye and Zhao, 1995), but it seems to have no or very little effect on the breeding success of *T. minor* (Långström, 1984). No information is available for the other *Tomicus* species.

The first consequence of intraspecific competition is a higher rate of earlier re-emergence of the parent females (Sauvard, 1989), and a reduction of their immediate fecundity sometimes so drastic that population replacement is not assured (Långström and Hellqvist, 1993b). It has been shown, however, that intraspecific competition occurs only above a certain density and an optimal density corresponding to a maximum brood production has been defined for *T. piniperda* (Nuorteva, 1954; Saarenmaa, 1983; Sauvard, 1989) and *T. destruens* (Faccoli, 2009). It has been estimated at 60 attacks/m^2 in Scots pine candles in the field (Nuorteva, 1954) and at 100 attacks/m^2 in laboratory logs (Sauvard, 1989). The effects of increasing attack densities on brood production are summarized in Figure 10.6. This density level seems also a critical value for the spatial distribution of attacks in logs (Saarenmaa, 1983): aggregative pattern when below 100 attacks/m^2, random between 100 and 200 attacks/m^2, and regular above 200 attacks/m^2.

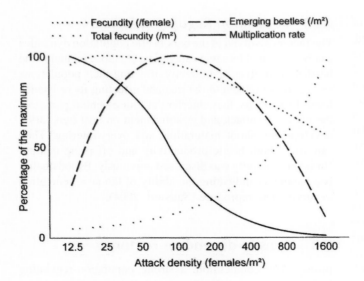

······· Fecundity (/female)   — — Emerging beetles (/m²)
· · · · Total fecundity (/m²) ——— Multiplication rate

FIGURE 10.6 Effect of intraspecific competition on the reproductive success of *Tomicus piniperda*. *From Sauvard (2004) with the permission of the publisher.*

Without any tree resistance to overcome, the beetle population would thus behave so that brood production is maximized. For *T. destruens* in the laboratory, the optimal attack density on *P. pinea* logs would be 50–75 attacks/m² (Faccoli, 2009). However in all cases, although brood production (number of offspring/m²) is maximized at the optimal density, brood productivity decreases constantly when attack density increases (Faccoli, 2009; Figure 10.6).

Brood quality is also strongly affected by intraspecific competition. Mean individual weight of emerging callow adults decreases constantly when density increases, even below the optimal density level defined above (Beaver, 1974; Sauvard, 1989; Amezaga and Garbisu, 2000). The offspring of early attacking adults would be less affected because their progeny have access to a fresher breeding substrate (Beaver, 1974; Sauvard, 1989). Moreover, only female weight seems to decrease (Amezaga and Garbisu, 2000). Intraspecific competition has thus very complex effects, mixing effect on brood productivity with that on brood quality, each varying according to different modalities. It has often been assumed that a lower weight of individual offspring means a lower level of their reproductive success (Botterweg, 1983; Anderbrandt, 1988; Birgersson *et al.*, 1988, among others). This is certainly true for bark beetles other than *Tomicus*, of which maturation of callow adults takes place on the same substrate as that used for larval development. However, *Tomicus* callow adults mature in shoots available in more or less large quantities in the forest, a situation corresponding to a much lower level of competition than in stems. This could allow them to recover, as demonstrated by Amezaga and Garbisu (2000) for *T. piniperda*. However, in particular situations, intraspecific competition can also occur in shoots, when high offspring populations emerge simultaneously with

little dispersal behavior. This is also the case for *T. yunnanensis*, whose callow adults seem to aggregate on the same trees for their maturation feeding (Ye and Lieutier, 1997). Considering the possibilities of recovering from the effects of intraspecific competition, the existence of sister broods must not be underestimated. Indeed, when parent females re-emerge earlier and at a higher rate under competition, sister broods may reduce the negative effects of competition (Sauvard, 2004).

Interspecific competition among bark beetles is generally very limited because the different species tend to segregate along the spatial, temporal or trophic axes of their niche. During trunk attack, bark thickness always plays an essential role in spatial segregation, in addition to host tree species, whereas the date of attack segregates along the temporal axis (Bakke, 1968; Långström, 1984; Haack and Lawrence, 1995; Amezaga and Rodríguez, 1998; Ye and Ding, 1999; Lu *et al.*, 2012b). Interspecific competition is also avoided during shoot attacks, mostly through spatial segregation (Långström, 1983b), possibly completed by temporal and trophic segregation for *T. yunnanensis* and *T. minor* (Chen, 2003) and *T. brevipilosus* (Lu *et al.*, 2014). Other xylophagous species can also be involved in the competition. The presence of the longhorn beetle *Acanthocinus aedilis* (L.) can considerably decrease (up to 78–84 %) the number of *T. piniperda* offspring (Nuorteva, 1962; Hellqvist, 1984; Schroeder and Weslien, 1994a).

Natural enemies are other biotic factors that play an important role in bark beetle population dynamics. However, their impact is difficult to quantify. Data on pathogens are often underappreciated in terms of their impact on beetles. Data on natural mortality by pathogenic fungi are scarce. More information exists regarding nematodes, predators, and parasitoids of *Tomicus* species. However, appreciating the impact of predators on population dynamics is difficult because predators are most often polyphagous. For nematodes and parasitoids, information on parasitism rates is available but is not sufficient to appreciate their impact on populations. Approaches using time series and life tables, and enemy exclusion experiments have been developed in a few cases.

Exclusion experiments in Sweden have demonstrated that the larvae of *T. formicarius*, *R. depressus*, and *R. ferrugineus* can jointly be responsible for decreasing the offspring production of *T. piniperda* by 81 to 90% (Schroeder and Weslien, 1994b). In another field experiment, offspring production was reduced by 81% when reared with *T. formicarius*, 41% when reared with *Rhizophagus*, and 89% when reared with both predators, demonstrating the essential role of *Thanasimus* (Schroeder, 1996). In Poland, Gidaszewski (1974) has also observed high densities of *Thanasimus* causing 48–82% mortality in *T. piniperda* broods. At the opposite end, Mazur (1975) has reported very low densities of *P. parallelipipedus* and *P. vulneratus* (Histeridae) in certain Polish stands and concluded that their role as a regulating factor was negligible. When predator densities reach high levels, intra- and interspecific competitions occur. Schroeder (1996) has shown that when both *Thanasimus* and *Rhizophagus* are present, the number of *Rhizophagus* larvae can be reduced by 49% while that of *Thanasimus* large larvae can be reduced by 34%. *Raphidia ophiopsis* has also been reported to drastically reduce *T. piniperda* populations (Pishchik, 1979). Fewer data are available on parasitoids and parasites. Detailed surveys of *T. piniperda* populations in relation to their insect associates in several pine species in North America suggested an inverse relation between *T. piniperda* brood productivity and both parasitism by native hymenopterans and predation by native dipterans (Ryall and Smith, 2000a). Microscopical investigations established that overall, 56% of *T. piniperda* individuals were parasitized in Polish forests (Gidaszewski, 1974).

Not much is known about natural enemies of *Tomicus* species other than *T. piniperda*. According to Gidaszewski (1974), *Thanasimus* could cause 11 to 14% mortality in *T. minor* broods, whereas 33% of the insects could be parasitized. Långström (1983b) found that ca. 10% of the exit holes of *T. minor* in fallen pine trees were attributable to unknown parasitoids. Using life tables, Ye and Zhao (1995) observed that *T. dubius* would prey on 1% of *T. yunnanensis* adults. However, after introducing *T. dubius* in caged trees, Ye and Liu (2006) have reported that this predator could kill 10.6 % of *T. yunnanensis* larvae and pupae.

6.4 Density-independent Factors

Availability of suitable breeding material is probably the most important factor responsible for pine shoot beetle population increase. Moreover, as *Tomicus* shoot feeding takes place in healthy pine trees, the abundance of breeding material is the only factor determining the extent of the shoot damage.

By suddenly providing large quantities of trees without defenses, storms are a major abiotic factor of *Tomicus* population dynamics, often responsible for the initiation of outbreaks. Following storms, mass attacks succeed on the broken trees, allowing an unlimited increase of population levels. This increase continues during sister broods, and attacks the subsequent year, which often take place on the remaining felled trees, finally lead the populations to exceed the attack density threshold of the living trees. Snow-broken pines can play a similar role as breeding material, but they occur on a more limted scale than storm-felled trees. Some silvicultural practices such as storage of fresh pulp wood or timber in or close to the forest, rough logging waste, thinning waste from early cleanings (pre-commercial thinning), as well as pine pulp wood stacks, can also contribute to bark beetle population increase (Långström *et al.*, 1984). Beginning in the 1960s, logging operations were mechanized and storage of pulp wood along forest roads became common practice, at least in northern Europe, and this led to a build of bark beetle populations as forest protection aspects were largely neglected (Nilsson, 1976). High *T. piniperda* and *I. typographus* population levels resulted in outbreaks in the 1970s, following the 1969 storms in which 37 million m^3 of trees were storm damaged (Nilsson, 1976). Similar scenarios have occurred in many parts of Europe in the last decades (Luitjes, 1976; Annila and Petäistö, 1978; Führer and Kerck, 1978; Winter and Evans, 1990; Gilbert *et al.*, 2005; Långström *et al.*, 2009).

As for most insects, temperature is suspected to influence population dynamics of *Tomicus*. It can affect both insects' survival and population increase. Population survival can be affected depending on the minimum and maximum temperatures tolerable by insects, whereas the duration of the development from eggs to adults can be affected depending on the preferred temperature range of the insects. The thermal preferenda of the different species have been presented previously. They reflect adaptation to local conditions but extreme tolerable temperatures can sometimes be largely exceeded, as in Scandinavia, for example, for winter temperatures. However, *T. piniperda* overwinters inside the bark at the base of trees and *T. minor* in the litter, a behavior which protects them efficiently from cold, especially in case of snow cover (Bakke, 1968). Thus, physiological and behavioral adaptations of the *Tomicus* species to the local conditions lead to low temperature effect on population survival. Only in particular years, when the snow cover is weak, for example, temperatures may cause mortality, especially for insects overwintering in thin bark.

Inside the tolerance intervals, the temperature must exceed a certain temperature threshold for development to occur. Above this temperature threshold, each species requires a certain thermal sum to complete its development. The temperature threshold is about 8°C for *T. piniperda* and *T. minor*, with thermal requirements (above 0°C) from trunk attack to emergence of about 1000 day-degrees for the former and 1300 day-degrees for the latter (Bakke, 1968; Långström, 1983b; Saarenmaa, 1985; Ryall and Smith, 2000b). Using the effective thermal sum (>8.5°C), Salonen (1973) reported that 503 day-degrees are required for development from egg to adult in *T. piniperda*. For *T. yunnanensis*, the temperature threshold is 7.2 to 8.2°C depending on insect stage, and the thermal requirement from oviposition to young adults is 570 degree-days (Ye, 1994). Temperatures thus can significantly impact the duration of the development with consequences on number and importance of sister broods. An increase of brood productivity is possible during favorable years but cold years with insufficient thermal sums can sometimes impede the completion of the life cycle, as for *T. minor* because of its high thermal sum requirement (Bakke, 1968).

Flight also depends on temperature. Through allowing more or less early flight, end of winter temperatures can affect the number of sister broods. Beetle localizations during overwintering can also have consequences for the possible completion of the life cycle and the number of sister broods. Insects located on the upper side of bolts (e.g., *T. piniperda*) can benefit from sun rays and early warming in spring, in contrast to those located on the lower side (e.g., *T. minor*; Bakke, 1968; Långström *et al.*, 1984). Bark thickness and color also influence the warming effect of sun rays. Winter itself can play an essential role in *T. piniperda* population dynamics. In temperate zones where adult maturation is terminated before overwintering, winter temperatures play a role in the synchronization of adult emergence in spring. In Nordic countries, snow cover plays a similar role, especially when after melting, temperatures increase slowly (Bakke, 1968). Such a synchronization of emergence also synchronizes attacks on trees, helping in reaching the critical attack density threshold.

6.5 Factors Affecting the Tree Resistance Level

Tree resistance to *Tomicus* attacks occurs only at the moment of trunk attacks. The mechanisms involved are presented in Chapter 5. Shoot defense mechanisms consist only in few preformed resin able to kill small quantities of beetles (Zhao and Långström, 2012). However, this resin flow becomes quickly inefficient because the hole of the beetle entrance dries up rapidly and because the beetle gallery is located in the upper part of the shoot, above the hole. Nevertheless, some factors may affect shoot attractiveness, which will be considered here as a "resistance" factor.

6.5.1 Biotic Factors

Genetic constitution of trees is an important factor of tree resistance to *Tomicus* attacks. Artificial beetle introductions or fungal inoculations in pine stems have demonstrated that resistance levels and defense parameters vary greatly from one tree to another among a same species, independently of environmental factors, as shown for *P. sylvestris* (Lieutier *et al.*, 1996; Bois and Lieutier, 1997, 2000) and for *P. halepensis* and *P. brutia* (Ben Jamâa *et al.*, 2007). The trees' ability to synthesize the phenolic compound pinosylvin in response to aggression has even been proposed as a predictor of Scots pine resistance to attacks by *T. piniperda* (Bois and Lieutier, 1997).

Indirect silvicultural aspects could play a role in resistance. In various models, the attack density threshold is positively correlated with tree vigor (Waring and Pitman, 1983, 1985; Mulock and Christiansen, 1986). Stressed Scots pines are more frequently attacked and more susceptible to *T. piniperda* than intermediate or dominant trees (Cedervind *et al.*, 2003). Similarly, after beetle attraction to Scots pines, trees surviving attacks by *T. piniperda* and *T. minor* had larger crown, diameter, and radial growth than killed trees (Långström and Hellqvist, 1993b). In southern China, there is a negative correlation between index of damage by *T. yunnanensis* and *P. yunnanensis* canopy density (Chen *et al.*, 2004). The length of Scots pine phloem reaction zone induced by single fungal inoculations was negatively correlated with productivity indexes (Lieutier *et al.*, 1993), a result interpreted as a lower resistance level to bark beetles in the low productive trees (Lieutier, 2004; Lieutier *et al.*, 2009). No data on the effect of stand density are available for *Tomicus* stem attacks, but shoot attacks would be favored by a low degree of thinning (Amezaga, 1997). On the other hand, trees located along a stand edge are less resistant to *Tomicus* stem attacks, as shown for Scots pine with *T. piniperda* (Långström and Hellqvist, 1993b) and *P. yunnanensis* with *T. yunnanensis* (Chen *et al.*, 2004).

Heavy phytosanitary problems always decrease tree resistance to bole attacks. There are several European examples of defoliator outbreaks rendering pine trees susceptible to pine shoot beetle attacks (Butovitsch, 1946; Crooke, 1959; Habermann and Geibler, 2001; Långström *et al.*, 2001a; Cedervind *et al.*, 2003). Quantitative evaluations have suggested that tree susceptibility to bark beetle attacks is significantly increased when at least 90% of the foliage is missing. This is the case for Scots pine attacked by *T. piniperda* and *T. minor* after heavy defoliations by *Diprion pini* L. and *Bupalus piniaria* L., tree susceptibility being highest 1 or 2 years after defoliation (Mihkelson, 1986; Annila *et al.*, 1999; Långström *et al.*, 2001a; Cedervind *et al.*, 2003). Similarly, 90% defoliation by *B. piniaria* was necessary to kill trees with *L. wingfieldii* mass inoculations (Långström *et al.*, 2001b). In the worst case, 2 years of severe to total defoliation lead to ca. 50% pine mortality over a 5-year period, half of which was attributed to *T. piniperda*, whereas stands sprayed with dilfubenzuron (Dimilin®) suffered 1 year of defoliation, no mortality, and modest growth losses (Långström *et al.*, 2001a).

Damage resulting from *Tomicus* maturation in shoots plays a similar role. The most spectacular example is *T. yunnanensis* of which intensive shoot feeding can predispose *P. yunnanensis* to lethal stem attacks (Långström *et al.*, 2002). The critical threshold of attack density by *T. yunnanensis* on *P. yunnanensis* stem decreases when the percentage of damaged shoots increases, and all stem attacks succeed above 60% damaged shoots (Lieutier *et al.*, 2003). This is a very low defoliation level, which is very easily reached especially if *T. yunnanensis* aggregates during shoot maturation (Ye and Lieutier, 1997). This low shoot damage threshold combined with shoot aggregation behavior would explain the dramatic damage in *P. yunnanensis* forests of southwestern China (Lieutier *et al.*, 2003). The stem attack density threshold corresponding to this shoot damage level is about 80 attacks/m² (Lieutier *et al.*, 2003). In Scandinavia, the same stem attack density by *T. piniperda* was necessary to kill defoliated Scots pine, but only in totally defoliated trees (Annila *et al.*, 1999). In a few cases, intensive shoot feeding by *T. piniperda* have also triggered stem attacks around sawmills or timber yards in England (Hanson, 1937) and the USA (Czokajlo *et al.*, 1997). Experimental pruning also increases tree susceptibility to *T. piniperda* bole attacks (Långström and Hellqvist, 1993a). Shoot attacks can be favored by defoliations, as in *P. radiata* and *P. sylvestris* stands attacked by *Thaumetopoea pityocampa* Denis and Schiff. in northern Spain (Amezaga, 1997). In southern France, a progressive destruction of the whole maritime pine forest of Maures-Estérel was observed from 1956 to 1987 for infestations of *T. destruens* and *Pissodes notatus* following tree decaying induced by the scale insect *Matsucoccus feytaudi* Duc. (Carle, 1974a; Abgrall and Soutrenon, 1991).

Parameters involved in tree defenses against *Tomicus* attacks are affected by defoliations. Stem resin flow rate decreases significantly in Scots pines heavily defoliated by *D. pini* (Annila *et al.*, 1999) or after artificial pruning (Långström *et al.*, 1993). Shoot pruning also considerably decreased the efficiency of phloem stem reactions induced by *T. piniperda* attacks (Långström *et al.*, 1992). However, defoliation or shoot pruning did not modify the size of the phloem-induced reaction zone or its total content of resin acids and phenolic compounds, after *T. piniperda* and *T. minor* attacks or after isolated inoculations with *L.*

wingfieldii (Långström *et al.*, 1993, 2001b; Croisé *et al.*, 1998b; Annila *et al.*, 1999). At the opposite end of a weakening caused by primary attacks, it has been shown that Scots pine mass inoculated with *L. wingfieldii* at sublethal densities become more resistant to subsequent mass inoculations (Krokene *et al.*, 2000).

Pine trees may also become susceptible to *Tomicus* attack by fungal diseases. For example, the root rot fungus *Heterobasidion annosum* (Fr.) Bref. may predispose pine trees to fatal attacks by *T. piniperda* (Jorgensen and Bejer-Petersen, 1951; Sierpinski, 1959; Bogdanova, 1998; Kolomiets and Bogdanova, 1998). In north Italy a large root infection of *H. annosum* occurring in a stone pine forest caused large tree mortality following outbreaks of *T. destruens* developing on the infected dying pines (Stergulc, 2002). Also in Tuscany, the occurrence of *T. destruens* infestations was correlated with the presence of fungal root rot (Sabbatini Peverieri *et al.*, 2005) and stem diseases (Villari *et al.*, 2008) in pine trees. Similarly, pine shoot beetle attacks have followed after outbreaks of the fungal disease *Gremmeniella abietina* (Lagernerg) M. Morelet in Fennoscandian pine forests (Kaitera and Jalkanen, 1994; Cedervind *et al.*, 2003; Sikström *et al.*, 2011).

6.5.2 Abiotic Factors

Drought as a factor in increasing tree susceptibility to bark beetle attacks has been popularized for a long time (Schwertfeger, 1944; Thalenhorst, 1958; Chararas, 1962). However, recent experiments aimed at testing this assertion have concluded that drought can have opposite effects on tree resistance, depending on water stress intensity. Other studies have examined pine susceptibility to *Tomicus* species and their associated fungi. After isolated inoculations of young Scots pines, *L. wingfieldii* was stopped by less extended reaction zones in moderately stressed trees than in unstressed ones (Croisé and Lieutier, 1993), allowing to conclude that defenses were more efficient and resistance higher in stressed trees. Similarly, the resistance level of Scots pines to mass inoculations was higher in trees submitted to several months of mild stress (predawn needle water potential $\psi_{wp} = -1.5$ to -1.8 MPa at its minimum) than in control trees (Dreyer *et al.*, 2002). Inversely, young Scots pines submitted to several cycles of severe stress ($\psi_{wp} < -2$ MPa at the peak drought intensity) were less resistant to mass inoculations with *L. wingfieldii* than control trees (Croisé *et al.*, 2001). In southern China, where dry and wet seasons alternate, *P. yunnanensis* water potential was followed during 18 months in two plots differing in soil water availability, which was -1 MPa at its minimum. Periodic inoculations with *L. yunnanense* showed that trees resistance to mass inoculations was higher and reactions to isolated inoculations more efficient during the dry season and in the dry plot than during the wet season and in the wet plot (Sallé *et al.*, 2008). Water stress can also affect host preference for shoot feeding. In an experiment with young *P. pinaster*, *T. destruens* preferred and had a better survival in unstressed ($-0.2 < \psi_{wp} < -0.7$ MPa) than in stressed plants ($-0.53 < \psi_{wp} < -2.1$ MPa) (Branco *et al.*, 2010).

Nutrient availability has been assayed for its effects on tree resistance to *Tomicus* attacks. Nitrogen fertilization did not affect the ability of *T. piniperda* to establish egg galleries and to cause stem damage in Scots pines (Löyttyniemi, 1978). It caused only a very slight decrease of the constitutive resin flow, whereas total concentration of phloem constitutive phenols was not affected (Kytö *et al.*, 1998, 1999; Viiri *et al.*, 1999).

In addition to providing large quantities of resources for bark beetles in the form of fallen trees, storms damage the remaining standing trees. Windthrown Scots pines with a declination higher than 50° from vertical are preferentially colonized by *T. piniperda* (Schlyter and Löfqvist, 1990). After fires, *Tomicus* is often the first arrival and the most abundant insect species on damaged Scots pines (Galaseva, 1976; Ehnström *et al.*, 1995; Bakke, 1996; Luterek, 1996; Långström *et al.*, 1999; Santolamazza-Carbone *et al.*, 2011), with attacks occurring mainly during the first 2 years (Ehnström *et al.*, 1995; Långström *et al.*, 1999). The success of *T. piniperda* stem attacks depends on the severity of fire injury to the crown. Attacks succeed in trees with less than 25% intact foliage and fail in trees with at least 40% full foliage (Långström *et al.*, 1999). Amezaga (1997) reported an increased number of shoot attacks by *T. piniperda* in *P. radiata* after fires. Industrial pollution may also render pine trees susceptible to pine shoot beetle attacks (Sierpinski, 1971; Oppermann, 1985; Kolomiets and Bogdanova, 1998). Kytö *et al.* (1998) indicated that heavy metal pollution increased Scots pine constitutive resin flow.

6.6 Conclusions on Population Dynamics

Numerous factors positively or negatively affect *Tomicus* population levels. All limiting factors certainly act in concert to maintain the populations at endemic levels, as well as to stop epidemics. However, it is difficult to determine their respective role due to several obstacles. One is the lack of reliable estimators of population levels. The aggregative behavior of the individuals makes them difficult to sample, and trapping methods are too much influenced by the local conditions. Another obstacle is the complexity of the interactions between trees, bark beetles, and their numerous and diversified associated organisms. In addition, particular aspects such as mortality during the two dispersal phases of *Tomicus* (shoot maturation and bole attacks) and the importance of sister broods are very poorly known.

Life tables and mathematical models could help in determining the relative importance of the factors potentially involved in population limitations, but they have been used in only a few cases and their conclusions vary. Moreover, the existence of the shoot-feeding phase makes it difficult to construct a life table. For endemic populations of *T. piniperda* in Lapland, it was concluded from models that in such an extreme situation, populations would be controlled rather by climatic disturbances than by regulating feedback mechanisms, thus giving preference to density independence over density dependence (Saarenmaa, 1985). Alternatively, for endemic *T. yunnanensis* populations in China, life tables gave the most important role to intraspecific competition and tree resistance mechanisms over natural enemies (Ye and Zhao, 1995). In southern France, an integrated approach of factors maintaining *T. destruens* populations at low endemic levels in *P. halepensis* stands did not reveal any density-dependent effect on larval survival. Tree resistance at the moment of stem attacks was proposed to be the main regulating factor (Durand-Gillmann, 2014).

7. DISPERSAL

7.1 Natural Dispersal

Pine shoot beetles have two different periods of dispersal during their life cycle. During their reproductive phase, the parent beetles have to locate and colonize suitable host material. This is the main period of dispersal as the host material where the beetles developed is destroyed and beetles therefore have to find new fresh breeding material, which may not be available in the vicinity. This reproductive flight takes place after shoot feeding and hibernation and is often referred to as the "spring flight," although it often occurs in November–December in the Mediterranean area and in southern China. After completing oviposition, the surviving parent beetles leave the breeding material and fly to crowns of living pine trees where they tunnel young pine shoots for their re-maturation feeding. The new brood emerges as callow and immature adults and they also fly to the pine shoots for their maturation feeding.

Adult *T. piniperda* are good flyers and may cover several kilometers during their spring flight, as indicated by flight mill studies (Forsse, 1989), which also indicate that the callow adults fly less when they leave the brood logs for the pine crowns. Field observations have confirmed that *T. piniperda* can fly long distances, even over water, in search for host material (Nuorteva and Nuorteva, 1968; Nilssen, 1978). However, most beetles do not fly far away as few beetles were captured beyond 250 m from the beetle source (Poland and Haack, 2000).

The maturation flight, i.e., when the beetles leave the host material where they have oviposited or developed, is normally a short flight to the crown of nearby pine trees. Many studies show that pine trees growing close to breeding material of *T. piniperda* display shoot attacks soon after emergence has begun (Långström, 1983a). At high attack levels and especially if shoot attacks occur over several years, nearby trees get heavily stunted by shoot feeding and then the attack levels decrease with increasing distance to the beetle source (Långström and Hellqvist, 1990, 1991). The largest observed dispersal distance during maturation flight is 1 km (Sauvard *et al.*, 1987; Långström and Hellqvist, 1990).

Regarding the other *Tomicus* species, no detailed studies have been made on their dispersal and flight behavior, but it is reasonable to assume that they disperse in a similar manner, except for *T. yunnanensis*. Studies in Yunnan revealed that shoot feeding in this species may predispose the tree under attack to subsequent stem attacks when the beetles move directly from the shoots for oviposition on the stem of the same tree (Lieutier *et al.*, 2003). This phenomenon makes *T. yunnanensis* the most aggressive species in the genus, but it remains to be demonstrated whether this mode of dispersal is the rule and to which extent normal dispersal by flight occurs in this phase of the life cycle of *T. yunnanensis*.

7.2 Anthropogenic Dispersal

As mentioned above, in North America, *T. piniperda* was first detected in Ohio in 1992, and the same year in several other states surrounding the Great Lakes. It was detected the following year in Ontario and by the year 2000 it had been recorded in 12 states in the USA and two provinces in Canada (Haack and Poland, 2001). *Tomicus piniperda* is believed to have been introduced in dunnage material by boat and is classified as an invasive species. However, damage surveys later indicated that *T. piniperda* must have been present at least in New York and Ontario long before the actual detection (Czokajlo *et al.*, 1997; Scarr *et al.*, 1999). In order to limit further spread of the species in North America, strict quarantine regulations were placed into effect in the USA and Canada (Scarr *et al.*, 1999; Haack and Poland, 2001). The lack of recent publications on this species in North America indicates that the situation is stable.

7.3 Climate Effects

The current paradigm of human-caused global warming is gaining increasing support in the scientific community and it will undoubtedly have profound effects on entire ecosystems. The distribution and dispersal of bark beetles will certainly be affected but so far there is little or no evidence for any such changes in the genus *Tomicus*. As *T. piniperda*

occurs in the entire geographic range of Scots pine and displays large ecological plasticity, changes in its distribution do not seem likely. A warmer climate may, however, alter the competition between *T. piniperda* and *T. destruens* in southern Europe where they occur sympatrically (Horn *et al.*, 2012). In the years following a massive storm in Sweden in 2005, *T. piniperda* was found to start flying ca. 3 weeks earlier than in the 1970s (Öhrn, 2012). This observation was supported by meteorological data showing that the first possible flight date (max temp $>12°C$) during this 30-year-period also occurred correspondingly earlier during the last decade than in the 1970s (Öhrn, 2012). Also, very few specimens of *T. minor* were captured in traps or found in fallen trees in the storm area (Öhrn *et al.*, in prep.). Compared to the distribution reported in Lekander *et al.* (1977), this could indicate the beginning of a retraction of the species in southern Sweden. Both these observations could be the first signs of a response to a warmer climate, but it remains to be seen if they were just accidental or the beginning of a trend.

8. DAMAGE AND ECONOMIC AND ECOLOGICAL IMPORTANCE

8.1 General Patterns

In principle, pine shoot beetles cause three different kinds of damage leading to economic losses: (1) growth losses caused by shoot feeding in the pine crowns; (2) tree mortality caused by stem attacks; and (3) deterioration of timber quality due to beetle-vectored blue-staining of saw logs and pulp wood. In early textbooks (Ratzeburg, 1839; Escherich, 1923), the hibernation galleries made by *T. piniperda* at the base of living pine trees was considered harmful, but no evidence for that has been found in later work. It was also said that shoot damage affects seed production in seed trees left for natural regeneration of pine stands, which might be a real problem, although it has not yet been studied.

The timber quality issue is the least of the three main problems, and mainly refers to *T. minor*, which is known to convey a deep and dark blue stain into its breeding material (Francke-Grosmann, 1952; Solheim *et al.*, 2001). The superficial wounding of log surface by the egg galleries of all *Tomicus* species does not cause any technical damage, but the blue-stain conveyed by the pine shoot beetles may cause some damage. Shoot feeding is a type of damage caused by all *Tomicus* species, but in *T. piniperda* this has long been considered the main damage caused by this species (Ratzeburg, 1839; Escherich, 1923; Postner, 1974). A special case of shoot damage is the esthetic and subsequently economic damage caused in Christmas tree plantations of Scots pine by *T. piniperda* in North America (Haack and Poland, 2001).

It has also long been known that both *T. piniperda* and *T. minor* may kill weakened trees during their stem attacks

(Ratzeburg, 1839), but there have been differing opinions about their aggressiveness. In old papers (e.g., Escherich, 1923), *T. minor* was considered more aggressive than *T. piniperda* mainly because of its horizontal gallery system, but this view has been challenged for several reasons. *Tomicus minor* often flies 1–2 weeks after *T. piniperda* (Långström, 1983b), and Annila *et al.* (1999) concluded that *T. minor* alone could not kill defoliated pine trees whereas many trees were killed by *T. piniperda* in the same stand. Finally, Solheim *et al.* (2001) demonstrated that the main blue-stain associate of *T. minor* had a very low virulence in living pine trees. However, it seems that *T. piniperda* becomes more aggressive in the southern parts of its distribution, where it often occurs with other primary bark beetles like *I. sexdentatus* and *I. acuminatus* (Lieutier *et al.*, 1989a; Fernández Fernández and Salgado Costas, 1999; Colombari *et al.*, 2012, 2013). Colombari *et al.* (2012, 2013) reported on the recent heavy outbreaks of *I. acuminatus* in Scots pine forests growing along the southern Alps. This species infests mainly the upper part of stressed pines and infestations occur in late spring, with a mean air temperature of about 16–18°C, i.e., after *Tomicus* emergence. Nevertheless, the lower part of the stem of trees infested by *I. acuminatus* keeps fresh phloem until next spring, when *T. piniperda* and *T. minor* as well as other large bark beetles species such as *I. sexdentatus* emerge finding new substrates suitable for reproduction. Full tree colonization may consequently be a joint effort by the species involved, although in this case *T. piniperda* and *T. minor* seem to have a secondary role in tree killing. After the re-establishment of *T. destruens* as a valid species, many of the Mediterranean studies concerning *T. piniperda* may in fact deal with *T. destruens*, which is considered more aggressive than *T. piniperda* (Ferreira and Ferreira, 1986, 1990). *Tomicus destruens* is rarely associated with other aggressive bark beetle species infesting Mediterranean pine species (Masutti, 1969; Stergulc, 2002). The large pine mortality observed in the last decades in many circum-Mediterranean countries is due mainly to *T. destruens*.

The Chinese situation differs drastically from that in Europe, in that large-scale tree mortality caused by *T. yunnanensis* has occurred in plantations of Yunnan pine during the last decades (Ye, 1991; Lieutier *et al.*, 2003). As there is no detailed information available on the damage by *T. brevipilosus*, *T. pilifer*, *T. puellus*, and *T. armandii*, the following sections will focus on the damage of the four most important species (*T. piniperda*, *T. minor*, *T. destruens*, and *T. yunnanensis*).

8.2 Shoot Damage

Although Ratzeburg (1839) pointed out shoot damage as the major forest protection problem with the pine shoot beetles in Europe, the first known attempt to clarify the impact of shoot damage on pine growth was made in Sweden after the outbreak described by Lagerberg (1911). Grönberg (1914) estimated that the damaged trees had lost 60–70% of their foliage and that the annual rings were very narrow. Trägårdh (1921) studied the shoot damage in detail and observed that not only current shoots but even twigs and leader shoots were killed, and pine crowns consequently stunted. Mattson-Mårn (1921) reported that an estimated foliage loss of 30% in 1 year yielded a radial growth loss of 22% the following year. The study was conducted in recently thinned young pine stands where the logging waste had been located and which produced high levels of *T. piniperda* that in turn had attacked the shoots of the remaining Scots pine trees.

The bark beetle outbreaks in the 1970s following storm damage in 1969 triggered renewed research interests in Sweden to clarify and quantify the growth losses following shoot feeding by pine shoot beetles. Andersson (1973) reported a volume growth loss of ca. 20% in a thinned pine stand with plentiful *T. piniperda* damage as compared to a similar undamaged stand. Nilsson (1974a) reported up to 40% loss in volume growth lasting for a decade after 1 year of attack originating from pulp wood stacks. He also concluded that ca. 20 damaged shoots per tree may cause losses and that high losses occurred at 100–150 lost shoots per tree (Nilsson, 1974b). In contrast, Elfving and Långström (1984) only recorded a 10% growth loss of short duration despite heavy shoot damage. A shoot-pruning experiment simulating low, medium, and heavy shoot damage on young Scots pines yielded no significant growth losses (Långström, 1980), but clipping current shoots with scissors did not fully represent pine shoot beetle attacks.

An experiment with caged *T. piniperda* beetles or clipped shoots showed that the damage levels of neither 25–50 beetles per tree nor clipped shoots did not cause any growth losses, whereas the higher levels of 100 and 200 resulted in small and transient growth losses, similar in cages and on cut trees (Ericsson *et al.*, 1985). In a subsequent experiment, the effect of early and late shoot damage, corresponding to the re-maturation and maturation feeding of *T. piniperda*, was studied on 20- and 50-year-old pine trees using cages (young trees only) and shoot clipping (both tree age groups) (Långström *et al.*, 1990). The damage level was set to either 200 caged beetles or clipped shoots per tree, and the result was a 30% loss in volume growth during 3 years for young caged trees and about half of that for the hand-pruned trees. In contrast, the old hand-pruned trees had not recovered from a 40% volume growth loss during the study period. There were no growth differences between damage in June or August (Långström *et al.*, 1990). Finally, Långström and Hellqvist (1991) demonstrated that 3 years of timber storage resulted in ca. 1000 lost shoots on nearby trees corresponding to more than half of the total needle biomass and a 65% loss in basal area growth during 6 years. Height growth was also affected and hence maximum volume growth loss was 75%. Damage levels and growth losses declined quickly with increasing distance to the timber yard, but could still be traced at a distance of 500 m. Borkowski (2001) found a similar pattern, with radial increment reduced to ca. 50% within 300 m from a sawmill, and the number of fallen shoots being more than five-fold in that area compared to more distant areas.

All these studies demonstrate a reduction in growth with increasing damage levels, but the former factor is easier to quantify than the latter. Nilsson (1974a) claimed that high growth losses occurred already at an estimated damage level of 100–150 lost shoots per tree, but he estimated the shoot damage several years after the damage had taken place, and it seems obvious that he substantially underestimated the true number of shoots under attack. This has great implications for his estimates on the total growth losses on the national level (Nilsson, 1976), which will be discussed in Section 8.4.

The growth losses caused by pine shoot beetles around sawmills and other permanent storage sites were first studied in Poland (Michalski and Witkowski, 1962), and later studies have confirmed the pattern of lasting and severe growth losses in surrounding pine forest up to 0.5 km and visible effects up to 1 km from the beetle source (Långström and Hellqvist, 1990; Czokajlo et al., 1997; Borkowski, 2001). Borkowski (2006) demonstrated that the growth losses might be permanent as long as new beetles originate from the timber yard.

In several of the studies mentioned above, fallen pine shoots with signs of *Tomicus* attack have been counted on the ground and used to estimate population size and/or the level of shoot damage. Knowing that one fallen shoot roughly corresponds to one attacking beetle, at least under Swedish conditions (Långström, 1979), the shoot numbers can be converted to beetle numbers and damage levels on standing trees. There are always pine shoot beetles in pine forests, even in well-managed ones, and in Sweden the baseline level seems to be below 0.1 shoots/m^2 and year corresponding to ca. 1000 beetles/ha and a few beetles per tree with a stem density ranging from ca. 2000 in pole-sized to ca. 500 in mature pine stands (Långström and Hellqvist, 1990; Ehnström et al., 1995). In a Polish study, the baseline figure was around 0.5 shoots/m^2 (Borkowski, 2001). In central France, the corresponding figure ranged from 0.2 to 0.4 shoots/m^2, but presence of any kind of brood material was directly reflected in elevated shoot and beetle numbers (Sauvard et al., 1987).

In southern Europe, the number of shoots infested by *T. destruens* may greatly vary according to population density and season. In a non-epidemic population, Stergulc (2002) found that in a *P. pinea* stand in northern Italy only 2.1% of the 285 examined shoots were infested by *T. destruens*. In the previous year in the same stand,

Faccoli (unpublished) found 13.3% of the 157 sampled shoots were infested. In samples of 500 pine shoots checked monthly in central Italy, 79 (15.8%) were found to be infested by *T. destruens* at the end of September 2003, but the number of adults in the shoots decreased progressively until mid-February 2004, when only six specimens were found (Sabbatini Peverieri et al., 2008).

In Christmas tree plantations in the USA and Canada, the problem is the presence of pine shoot beetles and not the extent of the damage. Reports vary from a few to ca. 50 (15%) damaged shoots per tree (Petrice et al., 2002). Shoot damage in natural and Scots pine stands has generally been low, but a survey in Ontario revealed considerable damage as shoot counts yielded 4 to 12 shoots/m^2 (Scarr et al., 1999) and even heavy tree mortality in one Scots pine stand. The shoot damage on Yunnan pine can be very severe as up to 80% of the current shoots can be attacked and the damage occurs all over the canopy (Långström et al., 2002). There are also signs of beetle aggregation during shoot feeding to certain tree individuals (Ye and Lieutier, 1997).

8.3 Tree Mortality

Tree mortality occurs when bark beetles and associated fungi are abundant enough to overcome the resistance in the trees under attack. In contrast to truly aggressive bark beetles in the genera *Dendroctonus*, *Ips*, and *Scolytus* (Chapters 8, 9, and 12, respectively), pine shoot beetles are generally not capable of overwhelming the resistance of healthy pine trees. The main reason for this is probably the lack of powerful aggregation pheromones leading to successful mass attacks (Byers, 2004). Another factor may be the lower virulence of the blue-stain fungi associated with the pine shoot beetles than that of more aggressive bark beetles (Lieutier, 2004), but this issue is discussed elsewhere both in this and in other chapters. The topic of host resistance and tree defensive reactions is very complex and covered in Chapter 5.

Considering that healthy pine trees are not available for stem attacks by the pine shoot beetles unless their resistance is substantially reduced, major tree mortality caused by pine shoot beetles are fairly rare episodes and always preceded by some predisposing factor(s). The role of predisposing factors to damage by *T. piniperda* was discussed in Section 6.2. There is one reported case of heavy tree mortality caused by *T. piniperda* in North America, which occurred in southern Ontario (Scarr et al., 1999). Surveys of eight damaged pine stands (Scots, white or jack pine) revealed almost total tree mortality in one Scots pine stand, none in an adjacent white pine stand, and up to 38% mortality in the other stands. As was mentioned earlier, fallen shoot numbers ranged from 4 to 12/m^2. In Eastern Europe, including Russia, there are many reports on the damage caused by *T. piniperda*: Estonia (Voolma and Luik, 2001), Poland

(Gidaszewski, 1974; Borkowski, 2001), Romania (Drugescu, 1980; Mihalciuc *et al.*, 2001), and Russia (Agafonov and Kuklin, 1979; Bogdanova, 1998; Kolomiets and Bogdanova, 1998; Gninenko and Vetrova, 2002).

Damage by *T. destruens* has been frequently reported from nearly all countries in southern Europe, e.g., Portugal (Ferreira and Ferreira, 1986, 1990), Spain (Amezaga, 1996; Fernández Fernández and Salgado Costas, 1999), France (Carle, 1975), Slovenia (Jurc, 2005), Greece (Kailides, 1964; Markalas, 1997; Avtzis and Gatzojannis, 2000), and Italy (Masutti, 1969; Triggiani, 1984; Boriani, 1998), where the beetle is listed among the most aggressive pests of the Mediterranean pine forests (Nanni and Tiberi, 1997). In addition, *T. destruens* is considered to be a major pest in Turkey, where the number of infested *P. brutia* has greatly increased in recent years (Sarikaya and Avci, 2010). In Tunisia, large and increasing damage to maritime pine forests has been recorded since 1972 (Hamza and Chararas, 1981) and is still occurring (Ben Jamâa *et al.*, 2000). Serious damage is reported also in Israel (Halperin, 1978; Halperin *et al.*, 1982; Mendel, 1987), Algeria (Chakali, 2003, 2005), and Morocco (Ghaioule *et al.*, 1998).

There are many Chinese papers reporting on *T. yunnanensis* damage. Most of these papers are published in Chinese, but starting with Ye (1991) there is an increasing number of papers in English (for references see Långström *et al.*, 2002 and Lieutier *et al.*, 2003). The damage situation in China differs drastically from that in Europe with *T. piniperda*, in that there has been large-scale tree mortality due to *T. yunnanensis* during the last decades (Ye, 1991; Lieutier *et al.*, 2003). Although these trees may suffer some drought stress from time to time (Ye, 1992), a more important explanation for this outbreak is that the intensive shoot damage itself may render the trees susceptible to further stem attacks, leading to a vicious self-perpetuating cycle (Lieutier *et al.*, 2003; Section 4.5). The extent of the damage is the largest ever observed for pine shoot beetle species, and the outbreak seemed to coincide with the maturation of large plantations of Yunnan pine that were established in the 1960s (for references see Långström *et al.*, 2002). This outbreak is discussed in more detail below.

8.4 Economic Impact of Attacks

8.4.1 Consequences of Growth Losses and Tree Mortality

Considering the large research interest devoted to the damage caused by the pine shoot beetles for almost 200 years, surprisingly few attempts have been made to quantify the economic damage caused by these beetles. In a joint European project called BAWBILT, *T. piniperda* was ranked as one of the "top-10" forest pests in Europe (Grégoire and Evans, 2004). During the 1990s, 14 million

ha were classified as the area threatened by *T. piniperda* and 13 million m³ were recorded as killed by the species in Europe, but most of that was reported from Poland alone and also included *I. acuminatus* and *Phaenops cyanea* F. as mortality causes (Grégoire and Evans, 2004). Hence, the specific role of pine shoot beetles in this exceptional damage estimate cannot be quantified. Despite a multitude of damage reports from Europe and Asia, little if anything can be concluded about the economic impact of pine shoot beetles.

During the large bark beetle outbreaks in Sweden following the 1969 storm when ca. 37 million m³ of pine and spruce was blown down, Nilsson (1976) estimated that the pine shoot beetles caused growth losses in Sweden of up to 10 million m³ per year in the early 1970s. These figures were based on nationwide surveys of fallen shoots that were converted to growth losses, but these losses were overestimated as they were based on the damage/loss ratios (Nilsson, 1974a; discussed above). Based on growth data compiled by the Swedish National Forest Inventory, Svensson (1980) concluded that there was a loss of 2.0–2.5 million m³ in the early 1970s that could be attributed to the pine shoot beetle damage. After a big storm in 2005, the Swedish National Forest Inventory again recorded fallen pine shoots that were attacked by pine shoot beetles in their nationwide survey program during 2005–2007 (Fransson, 2010). There was a clear increase in pine shoot beetle damage in the storm area in the years following the storm, but the levels were far lower than the levels recorded in the 1970s. Swedish National Forest Inventory statistics confirm that there was no difference in pine growth between the 5-year periods before and after the storm in 2005. A deeper comparison of the pine shoot beetle damage between the two periods is in progress (Långström and Wulff, in prep.), but it appears that shoot feeding caused by the pine shoot beetles does not cause major economic losses on the regional or national scale, at least not in Sweden. For the landowner with pine stands close to a large-scale storage site for pine timber, substantial growth losses can be expected and have been repeatedly demonstrated (Långström and Hellqvist, 1991; Borkowski, 2006). In fact, this damage could be seen as degradation in the site quality for production pine.

In the US and Canada, *T. piniperda* is causing substantial economic losses to the Christmas tree and nursery industry, but no overall loss figures are available. The quarantine rules imposed in order to restrict the further spread of the pest are restricting the market and the compliance management program makes the market risky for both Christmas tree producers and buyers (Haack and Poland, 2001). In China, the outbreak of *T. yunnanensis* on Yunnan pine has affected ca. 1.5 million ha (Ye, 1991; Långström *et al.*, 2002, Lieutier *et al.*, 2003) and devastated over 200,000 ha of pine forests (Ye and Ding, 1999).

Quantitative data on economic damage caused by *T. destruens* in the Mediterranean countries are scarce and mainly refer to single specific events. For instance, the more than 37,000 m^3 of pine timber harvested from 1988 to 2000 in Algeria following *T. destruens* infestations occurred in pine forests spread in semi-arid areas, over an area of about 20,000 ha (Chakali, 2003, 2005). Monleón *et al.* (1996) reported about 4000 trees killed close to Barcelona (Spain), while Stergulc (2002) assessed tree mortality within a stone pine forest of the northern Adriatic Sea of about 6% of the whole forest area (52 ha). In southern Europe, 900,000 m^3 of maritime pine was prematurely harvested after the devastating storm in 1999 (Nageleisen, 2004), but the role of *T. destruens* among the attacking bark beetles was not described.

8.4.2 Consequences of Degradation of Timber Quality

The degradation of timber quality caused by *T. piniperda* is mainly a problem in standing trees killed by the beetle, which often (maybe always) display heavy blue-staining of the whole sapwood (Långström and Hellqvist, 1993a). Compared to *T. minor*, the blue-stain caused by *T. piniperda* occurs less frequently in timber and is often more superficial but it is potentially a major problem as indicated by Löyttyniemi and Uusvaara (1978). Sapwood staining of Korean pine logs due to *T. piniperda* attack was also reported as a timber storage problem in Korea (Kim *et al.*, 2002). *Tomicus destruens* shares most of its associated fungi with the other European *Tomicus* species (Sabbatini Peverieri *et al.*, 2004, 2006; Section 5.2.1), causing similar timber alteration. No estimates of the economic losses due to blue-staining caused by any of the *Tomicus* species are available, but presence of any blue-stain in conifer saw logs reduces the market value from prime timber to lower grades or even to that of pulp wood. Saw mills have hence developed timber handling practices that minimize insect damage in general and especially that of *Trypodendron lineatum* (Olivier), which is the main problem and often occurs together with *T. piniperda* and *T. minor* in pine timber (Lekander and Rennerfelt, 1955). Although some blue-stain is acceptable in pulpwood, there is a cost connected to the processing of blue-stained timber, as more chemicals are needed for bleaching (Löyttyniemi *et al.*, 1978).

8.5 Ecological Consequences

As the pine shoot beetles generally are well adapted to their native environment and as outbreaks are limited in time and space, there are no foreseeable ecological consequences of their activities except for the situation in North America, where the introduction of *T. piniperda* may affect the pine ecosystem. Being an early flyer, *T. piniperda* may outcompete native bark beetles occupying the same part of the trees if it becomes common enough to compete with these species. Another cause of concern would be that the species is lacking its native complex of natural enemies and have a high rate of reproduction. Haack and Poland (2001) have raised questions about the possible impact of this exotic insect pest on the community structure of the native pine-infesting bark beetles and their natural enemies, but they conclude that so far there is no evidence for any changes due to the rather limited impact *T. piniperda* has had in natural or planted pine stands in the USA and Canada.

Regarding the long-lasting outbreak of *T. yunnannensis* in China, there is a possibility that the forest ecosystem will be affected if management strategies have to be changed, but it is essential first to understand the causes of this outbreak. Concerning *T. destruens*, extensive tree mortality in pine forests could alter species composition. In addition, since wildfire occurrence is high in Mediterranean pine forests, tree mortality caused by bark beetle creates higher fuel levels resulting in fires of greater intensity and frequency. In Cyprus, tree mortality caused by *T. destruens* often occurs on steep slopes with highly erosive soils. Under these conditions, high tree mortality could result in increased soil erosion (Ciesla, 2004), especially in sand dunes exposed to see winds along the Mediterranean coast.

9. MANAGEMENT

9.1 Detection and Survey

Presence of pine shoot beetles in the forest is revealed by fallen and tunneled pine shoots on the ground as well as typically stunted pine crowns in cases with high population levels. It is not possible to separate the shoot damage done to Scots pine by *T. piniperda* from that of *T. minor* or *T. destruens*, unless the beetles are outside the shoots. In southern Europe as well as in China, *Tomicus* species are rather host specific and hence the host under attack facilitates identifying the attacker. Shoot survey represents one of the possibilities to estimate population levels both locally and on a landscape level. Although the number of damaged shoots per beetle may vary with the local conditions, the number of fallen Scots pine shoots correlated well with the estimated beetle population in Sweden, showing a relationship of close to one shoot per beetle (Långström, 1979, 1983a). During the 1970s, nationwide surveys of fallen pine shoots were conducted by the Swedish National Forest Inventory, showing clearly elevated populations of pine shoot beetles in the early 1970s, which were linked to observed growth losses (see Section 8.4).

The brood of *Tomicus* species (except *T. puellus*, living on spruce) develops under the bark of fresh pine timber or in standing weakened trees. After the spring flight, the adults of *T. piniperda* disclose their presence in fresh pine timber or weakened trees by the typical boring dust containing brown bark and white wood grain (unique for this species) that is visible in bark crevices adjoining the entrance holes. In late summer, clusters of exit holes (ca. 1.5 mm in diameter) reveal successful brood emergence. Peeling off the bark will show the typical longitudinal egg galleries with egg niches with the developing brood. The same symptoms can be seen on successfully attacked standing trees, whereas failed attacks are visible as a white resin flow (or crystalized resin) in bark crevices on the bark. In Europe, *T. piniperda* normally attacks the lower part of the tree covered with rough bark (Långström, 1984). It is difficult to separate the egg galleries of *T. piniperda* from those of *T. destruens*, but the host species will mostly disclose the attacking species. It is easy to recognize the presence of *T. minor* in logs or standing trees from the typical egg galleries that are found under the thin bark, often on the underside of fallen trees. In contrast, attacks in China often occur on the lower stem under the thick bark (Ye and Ding, 1999). The galleries are unmistakable as they are oriented across the wood grain, and normally two-armed. The exit holes of the callow beetles are also visible in the xylem, as the larval galleries enter the sapwood.

After storm damage or other events producing abundant breeding material, early surveys of pine shoot beetle attacks in fallen pine trees yield information about the risk for beetle propagation and subsequent damage (Gilbert *et al.*, 2005). Such surveys have revealed highly variable rates of pine shoot beetle attacks ranging from a few to more than 50% of the fallen pine trees, and also that uprooted trees may sustain beetle attacks (Annila and Petäistö, 1978). This information is of vital importance for planning of salvage logging operations in order to save the timber quality and to prevent beetle propagation.

Another way of monitoring the occurrence of pine shoot beetles is the use of traps baited with host odors (Byers *et al.*, 1985; Lindgren, 1997; Czokajlo and Teale, 1999; Poland *et al.*, 2003, 2004). The trap technique can give information about the presence and the flight activity of the beetles, but trap catches give only relative estimates of the beetle abundance as they are biased by other factors like the availability of competing odor sources, e.g., fresh pine timber. Comparing trap catches captured under similar circumstances may, however, reveal something about population levels as well. In addition, permanent monitoring performed by traps baited with host volatiles may give indications about population trend between years. Similar survey protocols were applied also in some Mediterranean countries against *T. destruens*. Despite the large number of investigations carried out in this field and some promising results (Carle, 1974b, 1978; Carle *et al.*, 1978; Hamza and Chararas, 1981), an aggregation pheromone specific for *T. destruens* has not been identified. The possible use of host volatiles emitted by stressed trees

suitable to *T. destruens* bark colonization has been tested in the forest (Faccoli *et al.*, 2008). Both in Italy (Sabbatini Peverieri *et al.*, 2004, 2005, 2007) and in Spain (Gallego *et al.*, 2008), the highest catches (a few hundreds of beetles per trap) were obtained in traps baited with (−) α-pinene and ethanol. Lures registered against *T. piniperda* gave similar or lower captures (Sabbatini Peverieri *et al.*, 2004, 2005). These generic host volatiles may hence be useful in population monitoring of *T. destruens*.

Detection and survey of *T. destruens* populations in southern Europe was also experimentally performed by innovative procedures of remote monitoring and diagnostics. Aerial black and white, color, and infrared photographs of stone pine stands in central Italy were compared with tree ground survey. Photographs allowed the identification of trees suffering for climatic stress (water stress, wind damage) or attacked by insect pests, including *T. destruens*. False color infrared photographs gave the best results (Tinelli and Catena, 1992). Similar results were found also by aerial image acquisition with a lidar fluorosensor, an instrument that is mainly composed of a frequency-tripled laser and a telescope that detects Raman scattering by water and laser-induced fluorescence by chromophoric dissolved organic matter (Barbini *et al.*, 1995). Infested or stressed trees have different concentrations of plant pigments inducing an altered light reflection of the canopies, which is detected by the fluorosensor.

9.2 Population Management Methods

As the pine shoot beetles normally are dependent on a continuous supply of suitable host material for their survival, silvicultural and logging practices may greatly affect the population density of these beetles, both locally and regionally. The concept of forest hygiene, i.e., keeping the amounts of brood material available for bark beetles during their flight period as low as possible, has long been a key strategy in forest protection (Escherich, 1923; Postner, 1974; Wainhouse, 2005). This means that forestry operations, like thinning and final felling, should be done with a minimum of suitable host material left in the forests, that snow-breaks and windfalls should be cleared up before beetle attack or at least before beetle emergence, and correspondingly that the timber should be taken out from the woods in due time. This applies fully to the pine shoot beetles as well, and since most of the work has been done to prevent damage by *T. piniperda*, management and control of this species will be discussed. For the other *Tomicus* species, related management or control options will be presented.

9.2.1 The Case of *T. piniperda*

An important approach to maintain low beetle population is the proper timing of silvicultural operations like cleaning (i.e., pre-commercial thinning) and thinning of pine stands (Hanson, 1937; Wainhouse, 2005). In general, late summer operations should be preferred as the waste wood is neither attacked in the year of cutting (beetle flight terminated) nor in the following spring (waste wood unsuitable), as Escherich (1923) pointed out. In northern Europe, June–September are considered to be "*Tomicus* safe" months (Långström, 1986; Annila and Heikkilä, 1991), whereas Postner (1974) recommends August for central Europe. Cleanings should preferentially be made before DBH (diameter at 1.3 m stem height) exceeds 3 cm, i.e., before thick bark starts to form on the lower stem and the trees become suitable for *T. piniperda* (Butovitsch, 1954; Långström, 1979). The problem with the cleaned stems as well as other logging waste as breeding material for bark beetles is that it may, however, disappear with the increasing demand for woody material for energy production (Schroeder, 2008).

Before the Second World War, the availability of brood material was kept down by clean forestry practice (Escherich, 1923; Hanson, 1940). Felled timber was debarked if it could not be taken out in time, and it was even recommended that pine stumps after thinning and final cutting should be debarked in order to prevent breeding of *T. piniperda* (Lagerberg, 1911; Escherich, 1923). Attacked standing trees were also swiftly removed both to save timber quality and to prevent beetle propagation. Timing of cutting operations to late summer reduced attack risks, as the logging waste was drying out before the spring flight. When a beetle outbreak occurred, deployment of trap trees was the only way of reducing beetle populations. The trap trees were felled before the spring flight, and removed or debarked before beetle emergence (Escherich, 1923; Postner, 1974). In the 1950s, a major change took place in forestry operations when the timber, which until then mainly had been debarked in the forests prior to transportation (often by floating), instead was stored un-barked along roadsides until taken to the sawmill by trucks. In contrast to earlier practices, forest operations were also conducted year-round. This created an entirely new situation with huge amounts of un-barked timber stored in the forests during beetle flight, especially in northern Europe where annual cuttings were large and roads were often inaccessible during the thawing period in spring. This phenomenon has caused permanently elevated bark beetle populations and it is not discussed much in the literature (Nilsson, 1976; Eidmann, 1985; 1992; Jääskelä *et al.*, 1997). Still, the main option to avoid bark beetle damage was, and still is, timely removal (i.e., prior to beetle flight in spring) of the saw logs and pulpwood from the forest to the industry, where it is stored until processed. At the timber yard of a sawmill, the valuable saw logs are mostly either debarked or sprinkled with water to avoid quality losses, but the pulpwood is mostly stored in large stacks at the pulpmills without any measures to prevent beetle propagation (Långström and Hellqvist, 1990; Borkowski, 2006).

If timely transportation is not feasible, the timber can be protected in different ways. The practice of debarking the timber was abandoned as being too expensive, and replaced by the increasing use of insecticides for timber protection. This practice started in the 1960s and there are many references mainly from the 1970s dealing with timber protection with insecticides. Originally DDT was used, followed by lindane, and in the 1980s synthetic pyrethroids like permethrin became the main option (Srot, 1968; Novak, 1972; Dowding, 1974; Dominik and Kinelski, 1979; Szmidt, 1983; Glowacka and Wajland, 1992). Due to increased environmental concerns, the use of insecticides for timber protection has greatly decreased, at least in Fennoscandia. Also, spraying against the early flying *Tomicus* species may be tricky under northern conditions, as snow may sometimes still cover part of the timber resulting in poor spray coverage.

Instead, other ways of protecting log piles against bark beetles have been explored. Covering pulp wood stacks with plastic or other coatings has given variable but sometimes satisfactory results in Sweden and Finland (Dehlen and Nilsson, 1976; Heikkilä, 1978; Jääskelä *et al.*, 1997). Partial debarking or removal of the upper layers in the pines also substantially reduced beetle populations (Dehlen *et al.*, 1982; Jääskelä *et al.*, 1997), as did sprinkling with water (Regnander, 1976). A more modern approach is based on deterring the beetles from attacking by spraying log piles with verbenone or other substances with deterrent properties (Baader and Vité, 1990, Kohnle *et al.*, 1992; McCullough *et al.*, 1998). However, none of these techniques has attained greater use so far. Baited traps containing host odors, especially α-pinene, attract large numbers of pine shoot beetles, and these are excellent for monitoring purposes but give little hope for beetle control. The recent finding that trans-verbenol may act as an aggregation pheromone in *T. piniperda* (Poland *et al.*, 2003) may provide an even better monitoring tool but it is hardly a viable control option. The use of non-host volatiles may also be an option for tree and log protection against *T. piniperda* (Poland and Haack, 2000; Schlyter *et al.*, 2000; Kohnle, 2004).

Some *T. piniperda* biological control attempts have been made using *B. bassiana*, with variable results (Bychawska and Swiezynska, 1979; Nuorteva and Salonen, 1980). Lutyk and Swiezynska (1984) obtained satisfactory results when logs were covered with plastic after spraying with *B. bassiana*. The introduction of the clerid beetle *T. formicarius* to North America has been seriously considered (Haack and Poland, 2001), but this approach may now be redundant as the native and closely related clerid *T. dubius* seems to have adapted to the new prey (Kennedy and McCullough, 2002).

After its detection in North America, *T. piniperda* was classified as an invasive species and strict quarantine regulations were enforced for all potentially infested pine material (Haack and Poland, 2001; Scarr et al., 1999). In order to minimize the problem, a compliance management program developed for Christmas tree growers and buyers was initiated in 1997 (Haack and Poland, 2001). The program includes sanitation of cut trees and stumps, use of trap logs to collect and destroy beetles, and insecticide spray to control shoot-feeding beetles, as described by McCullough and Sadof (1998) and McCullough et al. (1998). They also reported a drop in infestation rates from 28–67% in unmanaged plantations to 0–4% in well-managed plantations. It is, however, the presence and not the magnitude of damage that is the problem for the trade, and although the quarantine rules have been questioned, as pine shoot beetle damage outside the plantations has not been as bad as feared, the US Animal and Plant Health Inspection Services decided to maintain the quarantine in 2001 (Haack and Poland, 2001).

Legal steps have also been taken in Europe in order to reduce bark beetle damage. The large bark beetle outbreaks of the 1970s (Nilsson, 1976) resulted in a 1979 change in the Swedish forest law, when a forest protection paragraph was included regulating the handling of pine and spruce wood that could be colonized by bark beetles (Eidmann, 1985, 1992). Finland (Jääskelä et al., 1997) and Norway followed a few years later, and a similar legislation probably exists in many European countries, although no references have been found. In Sweden, the nationwide surveys of fallen Tomicus-damaged shoots done by the Swedish National Forest Inventory show a substantial drop in shoot damage following the storm in 2005 as compared to the situation in the 1970s, after the 1969 storm (Nilsson, 1976; Wulff, 2008; Långström and Wulff, in prep.). This indicates that the improved forest protection strategy during the last decades has paid off in terms of a slower buildup of beetle populations, and consequently in lower damage levels, although the 2005 storm damage was twice as strong as the 1969 storm (75 versus 37 million m^3).

Keeping the bark beetle populations down is a sound forest protection strategy, but since the 1990s there is an increasing conflict between forest protection and nature conservation, as storm-felled trees and other woody debris are also important for maintaining a high biodiversity in the forest ecosystem. This environmental concern, which per se is sound, has also resulted in emergence of certification schemes in the 1990s for forests and forest products, which has changed forest protection policies and strategies in many countries (Wainhouse, 2005). The main principle of the Forest Stewardship Council (FSC) is to promote non-chemical methods in forest pest management, but it also imposes restrictions on forest management. Compliance with the FSC rules has become an important policy issue for forest companies. During the last decade, there has been considerable debate about storm-felled trees in natural reserves acting as "pest sources" that produce a lot of

beetles (Komonen et al., 2011), but the importance of those beetles on the landscape level remains to be determined. Regarding T. piniperda, it has been shown that green tree retention and prescribed burning lead to modestly increased shoot damage within 100 m from the stand edges (Martikainen et al., 2006) and a similar result was found after unsalvaged storm felling (Komonen et al., 2009).

9.2.2 Other Tomicus Species

There are no specific countermeasures known against T. minor that differ from the general control measures used and described above for T. piniperda. For T. minor, the main focus should, however, be on preventing damage to saw logs by timely transportation and debarking, or storage under water, or, as the last option when none of the previous options are available, spraying with insecticides. It is also noteworthy that T. minor mainly breeds in storm-felled trees (Långström, 1984), seldom occurs in log piles (Långström et al., 1984), and never attacks stumps (Hellqvist, 1984). Therefore, less attention needs to be paid to this species when planning silvicultural and logging operations.

As T. destruens and T. piniperda were mixed up for a long time, there is very little information on management and control that refers with certainty to T. destruens. According to Faccoli et al. (2005a), all Mediterranean Tomicus papers dealing with Mediterranean pine species should be attributed to T. destruens, which never has been recorded on Scots pine. The general strategies available to control T. destruens populations in the circum-Mediterranean countries are very similar to those applied against T. piniperda in central-northern Europe. A sound silvicultural management of the pine stands is probably the most feasible approach to prevent infestations. Pine clearing, thinning, and harvesting should be carried out in early spring to reduce risk of attacks, as the logging waste and timber is drying out during summer before the autumn flight of the T. destruens adults. Otherwise, when harvested logs are often left in the forest for extended periods, bark beetle breeding attacks may be prevented by log debarking or exposure to solar radiation, as suggested by Abgrall and Soutrenon (1991) for southern France.

Infestation control and containment of damage may be instead achieved by sanitation felling based on bark peeling or removal of infested trees (Capretti et al., 1987) associated with the use of trap-trees (Braquehais, 1973). Control of T. destruens using trap-trees or trap-logs was tested also by Capretti et al. (1987) and Triggiani (1984), which suggested the use of trap-trees set up vertically in September or October in pine stand clear-cuts and removed or debarked in February when pupae are still inside, as the most effective mean of T. destruens control. When log debarking is not possible or too expensive, emerging insects can be killed by spraying trap-trees with synthetic pyrethroids (Abgrall and Soutrenon, 1991). The use of insecticides in forests was related also to the chemical protection of live trees. Effectiveness, phytotoxicity, and residues of methomyl, acephate, and monocrotophos sprayed against *T. destruens* were evaluated in a field experiment conducted in central Italy. The trial consisted of a stem infusion of insecticides with pressured equipment through holes bored at the base of stone pine trees. Phytotoxicity was assessed based on the degree of closing of the holes used for the stem infusion, and presence of residues in pine needles, shoots, and pinoles. The results showed that methomyl was the most effective insecticide, monocrotophos had a phytotoxic effect, and acephate had a low degree of effectiveness (Paparatti *et al.*, 2000). Although effective, this protocol is very expensive and it could be adopted only for protection of trees of relevant economic, cultural, or historical value.

Blends of host volatiles active against *T. destruens* adults are now available, but their use in outbreak control or prevention does not represent a suitable solution due to their low attractiveness. Nevertheless, semiochemicals may be applied not only for insect trapping but also for tree and log protection. Recent applications of non-host volatiles (NHV) and green-leaf volatiles (GLV) gave positive results both in Spain (Guerrero *et al.*, 1997) and Italy (Sabbatini Peverieri *et al.*, 2004, 2005). Fresh timber and logs may be effectively protected from *T. destruens* bark colonization by the application of dispensers releasing NHV or GLV inducing an inhibition of the adult attack behavior (Guerrero *et al.*, 1997). Trap-logs treated with various blends of repellents (octanol, verbenone, limonene, hexenol, hexanol) in some cases exhibited a colonization density lower than control (untreated logs) (Sabbatini Peverieri *et al.*, 2004, 2005). Similar studies in the USA show that not only GLV but also verbenone have a potential for deterring pine shoot beetles and protecting timber (Poland and Haack, 2000; Haack *et al.*, 2004). Natural enemies are able to control *T. destruens* populations only if appropriate silvicultural measures are applied (Mendel *et al.*, 1986; Mendel, 1987). In conclusion, integrated pest management is probably the best strategy to reduce the risk of *T. destruens* infestations and manage the ongoing outbreaks (López Pantoja *et al.*, 2000).

In China, the same principles as described above for *T. piniperda* apply for the management and control of *T. yunnanensis*, although Wang *et al.* (1987) particularly stress the importance of silvicultural means of reducing tree susceptibility of Yunnan pine stands to beetle attacks. Early detection of active infestations as well as prompt action to remove infested trees in order to suppress populations is also emphasized. In the 1990s, an integrated management program developed to reduce the damage caused to Yunnan pine by *T. yunnanensis* included prompt felling of attacked trees as the main option, with the use of insecticides as a

supplement. Regulation of stem density to maintain healthy forests was recommended as well as the possible use of biological control and semiochemicals in monitoring and with trap-trees. Recently, a monitoring tool based on bundles of cut pine bolts was developed, and it was found that peeled bolts exposing the phloem were more attractive than intact bolts (Lu *et al.*, 2012a). The finding that a shoot damage exceeding 60% of the shoots is critical for successful stem attacks stresses the need for a management technique aimed at keeping the shoot damage below this critical level (Lieutier *et al.*, 2003). Quarantine procedures in order to prevent further dispersal of this serious pest are needed (Lieutier *et al.*, 2003). Regarding the newly described *T. armandii*, its biology and damage levels need to be clarified before the need for control measures can be estimated.

10. CONCLUSION

Among the eight known species of *Tomicus*, *T. piniperda* and *T. minor* have a very large distribution and cover both Europe and Asia. *Tomicus piniperda* has even been introduced in North America, giving it an almost Holarctic distribution. Among the five species endemic to Asia, *T. armandii* and *T. yunnanensis* seem much localized and until now have been described only from Yunnan, whereas *T. puellus*, *T. pilifer*, and certainly *T. brevipilosus* are largely distributed in Asia. *Tomicus destruens* is endemic to the Mediterranean Basin. However, whereas a considerable amount of research has been developed on the European species, especially on *T. piniperda*, the biology of the Asian endemic species, except *T. yunnanensis*, is very poorly known. No information is available on their biotic associations and population dynamics, as well as on their mechanisms of establishment on trees. Consequently, there is a considerable need for research on the Asian *Tomicus* species in the fields of basic biology, economic impact, relation with host tree, and population dynamics.

Although a large amount of data has been accumulated on the European species, there are still several aspects that need to be explored for all *Tomicus* species. For example, a reliable sampling method is still lacking, therefore making quantification and detailed survey of populations very difficult. Sampling during larval development is a methodological problem shared with all bark beetle species. Shoot feeding may offer an original and easy way to sample *Tomicus* populations, particularly after shoots have fallen on the ground, but the quantity of fallen shoots also depends greatly on climatic conditions. Very little information is available on the number and role of sister broods in the field, although they may be an important way to compensate intraspecific competition, possibly giving them a crucial role in population dynamics. Studies are also needed on the respective role of the various limiting factors in population dynamics, especially the density-dependent factors and their variations, tree resistance, and the impact of natural enemies.

Moreover, in the present climatic context, we urgently need to evaluate the possible consequences of climate change on *Tomicus* damage and geographic extension of their populations. For example, in Asia the localized ranges of species such as *T. yunnanensis* and *T. armandii* could expand if climate is the reason for this localization. In Europe, climate (combined with host tree distribution) is effectively the main reason for the localization of *T. destruens* in the Mediterranean area (Horn *et al.*, 2012). It is thus essential to clarify the relationships between abiotic factors on the one hand and beetle populations, tree resistance, and other regulating factors on the other. Studies of the relationships between water stress and tree resistance are also needed. Temperature has direct effects on insects through optimal developmental conditions and number of generations, especially through number of sister broods for *Tomicus* species. In addition, in the context of a geographic extension, host tree preference must also be considered. However, the various climate factors act in concert and their effects should be analyzed in combination on trees, beetles, and their associated organisms (natural enemies as well as fungi). In addition, as soil quality also conditions the ability of trees to respond to both biotic and abiotic aggressions, soil fertility should also be taken into account. Finally, since climate is a multivariate component, its possible effect on the multivariate *Tomicus* biological system could certainly be studied only through a global approach based on models.

ACKNOWLEDGMENTS

We thank Annie Yart (INRA, Orléans, France) for her efficient literature search. We are also grateful to Marc Kenis (CABI, Delémont, Switzerland) and Beat Wermelinger (WSL, Birmensdorf, Switzerland) for providing very useful information on natural enemies. Thanks also to Karin Eriksson for translating Russian, to Tao Zhao for help with Chinese papers, and to Claes Hellqvist for providing pictures.

REFERENCES

Abgrall, J.F., Soutrenon, A., 1991. La forêt et ses ennemis. CEMAGREF, Paris.

Agafonov, A.F., Kuklin, L.V., 1979. Stem pests on Scots pine on burns. (In Russian.). Lesnoe-Khozyaistvo 10, 55–57.

Alonso-Zarazaga, M.A., Lyal, C.H.C., 2009. A catalogue of family and genus group names in Scolytinae and Platypodinae with nomenclatural remarks (Coleoptera: Curculionidae). Zootaxa 2258, 1–134.

Amezaga, I., 1996. Monterrey pine (*Pinus radiata* D. Don) suitability for the pine shoot beetle (*Tomicus piniperda* L.) (Coleoptera: Scolytidae). For. Ecol. Manage. 86, 73–79.

Amezaga, I., 1997. Forest characteristics affecting the rate of shoot pruning by the pine shoot beetle (*Tomicus piniperda* L.) in *Pinus radiata* D. Don and *P. sylvestris* L. plantations. Forestry 70, 129–137.

Amezaga, I., Garbisu, C., 2000. Effect of intraspecific competition on progeny production of *Tomicus piniperda* (Coleoptera: Scolytidae). Environ. Entomol. 29, 1011–1017.

Amezaga, I., Rodríguez, M.Á., 1998. Resource partitioning of four sympatric bark beetles depending on swarming dates and tree species. For. Ecol. Manage. 109, 127–135.

Anderbrandt, O., 1988. Survival of parent and brood adult bark beetles, *Ips typographus*, in relation to size, lipid content and re-emergence day. Physiol. Entomol. 13, 121–129.

Andersson, S.O., 1973. Increment losses after thinning [in Scots Pine] caused by *Myelophilus piniperda*. Sveriges Skogsvardsförbunds Tidskrift 71, 359–379.

Andreev, E.A., 1988. On the fauna and ecology of mites of the genus *Procto-laelaps* (Aceosejidae) from galleries of bark beetles in the Moscow region. Nauchnye Doklady Vysshei Shkoly, Biolohicheskie Nauki 10, 34–37.

Annila, E., Heikkilä, R., 1991. Breeding efficiency in *Tomicus piniperda* and shoot damage after late autumn thinning of young *Pinus sylvestris* stands. Scandinavian J. For. Res. 6, 197–207.

Annila, E., Petäistö, R.-L., 1978. Insect attack on windthrown trees after the December 1975 storm in western Finland. Metsantutkimuslaitoksen Julkaisuja 94, 1–24.

Annila, E., Långström, B., Varama, M., Hiukka, R., Niemela, P., 1999. Susceptibility of defoliated Scots pine to spontaneous and induced attacks by *Tomicus piniperda* and *Tomicus minor*. Silva Fennica 33, 93–106.

Anonymous, 1997. Memoir of *Tomicus piniperda* L. research. Yunnan Forest Science and Technology 2, 1–118.

Astiaso, J.F., Leyva, E., 1970. Contribución al conocimiento de la biología y métodos de combate de *Blastophagus* sp. y *Pissodes notatus* F. Boletin del Servicio de Plagas Forestales 13, 203–211.

Avtzis, N., Gatzojannis, S., 2000. Auftreten der Borkenkaferart *Blastophagus piniperda* (L.) im Erholungswald von Thessaloniki und Waldbewirtschaftung. Mitteilungen Deutsche Gesellschaft Allgemeine Angewandte Entomologie 12, 29–32.

Baader, E.J., Vité, J.P., 1990. Ablenkstoffe gegen Borkenkafer. Allgemeine Forst- und Jagdzeitung 161, 145–148.

Bakke, A., 1968. Ecological studies on bark beetles (Coleoptera: Scolytidae) associated with Scots pine (*Pinus sylvestris* L.) in Norway with particular reference to the influence of temperature. Meddelelser fra det Norske Skogforsøksvesen 21, 443–602.

Bakke, A., 1996. Influence of forest fire in the beetle fauna. Rapport fra Skogforsk. 20, pp.

Barbini, R., Colao, F., Fantoni, R., Palucci, A., Ribezzo, S., Cinti, S., Paparatti, B., 1995. Monitoraggio mediante lidar fluorosensore di piante di pino domestico attaccate dal Coleottero Scolitide Blastofago distruttore (*Blastophagus destruens* Woll.) al centro ENEA di Frascati. Atti ENEA, Dipartimento Innovazione, RT/INN/95/27. Frascati, Italy, 38 pp.

Beaver, R.A., 1974. Intraspecific competition among bark beetle larvae (Coleoptera: Scolytidae). J. Anim. Ecol. 43, 445–467.

Ben Jamâa, M.L., Lieutier, F., Jerraya, A., 2000. Les Scolytids ravageurs des pins en Tunisie. Annales de l'INGREF 4, 27–39.

Ben Jamâa, M.L., Lieutier, F., Yart, A., Jerraya, A., Khouja, M.L., 2007. The virulence of phytopathogenic fungi associated with the bark beetles *Tomicus piniperda* and *Orthotomicus erosus* in Tunisia. For. Pathol. 37, 51–63.

Berryman, A.A., 1972. Resistance of conifers to invasion by bark beetle fungus associations. BioScience 22, 598–602.

Berryman, A.A., 1976. Theoretical explanation of mountain pine beetle dynamics in lodgepole pine forests. Environ. Entomol. 5, 1225–1233.

Berryman, A.A., 1982. Population Dynamics of Bark Beetles. In: Mitton, J.B., Sturgeon, K.B. (Eds.), Bark Beetles in North American Conifers. University of Texas, Austin, pp. 264–314.

Bevan, D., 1962. Pine shoot beetles. Forestry Commission Leaflet. No, 3, London, 8 pp.

Birgersson, G., Schlyter, F., Bergström, G., Löfqvist, J., 1988. Individual variation in aggregation pheromone content of bark beetle *Ips typographus*. J. Chem. Ecol. 14, 1737–1761.

Blinova, S.L., Gurando, E.V., 1974. *Parasitorhabditis fuchsi* n. sp. (Nematoda, Rhabditidae) a parasite of *Blastophagus minor*. Vestnik Zoologii 2, 50–55.

Bois, E., Lieutier, F., 1997. Phenolic response of Scots pine clones to inoculation with *Leptographium wingfieldii*, a fungus associated with *Tomicus piniperda*. Plant Physiol. Biochem. 35, 819–825.

Bois, E., Lieutier, F., 2000. Resistance level in Scots pine clones and artificial introductions of *Tomicus piniperda* (Col., Scolytidae) and *Leptographium wingfieldii* (Deuteromycetes). J. Appl. Entomol. 124, 163–167.

Bogdanova, D.A., 1982. The large pine pitch-borer in pine plantings of western Siberia. Izvestiya Sibirskogo Otdelenyia Akademii Nauk SSSR, Biologicheskikh Nauk 15, 109–113.

Bogdanova, D.A., 1998. Foci of *Heterobasidion annosum* and xylophagous insects in Scots pine forests of the Upper Ob region. Lesovedenie 2, 80–85.

Boriani, M., 1998. Il genere *Tomicus* (Latreille, 1802-3) (Coleoptera Scolytidae). Quaderni di ricerca e sperimentazione. Azienda Regionale delle Foreste, Regione Lombardia. Ufficio ricerca e sperimentazione, Difesa fitosanitaria, Riserve naturali, 68–71.

Borkowski, A., 2001. Threats to pine stands by the pine shoot beetles *Tomicus piniperda* (L.) and *Tomicus minor* (Hart.) around a sawmill in southern Poland. J. Appl. Entomol. 125, 489–492.

Borkowski, A., 2006. Spatial distribution of losses in growth of trees caused by the feeding of pine shoot beetles *Tomicus piniperda* and *T. minor* (Col., Scolytidae) in Scots pine stands growing within range of the influence of a timber yard in southern Poland. J. For. Sci. 52, 130–135.

Botterweg, P.F., 1983. The effect of attack density on size, fat content and emergence of the spruce bark beetle *Ips typographus* L. Z. Angew. Entomol. 96, 47–55.

Branco, M., Pereira, J.S., Mateus, E., Tavares, C., Paiva, M.R., 2010. Water stress affects *Tomicus destruens* host pine preference and performance during the shoot feeding phase. Ann. For. Sci. 67, 608.

Braquehais, F., 1973. Trap trees as an integral part of the control of beetle pests. Boletín de la Estación Central de Ecología 2, 65–70.

Braasch, H., Metge, K., Burgermeister, W., 1999. *Bursaphelenchus*-Arten (Nematoda, Parasitaphelenchidae) in Nadelgeholzen in Deutschland und ihre ITS-RFLP-Muster. Nachrichtenbl. Dtsch. Pflanzenschutzdienstes 51, 312–320.

Bright, D.E., 1996. Notes on the native parasitoids and predators of the larger pine shoot beetle, *Tomicus piniperda* (L.) in the Niagara region of Canada (Coleoptera: Scolytidae). Proc. Entomol. Soc. Ont. 127, 57–62.

Browne, F.G., 1968. Pests and Diseases of Forest Plantation Trees. An annotated list of the principal species occurring in the British Commonwealth. The Clarendon Press, Oxford.

Burjanadze, M., 2010. Efficacy of *Beauveria bassiana* Isolate against pine shoot beetle *Tomicus piniperda* L. (Coleoptera: Scolytidae) in laboratory. Bull. Georgian Natl. Acad. Sci 4, 119–122.

Burjanadze, M., Lortkipanidze, M., Supatashvili, A., Gorgadze, O., 2011. Occurrence of pathogens and nematodes of bark beetles (Coleoptera, Scolytidae) from coniferous forest in different region of Georgia. IOBC/WPRS Bull 66, 351–354.

Butovitsch, V., 1946. Redogörelse för flygbekämpningskampanjen mot tallmätaren under åren 1944–1945. Meddelanden från Statens skogsforskninginstitut 35, 1–108.

Butovitsch, V., 1954. Die Einwirkung der Läuterungszeit auf die Vermehrung des grossen Waldgärtners (*Blastophagus piniperda* L.). Berichte. 11. Kongress. Internationalen Verbandes Forstlicher Forschungsanstalten, Rom 1953, pp. 645–649. Firenze.

Bychawska, S., Swiezynska, H., 1979. Attempts to control *Myelophilus piniperda* by means of the entomopathogenic fungus *Beauveria bassiana* (Bals.) Vuill. (In Polish.). Sylwan 123, 59–64.

Byers, J.A., 2004. Chemical ecology of bark beetles in a complex olfactory landscape. In: Lieutier, F., Day, K.R., Battisti, A., Grégoire, J.-C., Evans, H.F. (Eds.), Bark and Wood Boring Insects in Living Trees in Europe, A Synthesis. Kluwer Academic Publishers, Dordrecht, pp. 89–134.

Byers, J.A., Lanne, B.S., Löfqvist, J., Schlyter, F., Bergström, G., 1985. Olfactory recognition of host-tree susceptibility by pine shoot beetles. Naturwissenschaften 72, 324–326.

Cao, H., Zhu, J.-Y., Yang, P., Zhao, N., Yang, B., 2012. Construction and analysis of enriched microsatellite library of *Tomicus brevipilosus*. Journal of Southwest Forestry University 32, 62–65.

Capretti, P., Panconesi, A., Parrini, C., 1987. Osservazioni sul deperimento del pino d'Aleppo e del pino marittimo in rimboschimenti dell'Alta Maremma. Monti e Boschi 1, 42–46.

Carle, P., 1974a. The decline of *Pinus pinaster* in Provence: role of insects in changing the biological equilibrium of forests invaded by *Matsucoccus feytaudi*. Ann. For. Sci. 31, 1–26.

Carle, P., 1974b. Mise en évidence d'une attraction secondaire d'origine sexuelle chez *Blastophagus destruens* Woll. (Col. Scolytidae). *Ann. Sci. Nat. Zool. Ecol. Anim.* 6, 539–550.

Carle, P., 1975. Problèmes posés par les ravageurs xylophages des conifères en forêt méditerranéenne. Rev. For. Fr. 27, 283–296.

Carle, P., 1978. Essais d'attraction en laboratoire et en foret de *Blastophagus* (*piniperda* L. et *destruens* Woll.). Colloques de l'INRA. 92–101.

Carle, P., Descoins, C., Gallois, M., 1978. Pheromones des *Blastophagus* (*piniperda* L. et *destruens* Woll.). Colloques de l'INRA. 87–91.

Cedervind, J., Pettersson, M., Långström, B., 2003. Attack dynamics of the pine shoot beetle, *Tomicus piniperda* (Col., Scolytinae) in Scots pine stands defoliated by *Bupalus pinaria* (Lep., Geometridae). Agric. For. Entomol. 5, 253–261.

Chakali, G., 2003. Influence climatique sur les populations de scolytes dans les peuplements de pin d'Alep en zone semi-aride (Djelfa). In: Seminaire impact des changements climatiques sur l'écologie des espèces animales, la santé et la population humaine Maghrébine., p. 10, GREDUR.

Chakali, G., 2005. L'Hylésine des Pins, *Tomicus destruens* Wollaston 1865 (Coleoptera-Scolytidae) en Zone Semi-Aride (Algérie). Silva Lusitana 13, 113–124.

Chakali, G., 2007. Stratégie d'attaque de l'hylésine *Tomicus destruens* (Wollaston 1865) (Coleoptera: Scolytidae) sur le pin d'Alep en zone semi-aride (Algérie, Djelfa). Ann. Soc. Entomol. Fr. 43, 129–137.

Chararas, C., 1962. Scolytides des conifères. Lechevalier, Paris.

Chararas, C., 1964. Le pin maritime. Dépérissement général dans le Var. Etude du role des Insectes, des conditions climatiques, des facteurs biologiques, Adaptation possible de certaines essences dans le Bassin Méditerranéen. Lechevalier, Paris.

Chen, P., 2003. Possible interspecific competition between *Tomicus yunnanensis* and *Tomicus minor* (Col. Scolytidae) during shoot-feeding and breeding period in China. M.S. thesis, Swedish University of Agricultural Sciences, Sweden.

Chen, P., Zhou, N., Zhao, T., Hu, G., Feng, Z., Li, L., 2004. Relationships between stand status of *Pinus yunnanensis* and damage of *Tomicus piniperda*. Journal of the Northwest Forestry University 32, 13–15.

Christiansen, E., Waring, R.H., Berryman, A.A., 1987. Resistance of conifers to bark beetle attack: searching for general relationships. For. Ecol. Manage. 22, 89–106.

Ciesla, W.M., 2004. Forests and forest protection in Cyprus. Forest. Chron. 80, 1–7.

Colombari, F., Battisti, A., Schroeder, L.M., Faccoli, M., 2012. Life history traits promoting outbreaks of the pine bark beetle *Ips acuminatus* (Coleoptera: Curculionidae, Scolytinae) in the south-eastern Alps. Eur. J. For. Res. 131, 553–561.

Colombari, F., Battisti, A., Schroeder, L.M., Faccoli, M., 2013. Spatial spot dynamics during an *Ips acuminatus* outbreak. Agric. For. Entomol. 15, 34–42.

Croisé, L., Lieutier, F., 1993. Effect of drought on the induced defence reaction of Scots pine to bark beetle-associated fungi. Ann. For. Sci. 50, 91–97.

Croisé, L., Lieutier, F., Dreyer, E., 1998a. Scots pine responses to number and density of inoculation points with *Leptographium wingfieldii* Morelet, a bark beetle associated fungus. Ann. For. Sci. 55, 497–506.

Croisé, L., Dreyer, E., Lieutier, F., 1998b. Effects of drought stress and severe pruning on the reaction zone induced by single inoculations with a bark beetle associated fungus (*Ophiostoma ips*) in the phloem of young Scots pines. Can. J. For. Res. 28, 1814–1824.

Croisé, L., Lieutier, F., Cochard, H., Dreyer, E., 2001. Effect of drought and high density stem inoculations with *Leptographium wingfieldii* on hydraulic properties of young Scots pine trees. Tree Physiol. 21, 427–436.

Crooke, M., 1959. Insecticidal control of the pine looper in Great Britain. I. Aerial spraying. Forestry 32, 166–196.

Czokajlo, D., Teale, S.A., 1999. Synergistic effect of ethanol to a-pinene in primary attraction of the larger pine shoot beetle, *Tomicus piniperda*. J. Chem. Ecol. 25, 1121–1130.

Czokajlo, D., Wink, R.A., Warren, J.C., Teale, S.A., 1997. Growth reduction of Scots pine, *Pinus sylvestris,* caused by the larger pine shoot beetle, *Tomicus piniperda* (Coleoptera, Scolytidae), in New York State. Can. J. For. Res. 27, 1394–1397.

de Beer, Z.W., Seifert, K.A., Wingfield, M.J., 2013. A nomenclator for ophiostomatoid genera and species in the Ophiostomatales and Microascales. *In* "Ophiostomatoid Fungi: Expanding Frontiers" (K. A. Seifert, Z. W. de Beer, and M. J. Wingfield, Eds.). CBS Biodiversity Series 12, 245–322.

Dehlen, R., Nilsson, S., 1976. Covering pine stacks with plastic to prevent attack by *Blastophagus piniperda*. Rapporter och Uppsatser, Institutionen for Skogsteknik 95, 1–36.

Dehlen, R., Herlitz, A., Johansson, I., Långström, B., Regnander, J., 1982. Through debarking of pulpwood—debarking results and protective effect against bark beetles, as well as some economic and ergonomic aspects. Rapport, Institutionen for Skogsteknik, Sveriges Lantbruksuniversitet 143, 1–40.

DeSalle, R., Freedman, T., Prager, E.M., Wilson, A.C., 1987. Tempo and mode of sequence evolution in mitochondrial DNA of Hawaiian *Drosophila*. J. Mol. Evol. 26, 157–164.

Dominik, J., Kinelski, S., 1979. Studies on the duration of effective protection of unpeeled Scots pine logs by several insecticides (in Polish). Sylwan 123, 13–19.

Dowding, P., 1974. Effect of felling time and insecticide treatment on the interrelationships of fungi and arthropods in Pine logs. Oikos 24, 422–429.

Drugescu, C., 1980. Coenological studies on scolytids (Coleoptera) on black pine (*Pinus nigra* var. banatica) in the Cerna Valley. *Studii si Cercetari de Biologie*. Biologie Animala 32, 155–162.

Dreyer, E., Guérard, N., Lieutier, F., Pasquier-Barré, F., Lung, B., Piou, D., 2002. Interactions between nutrient and water supply to potted *Pinus sylvestris* trees and their susceptibility to several pests and pathogens. In: Lieutier, F. (Ed.), Effects of Water and Nutrient Stress on Pine Susceptibility to Various Pest and Disease Guilds. pp. 63–81, Final Scientific Report of the EU Project FAIR 3 CT96-1854.

Duan, Y., Kerdelhue, C., Ye, H., Lieutier, F., 2004. Genetic study of the forest pest *Tomicus piniperda* (Col., Scolytinae) in Yunnan province (China) compared to Europe: new insights for the systematics and evolution of the genus *Tomicus*. Heredity 93, 416–422.

Durand-Gillmann, M., 2014. Interactions plantes-insectes dans deux écosystèmes forestiers méditerranéens contrastés: le cas des scolytes (Coleoptera: Curculionidae: Scolytinae) en région méditerranéenne. Thesis, Aix-Marseille Université, France.

Eidmann, H.H., 1985. Silviculture and insect problems. Z. Angew. Entomol. 99, 105–112.

Eidmann, H.H., 1992. Impact of bark beetles on forests and forestry in Sweden. J. Appl. Entomol. 114, 193–200.

Ehnström, B., Långström, B., Hellqvist, C., 1995. Insects in burned forests—forest protection and faunal conservation (preliminary results). Entomol. Fennica 6, 109–117.

Elfving, B., Långström, B., 1984. Crown damage and growth response in a pine stand attacked by *Tomicus piniperda*. Sveriges Skogsvardsförbunds Tidskrift 82, 49–56.

Ericsson, A., Hellqvist, C., Långström, B., Larsson, S., Tenow, O., 1985. Effects on growth of simulated and induced shoot pruning by *Tomicus piniperda* as related to carbohydrate and nitrogen dynamics in Scots pine. J. Appl. Ecol. 22, 105–124.

Escherich, K., 1923. Forstinsekten Mitteleuropas. Band II. Paul Parey, Berlin.

Faccoli, M., 2006. Morphological separation of *Tomicus piniperda* and *T. destruens* (Coleoptera: Curculionidae: Scolytinae): new and old characters. Eur. J. Entomol. 103, 433–442.

Faccoli, M., 2007. Breeding performances and longevity of *Tomicus destruens* on Mediterranean and continental pine species. Entomol. Exp. Applic. 123, 263–269.

Faccoli, M., 2009. Breeding performance of *Tomicus destruens* at different densities: the effect of intraspecific competition. Entomol. Exp. Appl. 132, 191–199.

Faccoli, M., Piscedda, A., Salvato, P., Simonato, M., Masutti, L., Battisti, A., 2005a. Phylogeography of the pine shoot beetles *Tomicus destruens* and *T. piniperda* (Coleoptera Scolytidae) in Italy. Ann. For. Sci. 62, 361–368.

Faccoli, M., Battisti, A., Masutti, L., 2005b. Phenology of *Tomicus destruens* (Wollaston) in northern Italian pine stands. In: Lieutier, F., Ghaioule, D. (Eds.), Entomological Research in Mediterranean Forest Ecosystems. INRA Editions, Paris, pp. 185–193.

Faccoli, M., Anfora, G., Tasin, M., 2008. Responses of the Mediterranean pine shoot beetle *Tomicus destruens* (Wollaston) (Coleoptera Curculionidae Scolytinae) to pine shoot and bark volatiles. J. Chem. Ecol. 34, 1162–1169.

Fernández Fernández, M.M., Salgado Costas, J.M.S., 1999. Susceptibility of fire-damaged pine trees (*Pinus pinaster* and *Pinus nigra*) to attacks by *Ips sexdentatus* and *Tomicus piniperda* (Coleoptera: Scolytidae). Entomologia Generalis 24, 105–114.

Fernández Fernández, M.M., Pajares Alonso, J.A., Salgado Costas, J.M., 1999a. Oviposition and development of the immature stages of *Tomicus minor* (Coleoptera, Scolytidae). Agric. For. Entomol. 1, 97–102.

Fernández Fernández, M.M., Pajares Alonso, J.A., Salgado Costas, J.M., 1999b. Shoot feeding and overwintering in the lesser pine shoot beetle *Tomicus minor* (Col., Scolytidae) in north-west Spain. J. Appl. Entomol. 123, 321–327.

Fernández Fernández, M.M., Pajares Alonso, J.A., Salgado Costas, J.M., 1999c. The seasonal development of the gonads and fat content of *Tomicus minor* (Coleoptera Scolytidae). Belgian. J. Entomol. 1–2, 311–324.

Ferreira, M.C., Ferreira, G.W.S., 1986. Pragas do pinheiro bravo em Portugal—escolitideos. Boletim Agricola 1–4, No. 36.

Ferreira, M.C., Ferreira, G.W.S., 1990. Pragas das Resinosas. Guia de Campo. Ministério de Agricultura, Pescas e Alimentação, Serie Divulgação, No. 3, Lisboa, 108 pp.

Forsse, E., 1989. Migration in bark beetles with special reference to the spruce bark beetle *Ips typographus*. PhD thesis, Swedish University of Agricultural Sciences, Sweden.

Fransson, J., 2010. Skogsdata 2010. Umeå, Department of Forest Resource Management, Swedish University of Agricultural Sciences, Sweden.

Francke-Grosmann, H., 1952. Über die Ambrosiazucht der beiden Kieferborkenkäfer *Myelophilus minor* Htg. und *Ips acuminatus* Gyll. Meddelanden Från Statens Skogsforskninginstitut 41, 1–52.

Francke-Grosmann, H., 1966. Über Symbiosen von xylo-mycetophagen und phloeophagen Scolytoidea mit holzbewohnenden Pilzen. Holz und Organismen, International Symposium, Berlin-Dahlen, pp. 503–522.

Francke-Grosmann, H., 1967. Ectosymbiosis in wood-inhabiting insects. In: Henry, S.M. (Ed.), SymbiosisVol. 2. Academic Press, New York, pp. 141–205.

Fuchs, G., 1930. Neue an Borkenkäfer und Rüsselkäfer gebundene Nematoden, halbparasitische und Wohnungseinmieter. Zoologische Jahrblatt, Jena, Abteil Systematik 59, 505–646.

Fuchs, G., 1937. Neue parasitische und halbparasitische Nematoden bei borkenkäfern und einige andere Nematoden. Zoologische Jahrblatt, Jena, Abteil Systematik 70, 291–380.

Führer, E., Kerck, K., 1978. Untersuchungen ueber Forstschutzprobleme in Kiefernschwachholz-Windwürfen in der Lüneburger Heide II. Die Gefährdung der Bestandesreste und Nachbarbestände durch rindenbrütende Insekten. Forstwiss.Centralbl 97, 156–167.

Galaseva, T.V., 1976. Life table for *Blastophagus piniperda* on burns in the Moscow region. Nauchnye Trudy, Moskovskii Lesotekhnicheskii Institut 90, 31–38.

Gallego, D., Galian, J., 2001. The internal transcribed spacers (ITS1 and ITS2) of the rDNA differentiate the bark beetle forest pests *Tomicus destruens* and *T. piniperda*. Insect Mol. Biol. 10, 415–420.

Gallego, D., Canovas, F., Esteve, M.A., Galian, J., 2004. Descriptive biogeography of *Tomicus* (Coleoptera: Scolytidae) species in Spain. J. Biogeogr. 31, 2011–2024.

Gallego, D., Galian, J., Diez, J.J., Pajares, J.A., 2008. Kairomonal responses of *Tomicus destruens* (Col., Scolytidae) to host volatiles a-pinene and ethanol. J. Appl. Entomol. 132, 654–662.

Gao, Y., Zhang, Y., Zhang, Z., Wu, W., 2012. Shoot feeding by Tomicus (Coleoptera: Scolytinae) on Pinus yunnanensis trees. Journal of the Northeastern Forestry University 40, 81–84.

Ghaioule, D., Abourouh, M., Bakry, M., Haddan, M., 1998. Insectes ravageurs des forets au Maroc. Ann. Rech. For. Maroc 31, 129–156.

Gibbs, J.N., Inman, A., 1991. The pine shoot beetle *Tomicus piniperda* as a vector of blue stain fungi to windblown pine. Forestry 64, 239–249.

Gidaszewski, A., 1974. An analysis of the occurrence and vigor of *Tomicus piniperda* (L.) and *T. minor* (Hartg.) in forest stands of Wielkopolski National Park during the years 1969–1970. Polskie Pismo Entomologiczne 44, 789–815.

Gilbert, M., Nageleisen, L.-M., Franklin, A., Grégoire, J.-C., 2005. Poststorm surveys reveal large-scale spatial patterns and influences of site factors, forest structure and diversity in endemic bark-beetle populations. Landsc. Ecol. 20, 35–49.

Glowacka, B., Wajland, M., 1992. Wood protection against secondary pests with new insecticides (in Russian). Prace Instytutu Badawczego Lesnictwa 738–745, 121–126.

Gninenko, Y.I., Vetrova, O.G., 2002. Protection of pines from pine bark beetles. Zashchita I Karantin Rastenii 7, 24.

Graf, P., Mzibri, M., 1994. Les Scolytes des pins. In: El Hassani, A., Graf, P., Hamdaoui, M., Harrachi, K., Messaoudi, J., Mzibri, M., Stiki, A. (Eds.), Ravageurs et maladies des forêts au Maroc. DPVCRTF, Morocco, pp. 33–47.

Greese, N.S., 1926. Zur frage über den Regenerationsfrass bei dem kleinen Waldgärtner (*Blastophagus minor* Hartig). Trudy po lisiviy dosivdniy spravi na ukraini. 3–25.

Grégoire, J.-C., Evans, H.F., 2004. Damage and control of BAWBILT organisms, an overview. In: Lieutier, F., Day, K.R., Battisti, A., Grégoire, J.-C., Evans, H.F. (Eds.), Bark and Wood Boring Insects in Living Trees in Europe, A Synthesis. Kluwer Academic Publishers, Dordrecht, pp. 19–38.

Grönberg, G., 1914. Märgborren, en fara för våra norrlandsskogar. Skogen 1, 185–198.

Grosmann, H., 1931. Beitrage zur Kenntniss der Lebensgemeinschaft zwischen Borkenkäfern und Pilzen. Z. Parasitenkd. 3, 56–102.

Guerrero, A., Feixas, J., Pajares, J., Wadhams, L.J., Pickett, J.A., Woodcock, C.M., 1997. Semiochemically induced inhibition of behaviour of *Tomicus destruens* (Woll.) (Coleoptera: Scolytidae). Naturwissenschaften 84, 155–157.

Gurando, E.V., 1974. Nematodes of *Blasophagus minor* Hartig. Relationships with their hosts in the forested areas of Kiev. Proc. thesis, Acad. Sci. Rep, Ukraine.

Gurando, E.V., 1979. The nematodes of *Blastophagus minor* Hartig. Vestnik Zoologii 4, 28–33.

Gurando, E.V., 1990. Ecofaunistic characterization of nematodes of the pine-shoot beetle. In: Sonin, M.D. (Ed.), Helminths of Insects. Brill, The Netherlands, pp. 53–58.

Gurando, O.V., Tsarichkova, D.B., 1974. Changes in the sex apparatus of females of the small pine bark beetle caused by parasitic nematodes. Zakhist Roslin 20, 35–40.

Haack, R.A., 2006. Exotic bark- and wood-boring Coleoptera in the United States: recent establishments and interceptions. Can. J. For. Res. 36, 269–288.

Haack, R.A., Lawrence, R.K., 1995. Attack densities of *Tomicus piniperda* and *Ips pini* (Coleoptera: Scolytidae) on Scotch pine logs in Michigan in relation to felling date. J. Entomol. Sci. 30, 18–28.

Haack, R.A., Kucera, D., 1993. New introduction—common pine shoot beetle, *Tomicus piniperda* (L.). *USDA Forest Service, North-Eastern Area, Pest Alert* NA-TP-05, 93.

Haack, R.A., Lawrence, R.K., Heaton, G.C., 2000. Seasonal shoot-feeding by *Tomicus piniperda* (Coleoptera: Scolytidae) in Michigan. The Great Lakes Entomologist 33, 1–8.

Haack, R.A., Poland, T.M., 2001. Evolving management strategies for a recently discovered exotic forest pest: the pine shoot beetle, *Tomicus piniperda* (Coleoptera). Biol. Invasions 3, 307–322.

Haack, R.A., Lawrence, R.K., Heaton, G.C., 2001. *Tomicus piniperda* (Coleoptera: Scolytidae) shoot-feeding characteristics and overwintering behavior in Scotch pine Christmas trees. J. Econ. Entomol. 94, 422–429.

Haack, R.A., Lawrence, R.K., Petrice, T.R., Poland, T.M., 2004. Disruptant effects of 4-allylanisole and verbenone on Tomicus piniperda (Coleoptera: Scolytidae) response to baited traps and logs. The Great Lakes Entomologist 37, 131–141.

Habermann, M., Geibler, A.V., 2001. Regenerationsfahigkeit von Kiefern (*Pinus sylvestris* L.) und Befall durch rindenbrutende Sekundarschadlinge nach Frass der Nonne (*Lymantria monacha* L.). Forst und Holz 56, 107–111.

Halperin, J., 1966. Forest insects in Israel. Proc. 6th World Forest Congress (Madrid) 2, 2100–2102.

Halperin, J., 1978. *Blastophagus piniperda* in Israel. La Yaaran 28, 20–28.

Halperin, J., Holzchuh, C., 1984. Contribution to the knowledge of bark beetles (Col., Scolytoidea) and associated organisms in Israel. Isr. J. Entomol. 18, 21–37.

Halperin, J., Mendel, Z., Golan, Y., 1982. On the damage caused by bark beetles to pine plantations. Preliminary report. La Yaaran 32, 31–38.

Hamza, M.H., Chararas, C., 1981. Etude de l'Attraction Primaire et de la Nutrition chez *Blastophagus piniperda* L. Annales de l'INRF 6, 1–30.

Hanson, H.S., 1937. Notes on the ecology and control of pine beetles in Great Britain. Bull. Entomol. Res. 28, 185–241.

Hanson, H.S., 1940. Further notes on the ecology and control of pine beetles in Great Britain. Bull. Entomol. Res. 30, 483–536.

Harrington, T.C., 1993. Biology and taxonomy of fungi associated with bark beetles. In: Schowalter, T.D., Filip, G.M. (Eds.), Beetle-Pathogen Interactions in Conifer Forests. Academic Press, New York, pp. 37–58.

Harrington, T.C., 2005. Ecology and evolution of mycetophagous bark beetles and their fungal partners. In: Vega, F.E., Blackwell, M. (Eds.), Insect-Fungal Associations: Ecology and Evolution. Oxford University Press, New York, pp. 257–289.

Hausner, G., Iranpour, M., Kim, J.J., Breuil, C., Davis, C.N., Gibb, E.A., et al., 2005. Fungi vectored by the introduced bark beetle *Tomicus piniperda* in Ontario, Canada, and comments on the taxonomy of *Leptographium lundbergii*, *L. terebrantis*, *L. truncatum*, and *L. wingfieldii*. Can. J. Bot. 83, 1222–1237.

Hedqvist, K.J., 1963. Die Feinde der Borkenkäfer in Schweden, 1. Erzwespen (Chalcidoidea). Stud. For. Suec. 11, 1–176.

Hedqvist, K.J., 1998. Bark beetle enemies in Sweden 2. Braconidae (Hymenoptera). *Entomologica Scandinavica*. Entomologica Scandinavica Suppl. 52, 1–86.

Heikkilä, R., 1978. Protection of pine pulpwood stacks against the common pine-shoot beetle in northern Finland. (In Finnish.). Folia Forestalia, Institutum Forestale Fenniae 351, 11.

Hellqvist, C., 1984. Produktion av större märgborre i tallstubbar (Production of *Tomicus piniperda* in pine stumps). Sveriges Skogsvardsförbunds Tidskrift 82, 37–47.

Hérard, F., Mercadier, G., 1996. Natural enemies of *Tomicus piniperda* and *Ips acuminatus* (Col., Scolytidae) on *Pinus sylvestris* near Orléans, France: temporal occurrence and relative abundance, and notes on eight predatory species. Entomophaga 41, 183–210.

Herting, B., 1973. A Catalogue of Parasites and Predators of Terrestrial Arthropods, Sect. A, Vol. III, Coleoptera to StrepsipteraCommonwealth Agricultural Bureaux, Farnham Royal.

Horn, A., Roux-Morabito, G., Lieutier, F., Kerdelhué, C., 2006. Phylogeographic structure and past history of the circum–Mediterranean species *Tomicus destruens* Woll. (Coleoptera: Scolytinae). Mol. Ecol. 15, 1603–1615.

Horn, A., Stauffer, C., Lieutier, F., Kerdelhué, C., 2009. Complex postglacial history of the temperate bark beetle *Tomicus piniperda* L. (Coleoptera, Scolytinae). Heredity 103, 238–247.

Horn, A., Kerdelhué, C., Lieutier, F., Rossi, J.-P., 2012. Predicting the distribution of the two bark beetles *Tomicus destruens* and *Tomicus piniperda* in Europe and the Mediterranean region. Agric. For. Entomol. 14, 358–366.

Humble, L.M., Allen, E.A., 2006. Forest biosecurity: alien invasive species and vectored organisms. Can. J. Plant Pathol. 28, 256–269.

Iljinsky, A., 1928. Gesetzmässigkeiten in der Vermehrung des kleinen Waldgärtners (*Blastophagus minor*, hartig) und teoretische Begründungen der Massnahmen zu seiner Bekämpfung im Walde. Mitteilungen Forstlichen Versuchswesen in der Ukraina 9, 54–91.

Jääskelä, M., Peltonen, M., Saarenmaa, H., Heliövaara, K., 1997. Comparison of protection methods of pine stacks against *Tomicus piniperda*. Silva-Fennica 31, 143–152.

Jacobs, K., Wingfield, M.J., 2001. *Leptographium* Species: Tree Pathogens, Insect Associates, and Agents of Blue Stain. American Phytopathological Society Press, St. Paul.

Jacobs, K., Wingfield, M.J., Coetsee, C., Kirisits, T., Wingfield, B.D., 2001a. *Leptographium guttulatum* sp. nov., a new species fromp spruce and pine in Europe. Mycologia 93, 380–388.

Jacobs, K., Wingfield, M.J., Uzunovic, A., Frisullo, S., 2001b. Three new species of *Leptographium* from pine. Mycol. Res. 105, 490–499.

Jacobs, K., Kirisits, T., Wingfield, M.J., 2003. Taxonomic re-evaluation of three related species of *Graphium*, based on morphology, ecology and phylogeny. Mycologia 95, 714–727.

Jacobs, K., Bergdahl, D.R., Wingfield, M.J., Halik, S., Seifert, K.A., Bright, D.E., Wingfield, B.D., 2004. *Leptographium wingfieldii* introduced in North America and found associated with exotic *Tomicus piniperda* and native bark beetles. Mycol. Res. 108, 411–418.

Jankoviak, R., 2006. Fungi associated with *Tomicus piniperda* in Poland and assessment of their virulence using Scots pine seedlings. Ann. For. Sci. 63, 801–808.

Jankoviak, R., 2008. Fungi associated with *Tomicus minor* on *Pinus sylvestris* in Poland and their succession into the sapwood of beetle-infested windblown trees. Can. J. For. Res. 38, 2579–2588.

Jankoviak, R., Bilanski, P., 2007. Fungal flora associated with *Tomicus piniperda* L. in an area close to a timber yard in southern Poland. J. Appl. Entomol. 131, 579–584.

Jorgensen, E., Bejer-Petersen, B., 1951. Attack of *Fomes annosus* (Fr.) Cke. and *Hylesinus piniperda* L. on *Pinus silvestris* (in Danish). Dan. Skovforen. Tidsskr. 36, 453–479.

Jurc, M., 2005. Gozna Zoologica. Univerza v Ljubljana, Slovenia.

Kaczmarek, S., Michalski, J., Ratajczak, E., 1992. Groups of mites (Acari, Gamasida) populating the feeding areas of some bark beetles. Sylwan 136, 51–59.

Kailides, D.S., 1964. Attacks by *Myelophilus piniperda* on *Pinus brutia* plantations. Dasos 33, 3–18.

Kaitera, J., Jalkanen, R., 1994. The history of shoot damage by *Tomicus* spp. (Col., Scolytidae) in a *Pinus sylvestris* L. stand damaged by the shoot-disease fungus *Gremmeniella abietina* (Lagerb.) Morelet. J. Appl. Entomol. 117, 307–313.

Kakuliya, G.A., 1966. New nematode *Micoletzkia cordovector* Kakuliya n. sp. (Nematoda: Diplogasteridae). Bulletin Academy Science Georgian SSR 41, 163–168.

Kakuliya, G.A., Devdariani, I.G., 1966. New species of nematode *Bursaphelenchus teratospicularis* Kakuliya and Devdariani, sp. nov. Nematoda: Aphelenchoïdea. Bulletin Academy Science Georgian SSR 38, 187–191.

Kangas, E., Perttunen, V., Oksanen, H., 1970. Responses of *Blastophagus minor* Hart. (Col., Scolytidae) to the pine phloem fraction known to be attractant to *Blastophagus piniperda* L. Entomol. Fennica 2, 120–122.

Kangas, E., Perttunen, V., Oksanen, H., 1971. Physical and chemical stimuli affecting the behavior of *Blastophagus piniperda* L. and *B. minor* Hart. (Col., Scolytidae). Acta Entomol. Fennica 28, 120–126.

Kauffman, W.C., Waltz, R.D., Cummings, R.B., 1998. Shoot feeding and overwintering behavior of *Tomicus piniperda* (Coleoptera: Scolytidae): implications for management and regulation. J. Econ. Entomol. 91, 182–190.

Kenis, M., Wermelinger, B., Grégoire, J.-C., 2004. Research on parasitoids and predators of Scolytidae—A review. In: Lieutier, F., Day, K.R., Battisti, A., Grégoire, J.-C., Evans, H.F. (Eds.), Bark and Wood Boring Insects in Living Trees in Europe, a Synthesis. Kluwer Academic Publishers, Dordrecht, pp. 237–290.

Kennedy, A.A., McCullough, D.G., 2002. Phenology of the larger European pine shoot beetle *Tomicus piniperda* (L.) (Coleoptera: Scolytidae) in relation to native bark beetles and natural enemies in pine stands. Environ. Entomol. 31, 261–272.

Kerdelhué, C., Roux-Morabito, G., Forichon, J., Chambon, J.-M., Robert, A., Lieutier, F., 2002. Population genetic structure of *Tomicus piniperda* L. (Curculionidae: Scolytinae) on different pine species and validation of *T. destruens* (Woll.). Mol. Ecol. 11, 483–494.

Kielcziewski, B., Moser, J.C., Wisniewski, J., 1983. Surveying the acarofauna associated with Polish Scolytidae. Bulletinde la Sociétédes Amis des Sciences et des Lettres de Poznan, Série D 22, 151–159.

Kim, G.H., Kim, J.J., Ra, J.B., 2002. Development of fungal sapstain in logs of Japanese red pine and Korean pine. Journal of the Korean Wood Science and Technology 30, 128–133.

Kim, J.J., Lim, Y., Breuil, C., Wingfield, M.J., Zhou, X., Kim, G., 2005. A new *Leptographium* species associated with *Tomicus piniperda* infesting logs in Korea. Mycol. Res. 109, 275–284.

Kirchhoff, J.-F., Führer, E., 1990. Experimentelle Analyse des Infektion und des Entwicklungszyklus von *Malamoeba scolyti* in *Dryocoetes autographus* (Coleoptera: Scolytidae). Entomophaga 35, 537–544.

Kirisits, T., 2004. Fungal associates of European bark beetles with special emphasis on the Ophiostomatoid fungi. In: Lieutier, F., Day, K.R., Battisti, A., Grégoire, J.-C., Evans, H.F. (Eds.), Bark and Wood Boring Insects in Living Trees in Europe, a Synthesis. Kluwer Academic Publishers, Dordrecht, pp. 181–235.

Kirisits, T., Grubelnik, R., Führer, E., 2000. Die ökologische Bedeutung von Bläuepilzen für rindenbrütende Borkenkäfer. In: Müller, F.

(Ed.), In: Mariabrunner Waldbautage 1999—Umbau sekundärer Nadelwälder, 111. Schriftenreihe der Forstlichen Bundesversuchsanstalt Wien, FBVA-Berichte, pp. 117–137.

Kirkendall, L.R., Faccoli, M., Ye, H., 2008. Description of the Yunnan shoot borer, *Tomicus yunnanensis* Kirkendall and Faccoli sp. n. (Curculionidae, Scolytinae), an unusually aggressive pine shoot beetle from southern China, with a key to the species of *Tomicus*. Zootaxa 1819, 25–39.

Kirschner, R., 2001. Diversity of filamentous fungi in bark beetle galleries in central Europe. In: Miskra, J.K., Horn, B.W. (Eds.), Trichomycetes and Other Fungal Groups: Professor Robert W. Lichtwardt Commemoration Volume. Science Publishers, New Hampshire, pp. 175–196.

Klepzig, K.D., Six, D.L., 2004. Bark beetle–fungal symbiosis: context dependency in complex associations. Symbiosis 37, 189–205.

Klepzig, K.D., Moser, J.C., Lombardero, M.J., Ayres, M.P., Hofstetter, R.W., Walkinshaw, C.J., 2001a. Mutualism and antagonism: ecological interactions among bark beetles, mites and fungi. In: Jeger, M.J., Spence, N.J. (Eds.), Biotic Interactions in Plant-Pathogen Associations. CAB International, Wallingford, pp. 237–267.

Klepzig, K.D., Moser, J.C., Lombardero, M.J., Hofstetter, R.W., Ayres, M.P., 2001b. Symbiosis and competition: complex interactions among beetles, fungi and mites. Symbiosis 30, 83–96.

Knížek, M., 2011. Subfamily Scolytinae. In: Löbl, I., Smetana, A. (Eds.), Catalogue of Palaearctic Coleoptera. Volume 7. Curculionoidea I. Apollo Books, Stenstrup, pp. 204–251.

Kohlmayr, B., 2001. Zum Auftreten von Krankheitserrengern in den Europäischen Waldgärtnerarten *Tomicus piniperda*, *Tomicus minor* und *Tomicus destruens* (Coleoptera; Scolytidae). PhD thesis, Universität für Bodenkultur, Austria.

Kohlmayr, B., Riegler, M., Wegensteiner, R., Stauffer, C., 2002. Morphological and genetic identification of the three pine pests of the genus *Tomicus* (Coleoptera, Scolytidae) in Europe. Agric. For. Entomol. 4, 151–157.

Kohlmayr, B., Weiser, R., Wegensteiner, R., Händel, U., Zizka, Z., 2003. Infection of *Tomicus piniperda* (Col.: Scolytidae) with *Canningia tomici* sp. n. (Microsporidia, Unikaryonidae). J. Pest Sci. 76, 65–73.

Kohnle, U., 2004. Host and non-host odour signals governing host selection by the pine shoot beetle, *Tomicus piniperda* and the spruce bark beetle, *Hylurgops palliatus* (Col., Scolytidae). J. Appl. Entomol. 128, 588–592.

Kohnle, U., Densborn, S., Duhme, D., Vité, J.P., 1992. Bark beetle attack on host logs reduced by spraying with repellents. J. Appl. Entomol. 114, 83–90.

Kolomiets, N.G., Bogdanova, D.A., 1998. Xylophagous insects of northern taiga forests in the oil extraction areas of Western Siberia. Lesovedenie 4, 34–42.

Komonen, A., Laatikainen, A., Similä, M., Martikainen, P., 2009. Ytimennävertäjien kasvainsyöntitrombin kaataman suojelumännikön ympäristössä Höytiäisen saaressa Pohjois-Karjalassa. Metsätieteen aikakauskirja 2, 127–134.

Komonen, A., Schroeder, L.M., Weslien, J., 2011. *Ips typographus* population development after a severe storm in a nature reserve in southern Sweden. J. Appl. Entomol. 135, 132–141.

Kotýnková-Sytchrová, E., 1966. The mycoflora of bark beetle galleries in Czecholslovakia. Ceska Mykologie 20, 45–53.

Krausse, A.M., 1920. Die Arten, Rassen und Varietäten des grossen Waldgärten (genus *Blastophagus* Eich.). Z. Forst- Jagdwes 52, 169–178.

Krokene, P., Solheim, H., Långström, B., 2000. Fungal infection and mechanical wounding induce disease resistance in Scots pine. Eur. J. For. Pathol. 106, 537–541.

Kytö, M., Niemela, P., Annila, E., 1998. Effects of vitality fertilization on the resin flow and vigour of Scots pine in Finland. For. Ecol. Manage. 102, 121–130.

Kytö, M., Niemela, P., Annila, E., Varama, M., 1999. Effects of forest fertilization on radial growth and resin exudation of insect defoliated Scots pines. J. Appl. Entomol. 36, 763–769.

Lagerberg, T., 1911. En märgborrshärjning i öfre Dalarna. Sveriges Skogs-vardsförbunds Tidskrift. 381–395.

Långström, B., 1979. Breeding of pine shoot beetles in cleaning residues of Scots pine and subsequent shoot damage on remaining trees. Skogsen-tomologiska Rapporter, Sveriges Lantbruksuniversitet, 1, 52 pp.

Långström, B., 1980. Distribution of pine shoot beetle (Tomicus piniperda, Tomicus minor) attacks within the crown of Scots pine (Pinus sylvestris). Stud. For. Suec. 154, 1–25.

Långström, B., 1983a. Life cycles and shoot-feeding of the pine shoot beetles. Stud. For. Suec. 163, 1–29.

Långström, B., 1983b. Within-tree development of Tomicus minor (Hart.) (Col., Scolytidae) in wind-thrown Scots pine. Acta Entomol. Fennica 42, 42–46.

Långström, B., 1984. Windthrown Scots pines as brood material for Tomicus piniperda and Tomicus minor. Silva Fennica 18, 187–198.

Långström, B., 1986. Attack density and brood production of Tomicus pini-perda in thinned Pinus sylvestris stems as related to felling date and latitude in Sweden. Scandinavian J. For. Res. 1, 351–357.

Långström, B., Hellqvist, C., 1985. Pinus contorta as a potential host for Tomicus piniperda L. and T. minor (Hart.) (Col., Scolytidae) in Sweden. Z. Angew. Entomol. 99, 174–181.

Långström, B., Hellqvist, C., 1990. Spatial distribution of crown damage and growth losses caused by recurrent attacks of pine shoot beetles in pine stands surrounding a pulp mill in southern Sweden. J. Appl. Entomol. 110, 261–269.

Långström, B., Hellqvist, C., 1991. Shoot damage and growth losses fol-lowing three years of Tomicus attacks in Scots pine stands close to a timber storage site. Silva Fennica 25, 133–145.

Långström, B., Hellqvist, C., 1993a. Scots pine susceptibility to attack by Tomicus piniperda (L) as related to pruning date and attack density. Ann. Sci. For. 50, 101–117.

Långström, B., Hellqvist, C., 1993b. Induced and spontaneous attacks by pine shoot beetles on young Scots pine trees: tree mortality and beetle performances. J. Appl. Entomol. 115, 25–36.

Långström, B., Hellqvist, C., Ehnström, B., 1984. Distribution and production of Tomicus piniperda in pine wood stacks. Sveriges Skogsvardsförbunds Tidskrift 82, 23–35.

Långström, B., Tenow, O., Ericsson, A., Hellqvist, C., Larsson, S., 1990. Effects of shoot pruning on stem growth, needle biomass, and dynamics of carbohydrates and nitrogen in Scots pine as related to season and tree age. Can. J. For. Res. 20, 514–523.

Långström, B., Hellqvist, C., Ericsson, A., Gref, D., 1992. Induced defense reaction in Scots pine following stem attacks by Tomicus piniperda L. Ecography 15, 318–327.

Långström, B., Solheim, H., Hellqvist, C., Gref, D., 1993. Effects of pruning young Scots pines on host vigour and susceptibility to Lepto-graphium wingfieldii and Ophiostoma minus, two blue-stain fungi associated with Tomicus piniperda. Eur. J. For. Pathol. 23, 400–415.

Långström, B., Lieutier, F., Hellqvist, C., Vouland, G., 1995. North American pines as hosts for Tomicus piniperda (L.) (Col., Scolytidae) in France and Sweden. In: Hain, F.P., Salom, S.M., Ravlin, W.F., Payne, T.L., Raffa, K.F. (Eds.), Behavior, Population Dynamics and Control of Forest Insects. Ohio State University Press, Wooster, pp. 547–557.

Långström, B., Hellqvist, C., Ehnström, B., 1999. Susceptibility of fire damaged Scots pine (Pinus sylvestris L.) to attack by Tomicus pini-perda L. In: Lieutier, F., Mattson, W.J., Wagner, M.R. (Eds.), Physi-ology and Genetics of Tree-Phytophage Interactions. INRA Editions, Versailles, pp. 299–311.

Långström, B., Annila, E., Hellqvist, C., Varama, M., Niemela, P., 2001a. Tree mortality, needle biomass recovery and growth losses in Scots pine following defoliation by Diprion pini (L.) and subsequent attacks by Tomicus piniperda (L.). Scandinavian J. For. Res. 16, 342–353.

Långström, B., Solheim, H., Hellqvist, C., Krokene, P., 2001b. Host resistance in defoliated Scots pine: effects of single and mass inoculations using bark beetle-associated blue-stain fungi. Agric. For. Entomol. 3, 211–213.

Långström, B., Li, L., Liu, H., Chen, P., Li, H., Hellqvist, C., Lieutier, F., 2002. Shoot feeding ecology of Tomicus piniperda and T. minor (Col. Scolytidae) in Southern China. J. Appl. Entomol. 126, 1–10.

Långström, B., Lindelöw, Å., Schroeder, M., Björklund, N., Öhrn, P., 2009. The spruce bark beetle outbreak in Sweden following the January-storms in 2005 and 2007. In: Kunca, A., Zubrik, M. (Eds.), Insects and Fungi in Storm Areas. Štrbské Pleso, Slovakia, pp. 13–19.

Lanne, B.S., Schlyter, F., Byers, J.A., Löfqvist, J., Leufvén, A., Bergström, G., et al., 1987. Differences in attraction to semiochem-icals present in sympatric pine shoot beetles, Tomicus minor and T. piniperda. J. Chem. Ecol. 13, 1045–1067.

Larroche, D., 1971. Importance et cycle biologique de Blastophagus pini-perda L. (Col. Scolytides) dans la foret de Bouconne. PhD thesis, Uni-versité Paul Sabatier, France.

Laumond, C., Carle, P., 1971. Nématodes associés et parasites de Blasto-phagus destruens Woll. (Col. Scolytidae). Entomophaga 16, 51–66.

Lawrence, R.K., Haack, R.A., 1995. Susceptibility of selected species of North American pines to shoot feeding by an Old World scolytid: Tomicus pini-perda. In: Hain, F.P., Salom, S.M., Ravlin, W.F., Payne, T.L., Raffa, K.F. (Eds.), Behavior, Population Dynamics and Control of Forest Insects. Ohio State University Press, Wooster, pp. 536–546.

Lekander, B., 1968. The number of larval instars in some bark beetle species. Entomologisk Tidskrkift 89, 25–34.

Lekander, B., 1971. On Blastophagus destruens Woll. and a description of its larva (Col. Scolytidae). Entomologisk Tidskrkift 92, 271–276.

Lekander, B., 1984. Tidpunkten for storre margborrens svarmning i olika delar av Sverige under aren 1970–79. Sveriges Skogsvardsförbunds Tidskrift 82, 7–21.

Lekander, B., Rennerfelt, E., 1955. Undersökningar över insekts- och blå-nadsskador på sågtimmer. Meddelanden Från Statens Skogsforsknin-ginstitut 45, 1–36.

Lekander, B., Bejer-Petersen, B., Kangas, E., Bakke, A., 1977. The distribution of bark beetles in the Nordic countries. Acta Entomol. Fennica 32, 1–37.

Li, L.S., Wang, H.L., Chai, X.S., Wang, Y.X., Shu, N.B., Yang, D.S., 1993. Study on the biological characteristics of Tomicus piniperda and its damage (in Chinese). For. Res. 6, 14–20.

Li, X., Zhang, Z., Wang, H., Wu, W., Cao, P., Zhang, P., 2010. Tomicus armandii Li and Zhang (Curculionidae, Scolytinae), a new pine shoot borer from China. Zootaxa 2572, 57–64.

Lieutier, F., 2002. Mechanisms of resistance in conifers and bark beetle attack strategies. In: Wagner, M.R., Clancy, K.M., Lieutier, F., Paine, T.D. (Eds.), Mechanisms and Deployment of Resistance in Trees to Insects. Kluwer, Dordrecht, pp. 31–75.

Lieutier, F., 2004. Host resistance to bark beetles and its variations. In: Lieutier, F., Day, K.R., Battisti, A., Grégoire, J.-C., Evans, H.F. (Eds.), Bark and Wood Boring Insects in Living Trees in Europe, a Synthesis. Kluwer Academic Publishers, Dordrecht, pp. 135–180.

Lieutier, F., Vallet, E., 1982. Observations sur les nématodes parasites et associés aux principaux Scolytidae ravageurs du pin sylvestre en forêts d'Orléans et de Sologne. Acta Oecol., Oecol. Appl 3, 131–148.

Lieutier, F., Yart, A., 1989. Preferenda thermiques des champignons associés à *Ips sexdentatus* Boern. et *Tomicus piniperda* L. (Coleoptera; Scolytidae). Ann. For. Sci. 46, 411–415.

Lieutier, F., Cheniclet, C., Garcia, J., 1989a. Comparison of the defense reactions of *Pinus pinater* and *Pinus sylvestris* to attacks by two bark beetles (Coleoptera: Scolytidae) and their associated fungi. Environ. Entomol. 18, 228–234.

Lieutier, F., Yart, A., Garcia, J., Ham, M.-C., Morelet, M., Lévieux, J., 1989b. Champignons phytopathogènes associés à deux coléoptères scolytidae du pin sylvestre (*Pinus sylvestris* L.) et étude préliminaire de leur agressivité envers l'hôte. Ann. For. Sci. 46, 201–216.

Lieutier, F., Garcia, J., Romary, P., Yart, A., Jactel, H., Sauvard, D., 1993. Inter-tree variability in the induced defense reaction of Scots pine to single inoculations by *Ophiostoma brunneo-ciliatum*, a bark-beetle-associated fungus. For. Ecol. Manage. 59, 257–271.

Lieutier, F., Garcia, J., Romary, P., Yart, A., 1995. Wound reactions of Scots pine (*Pinus sylvestris* L.) to attacks by *Tomicus piniperda* L. and *Ips sexdentatus* Boern. (Coleoptera: Scolytidae). J. Appl. Entomol. 119, 591–600.

Lieutier, F., Långström, B., Solheim, H., Hellqvist, C., Yart, A., 1996. Genetic and phenotypic variation in the induced reaction of Scots pine to *Leptographium wingfieldii*: reaction zone length and fungal growth. In: Mattson, W.J., Niemela, P., Rousi, M. (Eds.), Dynamics of Forest Herbivory: Quest for Pattern and Principle, pp. 166–177, USDA Forest Service, General Technical Report NC-183.

Lieutier, F., Ghaioule, D., Yart, A., Sauvard, D., 2002. Attack behavior of pine bark beetles in Morocco and association with phytopathogenic fungi. Ann. Rech. For. Maroc 35, 96–109.

Lieutier, F., Ye, H., Yart, A., 2003. Shoot damage by *Tomicus* sp. (Coleoptera: Scolytidae) and effect on *Pinus yunnanensis* resistance to subsequent reproductive attacks on the stem. Agric. For. Entomol. 5, 227–233.

Lieutier, F., Yart, A., Ye, H., Sauvard, D., Gallois, V., 2004. Between-isolate variations in the preformances of *Leptographium wingfieldii* Morelet, a fungus associated with the bark beetle *Tomicus piniperda* L. Ann. For. Sci. 61, 45–53.

Lieutier, F., Yart, A., Sallé, A., 2009. Stimulation of tree defenses by Ophiostomatoid fungi can explain attack success of bark beetles on conifers. Ann. For. Sci. 66, 801.

Lindgren, B.S., 1997. Optimal release rate of the host monoterpene a-pinene for trapping the European pine shoot beetle, *Tomicus piniperda* (Coleoptera: Scolytidae). Proc. Entomol. Soc. Ont. 128, 109–111.

Lindquist, E.E., 1970. Relationships between mites and insects in forest habitats. Can. Entomol. 102, 978–984.

Linnaeus, C., 1758. Systema naturae per regna tria naturae, secundum classes, ordines, genera, species, cum characteribus, differentiis, synonymis, locis, Tomus 1, 10, Holmiae, p. 563.

Liu, H., Zhang, Z., Ye, H., Wang, H., Clarke, S.R., Lu, J., 2010. Response of Tomicus yunnanensis (Coleoptera: Scolytinae) to infested and uninfested Pinus yunnanensis Bolts. J. Econom. Entomol. 103, 95–100.

López Pantoja, G., Sánchez Callado, F.M., Gómez de Dios, M.A., Jerez Fernández, A., 2000. Prototipo de control integrado para *Tomicus piniperda* ecotipo *destruens* (Woll.) y *Orthotomicus erosus* (Woll.) en los pinares litorales de la provincia de Huelva. 1st Symposium on Stone Pine (*Pinus pinea* L.). Valladolid, Spain, pp. 319–326.

Löyttyniemi, K., 1978. Metsanlannoituksen vaikutuksesta ytimennavertajiin (*Tomicus* spp., Col., Scolytidae). Folia Forestalia 348, 1–19.

Löyttyniemi, K., Uusvaara, O., 1978. Insect attack on pine and spruce sawlogs felled during the growing season. Commun. Inst. For. Fenn. 89, 1–48.

Löyttyniemi, K., Pekkala, O., Uusvaara, O., 1978. Deterioration of pine and spruce pulpwood stored during the growing season and its effects on sulphite pulping. Commun. Inst. For. Fenn. 92, 16.

Löyttyniemi, K., Heliövaara, K., Repo, S., 1988. No evidence of a population pheromone in *Tomicus piniperda* (Coleoptera: Scolytidae): a field experiment. Ann. Entomol. Fenn. 54, 93–95.

Lu, R.C., Wang, H.B., Zhang, Z., Byers, J.A., Jin, Y.J., Wen, H.F., Shi, W.J., 2012a. Attraction of *Tomicus yunnanensis* (Coleoptera: Scolytidae) to Yunnan pine logs with and without periderm or phloem: An effective monitoring bait. Psyche Articl, id 794683.

Lu, R.C., Wang, H.B., Zhang, Z., Byers, J.A., Jin, Y.J., Wen, H.F., Shi, W.J., 2012b. Coexistence and competition between *Tomicus yunnanensis* and *T. minor* (Coleoptera: Scolytinae) in Yunnan pine. Psyche Article, id 185312.

Lu, J., Zhao, T., Ye, H., 2014. The shoot-feeding ecology of three *Tomicus* species in Yunnan province, southwestern China. J. Insect Sci. 14, 37.

Luitjes, J., 1976. The development of insects in timber windthrown by the gales of 1972 and 1973 in the Netherlands. Zeitschrift für Pflanzenkrankheiten, Pflanzenpathologie und Pflanzenschutz 83, 87–95.

Luterek, R., 1996. Primary insect invaders of post-fire stands in the virgin Notecka Forest. Prace z Zakresu Nauk Lesnych 82, 103–110.

Lutyk, P., Swiezynska, H., 1984. Trials to control the larger pine-shoot beetle (*Tomicus piniperda* L.) by means of the fungus *Beauveria bassiana* (Bals.) Vuill. on piled wood. Sylwan 128, 41–45.

MacCallum, B.D., 1922. Some wood-staining fungi. Trans. Br. Mycol. Soc. 7, 231–236.

Markalas, S., 1997. Frequency and distribution of insect species on trunks in burnt forests of Greece. Mitt. Schweitz. Entomol. Ges. 70, 57–61.

Martikainen, P., 2001. Non-target beetles (Coleoptera) in *Trypodendron* pheromone traps in Finland. Anzeiger für Schädlingskunde 74, 150–154.

Martikainen, P., Kouki, J., Heikkala, O., Hyvärinen, E., Lappalainen, H., 2006. Effects of green tree retention and prescribed burning on the crown damage caused by the pine shoot beetles (Tomicus spp.) in pine-dominated timber harvest areas. J. Appl. Entomol. 130, 37–44.

Maslov, A.D., Verednikov, G.I., Andreyeva, I., 1988. The protection of forests against pests and diseases (in Russian), 2 Agropromizdat, Moskva, 139–141.

Masutti, L., 1959. Reperti sull'entomofauna del *Pinus nigra* Arn. Var. *Austriaca* Hoess nelle Prealpi Giulie. Ann. Accad. Ital. Sci. For 8, 263–308.

Masutti, L., 1969. Pinete dei litorali e *Blastophaus piniperda* L.—una difficile convivenza. Monti e Boschi 3, 15–27.

Masuya, H., Kaneko, S., Yamaoka, Y., 1998. Blue stain fungi associated with *Tomicus piniperda* (Coleoptera: Scolytidae). J. For. Res. 3, 15–17.

Mathiesen, A., 1950. Über einige mit Borkenkäfern assoziierte Bläuepilze in Schweden. Oikos 2, 275–308.

Mathiesen, A., 1951. Einige neue *Ophiostoma*-Arten in Schweden. Svensk Botanisk Tidskrift 45, 203–232.

Mathiesen-Käarik, A., 1953. Eine Übersicht über die gewöhnlichsten mit Borkenkäfern assoziierten Bläuepilze in Schweden und einige für Schweden neue Bläuepilze. Meddelanden frän Statens Skogforskningsinstitut 43, 1–74.

Mathiesen-Käarik, A., 1960. Studies on the ecology, taxonomy and physiology of Swedish insect-associated blue stain fungi, especially the genus *Ceratocystis*. Oikos 11, 1–24.

Mattson-Mårn, L., 1921. Märgborrens kronskadegörelse och dess inverkan på tallens tillväxt (Die Kronenbeschädigung des grossen Waldgärtners und deren Einfluss auf den Zuwachs der Kiefer). Meddelanden från Statens Skogsförsöksanstalt 18, 81–101.

Mazur, S., 1973. Contribution to the knowledge of the fauna of predatory beetles inhabiting feeding-places of *Tomicus* = *Blatophagus piniperda* L. Sylwan 117, 53–59.

Mazur, S., 1975. Economic significance of the predators *Paromalus Parallelipipedus* and *Plegaderus vulneratus* (Col.: Histeridae) in the control of *Tomicus piniperda*. Sylwan 119, 57–60.

Mazur, S., 1979. Beetle succession in feeding sites of the pine shoot beetle (*Tomicus piniperda* L., Coleoptera: Scolytidae) in one-species and mixed pine stands. Memorabilia Zoologica 30, 63–87.

Mazur, S., 1985. Remarks on the occurrence of some subcortical beetles in the feeding sites of *Tomicus piniperda* L. Sylwan 129, 35–42.

McCullough, D.G., Sadof, C.S., 1998. Evaluation of an integrated management and compliance program for *Tomicus piniperda* (Coleoptera: Scolytidae) in pine Christmas tree fields. J. Econ. Entomol. 91, 785–795.

McCullough, D.G., Haack, R.A., McLane, W.H., 1998. Control of *Tomicus piniperda* (Coleoptera: Scolytidae) in pine stumps and logs. J. Econ. Entomol. 91, 492–499.

Mendel, Z., 1985. Predation of *Orthotomicus erosus* (Col., Scolytidae) by the Syrian woodpecker (*Picoides syriacus*, Aves, Picidae). Zeitschift für Angewandte Entomologie 100, 355–360.

Mendel, Z., 1986. Hymenopterous parasitoids of bark beetles (Scolytidae) in Israel: host relation, host plant, abundance and seasonal history. Entomophaga 31, 113–125.

Mendel, Z., 1987. Major pests of man-made forests in Israel: origin, biology, damage and control. Phytoparasitica 15, 131–137.

Mendel, Z., Halperin, J., 1981. Parasites of bark beetles (Col.: Scolytidae) on pine and cypress in Israel. Entomophaga 26, 375–379.

Mendel, Z., Mandar, Z., Golan, Y., 1985. Comparison of the seasonal occurrence and behavior of seven pine bark beetles (Coleoptera Scolytidae) in Israel. Phytoparasitica 13, 21–32.

Mendel, Z., Mandar, Z., Golan, Y., 1986. Hymenopterous parasitoids of pine bark beetles in Israel. Hassadeh 66, 1899–1901.

Mendel, Z., Podoler, H., Livne, H., 1990. Interactions between *Aulonium ruficorne* (Coleoptera: Colydiidae) and other natural enemies of bark beetles (Coleoptera: Scvolytidae). Entomophaga 35, 99–105.

Merlin, J., Parmentier, C., Grégoire, J.-C., 1986. The feeding habits of *Rhizophagus dispar* (Col., Rhizophagidae), an associate of bark beetles. Mededelingen Faculteit Landbouwwetenschapen, Rijksunivesiteit Gent 51, 915–923.

Michalski, J., Witkowski, Z., 1962. Untersuchungen uber den Einfluss des Regenerations- und Reifungsfrasses von *Blastophagus piniperda* L. jungen Kiefernbeständes. XI International Congress of Entomology, Vienna 9, 258–261.

Mihalciuc, V., Danci, A., Lupu, D., Olenici, N., 2001. Situation of the main bark and wood boring insects which damaged conifer stands in the last 10 years in Romania (in Romanian). Anale Institutul de Cercetari si Amenajari Silvice 1, 48–53.

Mihkelson, S., 1986. Massive outbreak of *Bupalus piniarius* in Estonia. Metsanduslikud Uurimused 21, 64–72.

Mills, N.J., 1983. The natural enemies of scolytids infesting conifer bark in Europe in relation to the biological control of *Dendroctonus* spp. in Canada. Biocontrol News and Information 4, 305–328.

Monleón, A., Blas, M., Riba, J.M., 1996. Biologia de *Tomicus destruens* (Wollaston, 1865) (Coleoptera Scolytidae) en los bosques mediterraneos. Elytron 10, 161–167.

Morelet, M., 1988. Observations sur trois Deutéropycètes inféodés aux pins. Annales de la Société de Sciences Naturelles et d'Archéologie de Toulon et du Var 40, 41–45.

Mulock, P., Christiansen, E., 1986. The threshold of successful attack by *Ips typographus* on *Picea abies*: a field experiment. For. Ecol. Manage. 14, 125–132.

Nageleisen, L.-M., 2004. Les insectes sous-corticaux des résineux en 2002: diminution inattendue des dommages dus aux scolytes. *Les Cahiers du DSF* 1-2003/2004, (La santé des forêts [France] en 2002), pp. 29–31.

Nanni, C., Tiberi, R., 1997. *Tomicus destruens* (Wollaston): biology and behaviour in central Italy. *USDA Forest Service, General Technical Report* NE-236, 131–134.

Nedelchev, S., Takov, D., Pilarska, D., 2011. *Prothallonema tomici* n. sp. (Tylenchida: Sphaerulariidae) parasitizing *Tomicus piniperda* (Coleoptera: Curculionidae: Scolytinae) in Bulgaria. Nematology 13, 741–746.

Nilssen, A.C., 1978. Spatial attack pattern of the bark beetle *Tomicus piniperda* L. (Col., Scolytidae). Norwegian J. Entomol 25, 171–175.

Nilsson, S., 1974a. Tillväxtförluster hos tall vid angrepp av märgborrar. Skogshögskolan, Institutionen för Skogsteknik, Rapporter och uppsatser 78, 1–64.

Nilsson, S., 1974b. Märgborreskador vid överlagring av tallvirke. Skogshögskolan, Institutionen för Skogsteknik, Rapporter och uppsatser 74, 1–35.

Nilsson, S., 1976. Rationalization of forest operations gives rise to insect attack and increment losses. Ambio 5, 17–22.

Novak, V., 1972. Chemical protection of timber against barkbeetles and wood-destroying insects. Drevo 27, 314–317.

Nuorteva, M., 1954. Versuche über den Einfluβ der Bevölkerungsdichte auf die Nachkommenzahl des Groβen Walldgärtners, *Blastophagus piniperda* L. Ann. Entomol. Fenn. 20, 184–189.

Nuorteva, M., 1956. Über den Fichtenstamm-Bastkäfer, *Hylurgops palliatus* Gyll., und seine Insektenfeinde. Acta Entomol. Fenn. 13, 1–116.

Nuorteva, M., 1957. Zur kenntnis der parasitischen Hymenopteren der Borkenkäfer Finnlands. Ann. Entomol. Fenn. 23, 118–121.

Nuorteva, M., 1959. Untersuchungen über einige in den Frassbilden der Borkenkäfer lebende *Medetera*-Arten (Dipt. Dolichopodidae). Suom. Hyonteistiet. Aikak. 25, 192–210.

Nuorteva, M., 1962. Über die Nützlichkeit der Zimmerbocklarven (*Acanthocinus aedilis*) im Walde. XI International Congress of Entomology, Vienna 2, 171–173.

Nuorteva, M., 1964. Über den Einfluss der Menge des Brutmaterials auf die Vermehrlichkeit und die natürlichen Feinde des Grassen Waldgärtners, *Blastophagus piniperda* L. (Col., Scolytidae). Ann. Entomol. Fenn. 30, 1–17.

Nuorteva, M., 1971. Die Borkenkäfer (Col., Scolytidae) und deren Insektenfeinde im Kirchspiel Kuusamo. Nordfinnland. Ann. Entomol. Fenn. 37, 65–74.

Nuorteva, M., Nuorteva, P., 1968. The infestation of timber by bark beetles (Col., Scolytidae) and their natural enemies in the Finnish southwestern archipelago. Ann. Entomol. Fenn. 34, 56–65.

Nuorteva, M., Salonen, K., 1980. Versuche mit *Beauveria bassiana* (Bals.) Vuill. gegen *Blastophagus piniperda* L. (Col., Scolytidae). Ann. Entomol. Fenn. 34, 49–55.

Oppermann, T.A., 1985. Rinden- und holzbrutende Insekten an immissionsgeschadigten Fichten und Kiefern. Holz-Zentralbl 111, 213–217.

Öhrn, P., 2012. Seasonal flight patterns of the spruce bark beetle (*Ips typographus*) in Sweden. Phenology, voltininsm and development Licentiate thesis, Swedish University of Agricultural Sciences, Sweden.

Paciura, D., Zhou, X.D., de Beer, Z.W., Jacobs, K., Ye, H., Wingfield, M.J., 2010. Characterisation of synnematous bark beetle-associated fungi from China, including *Graphium carbonarium* sp. nov. Fungal Divers 40, 75–88.

Paine, T.D., Raffa, K.F., Harrington, T.C., 1997. Interactions among scolytids bark beetles, their associated fungi, and live host conifers. Annu. Rev. Entomol. 42, 179–206.

Paparatti, B., Cinti, S., Leandri, A., Pompi, V., Forchielli, L., 2000. Esperienze sul controllo di *Tomicus destruens* (Wollaston) (Coleoptera, Scolytidae). Atti Giornate Fitopatologiche 1, 433–440.

Park, K.N., Lee, S.O., 1972. Studies on ecology and control of the pine barl beetle *Blastophagus piniperda* L. Research Report Forest Research Institut, Seoul 19, 65–70.

Park, J.D., Byun, B.H., 1988. Trapping the overwintered pine bark beetle, *Tomicus piniperda* L. (Coleoptera: Scolytidae), by turpentine. Research Report Forest Research Institut, Seoul 36, 126–129.

Penas, A.C., Bravo, M.A., Naves, P., Bonifácio, L., Sousa, E., Mota, M., 2006. Species of *Bursaphelenchus* Fuchs, 1937 (Nematoda: Parasitaphelenchidae) and other nematode genera associated with insects from *Pinus pinaster* in Portugal. Ann. Appl. Biol. 148, 121–131.

Pettersen, H., 1976. Parasites (Hym., Chalcidoidea) associated with bark beetles in Norway. Norweg. J. Entomol. 23, 75–78.

Petrice, T.R., Haack, R.A., Poland, T.M., 2002. Selection of overwintering sites by *Tomicus piniperda* (Coleoptera: Scolytidae) during fall shoot departure. J. Entomol. Sci. 37, 48–59.

Pfeffer, A., 1995. Zental- und westpaläarktische Borken- und Kernkäfer (Coleoptera: Scolytidae, Platypodidae). Entomologica Brasiliensia 17, 5–310.

Piou, D., Lieutier, F., 1989. Observations symptomatologiques et roles possibles d'*Ophiostoma minus* Hedgc. (Ascomycete: Ophiostomatales) et de *Tomicus piniperda* L. (Coleoptera: Scolytidae) dans le dépérissement du pin sylvestre en forêt d'Orléans. Ann. For. Sci. 46, 39–53.

Pishchik, A.A., 1979. The effect of the common *Raphidia* on pest numbers. Lesnoe Khozyaistvo 2, 71–74.

Pischchik, A.A., 1980. An insect predator of *Blastophagus* (*Tomicus*) *piniperda* and *B.* (*T.*) *minor*. Lesnoe Khozyaistvo 11, 55–57.

Poinar Jr., G.O., 1972. Nematodes as facultative parasites of insects. Annu. Rev. Entomol. 17, 103–122.

Poland, T.M., Haack, R.A., 2000. Pine shoot beetle, *Tomicus piniperda* (Col., Scolytidae), responses to common green leaf volatiles. J. Appl. Entomol. 124, 63–69.

Poland, T.M., Haack, R.A., Petrice, T.R., 2002. *Tomicus piniperda* (Coleoptera: Scolytidae) initial flight and shoot departure along a north-south gradient. J. Econ. Entomol. 95, 1195–1204.

Poland, T.M., De Groot, P., Burke, S., Wakarchuk, D., Haack, R.A., Nott, R., Scarr, T., 2003. Development of an improved attractive lure for the pine shoot beetle, *Tomicus piniperda* (Coleoptera: Scolytidae). Agric. For. Entomol. 5, 293–300.

Poland, T.M., De Groot, P., Haack, R.A., Czokajlo, D., 2004. Evaluation of semiochemicals potentially synergistic to a-pinene for trapping the larger European pine shoot beetle, *Tomicus piniperda* (Col., Scolytidae). J. Appl. Entomol. 128, 639–644.

Postner, M., 1974. Scolytidae (=Ipidae), Borkenkäfer. In: Schwenke, W. (Ed.), Die forstschdlinge Europas. Band 2. Käfer. Hamburg-Berlin, Germany, pp. 334–482.

Raffa, K.F., Berryman, A.A., 1983. The role of host plant resistance in the colonization behavior and ecology of bark beetles (Coleoptera: Scolytidae). Ecol. Monogr. 53, 27–49.

Ratzeburg, J.T.C., 1839. Die Forst-Insekten. Erster Theil, Die Käfer, Berlin, 247 pp.

Razzauti, A., 1921. Contributi alla conoscenza faunistica delle isole toscane III. Coleotteri delle isole d'Elba, di Capraia e di Gorgona. Atti Soc. Ital. Sci. Nat. (Pisa) 31, 100–122.

Regnander, J., 1976. Vattenbegjutning av insektsangripet virke - ett viktigt led i bekampningen av barkborrar pa skog. Sveriges Skogsvardsförbunds Tidskrift 74, 497–504.

Reitter, E., 1913. Bestimmungstabelle der Borkenkäfer (Scolytidae) aus Europa und den angrenzenden Ländern. Wien Entomologie Zeitung 32, 1–116.

Rennerfelt, E., 1950. Über den Zusammenhang zwischen dem Verblauen des Holzes und den Insekten. Oikos 2, 120–137.

Ritchie, W., 1917. The structure, bionomics and forest importance of *Myelophilus minor* Hart. Trans.—R. Soc. Edinburgh 53, 213–234.

Ritzerow, S., Konrad, H., Stauffer, C., 2004. Phylogeography of the Eurasian pine shoot beetle *Tomicus piniperda* L. (Coleoptera, Scolytidae). Eur. J. Entomol. 101, 13–19.

Rollins, F., Jones, K.J., Krokene, P., Solheim, H., Blackwell, M., 2001. Phylogeny of asexual fingi associated with bark and ambrosia beetles. Mycologia 93, 991–996.

Romanyk, N., 1972. Daños de insectos perforadores en repoblaciones de *Pinus pinaster* Ait. Sugerencias para su prevención y combate. Boletín de la Estación Central de Ecología 1, 15–27.

Roversi, P.F., Sabbatini Peverieri, G., Pennacchio, F., Tiberi, R., 2004. Gli scolitidi del genere *Tomicus* Latreille in Italia centrale. XIX Congresso Nazionale Italiano di Entomologia, Catania pp. 927–930.

Rühm, W., 1956. Die nematoden der Ipiden. Parasitologische Schriftenreihe 6, 1–437.

Russo, G., 1940. Il blastofago del pino (*Blastophagus* (*Myelophilus*) *piniperda* L. *var. rubripennis* Reitter). Bollettino Regio Laboratiorio di Entomologia Agraria, Facoltà di Agraria di Portici 19, 1–13.

Russo, G., 1946. Scolitidi del pino del litorale toscano. *Bollettino Istituto Entomologia "G. Grandi" Univ.* Bologna 15, 297–314.

Ryall, K.L., Smith, S.M., 1997. Intraspecific larval competition and brood production in *Tomicus piniperda* (L.) (Col., Curculionidae, Scolytinae). Proc. Entomol. Soc. Ontario 128, 19–26.

Ryall, K.L., Smith, S.M., 2000a. Reproductive success of the introduced pine shoot beetle, *Tomicus piniperda* (L.) (Coleoptera, Scolytidae) on selected North American and European conifers. Proc. Entomol. Soc. Ontario 131, 113–121.

Ryall, K.L., Smith, S.M., 2000b. Brood production and shoot feeding by *Tomicus piniperda* (Coleoptera: Scolytidae). Can. Entomol. 132, 939–949.

Saarenmaa, H., 1983. Modeling the spatial pattern and intraspecific competition in *Tomicus piniperda* (Coleoptera, Scolytidae). Commun. Inst. For. Fenn. 118, 40.

Saarenmaa, H., 1985. Within tree population dynamics models for integrated management of *Tomicus piniperda* (Coleoptera, Scolytidae). Commun. Inst. For. Fenn. 128, 1–56.

Sabbatini Peverieri, G., Faggi, M., 2005. Determination of age in larvae of *Tomicus destruens* (Wollaston, 1865) (Coleoptera Scolytidae) based on head capsule width. Redia 88, 115–117.

Sabbatini Peverieri, G., Francardi, V., 2010. First record of *Cercoleipus coelonotus* Kinn (Acari Mesostigmata Cercomegistidae) from Italy. Redia 93, 79–81.

Sabbatini Peverieri, G., Faggi, M., Marziali, L., Panzavolta, T., Bonuomo, L., Tiberi, R., 2004. Use of attractant and repellent substances to control *Tomicus destruens* (Coleoptera: Scolytidae) in *Pinus pinea* and *P. pinaster* pine forests of Tuscany. Entomologica 38, 91–102.

Sabbatini Peverieri, G., Tiberi, R., Triggiani, O., Tarasco, E., 2005. Impiego di semiochimici nel monitoraggio e nel controllo di *Tomicus destruens* (Woll.) (Coleoptera: Scolytidae) in Toscana e in Puglia. Entomologica 39, 169–182.

Sabbatini Peverieri, G., Capretti, P., Tiberi, R., 2006. Associations between *Tomicus destruens* and *Leptographium* spp. in *Pinus pinea* and *P. pinaster* stands in Tuscany, central Italy. For. Pathol. 36, 14–20.

Sabbatini Peverieri, G., Pennacchio, F., Tiberi, R., 2007. Monitoraggio degli scolitidi indigeni delle conifere con l'impiego di sostanze naturali volatili. XXI Congresso Nazionale Italiano di Entomologia, Campobasso, p. 254.

Sabbatini Peverieri, G., Faggi, M., Marziali, L., Tiberi, R., 2008. Life cycle of *Tomicus destruens* in a pine forest of central Italy. Bulletin of Insectology 61, 337–342.

Sadof, C.S., Waltz, R.D., Kellam, C.D., 1994. Differential shoot feeding by adult Tomicus piniperda (Coleoptera: Scolytidae) in mixed stands of native and introduced pines in Indiana. The Great Lakes Entomologist 27, 223–228.

Sallé, A., Ye, H., Yart, A., Lieutier, F., 2008. Seasonal water stress and the resistance of *Pinus yunnanensis* to a bark beetle associated fungus. Tree Physiol. 28, 679–687.

Salonen, K., 1973. On the life cycle, especially on the reproduction biology of *Blastophagus piniperda* (Col., Scolytidae). Acta For. Fenn. 127, 1–72.

Santolamazza Carbone, S., Pestaña, M., Vega, J.A., 2011. Post-fire attractiveness of maritime pine (*Pinus pinaster* Ait.) to xylophagous insects. J. Pest. Sci. 84, 343–353.

Santos, A.M., Vasconcelos, T., Mateus, E., Paiva, M.R., Branco, M., 2005. Phloem monoterpene contents of four pine species and relative susceptibility to the attack of *Tomicus* spp. Proceedings, 21st Annual Meeting of the International Society of Chemical Ecology, Washington DC, p. 128.

Sarikaya, O., Avci, M., 2009. Predators of Scolytinae (Coleoptera: Curculionidae) species of the coniferous forests in the Western Mediterranean Region, Turkey. Türkiye Entomoloji Dergisi 33, 253–264.

Sarikaya, O., Avci, M., 2010. Distribution and biology of the Mediterranean pine shoot beetle *Tomicus destruens* (Wollaston, 1865) in the western Mediterranean region of Turkey. Türkiye Entomoloji Dergisi 34, 289–298.

Sauvard, D., 1989. Capacités de multiplication de *Tomicus piniperda* L. (Col., Scolytidae). 1. Effets de la densité d'attaque. J. Appl. Entomol. 108, 164–181.

Sauvard, D., 1993. Reproductive capacity of *Tomicus piniperda* L. (Col.: Scolytidae). 2. Analysis of the various sister broods. J. Appl. Entomol. 116, 25–38.

Sauvard, D., 2004. General biology of bark beetles. In: Lieutier, F., Day, K.R., Battisti, A., Grégoire, J.-C., Evans, H.F. (Eds.), Bark and Wood Boring Insects in Living Trees in Europe, a Synthesis. Kluwer Academic Publishers, Dordrecht, pp. 63–88.

Sauvard, D., Lieutier, F., Lévieux, J., 1987. Repartition spatiale et dispersion de *Tomicus piniperda* L. (Coleoptera: Scolytidae) en fôret d'Orléans. Ann. For. Sci. 44, 417–434.

Sauvard, D., Lieutier, F., Levieux, J., 1988. L'Hylesine du pin (*Tomicus piniperda* L.) en foret d'Orleans: repartition, degats, lutte. Rev. For. Fr. 40, 13–19.

Skarmoutsos, G., Braasch, H., Michalopoulou, H., 1998. *Bursaphelenchus hellenicus* sp. n. (Nematoda, Aphelenchoïdidae) from Greek pine wood. Nematologica 44, 623–629.

Scarr, T.A., Czerwinski, E.J., Howse, G.M., 1999. Pine shoot beetle damage in Ontario. In: Fosbroke, S.L.C., Gottschalk, K.W. (Eds.), Proc. USDA Interagency Research Forum on Gypsy Moth and Other Invasive Species.p. 56, USDA Forest Service General Technical Report NE-266.

Schedl, K.E., 1932. Scolytidae, Platypodidae. In: Winkler, A. (Ed.), Catalogus Coleopterorum Regionis Palaearcticae. Vienna, Austria, pp. 1632–1647.

Schedl, K.E., 1946. Bestimmungstabellen der palaearktischen Borkenkäfer, II. Die Gattung Blastophagus Eichh. Zentralblatt für das Gesamtgebiet der Entomologie 1, 50–58.

Schlyter, F., Löfqvist, J., 1990. Colonization pattern in the pine shoot beetle, *Tomicus piniperda*: effects of host declination, structure and presence of conspecifics. Entomol. Exp. Applic. 54, 163–172.

Schlyter, F., Zhang, Q.-H., Anderson, P., Byers, J.H., 2000. Electrophysiological and behavioural responses of *Tomicus piniperda* and *Tomicus minor* (Coleoptera: Scolytidae) to non-host leaf and bark volatiles. Can. Entomol. 132, 965–981.

Schönherr, J., 1972. Pheromon beim Kiefern-Borkenkäer "Waldgätner", *Myelophilus piniperda* L. (Coleopt., Scolytidae). Zeitschift für Angewandte Entomologie 71, 410–413.

Schroeder, L.M., 1987. Attraction of the bark beetle *Tomicus piniperda* to Scots pine trees in relation to tree vigor and attack density. Entomol. Exp. Applic. 44, 53–58.

Schroeder, L.M., 1988. Attraction of the bark beetle *Tomicus piniperda* and some other bark- and wood-living beetles to the host volatiles α-pinene and ethanol. Entomol. Exp. Appl. 46, 203–210.

Schroeder, L.M., 1996. Interactions between the predators *Thanasimus formicarius* (Col., Cleridae) and *Rhizophagus depressus* (Col.: Rhizophagidae), and the bark beetle *Tomicus piniperda* (Col.: Scolytidae). Entomophaga 41, 63–75.

Schroeder, L.M., 1999. Prolonged development time of the bark beetle predator *Thanasimus formicarius* (Col.: Cleridae) in relation to its prey species *Tomicus piniperda* (L.) and *Ips typographus* (L.) (Col.: Scolytidae). Agric. For. Entomol. 1, 127–135.

Schroeder, L.M., 2003. Differences in response to alpha-pinene and ethanol, and flight periods between the bark beetle predators *Thanasimus femoralis* and *T. formicarius* (Col.: Cleridae). For. Ecol. Manage. 177, 301–311.

Schroeder, L.M., 2008. Insect pests and forest energy. In: Röser, D., Asikainen, A., Raulund-Rasmussen, K., Stupak, I. (Eds.), Sustainable Use of Forest Biomass for Energy—A Synthesis with Focus on the Baltic and Nordic Region. Springer, Dordrecht, pp. 109–127.

Schroeder, L.M., Risberg, B., 1989. Establishment of a new brood in *Tomicus piniperda* (L.) (Col., Scolytidae) after a second hibernation. J. Appl. Entomol. 108, 27–34.

Schroeder, L.M., Weslien, J., 1994a. Interactions between the phloem-feeding species *Tomicus piniperda* (Col.: Scolytidae) and *Acanthocinus aedilis* (Col.: Cerambycidae), and the predator *Thanasimus formicarius* (Col.: Cleridae) with special reference to brood production. Entomophaga 39, 149–157.

Schroeder, L.M., Weslien, J., 1994b. Reduced offspring production in bark beetle *Tomicus piniperda* in pine bolts baited with ethanol and α-pinene, which attract antagonist insects. J. Chem. Ecol. 20, 1429–1444.

Schwertfeger, F., 1944. Die Waldkrankheiten. Ein Lehrbuch der Forstpathologie und des Forstschutzes. P. Parey, Berlin.

Siemaszko, W., 1939. Fungi associated with bark beetles in Poland. Planta Polonica 7, 1–54.

Sierpinski, Z., 1959. Hibernation places of the insect (*Blastophagus = Myelophilus minor* Hart.). Instituti Bawadczy Lesnictwa 1999, 37–48.

Sierpinski, Z., 1971. Secondary insect pests of Scots pine in stands affected by industrial air pollution with nitrogenous compounds. Sylwan 115, 11–18.

Sikström, U., Jacobson, S., Pettersson, F., Weslien, J., 2011. Crown transparency, tree mortality and stem growth of Pinus sylvestris, and colonization of Tomicus piniperda after an outbreak of Gremmeniella abietina. For. Ecol. Manage. 262, 2108–2119.

Six, D.L., 2003. Bark beetle fungal symbiosis. In: Miller, T., Kourtzis, K. (Eds.), Insect Symbiosis. CRC Presa, Boca Raton, pp. 97–114.

Six, D.L., Klepzig, K.D., 2004. *Dendroctonus* bark beetles as model systems for studies on symbiosis. Symbiosis 37, 207–232.

Six, D.L., Wingfield, M.J., 2011. The role of phytopathogenicity in bark beetle fungal symbioses: a challenge to the classical paradigm. Annu. Rev. Entomol. 56, 255–272.

Slobodyanyuk, O.V., 1973. On the biology of the nematode *Parasitaphelenchus papillatus* (Aphelenchoididae) from the beetle *Blastophagus piniperda*. Trudy Gel'mintologicheskoi Laboratorii 23, 155–159.

Slobodyanyuk, O.V., 1974. Biological features of the genus *Parasitorhabditis* (Rhabditidae). Trudy Gel'mintologicheskoi Laboratorii 24, 160–168.

Solheim, H., Långström, B., 1991. Blue-stain fungi associated with *Tomicus piniperda* in Sweden and preliminary observations on their pathogenicity. Ann. For. Sci. 48, 149–156.

Solheim, H., Långström, B., Hellqvist, C., 1993. Pathogenicity of the blue stain fungi *Leptographium wingfieldii* and *Ophiostoma minus* to Scots pine: effect of tree pruning and attack density. Can J. For. Res. 23, 1438–1443.

Solheim, H., Krokene, P., Långström, B., 2001. Effects of growth and virulence of associated blue-stain fungi on host colonization behaviour of the pine shoot beetles *Tomicus piniperda* and *T. minor*. Plant Pathol. 50, 111–116.

Srot, M., 1966. New knowledge on the establishment of a sister generation of pine beetle (*Myelophilus piniperda* L.) in Scots pine stands in Czeck regions. Lesnoje Casopis 12, 563–576.

Srot, M., 1968. The bionomics of *Myelophilus piniperda*, and new methods of chemical pest control. Lesnoje Casopis 14, 375–390.

Stark, V.N., 1952. Fauna SSSR 31. Koroedi (Scolytidae). Akademii Nauk SSSR, Moskva-Leningrad.

Stergulc, F., 2002. Studio sulle condizioni fitosanitarie della pineta di Valle Vecchia (Caorle, VE). Proposte per interventi di controllo del blastofago dei pini (*Tomicus destruens* Wollaston) e del marciume radicale da *Heterobasidion annosum* (Fr.). Bref. Veneto Agricoltura, Settore Ricerca e Sperimentazione Forestale e Fuori Foresta. Padova, 31 pp.

Sun, J.H., Clarke, S.R., Kang, L., Wang, H.B., 2005. Field trials of potential attractants and inhibitors for pine shoot beetles in the Yunnan province. China. Ann. For. Sci. 62, 9–12.

Svensson, S.A., 1980. Riksskogstaxeringen 1973–77. *Skogstillstånd, tillväxt och avverkning*. SLU, Institutionen för skogstaxering. Rapport 30, 1–167.

Szmidt, A., 1983. Possible control of *Blastophagus [Tomicus] piniperda* by spraying of unbarked Scots pine roundwood. Folia For. Pol. 25, 243–251.

Takov, D.I., Doychev, D.D., Linde, A., Draganova, S.A., Pilarska, D.K., 2012. Pathogens in bark beetles (Curculionidae: Scolytinae) and other beetles in Bulgaria. Biologia 67, 966–972.

Targioni Tozzetti, A., 1886. Invasione di *Myelophylus piniperda* nel territorio di Pisa. L'Agricoltore Italiano 734, 145–147.

Thalenhorst, W., 1958. Grundzüge der populationsdynamik des grössen Fichten-borkenkäfer *Ips typographus* L. Shriftenreihe Forst Fakültat Univität Göttingen 21, 1–126.

Thompson, W.R., 1943. A catalogue of the parasites and predators of the insect pests. Sect. I Parasite host catalogue. Part I Parasites of the arachnida and Coleoptera. Belleville, Canada.

Tiberi, R., Faggi, M., Panzavolta, T., Sabbatini Peverieri, G., Marziali, L., Niccoli, A., 2009. Feeding preference of *Tomicus destruens* progeny adults on shoots of five pine species. Bull. Insectology 62, 261–266.

Tinelli, A., Catena, G., 1992. Applicazione della fotografia aerea per indagini fitosanitarie della pineta. Monti e Boschi 43, 26–29.

Titovšek, J., 1988. Podlubniky (Scolytidae) Slovenije obvladovanje podlubnikov. Gozdarska zalozda, Ljubljana.

Tomalak, M., Michalski, J., Grocholski, J., 1984. The influence of nematodes on the structure of genitalia of *Tomicus piniperda* (Coleoptera: Scolytidae). J. Invertebr. Pathol. 43, 358–362.

Trägårdh, I., 1921. Untersuchungen uber den grossen Waldgärtner *Myelophilus piniperda*. Meddelanden från Statens Skogsförsöksanstalt 18, 1–80.

Triggiani, O., 1983. Sensibilita del *Tomicus (Blastophagus) piniperda* L. (Col. Scolytidae) ai nematodi della famiglia Steinernematidae e Heterorhabditidae. Entomologica 18, 215–223.

Triggiani, O., 1984. *Tomicus (Blastophagus) piniperda* (Coleoptera, Scolytidae Hylesininae): biologia, Danni e controllo nel litorale ionico. Entomologica 19, 5–21.

Vasconcelos, T., Branco, M., Goncalves, M., Cabral, M.T., 2005. Periods of flying activity of *Tomicus* spp. in Portugal. In: Lieutier, F., Ghaioule, D. (Eds.), Entomological Research in Mediterranean Forest Ecosystems. INRA Editions, Paris, pp. 177–184.

Vasconcelos, T., Horn, A., Lieutier, F., Branco, M., Kerdelhué, C., 2006. Distribution and population genetic structure of the Mediterranean pine shoot beetle *Tomicus destruens* in the Iberian Peninsula and Southern France. Agric. For. Entomol. 8, 103–111.

Viiri, H., Kytö, M., Niemela, P., 1999. Resistance of fertilized Norway spruce (*Picea abies* (L.) Karst.) and Scots pine (*Pinus sylvestris* L.). In: Lieutier, F., Mattson, W.J., Wagner, M.R. (Eds.), Physiology and Genetics of Tree-Phytophage Interactions. INRA Editions, Versailles, pp. 337–342.

Villari, C., Sabbatini Peverieri, G., Tiberi, R., Capretti, P., 2008. The occurrence of fungal diseases on pine trees and their relationship with bark beetles. In: Forster, B., Knízek, M., Grodzki, W. (Eds.), Proceedings of Workshop on Methodology of Forest Insect and Disease Survey in Central Europe (IUFRO Working Party 7.03.10, 2006. Federal Research and Training Centre for Forests, Forest Training Centre Gmunden, Austria, pp. 310–312.

Vité, J.P., Volz, H.A., Paiva, M.R., 1986. Semiochemicals in host selection and colonization of pine trees by the pine shoot beetle *Tomicus piniperda*. Naturwissenschaften 73, 39–40.

Voolma, K., Luik, A., 2001. Outbreaks of *Bupalus piniaria* (L.) (Lepidoptera, Geometridae) and *Pissodes piniphilus* (Herbst) (Coleoptera, Curculionidae). J. For. Sci. 47, 171–173.

Wainhouse, D., 2005. Ecological Methods in Forest Pest Management. Oxford University Press, New York.

Wang, C.S., 1981. A study of the Korean pine bark beetle (Blastophagus pilifer Spess). Entomol. Knowl. 18, 165–167.

Wang, H.L., Chen, S.W., Wu, Y., Pu, M.G., 1987. Preliminary studies on the bionomics and management of the pine bark beetle (*Blastophagus* piniperda L.) in Kunming district, China (in Chinese). Journal of Southwest Forestry College 2, 33–42.

Wang, P.Y., Zhang, Z., Kong, X., Wang, H., Zhang, S., Gao, X., Yuan, S., 2012. Antennal morphology and sensilla ultrastructure of three Tomicus species (Coleoptera: Curculionidae, Scolytinae). Microsc. Res. Tech. 75, 1672–1681.

Waring, R.H., Pitman, G.B., 1983. Physiological stress in lodgepole pine as a precursor for mountain pine beetle attack. Z. Angew. Entomol. 96, 265–270.

Waring, R.H., Pitman, G.B., 1985. Modifying lodgepole pine stands to change susceptibility to mountain pine beetle attack. Ecology 66, 889–897.

Wegensteiner, R., 2004. Pathogens in bark beetles. In: Lieutier, F., Day, K.R., Battisti, A., Grégoire, J.-C., Evans, H.F. (Eds.), Bark and Wood Boring Insects in Living Trees in Europe, a Synthesis. Kluwer Academic Publishers, Dordrecht, pp. 291–313.

Whitney, H.S., 1982. Relationships between bark beetles and symbiotic organisms. In: Mitton, J.B., Sturgeon, K.B. (Eds.), Bark Beetles in North American Conifers. University of Texas Press, Austin, pp. 183–211.

Wingfield, M.J., Gibbs, J.N., 1991. Leptographium and Graphium species associated with pine-infesting bark beetles in England. Mycol. Res. 95, 1257–1260.

Winter, T.G., Evans, H.F., 1990. Insects and storm-damaged conifers. Research Information Note, Forestry Commission, Research Division 173, 3.

Wisniewski, J., 1980. Four new species of heteromorphic Dendrolaelaps males (Acarina: Rhodocaridae) in bark-beetle galleries in Poland. Acarologia 21, 149–162.

Wollaston, T.V., 1865. Coleoptera Atlantidum, Being an Enumeration of the Coleopterous Insects of the Madeiras, Salvages, and Canaries. Voorst, London.

Wulff, S., 2008. Insekters utnyttjande av stormfällda träd och märgborreangripna tallskott på marken inom RIS/RT 2006–2007. Report to the Swedish Forestry Board 3.

Xu, F.Y., Yang, B.J., Ge, M.H., 1993. Investigation on insect vectors to carry the pine wood nematode. For. Pest Disease 2, 20–21.

Ye, H., 1991. On the bionomy of Tomicus piniperda (L.) (Col., Scolytidae) in the Kunming region of China. J. Appl. Entomol. 112, 366–369.

Ye, H., 1992. Approach to the reasons of Tomicus piniperda (L.) population epidemic. Journal of the Yunnan University 14, 211–215.

Ye, H., 1993. A preliminary study of chemical compounds inducing aggregation of Tomicus piniperda L. Sci. Silvae Sin. 29, 463–467.

Ye, H., 1994. Influence of temperature on the experimental population of the pine shoot beetle, Tomicus piniperda (L.) (Col., Scolytidae). J. Appl. Entomol. 117, 190–194.

Ye, H., 1996. Studies on the biology of Tomicus piniperda (L.) (Col., Scolytidae) in shoot-feeding period. Acta Entomol Sinica 39, 58–62.

Ye, H., Li, L., 1994. The distribution of Tomicus piniperda (L.) (Col., Scolytidae) population in the crown of Yunnan pine during the shoot feeding period (in Chinese). Acta Entomologica Sinica 37, 311–316.

Ye, H., Li, L., 1995. Preliminary observations on the trunk attacks by Tomicus piniperda (L.) (Col., Scolytidae) on Yunnan pine in Kunming, China. J. Appl. Entomol. 119, 331–333.

Ye, H., Zhao, Z., 1995. Life table of Tomicus piniperda (L.) (Col., Scolytidae) and its analysis. J. Appl. Entomol. 119, 145–148.

Ye, H., Lieutier, F., 1997. Shoot aggregation by Tomicus piniperda (Col; Scolytidae) in Yunnan, southwestern China. Ann. Sci. For. 54, 635–641.

Ye, H., Ding, X., 1999. Impacts of Tomicus minor on distribution and reproduction of Tomicus piniperda (Col., Scolytidae) on the trunk of the living Pinus yannanensis trees. J. Appl. Entomol. 123, 329–333.

Ye, H., Liu, H., 2006. Studies on predation of Thanasimus dubius (Col.: Cleridae) on Tomicus piniperda (Col.: Scolytidae). Forest Research, Beijing 19, 289–294.

Ye, H., Zhou, X., Lu, J., Yang, L., Ding, H., 2000. A preliminary study on the occurrence and pathogenicity of the fungi associated with Tomicus. Forest Research, Beijing 13, 451–454.

Ye, H., Mu, Q., Lu, J., Zhang, D., 2004. Observations on the frequency of the symbiotic fungus Leptographium yunnanense associated with Tomicus pinperda. Entomol. Knowl. 41, 555–558.

Zhao, T., 2003. Performance of the pine shoot beetles, Tomicus piniperda L., and . minor (Hart.) (Coleoptera: Scolytidae) on their principal and secondary hosts, Pinus yunnanensis and Pinus armandii in Yunnan, ChinaM.S. thesis, Swedish University of Agricultural Sciences, Sweden.

Zhao, T., Långström, B., 2012. Performance of Tomicus yunnanensis and Tomicus minor (Col., Scolytinae) on Pinus yunnanensis and Pinus armandii in Yunnan, southwestern China. Psyche, Article id 363767.

Zhou, N., Li, L., Jiang, Z., Liu, H., 1997. Study on effect of chemical attractants to Tomicus piniperda. Yunnan Forest Sciences Technics 2, 20–38.

Zhou, X.D., Jacobs, K., Morelet, M., Ye, H., Lieutier, F., Wingfield, M.J., 2000. A new Leptographium species associated with Tomicus piniperda in south-western China. Mycoscience 41, 573–578.

Zhou, X.D., de Beer, Z.W., Harrington, T.C., McNew, D., Kirisits, T., Wingfield, M.J., 2004. Epitypification of Ophiostoma galeiforme and phylogeny of species in the O. galeiforme complex. Mycologia 96, 1306–1315.

Chapter 11

The Genus *Hypothenemus*, with Emphasis on *H. hampei*, the Coffee Berry Borer

Fernando E. Vega[1], Francisco Infante[2], and Andrew J. Johnson[3]

[1] *Sustainable Perennial Crops Laboratory, United States Department of Agriculture, Agricultural Research Service, Beltsville, MD, USA,* [2] *El Colegio de la Frontera Sur (ECOSUR), Carretera Antiguo Aeropuerto Km. 2.5, Tapachula, Chiapas, Mexico,* [3] *School of Forest Resources and Conservation, University of Florida, Gainesville, FL, USA*

1. INTRODUCTION

Hypothenemus is one of the most speciose genera of Scolytinae, common in all tropical and subtropical areas (Wood, 1986). Most *Hypothenemus* species are very small (<2 mm long), poorly described, and difficult to distinguish. Several species are globally distributed, undoubtedly aided by human activities. Although the vast majority of *Hypothenemus* species live innocuously in twigs, some have become important pests, most notably the coffee berry borer *Hypothenemus hampei* (Ferrari), which lives inside the coffee berry and consumes the seeds, and the tropical nut borer *Hypothenemus obscurus* (F.), which attacks a range of seeds and fruits. This chapter will introduce the reader to taxonomic characters useful in identifying members of the genus, followed by some of the most important species, and concluding with an in-depth review of the vast body of multilingual literature on the coffee berry borer.

2. THE GENUS *HYPOTHENEMUS*

2.1 Key Characters for Identification to Genus

The majority of the 181 described *Hypothenemus* species are poorly known (Wood and Bright, 1992; Bright and Skidmore, 1997, 2002; Bright, 2014) and most species are not distinguishable when using the original published descriptions. The book *The Bark and Ambrosia Beetles of South America* (Wood, 2007) is the most extensive work to date and includes the 46 species reported from South America, in addition to others recorded from Central and North America.

Species are distinguished by details of vestiture (often lost from abrasion), frontal sculpture, and surface texture. However, the combination of some characters described below and illustrated in Figures 11.1–11.4 can be used to distinguish *Hypothenemus* from all other bark beetles.

The antennae have three to five funicular segments. The antennal club has sutures marked with setae and a partial septum, visible as a dark line. The eye in the female is emarginate, although in the smaller species this might be as slight as a few facets missing. There are one to 10 marginal asperities on the anterior margin of the pronotum, and usually more than 10 asperities on the pronotal declivity. A raised line that partially extends forward along the lateral margins marks the posterior edge of the pronotum. The males are smaller than females, often appearing deformed, and although in most keys they are described as wingless, they actually have vestigial wings and are effectively flightless and have reduced compound eyes by comparison with females (Vega *et al.*, 2014). Most species have prominent setae, particularly the interstrial bristles, which are in rows and usually flattened. With a few exceptions, the elytra are rounded without distinctive sculpturing.

2.2 Taxonomy

Hypothenemus was first established with the description of the species *H. eruditus* Westwood (Westwood, 1836; Figures 11.1 and 11.2A). The genus name was derived from "υπο subtus, εν, and νεμω pasco" referring to the downward facing mouthparts (Westwood, 1836). The species contained within *Hypothenemus* have been described in and moved from 23 other genera (Wood and Bright, 1992).

The genus *Stephanoderes* was first described for a number of species by Eichhoff (1871). This later encompassed Westwood's *Hypothenemus* genus (Eichhoff and Schwarz, 1896), giving priority to *Stephanoderes*, and accounting for the erroneous generic description by Westwood in which *H. eruditus* was described as having three funicular segments, when the specimens had four

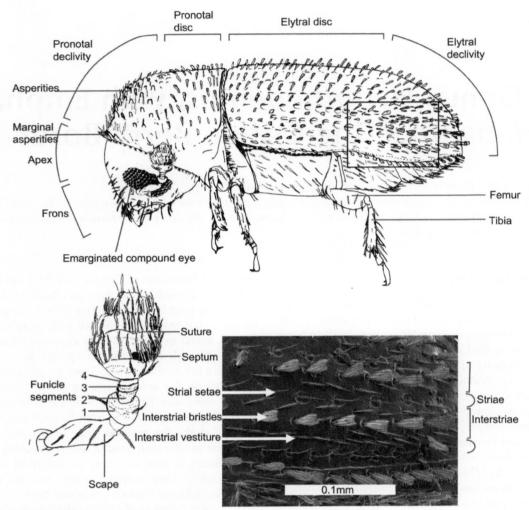

FIGURE 11.1   Top: Morphology of *Hypothenemus eruditus*, with terminology used in the chapter. Square box on lateral region of elytron is shown in detail in scanning electron microscopy photograph on bottom right. Bottom left: Enlarged diagram of an antenna of *H. eruditus*. In the text, "vestiture" is used to globally encompass all types of setae, and these can be divided into three types: strial setae, interstrial bristles, and interstrial vestiture.

(Figure 11.1). However, Swaine (1909) treated *Stephano-deres* as a redundant genus while Hopkins (1915a) listed both *Hypothenemus* and *Stephanoderes* as separate genera, with *Stephanoderes chapuisii* Eichhoff (current name: *H. dissimilis* (Zimmermann)) chosen as the generic type. Hopkins (1915b) also described the differences between the two genera, in that the antennal funicle of *Stephano-deres* is five segmented with the last segment widened, in contrast to only four segments for *Hypothenemus*. This distinction, however, is unreliable, especially with species such as *H. birmanus* (Eichhoff) having a range from three to five funicular segments. *Stephanoderes* was finally moved back into synonymy by Browne (1963). The genus has remained in occasional use since, especially for the coffee berry borer.

The genus *Trischidias* was also described by Hopkins (1915b). This genus of minute beetles shares many characters with *Hypothenemus* and differs only in lacking the septum of the antennal club, and a lack of emargination of the eye. Hopkins (1915b) also described several *Hypothenemus* species that have since been moved into *Trischidias*, highlighting the similarity or the two genera. Wood (1954) even comments that *Trischidias* "obviously were derived from" *Hypothenemus*. *Trischidias* has the unusual habit of breeding in wood infected with decaying fungi (Deyrup, 1987) and may just be a specialized group within *Hypothenemus*.

Other genera similar and likely to be phylogenetically close to *Hypothenemus* are *Cryptocarenus* and *Perioc-ryphalus*, which share many morphological characters and mating systems. The antennal club, however, is distinctly different, lacking a septum and sutures. These genera are naturally distributed in the Americas (Wood, 2007).

FIGURE 11.2 (A) *Hypothenemus eruditus* female and male. Females of (B) *H. areccae*; (C) *H. javanus*; (D) *H. birmanus*; (E) *H. dissimilis*; (F) *H. opacus*; (G) *H. curtipennis*; (H) *H. concolor*; and (I) *H. distinctus*. All photos by A. J. Johnson.

2.3 Typical *Hypothenemus* Life Cycle

A gallery is started by a single, mated female, referred to as the foundress, and in coffee berry borer literature as the colonizing female. The cues used to locate hosts are poorly known. Many species, however, are attracted to traps baited with ethanol, which is produced by plants under stress (Kimmerer and Kozlowski, 1982). The foundress initiates the gallery with a single entrance hole, usually located at stem or leaf nodes, or in a coffee berry for the coffee berry borer. A gallery may have just one hole, out of which frass and debris are pushed. Often, there are multiple galleries on the twig, which merge as they expand, becoming "inextricably intermingled" (Browne, 1961). Different species may be found together in the same gallery system after merging of original independent galleries (Wood, 1954). Galleries in twigs may also extend into leaf petioles or fruits.

The eggs are large relative to the size of the female, and the larvae are found within or very close to the parental gallery. Approximate development time in the field is 28 days (Browne, 1961). Males are produced at a much lower ratio, and for all *Hypothenemus* species, sex is believed to be determined by pseudo-arrhenotoky (see below). Adult females mate with their brothers (sibling mating), or perhaps non-sibling males if the gallery is merged with that of other families. The adult females remain in the galleries for some time, probably waiting for suitable weather conditions to disperse. When they disperse, the females may leave via the single entrance hole, or make new exit holes.

2.4 Host Plants

The diversity of plants any one *Hypothenemus* species can be found on is remarkable and many species are highly polyphagous (Wood, 1954; Wood and Bright, 1992). *Hypothenemus javanus* (Eggers) has been recorded from 32 families of plants (Atkinson, 2014). The most extreme, however, is *H. eruditus* (Figures 11.1 and 11.2A) which has been found breeding in innumerable hosts, in nearly any part of the plants, sometimes within the galleries of other insects including active and abandoned *Hypothenemus* galleries, fruiting bodies of fungi (Browne 1961; Deyrup, 1987), and manufactured products such as drawing boards (Browne, 1961) and books (Westwood, 1836). Some species, however, are much more restricted in their host range, such as *H. pubescens* Hopkins, which is only known from coastal grasses (Wood, 2007).

The plant host substrates for most *Hypothenemus* species, such as dead twigs, are nutritionally poor, and undoubtedly the microbial communities, inside or outside the beetle, play a role in allowing the beetles to thrive in such environments. These may also mediate the extreme polyphagy, especially since feeding on decomposed material may reduce plant species-specific defenses faced by insects that feed on live plant tissues. At least one species, *H. curtipennis* (Schedl), has adaptations associated with inoculation and cultivation of fungi (Beaver, 1986), which is known to permit a wide diversity of host plant substrates.

2.5 An Introduction to Some *Hypothenemus* Species

The diversity of *Hypothenemus* species is briefly described with figures depicting some of the most commonly encountered globally distributed species, morphological extremes, and economically significant species. The descriptions are focused on females since males are smaller, harder to find, and have fewer differentiating characters.

Hypothenemus areccae (Hornung) (Figure 11.2B) is found in all tropical regions, with origins in Southeast Asia. The body shape is slender, 1.2–1.4 mm long, and usually has eight marginal asperities and a distinctly concave frons. It is found on a broad range of plants, and is an occasional pest of transplants and seedlings.

Hypothenemus birmanus (Eichhoff) (Figure 11.2D) is another widely distributed, common species, found in every tropical region. Females can have three to five funicular segments. The marginal asperities are distinctive: they are narrowly separated, the median pair are large, and the outer pair are small. The flattened interstrial bristles are much denser on the elytral declivity than on the disc, and the interstrial vestiture on the declivity is made of rows of slightly flattened setae.

Hypothenemus concolor Hagedorn (Figure 11.2H) is one of the largest, most robust of all *Hypothenemus* species, with some specimens reaching 2.9 mm in length. The elytral disc is reduced, and there is an obvious summit between the elytral disc and declivity. It is likely that this species also has a fungus farming lifestyle, since the galleries are inside woody tissue and only extend to 20 mm (Schedl, 1961). Conversely, *H. distinctus* Wood (Figure 11.2I) is one of the smallest species, usually just 0.9 mm, seldom collected, and its biology is unknown.

Hypothenemus crudiae (Panzer) (Figures 11.3A and 11.4A) is widespread and common across much of the tropics, except Australia, with probable origins in the Americas (Wood, 2007). It is very similar and sometimes indistinct from *H. seriatus* (Eichhoff) (see below). The key character reported to distinguish *H. crudiae* from similar species is the presence of a frontal tubercle on the frons, often with a short central groove extending into the tubercle. However, the tubercle is sometimes reduced (Wood, 2007). There are usually six marginal asperities of a similar size and spacing, and the anterior margin is

FIGURE 11.3 Lateral and dorsal view of female (and lateral view of male, when available, in last column) for (A) *Hypothenemus crudiae*; (B) *H. seriatus*; (C) *H. interstitialis*; (D) *H. obscurus*; and (E) *H. hampei*. *All photos by A. J. Johnson.*

FIGURE 11.4   From left to right: Close-up of head of multiple individuals to show variation and enlarged dorsal view of eltyron showing vestiture and texture. (A) *H. crudiae*; (B) *H. seriatus*; (C) *H. interstitialis*; (D) *H. obscurus*; and (E) *H. hampei*. *All photos by A. J. Johnson.*

broadly rounded. The texture of the lateral regions of the pronotum is variable from smooth to rugose. The elytral declivity is rounded and quite steep, slightly steeper than *H. seriatus* (Wood, 2007). The interstrial bristles are strongly flattened, and usually have a square tip (sometimes curved along its length making the tip appear recurved). The interstrial bristles are only slightly longer on the declivity than on the elytral disc and the interstrial vestiture is usually completely absent, even on the lateral regions of the declivity. The hair-like strial setae arch over the strial punctures, which are prominent and distinct, rugose in the center. The interstrial areas are variable, from almost entirely smooth and shining (as in Figure 11.4A), to mostly rugose, but not with an irregular micropunctate surface as in *H. obscurus*. Usually at least some of the elytra in the discal region are smooth. The adult color is also variable, mostly monotonous, brown to black, sometimes red-brown at the elytral summit, and the setae are often light brown. This species inhabits twigs, fruits, and seeds of a wide range of plants and is rarely reported as a pest. García Martell (1980) illustrated this species breeding in the fruits of cocoa (*Theobroma cacao* L.).

Hypothenemus curtipennis (Schedl) (Figure 11.2G) has undoubtedly developed cuticular mycangia for an ambrosia fungus-farming lifestyle (Beaver, 1986). Mycangia (sing. mycangium) are pits or recesses on the cuticle or in the mandibles that serve to carry fungal spores (Batra, 1963; Crowson, 1981). In *H. curtipennis*, mycangia are evident as wide, abrupt pits on the sides of the pronotum (visible in Figure 11.2G), lined with plumose setae, which fill with fungal spores before the young beetles disperse. The behavior of this species is also modified for fungus cultivation. The foundress makes a gallery much shorter than those made by typical *Hypothenemus* species, which is followed by a period where the beetle waits, blocking the gallery entrance with the steep elytral declivity. The larvae develop in the gallery, sometimes extending it no more than 15 mm, and then wall themselves off for pupation (Beaver, 1986).

Hypothenemus dissimilis (Zimmermann) (Figure 11.2E) is a large, robust species from the Americas. This species has only two marginal asperities, and hair-like interstrial bristles among short, flattened interstrial vestiture. It has a typical habit of living in the pith or under the bark of a range of woody plants.

Hypothenemus eruditus Westwood (Figures 11.1 and 11.2A) is a widely distributed *Hypothenemus* species, present in every tropical and subtropical region. In the Americas, the range extends from Michigan (USA) to Argentina (Wood, 2007). This species is also remarkable for the extreme diversity of habits, recorded from hundreds of host plants and even fungal fruiting bodies, from all sorts of plant material including leaf petioles, twigs, seeds, fruits, and from manufactured products (Browne, 1961; Wood, 1982).

The type specimen was found living in the bindings of a book (Westwood, 1836); therefore, the name *eruditus* (i.e., erudite). This species has also been reported killing seedlings of cocoa and transplants of trees (Browne, 1961).

Hypothenemus eruditus is 1.1–1.3 mm long, usually with six marginal asperities, the median pair usually narrowly separated. The vestiture is variable in the shape and size of the interstrial bristles (Wood, 1954), and in the abundance of interstrial vestiture. The coloration is variable; the female and male depicted in Figure 11.2A are distinctly bicolored, but many are entirely dark brown to black. It is common to find multiple, distinct variants at any one location, which may represent different species, yet when specimens are compared to the global diversity, the distinctions are unclear. The distribution, variation, profound polyphagy, and minute size may explain the array of junior synonyms, with 71 recognized (Wood and Bright, 1992). It is very likely that a species complex is involved.

Hypothenemus interstitialis (Hopkins) (Figures 11.3C and 11.4C) is a slightly larger species (1.4–1.7 mm long) that is similar to *H. seriatus* and *H. hampei*. It is restricted to the Americas. The frons is variable, often with a frontal groove. There are four to six marginal asperities and the lateral regions of the pronotum are micropunctate. There is often a difference between the color of the pronotum and elytra. The interstrial bristles are prominent and as in *H. seriatus*, they are short on the disc, but on the declivity, the interstrial bristles are long and narrow, with the length being eight times the width or more (Wood, 2007). The strial setae are hair-like and arch over the strial punctures. The interstrial vestiture is absent. The elongate interstrial bristles on the declivity could cause confusion with *H. hampei*, especially since this species has been found on twigs of *Coffea* sp. (Wood, 2007). However, the interstrial bristles are short and flattened on the disc, whereas for *H. hampei*, the interstrial bristles are similar in length over all the elytra.

Hypothenemus javanus (Eggers) (Figure 11.2C) is a species with pantropical distribution with likely origins in Africa (Wood, 1977). This species is larger (1.4–1.7 mm long) and much more robust than *H. areccae*, with four marginal asperities, a concave frons, and hair-like interstrial vestiture (Wood, 2007).

Hypothenemus obscurus (F.) (Figures 11.3D and 11.4D), the tropical nut borer, is the second most economically damaging species in the genus (after the coffee berry borer), attacking a range of seeds and fruits. Originally from the Americas, this species is now found across the tropical world, although not in Australia. It is frequently intercepted in Brazil nuts (*Bertholletia excelsa* Bonpl.; Wood, 1982). *Hypothenemus obscurus* was first described in the genus *Hylesinus* by Fabricius (1801), briefly and non-specifically, mentioning only the color, minute size, and shape of the pronotum. In some reviews (e.g., Wood, 1954), this species' name has been used to describe specimens of *H. seriatus* and *H. crudiae*. *Hypothenemus obscurus* should not be confused with *Cryphalus obscurus* (current name: *H. eruditus*, synonymy by Wood, 1975), described by Ferrari (1867) alongside *Cryphalus hampei* (current name: *H. hampei*), or *Stephanoderes obscurus* as described by Eichhoff (1871) (current name: *H. setosus*, synonymy by Wood, 1975).

Hypothenemus obscurus usually has a narrow frontal groove although this is variable, sometimes just partial or absent, even within families in a gallery (Figure 11.4D). There are four to six marginal asperities. The pronotum and elytra is entirely very finely textured with irregular pits throughout (*H. seriatus* and *H. crudiae* usually have some smooth areas of interstriae). The interstrial bristles are flattened, about four to six times long as wide, and longer on the declivity. The apex of the interstrial bristles is usually rounded. There does not seem to be the amount of variation in setae between specimens as seen in *H. seriatus* and *H. crudiae*.

Hypothenemus obscurus has some interstrial vesture on the declivity (Mitchell and Maddox, 2010), often overlooked, which contradicts many keys (e.g., Wood, 1982). This can be seen as a few coarse, pointed setae lying flat among the erect interstrial bristles, restricted to the lateral regions of the declivity. Such setae are absent or rare in *H. crudiae* and *H. seriatus*. The described size of female *H. obscurus* ranges from 1.2–1.4 mm (Wood, 1982) to 1.4–1.6 mm (Wood, 2007), and specimens in Hawaii were 1.5–1.8 mm (Mitchell and Maddox, 2010), while insects reared in artificial diet in Colombia were on average 1.75 mm (Constantino *et al.*, 2011). The males are also variable; within one gallery in a tamarind seed (*Tamarindus indica* L.) collected in Florida, the males ranged from 0.8 to 1.2 mm (A. J. Johnson, unpubl.).

Hypothenemus obscurus feeds on a remarkably diverse array of seeds and fruits. Commercially important products include nutmeg (*Myristica fragrans* Houtt.), macadamia nuts (*Macadamia* sp.), cocoa, tamarind, longan (*Dimocarpus longan* Lour.), *Melicoccus bijugatus* Jacq., and jackfruit (*Artocarpus heterophyllus* Lam.) (Beardsley, 1990). It has also been collected in coffee berries, although it does not complete its development (Constantino *et al.*, 2011). The worldwide cost to the industry is unknown. Between 1998 and 2012, *H. obscurus* caused from 0.8 to 4.6% of harvested macadamia nuts in Hawaii to be rejected, equivalent to $0.3–$1.8 million per year (calculated from NASS reports; NASS 1998–2012). This does not account for the costs associated with mitigating the damage though management strategies. In particular, regular harvesting of fallen nuts avoids high insect prevalence, which otherwise could result in infestations as high as 30% (Jones, 2002). Working in Hawaii, Jones *et al.* (1992) found that continuous harvest and

processing within 3 weeks after *H. obscurus* infestation could result in decreased damage. At this time the insect was still in the husk and had not yet started consuming the kernel. Unfortunately, this recommendation is not practical among growers and processors. Unharvested nuts, including those that do not fall, could be a source for reinfestation, and reducing these through cultural practices and cultivar selection may reduce damage.

Hypothenemus opacus (Eichhoff) (Figure 11.2F) has been collected in Central and South America. This species, and a few others, has deep circular pits at the summit of the pronotum. These pits are often filled with debris, which led Wood (2007) to speculate that they could be mycangia.

Hypothenemus seriatus (Eichhoff) (Figures 11.3B and 11.4B) is widely distributed across the tropical regions. It is very similar to *H. crudiae*, usually distinguished by the absence of the tubercle. The frontal groove may be partial or absent, and in some specimens there is an area of shining cuticle in the place of the frontal tubercle. Beardsley (1990) and Mitchell and Maddox (2010) report than the frontal groove of *H. seriatus* in Hawaii is absent, whereas Beaver and Maddison (1990) report it is normally present in specimens from Niue (Polynesia). In Florida, the presence of the frontal groove is variable, even among specimens within a single gallery (A. J. Johnson, unpubl.).

There are usually six marginal asperities (sometimes five to eight), of approximately equal size, although the outermost pair may be reduced. The texture of the lateral regions of the pronotum is smooth to rugose or weakly micropunctate. Compared with *H. crudiae*, the overall shape of the pronotum and elytra tends to be narrower in *H. seriatus* than in *H. crudiae*. The declivity is usually less steep, and the strial rows are often more prominently impressed on the declivity.

The setae on the elytra are similar to *H. crudiae*, although the interstrial bristles are noticeably longer on the declivity than the elytral disc (typically with a length three times the width on the disc and six times the width on the declivity). The interstrial vesture is absent or one or two setae near the apex of the elytra. The sculpturing of the elytra is similar to *H. crudiae*, but *H. seriatus* usually have less prominent strial punctures. Like *H. crudiae*, usually at least some areas of the interstriae are smooth, and if rugose, the cuticle is not covered by small irregular pits. Fonseca (1937) presented an illustration of *H. seriatus* galleries inside a coffee berry. This species is also known to damage cocoa seedlings (García Martell, 1980).

Hypothenemus crudiae, *H. interstitialis*, *H. obscurus*, and *H. seriatus* are very similar. The differentiating characters appear variable within each species to the point where there is a large overlap and many specimens in collections cannot be assigned to a particular species. Wood (1982) suggested that some of the intermediate forms could be from hybridization between different species. Both

H. crudiae and *H. seriatus* have a worldwide distribution, and are undoubtedly still being transported within and between continents. If they once were regionally distinct, it is possible that this has since been lost. They also have similar habits and biology, and neither are significant pests, so conclusive differentiation between the four species is not always necessary.

2.6 Molecular Phylogenetics of *Hypothenemus* Species

Despite the difficulties in identification, *Hypothenemus* species are yet to receive much attention using molecular techniques. Jordal and Cognato (2012) included two species (*Hypothenemus* sp. 1 and *H. birmanus*) in a molecular phylogeny of many bark and ambrosia beetles, placing *Hypothenemus* nearest to *Ptilopodius*.

Within-species variation of the mitochondrial DNA (mtDNA) cytochrome oxidase subunit I gene (*COI*), the typical insect "barcode" region, has been studied for several species, finding variation within *H. obscurus*, *H. seriatus*, and *H. hampei* as 2.9%, 1.9%, and 1.9%, respectively (Mitchell and Maddox, 2010). More detailed work on *H. hampei* found variation to be as high as 11.8% (Gauthier, 2010), and another study examining specimens identified as *H. eruditus* found a remarkable level of variation as high as 20.1% (Kambestad, 2011). Such deep divergence suggests the presence of many distinct, cryptic species units present within *Hypothenemus* species. With very few distinct morphological characters between species, molecular tools are invaluable for species identification. However, the effects of routine inbreeding on species boundaries may remain unclear, and linking genetically determined species to traditional classification will remain a challenge.

No phylogenetic analyses have yet been done with a focus on between-species relationships. Further study could result in a better understanding of some of the unusual traits that vary within the genus, including the evolution of seed feeding, as well as the ambrosial habit. Molecular phylogenetics also could resolve taxonomic issues such as the questionable monophily of *Hypothenemus* with respect to *Trischidias*.

3. COFFEE AND THE COFFEE BERRY BORER

The genus *Coffea* (Rubiaceae) comprises 123 species (Davis *et al.*, 2006), of which only two are commercially traded: *C. arabica* L. and *C. canephora* Pierre ex A. Froehner (commonly referred to as robusta) (Vega, 2008a). In their natural habitat, both species were inhabitants of the humid, evergreen forests of Africa (Davis *et al.*, 2006), with *C. arabica* being a high altitude species (900–2000 m) occurring in southwestern Ethiopia and surrounding regions, and *C. canephora* being a predominantly lowland plant (50–1500 m) found throughout much of tropical Africa, west of the Rift Valley (Davis *et al.*, 2006). Coffee is grown in more than 10 million hectares in ca. 80 countries (FAOSTAT, 2014), with ca. 20 million families dependent on this plant for their subsistence (Osorio, 2002; Gole *et al.*, 2002; Vega *et al.*, 2003a, 2008a; Lewin *et al.*, 2004). The theoretical yearly gross earnings in coffee producing countries for 2000–2012 was US$11.6 billion, while in 2012 the value of the entire coffee industry was estimated at US$173.4 billion (ICO, 2014).

Of all *Hypothenemus* species, the coffee berry borer *Hypothenemus hampei* (Ferrari) (Figures 11.3E, 11.4E, and 11.5) is without doubt the most studied as a result of the losses in yield and quality that it causes in coffee plantations worldwide. Nevertheless, a recent review of the literature published on the coffee berry borer from 1910 to 2013 (Infante *et al.*, 2014) revealed that of 1603 papers published on the insect, only ca. 602 were peer-reviewed, equivalent to ca. six papers per year. This figure was contrasted with a total of 75 peer-reviewed papers per year published on the Mediterranean fruit fly, *Ceratitis capitata* (Weidemann) from 1990 to 2012. These figures indicate that coffee berry borer research outputs are not what would be expected for such an economically important commodity as coffee.

3.1 Taxonomy and Synonymies

The coffee berry borer was described in Austria by Count Johann Angelo Ferrari as *Cryphalus hampei* from coffee seeds imported into France from an unknown origin, and named after Dr. Clemens Hampe, who provided the samples (Ferrari, 1867). The species was later moved to *Stephanoderes* with Eichhoff's (1871) description of the genus. Swaine (1909) treated *Stephanoderes* as a synonym of *Hypothenemus*, although *hampei* was not specifically listed, and the genus *Stephanoderes* continued to be used widely by others.

Hagedorn (1910) described what is now a synonym of *H. hampei*, *Stephanoderes coffeae*, arguing that it was not the same as *hampei*, based on some morphological differences. The same year, van der Weele (1910) described *Xyleborus coffeivorus* from coffee plantations in Java (Indonesia), recognizing the species' potential as a pest to coffee production, as well as the life history, and the difficulties controlling it. Strohmeyer (1910), however, quickly recognized *X. coffeivorus* to be a synonym of *S. hampei*, yet acknowledged *S. coffeae* as a related but distinct species. Meanwhile, Hopkins (1915a) relisted *Stephanoderes* as a genus separate from *Hypothenemus*, and soon after described it (Hopkins, 1915b), along with *Stephanoderes cooki* Hopkins, which also became a synonym of *S. hampei* (Schedl, 1959).

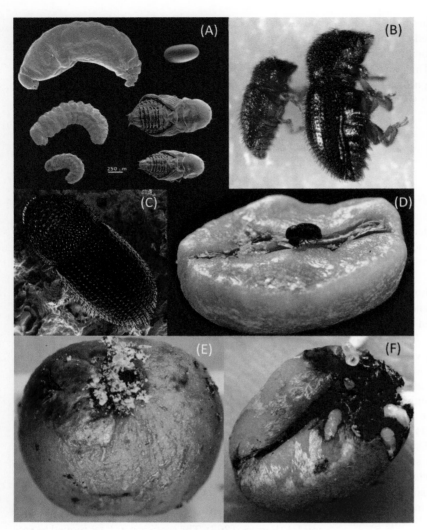

FIGURE 11.5 (A) Immature stages of the coffee berry borer. Clockwise starting in upper right: egg, female pupa, male pupa, first instar, second instar, and prepupa. There are two instars for females and only one for males; the first instar between sexes cannot be differentiated. (B) Male (left) and female (right) adults. (C) Dorsal view of an adult female. (D) Adult female on coffee seed. (E) Hole bored by colonizing female. (F) Damage caused to the seed. *Photos by: (A) Francisco Infante; (B) Ann Simpkins (USDA); (C) Eric Erbe (USDA); (D) Peggy Greb (USDA); (E) Guy Mercadier (USDA); and (F) Jaime Gómez (ECOSUR).*

Stephanoderes coffeae was synonymized with *S. hampei* by Roepke (1919). Eggers (1923) argued against the synonymy, suggesting that *hampei* had much broader interstrial bristles than *S. coffeae*. Sampson (1923), however, agreed with the synonymy, noting the specimens Eggers used for comparison were incorrectly identified as *S. hampei*, when instead they were *Stephanoderes cassiae* Eichhoff (current name: *Hypothenemus obscurus*).

Another synonym of *hampei* was described in a different genus as *Xyleborus cofeicola* (de Campos Novaes, 1922), synonymized 2 years later (Costa Lima, 1924a). Two further names, *Stephanoderes punctatus* Eggers (Eggers, 1924) and *Stephanoderes glabellus* Schedl (Schedl, 1952), were described and subsequently synonymized by Wood (1972 and 1989, respectively). The genus *Stephanoderes* was moved to *Hypothenemus* by Browne (1963), but the name *Stephanoderes hampei* continued to be used for some time.

Other *Hypothenemus* species have been recorded on coffee in Africa, including *H. areccae* (Ghesquière, 1933), *H. crudiae* (LePelley, 1968), *H. eruditus* (Schedl, 1960, 1961; Mayné and Donis, 1962), *H. grandis* Schedl (Schedl, 1961), *H. liberiensis* Hopkins (Schedl, 1961), *H. obscurus* (Le Pelley, 1968), *H. plumeriae* (Nordlinger) (Schedl, 1960), *H. seriatus* (Schedl, 1960; Le Pelley, 1968), and *H. solitarius* (Schedl) (Wood and Bright, 1992).

The term "falsa broca del café" or "false coffee berry borer" has been used to describe some of the *Hypothenemus* species that are similar in their appearance or habits to the coffee berry borer. It may be used in reference to a specific species such as *H. seriatus* (Fonseca, 1937; Decazy, 1987; Vega *et al.*, 2002b), *H. obscurus* (García Martell, 1980; Constantino *et al.*, 2011), or *H. crudiae* and *H. eruditus* (García Martell, 1980). Fonseca (1937) reported that *H. seriatus* never bores into green berries but that it bores into drier berries, where it consumes the pulp and reproduces, although it never attacks the seeds. In Mexico, *H. crudiae*, *H. eruditus*, and *H. obscurus* have been reported attacking coffee berries, but they never consume the seed, although all the life stages of *H. crudiae* and *H. eruditus* have been found inside the berry (García Martell, 1980).

In laboratory experiments, Constantino *et al.* (2011) found that *H. obscurus* will bore into coffee berries but only a small percentage causes superficial damage of the seed; the insect can complete its life cycle after larvae feeds on the pulp. *Hypothenemus obscurus* has been artificially reared in a coffee seed-based diet, with a reduced fecundity when compared to a macadamia-based artificial diet (Constantino *et al.*, 2011). In Hawaii, Greco and Wright (2012) noted that *H. obscurus* were usually only found in coffee berries when macadamia plantations are nearby, which suggests that they are unlikely to be, or become, a problem for coffee production.

3.2 Taxonomic Characters (Figures 11.3E and 11.4E)

The frons of *H. hampei* may have a broad, indistinct frontal groove, or no groove at all. There are usually four marginal asperities. The setae on the pronotum are mixed, with some slightly flattened. The shape of the pronotum, viewed from above, is slightly more narrowly rounded (i.e., more triangular) than the similar *Hypothenemus* species.

The elytral declivity of *H. hampei* is much more broadly rounded than in the similar species, without a distinct transition from the elytral disc. When viewed laterally, the declivity takes up more than half of the length of the elytra, whereas in the similar species, the elytral disc takes up more than half of the length.

As with most *Hypothenemus*, the interstrial bristles are prominent and in almost perfectly uniseriate rows. The shape of the interstrial bristles, however, is distinctive, and differentiates the coffee berry borer from most other *Hypothenemus* species. The bristles are long, narrow, and slightly flattened. The tip of each bristle is square, and not much wider than the rest of its length. The bristles on the elytral disc are not much shorter than those on the declivity. Males are smaller with reduced eyes (Vega *et al.*, 2014). The interstrial bristles are relatively long, and often not in distinct rows.

3.3 Distribution

The coffee berry borer is endemic to Africa (Vega *et al.*, 2009; Gauthier, 2010) and has disseminated to most of the coffee producing world (Table 11.1). It was first collected in the field in 1897 in Mount Coffee, Liberia, and reported as *S. cooki* (Hopkins, 1915b). In 1901 the insect was reported as a pest of *C. canephora* in the Republic of Congo (Fleutiaux, 1901). The insect was found in Indonesia

in 1908 (Hagedorn, 1910) and in 1913 was accidentally taken to Brazil (Table 11.1) in seeds imported from the Democratic Republic of the Congo (Berthet, 1913). Ihering (1924) proposed, to no avail, pruning to the ground all 900 million coffee plants in the state of São Paulo to completely eliminate all possibilities for survival of the insect.

Molecular methods have been used to elucidate the dissemination of the coffee berry borer throughout coffee-producing countries (Breilid *et al.*, 1997; Andreev *et al.*, 1998; Benavides *et al.*, 2005, 2006, 2007; Gauthier, 2010). Use of the mtDNA COI gene to track possible dispersal routes led to the identification of three clades: (1) Colombia, Honduras, and Mexico; (2) Fiji, Indonesia, Ivory Coast, Jamaica, New Caledonia, the Philippines, and Thailand; and (3) Kenya (Breilid *et al.*, 1997; Andreev *et al.*, 1998). Clades 1 and 2 imply introduction of two separate inbreeding lines, none of which is related to Kenyan specimens. It is noteworthy that specimens from Jamaica, where the insect was first reported in 1978 (Table 11.1), were more closely related to distant countries (clade 2, above) than to specimens from close-by countries, i.e., Honduras, Mexico, or Colombia, where the insect was first reported in 1977, 1978, and 1988, respectively (Table 11.1). This finding was also supported by Gauthier (2010), but not by Benavides *et al.* (2005).

Based on amplified length fragment polymorphism (AFLP) fingerprinting, Benavides *et al.* (2005, 2006, 2007) concluded that there were multiple coffee berry borer introductions into Colombia, with Brazil, Ecuador, and Peru being likely sources, and that introduction of the insect into Costa Rica most likely originated from Colombian insects. Benavides *et al.* (2005, 2007) also used AFLP fingerprinting to analyze genetic variability and biogeography in specimens originating in 17 countries. Results suggest the possibility of three separate introductions to the Americas and that West Africa was the origin of introductions into America and Asia. Genetic variability was low among specimens, as is to be expected for a species with extreme inbreeding (Andreev *et al.*, 1998), although using microsatellite markers developed by Gauthier and Rasplus (2004) as well as mtDNA, Gauthier (2010) reported low genetic variability within countries, but "considerable variation among groups of *H. hampei* specimens," which was presented as evidence for a species complex within *H. hampei*.

3.4 Damage and Losses

Damage caused by the coffee berry borer commences after adult females bore a hole in the coffee berry (Figure 11.5E) and lay their eggs in galleries built in the endosperm (i.e., the coffee seed, which is the marketable product), followed by larval feeding within the galleries (Figure 11.5F). Consequences of infestation include abscision of berries, loss in

TABLE 11.1 Reports for First Detection of the Coffee Berry Borer in Various Countries, in Chronological Order

Country and Year	Reference
Liberia 1897	Hopkins, 1915b
Gabon 1901	Beille, 1925
Republic of Congo 1901	Fleutiaux, 1901
Central African Republic and Republic of Chad 1902–1904	Chevalier, 1947
Democratic Republic of the Congo 1903	Leplae, 1928
Uganda 1908	Gowdey, 1911
Indonesia—Java 1908	Hagedorn, 1910
Angola 1912	Morstatt, 1912
Brazil 1913	Berthet, 1913; Neiva, 1928
Indonesia—Borneo and Sumatra 1919	Corporaal, 1921; Corbett, 1933
Côte d'Ivoire 1922	Beille, 1925
Cameroon 1924	Mbondji, 1988
United Republic of Tanzania 1924–1925	Ritchie, 1925
Benin 1925	Hesse, 1925
Kenya 1928	Wilkinson, 1928, 1929
Malaysia 1928	Corbett, 1933
Democratic Republic of São Tomé and Príncipe, 1929	Kaden, 1930
Togo 1930	Wegbe, 2012
Sri Lanka 1935	Hutson, 1936
Mariana Islands (Micronesia) 1945	Wood, 1960
New Caledonia 1948	Bugnicourt, 1950
Surinam 1951	van Dinther, 1960
Pohnpei (Micronesia) 1953	Wood, 1960
Peru 1962	de Ingunza, 1964
Tahiti 1963	Johnston, 1963
Philippines 1963	Gandia and Boncato, 1964
Ethiopia 1967	Davidson, 1967
Guatemala 1971	Hernández Paz, 1972
Honduras 1977	Muñoz, 1985
Jamaica 1978	McPherson, 1978; Reid 1983
Bolivia 1978	Rogg, 1997

TABLE 11.1 Reports for First Detection of the Coffee Berry Borer in Various Countries, in Chronological Order—cont'd

Country and Year	Reference
Mexico 1978	Baker, 1984
Fiji 1979	Anonymous, 1979
Ecuador 1981	Klein-Koch, 1990
El Salvador 1981	Vega Rosales and Romero, 1985
Colombia 1988	Cárdenas M. and Bustillo, 1991
Nicaragua 1988	Monterrey, 1991
India 1990[1]	Kumar et al., 1990
Cuba 1994	Hernández, 2002
Dominican Republic 1994	Serra, 2006
Bolivarian Republic of Venezuela 1995	Rosales Mondragón et al., 1998
Costa Rica 2000	Staver et al., 2001
Lao People's Democratic Republic 2004	CABI, 2008
Panama 2005	Inwood, 2005
Vietnam 2007	Beaver and Liu, 2010
Puerto Rico (USA) 2007[2]	Osorio, 2007; NAPPO, 2007
Hawaii (USA) 2010	Burbano et al., 2011
Martinique 2012	Dufour, 2013

[1]A 1930 report on the presence of the coffee berry borer in India (Anonymous, 1930) became a matter of proper identification (Coleman, 1931; Kannan, 1931) and turned out to be a misidentification (Coleman, 1931; Thomas, 1949).
[2]Earlier reports on the presence of H. hampei in Puerto Rico are incorrect and were based on a misidentification (Vega et al., 2002b).

seed weight, and loss in quality (Duque O., 2000; Duque-Orrego et al., 2002; Duque O. and Baker, 2003). To the best of our knowledge, of more than 3000 insect and mite species associated with coffee (Waller et al. 2007), the coffee berry borer is the only insect that completes its development after consuming the seeds inside coffee berries in the field, even though other insects have been reported to feed on the berry or as seed feeders in storage conditions (Bigger, 2008).

Yearly losses caused by the coffee berry borer have been estimated at US$500 million (Vega et al. 2002b), although a recent paper by Oliveira et al. (2013), which estimates losses for Brazil at US$215–358 million, indicates that the US$500 million is very conservative. In Colombia, more than 715,000 ha had been infested by 1998, equivalent to ca. 82% of the total coffee-growing area, and associated losses were estimated at ca. US$100 million in 2002 (Duque O., 2000). By 2002, over 800,000 ha were infested, equivalent to 90% of the total coffee-growing area (Duque-Orrego *et al.*, 2002). Costs associated with managing the insect in Colombia have been estimated at 5.5–11% of total production costs (Duque-Orrego *et al.*, 2002). To illustrate the magnitude of the problem in Brazil, in 1924 ca. 8,000,000 trees were infested in Campinas, out of a total of ca. 24,700,000 trees. To educate growers, a coffee berry borer film describing the biology of the insect, the damage it causes, how to differentiate it from other insects, and available control methods was produced in 1925 and seen by 104,634 people after being shown in cinemas 232 times (Pamplona, 1927).

3.5 Biology

Several extraordinary papers on the basic biology of the coffee berry borer and possible control methods were published in the early 1900s by scientists in Africa (Mayné, 1914; Beille, 1925; Hargreaves, 1926, 1935; Ghesquière, 1933), Brazil (Costa Lima, 1924a, b; Neiva *et al.*, 1924a, b; Neiva, 1928; Oliveira Filho, 1927; Fonseca, 1937, 1939; Fonseca and Araujo, 1939; Bergamin, 1943a), and Asia (Hagedorn, 1910; Roepke, 1919; Leefmans, 1920, 1923, 1924; Friederichs, 1922a, b, 1923, 1924a, b; Friederichs and Bally, 1922, 1923; Corbett, 1933). A common theme in these papers, mostly published in French, Portuguese, or Dutch, is that like many other Scolytinae, the coffee berry borer has female colonizers, males are smaller than females, sibling and pre-dispersal mating occurs, females are preponderant, and males cannot fly. These topics are discussed in detail below.

3.5.1 Boring into the Berry

Even though the term "berry" is commonly used when referring to the coffee fruit (and will be used throughout the chapter), the correct botanical name is drupe (Wrigley, 1988). Females usually bore the coffee berry through the disc, originally the floral disc of the flower, located on the upper part of the berry; at the other end of the fruit is the pedicel, which is the part attaching the fruit to the stem, via the infloresence (called the "infructescence" when the plant is in fruit) (Figures 11.5E and 11.8A, B). The style passes through the floral disc in the flower stage; during fruit development the hole closes up as the style dies back (Wrigley, 1988). It has been assumed that the disc is the preferred area for boring, as it provides a non-smooth surface for the insect to hold on while initiating the boring process (Costa and Faria, 2001; Cárdenas Murillo and Posada Flórez, 2001). The rough surface area on the disc presents a contrast to the smooth surfaces elsewhere on the berry.

The first step in the infestation process is the entrance of a colonizing female into the berry (Figure 11.5E). The entrance hole is circular, 0.6–0.8 mm (Varón *et al.*, 2004) to 1 mm in diameter (Wilkinson, 1928), which is very close to the width of a female insect (0.7 mm; Roepke, 1919; Bergamin, 1943a). In a laboratory study, Penagos Dardón and Flores (1974) placed female coffee berry borers on each one of eight green berries within four different branches removed from plants in the field and determined boring time until the insect "disappeared" inside the berry. The experiment was repeated several times and a total of 146 females were evaluated. The minimum time to enter the berry was 2 h, with a maximum time of 7 h 20 min. The average time to enter the berry was 4 h 16 min.

In a laboratory study in Ethiopia, the boring process until the insect is partially inside the berry took 8 h in green berries, 5.5 h in ripe berries, and 4 h in dry berries (Mendesil *et al.*, 2004). Wrigley (1988) stated that (1) it takes 2.5–4 h for the insect to enter the berry; (2) after 24 h the insect can no longer be seen through the entrance hole; and (3) it takes a minimum of 2 days before the insect starts building galleries in the seed. In a field study in Guatemala, Campos Almengor (1982) observed insects boring into berries for a 12-h period (6:00 AM–6:00 PM) and reported that highest boring activity occurred between 12:00 and 5:00 PM.

Boring in the field usually commences when berries are in the green stage and the determining factor for the progress of the penetration is the dry content of the berry, which has to be 20% or higher (Baker, 1984, 1999; Bustillo *et al.*, 1998). The 20% dry weight stage is reached ca. 120–150 days after flowering (Baker, 1999; Ruiz-Cárdenas and Baker, 2010; Arcila Moreno, 2011), with harvesting occurring 200–250 days after flowering (Baker *et al.*, 1992a). In a laboratory study, Ticheler (1961, 1963) found that there was no insect development in berries with moisture content of 75% or higher.

In some studies, four different positions for the colonizing female are described using letters: (A) female searching for a berry or initiating perforation; (B) female boring into berry, with part of the abdomen visible on the external part of the berry; (C) female inside the berry, boring into the seed; and (D) female and progeny inside seeds (Bustillo P. *et al.*, 1998). Camilo *et al.* (2003) assessed the position of colonizing females in berries starting at 77 days after flowering. Between 70 and 86% of females were observed to initiate the boring process and remained in position B until 112 days after flowering; at 161 days after flowering less than 10% remained in position B, having therefore completely entered the berry.

There is usually one hole per berry, unless infestations are high (Neiva *et al.*, 1924a; Hargreaves, 1940; Mendesil *et al.*, 2004; Vega *et al.*, 2011). As stated by Wrigley (1988), "during periods of intense infestation more than one female may bore into a single berry, each with its own entrance."

When suitable berries are absent, the insect could bore into the peduncles of young berries (Friederichs, 1922a) and after severe pruning, insects can bore into branches as well as older wood (Friederichs, 1923). Reproduction was not observed in any of these sites.

3.5.2 Oviposition

Females build galleries within the seed, where they oviposit. The number of eggs per female and oviposition period varies greatly. Friederichs (1924a) reported an average of 56 eggs per female and an oviposition period extending to 40 days. At 27°C in the laboratory, Bergamin (1943a) reported 24–63 eggs per female (with a maximum 119 eggs) and an oviposition period of 11–15 days. Leefmans (1923) obtained 164 insects from one coffee berry and Jaramillo et al. (2009a) recorded up to 288 eggs by an individual female in a berry.

Corbett (1933) found that after removing a female from a coffee berry it would not stop ovipositing, indicating they only need to be fertilized once. This also implies that "as in most female insects" (Chapman, 2003), coffee berry borer females must have a spermatheca, which by definition is "used for the storage of sperm from the time the female is inseminated until the eggs are fertilized" (Chapman, 2003). Rubio-Gómez et al. (2007) and Rubio G. et al. (2008) present details on the reproductive system of male and female coffee berry borers, including the spermatheca.

3.5.3 Larval Instars, Life Cycle, Adult Size, and Mating

Upon hatching, female coffee berry borers exhibit two larval instars in contrast to males, which only have one (Bergamin, 1943a; Figure 11.5A). The average time to complete the life cycle (egg to adult), as well as longevity, will depend on temperature and on how the assessment is made, i.e., natural conditions in the field, insects reared on berries in the lab, artificial diet, etc. Several papers have reported on these two parameters and only a few will be mentioned for illustrative purposes.

At constant temperatures of 19.2, 24.6, and 27°C, it takes 63, 27.5, and 21 days, respectively, to complete the life cycle (Bergamin, 1943a). In the field in Java, the life cycle takes about 1 month (Roepke, 1919) or 20–36 days, with an average of 25 days (Leefmans, 1923). In the Ivory Coast, Ticheler (1961, 1963) reported completion of the life cycle in 40.5 days at an average field temperature of 26°C, while in the Democratic Republic of the Congo, Steyaert (1935) reported completion of the life cycle in 36 days. In the laboratory, Muñoz (1989) reported completion of the life cycle in 35.8 days at 23°C, while in

Ethiopia, completion of the life cycle took 24–43 days at 25°C and 60% relative humidity (RH) (Mendesil et al., 2004). In artificial diet it takes 5–6 weeks (López-Pazos et al., 2009).

Adult females are larger (1.6–1.9 mm) than males (0.99–1.3 mm) (Roepke, 1919; Hargreaves, 1926; Corbett, 1933; Bergamin, 1943a) (Figures 11.3E and 11.5B–D). The width of males and females is ca. 0.6 and 0.7 mm, respectively (Roepke, 1919). Sibling mating occurs inside the berry (Leefmans, 1923; Sladden, 1934; Bergamin, 1943a) and males never leave the berry (Friederichs, 1922a, 1924a; Wilkinson, 1928; Corbett, 1933; Mathieu et al., 1997a). According to Bergamin (1943a), males hatch first and also reach the adult stage first in order to be sexually active once females become sexually active, ca. 3–4 days after molting into adults. Borsa and Kjellberg (1996a) also found that there is a tendency for earlier maturity in males than in females. In contrast, Dias Silva et al. (2012) reported that in the laboratory, both males and females take less than 2 days to reach sexual maturity. Similarly, Baker et al. (1992a) found that male and female offspring appear at the same time. Because there is a skewed sex ratio favoring females (discussed below; Table 11.2), males mate with multiple females (Brun et al., 1995a). Leefmans (1923) reported one male could mate with 12 females while Giordanengo (1992) recorded one male inseminating 128 females and four others individually inseminating 70–121 females. Multiple matings by females were observed by Dias Silva et al. (2012) in the laboratory, although it remains unknown whether this occurs in natural conditions in the field.

According to Roepke (1919), "The mother-beetle does not appear to leave the infested berry until the larvae are full grown; it then leaves with them, probably through the entrance hole." Various papers contradict this finding. Ticheler (1961, 1963) and López-Guillén et al. (2011) found that once females enter the berry and start ovipositing, their muscles degenerate; therefore, it is not clear why the colonizing female would leave the berry, as stated by Roepke (1919), if it cannot fly. Baker et al. (1992a) reported that the colonizing female remains with the developing progeny.

3.5.4 Generations per Year

Using seeds in the laboratory and not controlling for temperatures, Bergamin (1943a) was able to rear seven generations in 1 year. Ticheler (1961, 1963) reported that depending upon mean temperatures, there is an average of nine generations in Ivory Coast, a maximum of 13, and a minimum of five to six. In New Caledonia, Giordanengo (1992) reported four to five generations per year. Using degree-days, Jaramillo et al. (2009c) estimated the possible number of coffee berry borer generations in

TABLE 11.2 Sex Ratios Reported for the Coffee Berry Borer

Sex ratio (♀:♂)	Reference
56:1[1]	Leefmans, 1920
14.5:1[2]	Corporaal, 1921
40:1	Leefmans, 1923
8:1	von Ihering, 1924
10:1	Hargreaves, 1926
21.5:1	Wilkinson, 1928
8.5:1.5 to 8.3:1.7[3]	Bemelmans, 1930
56:1; 9.8:1; 13:1; 15.5:1[4]	Corbett, 1933
9.2:0.8[5]	Leroy, 1936
9.75:1	Bergamin, 1943a
11.4:1; 11.6:1; 15.9:1; 8.2:1	Ticheler 1961, 1963
10.3:1; 5.9:1; 10.3:1[6]	Ticheler 1961, 1963
5:1 to 20:1	Morallo-Rejesus and Baldos, 1980
231:1; 257:1; 494:1	Bautista Martínez and Atkinson Martín, 1988
10:1	Baker et al., 1992a
5.8:1; 113:1; 33.9:1; 11:1	Baker and Barrera, 1993
5.2:1 (artificial diet)	Pérez López et al., 1995
11.2:1; 6.8:1; 5:1 (artificial diet)[7]	Borsa and Kjellberg, 1996b
20:1 to 30:1	Mendesil et al., 2004
7.4:1 (artificial diet)	Portilla R. and Street, 2006
6.7:1[8]	Portilla R. and Street, 2006
5.7:1 (artificial diet)	López- Pazos et al., 2009
8.8:1.2	Jaramillo et al., 2009a
8.4:1.6; 8.5:1.5; 9:1	Jaramillo et al., 2009c

[1] *The percentage males was reported as 0.23 to 5%, with an average of 1.7% (427 males out of 23,842 females).*
[2] *The percentage males was reported as 0 to 50%, with an average of 6.87% (192 males out of 2793 females).*
[3] *Reported as 15 to 17% males.*
[4] *Ratios for insect collected from ripe berries, black berries from the ground, pupae from ripe and black berries, and from one black berry from the ground, respectively.*
[5] *Reported as 92% females.*
[6] *Sex ratio for 1, 2, and 4 females per berry.*
[7] *Sex ratios for 1, 2 or 3 females, respectively, in artificial diet.*
[8] *Sex ratio for progeny of females from the F_{48} to F_{64} generation in artificial diet after transferring to parchment coffee.*

four locations as follows: 2.4–4.7 in Tanzania; 2.9–4.3 in Colombia; 2–3.1 in Kenya; and 0–2 in Ethiopia. According to Baker (1999), there could be up to three generations within a berry.

3.5.5 Sex Ratio

The coffee berry borer is spanandrous, i.e., "females greatly preponderate" (see Hamilton, 1967). Even though most of the recent literature gives a 10:1 sex ratio for the coffee berry borer, in actuality a wide range of sex ratios favoring females (5.1:1 to 494:1) have been reported (Table 11.2). This range was recognized by Corbett (1933) 80 years ago: "The variation in the proportion of the sexes is significant."

The skewed sex ratio favoring females might be due to the presence of *Wolbachia* (Vega *et al.*, 2002a), a maternally inherited cytoplasmic α-proteobacterium that infects gonads and somatic tissues and which is quite common in insects (Jeyaprakash and Hoy, 2000; Hilgenboecker *et al.*, 2008). *Wolbachia* manipulates host reproduction via various mechanisms, including feminization (sex conversion), parthenogenesis, cytoplasmic incompatibility, and male killing (O'Neill *et al.*, 1998; Vega *et al.*, 2002a; Kageyama *et al.*, 2012).

The four studies reporting sex ratios for insects reared in artificial diet (Table 11.2) show a lower proportion of females than in all other studies, which report sex ratios in berries. A possible explanation is that among the ingredients included in the artificial diets are antimicrobial agents (e.g., sorbic acid, tetracycline, methyl paraben, benzoic acid, formol), some of which might be having an inhibitory effect on *Wolbachia*.

Borsa and Kjellberg (1996a) determined that competition among female coffee berry borers in artificial diet influenced the brood sex ratio, with the number of males significantly increasing as the number of competing females increased from one to two or three (Table 11.2). The experiment also revealed that competition among females led to fighting and occasional mutilation. Removal of dead immature stages from berries by the colonizing female, referred to as brood hygiene, has been reported by Baker *et al.* (1992a, 1994).

3.5.6 Longevity

Reports for female longevity vary widely, e.g., 55 days (Friederichs, 1924a); 96 days (Corbett, 1933); 102 days (Leefmans 1923); 81–282 days with an average of 156 (Bergamin 1943a); and 131 days (Muñoz, 1989). Brun *et al.* (1993) reported one female living 380 days in artificial diet. Longevity of males in the laboratory ranges from 21–43 days (Oliveira Filho, 1927) to 24–52 days (Giordanengo, 1992). Bergamin (1943a) reports that overall, males do not live more than 40 days, although three males he separately assessed lived to 78, 80, and 103 days. For insects reared in berries in the laboratory, Bautista Martínez and Atkinson Martín (1988) noted that adults could survive 6 months.

According to Corbett (1933), females can live without food for 81 days. This is in contrast to Mathieu *et al.*'s (1997a) finding that once females emerge from dry berries they can live for up to 11 days without food. Baker (1999) reported survival of more than 3, and even up to 8 months on dried or overripe berries.

3.5.7 Parthenogenesis

The production of fertile offspring from unfertilized eggs is known as parthenogenesis. Even though various papers have reported that the coffee berry borer does not reproduce parthenogenetically (Leefmans, 1923; Hargreaves, 1926; Bergamin, 1943a; Browne, 1961; Entwistle, 1964), Muñoz (1989) and Trejo et al. (2000) reported parthenogenetic reproduction in the coffee berry borer. Subsequent experiments have not been able to confirm parthenogenetic reproduction (Barrera et al., 1995; Alvarez Sandoval and Cortina Guerrero, 2004; Berrio E. and Benavides M., 2008; Constantino et al., 2011). Thus, virgin coffee berry borers can lay eggs, but they are not fertile (Bergamin, 1943a).

3.5.8 Functional Haplodiploidy

The coffee berry borer exhibits functional haplodiploidy (see Chapter 3). The concept is best understood by first presenting individual definitions for various terms. Haplodiploidy means that reproduction is arrhenotokous, i.e., males are only produced from unfertilized eggs and females are produced from fertilized eggs, resulting in haploid males and diploid females. The use of "functional" to define haplodiploidy in the coffee berry borer means that reproduction is pseudoarrhenotokous, i.e., male eggs are fertilized and even though the male is a diploid, it is functionally a haploid because the paternal set of chromosomes condenses into a mass of chromatin, resulting in (1) failure to be incorporated into semen during spermatogenesis or (2) inactivation in the somatic cells (Brun et al., 1995a, b; Borsa and Kjellberg, 1996b). It is possible that *Wolbachia* is involved in this process of chromosome condensation (see Vega et al., 2002a).

The fact that the eggs of unfertilized females do not hatch (Bergamin, 1943a) also serves as evidence for functional haplodiploidy (Brun et al., 1995b; Borsa and Kjellberg, 1996b), i.e., males would hatch from unfertilized eggs if the insect were truly haplodiploid. Further evidence for diploid males was presented by Borsa and Coustau (1996) after finding heterozigozyty in a cyclodiene resistance locus (*Rdl*) in male and female coffee berry borers. Cytogenetic evidence for diploid coffee berry borer males has been published by Bergamin and Kerr (1951), Brun et al. (1995a, b), and Constantino et al. (2011).

As mentioned above, once a colonizing female lays eggs within galleries in the berry the progeny has a skewed sex ratio favoring females and there is sibling mating, which by definition implies inbreeding. High inbreeding and low genetic variability has been demonstrated in various coffee

berry borer studies (Borsa and Gingerich, 1995; Gingerich et al., 1996; Borsa and Coustau, 1996; Andreev et al., 1998; Gauthier and Rasplus, 2004). When infestation levels are high, it is not unusual to find a berry attacked by more than one colonizing female as evidenced by several entrance holes (Leefmans, 1923; Neiva et al., 1924b; Wilkinson, 1928; Sladden, 1934; Leroy, 1936). Leefmans (1923) found that 83% of the black berries remaining on the plants and 8% of ripe berries were infested with more than one colonizing female. This situation could serve as a mechanism for outbreeding, although in such situations fecundity is hindered (Friederichs, 1924a; see Vega et al. (2011) for similar results in artificial diet). Based on a laboratory experiment, Mathieu et al. (1997a) proposed that females emerging from dry berries in the field during the interseason (i.e., between harvests) might enter dry berries again as a result of not finding suitable berries on trees. Such situations could theoretically result in outbreeding. The genetic relatedness among progenies where more than one colonizing female is present (as described above) has not been experimentally determined.

3.5.9 Pheromones

No sexual pheromones have been reported for the coffee berry borer. The biology of the female, which is inseminated by its sibling inside the berry, makes the production of a sex pheromone unlikely (de Kraker, 1988). According to Wood (1982) pheromones "apparently have not been reported from any species with the habit of consanguineous polygyny."

3.5.10 Vision

An important feature in the basic biology of an insect is the responses to movement and color. These responses will depend on visual acuity, and in the case of the coffee berry borer there is a marked sexual dimorphism in terms of the development of the compound eyes. Wood (2007) reported that males in 14 of 51 *Hypothenemus* species from South America have reduced eyes when compared to females. Using an optomotor response apparatus, Vega et al. (2014) demonstrated that in contrast to females, male coffee berry borers do not respond to movement, most likely as a result of the biology of males, whereby they are born inside the coffee berry and never leave it, thus not having need for vision. Male coffee berry borers have rudimentary eyes, with a significantly lower number of facets in males (19.1 ± 4.10) than in females ($127.5 \pm v3.88$) (Vega et al., 2014).

Color preferences by females have been examined using both artificial and field-collected berries. Ticheler (1961, 1963) compared female preference for black, red, yellow, and green artificial berries (painted cotton balls imbibed with paraffin and shaped like a coffee berry) and found preference for black berries, followed by red. Mathieu *et al.* (2001) tested visual responses to red and green artificial berries made with paraffin wax and found that red berries were more attractive to females. Mendoza *et al.* (2000) determined female preference for green, yellow, red, and black berries, as well as berries made with polystyrene. Females preferred red and black, both in real berries and polystyrene berries. The studies involving artificial berries eliminate olfactory cues associated with real berries and demonstrate that females can detect color.

In a choice test laboratory study using field-collected red, green, and dry berries placed in Petri dishes, females showed preference for red berries over mature green berries, for red berries over dry berries, and for dry berries over immature green berries (Giordanengo *et al.*, 1993). It is important to note that even though females can colonize red berries in the field, it is likely that development of the progeny into adults will not be completed before harvest (Baker, 1999).

3.5.11 Microbiota and its Role in Caffeine Detoxification

Caffeine (1,3,7-trimethylxanthine) is a purine alkaloid present in many plants, including coffee. The seeds of *C. arabica*, on which the coffee berry borer feeds, contain ca. 1.0% caffeine on a dry weight basis, while *C. canephora* contains ca. 1.7% (Lean *et al.*, 2012). The presence of caffeine and other alkaloids in plants have been proposed to have anti-herbivore roles (Levinson, 1976; Nathanson, 1984), and several studies have shown the negative effects of caffeine on insects (Vega *et al.*, 2003b and references therein).

Guerreiro Filho and Mazzafera (2003) tested adult female coffee berry borer feeding responses to seeds from 12 *Coffea* species containing different caffeine contents. The authors could not detect a significant negative correlation between caffeine content and damage caused by the insect. The same occurred when seeds were imbibed in aqueous solutions of caffeine to increase caffeine content. There was also no significant correlation in attraction to mature berries with different caffeine content. Guerreiro Filho and Mazzafera (2003) conclude that caffeine levels are not involved in resistance and that the insect "has evolved an adaptation to avoid the toxic effects of caffeine."

The adaptation required to survive caffeine consumption must involve a caffeine metabolizing mechanism. Based on the role of yeasts in detoxifying allelochemicals in insects (Vega and Dowd, 2005), Vega *et al.* (2003b) tested a yeast present in the coffee berry borer for caffeine breakdown properties, with negative results. Subsequent work has focused on the role of the gut microbiota. The terms microbiota and microbiome have been used to describe the "organisms that live inside and on humans" and "the genomes of these microbial symbionts," respectively (Turnbaugh *et al.*, 2007). The terms are no longer used just for humans and are applicable to all types of organisms.

The first step in elucidating the role of gut microbes in caffeine breakdown was to develop a technique to dissect the ca. 3.5-mm-long alimentary canal of live female coffee berry borers, in order to subsequently isolate and identify the associated microbiota (Ceja-Navarro *et al.*, 2012). Ceja-Navarro *et al.* (submitted) isolated and identified 13 bacterial species (*Pseudomonas fulva*, *P. fluorescens*, *Pantoea vagans*, *P. septica*, *P. eucalypti*, *Ocrobactrum* sp., *Enterobacter* sp., *Kosakonia cowanii*, *Brachybacterium rhamnosum*, Jonesiae, *Microbacterium binotii*, *Novosphigobium* sp., and *Stenotrophomonas maltophilia*) that subsist on caffeine as the sole source of carbon and nitrogen. Addition of antibiotics to the coffee berry borer artificial diet eliminated caffeine degradation, thus demonstrating the involvement of the microbes in the process. The caffeine demethylase gene (*ndmA*) was expressed *in vivo* in field specimens as well as in *P. fulva* isolated from the alimentary canal. Diet inoculation with *P. fulva* restored the ability to degrade caffeine. Elucidating the mechanism for caffeine detoxification in the coffee berry borer presents new research options, as well as challenges, for managing the insect. For example, the use of bacteriophages might result in interference with the microbes involved in caffeine detoxification, although their introduction and survival in the field might not be feasible due to the biology of the insect, i.e., the introduced trait, and by definition would cause death of the insect and would not be able to compete with wild populations.

3.5.12 Association with Fungi

The mycobiota associated with the coffee berry borer has been shown to be quite extensive. Pérez *et al.* (2003) isolated 38 fungal species in 21 genera from the insect cuticle (29 species), alimentary canal (18 species), and feces (10 species), and four genera from galleries (five species). Carrión and Bonet (2004) identified 12 fungal species associated with the insect and seven with the galleries, while Gama *et al.* (2005, 2006) identified 10 fungal genera associated with the insect and five with the galleries.

Rojas *et al.* (1999) isolated *Fusarium solani* (Mart.) Sacc. (current name: *Haematonectria haematococca* (Berk. and Broom) Samuels and Rossman; Hypocreales: Nectriaceae) from adult female coffee berry borers reared in artificial diet as well as from insects collected in the field in Mexico and Benin. This association led them to propose this as a "close association." In a subsequent study (Morales-Ramos *et al.*, 2000) concluded that this was a symbiotic association, and that insects reared in beans infected with *F. solani* had significantly higher fecundity than insects reared in sterile beans. This was ascribed to higher ergosterol levels in beans infected with the fungus. Such symbiotic association would require the presence of mycangia. Even though Morales-Ramos *et al.* (2000) were unable to find mycangia, they proposed that the asperities and setae on the pronotum might serve a similar function. One issue with this proposal is that the exposed asperities would have to be selective for *H. haematococca* spores, and it is not clear how this could happen. Another issue is that based on size, the photograph included in Morales-Ramos *et al.* (2000) depicts bacteria and not *H. haematococca* conidia, whose size is 6–24 × 2.5–5 μm (Rossman *et al.*, 1999). Pérez *et al.* (2005) pursued the topic of mutualistic fungi using three different fungi (including *H. haematococca*) and did not find any evidence for beneficial effects on the insect. To our knowledge, *H. curtipennis* (Figure 11.2G) remains the first and only cryphaline ambrosia beetle and the only *Hypothenemus* species for which mycangia have been identified (Beaver, 1986).

3.5.13   Genome

Very little is known about the coffee berry borer genome. Nuñez *et al.* (2012) reported an estimated genome size of 170–180 Mb with 20,653 unigenes while Benavides *et al.* (2014) reported a 194 Mb genome, with ≈ 20,500 unigenes. A gene of bacterial origin encoding for mannanase (discussed in Section 3.12), has been identified in the coffee berry borer genome (Acuña *et al.*, 2012).

3.6   Ecology

3.6.1   Host Plants

Vega *et al.* (2012a) presented evidence for the possible polyphagous nature of the coffee berry borer, based on a study by Schedl (1960), who collected the insect in 20 genera other than *Coffea* in forests in the Democratic Republic of Congo. Schedl (1960) suggested that the coffee berry borer might be polyphagous: "...there exists inside the rainforest a series of natural hosts for the parasite that give it the possibility to develop independently from coffee plantations" (translated from the French original). The paper by Schedl (1960) as well as two additional papers (Beille, 1925; Ghesquière, 1933) bring to the forefront the possibility that the insect originates in the humid, evergreen forests and that it feeds on various plants, perhaps including wild *Coffea* species. This scenario would make the insect movement to cultivated coffee more plausible as coffee cultivation increased in deforested areas.

Ghesquière (1933) unequivocally stated that he obtained different life stages of the coffee berry borer from the legume *Dialium lacourtianum* Vermoesen (current name: *D. englerianum* Henriq.; Leguminosae). Based on these

results, he proposed that the plant could be used as a trap. Gumier-Costa (2009) reported the insect breeding in Brazil nuts (Lecythidaceae) collected in the field.

Even though several papers have reported the coffee berry borer in many different plants, none has shown that the insect completes its life cycle in these plants (Eggers, 1922; Oliveira Filho, 1927; Hargreaves, 1935, 1940; Baguena Corella, 1941; Viana, 1965; Morallo-Rejesus and Baldos, 1980; Campos Almengor, 1981; Quezada and Urbina, 1987; Vijayalakshmi *et al.*, 1994; Messing, 2012). The presence of the insect in these plants is either temporary, perhaps as a result of seeking shelter, or possibly a misidentification of the insect (Oliveira Filho, 1927; Le Pelley, 1968; Wrigley, 1988).

3.6.2   Host Finding

Insect attraction to plants can be influenced by kairomones, as well as by plant shape and color, among others factors (Vinson, 1976; Prokopy and Owens, 1983; Miller and Strickler, 1984; Vet and Dicke, 1992; Vet, 1999). A kairomone is a chemical signal produced by one organism, in this case the coffee plant, which "evokes in the receiver" (the coffee berry borer), "a behavioral or physiological reaction which is adaptively favorable to the receiver, but not to the emitter" (Price *et al.*, 2011). Thus, a search for signals produced by coffee plants that end up attracting the coffee berry borer has been an area of research interest for many years.

The pioneer studies on coffee plant kairomones were based on extracts from coffee berries, without focusing on the identification of the particular components in the extract. For example, Prates (1969) conducted a laboratory experiment in which extracts of coffee berries (the solvent is not mentioned) were tested for their attraction to the coffee berry borer. Pure extracts and 50% diluted extracts were significantly more attractive to the insect than water or extracts diluted at 25%. In Mexico, Velasco Pascual *et al.* (1997a, b) collected berries from two varieties of *C. arabica* and from *C. canephora* and homogenized them in 80 ml of methanol and 120 ml of ethanol, followed by placement of traps with the extracts in a *C. canephora* plantation. There were significant differences in insect capture based on the coffee variety used for preparing the extracts, and traps containing methanol, ethanol, or a methanol: ethanol mixture were also shown to capture insects, in some cases at levels not significantly different than those in the traps with the extracts. The results indicate there might be one or more components in the berry extract that increases attraction.

In similar studies, Gutiérrez-Martínez *et al.* (1990) and Gutiérrez-Martínez and Ondarza (1996) used six different solvents to extract different *C. canephora* parts (flower, leaves, berries, branches, roots, etc.). The extracts were tested in the field and the laboratory for their attraction to the coffee berry borer. The highest insect capture in the field resulted from a methylene chloride/ethanol extract of ripe coffee berries. Giordanengo *et al.* (1993) used a Y-shaped olfactometer and reported significantly higher female attraction towards red berry volatiles than towards green berry volatiles. Females also responded to unidentified volatiles obtained from green berry acetone extracts, but not to hexane or ethanol extracts.

An ingenious experiment conducted by Ticheler (1961, 1963) used groups of green or red berries, each placed inside one half of a divided Petri dish. Females were allowed to walk on a mesh placed over the dish, and preferred walking over the area enclosing the green berries at a significantly higher rate than over red berries. The results suggest that volatiles emanating from green berries are more attractive to the insect than volatiles from red berries. Other experiments by Ticheler (1961, 1963) showed that females were significantly more attracted towards areas in Petri dishes where red or green berries were not covered with cellophane. Thus, the entire arena over which the insects were walking was the same color (red or green), and the only difference was the presumed volatiles emanating from the area not covered with cellophane (in separate experiments, cellophane had been shown not to influence selection). In an interesting twist, Ticheler (1961, 1963) removed the antennas and repeated the experiment using red berries and covered one half of the dish with cellophane. Females preferred the area over the red berries without cellophane at a significantly higher rate. He concluded that the antennae are not the only sensory organs involved in volatile detection in the coffee berry borer.

In a laboratory experiment in which red berries were placed on one half of a platform over which the insect release arena was located, Mathieu *et al.* (2001) found that colonizing females (those that emerged from berries or artificial diet) were significantly more attracted to the area above the berries than to the other half of the platform, under which no berries were present. In contrast, non-colonizing females, defined as virgin and mated nulliparous females still within the artificial diet, avoided the area above the berries, and the negative response became stronger with age. The results are noteworthy because it separates visual from olfactory cues and demonstrates that olfactory responses are influenced by the physiological state of the female.

Starting in the early 1990s, various studies focused on the identification of volatiles emanating from coffee berries with the goal of eventually identifying candidates that might be serving as coffee berry borer attractants (Mathieu *et al.* 1991, 1996, 1998; Ortiz *et al.*, 2004). Ortiz *et al.* (2004) identified 27, 34, 41, and 68 compounds in green, half-ripe, ripe, and overripe berries, respectively.

Mendesil *et al.* (2009) ran olfactometer bioassays using green, ripe, and dry berries and found positive responses to volatiles emanating from ripe and dry berries, but not to volatiles from green berries. They chose to examine the volatile profile from dry berries. Using gas chromatography-mass spectrometry (GC-MS), six compounds with positive electroantennogram responses were identified: ethylbenzene, limonene, methylcyclohexane, nonane, 1-octen-3-ol, and 3-ethyl-4-methylpentanol. Only one of these, methycyclohexane, was present in the volatile profile of green berries. All six were present in ripe berries but at lower concentrations than in dry berries. Three of the six compounds (nonane, 3-ethyl-4-methylpentanol, and 1-octen-3-ol) were also identified by Ortiz *et al.* (2004) as volatiles from coffee berries. Four of the electroantennogram-positive compounds elicited attraction in the olfactometer bioassay: 3-ethyl-4-methylpentanol, nonane, methylcyclohexane, and ethylbenzene. A blend of these four compounds also elicited attraction. Because these compounds are not specific to *C. arabica*, Mendesil *et al.* (2009) presume that as single compounds, they are not used for host recognition by the coffee berry borer, but that a blend might be more likely for host recognition.

Jaramillo *et al.* (2013) identified 49 volatile compounds in ripe berries (defined as having a yellow-orange exocarp) and 26 in green berries. Four compounds in ripe berries (conophthorin, chalcogran, frontalin, and sulcatone) serve as pheromones for some conifer bark beetles (e.g., *Dendroctonus*, *Pityogenes*, *Gnathotrichus*). Only one of these compounds (conophthorin) was present in green berries. In Y-tube olfactometer bioassays, female coffee berry borers were attracted to conophthorin and chalcogran.

3.6.3 Dispersal

Females are the dispersal unit in the coffee berry borer. Females living inside berries left on trees after harvest or in berries that have fallen on the ground serve as a source for subsequent dispersal and infestation of newly formed berries. For example, Corbett (1933) counted 93 females and six male coffee berry borers in one robusta berry collected from the ground. Leefmans (1924) calculated that ca. nine million coffee berry borers could be present in fallen berries in 875 acres, equivalent to 10,285 insects/acre. He also found that covering the infested berries in the soil with 8, 12, or 20 inches of loose soil would not prevent the insects from emerging, a finding also supported by Friederichs (1922a). Baker (1984) estimated that over 500,000 insects/ha could remain in berries that have fallen on the ground. Bustillo *et al.* (1999) and Vera *et al.* (2011) have reported on the use of fungal entomopathogen conidial suspension sprays on berries on the ground to reduce numbers of females emerging from these berries.

Another source of insects are the pruned branches left on the ground in the process of rejuvenation of plantations (Baker, 1999). In Colombia, insects emerging from berries on branches left on the ground after pruning plantations have been estimated at 1.5–2 million/ha (Baker, 1999) and 2–3.5 million flying females/ha from an estimated seven to nine million total number of insects (Benavides M., 2010).

Various factors influence female emergence from the native berry, such as high light intensity, and the presence of green berries nearby (Mathieu *et al.*, 1997a), although Steyaert (1935) found beetles were more active in cloudy weather than on bright days. Low (<60%) and high (>90%) RH occurring after rains also induces emergence from the berry (de Kraker, 1988; Baker *et al.*, 1992b; Baker and Barrera, 1993), as well as increases in temperature (Baker *et al.*, 1992b; Jaramillo *et al.*, 2010a). Baker *et al.* (1994) showed that the insect is very sensitive to RH, with survival time at 93.5% RH being twice that at 84%.

Acevedo-Bedoya *et al.* (2009) examined the use of a DayGlo® fluorescent pigment to mark coffee berry borers for dispersal studies. The pigment could only be detected for 5 days and examinations had to be conducted in the dark using a black light and a stereoscope. Therefore, the use of these pigments is of limited value for dispersal studies. On the other hand, a molecular marker developed by the same authors showed more potential for dispersal studies.

3.6.4 Flight

Seven different types of flight muscles have been identified in female coffee berry borers (López-Guillén *et al.*, 2011), and the flight muscle surface area in flying females is larger than that in ovipositing females, indicating (as mentioned above) that the muscles degenerate after oviposition commences (Ticheler, 1961, 1963; López-Guillén *et al.*, 2011).

In Java, Roepke (1919) observed females emerging from the berry and attempting to fly at sunset. Also in Java, Leefmans (1923) noted that flight was common between 4:00 and 6:00 PM and that females could cover distances extending up to 348 m. In Uganda, Hargreaves (1926) observed females flying late in the afternoon. Highest flight activity in Nicaragua was between 1:30 and 3:30 PM (Borbón-Martínez *et al.*, 2000). In the Democratic Republic of the Congo, Leroy (1936) observed that females exit the berry from 4:00 to 5:00 PM, flying at 4–5 m in elevation and covering up to 300–400 m. Bemelmans (1930) reported that females fly up to 5 m and more than that when pushed by the wind. In Mexico, Baker (1984) reported 22 min as the longest free flight observed in the laboratory and when tethered the longest continuous flight lasted 100 min. In New Caledonia, Giordanengo (1992) reported that colonizing female activity in lab and field peaks at 2:00 PM and noted that unmated females will not fly and therefore disperse. According to Cárdenas Murillo and Posada Flórez (2001) insects can fly in a spiral pattern for 1–2 h, which allows for wide dispersal if winds are prevalent.

The wing length of females is ca. 2.2 mm in contrast to 0.34 mm in males (Corbett, 1933). According to Leefmans (1923), "the wings in the male sex are so much reduced that the males certainly cannot fly." Corbett (1933) reached a similar conclusion when he stated, "the wings in the male do not appear to be sufficiently large for sustained flight." Hargreaves (1926) noted, "males seem incapable of flight" and Bemelmans (1930) stated, "The male cannot fly, being devoid of useful wings" (translated from the French original). Similarly, Leroy (1936) wrote, "The wings seem stunted and consist of small membranous appendages that do not allow the insect to fly" (translated from the French original).

3.7 Shade

As discussed above, in their natural habitat the two commercial coffee species were inhabitants of the humid, evergreen forests in Africa (Davis *et al.*, 2006). One of the first reports on the effects of deforestation on an insect is a 1925 paper by the French botanist Lucien Beille (1862–1946), in which he relates how the coffee berry borer became a pest of coffee (Beille, 1925): "The onset of the trouble coincided with the destruction of the forest; it appears that *Stephanoderes*,[1] lacking the plants that it frequents, has found in the new coffee plantations, conditions that are favorable to his evolution" (translated from the French original). Thus, from its natural habitat as an understory plant in forests, coffee plantations can now be found under a wide range of conditions, from full-sun (i.e., no shade trees) to variable shade levels provided by different types of plants (e.g., *Inga*, *Gliricidia*, avocados, bananas, etc.; Figure 11.6).

Shade trees and the ensuing shade levels provide several benefits to coffee plantations: (1) prevent soil erosion; (2) serve as windbreakers; (3) promote biodiversity; (4) increase nutrient recycling and organic matter from fallen leaves; (5) buffer temperatures (see below); (6) maintain higher RH; and (7) provide additional sources of food as well as income for the grower, e.g., wood for burning or for sale, or if shade is provided by bananas, avocados, etc., availability of fruits for family consumption or for sale (Muschler, 2004; Somarriba *et al.*, 2004; Albertin and Nair, 2004; Rice, 2011; Tscharntke *et al.*, 2011). On the other hand, yields in shaded plantations can be lower than in plantations at full sun (Brun *et al.*, 1989a; Gobbi, 2000; Soto-Pinto *et al.*, 2000; Muschler, 2004; Haggar *et al.*, 2011) while coffee berry borer

---

1. Old genus for the coffee berry borer.

FIGURE 11.6   Coffee grown at full sun (left) and under shade (right) in Puerto Rico. *Photos by Fernando E. Vega.*

infestation levels have been reported to be higher in shaded plantations (discussed below). These two issues (yields and infestation levels) have been used as justification for eliminating shade in coffee plantations.

Two main reasons have been proposed to explain why shaded plantations have higher coffee berry borer population levels than non-shaded plantations: (1) since the insect evolved in the shade of forests, it is better adapted to that environment and not to the lower RH of sun-exposed plantations, and (2) shade has a negative effect on parasitoids (discussed below). Adult coffee berry borers are very sensitive to RH with an optimum range for survival and development of 90–95% RH at 25°C (Baker *et al.*, 1994). These high humidity conditions would be more likely to be encountered in shaded plantations.

The first papers reporting on the effects of shade on the coffee berry borer were conducted by Hargreaves (1926, 1935, 1940) in Uganda and by Jervis (1939) in present-day Tanzania. We have chosen to include relevant quotes from these and other papers to accurately reflect the understanding of the situation at various points in time (Table 11.3). Hargreaves (1926, 1935, 1940) ascribed the reduced damage in the plantations growing at full sun to what he calls parasite preferences for this habitat, although no evidence for this was presented. Hargreaves (1935) further expanded on this topic and stated that shade "is distinctly favourable to the berry-borer" and explains that coffee berry borer damage is higher in unpruned trees and when large trees provide dense shade to the coffee plant (Table 11.3). Similarly, Jervis (1939) reported (Table 11.3) on the sanctuary provided to the insect by a dense plant canopy in Bukoba (Tanzania), where the

insect had caused a 73% reduction in value in 1931, and presents branch thinning as a solution to the problem. Jervis (1939) also mentions that the heaviest insect infestations occurred in areas of extreme humidity. It is important to point out that the papers by Hargreaves and by Jarvis (cited above) are based on observations and not on actual experiments, i.e., no data are presented.

From 1939 to 1945, 11 papers dealing with shade and the coffee berry borer were published in Brazil (Mendes, 1938; Fonseca, 1939; Mendes, 1939, 1940a, b; Bergamin, 1943b, 1944b, c, 1945a, b; Rocha Lima, 1945). The goal was to reach a recommendation for the state of São Paulo on whether plantations should be shaded or not. Mendes (1938) presents information from two farms that had shade-grown coffee in which coffee berry borer infestation levels were much higher than in sun-grown coffee, and two farms in which problems with the insect became so severe that the owners in one farm decided to abandon the crop while the owners in the other decided to cut down the entire plantation. Fonseca (1939) notes that during a trip to Uganda he observed higher infestation levels in shaded coffee and that in Brazil, dry and well-aired coffee plantations do not provide favorable conditions to the insect, when compared to more humid and wind-protected shaded plantations. Mendes (1939) presents data showing higher infestation levels in shade-grown coffee and observes that when shade is used, coffee takes longer to mature, which is favorable for development of the coffee berry borer. He concludes by stating that, based on the knowledge available at the time, shade cannot be recommended for growers in São Paulo. In a second study, Mendes (1940a) found higher

TABLE 11.3 Statements in the Literature Related to the Effects of Shade on the Coffee Berry Borer

	Reference
"We have noticed that shaded Coffee is more susceptible to attack than unshaded. In fact, we have seen unshaded Coffee entirely escape, whilst a few yards away, shaded trees were badly infested. It would be of great interest to learn if the beetle is a shade-loving insect."	Brown and Hunter, 1913
"Observations indicate that damage to coffee growing under shade is more extensive than is the case with unshaded coffee. It is the writer's opinion that this is due to preference by the parasites[1] for a sunny habitat, and not, as might be inferred, to any preference by the beetle for shaded conditions."	Hargreaves, 1926
"Damage in Uganda appears to be greatest to coffee under shade."	McDonald, 1930
"The shading of coffee berries, either by overhead shade or by dense foliage of the coffee tree themselves, is distinctly favourable to the berry-borer. In this case, however, it appears that the influence is indirect, by making the habitat less suitable for the parasites. Instances have occurred where estates or parts of estates under heavy shade have suffered intense borer damage continuously, and thinning or complete removal of shade has greatly reduced the incidence of the pest." "Again, large natural trees, providing dense shade, left growing in coffee have been proved to be centres of infestation; the further away from such trees the coffee, the less intense the infestation, until at 50 yards distance the borer could scarcely be found."	Hargreaves, 1935
"Under the methods of cultivation practised by the natives these trees become enormous botanical candelabra with their branches spreading to the ground and from which sprang innumerable shoots, forming a dense canopy and affording complete sanctuary to the coffee berry borer (*Stephanoderes hampei*)..."	Jervis, 1939
"Shade for coffee was, and still remains, the subject of much controversy in Uganda. In the past excessive shade was provided and this depressed yields and encouraged insect pests, especially *Stephanoderes*; later there was a movement for removal of all shade, but recently the consensus of opinion among planters has been more in favour of controlled shade. *Robusta* is essentially a forest plant and is more susceptible to lack of shade than is *arabica*."	Thomas, 1940
"The beetle is capable of causing very serious losses of crop during intense infestation, which occur normally under conditions of dense shade..." "A second factor of great importance[2] is that of shade, both overhead and that provided by the foliage of the coffee itself in closely planted or unpruned trees. Many examples of intense attack, due primarily to forest-like conditions, have been observed by the writer..." "In one instance, where spacing of plants was normal, an almost immediate reduction of the infestation followed thinning of the shade trees, and further removal of shade reduced the borer incidence to almost negligible proportions. In another instance, a reduction in density of the unusually close stand of coffee was necessary in addition to shade-thinning. A striking example of the influence of shade was observed on one plantation where a very large natural tree with immense spread was left among the coffee (otherwise unshaded): near the trunk of this shade tree the berries were intensely infested, and the infestation gradually became less intense in berries more and more distant, until at 50 yards from the trunk the borer could scarcely be found." "The effect of shade appears to be double: direct, as shown above; and indirect, because the parasites prefer better-lighted conditions, or because dense coffee growth makes the search for hosts by the parasites more difficult and more hazardous." "The fruits of wild coffee plants growing in dense forest are almost invariably intensely infested by *S. hampei*."	Hargreaves, 1940
"It is recommended to maintain the shade on the plantations to foster the development of fungal entomopathogens..."[3]	Chevalier, 1947
"Several authorities state that the pest is usually more troublesome in dense shade." "Heavy shade and close planting appear to favour the insect, possibly because of a moisture complex, since there would be too great a competition for the moisture in the soil. Wider spacing and the reduction of shade has immediately reduced an infestation. Dense shade may possibly reduce the population of the controlling parasites."	Haarer, 1962
"Heavy shade, from either shaded trees or inadequately pruned coffee causes conditions unsuitable for the natural enemies of the borer and should be removed."	Crowe and Gebremedhin, 1984
"It is worth remembering that originally coffee was an understory plant in tropical forest and that therefore the broca might find strong direct sunlight in dry season conditions inimical to survival."	Baker, 1984
"*Hypothenumus hampeii* [sic] prefers shaded coffee, and removal of shade trees is beneficial."	Bardner, 1985
"Attacks are also more severe where the coffee is grown under heavy shade or is closely planted and unpruned. A single, very large, dense shade tree can cause a serious local infestation. The fruits of wild coffee growing in the	Wrigley, 1988

TABLE 11.3 Statements in the Literature Related to the Effects of Shade on the Coffee Berry Borer—cont'd

	Reference
dense forest are frequently heavily infested. In Brazil the infestation is greater in damp, shaded plantations than in dry, open areas (da Fonseca 1939)."	
"The following cultural measures, if conscientiously applied, do much to reduce the infestation: 1) Reduce heavy shade. 2) Prune the coffee to keep the bush as open as possible..." "Heavy shading brought by inefficient pruning favors the survival of CBB, and is unfavorable to natural enemies. Proper pruning is, therefore, necessary for direct or indirect control of CBB."	Crowe, 2004
"At altitudes where berry borer is a problem, the incidence of damage caused by the pest can be reduced by thinning of shade trees and pruning the coffee bushes to open the canopy."	Waller *et al.*, 2007

[1] *What Hargreaves (1926, 1935) calls parasites are nowadays referred to as parasitoids, i.e., insects that oviposit on the coffee berry borer, causing its eventual death.*
[2] *In addition to altitude.*
[3] *Translated from the French: "On recommande de maintenir l'ombrage des plantations pour favoriser le développement des champignons entomophages..."*

infestation levels in shaded plots, while in a third set of experiments, Mendes (1940b) found lower infestation levels than in previous years, but these were still higher in shaded plantations. Bergamin (1943b) stated that it is impossible to mimic the shade conditions occurring in the natural habitat of coffee plants, and that berries produced in such conditions are slender with almost no body, and therefore production is null. He also states that depending upon the trees used to provide shade, leaves falling on the ground might decompose slowly, providing a hiding place for berries that have fallen from the plant and which can harbor the insect. Bergamin (1944b) reports infestation levels in shaded coffee in 1943 were 44% vs. 5% in coffee grown at full sun; in 1944, infestation levels were 89.5% and 13.9%, respectively. Bergamin (1944c) discusses that even though there are ideal goals for shading plantations (e.g., providing organic matter, reducing cold winds, reducing thermal variations, etc.), these theoretical observations do not account for the positive biological responses of the insect to shade conditions. Bergamin (1945a, b) and Rocha Lima (1945) conclude that after years of field research it is evident that the coffee berry borer prefers shaded plantations because they provide better conditions for their development, and that shading should not be recommended for coffee plantations in São Paulo.

Another series of experiments in Brazil, starting in 1953 and ending in 1971, in which one of the parameters assessed was coffee berry borer infestation levels in shade and no shade, revealed that in all instances infestations were higher in shaded plantations (Graner and Godoy Junior, 1959, 1962, 1971; Godoy Junior and Graner, 1961, 1967). In Honduras, Muñoz *et al.* (1987) reported higher infestation levels in "half shaded" plantations with shade provided by pruned *Erythrina* (Leguminosae), followed by unshaded plantations, and finally, plantations growing at "full shade" under *Erythrina*. Unfortunately, statistical analyses were not conducted.

In Brazil, Passos *et al.* (2005) sampled coffee berry borers in volatile-containing traps installed at different distances (3, 6, 12, 15, 24, and 36 m) from *Grevillea robusta* A. Cunn. ex R. Br. (Proteaceae), a tree used to provide shade in coffee plantations. The mean number of insects collected 3 m from the tree was 24 times higher (1506 insects) than the number collected at 36 m (63 insects), a statistically significant difference. They recommend placing emphasis—in terms of pest management strategies against the coffee berry borer—in areas that are proximate to trees that provide shade in order to reduce the number of insects that could infest the next crop.

Féliz Matos (2003) and Féliz Matos *et al.* (2004) examined coffee berry borer infestation levels under three shade levels in Nicaragua: no shade, medium shade (40–50%) using *Gliricidia sepium* (Jacq.) Walp. (Leguminosae), and dense shade (60–70%) using *Eugenia jambos* L. (current name: *Syzygium jambos* (L.) Alston; Myrtaceae). Percent infestation was significantly higher (17–25%) in dense shade compared to <2% under no shade and medium shade. Infestation levels for no shade and medium shade were not significantly different. Wegbe *et al.* (2007) in Togo have also reported significantly higher coffee berry borer infestation levels in densely shaded coffee plantations. In Colombia, Bosselmann *et al.* (2009) reported a trend towards higher infestation levels under shade.

In contrast to all the papers cited above, Baker *et al.* (1989) found no significant differences in coffee berry borer infestation levels at two different elevations in Mexico based on four different shade levels (no shade, light, medium, and high shade). Also working in Mexico, Soto-Pinto *et al.* (2002) found no correlation between shade cover and coffee berry borer levels. In Kenya, Jaramillo *et al.* (2013) reported lower infestation levels in shaded plantations.

In terms of abundance of parasitoids, Mendes (1940a) found lower *Prorops nasuta* population levels in shaded plots, while de Toledo (1948) found that shade and the parasitoid *P. nasuta* were perfectly compatible. Ticheler (1961, 1963) was able to find *Cephalonomia stephanoderis* in full sun plantations but not in shaded plantations. Parasitoids are discussed in detail below.

Shade has been shown to buffer temperatures in coffee plantations. For example, Vaast *et al.* (2006) found that coffee shading in Costa Rica reduced temperatures by 2°C and 4°C in outer and inner leaves, respectively, when compared to unshaded plantations. In New Caledonia, Brun and Suckling (1992) measured differences of 3°C between sun grown and shaded plantations. Using *Inga densiflora* Benth. (Leguminosae) to shade plantations, Siles *et al.* (2010) reported a reduction of up to 5°C in shaded coffee leaves vs. leaves in monocultures. Average maximum temperatures in non-shaded plantations in Mexico were 5.4°C higher that in shaded plantations (Barradas and Fanjul, 1986). In Kenya, Kirkpatrick (1935) found that between 11:00 AM and 1:00 PM, the temperatures in non-shaded coffee were up to 5–6°C higher than in shaded coffee. Also in Kenya, Jaramillo *et al.* (2013) reported that mean temperatures in shaded plantations were 2°C lower than in coffee growing at full sun. It has not been determined how these temperature differences translate to temperatures inside the berry, and consequently, on the coffee berry borer life history parameters.

3.7.1 Shade and Fungal Entomopathogens

One of the impediments towards the success of fungal entomopathogens is exposure to ultraviolet light and low humidity levels (Vega *et al.*, 2012b). Therefore, it is to be expected that under shade, fungal entomopathogens would have a higher chance of success than in unshaded plantations. Experimental results on this topic vary. For example, Chevalier (1947) recommended maintaining shade over coffee plantations in order to favor the development of fungal entomopathogens, a recommendation based on work by Friederichs and Bally (1923) in Java. Pascalet (1939) states that insect infection with *B. bassiana* will be much easier in shaded than in non-shaded plantations. Similarly, Féliz Matos (2003) recommended maintaining shade levels of 40–50% to encourage *B. bassiana* sporulation, although infection was almost non-existent in his study and not significantly different from levels encountered in dense shade or no shade. Vélez Arango (1997) reported no significant differences in *B. bassiana* recovery percentages (based in colony forming units) up to 14 days after spraying unshaded and shaded coffee plantations in Colombia. One interesting aspect of the study is that self-shading, due to dense foliage within individual plants, is mentioned as a factor that could have contributed to fungal survival due to a decrease in solar radiation reaching the fungal propagule.

3.7.2 Effects of Shade on the Effectiveness of Insecticides

Brun *et al.* (1990) found that resistance to endosulfan was significantly lower in shaded plantations, possibly due to (1) the lower temperatures (average of 3°C—see Brun and Suckling, 1992), which reduce the effectiveness of the insecticide and consequently the selective pressure for resistance, and (2) interruption of insecticidal sprays penetration by plant canopy in shaded plantations. Related to the last point, Parkin *et al.* (1992) reported more even deposition of insecticidal sprays in shaded plantations, possibly due to factors related to laminar flow whereby in non-shaded plantations the insecticidal mist can drift outside of the field due to unimpeded wind patterns.

3.7.3 Shade and Ants

Several papers have reported on the effects of shade on different ant species. For example, Armbrecht and Perfecto (2003) reported significantly different levels of litter and twig-nesting ants (e.g., *Pheidole*, *Solenopsis*, *Hypoponera*, *Wasmannia*, etc.) in Mexico when distance from the forest was compared for shaded monocultures (i.e., coffee under *Inga*) and shaded polycultures (coffee shaded with various tree species). For the shaded monoculture, ant species decreased with increased distance from the forest, while an increase in ant species was reported for the shaded polyculture with increased distance from the forest. Thus, even within one system (i.e., shaded coffee), various levels of different ant species can be found. This has important implications for the coffee berry borer because one particular shaded habitat may be more favorable towards ant species that might potentially prey on the insect when compared to a different habitat. Perfecto and Vandermeer (1996), Roberts *et al.* (2000), and Philpott and Armbrecht (2006) have also reported increased ant diversity in shaded coffee habitats.

In Colombia, Gallego Ropero and Armbrecht (2005) examined ant predation using coffee berry borer-infested parchment seeds placed under no shade or shade provided by different species (e.g., *Cedrela odorata* L., *Cordia alliodora* (Ruiz and Pavon) Oken, *Erythrina rubrinervia* Kunth, *Inga edulis* Mart., *Persea americana* Miller, etc.). Seven ant species in four genera (*Solenopsis*, *Tetramorium*, *Pheidole*, and *Myrmelachista*) were found inside the parchment seeds and a lower number of adult *H. hampei* were found when ants had access to infested seeds, and the number of adults was even lower in plantations under diverse shade. In a subsequent study in the same locations, Armbrecht and Gallego (2007) reported higher adult coffee berry borer predation in shaded coffee farms when a glass spiral trap was used. Ant predation is discussed in further detail below.

3.8 Rearing

Rearing large numbers of coffee berry borers is essential for obtaining large numbers of healthy insects of known age and developmental stage that can be used in controlled experiments, both in the field and in the laboratory. The use of coffee berries for coffee berry borer rearing presents many problems, including availability of berries and possible presence and development of fungi that could cause insect mortality (Villacorta, 1985; Pérez López et al., 1995). To avoid these problems, artificial diets have been developed to rear insects in the laboratory.

The first to develop an artificial diet for the coffee berry borer was Amador Villacorta (1985), a scientist at the Instituto Agronõmico do Paraná (IAPAR) in Brazil. Even though the main component in the 19-ingredient diet was cotton and not a single component included coffee, the diet was successful in rearing 30,000 insects per month and 15 generations of coffee berry borers. In contrast to Villacorta's (1985) diet, Bautista Martínez and Atkinson Martín (1988) used a diet consisting only of three components (ground coffee, distilled water, sorbic acid or methyl paraben) and observed insect burrowing into galleries, oviposition and hatching, but no pupae were formed.

The Villacorta (1985) diet was modified by Villacorta and Barrera (1993) and by Brun et al. (1993) to include ground coffee among its many ingredients. Subsequently, diet modifications and/or evaluations have been reported by Pérez López et al. (1995), Villacorta and Barrera (1996), Ruiz S. et al. (1996), Portilla-Reina (1999), Villacorta et al. (2000), Cirerol et al. (2002), Portilla R. and Streett (2006), and López-Pazos et al. (2009). Portilla R. and Streett (2006) developed an automated rearing system with the goal of rearing massive amounts of coffee berry borers and were able to rear ca. 900,000 females and males in 20 liters of diet.

The use of artificial diets for rearing coffee berry borers has also been instrumental for mass rearing parasitoids (Villacorta and Barrera, 1996; Portilla, 1999; Villacorta and Torrecillas, 2000). In lieu of using an artificial diet, Bautista Martínez and Atkinson Martín (1988), Benavides-G. and Portilla-R. (1990), Hirose and Neves (2002), and Jaramillo et al. (2009a) have developed rearing methodologies based on the use of coffee berries in the laboratory. Green coffee seeds and parchment coffee (i.e., coffee seeds removed from the berry and dried) have also been used for rearing the coffee berry borer (Benavides-G. and Portilla-R., 1990; Bustillo-Pardey et al., 1996; Portilla-Reina, 1999; Priyono et al., 2004). Finally, Friederichs (1924a) observed insects biting each other's legs off when crowded in a glass tube. This has occasionally been observed in artificial diet (Vega, unpubl.).

3.9 Sampling

Numerous sampling methods have been developed for estimating coffee berry borer population levels in the field (de Toledo, 1945; Decazy et al., 1989; Barrera et al., 1993a, 2004; Bustillo et al., 1998; Baker, 1999; Ruiz et al., 2000; Segura et al., 2004; Trujillo E. et al., 2006), including methods that account for losses due to infested berries that fall off the plant (Wegbe et al., 2003), as well as methods that correlate ethanol:methanol trap captures with infestation levels in the field (Pereira et al., 2012).

In a sampling method developed in Mexico (Barrera et al., 1993a), a plantation is divided into plots of approximately 4 ha/each (if the plot is <4 ha, then it does not need to be divided further). In each plot, 20 sampling sites representative of areas throughout the entire plot sites are selected. In each site, five coffee plants in a row are selected and from each plant a branch in the central part of the plant is selected. On this branch, all the berries are examined and the numbers that are infested and non-infested are counted. This method allows to sample different plots without bias, and to determine infestation levels in each plot. Another sampling method involves selecting 20 sites in a 1–5 ha plantation and in each site five plants in a row are selected and 20 berries per plant examined for infestation (Barrera, 2008).

According to Leefmans (1923), "Experiments with light traps did not give appreciable results; the beetles are practically not attracted to light." Subsequent studies have shed more light on this topic. Giordanengo (1992) found that 1, 7, and 14-day-old virgin females are negatively phototropic, although the level of negative phototropism diminishes as the insect becomes older. For mated colonizing females, phototropic responses ranged from 62% at 1 h, to 40% at 3 h, and 53% at 24 h. Giordanengo et al. (1993) also used the phototropic response to collect females as they emerge from berries. Infested berries were placed inside a black container to which an empty plastic tube was connected and into which the females would walk, attracted by the light.

More detailed light-related experiments were conducted by Chong et al. (2006), who tested 14 wavelengths ranging form 340 to 670 nm. Six wavelengths, between 400 and 540 nm, resulted in the highest attraction percentage for females. In addition, at 460, 490 and 520 nm, 90-day-old females had a stronger response than 45-day-old females. Therefore, mating status and age influence female coffee berry borer responses to light. We could not find any papers on the use of light traps to attract coffee berry borers in the field.

3.10 Traps and Attractants

Several different coffee berry borer traps have been developed and tested in various coffee producing countries, e.g., the IAPAR trap in Brazil (Villacorta *et al.*, 2001; Pereira *et al.*, 2012); the Ecobroca, ECOIAPAR and ETOTRAP in Mexico (Velasco Pascual *et al.*, 1997a; Barrera *et al.*, 2008); the tip CENICAFE trap in Colombia (Cárdenas M., 2000); and the Fiesta trap in Costa Rica (Borbón-Martínez *et al.*, 2000; Barrera *et al.*, 2006), among others (Figure 11.7). Most of these traps are made by hand (and referred to as artisanal traps) using empty 2-liter plastic soda bottles (IAPAR, ECOIAPAR, ETOTRAP, tip CEN-ICAFE) or with plastic cups (Fiesta). One trap, designed by scientists at PROCAFE (El Salvador) and CIRAD (France) (González and Dufour, 2000; Dufour *et al.*, 2002), is commercially available under the name BROCAP® and has been used in many countries throughout Latin America.

Many factors will influence the efficacy of coffee berry borer trapping devices, including trap color, shape, placement in the field, and attractant used (Dufour, 2002; da Silva *et al.*, 2006; Barrera *et al.*, 2006). The literature dealing with these topics is quite extensive, and because there are so many different trapping devices it is necessary to cover some topics briefly. For example, in terms of trap color, Mathieu *et al.* (1997b), Saravanan and Chozhan (2003), and Dufour and Frérot (2008) reported that red traps result in higher insect capture than the other colors tested in contrast to Borbón-Martínez *et al.* (2000) finding higher captures in white traps. As for vertical placement of the traps in the field, Fernández and Cordero (2005) found no significant differences in trap captures when traps were placed at 0.2 and 1 m in height, while Uemura-Lima *et al.* (2010) obtained significantly higher captures in traps placed at 0.5 m in height, when compared to 1 and 1.5 m. Dufour and Frérot (2008) captured three times as many insects when the traps were 1.2 m high, than when they were placed at ground level.

Ethanol, which is produced by plants under stress (Kimmerer and Kozlowski, 1982), has been shown to serve as an attractant for bark beetles in general (Cade *et al.*, 1970; Moeck, 1970; Montgomery and Wargo, 1983; Klimetzek *et al.*, 1986; Chénier and Philogène, 1989; Byers, 1992; Miller and Rabaglia, 2009; Gandhi *et al.*, 2010; Kelsey *et al.*, 2013). For the most part, all coffee berry borer traps use a mixture of methanol and ethanol as the attractant (Brun and Mathieu, 1997; Mathieu *et al.*, 1997b, 1999; Cárdenas M., 2000; Borbón-Martínez *et al.*, 2000; Saravanan and Chozhan, 2003; Fernández and Cordero, 2005; Dufour and Frérot, 2008; Barrera *et al.*, 2008; Agramont *et al.*, 2010; Fernandes *et al.*, 2011; Pereira *et al.*, 2012; Messing, 2012; Suárez *et al.*, 2013). Mendoza-Mora (1991) was the first to demonstrate the synergistic effect of a 3:1 mixture of methanol to ethanol in attracting the coffee berry borer.

As mentioned in Section 3.6.3, coffee berry borer levels per hectare can reach the millions. The main problem with the coffee berry borer traps is that they only capture a low

FIGURE 11.7   Coffee berry borer traps. (A) BROCAP® trap; (B) IAPAR trap (Brazil); (C) Fiesta trap (Costa Rica); and (D) Trampa Brocap casera (Mexico). *Photos by: (A) Bernard Dufour (CIRAD); (B) E. F. da Silva (IAPAR); (C) Fernando E. Vega; (D) Jaime Gómez (ECOSUR).*

percentage of the population. For example, insect captures per trap in various countries using methanol:ethanol mixtures follow: (1) in Mexico, weekly captures ranged from 83 to 1484 (Barrera *et al.*, 2008); (2) in Brazil, captures ranged from 11 to 87 insects per day, depending on trap height (Uemura-Lima *et al.*, 2010); (3) in Bolivia, 10 day insect captures averaged 3414 insects (Agramont *et al.*, 2010); (4) in Venezuela, weekly captures averaged 432 insects (Fernández and Cordero, 2005); (5) in Cuba, weekly captures averaged 205 insects (Moreno Rodríguez *et al.*, 2010); (6) in India, 2 week captures ranged from 18 to 303 insects (Saravanan and Chozhan, 2003); (7) in Costa Rica, daily captures ranged from ca. 18–33 insects (Borbón-Martínez *et al.*, 2000); and (8) in El Salvador, daily BROCAP® trap captures for different concentrations and composition of attractants tested ranged from 6 to 111 insects (González and Dufour, 2000). In contrast, Dufour (2002) states that when infestations are high and during periods of high migration, the BROCAP® trap can capture more than 10,000 coffee berry borers per trap per day. Barrera *et al.* (2006) compared the BROCAP® trap to the ECOIAPAR and Fiesta traps, with the BROCAP® trap capturing 2653 insects per trap per week, corresponding to 2.4 and 3.2 the captures obtained in the ECOIAPAR and Fiesta traps, respectively.

In addition to low captures, a more significant issue is that with few exceptions (Mathieu *et al.*, 1999; Fernandes *et al.*, 2011; Pereira *et al.*, 2012) trap captures are not correlated to actual coffee berry borer infestation levels in the field. Therefore, growers do not have any idea as to how effective trap captures are in reducing insect numbers and in increasing yields. It is clear that traps could be useful for monitoring the presence and dispersal of the coffee berry borer, especially when it is first reported in an area. Traps can also help monitor movement and increase in numbers of colonizing females throughout the season (as described by Pereira *et al.*, 2012), but at present trapping devices are not a practical pest management strategy against the coffee berry borer.

Another issue to consider is that, as mentioned before, alcohol-based traps are not specific to the coffee berry borer and, therefore, other bark beetles will also be trapped. Pereira *et al.* (2012) reported from 2850 to 9048 non-coffee berry borers collected in traps placed in four different fields. Consequently, insects will need to be manually sorted and identified, a tedious process that increases labor costs (Messing, 2012). Another cost involves servicing the traps to replenish the attractant. Messing (2012) obtained higher captures when the attractant was placed in a plastic pouch that required less servicing, but the plastic pouch on its own involves additional costs.

Based on finding various coffee berry borer life stages in *D. lacourtianum* (discussed above), Ghesquière (1933) suggested the use of this plant as a trap crop in the field, an experiment that has not been conducted in the 80 years that have passed since the publication of the paper.

There is an urgent need to identify and deploy attractants specific to the coffee berry borer. These should have a reasonable cost, be easy to use, and must result in dramatically higher captures than the currently used trapping methods.

3.11 Repellents

Some papers have provided field and laboratory evidence suggesting that coffee berry borers infesting berries might produce a female deterrent chemical. Wilkinson (1928) was the first to observe reduced offspring when a high number of coffee berry borers were infesting a limited number of berries. In Ivory Coast, Ticheler (1961, 1963) found that increasing the number of females per berry in the laboratory from one, to two, and to four, greatly reduced fecundity per female, with an average progeny number, 32 days post-infestation, of 44, 12, and 6, respectively. A decline in per capita fecundity with increased female density per berry in the laboratory was also reported by Moore *et al.* (1990). In a field experiment in Mexico, de Kraker (1988) obtained data that "suggests that already infested berries repel attacking borers." The possible repellency was further expanded upon by Gutiérrez-Martínez and Ondarza (1996), who hypothesized that a "chemical signal" deposited on the berry by a colonizing female might function as a deterrent for subsequent colonization by other females. Such a "chemical signal" could be a marking pheromone, as proposed by Vega *et al.* (2009). Marking pheromones can affect fecundity, as has been shown for many insects, including bark beetles (see Vega *et al.*, 2011).

A laboratory study using coffee berry borers in artificial diet revealed a reduction in fecundity as the number of females increased (Vega *et al.*, 2011), confirming the findings of Ticheler (1961, 1963) and Moore *et al.* (1990). The artificial diet study was expanded to include infested berries, leading to the identification of a sesquiterpene, which acts as a female deterrent in the laboratory (Vega and Cossé, unpubl.). This latter finding provides support to the repellency hypotheses presented by de Kraker (1988) and Gutiérrez-Martínez and Ondarza (1996).

Borbón-Martínez *et al.* (2000) reported various levels of repellency towards the coffee berry borer by verbenone (4,6,6-trimethylbicyclo [3.1.1]hept-3-en-2-one), (Z)-3-hexenol, and methylcyclohexenone (3-methylcyclohex-2-en-1-one). These compounds have been identified as anti-aggregation pheromones in other bark beetles (Byers, 1995; Zhang and Schlyter, 2004; Reddy and Guerrero, 2010; Strand *et al.*, 2012; Fettig *et al.*, 2012). Jaramillo *et al.* (2013) reported coffee berry borer avoidance of verbenone and α-pinene and recommended "cultivating coffee intercropped with plants producing conifer monoterpenes compounds that are repellent to *H. hampei*."

Góngora *et al.* (2012) identified high overexpression of the isoprene synthase gene in *Coffea liberica* Hiern berries exposed to the coffee berry borer, but not in *C. arabica* berries. Addition of isoprene to artificial diet had negative effects on survival and development of the coffee berry borer and the authors suggest that isoprene might have repellent properties against the insect. A concern with this concept is that isoprene is highly reactive and is produced and emitted by many species of trees in large amounts (Monson and Fall, 1989; Hewitt *et al.*, 2011). Therefore, with so much isoprene in the atmosphere it is unlikely that it could be used as a repellent against the coffee berry borer. Finally, isoprene is very volatile, highly flammable, and, consequently, not easy to handle.

The identification and field deployment of highly repellent volatile compounds could become another useful tool in the arsenal of strategies used to manage the coffee berry borer.

3.12 Plant Resistance

No resistance to the coffee berry borer has been reported in commercially traded coffee varieties (Friederichs, 1924b; Vuillet, 1925; Romero and Cortina-Guerrero, 2004a; Sera *et al.*, 2010). Two studies have reported some level of resistance in laboratory experiments using parchment coffee as the experimental unit, but it is not clear how the results translate to the field when insects have to select a berry (Romero and Cortina-Guerrero, 2004b; Romero and Cortina G., 2007).

In an attempt to elucidate differences in plant responses to coffee berry borer infestation, Idárraga *et al.* (2012) identified metabolic pathways induced in *C. arabica* var. Caturra and *C. liberica* after plants had been artificially infested with the insect for 24 h. *Coffea arabica* responded with an increased expression of stress-related proteins while *C. liberica* had increased expression of proteins related to insect defense. This information is useful for developing possible resistance strategies.

One area related to plant resistance to the coffee berry borer involves the digestive enzymes used by the insect and developing transgenic coffee plants with genes coding for inhibitors of these enzymes. About 50% of the dry weight in green coffee seeds is constituted by polysaccharides, including mannan, arabinogalactan (arabinose and galactose), and cellulose (a linear chain of glucose units) (Bradbury and Halliday, 1990; Redgwell *et al.*, 2003; Redgwell and Fischer, 2006). In order to metabolize these polymers as they move through the alimentary canal, the coffee berry borer needs to use enzymes such as amylases, mannanases, and galactosidases, among others. For a detailed description of the coffee berry borer alimentary canal see Rubio G. *et al.* (2008) and Ceja-Navarro *et al.* (2012).

The α-amylases are a family of carbohydrate-metabolizing enzymes common in plants, microorganisms, and animals (Grossi-de-Sá and Chrispeels, 1997).

Extensive research has been published on α-amylase inhibitors in insects (Grossi-de-Sá and Chrispeels, 1997; Carlini and Grossi-de-Sá, 2002; Strobl *et al.*, 1998; Zeng *et al.*, 2013; and references therein) and various papers have elucidated basic aspects of amylase presence in the midgut of the coffee berry borer (Valencia-Jiménez *et al.*, 1994; Martínez D. *et al.*, 2000; Martínez Díaz *et al.*, 2000; Valencia *et al.*, 2000; Valencia-Jiménez, 2000). A common bean (*Phaseolus vulgaris* L.) seed protein crude extract was shown to have amylase inhibitory properties when it reduced the α-amylase activity in whole coffee berry borer extracts by ca. 80% (Valencia *et al.*, 2000). An α-amylase inhibitory gene from *Phaseolus coccineus* L. expressed in tobacco plants inhibited α-amylase activity in whole coffee berry borer insect extracts by 65% (de Azevedo Pereira *et al.*, 2006).

Acuña *et al.* (2012) presented evidence for a horizontally transferred bacterial gene encoding mannanase (*HhMAN1*) in the genome of the coffee berry borer. The gene, which presumably allows the coffee berry borer to use seed galactomannans (a type of mannan, consisting of galactose and mannose), apparently originated from bacteria inhabiting the alimentary canal and could be detected in specimens originating in 16 countries. The horizontal transfer of this gene serves as an example of the biological complexity of the coffee berry borer, including the importance of the microbiota. Aguilera-Gálvez *et al.* (2013) cloned and characterized the mannanase gene (*HhMAN1*) from the midgut of the coffee berry borer, while Padilla-Hurtado *et al.* (2012) identified a xylanase gene (*HhXyl*) in the alimentary canal of the coffee berry borer; xylanases metabolize arabinoxylans (a copolymer of arabinose and xylose) present in the coffee seed.

In addition to polysaccharides, green coffee beans contain approximately 8.5–12% crude protein content (Rawel *et al.*, 2005). Proteases, also referred to as proteinases, are essential in order for the coffee berry borer to metabolize coffee seed proteins as they move through the alimentary canal. Proteases, including serine, cysteine, and aspartic, can be inhibited with plant protease inhibitors, which can be a valuable source of resistance against insect pests (Grossi-de-Sá and Chrispeels, 1997; Carlini and Grossi-de-sá, 2002; Bode *et al.*, 2013; da Silva *et al.*, 2014; and references therein). The first paper focused on understanding the role of proteases in the digestive system of the coffee berry borer was published by Valencia-Jiménez *et al.* (1994), who reported a high activity of trypsin and chymotrypsin (two serine proteases) in coffee berry borer larvae, and only a small amount of trypsin activity in adults. Ruiz Serna *et al.* (1995) ran bioassays with various commercial trypsins, chitinases, and trypsin-chymotrypsin inhibitors incorporated into coffee berry borer artificial diet and found significant differences in mortality levels when compared to the control, but not in the time from egg hatch to adult emergence. Preciado-Rodríguez *et al.* (2000) identified an aspartic protease in the coffee berry borer midgut. An aspartic protease inhibitor from *Lupinus bogotensis* Benth. was "highly effective" in inhibiting coffee berry borer midgut aspartic proteases in *in vitro* experiments (Molina *et al.*, 2014). In addition, two concentrations of the inhibitor separately mixed into artificial diet resulted in significant differences in larval mortality when compared to the control, depending on the concentration. Similarly, *in vitro* studies showed that a serine protease inhibitor from *P. coccineus* inhibited trypsin-like enzymes in the coffee berry borer (Azevedo Pereira *et al.*, 2007).

What this type of enzyme-related research is aiming for is transgenic coffee plants expressing genes codifying for amylase or protease inhibitors (Valencia-Jiménez *et al.*, 1994; Valencia *et al.*, 2000; Martínez D. *et al.*, 2000; de Azevedo Pereira *et al.*, 2006; Barbosa *et al.*, 2010). One such example is transgenic *C. arabica* with an α-amylase inhibitor-1 gene from *P. vulgaris* (Barbosa *et al.*, 2010). Use of transgenic seed extracts fed to the coffee berry borer resulted in up to 88% inhibition of α-amylase enzymatic activity. Even though the idea of using amylase inhibitors is straightforward, transfer of this type of technology to the field remains a challenge.

3.13 Endosulfan Resistance

Endosulfan ($C_9H_6Cl_6O_3S$), a broad-spectrum chlorinated cyclodiene insecticide, first entered the market in the mid-1950s. It has been used in many countries under the trade name Thiodan® as an insecticide against the coffee berry borer. Due to human and environmental hazards related to its use, including bioaccumulation, endosulfan has now been banned in at least 70 countries (Lubick, 2010; Janssen, 2011).

Ingram (1968) experimented with endosulfan in Uganda and found that it could have fumigant effects, i.e., contact toxicity did not appear to be essential. After 10 years of biannual applications, coffee berry borer resistance to endosulfan was reported in New Caledonia (Brun *et al.*, 1989a), with up to 1000-fold resistance detected in five localities. The development of resistance could have been due to a higher selective pressure, based on the higher levels of active ingredient used when compared to other countries (Brun *et al.*, 1989a). Another possibility the authors discuss is that fumigant action could have enhanced the development of resistance in all life stages of the insect inside the coffee berry. Finally, Brun *et al.* (1989a) hypothesized that selective pressure for resistance would be higher in sun grown plantations in contrast to shaded plantations, due to better spray coverage and higher temperatures, which would increase the fumigant action. In a subsequent paper, Brun *et al.* (1990) reported a significantly higher percentage

of resistant insects in sun grown plantations vs. shade plantations.

In order to assess and subsequently manage resistance in the field, Brun *et al.* (1989b) developed three methods for detecting endosulfan resistance, of which a method based on vapor action was the most convenient due to its low cost, reproducibility, and ease of use. This method was further developed by exposing coffee berry borers to endosulfan vapors at five different temperatures (Brun *et al.*, 1991). Another method involves the molecular detection of the cyclodiene resistance gene *Rdl*, which is present in coffee berry borers from New Caledonia (ffrench-Constant *et al.*, 1994; Borsa and Kjellberg, 1996b; Andreev *et al.*, 1998) and Colombia (Góngora *et al.*, 2001; Navarro *et al.*, 2010).

With resistance management in mind, Brun and Suckling (1992) conducted a study involving the traditional endosulfan application method in New Caledonia, which is based on roadside applications using sprayers mounted on vehicles. The findings showed reduced resistance frequency away from the road in both sun grown and shaded coffee. Two applications of endosulfan in a sun grown field resulted in a 61% increase in the frequency of resistance, implying it would be unwise to apply endosulfan in areas where resistance frequency is low (Brun and Suckling, 1992). The roadside vehicle-mounted sprayer results in most of the spray being deposited within 20 m of the point of application (Parkin *et al.*, 1992), thus confirming a higher selective pressure close to the road.

In Nicaragua, Pérez *et al.* (2000) used the bioassay method developed by Brun *et al.* (1991) and could not find evidence for endosulfan resistance in the coffee berry borer. Specimens from the Philippines, Guatemala, Brazil, and Cameroon also tested negative for endosulfan resistance (Kern *et al.*, 1991).

Endosulfan resistance has been used to study the segregation of resistance phenotypes, functional haplodiploidy, pseudoarrhenotoky, and extreme inbreeding (Brun *et al.*, 1995b; Borsa and Kjellberg, 1996b, Borsa and Coustau, 1996; Gingerich *et al.*, 1996; Andreev *et al.*, 1998).

3.14 Biological Control

3.14.1 Bacteria

A well-known entomopathogen is *Bacillus thuringiensis* (Bt), a Gram-positive bacterium that during sporulation produces crystal proteins known as delta endotoxins (δ-endotoxins; also referred to as Cry toxins; Jurat-Fuentes and Jackson, 2012). The toxin becomes activated in the midgut and disrupts the midgut epithelial cells, causing death of the insect.

After the coffee berry borer was first reported in Costa Rica in 2002, Arrieta *et al.* (2004) reported 202 Bt isolates in environmental samples (soil, leaf litter, leaves, coffee berries, coffee berry borers) collected in coffee berry borer-infested coffee plantations. Even though no laboratory bioassays were conducted with any of the isolates, the study reveals the widespread presence of a potential biological control agent in coffee agroecosystems.

Most laboratory bioassays examining toxicity of Bt to the coffee berry borer have used first instar larvae feeding on artificial diet surface contaminated with spore-crystal suspensions. Using Bt isolates from a Mexican collection, Méndez-López et al. (2003) found that out of 170 isolates, only the mosquitocidal strains, especially Bt serovar israelensis, exhibited significant toxicity. They also reported high toxicity when using four out of nine mosquitocidal strains from the Institut Pasteur collection. De la Rosa et al. (2005) used 61 Mexican Bt isolates from the same collection used by Méndez-López et al. (2003) and found toxicity levels ranging from 8 to 83%. López-Pazos et al. (2009) reported moderate levels of toxicity of recombinant Cry1Ba (from Bt serovar aizawai) and Cry3Aa (from Bt serovar san diego) proteins. López-Pazos et al. (2010) reported no toxicity of a hybrid Cry protein in contrast to the parental toxins Cry1B and Cry1I, which caused 60% and 52% mortality, respectively. In contrast to results published by Méndez- López et al. (2003), Naidu et al. (2001) reported Bt serovar sumiyoshiensis was toxic to larvae of the coffee berry borer.

The main issue with using Bt as a biological control strategy against the coffee berry borer is that the insect needs to ingest the bacterium or toxin for an effect to occur, and, furthermore, epidemics would need to be induced to result in significant reductions in population levels. Thus, the use of Bt in the field presents a formidable challenge, i.e., reaching the insects feeding inside the berry. It is also important to consider that use of commercial formulations of Bt would require spraying, a non-feasible option in most of the coffee-producing world due to cost and storage of the product, difficulty in properly spraying the entire plantation (e.g., steep hills), and easy access to water.

One obvious alternative strategy to spraying is the use of transgenic coffee plants expressing the Bt toxin protein, but whether this is a viable option for the coffee industry, in terms of grower and consumer acceptance, remains a debatable question. For example, a French team had developed transgenic C. canephora plants expressing the Bt toxin protein (Leroy et al., 1997) and planted them in French Guiana to test their effectiveness against the coffee leaf miner (Leucoptera coffeella Guérin-Mèneville and Perrottet) (Perthuis et al., 2005). Even though resistance to the coffee leaf miner was demonstrated in the field (Perthuis et al., 2005), the plantation was vandalized in 2004, bringing the project to an early demise (Coghlan, 2005).

As for other bacteria, Cárdenas (1995) mentions Serratia marcescens as an "antagonist" (i.e., opportunistic or potential pathogen) of coffee berry borer larvae and pupae. Bustillo et al. (1998) includes Serratia sp. and Bacillus sp. as uncommon natural enemies of larvae only found while dissecting infested berries.

3.14.2 Fungal Entomopathogens

The mode of action of fungal entomopathogens involves spore attachment to the insect cuticle, followed by germination and cuticle penetration (Vega et al., 2012b). Insect death is caused by hyphal growth and proliferation throughout the hemocoel, a process that depletes nutrients used by the insect and disrupts internal tissues; production of secondary metabolites could also contribute to death (Vega et al., 2012b).

In recent years, molecular research has vastly altered the phylogenetics and systematics of fungal entomopathogens, with major changes for Beavueria (Rehner et al., 2011) as well as Metarhizium (Bischoff et al., 2009), the two best-known fungal entomopathogens. In addition to these changes, the reader should become familiar with the new "one fungus, one name" concept (Taylor, 2011) that no longer uses two different scientific names for the same fungus based on sexual (teleomorphic) or asexual (anamorphic) reproduction (Gams et al., 2012). Being familiar with this change will help figure out what the new—and correct—scientific names are, and how they relate to the previously used names.

With one exception, all fungal entomopathogens attacking the coffee berry borer belong to the Phylum Ascomycota, Class Sordariomycetes, Order Hypocreales, Family Cordycipitaceae: Beauveria bassiana (Balsamo-Crivelli) Vuillemin, Metarhizium anisopliae (Metschn.) Sorokin, Isaria farinosa (Holmsk.) Fr. (formerly known as Paecilomyces farinosus (Holmsk.) Brown and Smith), Isaria fumosorosea Wize (formerly known as Paecilomyces fumosoroseus (Wize) Brown and Smith), and Lecanicillium lecanii (Zimm.) Zare and Gams (formerly known as Verticillium lecanii (Zimm.) Viégas). The exception is Ophiocordyceps entomorrhiza (Dicks.) Sung, Sung, Hywel-Jones and Spatafora (formerly known as Hirsutella eleutheratorum (Nees) Petch), which belongs to the Family Ophiocordycipitaceae (Bustillo et al., 1998, 2002; Vega et al., 1999).

Some papers list Nomuraea rileyi (Farlow) Samson (Clavicipitaceae) as a fungal entomopathogen of the coffee berry borer (Moore and Prior, 1988; Waterhouse and Norris, 1989; Klein-Koch, 1989; Murphy and Moore, 1990; Waterhouse, 1998; Damon, 2000). The reports are likely based on Le Pelley (1968), who stated that N. rileyi occurs in Brazil, while citing Averna-Saccá (1930). In actuality, what Averna-Saccá (1930) reported was the isolation of the fungus, at the time known as Botrytis rileyi (Farlow), from a twig borer attacking Melia azedarach L., followed by attempts to inoculate the coffee berry borer. Finding *N. rileyi* attacking the coffee berry borer would be highly unusual, as it is a common pathogen of Lepidoptera (Humber, 2012).

Beauveria bassiana has been the most commonly reported fungal entomopathogen infecting the coffee berry borer worldwide (Figure 11.8A–C): Brazil (Averna-Saccá, 1930; Drummond-Gonçalves, 1940; Mesquita, 1944; Villacorta, 1984; Costa *et al.*, 2002); Cameroon (Pascalet, 1939); Colombia (Vélez-Arango and Benavides-Gómez, 1990); Costa Rica (Echeverría Beirute, 2006); Cuba (Pérez León *et al.*, 2009); Democratic Republic of the Congo (Steyaert, 1935); Ecuador (Klein-Koch *et al.*, 1988); Guatemala (Monterroso Mayorga, 1981); Honduras (Lazo A., 1990); India (Haraprasad *et al.*, 2001); Indonesia (Friederichs, 1922b; Friederichs and Bally, 1922, 1923); Mexico (Méndez-López, 1990; Sampedro-Rosas *et al.*,

2008); Nicaragua (Monzón *et al.*, 2008); Puerto Rico (Gallardo-Covas *et al.*, 2010); and Venezuela (Bautista, 2000), among others.

In the first paper related to *Beauveria* and the coffee berry borer, Friederichs (1922b) reported a large natural epidemic in the field in Java, which greatly reduced infestation levels. In the Democratic Republic of the Congo, Steyaert (1935) found much higher *B. bassiana*-induced mortality in insects originating in green berries (highest monthly level: 64%) than in red berries (highest monthly level: 29%), a result he ascribes to females becoming infected while walking on the surface of the green berry, and moving among these, before eventually selecting a suitable berry. Pascalet (1939) concurs in that infection only occurs while females are outside the berry and while in the process of boring into the berry. In contrast to Steyaert's (1935) results, *B. bassiana* infection in green berries in

FIGURE 11.8 (A) Coffee berries, some of which show coffee berry borers infected with the fungal entomopathogen *Beauveria bassiana* (white areas on disc). (B) Red berry showing posterior part of the coffee berry borer abdomen infected with *B. bassiana*. (C) *B. bassiana* growing in culture. (D) Third-stage juveniles of the nematode *Metaparasitylenchus hypothenemi* emerging from an adult coffee berry borer. *Photos by: (A) Fernando E. Vega; (B) Aixa Ramirez Lluch (Departamento de Agricultura, Puerto Rico); (C) Keith Weller (USDA), and (D) Alfredo Castillo (ECOSUR).*

Ecuador was 15.9% vs. 13.1% in red berries, and 5% in black berries (Molinari, 1988).

Steyaert (1935) also found highest *B. bassiana*-induced mortality in insects in shaded (8.2–13.6%) vs. unshaded plots (1.8–3%), which is in agreement with Friederichs (1922b) and Friederichs and Bally (1923). Pascalet (1939) observed that *B. bassiana* was more widespread in forest areas than in the savanna and that shade was favorable for the development of the fungus. This is likely a result of higher fungal survival due to reduced ultraviolet light and higher moisture.

Coffee berry borer mortality levels caused by natural occurrences of *B. bassiana* in the field can vary widely (percentages given are highest levels reported in each study): 71% in Cameroon (Mbang *et al.*, 2012); 63% in the Democratic Republic of the Congo (Steyaert, 1935); 60% in India (Balakrishnan *et al.*, 1994); 44% in Nicaragua (Monzón *et al.*, 2008); 42% in Colombia (Posada-Flórez *et al.*, 1993); 30% in Ecuador (Klein-Koch *et al.*, 1988); <10% in Mexico (Méndez-López, 1990; Córdova-Gámez, 1995); and <1% in Brazil (Costa *et al.*, 2002) and Puerto Rico (Gallardo-Covas *et al.*, 2010).

In the field, *B. bassiana*-infected females can often be seen covered with white sporulating mycelia while fixed at the entrance hole in the berry, with the anterior part of the body sticking out of the hole (Pascalet, 1939; Villacorta, 1984; Figure 11.8A, B). Several possibilities exist for this situation. It is possible that females might have become infected while selecting a berry into which they can oviposit, and that they die in the process of boring the hole, followed by fungal sporulation on the cadaver. Another possibility is that described by Pascalet (1939). He mentions that in some cases, depending upon stage of infection, females move to the entrance hole and die, while the progeny continues its development inside the berry, free of infection. This argument presents the quandary that the progeny has to emerge from the berry through the entrance hole, which would then be blocked. A third possibility is that this phenomenon is the result of fungal manipulation of the insect to an area where the spores have a higher chance of finding a host, as has been reported for other fungal entomopathogens (Krasnoff *et al.*, 1995).

Several papers have reported results of laboratory bioassays aimed at testing the pathogenicity of *B. bassiana*: Fernandes *et al.* (1985); Jiménez-Gómez (1992); González G. *et al.* (1993); de la Rosa *et al.* (1997); Varela Ramírez (1997); Bustillo *et al.* (1999); Haraprasad *et al.* (2001); Samuels *et al.* (2002); Posada and Vega (2005); Neves and Hirose (2005); Sampedro-Rosas *et al.* (2008); and Vera *et al.* (2011). A common goal in these bioassay studies is to select the most virulent isolates for subsequent spraying and testing in the field, even though the bioassays are conducted under ideal conditions (e.g., constant temperatures, 100% RH, protection from ultraviolet light) that do not mimic field conditions. Thus, it is questionable whether laboratory results will be similar in the field. In addition, there are many parameters that are not considered in most laboratory bioassays. For example, Posada and Vega (2005) conducted bioassays using 50 different *B. bassiana* isolates from coffee berry borers collected in Cameroon, Ivory Coast, Kenya, Togo, Brazil, Mexico, and Nicaragua. In addition to determining normal parameters assessed in usual bioassay studies (percent mortality, survival time), they assessed spore germination, length of duration of the fungal life cycle in the insect, and spore production in the insect cadaver.

Other studies have focused on the enzymes produced by *B. bassiana* infecting the coffee berry borer (Rivera M. *et al.*, 1997; Varela-Ramírez, 1997; Castellanos Domínguez, 1997; Ito *et al.*, 2007; Dias *et al.*, 2008; Sassá *et al.*, 2008, 2009; Varéa *et al.*, 2012). Enzymatic action by the fungus is needed to breach the cuticle and enter the insect. In one study (Bridge *et al.*, 1990), enzymatic band patterns were used in an attempt to determine if 16 *B. bassiana* isolates attacking the coffee berry borer in 10 countries had disseminated with the insect, or whether isolates attacking the insect throughout the world were similar, with inconclusive results. In a similar study using molecular markers, Rehner *et al.* (2006) assessed the phylogenetic diversity of 34 coffee berry borer-infecting *B. bassiana* isolates from four African countries and five Neotropical countries. Results revealed that *B. bassiana sensu lato* is comprised by cryptic species, with four distinct lineages infecting the insect in Africa and the Neotropics. Also using molecular markers, Gaitan *et al.* (2002) detected low genetic variability among 49 *B. bassiana* isolates from coffee berry borers from Colombia, the Philippines, Brazil, Ecuador, and Guatemala.

Various papers have examined coffee berry borer infection levels in the field after spraying *B. bassiana* conidial suspensions (Vélez-Arango and Benavides-Gómez, 1990; Bustillo *et al.*, 1991, 1999; Tobar H. *et al.*, 1998; de la Rosa *et al.*, 2000; Haraprasad *et al.*, 2001; Montilla *et al.*, 2006). Unfortunately, these studies did not include cost to benefit analyses. Obtaining answers to the following questions is essential in order to determine whether a recommendation to the grower is warranted: (1) Is there increased insect mortality in sprayed plots compared to plots that were not sprayed?; (2) Are yields higher in sprayed plots?; (3) What costs, in terms of labor and material, are incurred when spraying?; and (4) Is spraying cost effective?

In a novel approach to using fungal entomopathogens, Cruz *et al.* (2006) developed the concept of using mixtures consisting of different *B. bassiana* strains. They conducted laboratory bioassays based on 10 *B. bassiana* strains from eight different hosts in four countries, all characterized for genetic diversity using different molecular methods (internally transcribed spacer region, β-tubulin gene, and

AFLP). Based on single strain bioassay results, five *B. bassiana* mixtures were designed to hypothetically take advantage of genetic diversity. Results revealed co-infections were prevalent when mixtures were used, and if strains in a mixture were genetically similar, virulence was similar to that obtained when single strains were used. In a very interesting twist, if strains in the mixture were not genetically similar but had caused similar virulence when individually tested, antagonism was observed if virulence was individually high and synergism if individual virulence was low. A field test using artificial infestations (i.e., insects were introduced into entomological sleeves) confirmed the synergism when a mixture of low virulence strains was used (Cárdenas-Ramírez *et al.*, 2007; Benavides *et al.*, 2012). Gene expression profiles of *B. bassiana* germinating conidia and growing hyphae on coffee berry borers was studied by Mantilla *et al.* (2012).

Coffee berry borer field and laboratory bioassays have also been conducted using *M. anisopliae*: D'Antonio and de Paula (1979); Lecuona *et al.* (1986); Bernal U. *et al.* (1994); de la Rosa-Reyes *et al.* (1995); de la Rosa *et al.* (2000); Bustillo *et al.* (1999); and Samuels *et al.* (2002). Pava-Ripoll *et al.* (2008) expressed a scorpion neurotoxin coding sequence in transformed *M. anisopliae*, resulting in higher virulence against the coffee berry borer.

Overall, there are many constraints to the effective use of fungal entomopathogens using traditional spraying methods. These include the inherent susceptibility of the fungus to low moisture levels and to UV light (Vega *et al.*, 2012b). Edgington *et al.* (2000) tested 22 substances as *B. bassiana* UV protectants, and two that were tested in the field did not improve coffee berry borer control. In addition, spraying fungal suspension requires ready access to water throughout the plantation, which can be difficult. Carrying a five-gallon (18.9 liters) back-sprayer over steep hills can quickly become a burden based on weight alone, i.e., 41.7 lb (18.9 kg). More importantly, spraying to reach an insect that has a cryptic life cycle is a great challenge and spraying must be done when the insect is boring the berry, i.e., ca. 90–120 days after flowering, but this is complicated by numerous flowering periods induced by rain.

The cost of a commercial product, which could be prohibitive for a grower, is another factor to consider, although artisanal production methods have been developed. These are usually based on using rice as a solid substrate inside glass bottles, where the fungus can be grown (Antía-Londoño *et al.*, 1992; Posada F. and Bustillo P., 1994). Production of large amounts of fungal entomopathogens would require a production facility, whose cost can be quite high (Grimm, 2001). Growing *B. bassiana* in liquid culture, followed by inoculating solid substrates such as cooked rice, has been demonstrated by Posada Flórez (2008), but the amounts of rice needed to produce high levels of inoculum for field application are too high to be practical.

3.14.3 Fungal Endophytes

A non-traditional method for using *B. bassiana*, as well as other fungal entomopathogens, is to attempt to establish them as fungal endophytes, i.e., as fungi that live internally in the plant (Posada and Vega, 2006; Posada *et al.* 2007; Vega *et al.* 2008b, c; Vega, 2008b). Various fungal entomopathogens have been reported as endophytes (see Vega *et al.*, 2008b) and three methods used to inoculate coffee plants with *B. bassiana* (spraying, injecting, and drenching the soil) were partially effective (Posada *et al.*, 2007), as recovery was confirmed but establishment was not long lasting. This lack of establishment was hypothesized to be due to the presence of other fungal endophytes that out-competed *B. bassiana* (Posada *et al.*, 2007). In the coffee-producing world, where seedlings are constantly grown in nurseries for subsequent transplant in the field, it would be ideal to develop a methodology effective in inoculating the seedlings with fungal entomopathogens.

The endophyte research also revealed the presence of bacterial endophytes in coffee plants (Vega *et al.*, 2005), and that coffee plants growing in the field in Hawaii, Colombia, Mexico, and Puerto Rico can harbor hundreds of fungal endophytes (Vega *et al.*, 2010), including *B. bassiana*.

3.14.4 Nematodes

Two vastly different groups of nematodes could be used as biological control agents against the coffee berry borer. The first group includes the entomopathogenic nematodes and the second group is the insect-parasitic ones. The entomopathogenic nematodes are in the genera *Steinernema* (Rhabditida: Steinernematidae) and *Heterorhabditis* (Rhabditida: Heterorhabditidae), and are mutualistically associated with bacteria in the genera *Xenorhabdus* and *Photorhabdus*, respectively. The mode of action of both nematode–bacteria complexes is similar. It involves infection of the hemocoel with infective juveniles followed by release of mutualistic bacteria that produce toxins that kill the insect (Lewis and Clarke, 2012). On the other hand, insect parasitic nematodes do not kill their hosts but reduce fecundity and/or sterilize females, and may reduce their longevity.

With entomopathogenic nematodes, laboratory bioassays using either coffee berry borer-infested coffee berries or specific insect stages exposed to *Steinernema* or *Heterorhabditis* infective juveniles have shown variable levels of mortality (Allard and Moore, 1989; Castillo and Marbán-Mendoza, 1996; Molina A. and López N., 2002; Molina Acevedo and López Núñez, 2003; Sánchez and Rodríguez, 2007, 2008; Manton *et al.*, 2012). In addition to laboratory bioassays, Manton *et al.* (2012) conducted field bioassays in Hawaii. Infested berries were placed on the soil surface around coffee plants, covered with leaf litter, followed by *S. carpocapsae* (Weiser) applications; resulting mortality was 4.7% in adults and 17% in larvae. In Colombia, Lara G. *et al.* (2004) also conducted field experiments using infested berries placed around coffee trees, followed by treatments consisting of various concentrations of *Steinernema* sp. or *Heterorhabditis* sp. (nematodes were not identified to species). Realpe-Aranda *et al.* (2007) developed a method for rearing *S. colombiense* López-Nuñez, Plichta, Góngora-Botero and Stock and *H. bacteriophora* Poinar in *Galleria mellonella* (L.) for use against the coffee berry borer. It remains unclear whether the use of entomopathogenic nematodes would be practical in the field, taking into consideration various constraints faced by coffee growers, such as the cost and proper storage of the product, labor costs, proper spray coverage, and ready access to water.

An area that might be more promising than the use of commercially available entomopathogenic nematode species is field sampling for parasitic nematodes infecting the coffee berry borer, followed by identification and subsequent studies aimed at determining their biocontrol potential. Varaprasad *et al.* (1994) reported on a *Panagrolaimus* species (Rhabditida: Panagrolamidae) attacking the coffee berry borer in India. However, most *Panagrolaimus* species are free-living nematodes and the parasitic nature of this species needs to be confirmed. In Mexico and Honduras, a new nematode species, *Metaparasitylenchus hypothenemi* Poinar, Vega, Castillo, Chavez and Infante (Tylenchida: Allantonematidae; Figure 11.8D) was found attacking the coffee berry borer in the field (Castillo *et al.*, 2002; Poinar *et al.*, 2004), the first such report in the Americas. *Metaparasitylenchus hypothenemi* appears to affect the female reproductive organs, reducing fecundity (Castillo *et al.* 2002).

3.14.5 Parasitoids

Murphy and Moore (1990) stated that parasitoids are probably the most promising biological control agents against the coffee berry borer. Approximately 12 species of parasitoids have been reported to attack the coffee berry borer (Morallo-Rejesus and Baldos, 1980; Benassi, 1995; Waterhouse, 1998; Pérez-Lachaud, 1998; Bustillo *et al.*, 2002), but only six species, all in the Hymenoptera, have been confirmed. This section will focus on the six parasitoids, four of them originating in Africa: (1) *Prorops nasuta* Waterston (Bethylidae); (2) *Cephalonomia stephanoderis* Betrem (Bethylidae); (3) *Phymastichus coffea* LaSalle (Eulophidae), and (4) *Heterospilus coffeicola* Schmiedeknecht (Braconidae). Two of the six parasitods originate in the Americas: (1) *Cryptoxilos sp.* Viereck (Braconidae), and (2) *Cephalonomia hyalinipennis* Ashmead (Bethylidae). A hyperparasitoid, *Aphanogmus dictynna* (Waterston) (Hymenoptera: Ceraphronidae), has been

reported in Kenya (Jaramillo and Vega, 2009; Buffington and Polaszek, 2009).

3.14.5.1 Prorops nasuta

Prorops nasuta, also known as the Uganda wasp, was the first reported natural enemy of the coffee berry borer. Even though Hargreaves (1926) states that it was "discovered early in 1923" attacking the coffee berry borer in Kampala, Uganda, the specimens described by Waterston (1923) as the new species *P. nasuta* were provided to him by Hargreaves in May of 1922. The species name, *nasuta*, denotes the elongated frontal process, described by Hargreaves (1926) as "a short median snout-like projection" (Figure 11.9D). This parasitoid appears to be indigenous to Uganda, Tanzania, and the Congo (Le Pelley, 1968); however, it has also been collected in Kenya, Cameroon, Ivory Coast, and Togo (Klein-Koch *et al.*, 1988; Barrera *et al.*, 1990a). This species has been used in several biological control programs throughout Latin America, the Caribbean, Asia, Madagascar, and the Pacific Islands (Klein-Koch *et al.*, 1988; Barrera *et al.*, 1990a; Infante, 1998; Baker, 1999; Waichert and Azevedo, 2012).

Females are minute wasps (ca. 2.3 mm long) with few body sculptures. They are blackish brown, with pale brown antennae and legs (Hargreaves, 1926). The males are similar to females but smaller (ca. 1 mm long). The head is nearly quadrate, with an elongated frontal process covering the clypeus and the antennal base (Figure 11.9D). The mandibles have three teeth and are strongly developed and especially large when compared to other bethylid species (Evans, 1964). The ocelli are arranged in an equilateral triangle and the antennae have 12 segments. The wings have long radial veins that lack closed cells with the exception of the subcostal vein. The forewings are faintly tinted and the hind wings are hyaline. The abdomen is smooth with a very short petiole. The females have a short ovipositor with about six to eight short bristles (Waterston, 1923; Hargreaves, 1926).

The immature stages of *P. nasuta* have been poorly studied. Eggs are comparatively large (0.53×0.18 mm), elongated, sausage-shaped, translucent, and white (Figure 11.9A). The larva is ca. 1.8 mm long, white and faintly segmented (Hargreaves, 1926) (Figure 11.9B, C). There are three instars in the larval stage (de Toledo, 1942). The pupa is initially white and gradually becomes dark brown as metamorphosis proceeds (Hargreaves, 1926).

Prorops nasuta is an idiobiont (stops development of the host after parasitizing it) solitary parasitoid. The female wasp enters an infested coffee berry, kills the adult borer and seals the entrance of the berry with the body of the insect, impeding the entry of other natural enemies (Hempel, 1933; Infante *et al.*, 2005). It usually spends the remainder of its life inside the berry. The preoviposition

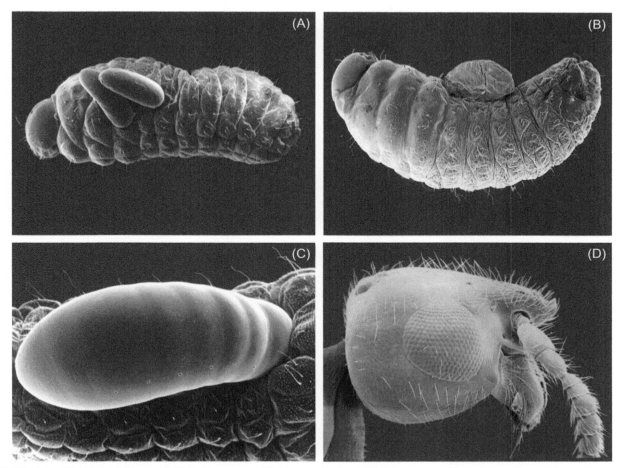

FIGURE 11.9 **Life stages of the parasitoid *Prorops nasuta*.** (A) An exceptional case of two eggs on a coffee berry borer larva; (B) a newly emerged larva on its host; (C) *P. nasuta* larva feeding on its host; and (D) the adult stage.

period ranges from 3 to 14 days. During this time, females feed on eggs and larvae, and paralyze fully grown larvae and pupae. The adult female is able to feed on all juvenile stages of the coffee berry borer (Infante *et al.*, 2005). Eggs are laid singly and externally on the insect cuticle, and in exceptional cases, two eggs can be laid (Figure 11.9A); when oviposition occurs on a pupa, the egg is placed in the dorso-abdominal region, while on larva, it is placed on the ventral surface (Figure 11.9B) (Hargreaves, 1926; de Toledo, 1942; Abraham *et al.*, 1990). The rate of egg laying varies between one and two eggs per day. Immediately after hatching, the larva starts to feed externally (Figure 11.9B, C) and slowly ingests the host fluids, leaving only a shriveled integument and cranial capsule (Abraham *et al.*, 1990). Each *P. nasuta* larva only consumes one host during its development. When completely mature, it spins a cocoon in which pupation occurs. Egg incubation lasts ca. 3 days and larval development ca. 4.5 days, passing through three instars. The prepupal stage lasts ca. 8 days and the pupal stage ca. 13 days. The life cycle from egg to adult lasts on average 28 days at a constant temperature of

22°C and at 75% RH, but adults remain in the berry for a few days in order to mate. In total, 297 degree-days are required to complete the development from egg to adult (Infante, 2000). As with other bethylids, males emerge 2–3 days before their sisters, with whom they mate (Hargreaves, 1926, 1935; Hempel, 1933; de Toledo, 1942; Abraham *et al.*, 1990). The proportion of sexes is one male to four females. Eggs laid by unfertilized females hatch and develop normally. In such instances, the resultant progeny are all males. The median longevity of *P. nasuta* adult females is 28 days when they feed on immature stages of the coffee berry borer but if no food is provided, longevity is drastically reduced to ca. 2.5 days (Infante *et al.*, 2005).

Rearing Methods The rearing method for *P. nasuta* in the laboratory is dependent on the availability of its host, which can be obtained from infested ripe coffee berries collected regularly in the field or from insects reared in artificial diet. Field-collected fresh berries are taken to the laboratory and placed for 3 days on trays lined with tissue paper to reduce humidity. *Prorops nasuta* cultures can then be established in circular plastic jars with ventilated mesh lids, at a ratio of one female parasitoid per 1.5 infested berries. Once inside the berry, the female parasitoid will feed and reproduce on its host. To avoid proliferation of saprophytic fungi, only a single layer of berries should be placed in each jar (approximately 200 berries). Three weeks after the culture has been initiated and just before progeny are expected to emerge, frass must be removed from the container to facilitate the emergence and collection of parasitoids. The jars are then placed under fluorescent lights and checked several times daily to collect and record the emerging adult parasitoids. In Mexico, *P. nasuta* has been reared for many years using fluctuating temperatures, ranging from 18 to 30°C (16 and 8 h, respectively), 60–85% RH, and a 12 h light:dark photoperiod (Barrera *et al.*, 1991; Infante *et al.*, 2005). In 10 years, more than three million parasitoids were produced in the laboratory (Infante *et al.*, 2005).

An alternative rearing method for *P. nasuta* was developed in Colombia using parchment coffee, which is rehydrated and treated with a fungicide and miticide to avoid contaminants. The parchment coffee is then placed in trays and infested in the laboratory with coffee berry borer females at a rate of two individuals per seed. Twenty-five days after infestation, 200 coffee seeds are placed in containers with ventilated lids and offered to 200 *P. nasuta* females. Parasitoid cultures are stored in the dark at 25°C and 70% RH. The progeny usually emerges 30 days later. This methodology results in an average of 3.7 wasps per seed and 20,000 wasps per month (Portilla and Bustillo, 1995).

Results of Cage Releases Cage exclusion or inclusion techniques are especially valuable because they provide a preliminary assessment of the impact of natural enemies upon pest populations, and also give quantitative information that can be used to understand the insect population dynamics (Luck *et al.*, 1999; Kidd and Jervis, 2005). To our knowledge, only one study has evaluated *P. nasuta* through the use of parasitoid inclusion field-cages techniques. In a preliminary evaluation in Ecuador in which 1605 adult parasitoids were released in sleeved cages, only 1430 individuals were recovered the following generation (Delgado and Sotomayor, 1991). Parasitism rates in the localities where the parasitoid was tested ranged from 2.3 to 38% in coffee berries on the plant, and 6% in coffee berries on the ground (Delgado *et al.*, 1990).

Results of Field Releases *Prorops nasuta* has been imported for biological control purposes to at least 14 coffee producing countries (Table 11.4). Unfortunately, classical biological control attempts using *P. nasuta* have, in almost all cases, not been satisfactory. There are various reasons for this. For instance, in Mexico (importation of 1988) and India (Table 11.4), the introductions failed due

TABLE 11.4 Introductions (in Chronological Order) of the Parasitoid *Prorops nasuta* to Countries outside Africa

Imported to:	Year	Exported from:	Reference
Indonesia	1924	Uganda	Le Pelley, 1968
Brazil	1929	Uganda	Hempel, 1933
Sri Lanka (Ceylon)	1938	Uganda	Le Pelley, 1968
Peru	1962	Brazil	de Ingunza, 1964
Ecuador	1987	Kenya and Togo	Klein-Koch *et al.*, 1988
Mexico	1988	Kenya and Togo	Barrera *et al.*, 1990c
Indonesia	1989	Togo	Murphy and Moore, 1990
Colombia	1990	Ecuador and Brazil	Baker, 1999; Bustillo Pardey, 2005
Mexico	1992	Brazil	Infante *et al.*, 2005
Guatemala	1993	Mexico	Infante, 1998
Honduras	1993	Mexico	Infante, 1998
El Salvador	1993	Mexico	Infante, 1998
India	1995	Mexico	Sujay *et al.*, 2010
Jamaica	1999	Honduras	Trejo S. and Fúnez, 2004
Costa Rica	2003	Colombia	Borbón-Martínez, 2007
Panama	2006	not stated	Contreras and Camilo, 2007

to problems with the rearing system. In these countries, *P. nasuta* was found to be difficult to maintain in a laboratory-rearing system and the colony collapsed. As a consequence, parasitoids could not be released (Barrera *et al.*, 1990c; Infante, 1998). In other cases, such as the introductions to Mexico (importation of 1992), Indonesia, Sri Lanka, Peru, and El Salvador (Table 11.4), the parasitoid never became established in the field after several releases (de Ingunza, 1964; Murphy and Moore, 1990; Infante *et al.*, 2001). It is assumed that in these cases the parasitoid only had a temporary effect on the coffee berry borer population, immediately after the release. *Prorops nasuta* has been recorded as established in Brazil, Ecuador, Colombia, Guatemala, and Honduras (Heinrich, 1965; Ruales, 1997; Trejo S. and Fúnez, 2004; Maldonado-Londoño and Benavides-Machado, 2007). However, parasitoid population levels have been barely perceptible and the parasitoid did not provide good control of the pest. In these countries, coffee growers typically use other methods to manage the coffee berry borer (Infante *et al.*, 2001).

It is apparent that *P. nasuta* is only able to maintain high populations in the field if there are multiple releases through the coffee season. In places where there was a constant rearing-release system, parasitism levels were acceptable but after several years with no releases, parasitoid populations decreased dramatically. In Brazil, reports following the introduction of *P. nasuta* introduction were very optimistic (Hempel, 1933; de Toledo, 1942, 1948; Yamamoto, 1948). Subsequently, it was demonstrated that although present in the country, *P. nasuta* populations were extremely low. Only 2% of the coffee berry borer infested berries were parasitized by *P. nasuta* (Heinrich, 1965; Ferreira and Bueno, 1995). Such low parasitism appears to have little impact on the pest population. A similar situation occurred in Ecuador. After its importation in 1987 (Table 11.4), average parasitism levels were 27% in berries on the plant and 25% in berries that had fallen on the ground (Cisneros and Tandazo, 1990). Unfortunately, 7 years later, the parasitoid was reported as barely present in some regions of the country (Ruales, 1997).

The best results obtained with *P. nasuta* have been reported in Colombia, where parasitoid releases started in 1991. The number of released individuals has been impressive. From 1994 to 2000, ca. 516 million *P. nasuta* were released in coffee plantations (Maldonado-Londoño and Benavides-Machado, 2007). Although parasitoid releases have decreased drastically in the recent years, the establishment of this species is evident in most coffee plantations (Bustillo Pardey, 2006). For example, Morales P. *et al.* (2011) reported the establishment of *P. nasuta* in 15 coffee farms 8 years after their release. The parasitoid was found in 80% of the farms evaluated, with parasitism between 0.2 and 11.6%. Another evaluation carried out 15 years after releases in 80 farms revealed that although

C. stephanoderis was not recovered, *P. nasuta* was recovered in 65% of the locations sampled, where the percentage of parasitism ranged from 0.25 to 50% (Maldonado-Londoño and Benavides-Machado, 2007). These results indicate that *P. nasuta* is well adapted to the environmental conditions of the Colombian coffee-growing areas, contributing in some degree to the control of the coffee berry borer. In fact, from the three African species introduced into Colombia for the biological control of the coffee berry borer, *P. nasuta* is considered the most promising species (Maldonado-Londoño and Benavides-Machado, 2007; Rivera-España *et al.*, 2010).

3.14.5.2 Cephalonomia stephanoderis

Ticheler (1961, 1963) discovered *Cephalonomia stephanoderis* (Figure 11.10) parasitizing the coffee berry borer by the end of 1950s in coffee plantations in Ivory Coast. He considered this species to be the most important natural enemy of the coffee berry borer in that country. Subsequently, Betrem (1961) described it as a new species. *Cephalonomia stephanoderis* is widely disseminated in Togo, Ivory Coast, Democratic Republic of the Congo, Burundi, Benin, and Cameroon (Koch, 1973; Damon, 1999; Barrera *et al.*, 2000). In a recent study, Jaramillo *et al.* (2009b) confirmed the presence of *C. stephanoderis* in coffee plantations in Kenya, although at low levels. It has been introduced to at least 16 countries (Table 11.5) in an attempt to use it as a biological control agent against the coffee berry borer (Klein-Koch *et al.*, 1988; Barrera *et al.*, 1990a; Bustillo *et al.*, 1998; Baker, 1999).

Cephalonomia stephanoderis is a macroptereous shiny black wasp (Figure 11.10F). Adult females are ca. 2 mm long and males ca. 1.4 mm long. The eggs are slightly curved and shiny white (Figure 11.10A, B). The eclosed larva is curved and looks like the egg because its segmentation is not apparent. The precise number of larval instars is unknown, but there are at least three larval instars (Infante *et al.*, 1994a). The pupal stage is similar to the adult in size and shape (Figure 11.10E). For a detailed description of the insect, see Betrem (1961) and Infante *et al.* (1994a).

Cephalonomia stephanoderis is a solitary ectoparasitoid of the coffee berry borer. A female wasp enters an infested coffee berry in search of potential hosts and will remain inside the berry for the rest of her life if there are enough hosts to feed on and parasitize (Koch, 1973). Females are synovigenic parasitoids that feed on all biological stages of the coffee berry borer, but have preference for eggs and adults (Koch, 1973; Lauzière *et al.*, 2001a). Lauzière *et al.* (2000) described in detail the behavior and daily parasitic activity, including host examination, adult feeding, paralysis, and oviposition. The preoviposition period usually takes 2 to 3 days and females feed during their entire lifetime (Lauzière *et al.*, 2001b). Ovipositing females

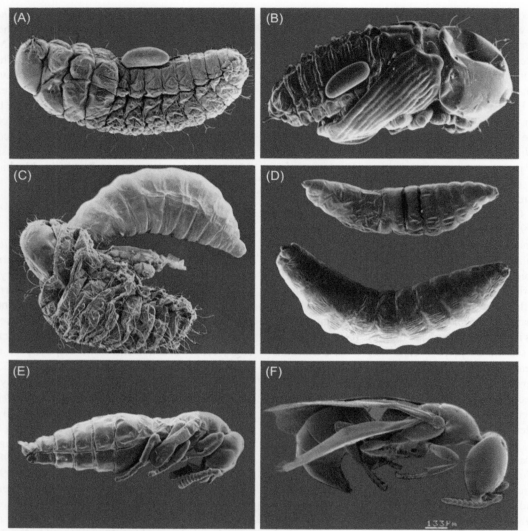

FIGURE 11.10   Life stages of *Cephalonomia stephanoderis.* (A) Parasitoid eggs laid on the prepupa and (B) pupa; (C) larva feeding on host; (D) fully grown male (top) and female (bottom) larvae; (E) pupa after removing the silk cocoon; and (F) the adult stage.

permanently paralyze full-grown coffee berry borer larvae, prepupae (Figure 11.10A), and pupae (Figure 11.10B) prior to oviposition and subsequently parasitoid eggs are laid singly and externally on these biological stages (Infante *et al.*, 1994a; Lauzière *et al.*, 2000). Individual wasps can parasitize two to three hosts per day for several weeks if there are enough hosts (Barrera *et al.*, 1989; Abraham *et al.*, 1990). Individual females can oviposit up to 63 eggs under optimal conditions in the laboratory (Infante and Luis, 1993). Upon hatching, the larva inserts its mouthpart into the host body and commences feeding externally until it has consumed the tissues of the host (Figure 11.10C). Once feeding is complete the larva detaches from the remains of the host and spins a cocoon. Adult parasitoid males emerge 1 day before females and sibling mating occurs inside the berry (Abraham *et al.*, 1990; Infante

et al., 1994a). The females exhibit arrhenotokous parthenogenesis (Koch, 1973; Infante *et al.*, 1993). In nearly all aspects, the biology and habits of *C. stephanoderis* is similar to *P. nasuta* (Abraham *et al.*, 1990). There is a skewed sex ratio favoring females: 4.8:1 (Ticheler, 1961, 1963); 3.5:1 (Koch, 1973); and 7:1 (Barrera *et al.*, 1993b). At 27°C, the developmental time for egg, larva, and pupa is 1.7, 4.2, and 12.7 days, respectively, while the development from egg to adult requires about 252 degree-days (Infante *et al.*, 1992a).

For a long time, both *C. stephanoderis* and *P. nasuta* were considered to be parasitoids specific to the coffee berry borer (Abraham *et al.*, 1990). However, Arcila *et al.* (1997) reported that *C. stephanoderis* also attacks *H. obscurus* in Colombia. In the laboratory, both parasitoid species can feed and reproduce on two curculionid species:

TABLE 11.5 Introductions (in Chronological Order) of the Parasitoid *Cephalonomia stephanoderis* to Countries outside Africa

Imported to:	Year	Exported from:	Reference
Ecuador	1988	Togo	Klein-Koch *et al.*, 1988
Mexico	1988	Togo	Barrera *et al.*, 1990a
Indonesia	1989	Togo	Murphy and Moore, 1990
New Caledonia	1989	Togo	Murphy and Moore, 1990
Colombia	1989	Ecuador	Benavides and Portilla, 1991
Guatemala	1990	Mexico	Barrera *et al.*, 1990b
El Salvador	1990	Mexico	Barrera *et al.*, 1990b
Honduras	1990	Mexico	Barrera *et al.*, 1990b
Nicaragua	1992	El Salvador	López *et al.*, 1993
Bolivia	1993	Ecuador	Sirpa-Roque, 1999
Brazil	1994	Colombia	Benassi, 1995
India	1995	Mexico	Sujay *et al.*, 2010
Dominican Rep.	1997	Honduras	Trejo S. and Fúnez, 2004
Jamaica	1999	Honduras	Trejo S. and Fúnez, 2004
Cuba	2003	Mexico	Peña *et al.*, 2006
Venezuela	2003	Dominican Rep.	Torrealba and Arcaya, 2005

Caulophilus oryzae (Gyllenhal) and *Sitophilus* sp. (Pérez-Lachaud and Hardy, 2001).

Rearing Methods Based on the similarity of their life cycles, the rearing methods for *P. nasuta* and *C. stephanoderis* are the same (Abraham *et al.*, 1990; Barrera *et al.*, 1991; Infante *et al.*, 2005). Therefore, no additional details for rearing *C. stephanoderis* will be presented in this section. The rearing methodology developed for *C. stephanoderis* in Mexico has been successful and this species has been reared uninterruptedly for over 25 years.

As in the case of *P. nasuta*, a rearing method based on parchment coffee (described above) has been developed in Colombia for *C. stephanoderis* (Portilla and Bustillo, 1995). This methodology has been reported as very successful, producing up to 10 million *C. stephanoderis* adults per month (Baker, 1999).

Results of Cage Releases Not much information is available on the field evaluation of *C. stephanoderis* in caged conditions. In Ecuador, Delgado *et al.* (1990) released *C. stephanoderis* in caged branches and caged plants after berries had been artificially infested with the coffee berry borer. Parasitization rates reached up to 86% in berries on the plant and 87% in berries on the ground.

In Mexico, field cages were placed individually on eight coffee plants and adult *C. stephanoderis* females were released inside the cages at a rate of 200 individuals per plant (Damon and Valle, 2002). One month later, berries were taken to the laboratory to estimate parasitism rates, with poor results. A high proportion of parasitoids were unable to find the infested berries, resulting in low levels of parasitism (4–37%). It became evident that *C. stephanoderis* has poor host searching capabilities and that a large numbers of parasitoids are needed to obtain high parasitism rates. Therefore, the use of *C. stephanoderis* in Mexico is not economically feasible (Damon and Valle, 2002).

Results of Field Releases Three papers from Africa have reported encouraging parasitism rates by *C. stephanoderis* (Ticheler, 1961, 1963; Koch, 1973; Borbón-Martinez, 1989). In Ivory Coast, the percentage of berries infested by the coffee berry borer and with presence of *C. stephanoderis* reached up to 50% at the end of the harvest (Ticheler, 1961, 1963). Also in Ivory Coast, Koch (1973) suggested that coffee harvesting considerably affects coffee berry borer population levels and, consequently, *C. stephanoderis* too. However, he mentioned that this parasitoid might be responsible for a 20–30% reduction in coffee berry borer population levels in berries remaining on the plant after harvest. In Togo, Borbón-Martinez (1989) reported 47% coffee berry borer mortality due to *C. stephanoderis*.

Cephalonomia stephanoderis has been introduced in at least 16 countries outside Africa for classical biological control purposes (Table 11.5). However, field evaluations have not been conducted in most countries, and therefore little information on parasitism levels is available in the literature. The parasitoid has been studied in Ecuador, Mexico, and Colombia and the results are quite different from those reported in Africa.

The first releases of *C. stephanoderis* in Ecuador were done in 1989 and resulted in parasitism rates of 9 to 52% (Delgado *et al.*, 1990). In another release, 10,000 parasitoids were released on four dates followed by monthly sampling of berries for parasitism assessments. The highest parasitism rate was 12% 3 months after release. The effect of the parasitoid decreased as time after release increased. Eight months after parasitoid release, parasitism levels were 1.3% (Delgado *et al.*, 1990). Also in Ecuador, Sponagel (1994) conducted an evaluation of *C. stephanoderis* after releasing 17,500 parasitoids in nine farms. Three months after release, the parasitoid was only detected in six farms. The parasitoid could not be detected 10 months after release. Sponagel (1994) concluded that *C. stephanoderis* is very susceptible to the humid conditions of the Ecuadorian Amazon region, where up to 25 days of rain per month are common.

In Mexico, a preliminary evaluation of the introduction of *C. stephanoderis* found that it could be detected in up to 81% of the infested berries; however, parasitism levels decreased to 3.2% after coffee was harvested (Barrera *et al.*, 1990c). In another study carried out in 26 localities in Chiapas, it was reported that the parasitoid was established in all of them, reaching parasitism rates between 0.5 and 19.6%, 3 years after being released (Barrera, 1994). Gómez *et al.* (2010) conducted a survey to evaluate the establishment of the three African parasitoids that have been released in Mexico. Sampling was conducted in 31 coffee plantations during the intercropping period. *Cephalonomia stephanoderis* was found in 67% of the plantations, with parasitism ranging from 0.3 to 26%. The highest level of parasitism was found in *C. canephora* plantations. The study confirmed the establishment of *C. stephanoderis* in Mexico 20 years after its first release. In Mexico, Dufour *et al.* (1999) reported a reduction of 22–56% in coffee berry borer infestation after releasing 35,000–40,000 wasps/ha during the intercropping period.

In Colombia, *C. stephanoderis* has become established in all the sites where it has been released. Parasitism rates between 2.2 and 13.8% have been reported (Portilla and Bustillo, 1995). Generally, field parasitism is lower than

10% and consequently not enough to reduce the pest population below the economic threshold (Bustillo *et al.*, 1998). Salazar and Baker (2002) conducted a field experiment to determine ensuing parasitoid infestation rates when different densities of the parasitoid were released based on the number of infested berries. A ratio of 100:1 (parasitoids:infested berries) resulted in an average of five coffee berry borer infested berries per tree. The 50:1 ratio had an average of 30 infested berries per tree, while the 10:1 ratio had 53 infested berries per tree. The control (no parasitoids) had 82 infested berries per tree (Salazar and Baker, 2002). In another experiment, Aristizábal *et al.* (1998) reported a significant reduction in the number of infested berries when using *C. stephanoderis*. The experimental plots, containing 2200 coffee plants, were treated with 30,000, 32,000, or 80,000 parasitoids, resulting in 3 to 28% parasitism. The conclusion was that very high numbers of parasitoids are required to improve parasitism levels (Aristizábal *et al.*, 1998). A similar conclusion was reached in Mexico by Damon (1999).

A tritrophic simulation model by Gutierrez *et al.* (1998), which included the coffee plant, the coffee berry borer, and their natural enemies, predicted that bethylid parasitoids, singly or in combination, have little impact on coffee berry borer population levels. Among other factors, poor control is predicted because these species have a low numerical response and their attack is limited to a single berry.

3.14.5.3 Phymastichus coffea

This parasitoid (Figure 11.11) was discovered parasitizing adult coffee berry borers in Togo in 1987 (Borbón-Martínez, 1989) and was described as a new genus and species in the family Eulophidae (LaSalle, 1990). *Phymastichus coffea* has been collected in Benin, Burundi, Cameroon, Ivory Coast, Kenya, and Togo (Infante *et al.*, 1992b) and is believed to be present in all African countries infested with the coffee berry borer (López-Vaamonde and Moore, 1998). *Phymastichus coffea* has been introduced in at least 12 countries (Table 11.6). It was introduced to Colombia in 1995 (Table 11.6) after being quarantined in England (López-Vaamonde and Moore, 1998). From Colombia it has been exported to Brazil, Ecuador, Honduras, Guatemala, India, and Costa Rica (Table 11.6).

Adults (Figure 11.11E) are dark brown wasps with reddish eyes and shiny wings (LaSalle, 1990; Vergara-Olaya *et al.*, 2001a). Adult females are ca. 1 mm long, and males are half that size (for a detailed description, see LaSalle, 1990). Sex in immature stages can be differentiated based on pupal size, with females being twice as large as males (Feldhege, 1992; Vergara-Olaya *et al.*, 2001a; Espinoza *et al.*, 2002, 2009).

Pymastichus coffea is a primary, gregarious, idiobiont endoparasitoid of coffee berry borer adults (Feldhege,

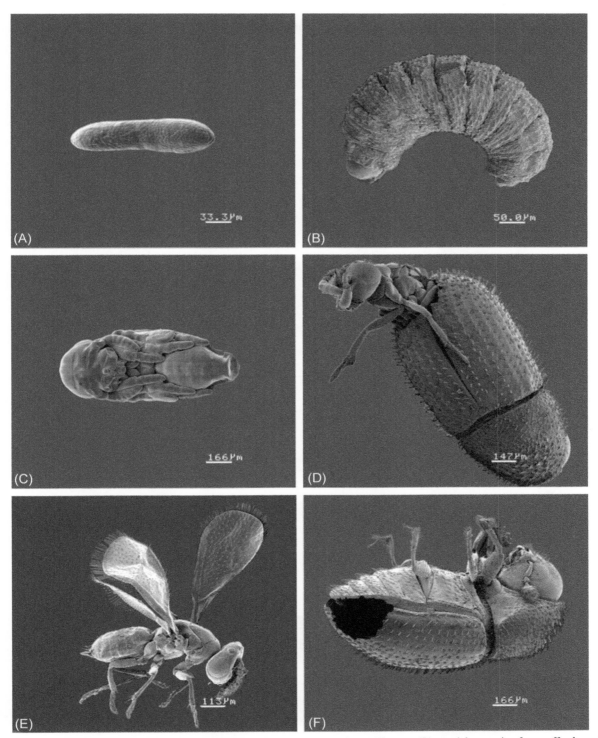

FIGURE 11.11 **Life stages of the parasitoid *Phymastichus coffea*.** (A) Egg; (B) larva; (C) pupa; (D) an adult emerging from coffee berry borer; (E) the adult female; and (F) a typical hole made by the female parasitoid, after emerging from its host.

1992; Infante *et al.*, 1994a; López-Vaamonde and Moore, 1998). Adult females start to search for hosts immediately after emergence and do not have a preoviposition period. Parasitization can occur within the first hours after the wasp reaches the adult stage (Infante *et al.*, 1994a). Females oviposit in the abdomen of coffee berry borer adults (Figure 11.12C), allocating two eggs per host, one of which will become a male and the other a female.

TABLE 11.6 Introductions (in Chronological Order) of the Parasitoid *Phymastichus coffea** to Countries outside Africa

Imported to:	Year	Exported from:	Reference
Colombia	1995	Kenya	López-Vaamonde and Moore, 1998; Baker, 1999
Brazil	1998	Colombia	Cantor *et al.*, 1999
Ecuador	1999	Colombia	Delgado A. *et al.*, 2002
Honduras	1999	Colombia	García, 2000
El Salvador	1999	Honduras	Baker *et al.*, 2002
Jamaica	1999	Honduras	Baker *et al.*, 2002
Guatemala	1999	Colombia	García, 2000
India	1999	Colombia	Bustillo Pardey, 2005; Sujay *et al.*, 2010
Mexico	2000	Guatemala	García, 2000
Cuba	?	Mexico	Vázquez-Moreno, 2005
Costa Rica	2003	Colombia	Borbón-Martínez, 2007
Panama	2006	?	Armuelles, 2007

*There is some evidence that P. coffea has also been introduced to Bolivia, Peru, Venezuela, Nicaragua, and the Dominican Republic. However, there are no reliable sources to support this information.

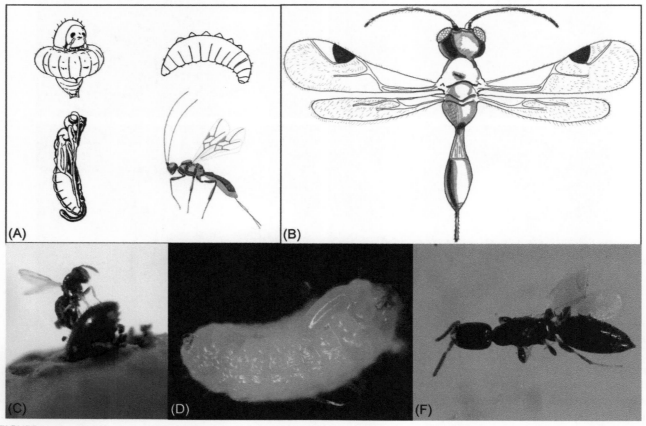

FIGURE 11.12 (A) Life stages of *Heterospilus coffeicola*: larva feeding on the pupal stage of the coffee berry borer (upper left); fully grown larva (upper right); pupa (lower left); and an adult female (lower right). No photographs are known for the biological stages of this species. *Redrawn from Fonseca and Araujo (1939)*. (B) Adult female of the Cryptoxilos sp. *Redrawn from Deyrup (1981)*. *No photographs are known for the biological stages of this parasitoid*. (C) An adult female Phymastychus coffea *at the moment of parasitizing the coffee berry borer, which is boring into a berry*. (D) An egg of the parasitoid Cephalonomia hyalinipennis *laid on a prepupa of the coffee berry borer and (E) adult female* C. hyalinipennis.

Sometimes it is possible to find more individuals in a single host, but only two will survive due to intense intraspecific competition (Castillo *et al.*, 2004a; Espinoza *et al.*, 2009). Upon hatching, parasitoid larvae feed on the internal tissues of the host's abdomen. After feeding for several days, the male larva migrates to the prothorax where it continues to feed, while the female larva remains feeding on the abdominal region. Pupation occurs inside the host without the formation of a cocoon. Males pupate in the prothorax, while females pupate in the abdomen. The female parasitoid makes an exit hole in the exoskeleton of the host in order to emerge (Figure 11.11D, F). The male comes out immediately afterwards, using the same hole made by the female. The sex ratio is close to 1:1, and presumably there is sibling mating just before emergence from the host (Espinoza *et al.*, 2009). The life cycle from egg to adult is ca. 30 days at 27°C and 70–80% RH (Feldhege, 1992). In the field, the life cycle from egg to adult is completed in 47 days at an average temperature of 23°C. The adult longevity is at most 3 days. Hosts parasitized by *P. coffea* do not live longer than 15 days (Espinoza *et al.*, 2009).

Some reports indicate that *P. coffea* can attack other scolytids in the laboratory. For example, in no-choice tests, *P. coffea* was able to parasitize and complete its development on *H. obscurus*, *H. seriatus*, and *Araptus* sp. (López-Vaamonde and Moore, 1998), three bark beetle species common in Colombian agroecosystems. In another laboratory host-specificity test, Castillo *et al.* (2004b) assessed parasitism on *H. crudiae*, *H. eruditus*, *H. plumeriae*, *Scolytodes borealis* Jordal, and *Araptus fossifrons* Wood. There were ovipositions attempts by *P. coffea* on all species tested, but parasitization and development of progeny was only completed in *H. crudiae* and *H. eruditus*. Both findings confirm the oligophagic behavior of *P. coffea*, although there are no field reports on parasitization of species other than the coffee berry borer.

Early studies suggested that attack by *P. coffea* occurred just while the coffee berry borer was initiating fruit perforation (Borbón-Martínez, 1989; Feldhege, 1992; López-Vaamonde and Moore, 1998). However, berries that had been infested with the coffee berry borer for 7 days were successfully parasitized by *P. coffea* (Echeverry-Arias, 1999; Jaramillo *et al.*, 2006; Espinoza *et al.*, 2002, 2009). Thus, it appears that *P. coffea* is able to attack the coffee berry borer at any time after fruit colonization. This could have important practical implications, as it would allow for field releases of the parasitoid at any vegetative period of the coffee plant, as long as infested berries are available.

Rearing Methods A method for rearing *P. coffea* was developed in Mexico by Infante *et al.* (1994b, 2003). Non-infested green coffee berries are collected in the field and taken to the laboratory. Berries are placed in a plastic container filled with water, in order to separate the floating berries, as they are not useful for rearing. Berries are washed and dried in trays for 2 or 3 days. To facilitate penetration by the coffee berry borer, the disc is slightly pierced with a dissecting needle. Immediately afterwards, berries are placed in a plastic container covered with a fine mesh lid. If the container has a one-liter capacity, a maximum of 50 berries should be used, distributed in a single layer. Adding more than 50 berries could lead to fungal proliferation due to increased moisture. A ratio of about 10 coffee berry borer females per berry are placed in the container and allowed to bore for ca. 2 hours, followed by introduction of parasitoids previously fed with honey. The container should not be moved to avoid injuring the parasitoids. At this stage, coffee berry borers will have about half of their bodies inside the berry, which makes parasitoid attack easier. Insects that do not bore the berries are also susceptible to attack. The containers can be kept at room temperature in the laboratory, keeping in mind that optimal ambient conditions for rearing are 80–90% RH, a 12:12 L:D photoperiod, and 26°C. Collection of the new generation of parasitoids should be done 25 days after initial set-up. Berries are brushed to remove dust and any fungal growth present, and transferred to a clean plastic container. Coffee berry borers that did not bore into berries should also be placed in these containers as they might have been parasitized. Generally, the parasitoids start emerging from the berries 1 month after set-up. The new generation of parasitoids will emerge on a daily basis for about 7 days. It is important to keep the containers under a light source to stimulate emergence (Infante *et al.*, 1994b, 2003).

As is the case for *P. nasuta* and *C. stephanoderis*, the rearing of *P. coffea* in Colombia is based on parchment coffee artificially infested with adult coffee berry borer (López-Vaamonde and Moore, 1998; Orozco Hoyos, 2002). Parasitoid colonies are held at 24°C, and 75% RH in total darkness. This methodology has been reported to be very successful, producing an average of one million parasitoids per month (Baker, 1999).

Results of Cage Releases Parasitoid inclusion experiments with *P. coffea* under field-cage conditions or entomological sleeves have been very promising for the control of the coffee berry borers. For instance, Echeverry-Arias (1999) evaluated releases of *P. coffea* in experimental plots in Colombia using different densities of parasitoids released on infested coffee berries inside entomological sleeves. Parasitism varied from 8 to 49% when releasing a 1:13 and 1:1 parasitoid:host ratio, respectively. Releasing parasitoids at a 1:1 ratio in entomological sleeves 5 days after the coffee berry borers had infested the berry resulted in up to 83% parasitism. Using a 1:1 parasitoid:host ratio, Jaramillo *et al.* (2005) observed that parasitism was significantly affected by the age of the berries at the time of infestation. They reported a maximum parasitism of 32% in

210-day-old berries and parasitism increased to 85% and 76% when berries were 150 and 90 days old, respectively.

Similar levels of parasitism were obtained in Mexico when evaluating *P. coffea* using different densities of parasitoid:host ratios in field-cage experiments. At the 1:30 parasitoid:host ratio, parasitism was ca. 12%. The pest population was significantly reduced when *P. coffea* was released at a 1:5 or 1:10 parasitoid:host ratio. The highest level of parasitism (62%) occurred at the 1:5 parasitoid:host ratio, with a significant reduction in damage to the seeds. These studies demonstrate that the use of *P. coffea* resulted in a 2.2–3.1-fold lower coffee berry borer damage to the coffee seeds' weight (Infante *et al.*, 2013).

Results of Field Releases Due to the short longevity of adult *P. coffea* (2–3 days), releases in Mexico have been carried out using the pupal stage. The entire rearing culture containing berries and coffee berry borers parasitized by *P. coffea* are taken to the field 3 days before the expected date of adult parasitoid emergence. Coffee berries are placed inside 10×15 cm metallic cages and hung on a coffee branch. The wire from which the cage is hung is covered with grease to avoid the interference of ants or other predators. The amount of individuals released is estimated by leaving 10% of the infested berries in the laboratory, to quantify the total emergence of parasitoids (Infante *et al.*, 2003).

Reports on the field performance of *P. coffea* have only been published in Colombia and Mexico. The first release in Colombia occurred in 1997 (Baker, 1999). The parasitoid was recovered from coffee plots the following year, with ensuing parasitism levels ranging between 41 and 67%. In a more extensive study comprising 33 coffee farms, ca. two million adult parasitoids were released, resulting in overall parasitism of 2–6% (Benavides *et al.*, 2002). Despite the low parasitism levels, the parasitoid appears to be adapted to the environmental conditions of the country (Benavides *et al.*, 2002). Echeverry-Arias (1999) carried out artificial infestations of coffee berries and released adult parasitoids at different densities. He reported parasitism rates of 47, 41, 22, 21, 13, 10, and 6% when using parasitoids:host ratios of 1:1, 1:3, 1:5, 1:7, 1:9, 1:11, and 1:13, respectively.

Vergara-Olaya *et al.* (2001b) reported on an augmentative release of 30,000 *P. coffea* adults in a 70×130 m plot with a 13% natural coffee berry borer infestation. Following the release of wasps in the center of the plot, samples were collected 25 days later at different distances from the center of the experimental plot. Results revealed that parasitism by *P. coffea* decreased gradually as the distance from the release point increased. There was a parasitism rate of 61, 63, 37, 25, and 27% at a distance of 0–10, 10.1–20, 20.1–30, 30.1–40, and 40.1–54 m from the release point, respectively. The overall mean parasitism was 47%. These results confirm

that high levels of parasitism can only be achieved if there are releases with high numbers of parasitoids. In Colombia, sampling for *P. coffea* to assess its establishment only resulted in detection for the first 3 years after its release (Bustillo Pardey, 2006).

In Mexico, *P. coffea* was released in 14 coffee farms at different elevations with the goal of having it established permanently (Galindo *et al.*, 2002). At least 9000 female parasitoids were released at each farm. Samples of 200 infested berries collected at random from each farm on a monthly basis resulted in parasitoid recovery from all farms for the first 6 months following release. The highest levels of parasitism occurred the first month (32–55%) following release. At 6 months, parasitism declined 10 to 28%. At eight to 12 months after release, no parasitoids were recovered. The following coffee season, parasitism was barely detected in three of the sites, and afterwards, the parasitoid was never recovered from any site (Galindo *et al.*, 2002). It is assumed that as is the case with the bethylid parasitoids, the coffee harvest has a severe effect on the survivorship of parasitoids and acts as a mortality factor. At present, *P. coffea* is not considered to have become established in Mexico.

3.14.5.4 Heterospilus coffeicola

Information on this species is very scarce and most data were published more than 80 years ago. *Heterospilus coffeicola* (Figure 11.12A) was discovered in Uganda in 1923 (Hargreaves, 1926, 1935) and has also been recorded in Tanzania, the Democratic Republic of the Congo, and Cameroon (Le Pelley, 1968). The adult female is a dark brownish wasp, ca. 2.5 mm long. There is sexual dimorphism, with males being smaller than females and having a small dark area (stigma) near the base of each hind wing (Fonseca and Araujo, 1939). The eggs are small and white, measuring approximately 0.39×0.13 mm, and can be confused with the eggs of the coffee berry borer. The size of first instar larvae is ca. 0.5 mm long, but fully developed larvae are 1.2 mm. The pupal stage can be found inside a white cocoon (Fonseca and Araujo 1939; Le Pelley, 1968).

In contrast with *P. nasuta* and *C. stephanoderis*, adult *H. coffeicola* spend little time inside infested berries (Fonseca and Araujo, 1939). A single egg is deposited per berry and the incubation period is about 6 days. After eclosion, the larva is able to feed on eggs and larvae of the coffee berry borer over a period of 18–20 days. The predatory rate of the parasitoid larva can reach 15 individuals during its entire life. By the end of the larval stage, a cocoon is formed in the gallery, where pupation occurs. Complete development from egg to adult takes about 40 days. Females prefer to oviposit in ripe berries that are attached to the plant. Berries that have fallen on the ground are not susceptible to be parasitized by *H. coffeicola*

(Hargreaves, 1926, 1935; Fonseca and Araujo, 1939; Le Pelley, 1968).

Hargreaves (1926, 1935) reported that, based on the large amount of individuals consumed by *H. coffeicola*, it could be the most important natural enemy of the coffee berry borer in Africa. In addition, *H. coffeicola* attacks the coffee berry borer shortly after it colonizes the berry, in contrast to the bethylids, which attack later. Fonseca and Araujo (1939) proposed that *H. coffeicola* could be complementary to *P. nasuta*. However, additional observations revealed that *H. coffeicola* larvae also feed on *P. nasuta* larvae, thereby reducing the efficacy of *P. nasuta* (Le Pelley, 1968). Thus, although *H. coffeicola* exerts some mortality over the coffee berry population in several countries in Africa, there is no evidence to suggest that its presence is sufficient to control the pest. Waterhouse (1998) stated that the biological control potential of *H. coffeicola* requires further study because it is not specific to the coffee berry borer. All attempts to rear *H. coffeicola* under laboratory conditions have been unsuccessful (Le Pelley, 1968).

3.14.5.5 Cryptoxilos sp

The genus *Cryptoxilos* (Figure 11.12B) belongs to the Braconidae subfamily Euphorinae. Species in this genus are characterized by brownish black color and small size (<2 mm) (Muesebeck, 1936). A wasp belonging to this genus was found parasitizing the coffee berry borer in a coffee plantation in Antioquía, Colombia (Cárdenas, 1995; Bustillo *et al.*, 2002). This finding constituted the first braconid parasitizing the coffee berry borer outside Africa. However, the species was never fully identified and there is very little information on this insect. According to Bustillo *et al.* (2002), *Cryptoxilos* enters coffee berries using the same hole made by the coffee berry borer. Female wasps are endoparasitoids of coffee berry borer adults. A single egg of the wasp is deposited internally in the coffee berry borer. After eclosion, the larva feeds and kills the adult host. When the larval stage is completed, the mature larva leaves the coffee berry borer cadaver and pupates in a gray cocoon that blocks the entrance tunnel bored into the berry by the colonizing female, presumably to avoid the entrance of potential predators. The adult wasp emerges through the entrance tunnel. Many individuals of this species have been collected in Colombia from coffee berry borers. Attempts to rear this species in the laboratory have been unsuccessful (Bustillo *et al.*, 2002).

Cryptoxilos has been recorded as parasitoids of adult scolytids (Deyrup, 1981; Jordal and Kirkendall, 1998; Kenis *et al.*, 2004). Therefore, it is possible that *Cryptoxilos* sp. is a native parasitoid of an unidentified scolytid inhabiting coffee plantations of Colombia and that it also attacks the coffee berry borer. Generally, this type of parasitism is considered incidental and with no important consequences in reducing pest population levels.

3.14.5.6 Cephalonomia hyalinipennis

Pérez-Lachaud (1998) reported *Cephalonomia* near *waterstoni* parasitizing the coffee berry borer in Chiapas, Mexico. The species was later confirmed to be *C. hyalinipennis* (Figure 11.12D, E) (Pérez-Lachaud and Hardy, 1999), which has been recorded attacking several coleopteran species in several countries throughout North America, South America, and Europe (Evans, 1978).

Females are dark wasps (Figure 11.12E), ca. 1.7 mm long, and males are smaller. This species feeds on all immature stages of the coffee berry borer and oviposits one to three eggs per host on the cuticle of last instar larva, prepupae (Figure 11.12D), or pupae. Most of these individuals are able to reach the adult stage, although the size of the progeny is reduced if there is more than one parasitoid on a host (Pérez-Lachaud, 1998). It takes ca. 20 days from egg to emergence of adults at 28°C. Emergence of males occurs earlier than females and there is sibling mating. Longevity of adult females is ca. 57 days and average fecundity per female is 88 eggs (Pérez-Lachaud, 1998; Pérez-Lachaud and Hardy, 1999).

In Brazil, Benassi (1989) reported the presence of an unidentified species of *Cephalonomia* attacking larvae of the coffee berry borer. Although there is no further information on this discovery, this species could be *C. hyalinipennis*, whose geographical range comprises all of South America (Evans, 1978). *C. hyalinipennis* is known to be a generalist parasitoid attacking larvae and pupae of several species of Coleoptera, especially Scolytinae and Anobiidae (Evans, 1964, 1978). Under laboratory conditions it is also capable of parasitizing several species of Curculionidae, Bostrichidae, and Bruchidae (Pérez-Lachaud and Hardy, 2001). Pérez-Lachaud *et al.* (2004) has shown that *C. hyalinipennis* can be a facultative hyperparasitoid of *C. stephanoderis* and *P. nasuta* and that its presence may have a negative effect on these parasitoids (Batchelor *et al.*, 2006). Consequently, it is not recommended for use in coffee berry borer biological control programs.

3.14.6 Predators
3.14.6.1 Ants

One of the first reports of ants (Hymenoptera: Formicidae) preying on the coffee berry borer was published by Leefmans (1923) in Java, where he showed 9.3% lower coffee berry borer infestation rates in coffee plants with *Dolichoderus bituberculatus* Mayr (current name: *D. thoracicus* (Smith)) than in plants without ants. In Brazil, Fonseca and Araujo (1939) concluded that *Crematogaster curvispinosus* Mayr, which uses dry berries as nests, was only an occasional predator of immature stages of the coffee berry borer. Almost 60 years later, Benassi (1995) found *C. curvispinosus* at low levels throughout coffee plantations in northern Espírito Santo (Brazil). The presence of the ant can be recognized based on the ant enlargement of the hole originally made by the coffee berry borer (Fonseca and Araujo, 1939; Benassi, 1995).

Varón *et al.* (2004) reported on *Solenopsis geminata* (F.), *Pheidole radoszkowskii* Mayr, and *Crematogaster torosa* Mayr as highly effective predators of the coffee berry borer in laboratory experiments with up to 100% predation, depending on insect growth stage provided. In contrast, predation levels in the field were much lower, not exceeding 25%, a result ascribed to the generalist diet of the ants and the lack of attraction towards coffee berry borer infested berries placed next to ant nests. These results point at the generalist nature of ants and the fact that in order for ants to be effective sources of mortality for the coffee berry borer they would have to enter the berry and consume life stages contained within. Even though Varón *et al.* (2004) observed the three ant species (as well as others) entering the berries placed near the ants nests, the size of the tunnel in the berry could be a limiting factor, impeding ant entrance based on ant size (see discussion in Varón *et al.*, 2004).

In Cuba, Vázquez Moreno *et al.* (2009) concluded that *Tetramorium bicarinatum* (Nylander) enters infested coffee berries and expands the galleries built by the colonizing female. The presence of immature ant stages inside the berry indicates that in addition to preying on the coffee berry borer, the ant uses infested berries for nesting.

Most papers reporting effects of ants on the coffee berry borer base their conclusions on observations of ants carrying coffee berry borers in their mandibles, predation of free living stages of the coffee berry borer offered in the field or laboratory, or on methods that allow or prevent ants to access the infested berries (bags with different mesh sizes) followed by subsequent sampling for assessment of coffee berry borer presence (Leefmans, 1923; Varón *et al.*, 2004; Gallego Ropero and Armbrecht, 2005; Vélez *et al.*, 2006; Vélez-Hoyos *et al.*, 2006; Perfecto and Vandermeer, 2006; Armbrecht and Gallego, 2007; Larsen and Philpott, 2010). Some studies have been conducted on coffee berry borers infesting parchment coffee, although it is not clear how the results relate to actual field situations involving infested coffee berries (Gallego Ropero and Armbrecht, 2005; Vélez *et al.*, 2006; Vélez-Hoyos *et al.*, 2006). For example, in a laboratory study Gallego Ropero and Armbrecht (2005) reported that *Solenopsis picea* Emery could enter coffee berry borer-infested parchment seeds but *Tetramorium simillimum* Smith did not do so, even though it had been observed to go into parchment seeds in the field. It is necessary to conduct molecular gut content analysis of ants (as done by Chapman *et al.* (2009) and Jaramillo *et al.* (2010b) for

thrips; discussed below) to determine if ants play an important role as coffee berry borer predators under natural field conditions involving infested berries on plants.

It is also important to consider that manipulating ants to increase coffee berry borer predation would be quite difficult, not to mention that ants can be a serious nuisance to coffee pickers (Vrydagh, 1940; Vandermeer *et al.*, 2002; Vera-Montoya *et al.*, 2007). Additionally, some ants tend other insects (e.g., coccids) that might create problems in coffee plantations (Leefmans, 1923; Vandermeer *et al.*, 2002). Leefmans (1923) found that even though a lower coffee berry borer infestation level was found in berries to which *D. bituberculatus* had access, greater damage was done to these berries by the ant-reared green scale of coffee, *Lecanium viride* (current name: *Coccus viridis* (Green); Coccidae), which prefers the stalks of the berries. He concluded, "…it is clear that by enticing ants to the coffee plants means more damage by the green coccids than is done by the borer." On the other hand, Perfecto and Vandermeer (2006) found a negative relationship between the number of ant-tended coccids and coffee berry borer damaged berries on a plant basis but not on a branch basis, presumably as a result of ant predation of the coffee berry borer. Ants were observed carrying coffee berry borers in their mandibles. For an in-depth discussion of this topic, see Philpott and Armbrecht (2006).

3.14.6.2   Birds

Coffee plantations, particularly when shaded, are important habitats for birds (Perfecto *et al.*, 1996; Moguel and Toledo, 1999; Sherry, 2000; Somarriba *et al.*, 2004). A few papers have examined bird predation on the coffee berry borer. Leefmans (1923) observed swallows (*Callocalia* sp.) feeding on swarming coffee berry borers in Java, and Sherry (2000) observed American redstarts (*Setophaga ruticilla* (L.)) eating coffee berry borers.

In Jamaica, Kellermann *et al.* (2008) identified 17 species of birds as potential coffee berry borer predators. Using coffee berry borer infested plants from which birds were excluded and contrasting them to plants to which birds had access resulted in a reduction in coffee berry borer infestation. They concluded: "These services likely result from avian predation of adult female borers as they search for an oviposition site or bore into the endosperm, which can take up to 8 h." Predation by birds as the insect searches for an oviposition site and starts boring the berry is possible, but not when the insect bores into the endosperm (seed), as this event occurs within the berry. Kellermann *et al.* (2008) cite an unpublished study that revealed the presence of coffee berry borers in the stomach contents of three bird species. They also mention that the bird exclosures do not exclude lizards, which were observed in both treatments. Johnson *et al.* (2010) conducted a similar experiment in Jamaica as the one conducted by Kellermann *et al.* (2008), with the exception that they examined responses in coffee plantations at full sun and in shaded plantations. Their results also show reduced coffee berry borer infestations in plants where birds were not excluded, and even though there were more birds in shaded plantations, their effect was not greater in shade than in sun grown coffee.

Using molecular methods for coffee berry borer DNA detection, Karp *et al.* (2013) confirmed the predatory status of five bird species after analyzing 469 fecal samples from 75 species of birds, and 53 samples from 13 bat species. Exclusion of birds from coffee plants increased coffee berry borer infestation levels from 4.6 to 8.5% in the wet season, and from 2.7 to 4.8% in the dry season. Presence or absence of bats did not have an effect on coffee berry borer infestation levels.

3.14.6.3 Thrips

Using molecular gut content analysis, Chapman *et al.* (2009) and Jaramillo *et al.* (2010b) confirmed that the black thrips, *Karnyothrips flavipes* (Jones) (Thysanoptera: Phlaeothripidae), serves as a predator of coffee berry borer eggs and larvae in Kenya. The study involved the collection of almost 18,000 coffee berry borer infested berries and of over 3000 thrips emerging from the berries. It serves as a model to confirm the predatory status of an organism based on detection of coffee berry borer DNA in the predator gut. The thrips is cosmopolitan and it could possibly be preying on the coffee berry borer in other countries.

3.14.6.4 Other predators

Leptophloeus sp. near *punctatus* Lefkovitch (Coleoptera: Laemophloeidae) collected in Togo and Ivory Coast has been observed preying on coffee berry borer larvae (Vega *et al.*, 1999). In Java, *Dindymus rubiginosus* (F.) (Hemiptera: Pyrrhocoridae), which preys on bark beetles in the forest, was observed feeding on coffee berry borers inside the berry as well as outside (Wurth, 1922; Sladden, 1934). Adults move among berries and insert their proboscis, which can be 9 mm long, in the galleries within the berry, sucking the contents of the insect (Wurth, 1922; Sladden, 1934). The insect can kill six coffee berry borers in an hour, but is not considered an important predator due to the small number of insects it can kill.

In sometimes-conflicting field observations and laboratory results in Colombia, nymphs and adults of unidentified species of *Calliodes*, *Scoloposcelis*, and *Xylocoris* (Hemiptera: Anthocoridae), *Cathartus quadricollis* (Guérin-Méneville) and *Monamus* sp. (Coleoptera: Cucujidae), and *Prometopia* sp. (Coleoptera: Nitidulidae) have been reported to feed on immature stages of the coffee berry borer (Cárdenas, 1995; Bustillo *et al.*, 2002; Vera-Montoya

et al., 2007). In laboratory studies in Costa Rica, *C. quadricollis*, *Ahasverus advena* (Waltl) (Coleoptera: Silvanidae) and *Lyctocoris* sp. (Hemiptera: Anthocoridae) preyed on various stages of the coffee berry borer (Rojas Barrantes, 2009; Rojas *et al.*, 2012).

Working in Mexico, Henaut *et al.* (2001) placed coffee berry borers on the webs of four spiders (*Cyclosa caroli* (Hentz), *Gasteracantha cancriformis* (L.), *Leucauge mariana* (Keyserling), and *L. venusta* (Walckenaer)) and concluded that preference for the insect was low when compared to other prey placed on the webs.

3.15 Cultural Control

Many authors have suggested that after the harvest, all berries in the field should be collected, including those left on trees and those that have fallen on the ground, to interrupt the life cycle of the insect. This method, known as rampassen in Dutch (Friederichs, 1922a), repasse in Portuguese (Bergamin, 1944a), repase in Spanish (Bustillo P. *et al.*, 1998), and re-picking in English, is the only method that would be guaranteed to eliminate the insect. Nevertheless, as discussed in Section 3.6.3, the number of insects that could be present on fallen berries is daunting; therefore, unless the insect presence is very limited, re-picking might be difficult to implement due to associated costs, the need for all growers to participate, and the requirement that absolutely all berries should be collected, both on the plant and on the ground. According to Pamplona (1927), finding an uninfested tree in the state of São Paulo (Brazil) was an unusual event in 1924, but after a large-scale re-picking effort just 228 infested berries were found in ca. 22,000 sampled trees.

Cultural control could also be used to reduce passive dispersal, which occurs when materials contaminated with the insect are moved from one place to another. These materials include coffee bags used during harvesting, agricultural implements, vehicles, workers clothing, and infested beans or coffee seeds for domestic use (Corporaal, 1921; Leefmans, 1923; Wilkinson, 1928). In Colombia, the dissemination of the insect from the southwest border with Ecuador to the main coffee growing areas has been ascribed to the movement of coffee pickers (Benavides *et al.*, 2006). Cultural control methods to avoid passive dispersal include tightly closing the bags containing harvested berries to prevent insect dispersal during transportation; placing screens covered with grease or other sticky substance over the areas where coffee is placed before initiating the wet processing; and properly managing pulp after depulping the berry to prevent insect dispersal (Bustillo P. *et al.*, 1998). The insect could also survive the drying process (Bustillo P. *et al.*, 1998). In coffee stores in Yemen, dead coffee berry borers were found in coffee beans imported from Ethiopia (Mahdi, 2006).

Another mechanism for passive dispersal is via animals. An experiment conducted by Leefmans (1923) found that 5.75 lb of coffee seeds in the feces of the Asian palm civet *Paradoxurus hermaphroditus* Pallas contained 17 coffee berry borer eggs, 11 newly hatched living adults, eight living old adults, seven dead adults, 33 living pupae, and 62 living larvae for a total of 114 live insects. Survival of coffee berry borers after being excreted by the palm civet was also confirmed by Gandrup (1922).

The use of cover crops could be an effective method to reduce insect levels inside berries that have fallen on the ground. According to Vázquez-Moreno (2005), coffee plantations in Cuba with *Zebrina pendula* Schnizl. (current name: *Tradescantia zebrina* var. *zebrina* Bosse; Commelinaceae) as a cover crop have a lower number of berries on the ground in the interseason period (i.e., between harvests). This is due to a faster decomposition of berries when the cover crop is present, which reduces survival of coffee berry borers within these berries. The decomposition of berries also makes dispersal of insects less successful due to the reduced number of suitable berries that they can find on the ground. In Mexico, Pohlan (2005) and Pohlan *et al.* (2008) found lower coffee berry borer levels in plantations with *Canavalia ensiformis* (L.) DC (Fabaceae) as a cover crop.

3.16 Climate Change

Jaramillo *et al.* (2009c) determined that 14.9 and 32°C were the upper and lower thresholds for coffee berry borer development, with 26.7°C being the optimal development temperature. According to Jaramillo *et al.* (2009c, 2011, 2013), increases in average daily temperature (global warming) in coffee-growing areas where temperatures have not reached 26.7°C could result in faster developmental time, increased number of generations, as well as an expanded distribution of the insect to elevations where it might not be able to otherwise survive. Bark beetles and climate change is covered in detail in Chapter 13.

4. CONCLUSIONS

There is very little information on the biology and ecology of the vast majority of *Hypothenemus* species. A reason for this is the difficulty in identifying the species and the difficulty in conducting the fieldwork that is essential to better understand them. The authors hope that this chapter will result in new research programs among entomologists and ecologists that will result in new and novel insights in the genus *Hypothenemus*.

As for the coffee berry borer, even though more than 100 years of research have been conducted on the insect, it remains the most economically important insect pest of coffee worldwide. This is likely due to a narrow focus on pest management approaches that have been repeatedly attempted in many different countries. For example, the use of methanol:ethanol traps has resulted in dozens of papers notwithstanding its lack of effectiveness in reducing population levels. Similarly, the use of parasitoids and fungal entomopathogens has been implemented in many countries, with mixed results, which have only slightly or temporarily alleviated the problem. In the same vein, it is unlikely that additional fungal entomopathogen field and laboratory bioassays will solve the problem, or that further studies on the predatory effects of ants or birds will suddenly reveal major insights over what has already been reported.

Currently available pest management strategies require collective action among coffee growers, a very difficult task. If one coffee grower implements one or several available strategies and the neighbor does not, then all the effort by the enterprising grower might be lost due to insect dispersal from the neighbor's field. A better understanding of basic biology issues related to the coffee berry borer that are just beginning to be elucidated, such as the genome and microbiota, might reveal novel strategies for pest management. Nevertheless, these possible strategies remain a distant dream. What is needed today is a novel strategy that dramatically reduces coffee berry borer population levels in the field. Such a strategy could involve the deployment of coffee berry borer-specific attractants and/or repellents. In order for these to be adopted, they need to be effective and repeatedly shown to reduce damage and consequently to increase yields. Another possibility is the use of fungal entomopathogens as endophytes. If a successful method can be developed to introduce fungal entomopathogens into coffee seedlings with long-term establishment or induced effects against insects, then they might become a feasible option for growers. After more than 100 years of dealing with the coffee berry borer, coffee growers deserve a novel pest management breakthrough that improves their economies and consequently their lives.

ACKNOWLEDGMENTS

For assistance with questions related to topics included in this chapter the authors wish to express their appreciation to M. C. Aime, T. H. Atkinson, M. Blackwell, E. L. Brodie, A. Castillo, J. Ceja-Navarro, A. Cossé, A. P. Davis, B. Dufour, F. Gallardo-Covas, J. Gómez, T. Halstead, R. A. Humber, M. A. Jackson, J. L. Jurat-Fuentes, M. J. Kern, E. Mendesil, A. M. Meneguim, J. Pérez, R. Pereira, A. Ramirez Lluch, H. W. Rogg, E. Schubert, D. Seudieu, D. I. Shapiro-Ilan, W. D. da Silva, S. R. Steiman, and J. M. Valdéz Carrasco. Special thanks to W. K. Olson at the National Agricultural Library and A. M. Galindo at the El Colegio de la Frontera Sur library for assistance in obtaining hard to find papers, and to Ann Simpkins for 14 years of excellent work at the Beltsville laboratory and for her inexhaustible good humor and camaraderie.

REFERENCES[2]

Abraham, Y.J., Moore, D., Godwin, G., 1990. Rearing and aspects of biology of *Cephalonomia stephanoderis* and *Prorops nasuta* (Hymenoptera: Bethylidae) parasitoid of the coffee berry borer, *Hypothenemus hampei* (Coleoptera: Scolytidae). Bull. Entomol. Res. 80, 121–128.

Acevedo-Bedoya, F.E., Gil-Palacio, Z.N., Bustillo-Pardey, A.E., Montoya-Restrepo, E.C., Benavides-Machado, P., 2009. Evaluación de marcadores físicos y moleculares como herramientas para el estudio de la dispersión de *Hypothenemus hampei*. Cenicafé 61, 72–85.

Acuña, R., Padilla, B.E., Flórez-Ramos, C.P., Rubio, J.D., Herrera, J.C., Benavides, P., et al., 2012. Adaptive horizontal transfer of a bacterial gene to an invasive insect pest of coffee. Proc. Natl. Acad. Sci. U. S. A. 109, 4197–4202.

Agramont, R., Cuba, N., Beltrán, J.L., Almanza, J.C., Loza-Murguía, M., 2010. Trampas artesanales con atrayentes alcohólicos una alternativa para el monitoreo y control de la broca del café, *Hypothenemus hampei* (Ferrari 1867). J. Selva Andina Res. Soc. 1, 2–12.

Aguilera-Gálvez, C., Vásquez-Ospina, J.J., Gutiérrez-Sanchez, P., Acuña-Zornosa, R., 2013. Cloning and biochemical characterization of an endo-1,4-β-mannanase from the coffee berry borer, Hypothenemus hampei. BMC Res. Notes 6, 333.

Albertin, A., Nair, P.K.R., 2004. Farmers' perspectives on the role of shade trees in coffee production systems: an assessment from the Nicoya Peninsula, Costa Rica. Hum. Ecol. 32, 443–463.

Allard, G.B., Moore, D., 1989. *Heterorhabditis* sp. nematodes as control agents for coffee berry borer, *Hypothenemus hampei* (Scolytidae). J. Invertebr. Pathol. 54, 45–48.

Alvarez Sandoval, J.E., Cortina Guerrero, H.A., 2004. ¿Presenta partenogénesis *Hypothenemus hampei* Ferrari (Coleoptera: Curculionidae: Scolytinae)? Fitotecnia Colombiana 4, 107–111.

Andreev, D., Breilid, H., Kirkendall, L., Brun, L.O., ffrench-Constant, R.H., 1998. Lack of nucleotide variability in a beetle pest with extreme inbreeding. Insect Mol. Biol. 7, 197–200.

Anonymous, 1930. Coffee borer beetle in India. The Times, London, 2nd July 1930. (Rev. Appl. Entomol., Ser. A, 1930, 18, 364).

Anonymous, 1979. New Records. Fiji. Coffee. Quarterly Newsletter. FAO Plant Protection Committee for the South East Asia and Pacific Region 22, 4.

Antía-Londoño, O.P., Posada-Florez, F., Bustillo-Pardey, A.E., González-García, M.T., 1992. Producción en finca del hongo *Beauveria bassiana* para el control de la broca del café. Cenicafé, Avances Técnicos No. 182, 12 pp.

Arcila Moreno, A., 2011. Período crítico del ataque de la broca del café. Brocarta 43.

Arcila, M.A., Cárdenas, M.R., Vélez, T.C., Bustillo, A.E., 1997. Registro de un nuevo hospedero de la avispita de Costa de Marfil *Cephalonomia stephanoderis* Betrem (Hymenoptera: Bethylidae). In: Memorias, XXIV Congreso de la Sociedad Colombiana de Entomología Pereira, Colombia, pp. 77–78.

Aristizábal, L.F., Bustillo, A.E., Orozco, J., Chaves, B., 1998. Efecto del parasitoide *Cephalonomia stephanoderis* (Hymenoptera: Bethylidae)

sobre las poblaciones de *Hypothenemus hampei* (Coleoptera: Scolytidae) durante y después de la cosecha. Rev. Colomb. Entomol. 24, 149–155.

Armbrecht, I., Perfecto, I., 2003. Litter-twig dwelling ant species richness and predation potential within a forest fragment and neighboring coffee plantations of contrasting habitat quality in Mexico. Agr. Ecosyst. Environ. 97, 107–115.

Armbrecht, I., Gallego, M.C., 2007. Testing ant predation on the coffee berry borer in shaded and sun coffee plantations in Colombia. Entomol. Exp. Appl. 124, 261–267.

Armuelles, H.P., 2007. Manejo de la broca del café en la República de Panamá. In: Barrera, J.F., García, A., Domínguez, V., Luna, C. (Eds.), La broca del café en América tropical: hallazgos y enfoques. pp. 33–36, Sociedad Mexicana de Entomología.

Arrieta, G., Hernández, A., Espinoza, A.M., 2004. Diversity of *Bacillus thuringiensis* strains isolated from coffee plantations infested with the coffee berry borer *Hypothenemus hampei*. Rev. Biol. Trop. 52, 757–764.

Atkinson, T.H., 2014. Bark and Ambrosia Beetles. Available online: http://www.barkbeetles.info. Last accessed: April 21, 2014.

Averna-Saccá, R., 1930. Os entomophagos cryptogamicos na broca do caffeiro (*Stephanoderes hampei* Ferr.) encontrados em S. Paulo. Boletim de Agricultura 31, 10–24, 195–213. (Rev. Appl. Entomol., Ser. A, 1930, 18, 640).

de Azevedo Pereira, R., Nogueira Batista, J.A., Mattar da Silva, M.C., de Oliveira Neto, O.S., Zangrando Figueira, E.L., Valencia Jiménez, A., Grossi-de-Sa, M.F., 2006. An α-amylase inhibitor gene from *Phaseolus coccineus* encodes a protein with potential for control of coffee berry borer (*Hypothenemus hampei*). Phytochemistry 67, 2009–2016.

de Azevedo Pereira, R., Valencia Jiménez, A., Picanço Magalhães, C., Prates, M.V., Taquita Melo, J.A., Maria de Lima, L., et al., 2007. Effect of a Bowman-Birk proteinase inhibitor from *Phaseolus coccineus* on *Hypothenemus hampei* gut proteinases in vitro. J. Agric. Food Chem. 55, 10714–10719.

Baguena Corella, L., 1941. El *Stephanoderes Hampei* [*sic*] Ferr. en los territorios españoles del Golfo de Guinea. Dirección General de Marruecos y Colonias, Dirección de Agricultura de los Territorios Españoles del Golfo de Guinea. Publicación 5, 117 pp.

Baker, P.S., 1984. Some aspects of the behaviour of the coffee berry borer in relation to its control in Southern Mexico (Coleoptera: Scolytidae). Folia Entomológica Mexicana 61, 9–24.

Baker, P.S., 1999. The coffee berry borer in Colombia. Final report of the DFID–Cenicafé–CABI Bioscience IPM for coffee project. Chinchiná (Colombia), DFID–CENICAFÉ. 144 pp. Also available in Spanish as: Baker, P. S. (1999). La broca del café en Colombia. Informe final del proyecto MIP para el café DFID–Cenicafé–CABI Bioscience (CNTR 93/1536A). Chinchiná (Colombia), DFID–CENICAFÉ. 148 pp.

Baker, P.S., Barrera, J.F., Valenzuela, J.E., 1989. The distribution of the coffee berry borer (*Hypothenemus hampei*) in Southern Mexico: a survey for a biocontrol project. Trop. Pest Manage. 35, 163–168.

Baker, P.S., Barrera, J.F., Rivas, A., 1992a. Life-history studies of the coffee berry borer (*Hypothenemus hampei*, Scolytidae) on coffee trees in southern Mexico. J. Appl. Entomol. 29, 656–662.

Baker, P.S., Ley, C., Balbuena, R., Barrera, J.F., 1992b. Factors affecting the emergence of *Hypothenemus hampei* (Coleoptera: Scolytidae) from coffee berries. Bull. Entomol. Res. 82, 145–150.

Baker, P., Barrera, J., 1993. A field study of a population of coffee berry borer, *Hypothenemus hampei* (Coleoptera; Scolytidae), in Chiapas, Mexico. Trop. Agric. (Trinidad) 70, 351–355.

---

2. Some of the references are appended with the *Review of Applied Entomology, Series A* citation in parenthesis, which might be easier to locate than the original article and which includes a summary of the article in English.

Baker, P.S., Rivas, A., Balbuena, R., Ley, C., Barrera, J.F., 1994. Abiotic mortality factors of the coffee berry borer (*Hypothenemus hampei*). Entomol. Exp. Appl. 71, 201–209.

Baker, P.S., Jackson, J., Murphy, S.T., 2002. Manejo integrado de la plaga de la broca del fruto del café. Informe del proyecto CFC/ICO/02. CABI Bioscience, Silwood Park, 16 pp.

Balakrishnan, M.M., Sreedharan, K., Krishnamoorthy Bhat, P., 1994. Occurrence of the entomopathogenic fungus *Beauveria bassiana* on certain coffee pests in India. J. Coffee Res. 24, 33–35.

Barbosa, A.E.A.D., Albuquerque, E.V.S., Silva, M.C.M., Souza, D.S.L., Oliveira-Neto, O.B., Valencia, A., et al., 2010. α-amylase inhibitor-1 gene from *Phaseolus vulgaris* expressed in *Coffea arabica* plants inhibits α-amylases from the coffee berry borer. BMC Biotechnol. 10, 44.

Bardner, R., 1985. Pest control. In: Clifford, M.N., Willson, K.C. (Eds.), Coffee: Botany, Biochemistry and Production of Beans and Beverage. Croom Helm, London, pp. 208–218.

Barradas, V.L., Fanjul, L., 1986. Microclimatic characterization of shaded and open-grown coffee (*Coffea arabica* L.) plantations in Mexico. Agric. For. Meteor. 38, 101–112.

Barrera, J.F., 1994. Dynamique des populations du scolyte des fruits du caféier, *Hypothenemus hampei* (Coleoptera: Scolytidae), et lutte biologique avec le parasitoide *Cephalonomia stephanoderis* (Hymenoptera: Bethylidae), au Chiapas, Mexique. Ph.D. thesis, Université Paul Sabatier, Toulouse, France, 301 pp.

Barrera, J.F., 2008. Coffee pests and their management. In: Capinera, J.L. (Ed.), Encyclopedia of Entomology, second ed. Springer, Dordrecht, pp. 961–998.

Barrera, J.F., Gómez, J., Infante, F., Castillo, A., de la Rosa, W., 1989. Biologie de *Cephalonomia stephanoderis* Betrem (Hymenoptera: Bethylidae) en laboratoire. I. Cycle biologique, capacité d'oviposition et émergence du fruit du caféier. Café, cacao, thé 33, 101–108.

Barrera, J.F., Baker, P.S., Schwarz, A., Valenzuela, J., 1990a. Introducción de dos especies de parasitoides africanos a México para el control biológico de la broca del cafeto *Hypothenemus hampei* (Ferr.) (Coleoptera: Scolytidae). Folia Entomológica Mexicana 79, 245–247.

Barrera, J.F., Infante, F., Vega, M., González, O., Carrillo, E., Campos, O., et al., 1990b. Introducción de *Cephalonomia stephanoderis* (Hymenoptera: Bethylidae) a Centroamérica para el control biológico de la broca del cafeto, *Hypothenemus hampei* (Coleoptera: Scolytidae). Turrialba 40, 570–574.

Barrera, J.F., Moore, D., Abraham, Y.J., Murphy, S.T., Prior, C., 1990c. Biological control of the coffee berry borer *Hypothenemus hampei* in Mexico and possibilities for further action. Brighton Crop Protection Conference—Pest and Diseases, pp. 391–396.

Barrera, J.F., Infante, F., Castillo, A., de la Rosa, W., Gómez, J., 1991. Cría y manejo de *Cephalonomia stephanoderis* y *Prorops nasuta*, parasitoides de la broca del café. Miscelánea de la Sociedad Colombiana de Entomología 18, 76–86.

Barrera, J.F., Infante, F., Gómez, J., Castillo, A., de la Rosa, W., 1993a. Guía practica. Umbrales económicos para el control de la broca del café. Centro de Investigaciones Ecológicas del Sureste, Tapachula, Chiapas, México, 50 pp.

Barrera, J.F., Infante, F., Alauzet, C., Gómez, J., de la Rosa, W., Castillo, A., 1993b. Biologie de *Cephalonomia stephanoderis* Betrem (Hymenoptera: Bethylidae) en laboratoire. II. Dureé de développement, sex-ratio, longévité et espérance de vie des adultes. Café, Cacao, Thé 37, 205–214.

Barrera, J.F., Gómez, G., Alauzet, C., 1994. Evidence for a marking pheromone in host discrimination by *Cephalonomia stephanoderis* (Hym.: Bethylidae). Entomophaga 39, 363–366.

Barrera, J.F., Gómez, J., Alauzet, C., 1995. Can the coffee berry borer (*Hypothenemus hampei*) reproduce by parthenogenesis. Entomol. Exp. Appl. 77, 351–354.

Barrera, J.F., Infante, F., de la Rosa, W., Castillo, A., Gómez, J., 2000. Control biológico de broca del café. In: Badii, M.H., Galán-Wong, J.L. (Eds.), Fundamentos y Perspectivas de Control Biológico. Universidad Autónoma de Nuevo León, México, pp. 211–229.

Barrera, J.F., Gómez, J., López, E., Herrera, J., 2004. Muestreo adaptivopara "la broca del café" (*Hypothenemus hampei*). Entomología Mexicana 3, 535–539.

Barrera, J.F., Herera, J., Villacorta, A., García, H., Cruz, L., 2006. Trampas de metanol-etanol para detección, monitoreo y control de la broca del café *Hypothenemus hampei*. In: Barrera, J.F., Montoya, P. (Eds.), Simposio sobre Trampas y Atrayentes en Detección, Monitoreo y Control de Plagas de Importancia Económica, pp. 71–83, Sociedad Mexicana de Entomología y El Colegio de la Frontera Sur.

Barrera, J.F., Herrera, J., Chiu, M., Gómez, J., Valle Mora, J., 2008. La trampa de una ventana (ECOIAPAR) captura más broca del café *Hypothenemus hampei* que la trampa de tres ventanas (ETOTRAP). Entomología Mexicana 7, 619–624.

Batchelor, T.P., Hardy, I.C.W., Barrera, J.F., 2006. Interactions among bethylid parasitoid species attacking the coffee berry borer, *Hypothenemus hampei* (Coleoptera: Scolytidae). Biol. Contr. 36, 106–118.

Batra, L.R., 1963. Ecology of ambrosia fungi and their dissemination by beetles. Trans. Kans. Acad. Sci. 66, 213–236.

Bautista, L., 2000. Hongos entomopatógenos parasitando en estado natural broca del café *Hypothenemus hampei* (Ferrari) (Coleoptera: Scolytidae) en el estado Táchira, Venezuela. VI Jornadas Científico Técnicas, Febrero 22–25, 2000, Universidad Nacional Experimental del Táchira, Venezuela.

Bautista Martínez, N., Atkinson Martín, T.T.H., 1988. Biología y respuesta a dietas semiartificiales de la broca del café *Hypothenemus hampei* (Ferr.) (Coleoptera: Scolytidae) bajo condiciones de laboratorio. Revista Chapingo 12, 26–30.

Beardsley, J.W., 1990. *Hypothenemus obscurus* (Fabricius) (Coleoptera: Scolytidae), a new pest of macadamia nuts in Hawaii. Proc. Hawaii. Entomol. Soc. 30, 147–150.

Beaver, R.A., 1986. The taxonomy, mycangia and biology of *Hypothenemus curtipennis* (Schedl), the first known cryphaline ambrosia beetle (Coleoptera: Scolytidae). Ent. Scand. 17, 131–135.

Beaver, R.A., Maddison, P.A., 1990. The bark and ambrosia beetles of the Cook Islands and Niue (Coleoptera: Scolytidae and Platypodidae). J. Nat. Hist. 24, 1365–1375.

Beaver, R.A., Liu, L.Y., 2010. An annotated synopsis of Taiwanese bark and ambrosia beetles, with new synonymy, new combinations and new records. Zootaxa 2602, 1–147.

Beille, L., 1925. Les *Stephanoderes* sur les Caféiers cultivés a la Côte d'Ivoire. Revue de Botanique Appliquée et d'Agriculture Tropicale 5, 387–388.

Bemelmans, J., 1930. Les ennemis du caféier. Ann. Gembloux 36, 418–424.

Benassi, V.L.R.M., 1989. Levantamento dos inimigos naturais da broca-do-café, *Hypothenemus hampei* (Ferr.) (Coleoptera: Scolytidae) no norte do Espírito Santo. Anais da Sociedade Entomológica do Brasil 24, 635–638.

Benassi, V.L.R.M., 1995. Introdução da espécie *Cephalonomia stephano-deris* Betrem 1961 (Hymenoptera: Bethylidae) parasitóide da broca-do-café, *Hypothenemus hampei* (Ferr., 1867) (Coleoptera: Scolytidae), Resumos, XV Congresso Brasileiro de Entomologia, p. 336.

Benavides, M., Portilla, M., 1991. Uso del café pergamino para la cría de la broca del café, *Hypothenemus hampei* y de su parasitoide *Cephalonomia stephanoderis* Betrem en Colombia. Miscelánea de la Sociedad Colombiana de Entomología 18, 87–90.

Benavides, P., Bustillo, A.E., Portilla, M., Orozco, J., 2002. Classical biological control of the coffee berry borer, *Hypothenemus hampei* (Coleoptera: Scolytidae) in Colombia with African parasitoids. Proceedings, First International Symposium on Biological Control of Arthopods, Honolulu, Hawaii, pp. 430–434.

Benavides, P., Vega, F.E., Romero-Severson, J., Bustillo, A.E., Stuart, J.J., 2005. Biodiversity and biogeography of an important inbred pest of coffee, coffee berry borer (Coleoptera: Curculionidae: Scolytinae). Ann. Entomol. Soc. Am. 98, 359–366.

Benavides, P., Stuart, J.J., Vega, F.E., Romero-Severson, J., Bustillo, A.E., Navarro, L., et al., 2006. Genetic variability of *Hypothenemus hampei* (Ferrari) in Colombia and development of molecular markers. Proceedings of the 21st International Scientific Colloquium on Coffee, Association Scientifique Internationale du Café (ASIC), Montpellier, France, pp. 1301–1315.

Benavides, P., Góngora, C., Bustillo, A., 2012. IPM program to control coffee berry borer *Hypothenemus hampei*, with emphasis on highly pathogenic mixed strains of *Beauveria bassiana*, to overcome insecticide resistance in Colombia. In: Perveen, F. (Ed.), Insecticides. Advances in Integrated Pest Management. InTech Europe, Croatia, pp. 511–540.

Benavides, P., Navarro, L., Acevedo, F., Acuña, R., O'Brochta, D., Stuart, J., et al., 2014. The genome of the coffee berry borer, *Hypothenemus hampei*, the major insect pest of coffee worldwide. Abstract W186, Plant & Animal Genome XXII, San Diego, CA.

Benavides-G. M., Portilla-R., M., 1990. Uso del café pergamino para la críade *Hypothenemus hampei* y de su parasitoide *Cephalonomia stephanoderis* en Colombia. Cenicafé 41, 114–116.

Benavides M., P., 2010. Manejo de la broca durante las renovaciones de café en la finca. Brocarta, 38.

Benavides Machado, P., Stuart, J.J., Vega, F.E., Romero-Severson, J., Bustillo, A.E., 2007. Biogeografía y aspectos genéticos de la broca del café *Hypothenemus hampei*. In: Anais Manejo da Broca-do-Café, Workshop Internacional, pp. 11–36, Londrina, Brasil.

Bergamin, J., 1943a. Contribuição para o conhecimento da biologia da broca do café "*Hypothenemus hampei* (Ferrari, 1867)" (Col. Ipidae). Arq. Inst. Biol., São Paulo 14, 31–72, (Rev. Appl. Entomol., Ser. A, 1945, 33, 203–204).

Bergamin, J., 1943b. A "broca" e as árvores de sombra. Revista do Departamento Nacional do Café 20, 31–32.

Bergamin, J., 1944a. O "repasse" como método de controle da broca do café "*Hypothenemus hampei* (Ferr., 1867)" (Col. Ipidae). Arq. Inst. Biol., São Paulo 15, 197–208, (Rev. Appl. Entomol., Ser. A, 1947, 35, 105).

Bergamin, J., 1944b. Sombreamento e broca. Revista do Departamento Nacional do Café 12, 181–184.

Bergamin, J., 1944c. Sombreamento e broca. Revista do Departamento Nacional do Café 12, 1009–1014.

Bergamin, J., 1945a. O sombreamento dos cafèzais e a "broca do café." Revista do Departamento Nacional do Café 13, 627–638.

Bergamin, J., 1945b. Broca do café. Revista de Agricultura 20, 427–430.

Bergamin, J., Kerr, W.E., 1951. Determinação do sexo e citologia da broca do café. Cienc. Cult. (Campinas, Brazil) 3, 117–121.

Bernal U., M.G., Bustillo P., A.E., Posada F., F.J., 1994. Virulencia de aislamientos de *Metarhizium anisopliae* y su eficacia en campo sobre *Hypothenemus hampei*. Rev. Colomb. Entomol. 20, 225–228.

Berrio E., A., Benavides M., P., 2008. Evaluación de la meiosis y del potencial de partenogénesis en la broca del café, *Hypothenemus hampei* (Coleoptera: Curculionidae: Scolyinae) mediante técnicas citológicas. Congreso de la Sociedad Colombiana de Entomología, Cali, p. 53.

Berthet, J.A., 1913. Caruncho do café. Informação prestada pelo Sr. Dr. Director do Instituto Agronômico a respeito de amostras de café vindas do Congo Belga. Boletim de Agricultura 14, 312–313.

Betrem, J.G., 1961. *Cephalonomia stephanoderis* nov. spec. (Hymenoptera: Bethylidae). Entomologische Berichten 21, 183–184.

Bigger, M., 2008. A geographical distribution of insects and mites associated with coffee, derived from literature published before 2008. Self-published.

Bischoff, J.F., Rehner, S.A., Humber, R.A., 2009. A multilocus phylogeny of the *Metarhizium anisopliae* lineage. Mycologia 101, 512–530.

Bode, R.F., Halitschke, R., Kessler, A., 2013. Herbivore damage-induced production and specific anti-digestive function of serine and cysteine protease inhibitors in tall goldenrod, *Solidago altissima* L. (Asteraceae). Planta 237, 1287–1296.

Borbón-Martínez, O., 1989. Bioécologie d'un ravageur des baies de caféier, *Hypothenemus hampei* Ferr. (Coleoptera: Scolytidae) et de ses parasitoides auTogo. Ph.D. thesis, Université Paul Sabatier, Toulousse, France, 185 pp.

Borbón-Martínez, O., 2007. Liberaciones, comportamiento y dispersión de los parasitoides de la broca del fruto del cafeto, *Hypothenemus hampei* (Ferrari, 1987), *Prorops nasuta* y *Phymastichus coffea* procedentes de Colombia, en Costa Rica. Anais Manejo da Broca-do-Café, Workshop Internacional. Londrina, Brasil, pp. 199–207.

Borbón-Martínez, O., Mora Alfaro, O., Cam Oehlschlager, A., González, L.M., 2000. Proyecto de trampas, atrayentes y repelentes para el control de la broca del fruto de cafeto, *Hypothenemus hampei* L. (Coleoptera: Scolytidae). Memoria, XIX Simposio Latinoamericano de Caficultura, San José, Costa Rica, pp. 331–348.

Borsa, P., Gingerich, D.P., 1995. Allozyme variation and an estimate of the inbreeding coefficient in the coffee berry borer, *Hypothenemus hampei* (Coleoptera: Scolytidae). Bull. Entomol. Res. 85, 21–28.

Borsa, P., Coustau, C., 1996. Single-stranded DNA conformation polymorphism at the *Rdl* locus in *Hypothenemus hampei* (Coleoptera: Scolytidae). Heredity 76, 124–129.

Borsa, P., Kjellberg, F., 1996a. Secondary sex ratio adjustment in a pseudoarrhenotokous insect, *Hypothenemus hampei* (Coleoptera: Scolytidae). C. R. Acad. Sci. Paris, Ser. III 319, 1159–1166.

Borsa, P., Kjellberg, F., 1996b. Experimental evidence for pseudoarrhenotoky in *Hypothenemus hampei* (Coleoptera: Scolytidae). Heredity 76, 130–135.

Bosselmann, A.S., Dons, K., Oberthur, T., Olsen, C.S., Ræbild, A., Usma, H., 2009. The influence of shade trees on coffee quality in small holder coffee agroforestry systems in Southern Colombia. Agr. Ecosyst. Environ. 129, 253–260.

Bradbury, A.G.W., Halliday, D.J., 1990. Chemical structures of green coffee bean polysaccharides. J. Agric. Food Chem. 38, 389–392.

Breilid, H., Brun, L.O., Andreev, D., ffrench-Constant, R.H., Kirkendall, L.R., 1997. Phylogeographic patterns of introduced populations of the coffee berry borer *Hypothenemus hampei* (Ferrari) (Coleoptera: Scolytidae) inferred from mitochondrial DNA sequences. Proceedings of the 17th International Scientific Colloquium on Coffee, Association Scientifique Internationale du Café (ASIC), Nairobi, Kenya, pp. 653–655.

Bridge, P.D., Abraham, Y.J., Cornish, M.C., Prior, C., Moore, D., 1990. The chemotaxonomy of *Beauveria bassiana* (Deuteromycotina: Hyphomycetes) isolates from the coffee berry borer *Hypothenemus hampei* (Coleoptera: Scolytidae). Mycopathologia 111, 85–90.

Bright, D.E., 2014. A catalog of Scolytidae and Platypodidae (Coleoptera), Supplement 3 (2000–2010), with notes on subfamily and tribal reclassifications. Insecta Mundi 0356, 1–336.

Bright, D.E., Skidmore, R.E., 1997. A catalog of Scolytidae and Platypodidae (Coleoptera), Supplement 1 (1990–1994). NRC Research Press, Ottawa.

Bright, D.E., Skidmore, R.E., 2002. A catalog of Scolytidae and Platypodidae (Coleoptera), Supplement 2 (1995–1999). NRC Research Press, Ottawa.

Brown, E., Hunter, H.H., 1913. Planting in Uganda. Coffee–Para Rubber–Cocoa. Longmans, Green and Co., Ltd, London.

Browne, F.G., 1961. The biology of Malayan Scolytidae and Platypodidae. Malayan Forest Records 22, 1–255.

Browne, F.G., 1963. Taxonomic notes on Scolytidae (Coleoptera). Entomologische Berichten 23, 53–59.

Brun, L.O., Suckling, D.M., 1992. Field selection for endosulfan resistance in coffee berry borer (Coleoptera: Scolytidae) in New Caledonia. J. Econ. Entomol. 85, 325–334.

Brun, L.O., Mathieu, F., 1997. Utilisation d'un piège à attractif kairomonal pour le suivi des populations du scolyte du café en champ. Proceedings of the 17th International Scientific Colloquium on Coffee, Association Scientifique Internationale du Café (ASIC), Nairobi, Kenya, pp. 714–717.

Brun, L.O., Marcillaud, C., Gaudichon, V., Suckling, D.M., 1989a. Endosulfan resistance in *Hypothenemus hampei* (Coleoptera: Scolytidae) in New Caledonia. J. Econ. Entomol. 82, 1311–1316.

Brun, L.O., Gaudichon, V., Marcillaud, C., 1989b. Provisional method for detecting endosulfan resistance in coffee berry borer, *Hypothenemus hampei* (Coleoptera: Scolytidae). FAO Plant Prot. Bull. 37, 125–129.

Brun, L.O., Marcillaud, C., Gaudichon, V., 1990. Monitoring of endosulfan and lindane resistance in the coffee berry borer *Hypothenemus hampei* (Coleoptera: Scolytidae) in New Caledonia. Bull. Entomol. Res. 80, 129–135.

Brun, L.O., Marcillaud, C., Gaudichon, V., Suckling, D.M., 1991. Evaluation of a rapid bioassay for diagnosing endosulfan resistance in coffee berry borer, *Hypothenemus hampei* (Ferrari) (Coleoptera: Scolytidae). Trop. Pest Manage. 37, 221–223.

Brun, L.O., Gaudichon, V., Wigley, P.J., 1993. An artificial diet for continuous rearing of the coffee berry borer, *Hypothenemus hampei* (Ferrari) (Coleoptera: Scolytidae). Insect Sci. Appl. 14, 585–587.

Brun, L.O., Borsa, P., Gaudichon, V., Stuart, J.J., Aronstein, K., Coustau, C., ffrench-Constant, R.H., 1995a. Functional haplodiploidy. Nature 374, 506.

Brun, L.O., Stuart, J., Gaudichon, V., Aronstein, K., ffrench-Constant, R.H., 1995b. Functional haplodiploidy: a mechanism for the spread of insecticide resistance in an important international insect pest. Proc. Natl. Acad. Sci. U. S. A. 92, 9861–9865.

Buffington, M.L., Polaszek, A., 2009. Recent occurrence of *Aphanogmus dictynna* (Waterston) (Hymenoptera: Ceraphronidae) in Kenya—an important hyperparasitoid of the coffee berry borer *Hypothenemus hampei* (Ferrari) (Coleoptera: Curculionidae). Zootaxa 2214, 62–68.

Bugnicourt, F., 1950. Le "scolyte du grain de café" en Nouvelle-Calédonie. Revue Agricole de la Nouvelle Calédonie 1–2, 3–4.

Burbano, E., Wright, M., Bright, D.E., Vega, F.E., 2011. New record for the coffee berry borer, *Hypothenemus hampei*, in Hawaii. J. Insect Sci. 11, 117.

Bustillo, A., Castillo, H., Villalba, D., Morales, E., Vélez, P., 1991. Evaluaciones de campo con el hongo *Beauveria bassiana* para el control de la broca del café, *Hypothenemus hampei* en Colombia, 14th International Scientific Colloquium on Coffee, San Francisco, California, pp. 679–686.

Bustillo, A.E., Bernal, M.G., Benavides, P., Chaves, B., 1999. Dynamics of *Beauveria bassiana* and *Metarhizium anisopliae* infecting *Hypothenemus hampei* (Coleoptera: Scolytidae) populations emerging from fallen coffee berries. Fla. Entomol. 82, 491–498.

Bustillo, A.E., Cárdenas, R., Posada, F.J., 2002. Natural enemies and competitors of *Hypothenemus hampei* (Ferrari) (Coleoptera: Scolytidae) in Colombia. Neotrop. Entomol. 31, 635–639.

Bustillo P., A.E., Cárdenas M., R., Villalba G., D.A., Benavides M., P., Orozco H., J., Posada, F., 1998. Manejo integrado de la broca del café *Hypothenemus hampei* (Ferrari) en Colombia. Centro Nacional de Investigaciones de Café (Cenicafé). Chinchiná, Colombia, 134 pp.

Bustillo Pardey, A.E., 2005. El papel del control biológico en el manejo integrado de la broca del café, *Hypothenemus hampei* (Ferrari) (Coleoptera: Curculionidae: Scolytinae). Revista de la Academia Colombiana de Ciencias Exactas Físicas y Naturales 29, 55–68.

Bustillo Pardey, A.E., 2006. Una revisión sobre la broca del café, *Hypothenemus hampei* (Coleoptera: Curculionidae: Scolytinae), en Colombia. Rev. Colomb. Entomol. 32, 101–116.

Bustillo-Pardey, A.E., Orozco-Hoyos, L., Benavides-Machado, P., Portilla-Reina, M., 1996. Producción masiva y uso de parasitoides para el control de la broca del café en Colombia. Cenicafé 47, 215–230.

Byers, J.A., 1992. Attraction of bark beetles, *Tomicus piniperda*, *Hylurgops palliatus*, and *Trypodendron domesticum* and other insects to short-chain alcohols and monoterpenes. J. Chem. Ecol. 18, 2385–2402.

Byers, J.A., 1995. Host-tree chemistry affecting colonization in bark beetles. In: Cardé, R.T., Bell, W.J. (Eds.), Chemical Ecology of Insects 2. Chapman and Hall, New York, pp. 154–213.

CABI, 2008. Final Report. CDF on Sustainable Pest Management for Coffee Small-holders in Laos and Vietnam, CABI Southeast and East Asia, Selangor, Malaysia, 19 pp.

Cade, S.C., Hrutfiord, B.F., Gara, R.I., 1970. Identification of a primary attractant for *Gnathotrichus sulcatus* isolated from western hemlock logs. J. Econ. Entomol. 63, 1014–1015.

Camilo, J.E., Olivares, F.F., Jiménez, H.A., 2003. Fenología y reproducción de la broca del café (*Hypothenemus hampei* Ferrari) durante el desarrollo del fruto. Agronomía Mesoamericana 14, 59–63.

Campos Almengor, O.G., 1981. El gandul *Cajanus cajan* como hospedero de la broca del fruto del café *Hypothenemus hampei* (Ferrari 1867) en Guatemala. IV Simposio Latinoamericano sobre Caficultura, Guatemala, Instituto Interamericano de Cooperación para la Agricultura, Ponencias, Resultados y Recomendaciones de Eventos Técnicos No. 322, pp. 155–159.

Campos Almengor, O.G., 1982. Estudio de hábitos de la broca del fruto del café (*Hypothenemus hampei*, Ferr. 1867) en el campo. V Simposio Latinoamericano sobre Caficultura, San Salvador, El Salvador,

Instituto Interamericano de Cooperación para la Agricultura, Ponencias, Resultados y Recomendaciones de Eventos Técnicos No. 323, pp. 38–49.

de Campos Novaes, J., 1922. Um broqueador do cafeeiro. *Xyleborus cofeicola* n. sp. Fam. Ipidae. Boletim de Agricultura, Secretaria da Agricultura, Commercio e Obras Publicas de S. Paulo 23, 67–70.

Cantor, R., Vilela, E.F., Bustillo, A.E., 1999. Primeira introdução no Brasil do endoparasitoide *Phymastichus coffea* LaSalle (Hymenoptera: Eulophidae), para o controle biológico da broca do café *Hypothenmus hampei* (Ferrari) (Coleoptera: Scolytidae). Anais III Seminário Internacional sobre Biotecnologia na Agroindústria Cafeeira, Londrina, Brazil, p. 305.

Cárdenas, R., 1995. Manejo de insectos que se alimentan del cafeto *Coffea arabica* L. (Rubiales: Rubiaceae) en Colombia. Memorias, XXII Congreso de la Sociedad Colombiana de Entomología, Santafé de Bogotá, pp. 86–90.

Cárdenas M., R., 2000. Trampas y atrayentes para monitoreo de poblaciones de broca del café *Hypothenemus hampei* (Ferrari) (Col., Scolytidae), Memorias del XIX Simposio Latinoamericano de Caficultura, San José, Costa Rica, pp. 369–379.

Cárdenas M., R., Bustillo, A.E., 1991. La broca del café en Colombia. In: Barrera, J.F., Castillo, A., Gómez, J., Malo, E., Infante, F. (Eds.), Resúmenes, I Reunión Intercontinental sobre Broca del Café Tapachula, Chiapas, México. pp. 42–44.

Cárdenas Murillo, R., Posada Flórez, F.J., 2001. Los insectos y otros habitantes de cafetales y platanales. Centro Nacional de Investigaciones de Café. Chinchiná, Colombia.

Cárdenas-Ramírez, A.B., Villalba-Guott, D.A., Bustillo-Pardey, A.E., Montoya-Restrepo, E.C., Góngora-Botero, C.E., 2007. Eficacia de mezclas de cepas del hongo *Beauveria bassiana* en el control de la broca del café. Cenicafé 58, 293–303.

Carlini, C.R., Grossi-de-Sá, M.F., 2002. Plant toxic proteins with insecticidal properties. A review on their potentialities as bioinsecticides. Toxicon 40, 1515–1539.

Carrión, G., Bonet, A., 2004. Mycobiota associated with the coffee berry borer (Coleoptera: Scolytidae) and its galleries in fruit. Ann. Entomol. Soc. Am. 97, 492–499.

Castellanos Domínguez, O.F., 1997. Importancia en la patogenicidad de la acción enzimática del hongo *Beauveria bassiana* sobre la broca del café. Rev. Colomb. Entomol. 23, 65–71.

Castillo, A., Marbán-Mendoza, N., 1996. Evaluación en laboratorio de nematodos entomopatógenos para el control biológico de la broca del café *Hypothenemus hampei*. Nematropica 26, 101–109.

Castillo, A., Infante, F., Barrera, J., Carta, L., Vega, F.E., 2002. First field report of a nematode (Tylenchida: Sphaerularioidea) attacking the coffee berry borer, *Hypothenemus hampei* (Ferrari) (Coleoptera: Scolytidae) in the Americas. J. Invertebr. Pathol. 79, 199–202.

Castillo, A., Infante, F., Vera-Graziano, J., Trujillo, J., 2004a. Host discrimination by *Phymastichus coffea* a parasitoid of the coffee berry borer. BioControl 49, 655–663.

Castillo, A., Infante, F., Lopez, G., Trujillo, J., Kirkendall, L.R., Vega, F. E., 2004b. Laboratory parasitism by *Phymastchus coffea* (Hymenoptera: Eulophidae) upon non-target bark beetles associated with coffee plantations. Fla. Entomol. 87, 274–277.

Ceja-Navarro, J., Brodie, E.L., Vega, F.E., 2012. A technique to dissect the alimentary canal of the coffee berry borer (*Hypothenemus hampei*), with isolation of internal microorganisms. J. Entomol. Acarol. Res. 44, e21.

Ceja-Navarro, J. A., Vega, F. E., Hao, Z., Lim, H. C., Kosina, P., Infante, F., Brodie, E. L. Gut microbiota mediate caffeine detoxification in the primary insect pest of coffee. Submitted.

Chapman, R.F., 2003. The Insects. Structure and Function, fourth ed. Cambridge University Press, New York.

Chapman, E.G., Jaramillo, J., Vega, F.E., Harwood, J.D., 2009. Biological control of coffee berry borer: the role of DNA-based gut-content analysis in assessment of predation. In: Mason, P.G., Gillespie, D.R., Vincent, C. (Eds.), Proceedings, 3rd International Symposium on Biological Control of Arthropods, Christchurch, New Zealand, pp. 475–484.

Chénier, J.V.R., Philogène, B.J.R., 1989. Field responses of certain forest Coleoptera to conifer monoterpenes and ethanol. J. Chem. Ecol. 15, 1729–1745.

Chevalier, A., 1947. Les Caféiers du Globe. Fascicule III. Systématique des Caféiers et Faux-Caféiers. Maladies et Insectes Nuisibles. Paul Lechevalier, Paris.

Chong, C., Chiu-Alvarado, P., Castillo, A., Barrera, J.F., Rojas, J., 2006. Preferencia espectral de la broca del café. *Hypothenemus hampei*. X Congreso International de Manejo Integrado de Plagas y Agroecología, El Colegio de la Frontera Sur, Mexico, pp. 162–164.

Cirerol, B., Infante, F., Castillo, A., 2002. Análisis químico de la nueva dieta merídica para *Hypothenemus hampei* (Ferrari) (Coleoptera: Scolytidae), con notas biológicas de su desarrollo en este substrato. Folia Entomológica Mexicana 41, 185–193.

Cisneros, P., Tandazo, A., 1990. Estudios preliminares del establecimiento del parasitoide *Prorops nasuta* W. en el sur del Ecuador. Sanidad Vegetal (Ecuador) 5, 43–50.

Coghlan, A., 2005. Coffee trial survives insects, but not vandals. New Scientist 2501, 14.

Coleman, L.C., 1931. Report on the coffee berry borer, *Stephanoderes hampei*, Ferr in Java, Department of Agriculture, Mysore State, General Series Bulletin 16, 1–20.

Constantino, L.M., Navarro, L., Berrio, A., Acevedo, F.E., Rubio, D., Benavides, P., 2011. Aspectos biológicos, morfológicos y genéticos de *Hypothenemus obscurus* e *Hypothenemus hampei* (Coleoptera: Curculionidae: Scolytinae). Rev. Colomb. Entomol. 37, 173–182.

Contreras, T., Camilo, J.E., 2007. Manejo de la broca del café en la República Dominicana. In: Barrera, J.F., García, A., Domínguez, V., Luna, C. (Eds.), La broca del café en América tropical: hallazgos y enfoques. Sociedad Mexicana de Entomología, pp. 43–55.

Corbett, G.H., 1933. Some preliminary observations on the coffee berry beetle borer, *Stephanoderes* (*Cryphalus*) *hampei* Ferr. Malayan Agricultural Journal 21, 8–22.

Córdova-Gámez, G., 1995. Plan de trabajo del Centro Reproductor de Entomopatógenos y Entomófagos (CREE), para 1995. Instituto Técnico Agropecuario de Oaxaca, No. 23, 15 pp.

Corporaal, J.B., 1921. De Koffiebesboorder op Sumatra's Oostkust en Atjeh. Mededeelingen van het Algemeen Proefstation der A.V.R.O.S. Algemeene Serie No. 12, 20 pp. (Rev. Appl. Entomol., Ser. A, 1922, 10, 571).

Costa, F.G., Faria, C.A., 2001. Por que fêmeas da broca do café perfuram preferencialmente a coroa dos frutos? Academia Insecta 1, 1–4.

Costa, J.N.M., da Silva, R.B., de Araújo Ribeiro, P., Garcia, A., 2002. Ocorrência de *Beauveria bassiana* (Bals.) Vuill. em broca-do-café (*Hypothenemus hampei*, Ferrari) no estado de Rondônia, Brasil. Acta Amazonica 32, 517–519.

da Costa Lima, A.M., 1924a. Sobre a broca do café (*Stephanoderes coffeæ*, Hag.). Chácaras e Quintaes (Brazil) 30, 316–319.

da Costa Lima, A.M., 1924b. Sobre a broca do café (*Stephanoderes coffeæ*, Hag). Chácaras e Quintaes (Brazil) 30, 413–416.

Crowe, T.J., 2004. Coffee pests in Africa. In: Wintgens, J.N. (Ed.), Coffee: Growing, Processing, Sustainable Production. Wiley-VCH, Weinheim, pp. 421–458.

Crowe, T.J., Gebremedhin, T., 1984. Coffee pests in Ethiopia. Their biology and control. Institute of Agricultural Research, Addis Abeba, Ethiopia.

Crowson, R.A., 1981. The Biology of the Coleoptera. Academic Press, London, 802 pp.

Cruz, L.P., Gaitan, A.L., Gongora, C.E., 2006. Exploiting the genetic diversity of *Beauveria bassiana* for improving the biological control of the coffee berry borer through the use of strain mixtures. Appl. Microbiol. Biotechnol. 71, 918–926.

D'Antonio, A.M., de Paula, V., 1979. Estudos preliminares de eficiência de *Metarhizium anisopliae* (Metsch) Sorokin no controle á broca do café, *Hypothenemus hampei* (Ferrari, 1867) em condições de laboratório. Resumos, 7° Congresso Brasileiro de Pesquisas Cafeeiras, Araxa, Brazil, p. 301.

Damon, A.A., 1999. Evaluation of current techniques and new approaches in the use of *Cephalonomia stephanoderis* (Hymenoptera: Bethylidae) as a biological control agent of the coffee berry borer, *Hypothenemus hampei* (Coleoptera: Scolytidae), in Chiapas, Mexico. Ph.D. thesis, University of London, 327 pp.

Damon, A., 2000. A review of the biology and control of the coffee berry borer, *Hypothenemus hampei* (Coleoptera: Scolytidae). Bull. Entomol. Res. 90, 453–465.

Damon, A.A., Valle, J., 2002. Comparison of two release techniques for the use of *Cephalonomia stephanoderis* (Hymenoptera: Bethylidae), to control the coffee berry borer, *Hypothenemus hampei* (Coleoptera: Scolytidae) in Soconusco, southeastern Mexico. Biol. Contr. 24, 117–127.

Davidson, A., 1967. The occurrence of coffee berry borer, *Hypothenemus* (*Stephanoderes*) *hampei* (Ferr), in Ethiopia. Café (Peru) 8, 1–3.

Davis, A.P., Govaerts, R., Bridson, D.M., Stoffelen, P., 2006. An annotated taxonomic conspectus of the genus *Coffea* (Rubiaceae). Bot. J. Linn. Soc. 152, 465–512.

de Ingunza, S.M.A., 1964. La "Broca del Café" *Hypothenemus hampei* (Ferrari, 1867) (Col.: Ipinae) en el Perú. Revista Peruana de Entomología 7, 96–98.

de Kraker, J., 1988. The coffee berry borer *Hypothenemus hampei* (Ferr.): factors affecting emergence and early infestation. Centro de Investigaciones Ecológicas del Sureste, Tapachula, Chiapas, Mexico, 57 pp.

de la Rosa, W., Alatorre, R., Trujillo, J., Barrera, J.F., 1997. Virulence of *Beauveria bassiana* (Deuteromycetes) strains against the coffee berry borer (Coleoptera: Scolytidae). J. Econ. Entomol. 90, 1534–1538.

de la Rosa, W., Alatorre, R., Barrera, J.F., Toriello, C., 2000. Effect of *Beauveria bassiana* and *Metarhizium anisopliae* (Deuteromycetes) upon the coffee berry borer (Coleoptera: Scolytidae) under field conditions. J. Econ. Entomol. 93, 1409–1414.

de la Rosa, W., Figueroa, M., Ibarra, J.E., 2005. Selection of *Bacillus thuringiensis* strains native to Mexico active against the coffee berry borer *Hypothenemus hampei* (Ferrari) (Coleoptera: Curculionidae: Scolytidae). Vedalia 12, 3–9.

de la Rosa-Reyes, W., Godinez-Aguilar, J.L., Alatorre-Rosas, R., 1995. Biological activity of five strains of *Metarhizium anisopliae* upon

the coffee berry borer *Hypothenemus hampei* (Col.: Scolytidae). Entomophaga 40, 403–412.

Decazy, B., 1987. Control de la broca del fruto del cafeto *Hypothenemus hampei*. Memorias, 1er Congreso sobre el Cultivo del Café, San Salvador, El Salvador, pp. 53–72.

Decazy, B., Ochoa, H., Lotodé, R., 1989. Indices de distribution spatiale et méthode d'échantillonnage des populations du scolyte des drupes du caféier, *Hypothenemus hampei* Ferr. Café, cacao, thé 33, 27–41.

Delgado, D., Sotomayor, I., 1991. Algunos resultados sobre la cría, adaptación y colonización de los entomófagos *Prorops nasuta* Waterston y *Cephalonomia stephanoderis* Betrem, en la regulación de poblaciones de *H. hampei* en Ecuador. Miscelánea de la Sociedad Colombiana de Entomología 18, 58–75.

Delgado, D., Sotomayor, I., Páliz, V., Mendoza, J., 1990. Cría, colonización y parasitismo de los entomófagos *Cephalonomia stephanoderis* Betrem y *Prorops nasuta* Waterston. Sanidad Vegetal (Ecuador) 5, 51–67.

Delgado A., P., Larco, A.M., García, C.E., Alcívar M., R., Chilán, W.P., Patiño, M.C., 2002. Café en Ecuador: Manejo de la Broca del Fruto (*Hypothenemus hampei* Ferrari). Informe de Terminación de Proyecto Manejo Integrado de la Broca del Café. Convenio CFC—OIC—CABI Commodities—ANECAFÉ, Manta, Ecuador.

Deyrup, M., 1981. A new species of *Cryptoxilos* (Hymenoptera: Braconidae) attacking adult *Lymantor decipens* LeConte (Coleoptera: Scolytidae). Entomol. News 92, 177–180.

Deyrup, M., 1987. *Trischidias exigua* Wood, new to the United States, with notes on the biology of the genus (Coleoptera: Scolytidae). The Coleopterists Bulletin 41, 339–343.

Dias, B.A., Neves, P.M.O.J., Furlaneto-Maia, L., Furlaneto, M.C., 2008. Cuticle-degrading proteases produced by the entomopathogenic fungus *Beauveria bassiana* in the presence of the coffee berry borer cuticle. Braz. J. Microbiol. 39, 301–306.

Dias Silva, W., Moura Mascarin, G., Manesco Romagnoli, E., Simões Bento, J.M., 2012. Mating behavior of the coffee berry borer, *Hypothenemus hampei* (Ferrari) (Coleoptera: Curculionidae: Scolytinae). J. Insect Behav. 25, 408–417.

Drummond-Gonçalves, R., 1940. Combate à broca. O Biológico (Brazil) 6, 269–270.

Dufour, B., 2002. Importance du piégeage pour la lutte intégrée contre le scolyte du café, *Hypothenemus hampei* Ferr. In: Berry, D. (Ed.), Recherche et caféiculture. CIRAD, Montpellier, pp. 108–113 (English translation, pp. 114–116).

Dufour, B.P., 2013. Le scolyte des baies du cafeier, *Hypothenemus hampei* (Ferr.), présent en Martinique. Cahiers Agricultures 22, 575–578.

Dufour, B.P., Frérot, B., 2008. Optimization of coffee berry borer, *Hypothenemus hampei* Ferrari (Col., Scolytidae), mass trapping with an attractant mixture. J. Appl. Entomol. 132, 591–600.

Dufour, B., Barrera, J.F., Decazy, B., 1999. La broca de los frutos del cafeto: ¿la lucha biológica como solución? In: Bertrand, B., Rapidel, B. (Eds.), Desafíos de la Caficultura Centroamericana. IICA PROMECAFE, San José, Costa Rica, pp. 293–325.

Dufour, B.P., Picasso, C., Gonzales, M.O., 2002. Contribution au développement d'un piège pour capturer le scolyte du café *Hypothenemus hampei* Ferr. en El Salvador. Proceedings of the 19th International Scientific Colloquium on Coffee, Association Scientifique Internationale du Café (ASIC), Trieste, Italy, 12 pp.

Duque O., H., 2000. Economics of coffee berry borer (*Hypothenemus hampei*) in Colombia. In: Workshop, Coffee Berry Borer: New

Approaches to Integrated Pest Management". Mississippi State University, Starkville, Mississippi, May 1–5, 2000, 14 pp.

Duque O., H., Baker, P.S., 2003. Devouring Profit. The Socio-Economics of Coffee Berry Borer IPM. The Commodities Press—CABI-CENICAFÉ, Chinchiná, 106 p.

Duque-Orrego, H., Márquez-Q., A., Hernández-S., M., 2002. Estudios de caso sobre costos de manejo integrado de la broca del café en el Departamento de Risaralda. Cenicafé 53, 106–118.

Echeverría Beirute, F., 2006. Caracterización biológica y molecular de aislamientos del hongo entomopatógeno *Beauveria bassiana* (Balsamo) Vuillemin. B.S. thesis, Escuela de Biología, Instituto Tecnológico de Costa Rica, 92 pp.

Echeverry-Arias, O.A., 1999. Determinación del impacto de *Phymastichus coffea* LaSalle (Hymenoptera: Eulophidae) sobre poblaciones de la broca del café *Hypothenemus hampei* (Ferrari) (Coleoptera: Scolytidae), en la zona cafetalera. B.S. thesis, Universidad Nacional de Colombia, Palmira, 94 pp.

Edgington, S., Segura, H., de la Rosa, W., Williams, T., 2000. Photoprotection of *Beauveria bassiana*: testing simple formulations for control of the coffee berry borer. Int. J. Pest Manag. 46, 169–176.

Eggers, H., 1922. Kulturschädliche Borkenkäfer des indischen Archipels. Entomologische Berichten 6, 84–88.

Eggers, H., 1923. Neue indomalayische Borkenkäfer (Ipidae). Zoologische Mededeelingen 7, 129–220.

Eggers, H., 1924. Neue Borkenkäfer (*Ipidae*) aus Afrika. Entomologische Blätter 20, 99–111.

Eichhoff, W.J., 1871. Neue exotische Tomiciden-Arten. Berliner entomologische Zeitschrift 15, 131–137.

Eichhoff, W.J., Schwarz, E.A., 1896. Remarks on the synonymy of some North American Scolytid beetles. Proc. Unit. States. Natl. Mus. 18, 605–610.

Entwistle, P.F., 1964. Inbreeding and arrhenotoky in the ambrosia beetle *Xyleborus compactus* (Eichh.) (Coleoptera: Scolytidae). Proceedings of the Royal Entomological Society of London Series A 38, 83–88.

Espinoza, J.C., Infante, F., Castillo, A., Barrera, J.F., Pinson, E., Galindo, V.H., 2002. Biología en campo de *Phymastichus coffea* (Hymenoptera: Eulophidae). Un parasitoide de adultos de la broca del café. In: Actas del XXV Congreso Nacional de Control Biológico, Hermosillo, Sonora, pp. 23–25.

Espinoza, J.C., Infante, F., Castillo, A., Pérez, J., Nieto, G., Pinson, E.P., Vega, F.E., 2009. The biology of *Phymastichus coffea* LaSalle (Hymenoptera: Eulophidae) under field conditions. Biol. Contr. 49, 227–233.

Evans, H.E., 1964. A synopsis of the American Bethylidae (Hymenoptera, Aculeata). Bull. Mus. Comp. Zool. 132, 1–222.

Evans, H.E., 1978. The Bethylidae of America north of Mexico. Memoirs of the American Entomological Institute 27, 1–332.

Fabricius, J.C., 1801. Systema eleutheratorum secundum ordines, genera, species: Adiectis synonymis, locis, observationibus, decriptionibus, vol II. Impensis Bibliopolii Academici Novi, Kiliae (Kiel, Germany), p. 395.

FAOSTAT, 2014. Food and Agriculture Organization of the United Nations,Statistics Division. Available online at: http://faostat.fao.org/. Last accessed: April 21, 2014.

Feldhege, M.R., 1992. Rearing techniques and aspects of biology of *Phymastichus coffea* (Hymenoptera: Eulophidae), a recently described endoparasitoid of the coffee berry borer, *Hypothenemus hampei* (Coleoptera: Scolytidae). Café, cacao, thé 31, 45–54.

Féliz Matos, D.A., 2003. Incidencia de la broca (*Hypothenemus hampei* Ferr. 1867) y sus controladores naturales en plantas de café bajo diferentes tipos de sombra en San Marcos, Nicaragua. M.S. thesis, Centro Agronómico Tropical de Investigación y Enseñanza, Turrialba, Costa Rica.

Féliz Matos, D., Guharay, F., Beer, J., 2004. Incidencia de la broca (*Hypothenemus hampei*) en plantas de café a pleno sol y bajo sombra de *Eugenia jambos* y *Gliricidia sepium* en San Marcos, Nicaragua. Agroforestería en las Américas 41/42, 56–61.

Fernandes, F.L., Picanço, M.C., Campos, S.O., Bastos, C.S., Chediak, M., Guedes, R.N.C., da Silva, R.S., 2011. Economic injury level for the coffee berry borer (Coleoptera: Curculionidae: Scolytinae) using attractive traps in Brazilian coffee fields. J. Econ. Entomol. 104, 1909–1917.

Fernandes, P.M., Lecuona, R.E., Alves, S.B., 1985. Patogenicidade *de Beauveria bassiana* (Bals.) Vuill. à broca-do-café, *Hypothenemus hampei* (Ferrari, 1867) (Coleoptera: Scolytidae). Ecossistema 10, 176–181.

Fernández, S., Cordero, J., 2005. Evaluación de atrayentes alcohólicos en trampas arteanales para el monitoreo y control de la broca del café, *Hypothenemus hampei* (Ferrari). Bioagro (Venezuela) 17, 143–148.

Ferrari, J.A.G., 1867. Die Forst- und Baumzuchtschädlichen Borkenkäfer (Tomicides Lac.) aus der Familie der Holzverderber (Scolytides Lac.), mit besonderer Berücksichtigung vorzüglich der europäischen Formen, und der Sammlung des k. k. zoologischen Kabinetes in Wien. Carl Gerold's Sohn, Vienna, 95 p.

Ferreira, A.J., Bueno, V.H.P., 1995. Ocorrência da vespa de Uganda, *Prorops nasuta* Waterston, 1923 (Hymenoptera: Bethylidae) na região de Lavras—MG. Ciência e Prática (Brazil) 19, 226–227.

Fettig, C.J., McKelvey, S.R., Dabney, C.P., Huber, D.P.W., 2012. Responses of *Dendroctonus brevicomis* (Coleoptera: Curculionidae) in behavioral assays: implications to development of a semiochemical-based tool for tree protection. J. Econ. Entomol. 105, 149–160.

ffrench-Constant, R.H., Steichen, J.C., Brun, L.O., 1994. A molecular diagnostic for endosulfan insecticide resistance in the coffee berry borer *Hypothenemus hampei* (Coleoptera: Scolytidae). Bull. Entomol. Res. 84, 11–16.

Fleutiaux, E., 1901. Un ennemi du café du Ronilon (Congo). La Nature—Revue des sciences et de leur application à l'art et à l'industrie 29, 4.

Fonseca, J.P., 1937. A broca verdadeira e a falsa broca do café. O Biológico (Brazil) 3, 366–368.

Fonseca, J.P. da, 1939. A "bróca" e o sombreamento dos cafezais. O Biologico (Brazil) 5, 133–136, (Rev. Appl. Entomol., Ser. A, 1940, 28, 61).

Fonseca, J.P. da, Araujo, R.L., 1939. Insetos inimigos do *Hypothenemus hampei* (Ferr.) ("Broca do Café"). Boletim Biológico 4, 485–503, (Rev. Appl. Entomol., Ser. A, 1941, 29, 292).

Friederichs, K., 1922a. De Bestrijding van de Koffiebessenboeboek op de Onderneming Karang Redjo. Mededeelingen van het Koffiebessenboeboek-Fonds 1, 7–21, (Rev. Appl. Entomol., Ser. A, 1922, 10, 506–507).

Friederichs, K., 1922b. Verslag van den Entomoloog over het Tijdvak, 1 Augustus 1921 t/m 31 December 1921. Mededeelingen van het Koffiebessenboeboek-Fonds 2, 21–26, (Rev. Appl. Entomol., Ser. A, 1922, 10, 507).

Friederichs, K., 1923. Kleine Mededeelingen omtrent de Koffiebessenboeboek. Mededeelingen van het Koffiebessenboeboek-Fonds 3, 55–61, (Rev. Appl. Entomol., Ser. A, 1922, 10, 601).

Friederichs, K., 1924a. Bionomische gegevens omtrent den Koffiebessen-boeboek. Mededeelingen van het Koffiebessenboeboek-Fonds 11, 261–286, (Rev. Appl. Entomol., Ser. A, 1925, 13, 11–12).

Friederichs, K., 1924b. In hoever bestaan er verschillen in de vatbaarheid der koffiesoorten voor den Koffiebessenboeboek? Mededeelingen Koffiebessenboeboek-Fonds 11, 315–358, (Rev. Appl. Entomol., Ser. A, 1925, 13, 12).

Friederichs, K., Bally, W., 1922. Resumé van een Publicatie over de parasitische schimmels van den bessenboeboek. Mededeelingen van het Koffiebessenboeboek-Fonds 5, 78–80, (Rev. Appl. Entomol., Ser. A, 1923, 11, 169).

Friederichs, K., Bally, W., 1923. Over de parasitische schimmels, die den Koffiebessenboeboek dooden. Mededeelingen van het Koffiebessenboeboek-Fonds 6, 103–147.

Gaitan, A., Valderrama, A.M., Saldarriaga, G., Velez, P., Bustillo, A., 2002. Genetic variability of Beauveria bassiana associated with the coffee berry borer Hypothenemus hampei and other insects. Mycol. Res. 106, 1307–1314.

Galindo, V.H., Infante, F., Castillo, A., Barrera, J.F., Pinson, E., González, G., Espinoza, J.C., 2002. Establecimiento preliminar del parasitoide Phymastichus coffea (Hymenoptera: Eulophidae) en Chiapas, México. Memorias del XXV Congreso Nacional de Control Biológico, Hermosillo, Sonora, México, pp. 44–46.

Gallardo-Covas, F., Hernández, E., Pagán, J., 2010. Presencia natural del hongo Beauveria bassiana (Bals.) Vuill. en la broca del fruto del café Hypothenemus hampei (Ferrari) en Puerto Rico. J. Agric. Univ. P. R. 94, 195–198.

Gallego Ropero, M.C., Armbrecht, I., 2005. Depredación por hormigas sobre la broca del café Hypothenemus hampei (Curculionidae: Scolytinae) en cafetales cultivados bajo dos niveles de sombra en Colombia. Manejo Integrado de Plagas y Agroecología 76, 32–40.

Gama, F. de C., Teixeira, C.A.D., Garcia, A., Costa, J.N.M., Lima, D.K.S., 2005. Influência do ambiente na diversidade de fungos associados a Hypothenemus hampei (Ferrari) (Coleoptera, Scolytidae) e frutos de Coffea canephora. Arq. Inst. Biol., São Paulo 72, 359–364.

Gama, F. de C., Teixeira, C.A.D., Garcia, A., Costa, J.N.M., Lima, D.K.S., 2006. Diversidade de fungos filamentosos associados a Hypothenemus hampei (Ferrari) (Coleoptera: Scolytidae) e suas galerias em frutos de Coffea canephora (Pierre). Neotrop. Entomol. 35, 573–578.

Gams, W., Humber, R.A., Jaklitsch, W., Kirschner, R., Stadler, M., 2012. Minimizing the chaos following the loss of Article 59: suggestions for a discussion. Mycotaxon 119, 495–507.

Gandhi, K.J.K., Cognato, A.I., Lightle, D.M., Mosley, B.J., Nielsen, D.G., Herms, D.A., 2010. Species composition, seasonal activity, and semiochemical response of native and exotic bark and ambrosia beetles (Coleoptera: Curculionidae: Scolytinae) in Northeastern Ohio. J. Econ. Entomol. 103, 1187–1195.

Gandia, I.M., Boncato, A.A., 1964. A note on the occurrence of the coffee borer in the Philippines. Coffee Cacao J. 7, 124–125, 138.

Gandrup, J., 1922. Over Boeboek in Loewak-Koffie. Mededeelingen van het Koffiebessenboeboek-Fonds 3, 53–54, (Rev. Appl. Entomol., Ser. A, 1922, 10, 184).

García, A., 2000. Proyecto manejo integrado de la broca del café CFC-OIC-IIBC-PROMECAFE. Memorias del XIX Simposio Latinoamericano de Caficultura, Costa Rica pp. 61–68.

García Martell, C., 1980. Falsas brocas del género Hypothenemus detectadas en frutos del cafeto en México. III Simposio Latinoamericano sobre Caficultura, Tegucigalpa, Honduras, pp 188–195.

Gauthier, N., 2010. Multiple cryptic genetic units in Hypothenemus hampei (Coleoptera: Scolytinae): evidence from microsatellite and mitochondrial DNA sequence data. Biol. J. Linn. Soc. 101, 113–129.

Gauthier, N., Rasplus, J.-Y., 2004. Polymorphic microsatellite loci in the coffee berry borer, Hypothenemus hampei (Coleoptera, Scolytidae). Mol. Ecol. Notes 4, 294–296.

Ghesquière, J., 1933. Rôle des Ipides dans la destruction des végétaux au Congo belge. Ann. Gembloux 39, 24–37.

Gingerich, D.P., Borsa, P., Suckling, D.M., Brun, L.-O., 1996. Inbreeding in the coffee berry borer, Hypothenemus hampei (Coleoptera: Scolytidae) estimated from endosulfan resistance phenotype frequencies. Bull. Entomol. Res. 86, 667–674.

Giordanengo, P., 1992. Biologie, eco-éthologie et dynamique des populations du Scolyte des grains de café, Hypothenemus hampei Ferr. (Coleoptera, Scolytidae), en Nouvelle-Calédonie. Ph.D. thesis, Université de Rennes I, Rennes, France.

Giordanengo, P., Brun, L.O., Frérot, B., 1993. Evidence for allelochemical attraction of the coffee berry borer Hypothenemus hampei, by coffee berries. J. Chem. Ecol. 19, 763–769.

Gobbi, J.A., 2000. Is biodiversity-friendly coffee financially viable? An analysis of five different coffee production systems in western El Salvador. Ecol. Econ. 33, 267–281.

Godoy Junior, C., Graner, E.A., 1961. Sombreamento dos cafèzais. II. Resultados do 4° biênio (1959/1960). Anais da Escola Superior de Agricultura "Luiz de Queiroz" 18, 61–75.

Godoy Junior, C., Graner, E.A., 1967. Sombreamento dos cafèzais. IV. Resultados de mais dois biênios: 1963/1964–1965/1966. Anais da Escola Superior de Agricultura "Luiz de Queiroz" 24, 1–17.

Gole, T.W., Denich, M., Teketay, D., Vlek, P.L.G., 2002. Human impacts on the Coffea arabica genepool in Ethiopia and the need for its in situ conservation. In: Engels, J.M.M., Ramanatha Rao, V., Brown, A.H.D., Jackson, M.T. (Eds.), Managing Plant Genetic Diversity. CABI Publishing, Oxon, pp. 237–247.

Gómez, J., Santos, A., Valle, J., Montoya, P.J., 2010. Determinación del establecimiento de parasitoides de la broca del café Hypothenemus hampei (Coleoptera: Curculionidae, Scolytinae) en cafetales del Soconusco, Chiapas, México. Entomotrópica 25, 25–36.

Góngora B., C.E., Posada F., F.J., Bustillo P., A.E., 2001. Detección molecular de un gen de resistencia al insecticida endosulfan en una población de broca, Hypothenemus hampei (Ferrari) (Coleoptera: Scolytidae) en Colombia. Resúmenes del 28vo Congreso de la Sociedad Colombiana de Entomología (SOCOLEN). Pereira, Colombia, August 8–10, 2001, pp. 47–48.

Góngora, C., Macea, E., Castro, A.M., Idarraga, S., Cristancho, M.A., Benavides, P., et al., 2012. Interaction between coffee plants and the insect coffee berry borer, Hypothenemus hampei. In: Proceedings of the 24th International Scientific Colloquium on Coffee, Association Scientifique Internationale du Café (ASIC), San José, Costa Rica, pp. 533–541.

González, M.O., Dufour, B.P., 2000. Diseño, desarrollo y evaluación del trampeo en el manejo integrado de la broca del café Hypothenemus hampei Ferr. en El Salvador. Memoria. XIX Simposio Latinoamericano de Caficultura, San José, Costa Rica pp. 381–396.

González G., M.T., Posada F., F.J., Bustillo P., A.E., 1993. Bioensayo para evaluar la patogenicidad de Beauveria bassiana (Bals.) Vuill. sobre la broca del café, Hypothenemus hampei (Ferrari). Rev. Colomb. Entomol. 19, 123–130.

Gowdey, C.C., 1911. Insects injurious to coffee. Report of the Government Entomologist of the year 1909–1910, Uganda Protectorate, Entebbe.

Graner, E.A., Godoy Junior, C., 1959. Sombreamento dos cafêzais. I. Resultados de três ciclos bienais (1953/1958) obtidos na Escola "Luiz de Queiróz.". Anais da Escola Superior de Agricultura "Luiz de Queiroz" 16, 139–165.

Graner, E.A., Godoy Junior, C., 1962. Sombreamento dos cafêzais. III. Resultados do 5° biênio (1961/1962). Anais da Escola Superior de Agricultura "Luiz de Queiroz" 19, 283–296.

Graner, E.A., Godoy Junior, C., 1971. Sombreamento dos cafêzais. V. Resultados de mais dois biênios: 1967/1968–1970/1971. Anais da Escola Superior de Agricultura "Luiz de Queiroz" 28, 153–164.

Greco, E.B., Wright, M.G., 2012. First report of exploitation of coffee beans by black twig borer (*Xylosandrus compactus*) and tropical nut borer (*Hypothenemus obscurus*) (Coleoptera; Curculionidae: Scolytinae) in Hawaii. Proc. Hawaii. Entomol. Soc. 44, 71–78.

Grimm, C., 2001. Economic feasibillty of a small-scale production plant for entomopathogenic fungi in Nicaragua. Crop Prot. 20, 623–630.

Grossi-de-Sá, M.F., Chrispeels, M.J., 1997. Molecular cloning of bruchid (*Zabrotes subfasciatus*) α-amylase cDNA and interactions of the expressed enzymes with bean amylase inhibitors. Insect Biochem. Mol. Biol. 27, 271–281.

Guerreiro Filho, O., Mazzafera, P., 2003. Caffeine and resistance of coffee to the berry borer *Hypothenemus hampei* (Coleoptera: Scolytidae). J. Agric. Food Chem. 51, 6987–6991.

Gumier-Costa, F., 2009. First record of the coffee berry borer, *Hypothenemus hampei* (Ferrari) (Coleoptera: Scolytidae), in Pará nut, *Bertholletia excelsa* (Lecythidaceae). Neotrop. Entomol. 38, 430–431.

Gutierrez, A.P., Villacorta, A., Cure, J.R., Ken Ellis, C., 1998. Tritrophic analysis of the coffee (*Coffea arabica*)–coffee berry borer [*Hypothenemus hampei* (Ferrari)]–parasitoid system. Anais da Sociedade Entomológica do Brasil 27, 357–385.

Gutiérrez-Martínez, A., Ondarza, R.N., 1996. Kairomone effect of extracts from *Coffea canephora* over *Hypothenemus hampei* (Coleoptera: Scolytidae). Environ. Entomol. 25, 96–100.

Gutiérrez-Martínez, A., Haro-González, R., Hernández Rivas, S., 1990. Kairomona responsable de la atracción de la broca del café *Hypothenemus hampei* Ferrari (Coleoptera: Scolytidae) al grano de café. Memorias del 4to Congreso Nacional MIP y 3er Congreso Internacional "Humberto Tapia Barquero" Managua, Nicaragua, October 1990, pp. 28–36.

Haarer, A.E., 1962. Modern Coffee Production. Leonard Hill [Books] Limited, London.

Hagedorn, M., 1910. Wieder ein neuer Kaffeeschädling. Entomologische Blätter 6, 1–4.

Haggar, J., Barrios, M., Bolaños, M., Merlo, M., Moraga, P., Manguia, R., et al., 2011. Coffee agroecosystems performance under full sun, shade, conventional and organic management regimes in Central America. Agroforestry Systems 82, 285–301.

Hamilton, W.D., 1967. Extraordinary sex ratios. Science 156, 477–488.

Haraprasad, N., Niranjana, S.R., Prakash, H.S., Shetty, H.S., Wahab, S., 2001. *Beauveria bassiana*—a potential mycopesticide for the efficient control of coffee berry borer, *Hypothenemus hampei* (Ferrari) in India. Biocontrol Sci. Technol. 11, 251–260.

Hargreaves, H., 1926. Notes on the coffee berry borer (*Stephanoderes hampei*, Ferr.) in Uganda. Bull. Entomol. Res. 16, 347–354.

Hargreaves, H., 1935. *Stephanoderes Hampei* [*sic*] Ferr., coffee berry-borer, in Uganda. The East African Agricultural Journal 1, 218–224.

Hargreaves, H., 1940. Coffee-pests. In: Tothill, J.D. (Ed.), Agriculture in Uganda. Oxford University Press, London, pp. 340–380.

Heinrich, W.O., 1965. Aspectos do combate biológico às pragas do café. O Biológico (Brazil) 31, 57–62.

Hempel, A., 1933. O combate á broca do café por meio da vespa de Uganda. Revista do Instituto do Café de São Paulo 16, 830–835.

Henaut, Y., Pablo, J., Ibarra-Nuñez, G., Williams, T., 2001. Retention, capture and consumption of experimental prey by orb-web weaving spiders in coffee plantations in Southern Mexico. Entomol. Exp. Appl. 98, 1–8.

Hernández, L.R., 2002. El bumerang maldito. Encuentro en la Red: Diario Independiente de Asuntos Cubanos, Año III, Edición 278, Lunes 14 de Enero de 2002. Available online: http://www.cubaencuentro.com/ecologia/2002/01/14.html. Last accessed: April 21, 2014.

Hernández Paz, M., 1972. Campaña nacional para el control y posible erradicación de la plaga "broca del cafeto". Revista Cafetalera (Guatemala) 116, 13–22.

Hesse, A., 1925. No. 389. Arrête Ministériel incorporant le Dahomey dans la liste des pays contaminés par le scolyte du grain de café (*Stephanoderes coffeae*). Bulletin Officiel du Ministere des Colonies, Paris, 25 August 1925.

Hewitt, C.N., Karl, T., Langford, B., Owen, S.M., Possell, M., 2011. Quantification of VOC emission rates from the biosphere. Trends Anal. Chem. 30, 937–944.

Hilgenboecker, K., Hammerstein, P., Schlattmann, P., Telschow, A., Werren, J.H., 2008. How many species are infected with *Wolbachia*? A statistical analysis of current data. FEMS Microbiol. Lett. 281, 215–220.

Hirose, E., Neves, P.M.O.J., 2002. Técnica para criação e manutenção da broca-do-café, *Hypothenemus hampei* (Ferrari) (Coleoptera: Scolytidae), em laboratório. Neotrop. Entomol. 31, 161–164.

Hopkins, A.D., 1915a. List of generic names and their type-species in the coleopterous superfamily Scolytoidea. Proc. Unit. States. Natl. Mus. 48, 115–136.

Hopkins, A.D., 1915b. Classification of the Cryphalinæ, with descriptions of new genera and species. United States Department of Agriculture, Contributions from the Bureau of Entomology, Report No. 99, 75 pp.

Humber, R.A., 2012. Identification of entomopathogenic fungi. In: Lacey, L.A. (Ed.), Manual of Techniques in Invertebrate Pathology, second ed. Academic Press, San Diego, pp. 151–187.

Hutson, J.C., 1936. The coffee berry-borer in Ceylon. Trop. Agric. (Trinidad) 87, 378–383.

ICO, 2014. International Coffee Organization—world coffee trade (1963–2013): a review of the markets, challenges and opportunities facing the sector. International Coffee Organization, London, ICC 111–115.

Idárraga, S.M., Castro, A.M., Macea, E.P., Gaitán, A.L., Rivera, L.F., Cristancho, M.A., Góngora, C.E., 2012. Sequences and transcriptional analysis of *Coffea arabica* var. Caturra and *Coffea liberica* plant responses to coffee berry borer *Hypothenemus hampei* (Coleoptera: Curculionidae: Scolytinae) attack. J. Plant Interact. 7, 56–70.

von Ihering, R., 1924. O caruncho da cereja do café. Chácaras e Quintaes (Brazil) 30, 111–114.

Infante, F., 1998. Biological control of *Hypothenemus hampei* (Coleoptera: Scolytidae) in Mexico, using the parasitoid *Prorops nasuta* (Hymenoptera: Bethylidae), Ph.D. thesis, University of London, 173 pp.

Infante, F., 2000. Development and population grow rates of *Prorops nasuta* (Hym., Bethylidae) at constant temperatures. J. Appl. Ent. 124, 343–348.

Infante, F., Luis, J.H., 1993. Estadísticos demográficos de *Cephalonomia stephanoderis* Betrem (Hymenoptera, Bethylidae) a temperaturas constantes. Folia Entomológica Mexicana 87, 61–72.

Infante, F., Luis, J.H., Barrera, J.F., Gómez, J., Castillo, A., 1992a. Thermal constants for preimaginal development of the parasitoid *Cephalonomia stephanoderis* Betrem (Hymenoptera, Bethylidae). Can. Entomol. 124, 935–941.

Infante, F., Barrera, J.F., Murphy, S.T., Gómez, J., Castillo, A., 1992b. Cría y cuarentena de *Phymastichus coffea* LaSalle (Hymenoptera: Eulophidae) un parasitoide de la broca del café introducido a México. Resúmenes, XV Simposio Latinoamericano de Caficultura, Xalapa, Veracruz, unpaginated.

Infante, F., Barrera, J.F., Castillo, A., de la Rosa, W., 1993. La capacidad reproductiva sexual y partenogenética del parasitoide *Cephalonomia stephanoderis* en laboratorio. Turrialba 42, 391–396.

Infante, F., Valdéz, J., Penagos, D.I., Barrera, J.F., 1994a. Description of the life stages of *Cephalonomia stephanoderis* (Hymenoptera, Bethylidae), a parasitoid of *Hypothenemus hampei* (Coleoptera: Scolytidae). Vedalia 1, 13–18.

Infante, F., Murphy, S.T., Barrera, G.J.F., Gómez, J., de la Rosa, W., Damon, A., 1994b. Cría de *Phymastichus coffea* parasitoide de la broca del café, y algunas notas sobre su historia de vida. Southwest. Entomol. 19, 313–315.

Infante, F., Mumford, J., Méndez, I., 2001. Non-recovery of *Prorops nasuta* (Hymenoptera: Bethylidae), an imported parasitoid of the coffee berry borer (Coleoptera: Scolytidae) in Mexico. Southwest. Entomol. 26, 159–163.

Infante, F., Castillo, A., Espinoza, J.C., Galindo, V.H., Ortíz, M. de J., Montes, R., et al., 2003. "*Phymastichus*", la avispita que parasita a los adultos de la broca del café. Manual Ilustrado. ECOSUR-SIBEJ. 14 pp.

Infante, F., Mumford, J., Baker, P., 2005. Life history studies of *Prorops nasuta*, a parasitoid of the coffee berry borer. BioControl 50, 259–270.

Infante, F., Castillo, A., Pérez, J., Vega, F.E., 2013. Field-cage evaluation of the parasitoid *Phymastichus coffea* as a natural enemy of the coffee berry borer, *Hypothenemus hampei*. Biol. Contr. 67, 446–450.

Infante, F., Pérez, J., Vega, F.E., 2014. The coffee berry borer: the centenary of a biological invasion in Brazil. Brazilian J. Biol. in press.

Ingram, W.R., 1968. Observations on the control of the coffee berry borer *Hypothenemus hampei* (Ferr.), with endosulfan in Uganda. Bull. Entomol. Res. 30, 539–547.

Inwood, J., 2005. Last to fall. The Panama News, Vol. 11, No. 12, June 19–July 2, 2005.

Ito, E.T., Varéa-Pereira, G., Miyagui, D.T., Pinotti, M.H.P., Neves, P.M.O. J., 2007. Production of extracellular protease by a Brazilian strain of *Beauveria bassiana* reactivated on coffee berry borer, *Hypothenemus hampei*. Braz. Arch. Biol. Tech. 50, 217–223.

Janssen, M.P.M., 2011. Endosulfan. A closer look at the arguments against a worldwide phase out. RIVM letter report 601356002/2011.National Institute for Public Health and the Environment. Ministry of Health, Welfare and Sport, The Netherlands.

Jaramillo, J., Vega, F.E., 2009. *Aphanogmus* sp. (Hymenoptera: Ceraphronidae): a hyperparasitoid of the coffee berry borer parasitoid *Prorops nasuta* (Hymenoptera: Bethylidae) in Kenya. Biocontrol Sci.Technol. 19, 113–116.

Jaramillo, J., Bustillo, A.E., Montoya, E.C., Borgemeister, C., 2005. Biological control of the coffee berry borer *Hypothenemus hampei* (Ferrari) (Coleoptera: Curculionidae) by *Phymastichus coffea* (Hymenoptera: Eulophidae) in Colombia. Bull. Entomol. Res. 95, 1–6.

Jaramillo, J., Borgemeister, C., Setamou, M., 2006. Field superparasitism by *Phymastichus coffea*, a parasitoid of adult coffee berry borer, *Hypothenemus hampei*. Entomol. Exp. Appl. 119, 231–237.

Jaramillo, J., Chabi-Olaye, A., Poehling, H.-M., Kamonjo, C., Borgemeister, C., 2009a. Development of an improved laboratory production technique for the coffee berry borer *Hypothenemus hampei*, using fresh coffee berries. Entomol. Exp. Appl. 130, 275–281.

Jaramillo, J., Chabi-Olaye, A., Borgemeister, C., Kamonjo, C., Poehling, H.-M., Vega, F.E., 2009b. Where to sample? Ecological implications of sampling strata in determining abundance and impact of natural enemies of the coffee berry borer, *Hypothenemus hampei*. Biol. Contr. 49, 245–253.

Jaramillo, J., Chabi-Olaye, A., Kamonjo, C., Jaramillo, A., Vega, F.E., Poehling, H.-M., Borgemeister, C., 2009c. Thermal tolerance of the coffee berry borer *Hypothenemus hampei*: predictions of climate change on a tropical insect pest. PLoS One 4 (8), e6487.

Jaramillo, J., Chabi-Olaye, A., Borgemeister, C., 2010a. Temperature-dependent development and emergence pattern of *Hypothenemus hampei* (Coleoptera: Curculionidae: Scolytinae) from coffee berries. J. Econ. Entomol. 103, 1159–1165.

Jaramillo, J., Chapman, E.G., Vega, F.E., Harwood, J.D., 2010b. Molecular diagnosis of a previously unreported predator-prey association in coffee: *Karnyothrips flavipes* Jones (Thysanoptera: Phlaeothripidae) predation on the coffee berry borer. Naturwissenschaften 97, 291–298.

Jaramillo, J., Muchugu, E., Vega, F.E., Davis, A., Borgemeister, C., Chabi-Olaye, A., 2011. Some like it hot: the influence and implications of climate change on coffee berry borer (*Hypothenemus hampei*) and coffee production in East Africa. PLoS One 6 (9), e24528.

Jaramillo, J., Setamou, M., Muchugu, E., Chabi-Olaye, A., Jaramillo, A., Mukabana, J., et al., 2013. Climate change or urbanization? Impacts on a traditional coffee production system in East Africa over the last 80 years. PLoS One 8 (1), e51815.

Jervis, T.S., 1939. The control of the coffee berry borer in Bukoba. The East African Agricultural Journal 5, 121–124.

Jeyaprakash, A., Hoy, M.A., 2000. Long PCR improves *Wolbachia* DNA amplification: *wsp* sequences found in 76% of sixty-three arthropod species. Insect Mol. Biol. 9, 393–405.

Jiménez-Gómez, J., 1992. Patogenicidad de diferentes aislamientos de *Beauveria bassiana* sobre la broca del café. Cenicafé 43, 84–98.

Johnson, M.D., Kellermann, J.L., Stercho, A.M., 2010. Pest reduction services by birds in shade and sun coffee in Jamaica. Anim. Conservat. 13, 140–147.

Johnston, A., 1963. *Stephanoderes hampei* in Tahiti. Plant Protection Committee for the South East Asia and Pacific Region, Information Letter No. 23. Food and Agriculture Organization of the United Nations Regional Office for Asia and the Far East, Bangkok, Thailand, 4 pp. (Rev. Appl. Entomol., Ser. A, 1964, 52, 535).

Jones, V.P., 2002. Macadamia integrated pest management: IPM of insects and mites attacking macadamia nuts in Hawaii. College of Tropical Agriculture and Human Resources, University of Hawaii, Manoa, 98 pp.

Jones, V.P., Burnam-Larish, L.L., Caprio, L.C., 1992. Effect of harvest interval and cultivar on damage to macadamia nuts caused by *Hypothenemus obscurus* (Coleoptera: Scolytidae). J. Econ. Entomol. 85, 1878–1883.

Jordal, B.H., Kirkendall, L.R., 1998. Ecological relationships of a guild of tropical beetles breeding in *Cecropia* petioles in Costa Rica. J. Trop. Ecol. 14, 153–176.

Jordal, B.H., Cognato, A.I., 2012. Molecular phylogeny of bark and ambrosia beetles reveals multiple origins of fungus farming during periods of global warming. BMC Evol. Biol. 12, 133.

Jurat-Fuentes, J.L., Jackson, T.A., 2012. Bacterial entomopathogens. In: Vega, F.E., Kaya, H.K. (Eds.), Insect Pathology, second ed. Academic Press, San Diego, pp. 265–349.

Kaden, O., 1930. Relatório Anual de 1929, Secção de Fitopatologia, Direcção dos Serviços de Agricultura. Imprensa Nacional, São Tomé e Príncipe, 56 pp.

Kageyama, D., Narita, S., Watanabe, M., 2012. Insect sex determination manipulated by endosymbionts: incidences, mechanisms and implications. Insects 3, 161–199.

Kambestad, M., 2011. Coexistence of habitat generalists in neotropical petiole-breeding bark beetles: molecular evidence reveals cryptic diversity, but no niche segregation. M.S. thesis, University of Bergen.

Kannan, K., 1931. Note on the berry borer situation in India. Department of Agriculture, Mysore State, General Series Bulletin 16, 21–22.

Karp, D.S., Mendenhall, C.D., Figueroa Sandí, R., Chaumont, N., Ehrlich, P.R., Hadly, E.A., Daily, G.C., 2013. Forest bolsters bird abundance, pest control and coffee yield. Ecol. Lett. 16, 1339–1347.

Kellermann, J.L., Johnson, M.D., Stercho, A.M., Hackett, S.C., 2008. Ecological and economic services provided by birds on Jamaican Blue Mountain coffee farms. Conserv. Biol. 22, 1177–1185.

Kelsey, R.G., Beh, M.M., Shaw, D.C., Manter, D.K., 2013. Ethanol attracts scolytid beetles to *Phytophthora ramorum* cankers on coast live oak. J. Chem. Ecol. 39, 494–506.

Kenis, M., Wermelinger, B., Grégoire, J.-C., 2004. Research on parasitoids and predators of Scolytidae—a review. In: Lieutier, F., Day, K.R., Battisti, A., Grégoire, J.-C., Evans, H.F. (Eds.), Bark and Wood Boring Insects in Living Trees in Europe. A Synthesis. Kluwer Academic Publishers, Dordrecht, pp. 237–290.

Kern, M.J., Beyhl, F.E., Braun, P., Grötsch, H., Knauf, W., 1991. Thiodan^R susceptibility in the coffee berry borer, *Hypothenemus hampei* Ferr. (Scolitidae: Coleoptera), from Brazil, Cameroon, Guatemala and the Philippines, documented by toxicological and physiological data. In: Proceedings of the First Asia-Pacific Conference of Entomology (APCE). The Entomology and Zoology Association of Thailand, Bangkok, pp. 468–486.

Kidd, N.A.C., Jervis, M.A., 2005. Population dynamics. In: Jervis, M.A. (Ed.), Insects as Natural Enemies: A Practical Perspective. Springer, The Netherlands, pp. 435–523.

Kimmerer, T.W., Kozlowski, T.T., 1982. Ethylene, ethane, acetaldehyde, and ethanol production by plants under stress. Plant Physiol. 69, 840–847.

Kirkpatrick, T.W., 1935. Studies on the ecology of coffee plantations in East Africa. I. The climate and eco-climates of coffee plantations. East African Agricultural Research Station, Amani, Kenya, 66 pp.

Klein-Koch, C., Espinoza, O., Tandazo, A., Cisneros, P., Delgado, D., 1988. Factores naturales de regulación y control biológico de la broca del café (*Hypothenemus hampei* Ferr.). Sanidad Vegetal (Ecuador) 3, 5–30.

Klein-Koch, C., 1989. Perspectivas en el control biotecnológico de la broca del café (*Hypothenemus hampei* Ferr.). Proceedings of the 13th International Scientific Colloquium on Coffee, Association Scientifique Internationale du Café (ASIC), Paipa, Colombia, pp. 717–725.

Klein-Koch, C., 1990. Natural regulation factors and classical biological control of the coffee berry borer (*Hypothenemus hampei* Ferrari) in Ecuador, Proceedings of the International DLG-Symposium on Integrated Pest Management in Tropical and Subtropical Systems, Bad Dürkheim, Federal Republic of Germany, February 8–15, 1989, pp. 331–344.

Klimetzek, D., Köhler, J., Vité, J.P., Kohnle, U., 1986. Dosage response to ethanol mediates host selection by "secondary" bark beetles. Naturwissenschaften 73, 270–272.

Koch, V.J.M., 1973. Abondance de *Hypothenemus hampei* Ferr., scolyte des graines de café, en fonction de sa plante-hôte et de son parasite *Cephalonomia stephanoderis* Betrem, en Côte d'Ivoire. Mededelingen Landbouwhogeschool Wageningen 73, 1–85.

Krasnoff, S.B., Watson, D.W., Gibson, D.M., Kwan, E.C., 1995. Behavioral effects of the entomopathogenic fungus *Entomophthora muscae* on its host *Musca domestica*: postural changes in dying hosts and gated pattern of mortality. J. Insect Physiol. 41, 895–903.

Kumar, P.K.V., Prakasan, C.B., Vijayalakshmi, C.K., 1990. Coffee berry borer *Hypothenemus hampei* (Coleoptera: Scolytidae): first record from India. J. Coffee Res. 20, 161–164.

Lara G., J.C., López N., J.C., Bustillo P., A.E., 2004. Effecto de entomonemátodos sobre poblaciones de la broca del café, *Hypothenemus hampei* (Coleoptera: Scolytidae), en frutos en el suelo. Rev. Colomb. Entomol. 30, 179–185.

Larsen, A., Philpott, S.M., 2010. Twig-nesting: the hidden predators of the coffee berry borer in Chiapas, Mexico. Biotropica 42, 342–347.

LaSalle, J., 1990. A new genus and species of Tetrastichinae (Hymenoptera: Eulophidae) parasitic on the coffee berry borer, *Hypothenemus hampei* (Ferrari) (Coleoptera: Scolytidae). Bull. Entomol. Res. 80, 7–10.

Lazo, A.R.R., 1990. Susceptibilidad de la broca del fruto del cafeto (*Hypothenemus hampei*) al hongo entomopatógeno *Beauveria bassiana*, y su tolerancia a oxicloruro de cobre. M.S. thesis. Centro Agronómico Tropical de Investigación y Enseñanza, Turrialba, Costa Rica, 61 pp.

Lauzière, I., Pérez-Lachaud, G., Brodeur, J., 2000. Behavior and activity pattern of *Cephalonomia stephanoderis* (Hymenoptera: Bethylidae) attacking the coffee berry borer, *Hypothenemus hampei* (Coleoptera: Scolytidae). J. Insect Behav. 13, 375–395.

Lauzière, I., Brodeur, J., Pérez-Lachaud, G., 2001a. Host stage selection and suitability in *Cephalonomia stephanoderis* Betrem (Hymenoptera: Bethylidae), a parasitoid of the coffee berry borer. Biol. Contr. 21, 128–133.

Lauzière, I., Pérez-Lachaud, G., Brodeur, J., 2001b. Importance of nutrition and host availability on oogenesis and oviposition of *Cephalonomia stephanoderis* (Hymenoptera: Bethylidae). Bull. Entomol. Res. 91, 185–191.

Le Pelley, R.H., 1968. Pests of Coffee. Longmans, Green and Co., Ltd, London, 590 pp.

Lean, M.E.J., Ashihara, H., Clifford, M.N., Crozier, A., 2012. Purine alkaloids: a focus on caffeine and related compounds in beverages. In: Crozier, A., Ashihara, H., Tomás-Barberán, F. (Eds.), Teas, Cocoa and Coffee: Plant Secondary Metabolism and Health. Wiley-Blackwell, Chichester, pp. 25–44.

Lecuona, R.E., Fernandes, P.M., Alves, S.B., Bleicher, E., 1986. Patogenicidade de *Metarhizium anisopliae* (Metsch.) Sorok., à broca-do-café, *Hypothenemus hampei* (Ferrari, 1867) (Coleoptera: Scolytidae). Anais da Sociedade Entomológica do Brasil 15, 21–27.

Leefmans, S., 1920. Voorloopige mededeelingen omtrent Koffiebessenboeboek. Publicaties van het Nederlandsch-Indisch Landbouw Syndicaat 12, 645–668, (Rev. Appl. Entomol., Ser. A, 1922, 10, 571).

Leefmans, S., 1923. De Koffiebessenboeboek (*Stephanoders hampei* Ferrari = *coffeae* Hagedorn). I. Levenswijze en oecologie. Mededeelingen van het Instituut voor Plantenziekten 57, 1–93, (Rev. Appl. Entomol., Ser. A, 1923, 11, 236–237).

Leefmans, S., 1924. De Koffiebessenboeboek. II. Bestrijding. Mededeelingen van het Instituut voor Plantenziekten 62, 1–99.

Leplae, E., 1928. Le scolyte des baies du caféier (*Stephanoderes*). Bulletin Agricole du Conge Belge 19, 271–276.

Leroy, J.V., 1936. Observations relatives à quelques insectes attaquant le Caféier. Publications de l'institut national pour l'étude agronomique du Congo Belge, Série Scientifique 8, 1–30.

Leroy, T., Royer, M., Paillard, M., Berthouly, M., Spiral, J., Tessereau, S., Legavre, T., Altosaar, I., 1997. Introduction de gènes d'intérêt agronomique dans l'espèce *Coffea canephora* Pierre par transformation avec *Agrobacterium* sp. Proceedings of the 19th International Scientific Colloquium on Coffee, Association Scientifique Internationale du Café (ASIC), Nairobi, Kenya, pp. 439–446.

Levinson, H.Z., 1976. The defensive role of alkaloids in insects and plants. Experientia 32, 408–411.

Lewin, B., Giovannucci, D., Varangis, P., 2004. Coffee Markets: New Paradigms in Global Supply and Demand. The International Bank for Reconstruction and Development, Agriculture and Rural Development Discussion Paper 3, 150.

Lewis, E.E., Clarke, D.J., 2012. Nematode parasites and entomopathogens. In: Vega, F.E., Kaya, H.K. (Eds.), Insect Pathology. second ed. Academic Press, San Diego, pp. 395–424.

López, M.N., Uriarte, S.L., López, L.N., Dufour, B., Sequeira, C.A., 1993. Cría de *Cephalonomia stephanoderis* (Hymenoptera: Bethylidae) en Nicaragua y pruebas preliminares de establecimiento en el campo. XVI Simposio sobre Caficultura Latinoamericana, Vol. 2, Managua, Nicaragua, pages unnumbered.

López-Guillén, G., Valdez Carrasco, J., Cruz-López, L., Barrera, J.F., Malo, E.A., Rojas, J.C., 2011. Morphology and structural changes in flight muscles of *Hypothenemus hampei* (Coleoptera: Curculionidae) females. Environ. Entomol. 40, 441–448.

López-Pazos, S.A., Cortázar Gómez, J.E., Cerón Salamanca, J.A., 2009. Cry1B and Cry3A are active against *Hypothenemus hampei* Ferrari (Coleoptera: Scolytidae). J. Invertebr. Pathol. 101, 242–245.

López-Pazos, S.A., Rojas Arias, A.C., Ospina, S.A., Cerón, J., 2010. Activity of *Bacillus thuringiensis* hybrid protein against a lepidopteran and a coleopteran pest. FEMS Microbiol. Lett. 302, 93–98.

López-Vaamonde, C., Moore, D., 1998. Developing methods for testing host specificity of *Phymastichus coffea* La Salle (Hym.: Tetrastichinae), a potential biological control agent of *Hypothenemus hampei* (Ferrari) (Col.: Scolytidae) in Colombia. Biocontrol Sci. Technol. 8, 397–411.

Lubick, N., 2010. Endosulfan's exit: U.S. EPA pesticide review leads to a ban. Science 328, 1466.

Luck, R.F., Merle-Shepard, B., Kenmore, P.E., 1999. Evaluation of biological control with experimental methods. In: Bellows, T.S., Fisher, T.W. (Eds.), Handbook of Biological Control. Academic Press, San Diego, pp. 225–242.

Mahdi, H.S.A., 2006. Survey of coffee insects under traditional storage conditions in Yemen. Ninth Annual Congress of Plant Protection. November 19–23, 2006, Damascus, Syria, p. E–19.

Maldonado-Londoño, C.E., Benavides-Machado, P., 2007. Evaluación del establecimiento de *Cephalonomia stephanoderis* y *Prorops nasuta*, controladores de *Hypothenemus hampei*, en Colombia. Cenicafé 58, 333–339.

Mantilla, J.G., Galeano, N.F., Gaitan, A.L., Cristancho, M.A., Keyhani, N.O., Góngora, C.E., 2012. Transcriptome analysis of the entomopathogenic fungus *Beauveria bassiana* grown on cuticular extracts of the coffee berry borer (*Hypothenemus hampei*). Microbiology 158, 1826–1842.

Manton, J.L., Hollingsworth, R.G., Cabos, R.Y.M., 2012. Potential of *Steinernema carpocapsae* (Rhabditida: Steinernematidae) against *Hypothenemus hampei* (Coleoptera: Curculionidae) in Hawai'i. Fla. Entomol. 95, 1194–1197.

Martínez D., C.P., González G., M.T., Bustillo P., A.E., Valencia J., A., 2000. Propiedades de amilasas provenientes de la broca del café *Hypothenemus hampei* (Coleoptera: Scolytidae). Rev. Colomb. Entomol. 26, 39–42.

Martínez Díaz, C.P., Valencia Jiménez, A., González García, M.T., Bustillo, A.E., 2000. Properties of amylases of coffee berries [*sic*] borer *Hypothenemus hampei* (Ferrari)—Coleoptera: Scolytidae. In: Sera, T., Soccol, C.R., Pandey, A., Roussos, S. (Eds.), Coffee Biotechnology and Quality. Kluwer Academic Publishers, Dordrecht, pp. 297–306, Proceedings, 3rd International Seminar on Biotechnology in the Coffee Agro-Industry, Londrina, Brazil.

Mathieu, F., Brun, L., Frérot, B., 1991. Preliminary results on comparative GC analyses of volatiles produced by the coffee berries. Proceedings of the 14th International Scientific Colloquium on Coffee, Association Scientifique Internationale du Café (ASIC), San Francisco, California pp. 687–689.

Mathieu, F., Malosse, C., Cain, A.H., Frérot, B., 1996. Comparative headspace analysis of fresh red coffee berries from different cultivated varieties of coffee trees. J. High Res. Chromatography 5, 298–300.

Mathieu, F., Brun, L.O., Frérot, B., 1997a. Factors related to native host abandonment by the coffee berry borer *Hypothenemus hampei* (Ferr.) (Col., Scolytidae). J. Appl. Entomol. 121, 175–180.

Mathieu, F., Brun, L.O., Marcillaud, C., Frérot, B., 1997b. Trapping of the coffee berry borer *Hypothenemus hampei* Ferr. (Col., Scolytidae) within a mesh-enclosed environment: interaction of olfactory and visual stimuli. J. Appl. Entomol. 121, 181–186.

Mathieu, F., Malosse, C., Frérot, B., 1998. Identification of the volatile components released by fresh coffee berries at different stages of ripeness. J. Agric. Food Chem. 46, 1106–1110.

Mathieu, F., Brun, L.O., Frérot, B., Suckling, D.M., Frampton, C., 1999. Progression of field infestation is linked with trapping of coffee berry borer, *Hypothenemus hampei* (Col., Scolytidae). J. Appl. Entomol. 123, 535–540.

Mathieu, F., Gaudichon, V., Brun, L.O., Frerot, B., 2001. Effect of physiological status on olfactory and visual responses of female *Hypothenemus hampei* during host plant colonization. Physiol. Entomol. 26, 189–193.

Mayné, R., 1914. Travau de l'Entomologiste de la Colonie. Bull. Agric. Congo Belge 5, 577–600.

Mayné, R., Donis, C., 1962. Hôtes entomologiques du bois. II. Distribution au Congo, au Rwanda et au Burundi. Observations éthologiques. Publications de l'institut national pour l'étude agronomique du Congo Belge, Brussels, Série Scientifique No. 100.

Mbang, J.A.à, Mounjouenpou, P., Mahob, R.J., Mbarga Amougou, M., Mouen Bedimo, J., Nyasse, S., et al., 2012. Evaluation naturelle de l'impact de *Beauveria bassiana*: champignon entomopathogène dans la dynamique de population de *Hypothenemus hampei*, scolyte de baies des cerises de *Coffea canefora* [*sic*]. Afr. Crop. Sci. J. 20, 443–451.

Mbondji, P.M., 1988. Étude épidémiologique d'*Hypothenemus hampei* (Coleoptera: Scolytidae), ravageur des baies du caféier, dans deux régions du Cameroun. Le Naturaliste Canadien 115, 245–249.

McDonald, J.H., 1930. Coffee Growing: With Special Reference to East Africa. East Africa, Ltd, London.

McPherson, G.I., 1978. Report on the presence of the coffee berry borer (*Hypothenemus hampei*) in Jamaica. Simposio sobre Caficultura, Ribeirão Preto, São Paulo, Brasil, October 24, 1978, Instituto Interamericano de Ciencias Agrícolas. Informes de Conferencias, Cursos y Reuniones No. 184, pp.15–24.

Mendes, C.T., 1938. A broca do café. Revista de Agricultura (Piracicaba) 13, 405–423.

Mendes, L.O.T., 1939. O sombreamento do cafeeiro e a "broca do café." Revista do Instituto de Café do Estado de S. Paulo 14, 874–891 (Rev. Appl. Entomol., Ser. A, 1940, 28, 514).

Mendes, L.O.T., 1940a. O sombreamento do cafeeiro e a "broca do café." Segunda contribuição. Revista do Instituto de Café do Estado de S. Paulo 15, 1578–1584.

Mendes, L.O.T., 1940b. O sombreamento do cafeeiro e a "Broca do café." Terceira contribuição. Revista do Instituto de Café do Estado de S. Paulo 15, 1817–1825.

Mendesil, E., Jembere, B., Seyoum, E., Abebe, M., 2004. The biology and feeding behavior of the coffee berry borer, *Hypothenemus hampei* (Ferrari) (Coleoptera: Scolytidae) and its economic importance in Southwestern Ethiopia. Proceedings of the 20th International Scientific Colloquium on Coffee, Association Scientifique Internationale du Café (ASIC), Bali, Indonesia, pp. 1209–1215.

Mendesil, E., Bruce, T.J.A., Woodcock, C.M., Caulfield, J.C., Seyoum, E., Pickett, J.A., 2009. Semiochemicals used in host location by the coffee berry borer, *Hypothenemus hampei*. J. Chem. Ecol. 35, 944–950.

Méndez-López, I., 1990. Control microbiano de la broca del fruto del cafeto *Hypothenemus hampei* Ferrari (Coleoptera: Scolytidae), con el hongo *Beauveria bassiana* (Bals.) Vuill. (Deuteromycetes) en el Soconusco, Chiapas. M.S. thesis, Colegio de Postgraduados, Chapingo, México, 135 pp.

Méndez-López, I., Basurto-Ríos, R., Ibarra, J.E., 2003. *Bacillus thuringiensis* serovar *israelensis* is highly toxic to the coffee berry borer, *Hypothenemus hampei* Ferr. (Coleoptera: Scolytidae). FEMS Microbiol. Lett. 26, 73–77.

Mendoza Mora, J.R., 1991. Resposta da broca-do-café, *Hypothenemus hampei*, a estímulos visuais e semioquímicos. M.S. thesis, Universidade Federal de Viçosa, Brasil, 44 pp.

Mendoza, J.R., Gomes de Lima, J.O., Vilela, E.F., Fanton, C.J., 2000. Atractividade de frutos à broca-do-café, *Hypothenemus hampei* (Ferrari): estímulos visuais e olfativos. Anais do Seminario Internacional sobre Biotecnologia na Agroindustria Cafeeira. 3. Londrina (Brasil), Maio 24–28, 1999. Londrina, Brasil, UFPR-IAPAR-IRD, pp. 313–315.

Mesquita, F. de C., 1944. A broca do café no Estado do Rio de Janeiro. Boletim Fitossanitário 1, 247–253.

Messing, R.H., 2012. The coffee berry borer (*Hypothenemus hampei*) invades Hawaii: preliminary investigations in trap response and alternate hosts. Insects 3, 640–652.

Miller, D.R., Rabaglia, R.J., 2009. Ethanol and (−)-α-pinene: attractant kairomones for bark and ambrosia beetles in the southeastern US. J. Chem. Ecol. 25, 435–448.

Miller, J.R., Strickler, K.L., 1984. Finding and accepting host plants. In: Bell, W.J., Cardé, R.T. (Eds.), Chemical Ecology of Insects. Chapman and Hall, London, pp. 127–157.

Mitchell, A., Maddox, C., 2010. Bark beetles (Coleoptera: Curculionidae: Scolytinae) of importance to the Australian macadamia industry: an integrative taxonomic approach to species diagnostics. Aust. J. Entomol. 49, 104–113.

Moeck, H.A., 1970. Ethanol as the primary attractant for the ambrosia beetle *Trypodendron lineatum* (Coleoptera: Scolytidae). Can. Entomol. 102, 985–995.

Moguel, P., Toledo, V.M., 1999. Biodiversity conservation in traditional coffee systems of Mexico. Conserv. Biol. 13, 11–21.

Molina A., J.P., López N., J.C., 2002. Desplazamiento y parasitismo de entomonematodos hacia frutos infestados con la broca del café *Hypothenemus hampei* (Coleoptera: Scolytidae). Rev. Colomb. Entomol. 28, 145–151.

Molina, D., Patiño, L., Quintero, M., Cortes, J., Bastos, S., 2014. Effects of the aspartic protease inhibitor from *Lupinus bogotensis* seeds on the growth and development of *Hypothenemus hampei*: an inhibitor showing high homology with storage proteins. Phytochemistry 98, 69–77.

Molina Acevedo, J.P., López Núñez, J.C., 2003. Supervivencia y parasitismo de nematodos entomopatógenos para el control de *Hypothenemus hampei* Ferrari (Coleoptera: Scolytidae) en frutos de café. Boletín de Sanidad Vegetal. Plagas (Spain) 29, 523–533.

Molinari, P.-A., 1988. Situación de la broca del café *Hypothenemus hampei* Ferrari (Coleoptera: Scolytidae) en Santo Domingo de los Colorados. Sanidad Vegetal (Ecuador) 3, 31–40.

Monson, R.K., Fall, R., 1989. Isoprene emission from aspen leaves. Plant Physiol. 90, 267–274.

Monterrey, J., 1991. La broca del café en Nicaragua. In: Barrera, J.F., Castillo, A., Gómez, J., Malo, E., Infante, F. (Eds.), Resúmenes, I Reunión Intercontinental sobre Broca del Café Tapachula, Chiapas, México, pp. 28–30.

Monterroso Mayorga, J.L., 1981. Incidencia del *Beauveria bassiana* sobre la broca del café y su reproducción en coco en Guatemala. Revista ANACAFE (Guatemala) 26, 10, 12.

Montgomery, M.E., Wargo, P.M., 1983. Ethanol and other host-derived volatiles as attractants to beetles that bore into hardwoods. J. Chem. Ecol. 9, 181–190.

Montilla, R., Camacho, B., Quintero, A., Cardozo, G., 2006. Parasitismo por *Beauveria bassiana* sobre la broca del café en el estado Trujillo, Venezuela. Agron. Trop. (Venezuela) 56, 183–198.

Monzón, A.J., Guharay, F., Klingen, I., 2008. Natural occurrence of *Beauveria bassiana* in *Hypothenemus hampei* (Coleoptera: Curculionidae) populations in unsprayed coffee fields. J. Invertebr. Pathol. 97, 134–141.

Moore, D., Prior, C., 1988. Present status of biological control of the coffee berry borer *Hypothenemus hampei*, Brighton Crop Protection Conference, Pests and Diseases—1988, Vol. 3, pp. 1119–1124.

Moore, D., Abraham, Y.J., Mills, N.J., 1990. Effects of competition in the coffee berry borer, *Hypothenemus hampei* (Ferrari) (Col., Scolytidae). J. Appl. Entomol. 109, 64–70.

Morales P., R., Bacca, T., Soto G., A., 2011. Establecimiento de los parasitoides de origen Africano de la broca del café en la zona cafetera del norte del Departamento de Nariño. Boletín Científico, Centro de Museos, Museo de Historia Natural 15, 81–93.

Morales-Ramos, J.A., Guadalupe Rojas, M., Sittertz-Bhatkar, H., Saldaña, G., 2000. Symbiotic relationship between *Hypothenemus hampei* (Coleoptera: Scolytidae) and *Fusarium solani* (Moniliales: Tuberculariaceae). Ann. Entomol. Soc. Am. 93, 541–547.

Morallo-Rejesus, B., Baldos, E., 1980. The biology of coffee berry borer *Hypothenemus hampei* (Ferr.) (Scolytidae, Coleoptera) and its incidence in the southern Tagalog provinces. The Philippine Entomologist 4, 303–316.

Moreno Rodríguez, D., Álvarez Núñez, A., Vázquez Moreno, L.L., Simonetti, J.A., 2010. Evaluación de atrayentes para la captura de hembras adultas de broca del café *Hypothenemus hampei* (Ferrari) con trampas artesanales. Fitosanidad (Cuba) 14, 177–180.

Morstatt, H., 1912. Die Schädlinge und Krankheiten des Kaffeebaumes in Ostafrika. Pflanzer. Zeitschrift für Land- und Forstwirtschaft in Deutsch-Ostafrika 2, 1–87.

Muesebeck, C.F.W., 1936. The genera of parasitic wasps of the braconid subfamily Euphorinae, with a review of the Nearctic species, United States Department of Agriculture, Miscellaneous Publication No. 241, 38 pp.

Muñoz, R., 1989. Ciclo biológico y reproducción partenogenética de la broca del fruto del cafeto *Hypothenemus hampei* (Ferr.). Turrialba 39, 415–421, Also published as: (1) Muñoz Hernández, R. (1988). Ciclo biológico y reproducción partenogenética de la broca del fruto del cafeto (Hypothenemus hampei Ferr.). XI Simposio de Caficultura Latinoamericana, San Salvador, El Salvador. IICA PROMECAFE, pp. 53–65; (2) Muñoz H., R. (1989). Ciclo biológico y reproducción partenogenética de la broca del fruto del cafeto (*Hypothenemus hampei* Ferr.). III Taller Regional de Broca, Antigua, Guatemala. IICA PROMECAFE, pp. 45–58.

Muñoz, R.I., 1985. Medidas de control de broca del fruto del cafeto efectuadas en Honduras. Memoria, Curso sobre Manejo Integrado de Plagas del Cafeto con énfasis en Broca del Fruto (*Hypothenemus hampei*, Ferr.), 15 al 19 de Julio de 1985. Instituto Interamericano de Cooperación para la Agricultura, Oficina en Guatemala, pp. 170–177.

Muñoz, R.I., Andino, A., Zelaya, R.R., 1987. Fluctuación poblacional de la broca del fruto del cafeto (*Hypothenemus hampei* Ferr.) en la zona del lago de Yojoa. Memoria, Taller Internacional sobre Manejo Integrado de la Broca del Café (*Hypothenemus hampei*, Ferr.) Tapachula, Mexico, IICA-PROMECAFE, pp. 75–99.

Murphy, S.T., Moore, D., 1990. Biological control of the coffee berry borer, *Hypothenemus hampei* (Ferrari) (Coleoptera: Scolytidae): previous programmes and possibilities for the future. Biocontrol News and Information 11, 107–117.

Muschler, R.G., 2004. Shade management and its effect on coffee growth and quality. In: Wintgens, J.N. (Ed.), Coffee: Growing, Processing, Sustainable Production. Wiley-VCH, Weinheim, pp. 391–418.

Naidu, M.M., Rang, C., Frutos, R., Sreenivasan, C.S., Naidu, R., 2001. Screening of *Bacillus thuringiensis* serotypes by polymerase chain reaction (PCR) for insecticidal crystal genes toxic against coffee berry borer. Indian J. Exp. Biol. 39, 148–154.

NAPPO, 2007. Detections of coffee berry borer, *Hypothenemus hampei*, in Puerto Rico—United States. North American Plant Protection Organization's Phytosanitary Alert System. Available online: http://www.pestalert.org/oprDetail.cfm?oprID=281. Last accessed: April 21, 2014.

NASS, 1998–2012. United States Department of Agriculture National Agricultural Statistics Service. Available online: http://www.nass.usda.gov/Statistics_by_State/Hawaii/Publications/Archive/. Last accessed April 21, 2014.

Nathanson, J.A., 1984. Caffeine and related methylxanthines: possible naturally occurring pesticides. Science 226, 184–187.

Navarro, L., Gongora, C., Benavides, P., 2010. Single nucleotide polymorphism detection at the *Hypothenemus hampei Rdl* gene by allele-specific PCR amplification with T_m-shift primers. Pestic. Biochem. Physiol 97, 204–208.

Neiva, A., 1928. Os trabalhos da Commissão de Estudo e Debellação da Praga Cafeeira desde seu inicio. Commissão de Estudo e Debellação da Praga Cafeeira. Publicação No. 21, 27 pp.

Neiva, A., da Costa Lima, A.M., Navarro de Andrade, Ed., 1924a. A praga do café. A Lavoura (Brazil) 15, 235–238.

Neiva, A., Navarro de Andrade, Ed, Queiroz Telles, A., 1924b. Instrucções para o combate á broca do café. Secretaria da Agricultura, Commercio e Obras Publicas. Serviço de Defesa do Café. Publicação No. 3, 1–15.

Neves, P.M.O.J., Hirose, E., 2005. Seleção de isolados de *Beauveria bassiana* para o controle biológico da broca-do-café, *Hypothenemus hampei* (Ferrari) (Coleoptera: Scolytidae). Neotrop. Entomol. 34, 77–82.

Nuñez, J., Hernández, E., Giraldo, W., Navarro, L., Gongora, C., Cristancho, M.A., et al., 2012. First draft genome sequence of coffee berry borer: the most invasive insect pest of coffee crops. Sixth Annual Arthropod Genomics Symposium and 15 k Community Workshop, Kansas City, MO, Abstract EG-12.

Oliveira, C.M., Auad, A.M., Mendes, S.M., Frizzas, M.R., 2013. Economic impact of exotic insect pests in Brazilian agriculture. J. Appl. Entomol. 137, 1–15.

de Oliveira Filho, M.L., 1927. Contribuição para o conhecimento da broca do café, *Stephanoderes hampei* (Ferrari 1867). Modo de comportar-se e ser combatida em S. Paulo-Brasil. Secretaria da Agricultura, Commercio e Obras Publicas. Comissão do Estudo e Debellação da Praga Caféeira 20, 1–95.

O'Neill, S.L., Hoffmann, A.A., Werren, J.H. (Eds.), 1998. Influential Passengers: Inherited Microorganisms and Arthropod Reproduction. Oxford University Press, New York.

Orozco Hoyos, J., 2002. Guía para la producción del parasitoide *Phymastichus coffea* para el control de la broca del café. A guide to the rearing of the parasitoid *Phymastichus coffea* for control of the coffee berry borer. The Commodities Press, Colombia, Published in Spanish and English.

Ortiz, A., Ortiz, A., Vega, F.E., Posada, F., 2004. Volatile composition of coffee berries at different stages of ripeness and their possible attraction to the coffee berry borer *Hypothenemus hampei* (Coleoptera: Curculionidae). J. Agric. Food Chem. 52, 5914–5918.

Osorio, I., 2007. Se cuela la broca en el café boricua. Oficina de Prensa, Noticias y Eventos, Universidad de Puerto Rico, Recinto Universitario de Mayagüez. Available online: http://www.uprm.edu/news/articles/as2007134.html. Last accessed: April 21, 2014.

Osorio, N., 2002. The global coffee crisis: a threat to sustainable development. International Coffee Organization, London. Available online: http://dev.ico.org/documents/globalcrisise.pdf. Last accessed: April 10, 2014.

Padilla-Hurtado, B., Flórez-Ramos, C., Aguilera-Gálvez, C., Medina-Olaya, J., Ramírez-Sanjuan, A., Rubio-Gómez, J., Acuña-Zornosa, R., 2012. Cloning and expression of an endo-1,4-β-xylanase from the coffee berry borer, *Hypothenemus hampei*. BMC Res. Notes 5, 23.

Pamplona, A., 1927. Divulgação, pelo cinema, dos methodos de combate á broca do café no estado de São Paulo, Secretaria da Agricultura, Commercio e Obras Publicas. Comissão para o Estudo e Debellação da Praga Cafeeira, Publicação No. 19, 101–104.

Parkin, C.S., Brun, L.O., Suckling, D.M., 1992. Spray deposition in relation to endosulfan resistance in coffee berry borer (*Hypothenemus hampei*) (Coleoptera: Scolytidae) in New Caledonia. Crop Prot. 11, 213–220.

Pascalet, P., 1939. La lutte biologique contre *Stephanoderes hampei* ou scolyte du caféier au Cameroun. Revue de Botanique Appliquée et d'Agriculture Tropicale 19, 753–764.

Passos, V. de J., Demoner, C.A., Morales, L., 2005. Efeito da arborização em cafeeiros na população da broca-do-café *Hypothenemus hampei* (Coleoptera: Scolytidae). Instituto Paranaense de Assistência Técnica e Extensão Rural—EMATER, Paraná, Brasil, 12 pp.

Pava-Ripoll, M., Posada, F.J., Momen, B., Wang, C., St. Leger, R., 2008. Increased pathogenicity against coffee berry borer, *Hypothenemus hampei* (Coleoptera: Curculionidae) by *Metarhizium anisopliae* expressing the scorpion toxin (AaIT) gene. J. Invertebr. Pathol. 99, 220–226.

Penagos Dardón, H., Flores, J.C., 1974. Hábito y tiempo de penetración de la broca del café *Hypothenemus hampei* (Ferrari) al fruto. Revista Cafetalera (Guatemala) 137, 5–15.

Peña, E., García, M., Blanco, E., Barrera, J.F., 2006. Introducción de la avispa de Costa de Marfil *Cephalonomia stephanoderis* Betrem (Hymenoptera: Bethyidae), parasitoide de la broca del fruto del cafeto *Hypothenemus hampei* (Ferrari) (Coleoptera: Scolytidae) en Cuba. Fitosanidad (Cuba) 10, 33–36.

Pereira, A.E., Vilela, E.F., Tinoco, R.S., de Lima, J.O.G., Fantine, A.K., Morais, E.G.F., França, C.F.M., 2012. Correlation between numbers captured and infestation levels of the coffee berry-borer, *Hypothenemus hampei*: a preliminary basis for an action threshold using baited traps. Int. J. Pest Manage 58, 183–190.

Pérez, C.J., Alvarado, P., Narváez, C., Miranda, F., Hernández, L., Vanegas, H., et al., 2000. Assessment of insecticide resistance in five insect pests attacking field and vegetable crops in Nicaragua. J. Econ. Entomol. 93, 1779–1787.

Pérez, J., Infante, F., Vega, F.E., Holguín, F., Macías, J., Valle, J., et al., 2003. Mycobiota associated with the coffee berry borer (*Hypothenemus hampei*) in Mexico. Mycol. Res. 7, 879–887.

Pérez, J., Infante, F., Vega, F.E., 2005. Does the coffee berry borer (Coleoptera: Scolytidae) have mutualistic fungi? Ann. Entomol. Soc. Amer. 98, 483–490.

Pérez-Lachaud, G., 1998. A new bethylid attacking the coffee berry borer in Chiapas (Mexico) and some notes on its biology. Southwest. Entomol. 23, 287–288.

Pérez-Lachaud, G., Hardy, I.C.W., 1999. Reproductive biology of *Cephalonomia hyalinipennis* (Hymenoptera: Bethylidae), a native parasitoid of the coffee berry borer, *Hypothenemus hampei* (Coleoptera: Scolytidae), in Chiapas, Mexico. Biol. Contr. 14, 152–158.

Pérez-Lachaud, G., Hardy, I.C.W., 2001. Alternative hosts for bethylid parasitoids of the coffee berry borer, *Hypothenemus hampei* (Coleoptera: Scolytidae). Biol. Contr. 22, 265–277.

Pérez-Lachaud, G., Batchelor, T.P., Hardy, I.C.W., 2004. Wasp eat wasp: facultative hyperparasitism and intra-guild predation by bethylid wasps. Biol. Contr. 30, 149–155.

Pérez León, R., Pérez Reyes, N., Pentón Valdivia, D., Mirabal Rodríguez, R., Cabrera Rodríguez, R., Galera Chongo, V.M., et al., 2009. Determinación de la cepa LLBB-11 de *Beauveria bassiana* para el contro de *Hypothenemus hampei* Ferrari en el municipio de Fomento. Centro Agrícola (Cuba) 36, 91–92.

Pérez López, E.J., Bustillo-Pardey, A.E., González-Garcia, M.T., Posada-Flórez, F.J., 1995. Comparación de dos dietas merídicas para la cría de *Hypothenemus hampei*. Cenicafé 46, 189–195.

Perfecto, I., Vandermeer, J., 1996. Microclimatic changes and the indirect loss of ant diversity in a tropical agroecosystem. Oecologia 108, 577–582.

Perfecto, I., Vandermeer, J., 2006. The effect of an ant-hemipteran mutualism on the coffee berry borer (*Hypothenemus hampei*) in southern Mexico. Agric. Ecosys. Environ. 117, 218–221.

Perfecto, I., Rice, R.A., Greenberg, R., van der Voort, M.E., 1996. Shade coffee: a disappearing refuge for biodiversity. BioScience 46, 598–608.

Perthuis, B., Pradon, J.L., Montagnon, C., Dufour, M., Leroy, T., 2005. Stable resistance against the leaf miner *Leucoptera coffeella* expressed by genetically transformed *Coffea canephora* in a pluriannual field experiment in French Guiana. Euphytica 144, 321–329.

Philpott, S.M., Armbrecht, I., 2006. Biodiversity in tropical agroforests and the ecological role of ants and ant diversity in predatory function. Ecol. Entomol. 31, 369–377.

Pohlan, H.A.J., 2005. Manejo de cenosis en cafetales y sus impactos sobre insectos, con especial énfasis en la broca del café. In: Barrera, J.F. (Ed.), Simposio sobre la Situación Actual y Perspectivas de la Investigación y Manejo de la Broca del Café en Costa Rica, Cuba, Guatemala y México. In: Sociedad Mexicana de Entomología y El Colegio de la Frontera Sur, Tapachula, México, pp. 22–30.

Pohlan, H.A.J., Janssens, M.J.J., Giesemann Eversbusch, B., 2008. Impact of *Canavalia* cover crop management in *Coffea arabica* L. on plant-invertebrate associations. Open Agric. J. 2, 84–89.

Poinar Jr., G., Vega, F.E., Castillo, A., Chavez, I.E., Infante, F., 2004. *Metaparasitylenchus hypothenemi* n. sp. (Nematoda: Allantonematidae), a parasite of the coffee berry borer, *Hypothenemus hampei* (Ferrari) (Curculionidae: Scolytinae). J. Parasitol. 90, 1106–1110.

Portilla, M., Bustillo, A.E., 1995. Nuevas investigaciones en la cría de *Hypothenemus hampei* y de sus parasitoides *Cephalonomia stephanoderis* y *Prorops nasuta*. Rev. Colomb. Entomol. 21, 25–33.

Portilla, M., 1999. Mass rearing technique for *Cephalonomia stephanoderis* (Hymenoptera: Bethylidae) on *Hypothenemus hampei* (Coleoptera: Scolytidae) developed using Cenibroca artificial diet. Rev. Colomb. Entomol. 25, 57–66.

Portilla-Reina, M., 1999. Desarrollo y evaluación de una dieta artificial para la cría de *Hypothenemus hampei*. Cenicafé 50, 24–38.

Portilla R., M., Streett, D., 2006. Nuevas técnicas de producción masiva automatizada de *Hypothenemus hampei* sobre la dieta artificial Cenibroca modificada. Cenicafé 57, 37–50.

Posada, F., Vega, F.E., 2005. A new method to evaluate the biocontrol potential of single spore isolates of fungal entomopathogens. J. Insect Sci. 5, 37.

Posada, F., Vega, F.E., 2006. Inoculation and colonization of coffee seedlings (*Coffea arabica* L.) with the fungal entomopathogen *Beauveria bassiana* (Ascomycota: Hypocreales). Mycoscience 47, 284–289.

Posada, F., Aime, M.C., Peterson, S.W., Rehner, S.A., Vega, F.E., 2007. Inoculation of coffee plants with the fungal entomopathogen *Beauveria bassiana* (Ascomycota: Hypocreales). Mycol. Res. 111, 748–757.

Posada F., F.J., Bustillo P., A.E., 1994. El hongo *Beauveria bassiana* y su impacto en la caficultura Colombiana. Agric. Trop. (Colombia) 31, 97–106.

Posada-Flórez, F.J., 2008. Production of *Beauveria bassiana* fungal spores on rice to control the coffee berry borer, *Hypothenemus hampei*, in Colombia. J. Insect Sci. 8, 41.

Posada-Flórez, F.J., Bustillo-Pardey, A.E., Saldarriaga-Correa, G., 1993. Primer registro del ataque de *Hirsutella eleutheratorum* sobre la broca del café, en Colombia. Cenicafé 44, 155–158.

Preciado-Rodríguez, D.P., Bustillo-Pardey, A.E., Valencia-Jiménez, A., 2000. Caracterización parcial de una proteinasa digestive proveniente de la broca del café (Coleoptera: Scolytidae). Cenicafé 51, 20–27.

Prates, H.S., 1969. Observações preliminares da atração da broca do café, *Hypothenemus hampei* (Ferrari, 1867), a extratos de frutos do cafeeiro (cereja e verde). Solo (Brazil) 61, 13–14.

Price, P.W., Denno, R.F., Eubanks, M.D., Finke, D.L., Kaplan, I., 2011. Insect Ecology: Behavior, Population and Communities. Cambridge University Press, New York.

Priyono, A., Rejesus, B.M., Reyes, S.G., 2004. Development of a mass-rearing technique for the coffee berry borer, *Hypothenemus hampei* Ferrari (Coleoptera: Scolytidae). The Philippine Entomologist 18, 176.

Prokopy, R.J., Owens, E.D., 1983. Visual detection of plants by insects. Annu. Rev. Entomol. 28, 337–364.

Quezada, J.R., Urbina, N.E., 1987. La broca del fruto del cafeto, *Hypothenemus hampei*, y su control. In: Pinochet, J. (Ed.), Plagas y Enfermedades de Carácter Epidémico en Cultivos Frutales de la Región Centroamericana, pp. 48–59, Centro Agronómico Tropical de Investigación y Enseñanza, Informe Técnico No. 110.

Rawel, H.M., Rohn, S., Kroll, J., 2005. Characterisation of 11s protein fractions and phenolic compounds from green coffee beans under special consideration of their interactions. A review. Dtsch. Lebensm.-Rundsch 101, 148–160.

Realpe-Aranda, F.J., Bustillo-Pardey, A.E., López-Núñez, J.C., 2007. Optimización de la cría de *Galleria mellonella* (L.) para la producción de nematodos entomopatógenos parásitos de la broca del café. Cenicafé 58, 142–157.

Reddy, G.V.P., Guerrero, A., 2010. New pheromones and insect control strategies. Vitam. Horm. 83, 493–519.

Redgwell, R.J., Curti, D., Rogers, J., Nicolas, P., Fischer, M., 2003. Changes to the galactose/mannose ratio in galactomannans during coffee bean (*Coffea arabica* L.) development: implications for in vivo modification of galactomannan synthesis. Planta 217, 316–326.

Redgwell, R., Fischer, M., 2006. Coffee carbohydrates. Braz. J. Plant Physiol. 18, 165–174.

Rehner, S.A., Posada, F., Buckley, E.P., Infante, F., Castillo, A., Vega, F. E., 2006. Phylogenetic origins of African and Neotropical *Beauveria bassiana s.l.* pathogens of the coffee berry borer, *Hypothenemus hampei*. J. Invertebr. Pathol. 93, 11–21.

Rehner, S.A., Minnis, A.M., Sung, G.-H., Luangsa-ard, J.J., Devotto, L., Humber, R.A., 2011. Phylogeny and systematics of the anamorphic, entomopathogenic genus *Beauveria*. Mycologia 103, 1055–1073.

Reid, J.C., 1983. Distribution of the coffee berry borer (*Hypothenemus hampei*) within Jamaica, following its discovery in 1978. Trop. Pest Manage. 29, 224–230.

Rice, R.E., 2011. Fruits from shade trees in coffee: how important are they? Agroforestry Systems 83, 41–49.

Ritchie, A.H., 1925. Entomological report, 1924–25. Tanganyika Terr. Rept. Dept. Agric. 1924–25, Dar-es-Salaam, pp. 41–44.

Rivera-España, P.A., Montoya-Restrepo, E.C., Benavides-Machados, P., 2010. Biología del parasitoide *Prorops nasuta* (Hymenoptera: Bethylidae) en el campo y su tolerancia a insecticidas. Cenicafé 61, 99–107.

Rivera, M.A., Bridge, P.D., Bustillo P., A.E., 1997. Caracterización bioquímica y molecular de aislamientos de *Beauveria bassiana* procedentes de la broca del café, *Hypothenemus hampei*. Rev. Colomb. Entomol. 23, 51–57.

Roberts, D.L., Cooper, R.J., Petit, L.J., 2000. Use of premontane moist forest and shade coffee agroecosystems by army ants in Western Panama. Conserv. Biol. 14, 192–199.

Rocha Lima, H. da, 1945. O sombreamento dos cafezais e o Instituto Biológico. O Biológico (Brazil) 6, 45–47.

Roepke, W., 1919. Gegevens omtrent de Koffiebessen-boeboek (*Stephanoderes hampei* Ferr. = *coffeae* Hgd. Mededeelingen van het Instituut voor Plantenziekten 38, 1–32 (Rev. Appl. Entomol., Ser. A, 1920, 8, 447–449).

Rogg, H.W., 1997. The coffee berry borer, *Hypothenemus hampei*, in Bolivia: distribution, incidence and control programs. Instituto Interamericano de Cooperación para la Agricultura, Bolivia, unpublished.

Rojas, M., Morales, J., Delgado, C., Marín, R., Torres-Murillo, L.C., 2012. Identification of natural enemies of the coffee berry borer (*Hypothenemus hampei*) in Costa Rica, Proceedings of the 24th International Scientific Colloquium on Coffee, Association Scientifique Internationale du Café (ASIC), San José, Costa Rica, pp. 1266–1269.

Rojas Barrantes, M., 2009. Ensayo Laboratorio de Entomología: estudio preliminar de enemigos naturales de broca identificados en Turrialba. Informe Anual de Investigaciones 2009. Centro de Investigaciones en Café, Instituto del Café de Costa Rica, pp. 67–70.

Rojas, M.G., Morales-Ramos, J.A., Harrington, T.C., 1999. Association between *Hypothenemus hampei* (Coleoptera: Scolytidae) and *Fusarium solani* (Moniliales: Tuberculariaceae). Ann. Entomol. Soc. Am. 92, 98–100.

Romero, J.V., Cortina-Guerrero, H., 2004a. Evaluación de germoplasma de café por antixenosis a *Hypothenemus hampei* (Ferrari) en condiciones controladas. Cenicafé 55, 341–346.

Romero, J.V., Cortina-Guerrero, H., 2004b. Fecundidad y ciclo de vida de *Hypothenemus hampei* (Coleoptera: Curculionidae: Scolytinae) en introducciones silvestres de café. Cenicafé 55, 221–231.

Romero, J.V., Cortina G., H.A., 2007. Tablas de vida de *Hypothenemus hampei* (Coleoptera: Curculionidae: Scolytinae) sobre tres introducciones de café. Rev. Colomb. Entomol. 33, 10–16.

Rosales Mondragón, M., Silva Acuña, R., Rodríguez González, G., 1998. Estrategias para el manejo integrado del minador de la hoja y la broca del fruto del cafeto. FONAIAP Divulga (Venezuela) 60, 19–24.

Rossman, A.Y., Samuels, G.J., Rogerson, C.T., Lowen, R., 1999. Genera of *Bionectriaceae*, *Hypocreaceae* and *Nectriaceae* (Hypocreales, Ascomycetes). Stud. Mycol. 42, 1–248.

Ruales, C., 1997. Aspectos generales sobre la broca del café en Ecuador. Café & Cacao: Noticias (Ecuador) 2, 7–11.

Rubio-Gómez, J.D., Bustillo-Pardey, A.E., Vallejo-Espinosa, L.F., Benavides-Machado, P., Acuña-Zornosa, J.R., 2007. Morfología del sistema reproductor femenino y masculino de *Hypothenemus hampei*. Cenicafé 58, 75–82.

Rubio G., J.D., Bustillo P., A.E., Vallejo E., L.F., Acuña Z., J.R., Benavides M., P., 2008. Alimentary canal and reproductive tract of *Hypothenemus hampei* (Ferrari) (Coleoptera: Curculionidae, Scolytinae). Neotrop. Entomol. 37, 143–151.

Ruiz Serna, L., López, J.C., Bustillo, A.E., 1995. Efecto de inhibidores comerciales de proteinasas sobre el ciclo de vida de la broca del café, *Hypothenemus hampei* (Ferrari), en dieta artificial. Rev. Colomb. Entomol. 21, 122–128.

Ruiz, T., Uribe, P.T., Riley, J., 2000. The effect of sample size and spatial scale on Taylor's power law parameters for the coffee berry borer (Coleoptera: Scolytidae). Trop. Agric. (Trinidad) 77, 249–261.

Ruiz-Cárdenas, R., Baker, P., 2010. Life table of *Hypothenemus hampei* (Ferrari) in relation to coffee berry phenology under Colombian field conditions. Sci. Agric. (Brazil) 67, 658–668.

Ruiz S., L., Bustillo-Pardey, A.E., Posada Flórez, F.J., González G., M.T., 1996. Ciclo de vida de *Hypothenemus hampei* en dos dietas merídicas. Cenicafé 47, 77–84.

Salazar, H.M., Baker, P.S., 2002. Impacto de las liberaciones de *Cephalonomia stephanoderis* sobre poblaciones de *Hypothenemus hampei*. Cenicafé 53, 306–316.

Sampedro-Rosas, L., Villanueva-Arce, J., Rosas-Acevedo, J.L., 2008. Aislamiento y validación en campo de *Beauveria bassiana* (Balsamo) contra *Hypothenemus hampei* (Ferrari) en la región cafetalera del municipio de Atoyac de Álvarez, Gro. México. Revista Latinoamericana de Recursos Naturales 4, 199–202.

Sampson, W., 1923. Notes on the nomenclature of the family Scolytidae. The Annals and Magazine of Natural History 11, 269–271.

Samuels, R.I., Pereira, R.C., Gava, C.A.T., 2002. Infection of the coffee berry borer *Hypothenemus hampei* (Coleoptera: Scolytidae) by Brazilian isolates of the entomopathogenic fungi *Beauveria bassiana* and *Metarhizium anisopliae* (Deuteromycotina: Hyphomycetes). Biocontrol Sci. Technol. 12, 631–635.

Sánchez, L., Rodríguez, M.G., 2007. Potencialidades de *Heterorhabditis bacteriophora* Poinar cepa HC1 para el manejo de *Hypothenemus hampei* Ferr. I. Parasitismo y capacidad de búsqueda. Revista de Protección Vegetal (Cuba) 22, 80–84.

Sánchez, L., Rodríguez, M.G., 2008. Potencialidades de *Heterorhabditis bacteriophora* Poinar cepa HC1 para el manejo de *Hypothenemus hampei* Ferr. II. Compatibilidad con *Beauveria bassiana* (Balsamo) Vuillemin y endosulfan. Revista de Protección Vegetal (Cuba) 23, 104–111.

Saravanan, P.A., Chozhan, K., 2003. Monitoring and management of coffee berry borer, *Hypothenemus hampei* Ferrari (Scolytidae: Coleoptera). Crop Res. 26, 154–158.

Sassá, D.C., Varéa-Pereira, G., Miyagui, D.T., Neves, P.M.O.J., Wu, J.I., Sugahara, V.I., et al., 2008. Avaliação de parâmetros cinéticos de quitinases produzidas por *Beauveria bassiana* (Bals.) Vuill. Semina: Ciências Agrárias (Brazil) 29, 807–814.

Sassá, D.C., Varéa-Pereira, G., Neves, P.M.O.J., Garcia, J.E., 2009. Genetic variation in a chitinase gene of *Beauveria bassiana*: lack of association between enzyme activity and virulence against *Hypothenemus hampei*. J. Entomol. 6, 35–41.

Schedl, K.E., 1952. Fauna Argentinensis, V. Acta Zoologica Lilloana 12, 443–463.

Schedl, K.E., 1959. A check list of the Scolytidae and Platypodidae (Coleoptera) of Ceylon with descriptions of new species and biological notes. Transactions of the Royal Entomological Society of London 111, 469–534.

Schedl, K.E., 1960. Insectes nuisibles aux fruits et aux graines. Publications de l'institut national pour l'étude agronomique du Congo Belge, Série Scientifique 82, 1–133.

Schedl, K.E., 1961. Scolytidae und Platypodidae Afrikas. Band I. Familie Scolytidae. Revista de Entomologia de Moçambique 4, 335–742.

Segura, H.R., Barrera, J.F., Morales, H., Nazar, A., 2004. Farmers' perceptions, knowledge, and management of coffee pests and diseases and their natural enemies in Chiapas, Mexico. J. Econ. Entomol. 97, 1491–1499.

Sera, G.H., Sera, T., Shiguer Ito, D., Ribeiro Filho, C., Villacorta, A., Seidi Kanayama, F., et al., 2010. Coffee berry borer resistance in coffee genotypes. Braz. Arch. Biol. Technol. 53, 261–268.

Serra, C.A., 2006. Manejo Integrado de Plagas de Cultivos. Estado Actual y Perspectivas para la República Dominicana. Centro para el Desarrollo Agropecuario y Forestal, Inc, (CEDAF), Santo Domingo, República Dominicana, 176 pp.

Sherry, T.W., 2000. Shade coffee: a good brew even in small doses. The Auk 117, 563–568.

Siles, P., Harmand, J.-M., Vaast, P., 2010. Effects of *Inga densiflora* on the microclimate of coffee (*Coffea arabica* L.) and overall biomass under optimal growing conditions in Costa Rica. Agrofores. Syst. 78, 269–286.

da Silva, F.C., Ursi Ventura, M., Morales, L., 2006. O papel das armadilhas com semioquímicos no manejo da broca-do-café, *Hypothenemus hampei*. Semina: Ciências Agrárias (Brazil) 27, 399–406.

da Silva, D.S., de Oliveira, C.F.R., Parra, J.R.P., Marangoni, S., Macedo, M.L.R., 2014. Short and long-term antinutritional effect of the trypsin inhibitor ApTI for biological control of sugarcane borer. J. Insect Physiol. 61, 1–7.

Sirpa-Roque, G., 1999. Comparación de métodos de cría de la avispita de Togo (*Cephalonomia stephanoderis*) parasitoide de la broca del café (*Hypothenemus hampei*). B.S. thesis, Universidad Mayor de San Andrés, Bolivia, 67 pp.

Sladden, G.E., 1934. Le *Stephanoderes Hampei* [*sic*] Ferr. Bull. Agricole Congo Belge 25, 26–77.

Somarriba, E., Harvey, C.A., Samper, M., Anthony, F., González, J., Staver, C., Rice, R.A., 2004. Biodiversity conservation in neotropical (*Coffea arabica*) plantations. In: Schroth, G., da Fonseca, G.A.B., Harvey, C.A., Gascon, C., Vasconcelos, H.L., Izac, A.-M.N. (Eds.), Agroforestry and Biodiversity Conservation in Tropical Landscapes. Island Press, Washington, pp. 198–226.

Soto-Pinto, L., Perfecto, I., Castillo-Hernandez, J., Caballero-Nieto, J., 2000. Shade effect on coffee production at the northern Tzeltal zone of the state of Chiapas, Mexico. Agr. Ecosyst. Environ. 80, 61–69.

Soto-Pinto, L., Perfecto, I., Caballero-Nieto, J., 2002. Shade over coffee: its effects on berry borer, leaf rust and spontaneous herbs in Chiapas, Mexico. Agroforestry Systems 55, 37–45.

Sponagel, K.W., 1994. La broca del café *Hypothenemus hampei* en plantaciones de café robusta en la amazonía ecuatoriana. Ph.D. thesis (traducida del alemán por Wilma Miranda-Sponagel), Universidad de Giessen, Alemania, 279 p.

Staver, C., Guharay, F., Monterroso, D., Muschler, R.G., 2001. Designing pest-suppressive multistrata perennial crop systems: shade-grown coffee in Central America. Agroforestry Systems 53, 151–170.

Steyaert, R.L., 1935. Un ennemi naturel du *Stephanoderes*. Publications de institut national pour l'étude agronomique du Congo Belge, Série Scientifique 2, 1–46, (Rev. Appl. Entomol., Ser. A, 1937, 25, 10–11).

Strand, T.M., Ross, D.W., Thistle, H.W., Ragenovich, I.R., Matos Guerra, I., Lamb, B.K., 2012. Predicting *Dendroctonus pseudotsugae* (Coleoptera: Curculionidae) antiaggregation pheromone concentrations using an instantaneous puff dispersion model. J. Econ. Entomol. 105, 451–460.

Strobl, S., Maskos, K., Wiegand, G., Huber, R., Gomis-Rüth, F.X., Glockshuber, R., 1998. A novel strategy for inhibition of α-amylases: yellow meal worm α-amylase in complex with the *Ragi* bifunctional inhibitor at 2.5 Å resolution. Curr. Biol. 6, 911–921.

Strohmeyer, H., 1910. Ueber Kaffeeschädlinge auf der Insel Java. Entomologische Blätter 6, 186–187.

Suárez, A., Arrieche, N., Paz, R., 2013. Monitoreo digitalizado de *Hypothenemus hampei* Ferrari 1867 (Coleoptera: Curculionidae) en el Parque Nacional Terepaima, Estado Lara, Venezuela. Bioagro (Venezuela) 25, 201–206.

Sujay, Y.H., Sattagi, H.N., Patil, R.K., 2010. Invasive alien insects and their impact on agroecosystem. Karnataka Journal of Agricultural Science 23, 26–34.

Swaine, J.M., 1909. Catalogue of the described Scolytidae of America, north of Mexico. 24th Report of the State Entomologist on Injurious and other Insects of the State of New York 1908, Appendix B., Education Department Bulletin no. 455, New York State Museum, Museum Bulletin 134, 76–159.

Taylor, J.W., 2011. One fungus = one name: DNA and fungal nomenclature twenty years after PCR. IMA Fungus 2, 113–120.

Thomas, A.S., 1940. Robusta coffee. In: Tothill, J.D. (Ed.), Agriculture in Uganda. Oxford University Press, London, pp. 289–313.

Thomas, K.M., 1949. The coffee berry-borer (*Stephanoderes hampei* Ferr). The Indian Coffee Board Monthly Bulletin 13, 83–88.

Ticheler, J.H.G., 1961. Étude analytique de l'épidémiologie du scolyte des graines de café, *Stephanoderes hampei* Ferr., en Côte d'Ivoire. Mededelingen van de Landbouwhogeschool te Wageningen, Nederland 61, 1–49 (Rev. Appl. Entomol., Ser. A, 1963, 51, 434–435).

Ticheler, J.H.G., 1963. Estudio analítico de la epidemiologia del escolítido de los granos de café *Stephanoderes hampei* Ferr., en Costa de Marfil. Cenicafé 14, 223–294 (Spanish translation of Ticheler, 1961).

Tobar H., P., Vélez A., P.E., Montoya R., E.C., 1998. Evaluación de campo de un aislamiento del hongo *Beauveria bassiana* seleccionado por resistencia a la luz ultravioleta. Rev. Colomb. Entomol. 24, 157–163.

de Toledo, A.A., 1942. Notas sobre a biologia da vespa de Uganda "*Prorops nasuta* Waterst". (Hym. Bethyl.) no estado de S. Paulo-Brasil. Arq. Inst. Biol., São Paulo 13, 233–260.

de Toledo, A.A., 1945. Estudos estatísticos da infestação num cafezal pela broca "*Hypothenemus hampei* (Ferr., 1867)" (Col. Ipidae). Arq. Inst. Biol., São Paulo 16, 27–39.

de Toledo, A.A., 1948. Comportamento da vespa de Uganda em cafezal sombreado. O Biológico (Brazil) 14, 189–191.

Torrealba R., G.E., Arcaya S., E.A., 2005. Cría masiva de *Cephalonomia stephanoderis* (Hymenoptera: Bethylidae) sobre la broca del café *Hypothenemus hampei* (Ferrari). Resúmenes, XIX Congreso Venezolano de Entomología. Entomotrópica 20, 132–133.

Trejo S., A.R., Fúnez C. R., 2004. Evaluación del establecimiento de los parasitoides *Cephalonomia stephanoderis* y *Prorops nasuta* sobre la broca del fruto del café (*Hypothenemus hampei*) en 14 años de liberación en Honduras. Memorias, IX Congreso Internacional de Manejo Integrado de Plagas, San Salvador, p. 72.

Trejo S., A.R., Muñoz H., R.I., Cabrera, L., 2000. Confirmación de la reproducción partenogenética de la broca del fruto del cafeto *Hypothenemus hampei* Ferr. en condiciones de laboratorio. Informes Técnicos de Investigaciones Realizadas en Honduras 1998–2000. Proyecto Manejo Integrado de la Broca del Fruto del Cafeto *Hypothenemus hampei* Ferrari. IICA-PROMECAFE-IHCAFE, Honduras, pp. 67–75.

Trujillo E., H.I., Ariztizábal A., L.F., Bustillo P., A.E., Jiménez Q., M., 2006. Evaluación de métodos para cuantificar poblaciones de broca del café, *Hypothenemus hampei* (Ferrari) (Coleoptera: Curculionidae: Scolytinae), en fincas de caficultores experimentadores. Rev. Colomb. Entomol. 32, 39–44.

Tscharntke, T., Clough, Y., Bhagwat, S.A., Buchori, D., Faust, H., Hertel, D., et al., 2011. Multifunctional shade-tree management in tropical agroforestry landscapes—a review. J. Appl. Ecol. 48, 619–629.

Turnbaugh, P.J., Ley, R.E., Hamady, M., Fraser-Liggett, C.M., Knight, R., Gordon, J.I., 2007. The Human Microbiome Project. Nature 449, 804–810.

Uemura-Lima, D.H., Ventura, M.U., Mikami, A.Y., da Silva, F.C., Morales, L., 2010. Responses of coffee berry borer, *Hypothenemus hampei* (Ferrari) (Coleoptera: Scolytidae), to vertical distribution of methanol:ethanol traps. Neotrop. Entomol. 39, 930–933.

Vaast, P., Bertrand, B., Perriot, J.-J., Guyot, B., Génard, M., 2006. Fruit thinning and shade improve bean characteristics and beverage quality of coffee (*Coffea arabica* L.) under optimal conditions. J. Sci. Food Agric. 86, 197–204.

Valencia-Jiménez, A., 2000. Amylase and protease inhibitors as alternative against herbivore insect. In: Sera, T., Soccol, C.R., Pandey, A., Roussos, S. (Eds.), Coffee Biotechnology and Quality. Kluwer Academic Publishers, Dordrecht, pp. 287–296. Proceedings, 3rd International Seminar on Biotechnology in the Coffee Agro-Industry, Londrina, Brazil.

Valencia-Jiménez, A., Ruiz-Serna, L., González-García, M.T., Riaño-Herrera, N.M., Posada Flórez, F., 1994. Efecto de inhibidores de proteinasas sobre la actividad tripsina y quimotripsina de *Hypothenemus hampei* (Ferrari). Cenicafé 45, 51–59.

Valencia, A., Bustillo, A.E., Ossa, G.E., Chrispeels, M.J., 2000. α-Amylases of the coffee berry borer (*Hypothenemus hampei*) and their inhibition by two plant amylase inhibitors. Insect Biochem. Mol. Biol. 30, 207–213.

van der Weele, H.W., 1910. *Xyleborus coffeivorus* nov. spec. een nieuwe koffieparasiet. Teysmannia 21, 308–316.

van Dinther, J.B.M., 1960. Insect pests of cultivated plants in Surinam. Landbouwproefstation in Suriname Bulletin 76, 1–159.

Vandermeer, J., Perfecto, I., Ibarra Nuñez, G., Phillpot, S., Garcia Ballinas, A., 2002. Ants (*Azteca* sp.) as potential biological control agents in shade production in Chiapas. Agroforestry Systems 56, 271–276.

Varaprasad, K.S., Balasubramanian, S., Diwakar, B.J., Rama Rao, C.V., 1994. First report of an entomogenous nematode, *Panagrolaimus* sp. from coffee berry borer, *Hypothenemus hampei* (Ferrari) from Karnataka. Plant Protect. Bull. 46, 42.

Varéa, G.S., Oliveira, J.A.Y., Sugahara, V.H., Ito, E.T., Pinto, J.P., Trevisan, D., et al., 2012. Identificação de proteases produzidas pelo fungo entomopatogênico *Beauveria bassiana* (Bals) Vuill. cepa CG432 previamente ativada em insetos vivos de broca do café (*Hypothenemus hampei*). Semina: Ciências Agrárias (Brazil) 33, 3055–3088.

Varela-Ramírez, A., 1997. Selección de aislamientos de *Beauveria bassiana* para el control de la broca del café. Rev. Colomb. Entomol. 23, 73–81.

Varón, E.H., Hanson, P., Borbón, O., Carballo, M., Hilje, L., 2004. Potencial de hormigas como depredadores de la broca del café (*Hypothenemus hampei*) en Costa Rica. Manejo Integrado de Plagas y Agroecología 73, 42–50.

Vázquez Moreno, L.L., 2005. Experiencia Cubana en el manejo agroecológico de plagas de cafeto y avances en la broca de café. In: Barrera, J. F. (Ed.), Simposio sobre la Situación Actual y Perspectivas de la Investigación y Manejo de la Broca del Café en Costa Rica, Cuba, Guatemala y México. Tapachula, Chiapas, México, pp. 46–57, Sociedad Mexicana de Entomología y El Colegio de la Frontera Sur.

Vázquez Moreno, L.L., Matienzo Brito, Y., Alfonso Simonetti, J., Moreno Rodríguez, D., Alvarez Núñez, A., 2009. Diversidad de especies de hormigas (Hymenoptera: Formicidae) en cafetales afectados por *Hypothenemus hampei* Ferrari (Coleoptera: Curculionidae: Scolytinae). Fitosanidad (Cuba) 13, 163–168.

Vega, F.E., 2008a. The rise of coffee. Am. Sci. 96, 138–145.

Vega, F.E., 2008b. Insect pathology and fungal endophytes. J. Invertebr. Pathol. 98, 277–279.

Vega, F.E., Dowd, P.F., 2005. The role of yeasts as insect endosymbionts. In: Vega, F.E., Blackwell, M. (Eds.), Insect-Fungal Associations:

Ecology and Evolution. Oxford University Press, New York, pp. 211–243.

Vega, F.E., Mercadier, G., Damon, A., Kirk, A., 1999. Natural enemies of the coffee berry borer, *Hypothenemus hampei* (Ferrari) (Coleoptera: Scolytidae) in Togo and Ivory Coast, and additional entomofauna associated with coffee beans. Afr. Entomol. 7, 243–248.

Vega, F.E., Benavides, P., Stuart, J., O'Neill, S.L., 2002a. *Wolbachia* infection in the coffee berry borer (Coleoptera: Scolytidae). Ann. Entomol. Soc. Am. 95, 374–378.

Vega, F.E., Franqui, R.A., Benavides, P., 2002b. The presence of the coffee berry borer, *Hypothenemus hampei*, in Puerto Rico: fact or fiction? J. Insect Sci. 2(13).

Vega, F.E., Rosenquist, E., Collins, W., 2003a. Global project needed to tackle coffee crisis. Nature 435, 343.

Vega, F.E., Blackburn, M.B., Kurtzman, C.P., Dowd, P.F., 2003b. Identification of a coffee berry borer-associated yeast: doest it break down caffeine? Entomol. Exp. Appl. 107, 19–24.

Vega, F.E., Pava-Ripoll, M., Posada, F., Buyer, J.S., 2005. Endophytic bacteria in *Coffea arabica* L. J. Basic Microbiol. 45, 371–380.

Vega, F.E., Ebert, A., Ming, R., 2008a. Coffee germplasm resources, genomics, and breeding. Plant Breed. Rev. 30, 415–447.

Vega, F.E., Posada, F., Aime, M.C., Peterson, S.W., Rehner, S.A., 2008b. Fungal endophytes in green coffee seeds. Mycosystema 27, 75–84.

Vega, F.E., Posada, F., Aime, M.C., Pava-Ripoll, M., Infante, F., Rehner, S.A., 2008c. Entomopathogenic fungal endophytes. Biol. Contr. 46, 72–82.

Vega, F.E., Infante, F., Castillo, A., Jaramillo, J., 2009. The coffee berry borer, *Hypothenemus hampei* (Ferrari) (Coleoptera: Curculionidae): a short review, with recent findings and future research directions. Terrestrial Arthropod Reviews 2, 129–147.

Vega, F.E., Simpkins, A., Aime, M.C., Posada, F., Peterson, S.W., Rehner, S.A., et al., 2010. Fungal endophyte diversity in coffee plants from Colombia, Hawai'i, Mexico, and Puerto Rico. Fungal Ecology 3, 122–138.

Vega, F.E., Kramer, M., Jaramillo, J., 2011. Increasing coffee berry borer (*Hypothenemus hampei*; Coleoptera: Curculionidae: Scolytinae) female density in artificial diet decreases fecundity. J. Econ. Entomol. 104, 87–93.

Vega, F.E., Davis, A.P., Jaramillo, J., 2012a. From forest to plantation? Obscure articles reveal alternative host plants for the coffee berry borer, *Hypothenemus hampei* (Coleoptera: Curculionidae). Biol. J. Linn. Soc. 107, 86–94.

Vega, F.E., Meyling, N.V., Luangsa-ard, J.J., Blackwell, M., 2012b. Fungal entomopathogens. In: Vega, F.E., Kaya, H.K. (Eds.), Insect Pathology, second ed. Academic Press, San Diego, pp. 171–220.

Vega, F.E., Simpkins, A., Bauchan, G., Infante, F., Kramer, M., Land, M.F., 2014. On the eyes of male coffee berry borers as rudimentary organs. PLoS One 9 (1), e85860.

Vega Rosales, M.I., Romero, C.E., 1985. La broca del fruto del cafeto (*Hypothenemus, hampei* Ferrari) en El Salvador. Memoria, Curso sobre Manejo Integrado de Plagas del Cafeto con énfasis en Broca del Fruto (*Hypothenemus hampei*, Ferr.), 15 al 19 de Julio de 1985. Instituto Interamericano de Cooperación para la Agricultura, Oficina en Guatemala, pp. 178–179.

Velasco Pascual, H., Beristain Ruiz, B., Díaz Cárdenas, S., Llavén Gómez, J.M., Velázquez Velázquez, A.F., 1997a. Respuesta de la broca del fruto *Hypothenemus hampei* Ferr. a extractos de cerezas de café utilizados como atrayentes en Tepatlaxco, Veracruz, México. Unpublished.

Velasco Pascual, H., Llavén Gómez, J.M., Velázquez Velázquez, A.F., 1997b. Respuesta a extractos de cerezas de café utilizados como atrayente para hembras intercosecha de la broca del fruto *Hypothenemus hampei* Ferr. Memorias XVIII Simposio Latinoamericano de Caficultura, San José, Costa Rica, pp. 349–352.

Vélez, M., Bustillo, A.E., Posada, F.J., 2006. Depredación de *Hypothenemus hampei* por hormigas durante el secado solar del café. Manejo Integrado de Plagas y Agroecología (Costa Rica) 77, 62–69. Also published as Vélez-Hoyos et al. (2006).

Vélez Arango, P.E., 1997. Evaluación de formulaciones en aceite y en agua del hongo *Beauveria bassiana* (Balsamo) Vuillemin en campo. Rev. Colomb. Entomol. 23, 59–64.

Vélez–Arango, P.E., Benavides-Gómez, M., 1990. Registro e identificación de *Beauveria bassiana* en *Hypothenemus hampei* en Anyuca, Departamento de Nariño, Colombia. Cenicafé 41, 50–57.

Vélez-Hoyos, M., Bustillo-Pardey, A.E., Posada-Flórez, F., 2006. Depredación de *Hypothenemus hampei* por hormigas, durante el secado solar del café. Cenicafé 57, 198–207. Also published as Vélez et al. (2006).

Vera, J.T., Montoya, E.C., Benavides, P., Góngora, C.E., 2011. Evaluation of *Beauveria bassiana* (Ascomycota: Hypocreales) as a control of the coffee berry borer *Hypothenemus hampei* (Coleoptera: Curculionidae: Scolytinae) emerging from fallen, infested coffee berries on the ground. Biocontrol Sci. Technol. 21, 1–14.

Vera-Montoya, L.Y., Gil-Palacio, Z.N., Benavides-Machado, P., 2007. Identificación de enemigos naturales de *Hypothenemus hampei* en la zona cafetalera central Colombiana. Cenicafé 58, 185–197.

Vergara-Olaya, J.D., Orozco-Hoyos, J., Bustillo-Pardey, A.E., Chaves-Córdoba, B., 2001a. Biología de *Phymastichus coffea* en condiciones de campo. Cenicafé 52, 97–103.

Vergara-Olaya, J.D., Orozco-Hoyos, J., Bustillo-Pardey, A.E., Chaves-Córdoba, B., 2001b. Dispersion de *Phymastichus coffea* en un lote de café infestado de *Hypothenemus hampei*. Cenicafé 52, 104–110.

Vet, L.E.M., 1999. From chemical to population ecology infochemical use in an evolutionary context. J. Chem. Ecol. 25, 31–49.

Vet, L.E.M., Dicke, M., 1992. Ecology of infochemical use by natual enemies in a tritrophic context. Annu. Rev. Entomol. 37, 141–172.

Viana, M.J., 1965. Datos ecológicos de Scolytidae argentinos (Coleoptera). Revista de la Sociedad Entomológica Argentina 27, 119–130.

Vijayalakshmi, C.K., Abdul Rahiman, P., Reddy, A.G.S., 1994. A note on the alternate shelters of coffee berry borer beetles. J. Coffee Res. 24, 47–48.

Villacorta, A., 1984. Ocorrência de *Beauveria* sp. infectando a broca do café—*Hypothenemus hampei* (Ferrari, 1867) (Coleoptera: Scolytidae) em lavouras no estado do Paraná. An. Soc. Entomol. Bras. 13, 177–178.

Villacorta, A., 1985. Dieta merídica para criação de sucessivas gerações de *Hypothenemus hampei* (Ferrari, 1867) (Coleoptera: Scolytidae). An. Soc. Entomol. Bras. 14, 315–319.

Villacorta, A., Barrera, J.F., 1993. Nova dieta merídica para criação de *Hypothenemus hampei* (Ferrari) (Coleoptera: Scolytidae). An. Soc. Entomol. Bras. 22, 405–409.

Villacorta, A., Barrera, J.F., 1996. Techniques for mass rearing of the parasitoid *Cephalonomia stephanoderis* (Hymenoptera: Bethylidae) on *Hypothenemus hampei* (Ferrari) (Coleoptera: Scolytidae) using an artificial diet. Vedalia 3, 45–48.

Villacorta, S., Torrecillas, S.M., 2000. New developments in mass production of parasitoids *Cephalonomia stephanoderis* (Hymenoptera:

Bethylidae) on *Hypothenemus hampei* (Coleoptera: Scolytidae) reared using artificial diet. In: Sera, T., Soccol, C.R., Pandey, A., Roussos, S. (Eds.), Coffee Biotechnology and Quality. Proceedings of the 3rd International Seminar on Biotechnology in the Coffee Agro-Industry, Londrina, Brazil. Kluwer Academic Publishers, The Netherlands, pp. 307–312.

Villacorta, A., Prela, A., Possagnolo, A., 2000. Redução dos custos na dieta artificial para a broca-do-café *Hypothenemus hampei* (Ferrari) para a criação de seus inimigos naturais, Simpósio de Pesquisa dos Cafés do Brasil, Poços de Caldas, Minas Gerais. Resumos expandidos, pp. 1273–1275.

Villacorta, A., Possagnolo, A.F., Silva, R.Z., Rodrigues, P.S., 2001. Um modelo de armadilha com semioquímicos para o manejo integrado da broca do café *Hypothenemus hampei* (Ferrari) no Paraná, Resumos Expandidos, II Simpósio de Pesquisa dos Cafés do Brasil, Vitória. Vol.2, pp. 2093–2098.

Vinson, S.B., 1976. Host selection by insect parasitoids. Annu. Rev. Entomol. 21, 109–133.

Vrydagh, J.M., 1940. Les fourmis du caféier (*Coffea robusta*). Courr. Agric. Afr. 4, 1–3.

Vuillet, J., 1925. Degré de sensibilité des differents Caféiers au *Stephanoderes coffeae*. Revue de Botanique Appliquée et d'Agriculture Coloniale 481, 601–604.

Waichert, C., Azevedo, C.O., 2012. The genus *Prorops* Waterson [*sic*], 1923 (Hymenoptera, Bethylidae) from Madagascar. European Journal of Taxonomy 16, 1–11.

Waller, J.M., Bigger, M., Hillocks, R.J., 2007. Coffee Pests, Diseases and their Management. CABI Publishing, Wallingford.

Waterhouse, D.F., 1998. Biological Control of Insect Pests: Southeast Asian Prospects, ACIAR Monograph No. 51, 548 pp.

Waterhouse, D.F., Norris, K.R., 1989. Biological Control: Pacific Prospects, Supplement 1, ACIAR Monograph No. 12, 125 pp.

Waterston, J., 1923. Notes on parasitic Hymenoptera. Bull. Entomol. Res. 14, 103–118.

Wegbe, K., 2012. Le scolyte des fruits du caféier (*Hypothenemus hampei* Ferr.) au Togo: etat actuel et perspectives. Proceedings of the 24th International Scientific Colloquium on Coffee, Association Scientifique Internationale du Café (ASIC), San José, Costa Rica, 5 pp.

Wegbe, K., Cilas, C., Decazy, B., Alauzet, C., Dufour, B., 2003. Estimation of production losses caused by the coffee berry borer (Coleoptera: Scolytidae) and calculation of an economic damage threshold in Togolese coffee plots. J. Econ. Entomol. 96, 1473–1478.

Wegbe, K., Cilas, C., Alauzet, C., Decazy, B., 2007. Impact des facteurs environnementaux sur les populations de scolytes (*Hypothenemus hampei* Ferrari) (Coleoptera: Scolytidae). Proceedings of the 21st International Scientific Colloquium on Coffee, Association

Scientifique Internationale du Café (ASIC), Montpellier, France, pp. 1349–1353.

Westwood, J.O., 1836. Description of a minute coleopterous Insect, forming the type of a new subgenus allied to *Tomicus*, with some observations upon the affinities of the Xylophaga. Trans. Entomol. Soc. London 1, 34–36.

Wilkinson, H., 1928. The coffee berry borer beetle *Stephanoderes hampei* (Ferr.). Printed by the Government Printer, Colony and Protectorate of Kenya, Nairobi, 10 pp.

Wilkinson, H., 1929. Annual report of the entomologist, 1928. Annual Report, Department of Agriculture, Kenya 1928, pp. 172–186. Nairobi.

Wood, S.L., 1954. A revision of North American Cryphalini (Scolytidae, Coleoptera). The University of Kansas Science Bulletin 36, 959–1089.

Wood, S.L., 1960. Coleoptera. Platypodidae and Scolytidae. Insects of Micronesia 18, 1–73.

Wood, S.L., 1972. New synonymy in the bark beetle tribe Cryphalini (Coleoptera: Scolytidae). Great Basin Nat. 32, 40–54.

Wood, S.L., 1975. New synonymy and new species of American bark beetles (Coleoptera: Scolytidae) Part II. Great Basin Nat. 35, 391–401.

Wood, S.L., 1977. Introduced and exported American Scolytidae (Coleoptera). Great Basin Nat. 37, 67–74.

Wood, S.L., 1982. The bark and ambrosia beetles of North and Central America (Coleoptera: Scolytidae), a taxonomic monograph. Great Basin Nat. Mem. 6, 1–1359.

Wood, S.L., 1986. A reclassification of the genera of Scolytidae (Coleoptera). Great Basin Nat. Mem. 10, 1–126.

Wood, S.L., 1989. Nomenclatural changes and new species of Scolytidae (Coleoptera), Part IV. Great Basin Nat. 49, 167–185.

Wood, S.L., 2007. Bark and Ambrosia Beetles of South America (Coleoptera, Scolytidae). Brigham Young University, Provo.

Wood, S.L., Bright Jr., D.E., 1992. A catalog of Scolytidae and Platypodidae (Coleoptera), Part 2: Taxonomic Index, Volume B. Great Basin Nat. Mem. 13, 835–1553.

Wrigley, G., 1988. Coffee. Longman Scientific & Technical, Essex, 639 pp.

Wurth, T., 1922. Een Vuurwants (*Dindymus rubiginosus*, F) die Jacht op de Bessenboeboek maakt. Mededeelingen van het Koffiebessenboeboek-Fonds 3, 49–52 (Rev. Appl. Entomol., Ser. A, 1922, 10, 601).

Yamamoto, K., 1948. Assim falou a vespa de Uganda. Guia pratico para o combate biologico á broca do café. Biblioteca Agropecuária Brasileira, São Paulo, 79 pp.

Zeng, F., Wang, X., Cui, J., Ma, Y., Li, Q., 2013. Effects of a new microbial α-amylase inhibitor protein on *Helicoverpa armigera* larvae. J. Agric. Food Chem. 61, 2028–2032.

Zhang, Q.-H., Schlyter, F., 2004. Olfactory recognition and behavioural avoidance of angiosperm nonhost volatiles by conifer-inhabiting bark beetles. Agr. Forest Entomol. 6, 1–19.

Chapter 12

Scolytus and other Economically Important Bark and Ambrosia Beetles

Sarah M. Smith[1] and Jiri Hulcr[2]

[1] *Department of Entomology, Michigan State University, East Lansing, MI, USA,* [2] *School of Forest Resources and Conservation and Department of Entomology, University of Florida, Gainesville, FL, USA*

1. INTRODUCTION TO DIVERSITY OF NORTH AMERICAN SPECIES

The overwhelming majority of bark and ambrosia beetle species are benign decomposers. Scolytines are among the first organisms to colonize woody debris and thus play an integral role in the decomposition of biomass in forest ecosystems by hastening the introduction of microbes and other xylophagous organisms (Stokland, 2012). Aside from the most destructive and notorious genera of *Dendroctonus*, *Ips*, *Hypothenemus*, and *Tomicus* described elsewhere in this volume, there are 11 genera covered in this chapter that can also cause significant destruction to North American forests, landscape trees, orchards, lumber, and even stored products. These genera are primarily secondary scolytines that under normal conditions colonize dead and dying host material. However, these genera have the potential to cause damage due to the beetle's association with pathogenic fungi and changes in abiotic conditions, such as rainfall and temperature, and biotic conditions, including host age and vigor (Wood, 1982; Breshers *et al.*, 2005; Raffa *et al.*, 2008). Other scolytine genera including *Phloeosinus* (Phloeosinini), *Hylastes* (Hylastini), and the recently revised *Hylurgops* (Hylastini) (Mercado-Vélez and Negrón, 2014) are not covered here but are economically and ecologically important to a lesser degree.

In this chapter we present information for the lesser known destructive Nearctic bark and ambrosia beetle genera. We provide a diagnosis, description, and taxonomic history of each genus and include a discussion of biology and ecology and highlight species of special importance. All common names given are those recognized by the Entomological Society of America.

2. *SCOLYTUS*

2.1 Overview

Scolytus Geoffroy (Scolytini) contains 127 species (Knížek, 2011; Petrov, 2013, Smith and Cognato, in press) distributed in the Nearctic, Palearctic, Oriental (Himalayan), and Neotropical regions. Twenty-one species are native to the Nearctic and four (*S. mali* (Bechstein), *S. multistriatus* (Marsham), *S. rugulosus* (Müller), and *S. schevyrewi* Semenov) are introduced from the Palearctic (Smith and Cognato, in press).

2.1.1 Diagnosis and Description (Modified from Smith and Cognato, in press) (Figure 12.1)

Diagnosis. *Scolytus* is easily distinguished by the unarmed protibia with a single curved process at the outer apical angle, flattened antennal club with 0–1 septate procurved sutures, seven-segmented funicle, the slightly declivous elytra, the depressed scutellar notch and scutellum and by the abruptly ascending abdominal sternites 2–5.

Description. *Scolytus* are stout to elongate, 1.7–2.9 times as long as wide and 1.7–6.0 mm in length. Mature color varies from red brown to black and teneral adults are often light brown.

Frons sexually dimorphic. Eye elongate, sinuate to shallowly emarginated and finely faceted. Antennal scape shorter than three funicle segments; funicle seven segmented; club large, flattened, oval to obovate, minutely pubescent and with strongly procurved sutures; suture 1 partially to completely septate and with or without a surface groove.

Pronotum large, head visible from above, lateral margins marked by a fine raised line. Scutellum large, depressed below level of elytra and triangular. Anterior coxae narrowly separated.

Elytra with a depressed scutellar notch and along the basal fifth to half of elytral suture; striate; slightly declivous apically. Abdomen ascending from posterior margin of sternite 1 to meet elytral apex, sternite 2 abruptly ascending (except *S. rugulosus*), often impressed, armed or both; abdomen dimorphic in most species (discussed above).

Bark Beetles. http://dx.doi.org/10.1016/B978-0-12-417156-5.00012-5

FIGURE 12.1 *Scolytus ventralis* male: (A) habitus lateral; (B) habitus dorsal; (C) venter posterior; (D) venter posterior oblique; (E) frons anterior.

Scolytus species are slightly to strongly sexually dimorphic. Sexually dimorphic structures typically exhibited are the frons, epistoma, and the abdominal sternites. Males typically have a flattened, impressed frons while the female frons is always more strongly convex. The male frons is more strongly and coarsely longitudinally aciculate than in the female and covered with longer, more abundant and dense erect setae. The epistomal process (when present) is more strongly developed in the male and less developed in the female. Venter armiture are more pronounced in males (except *S. multistriatus* and *S. piceae* (Swaine)).

2.1.2 Taxonomic History

Scolytus was the first bark beetle genus described and was the type for the family Scolytidae, now Curculionidae: Scolytinae. Blackman (1934) provides a detailed account of the taxonomic history of *Scolytus* and the

Scolytus/Eccoptogaster controversy, which occurred from the late 1800s to the early 1900s that resulted from intense debate on whether Geoffroy's (1762) description and drawing of *Scolytus* was a sufficient generic description. The type *Bostrichus scolytus* F. was subsequently designated as the type (Wood, 1982). China (1962, 1963) outline the International Commission on Zological Nomenclature ruling that preserved the name *Scolytus* over *Eccoptogaster*.

Twenty-nine species have been described from the Nearctic of which 21 indigenous species are recognized and two which were later confirmed as Palearctic exotics (Smith and Cognato, in press). Both species, *S. californicus* LeConte and *S. sulcatus* LeConte, were later identified as *S. scolytus* and *S. mali*, respectively. *Scolytus scolytus* is not established in the Nearctic and it is suspected that the locality was the result of a labeling error (Blackman, 1934). All Nearctic *Scolytus* were originally designated as *Scolytus*. However, Swaine was a vigorous opponent of Geoffroy's name and described all three of his species in *Eccoptogaster*.

There has been a great deal of taxonomic uncertainty regarding the status of several *Scolytus* species, particularly: *S. monticolae* (Swaine) and *S. tsugae* (Swaine); *S. abietis* Blackman and *S. opacus* Blackman; *S. reflexus* Blackman and *S. wickhami* Blackman; *S. fiskei* Blackman and *S. unispinosus* LeConte, with different authors presenting dramatically different opinions (McMullen and Atkins, 1959; Wood, 1966; Bright, 1976; Wood, 1977, 1982; Equihua-Martinez and Furniss, 2009). These differences likely resulted from overreliance on variable morphological features including the shape of the spine on the male second sternite and differences in host species rather than host genus.

Nearctic *Scolytus* have been revised four times: Blackman (1934), Edson (1967), Wood (1982), and most recently by Smith and Cognato (in press). In their monograph of Nearctic *Scolytus*, Smith and Cognato (in press) produced the first modern taxonomic treatment of *Scolytus*. Their investigation incorporated both molecular and morphological data in a phylogenetic analysis, a thorough review of taxonomic characters, assessment of intraspecific variation, and species boundaries tested using the phylogenetic species concept. The monograph is meant to be the definitive resource on the taxonomy and biology of Nearctic *Scolytus*. It also provides a glossary of terminology, fully illustrates all species, and includes the first key to both sexes of all Nearctic species. They recognize 25 *Scolytus* species including *S. monticolae*, *S. tsugae*, *S. reflexus* (=*S. wickhami*, =*S. virgatus* Bright), *S. praeceps* LeConte (=*S. abietis*, =*S. opacus*), and *S. fiskei* and *S. unispinosus*. Smith and Cognato (2010) have also produced a key to eastern North American species.

Palearctic species have been reviewed multiple times (Schedl, 1948; Balachowsky, 1949; Stark, 1952; Pfeffer,

1994b; Krivolutskaya, 1996; Knížek, 2011; Petrov, 2013) and keys have been produced for Russia (Stark, 1952), Japan (Nobuchi, 1973), Korea (Choo, 1983), China (Yin *et al.*, 1984), Spain (Lombardero and Novoa, 1994), Europe (Pfeffer, 1994a, b), Estonia (Voolma *et al.*, 1997), Italy (Faccoli *et al.*, 1998), Russian Far East (Krivolutskaya, 1996), France (de Laclos *et al.*, 2004), and Malta (Mifsud and Knížek, 2009) and for larvae of Scandinavian species (Lekander, 1968). Himalayan species were recently reviewed by Maiti and Saha (2009) and Mandelshtam and Petrov (2010a), both of which provide keys to the Indian fauna. The Central and South American fauna was revised by Schedl (1937) and Wood (1982, 2007). The Peruvian fauna was revised by Petrov and Mandelshtam (2010) and the publication provides a key that works for most western Amazonian species. It is quite probable that many South American and Chinese species are awaiting description. The Neotropical fauna is in need of further revision; however, some progress has recently been made (Smith and Cognato, 2013) and a synopsis of the Central American fauna is currently being prepared (Atkinson, in prep.).

A tremendous amount of progress toward revising *Scolytus* and the Scolytini has occurred within the past 10 years. The genus is rapidly becoming one of the best-known scolytine genera despite difficult morphology and strong sexual dimorphism. A molecular phylogeny of not only Nearctic species, but of species representing half of the genus has been reconstructed (Smith, 2013) and will serve as a framework toward species delimitation.

2.1.3 Biology

Scolytus are phloeophagous (phloem feeding) and specialized to either broadleaved host plants including Ulmaceae, Rosaceae, Fagaceae, Betulaceae, Juglandaceae, Fabaceae, and Oleaceae or Pinaceae conifers (*Abies*, *Cedrus*, *Larix*, *Picea*, *Pseudotsuga*, and *Tsuga*) (Wood and Bright, 1992). Seven species infest hardwoods with native species found in Fagaceae (*Fagus*, *Quercus*), Juglandaceae (*Carya*), Cannabaceae (*Celtis*), and exotic species colonizing Ulmaceae and Rosaceae. In North America about two-thirds of species infest conifer hosts including *Abies*, *Larix*, *Picea*, *Pseudotsuga*, and *Tsuga* (Wood and Bright, 1992; Smith and Cognato, in press).

North American conifer-feeding *Scolytus* species are distributed from the Atlantic to Pacific oceans and from the boundary of the Neotropical region to within the Arctic Circle. Native hardwood feeders are generally found from the Atlantic coast to Texas and west to the foothills of the Rocky Mountains. Typically, conifer-feeding *Scolytus* are restricted to the occurrence of host trees in western mountain ranges. However, *S. piceae* has an expansive range from the east and west coasts and from northern California and Colorado north to the Arctic Circle. Invasive *Scolytus* species are found throughout the USA, northern Mexico, and southern Canada (Smith and Cognato, in press).

Scolytus is primarily composed of secondary bark beetles but contains six potential tree-killing species in North America that can cause significant mortality of both conifers (Cibrián Tovar *et al.*, 1995) and *Carya* spp. (Furniss and Carolin, 1977). *Scolytus multistriatus* and *S. schevyrewi* Semenov are the primary vectors of Dutch elm disease, a fungal pathogen that has killed millions of *Ulmus* L. spp. trees in forest and urban areas across much of the USA and Canada (Furniss and Carolin, 1977; Jacobi *et al.*, 2013). Mortality caused by *Scolytus* species is often sporadic and short term, although some outbreaks locally affect thousands of acres a year. Damage is most severe in times of environmental stress, such as drought, fungal infections, and other insect infestations (Furniss and Carolin, 1977).

Scolytus ventralis LeConte and probably all conifer-feeding *Scolytus* exhibit primary attraction to host volatiles (Macías-Sámano *et al.*, 1998a). Attraction of Nearctic hardwood-feeding species is poorly investigated but primary attraction seems probable for *S. quadrispinosus* Say, *S. rugulosus*, and *S. schevyrewi* (Goeden and Norris, 1964a; Kovach and Gorsuch, 1985; Lee *et al.*, 2010). *Scolytus multistriatus* exhibits secondary attraction to 4-methyl-3-heptanol and multistriatin in combination with alpha-cubene (Lanier *et al.*, 1977). Typically, *Scolytus* infest overmature, unthrifty or weakened standing trees, shaded-out branches, fresh logging slash, fallen branches, and windthrown trees. During outbreaks, vigorous trees may be colonized by the more aggressive species *S. quadrispinosus*, *S. mundus* Wood, and *S. ventralis* (Edson, 1967; Furniss and Carolin, 1977; Cibrián Tovar *et al.*, 1995) and the secondary species *S. monticolae* (Swaine), *S. reflexus* (reported as *S. monticolae*) (USDA, 2004), and *S. unispinosus* (McMullen and Atkins, 1959).

All Holarctic *Scolytus* species are monogamous and several Neotropical species are bigamous or polygamous (Wood, 1982). In monogamous species, females select brood material, begin galley construction, and are subsequently joined by males, in contrast to bigamous and polygamous species, in which brood material is selected by the male. Males of monogamous species walk across the host in search of females. The female creates an entrance tunnel at a 45° angle, boring through the bark to the cambium. From the entrance tunnel, she then excavates a nuptial chamber and one or two egg galleries in either direction of the entrance tunnel. The female remains in the egg gallery and the male plugs the entrance tunnel with his abdomen. The nuptial chamber and galleries are excavated in the cambium and variously etch the sapwood. Females excavate egg niches on each side of the egg galleries and a single egg is deposited in each niche and covered with boring dust. Adult males assist in removing frass and typically stay with the female until egg gallery construction is complete. The male then leaves the gallery and the female dies in the entrance tunnel with her abdomen projecting onto the bark surface. Larval galleries radiate away from the egg tunnels as larvae feed on phloem, also variously etching the sapwood. Once larvae mature, the prepupae burrow into the outer sapwood and pupate. The brood overwinters as pupae with adults emerging in the spring (Edson, 1967). Upon emergence, several species including *S. mali*, *S. multistriatus*, *S. quadrispinosus*, *S. rugulosus*, and *S. schevyrewi* are known to engage in maturation feeding at twig crotches and/or leaf petioles (Hoffman, 1942; Baker, 1972; Negrón *et al.*, 2005). *Scolytus fiskei* feeds within small twigs (reported as *S. unispinosus* in McMullen and Atkins, 1962).

2.2 Economically Important Species

2.2.1 *Scolytus ventralis* LeConte—Fir Engraver

Scolytus ventralis is native to North America and is distributed in *Abies* spp. (fir) forests from Baja California and New Mexico north to British Columbia and east to Colorado and Montana (Smith and Cognato, in press). This species can cause significant fir mortality and is the most destructive conifer-feeding *Scolytus* (Keen, 1938; Bright and Stark, 1973). During a period between 1924 and 1936, *S. ventralis* killed 15% and damaged an additional 25% of the merchantable fir in California and more than 3000 acres of white fir (*Abies concolor* (Gordon and Glend.) Lindl. ex Hildebr.) in the Sandia Mountains of New Mexico during the 1950s (Massey, 1964). *Scolytus ventralis* is associated with a symbiotic stain fungus, *Trichosporium symbioticum* Wright, found in enlarged punctures on the vertex and gena of both sexes. Females introduce the fungus into the host during excavation of the adult gallery and the fungal hyphae spread out around the gallery system (Livingston and Berryman, 1972; Bright and Stark, 1973). Due to the aggressive habit of *S. ventralis*, this species is the most well-studied native *Scolytus* in North America. Attacks usually occur on the bole of weakened and stressed standing trees from a few feet above the base to the top of the tree, but can also occur in large slash, fallen, and recently dead trees (Chamberlin, 1958; Edson, 1967; Furniss and Johnson, 2002). Attacks at tree tops are more common on overmature standing trees during drought. Healthy, vigorous trees are not preferred (Chamberlin, 1958; Raffa and Berryman, 1987). Trees can also become successively attacked over a period of years and slowly die. Healthy trees may survive the attacks but can develop rots and defects that reduce timber value (Struble, 1937). Unlike most primary bark beetles, *S. ventralis* utilizes primary attraction to host volatiles rather than secondary attraction via an aggregation pheromone to aggregate conspecifics to a suitable host tree (Macías-Sámano *et al.*, 1998a, b).

Adult galleries of *S. ventralis*, consisting of two egg galleries with a central nuptial chamber (Edson, 1967), are constructed perpendicular to the grain of the wood and deeply score the sapwood and lightly score the cambium. The nuptial chamber is typically short and at a right angle to the egg galleries (Edson, 1967). Eggs are deposited singly in triangular niches spaced 1.0–1.5 mm apart on each side of the egg gallery with 80–300 niches per gallery. Galleries range in size from 8 to 30 cm in length (Chamberlin, 1958; Edson, 1967; Bright and Stark, 1973; Furniss and Johnson, 2002). Larval mines are perpendicular to the egg gallery and parallel with the grain. Larval mines are parallel to each other both above and below the egg gallery, giving the gallery a diamond-shaped appearance (Keen, 1938; Edson, 1967). Larval mines lightly score the sapwood and deeply score the cambium. Larvae pupate in the phloem or outer bark (Edson, 1967) and overwinter either as larvae or adults (Bright and Stark, 1973). The number of generations per year varies both geographically and with elevation. Development time can range from as little as 41 days at low latitudes and elevations to as many as 380 days at high latitudes and elevations (Bright and Stark, 1973). There is typically one generation per year (Bright and Stark, 1973). In Idaho pupation occurs from June to July and peak flight occurs in July (Furniss and Johnson, 2002).

2.2.2 *Scolytus multistriatus* (Marsham)— Smaller European Elm Bark Beetle, European Elm Bark Beetle

Scolytus multistriatus is a Palearctic species with a native range extending from Western Europe and North Africa east to Turkey and Russia (Michalski, 1973). It has not only been introduced to the Nearctic but also Australia, New Zealand, and temperate South America (Rosel and French, 1975; Bain, 1990; Wood and Bright, 1992; Wood, 2007; Smith and Cognato, 2013). The species was first encountered in Massachusetts in 1909 (Chapman, 1910) and is now distributed across the USA and temperate regions of Canada and Mexico (Smith and Cognato, in press). *Scolytus multistriatus* is the principal vector of the Dutch elm disease fungus *Ophiostoma ulmi* (Buisman) Melin and Nannf in North America, which killed 50–75% of the elm population in northeastern North America prior to the 1930s (Bloomfield, 1979). Adults become covered in fungal spores upon emergence from brood material and inoculate elms with the fungus as they perform maturation feeding in twig crotches. The feeding activity creates wounds in the bark that allow spores to be transferred from the beetle's cuticle to the tree tissues (Bright, 1976).

Scolytus multistriatus colonizes cut, stressed, weakened, and diseased elm trees (*Ulmus* spp.) (Wood, 1982). It seldom attacks healthy and vigorous trees (Bright, 1976). All native and introduced *Ulmus* spp. including *U. americana* L. (American elm) and *Zelkova serrata* (Thunb.) Makino are used as hosts. Females produce an aggregation pheromone to aggregate conspecifics to suitable hosts. The pheromone bouquet is composed of three components: (−)-4-methyl-3-heptanol, (−)-2,4-dimethyl-5-ethyl-6,8-dioxabicylo[3.2.1]octane (α)-multistriatin and (−)-α-cubebene (Pearce *et al.*, 1975). The adult gallery is excavated parallel to the grain of the wood and consists of a single egg gallery without a nuptial chamber. The adult gallery ranges in size from 2.5 to 5.0 cm in length. Egg niches are constructed along the gallery and score the sapwood. Twenty-four to 96 eggs may be singly laid along the egg gallery. Larval mines lightly score the sapwood and radiate perpendicular to the egg gallery. The larval galleries later meander often at an oblique angle to the grain, forming a fan-shaped pattern. Larvae construct pupal chambers in the bark (Bright, 1976). There are one and one-half generations per year in Canada and up to three in the southern USA (Furniss and Johnson, 2002). In Canada, adults emerge in June and July and maturation feed at twig crotches of healthy trees for 7–10 days prior to selecting a brood host (Chamberlin, 1958; Baker, 1972). The brood from these early summer adults emerges either in August or September or overwinters as larvae.

2.2.3 *Scolytus schevyrewi* Semenov—Banded Elm Bark Beetle

Scolytus schevyrewi is a Palearctic species with a native range extending from western Russia and Uzbekistan and east to China, Mongolia, and Korea (Michalski, 1973). It was first detected in North America in 2003 from Colorado and Utah. By 2005, it was recorded in 21 states suggesting that it had been present for many years before its initial detection (Negrón *et al.*, 2005; LaBonte, 2010). Subsequent examination of invasive bark beetle survey collections revealed an earlier occurrence of this species in Colorado in 1994 and New Mexico in 1998 (Lee *et al.*, 2006). It is currently distributed from British Columbia to Ontario in Canada and most of the continental USA (Smith and Cognato, in press).

The adult gallery solely consists of a single egg gallery constructed parallel with the grain of the wood and strongly etches the sapwood (Lee *et al.*, 2006). Egg niches are constructed along the gallery and score the sapwood. Twenty to 120 eggs are laid along the egg galleries (Lee *et al.*, 2010). Larval mines lightly score the sapwood and radiate perpendicular to the egg gallery. The larval galleries later meander usually at an oblique angle to the grain, forming a fan-shaped pattern. Pupation occurs in the outer bark and broods overwinter as mature larvae or pupae (Lee *et al.*, 2006). In California, adult flight occurs from April to September or October. In Nevada, Utah, Wyoming, Colorado, Kansas, and Utah, flight occurs from May to September (Lee *et al.*, 2011). Development from egg to adult takes 30–45 days (Negrón *et al.*, 2005). There are two to three generations per year (Lee *et al.*, 2011). Upon emergence, adults feed at twig crotches before selecting host material via attraction to host volatiles, showing preference for drought stressed *Ulmus* spp. (Negrón *et al.*, 2005; Lee *et al.*, 2010). *Scolytus schevyrewi* is a less effective vector of the Dutch elm disease fungus than *S. multistriatus* in North America (Jacobi *et al.*, 2013).

In areas where populations of *S. schevyrewi* and *S. multistriatus* co-occur, the abundance of *S. multistriatus* is decreasing to the point where this once abundant species is rare (Negrón *et al.*, 2005; Lee *et al.*, 2010). This competitive displacement is likely the result of differences in fecundity, generation time, and emergence date. *Scolytus schevyrewi* produces larger broods that may overwinter as pupae, have a quicker development period, have an earlier flight, and exhibit more rapid, strong aggregation to host kairomones as compared to *S. multistriatus* (Lee *et al.*, 2010).

2.2.4 *Scolytus quadrispinosus* Say—Hickory Bark Beetle

Scolytus quadrispinosus is a native species found east of the Rocky Mountains in the USA and Canada (Smith and Cognato, in press). It is one of the most destructive pests of hardwoods in North America and the most important pest of *Carya* spp. (hickory) (Juglandaceae) (Doane *et al.*, 1936; Baker, 1972). The species generally infests and kills single trees or treetops. However, outbreaks can develop during periods of drought, killing large stands. *Scolytus quadrispinosus* kills its host by a mass attack in which a multitude of broods develop under the bark, effectively girdling the host (Blackman, 1922). The primary hosts are *Carya* spp., including *C. illinoinensis* (Wangenh.) K. Koch (pecan) but the species also occurs in *Juglans cinerea* L. (butternut) (Wood and Bright, 1992).

Adult galleries are constructed parallel with the grain of the wood and deeply etch the sapwood. The adult gallery is short (2.5–5.0 cm), and consists of a single egg gallery (Blackman, 1922). Eggs are deposited singly in niches on each side of the egg gallery with 20–60 niches per gallery (Blackman, 1922). Larval mines are confined to the cambium. From the egg gallery, the mines are first perpendicular to the grain of the wood and then gradually turn and diverge creating a fan-shaped appearance. Larvae bore into the inner bark to overwinter, pupate the following spring, and emerge as adults the following summer (Blackman, 1922). Upon emergence, adults maturation feed at twig crotches and leaf petioles prior to selecting a host (Baker, 1972; Goeden and Norris, 1964b). There is one generation per year in the north with larvae completing their development in March and April and emergence in May. There are two generations per year in the south with the brood overwintering as larvae (Doane *et al.*, 1936). See Goeden and Norris (1964a, b, 1965a, b) for more information regarding the biology of *S. quadrispinosus*.

3. POLYGRAPHUS

3.1 Overview

Polygraphus Erichson (Polygraphini) contains 101 species in the Holarctic, Oriental, and Ethiopian regions (Wood and Bright, 1992; Alonso-Zarazaga and Lyal, 2009). Three species occur in the New World, all within coniferous forests of the USA and Canada.

3.1.1 Diagnosis and Description (Modified from Bright (1976) and Wood (1982)) (Figure 12.2A–C)

Diagnosis. *Polygraphus* are easily distinguished from other North American genera by the almost completely divided eye, unsegmented antennal club that lacks sutures and absent scutellum.

Description. Nearctic *Polygraphus* are stout to elongate, 2.0–2.4 times as long as wide and 1.8–3.2 mm in length. Mature color almost black; surface abundantly covered in short pale scales.

Frons sexually dimorphic, male impressed on lower half from epistoma to summit and convex above; summit slightly above midpoint and armed with a pair of small tubercles. Female frons convex and flat or impressed, unarmed, finely punctate and covered with fine hair-like setae. Eyes almost completely divided into dorsal and ventral halves, a few facets sometimes scattered between them. Antennal scape long, reaching posterior margin of eye; funicle 5–6 segmented; club asymmetrical, moderately flattened, aseptate, finely, uniformly pubescent, lacking sutures.

Pronotum unarmed, smooth, finely, closely punctate. Scutellum absent.

Elytral bases armed by 11–12 coarse crenulations, submarginal crenulations occasionally present from interstriae 2–4. Striae not impressed, finely punctate. Interstriae densely punctate. Declivity broadly convex, conservatively sculptured.

3.1.2 Taxonomic History

Six North American *Polygraphus* species have been described of which three are currently recognized. *Polygraphus rufipennis* Kirby was the first species described from the New World. The species and two of its synonyms were

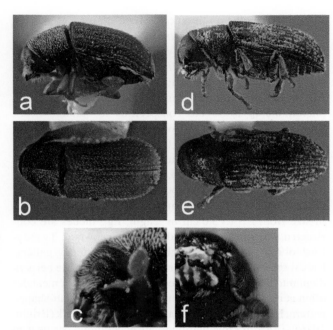

FIGURE 12.2 *Polygraphus rufipennis*: (A) habitus lateral; (B) habitus dorsal; (C) antennal club. *Pseudohylesinus* sp.: (D) habitus lateral; (E) habitus dorsal; (F) antennal club.

described from the same locality by Kirby (1837) in the same publication. Kirby (1837) described each sex as a distinct species as well as a recently emerged male that had all of its vesture intact. Mannerheim (1853) described an additional synonym. The remaining species were later described by Swaine (1925) and Wood (1951). The New World species have been revised twice, by LeConte (1868) and Wood (1982). Wood (1982) provides a key to the genus. Regional keys are available for Canada and Alaska (Bright, 1976), France (Balachowsky, 1949), the Palearctic (Schedl, 1955), Japan (Murayama, 1956; Nobuchi, 1979), Korea (Choo, 1983), Russian Far East (Krivolutskaya, 1996), Estonia (Voolma *et al.*, 1997), China (Yin and Hunag, 1996), greater Moscow, Russia (Chilahsaeva, 2010), India (Maiti and Saha, 2009), and to larvae of Scandinavian species (Lekander, 1968).

Wood (1982) noted that Schedl treated *P. rufipennis* as a synonym of *P. poligraphus* (L.) and remarked that the species are clearly separable. This distinction was recently supported by Jordal and Kambestad (2014) who sampled the mitochondrial gene for cytochrome oxidase I (COI) from both species and found that each formed unique lineages.

3.1.3 Biology

Polygraphus species are phloeophagous and polygamous scolytines. Hosts are either Pinaceae conifers, hardwoods, or in one unique European species, *P. grandiclava*

(Thomson), both conifers and cherry (*Prunus* L. spp.) (Pfeffer, 1994b; Avtzis *et al.*, 2008). Nearly all Holarctic and Oriental species occur on Pinaceae genera including *Abies*, *Cedrus*, *Larix*, *Picea*, and *Pinus* while African species occur on a diversity of hardwoods (Wood and Bright, 1992).

The biology of Holarctic and Oriental species is very similar to that of *P. rufipennis* discussed in detail below.

3.2 Economically Important Species

3.2.1 *Polygraphus rufipennis* Kirby—Foureyed Spruce Bark Beetle

Polygraphus rufipennis is transcontinental across America north of Mexico (Hilton, 1968). The species is polyphagous but exhibits preference to spruce (*Picea* spp.), particularly *P. glauca* (Moench) Voss (white spruce). It is also found on a wide diversity of Pinaceae conifers including: *Picea canadensis* (Mill.) Link, *P. engelmannii* Parry ex Engelm., *P. mariana* (Mill.) Britton, Sterns and Poggenb., *P. pungens* Engelm., *P. rubens* Sarg., *Abies concolor*, *A. fraseri* (Pursh) Poir., *A. lasiocarpa* Sarg., *Larix* spp., *Pinus banksiana* Lamb., *P. contorta* Douglas ex Loudon, *P. monticola* Douglas ex D. Don, *P. ponderosa* P. Lawson and C. Lawson, *P. resinosa* Torr., *P. strobus* L., *P. sylvestris* L., *Pseudotsuga menziesii* (Mirb.) Franco, and *Tsuga heterophylla* (Raf.) Sarg. (Hilton, 1968).

Polygraphus rufipennis is a common secondary species that excavates galleries under the bark of the smaller and drier regions of the bole of dead and moribund spruce (*Picea*) (Furniss and Carolin, 1977). Broods overwinter as larvae, pupae or adults. Adults emerge in May in June with individuals overwintering as larvae or pupae emerging 2–6 weeks later (Hilton, 1968). There is one generation per year; however, the female emerges in mid-summer and establishes a second brood in a different portion of the same tree or in another tree (Hilton, 1968; Furniss and Carolin, 1977). *Polygraphus rufipennis* attack stumps, boles, branches, and tops of weakened and fallen trees. Colonized trees are typically weakened from attack by *Dendroctonus rufipennis* Kirby (spruce bark beetle) or large populations of spruce budworm (*Choristoneura* spp.) (Simpson, 1929). The gallery initiating sex has historically been controversial with many authors reporting conflicting accounts (Hopkins, 1899; Blackman and Stage, 1918; Simpson, 1929; Hilton, 1968; Bright, 1976) and tested by Rudinsky *et al.* (1978). Galleries are either male or female initiated, an anomalous behavior among scolytines (Rudinsky *et al.*, 1978), but are typically male initiated with the male excavating the nuptial chamber in the phloem. The male stays in the nuptial chamber and is responsible for boring dust removal and admittance of additional females. The species is polygamous with each gallery containing 2–4 females (Simpson, 1929) that are attracted to the male-produced aggregation pheromone 3-methyl-3-buten-l-ol (Bowers *et al.*, 1991). Each female excavates her own egg gallery radiating away from the nuptial chamber with the direction varying with the numbers of females present. Galleries with 1–2 females have egg galleries perpendicular to the grain of the wood and those with 3–4 females have some egg galleries that are with the grain of the wood (see Hilton, 1968 for drawings). Egg galleries are elongated to a length of 7–10 cm over a 4-week period during which the female deposits eggs in niches on both sides of the egg gallery. After 4 weeks, the female bores an exit hole at the end of the egg gallery and wither, leaves to create a new brood during summer or stays at the end of the gallery, hibernates during late fall, and creates a second brood the following year (Hilton, 1968).

As a secondary bark beetle, *P. rufipennis* is of concern because of its propensity to attack and cause significant mortality of spruce trees weakened by cumulative spruce budworm damage (Raske and Sutton, 1986). The compromised host defense system favors colonization by the beetle and its associated *Ophiostoma piceaperdum* (Rumbold) Arx (Bowers *et al.*, 1991). In Newfoundland, *P. rufipennis* was found to attack 33% of severely damaged black spruce (*P. mariana*) weakened by eastern spruce budworm (*Choristoneura fumiferana* (Clemens)) and was aggressive enough to kill trees with little defoliation (Bowers *et al.*, 1996). In addition to causing tree mortality, *P. rufipennis* also reduces the time period that dead wood is useful because of decreased wood moisture content, resulting in decreased hardness and paper brightness. This reduced time period results in less time for salvage logging, and thus increased salvage costs (Bowers *et al.*, 1996).

4. *PSEUDOHYLESINUS*

4.1 Overview

Pseudohylesinus Swaine (Hylurgini, formerly Tomicini) is endemic to North America and contains 13 species and subspecies (Wood and Bright, 1992). Species are distributed in coniferous forests from Alaska, the western USA and Canada, and south into Oaxaca, Mexico (Bright, 1969; Wood, 1982; Wood and Bright, 1992).

4.1.1 Diagnosis and Description (Modified from Bright (1969, 1976) and Wood (1982)) (Figures 12.2D–F)

Diagnosis. *Pseudohylesinus* is distinguished from other Nearctic scolytines by the pronotal and elytral surface clothed in a combination of recumbent scales and erect hair-like setae, the seven-segmented funicle, antennal club with segment 1 occupying 2/3 of total length, the lateral areas of pronotum smooth, the arcuately impressed frons and by the nearly contiguous procoxae.

Description. *Pseudohylesinus* are stout to elongate, 2.1–2.5 times as long as wide and 2.4–6.0 mm in length. Mature color dark brown to black; surface shining, covered in scales and setae, often forming a variegated pattern of white to pale to dark brown scales.

Frons convex, punctate to granulate with an arcuate, transverse impression at or just below midpoint; area below impression divided by a transverse median carina. Epistomal margin broad, smooth and shining with a median bilobed epistomal process; epistoma broadly impressed, punctate. Antennal scape elongate, shorter than funicle; funicle seven-segmented, segments becoming wider and shorter distally; club conical to slightly flattened. Eyes entire, elongate.

Pronotum 0.7–1.3 times as long as wide; sides arcuate, strongly constricted on anterior fourth; surface shining, distinctly punctate, covered with recumbent scales and erect setae; median line visible. Scutellum small, longer than wide, slightly depressed.

Elytra 1.2–1.9 times as long as wide; sides straight and almost parallel on basal two-thirds, rather nearly rounded apically; bases arched, bearing 10–15 sharp to blunt overlapping crenulations; striae narrow to wide and distinctly punctate; interstriae flat to weakly convex, confusedly tuberclate at base, uniserially tuberculate and setose on apical three-quarters. Declivity evenly convex; interstriae 1, 3, 5, and 7 narrowed and weakly elevated, each bearing a row of tubercles and erect setae; interstria 2 depressed below level of interstriae 1 and 3, unarmed and glabrous; interstria 9 variably raised and serrate. Venter with abundant tufts of yellow scales.

The shape and sculpturing of the frons and the elytral vestiture are sexually dimorphic. The male frons is narrower between the eyes, the male frontal triangle is longer, the median carina is more strongly developed, the arcuate transverse depression is deeper and wider and the surface is rougher than that of the female. The male elytra is covered with both longer and coarser setae and broader and more abundant scales than the female (Blackman, 1942).

4.1.2 Taxonomic History

Swaine (1917) erected the genus to accommodate *Hylurgus sericeus* Mannerheim, *Hylastes granulatus* LeConte, *Hylesinus nebulosus* LeConte, and five new species that he described. An additional seven species were added by Blackman (1942) in the first revision of the genus. In a subsequent revision by Bright (1969), *P. nebulosus* was treated as two subspecies, *P. nebulosus nebulosus* and *P. nebulosus serratus* Bruck, and *P. dispar* Blackman as *P. dispar dispar* Blackman and *P. dispar pullatus* Blackman. Subspecies

were designated based on geographic differences in size, development of the first and third declivital interstriae and scale color and patterning (Bright, 1969). The genus was later revised by Wood and Bright (1992), who recognized 12 species. The Nearctic species are considered fairly well known but further investigation is necessary to determine the validity of Bright's subspecies and the status of *P. grandis* Swaine and *P. pini* Wood. There was a strong sustained disagreement between Wood and Bright regarding the status of these species. Bright (1969, 1970, 1976; Bright and Stark, 1973) treated *P. grandis* as a distinct species and *P. pini* as a synonym of *P. sericeus*, while Wood (1969, 1982; Wood and Bright 1992) treated *P. grandis* as a synonym of *P. sericeus* and *P. pini* as a distinct species. Future investigation is required to review the issue and clarify this ambiguity.

Wood (1982) is the most complete key to *Pseudohylesinus*. Numerous keys are also available, including Bright (1969), Bright and Stark (1973) for California, Bright (1976), and Furniss and Johnson (2002) for Idaho.

4.1.3 Biology

Pseudohylesinus are phloeophagous and are restricted to Pinaceae conifers (*Abies*, *Picea*, *Pinus*, *Pseudotsuga*, and *Tsuga*). Most species are specialized to *Abies* and a single species occurs on each of the other genera listed above.

Pseudohylesinus nebulosus is attracted to and colonizes a host tree via primary attraction to host compounds. It is quite probable this is characteristic of the genus and all species in the genus utilize primary attraction (Ryker and Oester, 1982). *Pseudohylesinus* are monogamous. The female initiates gallery construction, males subsequently join the female as she completes the entrance tunnel and both sexes work together to excavate the egg galleries, which have 1–2 branches (McGhehey and Nagel, 1969; Bright, 1969). Males keep the gallery clear of boring dust (Bright, 1969). Females deposit single eggs in individual niches along the egg gallery and cover it in boring dust. Galleries are transverse in *Abies*-feeding species and longitudinal in other host genera. Larvae mine their galleries in the phloem perpendicular to the egg gallery. There are three larval instars (Stoszek and Rudinsky, 1967; Wood, 1982). Pupation occurs either in the outer phloem (Stoszek and Rudinsky, 1967) or sapwood (McGhehey and Nagel, 1969).

Several species are known to exhibit maturation feeding. In Oregon, *P. nebulosus nebulosus* newly emerged teneral adults fly to tops of *P. menziesii* (Douglas fir) trees and bore into twigs slightly wider in diameter than the beetle. This boring activity occurs in the xylem and results in a hollow tube with a thin bark covering. The hollowed-out twig can damage the host tree resulting in the loss and breakage of the twig (Stoszek and Rudinsky, 1967).

Pseudohylesinus tsugae creates short feeding galleries in the phloem of live trees for 3–11 days until sexually mature. *Pseudohylesinus* may overwinter as adults in feeding galleries or as larvae. Adults abandon broods after laying eggs and may create 2–3 broods, depending on the species, before dying in the gallery of the last brood (McGhehey and Nagel, 1969). There is one generation per year in northern species and one to two and a partial third in southern species (Stoszek and Rudinsky, 1967; Wood, 1982).

Most *Pseudohylesinus* species are secondary bark beetles of minor economic importance. *Pseudohylesinus* species infest limbs, bole, and roots of weakened, injured, windthrown, and felled hosts (Bright and Stark, 1973; Bright, 1976; Wood, 1982). *Pseudohylesinus nebulosus nebulosus* is capable of killing small diameter weakened and suppressed *P. menziesii* and saplings and tops of healthy *P. menziesii* (Bright, 1976). *Pseudohylesinus granulatus* in conjunction with *P. sericeus* is the most destructive species in the genus and is of significant economic importance to *Abies* spp. especially in mature and overmature stands (Bright, 1976; Furniss and Carolin, 1977; Carlson and Ragenovich, 2012).

4.2 Economically Important Species

4.2.1 *Pseudohylesinus granulatus* LeConte— Fir Root Bark Beetle

Pseudohylesinus granulatus is distributed from southern British Columbia to California, east to Montana. The primary host is *Abies amabalis* Douglas ex J. Forbes (Pacific silver fir) but *P. granulatus* will also infest other *Abies* species including *A. grandis* (Douglas ex D. Don) Lindl., *A. concolor*, *A. magnifica* A. Murray bis, *A. lasiocarpa*, *A. procera* Rehder, and *Tsuga heterophylla* (Bright, 1969; Wood, 1982; Carlson and Ragenovich, 2012).

Pseudohylesinus granulatus is the most destructive species of the genus. The species is somewhat aggressive and is capable of killing overmature and unthrifty trees but will also colonize windblown and fallen trees (Bright, 1976; Carlson and Ragenovich, 2012). The brown-staining fungus *Ophiostoma subannulatum* (Livingston and Davidson) is associated with the beetle and is suspected to contribute to tree mortality (Carlson and Ragenovich, 2012). Top-kill can occur within a year; however, tree death occurs after at least a year of attack. Attacks may also be confined to patches that do not girdle the tree. Such attacks can heal and leave scars or the tree may be subjected to repeated attacks over several years and result in death. Feeding scars may also provide a path for fungi such as *Armillaria ostoyae* (Romagnes) Herink and *Phellinus weirii* (Murrill) Gilbertson to infect the tree (Carlson and Ragenovich, 2012) and aid the beetle in killing the host

(Furniss and Carolin, 1977). Attacks are typically confined to individual trees or small groups of trees (Carlson and Ragenovich, 2012). Outbreaks of *P. granulatus* are rare and occur after periods of drought stress, large blowdowns, and stands defoliated by sawflies and adelgids (McMullen *et al.*, 1981; Carlson and Ragenovich, 2012). A large drought-induced outbreak in Oregon between 1947 and 1955 resulted in the death of more than a million acres of *A. amabalis* (Carlson and Ragenovich, 2012).

The biology of *P. granulatus* is poorly understood, but is similar to the general habit described for *Pseudohylesinus* above. Transverse egg galleries are constructed at the base or roots of standing trees and large fallen or windthrown trees between May and August (Wood, 1982; Carlson and Ragenovich, 2012). Galleries are found a few centimeters below ground to 5 m above ground (Bright, 1976; Carlson and Ragenovich, 2012). Like other species, *P. granulatus* adults emerge in late summer and create maturation feeding tunnels in the phloem of the base of a tree in which they will overwinter (Carlson and Ragenovich, 2012). Adults emerge in late spring and excavate brood galleries. The larval stage lasts 12–14 months and the species has a 2-year life cycle. Beetles overwinter either as larvae or as adults in feeding galleries. Pupation occurs in the inner bark (Carlson and Ragenovich, 2012).

5. *DRYOCOETES*

5.1 Overview

Dryocoetes Eichhoff (Dryocoetini) contains 46 species (Alonso-Zarazaga and Lyal, 2009; Mandelshtam and Petrov, 2010b; Beaver and Liu, 2010) distributed in the Holarctic and Oriental regions. Seven species occur in North America (Bright, 1963; Wood, 1982; Alonso-Zarazaga and Lyal, 2009).

5.1.1 *Diagnosis and Description (Modified from Bright (1963) and Wood (1982))* (Figure 12.3)

Diagnosis. *Dryocoetes* can be differentiated from other Nearctic genera by the five-segmented funicle, obliquely subtruncate antennal club with its basal portion corneus and occupying more than half the length of club and bearing 1–2 transverse or recurved sutures on the pubescent anterior face, the evenly convex pronotum that is about as wide as long, large scutellum, and the short, steep, unarmed and granulate declivity on the apical fourth of the elytra.
Description. Nearctic *Dryocoetes* are elongate, 2.3–2.6 times as long as wide and 2.0–4.8 mm in length. Mature color dark reddish brown to black.

Frons convex, punctate to granulate; vestiture hair-like, varying from sparse in males to a dense brush in females of some species. Eye oval, emarginated. Antennal scape slightly longer than funicle, widened on distal half; funicle

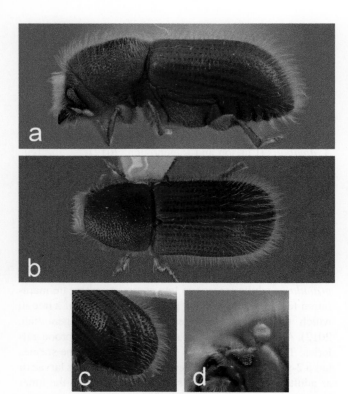

FIGURE 12.3 *Dryocoetes confusus* female: (A) habitus lateral; (B) habitus dorsal; (C) venter posterior; (D) antennal club.

five-segmented, segments increasing in size distally; club obliquely subtruncate, basal portion corneus, occupying more than half the length of club, anterior face pubescent with 1–2 transverse or recurved sutures.

Pronotum evenly convex, apical margin unarmed; surface granulate to finely asperate more strongly in the apical portion, occasionally punctured on the basal half. Scutellum large, longer than wide. Meso- and metathoracic tibiae slender, flattened and truncate distally, armed with 3–7 socketed teeth along the apical margin.

Elytral striae feebly to not impressed; strial punctures large, usually impressed, in rows. Interstriae slightly wider than striae, smooth; interstrial punctures much smaller than those of striae and slightly more abundant, each puncture bears a long hair-like seta. Declivity steep, convex or flattened, interstriae often granulate.

The frons and declivity of *Dryocoetes* are sexually dimorphic. The female frons has more pubescence than that of the male; frons shape and sculpturing (including granules) also vary. The male declivity also varies in the number and size of granules and interstrial impression and elevation.

5.1.2 Taxonomic History

Seventeen North American species have been described of which seven are currently recognized. The first Nearctic *Dryocoetes* were described from Alaska by Mannerheim

(1843, 1852). Additional species were described by LeConte, Hopkins, and Swaine. Many species were described based on the host plant from which it was collected and perceived geographic distributional differences, especially by Hopkins (1915) and Swaine (1915). The Nearctic species were first revised by Bright (1963) and have not been revised since. However, a recent study examining mitochondrial COI differences in Palearctic species by Jordal and Kambestad (2014) suggests that North American *D. autographus* (Ratzeburg) is distinct and distantly related to Palearctic *D. autographus*. Nearctic *D. autographus* formed a clade with other Nearctic species including *D. betulae* Hopkins and *D. confusus* Swaine. Further investigation is needed to determine species limits, diagnostic characters, and priority. Numerous keys are available to North American species including Bright (1963), Bright (1976), Wood (1982), and Furniss and Johnson (2002) for Idaho. Additional keys exist for France (Balachowsky, 1949), Japan (Murayama, 1957), Korea (Choo, 1983), China (Yin *et al.*, 1984), Europe (Pfeffer, 1994a, b), Russian Far East (Krivolutskaya, 1996), and to larvae of Scandinavian species (Lekander, 1968).

5.1.3 Biology

Dryocoetes are phloeophagous except for one Russian species (*D. krivolutzkajae* Mandelshtam) that feeds within roots of *Rhodiola rosea* L., a herbaceous plant. *Dryocoetes* are specialized to either broadleaved host plants including Betulaceae (*Alnus, Betula, Carpinus,* and *Corylus*), Juglandaceae (*Juglans*), Rosaceae (*Prunus*), Sapindaceae (*Acer*), Crassulaceae (*Rhodiola*) or Pinaceae conifers (*Abies, Larix, Picea, Pinus, Tsuga*) (Wood and Bright, 1992; Mandelshtam, 2001). In North America only one species, *D. betulae*, the birch bark beetle, infests hardwood hosts. The remaining six species infest conifers, especially *Picea* and *Abies*.

Overall, the biology of North American species is poorly understood and most investigations have focused on *D. confusus* Swaine, the most destructive species. *Dryocoetes* are polygamous. Males initiate gallery construction by excavating the entrance hole and nuptial chamber in the phloem. The male waits in the nuptial chamber until 2–4 females subsequently join him, depending on the species (Bright, 1976; Furniss and Kegley, 2006). However, Chamberlain (1939) reports that there may be 3–6 or more females per male. Each female excavates an egg gallery in the phloem radiating from the nuptial chamber, depositing eggs in niches and covered in frass along the margins. Galleries do not score the xylem (Bright and Stark, 1973; Wood, 1982; Furniss and Kegley, 2006). Females keep the egg gallery free of frass and males remove the frass out through the entrance hole (Bright, 1963; Furniss and Kegley, 2006). Larvae construct short, meandering galleries in the phloem. Development times are directly related to climate and altitude. North American species have one generation every 1–2 years while in Europe development may take 2 years to complete (Balachowsky, 1949; Bright and Stark, 1973; Bright, 1976; Furniss and Kegley, 2006). *Dryocoetes* spend winter either as larvae or as brood adults in hibernation galleries in the bark (Mathers, 1931; Bright, 1963; Bright and Stark, 1973; Negrón and Popp, 2009).

Most *Dryocoetes* are secondary pests and have no to minor economic importance. *Dryocoetes* species infest standing dead, injured, moribund, and felled or windthrown hosts. Species colonize the lower bole and roots of the host (Bright, 1963; Bright and Stark, 1973; Bright, 1976; Wood, 1982; Furniss and Kegley, 2006). *Dryocoetes betulae* prefers hosts that lack a live crown and have root-rot fungus (Furniss and Kegley, 2006). *Dryocoetes confusus* is the most destructive species in the genus and is of significant economic importance to *Abies lasiocarpa* (subalpine fir) (Bright, 1976; Furniss and Carolin, 1977; Borden *et al.*, 1987; Negrón and Popp, 2009).

5.2 Economically Important Species

5.2.1 *Dryocoetes confusus* Swaine—Western Balsam Bark Beetle (Figure 12.3)

Dryocoetes confusus is distributed from New Mexico and Arizona to British Columbia and Alberta. Specimens have also been collected from the Blue Mountains in Oregon (Bright, 1963). The primary host is *A. lasiocarpa* but *D. confusus* will less commonly infest other *Abies* spp. including *A. concolor* (white fir), and *Picea engelmannii* (Engelmann spruce) (Bright, 1963; Wood, 1982).

As previously mentioned, *D. confusus* is the most destructive *Dryocoetes* species. It is an aggressive species capable of killing overmature and unthrifty trees, and can also colonize windblown and fallen trees (Swaine, 1918; Chamberlain, 1939; Wood, 1982; Harris *et al.*, 2001; Negrón and Popp, 2009). *Dryocoetes confusus* is associated with the blue stain pathogenic fungus *Grosmannia dryocoetis* (W. B. Kendr. and Molnar) Zipfel, Z. W. de Beer and M. J. Wingf. both sexes of which transmit from paired mandibular mycangia (Kendrick and Molnar, 1965; Farris, 1969).

Total economic loss from the western balsam bark beetle species is unknown due to the inaccessibility of *A. lasiocarpa* at high elevation sites and thus reduced timber value (Bright, 1976; Negrón and Popp, 2009). Under normal conditions *D. confusus* selectively infests and kills small clusters of *A. lasiocarpa* in infested stands (Bleiker *et al.*, 2005). However, cumulative mortality may be significant in highly infested stands such as during the recent high mortality of *A. lasiocarpa* observed across the western USA

including Wyoming (McMillin *et al.*, 2003) and Colorado (Harris *et al.*, 2001).

Males produce a pheromone-based secondary attraction (Stock and Borden, 1983) and release *exo*-brevicomin and myrtenol as aggregation pheromones (Borden *et al.*, 1987). *Dryocoetes confusus* prefers weakened trees. Slow growing *A. lasiocarpa* are more likely to be attacked and killed than fast-growing trees (Bleiker *et al.*, 2005). Successfully attacked trees also exhibit a lower percentage of the bole with a constant crown, lower crown volume, reduced recent radial growth, and older age. Successfully attacked trees also have lower amounts of resin (Bleiker *et al.*, 2003).

Dryocoetes confusus may exhibit two flights, a peak flight in mid-June and early July (Utah), mid-late June (British Columbia), early-mid July (Colorado), and a flight in late August that is not consistently observed (Stock, 1991; Hanson, 1996; Negrón and Popp, 2009). The occurrence of the second flight is more likely at low to mid-elevation sites than at high elevations and has been suggested to be re-emerging parental beetles. Flight generally begins when temperatures are above 19–21°C in British Columbia and Colorado and peak flight occurs between 16 and 26°C (Negrón and Popp, 2009).

6. *XYLEBORUS*

6.1 Overview

Xyleborini is the most species-rich tribe within Scolytinae, with 1177 currently recognized species in 36 genera. The tribe also includes some of the most abundant and widespread scolytine species, as well as some of the most widespread animals on the planet. On the other hand, the tribe also includes many little known and rare species; approximately half of the described species are known from a single individual, the holotype. Most xyleborine species live in tropical rainforests, but many also occur in the temperate zone, wherever trees are present. *Xyleborus* Eichhoff is the largest genus with 416 recognized species (Wood and Bright, 1992; Chapter 2).

6.1.1 Diagnosis and Description (from Hulcr and Smith, 2010) (Figure 11.4A, B)

The tribe Xyleborini is defined by a combination of morphology, haplo-diploidy with arrhenotokous inbreeding, and obligate symbiosis with fungi. Diagnostic morphological characters shared by most taxa include emarginate eyes; strongly convex pronotum; flat and broad tibiae with convex outer edge, the tibial edge possesses numerous socketed denticles (which are actually fused setae); and mandibular, mesonotal or elytral mycangium. However, exceptions to these shared characters are frequent. Males are haploid clones of their mothers, dwarfed, with fused elytra. Male head, eyes, and antennae are also reduced compared to females (Kirkendall, 1993).

Diagnosis. Middle-sized, elongated, light-colored. Antennal club type 2 (Hulcr *et al.*, 2007b), with segment 1 dominant on posterior side, but not covering it all. Protibia obliquely triangular. Prosternal posterocoxal process inflated.

Description. Uniformly beige, brown or reddish, never black. Length: 2.1–3.0 mm. Eyes shallowly or deeply emarginate, upper part smaller than lower part. Antennal club approximately circular, club type 2 (obliquely truncated, pubescent segment 2 partly visible on posterior side). Segment 1 prominent, covering most of anterior and nearly all of the posterior side, its margin concave, and fully costate. Segment 2 visible on both sides, soft, or partly corneous on anterior side. Segment 3 absent from or partly visible on posterior side. Segment 1 of antennal funicle shorter than pedicel, funicle four-segmented, scapus regularly thick. Frons above epistoma rugged, coarsely punctate. Submentum deeply impressed, shaped as very narrow triangle. Anterior edge of pronotum with no conspicuous row of serrations (serrations no different than on pronotal slope).

Pronotum from lateral view of basic shape (type 0) (Hulcr *et al.*, 2007b), or elongated, with low summit (type 7) (Hulcr *et al.*, 2007b), from dorsal view basic, short, parallel-sided, rounded frontally (type 2) (Hulcr *et al.*, 2007b). Pronotal disc shining or smoothly alutaceous, with small punctures, lateral edge of pronotum obliquely costate. Procoxae contiguous, prosternal posterocoxal process inflated, rounded, not pointed. *Xyleborus sensu stricto* have mandibular mycangia (Hulcr, unpubl.), thus no tuft on pronotal base associated with mesonotal mycangium, and no setae on elytral bases associated with elytral mycangium.

Scutellum flat, flush with elytra. Elytral bases straight, with oblique edge, elytral disc longer than declivity, flat; elytral disc with distinct lines of strial punctures. Lateral profile of elytral declivity mostly flat or slightly convex, steep in some species, especially towards apex, dorsal profile of elytral apex rounded, not broadened posterolaterally. Posterolateral declivital costa absent or very short. Elytral declivity with few setae, not conspicuously pubescent, inner part of declivity with several tubercles on interstriae 1, 3, and beyond, and no tubercles on interstria 2. Tubercles on interstria 1 sometimes reduced to minor granules or absent, tubercles on interstria 3 always prominent. First interstria parallel, or parallel on disc but broadened towards apex of elytra.

Protibiae obliquely triangular, broadest at 2/3 of length, or distinctly triangular, slender on upper part, broad and denticulate on lower part. Posterior side of protibia flat, no granules, only setae. Protibial denticles large, distinctly longer than wide, bases of denticles slightly or distinctly enlarged, conical; fewer than six protibial denticles present.

6.1.2 Taxonomic History

Eichhoff (1864) described *Xyleborus* and designated *Bostrichus monographus* F. as the type species. LeConte (1876) was the first to recognize Xyleborini as a separate group of scolytine beetles and established the subtribe Xylebori. The first evolution-based classification of Xyleborini genera was that of Wood (1989). The classification of Xyleborini genera has recently been revised by several authors using both morphological and molecular data, and can be considered increasingly stable (Hulcr *et al.*, 2007b; Hulcr and Cognato, 2010; Cognato *et al.*, 2011). Xyleborine species were traditionally described as *Xyleborus* (Wood and Bright, 1992). Several authors including Blandford, Hagedorn, Hopkins, Reitter, and Sampson, began erecting other genera in the late 19th and early 20th centuries, transferring species out of *Xyleborus*. Recent phylogenetic research and reassessment of phylogenetic characters has demonstrated that *Xyleborus* is polyphyletic and thus several new genera were described for the Oriental and Oceanian faunas (Hulcr *et al.*, 2007a; Hulcr and Cognato, 2010, 2013). Given the current xyleborine phylogeny, additional genera will be described in the future (Cognato *et al.*, unpubl.).

Regional keys are available for Papua New Guinea (Hulcr and Cognato, 2013), durian orchards of southern Thailand (Sittichaya, 2013), India (Maiti and Saha, 2004), China (as a composite of multiple genera) (Yin *et al.*, 1984)), France (de Laclos *et al.*, 2004), Malta (Mifsud and Knížek, 2009), America north of Mexico (Rabaglia *et al.*, 2006; Hoebeke and Rabaglia, 2008), North and Central America (Wood, 1982), South America (Wood, 2007), and Puerto Rico (Bright and Torres, 2006). A key to the West Indies fauna is being developed by Bright (in prep.).

6.1.3 Biology

Ecologically, Xyleborini and *Xyleborus* are significant because of the combination of fungal farming and haplodiploid, inbred reproduction. Fungus farming freed most xyleborine species from strict specificity to particular tree hosts: the beetles feed on fungus, not on host tree tissue, and thus are not limited in their host selection by the host secondary metabolites. Historically, it has been thought that there is a relatively strict specificity between a particular xyleborine species or genus and their fungal symbionts. Increasingly, it is shown that this model applies in some xyleborine clades but not others. For example, beetle species from the clade that possesses mesonotal mycangia appear to be relatively fungus specific, and typically carry a single species of *Ambrosiella* (Microascales) (Harrington *et al.*, 2014; Kostovcik *et al.*, 2014). On the other hand, many *Xyleborus sensu stricto* with mandibular mycangia have been shown to be largely promiscuous in regards to the fungi they carry. Although their most common symbionts are typically *Raffaelea* spp. (Ophiostomatales), their mycangia frequently harbor multiple unrelated species (Carrillo *et al.*, 2014). Other clades, such as all *Euwallacea* spp. examined to date, carry several ambrosial species of *Fusarium* but can also harbor *Raffaelea* spp. (Kasson *et al.*, 2013).

The other xyleborine feature that confers a significant ecological advantage is sib-mating, or arrhenotokous inbreeding. Since females mate with their brother(s) in the native gallery, they do not need to seek mates in order to establish a new gallery in a new tree. This limits the impact of the Allee effect (an increased likelihood of extinction of small populations) on incipient populations of xyleborine species in new habitats. Consequently, Xyleborini are some of the most successful colonizers of new islands and continents (Jordal *et al.*, 2001). This feature also occurs in haplo-diploid and inbred Cryphalini, especially *Hypothenemus*, which similarly include some of the most widely distributed species (Chapter 11).

As a result of the capacity to colonize new environments and utilize a wide range of hosts, many xyleborine species become the world's most invasive and pestiferous insect species. In the USA alone, at least 25 exotic species of Xyleborini are now established, and some of them are causing unprecedented damage to the local ecosystems and tree-dependent industries (Bright, 2014).

6.1.4 Tree-killing Xyleborini

In the last several decades there has been a notable increase in the reports of tree-killing ambrosia beetles (Hulcr and Dunn, 2011). Most of these ambrosia species belong to Xyleborini. Several Platypodinae ambrosia beetles have also displayed outbreaks, but the discussion here is restricted to Xyleborini. There appear to be three mechanisms of xyleborine attack on living trees, each displayed by different beetle–fungus complexes:

1. Systemic virulence: The first type is a combination of a fungus that causes a systemic lethal response in the trees, and a beetle that routinely colonizes living trees. This type of attack has only been seen in a single case, that of *Xyleborus glabratus* Eichhoff (the beetle) and *Raffaelea lauricola* T. C. Harr., Fraedrich and Aghayeva (the fungus), which together cause laurel wilt disease (see below). The American population of the beetle is attracted to North American Lauraceae even if they are alive, for reasons currently unknown. The symbiotic fungus *R. lauricola* causes a collapse of the tree vascular system, but does not appear to secrete virulence factors (J. A. Smith, unpubl.) and appears to grow very slowly in the host (Ploetz *et al.*, 2012). Thus, the systemic pathogenicity in this case is likely due to the tree's exaggerated response, rather than the fungus

behavior. This is the only known case of this tree killing mechanism by ambrosia beetles.

2. Mass attack and weak pathogen: There are several cases of tree-killing Xyleborini aggregating on their host trees in large numbers before the tree succumbs. Examples include *Euwallacea* aff. *fornicatus* Eichhoff in Israel and California, *E. destruens* Blandford in Southeast Asia (Browne, 1958), and *Coptoborus ochromactonus* Smith and Cognato (Stilwell *et al.*, 2014). The fungal symbionts of those beetles appear to be weak tree pathogens causing localized necrosis, and become a significant agent of tree damage only if repeatedly inoculated by many attacking beetles (Freeman *et al.*, 2013).

3. Attack on water-stressed trees: The largest number of cases where ambrosia beetles have been reported attacking living trees seems to be restricted to trees in cultivated settings and typically overwatered. Flooding the root systems triggers the production of ethanol in the root system and trunk, attracting many Xyleborini (Ranger *et al.*, 2010). The beetles then bore into the trunk and repeated attacks eventually kill the tree. This is almost universally seen in tree nurseries and urban landscapes, where generous watering may cause temporary stress in the root system or even anoxia. Many ambrosia beetles have been observed to attack water-stressed trees including *Xylosandrus germanus* (Blandford) (Ranger *et al.*, 2012), *X. crassiusculus* (Motschulsky) (Hulcr, unpubl.), *Xyleborinus saxesenii* (Ratzeburg), and *Theoborus ricini* (Eggers) (Hulcr., unpubl.).

6.2 Economically Important Species

6.2.1 *Xyleborus glabratus* Eichhoff—Redbay Ambrosia Beetle (Figure 12.4A, B)

Xyleborus glabratus has been historically distributed throughout subtropical Southeast Asia, from India to Taiwan. In its native range, the species is rare and has no economic impact. It is one of the few xyleborine species that is relatively host specific, in this case to the family Lauraceae (Hulcr and Lou, 2013). Currently, the species is restricted to the southeastern USA (Formby *et al.*, 2012).

After its introduction to North America around the year 2000, the redbay ambrosia beetle became one of the most destructive wood-boring insects on the continent. The beetle is associated with several ambrosia fungi, most of which are harmless (Harrington *et al.*, 2010). However, one species, *Raffaelea lauricola*, is able to trigger a massive self-defense reaction in American Lauraceae trees. Susceptible tree species produce large amounts of tyloses that constrict water flow throughout xylem (Inch *et al.*, 2012). This reaction to the fungus is unfortunately further combined with the tendency of the beetle to attack live, healthy trees.

FIGURE 12.4 *Xyleborus glabratus* female: (A) habitus lateral; (B) habitus dorsal. *Euwallacea fornicatus* female: (C) habitus lateral; (D) habitus dorsal.

Such attacks introduce the fungus into the vascular tissues of the trees. An attack on susceptible tree species, even an unsuccessful attack of a single beetle, typically delivers a sufficient amount of fungal propagules to trigger the tree's reaction, and cause rapid death. The resulting disease has been termed laurel wilt disease (Fraedrich *et al.*, 2008).

It is not currently known why the beetle attacks healthy, living trees. This is a highly unusual behavior among ambrosia beetles, which are normally attracted to dead or dying trees. Attacking live trees has never been reported for *X. glabratus* in its native habitat. One of the hypotheses proposed for this apparent shift is called olfactory mismatch, whereby beetle olfactory response only partially matches to the profile of the new hosts (Hulcr and Dunn, 2011). In some cases, this mismatch may result in beetles still being attracted to host tree-specific volatiles, but not recognizing volatiles indicating that the trees are alive. *Xyleborus glabratus* is highly attracted to volatiles of American Lauraceae (Kendra *et al.*, 2011), but is not repelled by them when they are alive. On the other hand, it does not readily attack camphor (*Cinnamomum camphora* (L.) J. Presl.), an Asian laurel introduced to southeastern USA.

Since its introduction to Georgia, laurel wilt disease has rapidly spread across the southeastern USA. The tree species currently known to be sensitive to *R. lauricola* include nearly all North American Lauraceae (Fraedrich *et al.*, 2008). The species that are most attractive to the beetle and also extremely susceptible to the disease include several *Persea* species, including redbay, a formerly foundational species in southeastern American forest ecosystems.

Laurel wilt disease has caused unprecedented damage to southeastern American ecosystems. It has resulted in the death of over half billion trees (F. Koch, USDA Forest Service, pers. commun.) and in functional elimination of tree species from local ecosystems. Should the disease

spread outside of the southeastern USA, the damage may be even greater. For example, the western bay laurel, *Umbellularia californica* (Hook. and Arn.) Nutt., a key species in western coastal forest ecosystems, is also very susceptible to the disease (Fraedrich, 2008).

7. EUWALLACEA

7.1 Overview

Euwallacea Hopkins is a genus of ambrosia beetles within the Xyleborini and contains 54 recognized species (Alonso-Zarazaga and Lyal, 2009). The genus was originally distributed throughout Africa, East Asia, Southeast Asia, and Oceania. Several species have been introduced and established to other non-native regions, specifically North America, Europe, and the Middle East.

7.1.1 Diagnosis and Description (Modified from Hulcr and Smith, 2010) (Figure 12.4C, D)

Diagnosis. All *Euwallacea* are dark-colored (usually black), robust beetles. Their main diagnostic characters include tall, subquadrate ("oblique square") pronotum, laterally broadened elytral declivity, and fewer than six protibial denticles and the lateral edges of the elytral declivity, which are broadened and make the declivity appear flat and laterally expanded in the dorsal view (Hulcr and Cognato, 2013).

Description. Color uniform, brown to black. Size variable, 2.8–5.7 mm.

The pronotum is tall in lateral aspect, and subquadrate in dorsal aspect. Anterolateral edges of the pronotum are usually inflated ("bulging") which provides the subquadrate appearance. In some smaller species, however, the pronotum appears more rounded than subquadrate (e.g., *Euwallacea fornicatus* (Eichhoff)). The anterior margin of the pronotum is smooth, devoid of serrations.

The declivity is always gently descending, never steep or truncated. The surface of the declivity is unarmed by any large projections, only covered with strial punctures and interstrial tubercles, granules and hairs. Only in a very few Melanesian species does the declivity bear large tubercles or deep strial furrows. The antennal club is never truncated as in most Xyleborini; instead it is rounded, with the second and third segments visible on both sides of the club. In large species, the scape is longer than pedicel. This used to be considered a genus-level diagnostic character (Wood, 1986) but it is present only in several large species (Hulcr and Cognato, 2013).

Protibiae are also characteristic, always with fewer than six denticles (as few as three in large species). The shape of protibiae spans a range from distinctly triangular and narrow to obliquely triangular, but the lateral margin of protibiae is never rounded.

7.1.2 Taxonomic History

Euwallacea was described by Hopkins (1915) with *Xyleborus wallacei* Blandford as the type species. Hopkins (1915) included three species all of which occurred in Japan and the Oriental region. The overwhelming majority of species were originally placed in *Xyleborus* and most were transferred to *Euwallacea* by Wood and Bright (1992). The genus is somewhat well known but further investigation is needed as *Xyleborus* species continue to be transferred to the genus (Hulcr and Cognato, 2013).

Regional keys are available for Papua New Guinea (Hulcr and Cognato, 2013), America north of Mexico (Rabaglia *et al.*, 2006), durian orchards of southern Thailand (Sittichaya, 2013), and the New World (Wood, 2007).

7.1.3 Biology

All Xyleborini, and thus all *Euwallacea* spp., are ambrosia beetles obligately associated with nutritional ambrosia fungi. The genus *Euwallacea* is unusual among Xyleborini in that it is predominantly associated with several species of highly evolved ambrosial species of *Fusarium* (Kasson *et al.*, 2013). Most other Xyleborini are associated with Ophiostomatales and Microascales, while fusaria are only secondary associates on the beetle cuticle (Kostovcik *et al.*, 2014). In terms of tree hosts, most *Euwallacea* spp. are generalists, although several species display distinct preferences. For example, *E. funereus* (Lea) is a Moraceae specialist (Hulcr and Cognato, 2013). The most interesting relationship with tree hosts can be seen within the *E. fornicatus* species complex. It appears that several clades within this complex occur in separate regions of Southeast Asia and specialize on different hosts (see below).

7.2 Economically Important Species

7.2.1 Euwallacea destruens (Blandford)

Euwallacea destruens is one of the largest species of the genus, 3.8–4.5 mm in length (Hulcr and Cognato, 2013). It occurs in humid tropical rainforests throughout Southeast Asia and Melanesia. This species can have a significant impact on tree plantations. It occasionally mass attacks live trees such as *Tectona* L. f. (teak) (Browne, 1958) or *Casuarina* L. (Hulcr, unpubl.). The beetles may not only injure or kill the trees, but also degrade the wood with their unusually long galleries, up to 1.4 meters in total length (Browne, 1961). *Euwallacea destruens* is not yet known to be established outside of its native range.

7.2.2 Euwallacea fornicatus (Eichhoff) (Figure 12.4C, D)

Euwallacea fornicatus is a complex of morphologically similar populations throughout Southeast Asia and Oceania, some of which probably deserve the status of separate species. Several taxa originally described under various names have been synonymized under the name *E. fornicatus*, due to their morphological similarity (Beaver, 1991; Wood and Bright, 1992). However, the pestiferous nature of at least two of these clades prompted a reanalysis of *E. fornicatus* using molecular approaches. The results, although not yet published at the time of writing this chapter, suggest the existence of subpopulations of the *E. fornicatus* complex that are geographically and ecologically distinct (R. Stouthamer, pers. commun.). Characterizing these subpopulations as species proved difficult, primarily because the concept of "species" is not well defined for the highly inbred Xyleborini (including *Euwallacea*), which display the dynamics of clonal, rather than outcrossing populations (Jordal *et al.*, 2002). However, the combination of molecular marker divergence and the increasing ecological and economic impact of some clades may justify their designation as separate species.

A particularly important clade is currently called the polyphagous shot hole borer (Eskalen *et al.*, 2012). This population, originally occurring from northern Thailand and Vietnam, to Taiwan, has recently been introduced to Israel and California and has become a major pest of trees. Particularly affected is avocado (Freeman *et al.*, 2013), but as many as 200 other tree species are also affected, including oaks, maples, poplars, and other common species (Eskalen *et al.*, 2012). Heavy local infestations have significantly impacted tree communities in Israel and Los Angeles, CA. The beetle is associated with a highly coevolved species of ambrosial *Fusarium* (Mendel *et al.*, 2012, Kasson *et al.*, 2013). The *Fusarium* is a weak, locally pathogenic necrotroph. The beetle–fungus complex kills trees by a steady accumulation of massive numbers of attacking beetles, each of which inoculates the pathogenic *Fusarium* symbiont.

Another population within the *E. fornicatus* complex that is a prominent pest is the tea shot hole borer, occurring in Southeast Asia wherever tea (*Camellia sinensis* (L.) Kuntze) is grown. The beetle acts as a twig borer, and most of the damage is due to branch dieback, although death of whole tea bushes also occurs (Walgama, 2012). The tea-specific population is said to be morphologically distinguishable (Browne, 1961), and has been described as *Xyleborus fornicatior* Eggers, which was later considered a synonym of *E. fornicatus* (Wood and Bright, 1992). An overview of damage by the tea shot hole borer and other populations of *E. fornicatus* in Asia was published by Browne (1961). This beetle is also associated with a mutualistic ambrosial *Fusarium* (Gadd and Loos, 1947), though a different strain than the polyphagous shot hole borer (Kasson *et al.*, 2013).

Several other studied populations of *E. fornicatus* are also relatively host specific, but they are specific to different hosts in each region. For example, the East Asian (e.g., Borneo) and Melanesian populations of *E. fornicatus* are mostly associated with Moraceae, particularly *Artocarpus* J. R. Forst. and G. Forst. (Hulcr *et al.*, 2007a).

Multiple populations of the *E. fornicatus* complex have been introduced elsewhere throughout the world, including Central America (Wood, 1980; Kirkendall and Ødegaard, 2007) and southeast USA (Rabaglia *et al.*, 2006). The latter population has been identified as the tea shot hole borer (Stouthamer, unpubl.). None of the populations has been reported to cause significant tree mortality.

7.2.3 *Euwallacea validus* (Eichhoff)

Euwallacea validus is native to subtropical and temperate regions of eastern Asia. In its native range, the species has no commercial impact, as it typically colonizes dead trees in advanced stages of decay, including highly moist trunks and branches on the ground, in mud, or partly submerged in water. However, the species has been also seen attacking living *Ficus* L. trees in greenhouses in Japan (H. Kajimura, pers. commun.).

Euwallacea validus was inadvertently introduced to the USA in the mid-1970s (Wood, 1980), and has since spread throughout most of the eastern USA. Although it has had a very limited commercial impact, it may have a significant ecological impact in regions where *Ailanthus altissima* (Mill.) Swingle, the "tree of heaven," occurs. *Ailanthus altissima* is an invasive tree species aggressively spreading throughout the eastern USA. The tree is increasingly infected with a verticillium wilt (Kasson *et al.*, 2014a), the spread and persistence of which has been partly attributed to *E. validus* (Kasson *et al.*, 2014b).

8. *XYLOSANDRUS*

8.1 Overview

Xylosandrus Reitter is a genus of ambrosia beetles within the Xyleborini and contains 39 recognized species (Dole and Cognato, 2010). The genus was originally distributed throughout Africa, East Asia, Southeast Asia, Oceania, and North America. Several species have been introduced and established to other non-native regions, specifically North, Central, and South America, and Europe (Wood, 1982; Kirkendall and Ødegaard, 2007; Kirkendall and Faccoli, 2010; Garonna *et al.*, 2012). These species include *X. amputatus* (Blandford), *X. compactus* (Eichhoff), *X. crassiusculus*, *X. compactus* (Eichhoff), *X. germanus*, and

C. mutilatus (Blandford). Only one species, *X. curtulus* (Eichhoff), is native to the New World.

8.1.1 Diagnosis and Description (Modified from Hulcr and Smith, 2010 and Dole and Cognato, 2010) (Figure 12.5A, B)

Diagnosis. All *Xylosandrus* are relatively robust, although in absolute terms they may be very small. All species possess a tuft of setae, sometimes faint, on the base of the pronotum associated with the opening to the mesonotal mycangium. The antennal club is truncated, with segment 1 covering the entire posterior side. Segments 2 and 3 are inconspicuous, pubescent, visible only on the anterior side of the club. One of the best diagnostic characters is the separation of procoxae, which is particularly obvious in smaller species.

Description. Moderately to densely hairy. Uniformly light brown to dark brown to black, or pronotum dark, but elytra white semitransparent with a whitish patch. Body length varies greatly, from minute and very short, almost globular species to large montane species (1.3–4.2 mm).

Eyes shallowly or deeply emarginate, upper part smaller than lower part. Antennal club approximately circular, rarely taller than wide, flat. Club type 1 (truncated, segment 1 covering posterior side), type 4 (Hulcr *et al.*, 2007b) in *X. mixtus* (Schedl). Segment 1 of club circular around club, covering entire posterior face, margin of segment 1 costate. Segment 2 narrow, pubescent, visible on anterior side only.

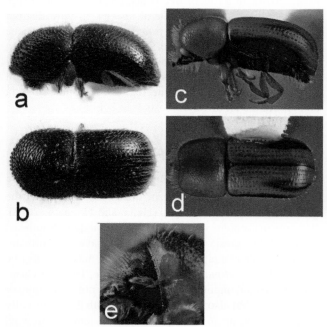

FIGURE 12.5 *Xylosandrus compactus* female: (A) habitus lateral; (B) habitus dorsal. *Trypodendron lineatum* male: (C) habitus lateral; (D) habitus dorsal; (E) antennal club.

Segment 3 absent from posterior side of club. Segment 1 of antennal funicle shorter than pedicel, funicle four-segmented, scapus regularly thick, or appears long and slender. Frons above epistoma mostly smooth, alutaceous, with minor punctures, or rugged, coarsely punctate. Submentum shaped as very narrow triangle, slightly impressed, or deeply impressed in small species, impression sometimes surrounded with costate edge or carina.

Anterior edge of pronotum with distinct row of serrations. Pronotum from lateral view low and rounded (type 1) (Hulcr *et al.*, 2007b), or tall (type 2) (Hulcr *et al.*, 2007b), or rounded and robust (type 5) (Hulcr *et al.*, 2007b), from dorsal view rounded (type 1). Pronotal disc shining or smoothly alutaceous, with small punctures, lateral edge of pronotum obliquely costate. Procoxae narrowly to widely separated, anterocoxal and posterocoxal prosternal processes merged into a process of variable shape, from short and conical, to flat and inconspicuous, to tall and pointed. Base of pronotum with small tuft of setae associated with mesonotal mycangium (sometimes very faint).

No elytral mycangium and no setae associated with it on elytral bases. Scutellum flat, flush with elytra. Elytral bases straight, with oblique edge, elytral disc longer or shorter than declivity, slightly to distinctly convex, punctures on elytral disc in strial lines (difficult to discern in some montane species). Lateral profile of elytral declivity slightly convex and gradually sloped to steep, especially towards apex, to obliquely truncated; dorsal profile of elytral apex rounded, or truncated, or rounded and broadened laterally. Posterolateral declivital costa ending in 7th interstria. Elytral declivity without armature, usually smooth or granulate-punctate (upper part of declivity in some Southeast Asian species with ridges and furrows). First interstria parallel.

Protibiae very slender, slightly broader only at distal end. Posterior side of protibia flat, no granules, only setae. Protibial denticles mostly large, distinctly longer than wide, bases of denticles usually distinctly enlarged and almost conical, in most species fewer than six protibial denticles present.

8.1.2 Taxonomic History

Xylosandrus was described by Reitter (1913) with *Xyleborus morigerus* Blandford as the type species. Complete taxonomic history can be found in Dole and Cognato (2010). The genus has been recently reclassified (Dole *et al.*, 2010; Dole and Cognato, 2010) using molecular-phylogenic and morphological data. The genus was made monophyletic by moving many species to their correct genera, particularly to *Anisandrus* and *Cnestus*.

A comprehensive key to the *Xylosandrus* is provided by Dole and Cognato (2010). Regional keys are available for

Australia (Dole and Beaver, 2008), Papua New Guinea (Hulcr and Cognato, 2013), durian orchards of southern Thailand (Sittichaya, 2013), India (Maiti and Saha, 2004), America north of Mexico (Rabaglia *et al.*, 2006), and the New World (Wood, 2007).

8.1.3 Biology

Most known *Xylosandrus* create a very short entrance tunnel that quickly expands into a cavity of variable shape. All *Xylosandrus* that have been studied are polyphagous, sometimes extremely so, and can colonize any woody material of appropriate size and moisture content, regardless of its taxonomic identity. Many *Xylosandrus*, particularly the globally invasive species, are capable of utilizing woody material which is drier than that required by most other ambrosia beetles. All studied species of *Xylosandrus*, as well as other genera that belong to the clade characterized by the mesothoracic mycangium, are associated with the mutualistic fungus genus *Ambrosiella*, a relative of the plant pathogens *Ceratocystis* (Ascomycota: Microascales). While other Xyleborini groups typically carry several different symbionts in their mycangia, each species of *Xyleborus* and the related genera usually carries only a single dominant *Ambrosiella* in its mycangium (Six *et al.*, 2009; Kostovcik *et al.*, 2014; Harrington *et al.*, 2014).

8.2 Economically Important Species

8.2.1 *Xylosandrus compactus* (Eichhoff) — Black Twig Borer (Figure 12.5A, B)

Xylosandrus compactus is one of the smallest (1.6–1.7 mm), but ecologically one of the most successful ambrosia beetle species in the world. It is a truly circumtropical species, common and expanding in most humid tropical and subtropical regions of the world (although in parts of East Asia and Melanesia it is often replaced by similar species such as *Xylosandrus morigerus* (Blandford) (Hulcr and Cognato, 2013)). It is introduced to the New World and occurs in the southeastern USA, Central America, West Indies, and South America (Rabaglia *et al.*, 2006; Wood, 2007). In many regions of tropical Africa and the New World, it is the most common scolytine beetle in twigs of hardwoods. It also frequently colonizes alternative habitats such as seeds and large leaf petioles (Jordal and Kirkendall, 1998).

Like most other Xyleborini, *X. compactus* can carry a whole community of fungal species, but its fungal community appears to be dominated by two fungi: *Fusarium* sp. from the *solani* complex (possibly several species) and *Ambrosiella* spp., also probably different species in different regions. *Ambrosiella* is probably the actual coevolved nutritional symbiont, since it is by far the most abundant fungus in the mycangium (Bateman and Hulcr, unpubl.). The fusaria, on the other hand, are almost universally carried on the cuticle, which may indicate a weak symbiotic bond, and potentially parasitic or commensal function of the fungus (Bateman and Hulcr, unpubl.). Both fungi probably grow in and around the gallery: the twig xylem discoloration can be bright red, which is typical for fusaria, or dark brown or black, which is more typical for Microascales, including *Ambrosiella* (Hulcr, unpubl.).

In most regions, *X. compactus* colonizes dead or dying substrates such as twigs or leaf petioles. In some regions, however, the species causes significant damages to live trees, particularly in coffee and cocoa plantations (Beaver and Browne, 1978). In the southeast USA and Italy, where *X. compactus* is considered invasive, the majority of the population attacks healthy twigs of living trees, including wild trees. The damage to mature trees can be so severe that the annual growth of new twigs is lost (Figure 12.6). Small tree seedlings may also be attacked and frequently die (Hulcr, unpubl.).

8.2.2 *Xylosandrus crassiusculus* (Motschulsky)—Granulate Ambrosia Beetle

Xylosandrus crassiusculus may be becoming one of the most abundant and widespread wood-boring beetles on the planet. Native to tropical and subtropical East Asia, it has recently colonized most warm and humid regions around the world, particularly in the subtropics and the tropics. It is now the most common ambrosia beetle in southeast USA and the rainforest of Africa (Hulcr, unpubl.), and is rapidly expanding in Papua New Guinea (Hulcr and Cognato, 2013). The species is also introduced to Central America and Europe (Kirkendall and Ødeggard, 2007; Kirkendall and Faccoli, 2010). In colder temperate zones with more extensive freezing, *X. crassiusculus* is often replaced by *X. germanus* (see below).

FIGURE 12.6 Damage to red maple by *Xylosandrus compactus*, the black twig borer, in Florida, visible in autumn after the leaves fall. Each clump of dead leaves that remains hanging on the tree is a result of an infestation in the twig that supported it. *Photo: J. Hulcr.*

In most cases, *X. crassiusculus* is a tree-generalist, although colonization of conifers is rare. It strongly prefers waterlogged, freshly dead wood, but at the same time it is capable of colonizing wood that is too dry for most other ambrosia beetles, such as small branches.

X. crassiusculus causes two distinct types of economic damage. First, it frequently colonizes and kills small trees in nurseries and urban settings. This damage appears to be restricted to trees that are watered or have been flooded (Ranger *et al.*, 2012). Similar damage in natural settings without flooding or irrigation is minimal to non-existent. In natural conditions *X. crassiusculus* only attacks dying or dead wood.

The second type of damage that *X. crassiusculus* causes is mass colonization of stored lumber. Beginning in the late 1990s (Atkinson *et al.*, 2000), but increasingly in recent years (Hulcr, unpubl.), mills and lumberyards in Florida and Georgia have reported losses of significant volumes of lumber, particularly hardwoods. Probably due to its prolific reproduction and the ability to thrive on relatively dry lumber, the species is able to produce enormous abundance within a single year given sufficient supply of wood, and create millions of holes and fungus-stained cavities in the wood.

8.2.3 *Xylosandrus germanus* (Blandford)

Xylosandrus germanus is equally prolific and invasive as the granulate ambrosia beetle *X. crassiusculus*, although it is much more abundant in temperate regions, and relatively rare in tropical regions. *Xylosandrus germanus* also originated in East Asia, and it is one of the most common ambrosia beetles there, particularly in Japan (Ito and Kajimura, 2008). It is introduced to eastern North America, the Pacific northwestern USA, and Hawaii (Cognato and Rubinoff, 2008; Atkinson, 2014).

Xylosandrus germanus has a relatively distinct morphology due to the entirely black, shining coloration, and a declivity that appears acuminated ("pointed") from the top. Similar species include the much smaller *X. compactus*, and several *Anisandrus*, particularly in North America and Europe, which do not have the separated procoxae. *Xylosandrus germanus* is one of the few scolytine beetles in which internal symbionts have been studied. Japanese populations turned out to be heavily colonized by multiple strains of *Wolbachia*, a widespread bacterial associate of insect whose roles range from parasitism to mutualism (Kawasaki *et al.*, 2010).

In North America, *X. germanus* has been a nuisance in nurseries for at least three decades (Weber, 1978). Just as in *X. crassiusculus*, it frequently attacks small trees, particularly flood-stressed or irrigated trees (Ranger *et al.*, 2012). It does not display any preference for any taxonomic lineage of trees, although it is only rarely found in conifers.

Xylosandrus germanus is currently one of the most commonly trapped scolytine beetles in the eastern USA (Oliver and Mannion, 2001). The species is also expanding in Europe (Kirkendall and Faccoli, 2010).

It is interesting to note that, despite the rapidly growing abundances of both *X. crassiusculus* and *X. germanus* in many parts of the world including North America and Europe, these species do not appear to have a discernible negative effect on the native ambrosia beetle faunas (Hulcr, unpubl. analysis of the USDA Forest Service Early Detection and Rapid Response Data (Rabaglia *et al.*, 2008)).

9. *TRYPODENDRON*

9.1 Overview

Trypodendron Stephens (Xyloterini) contains 13 species distributed in the Holarctic region. Six species occur in North America, all in the USA and Canada, five of which are indigenous and one, *T. domesticum* (L.), is exotic (Bright, 1976; Wood, 1982; Humble and Allen, 2006; Alonso-Zarazaga and Lyal, 2009). *Trypodendron lineatum* (Olivier) occurs throughout the Holarctic.

9.1.1 Diagnosis and Description (Modified from Chamberlain (1939, 1958) and Wood (1957b, 1982) (Figure 12.5C–E)

Diagnosis. *Trypodendron* can be recognized by the completed divided eyes, four-segmented funicle, antennal club with subcorenous basal area, strongly, narrowly procurved and contiguous procoxae.

Description. Nearctic *Trypodendron* are elongate, 2.5–2.6 times as long as wide and 2.7–4.6 mm in length. Mature color is solid brown to black, the pronotum and elytra may be differently colored and some species are bicolored, appearing striped. Surface smooth, shining.

Frons sexually dimorphic, convex in female, broadly, deeply excavated in male; eyes completely divided. Antennal scape elongate; funicle four-segmented; club with subcorenous basal area strongly, narrowly procurved.

Pronotum dimorphic, subcircular in female, subquadrate in male; anterior margin strongly procurved and asperate in female, almost straight to slightly recurved and unarmed in male. Female with a conspicuous narrow, longitudinal proepimeral excavation that is usually ornamented with setae. Scutellum present.

Elytra weakly striate, interstrial punctures typically, obsolete. Declivity convex, usually feebly sulcate; unarmed; subapical costal margin sharply elevated.

In all Xyloterini the female has a conspicuous longitudinal proepimeral excavation that is usually ornamented with setae. The proepimeral excavation is just above the procoxae and below the carinate pronotum basal lateral margin. This structure is absent in males. In *Trypodendron* the frons is deeply excavated from the epistoma to the vertex in males and convex in females. Females are larger than males and have a larger, subcircular pronotum as opposed to a shorter, subquadrate pronotum. The protibia of females are subinflated and tuberclate and those of the males are flattened and smooth.

9.1.2 Taxonomic History

Trypodendron was described by Stephens (1830) and included two species, *Dermestes domesticus* L. and *Bostrichus dispar* F., of which neither was designated as the type. Westwood (1838) subsequently designated *D. domesticus* as the type. Shortly thereafter, Erichson (1836) described *Xyloterus* to accommodate *D. domesticus*, *Bostrichus lineatus* Olivier, and *Bostrichus quinquelineatum* Adams. Thomson (1859) subsequently designated *B. lineatus* as the type. Ferrari (1868) placed *Xyloterus* in synonymy with *Trypodendron*. However, numerous authors including Balachowsky (1949) viewed *Trypodendron* as a synonym of *Xyloterus* due to a perceived inadequacy of the *Trypodendron* diagnosis despite priority. Wood (1957a) treated *Xyloterus* as a synonym of *Trypodendron*. Thirteen species were described between 1795 and 1917 by numerous authors including LeConte, Kirby, Eichhoff, Mannerheim, and Swaine (Wood and Bright, 1992).

The North American fauna has been revised a single time by Wood (1957b) who recognized five Nearctic species. Species were diagnosed primarily based on differences in color and color pattern, elytral luster, and male genitalia. Nearctic *Trypodendron* are in need of revision as several synonyms are likely valid species (Cognato, pers. commun.). Regional keys are available for California (Bright and Stark, 1973), Canada and Alaska (Bright, 1976), and Idaho (Furniss and Johnson, 2002). Additional keys exist for France (Balachowsky, 1949 as *Xyloterus*), Europe and Asia (Schedl, 1951), Japan (Murayama, 1957), China (Yin *et al.*, 1984), northwestern Spain (Lombardero and Novoa, 1994), Europe (Pfeffer, 1994a, b), Russian Far East (Krivolutskaya, 1996), Estonia (Voolma *et al.*, 1997), and to larvae of Scandinavian species (Lekander, 1968).

9.1.3 Biology

Trypodendron are monogamous ambrosia beetles with species either generalists or specialists (Kühnholz *et al.*, 2001). In North America, two species occur on hardwoods: *T. betulae* Swaine and *T. retusum* LeConte. *Trypodendron betulae* is found on Betulaceae hosts, primarily *Betula*, and rarely on *Alnus* (Kühnolz *et al.*, 2001) while *T. retusum* (LeConte) is found on *Populus* L. (Saliceae). Three species occur on conifers, *T. rufitarsus* Kirby on *Pinus*, *T. scabricollis* (LeConte) on *Pinus* and *Tsuga*, and *T. lineatum*

(Olivier) on all genera of Pinaceae, but exhibiting an apparent preference for *Pseudotsuga menziesii* (Prebble and Graham, 1957; Wood, 1982; Borden, 1988).

Beetles locate hosts via primary attraction to host ethanol (Moeck, 1970) and aggregate by means of secondary attraction to aggregation pheromones (Kühnolz et al., 2001). Galleries are female initiated and females possess mycangia (Kirkendall, 1983). In *Trypodendron*, the mycangia is paired tubes opening on proepimeron (Francke-Grossmann, 1959; Abrahamson et al., 1967; French and Roeper, 1972). Each species has its own specific associated microbiota primarily consisting of *Ophiostoma* spp. (Kühnolz et al., 2001).

Eggs are deposited in niches above and below each egg gallery. Larvae hatch and expand the niche into a larval cradle parallel with the grain of the wood, consuming wood to enlarge the cell and feeding upon fungi. Pupation occurs in the cradle with the head oriented toward the egg gallery. Males stay in the entrance tunnel and the female stays in the egg galleries. *Trypodendron* exhibits a high degree of parental care with the female paying constant attention to her larvae. The female will place a fungus plug in the junction of the egg gallery and the larval cradle and replace it as soon as it is consumed. The female will also collect the larval frass and move it to the entrance tunnel where it is expelled by the male. The male guards the entrance tunnel by plugging it with his body. Adults leave the gallery once the last larva has pupated (Borden, 1988).

Wood (1982) describes the genus as comprised of secondary species, attacking unthrifty hosts and logs; however, recently several species have begun to attack apparently healthy trees both in Europe (*T. domesticum* (L.) and *T. signatum* (F.)) and North America (*T. betulae* and *T. retusum*) (Kühnolz et al., 2001). *Trypodendron* pinholes and fungal staining are located in the outer sections of logs and thus can cause extensive economic loss through the degradation of products produced from infested logs such as lumber and plywood. Such ambrosia beetle infested wood suffers severe quality degradation, sells for a lower price, and is not acceptable for export. Additional costs are incurred to remove the damaged outer layer of wood, alteration of forest harvesting methods to remove vulnerable material, and implementation of pest management programs to reduce the impact of *Trypodendron* (Bright and Stark, 1973; Borden, 1988).

9.2 Economically Important Species

9.2.1 *Trypodendron lineatum* (Olivier) — Striped Ambrosia Beetle (Figure 12.5C–E)

Trypodendron lineatum occurs throughout Holarctic coniferous forests infesting almost any conifer species as well as hardwoods of the families Betulaceae (*Alnus* and *Betula*),

Rosaceae (*Malus*), and Sapindaceae (*Acer*) (Borden, 1988). *Trypodendron lineatum* is the most abundant and important ambrosia beetle of North American conifers (Furniss and Carolin, 1977) and can reach densities of up to 261 galleries per square foot (Prebble and Graham, 1957). Due to its abundance and propensity to colonize lumber, the species has an increased risk of being transported around the world and is the second most commonly intercepted species in New Zealand (Brockerhoff et al., 2006).

Borden (1988) provides a comprehensive review on the biology and ecology of this species. Initial flight occurs from late March to early May depending on the temperature, and a second flight occurs from mid-June until August. In both North America and Europe, initial flight occurs when temperatures reach at least 15.5–16°C with individuals exhibiting a midday or early afternoon diurnal flight pattern (Daterman et al., 1965). *Trypodendron lineatum* prefers aged timber, including windthrown, broken trees and 3–5-month-old logs. These older wood products have undergone anaerobic respiration for a longer period of time and consequently have an increased concentration of ethanol that serves as a primary attractant. Once a suitable host is found, pioneer beetles produce (+)-lineatin (3,3,7-trimethyl-2,9-dioxatricyclo[3.3.1^{4,7}]nonane, a powerful aggregation pheromone that attracts both sexes (Borden, 1988).

Galleries are initiated near bark crevices by females and construction lasts up to 3 weeks. Galleries are excavated into the sapwood and perpendicular to the grain (see Borden, 1988). Each gallery consists of an entrance tunnel and 2–3 radial egg galleries that are extended obliquely across the annual rings. Oviposition occurs during the first 2 weeks of gallery construction (Prebble and Graham, 1957). Gallery structure and the roles of males and females are as described above for the genus. Generation time is 9–10 weeks (Borden, 1988).

10. *PITYOPHTHORUS*

10.1 Overview

Pityophthorus Eichhoff (Corthylini: Pityophthorina) is distributed in the New World, Palearctic, Ethiopian, and Oriental regions. The genus contains 386 recognized species, of which more than half occur in North and Central America (Alonso-Zarazaga and Lyal, 2009; Chapter 2) and approximately 75% of the diversity is located in the New World (Bright, 1981; Wood and Bright, 1992; Alonso-Zarazaga and Lyal, 2009). An American species, *P. juglandis* Blackman, 1928 has been introduced to Europe (Montecchio and Faccoli, 2013).

10.1.1   Diagnosis and Description (Modified from Bright (1976, 1981) and Wood (1982)) (Figure 12.7)

Diagnosis. *Pityophthorus* is distinguished from other Nearctic genera by the costal margin of the declivity descending towards the apex, metepisternum almost completely covered by elytra, five-segmented funicle, pronotal asperities on anterior half, the transition from asperate to punctate area abrupt, pronotum with a distinct transverse impression behind summit, both antennal sutures marked by sclerotized septa at least on the lateral margins of the antennal club, sutures transverse to arcuate, and by the club less than 1.5 times as long as funicle.

Description. *Pityophthorus* are elongate, cylindrical, 2.0–3.4 times as long as wide and 1.0–4.5 mm in length. Mature color reddish brown to dark brown. Frons usually sexually dimorphic, male convex to variously impressed, female convex to concave; variously modified by longitudinal and/or transverse carina or setae or lacking modifications. Eyes elongate, emarginated at antennal insertion. Antennal scape club-shaped, bearing several plumose setae; funicle five-segmented (three-segmented in *P. costatus* Wood), about as long as scape; club flattened, broadly oval to circular, usually four-segmented (except *Hypopityophthorus*), sutures 1–2 distinct, transverse to arcuate, septate (partially to entirely) to lateral margins, suture 3 indicated only by an arcuate row of setae.

Pronotum slightly longer than wide; sides arcuate to nearly straight and parallel, constricted just before anterior margin; anterior margin broadly to narrowly and evenly rounded, bearing two or more erect serrations (serrations absent in costatus group *sensu* Bright (1981)); anterior slope with numerous asperities, either scattered or arranged in two or more concentric rows; dorsal surface convex or with an elevated summit at middle; usually a transverse impression behind summit; posterior area smooth or reticulate and usually bearing distinct punctures; basal and posterior half of margins bearing a fine raised line; lateral margins grooved in confusus group (Bright, 1981). Scutellum visible, large, flat. Protibia triangular, bearing one to four socketed teeth sinistral margin and two or three socketed teeth on apex, terminal mucro not socketed.

Elytra longer than wide; basal margins smooth, rounded; sides parallel or subparallel on basal three-quarters; apex rounded or acuminate; first stria distinctly impressed, other striae not impressed; striae and interstriae variously punctured. Declivity moderately to very steep; very variable, convex to strongly bisulcate with interstriae 1 and 3 variously elevated and granulate (several exceptions).

Pityophthorus sexual dimorphism is exhibited mainly in the shape and sculpturing of the frons and pubescence. The female is convex to concave, unarmed, glabrous to densely pubescent. The male frons is convex to variously impressed, sometimes armed by longitudinal and/or transverse carinae and sparser pubescence.

10.1.2   Taxonomic History

Eichhoff (1864) described the genus to accommodate three European species, *Bostrichus lichtensteini* Ratzeburg, *Dermestes micrographus* L., and *Pityophthorus exsculptus* Ratzeburg. Eichhoff (1864) failed to designate a type species, which was subsequently remedied by Hopkins (1914) who designated *B. lichtensteini*. Eichhoff described additional North American and European species between 1872 and 1881 (Wood and Bright, 1992). Since then, the generic concept of *Pityophthorus* has been quite variable with different authors expressing often very different opinions.

LeConte (1876) had a very board concept of *Pityophthorus* and created three species groups (A, B, and C) and included species that he previously considered as *Cryphalus* and recognized 18 species in the genus. Species group A included species that are now recognized as *Gnathotrichus*; group B included species now recognized as *Pseudopityophthorus*; and group C included species of *Pityoborus* Blackman, *Pityogenes* Bedel (Ipini), and *Pityophthorus*. LeConte (1883) recognized species group A (*Gnathotrichus*) as a distinct genus.

The concept of *Pityophthorus* began to stabilize and takes its present form in Swaine's (1918) revision. Swaine (1918) treated 26 species and removed all species

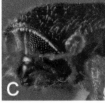

FIGURE 12.7 *Pityophthorus juglandis* male: (A) habitus lateral; (B) habitus dorsal; (C) antennal club.

(except two) that were not congeneric with the type. The exception being *P. ramiperda* Swaine and *P. boycei* Swaine, from which Blackman (1928) described *Myleborus* (type species = *P. ramiperda*). *Myleborus* was later synonymized with *Pityophthorus* by Bright (1977). In his monumental revision Blackman (1928) divided the genus into seven species groups and described two new genera (*Myleborus* and *Gnatholeptus*) to accommodate species. Blackman described 76 new species and included 35 previously described species increasing the known North American diversity to 106 species (Bright, 1981). Bright (1981) later revised the genus and placed Blackman's genera in synonymy, treating *Gnatholeptus* as a subgenus. In his revision, Bright incorporated over 90 new species from Mexico, USA, Canada, and Central America that were described by Bright between 1976 and 1978 and Wood between 1971 and 1977 (Bright, 1981). The genus more than doubled in size to include 220 species. To ease identification, Bright (1981) divided the genus into three subgenera: *Hypopityophthorus*, *Gnatholeptus*, and *Pityophthorus*. He further divided the largest subgenus *Pityophthorus* into 47 species groups (Bright, 1981). Almost concurrently, Wood (1982) produced his own revision, and recognized 218 species in North and Central America. The two authors reviewed each other's manuscripts but differences of opinion include the status of *Gnatholeptus*, which Wood (1982) treated as a genus and it is currently recognized as such (Wood, 1982; Wood and Bright, 1992; Alonso-Zarazaga and Lyal, 2009). *Pityophthorus* diversity is currently second only to *Xyleborus* and much of the diversity remains to be described, especially in South America, Central America, and Mexico (Alonso-Zarazaga and Lyal, 2009; Bright, pers. commun.). Further investigation using combined morphological and molecular phylogenetics would be useful to determine the monophyly of *Pityophthorus* and elucidate the generic boundary between *Pityophthorus* and *Araptus* (Wood, 1982).

Regional keys include Bright (1976) for Canada and Alaska, Bright and Stark (1973) for California, Furniss and Johnson (2002) for Idaho, and Bright and Torres (2006) for Puerto Rico. Additional keys exist for France (Balachowsky, 1949), China (Yin *et al.*, 1984), Europe (Pfeffer, 1976, 1994a, b), Russian Far East (Krivolutskaya, 1996), Estonia (Voolma *et al.*, 1997), and to larvae of Scandinavian species (Lekander, 1968). A key to the West Indies fauna is being developed (Bright, in prep.).

10.1.3 Biology

The genus is collectively referred to as the "twig beetles" (Furniss and Carolin, 1977) as they typically infest twigs and branches of live trees and branches and stems of trees

infested with other scolytines (Baker, 1972; Bright, 1981; Furniss and Johnson, 2002). *Pityophthorus* are phloeophagous or myelophagous, constructing their galleries either in the phloem or pith of the host, respectively. *Pityophthorus* species are specialized to coniferous hosts, woody shrubs, vines, hardwood trees, and even herbaceous plants (Bright, 1981). More than half of the Nearctic species occur on coniferous hosts with the overwhelming majority occurring on *Pinus* and a few species on *Abies*, *Picea*, *Pseudotsuga*, and *Larix* (Bright, 1981). Species do not occur on Taxodiaceae, Cupressaceae or Taxaceae. In general the biology of the genus is poorly known beyond host records.

The genus exhibits a diversity of mating systems including harem polygamy, parthogenesis (thelytoky), and monogamy (Bright, 1981; Deyrup and Kirkendall, 1983; Wood, 1986). All Nearctic species are polygamous and *P. puberulus* is parthenogenic (Deyrup and Kirkendall, 1983). The Palearctic species *P. henscheli* Seitner, *P. traegardhi* Spessivtseff, *P. carniolicus* Wichmann, and *P. morosovi* Spessivtseff are monogamous (Pfeffer, 1976). Galleries are male initiated (except *P. puberulus*) (Kirkendall, 1983). It is suspected that most species use aggregation pheromones to attract mates but pheromones have only been determined for a few species (Bright, 1981; Dallara *et al.*, 2000; López *et al.*, 2011; Seybold *et al.*, 2012b). In phloeophagous species, males excavate a nuptial chamber and are subsequently joined by 3–5 or more females. After mating, each female excavates an egg gallery from the central nuptial chamber, giving the adult gallery a star-like appearance. Eggs are deposited in niches along the length of each egg gallery and larval mines radiate away from the egg gallery in the cambium. Pupation occurs in enlarged cells at the end of the larval mine. There are 2–3 larval instars (Bright, 1981). In myelophagous species, the gallery is constructed in the pith of a small twig. The male and females bore into the pith and the females continue to excavate egg tunnels, depositing eggs along the wall. Larvae hatch and eventually mine all the tissue in the twig. In both types of galleries, the male stays in the entrance (Bright, 1981). No species have been reported to have maturation feeding (Sakamoto *et al.*, 2007).

Nearly all species are secondary bark beetles and are of almost no economic importance (Furniss and Carolin, 1977; Bright, 1981). However, some species may kill small Christmas trees and others may transmit fungal pathogens into healthy hosts (Bright, 1981; Storer *et al.*, 2004; Graves *et al.*, 2009). Several California *Pityophthorus* spp. (*P. setosus* Blackman, *P. nitidulus* (Mannerheim), *P. carmeli* Swaine, and *P. tuberculatus* Eichhoff), along with *Conophthorus radiatae* Hopkins, *Ips paraconfusus* Lanier, and the anobiid beetle *Ernobius punctulatus* (LeConte), can vector the pitch canker fungus *Fusarium circinatum* (Nirenberg and O'Donnell) in California (Dallara, 1997;

Storer *et al.*, 2004). Pitch canker is a conifer pathogen that primarily affects *Pinus radiata* D. Don (Monterey pine), other *Pinus* species, and *Pseudotsuga menziesii*. Symptoms of infected trees include infected branch tips, progression to multiple infected branch tips and to cankers on large branches and the trunk. Tree mortality is associated with bark beetle infestations (Storer *et al.*, 2004).

10.2 Economically Important Species

10.2.1 *Pityophthorus juglandis* Blackman — Walnut Twig Beetle (Figure 12.7)

The walnut twig beetle *Pityophthorus juglandis* is native to the southwestern USA and northern Mexico. The species was described from specimens collected from Lone Mountain (near Silver City in Grant County), New Mexico, and Portal, Arizona. The species was rarely encountered and by 1981 additional localities were known from southern Arizona (Chiricahua Mountains, Miller Canyon), northern Arizona (Oak Creek Canyon), southern California (San Fernando and Tarzana in Los Angeles County), and northern Chihuahua, Mexico (Bright, 1981; Cranshaw, 2011). Cranshaw (2011) examined museum collections and reported additional records from the California Central Valley and San Francisco Bay area collected during the 1970s. In 2001, the species was collected from diseased walnut (*Juglans*) in northern New Mexico and was subsequently found along the Front Range of eastern Colorado, where it was associated with black walnut (*Juglans nigra* L.) mortality and suspected of vectoring a pathogenic fungus (Cranshaw, 2011). Surveys were then initiated to delimit the range of *P. juglandis*. Additional records were reported from Utah (1996, 1998), Idaho (2003), Oregon (2004), and Washington (2008). The distribution was recently reviewed by Seybold *et al.* (2012c), who expanded the range to include Tennessee (2010), Nevada (2011), Virginia (2011), Pennsylvania (2012), North Carolina (2012), and Ohio (2012) (Thousand cankers disease, 2014). It is unknown why or how this rare species apparently expanded its range so rapidly between 1992 and 2011. Hypotheses include a natural dispersal event or anthropogenic transport and improved forest pest surveys (Cranshaw, 2011).

Prior to 1981, *P. juglandis* exhibited a distribution nearly identical to that of Arizona walnut (*Juglans major* (Torr.) A. Heller). However, specimen labels indicate that host tree was never identified to the species level, except *Juglans californica* S. Watson (southern California walnut) in California (Bright, 1981; Cranshaw, 2011). Since 1981, most host records exist from black walnut, a native of the eastern USA from New England west to the Great Plains, and from the Canadian border south to Texas and the Florida panhandle. It is also widely planted as an ornamental and nut tree and to a lesser extent for timber across the USA

(Tisserat *et al.*, 2011). Black walnut is an extremely valuable timber resource and its wood is prized for use in finished wood products (Utley *et al.*, 2013). The species has also been reported on other walnut species including *J. hindsii* Jeps. ex R. E. Sm. and *J. regia* L. (English walnut), as well as *Pterocarya* Kunth sp. (wingnut) (Seybold *et al.*, 2011, 2013)

Pityophthorus juglandis is the most economically important *Pityopthorus* species in the world, and vectors a pathogenic fungus, *Geosmithia morbida* M. Kolařík, E. Freeland, C. Utley, and N. Tisserat (Kolařík *et al.*, 2011). Tree mortality occurs as a direct result of aggressive feeding on the healthy branches and trunk in which hundreds or thousands of beetles colonize a host and cause subsequent canker development in the phloem around the galleries. Infected trees exhibit yellowing, flagging, and dieback and succumb to the pathogen after 3–4 years (Kolařík *et al.*, 2011). This insect–fungal complex is threatening the loss of black walnut forests worth an estimated $500 billion in the eastern USA (Newton and Fowler, 2009). Both the beetle and the fungus are capable of colonizing and infecting walnut and wingnut (Seybold *et al.*, 2013).

The insect is suspected to have 2–3 generations per year in California with an initial flight from April to May and a second flight from mid-July to mid-September (Graves *et al.*, 2009). Beetles are crepuscular and were found to be most active at 23–24°C in northern California (Seybold *et al.*, 2012b). Galleries are constructed in a pattern similar to other *Pityophthorus* species in which the male initiates gallery construction and releases an aggregation pheromone (Seybold *et al.*, 2012b) to attract 2–3 females and other conspecifics to the host tree (Graves *et al.*, 2009). Pheromone details are not yet available (Seybold *et al.*, 2012a). Females excavate a single egg gallery in the phloem that is against the grain of the wood and etch the xylem. Larvae create galleries radiating away from the maternal egg gallery (Graves *et al.*, 2009). There are three larval instars (Dallara *et al.*, 2012). After pupation, newly emerged adults either recolonize the brood host or fly to a new host (Graves *et al.*, 2009).

11. *CONOPHTHORUS*

11.1 Overview

Conophthorus Hopkins (Corthylini: Pityophthorina) is a Nearctic endemic and contains 13 species distributed in pine forests from southern Canada to Guatemala (Alonso-Zarazaga and Lyal, 2009).

11.1.1 Diagnosis and Description (Modified from Hopkins (1915) and Wood (1982)) (Figure 12.8)

Diagnosis. Costal margin of declivity descending toward the apex, metepisternum almost completely covered by

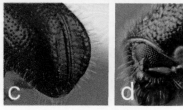

FIGURE 12.8 *Conophthorus ponderosae*: (A) habitus lateral; (B) habitus dorsal; (C) venter posterior; (D) antennal club.

elytra, antennal club sutures aseptate, antennal funicle five-segmented, pronotal asperities reaching past middle, the transition from asperate to punctate area gradual, pronotum without transverse impression behind an indefinite summit, and by the spermatophagous habit of breeding in pine cones.

Description. *Conophthorus* exhibits very limited taxonomic characters and a high degree of morphological similarity. *Conophthorus* are stout, 2.3–2.4 times as long as wide and 2.2–4.1 mm in length. Mature color is brown to black.

Frons simple, convex, sparsely punctate, vesiture sparse, hair-like. Eye emarginated, finely faceted. Antennal scape long, slender; funicle five-segmented; club oval, compressed, three sutures on anterior and two on posterior face, sutures 1 and 2 aseptate, conspicuously marked by grooves and setae.

Pronotum longer than wide, sides broadly rounded from near base to apex, slightly constricted beyond middle with base margined; indefinite summit on basal third; anterior slope asperate, asperities reaching past middle, the transition from asperate to punctate area gradual, pronotum without transverse impression behind an indefinite summit.

Elytral punctures striate. Declivity sulcate, simple.

Sexual differences among *Conophthorus* are seen in frons shape, sculpturing, and punctation and declivital tubercle size. See Wood (1982) for further details regarding each species.

11.1.2 Taxonomic History

Hopkins (1915) erected the genus to accommodate *Pityophthorus coniperda* Schwarz and 14 new species, all of which shared the species epithet with that of the *Pinus* host, a pattern followed by subsequent authors. There have been 22 species described of which 13 are currently recognized. Wood added six species, mostly Mexican, between 1962 and 1980 (Wood and Bright, 1992). McPherson (1970) described *C. banksianae* breeding in *P. banksiana* twigs and Flores and Bright (1987) described the most recent species. Wood's (1977) revision of the genus resulted in many species described by Hopkins (1915) being placed in synonymy with *C. coniperda* (1), *C. ponderosae* (5), and *C. resinosae* (1). Studies involving ecological, morphological, karyological, and molecular data suggested that *C. banksianae* is a synonym of *C. resinosae* (Wood, 1989; de Groot *et al.*, 1992; Cognato *et al.*, 2005). Cognato *et al.* (2005) presented the first phylogeny of the genus using mitochondrial COI sequences. Geographic isolation, rather than host species appears to be the important factor in diversification among *Conophthorus* species (Cognato *et al.*, 2005). The genus is in need of revision and at least one species, *C. ponderosae*, may be polyphyletic (Cognato *et al.*, 2005). *Conophthorus* will be especially difficult to revise due to the paucity of morphological characters. Bright and Stark (1973) provide a key to California species and Furniss and Johnson (2002) provide a key to Idaho species. Wood (1982) presents a Nearctic key but caution should be used identifying Mexican material as *C. terminalis* Flores and Bright (1987) is not included.

11.1.3 Biology

Conophthorus are spermatophagous (seed feeding), with adults and larvae feeding within pine cones. *Conophthorus resinosae* Hopkins and *C. terminalis* Flores and Bright can also infest twig terminals and growing tips and buds (Wood, 1982; Flores and Bright, 1987; Cibrián Tovar *et al.*, 1995).

Conophthorus are monogamous and galleries are established by females who bore into the cone near the base or cone stalk of second year cones in late spring to early summer (Hedlin *et al.*, 1980; Wood, 1982; Kirkendall, 1983). Boring into the cone results in the production of a pitch tube, the odor of which combined with sex pheromones attracts males to the cone (Hedlin *et al.*, 1980; Miller *et al.*, 2000; Trudel *et al.*, 2004). Infested cones typically contain a single mating pair (Trudel *et al.*, 2004). Females construct the egg gallery in the central stalk and deposit eggs in individual niches along the length of the gallery and close to the developing seeds. Larvae have two instars and feed upon the seeds and bracts within the cone. Larvae pupate at the end of summer and new adults overwinter within the cone emerging in the following spring (Miller, 1915; Hedlin *et al.*, 1980; Cibrián

Tovar *et al.*, 1995). There is typically one generation per year (Bright and Stark, 1973; Hedlin *et al.*, 1980). Emerging brood adults of some species overwinter in twig tips (McPherson *et al.*, 1970; Bright and Stark, 1973). Other species including *C. terminalis* and *C. resinosae* create egg galleries in growing tips and buds of the host (McPherson *et al.*, 1970; Wood, 1982; Flores and Bright, 1987). Emerging adults were believed to infest cones of the same or nearby tree (Henson, 1962; Kinzer *et al.*, 1970). However, Menard and Cognato (2007) demonstrated a variable haplotype diversity from beetles collected from cones produced by a tree, suggesting that beetles disperse from their natal tree in search of cones to lay eggs.

Adults and larvae primarily feed within cones, causing serious economic loss to pine seed crops as well as in forest ecosystems where they decrease natural stand regeneration (Godwin and Odell, 1965; Cibrián Tovar *et al.*, 1986, 1995; Hedlin *et al.*, 1980). When the female bores into the cone, the phloem and xylem of the cone stalk are girdled, killing the cone (Kinzer *et al.*, 1970; Hedlin *et al.*, 1980; Furniss, 1997). *Conophthorus* infestations can cause high cone mortality of up to 100% (Schaefer, 1962), although typical mortality is between 25 and 75% (Furniss and Carolin, 1977; Cibrián Tovar *et al.*, 1995). However, damage from all species is not confined to cones. Heavy attack by *C. terminalis* caused a significant amount of damage to a stand of *P. cembroides* with 60–90% of all lateral and terminal shoots infected and killed. In addition, *C. radiatae* has been implicated as a potential vector of the pitch canker fungus to *P. radiata* in California (Storer *et al.*, 2004).

11.2 Economically Important Species

11.2.1 *Conophthorus ponderosae* Hopkins — Lodgepole Cone Beetle, Ponderosa Pine Cone Beetle, Sugar Pine Cone Beetle (Figure 12.8)

Cognato *et al.* (2005) found *C. ponderosae* to be polyphyletic with species forming multiple clades, including a Cascade and northern Rocky Mountain clade and a southern Rocky Mountain clade. Based on these findings, it appears that the southern Rocky Mountain populations of *C. ponderosae*, including New Mexico, Arizona and Mexican populations, are more closely related to *C. resinosae* and *C. coniperda* than to *C. ponderosae*, which is from Ashland, Oregon. This southern population potentially represents a different species.

Conophthorus ponderosae is distributed from southern British Columbia and Montana to Puebla and Veracruz, Mexico (Wood, 1982; Bright and Skidmore, 2002) and occurs in many *Pinus* species including *P. aristata* Engelm., *P. contorta* Douglas ex. Loudon, *P. douglasiana* Martínez, *P. durangensis* Martínez, *P. flexilis* E. James, *P. hartwegii* Lindl., *P. jeffreyi* Balf., *P. lambertiana*

Douglas, *P. leiophylla* Schiede ex Schltdl. and Cham., *P. montezumae* Gordon and Glend., *P. monticola* Douglas ex D. Don, *P. rudis* Endl., *P. strobiformis* Engelm., *P. washoensis* H. Mason and Stockw. (Wood, 1982; Atkinson and Equihua, 1985).

Conophthorus ponderosae infestations can cause high cone mortality of 75% in the northern Rockies (Keen, 1958) and 76% in New Mexico (Kinzer *et al.*, 1970), where up to 90% has been reported (Bennett, 2000). Seed mortality within each infested cone is 100%. The biology and life history of *C. ponderosae* is as described for the genus. Females produce a sex pheromone, (+)-pityol, to attract males to cones (Miller *et al.*, 2000). Miller (1915) and Furniss (1997) provide detailed life history studies for Ashland, Oregon (the type locality), and Moscow, Idaho, respectively. In Oregon, flight occurs from early to late May, pupae develop early June to mid-July, and overwintering adults can be found from late July to the end of April (Miller, 1915). The number of overwinter adults ranges from 1 to 20 per cone, 5 to 8 on average in Oregon, and 2 to 15 per cone, 5.3 on average in Idaho (Miller, 1915; Furniss, 1997).

12. *GNATHOTRICHUS*

12.1 Overview

Gnathotrichus Eichhoff (Corthylini: Corthylina) is a New World endemic and contains 17 species in North and Central America. Fifteen species occur in the Nearctic (Alonso-Zarazaga and Lyal, 2009). *Gnathotrichus materiarius* (Fitch) is widely distributed across western Europe (Kirkendall and Faccoli, 2010).

12.1.1 Diagnosis and Description (Modified from Blackman (1931) and Wood (1982)) (Figure 12.9)

Diagnosis. *Gnathotrichus* can be differentiated from other Nearctic genera by the costal margin of the declivity descending toward the apex, metepisternum almost completely covered by elytra, antennal funicle five-segmented, club symmetrical, bearing two sutures, sutures straight to moderately procurved, segment 1 not reduced in size, nearly glabrous elytra and pronotum, protibia widest at apex, prosternal intercoxal piece obtuse, subapical margin near apex acutely elevated, sutural apex entire, rather narrowly rounded behind and the xylomycetophagous habit.

Description. *Gnathotrichus* are elongate, 2.9–3.3 times as long as wide and 2.0–3.7 mm in length. Mature color is light brown to nearly black. Surface smooth, finely reticulate, varying from subopaque to shining, finely to minutely punctate. Vestiture sparse, hair-like.

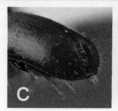
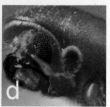

FIGURE 12.9 Gnathotrichus sulcatus female: (A) habitus lateral; (B) habitus dorsal; (C) venter posterior; (D) antennal club.

Frons weakly to strongly convex, either punctate with median area elevated or aciculate with spare hair-like setae. Eyes oval, emarginated. Antennal scape elongate; funicle five-segmented; club more than 1.5 times as long as funicle, sparsely pubescent, segment 1 not reduced, segments 2 and 3 subequal in length, and with two straight to moderately procurved, septate sutures.

Pronotum elongate, sides not constricted before the middle; anterior slope finely asperate, asperites broad, low; summit on anterior half, marked by a sharply elevated transverse carina. Basal margin usually not carinate. Scutellum large, flat. Tibiae widened apically, two or more socketed teeth on lateral margin near apex, lateral margin armed by a row of serrations.

Elytra elongate, finely rugulose, striate or not; finely to minutely punctate. Disc glabrous to subglaborus. Declivity moderately steep, convex, weakly to strongly sulcate or bisulcate; conservatively sculptured; sutural apex entire, costal margin near apex moderately to strongly acutely elevated.

Like several other Corthylina genera, the female antennal funicle and club bears a few very long curved hair-like setae on posterior face. The frons and/or declivity may also be sexually dimorphic in some species with variation observed in the convexity and sculpturing of the frons and shape and sculpturing of the declivity. See Wood (1982) for further details.

12.1.2 Taxonomic History

Twenty North American species have been described and 17 are currently recognized. Almost all of the Mexican and Central American species were described by Wood (Wood and Bright, 1992). Other authors including Fitch, LeConte, Blandford, Blackman, and Bright described species (Wood and Bright, 1992). There has been very little conflict regarding species limits; however, four genera have been described for *Gnathotrichus* species. *Gnathotrichus materiarus* (Fitch) was described as a new genus (*Paraxyleborus*) and species (*Xyleborus duprezi* Hoffman) when it was first found in France in 1933. Balachowsky (1949) recognized that the species were the same and placed *X. duprezi* in synonymy. *Gnathotrichus* has been revised by Blackman (1931) and Wood (1982). Wood (1982) provides a key to the entire genus. Regional keys are available for California (Bright and Stark, 1973), Canada and Alaska (Bright, 1976), and Idaho (Furniss and Johnson, 2002).

12.1.3 Biology

Gnathotrichus are monogamous ambrosia beetles. Males are attracted to host trees by primary attraction to ethanol (Cade *et al.*, 1970), initiate gallery construction (Kirkendall, 1983), and possess procoxal mycangia (Farris, 1963; Schneider and Rudinsky, 1969). A radial gallery is bored into the sapwood, consisting of a short entrance tunnel and usually four egg galleries that curve along growth rings. Each gallery is about 15–25 cm in length. The cradles and biology are essentially as described for *Monarthrum* (Doane and Gilliland, 1929; Prebble and Graham, 1957). Eggs are deposited in niches above and below each egg gallery. Larvae hatch and expand the niche into a larval cradle parallel with the grain of the wood, consuming wood to enlarge the cell and feeding upon fungi. Pupation occurs in the cradle with the head oriented toward the egg gallery. Males stay in the entrance tunnel and the female stays in the egg galleries. *Gnathotrichus* exhibits a high degree of parental care with the female paying constant attention to her larvae. The female will place a fungus plug at the junction of the egg gallery and the larval cradle and replace it as soon as it is consumed. The female will also collect the larval frass and move it to the entrance tunnel where it is expelled by the male. The male guards the entrance tunnel by plugging it with his body (Doane and Gilliland, 1929).

Gnathotrichus consists of species either specialized on conifers or hardwoods. *Gnathotrichus retusus* (LeConte) has the ability to live in either hardwoods or conifers including *Alnus*, *Betula*, *Populus*, *Abies*, *Larix*, *Pinus*, *Pseudotsuga*, and *Tsuga* (Wood, 1982; Furniss and Johnson, 2002). Hardwood feeding species occur strictly on *Quercus* (oak) from Arizona to southern Mexico. Central American species occur at high elevation sites on *Oreopanax* (Araliaceae) (Wood, 1982).

12.2 Economically Important Species

12.2.1 *Gnathotrichus sulcatus* LeConte
(Figure 12.9)

Gnathotrichus sulcatus is distributed in conifer forests from British Columbia and South Dakota to Honduras (Wood, 1982). The primary hosts are *Pseudotsuga menziesii* and *Abies*, including *A. concolor*, *A. magnifica*, and *A. religiosa* (Kunth) Schltdl. and Cham., *Pinus* spp., and *Tsuga heterophylla* (Raf.) Sarg. (Wood, 1982; Furniss and Johnson, 2002). *Gnathotrichus sulcatus* is considered the second most important conifer ambrosia beetle in North America, after *T. lineatum* (Furniss and Carolin, 1977).

The natural history of the species is as outlined for the genus. Males produce the aggregation pheromone sulcatol (65/35 mixture of the (S)-(+) and the (R)-(−) enantiomers of 6-methyl-s-hepten-2-ol) to attract mates and conspecifics to suitable hosts (Byrne *et al.*, 1974). Males initiate gallery construction and inoculate galleries with fungal spores of *Ambrosiella sulcati* Funk and *Raffaelea sulcati* Funk (Funk, 1970). Up to 60 eggs are laid per gallery system and all stages overwinter within the brood log. Overwintered brood adults engage in maturation feeding on ambrosia fungus before emergence in the spring. Flight begins from early April to end of May, when temperatures reach 14–16°C, with peak activity in late June and early July, and subside in late September when another peak occurs. *Gnathotrichus sulcatus* is crepuscular and some individuals fly in the morning and early afternoon (Daterman *et al.*, 1965). One generation per year has been reported in Canada and it is suspected that there are two generations per year with overlapping broods in California (Bright and Stark, 1973; Bright, 1976).

The species is a secondary bark beetle and colonizes moribund trees and recently felled timber. Infestation by ambrosia beetles such as *G. sulcatus* degrade lumber and thus reduce profits for forestry companies. *Gnathotrichus sulcatus* can penetrate up to 3 cm deep in *P. menziesii*, not extending beyond the sapwood and 8 cm deep in *T. heterophylla* because of higher moisture content and can degrade up to 64% of *T. herterophylla* total log volume (McLean, 1985). Doane and Gilliland (1929) reported high densities of *G. sulcatus* in recently felled *P. menziesii*, infesting the top first, then trunk and persisting in stumps for several generations. The species can survive and reproduce in sawn lumber and thus poses an introduction threat to lumber-importing countries and has been intercepted in New Zealand on numerous occasions (Milligan, 1970; Bain, 1974; McLean and Borden, 1975; Brockerhoff *et al.*, 2006). Beetles can survive in green lumber for at least 2 months, long enough to reach anywhere in the world alive (McLean and Borden, 1975).

13. *MONARTHRUM*

13.1 Overview

Monarthrum Kirsch (Corthylini: Corthylina) is a New World endemic and contains 145 species in both temperate and tropical forests. The vast majority of species occur in the Neotropics but 22 occur in the Nearctic. *Monarthrum mali* (Fitch) has become established in Italy (Kirkendall *et al.*, 2008; Kirkendall and Faccoli, 2010).

13.1.1 Diagnosis and Description (Modified from Bright (1976) and Wood (1982, 2007))
(Figure 12.10)

Diagnosis. *Monarthrum* can be distinguished by the costal margin of the declivity descending toward the apex, metepisternum almost completely covered by elytra, body slender, elongate, antennal funicle two-segmented, elytral apex usually emarginated, lateral margin of pronotum finely raised and by the prosternal intercoxal piece large, posteriorly angulate and projecting between procoxae.

Description. Nearctic *Monarthrum* are elongate 2.5–3.4 times as long as wide and 1.4–4.7 mm in length. Mature color yellow brown to very dark brown, species often bicolored.

Frons often sexually dimorphic, convex and variously conservatively sculptured in male, females often similar

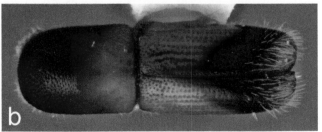
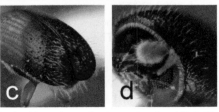

FIGURE 12.10 *Monarthrum fasciatum* female: (A) habitus lateral; (B) habitus dorsal; (C) venter posterior; (D) antennal club.

to male but may be concave to elaborately impressed, glabrous to elaborately ornamented by hair-like setae. Eye oval, deeply emarginated, finely faceted. Antennal scape slender to triangular, usually finely pubescent in females; funicle two-segmented; club slender to broadly oval or strongly triangular, females with very long curved hair-like setae on posterior face; two sutures straight to moderately procurved.

Pronotum longer than wide, summit at or on apical half; anterior slope asperate; anterior margin usually unarmed in female, asperate in male; posterior half very finely sculptured. Scutellum large, flat. Procoxae contiguous, prosternal intercoxal piece large, posteriorly angulate and projecting between procoxae; tibiae dimorphic, posterior face subinflated and armed by numerous confused tubercles in female, not inflated or tuberclate in male.

Elytra very finely sculptured, punctures confused; typically glabrous. Declivity very gradual to subvertical, convex to elaborately excavated, unarmed to armed by spines; sutural apex entire to conspicuously emarginated; glabrous to subglabrous.

Like several other Corthylina genera, the female antennal funicle and club bears a few very long curved hair-like setae on posterior face. The frons and/or declivity may also be sexually dimorphic in some species with variation observed in the convexity, sculpturing and vesture abundance of the frons and shape and sculpturing of the declivity. See Wood (1982) for further details.

13.1.2 Taxonomic History

Thirty-four Nearctic species have been described of which 27 are currently recognized. The majority of species were described by Wood. Other authors, including Bright, Schedl, Blandford, Eichhoff, LeConte, Say and Fitch, described species (Wood, 1982, 2007). Almost all species described prior to 1940 were described as *Ptercyclon* Eichhoff. Swaine (1918) considered the Nearctic species *M. dentigerum* (LeConte), *M. fasciatum* (Say), *M. mali* (Fitch), and *M. scutellare* (LeConte) as *Pterocyclon* because of what he viewed as an inadequate description of *Monarthrum* by Kirsch. Chamberlain (1939, 1958) rejected this because of the widespread use of *Monarthrum* in the literature. Wood (1966) placed *Pterocyclon* in synonymy with *Monarthrum*. *Monarthrum* has been revised twice (Wood, 1982, 2007).

The generic limits and species need revision, especially for the Neotropical fauna. The genus is very morphologically varied with much of its diversity awaiting description. Generic and species limits are in need of revision, especially in South America (Wood, 2007, Smith. pers. obs.). Wood (2007) provides a key to the entire genus. In his monograph of South America, Wood (2007) describes four new Nearctic Mexican species, thus caution should be used

when using Wood (1982) as it does not contain the entire Nearctic diversity. Regional keys are available for California (Bright and Stark, 1973), Canada, and Alaska (Bright, 1976). A key to the West Indies fauna is also being developed (Bright, in prep.).

13.1.3 Biology

Monarthrum are monogamous ambrosia beetles although some species may be polygynous (Hubbard, 1897; Wood, 1982, 2007; Kirkendall, 1983; T. H. Atkinson, pers. commun.). The vast majority of *Monarthrum* species are found in montane Neotropical forests and are known from a wide diversity of hosts at these sites (Wood, 2007). Most Nearctic species occur on *Quercus* but several species occur on *Alnus* and the two eastern species, *M. mali* and *M. fasciatum*, are polyphagous with records from numerous families of hardwoods. Species attack moribund, weakened, diseased or recently dead trees and branches.

There has been a paucity of studies on the natural history of any *Monarthrum* (Hubbard, 1897; Blackman, 1922; Doane and Gilliland, 1929; Roling and Kearby, 1974, 1975, 1977), which were recently summarized by Kirkendall *et al.* (2008). Males initiate galleries by boring into the xylem, creating an entrance tunnel (Roling and Kearby, 1974; Wood, 1982, 2007; Kirkendall, 1983). The entrance tunnel leads to a star-like nuptial chamber from which egg galleries radiate. Eggs are deposited in niches above and below each egg gallery. Larvae hatch and expand the niche into a larval cradle parallel with the grain of the wood, consuming wood to enlarge the cell and feeding upon fungi. Pupation occurs in the cradle with the head oriented toward the egg gallery. Males stay in the entrance tunnel and the female stays in the egg galleries (Hubbard, 1897; Roling and Kearby, 1974; Smith, pers. obs.). *Monarthrum* exhibits a high degree of parental care with the female paying constant attention to her larvae. The female will place a fungus plug at the junction of the egg gallery and the larval cradle and replace it as soon as it is consumed. The female will also collect the larval frass and move it to the entrance tunnel where it is expelled by the male. The male guards the entrance tunnel by plugging it with his body (Hubbard, 1897).

Females have procoxal mycangia in *M. dentiger*, *M. fasciatum*, *M. mali*, *M. scutellare*, and the Neotropical *M. nudum* (Schedl) (Schedl, 1962; Batra, 1963; Farris, 1965; Lowe *et al.*, 1967; Roeper and French, 1978). Only one species, *M. bicallosum*, is known to possess pregular mycangia (Schedl, 1962). *Monarthrum* is quite unusual because in other scolytines the sex that initiates the gallery possesses the mycangia (Schedl, 1962; Beaver, 1989). Species are highly attracted to ethanol (Roling and Kearby, 1975; Montgomery and Wargo, 1983).

Monarthrum are of almost no economic importance and can be considered beneficial to forest ecosystems by hastening the decay of dead wood (Roling and Kearby, 1974). However, *M. fasciatum* and *M. mali* are highly destructive pests of green lumber and fresh logs of *Liquidambar styracifluia* L. (gum) in the Gulf States (Baker, 1972). In addition, these two species have been known to bore into wine casks, causing leaking and loss (Hubbard, 1897; Swaine, 1909; Cognato, pers. commun.) and beetle galleries and associated fungal staining can decrease the value of wood for finish or structural material (Blackman, 1922).

13.2 Economically Important Species

13.2.1 *Monarthrum fasciatum* (Say)
(Figure 12.10)

Monarthrum fasciatum occurs from southern Ontario and Quebec, Canada, and the eastern USA south to Florida (Bright, 1976; Wood, 1982). The species is polyphagous with most individuals colonizing the Fagaceae hosts *Quercus*, *Carya* (hickory), and *Castanea* (chestnut), as well as Sapindaceae (*Acer*; maple), Altingiaceae (*Liquidambar*), Fabaceae (*Mimosa*), and Rosaceae (*Prunus*) (Chamberlain, 1939; Bright, 1976; Wood, 1982).

The biology is essentially as outlined for the genus with the female having procoxal mycangia (Batra, 1963; Lowe *et al.*, 1967). Two studies demonstrated that the ascomycete fungus *Monilia brurnnea* J.C. Gilman and E.V. Abbott is commonly present in galleries (Verrall, 1943; Batra, 1963). Roling and Kearby (1974, 1975, 1977) provide the most detailed account of the biology of any species of the genus based on observations from Missouri, where there are 2–3 generations per year and a 2:1 female to male ratio. Flight periods occur from March to mid-May, early June to late July, and late August to early September (Roling and Kearby, 1974, 1977). Beetles overwinter as pupae in cradles or as adults emerging from parental brood gallery in spring or overwinter in a partially completed gallery. The species prefers larger diameter trees greater than 24 cm DBH (diameter at breast height) (Roling and Kearby, 1977). The average gallery was found to be 2.5 cm and complete measurements for each gallery component are given (Roling and Kearby, 1974).

13.2.2 *Monarthrum mali* (Fitch)

Monarthrum mali is naturally distributed from southern Ontario to New Brunswick and in the USA east of the Rocky Mountains and California (Bright, 1976; Wood, 1982). The species has been recently established in Italy (Kirkendall *et al.*, 2008; Kirkendall and Faccoli, 2010). It commonly occurs on Fagaceae hosts, especially *Quercus* and *Fagus* (beech) but can also be found infesting *Acer*

(maple), *Betula* (birch), *Liquidambar* (gum), and *Tilia* (linden) (Wood, 1982). Even though records also exist for two conifer genera, *Pinus* and *Tsuga*, these records are rare and suggest that conifers are not typical hosts (Wood, 1982; Kirkendall *et al.*, 2008). *Monarthrum mali* is also considered an orchard pest of apple, plum, cherry, and orange and was a serious pest of apple in Massachusetts during the 1860s (Brooks, 1916).

The biology is essentially as outlined for the genus with the female having procoxal mycangium (Batra, 1963). The species also displays a preference for larger diameter trees greater than 24 cm DBH (Roling and Kearby, 1977).

13.3.3 *Monarthrum scutelllare* (LeConte)

Monarthrum scutellare is distributed along the Pacific Coastal region of North America ranging from British Columbia, Canada, to Baja California, Mexico (Bright, 1976; Wood, 1982) and occurs on Fagaceae hosts especially *Quercus* spp. and also *Chrysolepis* and *Lithocarpus densiflorus* (Hook. and Arn.) Rehder (Bright, 1976; Wood, 1982). The biology is essentially as outlined for the genus with male initiated galleries (Doane and Gilliland, 1929) and females with procoxal mycangia that contain the black staining ambrosia fungus, *Ambrosiella brunnea* (Verrall) Batra (Farris, 1965). The biology was elucidated by Doane and Gilliland (1929) who examined the species in California. The species prefers moribund or severely weakened coast live oaks and also diseased tissues of otherwise healthy trees. There are two generations per year, with egg laying activity occurring in late March and in October.

Monarthrum scutellare is one of the most abundant scolytine species found colonizing *Quercus* species infected with sudden oak death, *Phytophthora ramorum* Werres, De Cock and Man in't Veld in California. This species selectively bores into stem cankers caused by the pathogen, thus considerably shortening the survival of infected oaks (McPherson *et al.*, 2013).

14. CONCLUSION

The 12 genera included in this chapter are arguably not as economically or ecologically important as North American *Ips* and *Dendroctonus*. These genera have received less funding and research attention, and as a consequence natural history and phylogenetic knowledge is minimal. For some genera with a pest species, there may be a great deal of known biological information. However, these habits may not be inferred for related benign species. Recent phylogenetic investigations have showed that some widespread pest species have been improperly delineated and contain cryptic species. This was especially apparent in the study of Holarctic scolytines by Jordal and Kambestad (2014) who demonstrated that *Trypodendron lineatum, Dryocoetes autographus,* and western North American *Orthotomicus caelatus* (Eichhoff) each formed separate lineages from their Palearctic relatives. Additional examples of taxonomic confusion are found in *Conophthorus ponderosae* (Cognato *et al.*, 2005), *Euwallacea fornicatus* and *Pseudohylesinus grandis.* Molecular and morphological phylogenetic studies are needed to elucidate species relationships within Nearctic genera. Only *Ips* DeGeer (Cognato and Sun, 2007), *Scolytus* (Smith and Cognato, in press), and *Xylosandrus* Reitter (Dole and Cognato, 2010) have received such a modern holistic taxonomic treatment. Further phylogenetic investigations of taxonomically challenging genera such as *Conophthorus*, *Pityophthorus*, and *Trypodendron* will test species limits, reveal species relationships, and provide inference to the biology of closely related species.

ACKNOWLEDGMENTS

We would like to thank David Adamski (USDA-SEL), Thomas H. Atkinson (University of Texas), Donald Bright, Anthony Cognato (Michigan State University), Lawrence Kirkendall (University of Bergen), and Robert Rabaglia (USDA-FS) for thoughtful discussions regarding this chapter. Funding for J. Hulcr was provided by NSF DEB 1256968 and the USDA-FS 11-CA-11330129-092 and 12-CA-11420004-042. S. M. Smith thanks the Michigan State University A. J. Cook Research Collection and the Smithsonian Museum of Natural History for providing desk space and access to resources.

REFERENCES

Abrahamson, L.P., Chu, H.-M., Norris Jr., D.M., 1967. Symbiotic interrelationships between microbes and ambrosia beetles: II. The organs of microbial transport and perpetuation in *Trypodendron betulae* and *Trypodendron retusum* (Coleoptera: Scolytidae). Ann. Entomol. Soc. Am. 60, 1107–1110.

Alonso-Zarazaga, M.A., Lyal, C.H.C., 2009. A catalogue of family and genus group names in Scolytinae and Platypodinae with nomenclatural remarks (Coleoptera: Curculionidae). Zootaxa 2258, 1–134.

Atkinson, T.H., 2014. Bark and ambrosia beetles of the US and Canada. Available online: www.barkbeetles.info. Last accessed: May 20, 2014.

Atkinson, T.H., Equihua, M.A., 1985. Lista comentada de los coleópteros Scolytidae y Platypodidae del Valle de México. Fol. Entomol. Mex. 65, 63–108.

Atkinson, T.H., Foltz, J.L., Wilkinson, R.C., Mizell, R.F., 2000. Granulate ambrosia beetle, *Xylosandrus crassiusculus* (Motschulsky) (Insecta: Coleoptera: Curculionidae: Scolytinae). Florida Department of Agriculture and Consumer Services, Division of Plant Industry, Entomology Circula 310.

Avtzis, D., Knížek, M., Hellrigl, K., Stauffer, C., 2008. *Polygraphus grandiclava* (Coleoptera: Curculionidae) collected from pine and cherry trees: a phylogenetic analysis. Eur. J. Entomol. 105, 789–792.

Bain, J., 1974. Overseas wood- and bark-boring insects intercepted at New Zealand ports. New Zealand Forest Service Technical Paper 61, 1–24.

Bain, J., 1990. Dutch Elm Disease: New Zealand. Proceedings of the Symposium on: does the elm have a future in Australia? Victorian College of Agriculture & Horticulture, Burnley.

Baker, W.L., 1972. Eastern Forest Insects. U.S. Dept. Agric. Misc. Publ. 1175, 642.

Balachowsky, A.S., 1949. Coléoptères: Scolytides. Faune de France 50, 320 p.

Batra, L.R., 1963. Ecology of ambrosia fungi and their dissemination by beetles. Trans. Kans. Acad. Sci. 66, 218–236.

Beaver, R.A., 1989. Insect-fungus relationships in the bark and ambrosia beetles. In: Wilding, N., Collins, N.M., Hammond, P.M., Webber, J. F. (Eds.), Insect-Fungus Interactions. Academic Press, London, pp. 121–143.

Beaver, R.A., 1991. New synonymy and taxonomic changes in Pacific Scolytidae (Coleoptera). Annalen des Naturhistorisches Museums Wien 92B, 87–97.

Beaver, R.A., Browne, F.G., 1978. The Scolytidae and Platypodidae (Coleoptera) of Penang, Malaysia. Oriental Insects 12, 575–624.

Beaver, R.A., Liu, L.Y., 2010. An annotated synopsis of Taiwanese bark and ambrosia beetles, with new synonymy, new combinations and new records (Coleoptera: Curculionidae: Scolytinae). Zootaxa 2602, 1–47.

Bennett, R., 2000. Management of cone beetles (Conophthorus ponderosae, Scolytidae) in blister rust resistant western white pine seed orchards in British Columbia. Seed and Seedlings Extension Topics 12, 16–18.

Blackman, M.W., 1922. Mississippi bark beetles. Miss. Agric. Exp. Stn., Tech. Bull. 11, 130.

Blackman, M.W., 1928. The genus *Pityophthorus* Eichh. in North America: a revisional study of the Pityophthori, with descriptions of two new genera and seventy-one new species. Bulletin of the New York State College of Forestry, Syracuse University, Technical Publication, 25, 183 p.

Blackman, M.W., 1931. A revisional study of the genus *Gnathotrichus* Eichhoff in North America. J. Wash. Acad. Sci. 21, 264–276.

Blackman, M.W., 1934. A revisional study of the genus *Scolytus* Geoffroy (*Eccoptogaster* Herbst) in North America. U.S. Dep. Agric., Tech. Bull. 431, 30 p.

Blackman, M.W., 1942. Revision of the bark beetles belonging to the genus *Pseudohylesinus* Swaine. U.S. Dept. Agric. Misc. Publ. 461, 32.

Blackman, M.W., Stage, H.H., 1918. Notes on insects bred from the bark and wood of the American larch. Tech. Publ.—State Univ. Coll. For. Syracuse Univ 10, 9–115.

Bleiker, K.P., Lindgren, B.S., Maclauchian, L.E., 2003. Characteristics of subalpine fir susceptible to attack by western balsam bark beetle (Coleoptera: Scolytidae). Can. J. Forest Res. 33, 1538–1543.

Bleiker, K.P., Lindgren, B.S., Maclauchian, L.E., 2005. Resistance of fast- and slow-growing subalpine fir to pheromone-induced attack by western balsam bark beetle (Coleoptera: Scolytinae). Agr. Forest Entomol. 7, 237–244.

Bloomfield, H., 1979. Elms for always. Am. For. 85, 24–26, 48, 50.

Borden, J.H., 1988. The striped ambrosia beetle. In: Berryman, A.A. (Ed.), Dynamics of Forest Insect Populations. Plenum, New York, pp. 579–596.

Borden, J.H., Pierce, A.M., Pierce Jr., H.D., Chong, L.J., Stock, A.J., Oehlschlager, A.C., 1987. Semiochemicals produced by the western balsam bark beetle, *Dryocoetes confusus* Swaine (Coleoptera: Scolytidae). J. Chem. Ecol. 13, 823–836.

Bowers, W.W., Gries, G., Borden, J.H., Pierce Jr., H.D., 1991. 3-methyl-3-buten-1-ol: an aggregation pheromone of the four-eyed spruce beetle, *Polygraphus rufipennis* (Kirby) (Coleoptera: Scolytidae). J. Chem. Ecol. 17, 1989–2002.

Bowers, W.W., Borden, J.H., Raske, A., 1996. Incidence and impact of *Polygraphus rufipennis* (Coleoptera: Scolytidae) in Newfoundland. For. Ecol. Manage. 89, 173–187.

Breshears, D.D., Cobb, N.S., Rich, P.M., Price, K.P., Allen, C.D., Balice, R.G., et al., 2005. Regional vegetation die-off in response to global-change-type drought. Proc. Natl. Acad. Sci. U.S.A. 102, 15144–15148.

Bright, D.E., 1963. Bark beetles of the genus *Dryocoetes* (Coleoptera: Scolytidae) in North America. Ann. Entomol. Soc. Am. 56, 103–115.

Bright, D.E., 1969. Biology and taxonomy of bark beetles in the genus *Pseudohylesinus* Swaine (Coleoptera: Scolytidae). Univ. Calif. Publ. Entomol. 54, 1–49.

Bright, D.E., 1970. A note concerning *Pseudohylesinus sericeus*. Can. Entomol. 102, 499–500.

Bright, D.E., 1976. The Insects and Arachnids of Canada, Part 2: The Bark Beetles of Canada and Alaska, Coleoptera: Scolytidae. Canada Department of Agriculture, Research Branch, Biosystematic Research Institute, Publication, 1576, 1–241.

Bright, D.E., 1977. New synonymy, new combinations, and new species of North American *Pityophthorus* (Coleoptera: Scolytidae), Part 1. Can. Entomol. 109, 511–532.

Bright, D.E., 1981. Taxonomic monograph of the genus *Pityophthorus* Eichhoff in North and Central America (Coleoptera: Scolytidae). Mem. Entomol. Soc. Can. 118, 1–378.

Bright, D.E., 2014. A catalog of Scolytidae and Platypodidae (Coleoptera), Supplement 3 (2000–2010), with notes on subfamily and tribal reclassifications. Insecta Mundi 0356, 1–336.

Bright, D.E., Stark, R.W., 1973. The bark and ambrosia beetles of California (Coleoptera: Scolytidae and Platypodidae). Bulletin of the California Insect Survey 16, 1–169.

Bright, D.E., Skidmore, R.E., 2002. A Catalog of Scolytidae and Platypodidae (Coleoptera): Supplement 2 (1995–1999). NRC Research Press, Ottawa.

Bright, D.E., Torres, J.A., 2006. Studies on West Indian Scolytidae (Coleoptera) 4 A review of the Scolytidae of Puerto Rico, U.S.A. with descriptions of one new genus, fourteen new species and notes on new synonymy. Koleopterologische Rundschau 76, 389–428.

Brockerhoff, E.G., Bain, J., Kimberley, M., Knížek, M., 2006. Interception frequency of exotic bark and ambrosia beetles (Coleoptera: Scolytinae) and relationship with establishment in New Zealand and worldwide. Can. J. Forest Res. 36, 289–298.

Brooks, F.E., 1916. Orchard bark beetles and pinhole borers, and how to control them. United States Department of Agriculture Farmers' Bulletin 763, 16.

Browne, F.G., 1958. Some aspects of host selection among ambrosia beetles in the humid tropics of South-East Asia. Malay. For. 21, 164–182.

Browne, F.G., 1961. The biology of Malayan Scolytidae and Platypodidae. Malayan Forest Records 22, 1–255.

Byrne, K.J., Swigar, A.A., Silverstein, R.M., Borden, J.H., Stokkink, E., 1974. Sulcatol—population aggregation pheromone in the scolytid beetle, Gnathotrichus sulcatus. J. Insect Physiol. 20, 1895–1900.

Cade, S.C., Hrutfiord, B.F., Gara, R.I., 1970. Identification of a primary attractant for *Gnathotrichus sulcatus* isolated from western hemlock logs. J. Econ. Entomol. 63, 1014–1015.

Carlson, D., Ragenovich, I., 2012. Silver fir beetle and fir root bark beetle. United States Department of Agriculture Forest Service, U.S. Forest Insect and Disease Leaflet, 60.

Carrillo, D., Duncan, R.E., Ploetz, J.N., Campbell, A.F., Ploetz, R.C., Peña, J.E., 2014. Lateral transfer of a phytopathogenic symbiont among native and exotic ambrosia beetles. Plant Pathol. 63, 54–62.

Chamberlin, W.J., 1939. The bark and timber beetles of North America north of Mexico. The taxonomy, biology and control of 575 species belonging to 72 genera of the superfamily Scolytoidea. Oregon State College Cooperative Association, Corvallis, Oregon, 513 p.

Chamberlin, W.J., 1958. The Scolytoidea of the Northwest: Oregon, Washington, Idaho, and British Columbia. Oregon State Monograph, Oregon State College, Corvallis, Studies in Entomology 2, 1–208.

Chapman, J.W., 1910. The introduction of a European scolytid (the smaller elm bark-beetle, *Scolytus multistriatus* Marsham) into Massachusetts. Psyche 17, 63–68.

Chilahsaeva, E.A., 2010. Genus *Polygraphus* Erichson, 1836 (Coleoptera, Scolytidae): species of Moscow region fauna survey. Bulletin of the Moscow Society of Naturalists 115, 48–50.

China, W.E., 1962. Scolytus Geoffroy, 1762 (Insecta, Coleoptera); proposed validation under the plenary powers. Z.N. (S) 81. Bulletin of Zoological Nomenclature 19, 3–8.

China, W.E., 1963. Opinion 683. *Scolytus* Geoffroy, 1762 (Insecta, Coleoptera); validated under the plenary powers. Bulletin of Zoological Nomenclature 20, 416–417.

Choo, H.Y., 1983. Taxonomic studies on the Platypodidae and Scolytidae (Coleoptera) from Korea. Dissertation, Seoul National University, Seoul, Korea, 128 p.

Cibrián Tovar, D., Ebel, B.H., Yates III., H.O., Mendéz Montiel, J.T., 1986. Cone and seed insects of the Mexican Conifers. USDA Forest Service Southeastern Station, Asheville, North Carolina, 110 p.

Cibrián Tovar, D., Mendéz Montiel, J.T., Campos Bolaños, R., Yates III., H.O., Flores Lara, J., 1995. Insectos forestales de Mexico (Forest insects of Mexico). North American Forestry Commission, FAO, Publication 6, 453.

Cognato, A.I., Gillette, N.E., Campos Bolaños, R., Sperling, F.A.H., 2005. Mitochondrial phylogeny of pine cone beetles (Scolytinae, *Conophthorus*) and their affiliation with geographic area and host. Mol. Phylogenet. Evol. 36, 494–508.

Cognato, A.I., Sun, J.H., 2007. DNA based cladograms augment the discovery of a new Ips species from China (Coleoptera: Curculionidae: Scolytinae). Cladistics 23, 539–551.

Cognato, A.I., Rubinoff, D., 2008. New exotic ambrosia beetles found in Hawaii (Curculionidae: Scolytinae: Xyleborina). Coleopterists Bulletin 62, 421–424.

Cognato, A.I., Hulcr, J., Dole, S.A., Jordal, B.H., 2011. Phylogeny of haplo-diploid, fungus-growing ambrosia beetles (Curculionidae: Scolytinae: Xyleborini) inferred from molecular and morphological data. Zool. Scr. 40, 174–186.

Cranshaw, W., 2011. Recently recognized range extensions of the walnut twig beetle, *Pityophthorus juglandis* Blackman (Coleoptera: Curculionidae: Scolytinae), in the Western United States. Coleopterists Bulletin 65, 48–49.

Dallara, P.L., 1997. Studies on the distribution, interspecific relationships, host range, and chemical ecology of *Pityophthorus* spp. (Coleoptera: Scolytidae) and selected insect associates, and their associations with *Fusarium subglutinans* f. sp. *pini* in central coastal California. PhD thesis, University of California, Berkeley.

Dallara, P.L., Seybold, S.J., Meyer, H., Tolasch, T., Francke, W., Wood, D.L., 2000. Semiochemicals from three species of *Pityophthorus* (Coleoptera: Scolytidae): identification and field response. Can. Entomol. 132, 889–906.

Dallara, P.L., Flint, M.F., Seybold, S.J., 2012. An analysis of the larval instars of the walnut twig beetle, *Pityophthorus juglandis* Blackman (Coleoptera: Scolytidae), in northern California black walnut, *Juglans hindsii*, and a new host record for *Hylocurus hirtellus*. Pan-Pac. Entomol. 88, 248–266.

Daterman, C.E., Rudinsky, J.A., Nagel, W.P., 1965. Flight patterns of bark and timber beetles associated with coniferous forests of Western Oregon. Oregon State University Agricultural Experiment Station Technical Bulletin 87, 46.

de Groot, P., Harvey, G.T., Roden, P.M., 1992. Genetic divergence among Eastern North American cone beetles, *Conophthorus* (Coleoptera: Scolytidae). Can. Entomol. 124, 189–199.

Deyrup, M., Kirkendall, L.R., 1983. Apparent parthenogenesis in *Pityophthorus puberulus* (Coleoptera: Scolytidae). Ann. Entomol. Soc. Am. 76, 400–402.

Doane, R.W., Gilliland, O.J., 1929. Three California ambrosia beetles. J. Econ. Entomol. 22, 915–921.

Doane, R.W., VanDyke, E.C., Chamberlin, W.J., Burke, H.E., 1936. Forest insects, a textbook for the use of students in forest schools, colleges, and universities, and for forest workers. McGraw-Hill Co., New York

Dole, S.A., Beaver, R.A., 2008. A review of the Australian species of *Xylosandrus* Reitter (Coleoptera: Curculionidae: Scolytinae). Coleopterists Bulletin 62, 481–492.

Dole, S.A., Jordal, B.H., Cognato, A.I., 2010. Polyphyly of *Xylosandrus* Reitter inferred from nuclear and mitochondrial genes (Coleoptera: Curculionidae: Scolytinae). Mol. Phylogenet. Evol. 54, 773–782.

Dole, S.A., Cognato, A.I., 2010. Phylogenetic revision of *Xylosandrus* Reitter (Coleoptera: Curculionidae: Scolytinae: Xyleborina). Series 4, Proc. Calif. Acad. Sci. 61, 451–545.

Edson, L.J., 1967. An annotated and illustrated key to the species of the genus *Scolytus* (Coleoptera, Scolytidae) attacking coniferous trees of the Nearctic region. Masters thesis, Humboldt State College, California.

Eichhoff, W.J., 1864. Uber die Mundtheile und die Fuhlerbildung der Europäischen Xylophagi sens strict. Dtsch. Entomol. Z. 8, 17–46.

Equihua-Martinez, A., Furniss, M.M., 2009. Taxonomic status of *Scolytus opacus* and *Scolytus abietis* (Coleoptera: Curculionidae: Scolytinae): a comparative study. Ann. Entomol. Soc. Am. 102, 597–602.

Erichson, W.F., 1836. Systematische Auseinandersetzung der familie der Borkenkäfer (Bostrichidae). Archiv für Naturgeschichte 2, 45–65.

Eskalen, A., Gonzalez, A., Wang, D.H., Twizeyimana, M., Mayorquin, J.S., Lynch, S.C., 2012. First report of a *Fusarium* sp. and its vector tea shot hole borer (*Euwallacea fornicatus*) causing *Fusarium* dieback on avocado in California. Plant Dis. 96, 1070.

Faccoli, M., Zanocco, D., Battisti, A., Masutti, L., 1998. Chiave semplificata per la determinazione degli *Scolytus* Geoffroy (Coleoptera Scolytidae) Italiani viventi sugli Olmi. Redia 81, 183–197.

Farris, S.H., 1963. Ambrosia fungus storage in two species of *Gnathotrichus* Eichhoff (Coleoptera: Scolytidae). Can. Entomol. 95, 257–259.

Farris, S.H., 1965. Repositories of symbiotic fungus in ambrosia beetle *Monarthrum scutellare* Lec (Coleoptera: Scolytidae). Proc. Entomol. Soc. B. C. 62, 30–33.

Farris, S.H., 1969. Occurrence of mycangia in the bark beetle *Dryocoetes confusus* (Coleoptera: Scolytidae). Can. Entomol. 101, 527–532.

Ferrari, J.A., 1868. Nachtrage, Berichtigungen und Aufklarungen uber zweifehalf gebiebene Arten. *In* "Die forst- und baumzuchtschädlichen Borkenkäfer (Tomicides Lac)" Dtsch. Entomol. Z. 12, 251–258.

Flores, L.J., Bright, D.E., 1987. A new species of *Conophthorus* from Mexico: descriptions and biological notes (Coleoptera: Scolytidae). Coleopterists Bulletin 41, 181–184.

Formby, J.P., Schiefer, T.L., Riggins, J.J., 2012. First records of *Xyleborus glabratus* (Coleoptera: Curculionidae) in Alabama and in Harrison County. Mississippi. Fla. Ent. 95, 192–193.

Fraedrich, S.W., 2008. California laurel is susceptible to laurel wilt caused by *Raffaelea lauricola*. Plant Dis. 92, 1469.

Fraedrich, S.W., Harrington, T.C., Rabaglia, R.J., Ulyshen, M.D., Mayfield III, A.E., Hanula, J.L., et al., 2008. A fungal symbiont of the redbay ambrosia beetle causes a lethal wilt in redbay and other Lauraceae in the southeastern United States. Plant Dis. 92, 215–224.

Francke-Grossmann, H., 1959. Beiträge zur kenntnis der übertragungsweise von planzenkrankheiten durch käfer. Proceedings of the International Congress of Botany 4, 805–809.

French, J.R.J., Roeper, R.A., 1972. Observations of *Trypodendron rufitarsis* (Coleoptera: Scolytidae) and its primary symbiotic fungus, *Ambrosiella ferruginea*. Ann. Entomol. Soc. Am. 65, 282.

Freeman, S., Sharon, M., Maymon, M., Mendel, Z., Protasov, A., Aoki, T., et al., 2013. *Fusarium euwallaceae* sp. nov—a symbiotic fungus of *Euwallacea* sp., an invasive ambrosia beetle in Israel and California. Mycologia 105, 1595–1606.

Funk, A., 1970. Fungal symbionts of the ambrosia beetle *Gnathotrichus sulcatus*. Can. J. Bot. 48, 1445–1448.

Furniss, M.M., 1997. *Conophthorus ponderosae* (Coleoptera: Scolytidae) infesting lodgepole pine cones in Idaho. Environ. Entomol. 26, 855–858.

Furniss, M.M., Johnson, J.B., 2002. Field guide to the bark beetles of Idaho and adjacent regions. University of Idaho, Moscow.

Furniss, M.M., Kegley, S.J., 2006. Observations on the biology of *Dryocoetes betulae* (Coleoptera: Curculionidae) in paper birch in Northern Idaho. Environ. Entomol. 35, 907–911.

Furniss, R.L., Carolin, V.M., 1977. Western forest insects. United States Department of Agriculture, Forest Service Miscellaneous Publication 1339, 654.

Gadd, C.H., Loos, C.A., 1947. The ambrosia fungus of *Xyleborus fornicatus* Eich. Trans. Br. Mycol. Soc. 30, 13–18.

Garonna, A.P., Dole, S.A., Saracino, A., Mazzoleni, S., Cristinzio, G., 2012. First record of the black twig borer *Xylosandrus compactus* (Eichhoff) (Coleoptera: Curculionidae, Scolytinae) from Europe. Zootaxa 3251, 64–68.

Geoffroy, E.L., 1762. Histoire abrégée des insects qui se trouvent aux environs de Paris, dans laquelle ces animaux sont ranges suivant un ordre méthodique. Paris. 523 p.

Godwin, P.A., Odell, T.M., 1965. The life history of the white-pine cone beetle, *Conophthorus coniperda*. Ann. Entomol. Soc. Am. 58, 213–219.

Goeden, R.D., Norris Jr., D.M., 1964a. Attraction of *Scolytus quadrispinosus* (Coleoptera: Scolytidae) to *Carya* spp. for oviposition. Ann. Entomol. Soc. Am. 57, 141–146.

Goeden, R.D., Norris Jr., D.M., 1964b. Some biological and ecological aspects of the dispersal flight of *Scolytus quadrispinosus* (Coleoptera: Scolytidae). Ann. Entomol. Soc. Am. 57, 743–749.

Goeden, R.D., Norris Jr., D.M., 1965a. The behavior of *Scolytus quadrispinosus* (Coleoptera: Scolytidae) during the dispersal flight as related to its host specificities. Ann. Entomol. Soc. Am. 58, 249–252.

Goeden, R.D., Norris Jr., D.M., 1965b. Some biological and ecological aspects of ovipositional attack in *Carya* spp. by *Scolytus quadrispinosus* (Coleoptera: Scolytidae). Ann. Entomol. Soc. Am. 58, 771–776.

Graves, A.D., Coleman, T.W., Flint, M.L., Seybold, S.J., 2009. Walnut twig beetle and thousand cankers disease: field identification guide. UC-IPM Website Publication. Available online: http://www.ipm.ucdavis.edu/PDF/MISC/thousand_cankers_field_guide.pdf. Last accessed: May 20, 2014.

Hansen, E.M., 1996. Western balsam bark beetle, *Dryocoetes confusus* Swaine, flight periodicity in northern Utah. Great Basin Nat. 56, 348–359.

Harrington, T.C., Aghayeva, D.N., Fraedrich, S.W., 2010. New combinations in *Raffaelea*, *Ambrosiella*, and *Hyalorhinocladiella*, and four new species from the redbay ambrosia beetle, *Xyleborus glabratus*. Mycotaxon 111, 337–361.

Harrington, T.C., McNew, D., Mayers, C., Fraedrich, S.W., Reed, S.E., 2014. *Ambrosiella roeperi* sp. nov. is the mycangial symbiont of the granulate ambrosia beetle, *Xylosandrus crassiusculus*. Mycologia 106, 835–845.

Harris, J.L., Frank, M., Johnson, S. (Eds.), 2001. Forest insect and disease conditions in the Rocky Mountain region 1997–1999. USDA Forest Service, Rocky Mountain Region, Renewable Resources Denver, CO.

Hedlin, A.F., Yates III., H.O., Cibrián Tovar, D., Ebel, B.H., Koerber, T.W., Merkel, E.P., 1980. Cone and seed insects of North American conifers. United States Department of Agriculture Forest Service, Washington DC, 122 p.

Henson, W.R., 1962. Laboratory studies on the adult behavior of *Conophthorus coniperda* (Coleoptera: Scolytidae). III. Flight. Ann. Entomol. Soc. Am. 55, 524–530.

Hilton, D.F.J., 1968. A review of the genus *Polygraphus*. Univ. Kans. Sci. Bull. 47, 21–44.

Hoebeke, E.R., Rabaglia, R.J., 2008. *Xyleborus seriatus* Blandford (Coleoptera: Curculionidae: Scolytinae), an Asian ambrosia beetle new to North America. Proc. Entomol. Soc. Wash. 110, 470–476.

Hoffman, C.H., 1942. Annotated list of elm insects in the United States. Misc. Publ. U.S. Dep. Agric. 466, 20 p.

Hopkins, A.D., 1899. Report on investigations to determine the cause of unhealthy conditions of the spruce and pine from 1880–1883. Bull. W. Va. Agric. Exp. Stn. 56, 197–461.

Hopkins, A.D., 1914. List of generic names and their type-species in the coleopterous superfamily Scolytoidea. Proc. U.S. Natl. Mus. 48, 115–136.

Hopkins, A.D., 1915. A new genus of Scolytoid beetles. J. Wash. Acad. Sci. 5, 429–433.

Hubbard, H.G., 1897. The ambrosia beetles of the United States. U.S. Dept. Agr. Div. Entomol. Bull. 7, 9–30.

Hulcr, J., Cognato, A.I., 2010. New genera of Palaeotropical Xyleborini (Coleoptera: Curculionidae: Scolytinae) based on congruence between morphological and molecular characters. Zootaxa 2717, 1–33.

Hulcr, J., Smith, S.M., 2010. Xyleborini ambrosia beetles. An identification tool to the world genera. Available online: http://idtools.org/id/wbb/xyleborini/index.htm. Last accessed: May 20, 2014.

Hulcr, J., Dunn, R.R., 2011. The sudden emergence of pathogenicity in insect-fungus symbiosis threatens naïve forest ecosystems. Proc. R. Soc. B 278, 2866–2873.

Hulcr, J., Cognato, A.I., 2013. Xyleborini (Curculionidae: Scolytinae) of New Guinea—taxonomic monograph. Thomas Say Publications in Entomology, Entomological Society of America, 176 p.

Hulcr, J., Lou, Q.-Z., 2013. The redbay ambrosia beetle (Coleoptera: Curculionidae) prefers Lauraceae in its native range: records form the Chinese National Insect Collection. Fla. Entomol. 96, 1595–1597.

Hulcr, J., Mogia, M., Isua, B., Novotny, V., 2007a. Host specificity of ambrosia and bark beetles (Col., Curculionidae: Scolytinae and Platypodinae) in a New Guinea rain forest. Ecol. Entomol. 32, 762–772.

Hulcr, J., Beaver, R., Dole, S., Cognato, A., 2007b. Cladistic review of xyleborine generic taxonomic characters (Coleoptera: Curculionidae: Scolytinae). Syst. Entomol. 32, 568–584.

Humble, L.M., Allen, E.A., 2006. Forest biosecurity: alien invasive species and vectored organisms. Can. J. Plant Pathol. 28, S256–S269.

Inch, S., Ploetz, R., Held, B., Blanchette, R., 2012. Histological and anatomical responses in avocado, *Persea americana*, induced by the vascular wilt pathogen, *Raffaelea lauricola*. Botany 90, 627–635.

Ito, M., Kajimura, H., 2008. Genetic structure of Japanese populations of an ambrosia beetle, *Xylosandrus germanus* (Curculionidae: Scolytinae). J. Entomol. Sci. 11, 375–383.

Jacobi, W.R., Koski, R.D., Negrón, J.F., 2013. Dutch elm disease pathogen transmission by the banded elm bark beetle *Scolytus schevyrewi*. Forest Pathol. 43, 232–237.

Jordal, B.H., Kirkendall, L.R., 1998. Ecological relationships of a guild of tropical beetles breeding in *Cecropia* petioles in Costa Rica. J. Trop. Ecol. 14, 153–176.

Jordal, B.H., Kambestad, M., 2014. DNA barcoding of bark and ambrosia beetles reveals excessive NUMTs and consistent east-west divergence across Palearctic forests. Mol. Ecol. Resour. 14, 7–17.

Jordal, B.H., Beaver, R.A., Kirkendall, L.R., 2001. Breaking taboos in the tropics: incest promotes colonization by wood-boring beetles. Global Ecology & Biogeography 10, 345–357.

Jordal, B.H., Beaver, R.A., Normark, B.B., Farrell, B.D., 2002. Extraordinary sex ratios and the evolution of male neoteny in sib-mating *Ozopemon* beetles. Biol. J. Linn. Soc. 75, 353–360.

Kasson, M.T., O'Donnell, K., Rooney, A., Sink, S., Ploetz, R., Ploetz, J.N., et al., 2013. An inordinate fondness for *Fusarium*: phylogenetic diversity of fusaria cultivated by ambrosia beetles in the genus *Euwallacea* on avocado and other plant hosts. Fungal Genet. Biol. 56, 147–157.

Kasson, M.T., Short, D.P., O'Neal, E.S., Subbarao, K.V., Davis, D.D., 2014a. Comparative pathogenicity, biocontrol efficacy, and multilocus sequence typing of *Verticillium nonalfalfae* from the invasive *Ailanthus altissima* and other hosts. Phytopathology 104, 282–292.

Kasson, M.T., O'Neal, E.S., Davis, D.D., 2014b. Expanded host range testing for *Verticillium nonalfalfae*: potential biocontrol agent against the invasive *Ailanthus altissima*. Plant Dis, submitted.

Kawasaki, Y., Ito, M., Miura, K., Kajimura, H., 2010. Superinfection of five *Wolbachia* in the alnus ambrosia beetle, *Xylosandrus germanus* (Blandford) (Coleoptera: Curculionidae). Bull. Entomol. Res. 100, 231–239.

Keen, F.P., 1938. Insect enemies of western forests. U.S. Dep. Agric., Misc. Publ. 273.

Keen, F.P., 1958. Cone and seed insects of western forest trees. U.S. Dep. Agric., Tech. Bull. 1169.

Kendra, P.E., Montgomery, W.S., Niogret, J., Peña, J.E., Capinera, J.L., Brar, G., et al., 2011. Attraction of redbay ambrosia beetle (Coleoptera: Curculionidae: Scolytinae) to avocado, lychee, and essential oil lures. J. Chem. Ecol. 37, 932–942.

Kendrick, W.B., Molnar, A.C., 1965. A new *Ceratocystis* and its *Verticicladiella* imperfect state associated with the bark beetle *Dryocoetes confusus* on *Abies lasiocarpa*. Can. J. Bot. 43, 39–43.

Kinzer, H.B., Ridgill, B.J., Watts, J.G., 1970. Biology and cone attack behavior on *Conophthorus ponderosae* in Southern New Mexico (Coleoptera: Scolytidae). Ann. Entomol. Soc. Am. 63, 795–798.

Kirby, W., 1837. Insects, Coleoptera. In: Richardson, J. (Ed.), In: Fauna Boreali-Americana: or the Zoology of the Northern Parts of British America"Vol. 4. Josiah Fletcher, Norwich, p. 249.

Kirkendall, L.R., 1983. The evolution of mating systems in bark and ambrosia beetles. J. Linn. Soc. London Zool. 77, 293–352.

Kirkendall, L.R., 1993. Ecology and evolution of biased sex ratios in bark and ambrosia beetles. In: Wrensch, D.L., Ebbert, M.A. (Eds.), Evolution and Diversity of Sex Ratio: Insects and Mites. Chapman and Hall, New York, pp. 235–345.

Kirkendall, L.R., Ødegaard, F., 2007. Ongoing invasions of old-growth tropical forests: establishment of three incestuous beetle species in Central America (Curculionidae, Scolytinae). Zootaxa 1588, 53–62.

Kirkendall, L.R., Faccoli, M., 2010. Bark beetles and pinhole borers (Curculionidae, Scolytinae, Platypodinae) alien to Europe. Zookeys 56, 227–251.

Kirkendall, L.R., Dal Cortivo, M., Gatti, E., 2008. First record of the ambrosia beetle, *Monarthrum mali* (Curculionidae, Scolytinae) in Europe. Journal of Pest Science 81, 175–178.

Knížek, M., 2011. Scolytinae. In: Löbl, I., Smetana, A. (Eds.), Catalogue of Palaearctic Coleoptera, Vol. 7, Curculionoidea I. Apollo Books, Stenstrup, pp. 204–251.

Kolařík, M., Freeland, E., Utley, C., Tisserat, N., 2011. *Geosmithia morbida* sp. nov., a new phytopathogenic species living in symbiosis with the walnut twig beetle (*Pityophthorus juglandis*) on *Juglans* in USA. Mycologia 103, 325–332.

Kostovcik, M., Bateman, C., Kolarik, M., Stelinski, L., Jordal, B.H., Hulcr, J., 2014. The ambrosia symbiosis is specific in some species and promiscuous in others: evidence from community pyrosequencing. ISME J. in press.

Kovach, D.A., Gorsuch, C.S., 1985. Survey of ambrosia beetle species infesting South Carolina orchards and a taxonomic key for the most common species. J. Agric. Entomol. 2, 238–247.

Krivolutskaya, G.O., 1996. 113. Cem. Scolytidae. In: Key to the Insects of the Russian Far East. Vol. 3, Coleoptera, Part 3. Dal'nauka, Vladivostok, pp. 312–373.

Kühnholz, S., Borden, J.H., Uzunovic, A., 2001. Secondary ambrosia beetles in apparently healthy trees: adaptations, potential causes and suggested research. Integr. Pest Manage. Rev. 6, 209–219.

LaBonte, J.R., 2010. The banded elm bark beetle, *Scolytus schevyrewi* Semenov (Coleoptera, Curculionidae, Scolytinae) in North America: a taxonomic review and modifications to the Wood (1982) key to the species of *Scolytus* Geoffroy in North and Central America. Zookeys 56, 207–218.

de Laclos, E., Mouy, C., Strenna, L., Agou, P., 2004. Les Scolytes de Bourgogne (Coléoptères: Scolytidae—Platypodidae). Société d'Histoire Naturelle et des Amis du Muséum d'Autun, 239 p.

Lanier, G.N., Gore, W.E., Pearce, G.T., Peacock, J.W., Silverstein, R.M., 1977. Response of the European elm bark beetle. *Scolytus multistriatus* (Coleoptera: Scolytidae), to isomers and components of its pheromone. J. Chem. Ecol. 3, 1–8.

LeConte, J.L., 1868. Notes and appendix. In: Zimmermann, C. (Ed.), In: Synopsis of the Scolytidae of America North of Mexico"Vol. 2. Transactions of the American Entomological Society, pp. 141–178.

LeConte, J.L., 1876. Family IX, Scolytidae. In: LeConte, J.L., Horn, G.H. (Eds.), The Rhynchophora of America North of Mexico.pp. 341–391, Proc. Am. Philos. Soc. 15.

LeConte, J.L., Horn, G.H., 1883. Classification of the Coleoptera of North America. Smithson. Misc. Collect. 507, 1–567.

Lee, J.C., Negrón, J.F., McElwey, S.J., Witcosky, J.J., Seybold, S.J., 2006. Banded elm bark beetle—*Scolytus schevyrewi*. *United States Department of Agriculture Forest Service,* Pest Alert R2-PR-01-06.

Lee, J.C., Hamud, S.M., Negrón, J.F., Witcosky, J.J., Seybold, S.J., 2010. Semiochemical-mediated flight strategies of two invasive elm bark beetles: a potential factor in competitive displacement. Environ. Entomol. 39, 642–652.

Lee, J.C., Negrón, J.F., McElwey, S.J., Williams, L., Witcosky, J.J., Popp, J.B., Seybold, S.J., 2011. Biology of the invasive banded elm bark beetle (Coleoptera: Scolytidae) in the western United States. Ann. Entomol. Soc. Am. 104, 705–717.

Lekander, B., 1968. Scandinavian bark beetle larvae, description and classification. *Institutionen for Skogszoologi, Skogshogskolan, Rapporter och Uppsater* 4, 186 p.

Livingston, R.L., Berryman, A.A., 1972. Fungus transport structure in the fir engraver, *Scolytus ventralis* (Coleoptera: Scolytidae). Can. Entomol. 104, 1793–1800.

Lombardero, M.J., Novoa, F., 1994. Datos faunísticos sobre escolítidos ibéricos (Coleoptera: Scolytidae). Boletín de la Asociación Española de Entomología 18, 181–186.

López, S., Quero, C., Iturrondobeitia, J.C., Guerrero, Á., Goldarazena, A., 2011. Evidence for (E)-pityol as an aggregation pheromone of *Pityophthorus pubescens* (Coleoptera: Curculionidae: Scolytinae). Can. Entomol. 143, 447–454.

Lowe, R.E., Giese, R.L., McManus, M.L., 1967. Mycetangia of the ambrosia beetle *Monarthrum fasciatum.* J. Invertebr. Pathol. 9, 451–458.

Macías-Sámano, J.E., Borden, J.H., Gries, R., Pierce Jr., H.D., Gries, G., King, G.G.S., 1998a. Lack of evidence for pheromone-mediated secondary attraction in the fir engraver, *Scolytus ventralis,* (Coleoptera: Scolytidae). J. Entomol. Soc. B. C. 95, 117–125.

Macías-Sámano, J.E., Borden, J.H., Gries, R., Pierce Jr., H.D., Gries, G., King, G.G.S., 1998b. Primary attraction of fir engraver, *Scolytus ventralis.* J. Chem. Ecol. 24, 1049–1075.

Maiti, P.K., Saha, N., 2004. Fauna of India and the adjacent countries. Scolytidae: Coleoptera (Bark and ambrosia beetles) Volume: I (Part 1), Introduction and tribe Xyleborini. Zool. Soc. India, 268 p.

Maiti, P.K., Saha, N., 2009. Fauna of India and the Adjacent Countries. Scolytidae: Coleoptera (Bark and ambrosia beetles) Volume: I (Part 2). Zool. Soc. India, 245 p.

Mandelshtam, M.Y., 2001. A new species of bark-beetles (Coleoptera: Scolytidae) from Russian Far East. Far Eastern Entomologist 105, 11–12.

Mandelshtam, M.Y., Petrov, A.V., 2010a. *Scolyus stepheni* sp. n.—a new species of bark-beetle (Coleoptera, Curculionidae, Scolytinae) from Northern India with a key to Indian *Scolytus* Geoffroy, 1762 species. Zookeys 56, 171–178.

Mandelshtam, M.Y., Petrov, A.V., 2010b. Description of new *Dryocoetes* (Coleoptera, Curculionidae, Scolytinae) species from Afghanistan and Northern India and redescription of *Scolytoplatypus kunala* Strohm. Zookeys 56, 179–190.

Mannerheim, C.G.v., 1843. Beitrag zur Käfer-fauna der Aleutischen Inseln, der Insel Sitkha und Neu-Californiens. Bull. Soc. Imp. Nat. Moscou 16, 175–314.

Mannerheim, C.G.v., 1852. Zweiter Nachtrag zur Käfer-fauna der Nord-Amerikanischen Laender des Russischen Reiches Bull. Soc. Imp. Nat. Moscou 25, 283–387.

Mannerheim, C.G.v., 1853. Dritter Nachtrag zur Käfer-Fauna der Nord-Amerikanischen Laender des Russischen Reiches. Bull. Soc. Imp. Nat. Moscou 26, 95–273.

Massey, C.L., 1964. The nematode parasites and associates of the fir engraver beetle, *Scolytus ventralis* LeConte, in New Mexico. J. Insect Pathol. 6, 133–155.

Mathers, W.G., 1931. The biology of Canadian bark beetles, the seasonal history of *Dryocoetes confusus* SW. Can. Entomol. 63, 247–248.

McGhehey, J.H., Nagel, W.P., 1969. The biologies of *Pseudohylesinus tsugae* and *P. grandis* (Coleoptera: Scolytidae) in western hemlock. Can. Entomol. 101, 269–279.

McLean, J.A., 1985. Ambrosia beetles: a multimillion dollar degrade problem of sawlogs in coastal British Columbia. Forest. Chron. 61, 295–298.

McLean, J.A., Borden, J.H., 1975. *Gnathotrichus sulcatus* attack and breeding in freshly sawn lumber. J. Econ. Entomol. 68, 605–606.

McMillin, J.D., Allen, K.K., Long, D.F., Harris, J.L., Negrón, J.F., 2003. Effects of western balsam bark beetle on spruce-fir forests of north-central Wyoming. West. J. Appl. Forest. 18, 259–266.

McMullen, L.H., Atkins, M.D., 1959. Life-history and habits of *Scolytus tsugae* (Swaine) (Coleoptera: Scolytidae) in the interior of British Columbia. Can. Entomol. 91, 416–426.

McMullen, L.H., Atkins, M.D., 1962. Life-history and habits of *Scolytus unispinosus* LeConte (Coleoptera: Scolytidae) in the interior of British Columbia. Can. Entomol. 94, 17–25.

McMullen, L.H., Fiddick, R.L., Wood, R.O., 1981. Bark beetles, *Pseudohylesinus* spp. (Coleoptera: Scolytidae), associated with amabalis fir defoliated by Neodiprion. J. Entomol. Soc. B. C. 84, 43–45.

McPherson, B.A., Erbilgin, N., Bonello, P., Wood, D.L., 2013. Fungal species assemblages associated with *Phytophthora ramorum*-infected coast live oaks following bark and ambrosia beetle colonization in northern California. For. Ecol. Manage. 291, 30–42.

McPherson, J.E., Stehr, F.W., Wilson, L.F., 1970. A comparison between *Conophthorus* shoot-infesting beetles and *Conophthorus resinosae* (Coleoptera: Scolytidae) II. Reciprocal host and resin toxicity tests; with description of a new species. Can. Entomol. 102, 1016–1022.

Menard, K.L., Cognato, A.I., 2007. Mitochondrial haplotypic diversity of pine cone beetles (Scolytinae: *Conophthorus*) collected on food sources. Environ. Entomol. 36, 962–966.

Mendel, Z., Protasov, A., Sharon, M., Zveibil, A., Yehuda, S.B., O'Donnell, K., et al., 2012. An Asian ambrosia beetle *Euwallacea fornicatus* and its novel symbiotic fungus *Fusarium* sp. pose a serious threat to the Israeli avocado industry. Phytoparasitica 40, 235–238.

Mercado-Vélez, J.E., Negrón, J.F., 2014. Revision of the new world species of *Hylurgops* LeConte, 1876 with the description of a new genus in the Hylastini (Coleoptera: Scolytinae) and comments on some Palearctic species. Zootaxa 3785, 301–342.

Michalski, J., 1973. Revision of the Palearctic species of the genus *Scolytus* Geoffroy (Coleoptera, Scolytidae). Polska Akademia Nauk Zaklad Zoologii Systematycznej i Doswiadczalnej, Warsaw.

Miller, D.R., Pierce Jr., H.D., deGroot, P., Jeans-Williams, N., Bennett, R., Borden, J.H., 2000. Sex pheromone of *Conophthorus ponderosae* (Coleoptera: Scolytidae) in a coastal stand of western white pine (Pinaceae). Can. Entomol. 132, 243–245.

Miller, J.M., 1915. Cone beetles: injury to sugar pine and western yellow pine. U.S. Dept. Agric. Bull 243, 12.

Milligan, R.H., 1970. Overseas wood- and bark-boring insects intercepted at New Zealand ports. N. Z. For. Serv. Tech. Pap. 57, 80 p.

Mifsud, D., Knížek, M., 2009. The bark beetles (Coleoptera: Scolytidae) of the Maltese Islands (Central Mediterranean). Bulletin of the Entomological Society of Malta 2, 25–52.

Moeck, H.A., 1970. Ethanol as the primary attractant for the ambrosia beetle *Trypodendron lineatum* (Coleoptera: Scolytidae). Can. Entomol. 102, 985–995.

Montecchio, L., Faccoli, M., 2013. First record of thousand cankers disease *Geosmithia morbida* and walnut twig beetle *Pityophthorus juglandis* on *Juglans nigra* in Europe. Plant Dis. 98, 696.

Montgomery, M.E., Wargo, P.M., 1983. Ethanol and other host-derived volatiles as attractants to beetles that bore into hardwood. J. Chem. Ecol. 9, 181–190.

Murayama, J.J., 1956. Polygraphinae (Coleoptera, Ipidae) from the northern half of the Far East. Yamaguti University, Faculty of Agriculture, Bulletin 7, 275–292.

Murayama, J.J., 1957. Studies in the scolytid-fauna of the northern half of the Far East, II: Xyloterinae. Yamaguti University, Faculty of Agriculture, Bulletin 8, 569–586.

Negrón, J.F., Witcosky, J.J., Cain, R.J., LaBonte, J.R., Durr II., D.A., McElwey, S.J., et al., 2005. The banded elm bark beetle: A new threat to elms in North America. Am. Entomol. 51, 84–94.

Negrón, J.F., Popp, J.B., 2009. The flight periodicity, attack patterns, and life history of *Dryocoetes confusus* Swaine (Coleoptera: Curculionidae: Scolytinae), the western balsam bark beetle, in north central Colorado. Great Basin Nat. 69, 447–458.

Newton, L., Fowler, G., 2009. Pathway assessment: *Geosmithia* sp. and *Pityophthorus juglandis* Blackman movement from the western into the eastern United States. United States Department of Agriculture, Animal and Plant Health Inspection Service. Available online: http://mda.mo.gov/plants/pdf/tc_pathwayanalysis.pdf. Last accessed: May 20, 2014.

Nobuchi, A., 1973. Studies on Scolytidae XI *Scolytus* Geoffroy of Japan (Coleoptera). Japan. Bulletin of the Government Forest Experiment Station 258, 13–27.

Nobuchi, A., 1979. Studies on Scolytidae, XVIII. Bark beetles of tribe Polygraphini in Japan (Coleoptera, Scolytidae). Bull. Forestry and Forest Prod. Res. Inst. 308, 1–16.

Oliver, J.B., Mannion, C.M., 2001. Ambrosia beetle (Coleoptera: Scolytidae) species attacking chestnut and captured in ethanol-baited traps in middle Tennessee. Environ. Entomol. 30, 909–918.

Pearce, G.T., Gore, W.E., Silverstein, R.M., Peacock, J.W., Cuthbert, R.A., Lanier, G.N., Simeone, J.B., 1975. Chemical attractants for the smaller European elm bark beetle *Scolytus multistriatus* (Coleoptera: Scolytidae). J. Chem. Ecol. 1, 115–124.

Petrov, A.V., Mandelshtam, M.Y., 2010. New data on Neotropical *Scolytus* Geoffroy, 1762 with descriptions of five new species from Peru (Coleoptera, Curculionidae, Scolytinae). Zookeys 56, 65–104.

Petrov, A.V., 2013. New data and synonymy of bark beetles of the genus *Scolytus* Geoffroy, 1792 from Russia and adjacent countries. Lesnoy Vestnik, Bulletin of Moscow State Forest University 98, 39–47.

Pfeffer, A., 1976. Revision der palaarktischen Arten der Gattung *Pityophthorus* Eichhoff (Coleoptera, Scolytidae). Acta Entomologica Bohemoslavaca 73, 324–342.

Pfeffer, A., 1994a. 91. Familie: Scolytidae. In: Lohse, G.H., Licht, W.H. (Eds.), Die Käfer Mitteleuropas. 3. Supplementband mit Katalogteil". Goecke and Evers, Krefeld, pp. 153–180.

Pfeffer, A., 1994b. Zentral- und Westpaläarktische Borken- und Kernkäfer (Coleoptera, Scolytidae, Platypodidae). Entomologica Basiliensia 17, 5–310.

Ploetz, R.C., Pérez-Martínez, J.M.J., Smith, A., Hughes, M., Dreaden, T.J., Inch, S.A., Fu, Y., 2012. Responses of avocado to laurel wilt, caused by *Raffaelea lauricola*. Plant Path. 61, 801–808.

Prebble, M.L., Graham, K., 1957. Studies of attack by ambrosia beetles in softwood logs on Vancouver Island, British Columbia. Forest Sci. 3, 90–112.

Rabaglia, R.J., Dole, S.A., Cognato, A.I., 2006. Review of American Xyleborina (Coleoptera: Curculionidae: Scolytinae) occurring north of Mexico, with an illustrated key. Ann. Ent. Soc. Am. 99, 1034–1056.

Rabaglia, R.J., Duerr, D., Acciavatti, R., Ragenovich, I., 2008. Early detection and rapid response for non-native bark and ambrosia beetles. US Department of Agriculture Forest Service, Forest Health Protection.

Raffa, K.F., Berryman, A.A., 1987. Interacting selective pressures in conifer-bark beetle systems: a basis for reciprocal adaptations? Am. Nat. 129, 234–262.

Raffa, K.F., Aukema, B.H., Bentz, B.J., Carroll, A.L., Hicke, J.A., Turner, M.G., Romme, W.H., 2008. Cross-scale drivers of natural disturbances prone to anthropogenic amplification: the dynamics of bark beetle eruptions. Bioscience 58, 501–517.

Ranger, C.M., Reding, M.E., Persad, A.B., Herms, D.A., 2010. Ability of stress-related volatiles to attract and induce attacks by *Xylosandrus germanus* (Coleoptera: Curculionidae, Scolytinae) and other ambrosia beetles. Agr. Forest Entomol. 12, 177–185.

Ranger, C.M., Reding, M.E., Schultz, P.B., Oliver, J.B., 2012. Influence of flood-stress on ambrosia beetle host-selection and implications for their management in a changing climate. Agr. Forest Entomol. 15, 56–64.

Raske, A.G., Sutton, W.J., 1986. Decline and mortality of black spruce caused by spruce budworm defoliation and secondary organisms. Can. For. Serv. Inf. Rep. 29, N-X-236.

Reitter, E., 1913. Bestimmungs-tabelle der Borkenkäfer (Scolytidae) aus Europa und den angrenzenden Ländern. Wiener Entomologische Zeitung 32, 1–116.

Roeper, R.A., French, J.R.J., 1978. Observations on *Monarthrum dentiger* (Coleoptera: Scolytidae) and its primary symbiotic fungus *Ambrosiella brunnea* (Fungi: Imperfecta) in California. Pan-Pac. Entomol. 54, 68–69.

Roling, M.P., Kearby, W.H., 1974. Life stages and development of *Monarthrum fasciatum* (Coleoptera: Scolytidae) in dying and dead oak trees. Can. Entomol. 106, 1301–1308.

Roling, M.P., Kearby, W.H., 1975. Seasonal flight and vertical distribution of Scolytidae attracted to ethanol. Can. Entomol. 107, 1315–1320.

Roling, M.P., Kearby, W.H., 1977. Influence of tree diameter, aspect, and month killed on behavior of scolytids infesting black oaks. Can. Entomol. 109, 1235–1238.

Rosel, A., French, J.R.J., 1975. Dutch elm disease beetle in Australia. Nature 253, 305.

Rudinsky, J.A., Oester, P.T., Ryker, L.C., 1978. Gallery initiation and male stridulation of the polygamous spruce bark beetle *Polygraphus rufipennis*. Ann. Entomol. Soc. Am. 71, 317–321.

Ryker, L.C., Oester, P.T., 1982. *Pseudohylesinus nebulosus* (LeConte) (Col., Scolytidae): aggregation by primary attraction. Zeitschrift für Angewandte Entomologie 94, 377–382.

Sakamoto, J.M., Gordon, T.R., Storer, A.J., Wood, D.L., 2007. The role of *Pityophthorus* spp. as vectors of pitch canker affecting *Pinus radiata*. Can. Entomol. 139, 864–871.

Schaefer, C.H., 1962. Life history of *Conophthorus radiatae* (Coleoptera: Scolytidae) and its principal parasite *Cephalonomia utahensis* (Hymenoptera: Bethylidae). Ann. Entomol. Soc. Am. 55, 569–577.

Schedl, K.E., 1937. Scolytidae und Platypodidae—Zentral und Südamerikanische arten. Archivos do Instituto de Biologia Vegetal, Rio de Janeiro 3, 155–170.

Schedl, K.E., 1948. Bestimmungstabellen der Palaearktischen borkenkäfer, Teil III. Die Gattung *Scolytus* Geoffr. Zentralblatt für das Gesamtgebiet der Entomologie, Monographie 1, 1–67.

Schedl, K.E., 1951. Bestimmungstabellen der Palaearktischen borkenkäfer, V. Tribus Xyloterini. 98 Beitrag. Mitteilungen der Forstlichen Bundes-Versuchsanstalt Maria-Brunn 47, 74–100.

Schedl, K.E., 1955. Bestimmungstabellen Palaearktischer borkenkäfer, VII. Gattung *Polygraphus* Er. 102 Beitrag. Mitteilungen Munchener Entomologischen Gesellschaft, 44–45, 3–25.

Schedl, W., 1962. Ein beitrag zur kenntnis der Pilzübertragungsweise bei xylomycetophagen Scolytiden. Österreichische der Akademie der Wissenschaften mathematisch-naturwissenschaftliche Klasse 171, 363–387.

Schneider, I., Rudinsky, J.A., 1969. Mycetangial glands and their seasonal changes in *Gnathotrichus retusus* and *G. sulcatus*. Ann. Entomol. Soc. Am. 62, 39–43.

Seybold, S.J., Graves, A.D., Coleman, T.W., 2011. Walnut twig beetle: update on the biology and chemical ecology of a vector of an invasive fatal disease of walnut in the western U.S. In: McManus, K.A., Gottschalk, K.W. (Eds.), Proceedings. 2010 U.S. Department of Agriculture Interagency Research Forum on Invasive Species. General Technical Report NRS-P-75, U.S. Department of Agriculture, Forest Service, Northern Research Station, Newtown Square, PA, pp. 55–57.

Seybold, S.J., Dallara, P.L., Nelson, L.J., Graves, D.A., Hishinuma, S.M., Gries, R., 2012a. Methods of monitoring and controlling the walnut twig beetle, *Pityophthorus juglandis*, U.S. Patent Application No. 13/548,319.

Seybold, S.J., King, J.A., Harris, D.R., Nelson, L.J., Hamud, S.M., Chen, Y., 2012b. Diurnal flight response of the walnut twig beetle, *Pityophthorus juglandis* Blackman (Coleoptera: Scolytidae), to pheromone-baited traps in two northern California walnut habitats. Pan-Pac. Entomol. 88, 231–247.

Seybold, S.J., Coleman, T.W., Dallara, P.L., Dart, N.L., Graves, A.D., Pederson, L.A., Spichiger, S.-E., 2012c. Recent collecting reveals new state records and geographic extremes in the distribution of the walnut twig beetle, *Pityophthorus juglandis* Blackman (Coleoptera: Scolytidae), in the United States. Pan-Pac. Entomol. 88, 277–280.

Seybold, S.J., Haugen, D., Graves, A., 2013. Thousand cankers disease. United States Department of Agriculture Forest Service, Pest Alert NA-PR-02-10.

Simpson, L.J., 1929. The seasonal history of *Polygraphus rufipennis* Kirby. Can. Entomol. 61, 146–151.

Sittichaya, W., 2013. Xyleborine ambrosia beetles (Coleoptera: Curculionidae; Scolytinae, Xyleborini) pest of durian trees in southern Thailand: taxonomic and some ecological aspects. Prince of Songkla University, Songkla, Thailand, Dissertation.

Six, D., Stone, D.W., de Beer, Z.W., Woolfolk, S.W., 2009. *Ambrosiella beaveri* sp. nov., associated with an exotic ambrosia beetle, *Xylosandrus mutilatus* (Coleoptera: Curculionidae, Scolytinae), in Mississippi, USA. Antonie Van Leeuwenhoek 96, 17–29.

Smith, S.M., 2013. Phlyogenetics of the Scolytini (Coleoptera: Curculionidae: Scolytinae) and host-use evolution. Michigan State University, East Lansing, Dissertation.

Smith, S.M., Cognato, A.I., 2010. Notes on *Scolytus fagi* Walsh 1867 with the designation of a neotype, distribution notes and a key to *Scolytus* Geoffroy of America east of the Mississippi River (Coleoptera: Curculionidae: Scolytinae: Scolytini). Zookeys 56, 35–43.

Smith, S.M., Cognato, A.I., 2013. A new species of *Scolytus* Geoffroy, 1762 and taxonomic changes regarding Neotropical Scolytini (Coleoptera: Curculionidae: Scolytinae). Coleopterists Bulletin 67, 547–556.

Smith, S. M., and Cognato, A. I. A taxonomic monograph of Nearctic *Scolytus* Geoffroy (Coleoptera, Curculionidae, Scolytinae). *Zookeys*, in press

Stark, V.N., 1952. Koredy. Fauna SSSR, Zhestkokrylye 31. Akademiia Nauk SSSR, Zoologischeskii Institut (N.S.) 49, 462.

Stephens, J.F., 1830. Illustrations of British entomology; or, A synopsis of indigenous insects: Containing their generic and specific distinctions. In: Mandibulata Coleoptera, Vol. 3. London 374 pp.

Stilwell, A.R., Smith, S.M., Cognato, A.I., Martinez, M., Flowers, R.W., 2014. *Coptoborus ochromactonus*, n. sp. (Coleoptera: Curculionidae: Scolytinae), an emerging pest of cultivated balsa (*Ochroma pyramidale*) in Ecuador. J. Econ. Entomol. 107, 675–683.

Stock, A.J., 1991. The western balsam bark beetle, *Dryocoetes confusus* Swaine: impact and semiochemical based management. Simon Fraser University, Canada, MS thesis.

Stock, A.J., Borden, J.H., 1983. Secondary attraction in the western balsam bark beetle, *Dryocoetes confusus* (Coleoptera: Scolytidae). Can. Entomol. 115, 539–550.

Stokland, J.N., 2012. The saproxylic food web. In: Stokland, J.N., Sitonen, J., Jonsson, B.G. (Eds.), Biodiversity in Dead Wood. Cambridge University Press, Cambridge, pp. 29–57.

Storer, A.J., Wood, D.L., Gordon, T.R., 2004. Twig beetles, *Pityophthorus* spp. (Coleoptera: Scolytidae), as vectors of the pitch canker pathogen in California. Can. Entomol. 136, 685–693.

Stoszek, K.J., Rudinsky, J.A., 1967. Injury of Douglas-fir trees by maturation feeding of the Douglas-fir Hylesinus, *Pseudohylesinus nebulosus* (Coleoptera: Scolytidae). Can. Entomol. 99, 310–311.

Struble, G.R., 1937. The fir engraver beetle, a serious enemy of white fir and red fir. United States Department of Agriculture, Bureau of Entomology, Circular 419. 15 p.

Swaine, J.M., 1909. Catalogue of the described Scolytidae of America, north of Mexico Pages 76 to 159, pls 3–17 in E. P. Felt, 24th report of the State Entomologist, 1908, Appendix B New York Education Department Bulletin 455 (New York State Museum Bulletin 134). New York State University, Albany, 206 p.

Swaine, J.M., 1915. Descriptions of new species of Ipidae (Coleoptera). Can. Entomol. 47, 355–369.

Swaine, J.M., 1917. Canadian bark-beetles, part 1. Descriptions of new species. Dominion of Canada Department of Agriculture, Entomological Branch, Technical Bulletin 14 (1), 32 p.

Swaine, J.M., 1918. Canadian bark-beetles, part 2. A preliminary classification with an account of the habits and means of control. Dominion of Canada Department of Agriculture, Entomological Branch, Technical Bulletin 14 (2), 143 p.

Swaine, J.M., 1925. A new species of *Polygraphus* (Coleoptera). Can. Entomol. 57, 51.

Thomson, G.C., 1859. Familia Tomicidae. In: Skandinaviens Coleoptera synoptiskt bearbetade, Vol. 1. Lund, pp. 145–147.

Thousand cankers disease, 2014. Available online: http://www.thousandcankers.com/about-us.php. Last accessed: May 20, 2014.

Tisserat, N., Cranshaw, W., Putnam, M., Pscheidt, J., Leslie, C.A., Murray, M., et al., 2011. Thousand cankers disease is widespread on black walnut, *Juglans nigra*, in the western United States. Available online: http://www.fs.fed.us/psw/publications/seybold/psw_2010_seybold008(tisserat).pdf. Last accessed May 20, 2014.

Trudel, R., Guertin, C., de Groot, P., 2004. Use of pityol to reduce damage by the white pine cone beetle, *Conophthorus coniperda* (Col., Scolytidae) in seed orchards. J. Appl. Entomol. 128, 403–406.

U.S. Department of Agriculture Forest Service, 2004. Forest Insect and Disease Conditions in the Southwestern Region, 2004. United States Department of Agriculture, Forest Service, Southwestern Region, Forestry and Forest Health R3-05-10. Available online: http://www.fs.usda.gov/Internet/FSE_DOCUMENTS/stelprdb5238440.pdf. Last accessed: May 20, 2014.

Utley, C., Nguyen, T., Roubtsova, T., Coggeshall, M., Ford, T.M., Grauke, L.J., et al., 2013. Susceptibility of walnut and hickory species to *Geosmithia morbida*. Plant Dis. 97, 601–607.

Verrall, A.F., 1943. Fungi associated with certain ambrosia beetles. J. Agric. Res. 66, 135–144.

Voolma, K., Ounap, H., Suda, I., 1997. Eesti urasklaste (Coleoptera, Scolytidae) maaraja. Eesti Pollumajandusulikool Metsandusteaduskond, Tartu, 43 p.

Walgama, R.S., 2012. Ecology and integrated pest management of *Xyleborus fornicatus* (Coleoptera: Scolytidae) in Sri Lanka. Journal of Integrated Pest Management 3, A1–A8.

Weber, B.C., 1978. *Xylosandrus germanus* (Blandf.) (Coleoptera: Scolytidae), a new pest of black walnut: a review of its distribution, host plants, and environmental conditions of attack. In: Proceedings, Walnut insects and diseases workshop. June. Carbondale, IL, pp. 63–68.

Westwood, J.O., 1838. Family Scolytidae. In: Synopsis of the genera of British insects. Longman, Orme, Brown, Green and Longmans, London, pp. 39–40.

Wood, S.L., 1951. Two new species and a new genus of Scolytidae (Coleoptera) from Utah. J. Kans. Entomol. Soc. 24, 31–32.

Wood, S.L., 1957a. Distributional notes on and synonymies of some North American Scolytidae (Coleoptera). Can. Entomol. 89, 396–403.

Wood, S.L., 1957b. Ambrosia beetles of the tribe Xyloterini (Coleoptera: Scolytidae) in North America. Can. Entomol. 89, 337–354.

Wood, S.L., 1966. New synonymy in the Platypodidae and Scolytidae (Coleoptera). Great Basin Nat. 26, 17–33.

Wood, S.L., 1969. New synonymy and records of Platypodidae and Scolytidae (Coleoptera). Great Basin Nat. 29, 113–128.

Wood, S.L., 1977. New synonymy and new species of American bark beetles (Coleoptera: Scolytidae), Part V. Great Basin Nat. 37, 383–394.

Wood, S.L., 1980. New American bark beetles (Coleoptera: Scolytidae), with two recently introduced species. Great Basin Nat. 40, 353–358.

Wood, S.L., 1982. The bark and ambrosia beetles of North and Central America (Coleoptera: Scolytidae), a taxonomic monograph. Great Basin Nat. Mem. 8, 1–1359.

Wood, S.L., 1986. A reclassification of the genera of Scolytidae (Coleoptera). Great Basin Nat. Mem. 10, 1–126.

Wood, S.L., 1989. Nomenclatural changes and new species of Scolytidae (Coleoptera), part IV. Great Basin Nat. 49, 167–185.

Wood, S.L., 2007. Bark and ambrosia beetles of South America (Coleoptera: Scolytidae). Brigham Young University, M.L. Bean Life Science Museum, Provo, 900 p.

Wood, S.L., Bright, D.E., 1992. A catalog of Scolytidae and Platypodidae (Coleoptera), Part 2: Taxonomic Index. Great Basin Nat. Mem. 13, 1–1553.

Yin, H.-F., Huang, F.-S., 1996. A taxonomic study on Chinese *Polygraphus* Erichson with descriptions of three news species and a new subspecies (Coleoptera: Scolytidae). Acta Zootaxonomica Sinica 21, 345–351.

Yin, H.-F., Huang, F.-S., Li, C.-L., 1984. Coleoptera: Scolytidae. Fasc. 29, In: Economic Insect Fauna of China. Science Press, Beijing, pp. 1–205.

Chapter 13

Modeling Bark Beetle Responses to Climate Change

Barbara J. Bentz[1] and Anna Maria Jönsson[2]

[1] USDA Forest Service, Rocky Mountain Research Station, Logan, UT, USA, [2] Department of Physical Geography and Ecosystem Science, Lund University, Lund, Sweden

1. INTRODUCTION

Climate change is happening, and at a quicker pace than previously predicted. Since the late 1970s there have been twice the number of days with a high temperature record as a cold record, and in the Northern Hemisphere, 1983–2012 was *likely* the warmest 30-year period in the last 1400 years (IPCC, 2013). These trends suggest that ecological effects of climate change are occurring sooner than anticipated and could potentially be more catastrophic. Forest and agriculture ecosystems are not immune to changing climatic conditions. An alarmingly large number of recent tree mortality events are a consequence of globally increasing temperatures and drought that directly influence tree survival (Allen *et al.*, 2010), and yields of some crops worldwide have declined in recent decades due in large part to increasing temperatures (Lobell *et al.*, 2011). In addition to direct effects of climate change on plant survival, herbivorous insects that contribute significantly to plant mortality are also being influenced. Due to their short generation time and tight connection between temperature and life history traits, insects are particularly sensitive to warming associated with climate change (Bale *et al.*, 2002).

Bark beetles (Coleoptera: Curculionidae, Scolytinae) comprise a large subfamily of insects, although only a small percentage of the more than 6000 bark beetle species found worldwide are capable of causing significant economic impacts. Bark beetles mostly feed in the phloem (true bark beetles) or on fungi in the sapwood (ambrosia beetles) of woody plants, although there are a few species that feed in seeds, annual plants, grasses, and other herbaceous vegetation (Wood, 1982). Bark beetles are among the few insects that burrow into their host material for egg deposition, and all parts of a plant may be used including twigs, cones, roots, and the main trunk, although each genus is generally restricted to a particular part. For example, larvae of the genera *Dendroctonus* and *Ips* feed in phloem, whereas *Conophthorus* and some *Hypothenemus* species feed on tissues within seeds. Consumption of live tissue results in death of the entire plant or the plant part that is fed on, and new host material is therefore required for each beetle generation. Bark beetles that feed on live tissue are major contributors to tree mortality globally (Schelhaas *et al.*, 2003; Meddens *et al.*, 2012; Thom *et. al.*, 2013), and can cause declines in crop systems (Jaramillo *et al.*, 2011).

As with all poikilotherms, many bark beetle life history traits that influence population success are temperature dependent (Danks, 2007). Climate change can therefore cause significant alterations to bark beetle population dynamics, both positive and negative. Important insect traits that will be affected by warming associated with climate change include thermal thresholds and rates of development, diapause, and cold hardening. These traits are important regulators of insect seasonality and synchrony, and ultimately the mean fitness of the population. For example, adult emergence synchrony is key to the mass attack strategy that is used by some bark beetle species to overwhelm well-defended live trees (Logan and Bentz, 1999). Additionally, all species must synchronize vulnerable life stages with appropriate seasons (i.e., seasonality), and thereby increase the probability of surviving adverse conditions such as extreme cold or heat. In northern habitats, winter and spring temperatures have warmed more than summer and fall temperatures, although some amount of warming has occurred year round (Vose *et al.*, 2005; Walsh *et al.*, 2014). Several bark beetle species have benefited through expansion into new habitats (Cudmore *et al.*, 2010; Jaramillo *et al.*, 2011; de la Giroday *et al.*, 2012), and sustained outbreaks in habitats that were previously too cold for outbreak continuity (Logan and Powell, 2001; Raffa *et al.*, 2013; Weed *et al.*, 2013). In some cases, climate extremes that influence susceptibility of the host plant also play a role in outbreak dynamics. For example, excessively warm and dry conditions that increase plant stress (McDowell *et al.*, 2008; Chapman *et al.*, 2012; Hart *et al.*, 2013), and severe storm events that result in blow-down (Schmid and Frye, 1977; Christiansen and Bakke, 1988) can increase vulnerability to bark beetle colonization. The positive population responses seen in recent years highlight the important role of weather in bark beetle outbreaks. Projected changes in climate will undoubtedly continue to manifest through alterations in bark beetle population outbreaks and shifts in species distributions.

To adequately manage future forest and agriculture ecosystems, knowledge of potential bark beetle impacts in a changing climate will be required. In this chapter, we review quantitative models that have been developed to predict the influence of temperature and precipitation on bark beetle population outbreaks. These include models developed to describe the direct effects of temperature and photoperiod on bark beetle populations, and the indirect effects of temperature and precipitation on host plant susceptibility to population outbreaks. We focus on bark beetle species with the potential for significant economic and ecological impacts, and where sufficient biological information regarding climate effects on population success has allowed model development (Table 13.1, Figure 13.1). Several approaches have been used to incorporate climate effects into models that describe bark beetle population growth. We partition the approaches based on the data (i.e., insect and host tree) and type of analyses (i.e., analytical, statistical, or a combination of the two) used in model development. Our discussion is focused on insect *phenology models*, *ecosystem models* that incorporate aspects of climate, insects and their host trees, and models that incorporate the ability to test *management* strategies for reducing

bark beetle impacts in a changing climate. Most bark beetle species that cause economic impact have been modeled extensively for predicting habitat conditions that increase the probability of outbreaks. We acknowledge the important role of host density and continuity across a landscape in bark beetle population outbreaks, but here only include models that also incorporate a climate component.

2. MODEL TYPES AND DATA REQUIREMENTS

2.1 Phenology Models

Insect phenology models provide a quantitative description of physiological processes that are influenced by temperature, and based on these processes predict how temperature will influence life cycle timing and ultimately population success. Describing the physiological responses of insects to temperature has been a topic of research for decades, focusing on the influence of temperature on development time and survival (Janisch, 1932; Sømme, 1964), and photoperiod and temperature on diapause (Bradshaw and Holzapfel, 2010; Saunders, 2014). Collectively, we refer to these processes as phenology, the study of the timing of recurring life cycle events and survival, and how they are influenced by seasonal changes in climate (Tauber and Tauber, 1976). The complex relationship between temperature and physiological processes that make up phenology affect multiple aspects of bark beetle population dynamics and each species' geographic distribution. Mechanisms that promote synchrony of vulnerable life states with appropriate seasons (i.e., seasonality), and survival

TABLE 13.1 Bark Beetle Species with Sufficient Information for Development of Models Describing Climate Effects on Population Outbreaks

Species	Voltinism	Major Host Tree Species	Latitudinal Range
Dendroctonus frontalis	multivoltine	*Pinus echinata, P. elliottii, P. engelmannii, P. leiophylla, P. palustris, P. ponderosae, P. taeda, P. virginiana,* others	15°N to 40°N
Dendroctonus ponderosae	univoltine, semivoltine	*Pinus albicaulis, P. aristata, P. contorta, P. flexilis, P. lambertiana, P. monticola, P. ponderosae,* others	31°N to 60°N
Dendroctonus rufipennis	univoltine, semivoltine	*Picea engelmannii, Pi. glauca, Pi. sitchensis*	32°N to 65°N
Hylobius abietis	semivoltine	*Pinus sylvestris, Picea abies, Pi. sitchensis, Pseudotsuga menziesii,* others	40°N to 54°N
Hypothenemus hampei	multivoltine	*Coffea arabica* and *C. canephora*	30°S to 20°N
Ips typographus	bivoltine, univoltine	*Picea abies*	43°N to 66°N

FIGURE 13.1 **Bark beetle species (adult size) with sufficient information for development of models that describe potential climate influences on population success.**(A) *Dendroctonus rufipennis* (photo: B. J. Bentz) and impact (photo: Getty Images); (B) *D. frontalis* (photo: Erich Vallery, Bugwood. org) and impact (photo: Paul Butts, Bugwood.org); (C) *D. ponderosae* (photo: Matt Ayres) and impact (photo: B. J. Bentz); (D) *Hylobius abietis* (photo: Claes Hellqvist) and damage (photo: Beat Forster, Bugwood.org); (E) *Hypothenemus hampei* (photo: Pest and Diseases Image Library, Bugwood.org) and unripe coffee berry with entering female (photo: Gonzalo Hoyos); (E) *Ips typographus* (photo: Maja Jurc) and damage (photo: A. M. Jönsson).

of adverse environmental conditions, are critical to bark beetle population growth and outbreak potential. Models developed to describe bark beetle phenology are typically analytical or process based and require detailed information on the response of individual beetles. Because they are driven by functional, rather than statistical, relationships between physiological processes and temperature, phenology models are important tools for describing life history events at multiple spatial and temporal scales. Due to their mechanistic nature, phenology models have the capacity to describe emergent processes in bark beetle population dynamics in a changing climate.

2.1.1 Development Time

Because temperature is such a significant driver of development time, its effects on development and life cycle timing are well studied for a number of bark beetle species (see references below for each species). Temperature-dependent development rates are typically estimated as the inverse of development time, rather than a direct measurement of physiological rate processes. In the laboratory, individuals are monitored at constant or fluctuating temperatures, and the time required to complete a particular life stage or phase is measured. Mathematical descriptions of the distribution of development time or rate data, as a function of temperature, can then be parameterized. Although linear approximations to temperature response (i.e., degree-days) have often been used because they are easy to derive and implement, this type of model only captures the response over a limited range of temperatures and often does not include critical upper and lower developmental thresholds. Information on response at the extremes is important, particularly when making climate change

predictions, and non-linear descriptions are critical (Régnière *et al.*, 2012). Examining low and high temperature thresholds is complicated, however, by non-linearities that occur at thresholds and reduced survival at temperature extremes. Due to these difficulties, low and high temperature thresholds are unclear for many bark beetle species.

Given a mathematical description of the temperature/development rate response, simulation models can be used to generate stage or age occurrences through time based on an input temperature file. A variety of model types have been used including distributed delays (Manetsch, 1976), cohort based (Sharpe *et al.*, 1977; Logan, 1988), and individual based (Cooke and Régnière, 1996; Régnière *et al.*, submitted). In all cases, development rates are integrated over short time steps, typically hourly or daily (Régnière and Powell, 2003). Because of this, the input temperature file must be no coarser than a daily time step, and have some relationship with temperatures within the habitat of the insect. For example, it is clear that tree phloem and air temperature can be different, particularly when sun exposure is a factor, and a variety of methods have been applied to describe the complex relationship (Harding and Ravn, 1985; Bolstad *et al.*, 1997; Baier *et al.*, 2007; Lewis 2011; Trân *et al.*, 2007; Wainhouse *et al.*, 2013).

2.1.2 Diapause

Diapause and its seasonal progression are also a critical aspect of phenology. Among insects, diapause is the most common strategy to synchronize individuals, and it enables survival of extreme climatic conditions. Diapause is a developmental arrest characterized by a reduction in metabolic activity and an increase in energy reserves and tolerance to stress. Although only a small number of species have been investigated, facultative and obligatory diapause has been demonstrated or suggested to occur in multiple bark beetles species in the larval/prepupal stage (Christiansen, 1971; Scott and Berryman, 1972; Hansen *et al.*, 2001), adult stage (Ryan, 1959; Birch, 1974; Clark, 1974; Langor and Raske, 1987; Doležal and Sehnal, 2007; Lester and Irwin, 2012), and both larval and adult stages (Safranyik *et al.*, 1990; Hansen *et al.*, 2001; Inward *et al.*, 2012; Wainhouse *et al.*, 2013). Although temperature and photoperiod are the most common diapause cues (Denlinger, 2002), in prepupal *Dendroctonus rufipennis* (Kirby) photoperiod was found not to be a significant driver, and was replaced with thermoperiod (Hansen *et al.*, 2011). Photoperiod and temperature, however, determine the length of diapause in adult *Ips typographus* (L.) (Doležal and Sehnal, 2007). Due to its critical role in synchronizing life cycles and determining the number of generations per year, diapause is critical to include in phenology models (Tobin *et al.*, 2008), particularly when making predictions based on a changing climate (Jönsson *et al.*, 2011). However, quantifying the appropriate cues for diapause, particularly facultative diapause, can be difficult. Experimental designs to tease out photoperiod and/or thermoperiod regimes necessary to induce diapause can be complex, and unlike development time, the response variable is not always straightforward. For example, the typical measure of reduced respiration to indicate diapause is confounded during the prepupal stage when reduced respiration occurs as a result of histolysis in preparation for pupation, regardless of the diapause state (Hansen *et al.*, 2011).

2.1.3 Cold Hardening

Diapause is not the only process insects use to survive adverse environmental conditions. Many insects, including bark beetles, rely on supercooling for overwintering survival. Supercooling allows individuals to cool below the freezing point, and the acclimation process is temperature dependent. The capacity to supercool has been found in all bark beetle species investigated. Bark beetles are freeze intolerant, meaning they cannot survive ice formation within their tissues (Lee, 1991). Supercooling to survive subfreezing temperatures is accomplished in bark beetles by accumulating antifreeze proteins and low molecular weight polyols and sugars, including glycerol (most reported), ethylene glycol, glucose, sorbitol, mannitol, dulcitol, and trehalose (Gehrken, 1984; Miller and Werner, 1987; Bentz and Mullins, 1999; Lombardero *et al.*, 2000; Koštál *et al.*, 2011). Supercooling points (SCP), the temperature at which crystallization of tissues and death occur, are often seasonally dynamic and varying among life stages. SCPs have been determined for a number of bark beetle species, but quantified into models for only a few.

2.2 Ecosystem Models

Ecosystem models describe the effect of climate variables and host plant conditions on bark beetle population abundance using statistical and analytical techniques, or a combination of the two. In contrast to the mechanistic nature of phenology models, *statistical ecosystem models* provide a statistical association between field-based conditions of host trees, insect population success, and weather. This type of model is often referred to as a risk or hazard model, and the goal is to describe forest and weather conditions that promote bark beetle population outbreaks. *Process-based ecosystem models* incorporate at least some ecosystem and bark beetle processes. In addition to describing how climate and ecosystem properties affect bark beetle outbreaks, they also describe how outbreaks and climate influence future ecosystem composition and structure. For these types of models, bark beetle population abundance is typically measured as plant impacts estimated from measures of the number of trees or plants killed derived from remotely sensed data, aerial surveys, pheromone traps or ground plots. Weather data used to develop the statistical associations are often at a monthly or annual temporal scale with varying spatial scale, and the processes that influence the transition between endemic and epidemic populations are not always captured. The models can be spatial or aspatial. Observed data on bark beetle-caused plant mortality is statistically correlated with weather variables that describe conditions during the mortality event. Output from mechanistic models, such as those describing phenological aspects, can be used as input variables in statistical ecosystem models. Analyses are often retrospective, describing specific events that occurred rather than the processes responsible for the plant mortality event.

A variety of statistical approaches have historically been used including logistic regression and discriminate analyses. More recently, species–environmental matching models, or niche models, have been used to map suitable climate habitat for a variety of bark beetle species. Ecosystem models provide a description of climate and stand metrics that have influenced forest insect population outbreaks in the past. Because these relationships may change in a future climate, ecosystem models do not have the capacity of mechanistic phenology models to describe emergent properties of a system in a changing climate.

2.3 Management Models

Management models integrate management options with either mechanistic or statistical descriptions of climate effects on insect and host plant populations. Models provide a quantitative assessment of insect effects on forest resources and provide a means to assess the benefit of various management strategies (Seidl *et al.*, 2011). The ability to test the influence of climate and specific management strategies on insect-caused tree mortality is a novel aspect of this type of model. In the context of bark beetles, management models provide tools for predicting the integrated effect of climate and management actions on bark beetle outbreaks. Management models integrate either mechanistic or statistical descriptions of climate and stand dynamics on bark beetle population success.

3. DEVELOPED MODELS

3.1 *Dendroctonus frontalis* (Zimmermann)

3.1.1 Phenology Models

Dendroctonus frontalis, the southern pine beetle, is found across the southern and southeastern USA, roughly coinciding with the distribution of its main host *Pinus taeda* (L.). *Dendroctonus frontalis* also occurs in Arizona, New Mexico, and on coastal facing slopes in Mexico and Central America, and in recent years outbreak populations have been found as far north as New Jersey (Weed *et al.*, 2013). Most *D. frontalis* infestations begin in the spring, completing three to six generations (depending on geographic location and weather) per year. The important role of temperature in *D. frontalis* life cycle timing motivated multiple early studies, mostly in uncontrolled field conditions (Thatcher and Pichard, 1967; Goldman and Franklin, 1977; Mizell and Nebeker, 1978). In the interest of developing a model, Wagner *et al.* (1984) measured development time of eggs, larvae, pupae, and teneral adults at constant temperatures in the laboratory and used a biophysical model based on enzyme kinetics (Sharpe *et al.*, 1977) to describe life cycle timing. They assumed that the distribution of development rates do not change with temperature when normalized, yielding a standard curve that was used to distribute development times across all life stages reared at constant temperatures. Although they did not capture the low and high thermal thresholds for larval development (larvae migrate to the outer bark during the fourth instar, which complicates monitoring), egg rates peaked at 30°C, and teneral adults and pupae at 31.1°C. Similar to *D. ponderosae* (Hopkins) (Bentz *et al.*, 1991), *D. frontalis* larvae fail to pupate at low temperatures (Wagner *et al.*, 1984; Trần *et al.*, 2007), thereby synchronizing as mature larvae. Comparisons of model-predicted development from eggs to emerged adults compared favorably with field data. Using the data described in Wagner *et al.* (1984) and Ungerer *et al.* (1999), Friedenberg *et al.* (2007) improved the process-based model by incorporating individual variation based on Gilbert *et al.* (2004). The model was used to test the role of emergence synchrony in *D. frontalis* population growth. Unlike other *Dendroctonus* species with strong selection for emergence synchrony (Logan and Bentz, 1999),

asynchrony in *D. frontalis* attacks among trees provides a continual plume of pheromones across generations, especially at small population sizes, thereby counteracting a strong Allee effect (Friedenberg *et al.*, 2007).

The influence of minimum winter temperature on *D. frontalis* population dynamics was recognized as early as 1899 (Beal, 1933). Observations of field populations suggested that temperatures below −12°C were detrimental to *D. frontalis* survival (see references in Ungerer *et al.*, 1999). Using individuals from a southern population in Louisiana in a controlled laboratory experiment, Ungerer *et al.* (1999) estimated that >90% mortality would be expected in adults exposed to −16°C and below. Also using individuals from populations in Louisiana, Lombardero *et al.* (2000) observed no seasonal change (between October and March) in adult SCP levels that ranged from −11 to −13°C. SCPs of adults from New Jersey were found to be significantly lower than adults from Alabama, although only prepupae overwintered successfully in New Jersey (Trần *et al.*, 2007). Of all life stages and populations sampled, prepupae from New Jersey, the northernmost population, were the most cold tolerant. Relative to all bark beetle species investigated, however, *D. frontalis* is the least cold tolerant.

Data on cold tolerance and the biophysical development model of Wagner *et al.* (1984) were combined to evaluate how temperature may limit the northern distribution of *D. frontalis* (Ungerer *et al.*, 1999). In climate change scenarios, they increased average minimum temperature, and increased and decreased variability around the mean. Their results supported the hypothesis that the northern distributional limit of *D. frontalis* is maintained by lethal winter temperatures, not by summer temperatures that dictate the number of generations per year. In climate change scenarios, they also observed that variability in annual minimum temperatures affected winter mortality more than average increases. The authors noted, however, that the cold tolerance used in their model was based on individuals from southern, not northern, populations. Using new information on increased cold tolerance capacity of prepupae from New Jersey, Trần *et al.* (2007) incorporated cold tolerance with a model for predicting phloem temperatures from air temperatures in trees that do not experience solar radiation, and a process-based model of climatic effects on interannual growth of populations. The addition of the new cold tolerance information increased model fit. They concluded that winter mortality was only a minor driver of *D. frontalis* dynamics over much of its range, but that the northward range expansion predicted by Ungerer *et al.* (1999) was ongoing in large part due to increased winter survival. Model predictions suggest that a minimum temperature of the coldest night of the year, that is −12 to −16°C, would produce about 50% mortality. Their model results also suggested that the best predictor of beetle abundance in one year is beetle abundance the previous year.

Like many *Dendroctonus* species, *D. frontalis* is associated with a wide array of community associates including beetle-mutualistic fungi, antagonistic fungi, and mites (Klepzig and Wilkens, 1997; Chapters 6 and 7). Growth of the associated fungi and mites are also temperature dependent. For example, one of the beetle-mutualistic fungi has maximal growth at cooler temperatures than other associates, resulting in varying seasonal abundance among the community associated with *D. frontalis* (Hofstetter *et al.*, 2007). Predicting the fungal species being carried by beetles emerging from trees is further complicated, however, by temperature dependency of mites that help to transport and propagate fungi within a tree (Lombardero *et al.*, 2003).

3.1.2 Ecosystem Models

Using county-level presence–absence data of *D. frontalis* as the response variable in a logistic regression model and a Classification and Regression Tree (CART) analyses, infestation either in the previous year or further in past was found to be the most important variable in predicting *D. frontalis* presence (Duehl *et al.*, 2011). Minimum winter temperature, influencing cold-induced mortality, and seasonal average temperature were the important temperature variables. High levels of fall precipitation lead to high levels of infestation the following year, potentially due to increased nutritional quality of the food resource. The authors acknowledge the significant variables in their analyses only explained a small amount of the variability in infestation occurrence, and highlighted the need for more research to capture other important aspects of the system including natural enemies. Using the same presence–absence data with a panel data modeling approach, Gan (2004) found that winter temperature and spring temperature were the most important climatic variables in predicting *D. frontalis* presence. Based on a marginal logistic regression modeling approach, Gumpertz and Pye (2000) found similar results. Average daily temperature and precipitation in the fall were the most significant predictors.

TAMBEETLE is a spatially explicit, stochastic model of population dynamics based on submodels that describe temperature-dependent development, fecundity, and survival to estimate the number of emerging beetles (Feldman *et al.*, 1981; Wagner *et al.* 1984; Coulson *et al.*, 1989). These values are used to determine the probability that emerging distributions will be adequate to overcome neighboring trees, and hence provide a measure of outbreak probability. Similarly, SPBMODEL is a simulation model that incorporates development and mortality of stage-specific cohorts of southern pine beetle to predict infestation growth in currently infested stands over a 3-month period (Lih and Stephen, 1989). The model includes the influence of stand conditions and predicts the number of infested trees through time and, if tree diameter distributions are included, the economic return from a salvage operation (Ghosh, 1983). The model was re-engineered using JAVA and is currently available on the web (Satterlee, 2002). Model projections with climate change scenarios have not been published.

3.1.3 Management Models

No models that include climate and options for evaluating the effect of management on *D. frontalis* outbreaks have been published.

3.2 *Dendroctonus ponderosa* (Hopkins)

3.2.1 Phenology Models

The historical distribution of *D. ponderosae*, the mountain pine beetle, follows that of its major *Pinus* host trees, spanning from Baja California Norte, Mexico, to central British Columbia, Canada. In recent years, likely as a result of increasing temperatures, sustained outbreak populations have been observed in northern British Columbia and the Northwest Territories and east across the Rocky Mountains into stands of a novel host tree, jack pine (*Pinus banksiana* Lamb.) in Alberta, Canada (Cudmore *et al.*, 2010; Cullingham *et al.*, 2011; de la Giroday *et al.* 2012). Life cycle timing is univoltine at low elevations and a mix of univoltine and semivoltine at high elevations (Bentz *et al.*, 2014). Brood adult emergence typically occurs in mid-summer, although parents may re-emerge to attack new trees in early summer. Offspring from early summer attacks by re-emerged parents can complete development by the following fall, although thermal requirements for bivoltinism are currently lacking across the *D. ponderosae* realized distribution (Bentz and Powell, in press).

Descriptions of life stage response to temperature were initially quantified as degree-hours or degree-days above particular thresholds (Reid and Gates, 1970; Safranyik and Whitney, 1985). This type of description captures the linear portion of the temperature response, but is less descriptive at the upper and lower temperature thresholds. Informed by the previous work, non-linear responses were later quantified for the egg (Logan and Amman, 1986), each larval instar, and pupal stage (Bentz *et al.*, 1991). In the Bentz *et al.* (1991) model, individual variation was incorporated based on the "same shape" approach of Sharpe *et al.* (1977) (see *D. frontalis*, above). A cohort-based approach was used to simulate movement of individuals from one life stage to the next based on advancement of physiological age, a function of temperature and time. Logan and Bentz (1999) later added rate curves for the teneral and ovipositional adult stages using a similar approach. Recognizing the need for individual variability in the model, Gilbert *et al.* (2004) expanded the existing model based on the age-structured McKendrick–von Foerester partial differential model to account for phenotypic variability in development rates.

Over a span of 10 years, additional development time data were collected on the larval stages, the time required for transformation of a teneral adult to a fully mature adult that occurs prior to emergence from a tree, and temperature-dependent oviposition. These new data, with the previously collected data, were used to re-parameterize development and oviposition rate curves. A single functional form was used in the re-parameterization and variation among individuals was described using a lognormal distribution (Régnière *et al.*, 2012). A novel aspect of the more recent data analyses was that censored data were used to provide better estimates of low and high temperature thresholds. Censored observations occur when development is slow and may end before the stage is completed, or when stressful conditions result in death before the stage is completed. The presence of censored data increased the ability to describe low temperature thresholds for development, particularly in the fourth instar, a stage believed to play an important role in this insect's life cycle synchrony (Bentz *et al.*, 1991). *Dendroctonus ponderosae* fourth instar does not pupate at temperatures below 15–17.5°C (Safranyik and Whitney, 1985; Régnière *et al.*, 2012), and preliminary investigations suggest some amount of facultative diapause in the prepupal life stage (BJ Bentz, unpublished data).

Data used to parameterize the *D. ponderosae* development time model were collected from populations in northern Utah and central Idaho. We know, however, that there is geographic variation in developmental response of this insect to temperature. Common garden experiments revealed that populations in southern latitudes develop significantly slower, and were less synchronized, than populations from northern latitudes at the same constant temperature (Bentz *et al.*, 2001, 2011; Bracewell *et al.*, 2013). Moreover, in the field, despite receiving significantly more thermal input, generation time for univoltine southern populations was similar to that of populations at more northern latitudes (Bentz *et al.*, 2014). These results suggest differences in development rates or thresholds among populations along a latitudinal cline. A stage-specific model has not been parameterized for *D. ponderosae* at southern latitudes.

Another aspect of *D. ponderosae* phenology is survival. The influence of cold temperatures on *D. ponderosae* was recognized early (Yuill, 1941), and predicted to be a major factor influencing the insect's historical distribution, particularly in Canada (Safranyik *et al.*, 1975). Eggs and pupae were considered the least cold tolerant life stages (Reid and Gates, 1970; Amman, 1973), whereas large larvae were considered the most tolerant (Sømme, 1964; Safranyik and Linton, 1998). The seasonality of larval cold tolerance

was evaluated at multiple field sites in the western USA by collecting individuals throughout the life cycle and measuring SCPs (Bentz and Mullins, 1999). Similarly, post-ovipositional adults collected in Washington were also examined (Lester and Irwin, 2012). Although no differences were found among larval instars, the data showed that, like other insects, *D. ponderosae* has the capacity to acclimate to decreasing temperatures by increasing production of cryoprotectants and thereby decreasing their supercooling point. Cold tolerance was dynamic and changed throughout the life cycle. The lowest supercooling point, and hence when individuals were most cold hardy, occurred in the middle of winter with the highest supercooling points in the fall and spring. Adults in Washington did not supercool to temperatures as low as observed for larvae, although individuals for the two studies were collected in different thermal regimes.

The observed dynamic seasonal nature of cold hardening suggested that a single threshold value was inappropriate for describing cold-induced mortality in *D. ponderosae*. Based on the assumption that supercooling is a function of temperature (Lee, 1991), a model was developed based on the field-collected SCPs for larvae and associated phloem temperature measurements (Bentz and Mullins, 1999; Régnière and Bentz, 2007). In general, the model is based on the changing proportion of individuals in three states: (1) a non-cold hardened feeding state (summer); (2) an intermediate state in which insects have ceased feeding (spring and fall); and (3) a fully cold hardened state where insects have accumulated a maximum concentration of cryoprotectants (winter). Shifts in the proportion of individuals in each state are determined by the influence of temperature on the gain and loss of cryoprotectants, modeled using logistic probability distribution functions and field-collected data. The proportion of individuals that die when temperatures drop below the median SCP for each state is calculated for a given daily temperature. In Alberta, Canada, Cooke (2009) found that the cold tolerance model predicted overwintering survival rates in close agreement with observed survival during years with sudden drops in temperature, although there was substantial unexplained variation.

The two aspects of *D. ponderosae* phenology, development time and survival, and the influence of precipitation on host suitability were combined to predict climatic suitability based on a conceptual modeling approach (Safranyik *et al.*, 1975; Carroll *et al.*, 2004). Development time was incorporated as degree-days required to complete a univoltine generation, winter mortality was considered 100% when temperatures reach −40°C, and mass attack potential was based on maximum August temperatures. The influence of drought on host tree suitability was incorporated by including precipitation in April and an index of water deficit (see Safranyik *et al.*, 2010 for additional details). Régnière *et al.* (submitted) took a more mechanistic approach, and used an individual-based, object-oriented model to combine the influence of temperature on development time and cold tolerance (as described above), and also includes cold-induced egg mortality based on information from Reid and Gates (1970).

3.2.1.1 Application of phenology models

The development time model proved useful in analyzing the important role of seasonality and synchronicity in *D. ponderosae* population dynamics, and the role of different development thresholds and rates in this process (Logan and Bentz, 1999; Jenkins *et al.*, 2001). When driven with hourly temperature, the model was also used to analyze the influence of weather and climate on *D. ponderosae* population success in historical and future climates. Basing their results on median adult emergence time, the Logan and Bentz (1999) version of the development time model was used by Logan and Powell (2001) to show that warming of a relatively small amount (i.e., 2°C) could cause a shift from semi- to univoltinism in *D. ponderosae* populations at high elevations, and that warming associated with climate change could result in a northward range expansion of the species. Northward expansion has indeed happened (Cullingham *et al.*, 2011; de la Giroday *et al.*, 2012), and populations are currently found in historical and novel hosts near the Yukon Territory border with British Columbia and in the Northwest Territories, and as far east as central Alberta (Nealis and Cooke, 2014).

The development time and cold tolerance models were integrated with topography and observed and projected hourly temperature using BioSim (Régnière and Saint-Amant, 2013) to analyze *D. ponderosae* population success in future climates in Canada and the USA. Population success was predicted using a variant of the development time model that was based on predicted timing of median adult emergence, and assumptions that univoltinism and emergence during a specific window of time in the summer were fundamental to population success (i.e., adaptive seasonality). When consecutive years of adaptive seasonality were a requisite, high probability was found in areas that have been under extensive attack in recent years (Bentz *et al.*, 2010; Safranyik *et al.*, 2010). The area suitable for adaptive seasonality and climatic suitability as temperatures rise throughout this century was predicted to be restricted to northern provinces in Canada and high elevations in the western USA (Hicke *et al.*, 2006; Bentz *et al.*, 2010; Safranyik *et al.*, 2010). Based on the cold tolerance model, cold survival probability was predicted to substantially increase at high elevations and across Canada, although in areas key to *D. ponderosae* migration in central Canada, the probability for low survival remained low.

Based on an individual-based model that incorporates both cold survival and development time, Régnière *et al.* (submitted) showed that recent climate trends in western North America affected *D. ponderosae* population growth rates, influencing the size and severity of recent outbreaks. Predictions based on climate projections suggest that by the middle of the century, probability of population success will be moderate to high in most of Alberta, although moderate to low in the northern and eastern Canadian Provinces where population growth actually declines in the future. These results highlight the differential effect of temperature on mountain pine beetle cold tolerance and population synchrony. Increasing minimum temperatures may result in higher overwinter survival, although univoltinism will be disrupted when temperatures are too warm (Logan and Bentz, 1999; Sambaraju *et al.*, 2012). With the exception of the highest elevation areas, much of the western USA will be highly suitable for transition from endemic to epidemic population levels, especially when additional factors that influence population size (i.e., stand conditions) are favorable.

Increased knowledge of the important role of flexibility in *D. ponderosae* life cycle timing suggested that the restrictions applied when using the median model and assumptions about adaptive seasonality were too strict. To improve on this, the developmental time model that includes individual variation, resulting in a distribution of emergence times rather than timing of a median individual, was employed to predict population growth rates at a landscape scale. The concept was again based on the notion of adaptive seasonality, in that developmental synchrony must occur that allows adult emergence to occur in a relatively short window of time, i.e., effective adults. The developmental time model was used to predict adult emergence distributions that were connected to a mathematical framework to describe population success at a landscape scale (Powell and Bentz, 2009). The model was parameterized with aerial detection data of *D. ponderosae*-killed trees over a span of years and observed phloem temperatures over the same time span. The resulting population growth model provides a direct connection between highly variable temperature data and predicted adult emergence distributions to estimate *D. ponderosae* population growth at a landscape scale. This model was then expanded to include a dispersal component that is conditioned on host tree density across a landscape (Powell and Bentz, 2014). Because both models are temperature driven, they can be used in forecasting *D. ponderosae* population success in future climates.

Dendroctonus ponderosae is associated with two species of mutualistic fungi, *Grosmannia clavigera* (Robinson-Jeffrey & Davidson) and *Ophiostoma montium* (Rumbold) Arx (Six, 2013). Temperature influences the growth rate of both fungal species, and optimal growth for *G. clavigera* is at cooler temperatures than *O. montium* (Rice *et al.*, 2008). Climate change could therefore impact the evolved multipartite symbiosis between fungi and beetle. To analyze this potential, temperature-driven models describing growth of the two mycangial fungal associates were developed and integrated with the median *D. ponderosae* development time model (Addison *et al.*, 2013). Preliminary evaluations suggest that thermal regimes that vary either intra- or interannually, in addition to interannual changes in density of attacking beetles and fungal inoculation sites, could maintain the symbiosis. Potential disruption of the symbiosis given climate change scenarios is under investigation.

3.2.2 Ecosystem Models

Statistical "risk" models were initially developed to evaluate susceptibility to *D. ponderosae* outbreaks, although they did not explicitly include a climate variable and instead used elevation, latitude, and longitude as a surrogate for climate effects on the insect (Amman *et al.*, 1977; Shore and Safranyik, 1992). More recently, several studies have taken a statistical approach to developing models that predict suitability for *D. ponderosae* outbreaks in the western USA and Canada using climate as a predictor variable. In all cases, the studies used some type of logistic regression model, were retrospective, and *D. ponderosae* impact data were based on aerially detected measures of *D. ponderosae*-killed trees during previous outbreaks (Aukema *et al.*, 2008; Preisler *et al.*, 2012; Sambaraju *et al.*, 2012; Creeden *et al.*, 2014). None of the studies included a measure of host density. All four studies found that some metric of winter minimum temperature and August maximum temperature, and simultaneously outbreaking populations in the vicinity were important predictors of outbreak progression. Sambaraju *et al.* (2012) found that sudden drops in daily temperature were particularly important, an indication of how non-acclimated individuals could be susceptible to cold-induced mortality (Régnière and Bentz, 2007). Reduced precipitation in the two prior years also significantly influenced the number of killed trees (Preisler *et al.*, 2012). A one-year lag of increased precipitation had a potentially positive effect on beetle reproduction, while reduced precipitation over two previous years had a negative effect on host susceptibility. Based on the same impact data as Preisler *et al.* (2012) and Creeden *et al.* (2014), Evangelista *et al.* (2011) used an ecological niche model (Maxent) and also found that precipitation during the warmest quarter of the year (data from 1998 to 2008 were included in the analyses) was the best predictor of suitable habitat for *D. ponderosae*. It is important to note that in 2001–2002, years included in these studies, one of the most severe droughts in the past 500 years occurred in many parts of the interior west of the USA (Pielke *et al.*, 2005).

In addition to climate variables, Preisler *et al.* (2012) and Creeden *et al.* (2014) tested for significance of output from the "median adaptive seasonality" model and cold tolerance model (see above) in describing phases of outbreaks (based on the number of trees killed in a given year). The cold tolerance model was found to be useful for predicting the transition of outbreaks to epidemics at regional scales in Washington and Oregon (Preisler *et al.*, 2012), and at times were associated with higher tree mortality in other western states (Creeden *et al.*, 2014), although the median-adaptive seasonality model was not found to be significant at any scale. As described above, predictions of the median individual and the strictness of the rule base for seasonality may be unrealistic, and other model variations have been developed for use in forecasts (Powell and Bentz, 2009, 2014). Only Sambaraju *et al.* (2012) evaluated their model predictions using climate scenarios (i.e., mean temperature increase between 1–4°C), and, similar to findings in Logan and Powell (2001), small shifts in temperature influenced population success at high elevations (1 or 2°C) and northern latitudes (4°C). Variability in mean temperature increase was found to be not important (Preisler *et al.*, 2012; Sambaraju *et al.*, 2012).

Coops *et al.* (2012) used a hybrid approach that incorporates process-based models of tree vigor and output describing climatic suitability for *D. ponderosae* (based on Safraynik *et al.*, 2010 as described above) to predict future vulnerability of lodgepole and jack pine in Canada. The model predicted areas of overlap where climate-driven processes predict trees will be more vulnerable to attack and where climatic suitability for the beetle will be at least moderate. These predictions can be used to focus future monitoring and management efforts.

3.2.3 Management Models

Safranyik *et al.* (1999) developed a simulation model of mountain pine beetle dynamics that includes a submodel of beetle biology driven by daily temperature that simulates host colonization, brood development and survival, and subsequent tree mortality. The beetle submodel is connected with a submodel of lodgepole pine growth and yield, and a submodel for invoking management. Safranyik's model was incorporated into a landscape simulation platform (SELES-MPB) to evaluate the effects of various management strategies, and results highlight the important role of weather and climate in outbreak growth rates (Riel *et al.*, 2004). Expanding on this approach by incorporating projected climatic suitability (Carroll *et al.*, 2004), Shore *et al.*, (2008) suggested that if *D. ponderosae* populations could be maintained at low levels through management until surrounding populations subside, beetle-killed trees could be minimized.

3.3 *Dendroctonus rufipennis* (Kirby)

3.3.1 Phenology Models

Dendroctonus rufipennis, the spruce beetle, is distributed across northern North America from Alaska to Newfoundland, and in the western USA throughout the Rocky Mountains south to Arizona and New Mexico. It attacks the two most abundant spruce species within its current range, *Picea glauca* (Moench) and *P. engelmannii* (Parry ex Engelm.). Life cycle timing is strongly associated with temperature, and life cycles of 1, 2, and 3 years have been observed across its range (Schmid and Frye, 1977; Werner et al., 2006). Adult emergence and subsequent attacks on new trees occurs in late spring. If temperatures the first summer are not warm enough to allow development to the adult stage by September, a facultative prepupal diapause is invoked and a semivoltine life cycle, or longer, occurs (Hansen et al., 2001). Conversely, warm temperatures the first summer increase the probability that development will progress to the pupal and adult stages prior to winter, avoiding the prepupal diapause, and emergence of new brood adults on a univoltine life cycle the following year. In addition to a facultative diapause in the prepupal phase, *D. rufipennis* has an obligate adult reproductive diapause (Safranyik et al., 1990) that requires cold temperature to satisfy. Therefore, adult emergence in the late spring is synchronized in both univoltine and semivoltine pathways.

Outbreak potential is greatest when some proportion of univoltine broods are present (Hansen and Bentz, 2003; Berg et al., 2006). Both univoltine and semivoltine beetles have been found developing in the same tree (Hansen et al., 2001), suggesting phenotypic plasticity for the trait and highlighting the role of temperature in diapause induction. Based on data from laboratory experiments, development rate of the fourth instar, which includes diapause as a continuous process, was modeled (Hansen et al., 2011). The model estimates developmental rates, as a function of hourly temperature, and the rates are continuously modified by diapause-inducing conditions. Development time information has been quantified for other spruce beetle life stages (Hansen et al., 2001), although a process-based, stage-specific phenology model has not been completed. Adult and larval spruce beetle are cold intolerant, and like other bark beetle species have a dynamic capacity to supercool to temperatures $< -30°C$ in the middle of winter (Miller and Werner, 1987). The cold hardening process has not been quantified in a predictive model.

3.3.2 Ecosystem Models

Because knowing the proportion of a population that will be univoltine is important to predicting the probability of an outbreak, Hansen et al. (2001) developed a temperature-based model that predicts the proportion of univoltine

beetles given air temperature. The proportion of univoltine beetles observed in field plots was used with associated air temperature in a linear mixed modeling framework. Data from additional locations and pheromone traps were subsequently analyzed, and a new model was developed that is based on cumulative hours above a threshold of 17°C that occurs from 40 to 90 days following peak adult trap captures to predict the proportion univoltine of beetles. The model was incorporated into the BioSim modeling framework to predict areas of vulnerability in current and future climates. During the historical period 1961–1990, spruce forests in Alaska and at high elevations in the contiguous western USA and northern latitudes of Canada were rated moderate to low probability of spruce beetle developing in a single year (Bentz et al., 2010). In 2001–2030 and again from 2071 to 2100, substantial increases in the spruce forest area with high probability of univoltine spruce beetle were predicted. The model does not factor in the influence of climate on spruce trees, yet reductions in Engelmann spruce habitat in the western USA is predicted to decline throughout the century (Rehfeldt et al., 2006).

Based on sustained growth releases observed in tree rings and reconstructed climate data, other studies also found significant relationships between *D. rufipennis*-killed trees and warm and drier-than-average late summer conditions and warm winter and fall temperatures (Hebertson and Jenkins, 2008; Sherriff et al., 2011). In the only study to include both climate and habitat variables, DeRose et al. (2013) found that while cool season minimum and warm season maximum temperatures were important to predicting *D. rufipennis* presence, spruce basal area and composition were more influential in the model. Interestingly, when projected in the future using Global Climate Models, habitat variables that characterized current spruce beetle susceptibility changed as future temperatures increased. Model predictions suggest that increased temperatures will allow the spruce beetle to be successful in stands of lower basal area and percent spruce, stand types that may not promote landscape-wide outbreaks despite a climatic release of spruce beetle.

3.3.3 Management Models

No models that include climate and options for evaluating the effect of management on *D. rufipennis* outbreaks have been published.

3.4 *Hylobius abietis* (L.)

3.4.1 Phenology Models

The range of *Hylobius abietis*, the large pine weevil, extends latitudinally from the Mediterranean area to northern Scandinavia. It is a polyphagous pest in conifer forests where larvae develop in the stumps of recently felled trees, and the adults feed on new seedlings of multiple conifer species (Table 13.1). Although predominantly semivoltine, its life cycle can range from 1 to 4 years, and can be influenced by the host tree species (Thorpe and Day, 2002). The majority of the life cycle is spent as larvae underground feeding within the phloem of root stumps. Adults overwinter in the soil or litter and emerge in the spring to feed on seedlings or twigs of mature trees before dispersing to new stumps where eggs are laid. Similar to other bark beetle species there is variability in life cycle timing among geographic locations, suggesting temperature is likely a dominant factor. The linear portion of the development–temperature relationship for multiple *Hyl. abietis* life stages was determined in laboratory conditions (Inward *et al.*, 2012). The length of time in the prepupal stage was highly variable with no discernible threshold, suggesting a potential facultative diapause (Christiansen, 1971; Inward *et al.*, 2012) similar to *Hyl. pales* (L.) (Salom *et al.*, 1987), *D. rufipennis* (Hansen *et al.*, 2001), and potentially *D. ponderosae* (Bentz, unpubl.). Although the adult stage was originally thought to have an obligatory reproductive diapause, recent evidence suggests some eggs can mature without adult overwintering (Tan *et al.*, 2010). Using the laboratory-derived data, a simulation model was developed (Wainhouse *et al.*, 2013). Degree-days were used to estimate development time for each of the life stages except the facultative diapause in prepupae, which was modeled with a non-linear log-logistic distribution bounded by data from laboratory experiments. When driven with climate change projections, by 2030 an increasing predominance of a 2-year life cycle, rather than the current 3-year life cycle, in northern and western United Kingdom was predicted. When the obligatory diapause in the adult stage was relaxed, life cycles of <2 years were produced (Wainhouse *et al.*, 2013).

Similar to other bark beetle species, *Hyl. abietis* is associated with ophiostomatoid fungi including *Leptographium proderum* (Kendr.) M. J. Wingf., and *O. quercus* (Georgev.) Nannf., although climate effects of these associates have not been investigated (Jankowiak and Bilański, 2013).

3.4.2 Ecosystem Models

No statistical ecosystem models that include climate variables to predict *Hyl. abietis* population outbreaks have been published.

3.4.3 Management Models

No models that include climate and options for evaluating the effect of management on *Hyl. abietis* outbreaks have been published.

3.5 *Hypothenemus hampei* (Ferrari)

3.5.1 Phenology Models

Hypothenemus hampei, the coffee berry borer, is endemic to Africa but has disseminated to most coffee growing countries worldwide (Chapter 11). The insect consumes the seeds within the berries of *Coffea arabica* L. and *C. canephora* Pierre ex A. Froehner and is the most economically important insect pest of coffee. Recently, *H. hampei* has been found attacking coffee plantations at higher elevations than historically observed (e.g., above 1500 m) (Jaramillo *et al.*, 2011). It has a multivoltine life cycle with overlapping developmental stages, and emergence is dictated by temperatures between 20 and 25°C (Baker *et al.*, 1992). As temperatures increase, female beetles lay more eggs, and disperse earlier (Jaramillo *et al.*, 2010).

A degree-day model was initially developed based on field-collected samples and temperatures in Brazil, and information from the literature (Gutierrez *et al.*, 1998). Later, based on data from controlled experiments in the laboratory using an *H. hampei* population from western Kenya, Jaramillo *et al.* (2009) found that no life stages developed at 15 and 35°C, suggesting these are the lower and upper development thresholds. Development rate peaked between 27 and 30°C depending on the life stage. These data were used to develop a linear, degree-day model of generation time. Using this model, the average number of generations per year in major coffee growing areas was estimated to vary from 1.3 in Ethiopia, to 3.4 in Colombia, to 3.1 in Kenya, and to 3.1 in Tanzania (Jaramillo *et al.*, 2009).

Degree-day information was incorporated into CLIMEX, an ecological niche modeling approach. Model projections suggest that a 1 to 2°C increase in temperature could lead to increased number of generations, and by 2050, *H. hampei* will be particularly damaging in major coffee growing areas that are currently considered marginal, including areas at higher elevations (Jaramillo *et al.*, 2011).

3.5.2 Ecosystem Models

There is a tight relationship between the reproductive phenology of berry ripening and *H. hampei* attack, in part due to changing conditions that favor beetle development. Ripening coffee berries release high quantities of volatile compounds, similar to compounds found to attract coniferous Scolytinae, that elicit responses from *H. hampei* (Jaramillo *et al.*, 2013; Rodríguez *et al.*, 2013). Given the importance of phenological matching between host plant and insect, Gutierrez *et al.* (1998) combined a temperature-dependent physiological model of *H. hampei* (based on the distributed delay concept of Manetsch, 1976) with a model of coffee fruiting dynamics that also included a model of three important parasitoids of *H. hampei*. Rodríguez *et al.* (2011, 2013) improved the model for coffee to better capture the effects of temperature on coffee fruiting, and also the influence of temperature and rainfall on *H. hampei* adult emergence using laboratory-derived data from Jaramillo *et al.* (2009). Model predictions compared favorably to field data (Rodríguez *et al.*, 2011). Although the influence of future climate on coffee and *H. hampei* using the coupled models has not been evaluated, coffee production is predicted to decrease by up to 10% due to effects on the plant alone (Gay *et al.*, 2006). Due to the effects of increasing temperature on both the plant and the insect, shading coffee plants is an important strategy that could lead to a decrease in temperature and potentially reduce the number of *H. hampei* generations (Jaramillo *et al.*, 2011).

3.5.3 Management Models

No models that include climate and options for evaluating the effect of management on *H. hampei* outbreaks have been published.

3.6 *Ips typographus* (L.)

3.6.1 Phenology Models

The range of *Ips typographus*, the European spruce bark beetle, is mainly determined by its principal host tree *Picea abies* (L.) H. Karst (Christiansen and Bakke, 1988). Life cycle timing is temperature dependent, and phenological models originated from a series of field studies and laboratory experiments in which the brood initiation and development were evaluated using thermal thresholds and temperature sums (degree-days) (Annila, 1969; Harding and Ravn, 1985; Netherer and Pennerstorfer, 2003; Wermelinger and Seifert, 1998). Later studies used information on temperature-dependent flight activity and development from egg to adult in combination with climate model projections to provide climate impact assessments (Lange *et al.*, 2006; Jönsson *et al.*, 2009). The PHENIPS model (Baier *et al.*, 2007) was developed to capture the seasonal development of *I. typographus* in mountainous regions. Digital elevation data are used in the model for interpolation of temperature and solar radiation to calculate bark temperature to simulate brood development. The phenological model by Jönsson *et al.* (2007), developed for Scandinavian boreal forest conditions, uses degree-day requirements calculated from gridded daily temperature data to simulate brood development in sun-exposed conditions (i.e., forest edges) and in shaded conditions (i.e., forest interior). The two approaches reflect regional differences in climate variability. Local topography can have a large influence on incoming solar radiation and air temperature, both influencing bark thermal conditions. Large-scale gridded climate data only provide useful approximations of temperature conditions in more homogeneous

landscapes, including most parts of Scandinavia. In the Scandinavian mountains, however, gridded climate data generally represents high altitude conditions, and are therefore not appropriate for model simulations of *I. typographus* in valleys at substantially lower altitudes (Jönsson *et al.*, 2011).

3.6.2 Ecosystem Models

In statistical ecosystem models, the risk of *I. typographus* damage is quantified by taking into account the interaction between the insect, climate, and host tree. These types of risk rating models have been developed based on statistical relationships between tree killing, predisposing site factors, and stand characteristics. For the high Tatra Mountains in central Europe, terrain, climate, soil, forest structure, tree species composition, vitality, and predisposition to storm and snow damage were identified as key factors (Netherer and Nopp-Mayr, 2005). For the European Alps, a discrete population model highlighted dry summers combined with warm temperatures as the main trigger of outbreaks (Marini *et al.*, 2012). Based on a PCR analysis of the spatio-temporal variation in Austria, Thom *et al.* (2013) concluded that predisposing factors such as species composition, climate conditions, and management had a larger influence on the risk of damage than inciting factors such as wind damage. Using a Poisson log-normal model in a Bayesian framework, Stadelmann *et al.* (2013) found a significant influence of temperature, volume of standing *P. abies*, and previous year infestation spots when analyzing the spatio-temporal dynamics of *I. typographus* in Switzerland. They also concluded that information about forest management of storm damage was not necessary for the model to provide accurate predictions. In contrast, a study of *I. typographus* population dynamics in Sweden, using a discrete population model and a multi-model inference approach, identified storm-felled trees as the main outbreak trigger, and the temperature-related metrics did not emerge as important drivers of population dynamics (Marini *et al.*, 2013). That is, biologically relevant spatial or temporal variation is needed to detect a significant influence. In the alpine region, voltinism is highly variable and dependent on altitude and exposure. In Sweden, however, temperature conditions generally allow for the production of only one generation per year and other driving factors, such as susceptible host trees, are often more important.

The relative strength of environmental variables influencing the occurrence of *I. typographus* damage is also depicted in ecosystem models developed for the different regions. The submodel of *I. typographus*-caused disturbances, developed and integrated in the hybrid forest patch model PICUS (Seidl *et al.*, 2007) and the large-scale forest scenario model EFISCEN (Seidl *et al.*, 2009), was set up to simulate *I. typographus* disturbances in alpine regions. The *I. typographus* submodel includes the effect of temperature by accounting for the number of generations as determined by the PHENIPS model (Baier *et al.*, 2007), in addition to the amount of host trees and occurrence of tree drought stress. The model does not explicitly consider the influence of brood material, created by windstorms or snow breakage, on the interannual development of *I. typographus* populations. These factors, however, are a main component in the *I. typographus* population model included in the ecosystem model LPJ-GUESS (Jönsson *et al.*, 2012). The model was set up to simulate Swedish conditions, and a sensitivity test indicated that tree drought stress had minor importance to the risk of damage in current climate conditions, similar to the findings of Marini *et al.* (2013). A sensitivity test suggested a major reduction in the risk of attacks on living trees could be achieved by timely salvage and cutting of infested trees. Temperli *et al.* (2013) concluded that climate change may shift the relative importance of the drivers of disturbances, including the distribution of host trees, by using a spatially explicit model of *I. typographus* dynamics that incorporates beetle phenology and forest susceptibility integrated with a climate-sensitive landscape model (LandClim; Seidl *et al.*, 2009).

3.6.3 Management Models

Inclusion of forest management in simulation models enables the user to evaluate management alternatives and policy options. Forest management can alter forest predisposition by shaping the forest stands and the landscape dynamics, and reactive strategies such as salvage and sanitary cutting can be of importance for modifying an outbreak pattern (Jönsson *et al.*, 2013). In a study in Slovenia, Ogris and Jurc (2010) predicted an increased need of sanitary felling due to *I. typographus* in response to climate change. Model simulations of adaptive management strategies (tree species change) revealed that there can be a considerable time lag between the start of adaptation measures and a decrease in bark beetle-caused damage (Seidl *et al.*, 2009). Fahse and Heurich (2011) developed a spatially explicit agent-based simulation model that takes into account individual trees and bark beetles to simulate the stand scale. The simulations provided a simple rule of thumb: if roughly 80% of individual beetles are killed by antagonists or foresters, outbreaks will rarely take place.

4. COMPARISONS AMONG BARK BEETLE SPECIES IN RESPONSE TO CLIMATE

4.1 Phenology Models

Phenology models incorporate thermally dependent traits that are important regulators of insect seasonality and

synchrony, and ultimately population fitness. Diapause provides a mechanism for synchronizing individuals and promoting seasonality (Denlinger, 2002). Stage-specific thresholds and rates of development can serve a similar role (Jenkins *et al.*, 2001). Of the bark beetle species examined in constant temperature experiments, pupation was delayed or did not occur when individuals were reared at constant temperatures between 10 and 15 to 17°C, either due to a facultative diapause or developmental threshold in the prepupal stage (Wagner *et al.*, 1984; Salom *et al.*, 1987; Wermelinger and Seifert 1998; Hansen *et al.*, 2011; Inward *et al.*, 2012; Régnière *et al.*, 2012). An exception is the tropical species *H. hampei* where none of the life stages developed at 15°C. In temperate species, this arrestment of pupation is hypothesized to be a mechanism that promotes the synchrony required for aggregated attacks on well-defended trees, in addition to reducing the probability of overwintering in the cold sensitive pupal life stage. Presence of this trait in multiple bark beetle species suggests it could be a derived trait due to common ancestry.

Incorporation of physiological processes into phenology models increases the probability that model predictions will capture seasonality, and hence emergent properties of population dynamics. The flexibility of a facultative prepupal diapause and developmental threshold for pupation allows species to shift among voltinism pathways depending on the timing and amount of thermal heat. For example, *D. rufipennis* and *D. ponderosae* often develop on a semivoltine life cycle in cool years, but can shift to univoltine in warm years (Hansen *et al.*, 2001; Werner *et al.*, 2006; Bentz *et al.*, 2014). Similarly, *I. typographus* can produce at least one extra generation in most parts of Europe in warm years (Jonsson *et al.*, 2011), and *Hyl. abietis* can complete a life cycle in 1 or 2 rather than 3 years (Tan *et al.*, 2010). Certain thermal regimes allow these shifts in voltinism yet also maintain seasonality due to the adaptive diapause and threshold processes evolved. Other thermal regimes, however, are predicted to cause shifts in voltinism that could also disrupt seasonality (Logan and Bentz, 1999; Hicke *et al.*, 2006). For example, warm summers could accelerate *D. ponderosae* development, but result in cold-sensitive life stages entering winter. Without adaptation, warmer climates could therefore lead to lower overall population fitness as a result of poor synchrony (Régnière *et al.*, submitted). Including these processes in analytic models, therefore, allows for predictions of thermal patterns that can result in both positive and negative species-specific population response in a changing climate. Without considering the evolved processes, misconceptions of increasing temperature on voltinism can occur (Bentz and Powell, in press). As quantitative data on evolutionary adaptive potential in changing thermal regimes are generated, mechanistic individual-based models provide a framework for including this important aspect of projecting population success in a changing climate (Régnière *et al.*, 2012).

Diapause in the adult state has also been found in multiple bark beetle species, and incorporated into models. For *I. typographus*, the timing of the spring swarming period is an important aspect of predicting forest damage. Only adults survive winter, in a diapause mediated by day length, and successful production of a second or third generation requires development to the adult stage prior to winter. Because photoperiod is considered a critical aspect of adult *I. typographus* diapause, the combined effects of day length and temperature were used by Jönsson *et al*. (2011) to show how diapause could limit bi- and trivoltinism in parts of this insect's range. Although adult diapause has been considered obligatory in the bark beetle species investigated (Safranyik *et al*., 1990; Doležal and Sehnal, 2007), genetic variation among individuals could allow for variability in diapause expression. When this occurs, warming temperatures could result in even shorter life cycles for some proportion of a population (Wainhouse *et al*., 2013).

For some species, only the linear portion of the development response (i.e., degree-days) is known and incorporated into mechanistic models. Specific upper and lower thresholds may therefore not be included. High temperature thresholds, in particular, play an important role in population response to a warming climate as temperatures exceed the high temperature threshold. A rise in development rate with increasing temperature up to some maximum, followed by a rapid decline, is common in insects (Janisch, 1932). Of the six species included here, *D. ponderosae*, *D. rufipennis*, and *Hyl. abietis* have thermal optimum less than 27°C, and the thermal optimum for *I. typographus*, *D. frontalis*, and *H. hampei* is around 30°C. Species already living at or near their thermal maximum (i.e., small thermal safety margin) may be more impacted by climate warming as temperatures exceed optimal thresholds. Populations in cooler environments typically have thermal optima higher than their current environment (i.e., large thermal safety margin), relative to populations in warmer environments (Deutsch *et al*., 2008), and therefore are initially responding positively to warming temperatures. For example, at many cool, high elevation sites, recent temperature increases remain in the thermal range of increasing development for *D. ponderosae* resulting in a decrease in generation time. At warm, low elevation sites, however, temperatures prior to warming were already at or near the threshold for optimal development and slight increases in warming have had little effect on generation time (Bentz *et al*., 2014). Faccolli (2009) found a similar result for *I. typographus* during a warm year (2003) in the southeastern Alps. Increases in generation time, rather than reductions, could occur as increasing temperatures exceed thermal optima, particularly those species with the highest thermal optima.

The multivoltine *H. hampei* was predicted to have significantly increased number of generations per year (i.e., up to 10 generations) by 2050 based on degree-day estimates (Jaramillo *et al*., 2011), yet the authors acknowledge that average daily temps >26°C could lead to a reduction in *H. hampei* population growth as temperatures exceed optimal thermal maxima. Incorporation of high temperature thresholds and mechanisms that promote seasonality into the model could result in more detailed estimates of generation time. It will also be important to adequately model the relationship between air temperature and the temperature of the beetle habitat. High temperatures in particular can be influenced by such factors as direct solar radiation (Harding and Ravn, 1985, Bolstad *et al*., 1997).

Phenology models that incorporate cold tolerance (i.e., *D. ponderosae*, *D. frontalis*) suggest higher probability of survival as winters warm, particularly in habitats further north of historical range boundaries (Ungerer *et al*., 1999; Bentz *et al*., 2010; Safranyik *et al*., 2010), and at higher elevations (Régnière *et al*., submitted). Models for species with an obligatory adult diapause (i.e., *D. rufipennis*, *I. typographus*, *Hyl. abietis*) have not considered the role of cold temperature on survival, and instead focus on predicting if summer temperatures and fall day length result in a developmental pathway that ensures progression to the adult life stage prior to winter.

4.2 Ecosystem and Management Models

The simplest form of model developed was based on a statistical association between past population outbreaks and the values of climate and host tree variables present during the years of outbreak. There were several commonalities in model results among the species. In each study that included population size in neighboring areas, this variable was found to be one of the most important for describing outbreak potential and severity (Aukema *et al*., 2008; Duehl *et al*., 2011; Preisler *et al*., 2012). Due to the need for aggregated attacks, bark beetle population growth is known to be highly density dependent (Berryman, 1982; Martinson *et al*., 2013). Some metric of temperature and/ or precipitation was also found to be important for all species.

Although in most cases precipitation does not influence bark beetle generation time directly, when rain events occur during bark beetle emergence and dispersal, precipitation can disrupt life cycle timing and beetle survival (Rodríguez *et al*., 2011). Years of high spring precipitation were correlated with lower damage due to *I. typographus* (Faccolli 2009), potentially due to disruption of adult emergence timing. Conversely, increased precipitation can have a positive effect on bark beetle population growth by providing a more nutritious food resource for developing larvae as was found for *D. frontalis* (Gumpertz and

Pye, 2000; Duehl *et al.*, 2011) and *D. ponderosae* (Preisler *et al.*, 2012). Reduced precipitation can also create susceptible habitat for bark beetles by stressing host plants (McDowell *et al.*, 2008; Gaylord *et al.*, 2013). Reduced precipitation in the current year and years leading up to *D. frontalis*, *D. ponderosae*, and *D. rufipennis* outbreaks were also found to be significant in explaining outbreak presence (Hebertson and Jenkins, 2008; Safranyik *et al.*, 1975; Evangelista *et al.*, 2011; Chapman *et al.*, 2012; Preisler *et al.*, 2012; Hart *et al.*, 2013). For *I. typographus*, however, the strength of the precipitation effect varied geographically. Dry summers were found to be important in the European Alps (Faccoli, 2009; Marini *et al.*, 2012), but in Sweden only one of four sites showed a correlation between low precipitation and *I. typographus* damage, and the presence of storm-felled trees was most important (Marini *et al.*, 2013). Whether a drought or storm event of significant magnitude occurs during the years and location of a statistically-based study could influence whether these variables are found to be significant drivers of population outbreaks. For example, recent statistical models developed for *D. ponderosae* and *D. rufipennis* included the years 2001–2002, one of the most severe droughts in the past 500 years in many parts of the interior west of the USA (Pielke *et al.*, 2005).

Warm summer temperatures, August maxima in particular, were found to be important predictors for *D. ponderosae* and *D. rufipennis*. Warm temperatures in late summer dictate the overwintering life stage, and in the case of *D. rufipennis* if larval diapause is invoked. Warm springs were important for *D. frontalis* and *I. typographus*, influencing the timing of adult emergence and the potential for additional generations. Variability in temperature increase was not found to be important for *D. ponderosae*, but was for *D. frontalis*, potentially due to the multivoltine nature of *D. frontalis*. Minimum temperature, either as a climatic variable or as output from a process-based cold tolerance model, was found to be an important predictor for *D. ponderosae*, and in more northern parts of the *D. frontalis* range. While it is clear that warm temperatures and reduced precipitation can directly and indirectly affect both bark beetles and their host trees, the relative role of insects in subsequent tree mortality is not always clear, and will differ among species. Species with mechanisms for positive feedback following a host stressing event such as drought (i.e., the species included in this review) are not as dependent on continuation of drought for outbreak continuation as species that are only capable of population buildup when trees are stressed (Raffa *et al.*, 2008). In general, warm summer temperatures positively influence bark beetle population success and the presence of a drought event will magnify the effect. Drought in the absence of warm temperatures will have lessened effect on bark beetle population outbreaks.

5. MODEL LIMITATIONS

Mechanistic-based phenology models can provide predictions of population response that incorporate the important role of seasonality and allow for population processes to emerge when driven by climate change projections. Because of this, these types of models, compared to purely statistical associations of climate variables and numbers of bark beetle-killed trees, will be important to the management of future ecosystems. Data describing specific physiological mechanisms, however, including diapause and stage-specific thresholds and development rates, are not available for many bark beetle species that can cause landscape-scale economic and ecological impacts. Many of these species have expansive ranges that follow the distribution of their host plants, yet their distributions are currently limited by climate, most likely due to a lack of sufficient thermal input or temperatures beyond evolved tolerances. To adequately predict range expansions and future impacts, a mechanistic understanding of thermal responses will be needed.

It is clear that developmental responses to temperature are highly evolved traits (Angilletta *et al.*, 2002) that can vary latitudinally among and within species as populations adapt to strong selection pressures exerted by local climates (Deutsch *et al.*, 2008). Yet data on such adaptations are only available for a single bark beetle species (Bentz *et al.*, 2001, 2011), and specific data required for region-specific model development are lacking. The potential for adaptation in specific thermal responses is also unclear and not currently included in any bark beetle phenology model. This type of advancement can be readily undertaken with individual-based models (Régnière *et al.*, 2012) and models that can project potential changes with future selection pressures (Jönsson *et al.*, 2011). To adequately address this challenge, however, quantitative data, including genomic data, on inheritance of thermal-response traits are needed.

Incorporation of the important effects of host tree size and stand composition at local and landscape scale is needed in both mechanistic and statistical models. While temperature may directly control bark beetle population dynamics, the presence of suitable habitat, and more importantly, the structure and composition of that habitat, can be more influential in determining whether bark beetles will occur at a particular location (DeRose *et al.*, 2013). Regardless of temperature conditions, suitable hosts must be available. Powell and Bentz (2014) connected a mechanistic model that incorporates temperature-dependent *D. ponderosae* population dynamics with remotely sensed data describing the available host tree landscape. Although the spread of *D. ponderosae* from established pockets was predicted with substantial accuracy, the model was unable to predict establishment of new infested pockets. Jönsson *et al.* (2012) also found that knowledge of initial beetle population size and location was important for modeling *I. typographus* migration across a landscape of hosts. A major hurdle to incorporating landscape-scale host tree information is adequate data at the appropriate scale. These results also highlight the need for quantitative measures of climate effects on tree vulnerability to bark beetles (e.g., host defenses) for predicting the establishment and growth of endemic populations that are unable to overwhelm tree defenses by mass attack. A major challenge for climate change impact studies is to quantify how a series of events that stress trees can influence vulnerability to bark beetle attack. An understanding of the connection between tree-level processes, climate, and increased susceptibility to insects will further the capacity to include bark beetles and other insect guilds into Dynamic Global Vegetation Models (DGVM) used to simulate changes in host distribution as an effect of climate change.

Although all bark beetle species have a wide array of community associates including fungi, insects, and microbes that could have different responses to climate, very little quantitative data on thermal responses are available for most species. Associated fungi can play a role in depleting host tree resources used in defense (Lahr and Krokene, 2013), and these captured resources can provide vital nutrients to developing larvae and adults (Ayres *et al.*, 2000; Bentz and Six, 2006). To adequately project bark beetle population success in future climates, knowledge on temperature response for the associated fungi is needed (Addison *et al.*, 2013). The same can be said for invertebrate natural enemies that are important factors in the dynamics of some bark beetles, each with its own complex relationship with temperature, including physiological processes such as diapause that control seasonality (Reeve, 2000).

When analyzing the outcome of model simulations that capture the impact of climate change, it is important to acknowledge uncertainties associated with both the driving climate data and the biological model. Although confidence in projections of future climate has increased, a wide range of potential changes from climate model projections exist because multiple options for future technological developments, human behavior, and demography are considered (IPCC, 2000; Walsh *et al.*, 2014). An array of new scenarios have been produced ranging from RCP 2.6, which assumes rapid reductions in emissions and a smaller amount of warming, to RCP 8.5, which assumes continued increases in emissions and a corresponding greater amount of warming. Likewise, global scenarios (SRES) are categorized into families (i.e., A1, A2, B1, and B2) that represent the influence of varying demographic and economic driving forces on greenhouse gas emissions (IPCC, 2000). Assessment of management strategies for future forests would benefit from a comparison of output from bark beetle models driven by a range of scenarios that are optimistic,

moderate, and pessimistic. Ensemble simulations using several climate model scenarios can also highlight particular aspects and parameters of bark beetle and ecosystem models that are particularly sensitive to climate shifts.

6. CONCLUSIONS

It is clear that bark beetles are directly and indirectly influenced by climate. Acknowledging this, a variety of modeling approaches have been used to describe climate effects on bark beetle population outbreak potential. Of the more than 30 bark beetle species found worldwide that are capable of causing landscape-scale plant mortality, only six species have sufficient information for development of climate-driven phenology and ecosystem models (Table 13.1). Bark beetles are major disturbance agents in some agriculture and most forest ecosystems, and an understanding of their species-specific community response is needed to adequately predict future distributions and potential impacts to the ecosystems they inhabit. In some regions, predicted increases in climate extremes will result in increased vulnerability of host trees. With a dramatically faster reproduction rate, however, bark beetles and their community associates will have the capacity to respond more quickly than their host trees to a rapidly change climate.

Phenology and ecosystem models have proven useful in evaluating the relative response of populations among current and potentially new habitats, and also provide a framework to assess management options, and vulnerability and sensitivity of ecosystems to climate. Because phenology models mechanistically describe the evolved response of insects to temperature, this type of model can highlight thermal regimes that both reduce and increase population success. Range expansion northward of several bark beetle species over the past decade emphasizes how insects can respond to fluctuating conditions through phenotypic plasticity. The variability among individuals in response to temperature that is included in most mechanistic phenology models accounts for this phenotypic plasticity. Phenotypic plasticity, however, is not a long-term solution to changing climate. Incorporation of the potential for evolutionary adaptation in thermally dependent processes and quantitative descriptions of intraspecific genetic variability will improve predictions of bark beetle response and range expansion in a changing climate.

The important role of host tree vulnerability, density, and continuity across a landscape in bark beetle population outbreaks is clear, but generally not currently included in modeling frameworks. As spatial representations of vegetation data become more available, and research on climate effects on host vulnerability is advanced, these critical aspects can be more thoroughly incorporated into models. Although species-specific thermal responses are apparent, there are several shared traits among bark beetle species that may facilitate a common process-based modeling framework that can be used to create climate-sensitive stochastic mortality functions in DGVM and forest growth models.

ACKNOWLEDGMENTS

AMJ was supported by FORMAS (2010-822). We thank Jacques Régnière for comments on an earlier draft.

REFERENCES

Addison, A.L., Powell, J.A., Six, D.L., Moore, M., Bentz, B.J., 2013. The role of temperature variability in stabilizing the mountain pine beetle-fungus mutualism. J. Theor. Biol. 335, 40–50.

Allen, C.D., Macalady, A.K., Chenchouni, H., Bachelet, D., McDowell, N., Vennetier, M., et al., 2010. A global overview of drought and heat-induced tree mortality reveals emerging climate change risks for forests. For. Ecol. Manage. 259, 660–684.

Amman, G.D., 1973. Population changes of the mountain pine beetle in relation to elevation. Environ. Entomol. 2, 541–548.

Amman, G.D., McGregor, M.D., Cahill, D.B., Klein, W.H., 1977. Guidelines for reducing losses of lodgepole pine to the mountain pine beetle in unmanaged stands in the Rocky Mountains, General Technical Report INT-36. United States Department of Agriculture Forest Service, Ogden, UT.

Angilletta Jr., M.J., Niewiarowski, P.H., Navas, C.A., 2002. The evolution of thermal physiology in ectotherms. J. Therm. Biol. 27, 249–268.

Annila, E., 1969. Influence of temperature upon the development and voltinism of *Ips typographus* L. (Coleoptera, Scolytidae). Ann. Zool. Fenn. 6, 161–208.

Aukema, B.H., Carroll, A.L., Zheng, Y., Zhu, J., Raffa, K.F., Dan Moore, R., et al., 2008. Movement of outbreak populations of mountain pine beetle: influences of spatiotemporal patterns and climate. Ecography 31, 348–358.

Ayres, M.P., Wilkens, R.T., Ruel, J.J., Lombardero, M.J., Vallery, E., 2000. Nitrogen budgets of phloem-feeding bark beetles with and without symbiotic fungi. Ecology 81, 2198–2210.

Baier, P., Pennerstorfer, J., Schopf, A., 2007. PHENIPS—a comprehensive phenology model of *Ips typographus* (L.) (Col., Scolytinae) as a tool for hazard rating of bark beetle infestation. For. Ecol. Manage. 249, 171–186.

Baker, P., Ley, C., Balbuena, R., Barrera, J., 1992. Factors affecting the emergence of *Hypothenemus hampei* (Coleoptera: Scolytidae) from coffee berries. Bull. Entomol. Res. 82, 145–150.

Bale, J.S., Masters, G.J., Hodkinson, I.D., Awmack, C., Bezemer, T.M., Brown, V.K., et al., 2002. Herbivory in global climate change research: direct effects of rising temperature on insect herbivores. Glob. Chang. Biol. 8, 1–16.

Beal, J.A., 1933. Temperature extremes as a factor in the ecology of the southern pine beetle. J. For. 31, 329–336.

Bentz, B.J., Mullins, D.E., 1999. Ecology of mountain pine beetle (Coleoptera: Scolytidae) cold hardening in the Intermountain West. Environ. Entomol. 28, 577–587.

Bentz, B.J., Six, D.L., 2006. Ergosterol content of fungi associated with *Dendroctonus ponderosae* and *Dendroctonus rufipennis* (Coleoptera: Curculionidae, Scolytinae). Ann. Entomol. Soc. Am. 99, 189–194.

Bentz, B. J., and Powell, J. A. Mountain pine beetle seasonal timing and constraints to bivoltinism: a comment on Mitton and Ferrenberg. Am. Nat. in press.

Bentz, B.J., Logan, J.A., Amman, G.D., 1991. Temperature-dependent development of the mountain pine beetle (Coleoptera: Scolytidae) and simulation of its phenology. Can. Entomol. 123, 1083–1094.

Bentz, B.J., Logan, J.A., Vandygriff, J.C., 2001. Latitudinal variation in *Dendroctonus ponderosae* (Coleoptera: Scolytidae) development time and adult size. Can. Entomol. 133, 375–387.

Bentz, B.J., Régnière, J., Fettig, C.J., Hansen, E.M., Hayes, J.L., Hicke, J. A., et al., 2010. Climate change and bark beetles of the western United States and Canada: direct and indirect effects. BioSci. 60, 602–613.

Bentz, B.J., Bracewell, R.R., Mock, K.E., Pfrender, M.E., 2011. Genetic architecture and phenotypic plasticity of thermally-regulated traits in an eruptive species, *Dendroctonus ponderosae*. Evol. Ecol. 25, 1269–1288.

Bentz, B., Vandygriff, J., Jensen, C., Coleman, T., Maloney, P., Smith, S., et al., 2014. Mountain pine beetle voltinism and life history characteristics across latitudinal and elevational gradients in the western United States. For. Sci. in press.

Berg, E.E., David Henry, J., Fastie, C.L., De Volder, A.D., Matsuoka, S.M., 2006. Spruce beetle outbreaks on the Kenai Peninsula, Alaska, and Kluane National Park and Reserve, Yukon Territory: relationship to summer temperatures and regional differences in disturbance regimes. For. Ecol. Manage. 227, 219–232.

Berryman, A.A., 1982. Biological control, thresholds, and pest outbreaks. Environ. Entomol. 11, 544–549.

Birch, M.C., 1974. Seasonal variation in pheromone-associated behavior and physiology of *Ips pini*. Ann. Entomol. Soc. Am. 67, 58–60.

Bolstad, P.V., Bentz, B.J., Logan, J.A., 1997. Modelling micro-habitat temperature for *Dendroctonus ponderosae* (Coleoptera: Scolytidae). Ecol. Model. 94, 287–297.

Bracewell, R.R., Pfrender, M.E., Mock, K.E., Bentz, B.J., 2013. Contrasting geographic patterns of genetic differentiation in body size and development time with reproductive isolation in *Dendroctonus ponderosae* (Coleoptera: Curculionidae, Scolytinae). Ann. Entomol. Soc. Am. 106, 385–391.

Bradshaw, W.E., Holzapfel, C.M., 2010. What season is it anyway? Circadian tracking vs. photoperiodic anticipation in insects. J. Biol. Rhythms 25, 155–165.

Carroll, A.L., Taylor, S.W., Régnière, J., Safranyik, L., 2004. Effects of climate change on range expansion by the mountain pine beetle in British Columbia. In: Shore, T.L., Brooks, J.E., Stone, J.E. (Eds.), Mountain Pine Beetle Symposium: Challenges and Solutions". Information Report BC-X-399, Canadian Forest Service, Victoria, British Columbia, pp. 223–232.

Chapman, T.B., Veblen, T.T., Schoennagel, T., 2012. Spatiotemporal patterns of mountain pine beetle activity in the southern Rocky Mountains. Ecology 93, 2175–2185.

Christiansen, E., 1971. Laboratory study on factors influencing preimaginal development in *Hylobius abietis* L. (Co., Curculionidae). Norweg. J. Entomol. 18, 1–8.

Christiansen, E., Bakke, A., 1988. The spruce bark beetle of Eurasia. In: Berryman, A.A. (Ed.), Dynamics of Forest Insect Populations. Plenum Publishing, New York, pp. 479–503.

Clark, E.W., 1974. Reproductive diapause in *Hylobius pales*. Ann. Entomol. Soc. Am. 68, 349–352.

Cooke, B.J., 2009. Forecasting mountain pine beetle-overwintering mortality in a variable environment. Natural Resources Canada, Canadian forest Service, Pacific Forestry Centre, Victoria BC, Mountain Pine Beetle Working Paper 2009–03.

Cooke, B.J., Régnière, J., 1996. An object-oriented, process-based stochastic simulation model of *Bacillus thuringiensis* efficacy against spruce budworm, *Choristoneura fumiferana* (Lepidoptera: Tortricidae). Int. J. Pest. Manage. 42, 291–306.

Coops, N.C., Wulder, M.A., Waring, R.H., 2012. Modeling lodgepole and jack pine vulnerability to mountain pine beetle expansion into the western Canadian boreal forest. For. Ecol. Manage. 274, 161–171.

Coulson, R.N., Feldman, R., Sharpe, P., Pulley, P., Wagner, T., Payne, T., 1989. An overview of the TAMBEETLE model of *Dendroctonus frontalis* population dynamics. Ecography 12, 445–450.

Creeden, E.P., Hicke, J.A., Buotte, P.C., 2014. Climate, weather, and recent mountain pine beetle outbreaks in the western United States. For. Ecol. Manage. 312, 239–251.

Cudmore, T.J., Björklund, N., Carroll, A.L., Lindgren, S.B., 2010. Climate change and range expansion of an aggressive bark beetle: evidence of higher beetle reproduction in naïve host tree populations. J. Appl. Ecol. 47, 1036–1043.

Cullingham, C.I., Cooke, J.E., Dang, S., Davis, C.S., Cooke, B.J., Coltman, D.W., 2011. Mountain pine beetle host-range expansion threatens the boreal forest. Mol. Ecol. 20, 2157–2171.

Danks, H., 2007. The elements of seasonal adaptations in insects. Can. Entomol. 139, 1–44.

de la Giroday, H.M.C., Carroll, A.L., Aukema, B.H., 2012. Breach of the northern Rocky Mountain geoclimatic barrier: initiation of range expansion by the mountain pine beetle. J. Biogeogr. 39, 1112–1123.

Denlinger, D.L., 2002. Regulation of diapause. Annu. Rev. Entomol. 47, 93–122.

DeRose, R.J., Bentz, B.J., Long, J.N., Shaw, J.D., 2013. Effect of increasing temperatures on the distribution of spruce beetle in Engelmann spruce forests of the Interior West, USA. For. Ecol. Manage. 308, 198–206.

Deutsch, C.A., Tewksbury, J.J., Huey, R.B., Sheldon, K.S., Ghalambor, C.K., Haak, D.C., Martin, P.R., 2008. Impacts of climate warming on terrestrial ectotherms across latitude. Proc. Natl. Acad. Sci. U. S. A. 105, 6668–6672.

Doležal, P., Sehnal, F., 2007. Effects of photoperiod and temperature on the development and diapause of the bark beetle *Ips typographus*. J. Appl. Entomol. 131, 165–173.

Duehl, A., Bishir, J., Hain, F.P., 2011. Predicting county-level southern pine beetle outbreaks from neighborhood patterns. Environ. Entomol. 40, 273–280.

Evangelista, P.H., Kumar, S., Stohlgren, T.J., Young, N.E., 2011. Assessing forest vulnerability and the potential distribution of pine beetles under current and future climate scenarios in the Interior West of the US. For. Ecol. Mgmt. 262, 307–316.

Faccoli, M., 2009. Effect of weather on *Ips typographus* (Coleoptera Curculionidae) phenology, voltinism, and associated spruce mortality in the southeastern Alps. Environ. Entomol. 38, 307–316.

Fahse, L., Heurich, M., 2011. Simulation and analysis of outbreaks of bark beetle infestations and their management at the stand level. Ecol. Model. 222, 1833–1846.

Feldman, R.M., Curry, G.L., Coulson, R.N., 1981. A mathematical model of field population dynamics of the southern pine beetle, *Dendroctonus frontalis*. Ecol. Model. 13, 261–281.

Friedenberg, N.A., Powell, J.A., Ayres, M.P., 2007. Synchrony's double edge: transient dynamics and the Allee effect in stage structured populations. Ecol. Lett. 10, 564–573.

Gan, J., 2004. Risk and damage of southern pine beetle outbreaks under global climate change. For. Ecol. Mgmt. 191, 61–71.

Gay, C., Estrada, F., Conde, C., Eakin, H., Villers, L., 2006. Potential impacts of climate change on agriculture: a case of study of coffee production in Veracruz, México. Clim. Change 79, 259–288.

Gaylord, M.L., Kolb, T.E., Pockman, W.T., Plaut, J.A., Yepez, E.A., Macalady, A.K., et al., 2013. Drought predisposes pinon-juniper woodlands to insect attacks and mortality. New Phytol. 198, 567–578.

Gehrken, U., 1984. Winter survival of an adult bark beetle *Ips acuminatus* Gyll. J. Insect Physiol. 30, 421–429.

Ghosh, J., 1983. Mathematical and simulation modeling of southern pine beetle infestations. Ph.D. thesis, University of Arkansas, Fayetteville.

Gilbert, E., Powell, J.A., Logan, J.A., Bentz, B.J., 2004. Comparison of three models predicting developmental milestones given environmental and individual variation. Bull. Math. Biol. 66, 1821–1850.

Goldman, S., Franklin, R., 1977. Development and feeding habits of southern pine beetle larvae. Ann. Entomol. Soc. Am. 70, 54–56.

Gumpertz, L., Pye, M., 2000. Logistic regression for southern pine beetle outbreaks with spatial and temporal autocorrelation. For. Sci. 46, 95–107.

Gutierrez, A.P., Villacorta, A., Cure, J.R., Ellis, C.K., 1998. Tritrophic analysis of the coffee (*Coffea arabica*)-coffee berry borer [*Hypothenemus hampei* (Ferrari)]-parasitoid system. An. Soc. Entomol. Bras. 27, 357–385.

Hansen, E.M., Bentz, B.J., 2003. Comparison of reproductive capacity among univoltine, semivoltine, and re-emerged parent spruce beetles (Coleoptera: Scolytidae). Can. Entomol. 135, 697–712.

Hansen, E.M., Bentz, B.J., Turner, D.L., 2001. Physiological basis for flexible voltinism in the spruce beetle (Coleoptera: Scolytidae). Can. Entomol. 133, 805–817.

Hansen, E.M., Bentz, B.J., Powell, J.A., Gray, D.R., Vandygriff, J.C., 2011. Prepupal diapause and instar IV development rates of spruce beetle, *Dendroctonus ruifpennis* (Coleoptera: Curculionidae, Scolytinae). J. Insect Physiol. 57, 1347–1357.

Harding, S., Ravn, H., 1985. Seasonal activity of *Ips typographus* L. (Col., Scolytidae) in Denmark. J. Appl. Entomol. 99, 123–131.

Hart, S.J., Veblen, T.T., Eisenhart, K.S., Jarvis, D., Kulakowski, D., 2013. Drought induces spruce beetle (*Dendroctonus rufipennis*) outbreaks across northwestern Colorado. Ecology 95, 930–939.

Hebertson, E.G., Jenkins, M.J., 2008. Climate factors associated with historic spruce beetle (Coleoptera: Curculionidae) outbreaks in Utah and Colorado. Environ. Entomol. 37, 281–292.

Hicke, J.A., Logan, J.A., Powell, J., Ojima, D.S., 2006. Changing temperatures influence suitability for modeled mountain pine beetle (*Dendroctonus ponderosae*) outbreaks in the western United States. J. Geophys. Res.: Biogeosci. 111, G02019.

Hofstetter, R., Dempsey, T., Klepzig, K., Ayres, M., 2007. Temperature-dependent effects on mutualistic, antagonistic, and commensalistic interactions among insects, fungi and mites. Community Ecol. 8, 47–56.

Inward, D.J.G., Wainhouse, D., Peace, A., 2012. The effect of temperature on the development and life cycle regulation of the pine weevil *Hylobius abietis* and the potential impacts of climate change. Agr. Forest Entomol. 14, 348–357.

IPCC, 2000. Intergovernmental Panel on Climate Change. Summary for Policymakers. Emissions Scenarios. A Special Report of IPPC Working Group III. Intergovernmental Panel on Climate Change.

IPCC, 2013, 2013. Summary for Policymakers. In: Stocker, T.F., Qin, D., Plattner, G.K., Tignor, M., Allen, S.K., Boschung, J., et al., (Eds.), Climate Change 2013: The Physical Science Basis, Contributions of Working Group I to the Fifth Assessment Report of the Intergovernmental Panel on Climate Change. Cambridge University Press, Cambridge.

Janisch, E., 1932. The influence of temperature on the life-history of insects. Transactions of the Royal Entomological Society of London 80, 137–168.

Jankowiak, R., Bilański, P., 2013. Diversity of ophiostomatoid fungi associated with the large pine weevil, *Hylobius abietis*, and infested Scots pine seedlings in Poland. Ann. For. Sci. 70, 391–402.

Jaramillo, J., Chabi-Olaye, A., Kamonjo, C., Jaramillo, A., Vega, F.E., Poehling, H.-M., Borgemeister, C., 2009. Thermal tolerance of the coffee berry borer *Hypothenemus hampei*: predictions of climate change impact on a tropical insect pest. PLoS One 4, e6487.

Jaramillo, J., Chabi-Olaye, A., Borgemeister, C., 2010. Temperature-dependent development and emergence pattern of *Hypothenemus hampei* (Coleoptera: Curculionidae: Scolytinae) from coffee berries. J. Econ. Entomol. 103, 1159–1165.

Jaramillo, J., Muchugu, E., Vega, F.E., Davis, A., Borgemeister, C., Chabi-Olaye, A., 2011. Some like it hot: the influence and implications of climate change on coffee berry borer (*Hypothenemus hampei*) and coffee production in East Africa. PLoS One 6, e24528.

Jaramillo, J., Torto, B., Mwenda, D., Troeger, A., Borgemeister, C., Poehling, H.-M., Francke, W., 2013. Coffee berry borer joins bark beetles in coffee klatch. PLoS One 8, e74277.

Jenkins, J.L., Powell, J.A., Logan, J.A., Bentz, B.J., 2001. Low seasonal temperatures promote life cycle synchronization. Bull. Math. Biol. 63, 573–595.

Jönsson, A.M., Harding, S., Bärring, L., Ravn, H.P., 2007. Impact of climate change on the population dynamics of *Ips typographus* in southern Sweden. Agr. Forest. Meteorol. 146, 70–81.

Jönsson, A.M., Appelberg, G., Harding, S., Bärring, L., 2009. Spatio-temporal impact of climate change on the activity and voltinism of the spruce bark beetle, *Ips typographus*. Glob. Chang. Biol. 15, 486–499.

Jönsson, A.M., Harding, S., Krokene, P., Lange, H., Lindelöw, Å., Økland, B., et al., 2011. Modelling the potential impact of global warming on *Ips typographus* voltinism and reproductive diapause. Clim. Change 109, 695–718.

Jönsson, A.M., Schroeder, L.M., Lagergren, F., Anderbrant, O., Smith, B., 2012. Guess the impact of *Ips typographus*—an ecosystem modelling approach for simulating spruce bark beetle outbreaks. Agr. Forest. Meteorol. 166–167, 188–200.

Jönsson, A., Lagergren, F., Smith, B., 2013. Forest management facing climate change—an ecosystem model analysis of adaptation strategies. Mitigation and Adaptation Strategies for Global Change. http://dx.doi.org/10.1007/s11027-013-9487-6.

Klepzig, K.D., Wilkens, R.T., 1997. Competitive interactions among symbiotic fungi of the southern pine beetle. Appl. Environ. Microbiol. 63, 621–627.

Koštál, V., Doležal, P., Rozsypal, J., Moravcová, M., Zahradníčková, H., Šimek, P., 2011. Physiological and biochemical analysis of overwintering and cold tolerance in two Central European populations of the spruce bark beetle, Ips typographus. J. Insect Physiol. 57, 1136–1146.

Lahr, E.C., Krokene, P., 2013. Conifer stored resources and resistance to a fungus associated with the spruce bark beetle *Ips typographus*. PLoS One 8, e72405.

Lange, H., Økland, B., Krokene, P., 2006. Thresholds in the life cycle of the spruce bark beetle under climate change. InterJournal 1648, 1–10.

Langor, D.W., Raske, A.G., 1987. Reproduction and development of the eastern larch beetle, *Dendroctonus simplex* Leconte (Coleoptera: Scolytidae) in Newfoundland. Can. Entomol. 119, 985–992.

Lee, R.E., 1991. Principles of insect low temperature tolerance. In: Lee, R.E., Denlinger, D.L. (Eds.), Insects at Low Temperature. Chapman and Hall, New York, pp. 17–46.

Lester, J.D., Irwin, J.T., 2012. Metabolism and cold tolerance of overwintering adult mountain pine beetles (*Dendroctonus ponderosae*): Evidence of facultative diapause? J. Insect Physiol. 58, 808–815.

Lewis, M., 2011. Modeling Phloem Temperatures Relative to Mountain Pine Beetle Development. M.S. thesis. Utah State University, Logan, UT. 58 p.

Lih, M.P., Stephen, F.M., 1989. Modeling southern pine beetle (Coleoptera: Scolytidae) population dynamics: methods, results and impending challenges. In: McDonald, L.L., Manly, B.F.J., Lockwood, J.A., Logan, J.A. (Eds.), Estimation and Analysis of Insect Populations. Springer, Heidelberg, pp. 256–267.

Lobell, D.B., Schlenker, W., Costa-Roberts, J., 2011. Climate trends and global crop production since 1980. Science 333, 616–620.

Logan, J.A., 1988. Toward an expert system for development of pest simulation models. Environ. Entomol. 17, 359–376.

Logan, J.A., Amman, G., 1986. A distribution model for egg development in mountain pine beetle. Can. Entomol. 118, 361–372.

Logan, J.A., Bentz, B.J., 1999. Model analysis of mountain pine beetle (Coleoptera: Scolytidae) seasonality. Environ. Entomol. 28, 924–934.

Logan, J.A., Powell, J.A., 2001. Ghost forests, global warming, and the mountain pine beetle (Coleoptera: Scolytidae). American Entomologist 47, 160.

Lombardero, M.J., Ayres, M.P., Ayres, B.D., Reeve, J.D., 2000. Cold tolerance of four species of bark beetle (Coleoptera: Scolytidae) in North America. Environ. Entomol. 29, 421–432.

Lombardero, M.J., Ayres, M.P., Hofstetter, R.W., Moser, J.C., Klepzig, K.D., 2003. Strong indirect interactions of *Tarsonemus* mites (Acarina: Tarsonemidae) and *Dendroctonus frontalis* (Coleoptera: Scolytidae). Oikos 102, 243–252.

Manetsch, T.J., 1976. Time-varying distributed delays and their use in aggregative models of large systems. IEEE Trans. Syst. Man Cybern. 6, 547–553.

Marini, L., Ayres, M.P., Battisti, A., Faccoli, M., 2012. Climate affects severity and altitudinal distribution of outbreaks in an eruptive bark beetle. Clim. Change 115, 327–341.

Marini, L., Lindelow, A., Jönsson, A.M., Wulff, S., Schroeder, L.M., 2013. Population dynamics of the spruce bark beetle: a long-term study. Oikos 122, 1768–1776.

Martinson, S.J., Ylioja, T., Sullivan, B.T., Billings, R.F., Ayres, M.P., 2013. Alternate attractors in the population dynamics of a tree-killing bark beetle. Pop. Ecol. 55, 95–106.

McDowell, N., Pockman, W.T., Allen, C.D., Breshears, D.D., Cobb, N., Kolb, T., et al., 2008. Mechanisms of plant survival and mortality during drought: why do some plants survive while others succumb to drought? New Phytol. 178, 719–739.

Meddens, A.J., Hicke, J.A., Ferguson, C.A., 2012. Spatiotemporal patterns of observed bark beetle-caused tree mortality in British Columbia and the western United States. Ecol. Appl. 22, 1876–1891.

Miller, L.K., Werner, R.A., 1987. Cold-hardiness of adult and larval spruce beetles *Dendroctonus rufipennis* (Kirby) in interior Alaska. Can. J. Zool. 65, 2927–2930.

Mizell, R., Nebeker, T., 1978. Estimating the developmental time of the southern pine beetle *Dendroctonus frontalis* as a function of field temperatures. Environ. Entomol. 7, 592–595.

Nealis, V.G., Cooke, B.J., 2014. Risk assessment of the threat of mountain pine beetle to Canada's boreal and eastern pine forests. Cat. No. Fo79-14/2014E, ISBN: 978-1-100-23301-7, Forest Pest Working Group on the Canadian Council of Forest Ministers, Natural Resources Canada.

Netherer, S., Pennerstorfer, J., 2003. Parameters relevant for modelling the potential development of *Ips typographus* (Coleoptera: Scolytidae). Integrated Pest Manag. Rev. 6, 177–184.

Netherer, S., Nopp-Mayr, U., 2005. Predisposition assessment systems (PAS) as supportive tools in forest management—rating of site and stand-related hazards of bark beetle infestation in the High Tatra Mountains as an example for system application and verification. For. Ecol. Manage. 207, 99–107.

Ogris, N., Jurc, M., 2010. Sanitary felling of Norway spruce due to spruce bark beetles in Slovenia: a model and projections for various climate change scenarios. Ecol. Model. 221, 290–302.

Pielke Sr., R.A., Doesken, N., Bliss, O., Green, T., Chaffin, C., Salas, J.D., et al., 2005. Drought 2002 in Colorado: an unprecedented drought or a routine drought. Pure Appl. Geophys. 162, 1455–1479.

Powell, J.A., Bentz, B.J., 2009. Connecting phenological predictions with population growth rates for mountain pine beetle, an outbreak insect. Landsc. Ecol. 24, 657–672.

Powell, J.A., Bentz, B.J., 2014. Phenology and density-dependent dispersal predict patterns of mountain pine beetle (*Dendroctonus ponderosae*) impact. Ecol. Model. 273, 173–185.

Preisler, H.K., Hicke, J.A., Ager, A.A., Hayes, J.L., 2012. Climate and weather influences on spatial temporal patterns of mountain pine beetle populations in Washington and Oregon. Ecology 93, 2421–2434.

Raffa, K.F., Aukema, B.H., Bentz, B.J., Carroll, A.L., Hicke, J.A., Turner, M.G., Romme, W.H., 2008. Cross-scale drivers of natural disturbances prone to anthropogenic amplification: the dynamics of bark beetle eruptions. BioSci. 58, 501–517.

Raffa, K.F., Powell, E.N., Townsend, P.A., 2013. Temperature-driven range expansion of an irruptive insect heightened by weakly coevolved plant defenses. Proc. Natl. Acad. Sci. U. S. A. 110, 2193–2198.

Reeve, J.D., 2000. Complex emergence patterns in a bark beetle predator. Agr. Forest Entomol. 2, 233–240.

Régnière, J., Bentz, B., 2007. Modeling cold tolerance in the mountain pine beetle, Dendroctonus ponderosae. J. Insect Physiol. 53, 559–572.

Régnière, J., Powell, J.A., 2003. Animal life cycle models. In: Schwartz, M. (Ed.), Phenology: An Integrative Environmental Science". Springer, Dordrecht, pp. 237–254.

Régnière, J., Saint-Amant, R., 2013. BioSIM: Optimizing Pest Control Efficacy in Forestry. Branching Out from the Canadian Forest Service, Laurentian Forestry Centre. No. 81.

Régnière, J., Powell, J., Bentz, B., Nealis, V., 2012. Effects of temperature on development, survival and reproduction of insects: experimental design, data analysis and modeling. J. Insect Physiol. 58, 634–647.

Régnière, J., Bentz, B. J., Powell, J. A., and St-Amant, R. Mountain pine beetle and climate change. In: Sturtevant, B., Buse, L. (Eds.), Modeling Forest Landscape Disturbances. Springer, submitted.

Rehfeldt, G.E., Crookston, N.L., Warwell, M.V., Evans, J.S., 2006. Empirical analyses of plant-climate relationships for the western United States. Int. J. Plant Sci. 167, 1123–1150.

Reid, R., Gates, H., 1970. Effect of temperature and resin on hatch of eggs of the mountain pine beetle (*Dendroctonus ponderosae*). Can. Entomol. 102, 617–622.

Rice, A., Thormann, M., Langor, D., 2008. Mountain pine beetle-associated blue-stain fungi are differentially adapted to boreal temperatures. Forest Pathol. 38, 113–123.

Riel, W., Fall, A., Shore, T.L., Safranyik, L., 2004. A spatio-temporal simulation of mountain pine beetle impacts on the landscape. In: Shore, T.L., Brooks, J.E., Stone, J.E. (Eds.), Mountain Pine Beetle Symposium: Challenges and Solutions". Information Report BC-X-399, Canadian Forest Service, Victoria, British Columbia, pp. 106–113.

Rodríguez, D., Cure, J.R., Cotes, J.M., Gutierrez, A.P., Cantor, F., 2011. A coffee agroecosystem model: I. Growth and development of the coffee plant. Ecol. Modell. 222, 3626–3639.

Rodríguez, D., Cure, J.R., Gutierrez, A.P., Cotes, J.M., Cantor, F., 2013. A coffee agroecosystem model: II. Dynamics of coffee berry borer. Ecol. Model. 248, 203–214.

Ryan, R.B., 1959. Termination of diapause in the Douglas-fir beetle, *Dendroctonus pseudotsugae* Hopkins (Coleoptera: Scolytidae), as an aid to continuous laboratory rearing. Can. Entomol. 91, 520–525.

Safranyik, L., Whitney, H., 1985. Development and survival of axenically reared mountain pine beetles, *Dendroctonus ponderosae* (Coleoptera: Scolytidae), at constant temperatures. Can. Entomol. 117, 185–192.

Safranyik, L., Linton, D.A., 1998. Mortality of mountain pine beetle larvae, *Dendroctonus ponderosae* in logs of lodgepole pine at constant low temperatures. J. Entomol. Soc. B. C. 95, 81–87.

Safranyik, L., Shrimpton, D., Whitney, H., 1975. An interpretation of the interaction between lodgepole pine, the mountain pine beetle and its associated blue stain fungi in western Canada. In: Baumgartner, D. M. (Ed.), Management of Lodgepole Pine Ecosystems. Washington State University, Pullman, pp. 406–428.

Safranyik, L., Simmons, C., Barclay, H.J., 1990. A conceptual model of spruce beetle population dynamics. Information Report BC-X-316, Forestry Canada, Victoria, British Columbia.

Safranyik, L., Barclay, H.J., Thomson, A., Riel, W., 1999. A population dynamics model for the mountian pine beetle, *Dencroctonus ponderosae* Hopk. (Coleoptera: Scolytidae). Information Report BC-X-386, Natural Resources Canada, Victoria, British Columbia.

Safranyik, L., Carroll, A., Régnière, J., Langor, D., Riel, W., Shore, T., et al., 2010. Potential for range expansion of mountain pine beetle into the boreal forest of North America. Can. Entomol. 142, 415–442.

Salom, S.M., Stephen, F.M., Thompson, L.C., 1987. Development rates and a temperature-dependent model of pales weevil, *Hylobius pales* (Herbst), development. Environ. Entomol. 16, 956–962.

Sambaraju, K.R., Carroll, A.L., Zhu, J., Stahl, K., Moore, R.D., Aukema, B.H., 2012. Climate change could alter the distribution of mountain pine beetle outbreaks in western Canada. Ecography 35, 211–223.

Satterlee, S.M., 2002. Evolution of the southern pine beetle legacy simulation model "SPBMODEL" using genetic algorithms. M.S. thesis Virginia Polytechnic Institute and State University, Blacksburg.

Saunders, D.S., 2014. Insect photoperiodism: effects of temperature on the induction of insect diapause and diverse roles for the circadian system in the photoperiodic response. J. Entomol. Sci. 17, 25–40.

Schelhaas, M.J., Nabuurs, G.J., Schuck, A., 2003. Natural disturbances in the European forests in the 19th and 20th centuries. Global Change Biol. 9 (11), 1620–1633.

Schmid, J.M., Frye, R.H., 1977. Spruce beetle in the Rockies. USDA Forest Service, Rocky Mountain Forest and Range Experiment Station, General Technical Report, RM-49.

Scott, B., Berryman, A., 1972. Larval diapause in *Scolytus ventralis* (Coleoptera: Scolytidae). J. Entomol. Soc. B. C. 69, 50–53.

Seidl, R., Baier, P., Rammer, W., Schopf, A., Lexer, M.J., 2007. Modelling tree mortality by bark beetle infestation in Norway spruce forests. Ecol. Model. 206, 383–399.

Seidl, R., Schelhaas, M.J., Lindner, M., Lexer, M.J., 2009. Modelling bark beetle disturbances in a large scale forest scenario model to assess climate change impacts and evaluate adaptive management strategies. Reg. Environ. Change 9, 101–119.

Seidl, R., Fernandes, P.M., Fonseca, T.F., Gillet, F., Jönsson, A.M., Merganičová, K., et al., 2011. Modelling natural disturbances in forest ecosystems: a review. Ecol. Model. 222, 903–924.

Sharpe, P.J., Curry, G.L., DeMichele, D.W., Cole, C.L., 1977. Distribution model of organism development times. J. Theor. Biol. 66, 21–38.

Sherriff, R.L., Berg, E.E., Miller, A.E., 2011. Climate variability and spruce beetle (*Dendroctonus rufipennis*) outbreaks in south-central and southwest Alaska. Ecology 92, 1459–1470.

Shore, T.L., Safranyik, L., 1992. Susceptibility and risk rating systems for the mountain pine beetle in lodgepole pine stands. Information Report BC-X-336, Forestry, Canada, Victoria, British Columbia.

Shore, T. L., Fall, A., and Riel, W. G. (2008). Incorporating present and future climatic suitability into decision support tools to predict geographic spread of the mountain pine beetle. Working Paper 2008-10, Canadian Forest Service, Victoria, British Columbia.

Six, D.L., 2013. The bark beetle holobiont: why microbes matter. J. Chem. Ecol. 39, 989–1002.

Sømme, L., 1964. Effects of glycerol on cold-hardiness in insects. Can. J. Zool. 42, 87–101.

Stadelmann, G., Bugmann, H., Wermelinger, B., Meier, F., Bigler, C., 2013. A predictive framework to assess spatio-temporal variability of infestations by the European spruce bark beetle. Ecography 36, 1208–1217.

Tan, J.Y., Wainhouse, D., Day, K.R., Morgan, G., 2010. Flight ability and reproductive development in newly-emerged pine weevil *Hylobius abietis* and the potential effects of climate change. Agr. Forest Entomol. 12, 427–434.

Tauber, M.J., Tauber, C.A., 1976. Insect seasonality: diapause maintenance, termination, and postdiapause development. Annu. Rev. Entomol. 21, 81–107.

Temperli, C., Bugmann, H., Elkin, C., 2013. Cross-scale interactions among bark beetles, climate change, and wind disturbances: a landscape modeling approach. Ecol. Monogr. 83, 383–402.

Thatcher, R., Pichard, L., 1967. Seasonal development of the southern pine beetle in east Texas. J. Econ. Entomol. 60, 656–658.

Thom, D., Seidl, R., Steyrer, G., Krehan, H., Formayer, H., 2013. Slow and fast drivers of the natural disturbance regime in Central European forest ecosystems. For. Ecol. Manage. 307, 293–302.

Thorpe, K., Day, K., 2002. The impact of host plant species on the larval development of the large pine weevil *Hylobius abietis* L. Agr. Forest Entomol. 4, 187–194.

Tobin, P.C., Nagarkatti, S., Loeb, G., Saunders, M.C., 2008. Historical and projected interactions between climate change and insect voltinism in a multivoltine species. Glob. Chang. Biol. 14, 951–957.

Trần, J.K. i, Ylioja, T., Billings, R.F., Régnière, J., Ayres, M.P., 2007. Impact of minimum winter temperatures on the population dynamics of *Dendroctonus frontalis*. Ecol. Appl. 17, 882–899.

Ungerer, M.J., Ayres, M.P., Lombardero, M.J., 1999. Climate and the northern distribution limits of *Dendroctonus frontalis* Zimmermann (Coleoptera: Scolytidae). J. Biogeogr. 26, 1133–1145.

Vose, R.S., Easterling, D.R., Gleason, B., 2005. Maximum and minimum temperature trends for the globe: an update through 2004. Geophys. Res. Lett. 32, L23822.

Wagner, T.L., Gagne, J.A., Sharpe, P.J.H., Coulson, R.N., 1984. A biophysical model of southern pine beetle, *Dendroctonus frontalis* Zimmermann (Coleoptera: Scolytidae), development. Ecol. Model. 21, 125–147.

Wainhouse, D., Inward, D.J., Morgan, G., 2013. Modelling geographical variation in voltinism of *Hylobius abietis* under climate change and implications for management. Agr. Forest Entomol. 16, 136–146.

Walsh, J., Wuebbles, D., Hayhoe, K., Kunkel, K., Stephens, B., Thorne, P., et al., 2014. Appendix 3: Climate Science Supplement. Climate Change Impacts in the United States: The Third National Climate Assessment. Available online: In: Melillo, J.M., Richmond, T.C., Yohe, G.W. (Eds.), U.S. Global Change Research Program", pp. 735–789. http://nca2014.globalchange.gov/report/appendices/climate-science-supplement, Last (accessed: 28.05.14.).

Weed, A.S., Ayres, M.P., Hicke, J.A., 2013. Consequences of climate change for biotic disturbances in North American forests. Ecol. Monogr. 83, 441–470.

Wermelinger, B., Seifert, M., 1998. Analysis of the temperature dependent development of the spruce bark beetle *Ips typographus* (L.) (Col., Scolytidae). J. Appl. Entomol. 122, 185–191.

Werner, R.A., Holsten, E.H., Matsuoka, S.M., Burnside, R.E., 2006. Spruce beetles and forest ecosystems in south-central Alaska: a review of 30 years of research. For. Ecol. Manage. 227, 195–206.

Wood, S.L., 1982. The bark and ambrosia beetles of North and Central America (Coleoptera: Scolytidae), a taxonomic monograph. Great Basin Nat. Mem. 6, 1–1359.

Yuill, J., 1941. Cold hardiness of two species of bark beetles in California forests. J. Econ. Entomol. 34, 702–709.

Chapter 14

Management Strategies for Bark Beetles in Conifer Forests

Christopher J. Fettig[1] and Jacek Hilszczański[2]

[1]*Invasives and Threats Team, Pacific Southwest Research Station, USDA Forest Service, Davis, CA, USA,* [2]*Department of Forest Protection, Forest Research Institute, Sękocin Stary, Raszyn, Poland*

1. INTRODUCTION

Bark beetles (Coleoptera: Curculionidae, Scolytinae) are important disturbance agents in conifer forests. The genera *Dendroctonus*, *Ips*, and *Scolytus* are well recognized in this regard (Table 14.1). For example, in western North America, the mountain pine beetle (*Dendroctonus ponderosae* Hopkins) colonizes several tree species, most notably lodgepole pine (*Pinus contorta* Dougl. ex Loud.), ponderosa pine (*Pinus ponderosa* Dougl. ex Laws.), and whitebark pine (*Pinus albicaulis* Engelm.). Recent outbreaks have been severe, long lasting, and well documented, with over 27 million hectares impacted (BC Ministry of Forests, Lands and Natural Resource Operations, 2012; USDA Forest Service, 2012). In British Columbia, Canada, alone 710 million m^3 of timber have been killed (BC Ministry of Forests, Lands and Natural Resource Operations, 2012). In Europe, the European spruce beetle (*Ips typographus* (L.)) is regarded as the most important pest of Norway spruce (*Picea abies* (L.) Karst.) (Christiansen and Bakke, 1988; Schelhaas *et al.*, 2003), an indigenous species also widely planted for commercial timber production outside its native range. It is estimated that 8% of all tree mortality that occurred in Europe between 1850 and 2000 was caused by bark beetles, primarily *I. typographus* (Schelhaas *et al.*, 2003). In Asia, the red turpentine beetle (*Dendroctonus valens* LeConte), an exotic invasive introduced from North America, has caused significant levels of tree mortality since being detected in China in 1998 (Yan *et al.*, 2005). Although considered a minor pest in its native range, more than 10 million Chinese red pine (*Pinus tabuliformis* Carr.), China's most widely planted pine species, have been killed by *D. valens*.

Over the last century, substantial basic and applied research has been devoted to the development of effective tools and tactics for mitigating undesirable levels of tree mortality attributed to bark beetles. There are two basic approaches. *Direct control* involves short-term tactics designed to address current infestations by manipulating beetle populations, and often includes the use of fire, insecticides, semiochemicals (i.e., chemicals released by one organism that elicit a response, usually behavior, in another organism), sanitation harvests, or a combination of these treatments. *Indirect control* is preventive, and designed to reduce the probability and severity of future bark beetle infestations within treated areas by manipulating stand, forest, and/or landscape conditions by reducing the number of susceptible hosts through thinning, prescribed burning, and altering age classes and species composition. Unlike direct control, the focus of indirect control is on the susceptibility of residual forest structure and composition to future infestations.

The purpose of this chapter is to synthesize information related to the management of bark beetles in conifer forests, and to present a case study on the management of *I. typographus* in central Europe. We concentrate on what some authors commonly refer to as *aggressive* species (i.e., they are capable of causing large amounts of tree mortality during certain circumstances) (Table 14.1), and draw heavily from research conducted and practical experience gained while working in North America and Europe. Our hope is that this synthesis provides a basic understanding of current and evolving strategies for reducing the negative impacts of bark beetles on forests. However, we stress that in most cases we concentrate on native species important to the proper functioning of forest ecosystems as they regulate certain aspects of primary production, nutrient cycling, and ecological succession (Romme *et al.*, 1986). In this context, some level of tree mortality is desirable and often results in a mosaic of age classes and species compositions that increases resilience to bark beetles and other disturbances. This differs from the negative impacts associated with outbreaks, which often merit intervention. We encourage the reader to delve deeper into the literature cited for more detailed information on specific bark beetle–host complexes.

Bark Beetles. http://dx.doi.org/10.1016/B978-0-12-417156-5.00014-9

TABLE 14.1 Bark Beetle Species Notable for Causing Substantial Levels of Tree Mortality in Conifer Forests within their Native Ranges

Common Name	Scientific Name	Common Host(s)
Arizona fivespined ips	*Ips lecontei*	*P. ponderosa*
California fivespined ips	*Ips paraconfusus*	*P. contorta, Pinus lambertiana, P. ponderosa*
Douglas-fir beetle	*Dendroctonus pseudotsugae*	*Pseudotsuga menziesii*
eastern fivespined ips	*Ips grandicollis*	*Pinus echinata, Pinus elliottii, Pinus taeda, Pinus virginiana*
eastern larch beetle	*Dendroctonus simplex*	*Larix laricina*
eastern six-spined engraver	*Ips calligraphus*	*P. echinata, P. elliotti, P. ponderosa, P. taeda, P. virginiana*
European spruce beetle	*Ips typographus*	*Pi. abies, Picea orientalis, Picea yezoensis,* occasionally *Pinus sylvestris*
fir engraver	*Scolytus ventralis*	*Abies concolor, Abies grandis, Abies magnifica*
Jeffrey pine beetle	*Dendroctonus jeffreyi*	*Pinus jeffreyi*
larger Mexican pine beetle	*Dendroctonus approximatus*	*P. ponderosa*
mountain pine beetle	*Dendroctonus ponderosae*	*P. albicaulis, P. contorta, Pinus flexilis, P. lambertiana, Pinus monticola, P. ponderosa*
northern spruce engraver	*Ips perturbatus*	*Picea glauca, Picea x lutzii*
pine engraver	*Ips pini*	*P. contorta, P. jeffreyi, P. lambertiana, P. ponderosa, Pinus resinosa*
pinyon ips	*Ips confusus*	*Pinus edulis, Pinus monophylla*
roundheaded pine beetle	*Dendroctonus adjunctus*	*Pinus arizonica, Pinus engelmannii, P. flexilis, Pinus leiophylla, P. ponderosa, Pinus strobiformis*
six-toothed bark beetle	*Ips sexdentatus*	*Pinus heldreichii, Pinus nigra, Pinus pinaster, P. sylvestris, Pi. orientalis*
southern pine beetle	*Dendroctonus frontalis*	*P. echinata, P. engelmannii, P. leiophylla, P. ponderosa, Pinus rigida, P. taeda, P. virginiana*
spruce beetle	*Dendroctonus micans*	*P. sylvestris, Pi. abies*
spruce beetle	*Dendroctonus rufipennis*	*Picea engelmannii, Pi. glauca, Picea pungens, Picea sitchensis*
western balsam bark beetle	*Dryocoetes confusus*	*Abies lasiocarpa*
western pine beetle	*Dendroctonus brevicomis*	*Pinus coulteri, P. ponderosa*

1.1   Bark Beetle Ecology

Some knowledge of bark beetle ecology and physiology is important to understanding the utility and proper implementation of control strategies. In brief, adult bark beetles maintain limited energy reserves (Atkins, 1966), and are highly susceptible to predation, starvation, and adverse weather conditions when searching for hosts. Beetles therefore must detect and locate the correct habitat, correct tree species, and the most susceptible trees within these species with efficiency (Byers, 1995; Borden, 1997; Schlyter and Birgersson, 1999). For example, the dominant theory of host finding and selection in *D. ponderosae* suggests pioneering females use a combination of random landings and visual orientations followed by direct assessment of hosts based on olfactory and/or gustatory cues (Raffa and Berryman, 1982, 1983; Wood, 1982). Given the cues received during this process and other factors, such as the beetle's internal physiology (Wallin and Raffa, 2000), the host is either rejected or accepted. If the host is accepted, gallery construction is initiated upon which many species release aggregation pheromones that enhance attraction of conspecifics to the target tree (Borden, 1985; Byers, 1995; Zhang and Schlyter, 2004) as successful colonization requires overcoming host tree defenses (Wood, 1972; Hodges *et al.*, 1979, 1985; Raffa *et al.*, 1993; Franceschi *et al.*, 2005). This can only be accomplished by recruitment of a critical minimum number of beetles to *mass attack* the tree and overwhelm its defenses.

Most conifers are capable of mobilizing large amounts of oleoresin following wounding, which constitutes their primary defense against bark beetle attack (Vité, 1961, Reid *et al.*, 1967, Franceschi *et al.*, 2005) (see Chapter 5); however, resin chemistry also plays an important role (Smith, 1966; Cook and Hain, 1988; Reid and Purcell, 2011). The development of a hypersensitive response, consisting mainly of secondary metabolites around points of attack, has also been demonstrated to be important (Lieutier, 2004). Beetles that initiate host selection are often killed by drowning or immobilization in resin (termed *pitch out*) especially when adequate moisture, flow, and oleoresin exudation pressure exist, such as in the case of vigorous hosts (Raffa and Berryman, 1983) or when beetle populations are low (Figure 14.1). The presence of pitch tubes and/or boring dust is commonly used to identify trees that have been attacked by bark beetles. Monoterpenes released from pitch tubes may enhance attraction to the host tree. However, for most aggressive species attraction to host volatiles has not been demonstrated in the absence of aggregation pheromone components (Borden, 1997). Many bark beetles introduce a variety of microbes into the tree upon colonization (see Chapter 6), which may have deleterious effects on tree health, but mortality occurs primarily through girdling of the phloem and cambium tissues. The resultant tree mortality may impact timber and fiber production, water quality and quantity, fish and wildlife populations, recreation, grazing capacity, real estate values, biodiversity, carbon storage, endangered species, and cultural resources (Coulson and Stephen, 2006), among other factors.

Following pupation, adult beetles of the next generation tunnel outward through the bark and initiate flight in search of new hosts. The life cycle may be repeated once every several years (e.g., the spruce beetle, *Dendroctonus rufipennis* Kirby) or several times a year (e.g., the western pine beetle, *Dendoctonus brevicomis* LeConte), which has obvious implications to their management. For example, mechanical fuel treatments (e.g., thinning of small-diameter trees) are commonly implemented in the western United

FIGURE 14.1 Beetles that initiate host colonization are often killed by drowning or immobilization in resin when hosts are vigorous, as depicted by this *Dendroctonus brevicomis*. This is usually considered the primary defense of conifers against bark beetle attack. Management strategies exist to increase tree vigor, and thus reduce the susceptibility of trees and forests to bark beetles. *Photo credit: C. Fettig, Pacific Southwest Research Station, USDA Forest Service.*

States to reduce the risk, severity, and extent of wildfires (Stephens *et al.*, 2012). However, much of the biomass removed is unmerchantable, and therefore cut and lopped (i.e., the boles are severed into short lengths and limbs removed) or chipped and redistributed on site. Chipping has been demonstrated to increase levels of tree mortality attributed to bark beetles, presumably due to the plumes of monoterpenes released, but conducting chipping operations in autumn (as compared to spring/early summer) after most species have become relatively inactive results in fewer trees being attacked and killed (Fettig *et al.*, 2006a; DeGomez *et al.*, 2008).

1.2 Development of Outbreaks

Mechanisms contributing to bark beetle outbreaks are complex and include density-dependent and density-independent factors (see Chapters 1, 4, and 7), but two requirements must be met for an outbreak to occur: (1) there must be several years of favorable weather conducive to beetle survival and population growth; and (2) there must be an abundance of susceptible host trees. In many cases, age–class structure and tree species composition will be dominant factors influencing the severity of outbreaks. However, many experts agree that anthropogenic-induced climate change has also contributed to some outbreaks due to shifts in temperature and precipitation that influence both the beetles and their hosts (Bentz *et al.*, 2010; Sambaraju *et al.*, 2012).

During endemic bark beetle populations, trees weakened or damaged by other agents (e.g., pathogens) are often colonized and killed by bark beetles. For example, endemic populations of northern spruce engraver (*Ips perturbatus* (Eichhoff)) infest forest debris, widely scattered individual trees or small groups of trees. However, natural (e.g., flooding, wildfire, and wind storms) and anthropogenic-induced (e.g., road building, construction of utility rights-of-way, and logging) disturbances may produce large quantities of damaged, dead, or dying spruce that serve as ideal hosts. If favorable climatic conditions coincide with large quantities of suitable host material, populations may erupt resulting in the mortality of apparently healthy trees over extensive areas (Holsten and Werner, 1987). Similarly, outbreaks of *I. typographus* in central Europe are often precipitated by large-scale blowdown events associated with severe storms (see Section 6). In the absence of such large-scale disturbances, damage to individual hosts from subcortical insects (Boone *et al.*, 2011), defoliators (Wallin and Raffa, 2001), drought (Fettig *et al.*, 2013a), lightning strikes (Hodges and Pickard, 1971), and root pathogens (Klepzig *et al.*, 1991) may reduce host resistance and facilitate successful colonization by bark beetles. Such hosts are thought to be important in maintaining localized populations between outbreaks.

Individual bark beetle species generally exhibit a preference for trees of certain sizes. For example, it is well established that *D. ponderosae* initially colonizes the largest trees within *P. contorta* forests (Shepherd, 1966; Rasmussen, 1972), with progressively smaller trees being attacked over time (Klein *et al.*, 1978; Cole and Amman, 1980; Amman and Cole, 1983). This is despite larger-diameter *P. contorta* having more pronounced defenses (Shrimpton, 1973; Boone *et al.*, 2011), but provide for a higher reproductive potential and probability of beetle survival (Amman, 1969, 1975; Reid and Purcell, 2011; Graf *et al.*, 2012) because of the greater quantity of food (phloem) available on which larvae feed. To that end, Safranyik *et al.* (1974) reported that *P. contorta* ≤25 cm dbh (diameter at breast height) (diameter at 1.37 m in height) serve as *D. ponderosae* sinks, whereas trees >25 cm dbh serve as sources producing more *D. ponderosae* than required to overcome host defenses. This has obvious implications to the population dynamics of *D. ponderosae*. In other species, a preference for smaller-diameter trees may be exhibited. For example, the pine engraver (*Ips pini* (Say)) most frequently colonizes trees 5–20 cm dbh, and attack rates are negatively correlated with tree diameter (Kolb *et al.*, 2006). Understanding host preferences and how these influence outbreak dynamics is critical to the proper implementation of management strategies.

A considerable amount of effort has been devoted to the identification of tree, stand, and landscape conditions associated with bark beetle infestations. Most aggressive species exhibit a preference for larger-diameter trees growing in high-density stands with a high percentage of host type (reviewed by Fettig *et al.*, 2007a for North America) (see Section 3). Furthermore, forested landscapes that contain little heterogeneity may result in large contiguous areas susceptible to bark beetles. It is clear that efforts to prevent undesirable levels of tree mortality attributed to bark beetles must account for these variables (see Section 5).

2. DETECTION AND SURVEY

Information on the intensity and extent of bark beetle infestations adequate to plan appropriate control strategies requires accurate detection and survey methods. Many methods have been developed to address different bark beetle species, host species, and spatial scales. These range from trapping programs to monitor populations, to simple ground-based surveys, to a broad array of aerial surveys using methods such as sketch mapping, to more sophisticated methods using remotely sensed data obtained from satellites (Wulder *et al.*, 2006a, b; Meigs *et al.*, 2011).

2.1 Aerial Survey

Research concerning the application of remote sensing methods for detection and survey was initiated in the mid-20th century. Aerial photography was frequently used in the 1970–1980s, including both true color and color-infrared photography (Puritch, 1981; Gimbarzevsky, 1984). Usually, these surveys were limited to detection of infestations followed by more detailed surveys to identify currently infested trees. Infestations were manually drawn (sketched) on maps, but such techniques have largely been replaced by more sophisticated methods, particularly in North America. For example, surveys using helicopters and/or fixed-wing aircraft with global positioning systems (GPS) and digital sketch-mapping equipment is one of the most precise and widely used methods today (Wulder *et al.*, 2005a) (Figure 14.2). In addition to showing your position on a digital map, sketch mapping allows real-time acquisition of geographic information system (GIS) data without being at the corresponding physical location, and is relatively inexpensive compared to other survey methods (often < $US1/ha). It also allows for quick processing of data and reporting compared to waiting weeks or months for quality aerial or satellite imagery. However, flying presents unique risks, and considerable variability has been observed in data reported from different observers (Figure 14.2).

The landscape scale of aerial survey (1:10,000–1:50,000) is often considered sufficient for control planning purposes (Wulder *et al.*, 2004), but requires survey methods that are accurate and provide spatially distinct data. Landsat data, as those derived from the analysis of enhanced wetness difference index (EWDI), are sufficient to detect

FIGURE 14.2 Digital sketch-mapping systems are now commonly used during aerial survey. The system, consisting of a tablet PC, external GPS receiver, and stylus, has the capability to display multiple types of background images for navigation and mapping (aerial imagery, topographic maps, etc.) and vector data (e.g., administrative and political boundaries, aerial hazards, etc.). *Photo credit: D. Wittwer, Forest Health Protection, USDA Forest Service.*

larger groups of trees, but not small or low-density infestations (Skakun *et al.*, 2003). However, it is hard to achieve sufficient accuracy within large areas, especially when infested trees or groups of trees are scattered across the landscape (Wulder *et al.*, 2006a, b). Some methods provide quite precise data on individual trees through imagery collected on multiple dates or spatial high resolution, but are expensive (Bone *et al.*, 2005). These methods enable detection of trees during the later stages of infestation when their foliage is fading and distinctly different from that of healthy trees or those previously killed by bark beetles or other agents (Figure 14.3). During surveys, a common method of estimating when trees died uses needle color and retention. For example, for *D. ponderosae* in *P. contorta* these stages are commonly referred to as the *green stage* (within 1 year of attack; green foliage or foliage just beginning to fade), *red stage* (1–3 years since death; red foliage), and *gray stage* (>3 years since death; gray, limited or no foliage). However, relationships between foliage characteristics and time since tree death vary considerably by bark beetle species and host species, among other factors. It is also important to emphasize that these are crude

estimates that may vary by several years from the actual time since tree death.

The identification of currently infested trees is critical to maximizing the effectiveness of direct control strategies such as sanitation (Niemann and Visintini, 2005) (see Section 4.3.1). As indicated, trees that have been dead for 1 or more years and which the beetles have vacated are detected based on patterns of crown fade, and currently infested trees (i.e., which exhibit little or no crown fade) are then detected by their proximity to faded trees (Wulder *et al.*, 2006a, 2009) and confirmed by the presence of pitch tubes and/or boring dust during ground-based surveys. Some experiments have shown that detection of currently infested trees (green stage) is possible with the use of thermal scanners (Heller, 1968), and on aerial photographs with the use of color-infrared film to improve contrasts between infested and uninfested trees (Arnberg and Wastenson, 1973). However, neither method has been widely adopted. At the local scale, detailed surveys of red stage trees can be performed with aerial photography or high-resolution satellite imagery such as IKONOS (White *et al.*, 2004).

2.2 Ground-based Surveys

Methods for identifying currently infested trees depend primarily on ground-based surveys. In North America, these surveys are supported by data from aerial surveys focused on detection of red-stage trees. In many European countries, currently infested trees are detected by trained field observers called *sawdusters* (see Section 6). During outbreaks, sawdusters are actively employed searching for currently infested trees on a systematic basis throughout the year. In well-organized management units, where one sawduster is operating on a scale of $\approx$ 1000 ha, the effectiveness of infested tree detections is very close to 100%. Once identified, infested trees are marked, numbered, and mapped.

FIGURE 14.3 An outbreak of *Dendroctonus ponderosae* in *Pinus albicaulis* forests in California, United States. During aerial survey, host and bark beetle signatures are often differentiated by crown color and pattern of mortality. *Photo credit: D. Cluck, Forest Health Protection, USDA Forest Service.*

Sometimes the date of detection is also placed on the tree. Usually after several days, these trees are cut and removed or debarked (see Section 6). During a recent outbreak of *I. typographus* in southern Poland, ≈1.5 million currently infested trees were identified during ground-based surveys and harvested (Szabla, 2013). In Europe, ground-based surveys using well-trained dogs to detect infested trees have been demonstrated to be effective, even when visible signs of attack were not evident on tree boles (Feicht, 2006).

Ground-based surveys may also be conducted to quantify the impact of bark beetles on forests. Sample designs vary widely depending on variables of interest, but often include collection of standard forest mensuration data. For example, in France infestations of the six-toothed bark beetle (*Ips sexdentatus* (Boern)) were located using color-infrared aerial photography and validated by ground-based survey. Assessments then concentrated on counting all dead and dying trees sighted within a fixed distance of roads (Samalens *et al.*, 2007). In the United States, the USDA Forest Service has installed a large network of plots in the Rocky Mountains to quantify the impacts of *D. ponderosae* outbreaks on forest fuels and other attributes (Fettig *et al.*, unpubl. data).

3. RISK AND HAZARD RATING

Risk and hazard rating systems have been developed for several species of bark beetles to provide land managers and others with means of identifying stands or forests that foster initiation and/or spread of infestations. In general, rating systems that estimate the probability of stand infestation define "risk," while those that predict the extent of tree mortality define "hazard," although conventions vary among authors resulting in confusion between differences in these systems. Some authors have reserved "risk" solely for rating systems in which measures of insect population pressure are included (Waters, 1985). Risk and hazard rating systems represent a critical step in forest planning,

especially where bark beetles are known to cause significant levels of tree mortality.

As indicate earlier, most bark beetle species capable of causing extensive levels of tree mortality exhibit a preference for larger diameter trees (often with declining radial growth) growing in high-density stands with a high percentage of host type (Table 14.2), and therefore such variables serve as a foundation for many risk and hazard rating systems (Table 14.3). In western North America, among the most commonly used is that of Shore and Safranyik (1992) for *D. ponderosae* in *P. contorta*. Susceptibility is calculated based on four factors: (1) percentage of susceptible basal area (trees ≥15 cm dbh); (2) average stand age of dominant and co-dominant trees; (3) stand density of all trees ≥7.5 cm dbh; and (4) the geographic location of the stand in terms of latitude, longitude, and elevation. *Dendroctonus ponderosae* population data, referred to as a beetle pressure index, incorporates the proximity and size of *D. ponderosae* populations (Table 14.4). The stand susceptibility index and beetle pressure index are then used to compute an overall stand risk index (Shore and Safranyik, 1992; Shore *et al.*, 2000). Due to the unique ability of *D. ponderosae* to cause extensive levels of tree mortality in several hosts, numerous risk and hazard rating systems have been developed for this species (reviewed by Fettig *et al.*, 2014a), but also for other bark beetle–host systems, particularly for the more aggressive bark beetle species. For example, several models have been developed to predict tree losses attributed to Douglas-fir beetle (*Dendroctonus pseudotsugae* Hopkins) (Weatherby and Thier, 1993; Negrón, 1998; Shore *et al.*, 1999); roundheaded pine beetle (*Dendroctonus adjunctus* Blandford) (Negrón, 1997); spruce beetle (Schmid and Frye, 1976; Reynolds and Holsten, 1994, 1996; Steele *et al.*, 1996); southern pine beetle (*Dendroctonus frontalis* Zimmermann) (Billings and Hynum, 1980; Reed *et al.*, 1981; Hedden, 1985; Stephen and Lih, 1985), most recently using GIS-based three-dimensional platforms (Chou *et al.*, 2013); *D. brevicomis* (Liebhold *et al.*, 1986; Steele *et al.*, 1996; Hayes

---

TABLE 14.2 Factors Characteristic of Stands Susceptible to *Dendroctonus frontalis* in Three Physiographic Regions of the Southern United Sates

Coastal Plain	Piedmont	Appalachian Mountains
Dense stocking	Dense stocking	Dense stocking, natural regeneration
Declining radial growth	Declining radial growth	Declining radial growth
Poorly drained soils	High clay content	Southern aspects
High proportion of *Pinus echinata* and *P. taeda*	High percentage of *P. echinata*	High percentage of *P. echinata* and/or *P. rigida*

(Modified from Belanger and Malac, 1980.)

TABLE 14.3 Rating the Probability of *Pinus ponderosa* Stands becoming Infested by *Dendroctonus ponderosae* in the Black Hills of South Dakota and Wyoming, United States

Variables	Probability on Infestation Classes		
	Low = 1	*Moderate = 2*	*High = 3*
Stand structure		Two-storied	Single-storied
Mean dbh[1] (cm)	<15.2	15.2–25.4	>25.4
Basal area (m²/ha)	<18.4	18.4–34.4	>34.4
Stand Value		**Overall Rating**	
2–6		Low	
8–12		Moderate	
18–27		High	

[1]*Diameter at breast height, 1.37 m.*
A number of rating systems use similar approaches of assigning values to model variables which are then multiplied (or added) to obtain an overall rating.
(Modified from Stevens et al., 1980.)

TABLE 14.4 Determination of the Relative Size of a *Dendroctonus ponderosae* Infestation (Small to Large, Top) and then the Bark Beetle Index (0.06–1.0, Bottom) based on the Relative Size of the Infestation

Number of Infested Trees outside Stand (within 3 km)		Number of Infested Trees within Stand		
		<10	*10–100*	*>100*
<900		Small	Medium	Large
900–9000		Medium	Medium	Large
>9000		Large	Large	Large
Distance to Nearest Infestation (km)				
Relative infestation size	In stand	0–1	1–2	2–3	3–4	>4
Beetle Pressure Index (B)				
Small	0.6	0.5	0.4	0.3	0.1	0.06
Medium	0.8	0.7	0.6	0.4	0.2	0.08
Large	1.0	0.9	0.7	0.5	0.2	0.1

Once the beetle pressure index (B) and stand susceptibility index (not presented here) are known, these values are used to compute an overall stand risk index.
(Modified from Shore and Safranyik, 1992.)

et al., 2009); and *I. typographus* (see Section 6), among others.

Risk and hazard rating systems are influenced by geographic location, site quality, and tree-diameter distributions. Measures of density are usually stand-level means, while differences in microtopography may create localized differences in productivity important to determining risk and hazard (Fettig, 2012), specifically in reference to the probability of infestation. As such, rating systems should primarily be used to identify areas most susceptible to bark beetles, as actual predictions may not be very accurate. Bentz *et al.* (1993) evaluated several *D. ponderosae* rating systems in *P. contorta* forests in Montana, and reported that none provided adequate predictions of tree losses. Alternatively, Shore *et al.* (2000) evaluated the Shore and Safranyik (1992) rating system in *P. contorta* forests in British Columbia, and reported most stands fell within the 95% prediction interval of the original model data. Finally, it is likely climate change will affect the predictive capacities of some systems due to the effects of projected changes on host-tree vigor, and on the temperature-dependent life history traits of bark beetles. We expect that the threshold values identified in many rating systems will require revision in the future (e.g., reductions in existing tree density thresholds associated with highly susceptible stands).

Other methods have been developed to predict tree losses attributed to bark beetles based on trap catches. For example, Billings (1988) developed a practical system for predicting risk of *D. frontalis* infestations in the southern United States based on captures of *D. frontalis* in attractant-baited multiple-funnel traps and the ratio of *D. frontalis* to one of its major predators, *Thanasimus dubius* (Fabricius) (Coleoptera: Cleridae). Traps are deployed on a county basis and monitored for several weeks in spring. Since its inception, this system has received widespread use and is generally regarded as an accurate means of forecasting *D. frontalis* population trends (i.e., increasing, declining, or static) and infestation levels (i.e., low, moderate, high, or outbreak). Similarly, Hansen *et al.* (2006) developed an effective method using attractant-baited multiple-funnel traps to estimate relative levels of tree mortality attributed to *D. rufipennis* in the central Rocky Mountain region. However, trap catches are regarded as poor indicators of future levels of tree mortality in some bark beetle–host systems. For example, Hayes *et al.* (2009) showed that monitoring of *D. brevicomis* populations through the use of attractant-baited multiple-funnel traps was ineffective for predicting levels of *D. brevicomis*-caused tree mortality. However, levels of tree mortality could be effectively predicted at large spatial scales (forests; ≈3000 to 14,000 hectares of contiguous host) by simply measuring stand density.

4. DIRECT CONTROL

Bark beetles have been the focus of direct control dating back to the 1700s. For example, in central Europe the Royal Society of Sciences at Göttingen, Germany, established an award to recognize the best proposal for bark beetle control in response to large-scale outbreaks of *I. typographus* in the mid-18th century. In response, Gmelin (1787) described two treatments, sanitation and burning of infested host material, that are still used today. In North America, the first documented use of large-scale direct control occurred in response to outbreaks of *D. ponderosae* in the Black Hills of South Dakota and Wyoming (Hopkins, 1905). Significant efforts have been undertaken since to develop effective direct control strategies for several species of bark beetles. Most target reducing localized populations, slowing the rate of infestation spread, and protecting individual trees or stands.

A successful direct control program requires prompt and thorough applications of the most appropriate strategies at a magnitude dictated by the bark beetle population and the spatial extent of the infested area. Treatments applied to areas adjacent to untreated areas where elevated populations occur are likely to be less successful due to immigration from untreated to treated areas. Coggins *et al.* (2011) found that mitigation rates of >50% (sanitation harvests) coupled with ongoing detection, monitoring, and treatment of infested trees within treated sites in British Columbia was sufficient to control *D. ponderosae* infestations. Alternatively, others have stressed that many large-scale, well-funded, and well-coordinated direct control programs (sanitation harvests) were largely ineffective (Wickman, 1987), and that resources would be better allocated to indirect control. Direct control is an expensive endeavor, and therefore decisions regarding its use and implementation are often dictated by more practical concerns such as resource availability (e.g., budget, time, personnel, and equipment), market conditions, logistical constraints (e.g., accessibility and ownership patterns), and environmental concerns.

4.1 Acoustics

Bark beetles use acoustics in a variety of behaviors, including territoriality (Rudinsky *et al.*, 1976), mate recognition (Rudinsky and Michael, 1973), and predator escape (Lewis and Cane, 1990). While applied research is in its infancy, Hofstetter *et al.* (2014) reported that applications of biologically derived acoustical signals disrupted behaviors in *D. frontalis* important to their reproductive performance, and therefore may have utility in the future management of this and other bark beetle species.

4.2 Biological Control

Natural enemies, such as predators and parasitoids, are important in regulating bark beetle populations at endemic levels, and have potential utility in biological control programs. In portions of China, successful classical biological control has been implemented in response to the introduction of *D. valens* by mass rearing and release of *Rhizophagus grandis* Gyllenhal (Coleoptera: Rhizophagidae), a predatory beetle native to Eurasia (Yang *et al.*, 2014). The use of *R. grandis* is also a common strategy for control of the great spruce beetle (*Dendroctonus micans* (Kugelann)). Native to Siberia, *D. micans* invaded Europe in the 19th century and its range is still expanding. Successful classical biological control efforts have been implemented using *R. grandis* in France (Grégoire *et al.*, 1985), Georgia (Kobakhidze *et al.*, 1970), the United Kingdom

(Fielding *et al.*, 1991), and Turkey (Yüksel, 1996). A common approach is to inundate stands with *R. grandis* at the leading edge of infested areas. Other research has indicated that conservation and supplemental feeding may be useful to enhance the effect of native biological control agents (Stephen *et al.*, 1997). For example, the parasitoid complex of *D. frontalis* in the southern United States consists of several species that may be important in regulating small infestations. Supplemental feedings of parasitoids in the laboratory and field with Eliminade™ (Entopath Inc., Easton, PA), an artificial diet consisting largely of sucrose, has been shown to increase longevity and fecundity (Mathews and Stephen, 1997, 1999; Stephen and Browne, 2000), but is not used operationally.

Synthetic formulations of entomopathogenic microorganisms, such as fungi, bacteria, and viruses, may also be useful for managing bark beetle populations. Efforts have focused largely on the fungus *Beauveria bassiana* (Bals.) Vuill. (Ascomycota: Hypocreales), which has been demonstrated to cause high levels of mortality in several species of bark beetles, including *I. typographus* (Wegensteiner, 1992, 1996; Kreutz *et al.*, 2000, 2004). One tactic being developed includes contaminating beetles collected in attractant-baited traps, and then releasing these individuals back into the field to contaminate the pest population (Vaupel and Zimmermann, 1996; Kreutz *et al.*, 2000). While this method has potential, additional research is needed to develop more practical methods of release and spread of *B. bassiana* in bark beetle populations as field studies have provided less conclusive evidence of mycosis than under laboratory conditions (Safranyik *et al.*, 2002). Related research is being conducted in the western United States to developed *B. bassiana* as a tool for protecting trees from colonization by bark beetles (Fettig *et al.*, unpubl. data). Other research has focused on bacteria. For example, Sevim *et al.* (2012) showed that strains of *Pseudomonas fluorescens* Flügge can be modified to express insecticidal toxins, and may represent a new method of control for *I. sexdentatus*, and perhaps other bark beetles. Chapter 7 presents detailed information on natural enemies of bark beetles.

4.3 Cultural

4.3.1 Sanitation

Sanitation involves the identification of trees infested by bark beetles, and subsequent felling and removal or treatment to destroy adults and brood beneath the bark, thereby reducing their populations. Where it is economically feasible, trees may be harvested and transported to mills where broods will be killed during processing. Otherwise, felled trees are burned, chipped, peeled, and

FIGURE 14.4 Log Wizard™ being used to peel bark from *Picea engelmannii* infested with *Dendroctonus rufipennis* in Utah, United States. This and other similar methods are often used in conjunction with sanitation and trap tree methods to destroy brood and adults beneath the bark. *Photo credit: S. Munson, Forest Health Protection, USDA Forest Service.*

debarked (Figure 14.4) or treated by solarization (i.e., placement of infested material in the direct sun, which is often sufficient to kill brood beneath the bark in warmer climates). In some cases, an emphasis is placed on sanitation of newly infested trees during the very early stages of colonization in order to also reduce the quantity of attractive semiochemicals (e.g., aggregation pheromones) released into the stand (see Section 4.6). However, reducing the level of attractive semiochemicals is difficult due to complications regarding the identification of newly attacked trees and the level of responsiveness required in their prompt removal. Identifying susceptible stands (see Section 3), coupled with the ability to address the infestation and resource values adversely affected, will determine where sanitation is most effective. Synthetic attractants may be used to concentrate existing infestations within small groups of trees prior to sanitation.

Sanitation, one of the oldest *D. frontalis* control tactics (St. George and Beal, 1929), continues to be the most recommended. Harvesting and utilizing currently infested trees, plus a buffer strip of uninfested trees, can halt infestation growth. *Dendroctonus frontalis* infests concentrated groups of trees (*spots*) creating infestations that can expand over time without intervention. These groups may range in size from a few trees to several thousand hectares. Timely sanitation is often not possible during large-scale outbreaks of *D. frontalis* due to limitations in labor, but in this case *cut-and-leave* (i.e., felling all freshly attacked and currently infested trees toward the center of an infestation) may be employed (Figure 14.5). Similar, sanitation is considered the most effective direct control method for *I. typographus*, and is widely implemented throughout central Europe (see Section 6). Depending on the scale and extent, sanitation

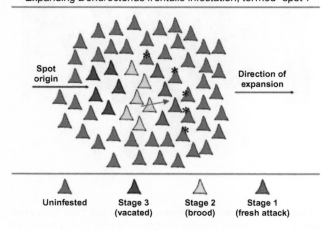

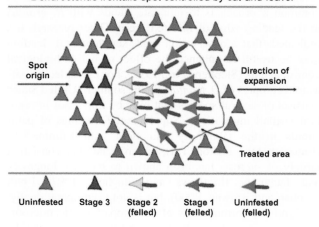

FIGURE 14.5 Illustrations of an expanding *Dendroctonus frontalis* "spot" (A) and one that has been controlled by cut-and-leave tactics (B). *Modified from Fettig et al. (2007a).*

may have the added benefit of reducing stand risk and hazard by influencing structure and composition.

4.3.2 Salvage

Salvage involves the harvest and removal of dead trees in order to recover some economic value that would otherwise be lost. Technically, salvage is not a direct control method as its implementation has no immediate effect on bark beetle populations (i.e., beetles have already vacated the trees). However, the term commonly appears in the bark beetle literature, particularly in Europe (see Section 6). In some cases, thinning (see Section 5) may be combined with sanitation and salvage in order to manipulate stand structure to reduce risk and hazard. Salvage or other treatment of hazardous trees may be necessary for safety concerns prior to accessing forests where high levels of tree mortality have occurred.

4.4 Insecticides

Insecticides are highly regulated by federal, provincial, state, and local governments, and therefore their use for protecting trees from mortality attributed to bark beetles varies accordingly. For example, hundreds of thousands of trees may be treated with insecticides during outbreaks of *D. ponderosae* in the western United States, yet their use for control of *I. typographus* is banned in most European

countries. A list of insecticides registered for protecting trees from bark beetle attack can usually be obtained online from regulatory agencies and/or cooperative extension offices, and should be consulted prior to implementing any treatment. It is important to note whether the product is registered for ornamental and/or forest settings, and to limit applications to appropriate sites using suitable application rates while carefully following label restrictions. Generally, only high-value, individual trees growing in unique environments are treated (e.g., developed campgrounds and wildland-urban environments). Tree losses in these environments result in undesirable impacts such as reduced shade, screening, aesthetics, and increased fire risk. Dead trees also pose potential hazards to public safety requiring routine inspection, maintenance, and eventual removal. In addition, trees growing in progeny tests, seed orchards, or those genetically resistant to forest diseases may be treated, especially when epidemic populations of bark beetles exist. Applied correctly, failures in insecticide efficacy are rare and often associated with inadequate coverage, improper mixing, improper storage, and/or improper timing (i.e., applying insecticides to trees already attacked). Remedial applications to kill adults and brood beneath the bark of infested trees are rarely used today (Fettig *et al.*, 2013b).

Most insecticide treatments involve topical sprays applied to the tree bole from the root collar to the mid-crown until runoff (Figure 14.6). It is important that all parts of the tree that are likely to be attacked are

FIGURE 14.7 Experimental injections of emamectin benzoate for protecting *Pinus ponderosa* from mortality attributed to *Dendroctonus brevicomis*. Small quantities [usually <500 ml tree (total volume) based on tree size] were injected with the Arborjet Tree IV™ microinfusion system (Arborjet Inc., Woburn, MA), and later trees were challenged by baiting. A single injection provided adequate protection for three field seasons spurring additional research and later registration of a commercial formulation. *Photo credit: C. Fettig, Pacific Southwest Research Station, USDA Forest Service.*

FIGURE 14.6 A common method of protecting conifers from bark beetle attack in the United States is to saturate all surfaces of the tree bole with an insecticide. Bole sprays are typically applied in late spring prior to initiation of the adult flight period for the target bark beetle species. Usually only high-value, individual trees growing in unique environments or under unique circumstances are treated. *Photo credit: C. Fettig, Pacific Southwest Research Station, USDA Forest Service.*

adequately protected. For some bark beetle species, such as *Ips*, this may require coverage of small limbs and branches. The amount of insecticide (product + carrier, usually water) applied varies considerably with tree species, bark beetle species, tree size, bark and tree architecture, equipment and applicator, among other factors (Fettig *et al.*, 2013b). However, application efficiency (i.e., the percentage of material applied that is retained on trees) is surprisingly high, generally ranging between 80 and 90% (Haverty *et al.*, 1983; Fettig *et al.*, 2008). Bole sprays are usually applied in late spring prior to initiation of the adult flight period for the target bark beetle species. Length of residual activity varies by active ingredient, formulation, bark beetle species, tree species, and location. In most cases, a minimum of one field season of efficacy is expected, but two field seasons is common in some bark beetle–host systems (Fettig *et al.*, 2013b). In rare cases, more than one application per year may be recommended, but this is usually not supported by the scientific literature (Fettig *et al.*, 2006b).

Researchers attempting to find safer, more portable and longer lasting alternatives to bole sprays have evaluated the

effectiveness of injecting small quantities of systemic insecticides directly into the tree bole with pressurized systems (Figure 14.7). These systems push adequate volumes of product (i.e., generally less than several hundred milliliters for even large trees) into the small vesicles of the sapwood. Following injection, the product is transported throughout the tree to the target tissue (i.e., the phloem where bark beetle feeding occurs). Injections can be applied at any time of year when the tree is actively translocating, but time is needed to allow for full distribution of the active ingredient within the tree prior to the tree being attacked by bark beetles. This takes at least several weeks (Fettig *et al.*, 2014b). Tree injections represent essentially closed systems that eliminate drift, and reduce non-target effects and applicator exposure. With the advent of systemic formulations specifically for tree injection, tree injections may become more common tools for protecting conifers from bark beetle attack (Fettig *et al.*, 2013b), particularly in areas where bole sprays are not practical.

4.5 Fire

Burning of infested host material may cause significant beetle mortality and provide some level of direct control (DeGomez *et al.*, 2008); however, attempts to burn standing infested trees have produced mixed results and are seldom used. The application of prescribed fire and/or broadcast burns to suppress bark beetle populations is largely ineffective and rarely practiced (Carroll *et al.*, 2006), but may be appropriate for some species. For example, the use of prescribed fire in the late summer and early autumn in interior Alaska is becoming more common to reduce hazardous fuels and improve wildlife habitat. Such burns may have the added benefit of killing adult *I. perturbatus* that overwinter in the litter layer (Burnside *et al.*, 2011), yet the effectiveness of this treatment has not been adequately explored.

4.6 Semiochemicals

After discovery of the first bark beetle pheromone in the mid-1960s (Silverstein *et al.*, 1966), several bark beetle species were among the first organisms investigated for pheromones, but it was not until years later that these and other semiochemicals were used in management. Utilization has centered on aggregation pheromones that attract the subject species for purposes of retention and later destruction, and antiaggregation pheromones that inhibit host finding and colonization success. The primary semiochemicals associated with most aggressive bark beetle species have been isolated and identified (Wood, 1982; Borden, 1997; Zhang and Schlyter, 2004), and combined with an integrated understanding of their context in the chemical ecology of forests have led to the development of several direct control strategies.

4.6.1 Attractants

The use of attractants in traps to detect or monitor bark beetles is common (Figure 14.8), and often used to survey for exotic, invasive species. For example, the United States Cooperative Agricultural Pest Survey (CAPS) for *I. typographus* includes use of *cis*-verbenol, ipsdienol and 2-methyl-3-buten-2-ol in multi-funnel traps. As mentioned earlier, attractants are also used in trapping programs to monitor population trends and to predict levels of tree mortality attributed to bark beetles (see Section 3), as well as to time deployment of direct control tactics with peak emergence or flight activity patterns. However, some caution should be exerted when interpreting trap catches. For example, Bentz (2006) showed that emergence of *D. ponderosae* from naturally infested trees occurred during a short period of time (30 days), while beetles were caught in attractant-baited traps over a much longer period (130 days). Furthermore, a large proportion of the total number of beetles caught in traps occurred prior to and following peak emergence from trees. In this case, trap catches were a poor representation of overall activity levels.

Attractants are also used in traps to collect and remove beetles through *mass trapping*, and to a lesser extent are placed on insecticide-treated trees to create *lethal trap trees* that induce mortality of beetles upon contact with the tree.

FIGURE 14.8 An attractant-baited multiple-funnel trap used for monitoring bark beetle populations. *Photo credit: C. Fettig, Pacific Southwest Research Station, USDA Forest Service.*

The tactic of controlling bark beetle populations by mass trapping has been attempted for several species in Europe (Bakke *et al.*, 1983, Weslien *et al.*, 1989, Hübertz *et al.*, 1991) and North America (Bedard *et al.*, 1979; Bedard and Wood, 1981; Borden and McLean, 1981; Shea and Neustein, 1995; Ross and Daterman, 1997; Bentz and Munson, 2000). Trapping efficiency varies by bait composition, placement and release rate, trap design and placement, stand structure and composition, and abiotic factors. As mentioned earlier, attractants may be used to induce attacks on individual trees or small groups of trees (termed *trap trees*) to induce colonization prior to sanitation (see Section 4.3.1). An alternative, known as *push pull*, combines the use of mass trapping or trap-tree methods with inhibitors to divert beetles from high-value stands to attractant-baited traps or trees. However, as with any method using attractants, some beetles may infest or *spill over* onto adjacent trees resulting in additional levels of tree mortality, a behavior exhibited in many *Dendroctonus* species. When using attractant-baited traps, placement in areas of non-host trees or in forest openings should limit spillover. Similarly, baiting trees that are widely separated from other hosts (e.g., by >10 m) should reduce the probability of spillover.

4.6.2 Inhibitors

Inhibitors, such as antiaggregation pheromones, are used to protect individual trees and forest stands. Verbenone has received considerable attention and is the primary antiaggregation pheromone of *D. ponderosae*, *D. frontalis*, and *D. brevicomis*, but also causes inhibition in several other species (Zhang and Schlyter, 2004). Production occurs by the beetles themselves (Byers *et al.*, 1984), by autooxidation of the host monoterpene α-pinene via the intermediary compounds *cis*- and *trans*-verbenol (Hunt *et al.*, 1989; Hunt and Borden, 1990), and by degradation of host material by microorganisms associated with bark beetles (Leufvén *et al.*, 1984). Lindgren *et al.* (1996) proposed that verbenone is an indicator of host tissue quality and that its quantity is a function of microbial degradation. Verbenone is presumed to reduce intra- and interspecific competition by altering adult beetle behavior to minimize overcrowding of developing brood within the host. Fettig *et al.* (2007b) showed that *Temnochila chlorodia* (Mannerheim) (Coleoptera: Trogositidae), a common bark beetle predator in western North America, is attracted to verbenone, and therefore its impact on beetle populations may be enhanced by verbenone treatments.

In North America, verbenone has been demonstrated effective for reducing tree mortality attributed to *D. ponderosae* and *D. frontalis*, but not *D. brevicomis*. During the 1990s, *D. frontalis* populations were epidemic in many areas of the southern United States, and research there led to the development and registration of a 5-g verbenone-releasing pouch (Clarke *et al.*, 1999). Subsequently, larger capacity pouches (7-g and 7.5-g) were evaluated and registered (Progar *et al.*, 2013). The effectiveness of verbenone varies with time and geographical area (Amman, 1994), outbreak intensity (Progar *et al.*, 2013), dose (Borden and Lindgren, 1988; Gibson *et al.*, 1991), tree species (Negrón *et al.*, 2006), and bark beetle species (Fettig *et al.*, 2009). Failures in efficacy are not uncommon, and have limited more widespread use (Table 14.5). Another antiaggregation pheromone, 3-methylcyclohex-2-en-l-one (MCH), which has been demonstrated effective for reducing colonization of Douglas-fir (*Pseudotsuga menziesii* (Mirb.) Franco) by *D. pseudotsugae*, has yielded more consistent efficacy (Ross *et al.*, 2001). However, research in North America has largely focused on verbenone, presumably due to the substantial impacts of recent outbreaks of *D. ponderosae*.

Recent research has concentrated on combining verbenone with other inhibitors to increase levels of inhibition (Zhang and Schlyter, 2004). In this context, a diverse array of chemical cues from con- and heterospecifics and nonhosts is likely to disrupt bark beetle searching more than high doses of a single semiochemical (e.g., verbenone) or even mixtures of semiochemicals intended to mimic one

TABLE 14.5 Barriers to Successful Development of Semiochemical-based Tools for Protecting Conifers from Mortality Attributed to Bark Beetles Based Largely on Experiences with Verbenone and *Dendroctonus ponderosae* and *D. brevicomis* in Western North America, but with Wider Applicability

Chemical stability of formulations in the forest environment	Little is known about chemical stability once released into the active airspace.
Complexity of semiochemical signals used in host finding, selection and colonization processes	Bark beetles use a variety of contextual cues during host finding, selection, and colonization. Insufficient reductions in tree mortality may be due, in part, to inadequate chemical or other signaling. For example, synthetic verbenone deployed alone without other beetle-derived or non-host cues may not provide sufficient levels of inhibition.
Costs and small market conditions	These factors are significant barriers to investment in research and development, specifically basic science.
Inconsistent release	Several authors have speculated that failures in effectiveness have resulted from problems associated with passive release, which is largely controlled by ambient temperature.
Levels of inhibition	Sensitivity varies among populations and among individuals within a population thus influencing effectiveness.
Managing expectations	Research is needed to determine what levels of efficacy are acceptable (e.g., based on reductions of negative impacts to forests), and under what conditions inhibitors are likely to be most effective.
Population size	Effectiveness declines with increasing population density. Higher levels of tree mortality are expected during severe infestations and with a declining proportion of preferred hosts when populations still exist at epidemic levels.

Continued

TABLE 14.5 Barriers to Successful Development of Semiochemical-based Tools for Protecting Conifers from Mortality Attributed to Bark Beetles Based Largely on Experiences with Verbenone and *Dendroctonus ponderosae* and *D. brevicomis* in Western North America, but with Wider Applicability—cont'd

Range of inhibition	Studies show that the maximum range of inhibition is quite limited. Higher densities of small, point-source releasers may provide for better dispersal patterns and greater reductions in tree mortality.
Ratio of inhibitors to attractants	Levels of inhibition vary based on this ratio.
Variation in stand structure, especially tree density	Concentrations of semiochemicals rapidly decrease with increasing distance from a point source, and in low density forests unstable layers and multi-directional traces (eddies) may dilute concentrations and reduce effectiveness.

FIGURE 14.9 An example of a 7-g verbenone pouch (Contech Inc., Delta, BC) applied to reduce the amount of mortality attributed to *Dendroctonus ponderosae* in *Pinus contorta* stands. Semiochemical release devices are typically stapled at maximum reach (≈2 m in height) to individual trees or applied in a gridded pattern to achieve more uniform coverage when stand protection is the objective. *Photo credit: C. Fettig, Pacific Southwest Research Station, USDA Forest Service.*

type of signal (e.g., antiaggregation pheromones), as they represent heterogeneous stand conditions to foraging insects. To that end, a bark beetle encounters several decision nodes during host searching that may be exploited by combining verbenone (or other antiaggregation pheromones) with non-host volatiles, including (1) habitat suitability (e.g., green leaf volatiles and angiosperm bark volatiles), (2) host presence (e.g., green leaf volatiles and angiosperm bark volatiles), and (3) host suitability and susceptibility (e.g., antiaggregation and aggregation pheromone components of con- and heterospecifics, and host volatiles that signal changes in host vigor and/or tissue quality) (Borden, 1997; Schlyter and Birgersson, 1999; Zhang and Schlyter, 2004; Progar *et al.*, 2014).

The most common method of applying inhibitors includes pouch release devices (Figure 14.9) stapled at maximum reach (≈2 m in height) to individual trees prior to beetle flight, or applied in a gridded pattern to achieve uniform coverage when stand protection is the objective. For some species, such as *D. frontalis*, a unique distribution of release points may be required (Clarke *et al.*, 1999). Bead, flake, and sprayable formulations have been evaluated but are not widely used. The release rates of passive releasers vary with changes in temperature and humidity, and since they dispense semiochemicals through a membrane, are strongly influenced by meteorological conditions making the amount of semiochemical released somewhat unpredictable (Holsten *et al.*, 2002). Puffers are small battery-activated reservoirs that emit frequent, measured puffs of aerosolized liquid, thus overcoming some of the problems associated with passive releasers, but are prohibitively expensive for forestry applications (Progar *et al.*, 2013). However, once the fluid is dispensed from puffers, evaporative properties and thus release rates are still heavily influenced by meteorological conditions. Mafra-Neto *et al.* (2013) have recently developed a novel matrix impregnated with verbenone (SPLAT® Verb, ISCA Technologies Inc., Riverside, CA) that has shown a high degree of efficacy for protecting *P. contorta* from *D. ponderosae*. Rather than a single release device, SPLAT is an amorphous, flowable controlled-release emulsion with chemical and physical properties that can be adjusted by small changes in composition and application. This formulation is also biodegradable, which has been an objective for the development of release devices as significant labor cost savings are achieved by not having to retrieve release devices from the field after use. A formulation of biodegradable flakes and a novel dispenser have also been developed and are being evaluated (Gillette *et al.*, 2012; Fettig, unpubl. data). In all cases, the fate of semiochemicals once released into the active airspace of forests is highly influenced by forest structure and meteorological conditions.

4.6.3 Future Semiochemical Research

Significant advances have been made concerning the molecular biology and biochemistry of pheromone production in bark beetles, the synthesis of semiochemicals in the laboratory, the deployment of semiochemicals in the field, and the fate of semiochemicals once released into the active airspace of forests. Despite this, significant research needs exist, including (1) improving the efficacy and cost effectiveness of blends and delivery systems, (2) redefining selection criteria for target areas where semiochemical-based treatments are likely to be most effective, (3) examining the effects of forest structure and abiotic factors on semiochemical plumes, (4) expansion of related research into understudied forest types, and (5) assessment of semiochemical performance at varied levels of beetle population and host availability (Progar *et al.*, 2014). The results of a recent meta-analysis demonstrating the effectiveness of semiochemicals to reduce levels of tree mortality attributed to bark beetles are encouraging (Schlyter, 2012) and should spur additional research and development.

5. INDIRECT CONTROL

5.1 Thinning

It is widely accepted that thinning is an effective means of increasing the resiliency of forests to bark beetle infestations and other disturbances (Fettig *et al.*, 2007a). However, it is important to stress that prescriptions vary widely and have different effects on structure and composition. For example, in the western United States many thinning treatments are implemented for fuels reduction, which concentrates on reducing surface fuels, increasing the height to live crown, decreasing crown density, and retaining large trees of fire-resistant species (Agee and Skinner, 2005). While such treatments may also reduce the susceptibility of forests to some bark beetle species, related prescriptions vary from those that might be implemented specifically for bark beetles. In the latter case, *crown* or *selection thinning* (i.e., removal of larger trees in the dominant and codominant crown classes) may be required to achieve target threshold densities, residual tree spacing, and significant reductions in the abundance of preferred hosts necessary to adequately reduce stand susceptibility (e.g., from *D. brevicomis*). Furthermore, thinning may have differential

effects among bark beetle species. In many systems, a suite of less aggressive species is attracted to logging residues (e.g., several *Ips* spp. in North America), but depending on the vigor of residual trees may result in little tree mortality. Thinning conducted in a careless manner may result in increases in other subcortical insects and root pathogens (Harrington *et al.*, 1985).

Fettig *et al.*, (2007a) used the concept of growing space as a mechanism to illustrate how changes in host tree vigor, among other factors, influence susceptibility of individual trees and forest stands to bark beetle attack following thinning. Trees utilize growth factors, such as sunlight, water, nutrients, temperature, oxygen, and carbon dioxide, until one or more factors become limiting (Oliver and Larson, 1996). Disturbances can make growing space available to some trees at the expense of others (e.g., herbivory), or alter the amount of growing space available to all trees (e.g., drought). For example, when soil moisture is limited, trees close their stomata to avoid excessive water loss, which inherently leads to reduced productivity as stomatal closure also prohibits uptake of carbon dioxide and therefore photosynthesis. A tree's photosynthates are allocated to different uses in an order of priorities (Oliver and Larson, 1996): (1) maintenance respiration; (2) production of fine roots; (3) reproduction; (4) primary (height) growth; (5) xylem (diameter) growth; and (6) insect and disease resistance mechanisms. While somewhat conceptual, this hierarchy illustrates how production of insect resistance mechanisms is compromised first when growing space becomes limited by one or more factors. Conversely, it demonstrates how cultural practices that release growing space through reductions in tree density influence the susceptibility of individual trees, stands, and forests by strengthening insect resistance mechanisms (Fettig *et al.*, 2007a).

Reductions in tree density also cause changes in microclimate that affect beetle fecundity and fitness, phenology, and voltinism (number of generations per year), as well as that of predators, parasitoids, and competitors. Changes in tree density may also cause turbulences that disrupt pheromone plumes used for recruiting conspecifics during initial phases of host tree colonization. Bartos and Amman (1989) suggested that changes in microclimate were the principal factors associated with reductions in stand susceptibility to *D. ponderosae* following thinning in *P. contorta*. Thinning increased light intensity, wind movement, insolation, and temperature in affected stands. Thistle *et al.* (2004) examined near field canopy dispersion of a tracer gas (SF_6), as a surrogate for bark beetle pheromones, within the trunk space of trees. They showed that when surface layers of air are stable (e.g., during low wind velocities), the tracer plume remained concentrated and directional because of suppression of turbulent mixing by the forest canopy. Lower density stands result in unstable layers of air and multi-directional traces (eddies) that diluted "pheromone" concentrations (Thistle *et al.*, 2004) and presumably reduce beetle aggregation, thus influencing host finding and colonization successes. Furthermore, the killing of groups of trees is fundamental to expansion of some infestations, and therefore some authors have suggested that residual spacing of leave trees is more important than reductions in overall tree density (Whitehead *et al.*, 2004; Whitehead and Russo, 2005). This is likely not the case in all bark beetle–host systems.

In North America, *thinning from above* or diameter-limit thinning, and *thinning from below* (Cole and Cahill, 1976; McGregor *et al.*, 1987) applied to reduce basal area (Amman *et al.*, 1977; Cahill, 1978; Bennett and McGregor, 1980), remove trees with thick phloem (Hamel, 1978), and/or increase residual tree spacing (Whitehead *et al.*, 2004; Whitehead and Russo, 2005; Table 14.6) have all been implemented to reduce the susceptibility of *P. contorta* forests to *D. ponderosae*. Schmidt and Alexander (1985) found that thinning from above was effective until residual trees grew to susceptible sizes; however, it left stands with reduced silvicultural value that were often vulnerable to windthrow or snow damage. Thinning from below may optimize the effects of microclimate, inter-tree spacing, and tree vigor (Whitehead and Russo, 2005; Coops *et al.*, 2008) even though residual trees are of diameter classes considered more susceptible to attack (Waring and Pitman, 1980; Mitchell *et al.*, 1983, but see Ager *et al.*, 2007). However, this practice may not be economically viable since only smaller diameter trees are removed. Recommended residual conditions include inter-tree spacings of at least 4 m (Whitehead *et al.*, 2004; Whitehead and Russo, 2005) or 400–625 trees/hectare (Whitehead and Russo, 2005). While thinning during endemic populations is most desirable, thinning may also be useful during an outbreak, specifically if combined with sanitation harvests and/or other direct control methods (Waring and Pitman, 1985).

Schmid and Mata (2005) monitored levels of tree mortality attributed to *D. ponderosae* in 1-hectare plots over a 17-year period in South Dakota. The authors concluded that the effectiveness of thinning *P. ponderosa* forests to residual densities between 18.4 and 27.6 m^2/hectare to reduce susceptibility was questionable. However, they suggested that their results were confounded by small study plots being surrounded by extensive areas of unmanaged forest where *D. ponderosae* populations were epidemic. Later, Schmid *et al.* (2007) reported thinning to 18.4 m^2/hectare in susceptible stands may not be sufficient to yield long-term reductions in susceptibility if not followed with subsequent thinning over time to maintain lower tree densities. These publications raise important issues that likely apply to other bark beetle species for which thinning and other management strategies (see Section 4) have been demonstrated effective. That is, it is critical that treatments are applied at a frequency, scale, and intensity dictated by the bark beetle population and the spatial extent of infested areas. Relatedly, Ager *et al.* (2007) simulated the impacts of thinning over 60 years, coupled with a *D. ponderosae* outbreak at 30 years, to examine how thinning might influence

TABLE 14.6 Cumulative Number of *Pinus contorta* Killed by *Dendroctonus ponderosae* 9–12 years after Thinnings were Conducted, British Columbia, Canada

Location (Year of Treatment)	Treatment	No. Trees Attacked/ha	Green: Red Attack Ratio[1]
Cranbook (1992)	Untreated	22	1.8
	Spaced to 4 m	2	0.3
	Spaced to 5 m	7	0.5
Parson (1993)	Untreated	56	2.9
	Untreated	15	0.3
	Spaced to 4 m	0	–
	Spaced to 5 m	0	–
Hall Lake (1994)	Untreated	158	1.8
	Thinned to 500 trees/ha	37	1.4
Quesnel (1991)	Untreated	452	3.3
	Spaced to 4 m	167	1.2

[1]*Ratios >1.0 indicate that infestations are building.*
(Modified from Whitehead *et al.*, 2004.)

bark beetle impacts in a 16,000-hectare landscape in eastern Oregon. They employed the Forest Vegetation Simulator and Westwide Pine Beetle Model (WPBM). The latter simulates beetle populations in terms of a "beetle kill potential" (BKP), where one unit of BKP is sufficient to kill $0.0929\ m^2$ of host tree basal area (Smith *et al.*, 2005). While not widely adopted, the model assumes that beetles emerge and disperse, and choose stands to attack based on distance and certain stand attributes. The authors reported that contrary to expectations, WPBM predicted higher levels of tree mortality from an outbreak in thinned versus unthinned scenarios. In this case, thinning favored retention of early seral tree species (e.g., *P. ponderosa*), leading to increases in the proportion and average diameter of preferred hosts.

5.2 Landscape Heterogeneity

Efforts to prevent undesirable levels of tree mortality must also account for the spatial distribution of cover types. In many areas, treatments should be implemented to increase heterogeneity (e.g., of age, size, and species compositions) as homogeneous forested landscapes promote creation of large contiguous areas susceptible to similar disturbances (Fettig *et al.*, 2007a). Studies have shown that insects tend to focus host searching in patches of high host concentrations (Root, 1973), which increases the probability of encounters with suitable hosts. In heterogeneous stands or landscapes this occurs with less efficiency (Jactel and Brockerhoff, 2007). For example, in North America several authors have suggested that shorter rotations and promotion of multiple age classes will minimize levels of tree mortality attributed to *D. ponderosae* (Safranyik *et al.*, 1974; Taylor and Carroll, 2004; Whitehead *et al.*, 2004).

5.3 Prescribed Fire

Prescribed fire is primarily used to reduce surface and ladders fuels in fire prone forests. Tree mortality resulting from prescribed fire may be immediate due to consumption of living tissue and heating of critical plant tissues, or can be delayed occurring over the course of several years. Levels of delayed tree mortality are difficult to predict, and depend on numerous factors including tree species, tree size, phenology, degree of fire-related injuries, initial and post-fire levels of tree vigor, the post-fire environment, and the frequency and severity of other predisposing, inciting, and contributing factors. Following prescribed fire, short-term increases in levels of bark beetle-caused tree mortality are often reported, primarily in the smaller-diameter classes (Stephens *et al.*, 2012). However, in the longer term burned areas may benefit from the positive impacts of prescribed fire on growing space and other factors that reduce forest susceptibility to bark beetles (Fettig *et al.*, 2006a; Fettig and McKelvey, 2010).

5.4 Social Acceptance of Management Strategies

Although public opinion is an important factor influencing the management of bark beetles (Wellstead *et al.*, 2006), few contemporary studies have evaluated the social acceptance of various direct and indirect control strategies. McFarlane *et al.* (2006) examined public attitudes relevant to management preferences for *D. ponderosae* in Banff and Kootenay National Parks, Canada. Data were collected by mail survey from a large pool of residents living in or near the parks. All groups agreed that "allowing the outbreak to follow its course without intervention" was not an acceptable option. Preferred options included "sanitation cutting to remove infested trees from small areas" and the "use of pheromones to attract beetles to one area." Other acceptable options included the use of prescribed burning, sanitation of large areas, and "thinning the forest to remove some of the uninfested but susceptible trees." Visitors to these parks had similar attitudes to the local residents in support of direct and indirect control (McFarlane and Watson, 2008). This differs from tourists' perception of an *I. typographus* outbreak in Bavarian Forest National Park, Germany, where respondents showed a neutral attitude toward the bark beetle, and were somewhat disinclined to support control measures within the park (Müller and Job, 2009). In Virginia, a survey of landowners indicated that those that were college educated were more willing to participate in the state's Southern Pine Beetle Prevention Program (Watson *et al.*, 2013), which concentrates on pre-commercial thinning to reduce forest susceptibility to *D. frontalis*. A better understanding of public perceptions towards outbreaks and proposed management strategies may help managers to better inform the public of the usefulness and consequences of different treatments.

6. CASE STUDY—MANAGEMENT OF *IPS TYPOGRAPHUS* IN CENTRAL EUROPE

As previously discussed, *I. typographus* is one of the most important forest insects in Europe due to its role in the dynamics of forest ecosystems and the profound impact of outbreaks on ecosystem goods and services (Grégoire and Evans, 2004; Stadelmann *et al.*, 2013). Outbreaks are usually precipitated by other disturbances such as windstorms, severe drought, or weakening of trees by pathogenic fungi (Wermelinger, 2004). Such trees attract beetles by releasing host volatiles (Lindelöw *et al.*, 1992), and provide abundant host material. In recent years, spectacular storms such as Vivian (1990), Lothar (1999), Gudrun (2005), and Kirill (2007) impacted huge areas of Europe and destroyed millions of trees, creating large quantities of susceptible host material and subsequently *I. typographus* outbreaks (Komonen *et al.*, 2011). Furthermore, the frequency and severity of outbreaks is expected to increase as a result of climate change (Schlyter *et al.*, 2006). During outbreaks, thousands of trees are attacked and killed within several weeks prompting large and well-coordinated direct control efforts. For example, during 1940–1951 one of the largest outbreaks in central Europe resulted in sanitation of 30 million m^3 of infested trees. In northeastern Poland alone, ≈100,000 hectares of infested spruce forest were harvested during 1945–1948, and decreased spruce inventories there by ≈50% (Puchniarski, 2008).

Several thousand scientific and popular papers have been published on various aspects of the biology, ecology, and management of *I. typographus*. In this regard, the species ranks among the best studied of forest insects. Since publication of the first forestry text that addressed bark beetles (Ratzeburg, 1839), control methods and strategies have been systematically developed based on the practical experiences of foresters and research executed by scientists. In many European countries, several methods of control are regularly implemented in response to outbreaks of *I. typographus*.

6.1 Sanitation Salvage

As previously discussed, salvage involves the harvest and removal of dead trees, but in the context of *I. typographus* has the added benefit of removing attractive host material that facilitates rapid population growth. In central Europe, "salvage" is traditionally regarded by foresters and other practitioners as removal of infested trees (sanitation) combined with the salvage of previously infested trees or those impacted by other disturbances, termed here *sanitation salvage*. This is considered the most effective direct control strategy for reducing levels of tree mortality attributed to *I. typographus* in Europe (Wermelinger, 2004). It is common practice that a large number of infested and susceptible downed, damaged, and standing trees are treated during outbreaks, especially in countries where responding agencies are well organized and technically prepared for such large and logistically complicated operations (Szabla, 2013) (Table 14.7).

The first step in sanitation-salvage operations is to locate and mark trees to be removed. While selection of such trees is obvious in some cases (e.g., on the basis of crown fade), the process is much more difficult in the case of newly infested trees. Infested trees are systematically marked by experienced and trained sawdusters, who are usually very adept at finding trees attacked by *I. typographus* (Table 14.7). Conducting sanitation salvage during proper periods is an important factor (Figure 14.10). In the case of infested trees, treatment is critical before emergence of the next generation of beetles (Jönsson *et al.*, 2012; Stadelmann *et al.*, 2013). Trees recently damaged by windstorms or other disturbances should be salvaged before

mid-summer (Göthlin *et al.*, 2000) or may be used as trap trees (Wichmann and Ravn, 2001). Unfortunately, timely sanitation salvage is often not possible during large-scale outbreaks due to limitations in labor and logistical constraints, but delay to after the flight activity period of the following year makes these tactics ineffective and perhaps even harmful. By mid-summer, infested trees are usually heavily colonized by a rich community of natural enemies that may be adversely impacted by sanitation salvage, particularly during latter stages of an outbreak. Furthermore, retention of some dead trees is beneficial for a variety of organisms including predators and parasitoids of *I. typographus*, and several endangered species (Weslien, 1992a; Siitonen, 2001; Jonsell and Weslien, 2003).

During large-scale outbreaks, sanitation-salvage operations may be carried into winter, but is rarely implemented simply because weather conditions and snow loads often preclude access. During winter, the role of *winter sawdusters* is also very important, and the method is highly effective for limiting numbers of attacked trees the following spring and summer (Kolk and Grodzki, 2013). However, winter operations may have a stronger negative impact on natural enemy communities than sanitation-salvage implemented during the spring and summer as many species overwinter in high numbers in and under the bark of trees colonized by *I. typographus* (Weslien, 1992a). Most cluster in the bottoms of trees beneath the snow, which enhances overwintering survival (Hilszczański, 2008).

6.2 Trap Trees

It is common practice to use broken and windthrown trees as trap trees for *I. typographus*. Such trees are attractive for at least two seasons, and have limited defensive mechanisms to deter attack (Eriksson *et al.*, 2005). Covering downed logs with the branches and foliage of spruce, while labor intensive, is supposed to protect them from rapid desiccation and is recommended in the Czech and Slovak Republics (Zahradník *et al.*, 1996). The idea of preparing special trap trees in the form of logs for control *I. typographus* and other bark beetles dates back to the first half of the 19th century (Skuhravý, 2002). Trap trees are often prepared a few times per year, usually twice, to adequately cover the most important generations of *I. typographus* (Figure 14.10). It is important to debark or otherwise destroy trap trees in a timely manner after oviposition otherwise a high proportion of adults may leave (emerge), colonize adjacent trees, and establish a sister generation (Bakke, 1983). In some countries, standing live trees are used as trap trees. In these cases, trees are usually baited with synthetic pheromones and are often several times more effective at trapping *I. typographus* than pheromone-baited traps (Raty *et al.*, 1995). For example, a study in Belgium

TABLE 14.7 Methods of Control for *Ips typographus* during Outbreaks in Southern Poland, 2007–2010 (Based on Szabla, 2013)

Method	2007	2008	2009	2010	Total
Trap trees (thousands)	8.4	5.2	4.4	3.7	21.7
Baited-trap trees (thousands m^3)	17.5	30.9	31.7	16.2	96.3
Marked "sawdust" trees (thousands)	510	424	272	158	1364
Sanitation (thousands m^3)	803	798	466	231	2298
Debarked-infested trees (thousands m^3)	297	254	17	23	591
Pheromone-baited traps (thousands)	11.6	12.2	11.8	10.6	46.2

FIGURE 14.10 Annual timing of direct control strategies used for management of *Ips typographus* in central Europe. Here, "salvage" is synonymous with sanitation or sanitation salvage (see Section 4).

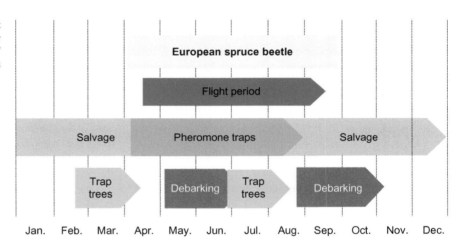

reported trap trees were 14 times more efficient than artificial traps (Drumont *et al.*, 1992). However, efficiency varies with population densities and during endemic populations more beetles are often collected in pheromone-baited traps (Król and Bakke, 1986). On the contrary, trap trees are more effective during outbreaks (Grodzki *et al.*, 2008).

6.3 Pheromone-baited Traps

Pheromone-baited traps were initially used for monitoring of *I. typographus*, but were quickly recognized as a cheaper alternative to trap trees. The number of *I. typographus* collected depends on many factors, including bait composition, placement and release rate, trap design and placement, stand structure and composition, competition from accessible host material such as windthrown trees, harvested logs and damaged trees, and abiotic factors (Bakke, 1992; Lobinger, 1995; Mezei *et al.*, 2012). For example, traps located on southern aspects are often several times more effective at capturing *I. typographus* than those

on northern aspects (Lobinger and Skatulla, 1996), presumably due to solar inputs. Mass trapping is regarded as an effective method for protecting stands of wind-damaged spruce (Grégoire *et al.*, 1997), but is considered rather ineffective during large-scale outbreaks (Dmitri *et al.*, 1992; Wichmann and Ravn, 2001). Estimates concerning the effectiveness of traps for reducing *I. typographus* populations range from 0.2 to 80% (Zahradník *et al.*, 1993), but most studies have shown that only a minor portion is captured (Weslien and Lindelow, 1990; Lobinger and Skatulla, 1996) despite substantial numbers being collected (Szabla, 2013) (Table 14.8). Mass trapping has been implemented during large-scale outbreaks in Sweden ($\approx$270,000 traps; Weslien, 1992b), and Poland ($\approx$50,000 traps; Szabla, 2013), but in the latter case was just one of several direct control methods employed (Table 14.7). Generally, high trap catches are not well correlated with activity on trees, but low catches usually coincided with little beetle activity (Weslien, 1992b; Lindelöw and Schroeder, 2001). Mass trapping could be effective as an additional method of control during outbreaks, especially in the context of

TABLE 14.8 Numbers of Individuals (in Millions) Collected at the Time of Mass Trapping Efforts for *Ips typographus* during Outbreaks in Southern Poland, 2007–2010 (Based on Szabla, 2013)

Species	2007	2008	2009	2010	Total
Ips typographus	194	189	170	72	625
Pityogenes chalcographus (I.)	156	122	162	72	512
Trypodendron lineatum (Olivier)	6	3	5	1	15

protecting living trees rather than reducing *I. typographus* populations (Dubbel *et al.*, 1995; Jakuš, 2001).

A potentially negative impact of mass trapping is that several members of the natural enemy community may be collected and killed. This is of specific concern for the European red-bellied clerid (*Thanasimus formicarius* (L.)) an important predator of *I. typographus* and other bark beetles. However, the proportion of trap catches represented by *T. formicarius* is usually <4% (Babuder *et al.*, 1996; Valkama *et al.*, 1997; Grodzki, 2007).

6.4 Push Pull

Advanced methods of semiochemical-based control have not been widely implemented for *I. typographus* despite numerous experiments being conducted. In the Slovak Republic, the use of verbenone and aggregation pheromones of *I. typographus* were ineffective (Jakuš and Dudova, 1999). Verbenone and non-host volatiles have been tested in several different countries with variable results. Promising results were obtained in the Šumava Mountains of the Czech Republic where push pull significantly decreased the probability of *I. typographus* attack on standing healthy trees by 60–80% (Jakuš *et al.*, 2003).

6.5 Debarking of Infested Host Material

Debarking has been implemented in response to concerns of transporting infested logs as beetles could emerge prior to processing (Drumont *et al.*, 1992). While debarking has been demonstrated to kill up to 93% of *I. typographus* beneath the bark, this may still be insufficient to achieve adequate levels of control during outbreaks (Dubbel, 1993). Furthermore, the method is time consuming and relatively expensive. Occasionally infested trees are cut and left untreated for conservation of biodiversity (Jonasova and Matejkova, 2007).

6.6 Biological Control

Several natural enemies of *I. typographus* have been extensively studied (see Chapter 7), specifically insect parasitoids and predators (Kenis *et al.*, 2004; Wermelinger,

2004), pathogens (Wegensteiner, 2004), and to a lesser extent woodpeckers (Fayt *et al.*, 2005). *Ips typographus* parasitoids and predators exhibit clear habitat preferences. Some species prefer standing trees or high stumps while others prefer open areas or shady conditions (Hedgren, 2004; Hilszczański *et al.*, 2005). Similarly, woodpeckers that commonly feed on *I. typographus*, such as the three-toed woodpecker (*Picoides tridactylus* (L.)), require certain habitat features such as dead standing trees for cavity nesting (Fayt *et al.*, 2005). In that context, habitat manipulation and forestry practice modification could be implemented as measures of natural enemy control enhancement. To date, biological control efforts have not been formally implemented for *I. typographus* (Wermelinger, 2004), but some recent experiments involving *B. bassiana* have proven promising (Vaupel and Zimmermann, 1996; Kreutz *et al.*, 2004; Landa *et al.*, 2008; Jakuš and Blaženec, 2011).

6.7 Insecticides

The use of insecticides for management of *I. typographus* is banned in most European countries. Treatments were more widely used in the late 20th century usually in the context of baited trap trees or as a means to protect timber (Drumont *et al.*, 1992; Lubojacký and Holuša, 2011).

6.8 Risk and Hazard Rating and Silviculture

Risk and hazard rating represents critical elements in the management of *I. typographus*. Optimally, both direct and indirect control strategies are prioritized, planned, and implemented based on predicted risks. In most countries where *I. typographus* creates serious problems, estimation of the dynamics of pest activity is based on monitoring of the volume of infested trees (Cech and Krehan, 1997; Knížek and Lubojacký, 2012) (Figure 14.11). Unfortunately, this method provides very little information about potential risks. Moreover the interpretation of other monitoring techniques, such as the use of pheromone-baited traps, is also difficult since the effectiveness of traps depends on so many environmental and technical factors

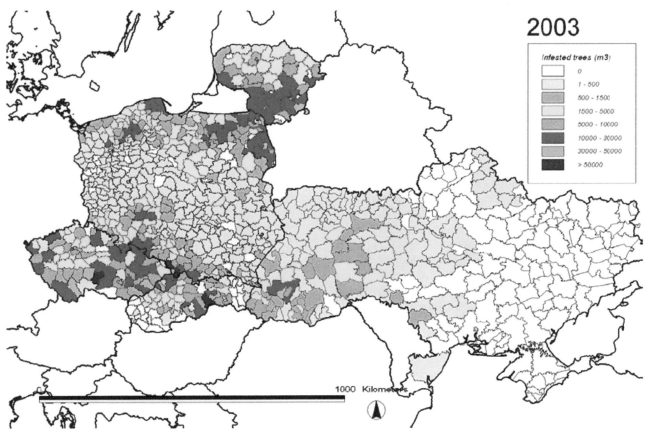

2003

Infested trees (m3)

	0
	1 - 500
	500 - 1500
	1500 - 5000
	5000 - 10000
	10000 - 30000
	30000 - 50000
	> 50000

1000 Kilometers

FIGURE 14.11 Distribution of bark beetle-killed trees in *Picea abies* stands of central Europe in 2003, expressed as the volume of infested trees cut in individual territorial units. *From Grodzki (2005).*

(Bakke, 1992; Lobinger, 1995; Grodzki, 2007; Mezei *et al.*, 2012). Nevertheless some tree, stand, and other environmental characteristics related to site, exposure (aspect), water supply, temperature, and co-occurring pathogens, among others, are known to have strong effects on the probability of *I. typographus* attack. Since the first study by Annila (1969), several attempts to develop models of *I. typographus* development and risk assessment have been made (Zumr, 1982; Schopf, 1985; Anderbrant, 1986; Coeln *et al.*, 1996; Wermelinger and Seifert, 1999; Netherer and Nopp-Mayr, 2005). More sophisticated models, such as PHENIPS, provide a tool for hazard rating at local and regional scales (Baier *et al.*, 2007). Implementation of the TANABBO model enables prediction of stand susceptibility to attack based on select environmental factors and their influence on *I. typographus* population density (Kissiyar *et al.*, 2005). Additionally, GIS techniques provide useful visualizations of outbreak dynamics (Jakuš *et al.*, 2005).

Stand characteristics that can be manipulated by silvicultural treatment to reduce the susceptibility of forests to *I. typographus* include age class diversity, stand density, host density, and stand composition (Wermelinger, 2004).

Key characteristics positively correlated with the severity of outbreaks include the proportion (density) and age of spruce trees (Becker and Schröter, 2000; Gilbert *et al.*, 2005; Hilszczański *et al.*, 2006; Grodzki, 2010). Susceptibility to *I. typographus* attack also increases with the so-called "edge effect" and sudden "opening" of the forest (Göthlin *et al.*, 2000; Grodzki *et al.*, 2006; Hilszczański *et al.*, 2006). Others factors affecting the susceptibility of trees include crown length, which is often related to stand density, and tree size, specifically diameter, which is relevant to both standing and downed trees (Lekander, 1972; Weslien and Regnander, 1990; Zolubas, 2003; Eriksson *et al.*, 2005). Silvicultural activities (e.g., thinning) that reduce the susceptibility of trees, stands, and forests to *I. typographus* in central Europe are for the long term the most acceptable both for environmental and economic reasons.

7. CONCLUSIONS

Bark beetle infestations will continue to occur as long as susceptible forests and favorable climatic conditions coincide. As discussed above, there are a wide variety of tactics available to reduce their severity and extent when properly applied at appropriate scales. However, the only long-term solution is to change forest structure and composition to increase resiliency. Experience has shown that even a course of no action is not without consequence, although this alternative may be the most appropriate under some circumstances. Several assessments have concluded that forests are increasingly vulnerable to tree mortality as a result of the direct and indirect effects of climate change (Fettig et al., 2013a), and that the use of sound, ecologically appropriate management strategies, and prioritization of their application to enhance resiliency is critical. Gillette et al. (2014) examined the various D. ponderosae treatment options available to land managers in North America, and described their long-term consequences in terms of risk of future outbreaks, wildfire, invasion by exotic weeds, loss of hydrological values, and carbon sequestration. They, like us, argue for the increased use of science-based indirect control, specifically thinning, to increase resiliency of forests to multiple stressors including bark beetle infestations.

REFERENCES

Agee, J.K., Skinner, C.N., 2005. Basic principles of forest fuel reduction treatments. For. Ecol. Manage. 211, 83–96.

Ager, A.A., McMahan, A., Hayes, J.L., Smith, E.L., 2007. Modeling the effects of thinning on bark beetle impacts and wildfire potential in the Blue Mountains of eastern Oregon. Landsc. Urban Plann. 80, 301–311.

Amman, G.D., 1969. Mountain pine beetle emergence in relation to depth of lodgepole pine bark. U.S. Department of Agriculture, Forest Service, Ogden, UT, Research Note INT-RN-96.

Amman, G.D., 1975. Insects affecting lodgepole pine productivity. In: Baumgartner, D.M. (Ed.), Management of Lodgepole Pine Ecosystems: Symposium and Proceedings. Washington State University Cooperative Extension Service, Pullman, pp. 310–341.

Amman, G.D., 1994. Potential of verbenone for reducing lodgepole and ponderosa pine mortality caused by mountain pine beetle in high-value situations. U.S. Department of Agriculture, Forest Service, Berkeley, CA, General Technical Report PSW-GTR-150.

Amman, G.D., Cole, W.E., 1983. Mountain pine beetle dynamics in lodgepole pine forests. Part 2. U.S. Department of Agriculture, Forest Service, Ogden, UT, General Technical Report INT-GTR-145.

Amman, G.D., McGregor, M.D., Cahill, D.B., Klein, W.H., 1977. Guidelines for reducing losses of lodgepole pine to the mountain pine beetle in unmanaged stands in the Rocky Mountains. U.S. Department of Agriculture, Forest Service, Ogden, UT, General Technical Report INT-GTR-36.

Anderbrant, O., 1986. A model for the temperature and density dependent reemergence of the bark beetle Ips typographus. Entomol. Exp. Appl. 40, 81–88.

Annila, E., 1969. Influence of temperature upon the development and voltinism of Ips typographus L. (Coleoptera; Scolytidae). Ann. Zool. Fenn. 6, 161–208.

Arnberg, W., Wastenson, L., 1973. Use of aerial photographs for early detection of bark beetle infestations of spruce. Ambio 2, 77–83.

Atkins, M.D., 1966. Behavioral variation among scolytids in relation to their habitat. Can. Entomol. 98, 285–288.

Babuder, G., Pohleven, F., Brelih, S., 1996. Selectivity of synthetic aggregation pheromones Linoprax and Pheroprax in the control of the bark beetles (Coleoptera, Scolytidae) in a timber storage yard. J. Appl. Entomol. 120, 131–136.

Baier, P., Pennerstorfer, J., Schopf, A., 2007. PHENIPS—a comprehensive phenology model of Ips typographus (L.) (Col., Scolytinae) as a tool for hazard rating of bark beetle infestation. For. Ecol. Manage. 249, 171–186.

Bakke, A., 1983. Host tree and bark beetle interaction during a mass outbreak of Ips typographus in Norway. Z. Angew. Entomol. 96, 118–125.

Bakke, A., 1992. Monitoring bark beetle populations: effect of temperature. J. Appl. Entomol. 114, 208–211.

Bakke, A., Saether, T., Kvamme, T., 1983. Mass trapping of the spruce bark beetle Ips typographus. Pheromone and trap technology. Medd. Nor. Inst. Skogforsk 38, 1–35.

Bartos, D.L., Amman, G.D., 1989. Microclimate: an alternative to tree vigor as a basis for mountain pine beetle infestations. U.S. Department of Agriculture, Forest Service, Ogden, UT, Research Paper INT-RP-400, Ogden, UT.

Becker, T., Schröter, H., 2000. Ausbreitung von rindenbrütenden Borkenkäfern nach Sturmschäden. Allg. Forsztg. 55, 280–282.

Bedard, W.D., Wood, D.L., 1981. Suppression of Dendroctonus brevicomis by using a mass-trapping tactic. In: Mitchell, E.R. (Ed.), Management of Insect Pests with Semiochemicals. Plenum, New York, pp. 310–341.

Bedard, W.D., Wood, D.L., Tilden, P.E., 1979. Using behavior modifying chemicals to reduce western pine beetle-caused tree mortality and protect trees. In: Waters, W.E. (Ed.), Current Topics in Forest Entomology. U.S. Department of Agriculture, Forest Service, Washington, DC, pp. 159–163, General Technical Report WO-GTR-8.

Belanger, R.P., Malac, B.F., 1980. Silviculture can reduce losses from the southern pine beetle. U.S. Department of Agriculture, Combined Forest Research and Development Program, Washington, DC, Agricultural Handbook 576.

Bennett, D.D., McGregor, M.D., 1980. A demonstration of basal area cutting to manage mountain pine beetle in second growth ponderosa pine. U.S. Department of Agriculture, Forest Service, Missoula, MT, Forest Pest Management Report FPM 88-16.

Bentz, B.J., 2006. Mountain pine beetle population sampling: inferences from Lindgren pheromone traps and tree emergence cages. Can. J. For. Res. 36, 351–360.

Bentz, B.J., Munson, A.S., 2000. Spruce beetle population suppression in northern Utah. West. J. Appl. For. 15, 122–128.

Bentz, B.J., Amman, G.D., Logan, J.A., 1993. A critical assessment of risk classification systems for the mountain pine beetle. For. Ecol. Manage. 61, 349–366.

Bentz, B.J., Régnière, J., Fettig, C.J., Hansen, E.M., Hayes, J.L., Hicke, J.A., et al., 2010. Climate change and bark beetles of the western United States and Canada: direct and indirect effects. Bioscience 60, 602–613.

Billings, R.F., 1988. Forecasting southern pine beetle infestation trends with pheromone traps. In: Payne, T.L., Saarenmaa, H. (Eds.), Proceedings on the Symposium on Integrated Control of Scolytid Bark Beetles. Virginia Polytechnic Institute and State University, Blacksburg, pp. 247–255.

Billings, R.F., Hynum, B., 1980. Southern pine beetle: guide for predicting timber losses from expanding spots in East Texas. Texas Forest Service, Lufkin, TX, Circular 249.

Bone, C., Dragicevic, S., Roberts, A., 2005. Integrating high resolution remote sensing, GIS and fuzzy set theory for identifying susceptibility areas of forest insect infestations. Int. J. Rem. Sens. 26, 4809–4828.

Boone, C.K., Aukema, B.H., Bohlmann, J., Carroll, A.L., Raffa, K.P., 2011. Efficacy of tree defense physiology varies with bark beetle population density: a basis for positive feedback in eruptive species. Can. J. For. Res. 41, 1174–1188.

Borden, J.H., 1985. Aggregation pheromones. In: Kerkut, G.A., Gilbert, L.I. (Eds.), Comprehensive Insect Physiology, Biochemistry and Pharmacology Vol. 9. Pergamon Press, Oxford, pp. 257–285.

Borden, J.H., 1997. Disruption of semiochemical-mediated aggregation in bark beetles. In: Cardé, R.T., Minks, A.K. (Eds.), Insect Pheromone Research, New Directions. Chapman & Hall, New York, pp. 421–437.

Borden, J.H., Lindgren, B.S., 1988. The role of semiochemical in IPM of the mountain pine beetle. In: Payne, T.L., Saarenmaa, H. (Eds.), Proceedings on the Symposium on Integrated Control of Scolytid Bark Beetles. Virginia Polytechnic Institute and State University, Blacksburg, pp. 247–255.

Borden, J.H., McLean, J.A., 1981. Pheromone-based suppression of ambrosia beetles in industrial timber processing areas. In: Mitchell, E.R. (Ed.), Management of Insect Pests with Semiochemicals. Plenum, New York, pp. 133–154.

British Columbia (BC) Ministry of Forests, Lands and Natural Resource Operations, 2012. Available online: www.for.gov.bc.ca/hfp/mountain_pine_beetle/Updated-Beetle-Facts_April2013.pdf, Last accessed: November 19, 2013.

Burnside, R.E., Holsten, E.H., Fettig, C.J., Kruse, J.J., Schultz, M.E., Hayes, C.J., et al., 2011. The northern spruce engraver, *Ips perturbatus*. U.S. Department of Agriculture, Forest Service, Portland, OR, Forest Insect and Disease Leaflet FIDL 180.

Byers, J.A., 1995. Host tree chemistry affecting colonization in bark beetles. In: Cardé, R.T., Bell, W.J. (Eds.), Chemical Ecology of Insects 2. Chapman & Hall, New York, pp. 154–213.

Byers, J.A., Wood, D.L., Craig, J., Hendry, L.B., 1984. Attractive and inhibitory pheromones produced in the bark beetle *Dendroctonus brevicomis* during host colonization: regulation of inter- and intraspecific competition. J. Chem. Ecol. 10, 861–878.

Cahill, D.B., 1978. Cutting strategies as control measures of the mountain pine beetle in lodgepole pine in Colorado. In: Berryman, A.A., Amman, G.D., Stark, R.W. (Eds.), Theory and Practice of Mountain Pine Beetle Management in Lodgepole Pine Forests: Symposium Proceedings. Washington State University, Pullman, pp. 188–191.

Carroll, A.L., Shore, T.L., Safranyik, L., 2006. Direct control: theory and practice. In: Safranyik, L., Wilson, B. (Eds.), The Mountain Pine Beetle—A Synthesis of Biology, Management, and Impacts on Lodgepole Pine. Natural Resources Canada, Canadian Forest Service, Victoria, BC, pp. 155–172.

Cech, T.L., Krehan, H., 1997. Important and new forest pests in Austria. In: Knížek, M., Zahradník, P., Diviš, K. (Eds.), Workshop on Forest Insect and Disease Survey, pp. 27–36, Pisek, Czech Republic, April 7–10, 1997, *VULHM Jiloviště-Strnady*.

Chou, C.-Y., Hedden, R.L., Song, B., Williams, T.M., 2013. Using a GIS-based spot growth model and visual simulator to evaluate the effects of silvicultural treatments on southern pine beetle-infested stands. In: Guldin, J.M. (Ed.), Proceedings of the 15th Biennial Southern Silvicultural Research Conference. U.S. Department of Agriculture, Forest Service, Asheville, NC, pp. 423–430, General Technical Report SRS-GTR-175.

Christiansen, E., Bakke, A., 1988. The spruce bark beetle of Eurasia. In: Berryman, A.A. (Ed.), Dynamics of Forest Insect Populations; Patterns, Causes, Implications. Plenum, New York, pp. 479–503.

Clarke, S.R., Salom, S.M., Billings, R.F., Berisford, C.W., Upton, W.W., McClellan, Q.C., Dalusky, M.J., 1999. A scentsible approach to controlling southern pine beetles: two new tactics using verbenone. J. For. 97, 26–31.

Coeln, M., Niu, Y., Führer, E., 1996. Entwicklung von Borkenkäfern in Abhängigkeit von thermischen Bedingungen verschiedener montaner Waldstufen (Coleoptera: Scolytidae). Entomologia Generalis 21, 37–54.

Coggins, S.B., Coops, N.C., Wulder, M.A., Bater, C.W., Ortlepp, S.M., 2011. Comparing the impacts of mitigation and non-mitigation on mountain pine beetle populations. J. Environ. Manage. 92, 112–120.

Cole, W.E., Amman, G.D., 1980. Mountain pine beetle dynamics in lodgepole pine forests. Part 1: Course of an infestation. U.S. Department of Agriculture, Forest Service, Ogden, UT, General Technical Report INT-GTR-89.

Cole, W.E., Cahill, D.B., 1976. Cutting strategies can reduce probabilities of mountain pine beetle epidemics in lodgepole pine. J. For. 74, 294–297.

Cook, S.P., Hain, F.P., 1988. Toxicity of host monoterpenes to *Dendroctonus frontalis* and *Ips calligraphus* (Coleoptera: Scolytidae). J. Entomol. Sci. 23, 287–292.

Coops, N.C., Timko, J.A., Wulder, M.A., White, J.C., Ortlepp, M.A., 2008. Investigating the effectiveness of mountain pine beetle mitigation strategies. Int. J. Pest Manag. 54, 151–165.

Coulson, R.N., Stephen, F.M., 2006. Impacts of insects in forest landscapes: implications for forest health management. In: Payne, T.D. (Ed.), Invasive Forest Insects, Introduced Forest Trees, and Altered Ecosystems: Ecological Pest Management in Global Forests of a Changing World. Springer-Verlag, New York, pp. 101–125.

DeGomez, T., Fettig, C.J., McMillin, J.D., Anhold, J.A., Hayes, C.J., 2008. Managing slash to minimize colonization of residual leave trees by *Ips* and other bark beetle species following thinning in southwestern ponderosa pine. University of Arizona Press, Tucson, AZ.

Dimitri, L., Gebauer, U., Lösekrug, R., Vaupel, O., 1992. Influence of mass trapping on the population dynamic and damage-effect of bark beetles. J. Appl. Entomol. 114, 103–109.

Drumont, A., Gonzalez, R., Dewindt, N., Grégoire, J.-C., Deproft, M., Seutin, E., 1992. Semiochemicals and integrated management of *Ips typographus* L. (Col., Scolytidae) in Belgium. J. Appl. Entomol. 114, 333–337.

Dubbel, V., 1993. Überlebensrate von Fichtenborkenkäfern bei maschineller Entrindung. Allg. Forsztg. 48, 359–360.

Dubbel, V., Dimitro, L., Niemeyer, H., Vaupel, O., 1995. Borkenkäferfallen—sinnlos bei Massenvermehrungen? Allgemeine Forst Zeitschrift für Waldwirtschaft und Umweltvorsorge 50, 258.

Eriksson, M., Pouttu, A., Roininen, H., 2005. The influence of windthrow area and timber characteristics on colonization of wind-felled spruces by *Ips typographus* (L.). For. Ecol. Manage. 216, 105–116.

Fayt, P., Machmer, M.M., Steeger, C., 2005. Regulation of spruce bark beetles by woodpeckers—a literature review. For. Ecol. Manage. 206, 1–14.

Feicht, E., 2006. Bark beetle hunting dogs. The suitability of dogs to find bark beetle infestations. Forst und Holz 61, 70–71.

Fettig, C.J., 2012. Forest health and bark beetles. In: North, M. (Ed.), Managing Sierra Nevada Forests. U.S. Department of Agriculture, Forest Service, Albany, CA, pp. 13–22, General Technical Report PSW-GTR-237.

Fettig, C.J., McKelvey, S.R., 2010. Bark beetle responses to stand structure and prescribed fire at Blacks Mountain Experimental Forest, California, USA: 5-year data. Fire Ecol. 6, 26–42.

Fettig, C.J., McMillin, J.D., Anhold, J.A., Hamud, S.M., Borys, R.R., Dabney, C.P., Seybold, S.J., 2006a. The effects of mechanical fuel reduction treatments on the activity of bark beetles (Coleoptera: Scolytidae) infesting ponderosa pine. For. Ecol. Manage. 230, 55–68.

Fettig, C.J., Allen, K.K., Borys, R.R., Christopherson, J., Dabney, C.P., Eager, T.J., et al., 2006b. Effectiveness of bifenthrin (Onyx®) and carbaryl (Sevin SL®) for protecting individual, high-value conifers from bark beetle attack (Coleoptera: Curculionidae: Scolytinae) in the western United States. J. Econ. Entomol. 99, 1691–1698.

Fettig, C.J., Klepzig, K.D., Billings, R.F., Munson, A.S., Nebeker, T.E., Negrón, J.F., Nowak, J.T., 2007a. The effectiveness of vegetation management practices for prevention and control of bark beetle infestations in coniferous forests of the western and southern United States. For. Ecol. Manage. 238, 24–53.

Fettig, C.J., McKelvey, S.R., Dabney, C.P., Borys, R.R., 2007b. The response of *Dendroctonus valens* (Coleoptera: Scolytidae) and *Temnochila chlorodia* (Coleoptera: Trogossitidae) to *Ips paraconfusus* (Coleoptera: Scolytidae) pheromone components and verbenone. Can. Entomol. 139, 141–145.

Fettig, C.J., Munson, A.S., McKelvey, S.R., Bush, P.B., Borys, R.R., 2008. Spray deposition from ground-based applications of carbaryl to protect individual trees from bark beetle attack. J. Environ. Qual. 37, 1170–1179.

Fettig, C.J., McKelvey, S.R., Borys, R.R., Dabney, C.P., Hamud, S.M., Nelson, L.J., Seybold, S.J., 2009. Efficacy of verbenone for protecting ponderosa pine stands from western pine beetle (Coleoptera: Curculionidae, Scolytinae) attack in California. J. Econ. Entomol. 102, 1846–1858.

Fettig, C.J., Reid, M.L., Bentz, B.J., Sevanto, S., Spittlehouse, D.L., Wang, T., 2013a. Changing climates, changing forests: a western North American perspective. J. For. 111, 214–228.

Fettig, C.J., Grosman, D.M., Munson, A.S., 2013b. Advances in insecticide tools and tactics for protecting conifers from bark beetle attack in the western United States. In: Trdan, S. (Ed.), Insecticides—Development of Safer and More Effective Technologies. InTech, Croatia, pp. 472–492.

Fettig, C.J., Gibson, K.E., Munson, A.S., Negrón, J.F., 2014a. Cultural practices for prevention and mitigation of mountain pine beetle infestations. For. Sci. 60, 450–463.

Fettig, C.J., Munson, A.S., Grosman, D.M., Bush, P.B., 2014b. Evaluations of emamectin benzoate and propiconazole for protecting individual *Pinus contorta* from mortality attributed to colonization by *Dendroctonus ponderosae* and associated fungi. Pest Manage. Sci. 70, 771–778.

Fielding, N.J., O'Keefe, T., King, C.J., 1991. Dispersal and host-finding capability of the predatory beetle *Rhizophagus grandis* Gyll. (Coleoptera, Rhizophagidae). J. Appl. Entomol. 112, 89–98.

Franceschi, V.R., Krokene, P., Christiansen, E., Krekling, T., 2005. Anatomical and chemical defenses of conifer bark against bark beetles and other pests. New Phytol. 167, 353–376.

Gibson, K.E., Schmitz, R.F., Amman, G.D., Oakes, R.D., 1991. Mountain pine beetle response to different verbenone dosages in pine stands of western Montana. U.S. Department of Agriculture, Forest Service, Ogden, UT, Research Paper INT-RP-444.

Gilbert, M., Nageleisen, L.M., Franklin, A., Grégoire, J.-C., 2005. Poststorm surveys reveal large-scale spatial patterns and influences of site factors, forest structure and diversity in endemic bark-beetle populations. Lands. Ecol. 20, 35–49.

Gillette, N.E., Mehmel, C.J., Mori, S.R., Webster, J.N., Wood, D.L., Erbilgin, N., Owen, D.R., 2012. The push-pull tactic for mitigation of mountain pine beetle (Coleoptera: Curculionidae) damage in lodgepole and whitebark pines. Environ. Entomol. 41, 1575–1586.

Gillette, N.E., Wood, D.L., Hines, S.J., Runyon, J.B., Negrón, J.F., 2014. The once and future forest: consequences of mountain pine beetle treatment decisions. For. Sci. 60, 527–538.

Gimbarzevsky, P., 1984. Remote sensing in forest damage detection and appraisal-selected annotated bibliography. Canadian Forestry Service, Victoria, BC, Information Report, BC-X-253.

Gmelin, J.F., 1787. Abhandlung über die Wurmtrocknis. Cruisiusschen, Leipzig.

Göthlin, E., Schroeder, L.M., Lindelöw, A., 2000. Attacks by *Ips typographus* and *Pityogenes chalcographus* on windthrown spruces (*Picea abies*) during the two years following a storm felling. Scand. J. For. Res. 15, 542–549.

Graf, M., Reid, M.L., Aukema, B.H., Lindgren, B.S., 2012. Association of tree diameter with body size and lipid content of mountain pine beetles. Can. Entomol. 144, 467–477.

Grégoire, J.-C., Evans, H.F., 2004. Damage and control of BAWBILT organisms, an overview in European bark and wood boring insects in living trees: a synthesis. In: Lieutier, F., Day, K.R., Battisti, A., Grégoire, J.-C., Evans, H.F. (Eds.), Bark and Wood Boring Insects in Living Trees in Europe, a Synthesis. Kluwer Academic Publishers, Dordrecht, pp. 19–37.

Grégoire, J.-C., Merlin, J., Pasteels, J.M., Jaffuel, R., Vouland, G., Schvester, D., 1985. Biocontrol of *Dendroctonus micans* by *Rhizophagus grandis* Gyll. (Coleoptera: Rhizophagidae) in the Massif Central (France). A first appraisal of the mass-rearing and release methods. Zeitschrift für Angewandte Entomologie 99, 182–190.

Grégoire, J.-C., Raty, L., Drumont, A., De Windt, N., 1997. Pheromone mass trapping: does it protect windfalls from attack by *Ips typographus* L. (Coleoptera: Scolytidae)? In: Grégoire, J.-C., Liebhold, A.M., Stephen, F.M., Day, K.R., Salom, S.M. (Eds.), Proceedings: Integrating Cultural Tactics into the Management of Bark Beetle and Reforestation Pests. U.S. Department of Agriculture, Forest Service, Newton Square, PA, pp. 1–8, General Technical Report NE-GTR-236.

Grodzki, W., 2005. GIS, spatial ecology and research on forest protection. In: Grodzki, W. (Ed.), GIS and Databases in the Forest Protection in Central Europe. Forest Research Institute, Warsaw, Poland, pp. 7–14.

Grodzki, W., 2007. Wykorzystanie pułapek feromonowych do monitoringu populacji kornika drukarza w wybranych parkach narodowych w Karpatach (Pheromone traps as a tool for monitoring of *Ips typographus* in selected national parks in Carpathian Mountains). Prace Instytutu Badawczego Leśnictwa, Rozprawy i Monografie 8, 1–127.

Grodzki, W., 2010. The decline of Norway spruce *Picea abies* (L.) Karst. stands in Beskid Śląski and Żywiecki: theoretical concept and reality. Beskydy 3, 19–26.

Grodzki, W., Jakuš, R., Lajzová, E., Sitková, Z., Maczka, T., Škvarenina, J., 2006. Effects of intensive versus no management strategies during an outbreak of the bark beetle *Ips typographus* (L.) (Col.: Curculionidae, Scolytinae) in the Tatra Mts. in Poland and Slovakia. Ann. Forest Sci. 63, 55–61.

Grodzki, W., Kosibowicz, M., Mączka, T., 2008. Skuteczność wystawiania pułapek feromonowych na kornika drukarza *Ips typographus* (L.) w sąsiedztwie wiatrowałów i wiatrołomów (Effectiveness of pheromone traps for ESB in the proximity of windthrows and windbreaks). Leśne Prace Badawcze 69, 365–370.

Hamel, D.R., 1978. Results of harvesting strategies for management of mountain pine beetle infestations in lodgepole pine on the Gallatin National Forest, Montana. In: Berryman, A.A., Amman, G.D., Stark, R.W. (Eds.), Theory and Practice of Mountain Pine Beetle Management in Lodgepole Pine Forests: Symposium Proceedings. Washington State University, Pullman, pp. 192–196.

Hansen, E.M., Vandygriff, J.C., Cain, R.J., Wakarchuk, D., 2006. Comparison of naturally and synthetically baited spruce beetle trapping systems in the central Rocky Mountains. J. Econ. Entomol. 99, 373–382.

Harrington, T.C., Cobb Jr., F.W., Lownsberry, J.W., 1985. Activity of *Hylastes nigrinus*, a vector of *Verticicladiella wageneri*, in thinned stands of Douglas-fir. Can. J. For. Res. 15, 579–593.

Haverty, M.I., Page, M., Shea, P.J., Hoy, J.B., Hall, R.W., 1983. Drift and worker exposure resulting from two methods of applying insecticides to pine bark. Bull. Environ. Contam. Tox. 30, 223–228.

Hayes, C.J., Fettig, C.J., Merrill, L.D., 2009. Evaluation of multiple funnel traps and stand characteristics for estimating western pine beetle-caused tree mortality. J. Econ. Entomol. 102, 2170–2182.

Hedden, R.L., 1985. Simulation of southern pine beetle associated timber loss using CLEMBEETLE. In: Branham, S.J., Thatcher, R.C. (Eds.), Integrated Pest Management Research Symposium, Proceedings. U.S. Department of Agriculture, Forest Service, New Orleans, LA, pp. 288–291, General Technical Report SO-GTR-56.

Hedgren, P.O., 2004. The bark beetle *Pityogenes chalcographus* (L.) (Scolytidae) in living trees: reproductive success, tree mortality and interaction with *Ips typographus*. J. Appl. Entomol. 128, 161–166.

Heller, R.C., 1968. Previsual detection of ponderosa pine trees dying from bark beetle attack. In: Forest Science Proceedings, 5th Symposium on Remote Sensing of Environment. University of Michigan, Ann Arbor, pp. 387–434.

Hilszczański, J., 2008. Kora zamarłych świerków jako miejsce zimowania owadzich drapieżników związanych z kambio i ksylofagami (Bark of dead infested spruce trees as an overwintering site of insect predators associated with bark and wood boring beetles). Leśne Prace Badawcze 69, 15–19.

Hilszczański, J., Gibb, H., Hjältén, J., Atlegrim, O., Johansson, T., Pettersson, R.B., et al., 2005. Parasitoids (Hymenoptera, Ichneumonoidea) of saproxylic beetles are affected by forest management and dead wood characteristics in boreal spruce forest. Biol. Conserv. 126, 456–464.

Hilszczański, J., Janiszewski, W., Negrón, J., Munson, A.S., 2006. Stand characteristics and *Ips typographus* (L.) (Col., Curculionidae, Scolytinae) infestation during outbreak in northeastern Poland. Folia For. Pol., Ser. A 48, 53–65.

Hodges, J.D., Pickard, L.S., 1971. Lightning in the ecology of the southern pine beetle, *Dendroctonus frontalis* (Coleoptera: Scolytidae). Can. Entomol. 103, 44–51.

Hodges, J.D., Elam, W.W., Watson, W.F., Nebeker, T.E., 1979. Oleoresin characteristics and susceptibility of four southern pines to southern pine beetle (Coleoptera: Scolytidae) attacks. Can. Entomol. 111, 889–896.

Hodges, J.D., Nebeker, T.E., DeAngelis, J.D., Karr, B.L., Blanche, C.A., 1985. Host resistance and mortality: a hypothesis based on the southern pine beetle-microorganism-host interactions. Bull. Entomol. Soc. Am. 31, 31–35.

Hofstetter, R.W., Dunn, D.D., McGuire, R., Potter, K.A., 2014. Using acoustic technology to reduce bark beetle reproduction. Pest Manage. Sci. 70, 24–27.

Holsten, E.H., Werner, R.A., 1987. Engraver beetles in Alaska forests. U.S. Department of Agriculture, Forest Service, Leaflet, Portland, OR.

Holsten, E.H., Webb, W., Shea, P.J., Werner, R.A., 2002. Release rates of methylcyclohexenone and verbenone from bubble cap and bead releasers under field conditions suitable for the management of bark beetles in California, Oregon, and Alaska. U.S. Department of Agriculture, Forest Service, Portland, OR, Research Paper PNW-RP-544.

Hopkins, A.D., 1905. The Black Hills beetle. U.S. Department of Agriculture, Washington, DC, Bureau of Entomology, Bulletin 56.

Hübertz, H., Larsen, J.R., Bejer, B., 1991. Monitoring spruce bark beetle (*Ips typographus* (L.)) populations under nonepidemic conditions. Scand. J. For. Res. 6, 217–226.

Hunt, D.W.A., Borden, J.H., 1990. Conversion of verbenols to verbenone by yeasts isolated from *Dendroctonus ponderosae* (Coleoptera: Scolytidae). J. Chem. Ecol. 16, 1385–1397.

Hunt, D.W.A., Borden, J.H., Lindgren, B.S., Gries, G., 1989. The role of autoxidation of α-pinene in the production of pheromones of *Dendroctonus ponderosae* (Coleoptera: Scolytidae). Can. J. For. Res. 19, 1275–1282.

Jactel, H., Brockerhoff, E.G., 2007. Tree diversity reduces herbivory by forest insects. Ecol. Lett. 10, 835–848.

Jakuš, R., 2001. Bark beetle (Coleoptera, Scolytidae) outbreak and system of IPM measures in an area affected by intensive forest decline connected with honey fungus (*Armillaria* sp.). J. Pest Sci. 74, 46–51.

Jakuš, R., Blaženec, M., 2011. Treatment of bark beetle attacked trees with entomopathogenic fungus *Beauveria bassiana* (Balsamo) Vuillemin. Folia For. Pol., Ser. A 53, 150–155.

Jakuš, R., Dudová, A., 1999. Pokusné použitie agregačných a antiagregačných feromónov proti lykožrútovi smrekovému (*Ips typographus*) v rozpadávajúcich sa smrekových porastoch so zníženým zakmenením. J. For. Sci. 45, 525–532.

Jakuš, R., Schlyter, F., Zhang, Q.-H., Blazenec, M., Vavercák, R., Grodzki, W., et al., 2003. Overview of development of anti-attractant based technology for spruce protection against *Ips typographus*: from past failures to future success. J. Pest Sci. 76, 89–99.

Jakuš, R., Ježík, M., Kissiyar, O., Blaženec, M., 2005. Prognosis of bark beetle attack in TANABBO model. In: Grodzki, W. (Ed.), GIS and Databases in the Forest Protection in Central Europe. Forest Research Institute, Warsaw, pp. 35–43.

Jonasova, M., Matejkova, I., 2007. Natural regeneration and vegetation changes in wet spruce forests after natural and artificial disturbances. Can. J. For. Res. 37, 1907–1914.

Jonsell, M., Weslien, J., 2003. Felled or standing retained wood—it makes a difference for saproxylic beetles. For. Ecol. Manage. 175, 425–435.

Jönsson, A.M., Schroeder, L.M., Lagergren, F., Anderbrant, O., Smith, B., 2012. Guess the impact of *Ips typographus*—an ecosystem modelling approach for simulating spruce bark beetle outbreaks. Agr. Forest. Meteorol. 166, 188–200.

Kenis, M., Wermelinger, B., Grégoire, J.-C., 2004. Research on parasitoids and predators of Scolytidae—a review. In: Lieutier, F., Day, K.R., Battisti, A., Grégoire, J.-C., Evans, H. (Eds.), Bark and Wood Boring Insects in Living Trees in Europe, a Synthesis. Kluwer Academic Publishers, Dordrecht, pp. 237–290.

Kissiyar, O., Blaženec, M., Jakuš, R., Willekens, A., Ježík, M., Baláž, P., et al., 2005. TANABBO model—a remote sensing based early warning system for forest decline and bark beetle outbreaks in Tatra Mts.—overview. In: Grodzki, W. (Ed.), GIS and Databases in the

Forest Protection in Central Europe. Forest Research Institute, Warsaw, pp. 15–34.

Klein, W.H., Parker, D.L., Jensen, C.E., 1978. Attack, emergence, and stand depletion trends of the mountain pine beetle in a lodgepole pine stand during an outbreak. Environ. Entomol. 112, 185–191.

Klepzig, K.D., Raffa, K.F., Smalley, E.M., 1991. Association of insect-fungal complexes with red pine decline in Wisconsin. For. Sci. 37, 1119–1139.

Knížek, M., Lubojacký, J., 2012. Podkorní hmyz (Bark beetles). In: Knížek, M., Modlinger, R. (Eds.), Výskyt Lesních Škodlivých Činitelů v Roce 2011 a Jejich Očekávaný Stav v Roce 2012. Zpravodaj ochrany lesa, Supplementum, pp. 20–30.

Kobakhidze, D.N., Tvaradze, M.S., Kraveishvili, I.K., 1970. Predvaritel'nye rezul'taty k introduktsii, izucheniyu bioekologii, razrabotke metodiki iskusstvennogo razvedeniya i naturalizatsii v elovykh nasazhdeniyakh Gruzii naibolee effektivnogo entomofaga Dendroctonus micans Kugel. Rhizophagus grandis Gyll. Soobshcheniya Akademii Nauk. Gruzinskoi SSR 60, 205–208.

Kolb, T.E., Guerard, N., Hofstetter, R.W., Wagner, M.R., 2006. Attack preference of Ips pini on Pinus ponderosa in northern Arizona: tree size and bole position. Agric. For. Entomol. 8, 295–303.

Kolk, A., Grodzki, W., 2013. Metody ograniczania liczebności kornika drukarza (Methods of ESB control). In: Grodzki, W. (Ed.), Kornik Drukarz i Jego Rola w Ekosystemach Leśnych.pp. 149–154, CILP.

Komonen, A., Schroeder, L.M., Weslien, J., 2011. Ips typographus population development after a severe storm in a nature reserve in southern Sweden. J. Appl. Entomol. 135, 132–141.

Kreutz, J., Zimmermann, G., Marohn, H., Vaupel, O., Mosbacher, G., 2000. Preliminary investigations on the use of Beauveria bassiana (Bals.) Vuill. and other control methods against the bark beetle, Ips typographus L. (Col., Scolytidae) in the field. IOBC/WPRS-Bull. 23, 167–173.

Kreutz, J., Vaupel, O., Zimmermann, G., 2004. Efficacy of Beauveria bassiana (Bals.) Vuill. against the spruce bark beetle, Ips typographus L. in the laboratory under various conditions. J. Appl. Entomol. 128, 384–389.

Król, A., Bakke, A., 1986. Comparison of trap trees and pheromone loaded pipe traps in attracting Ips typographus L. (Col., Scolytidae). Pol. Pismo Entomol. 56, 437–445.

Landa, Z., Kalista, M., Křenová, Z., Vojtěch, O., 2008. Praktické využití entomopatogenní houby Beauveria bassiana proti lýkožroutu smrkovému Ips typographus. In: Vojtech, O., Šustr, P. (Eds.), Ekologické Metody Ochrany Lesa Před Podkorním Hmyzem". Sborník referátů ze semináře, Sborníky z výzkumu na Šumavě—sešit 1. Správa NP a CHKO Šumava, Vimperk, pp. 62–71.

Lekander, B., 1972. A mass outbreak of Ips typographus in Gästrikland, Central Sweden, in 1945–1952. Department of Forest Zoology, Royal College of Forestry, Stockholm, Research Notes, 10.

Leufvén, A., Bergström, G., Falsen, E., 1984. Interconversion of verbenols and verbenone by identified yeasts isolated from the spruce bark beetle Ips typographus. J. Chem. Ecol. 10, 1349–1361.

Lewis, E.E., Cane, J.H., 1990. Stridulation as a primary anti-predator defence of a beetle. Anim. Behav. 40, 1003–1004.

Liebhold, A.M., Berck, P., Williams, N.A., Wood, D.L., 1986. Estimating and valuing western pine beetle impacts. For. Sci. 32, 325–338.

Lieutier, F., 2004. Host resistance to bark beetles and its variations. In: Lieutier, F., Day, K.R., Battisti, A., Grégoire, J.-C., Evans, H.F. (Eds.), Bark and Wood Boring Insects in Living Trees in Europe, a Synthesis. Kluwer Academic Publishers, Dordrecht, pp. 135–180.

Lindelöw, Å., Schroeder, M., 2001. Spruce bark beetle (Ips typographus L.) in Sweden: monitoring and risk assessment. J. For. Sci. 47, 39–41.

Lindelöw, Å., Risberg, B., Sjodin, K., 1992. Attraction during flight of scolytids and other bark and wood-dwelling beetles to volatiles from fresh and stored spruce wood. Can. J. For. Res. 22, 224–228.

Lindgren, B.S., Nordlander, G., Birgersson, G., 1996. Feeding deterrency and acute toxicity of verbenone to the pine weevil, Hylobius abietis. J. Appl. Entomol. 120, 397–403.

Lobinger, G., 1995. Einsatzmöglichkeiten von Borkenkäferfallen. Allg. Forst. Z. 50, 198–201, Waldwirtsch. Umwelvorsorge.

Lobinger, G., Skatulla, U., 1996. Untersuchungen zum Einfluss von Sonnenlicht auf das Schwärmverhalten von Borkenkäfern. Anz. Schädl., Pflanzens. Umwelts. 69, 183–185.

Lubojacký, J., Holuša, J., 2011. Comparison of spruce bark beetles (Ips typographus) catches between treated trap logs and pheromone traps. Šumarski list br. 5–6 (135), 233–242.

Mafra-Neto, A., de Lame, F.M., Fettig, C.J., Munson, A.S., Perring, T.M., Stelinski, L.L., et al., 2013. Manipulation of insect behavior with Specialized Pheromone & Lure Application Technology (SPLAT®). In: Beck, J., Coats, J., Duke, S., Koivunen, M. (Eds.), Natural Products for Pest Management. ACS Publications, Washington, DC, pp. 31–58.

Mathews, P.L., Stephen, F.M., 1997. Effect of artificial diet on longevity of adult parasitoids of Dendroctonus frontalis (Coleoptera: Scolytidae). Environ. Entomol. 26, 961–965.

Mathews, P.L., Stephen, F.M., 1999. Effects of an artificial diet and varied environmental conditions on longevity of Coeloides pissodis (Hymenoptera: Braconidae), a parasitoid of Dendroctonus frontalis (Coleoptera: Scolytidae). Environ. Entomol. 28, 729–734.

McFarlane, B.L., Watson, D.O., 2008. Perceptions of ecological risk associated with mountain pine beetle (Dendroctonus ponderosae) infestations in Banff and Kootenay National Parks of Canada. Risk Anal. 28, 203–212.

McFarlane, B.L., Stumpf-Allen, R.C.G., Watson, D.O., 2006. Public perceptions of natural disturbance in Canada's national parks: the case of the mountain pine beetle (Dendroctonus ponderosae Hopkins). Biol. Conserv. 130, 340–348.

McGregor, M.D., Amman, G.D., Schmitz, R.F., Oakes, R.D., 1987. Partial cutting lodgepole pine stands to reduce losses to the mountain pine beetle. Can. J. For. Res. 17, 1234–1239.

Meigs, G.W., Kennedy, R.E., Cohen, W.B., 2011. A Landsat time series approach to characterize bark beetle and defoliator impacts on tree mortality and surface fuels in conifer forests. Remote Sens. Environ. 115, 3707–3718.

Mezei, P., Jakuš, R., Blazenec, M., Belanova, S., Smidt, J., 2012. The relationship between potential solar radiation and spruce bark beetle catches in pheromone traps. Ann. For. Sci. 55, 243–252.

Mitchell, R.G., Waring, R.H., Pitman, G.B., 1983. Thinning lodgepole pine increases the vigor and resistance to mountain pine beetle. For. Sci. 2, 204–211.

Müller, M., Job, H., 2009. Managing natural disturbance in protected areas: tourists' attitude towards the bark beetle in a German national park. Biol. Conserv. 142, 375–383.

Negrón, J., 1997. Estimating probabilities of infestation and extent of damage by the roundheaded pine beetle in ponderosa pine in the Sacramento Mountains, New Mexico. Can. J. For. Res. 27, 1634–1645.

Negrón, J., 1998. Probability of infestation and extent of mortality associated with the Douglas-fir beetle in the Colorado Front Range. For. Ecol. Manage. 107, 71–85.

Negrón, J.F., Allen, K., McMillin, J., Burkwhat, H., 2006. Testing verbenone for reducing mountain pine beetle attacks in ponderosa pine in the Black Hills, South Dakota. U.S. Department of Agriculture, Forest Service, Fort Collins, CO, Research Note RMRS-RN-31.

Netherer, S., Nopp-Mayr, U., 2005. Predisposition assessment systems (PAS) as supportive tools in forest management-rating of site and stand-related hazards of bark beetle infestation in the High Tatra Mountains as an example for system application and verification. For. Ecol. Manage. 207, 99–107.

Niemann, K.O., Visintini, F., 2005. Assessment of potential for remote sensing detection of bark beetle-infested areas during green attack: a literature review. *CAB Abstracts Working Paper*—Canadian Forest Service, (2), iii + 14 p., Victoria.

Oliver, C.D., Larson, B.C., 1996. Forest Stand Dynamics. John Wiley & Sons Inc., New York.

Progar, R.A., Blackford, D.C., Cluck, D.R., Costello, S., Dunning, L.B., Eager, T., et al., 2013. Population densities and tree diameter effects associated with verbenone treatments to reduce mountain pine beetle-caused mortality of lodgepole pine. J. Econ. Entomol. 106, 221–228.

Progar, R.A., Gillette, N., Fettig, C.J., Hrinkevich, K., 2014. Applied chemical ecology of the mountain pine beetle. For. Sci 60, 414–433.

Puchniarski, T.H., 2008. Świerk pospolity, hodowla i ochrona (Norway spruce, silviculture and protection). In: Świerk Pospolity w Praktyce Ochrony Lasu.pp. 163–206, PWRiL Warszawa.

Puritch, G.S., 1981. Nonvisual remote sensing of trees affected by stress: a review. Natural Resources Canada. Canadian Forest Service, Forestry Technical Report, Victoria, BC, BC-X- 30.

Raffa, K.F., Berryman, A.A., 1982. Gustatory cues in the orientation of *Dendroctonus ponderosae* (Coleoptera: Scolytidae) to host trees. Can. Entomol. 114, 97–104.

Raffa, K.F., Berryman, A.A., 1983. The role of host plant resistance in the colonization behavior and ecology of bark beetles (Coleoptera: Scolytidae). Ecol. Monogr. 53, 27–49.

Raffa, K.F., Phillips, T.W., Salom, S.M., 1993. Strategies and mechanisms of host colonization by bark beetles. In: Schowalter, T.D., Filip, G.M. (Eds.), Beetle Pathogen Interactions in Conifer Forests. Academic Press Inc., San Diego, CA, pp. 103–128.

Rasmussen, L.A., 1972. Attraction of mountain pine beetle to small diameter lodgepole pines baited with *trans*-verbenol and alpha-pinene. J. Econ. Entomol. 65, 1396–1399.

Raty, L., Drumont, A., De Windt, N., Grégoire, J.-C., 1995. Mass trapping of the spruce bark beetle *Ips typographus* L.: traps or trap trees? For. Ecol. Manage. 78, 191–205.

Ratzeburg, J.T.C., 1839. Die Forst-Insecten. 1. Die Käfer. Nicolai, Berlin.

Reed, D.D., Burkhart, H.E., Leushner, W.A., Hedden, R.L., 1981. A severity model for southern pine beetle infestations. For. Sci. 27, 290–296.

Reid, M.L., Purcell, J.R.C., 2011. Condition-dependent tolerance of monoterpenes in an insect herbivore. Arthropod Plant Interactions 5, 331–337.

Reid, R.W., Whiney, H.S., Watson, J.A., 1967. Reactions of lodgepole pine to attack by *Dendroctonus ponderosae* Hopkins and blue-stain fungi. Can. J. Bot. 45, 1115–1126.

Reynolds, K.M., Holsten, E.H., 1994. Classification of spruce beetle hazard in Lutz spruce (*Picea* X *lutzii*) stands on the Kenai Peninsula, Alaska. Can. J. For. Res. 24, 1015–1021.

Reynolds, K.M., Holsten, E.H., 1996. Classification of spruce beetle hazard in Lutz and Sitka spruce stands on the Kenai Peninsula, Alaska. For. Ecol. Manage. 84, 251–262.

Romme, W.H., Knight, D.H., Yavitt, J.B., 1986. Mountain pine beetle outbreaks in the Rocky Mountains: regulators of primary productivity? Am. Nat. 127, 484–494.

Root, R.B., 1973. Organization of a plant-arthropod association in simple and diverse habitats: the fauna of collards (*Brassica oleracea*). Ecol. Monogr. 43, 95–124.

Ross, D.W., Daterman, G.E., 1997. Using pheromone-baited traps to control the amount and distribution of tree mortality during outbreaks of the Douglas-fir beetle. For. Sci. 43, 65–70.

Ross, D.W., Gibson, K.E., Daterman, G.E., 2001. Using MCH to protect trees and stands from Douglas-fir beetle infestation. U.S. Department of Agriculture, Forest Service, Morgantown, WV, Forest Health Technology Enterprise Team Report FHTET-2001-09.

Rudinsky, J.A., Michael, R.R., 1973. Sound production in Scolytidae: stridulation by female *Dendroctonus* beetles. J. Insect Physiol. 19, 689–705.

Rudinsky, J.A., Ryker, L.C., Michael, R.R., Libbey, L.M., Morgan, M.E., 1976. Sound production in Scolytidae: female sonic stimulus of male pheromone release in two *Dendroctonus* beetles. J. Insect Physiol. 22, 167–168.

Safranyik, L., Shrimpton, D.M., Whitney, H.S., 1974. Management of lodgepole pine to reduce losses from the mountain pine beetle. Natural Resources Canada, Canadian Forest Service, Victoria, BC, Pacific Forest Research Centre, For. Tech. Rep. 1.

Safranyik, L., Shore, T.L., Moeck, H.A., Whitney, H.S., 2002. *Dendroctonus ponderosae* Hopkins, mountain pine beetle (Coleoptera: Scolytidae). In: Mason, P.G., Huber, J.T. (Eds.), Biological Control Programmes in Canada 1981–2000. CABI, Wallingford, pp. 104–110.

Samalens, J.C., Rossi, J.P., Guyon, D., Van Halder, I., Menassieu, P., Piou, D., Jactel, H., 2007. Adaptive roadside sampling for bark beetle damage assessment. For. Ecol. Manage. 253, 177–187.

Sambaraju, K.R., Carroll, A.L., Zhu, J., Stahl, K., Moore, R.D., Aukema, B.H., 2012. Climate change could alter the distribution of mountain pine beetle outbreaks in western Canada. Ecography 35, 211–223.

Schelhaas, M.-J., Nabuurs, G.-J., Schuck, A., 2003. Natural disturbances in the European forests in the 19th and 20th centuries. Global Change Biol. 9, 1620–1633.

Schlyter, F., 2012. Semiochemical diversity in practice: antiattractant semiochemicals reduce bark beetle attacks on standing trees—a first meta-analysis. Psyche. 268621.

Schlyter, F., Birgersson, G., 1999. Forest beetles. Pheromones of non-lepidopteran insects. In: Hardie, R.J., Minks, A.K. (Eds.), Associated with Agricultural Plants. CABI, Wallingford, pp. 113–148.

Schlyter, P., Stjernquist, I., Barring, L., Jönsson, A.M., Nilsson, C., 2006. Assessment of the impacts of climate change and weather extremes on boreal forests in northern Europe, focusing on Norway spruce. Clim. Res. 31, 75–84.

Schmid, J.M., Frye, R.H., 1976. Stand ratings for spruce beetles. U.S. Department of Agriculture, Forest Service, Fort Collins, CO, Research Note RM-RN-309.

Schmid, J.M., Mata, S.A., 2005. Mountain pine beetle-caused tree mortality in partially cut plots surrounded by unmanaged stands. U.S. Department of Agriculture, Forest Service, Fort Collins, CO, Research Paper RM-RP-54.

Schmid, J.M., Mata, S.A., Kessler, R.R., Popp, J.B., 2007. The influence of partial cutting on mountain pine beetle-caused tree mortality in Black Hills ponderosa pine stands. U.S. Department of Agriculture, Forest Service, Fort Collins, CO, Research Paper RMRS-RP-68.

Schmidt, W.C., Alexander, R.R., 1985. Strategies for managing lodgepole pine. In: Baumgartner, D.M., Krebill, R.G., Arnott, J.T., Weetman, G. F. (Eds.), Lodgepole Pine: The Species and Its Management, Symposium Proceedings. Washington State University, Spokane, pp. 201–210.

Schopf, A., 1985. Zum Einfluß der Photoperiode auf die Entwicklung und Kälteresistenz des Buchdruckers, *Ips typographus* L. (Col., Scolytidae). Anzeiger für Schädlingskunde, Pflanzenschutz, Umweltschutz 58, 73–75.

Sevim, A., Gokce, C., Erbas, Z., Ozkan, F., 2012. Bacteria from *Ips sexdentatus* (Coleoptera: Curculionidae) and their biocontrol potential. J. Basic Microbiol. 52, 695–704.

Shea, P.J., Neustein, M., 1995. Protection of a rare stand of Torrey pine from *Ips paraconfusus*. In: Salom, S.M., Hobson, K.R. (Eds.), Application of Semiochemicals for Management of Bark Beetle Infestations: Proceedings of an Informal Conference. U.S. Department of Agriculture, Forest Service, Ogden, UT, pp. 39–43, General Technical Report INT-GTR-318.

Shepherd, R.F., 1966. Factors influencing the orientation and rates of activity of *Dendroctonus ponderosae* Hopkins (Coleoptera: Scolytidae). Can. Entomol. 98, 507–518.

Shore, J.L., Safranyik, L., 1992. Susceptibility and risk rating stands for the mountain pine beetle in lodgepole pine stands. Natural Resources Canada, Canadian Forest Service, Victoria, BC, Information Report BC-X-336.

Shore, T.L., Safranyik, L., Riel, W.G., Ferguson, M., Castonguay, J., 1999. Evaluation of factors affecting tree and stand susceptibility to the Douglas-fir beetle (Coleoptera: Scolytidae). Can. Entomol. 131, 831–839.

Shore, T.L., Safranyik, L., Lemieux, J.P., 2000. Susceptibility of lodgepole pine stands to the mountain pine beetle: testing of a rating system. Can. J. For. Res. 30, 44–49.

Shrimpton, D.M., 1973. Age- and size-related response of lodgepole pine to inoculation with *Europhium clavigerum*. Can. J. Bot. 51, 1155–1160.

Siitonen, J., 2001. Forest management, coarse woody debris and saproxyllic organisms: Fennoscandian boreal forests as an example. Ecol. Bull. 49, 11–41.

Silverstein, R.M., Rodin, J.O., Wood, D.L., 1966. Sex attractants in frass produced by male *Ips confusus* in ponderosa pine. Science 154, 509–510.

Skakun, R.S., Wulder, M.A., Franklin, S.E., 2003. Sensitivity of the thematic mapper enhanced wetness difference index to detect mountain pine beetle red-attack damage. Rem. Sens. Environ. 86, 433–443.

Skuhravý, V., 2002. Lýkožrout smrkový (*Ips typographus* L.) a jeho calamity, Agrospoj, Praha.

Smith, R.H., 1966. Resin quality as a factor in the resistance of pines to bark beetles. In: Gerhold, H.D., McDermott, R.E., Schreiner, E.J., Winieski, J.A. (Eds.), Breeding Pest-resistant Trees. Pergamon Press, New York, pp. 189–196.

Smith, E.L., McMahan, A.J., David, L., Beukema, S.J., Robinson, D.C., 2005. Westwide Pine Beetle Model version 2.0: detailed description. U.S. Department of Agriculture, Forest Service, Forest Health Protection, Fort Collins, CO, Forest Health Technology Enterprise Team Report FHTET 05-06.

St. George, R.A., Beal, J.A., 1929. The southern pine beetle: a serious enemy of pines in the South. U.S. Department of Agriculture, Washington, DC, Farmers' Bulletin 1586.

Stadelmann, G., Bugmann, H., Meier, F., Wermelinger, B., Bigler, C., 2013. Effects of salvage logging and sanitation felling on bark beetle (*Ips typographus* L.) infestations. For. Ecol. Manage. 305, 273–281.

Steele, R., Williams, R.E., Weatherby, J.C., Reinhardt, E.D., Hoffman, J. T., Thier, R.W., 1996. Stand hazard rating for central Idaho forests. U. S. Department of Agriculture, Forest Service, Ogden, UT, General Technical Report INT-GTR-332.

Stephen, F.P., Browne, L.E., 2000. Application of Eliminade™ parasitoid food to boles and crowns of pines (Pinaceae) infested with *Dendroctonus frontalis* (Coleoptera: Scolytidae). Can. Entomol. 132, 983–985.

Stephen, F.M., Lih, M.P., 1985. A *Dendroctonus frontalis* infestation growth model: organization, refinement, and utilization. In: Branham, S.J., Thatcher, R.C. (Eds.), Integrated Pest Management Research Symposium, Proceedings. U.S. Department of Agriculture, Forest Service, New Orleans, LA, pp. 186–194, General Technical Report, SO-GTR-56.

Stephen, F.M., Lih, M.P., Browne, L.E., 1997. Augmentation of *Dendroctonus frontalis* parasitoid effectiveness by artificial diet. In: Grégoire, J.-C., Liebhold, A.M., Stephen, F.M., Day, K.R., Salom, S.M. (Eds.), Proceedings: Integrating Cultural Tactics into the Management of Bark Beetle and Reforestation Pests. U.S. Department of Agriculture, Forest Service, Newton Square, PA, pp. 15–22, General Technical Report NE-GTR-236.

Stephens, S.L., McIver, J.D., Boerner, R.E.J., Fettig, C.J., Fontaine, J.B., Hartsough, B.R., et al., 2012. The effects of forest fuel-reduction treatments in the United States. Bioscience 62, 549–560.

Stevens, R.E., McCambridge, W.F., Edminster, C.B., 1980. Risk rating guide for mountain pine beetle in Black Hills ponderosa pine. U.S. Department of Agriculture, Forest Service, Fort Collins, CO, Research Note RM-RN-385.

Szabla, K., 2013. Praktyczna realizacja strategii ograniczania liczebności kornika drukarza na przykładzie świerczyn Beskidu Śląskiego i Żywieckiego. In: Grodzki, W. (Ed.), Kornik Drukarz i Jego Rola w Ekosystemach Leśnych. Centrum Informacyjne Lasów Państwowych, Warsaw, pp. 161–175.

Taylor, S.W., Carroll, A.L., 2004. Disturbance, forest age, and mountain pine beetle outbreak dynamics in BC: a historical perspective. In: Shore, T.L., Brooks, J.E., Stone, J.E. (Eds.), Mountain Pine Beetle Symposium: Challenges and Solutions. Natural Resources Canada, Canadian Forest Service, Victoria, BC, pp. 41–51, Information Report BC-X-399.

Thistle, H.W., Peterson, H., Allwine, G., Lamb, B.K., Strand, T., Holsten, E.H., Shea, P.J., 2004. Surrogate pheromone plumes in three forest trunk spaces: composite statistics and case studies. For. Sci. 50, 610–625.

USDA Forest Service, 2012. Areas with tree mortality from bark beetles: summary for 2000–2011, Western U.S.

Valkama, H., Räty, M., Niemelä, P., 1997. Catches of *Ips duplicatus* and other non-target Coleoptera by *Ips typographus* pheromone trapping. Entomol. Fennica 8, 153–159.

Vaupel, O., Zimmermann, G., 1996. Orientierende Versuche zur Kombination von Pheromonfallen mit dem insektenpathogenen Pilz *Beauveria bassiana* (Bals.) Vuill. gegen die Borkenkäferart *Ips typographus* L. Anz. Schädl., Pflanzens., Umwelts. 69, 175–179.

Vité, J.P., 1961. The influence of water supply on oleoresin exudation pressure and resistance to bark beetle attack in *Pinus ponderosa*. Contr. Boyce Thompson Inst. 21, 37–66.

Wallin, K.F., Raffa, K.F., 2000. Influence of host chemicals and internal physiology on the multiple steps of postlanding host acceptance behavior of *Ips pini* (Coleoptera: Scolytidae). Environ. Entomol. 29, 442–453.

Wallin, K.F., Raffa, K.F., 2001. Host-mediated interactions among feeding guilds: incorporation of temporal patterns can integrate plant defense theories to predict community level processes. Ecology 82, 1387–1400.

Waring, R.H., Pitman, G.B., 1980. A simple model of host resistance to bark beetles. Oregon State University, Corvallis, School of Forestry Research Note OSU-RN-65.

Waring, R.H., Pitman, G.B., 1985. Modifying lodgepole pine stands to change susceptibility to mountain pine beetle attack. Ecology 66, 889–897.

Waters, W.E., 1985. Monitoring bark beetle populations and beetle-caused damage. In: Waters, W.W., Stark, R.W., Wood, D.L. (Eds.), Integrated Pest Management in Pine-bark Beetle Ecosystems. John Wiley & Sons, New York, pp. 141–175.

Watson, A.C., Sullivan, J., Amacher, G.S., Asaro, C., 2013. Cost sharing for pre-commercial thinning in southern pine plantations: willingness to participate in Virginia's pine bark beetle prevention program. Forest Pol. Econ. 34, 65–72.

Weatherby, J.C., Thier, R.W., 1993. A preliminary validation of a Douglas-fir beetle hazard rating system, Mountain Home Ranger District, Boise National Forest, 1992. U.S. Department of Agriculture, Forest Service, Ogden, UT, Forest Pest Management Report FPM 93-05.

Wegensteiner, R., 1992. Untersuchungen zur Wirkung von Beauveria-Arten auf *Ips typographus* (Col., Scolytidae). Mitteilungen der Deutschen Gesellschaft für Allgemeine und Angewandte Entomologie 8, 104–106.

Wegensteiner, R., 1996. Laboratory evaluation of *Beauveria bassiana* (Bals.) Vuill. against the bark beetle, *Ips typographus* (L.) (Coleoptera, Scolytidae). IOBC/WPRS Bull 19, 186–189.

Wegensteiner, R., 2004. Pathogens in bark beetles. In: Lieutier, F., Day, K. R., Battisti, A., Grégoire, J.-C., Evans, H.F. (Eds.), Bark and Wood Boring Insects in Living Trees in Europe, a Synthesis. Kluwer Academic Publishers, Dordrecht, pp. 291–313.

Wellstead, A.M., Davidson, D.J., Stedman, R.C., 2006. Assessing approaches to climate change-related policy formulation in British Columbia's forest sector: the case of the mountain pine beetle epidemic. BC J. Ecosyst. Manag. 7, 1–9.

Wermelinger, B., 2004. Ecology and management of the spruce bark beetle *Ips typographus*—a review of recent research. For. Ecol. Manage. 202, 67–82.

Wermelinger, B., Seifert, M., 1999. Temperature dependent reproduction on the spruce bark beetle *Ips typographus*, and analysis of the potential population growth. Ecol. Ent. 24, 103–110.

Weslien, J., 1992a. The arthropod complex associated with *Ips typographus* (L.) (Coleoptera, Scolytidae): species composition, phenology, and impact on bark beetle productivity. Entomol. Fennica 3, 205–213.

Weslien, J., 1992b. Monitoring *Ips typographus* (L.) populations and forecasting damage. J. Appl. Entomol. 114, 338–340.

Weslien, J., Lindelöw, Å., 1990. Recapture of marked spruce beetle *Ips typographus* (L.) in pheromone traps using area-wide mass trapping. Can. J. For. Res. 20, 1786–1790.

Weslien, J., Regnander, J., 1990. Colonization densities and offspring production in the bark beetle *Ips typographus* (L.) in standing spruce trees. J. Appl. Entomol. 109, 358–366.

Weslien, J., Annila, E., Bakke, A., Bejer, B., Eidmann, H.H., Narvestad, K., et al., 1989. Estimating risks for spruce bark beetle (*Ips typographus* (L.)) damage using pheromone-baited traps and trees. Scand. J. For. Res. 4, 87–98.

White, J.C., Wulder, M.A., Brooks, D., Reich, R., Wheate, R., 2004. Mapping mountain pine beetle infestation with high spatial resolution imagery. For. Chron. 80, 743–745.

Whitehead, R.J., Russo, G.L., 2005. "Beetle-proofed" lodgepole pine stands in interior British Columbia have less damage from mountain pine beetle. Natural Resources Canada, Canadian Forest Service, Victoria, BC, Information Report BC-X-402.

Whitehead, R.J., Safranyik, L., Russo, G., Shore, T.L., Carroll, A.L., 2004. Silviculture to reduce landscape and stand susceptibility to the mountain pine beetle. In: Shore, T.L., Brooks, J.E., Stone, J.E. (Eds.), Mountain Pine Beetle Symposium: Challenges and Solutions. Natural Resources Canada, Canadian Forest Service, Victoria, BC, pp. 233–244, Information Report BC-X-399.

Wichmann, L., Ravn, H.P., 2001. The spread of *Ips typographus* (L.) (Coleoptera, Scolytidae) attacks following heavy windthrow in Denmark, analysed using GIS. For. Ecol. Manage. 148, 31–39.

Wickman, B.E., 1987. The battle against bark beetles in Crater Lake National Park: 1925–34. U.S. Department of Agriculture, Forest Service, Portland, OR, General Technical Report PNW-GTR-259.

Wood, D.L., 1972. Selection and colonization of ponderosa pine by bark beetles. In: van Emden, H.F. (Ed.), Insect/Plant Relationships. Blackwell Scientific Publications, Oxford, pp. 101–117.

Wood, D.L., 1982. The role of pheromones, kairomones and allomones in the host selection and colonization behavior of bark beetles. Ann. Rev. Entomol. 27, 411–446.

Wulder, M.A., Dymond, C.C., Erickson, B., 2004. Detection and monitoring of the mountain pine beetle. Natural Resources Canada, Canadian Forest Service, Victoria, BC, Information Report BC-X-398.

Wulder, M.A., Skakun, R.S., Dymond, C.C., Kurz, W.A., White, J.C., 2005. Characterization of the diminishing accuracy in detecting forest insect damage over time. Can. J. Rem. Sens. 31, 421–431.

Wulder, M.A., Dymond, C.C., White, J.C., Erickson, B., 2006a. Detection, mapping, and monitoring of the mountain pine beetle. In: Safranyik, L., Wilson, B. (Eds.), The Mountain Pine Beetle—A Synthesis of Biology, Management, and Impacts on Lodgepole Pine. Natural Resources Canada, Canadian Forest Service, Victoria, BC, pp. 123–154.

Wulder, M.A., Dymond, C.C., White, J.C., Leckie, D.G., Carroll, A.L., 2006b. Surveying mountain pine beetle damage of forests: a review of remote sensing opportunities. For. Ecol. Manage. 221, 27–41.

Wulder, M.A., White, J.C., Carroll, A.L., Coops, N.C., 2009. Challenges for the operational detection of mountain pine beetle green attack with remote sensing. For. Chron. 85, 32–38.

Yan, Z.-G., Sun, J., Owen, D., Zhang, Z., 2005. The red turpentine beetle, *Dendroctonus valens* LeConte (Scolytidae): an exotic invasive pest of pine in China. Biodivers. Conserv. 14, 1735–1760.

Yang, Z.-Q., Wand, X.-Y., Zhang, Y.-N., 2014. Recent advances in biological control of important native and invasive forest pests in China. Bio. Control 68, 117–128.

Yüksel, B., 1996. Mass-rearing of *Rhizophagus grandis* (Gyll.) for the biological control of *Dendroctonus micans* (Kug.). Orman Fakültesi Seminer Trabzon 1, 133–141.

Zahradník, P., Knížek, M., Kapitola, P., 1993. Zpětné odchyty značených lýkožroutu smrkových (*Ips typographus* L.) do feromonových lapačů v podmínkach smrkového a dubového porostu. Zpravy Lesnickeho Vyzkumu 38, 28–34.

Zahradník, P., Švestka, M., Novák, V., Knížek, M., 1996. Podkorní škůdci. In: Švestka, M., Hochmut, R., Jančařík, V. (Eds.), In: Praktické Metody v Ochraně Lesa"2, Silva Regina, Praha, pp. 118–144.

Zhang, Q.-H., Schlyter, F., 2004. Olfactory recognition and behavioural avoidance of angiosperm nonhost volatiles by conifer-inhabiting bark beetles. Agric. For. Entomol. 6, 1–19.

Zolubas, P., 2003. Spruce bark beetle (*Ips typographus* L.) risk based on individual tree parameters. In: IUFRO Kanazawa 2003 Forest Insect Population Dynamics and Host Influences, Proceeding. pp. 92–93.

Zumr, V., 1982. The data for the prognosis of spring swarming of main species of bark beetles (Coleoptera, Scolytidae) on the spruce (*Picea excelsa*). Z. Angew. Entomol. 93, 305–320.

Chapter 15

Economics and Politics of Bark Beetles

Jean-Claude Grégoire[1], Kenneth F. Raffa[2], and B. Staffan Lindgren[3]

[1]*Biological Control and Spatial Ecology Laboratory, Université Libre de Bruxelles, Bruxelles, Belgium,* [2]*Department of Entomology, University of Wisconsin-Madison, Madison, WI, USA,* [3]*Natural Resources and Environmental Studies Institute, University of Northern British Columbia, Prince George, BC, Canada*

1. INTRODUCTION—ECOSYSTEMS, HUMANS, AND BARK BEETLES

Large bark beetle outbreaks are regarded as major forest disturbances. In the United States, Dale *et al.* (2001) ranked them first, before hurricanes, tornadoes, and fire, with a 20,400,000 ha average annual impact area and annual average costs (shared with pathogens) above US$2 billion per year. In Europe, over the period 1950–2000, Schelhaas *et al.* (2003) ranked them third (8% of the total damage), after storms (53%) and fire (16%), with 2.88 million m^3 per year between 1958 and 2001 (Seidl *et al.*, 2011). The recent major outbreak of the mountain pine beetle *Dendroctonus ponderosae* Hopkins in British Columbia and neighboring areas has certainly promoted bark beetles even higher on these scales.

The major direct economic consequences of these outbreaks have been widely analyzed, various mitigation methods have been designed and implemented, and diverse political, industrial, and commercial initiatives have been developed to salvage the remains of the devastated forests. At this point, however, the many other, environmental and sociological, consequences of these disturbances are still largely unexplored, although significant progress has been made since Stark and Waters (1987) stressed the importance of understanding the ecological impact of bark beetle damage, regretting the paucity of the information available. A substantial amount of research is now filling this gap. Progar *et al.* (2009) provide a comprehensive review of the progress in this direction. They outline the multi-scale positive influence of bark beetle activity, from the landscape to stand levels, as well as the various socioeconomic changes brought by bark beetle outbreaks.

2. ECONOMICS

2.1 Damage

Aerial surveys for British Columbia (2001–2010) and western conterminous USA (1997–2010) estimate total mortality area (i.e., the area covered by all the dead trees) to be 5.46 million ha (Mha) and 0.47–5.37 Mha, respectively (Meddens *et al.*, 2012). Total bark beetle damage in Europe from 1958 to 2001 was estimated from forest inventory data at about 124 million m^3 (Seidl *et al.*, 2011). These striking figures are difficult to compare as they appear in different units (forest areas vs. log volumes), which also illustrates the difficulty to collate and compare damage information.

2.1.1 Silvicultural Consequences

Stand composition and structure are modified by the selective choices of bark beetles. For example, Dymerski *et al.* (2001) surveyed stands of Engelmann spruce (*Picea engelmannii* Parry ex Engelm.) in central Utah during a large outbreak of the spruce beetle (*Dendroctonus rufipennis* Kirby). They found that basal area had decreased by an average 78% in trees larger than 13 cm in diameter at breast height (DBH) in 1996 and by 90% in 1998. Spruce mortality for trees the same size as above averaged 53% in 1996 and 73% in 1998. Stand composition markedly changed, with subalpine fir (*Abies lasiocarpa* (Hooker) Nuttall) dominating the overstory.

In lodgepole pine (*Pinus contorta* var. *latifolia* Engelm. ex S. Watson) stands in Rocky Mountain National Park, Colorado, 47% of the stems were killed and basal area was reduced by 71% by a *D. ponderosae* outbreak (Nelson *et al.*, 2014). Average DBH decreased from 17.4 to 11.0 cm, and density decreased from 1393 to 915 stems/ha, while the proportion of non-host species grew from 10.6 to 23.1%. The preferential attack of larger trees suggested above was analyzed and discussed further by Boone *et al.* (2011), who surveyed lodgepole stands attacked by *D. ponderosae* in British Columbia. At increasing population densities, the beetles increasingly selected larger trees, despite their stronger defenses.

The gaps created by mortality to the largest trees can make stands more vulnerable to wind, hence increasing

Bark Beetles. http://dx.doi.org/10.1016/B978-0-12-417156-5.00015-0

the chance of new attacks on windthrows by species such as *Ips typographus* (L.), *D. rufipennis*, or *D. pseudotsugae* Hopkins. Under favorable weather conditions, there may also be increased fire risk during the period when red needles remain in the crown (Kulakowski and Jarvis, 2011). Lynch *et al.* (2006) analyzed historical records from Yellowstone National Park, Wyoming, for the 25-year period before the 1988 Yellowstone fires and developed a model in which mountain pine beetle activity in the period 1972–1975 increased the likelihood of fire in 1988 by 11% over unaffected areas. From data collected in endemic, epidemic, and post-epidemic Douglas-fir *Pseudotsuga menziesii* (Mirb.) Franco, lodgepole pine and Engelmann spruce stands, Jenkins *et al.* (2008) found that changes in fuels over the course of an epidemic either increase or decrease the potential for fire. Globally, bark beetle epidemics result in substantial changes in species composition and altered fuel complexes. Hoffmann *et al.* (2012) used a fire risk model, the Wildland-Urban Interface Fire Dynamics Simulator, and field data at the tree scale to investigate how tree spatial arrangements and *D. ponderosae*-caused mortality influenced fire hazard after outbreak. They found a positive link between beetle-caused tree mortality and the intensity of crown fires, while dead needles remained in the crowns. This relationship varied according to stand structure and other factors. For example, linkages between bark beetle outbreaks and fire can also be quite weak (Simard *et al.*, 2011). DeRose and Long (2009), using another simulator, assessed potential wildfire behavior after a massive *D. rufipennis* outbreak in southern Utah and found a reduced probability of active crown fire for 10 or 20 years, due to a reduction of crown fuel after beetle attack. The host trees species seems thus to influence fire hazards. Page *et al.* (2014) provides a comprehensive review of the research on effects of *D. ponderosae* outbreaks on fire.

Linkages between fire and bark beetles can potentially work both ways. Surveying ponderosa pine stands attacked by several bark beetle species in the southwestern USA after two wildfires and a prescribed fire, McHugh *et al.* (2003) found that tree colonization by several *Dendroctonus* and *Ips* species was promoted by heavy crown fire damage. Wildfire injury reduces inducible defenses of lodgepole pine against mountain pine beetle (Powell and Raffa, 2011). However, the increased but localized colonization of fire-injured trees is unlikely to cause a transition into outbreaks, unless there is an additional region-wide factor such as severe drought or high temperatures (Hood and Bentz, 2007; Powell *et al.*, 2012).

2.1.2 Environmental Consequences

Ecosystem services cover many aspects (Krieger, 2001): watershed services (water quantity and quality; soil stabilization; air quality; climate regulation and carbon sequestration; biological diversity); recreation (economic impact; wilderness recreation; hunting and fishing; non-timber products); and cultural values (aesthetic and passive use; endangered species; cultural heritage). There have been increased efforts to assign monetary value to ecosystem services, as this can potentially allow for more objective choices in priorities and resource allocation (Costanza *et al.*, 1997; Krieger, 2001). One difficult issue is that the ultimate cost of many decisions that could affect ecosystem services (e.g., the planning of forest operations) is often delayed and, thus, those who benefit in the short term from these decisions are not those who will face their costs. Many ecosystem services can be seriously affected by bark beetle outbreaks (Embrey *et al.*, 2012; see also Chapter 1). Likewise, the costs and benefits of policy decisions are often spatially segregated. For example, high profits can be derived during global trade, while the costs of invasive species are disproportionately high at the local level (Aukema *et al.*, 2011).

The water and soil nutrient balance can be affected after an outbreak, before vegetation regrowth (Bosch and Hewlett, 1982; Stednick, 1996; Zimmermann *et al.*, 2000; Brown *et al.*, 2005). Bark beetle damage can result in reduced cover and the reduction of small roots, leading to an increase in ground moisture, an increase in water discharge and recharge and in nutrient uptake. Enhanced insolation leads to increased soil temperature, which in combination with increased moisture leads to faster decomposition and mineralization of the dead biomass. Under conditions of reduced nutrient uptake, there is a higher nitrate concentration in the seepage water, and percolation increases, at least before regrowth occurs. This could increase soil acidity, depending on local conditions, which could lead to increased cation leaching. Increased acidity and aluminum leaching can endanger river ecosystems. A 25 to 40 mm increase in annual water yield per 10% cover change is observed for pine and hardwood forests, respectively (Bosch and Hewlett, 1982), although, according to Stednick (1996), these changes are not noticeable below 20–30% deforestation. Beetle-infested plots have lower C:N mass ratios of pine needlefall than uninfested plots, with higher nitrification rates in the mineral soils from infested plots (Morehouse *et al.*, 2008).

The timing and amount of snow melt can be affected by bark beetle activity (Logan *et al.*, 2010; Edburg *et al.*, 2012; Perrot *et al.*, 2014), with earlier snow disappearance under attacked trees. Tree death may reduce protection against avalanches although, according to Kupferschmid Albisetti *et al.* (2003), spruce snags and dead wood on the ground can still provide some protection for several decades.

The present mountain pine beetle outbreak (2000–2014, and continuing) in British Columbia has affected the global carbon balance, converting the forest from a small carbon sink to a large source during a long period (2000–2020) (Kurz *et al.*, 2008). However, globally, Canada's managed forests remain a carbon sink (Stinson *et al.*, 2011).

From a conservation perspective, large outbreaks have been shown to increase biodiversity by opening closed conifer stands. For instance, unmanaged outbreak of *I. typographus* in the German National Park "Bavarian Forest" has favored large numbers of arthropod and plant species with a preference for open habitats (Müller *et al.*, 2008, 2010; Lehnert *et al.*, 2013).

2.1.3 Economic Consequences

As suggested above, many consequences of a bark beetle outbreak incur costs: wood losses or downgrading, changes in the ecosystem services, public health consequences, and changes in the landscape aesthetic value. The costs in each of these categories are estimated following different rules.

Wood colonized by the fungi associated with tree-killing bark beetles often results in large areas of blue-stained wood, which reduces the market value of the wood. For example, Chow and Obermajer (2007) found that the volume of bluestain in lodgepole pine wood increased with time since mountain pine beetle attack, with maximum discoloration at about 3 m above ground. Chow and Obermajer (2007) measured the economic implication of this staining by analyzing the percentage of Japanese grade (J-grade) lumber produced, and showed a decrease in J-grade production with increasing time since beetle attack. They recommended early harvest and processing of attacked trees and predicted a reduced supply to the Japanese J-grade market, with an estimated loss of sales of about US$400 million in the following 10 years.

Patriquin *et al.* (2007) used a computable general equilibrium framework to investigate the regional economic impact sensitivity to the current mountain pine beetle infestation in British Columbia and analyzed the short- and long-term changes in timber supply. They concluded that, in the short term, an increased timber supply would favor the regional economies, but that, in the longer-term, the decreasing timber supply would negatively impact regional economies. This raises the concern that severe outbreaks can cause sustainable resource-based economies to behave more like boom-and-bust mining economies. The model can help local decision-makers to develop policies and priority areas for mitigation planning in response to the anticipated fluctuations in timber supply. Fluctuations in employment have strongly impacted local communities in the Alaskan Kenai Peninsula, where timber harvesting developed after the 1989–2004 *D. rufipennis* outbreak but collapsed when the local wood chip facility closed down in 2004 following the decline in quality of the salvaged wood (Flint *et al.*, 2009).

Starting from the global estimates of Costanza *et al.* (1997), Krieger (2001) estimated the annual value of ecosystem services provided by temperate/boreal forests to be US$63.6 billion (see also Chapter 1). Price *et al.* (2010) applied hedonic analysis to property value. They estimated willingness-to-pay to prevent mountain pine beetle damage in Grand County, Colorado. According to their results, property values decline by $648, $43, and $17, respectively, for every tree killed within a 0.1, 0.5, and 1.0 km buffer.

2.1.4 Social Dimensions

The recent outbreaks in North America and Europe triggered a set of studies centered on public health consequences of bark beetle damage, their impact on the standards of living and on employment, the social perception of forest changes and public acceptance of their social, economic, and aesthetic consequences.

Embrey *et al.* (2012) reviewed direct and indirect health impacts in the broader context of ecosystem services and climate changes. They mention increased gastrointestinal disorders brought by higher water turbidity, psychological issues linked to unemployment or loss of property value and, from a more long-term forecasting standpoint, heat-related mortality and morbidity due to climate change. They also discuss possible prevention strategies and argue that the mountain pine beetle outbreak highlights the need for adopting an ecological, systems-oriented public health approach, able to anticipate all potential health impacts.

Flint (2006) analyzed the response of people and communities to a *D. rufipennis* outbreak on the Kenai Peninsula, by interviews and mail surveys. She observed differences in perception of the impacts of changing forest conditions (fire, falling trees, declining watershed quality and wildlife habitat, economic fluctuations, landscape change, emotional loss). Some communities benefited from increased timber harvesting, others suffered from the loss of the spruce forest, which profoundly affected quality of life, and led to community conflict and economic challenges. She discusses how these different perceptions present both opportunities and difficulties for forest management. In a wider context, Flint *et al.* (2009) offer a seminal international approach of the human context of forest disturbances by insects. They review four cases of bark beetle forest disturbance: the *D. ponderosae* outbreaks in British Columbia and north central Colorado, the *I. typographus* epidemics in the Bavarian Forest National Park, and the *D. rufipennis* outbreak in the Kenai Peninsula. The diverse communities in these case studies varied in their concerns for different issues (employment, security, changing environment). Findings and lessons learned from these studies are outlined along with their implications for managing forest disturbances by insects in general. Conclusions focus on the need to assess the broad array of impacts and risks perceived by local residents and the capacity for local action and involvement in managing forest disturbances. From various examples, the study also highlights the variability in cohesiveness of many local communities, and the high need to involve local shareholders in the decision-making processes. Müller (2011) proposes a comprehensive review of studies concerned with the social dimensions of natural disturbances in forests (wildfires and insects). He also discusses the case of the Bavarian Forest National Park in Germany, which will be examined further in this chapter (see Section 4.1).

From surveys among residents and land managers responsible for forest health management in three regions of Alberta suffering a *D. ponderosae* outbreak, McFarlane *et al.* (2012) analyze regional variation in the public perceptions of risk, and compare the perception of the residents and the land managers. Residents were not well informed about the mountain pine beetle issue and showed little trust that the provincial government and forest industry would satisfactorily manage the outbreak. Land managers were less concerned about non-timber effects.

2.2 Salvage

Salvaging the wood and restoring the land are also needed when sporadic and local damage occurs due to endemic bark beetle populations. Scale is important, since it influences the market value of the salvaged timber, the technical, administrative and commercial feasibility of silvicultural operations, as well as their overall economic, environmental, and human impacts.

2.2.1 Silvicultural Salvage

Sanitary thinning, felling, pest control, and an economic component are commonly performed operations when it comes to restoring forest health (Figure 15.1) (Carroll *et al.*, 2006; Coggins *et al.*, 2011). Removing attacked trees in time prevents damage from new pest generations and, at the same time, preserves lumber value by preventing further wood deterioration by fungal agents or insects. Cost-effectiveness analyses can be applied to determine optimal options, e.g., between salvage, quarantine, or biological control (O'Neill and Evans, 1999). Ground surveys, aerial surveys, and satellite image analysis provide foresters with spatially referenced quantitative estimates and, depending on local rules, the local forest services or private companies proceed to salvage logging. For quantitative accuracy and spatial precision, large-scale satellite (Meigs *et al.*, 2011) or aerial monitoring can be complemented by targeted helicopter surveys, followed by ground surveys (Coggins *et al.*, 2011). Ground surveys can be tailored to fit local constraints. For example, in southwestern France, Samalens

FIGURE 15.1 Logging operation near Prince George, British Columbia. Each bunch of logs corresponds to one or several truckloads. *Photo: J.-C. Grégoire.*

et al. (2007) designed an adaptive technique of road sampling for assessing the damage of *Ips sexdentatus* (Boerner) in plantations of *Pinus maritima* Mill. Sometimes, indirect ways can be used (Meurisse *et al.*, 2008).

The use of thinning to mitigate bark beetle outbreaks has always generated contention and controversy, and components of these arguments remain. For example, Six *et al.* (2014) argue that while thinning may decrease the likelihood of outbreaks erupting, and so may have some benefit as a proactive tool, its ability to reduce outbreaks already under way is not supported by evidence. Fettig *et al.* (2014) emphasize that it is important to distinguish these functions, as well as additional management intentions of thinning operations, and that understating such distinctions can have negative policy implications through lost opportunities. Likewise, Black (2005) argued that the evidence for deleterious effects on non-target invertebrates was stronger than that for effective pest management, a view disputed by Fettig *et al.* (2007).

Under outbreak conditions, these operations are complicated by decreasing prices for the wood, competition for loggers, machinery and logging trucks, and the availability of an extensive road network. In many cases, it then becomes difficult or impossible to remove the killed trees soon enough to prevent damage to the wood. Strategic choices must be made (e.g., to increase harvest) when infestation levels exceed a certain threshold (Bogle and van Kooten, 2012). In the southern Rocky Mountains, it is foreseen that terrain, economic, and administrative limitations will limit salvage logging to a small fraction (<15%) of the forests killed by *D. ponderosae* (Collins *et al.*, 2011).

Several critical issues arise from salvage logging. One of them is the destination and use of the salvage timber. Flint *et al.* (2009) analyzed this issue in Colorado in the context of the *D. ponderosae* outbreak. Communities highly dependent on tourism and recreation were less supportive to large-scale forest industry, which, on the other hand, received more support in areas with an existing tradition in resource extraction. There was a wide cross-community support for small-scale niche markets (e.g., posts and poles and furniture) and biofuel energy production.

From the biodiversity standpoint, Foster and Orwig (2006) concluded from a study in New England that leaving the forest alone brought more ecological benefits than salvage logging. This debate has also been widely explored around *I. typographus* outbreaks in Germany (see Section 4.1).

Carbon budgets can also be influenced by salvage logging. A study of harvested vs. unharvested stands in British Columbia after the mountain pine beetle outbreak showed carbon release in the harvested stands even 10 years after harvesting, while the unharvested stands were still carbon sinks (Brown *et al.*, 2010).

Tree regeneration was compared in the southern Rocky Mountains between paired harvested and untreated lodgepole pine stands that had suffered more than 70% mortality due to *D. ponderosae* (Collins *et al.*, 2011). In harvested stands, the density of new seedlings was four times higher than in the non-harvested stands. Growth simulations suggested that lodgepole pine will remain the dominant species in harvested stands, while *A. lasiocarpa* will become the most abundant species in untreated areas.

2.2.2 Industrial Salvage

Bark beetle outbreaks generate changes in different directions in the market value of timber and wood derived products. At first, the sudden increase of raw materials reduces the market value of wood. Later on, yearly harvests can be regulated to compensate for the depletion caused by the insects, affecting global log and lumber prices. Abbott *et al.* (2009) reported that the 2006 harvest in British Columbia was 8.7 million m^3 above the pre-outbreak annual allowable cut, and that, after 2009, the allowable annual cut would be reduced by 12 million m^3. Other products can then be developed, such as bioenergy (Pan *et al.*, 2007, 2008; Stennes *et al.*, 2010; Luo *et al.*, 2010; Zhu *et al.*, 2011). More information can be found in Section 4.4.

3. POLITICS

3.1 Management

A global approach is needed for large area-wide outbreaks. Abbott *et al.* (2008, 2009) consider the economics of the mountain pine beetle outbreak in British Columbia in an international context, where province-wide forest management policies and the international market for timber and wood products must be simultaneously taken into account.

Public policies also bear on forest management and vary worldwide, in terms of regulatory constraints and financial incentives, with variable weight on either public support or private initiative. Brunette and Couture (2008) analyzed how some European governments compensate forest owners for windstorm damage. They concluded that this policy is likely to interfere with the propensity of private forest owners to purchase a personal insurance policy for the coverage of natural disturbances and develop a proactive attitude towards prevention. Sims *et al.* (2010) developed a bioeconomic model of tree harvesting after mountain pine beetle damage, to measure the consequences of alternative public management strategies. They suggest that the commonly practiced procedures increase the severity of mountain pine beetle cycles, while more centrally coordinated management could eliminate mountain pine beetle cycles and lessen their impacts with only small reductions in the long-run stock of wood. Watson *et al.* (2013) were interested in cost sharing for pre-commercial thinning (PCT) in pine plantations in Virginia, in view of reducing southern pine beetle (*Dendroctonus frontalis* Zimmermann) risks. PCT has a cost and delayed impact; therefore, it is not always seen positively by landowners. The Virginia Pine Bark Beetle Prevention Program attempts to reconcile differing public attitudes by partly reimbursing PCT costs. A survey sent to landowners indicated a significant, positive effect of cost sharing on willingness to participate, with a 50% upper limit of reimbursement beyond which participation is unlikely to increase substantially.

Management policy also includes the choice of tree species for replanting. In a meta-analysis, Bertheau *et al.* (2010) showed that, in some limited instances, some pest species have a higher fitness on exotic trees. It has been frequently observed that native trees are more sensitive to exotic pests. For example, *Dendroctonus valens* LeConte is a tree killer in China, while it is much less aggressive in North America (Sun *et al.*, 2013; Chapter 8). Forest biodiversity has also been shown to increase stand resilience (Jactel and Brokerhoff, 2007), and replantation policies could be designed along these lines.

3.2 Anticipating Trouble

3.2.1 Predictive Models

Various types of models are used for risk prediction, phenology planning, anticipation of spatio-temporal population changes or, more prospectively, for anticipating the effects of climate change (Hansen *et al.*, 2001a; Williams and Liebhold, 2002; Gan, 2004; Jönsson *et al.*, 2007; Seidl *et al.*, 2008; Waring *et al.*, 2009; Bentz *et al.*, 2010; Evangelista *et al.*, 2011; Temperli *et al.*, 2013). A synoptic presentation of such models developed to date for *I. typographus* is given in Table 15.1.

3.2.2 Exotic Species

Exotic bark beetles are frequently intercepted with imported goods and materials. Haack (2001) reported 6825 records of bark and ambrosia beetles from countries outside of North America that had been intercepted during the 1985–2000 period at US ports of entry. Similar information for New Zealand is provided by Brockerhoff *et al.* (2006). Most of the insects come in wood packaging material containing various goods (tiles, marble, machinery, steel, ironware, granite, slate, etc.) (Haack, 2001). An International Standard for Phytosanitary Measures (ISPM 15) has since been established (FAO, 2009), which requires that the wood is debarked, and either heat treated (at a minimum temperature of 56°C for a minimum duration of 30 minutes) or fumigated with methyl bromide, although this latter treatment is being phased out. However, ISPM 15 does not totally guarantee bark beetle-free importations. Analyzing importation data for goods entering into the US, Haack *et al.* (2014) found only a small reduction in contaminated wood packaging material following implementation of ISPM15, from about 0.2% (for the 2 years pre-ISPM) to about 0.1% (for the 4 years following ISPM15). There are numerous examples of introductions of exotic bark beetles, e.g., *Dendroctonus micans* Kug. in Britain (Bevan and King, 1983), *Tomicus piniperda* L. in North America (Haack and Lawrence, 1994), and *D. valens* in China (Sun *et al.*, 2013). A recent occurrence is the discovery of the thousand cankers disease pathogen *Geosmithia morbida* Kolařik (Ascomycota: Hypocreales) and its vector *Pityophthorus juglandis* Blackman on an infected walnut tree in Italy in 2013 (Montecchio and Faccoli, 2014). International regulations such as ISPM15, national quarantine regulations implemented by each national plant protection organization, and inspections at the national borders

are methods developed to mitigate this threat, but they are obviously not 100% effective.

4. A DIVERSITY OF PATTERNS— ILLUSTRATIVE CASE STUDIES

Five specific cases are illustrated by a more thorough presentation: the eight-toothed spruce bark beetle in Eurasia; secondary ambrosia beetles attacking living beech in Europe; the spruce beetle in North America; the mountain pine beetle crisis in British Columbia; and the Eastern pine engraver across the continent.

4.1 Fallen and Standing Alike—The Eight-toothed Spruce Bark Beetle in Eurasia

Ips typographus is a pest of spruce (*Picea* spp.) throughout Eurasia. At endemic levels, it attacks windthrows or snowbreaks, but when populations grow, it can kill standing trees (Christiansen and Bakke, 1988). Wermelinger (2004) provides a comprehensive review of the biology and management of this species. Depending on competition and natural enemies, each attacked tree produces 25,000 to 70,000 individuals (Fahse and Heurich, 2011 and references therein; Gonzalez *et al.*, 1996). Grégoire *et al.* (1997) counted on average 4800 to 6400 beetles of both genders attacking new host trees in about 1.5 m^3 in volume. Comparing these two sets of figures suggests that, when a substantial proportion of the emerging insects can find a susceptible host (e.g., after a storm), catastrophic outbreaks can easily develop, especially in multivoltine populations.

Ips typographus is the most damaging of the bark- and wood-boring insects attacking living trees in Europe (Grégoire and Evans, 2004). According to Carpenter (1940), 15 outbreak episodes occurred between 1769 and 1931. Bark beetles (mainly *I. typographus*) have been responsible for losses of about 2.9 million m^3 of spruce timber per year in Europe during the period 1950–2000 (Schelhaas *et al.*, 2003). Recent detailed data for Switzerland, France, Austria, and Sweden are provided in Table 15.2. Dale *et al.* (2001) listed seven major disturbances affecting forests, including insects and hurricanes. In the case of *I. typographus*, both disturbances are combined, as outbreaks usually occur after storms (Table 15.1), in particular when the windthrown host trees are not removed fast enough (Schroeder and Lindelöw, 2002; Schelhaas *et al.*, 2003). Insect success is increased by dry and hot springs and summers (another combination of disturbances), and the value of the attacked timber decreases as infestation time increases. The wood is first stained by the pathogenic fungi (e.g., *Ophiostoma* sp.) associated with the beetles (Figure 15.2). This does not affect its structural properties (Chow and Obermajer, 2007) but can

TABLE 15.1 Risk Models Developed for the Management of *Ips typographus* in Europe

References	Type of Model/Objectives	Data	Method	Main Results
Faccoli (2009)	Risk model Investigate the possible weather effect on the biology of and damage caused by *I. typographus* in the southeastern Alps.	Temperature records (1962–2007), precipitation data (1922–2007), damage caused by *I. typographus* (1993–2007). Data from pheromone-baited traps (1996–2005) in the southeastern Alps.	Statistical model - multiple regressions.	• Damage caused by *I. typographus* was inversely correlated with March–July precipitation from the previous year but not correlated with temperature. • Spring drought increased damage caused by *I. typographus* in the following year, whereas warmer spring affected insect phenology.
Fahse and Heurich (2011)	Risk model • Analyze the spatial and temporal aspects of bark beetle outbreaks at the stand scale. • Assess the impact of both antagonists and management. • Predict outbreak probabilities under different conditions.	Data (1994 to 2009) from the recent outbreak in the Bavarian Forest National Park (Germany).	Spatially explicit agent-based bottom-up simulation model taking into account individual trees and beetles (SAMBIA).	• Distinct threshold above a certain level of impact from natural enemies or silvicultural management. • Also validated by the model: anisotropic growth of infestation spots; abrupt collapse of attacks even in the presence of potential host trees.
Jakuš *et al.* (2011)	Risk model Define the characteristics of individual Norway spruces that survived a massive bark beetle outbreak.	Measurements made in the Šumava National Park (Czech Republic).	Statistical model, based on parameters related to crown geometry, stand conditions and distances between trees.	• Trees with a longer crown length tended to survive. • Attacked trees usually located in the south aspects of areas with larger basal areas. • Probability of additional attack inversely proportional to distance to a previously attacked tree.
Jönsson *et al.* (2012)	Risk model Analyze the influence of multiple environmental factors on the risk for *I. typographus* outbreaks.	Gridded daily climate data covering Sweden (spatial resolution: 0.5°). Data on storm damage and *I. typographus* outbreak in 1960–2009.	Ecosystem modeling approach. "Model calculations of *I. typographus* phenology and population dynamics as a function of weather and brood tree availability were developed and implemented in the LPJ-GUESS ecosystem modeling framework." Sensitivity analysis.	• Good fit between the model simulations and the observed pattern in outbreak frequency. • Higher risk for attacks on living trees under a warmer climate allowing multivoltinism. • Timely salvage cutting and removing of infested trees leads to a major reduction in the risk of attacks on living trees.
Kärvemo *et al.* (2014)	Risk model Locate areas of high risk for tree mortality across forest landscapes.	Calibration and validation data each from a different set of 130,000 ha of managed lowland forest in southern Sweden in 2007–2009, at a 100 × 100 m resolution.	Statistical model based on boosted regression trees.	• Host tree volume ha⁻¹ (up to 200 m³ ha⁻¹) was the most important predictor of beetle attack. • Birch volume of (up to 25 m³ ha⁻¹) also positively correlated with infestation risk. • Tree height (above 10–15 m) associated with increased infestation risk. • The attacked trees are distributed in many small spots spread out over the landscape.

Continued

TABLE 15.1 Risk Models Developed for the Management of *Ips typographus* in Europe—cont'd

References	Type of Model/Objectives	Data	Method	Main Results
Lausch *et al.* (2011)	Risk model Identify key habitat variables (topography; climate; soil type; forest stage; biological/structural characteristics of the patch) influencing attack risk.	Annual color-infrared aerial photographs (1:10,000–1:15,000) of deadwood areas (100% mortality due to *I. typographus*) taken from 1990 to 2007 in the Bavarian Forest National Park (Germany).	Ecological Niche Factor Analysis (ENFA) models calculated yearly from a spatially explicit database.	• No single causal factor was identified over the entire model period. • The distance from the previous year's infestation and the area and perimeter of the previous year's infestation patch influenced the probability of a new attack, but not across all years.
Netherer and Nopp-Mayr (2005)	Risk model Identify mechanisms of disturbance agents and establish spatial distribution of predisposed stands.	Forest inventory data from the Slovak and Polish High Tatras National Parks, combined to a digital elevation model.	A spatially explicit predisposition assessment system was developed, scoring abiotic and biotic factors based on the literature and expert advice. The resulting predisposition scores (11 sites and nine stand criteria) were compared to the distribution patterns of damaged and undamaged forest stands.	• At the site level, the distribution of sound and attacked forest units was significantly different between low–medium and high scores of "Radiation" and between the categories of "Slope Position." • At the stand level, higher values for the criteria "Proportion of Spruce," "Age Class," "Predisposition to Storm Damage," and "Stand Density" significantly characterized attacked forest units.
Pasztor *et al.* (2014)	Risk model Develop tools to assess the risks of damage from bark beetle disturbances at the operational scale of forest stands.	Ten-year forest management plans and related harvest records of four management units of the Austrian Federal Forests (40,000 ha) within the 1992–2010 period; gridded climate data set provided by the Austrian Central Institution for Meteorology and Geodynamics.	Statistical binomial generalized linear mixed models were used to assess the effects of site, stand, and climate conditions on the probability of bark beetle disturbance events at forest stand level, and linear mixed models to assess the intensity of these events.	• Increases in some of the predictor variables increased probability of damage substantially, mainly previous bark beetle damage during the four previous years and current timber stock. • Potential bark beetle generations estimated from a beetle phenology model were also a useful predictor. • The model of disturbance probability correctly classified 90% of all cases in the dataset (specificity 95%, sensitivity 29%). • The model for damage intensity explained only low shares of the variation in the recorded damage data.
Schmidt *et al.* (2010)	Risk model Storm damage risk of for individual trees.	Individual tree damage data from the storm "Lothar" (1999) in Baden-Württemberg (Germany).	Statistical model inferring probability of damage, and separating the effects of tree-dependent variables, topography, site conditions, and flow field related effects.	• Good validation of predicted geographical location of risk hotspots using forest service data. • Tree height (but not height to DBH ratio) influences damage. • *Picea abies* has the highest damage potential. • Higher risks for west- to south-exposed locations and waterlogged soils show an increased risk.

Stadelmann et al. (2013b)	Risk model Assess the impact of drivers influencing bark beetle infestations at the forest district level, in particular salvage logging and sanitation felling.	Annual survey dataset covering nine years and 487 Swiss forest districts (82% of the forested area).	Statistical Poisson log-normal models.	• Bark beetle damage proportional to storm damage, heat sum, volume of Norway spruce stock, and the number of infestation spots in the previous year. • Damage inversely proportional to sanitation felling relative to the total volume of infested spruce, and to proportions of salvaged windthrows.
Zolubas et al. (2009)	Risk model	Fixed-radius plots around attacked trees or controls in 80–100-year-old *Picea abies* pure stands. Ninety-two paired plots in 2000–2002. Characteristics under endemic bark beetle densities.	Statistical model—classification and regression trees.	• Most significant variable: spruce basal area, positively correlated with risk. • Lack of sensitivity for decision-making (70% of the non-attacked (control) plots and 88% of the attacked plots are in the "high risk" category).
Marini et al. (2012)	Population dynamics model Characterize the combined effects of climatic factors and density-dependent feedbacks on damage; test whether climate modify the species' altitudinal outbreak range.	Sixteen-year time-series of *P. abies* timber loss due to *I. typographus* attacks and abiotic events in the Friuli-Venezia Giulia region (Italy). Annual time series (1994–2009) from daily climatic data from eight meteorological stations distributed over the region.	Discrete population dynamics model and information theoretic approach.	• Dry summers combined with warm temperatures appeared as the main abiotic triggers of severe outbreaks. • Endogenous negative feedback with a 2-year lag suggesting a potential important role of natural enemies. • Forest damage would be on the average sevenfold higher in warmer sites than in spruce's historical climatic range. • Dry summers (not temperature) influence upward altitudinal shifts of the outbreaks.
Marini et al. (2013)	Population dynamics model Quantify and compare the relative importance of predation, negative density feedback, and abiotic factors as drivers of *I. typographus* population dynamics.	Pheromone-baited traps from 1995 to 2011 in two 60,000 ha areas in central Sweden, and two areas in southern Sweden. Temperature and rainfall data. Annual amount of timber loss due to snow and wind.	Discrete population dynamics model; multi-model inference approach.	• The main outbreak trigger was the availability of breeding substrates (windthrows). • The main endogenous regulating factor was a strong intraspecific competition for host trees. • Temperature-related metrics did not significantly influence population dynamics, even though they are known to influence voltinism. • Predator (*Thanasimus formicarius*) density did not exert any important regulating impact.
Økland and Berryman (2004)	Population dynamics model Identify the role played by resource dynamics in regional population changes.	Time series of pheromone trap catches from 1979 to 2000 in approximately 100 localities throughout southeast Norway.	Statistical model at two spatial scales (whole area and 12 subregions); additional analyses of time-series before and after a large windfelling in 1987.	• The endogenous dynamics were dominated by lag 1 density dependence. • Windfelling appears to be an important predictor of the dynamics; uncertainty due to only one large windfall event in the time series. • Weak influence of drought stress; uncertainty linked to the absence of severe droughts within the time series.

Continued

TABLE 15.1 Risk Models Developed for the Management of *Ips typographus* in Europe—cont'd

References	Type of Model/Objectives	Data	Method	Main Results
Baier *et al.* (2007)	Phenology model Spatio-temporal simulation of *I. typographus'* seasonal development in Kalkalpen National Park, Austria.	Digital elevation model for interpolating temperature and solar radiation, as well as air and bark temperature measurements.	Phenology model (PHENIPS), using a flight initiation lower threshold of 16.5°C and thermal accumulation of 140 degree-days (dd) from April 1st onward (upper and lower thresholds: 38.9 and 8.3°C, respectively). Thermal sum for total development: 557 dd. Re-emergence of parental beetles when 49.7% of this sum is reached. Discontinuance of reproductive activity at a day length <14.5 h.	Spatially explicit estimate of local maximum number of generations, allowing to predict the potential impact of bark beetle outbreaks.
Jönsson *et al.* (2009)	Phenology model Describe the temperature thresholds for swarming and temperature requirements for development from egg to adult for three future climate change scenarios during the period 1961–2100.	Daily climatic data (1961–2100) from three climate change scenarios obtained from the Rossby Centre Regional Climate Model RCA3 with different adjustments.	Phenology model. The model of Jönsson *et al.* (2007) was used, with some adjustments.	• *I. typographus* able to initiate a second generation in south Sweden during 50% of the years around the mid-century. • By the end of the century, a second generation will be initiated in south Sweden in 63–81% of the years; and less frequently in the rest of the country. • Later, 1–2 generations per year are predicted, and the northern distribution limit for the second generation will vary.
Jönsson *et al.* (2011)	Phenology model Extends the existing model of Jönsson *et al.* (2007), based on temperature only by including reproductive diapause initiated by photoperiodic and thermal cues; use this extended model to assess the impact of global warming on voltinism in *I. typographus*.	Three different climate datasets (1950–2010) including climate change scenarios. Monitoring data from trap catches over various periods between 1979 and 2007, according to country (Sweden, Norway, Denmark).	Phenology model based on several steps: (1) comparison of the output of a phenology model and monitoring data; (2) development and parameterization of a diapause model; (3) analysis of model sensitivity; (4) inclusion of climatic scenarios in the model.	• Higher temperatures can result in increased frequency and length of late summer swarming (producing a second generation in southern Scandinavia and a third generation in lowland parts of central Europe). • Reproductive diapause will not prevent the occurrence of an additional generation per year. • However, day length could restrict late summer swarming.
Gilbert *et al.* (2005)	Spatial model Study large-scale patterns in bark beetle populations that would benefit from the abundant breeding material provided by the 1999 storm in France.	Large-scale survey in the spring and in the autumn of 2000, after the December 1999 storm, in 898 locations distributed throughout wind-damaged areas in France. Local abundance of four conifer bark beetle species scored on a 0 to 5 scale.	Geostatistical estimators to explore the extent and intensity of spatial autocorrelation. Statistical analysis to correlate results with site, stand, and neighborhood landscape metrics of the forest cover.	• Large-scale spatial dependence and regional variations in abundance. • Significant relationships with the number of coniferous patches.

Reference	Model / Objectives	Data	Methods	Results
Kautz et al. (2011)	Spatial model. Quantify the spatio-temporal dispersion of I. typographus: • Parameterize the size and shape of infestation patches. • Model an infestation gradient. • Assess the risk of subsequent infestations at the landscape scale.	Analysis of attacked patches (≥ 5 trees), based on a 22-year time series of annual color-infrared images (1:10,000 to 1:15,000) of a 130 km³ area in the Bavarian Forest National Park (Germany).	GIS-based spatial correlations between successive patches calculated by a distance ring approach based on nearest distance relations. Overlaying this distribution with the distribution of potential hosts.	• The infestation spread was strongly distance dependent, following an inverse power law. • On average, 65% of new infestations occurred within a 100 m radius of the previous year's infestations, and 95% within 500 m. • During outbreak periods within the study's time series, a higher percentage of infestations within short distance (<100 m) were observed. • Larger patches tended to have more complex shapes.
Lausch et al. (2013)	Spatial model. • Describe the long-term spatio-temporal infestation patterns of I. typographus in the Bavarian Forest National Park (Germany), at the landscape scale. • Analyze the spatio-temporal movements of infestation patches.	Color-infrared aerial photographs (1:10,000–1:15,000) of deadwood areas (100% mortality due to I. typographus) taken twice a year (June–July and September–October) from 1988 to 2010 in the Bavarian Forest National Park (Germany).	Spatially explicit variables seen as meaningful for structure and pattern analysis were calculated using the structural analysis program FRAGSTATS. Comparison with, and incorporation to, an agent-based simulation model (SAMBIA) (Fahse and Heurich, 2011).	• Non-directional movements of the centroid of the deadwood patches from 1988 to 2001. • Northeast-southwest movement during the 2001–2007 period. • The mean Euclidean nearest neighbor distance of dead wood patches over the whole period was 116 m (±143), the minimum was 22 m.
Wichmann and Ravn (2001)	Spatial model. Analysis of dispersion patterns of infestation spots after an outbreak.	Field collected data from the forest of Rold Skov (7280 ha; Denmark) in 1982–1983: ground surveys of windthrown areas and infestation patches, salvage harvests and pheromone trap catches.	GIS and statistical analyses.	• Attack densities were not spatially correlated with trap catches. • Attack densities were correlated with the timing of salvage harvests (the later the harvest, the higher the attacks). • Nearly 90% of the new attacks occur within 100 m from an old attack, nearly 80% within 50 m, and 50% of the new attacks occur within 20 m from an old attack.
Jönsson et al. (2007)	Climate change model. Evaluate the effect of regional (southern Sweden) climate change scenarios for the period 2070–2099.	Temperatures data (1961–1990). Bark beetle activity monitored in 1980, 1981, 1984, and 1985 used for validating the model.	Phenology model based on the relationship between thermal conditions and phenology models of I. typographus presented in the literature.	• Step-wise effect of temperature increase on the population dynamics. • Earlier spring swarming and faster development increase the probability of a second swarming during summer. • Because immature stages die during the winter, the autumn temperature will have a decisive impact on the population size of the following spring.
Seidl et al. (2008)	Climate change model. Effects of bark beetle disturbance on timber production and carbon sequestration over 100 years.	Norway spruce management unit in Austria.	Simulation under two scenarios of climatic change, including a submodule of bark beetle-induced tree mortality, under four management strategies (no management; three active management strategies).	• Strong increase in bark beetle damage under climate change scenarios. • Reduced C storage in the actively managed strategies. • Under some scenarios: increased C sequestration in unmanaged control (stand density effect).

Continued

TABLE 15.1 Risk Models Developed for the Management of *Ips typographus* in Europe—cont'd

References	Type of Model/Objectives	Data	Method	Main Results
Temperli et al. (2013)	Climate change model Identify and assess the mechanisms and feedbacks driving short-term and long-term interactions between beetle disturbance, climate change, and windthrow. Predict how they may change in the future.	Model parameterization using measurements of infested area from the recent outbreak in the Bavarian Forest, Germany (Kautz et al., 2011) and monthly climate data from the Black Forest (1950–2000).	Spatially explicit model incorporating beetle phenology and forest susceptibility, and integrated in a climate-sensitive fine-grain landscape model (*LandClim*) Four spatiotemporal scales: short-term, patch scale; short-term, landscape scale; long-term, patch scale; long-term, landscape-scale Baseline climate compared to a weak and a strong climate change scenario.	• *Short-term, patch scale*: spruce age, spruce share, drought index, and windthrown spruce biomass positively correlated alone and in combinations with tree susceptibility; increased infestation probabilities occurred in decades with large windthrow events. • *Short-term, landscape scale*: windthrow had a comparatively weak influence on bark beetle damage because it affected only a small fraction of the landscape, whereas changes in temperature and drought affected trees throughout the landscape. • *Under climate change scenarios*, beetle activity combined with warmer and dryer conditions at the drier-warmer parts of the new range, generating a negative feedback for the beetles by suppressing the host trees.

TABLE 15.2 Storm Damage and Subsequent Damage caused by *Ips typographus* in Germany, Switzerland, France, Sweden, and Austria

Country *(storm)*	Storm Damage (m³)		*I. typographus* Damage (m³)		References
	Date	Damage (spruce)	Date	Damage	
Germany	1972	9,200,000 m³	1972	>8000	Abgrall (2000)
			1973–1975	>567,450	
			1976–1978	>134,920	
Switzerland (*Vivian*)	1990	5,000,000 m³	1990	60,000	Abgrall (2000)
			1991	140,000	
			1992	500,000	
			1993	480,000	
			1995	300,000	
			1995	135,000	
			1996	289,000	
			1997	90,000	
			1998	87,000	
			1999	86,000	
			Total	**2,167,000**	
Switzerland (*Lothar*)	1999	8,000,000 m³ (spruce)	2000	162,000 NB—warm spring and summer	Meier *et al.* (2013) WSL—Forest Protection Overviews, 2014
			2001	1,300,000	
			2002	1 100,000	
			2003	2,067,000 NB—extremely hot summer	
			2004	1,350,000	
			2005	1,015,000	
			2006	727,000	
			2007	285,000	
			2008	85,000	
			2009	100,000	
			Total	**8,191,000**	
France (*Lothar*)	1999	87,600,000 m³ (all conifers, North-Eastern France)	1999	24,500	Nageleisen (2006, 2007) (partial reports, for North-Eastern France)
			2000	–	
			2001	514,000	
			2002	295,000	
			2003	308,700 NB—extremely hot summer	
			2004	378,000	
			2005	453,000	
			Total	**1,948,700**	

Continued

TABLE 15.2 Storm Damage and Subsequent Damage caused by *Ips typographus* in Germany, Switzerland, France, Sweden, and Austria—cont'd

Country (storm)	Storm Damage (m³)		*I. typographus* Damage (m³)		References
	Date	Damage (spruce)	Date	Damage	
Sweden (Gudrun)	2005	75,000,000 m³	2006	1,500,000	Lindelöw and Schroeder (2008)
	2007	12,000,000 m³	2007	>500,000	
Austria	2002	4,000,000 m³ (all tree species)	2002	545,762	Steyrer and Krehan (2009); Krehan *et al.* (2012); Bundesforschungszentrum für Wald. (2014)
			2003	1,485,421	
			2004	1,945,001	
			2005	2,148,970	
			2006	1 953,765	
	2007–2008	Ca. 18,700,000 m³ (all tree species)	2007	1,738,468	
			2008	1,563,216	
			2009	2,470,772	
			2010	2,350,623	
			2011	1,375,634	
			2012	702,126	
			Total	**18,279,758**	

FIGURE 15.2 Blue staining (*Ophiostoma* sp.) and symptoms of lignivorous fungi, *Stereum sanguinolentum* (Alb. and Schwein.) Fr. (brown staining) on a standing spruce mass attacked and killed six months previously. Right figure: close up of the blue and brown staining. *Photo courtesy Emmanuel Defay.*

reduce its value by 50%. Later on, lignivorous fungi (e.g., *Stereum sanguinolentum* (Alb. and Schwein.) Fr.) may colonize the wood, which then loses most of its remaining value. The damage caused by *I. typographus* is not restricted to timber losses and changes in silvicultural planning. In mountainous areas, losing the trees represents a reduced protection against avalanches (Bebi *et al.*, 2012). However, leaving the snags on the slope can still provide effective protection for about 30 years (Kupferschmid Albisetti *et al.*, 2003).

The environmental and social impacts of *I. typographus* outbreaks have been only partially investigated. The Bavarian Forest National Park in Germany is an extremely rich source of information regarding the multiple features of a large and long-lasting *I. typographus* outbreak, because of its unique beetle management rules. The National Park was established in 1970 and now covers more than 240 km^2. Its large forests have been allowed to develop free of human interference. In direct continuity, on the other side of the border with the Czech Republic, the 690 km^2-wide Šumava National Park is also a protected area. These protected zones are only a part of the Bavarian Forest Nature Park (3070 km^2) and the Šumava Protected Landscape Area (1000 km^2), respectively. The entire area is known as the "Greater Bohemian Forest Ecosystem" (Heurich *et al.*, 2011). Large outbreaks of *I. typographus* occurred in the park in the 1980s after several windthrow events. A decision was made at that time to exert no control on the beetles in the natural zone of the national park. This decision is still under effect and as a result the outbreak is still ongoing. Consequently, the spruce forest has been killed in over 6000 ha (Lausch *et al.*, 2011). This exceptional situation of an undisturbed beetle population over a vast territory has allowed in-depth investigations regarding various negative or positive impacts of the bark beetles, as well as extensive modeling.

Some aspects of the environmental impacts of *I. typographus* have been measured in the Bavarian Forest National Park. Measurements on a 110 ha water catchment characterized by 81% dead trees at the end of a 1989–1999 observation period showed a steep increase of the runoff/precipitation ratio (0.84 in 1997–1999 vs. 0.64 in 1989–1996), correlated with deforestation. Nitrate concentration in the soil solution peaked (up to 60 mg/l) at 50 and 100 cm depth during the first 4 years of beetle activity, then decreased with the regrowth of the vegetation. Nitrate leaching was important, with peak values temporarily exceeding 50 mg/l in seepage water and 25 mg/l in springs and streams (Zimmermann *et al.*, 2000). Similar observations were made by Huber (2005), who also found spatial heterogeneity in nitrate leaching, which he attributed to different patterns of vegetation regrowth.

Ips typographus can also have positive effects. In the Bavarian Forest National Park, the recolonization dynamics of the 5800 ha of naturally occurring Norway spruce stands killed by *I. typographus* from 1988 to 2010 (Lausch *et al.*, 2013) was studied by Lehnert *et al.* (2013) and Müller *et al.* (2008, 2010), who concluded that *I. typographus* is a "keystone species" for the maintenance or improvement of forest biodiversity, because its activities open the stands, and the deadwood it creates favors endangered saproxylic beetles. In Switzerland, salvage logging resulted in considerable amounts of deadwood, providing a key resource for biodiversity (Priewasser *et al.*, 2013). In Sweden, Schroeder (2007) followed six reserves hit by a storm in 1995 and found that 81% of the snags remaining in 2006 were from bark beetle-killed trees (19% were felled by the storm) and argued that preserving bark beetle-killed trees would be a cost-effective means to increase the amounts of coarse woody debris in the forest, hence favoring endangered saproxylic species.

Public perception of *I. typographus* outbreaks has also been analyzed in the Bavarian Forest National Park. As central stakeholders in the national park, tourists are facing scenes of utter devastation, which do not correspond to their expectations. Müller and Job (2009) used structural equation modeling to compare three models explaining their "bark beetle attitude." They found that, globally, tourists have a neutral attitude towards bark beetles but that tourists with a higher familiarity with the park have a more positive attitude towards the park's policy. Müller (2011) analyzed political conflicts going on for 20 years within and around the park regarding bark beetle management. He describes two diverging attitudes, one of them hostile to the present policy, seen as imposed by external forces, the other more favorable.

"Sauber Forstwirtschaft" (clean forestry) has long been a practical or legal rule in European countries, prescribing that felled conifers must be peeled, and that bark beetle-attacked trees must be immediately removed from the stands. When comparing tree mortality during the years following a storm in Sweden, fewer trees were killed by the beetles during the first year in unmanaged stands (windthrows not removed) than in managed stands, probably because the windthrows left in place captured most of the insects the year following the storm (Schroeder and Lindelöw, 2002). This trend reversed in subsequent years, however, and in the 4-year period after the storm, twice as many trees were killed per ha in the unmanaged stand as compared to the managed stands. In a recent study covering 9 years and 487 forest districts in Switzerland, Stadelmann *et al.* (2013a) provide quantitative arguments in favor of salvage-logging, stressing the priority of salvage logging after a storm. Jönsson *et al.* (2012) reached similar conclusions from a modeling approach in Sweden. As discussed above, the situation of national parks is particular, because their priorities focus on biodiversity. In the Šumava National Park (Czech Republic), salvage logging had a

FIGURE 15.3 Stocks of windthrows stored under sprinkling water in the Vosges (France) after the Lothar storm (December 1999). The pictures were taken in June 2002. *Photos: J.-C. Grégoire.*

detrimental effect on forest recovery, compared to leaving the dead trees on site (Jonášová and Prach, 2008).

Salvage logging often requires careful planning, in particular when very large amounts of timber become suddenly available, with markets plummeting and with logging personnel and equipment in very high demand. Often, the vast amounts of timber salvaged after a storm cannot be processed immediately and need to be safely stored. Millions of m^3 were thus kept under water sprinkling after the recent storms in Europe (Björkhem *et al.*, 1977; Jonsson, 2004). For example, in Sweden, the Byholma site sheltered one million m^3 in 2007 (Lindelöw and Schroeder, 2008), and more than 6.5 million m^3 were stored under sprinkling water in France between 1999 and 2001 (Figure 15.3) (Flot and Vautherin, 2002; Moreau *et al.*, 2006). These massive salvage-logging and storage operations require equally massive logistics. A European lorry carries on average 30 m^3 of timber. To fill a one million m^3 storage unit such as the Byholma site mentioned above, more than 33,300 such lorry loads are necessary. Supposing that 50 lorries could be operated daily, about 670 days (almost 2 years) of uninterrupted work (logging and transportation) are necessary, with the consequence that timely removal of vulnerable material from the stands is not always feasible. In addition, the water storage of conifer logs raises environmental problems, as phenols and diterpene resin acids leak into the soil or the aquatic ecosystems (Jonsson, 2004; Hedmark *et al.*, 2009).

Spatially explicit damage assessment is an extremely important issue regarding *I. typographus*, since salvage logging is the preferred option to prevent further damage. Pest monitoring is intensively carried out in many countries. For example, in Switzerland (WSL—Forest Protection Overviews, 2014) and Austria (Bundesforschungszentrum für Wald, 2014), yearly damage reports fed by a network of local observers are available online. In France, a similar database is kept centrally (Département de la Santé des Forêts, 2014). The same situation exists in the Belgian Walloon region (Observatoire wallon de la Santé des Forêts, 2014). One difficulty in these assessments is that they rely upon forest inventories, which even in the best cases are not totally accurate because some forest officers tend to overestimate or underestimate damage (Franklin *et al.*, 2004).

Based on the issues discussed above, risk planning is an extremely important component in the politics of *I. typographus* management. Enormous progress has been made recently in risk modeling (Table 15.2), for immediate and local use, as well as on a more prospective level, for long-term planning in view of climate change. Among the drivers that are recurrently identified in these models are the previous year's volume of windthrows and volume of attacked timber, as well as the local volume of standing trees.

A second "political" element of long-term planning could concern, whenever possible, the choice of the species selected for reforestation. In general terms, forest tree diversity reduces herbivory in oligophagous animal species (Jactel and Brockerhoff, 2007). More specifically, Warzée *et al.* (2006), analyzing the relationships between *I. typographus* and the predatory clerid beetle, *Thanasimus formicarius* L. in northeastern France, caught a much higher predator/prey ratio (many more predators and less prey) in spruce stands mixed with pines than in pure spruce stands. They attributed this difference to the higher reproductive success of *T. formicarius* when it can pupate in the thicker bark of pines.

A third aspect that relates to politics is the quarantine dimension, i.e., the set of rules and practices designed to prevent the pest from entering new areas. Within Europe, *I. typographus* outbreaks seem to happen only in areas where the insects have been long established. Recolonizing Eurasia after the glaciations, Norway spruce (*Picea abies* (L.)) has spread naturally only in higher elevations and latitudes (Taberlet *et al.*, 1998; EUFORGEN, 2009). During the last 150 years, however, it has been widely planted outside of this limited range. At the same time, *Picea sitchensis* (Bong.) Carrière was also introduced in Europe from northwestern America. *Ips typographus* has followed its ancient and new hosts into these new territories, but with a time lag. In Belgium, for example, *P. abies* plantations started around 1885 (Scheepers *et al.*, 1997), and *I. typographus* colonized the country quite slowly afterwards. In the early 1970s, it was largely established in the country (Dourojeanni, 1971), but at densities too low for causing outbreaks. The first outbreaks only appeared in 1976 (J.C. Grégoire, pers. observ.), during an exceptionally hot, dry summer (IRM, 2014). Some northwestern, lower elevation parts of France (Normandy, Brittany), also recently planted, are still under colonization. The beetles are present and occasionally colonize windthrows (Gilbert *et al.*, 2005), but never reach outbreak level, suggesting that an Allee threshold has not yet been reached (Liebhold and Tobin, 2008), in spite of the heavy commercial movements of spruce roundwood, sometimes infested, within Europe or from outside the European Union (Piel *et al.*, 2006, 2008). This is probably also an important reason why it has never established in the USA or New Zealand, where it is listed as a quarantine pest, although it is regularly intercepted. *Ips typographus* was found 286 times at US ports between 1985 and 2000 (Haack, 2001), and constituted 6% of all bark beetles intercepted in New Zealand between 1950 and 2000 (Brockerhoff *et al.*, 2006). In the European Union, special provisions in the phytosanitary rules (Commission Directive, 2008) grant to Great Britain and Ireland the status of "Protected Zones" that allow these countries, free so far from *I. typographus*, to restrict intra-European Union commercial movements of coniferous logs and timber.

4.2 A Deadly Mistake, but for Which Party?—Secondary Ambrosia Beetles Attacking Living Beech in Europe

The ambrosia beetles *Trypodendron domesticum* (L.) and *Trypodendron signatum* (F.) are known as secondary species, attacking dying or dead broadleaved trees to which they are attracted by volatiles (e.g., ethanol) produced in the trees' fermenting tissues (Kerck, 1972; Holighaus and Schütz, 2006). In the early 2000s, however, these species attacked standing beech trees (*Fagus sylvatica* L.) in Belgium, Germany, France, and Luxemburg, affecting more than 1.8 million m³ (Eisenbarth *et al.*, 2001; Huart *et al.*, 2003; Arend *et al.*, 2006). The first symptoms, observed in 1999, were not very surprising, as they were concentrated around necrotic areas dating from the winter 1998–1999 and probably related to frost damage (Huart *et al.*, 2003). In the following years, however, these insects attacked areas on apparently healthy trees. Three

Ophiostoma species were found in the galleries: *O. quercus* (Georgev.) Nannf., *O. bacillisporum* (Butin and G. Zimm.) de Hoog and R. J. Scheff., and a new species, *O. arduennense* F.-X. Carlier, Decock, K. Jacobs and Maraite (Carlier *et al.*, 2006). Several secondary fungi rapidly colonized the stems of some of the attacked trees. These included *Fomes fomentarius* (L.) Fr., *Fomitopsis pinicola* (Sw.) P. Karst., *Stereum hirsutum* (Willd.) Pers., and *Trametes versicolor* (L.) Lloyd) (La Spina *et al.*, 2013). The breaking of whole trees or large branches raised serious safety issues for foresters, forest workers, hikers, and hunters, but the main consequence of this outbreak was economic. Valuable bolts that had been reserved for slicing and that were to be exported to China were embargoed because of the staining of the wood. It was even difficult to sell the wood for pulp, because of the rapid development of lignivorous fungi. As the timber market had decreased after the Lothar storm in December 1999, many owners had preferred to keep their stock standing, postponing any sale in hopes of an improved market, and found their assets seriously diminished. After 2002, the outbreak subsided and many trees recovered, sealing the often-aborted galleries under new wood. However, the remaining stands are still marked in the memory of sawyers, and very low prices are offered for the local timber. The losses were high for the forest market, as the remaining beetles attacked the standing trees, perished in their attempt to colonize these trees, but left stains in the wood.

Similar, smaller outbreaks were observed in the past in Belgium (1929 and 1942) and in neighboring countries (Zycha, 1943; Prieels, 1961; Poncelet, 1965; Nageleisen, 1993), and other secondary ambrosia bark beetles have been observed to attack living trees in other parts of the world (Kühnholz *et al.*, 2001; Coyle *et al.*, 2005). The causes of this phenomenon are still unknown. One hypothesis regarding the Belgian outbreak is that the early frost in the winter of 1998–1999 affected trees that were still physiologically unprepared. The trees were affected in two ways: direct frost necroses (explaining the first insect attacks in 1999) and longer-term damage (explaining the subsequent attacks on apparently healthy trees). La Spina *et al.* (2013) explored this hypothesis further by analyzing the Belgian weather records, and by direct experiments where they inflicted frost wounds to mature trees using dry ice. The meteorological records showed "very exceptional" (one occurrence in 56 years) cold waves 1 year before the outbreaks in 1929 and 1942, and an "exceptional" (one occurrence in 30 years) one in 1998. The field experiments carried out by La Spina *et al.* (2013) resulted in beetle attacks but, contrary to the field observations in 1999–2002, the galleries were limited to a heavily necrotic zone in the sapwood.

Other hypotheses could be developed, such as adverse soil conditions (drought, waterlogging) and/or sublethal attacks of pathogenic fungi could have weakened the trees and made them attractive to the insects. Ranger *et al.* (2010, 2013) induced ethanol production in potted trees submitted to flood-stress, which made them attractive to *Xylosandrus germanus* (Blandford). McPherson *et al.* (2001, 2008) observed ambrosia beetles attracted to and colonizing oaks infected by *Phytophthora ramorum* Werres, De Cock and Man in't Veld, and Kelsey *et al.* (2013) showed that the infected trees produced ethanol that was attractive to ambrosia beetles. In Europe, Jung (2009) linked beech decline to infection by *Phytophthora* spp. In a survey of 49 sites in southern Belgium, Schmitz *et al.* (2009) found *P. cambivora* (Petri) Buisman and *P. gonapodyides* (H. E. Petersen) Buisman infecting living trees in 19 sites. Waterlogging and fungal infection could act jointly, as flood-stress on the one hand reduces tree resistance and induces ethanol production, and on the other hand favors the production and propagation of the *Phytophthora* zoospores.

4.3 Fallen and Standing Alike— The Spruce Beetle in North America

The spruce beetle *D. rufipennis* is widely distributed across North America (Chapter 8). It extends from central Alaska to Newfoundland and down the Rocky Mountains almost to Mexico (Wood, 1982). This beetle breeds in all *Picea* species within its range, although black spruce, *P. mariana* (Mill.) Britton, Sterns and Poggenburg, is rarely attacked, and susceptibility and suitability vary among other species (Werner *et al.*, 2006a). The spruce beetle is associated with several species of fungi, most commonly *Leptographium abietinum* (Peck) M. J. Wingf. (Six and Bentz, 2003). This insect shows markedly different population dynamics in different regions, and hence poses very different levels and types of management concerns. Appropriate management tactics, and accompanying policy issues, vary accordingly. This diversity of impacts and behaviors also makes spruce beetle a useful model for understanding management options for bark beetles in general.

Throughout much of its range, *D. rufipennis* is a truly eruptive species, capable of undergoing intermittent large-scale outbreaks. Landscape-scale outbreaks occur throughout coastal Alaska, British Columbia, and the northwestern USA and Rocky Mountains (Safranyik, 1988; Eisenhart and Veblen, 2000; Werner *et al.*, 2006a). A large outbreak in central British Columbia resulted in mortality over 175,000 ha (Cozens, 1997). From 1920 to 1989, 847,000 ha of spruce forest in Alaska were impacted (Holsten, 1990). An outbreak from 1989 to 2004 in Alaska resulted in 1.2 million ha of affected spruce forests, with an estimated 30 million trees killed per year. More than 90% of trees >11 cm were killed in some stands (Werner *et al.*, 2006a). These outbreaks completely transform the structures and compositions of forests, converting them from predominantly spruce to either angiosperms, such as birch and aspen, conifers such as pine, hemlock or fir, and sometimes grasses, depending on location and the heterogeneity of the forest (Lewis and Lindgren, 2000; Boucher and Mead 2006; Werner *et al.*, 2006a). Impacts include economic losses, wildlife habitat effects, hydrological changes, aesthetic value loss, and increased risk to humans due to danger trees and possibly catastrophic fire (Werner *et al.*, 2006a). Outbreaks are commonly followed by increased populations of secondary bark beetles such as *Ips* spp., which can be problematic in residential areas. Spruce trees that are too small to support beetle development survive these outbreaks, forming the basis for an eventual return to spruce.

Like other eruptive bark beetles, *D. rufipennis* outbreaks are relatively uncommon, and there are long intervening periods during which populations are low. During these endemic periods, populations are held within a relatively stable range by a combination of tree defense, resource availability, interspecific competition, predators, weather, and their interactions (Chapter 1). The release of *D. rufipennis* populations from endemic to eruptive dynamics is often associated with scattered windthrow events (Safranyik, 1985), in combination with stresses on host tree defense, such as drought (Hart *et al.*, 2014), and abnormally high temperatures that increase the beetles' overwintering survival and accelerate development (Werner *et al.*, 1977; Werner and Holsten, 1985; Hansen and Bentz, 2003). This insect has a facultative diapause, and in regions where semivoltinism is common, responds to warm conditions by shifting to a univoltine life cycle, greatly increasing the likelihood of outbreaks (Hansen *et al.*, 2001b). At high densities, however, populations of this insect become self-amplifying, as they successfully enter and overwhelm vigorous trees regardless of their defensive capabilities (Lewis and Lindgren, 2002; Wallin and Raffa, 2004; Raffa *et al.*, 2008).

Throughout much of its range, the spruce beetle never or rarely undergoes major outbreaks. The reasons vary with region, but further demonstrate how a combination of factors is required for an outbreak to occur (Raffa *et al.*, 2008). In interior Alaska, populations are typically univoltine and the habitat consists of extensive tracts of *Picea*, two ingredients that foster outbreaks. However, spruce beetles there are almost entirely limited to windthrown or otherwise stressed trees. The reasons are not entirely clear, but interspecific competition appears much higher in interior than coastal Alaska, probably arising from the drier conditions, and hence drier phloem, that favor *Ips* spp. (Werner *et al.*, 2006b). Likewise, throughout the Great Lakes region, *D. rufipennis* is almost entirely limited to highly stressed trees. Although populations there are univoltine, the forests are much more diverse than in the west, which can dampen population responses to environmental perturbations. Further, interspecific competition, especially from *Dryocoetes* spp., is much more intense in the Midwest than west (Haberkern and Raffa, 2003; Raffa *et al.*, in press). For example, a huge blowdown event in the Boundary Waters region of northern Minnesota raised concerns about subsequent spruce beetle outbreaks. Instead, a highly diverse subcortical community emerged (Gandhi *et al.*, 2009) and large-scale outbreaks did not follow. Further, ratios of predator to tree-killing bark beetle populations appear to be relatively higher in midwestern than western forests (Raffa *et al.*, in press), and predacious beetles in this region are strongly attracted to frontalin (Haberkern and Raffa, 2003), a component of the spruce beetle's pheromone plume (Dyer, 1973). In this region, therefore, protection from spruce beetles need not be as proactive as in western North America, and remedial responses can be initiated after events such as spruce budworm, *Choristoneura fumiferana* (Clemens), outbreaks. In the eastern provinces of Canada and New England, spruce beetle appears more aggressive than in the Midwest, but less than in the west. Small, localized outbreaks may follow drought or outbreak by spruce budworm. However, these outbreaks do not appear to become self-sustaining on a landscape scale. The optimal strategy, then, is again one of careful monitoring, followed by sanitation where potential losses appear imminent, coupled with landscape-scale management of *C. fumiferana*.

4.4 A Political, Economic and Ecological Challenge—The Mountain Pine Beetle in British Columbia

The mountain pine beetle *D. ponderosae* is without a doubt the most destructive bark beetle in North America (Safranyik and Carroll, 2006), and possibly the world. Large-scale eruptions have occurred with semi-regular frequency in western North America, averaging about 40 years in British Columbia (Alfaro *et al.*, 2010). During the past decades, several outbreaks have occurred that have been characterized by increasing intensity and scale, with the most recent surpassing all others by a large margin (Westfall and Ebata, 2014; Petersen and Stuart, 2014). The result has been substantial ecological and socioeconomic impacts, even including a notable occurrence of allergies to air-borne mountain pine beetle allergens (Stark and Li, 2009), and an international court challenge to Canadian lumber pricing practices under the US-Canada Softwood Lumber Agreement (Woo, 2012; Petersen and Stuart, 2014). The impacts have been particularly notable in British Columbia, where the main host tree, lodgepole pine (*Pinus contorta* var. *latifolia* Engelm.), is one of the primary commercial conifer species. As of 2012, well over

one billion US dollars had been invested by provincial and federal governments to mitigate the impact of the outbreak, and about 700 million m³ of lodgepole pine had been killed (Ministry of Forests, Lands and Natural Resources Operations, 2012).

Lodgepole pine is a highly adaptable species with a wide distribution in western North America (Forrest, 1980). It is a fast-growing seral conifer that occupies vast areas of the Central Plateau of British Columbia as a leading species. In areas with frequent fires, this species often persists as a climax species in even-aged monocultures because a high proportion of the population has serotinous cones that persist on trees for many years, only opening and releasing seed after exposure to heat (Lotan and Critchfield, 1990). Homogeneous, mature lodgepole pine stands in Tweedsmuir and Entiako Provincial Parks (British Columbia), and the lack of management in these areas, are frequently pointed to as the initial cause of the current outbreak (Gawalko, 2004), but Aukema *et al.* (2006) showed that outbreaks also started concurrently in many parts of British Columbia. In the absence of fire, lodgepole pine is susceptible to mountain pine beetle outbreaks, leading to complex, multi-layered stand structures (Axelson *et al.*, 2009). In mixed species stands, lodgepole pine is eventually displaced by long-lived species like *Picea*, *Pseudotsuga*, and *Abies* spp., in part due to mountain pine beetle-caused mortality of pines older than 80 years. Taylor *et al.* (2006), using projections from 1990 inventory data, showed that the average age of lodgepole pine stands changed from 51 years in 1910 to 114 years in 2010, and the age distribution from 17% mountain-pine beetle susceptible trees to 56% (Figure 15.4). This change was due to increasingly effective fire protection and relatively low harvesting rates until the 1960s. The increase in availability of

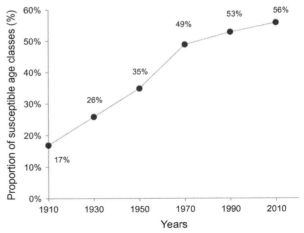

FIGURE 15.4 Change over time of the proportion of susceptible age classes (90 to 150 cm DBH), following increasingly successful fire control. *Redrawn from Taylor et al. (2006).*

susceptible hosts played a major role in driving the development of the current outbreak.

Climate plays an important role in the population dynamics of mountain pine beetle (Chapter 13). Cold fall and winter temperatures, rather than host availability, have limited the latitudinal and elevational range of mountain pine beetle (Carroll et al., 2004). Significant increases in mean temperatures over the past few decades have reduced the occurrence of population-limiting cold events. Petersen and Stuart (2014) cite data indicating a 1.5°C increase of mean annual temperature from the mid-20th century and 10.4°C higher spring minimum temperatures from 1943 to 2008. Consequently, previously climatically unsuitable habitat for mountain pine beetle has now become suitable, generating conditions conducive to range expansion (Carroll et al., 2006; Cudmore et al., 2010; Safranyik et al., 2010; Cullingham et al., 2011), threatening sensitive whitebark pine (Pinus albicaulis Engelm.) ecosystems (Logan et al., 2010; Raffa et al., 2013), and impacting management (Konkin and Hopkins, 2009; Petersen and Stuart, 2014).

After a large outbreak in the late 1980s, general strategies and tactics for bark beetle management (https://www.for.gov.bc.ca/tasb/legsregs/fpc/fpcguide/beetle/chap2.htm) were implemented under the Forest Practices Code of British Columbia Act. As of January 1, 2004, this prescriptive piece of legislation was replaced by the results-based Forest and Range Practices Act. Further legislative changes at the time of the initial population buildup of the current outbreak had a potentially negative impact on the ability of the British Columbia government to manage the most recent outbreak (Petersen and Stuart, 2014). Government oversight and staffing were reduced both in response to legislative changes and budget cuts, and this included the closure of many Regional Forest Service Offices in a number of small, resource-dependent communities, causing significant socioeconomic impact, which was further exacerbated in areas affected by the outbreak (Parfitt, 2010; Petersen and Stuart, 2014). The Federal Government of Canada provided funding for research and limited management activities through the Mountain Pine Beetle Initiative (Ministry of Forests, Lands and Natural Resources Operations, 2012; Petersen and Stuart, 2014). However, it is unlikely that the outbreak would have been stopped in a more favorable economic climate.

Due to the uplift in annual allowable cut (AAC) and a focus on salvage of low-value beetle-killed pine (Burton, 2006; Petersen and Stuart, 2014) there has been a need to find alternate uses for the harvested pine. Dead pine may remain standing for many years, but loses value due to checking, staining, and decay (Lewis and Hartley, 2006; Lewis and Thompson, 2011). Consequently, other markets have been sought, e.g., utilization of dead wood for bioenergy (Kumar et al., 2008; Mahmoudi et al., 2009) or for innovative products such as "Denim Pine" (Byrne et al., 2006) and "Beetlecrete" (Hopper, 2010). Further economic impact has been on recreation (McFarlane and Watson, 2008). There has also been concern about the potential impact on carbon sequestration (Kurz et al., 2008), although some studies in areas with advanced regeneration have indicated that they serve as carbon sinks due to increased sequestration by remaining live vegetation (Brown et al., 2010; Hansen, 2014). A pervasive paradigm has been that bark beetle outbreaks result in an increased risk of fire. However, recent studies indicated that climate is the primary determinant factor of wildfire frequency and intensity (Kulakowski and Jarvis, 2011; Simard et al., 2011). In a review, Page et al. (2014) questioned the validity of conclusions based on modeling, however.

Due to the magnitude of the outbreak, there has been considerable concern over impacts on wildlife (Chan-McLeod, 2006; Ritchie, 2008; Saab et al., 2014). Bark insectivores, and particularly cavity nesters, appear to benefit initially (Martin et al., 2006; Norris and Martin, 2008; Drever et al., 2009; Saab et al., 2014), whereas populations of some other guilds, e.g., flickers (Colaptes auratus (L)) and red-naped sapsuckers (Sphyrapicus nuchalis Baird), decreased (Martin et al., 2006). Impacts on mammals varied (Saab et al., 2014). Direct impact on American marten (Martes americana (Turton)) populations may depend on management scenario (Steventon and Daust, 2009), although loss of primary prey species like American red squirrel (Tamiasciurus hudsonicus (Erxleben)) had a ripple effect, and fragmentation was negative for both marten and fisher (Martes pennanti (Erxleben)) populations due to their poor dispersal ability (Chan-McLeod, 2006). Ungulates are affected in different ways depending on habitat requirements (Chan-McLeod, 2006), with loss of cover being a primary driver. There is particular concern for impacts on caribou (Rangifer tarandus caribou (Gmelin)) with potential loss of habitat (Cichowski and Williston, 2005; McNay et al., 2008; Ritchie, 2008), including loss of terrestrial lichens and changes in snow accumulation patterns. Pacific salmon are indirectly impacted by the outbreak due to increasing water temperatures and altered hydrological cycles (Pacific Fisheries Resource Conservation Council, n.d.; Bewley et al., 2010).

4.5 A Chronic Presence—The Pine Engraver Across the Continent

The pine engraver Ips pini (Say) has a transcontinental distribution across North America, and can utilize almost all Pinus and sometimes other genera within its range (Wood, 1982). With such a broad geographic and host species range, I. pini exemplifies the adaptive plasticity that bark beetles can show in their ecology, behavior, and physiology with local biotic and abiotic conditions. Because of this plasticity, the socioeconomic impacts of this insect and optimal management approaches vary widely.

In the western United States and Canada, *I. pini* coincides with several outbreak pine-killing species, such as *D. ponderosae* and *D. brevicomis* LeConte. In these regions, *I. pini* is largely a secondary insect, orienting to plant and insect volatiles emitting from trees attacked by the more aggressive species, or colonizing severely stressed trees either alone or in a scramble competition (Rankin and Borden, 1991; Safranyik *et al.*, 1996). Throughout much of the west, *I. pini* is at least partially beneficial to humans, because it reduces reproductive success of primary bark beetles. For example, when *D. ponderosae* colonize fire-injured trees, competition with *I. pini* is one of the factors that limit its population increase (Powell *et al.*, 2012). However, this competitive effect can be reduced somewhat, by vertical partitioning of the resource, whereby *I. pini* is often concentrated in the upper stems. Under some conditions, *I. pini* can be a pest in western forests, particularly during drought years or in highly dense stands (Kegley *et al.*, 1997). During a chronic outbreak in Montana from 1974 to 1994 in ponderosa pine (Gara *et al.*, 1999), slash management to promote rapid drying of host material was shown to be important. For example, how slash is distributed and treated, and how the equipment is used affects colonization rate, rate of drying, and prevalence of natural enemies (Six *et al.*, 2002). The timing of thinning operations can be optimized to minimize population buildup (Gara *et al.*, 1999). Finally, providing a "green chain," i.e., providing a continuous supply of fresh slash during beetle flight, was recommended to prevent spillover attacks into live trees (Kegley *et al.*, 1997)

In the midwestern and northeastern portions of North America, there are no landscape-scale aggressive bark beetles that attack pine. In these regions, *I. pini* fills the niche of a primary, tree-killing species. However, the live trees this beetle selects almost always show at least moderate acute or chronic stress prior to attack. In the Great Lakes region, stress caused by belowground herbivory and accompanying root infection provide a continuous but limited source of susceptible trees (Klepzig *et al.*, 1991). In plantations having high populations of these predisposing agents, *I. pini* can be problematic and sometimes requires direct control by sanitation or pheromone-based mass trapping. However, unlike aggressive species, *I. pini* populations do not become self-sustaining and encompass entire landscapes after an initial population increase. For example, during drought years, both the numbers of *I. pini* and the proportion of trees it kills that did not have prior root infection increase markedly. Unlike species such as *D. ponderosae* and *D. rufipennis*, after the drought subsides, *I. pini* populations again become restricted to trees with previously colonized roots or lower stems, and populations decline (Aukema *et al.*, 2010). Reliance on such

a predictably and spatially concentrated resource as root-infested trees appears to facilitate predator impacts (Erbilgin *et al.*, 2002), but the spatial separation of plantations can inhibit predator dispersal to new infestations (Ryall and Fahrig, 2005).

Ips pini also shows high plasticity in its pheromone chemistry. All *I. pini* produce ipsdienol, but local populations vary in the stereochemistry of their signals (Lanier *et al.*, 1972, 1975; Miller *et al.*, 1989). Most western populations produce almost entirely (−)-ipsdienol. In contrast, midwestern and eastern populations produce blends that are either racemic or biased toward (+)-ipsdienol. Some areas of western Canada produce substantial amounts of (+)-ipsdienol. In addition to enantiomeric differences, midwestern and eastern populations produce lanierone, which is not attractive by itself, but greatly increases attraction to ipsdienol (Teale and Lanier, 1991; Miller *et al.*, 1997). Western populations do not produce lanierone, although it is weakly to strongly synergistic to (−)-ipsdienol in Arizona and Montana, and British Columbia, Canada (Miller *et al.*, 1997; Steed and Wagner, 2008). Furthermore, populations in both Arizona and Montana had seasonal shifts in preference (Steed and Wagner, 2008). These patterns appear to arise from local selective pressures, specifically avoidance of interspecific competition with sympatric *Ips* (Birch *et al.*, 1980; Borden *et al.*, 1992), and escape from predators that exploit beetle pheromones in prey finding (Raffa and Dahlsten, 1995). That is, pheromone blends produced by *I. pini* differ from those of sympatric congenerics, which in the west include other species producing ipsdienol but in the midwest produce ipsenol. Likewise, local predators show mismatches from their prey in preferences for stereochemistry and derived components, suggesting time-lagged coevolution (Raffa *et al.*, 2007). Regardless of their evolutionary origins, these variable mixtures, and the distinctions between local predator and prey preferences, provide opportunities to greatly improve both the efficacy and selectivity of pheromonally-based population monitoring and control methods (Dahlsten *et al.*, 2003), but at the same time cause commercial challenges due to the need for locally specific blends, which increase cost of management.

In addition to the above distinctions between eastern and western populations, *I. pini* also shows plasticity in its life history with latitude and elevation. It has an apparently facultative diapause, showing variation in cold tolerance and voltinism, both between regions and between years within regions (Lombardero *et al.*, 2000). The number of generations per year can range from one to five depending on regional temperatures. This insect also shows plasticity in its overwintering behavior. In northern regions of the Midwest, it overwinters as adults in the soil. In other regions, it overwinters both during various life stages under the bark and as adults in soil.

From a management perspective, knowledge of the population dynamics of *I. pini*, i.e., its responding to a resource pulse but not becoming self-driving, can be used to guide control strategies. Specifically, losses to this insect can be reduced by controlling the predisposing agents, by controlling *I. pini* directly, or both. This can involve tactics such as seasonally timing thinning operations to avoid infestation by lower-stem colonizing beetles such as *D. valens*, removing slash, sanitation clearing, or localized application of semiochemicals (Kegley *et al.*, 1997). This contrasts with the eruptive bark beetle species, which can only be successfully managed by preventing initial population increases beyond a critical threshold. Thus, from a policy standpoint, *I. pini* does not necessitate coordinated actions on a large scale, but instead can be effectively addressed as needed by local private or government land managers.

5. CONCLUSIONS

Bark beetles are important forest disturbance agents, reshaping whole landscapes and exerting a large variety of economic, environmental, and social impacts. Some of these impacts incur very high socioeconomic costs, while others exert positive influences on species richness and biodiversity. Although a substantial amount of information is available, and much practical and political knowledge has been developed, we are very far from mastering the "bark beetle ecosystem." In the best of cases, we can to some extent anticipate or mitigate bark beetle impact where such actions are consistent with management objectives. Climate change and biological invasions are important threats against which satisfactory solutions, if any, remain to be found.

REFERENCES

Abbott, B., Stennes, B., van Kooten, C.G., 2008. An Economic Analysis of Mountain Pine Beetle Impacts in a Global Context. Resource Economics and Policy Analysis (REPA) Research Group, Department of Economics, University of Victoria, Victoria, BC, Canada.

Abbott, B., Stennes, B., van Kooten, G.C., 2009. Mountain pine beetle, global markets, and the British Columbia forest economy. Can. J. For. Res. 39, 1313–1321.

Abgrall, J.-F., 2000. La tornade de décembre 1999. Risques sanitaires et stratégies de gestion. Forêt et tempête. Expertise collective sur les tempêtes, la sensibilité des forêts et sur leur reconstitution. Dossier de l'environnement de l'INRA n°20, INRA-ME&S. Available online: http://www7.inra.fr/dpenv/do20.htm (Last accessed 06.06.14.).

Alfaro, R.I., Campbell, E., Hawkes, B.C., 2010. Historical Frequency, Intensity and Extent of Mountain Pine Beetle Disturbance in British Columbia. Natural Resources Canada, Canadian Forest Service, Pacific Forestry Centre, Victoria, BC, Mountain Pine Beetle Working Paper 2009–30.

Arend, J.-P., Eisenbarth, E., Petercord, R., 2006. Buchenkomplexkrankheit in Luxemburg und Rheinland-Pfalz—Schadsymptome, Ausmass und Entwicklung der Schäden. Mitteilungen aus der Forschungsanstalt für Waldökologie und Forstwirtschaft Rheinland- Pfalz 59, 11–22.

Aukema, B.H., Carroll, A.L., Zhu, J., Raffa, K.F., Sickley, T.A., Taylor, S.W., 2006. Landscape level analysis of mountain pine beetle in British Columbia, Canada: spatiotemporal development and spatial synchrony within the present outbreak. Ecography 29, 427–441.

Aukema, B.H., Zhu, J., Moeller, J., Rasmussen, J., Raffa, K.F., 2010. Interactions between below- and above-ground herbivores drive a forest decline and gap-forming syndrome. For. Ecol. Manage. 259, 374–382.

Aukema, J.E., Leung, B., Kovacs, K., Chivers, C., Britton, K.O., Englin, J., et al., 2011. Economic impacts of non-native forest insects in the continental United States. PLoS One 6, e24587.

Axelson, J.N., Alfaro, R.I., Hawkes, B.C., 2009. Influence of fire and mountain pine beetle on the dynamics of lodgepole pine stands in British Columbia, Canada. For. Ecol. Manage. 257, 1874–1882.

Baier, P., Pennerstorfer, J., Schopf, A., 2007. PHENIPS—a comprehensive phenology model of *Ips typographus* (L.) (Col., Scolytinae) as a tool for hazard rating of bark beetle infestation. For. Ecol. Manage. 249, 171–186.

Bebi, P., Teich, M., Hagedorn, F., Zurbriggen, N., Brunner, S.H., Grêt-Regamey, A., 2012. Veränderung von Wald und Waldleistungen in der Landschaft Davos im Zuge des Klimawandels. Schweiz. Z. Forstwes. 163, 493–501.

Bentz, B.J., Régnière, J., Fettig, C.J., Hansen, E.M., Hayes, J.L., Hicke, J.A., et al., 2010. Climate change and bark beetles of the western United States and Canada: direct and indirect effects. BioScience 60, 602–613.

Bertheau, C., Brockerhoff, E.G., Roux-Morabito, G., Lieutier, F., Jactel, H., 2010. Novel insect-tree associations resulting from accidental and intentional biological "invasions": a meta-analysis of effects on insect fitness. Ecol. Lett. 13, 506–515.

Bevan, D., King, C.J., 1983. *Dendroctonus micans* Kug., a new pest of spruce in U.K. Emp. For. Rev. 62, 41–51.

Bewley, D., Alila, Y., Varhola, A., 2010. Variability of snow water equivalent and snow energetics across a large catchment subject to mountain pine beetle infestation and rapid salvage logging. J. Hydrol. 388, 464–479.

Birch, M.C., Light, D.M., Wood, D.L., Browne, L.E., Silverstein, R.M., Bergot, B.J., et al., 1980. Pheromonal attraction and allomonal interruption of *Ips pini* in California by the two enantiomers of ipsdienol. J. Chem. Ecol. 6, 703–717.

Björkhem, U., Dehlén, R., Lundin, L., Nilsson, S., Olsson, M.T., Regnander, J., 1977. Storage of Pulpwood Under Water Sprinkling—Effects on Insects and the Surrounding Area. Royal College of Forestry, Department of Operational Efficiency, Research Notes 107.

Black, S.H., 2005. Logging to Control Insects: The Science and Myths behind Managing Forest Insect "pests". A Synthesis of Independently Reviewed Research. The Xerces Society for Invertebrate Conservation, Portland.

Bogle, T., van Kooten, G.C., 2012. Why mountain pine beetle exacerbates a principal–agent relationship: exploring strategic policy responses to beetle attack in a mixed species forest. Can. J. For. Res. 42, 621–630.

Boone, C.K., Aukema, B.H., Bohlmann, J., Carroll, A.L., Raffa, K.F., 2011. Efficacy of tree defense physiology varies with bark beetle population density: a basis for positive feedback in eruptive species. Can. J. For. Res. 41, 1174–1188.

Borden, J.H., Devlin, D.R., Miller, D.R., 1992. Synomones of two sympatric species deter attack by the pine engraver, *Ips pini* (Coleoptera: Scolytidae). Can. J. For. Res. 22, 381–387.

Bosch, J.M., Hewlett, J.D., 1982. A review of catchment experiments to determine the effect of vegetation changes on water yield and evapotranspiration. J. Hydrol. 55, 3–23.

Boucher, T.V., Mead, B.R., 2006. Vegetation change and forest regeneration on the Kenai Peninsula, Alaska following a spruce beetle outbreak, 1987–2000. For. Ecol. Manage. 227, 233–246.

Brockerhoff, E.G., Bain, J., Kimberley, M., Knizek, M., 2006. Interception frequency of exotic bark and ambrosia beetles (Coleoptera: Scolytinae) and relationship with establishment in New Zealand and worldwide. Can. J. For. Res. 36, 289–298.

Brown, A.E., Zhang, L., McMahon, T.A., Western, A.W., Vertessy, R.A., 2005. A review of paired catchment studies for determining changes in water yield resulting from alterations in vegetation. J. Hydrol. 310, 28–61.

Brown, M., Black, T.A., Nesic, Z., Foord, V.N., Spittlehouse, D.L., Fredeen, A.L., et al., 2010. Impact of mountain pine beetle on the net ecosystem production of lodgepole pine stands in British Columbia. Agric. For. Meteorol. 150, 254–264.

Brunette, M., Couture, S., 2008. Public compensation for windstorm damage reduces incentives for risk management investments. Forest Policy and Economics 10, 491–499.

Bundesforschungszentrum für Wald, 2014. Available online: http://bfw.ac. at/rz/bfwcms.web?dok=9605 (Last accessed 17.06.14.).

Burton, P.J., 2006. Restoration of forests attacked by mountain pine beetle: misnomer, misdirected, or must-do? BC J. Ecosyst. Manag. 7, 1–10.

Byrne, T., Stonestreet, C., Peter, B., 2006. Characteristics and utilization of post mountain pine beetle wood in solid wood products. In: Safranyik, L., Wilson, B. (Eds.), The Mountain Pine Beetle: A Synthesis of Biology, Management, and Impacts on Lodgepole Pine. Natural Resources Canada, Canadian Forest Service, Pacific Forestry Centre, Victoria, B.C., Canada, pp. 236–254.

Carlier, F.-X., Decock, C., Jacobs, K., Maraite, H., 2006. *Ophiostoma arduennense* sp. nov. (Ophiostomatales, Ascomycota) from *Fagus sylvatica* in southern Belgium. Mycol. Res. 110, 801–810.

Carpenter, J.R., 1940. Insect outbreaks in Europe. J. Anim. Ecol. 9, 108–147.

Carroll, A.L., Taylor, S., Régnière, J., Safranyik, L., 2004. Effects of climate change on range expansion by the mountain pine beetle in British Columbia. In: Shore, T.L., Brooks, J.E., Stone, J.E. (Eds.), Mountain Pine Beetle Symposium: Challenges and Solutions. Natural Resources Canada, Canadian Forest Service, Pacific Forestry Centre, Victoria, B.C., Canada, pp. 223–232.

Carroll, A.L., Shore, T.L., Safranyik, L., 2006. Direct control: theory and practice. In: Safranyik, L., Wilson, B. (Eds.), The Mountain Pine Beetle: A Synthesis of Biology, Management, and Impacts on Lodgepole Pine. Natural Resources Canada, Canadian Forest Service, Pacific Forestry Centre, Victoria, BC, Canada, pp. 155–172.

Chan-McLeod, A.C.A., 2006. A review and synthesis of the effects of unsalvaged mountain-pine-beetle-attacked stands on wildlife and implications for forest management. BC J. Ecosyst. Manag. 7, 119–132.

Chow, S., Obermajer, A., 2007. Moisture and blue stain distribution in mountain pine beetle infested lodgepole pine trees and industrial implications. Wood Sci. Technol. 41, 3–16.

Christiansen, E., Bakke, A., 1988. The spruce bark beetle of Eurasia. In: Berryman, A.A. (Ed.), Dynamics of Forest Insect Populations: Patterns, Causes, Implications. Plenum Press, New York, pp. 479–503.

Cichowski, D., Williston, P., 2005. Mountain pine beetle and emerging issues in the management of woodland caribou in westcentral British Columbia. Rangifer 16, 97–103.

Coggins, S.B., Coops, N.C., Wulder, M.A., Bater, C.W., Ortlepp, S.M., 2011. Comparing the impacts of mitigation and non-mitigation on mountain pine beetle populations. J. Environ. Manage. 92, 112–120.

Collins, B.J., Rhoades, C.C., Hubbard, R.M., Battaglia, M.A., 2011. Tree regeneration and future stand development after bark beetle infestation and harvesting in Colorado lodgepole pine stands. For. Ecol. Manage. 261, 2168–2175.

Commission Directive, 2008. Official Journal of the European Union. Commission Directive 2008/61/EC of 17 June 2008. Available online: http://eur-lex.europa.eu/search.html?qid=1403114180582& text=directive%202000/29/ec&scope=EURLEX&type=quick& lang=en (Last accessed 17.06.14.).

Costanza, R., d'Arge, R., de Groot, R., Farber, S., Grasso, M., Hannon, B., et al., 1997. The value of the world's ecosystem services and natural capital. Nature 387, 253–260.

Coyle, D.R., Booth, D.C., Wallace, M.S., 2005. Ambrosia beetle (Coleoptera: Scolytidae) species, flight, and attack on living eastern cottonwood trees. J. Econ. Entomol. 98, 2049–2057.

Cozens, R.D., 1997. The Upper Bowron spruce beetle outbreak: a case study. In: Proceedings, Western Forest Insect Work Conference, April 14–17, 1997, Prince George, B.C., Canada.

Cudmore, T.J., Björklund, N., Carroll, A.L., Lindgren, B.S., 2010. Climate change and range expansion of an aggressive bark beetle: evidence of higher reproductive success in naïve host tree populations. J. Appl. Ecol. 47, 1036–1043.

Cullingham, C.I., Cooke, J.E.K., Dang, S., Davis, C.S., Cooke, B.J., Coltman, D.W., 2011. Mountain pine beetle host-range expansion threatens the boreal forest. Mol. Ecol. 20, 2157–2171.

Dahlsten, D.L., Six, D.L., Erbilgin, N., Raffa, K.F., Lawson, A.B., Rowney, D.L., 2003. Attraction of *Ips pini* (Coleoptera: Scolytidae) and its predators to various enantiomeric ratios of ipsdienol and lanierone in California: implications for the augmentation and conservation of natural enemies. Environ. Entomol. 32, 1115–1122.

Dale, V.H., Joyce, L.A., McNulty, S., Nielson, R.P., Ayres, M.P., Flannigan, M.D., et al., 2001. Climate change and forest disturbances. BioScience 9, 723–734.

Département de la Santé des Forêts, 2014. Available online: http://agriculture.gouv.fr/departement-de-la-sante-des-forets (Last accessed 17.06.14.).

DeRose, J., Long, J., 2009. Wildfire and spruce beetle outbreak: simulation of interacting disturbances in the central Rocky Mountains. Ecoscience 16, 28–38.

Dourojeanni, M.J., 1971. Catalogue des Coléoptères de Belgique, fasc. V, 100–101: Catalogue raisonné des Scolytidae et Platypodidae. Société Royale d'Entomologie de Belgique, Bruxelles (BE).

Drever, M.C., Goheen, J.R., Martin, K., 2009. Species–energy theory, pulsed resources, and regulation of avian richness during a mountain pine beetle outbreak. Ecology 90, 1095–1105.

Dyer, E.D.A., 1973. Spruce beetle aggregated by the synthetic pheromone frontalin. Can. J. For. Res. 3, 486–494.

Dymerski, A.D., Anhold, J.A., Munson, A.S., 2001. Spruce beetle (*Dendroctonus rufipennis*) outbreak in Engelmann spruce (*Picea engelmannii*) in central Utah, 1986–1998. West. N. Am. Naturalist 61, 19–24.

Edburg, S.L., Hicke, J.A., Brooks, P.D., Pendall, E.G., Ewers, B.E., Norton, U., et al., 2012. Cascading impacts of bark beetle-caused tree mortality on coupled biogeophysical and biogeochemical processes. Front. Ecol. Environ. 10, 416–424.

Eisenbarth, E., Wilhelm, G.J., Berens, A., 2001. Buchen-Komplexkrankheit in der Eifel und den angrenzenden Regionen. Allgemeine Forst Zeitschrift für Wald und Forstwirtschaft 56, 1212–1217.

Eisenhart, K.S., Veblen, T.T., 2000. Dendroecological detection of spruce bark beetle outbreaks in northwestern Colorado. Can. J. For. Res. 30, 1788–1798.

Embrey, S., Remais, J.V., Hess, J., 2012. Climate change and ecosystem disruption: the health impacts of the North American Rocky Mountain pine beetle infestation. Am. J. Public Health 102, 818–827.

Erbilgin, N., Nordheim, E.V., Aukema, B.H., Raffa, K.F., 2002. Population dynamics of *Ips pini* and *Ips grandicollis* in red pine plantations in Wisconsin: within- and between-year associations with predators, competitors, and habitat quality. Environ. Entomol. 31, 1043–1051.

EUFORGEN, 2009. Distribution map of Norway spruce (*Picea abies*). Available online, www.euforgen.org, (Last accessed 15.06.14.).

Evangelista, P.H., Kumar, S., Stohlgren, T.J., Young, N.E., 2011. Assessing forest vulnerability and the potential distribution of pine beetles under current and future climate scenarios in the interior west of the US. For. Ecol. Manage. 262, 307–316.

Faccoli, M., 2009. Effect of weather on *Ips typographus* (Coleoptera Curculionidae) phenology, voltinism, and associated spruce mortality in the southeastern Alps. Environ. Entomol. 38, 307–316.

Fahse, L., Heurich, M., 2011. Simulation and analysis of outbreaks of bark beetle infestations and their management at the stand level. Ecol. Model. 222, 1833–1846.

FAO, 2009. International Standards for Phytosanitary Measures: Revision of ISPM No. 15, Regulating Wood Packaging Material in International Trade, Draft Publ. No. 15. Food and Agriculture Organization of the United Nations, Rome, Italy.

Fettig, C.J., Klepzig, K.D., Billings, R.F., Munson, A.S., Nebeker, T.E., Negrón, J.F., Nowak, J.T., 2007. The effectiveness of vegetation management practices for prevention and control of bark beetle infestations in coniferous forests of the western and southern United States. For. Ecol. Manage. 238, 24–53.

Fettig, C.J., Gibson, K.E., Munson, A.S., Negrón, J.F., 2014. Cultural practices for prevention and mitigation of mountain pine beetle infestations. For. Sci. 60, 450–463.

Flint, C.G., 2006. Community perspectives on spruce beetle impacts on the Kenai Peninsula, Alaska. For. Ecol. Manage. 227, 207–218.

Flint, C.G., McFarlane, B., Müller, M., 2009. Human dimensions of forest disturbance by insects: an international synthesis. Environ. Manag. 43, 1174–1186.

Flot, J.-L., Vautherin, P., 2002. Des stocks de bois à conserver en forêt ou hors forêt. Revue Forestière Française 136–144, (special issue) "Après les tempêtes".

Forrest, G.I., 1980. Geographical variation in the monoterpenes of *Pinus contorta* oleoresin. Biochem. Syst. Ecol. 8, 343–359.

Foster, D.R., Orwig, D.A., 2006. Preemptive and salvage harvesting of New England forests: when doing nothing is a viable alternative. Conserv. Biol. 20, 959–970.

Franklin, A., De Cannière, C., Grégoire, J.-C., 2004. Using sales of infested timber to quantify attacks by bark beetles: testing a cost-effective method for the assessment of population levels. Ann. For. Sci. 61, 477–480.

Gan, J., 2004. Risk and damage of southern pine beetle outbreaks under global climate change. For. Ecol. Manage. 191, 61–71.

Gandhi, K.J.K., Gilmore, D.W., Haack, R.A., Katovich, S.A., Krauth, S.J., Mattson, W.J., et al., 2009. Application of semiochemicals to assess the biodiversity of subcortical insects following an ecosystem disturbance in a sub-boreal forest. J. Chem. Ecol. 35, 1384–1410.

Gara, R.I., Millegan, D.R., Gibson, K.E., 1999. Integrated pest management of *Ips pini* (Col., Scolytidae) populations in south-eastern Montana. J. Appl. Entomol. 123, 529–534.

Gawalko, L., 2004. Mountain pine beetle management in British Columbia parks and protected areas. In: Shore, T.L., Brooks, J.E., Stone, J.E. (Eds.), Mountain Pine Beetle Symposium: Challenges and Solutions. Natural Resources Canada, Canadian Forest Service, Pacific Forestry Centre, Victoria, B.C., Canada, pp. 79–86.

Gilbert, M., Nageleisen, L.-M., Franklin, A., Grégoire, J.-C., 2005. Poststorm surveys reveal large-scale spatial patterns and influences of site factors, forest structure and diversity in endemic bark-beetle populations. Landsc. Ecol. 20, 35–49.

Gonzalez, R., Grégoire, J.-C., Drumont, A., Windt, N., 1996. A sampling technique to estimate within-tree populations of pre-emergent *Ips typographus* (Col., Scolytidae). J. Appl. Entomol. 120, 569–576.

Grégoire, J.-C., Evans, H.F., 2004. Damage and control of BAWBILT organisms—an overview. In: Lieutier, F., Day, K.R., Battisti, A., Grégoire, J.-C., Evans, H.F. (Eds.), Bark and Wood Boring Insects in Living Trees in Europe, a Synthesis. Kluwer, Dordrecht, pp. 19–37.

Grégoire, J.-C., Raty, L., Drumont, A., De Windt, N., 1997. Pheromone mass-trapping: does it protect windfalls from attack by *Ips typographus* (Coleoptera: Scolytidae)? In: Grégoire, J.-C., Liebhold, A.M., Stephen, F.M., Day, K.R., Salom, S.M. (Eds.), Integrating Cultural Tactics into the Management of Bark Beetle and Reforestation Pests. Proceedings of the IUFRO Conference, 1–4 September. 1996. USDA, Forest Service General Technical Report NE-236, Vallombrosa, Italy, pp. 1–8.

Haack, R.A., 2001. Intercepted Scolytidae (Coleoptera) at US ports of entry: 1985–2000. Integrated Pest Manag. Rev. 6, 253–282.

Haack, R.A., Britton, K.O., Brockerhoff, E.G., Cavey, J.F., Garrett, L.J., Kimberley, M., et al., 2014. Effectiveness of the International Phytosanitary Standard ISPM No. 15 on reducing wood borer infestation rates in wood packaging material entering the United States. PLoS One 9, e96611.

Haack, R.A., Lawrence, R.K., 1994. Geographic distribution of *Tomicus piniperda* in North America: 1992–1994. Newslett. Mich. Entomol. Soc. 39, 10–11.

Haberkern, K.E., Raffa, K.F., 2003. Phloeophagous and predaceous insects responding to synthetic pheromones of bark beetles inhabiting white spruce stands in the Great Lakes Region. J. Chem. Ecol. 29, 1651–1663.

Hansen, E.M., 2014. Forest development and carbon dynamics after mountain pine beetle outbreaks. For. Sci. 60, 476–488.

Hansen, E.M., Bentz, B.J., 2003. Comparison of reproductive capacity among univoltine, semivoltine, and re-emerged parent spruce beetles (Coleoptera: Scolytidae). Can. Entomol. 135, 697–712.

Hansen, E.M., Bentz, B.J., Turner, D.L., 2001a. Temperature-based model for predicting univoltine brood proportions in spruce beetle (Coleoptera: Scolytidae). Can. Entomol. 133, 827–841.

Hansen, E.M., Bentz, B.J., Turner, D.L., 2001b. Physiological basis for flexible voltinism in the spruce beetle (Coleoptera: Scolytidae). Can. Entomol. 133, 805–817.

Hart, S.J., Veblen, T.T., Eisenhart, K.S., Jarvis, D., Kulakowski, D., 2014. Drought induces spruce beetle (*Dendroctonus rufipennis*) outbreaks across northwestern Colorado. Ecology 95, 930–939.

Hedmark, Å., Scholz, M., Elowson, T., 2009. Treatment of log yard runoff impacted by aged logs in a free water surface constructed wetland. Environ. Eng. Sci. 26, 1623–1632.

Heurich, M., Baierl, F., Günther, S., Sinner, K.F., 2011. Management and conservation of large mammals in the Bavarian Forest National Park. Silva Gabreta. 17, 1–18.

Hoffmann, C., Morgan, P., Mell, W., Parsons, R., Strand, E.K., Cook, S., 2012. Numerical simulation of crown fire hazard immediately after bark beetle-caused mortality in lodgepole pine forests. For. Sci. 58, 178.

Holighaus, G., Schütz, S., 2006. Odours of wood decay as semiochemicals for *Trypodendron domesticum* L. (Col., Scolytidae). Mitteilungen der Deutschen Gesellschaft für Allgemeine und Angewandte Entomologie 15, 161–165.

Holsten, E.H., 1990. Spruce beetle activity in Alaska: 1920–1989. USDA Forest Service, Anchorage, Alaska, For. Pest. Manage. Tech. Rep., R10-90-18.

Hood, S., Bentz, B., 2007. Predicting postfire Douglas-fir beetle attacks and tree mortality in the northern Rocky Mountains. Can. J. For. Res. 37, 1058–1069.

Hopper, T., 2010. Pine beetle wood spurs innovation. BC Business. February 3, 2010. Available online, http://www.bcbusiness.ca/natural-resources/pine-beetle-wood-spurs-innovation (Last accessed 19.06.14.).

Huart, O., De Proft, M., Grégoire, J.-C., Piel, F., Gaubicher, B., Carlier, F. X., et al., 2003. Le point sur la maladie du hêtre en Wallonie. Forêt Wallonne 64, 2–20.

Huber, C., 2005. Long lasting nitrate leaching after bark beetle attack in the highlands of the Bavarian Forest National Park. J. Environ. Qual. 34, 1772–1779.

IRM (Institut Royal Météorologique de Belgique), 2014. Evénements marquants depuis 1901. Available online: http://www.meteo.be/meteo/view/fr/1103100-Canicule.html (Last accessed 15.06.14.).

Jactel, H., Brockerhoff, E.G., 2007. Tree diversity reduces herbivory by forest insects. Ecol. Lett. 10, 835–848.

Jakuš, R., Edwards-Jonášová, M., Cudlín, P., Blaženec, M., Ježík, M., Havlíček, F., Moravec, I., 2011. Characteristics of Norway spruce trees (*Picea abies*) surviving a spruce bark beetle (*Ips typographus* L.) outbreak. Trees 25, 965–973.

Jenkins, M.J., Hebertson, E., Page, W., Jorgensen, C.A., 2008. Bark beetles, fuels, fires and implications for forest management in the Intermountain West. For. Ecol. Manage. 254, 16–34.

Jonášová, M., Prach, K., 2008. The influence of bark beetles outbreak vs. salvage logging on ground layer vegetation in central European mountain spruce forests. Biol. Conserv. 141, 1525–1535.

Jönsson, A.M., Harding, S., Bärring, L., Ravn, H.P., 2007. Impact of climate change on the population dynamics of *Ips typographus* in southern Sweden. Agric. For. Meteorol. 146, 70–81.

Jönsson, A.M., Appelberg, G., Harding, S., Bärring, L., 2009. Spatio-temporal impact of climate change on the activity and voltinism of the spruce bark beetle, *Ips typographus*. Glob. Change Biol. 15, 486–499.

Jönsson, A.M., Harding, S., Krokene, P., Lange, H., Lindelöw, Å., Økland, B., et al., 2011. Modelling the potential impact of global warming on *Ips typographus* voltinism and reproductive diapause. Clim. Change 109, 695–718.

Jönsson, A.M., Schroeder, L.M., Lagergren, F., Anderbrant, O., Smith, B., 2012. Guess the impact of *Ips typographus*—an ecosystem modelling approach for simulating spruce bark beetle outbreaks. Agric. For. Meteorol. 166–167, 188–200.

Jonsson, M., 2004. Wet storage of roundwood—effects on wood properties and treatment of run-off water. PhD thesis, Swedish University of Agricultural Sciences, Uppsala.

Jung, T., 2009. Beech decline in Central Europe driven by the interaction between *Phytophthora* infections and climatic extremes. For. Pathol. 39, 73–94.

Kärvemo, S., Van Boeckel, T.P., Gilbert, M., Grégoire, J.-C., Schroeder, M., 2014. Large-scale risk mapping of an eruptive bark beetle—importance of forest susceptibility and beetle pressure. For. Ecol. Manage. 318, 158–166.

Kautz, M., Dworschak, K., Gruppe, A., Schopf, R., 2011. Quantifying spatio-temporal dispersion of bark beetle infestations in epidemic and non-epidemic conditions. For. Ecol. Manage. 262, 598–608.

Kegley, S.J., Livingston, L., Gibson, K.E., 1997. Pine Engraver *Ips pini* (Say) in the Western United States. USDA Forest Service, Forest Insect and Disease Leaflet 122.

Kelsey, R.G., Beh, M.M., Shaw, D.C., Manter, D.K., 2013. Ethanol attracts scolytid beetles to *Phytophthora ramorum* cankers on coast live oak. J. Chem. Ecol. 39, 494–506.

Kerck, K., 1972. Aethylalkohol und Stammkontur als Komponenten der Primäranlockung bei *Xyloterus domesticus* L. (Col.: Scolytidae). Naturwissenschaften 59, 1–2.

Klepzig, K.D., Raffa, K.F., Smalley, E.B., 1991. Association of insect-fungal complexes with red pine decline in Wisconsin. For. Sci. 37, 1119–1139.

Konkin, D., Hopkins, K., 2009. Learning to deal with climate change and catastrophic forest disturbances. Unasylva 60, 17–23.

Krehan, H., Steyrer, G., Hoch, G., 2012. Borkenkäfer-Situation 2011: Schäden deutlich geringer. Forstschutz Aktuell 56, 11–15.

Krieger, D.J., 2001. The Economic Value of Forest Ecosystem Services: A Review. Wilderness Society, Washington, DC.

Kühnholz, S., Borden, J.H., Uzunovic, A., 2001. Secondary ambrosia beetles in apparently healthy trees: adaptations, potential causes and suggested research. Integrated Pest Manag. Rev. 6, 209–219.

Kulakowski, D., Jarvis, D., 2011. The influence of mountain pine beetle outbreaks and drought on severe wildfires in northwestern Colorado and southern Wyoming: a look at the past century. For. Ecol. Manage. 262, 1686–1696.

Kumar, A., Flynn, P., Sokhansanj, S., 2008. Biopower generation from mountain pine infested wood in Canada: an economical opportunity for greenhouse gas mitigation. Renew. Energy 33, 1354–1363.

Kupferschmid Albisetti, A.D., Brang, P., Schönenberger, W., Bugmann, H., 2003. Decay of *Picea abies* snag stands on steep mountain slopes. For. Chron. 79, 247–252.

Kurz, W.A., Dymond, C.C., Stinson, G., Rampley, G.J., Neilson, E.T., Carroll, A.L., et al., 2008. Mountain pine beetle and forest carbon feedback to climate change. Nature 452, 987–990.

La Spina, S., De Cannière, C., Dekri, A., Grégoire, J.-C., 2013. Frost increases beech susceptibility to scolytine ambrosia beetles. Agric. For. Entomol. 15, 157–167.

Lanier, G.N., Wood, D.L., 1975. Specificity of response to pheromones in the genus *Ips* (Coleoptera: Scolytidae). J. Chem. Ecol. 1, 9–23.

Lanier, G.N., Birch, M.C., Schmitz, R.F., Furniss, M.M., 1972. Pheromones of *Ips pini* (Coleoptera: Scolytidae): variation in response among three populations. Can. Entomol. 104, 1917–1923.

Lausch, A., Fahse, L., Heurich, M., 2011. Factors affecting the spatiotemporal dispersion of *Ips typographus* (L.) in Bavarian Forest National Park: a long-term quantitative landscape-level analysis. For. Ecol. Manage. 261, 233–245.

Lausch, A., Heurich, M., Fahse, L., 2013. Spatio-temporal infestation patterns of *Ips typographus* (L.) in the Bavarian Forest National Park, Germany. Ecol. Indic. 31, 73–81.

Lehnert, L.W., Bässler, C., Brandl, R., Burton, P.J., Müller, J., 2013. Conservation value of forests attacked by bark beetles: highest number of indicator species is found in early successional stages. J. Nat. Conserv. 21, 97–104.

Lewis, K.J., Lindgren, B.S., 2000. A conceptual model of biotic disturbance ecology in the central interior of B.C.: how forest management can turn Dr. Jekyll into Mr. Hyde. For. Chron. 76, 433–443.

Lewis, K.J., Lindgren, B.S., 2002. Relationship between spruce beetle and tomentosus root disease: two natural disturbance agents of spruce. Can. J. For. Res. 32, 31–37.

Lewis, K.J., Hartley, I.D., 2006. Rate of deterioration, degrade, and fall of trees killed by mountain pine beetle. BC J. Ecosyst. Manag. 7, 11–19.

Lewis, K.J., Thompson, D., 2011. Degradation of wood in standing lodgepole pine killed by mountain pine beetle. Wood Fiber Sci. 43, 130–142.

Liebhold, A.M., Tobin, P.C., 2008. Population ecology of insect invasions and their management. Annu. Rev. Entomol. 53, 387–408.

Lindelöw, A., Schroeder, M.L., 2008. The storm "Gudrun" and the spruce bark beetle in Sweden. Forstschutz Aktuell 44, 5–7.

Logan, J.A., MacFarlane, W.W., Willcox, L., 2010. Whitebark pine vulnerability to climate-driven mountain pine beetle disturbance in the Greater Yellowstone Ecosystem. Ecol. Appl. 20, 895–902.

Lombardero, M.J., Ayres, M.P., Ayres, B.D., Reeve, J.D., 2000. Cold tolerance of four species of bark beetle (Coleoptera: Scolytidae) in North America. Environ. Entomol. 29, 421–432.

Lotan, J.E., Critchfield, W.B., 1990. *Pinus contorta* Dougl. ex. Loud. lodgepole pine. In: Burns, R.M., Honkala, B.H. (Eds.), Silvics of North America, Volume 1 Conifers. U.S. Department of Agriculture, Forest Service, Washington, DC, pp. 302–315, Agriculture Handbook 654.

Luo, X., Gleisner, R., Tian, S., Negrón, J., Zhu, W., Horn, E., et al., 2010. Evaluation of mountain beetle-infested lodgepole pine for cellulosic ethanol production by sulfite pretreatment to overcome recalcitrance of lignocellulose. Ind. Eng. Chem. Res. 49, 8258–8266.

Lynch, H.J., Renkin, R.A., Crabtree, R.L., Moorcroft, P.R., 2006. The influence of previous mountain pine beetle (*Dendroctonus ponderosae*) activity on the 1988 Yellowstone sires. Ecosystems 9, 1318–1327.

Mahmoudi, M., Sowlati, T., Sokhansanj, S., 2009. Logistics of supplying biomass from a mountain pine beetle-infested forest to a power plant in British Columbia. Scand. J. For. Res. 24, 76–86.

Marini, L., Ayres, M.P., Battisti, A., Faccoli, M., 2012. Climate affects severity and altitudinal distribution of outbreaks in an eruptive bark beetle. Clim. Change 115, 327–341.

Marini, L., Lindelöw, Å., Jönsson, A.M., Wulff, S., Schroeder, L.M., 2013. Population dynamics of the spruce bark beetle: a long-term study. Oikos 122, 1768–1776.

Martin, K., Norris, A., Drever, M., 2006. Effects of bark beetle outbreaks on avian biodiversity in the British Columbia interior: implications for critical habitat management. BC J. Ecosyst. Manag. 7, 10–24.

McFarlane, B.L., Watson, D.O., 2008. Perceptions of ecological risk associated with mountain pine beetle (*Dendroctonus ponderosae*) infestations in Banff and Kootenay National Parks of Canada. Risk Anal. 28, 203–212.

McFarlane, B.L., Parkins, J.R., Watson, D.O.T., 2012. Risk, knowledge, and trust in managing forest insect disturbance. Can. J. For. Res. 42, 710–719.

McHugh, C.W., Kolb, T.E., Wilson, J.L., 2003. Bark beetle attacks on ponderosa pine following fire in northern Arizona. Environ. Entomol. 32, 510–522.

McNay, R.S., Sulyma, R., Voller, J., Brumovsky, V., 2008. Potential Implications of Beetle-Related Timber Salvage on the Integrity of Caribou Winter Range. Wildlife Infometrics Inc, Mackenzie, British Columbia, Canada, Wildlife Infometrics Inc. Report No. 273.

McPherson, B.A., Wood, D.L., Storer, A.J., Kelly, N.M., Standiford, R.B., 2001. Sudden Oak Death, a new forest disease in California. Integrated Pest Manag. Rev. 6, 243–246.

McPherson, B.A., Erbilgin, N., Wood, D.L., Svihra, P., Storer, A.J., Standiford, R.B., 2008. Attraction of ambrosia and bark beetles to coast live oaks infected by *Phytophthora ramorum*. Agric. For. Entomol. 10, 315–321.

Meddens, A.J.H., Hicke, J.A., Ferguson, C.A., 2012. Spatiotemporal patterns of observed bark beetle-caused tree mortality in British Columbia and the western United States. Ecol. Appl. 22, 1876–1891.

Meier, F., Engesser, R., Forster, B., Odermatt, O., Angst, A., 2013. Protection des forêts—vue d'ensemble 2012. Institut fédéral de recherches sur la forêt, la neige et le paysage WSL, Birmensdorf.

Meigs, G.W., Kennedy, R.E., Cohen, W.B., 2011. A landsat time series approach to characterize bark beetle and defoliator impacts on tree mortality and surface fuels in conifer forests. Remote Sens. Environ. 115, 3707–3718.

Meurisse, N., Couillien, D., Grégoire, J.-C., 2008. Kairomone traps: a tool for monitoring the invasive spruce bark beetle *Dendroctonus micans* (Coleoptera: Scolytinae) and its specific predator, *Rhizophagus grandis* (Coleoptera: Monotomidae). J. Appl. Ecol. 45, 537–548.

Miller, D.R., Borden, J.H., Slessor, K.N., 1989. Inter- and intrapopulation variation of the pheromone, ipsdienol, produced by male pine engravers, *Ips pini* (Say) (Coleoptera: Scolytidae). J. Chem. Ecol. 15, 233–247.

Miller, D.R., Gibson, K.E., Raffa, K.F., Seybold, S.J., Teale, S.A., Wood, D.L., 1997. Geographic variation in response of pine engraver, *Ips pini*, and associated species to pheromone, lanierone. J. Chem. Ecol. 23, 2013–2031.

Ministry of Forests, Lands and Natural Resources Operations, 2012. A history of the battle against the mountain pine beetle: 2000 to 2012. Available online, https://www.for.gov.bc.ca/hfp/mountain_pine_beetle/Pine%20Beetle%20Response%20Brief%20History%20May%2023%202012.pdf (Last accessed 06.06.14.).

Montecchio, L., Faccoli, M., 2014. First record of thousand cankers disease *Geosmithia morbida* and walnut twig beetle *Pityophthorus juglandis* on *Juglans nigra* in Europe. Plant Dis. 98, 696.

Moreau, J., Chantre, G., Vautherin, P., Gorget, Y., Ducray, P., Léon, P., 2006. Conservation de bois sous aspersion. Revue Forestière Française 58, 377–387.

Morehouse, K., Johns, T., Kaye, J., Kaye, M., 2008. Carbon and nitrogen cycling immediately following bark beetle outbreaks in southwestern ponderosa pine forests. For. Ecol. Manage. 255, 2698–2708.

Müller, J., Bussler, H., Gossner, M., Rettelbach, T., Duelli, P., 2008. The European spruce bark beetle *Ips typographus* in a national park: from pest to keystone species. Biodiv. Conserv. 17, 2979–3001.

Müller, J., Noss, R.F., Bussler, H., Brandl, R., 2010. Learning from a "benign neglect strategy" in a national park: response of saproxylic beetles to dead wood accumulation. Biol. Conserv. 143, 2559–2569.

Müller, M., 2011. How natural disturbance triggers political conflict: bark beetles and the meaning of landscape in the Bavarian forest. Global Environ. Change 21, 935–946.

Müller, M., Job, H., 2009. Managing natural disturbance in protected areas: tourists' attitude towards the bark beetle in a German national park. Biol. Conserv. 142, 375–383.

Nageleisen, L.-M., 1993. Les dépérissements d'essences feuillues en France. Revue Forestière Française 45, 605–620.

Nageleisen, L.-M., 2006. Insectes sous-corticaux des résineux en 2005: poursuite de la pullulation de typographe de l'épicéa dans les montagnes de l'Est. La santé des forêts [France] en 2005. Min. Agri. et Pêche. (DGFAR), Paris, France.

Nageleisen, L.-M., 2007. Insectes sous-corticaux des résineux en 2006: le typographe de l'épicéa reste le principal ravageur des forêts de conifères. La santé des forêts [France] en 2006. Min. Agri. et Pêche (DGFAR), Paris, France.

Nelson, K.N., Rocca, M.E., Diskin, M., Aoki, C.F., Romme, W.H., 2014. Predictors of bark beetle activity and scale-dependent spatial heterogeneity change during the course of an outbreak in a subalpine forest. Landsc. Ecol. 29, 97–109.

Netherer, S., Nopp-Mayr, U., 2005. Predisposition assessment systems (PAS) as supportive tools in forest management—rating of site and stand-related hazards of bark beetle infestation in the high Tatra Mountains as an example for system application and verification. For. Ecol. Manage. 207, 99–107.

Norris, A.R., Martin, K., 2008. Mountain pine beetle presence affects nest patch choice of red-breasted nuthatches. J. Wildlife Manage. 72, 733–737.

O'Neill, M., Evans, H.F., 1999. Cost-effectiveness analysis of options within an integrated crop management regime against great spruce bark beetle, *Dendroctonus micans* Kug. (Coleoptera: Scolytidae). Agric. For. Entomol. 1, 151–156.

Observatoire wallon de la Santé des Forêts, 2014. Available online: http://environnement.wallonie.be/cgi/dgrne/plateforme_dgrne/Visiteur/V2/FrameSet.cfm?page=http://environnement.wallonie.be/sante-foret/ (Last accessed 17.06.14.).

Økland, B., Berryman, A., 2004. Resource dynamic plays a key role in regional fluctuations of the spruce bark beetles *Ips typographus*. Agric. For. Entomol. 6, 141–146.

Pacific Fisheries Resource Conservation Council, No date. Mountain pine beetle: salmon are suffering too. Available online: http://www.fish.bc.ca/mountain-pine-beetle-salmon-are-suffering-too (Last accessed 16.06.14.).

Page, W.G., Jenkins, M.J., Alexander, M.E., 2014. Crown fire potential in lodgepole pine forests during the red stage of mountain pine beetle attack. Forestry. http://dx.doi.org/10.1093/forestry/cpu003.

Pan, X., Xie, D., Yu, R.W., Lam, D., Saddler, J.N., 2007. Pretreatment of lodgepole pine killed by mountain pine beetle using the ethanol

organosolv process: fractionation and process optimization. Ind. Eng. Chem. Res. 46, 2609–2617.

Pan, X., Xie, D., Yu, R.W., Saddler, J.N., 2008. The bioconversion of mountain pine beetle-killed lodgepole pine to fuel ethanol using the organosolv process. Biotechnol. Bioeng. 101, 39–48.

Parfitt, B., 2010. Axed: A Decade of Cuts to BC's Forest Service. Sierra Club BC and Canadian Centre for Policy Alternatives, Vancouver, B.C., Canada. Available online, https://www.policyalternatives.ca/sites/default/files/uploads/publications/BC%20Office/2010/12/CCPA_BTN_forest_service_web.pdf (Last accessed 16.06.14.).

Pasztor, F., Matulla, C., Rammer, W., Lexer, M.J., 2014. Drivers of the bark beetle disturbance regime in alpine forests in Austria. For. Ecol. Manage. 318, 349–358.

Patriquin, M.N., Wellstead, A.M., White, W.A., 2007. Beetles, trees, and people: regional economic impact sensitivity and policy considerations related to the mountain pine beetle infestation in British Columbia, Canada. Forest Pol. Econ. 9, 938–946.

Perrot, D., Molotch, N.P., Musselman, K.N., Pugh, E.T., 2014. Modelling the effects of the mountain pine beetle on snowmelt in a subalpine forest. Ecohydrology 7, 226–241.

Petersen, B., Stuart, D., 2014. Explanations of a changing landscape: a critical examination of the British Columbia bark beetle epidemic. Environ. Plann. A 46, 598–613.

Piel, F., Grégoire, J.-C., Knížek, M., 2006. New occurrence of *Ips duplicatus* Sahlberg in Herstal (Liege, Belgium). EPPO/OEPP Bulletin 36, 529–530.

Piel, F., Gilbert, M., De Cannière, C., Grégoire, J.-C., 2008. Coniferous round wood imports from Russia and Baltic countries to Belgium. A pathway analysis for assessing risks of exotic pest insect introductions. Diversity and Distributions 14, 318–328.

Poncelet, J., 1965. Eclaircies. Bulletins de la Société Royale Forestière de Belgique 7, 293–300.

Powell, E.N., Raffa, K.F., 2011. Fire injury reduces inducible defenses of lodgepole pine against mountain pine beetle. J. Chem. Ecol. 37, 1184–1192.

Powell, E.N., Townsend, P.A., Raffa, K.F., 2012. Wildfire provides refuge from local extinction but is an unlikely driver of outbreaks by mountain pine beetle. Ecol. Monogr. 82, 69–84.

Price, J.I., McCollum, D.W., Berrens, R.P., 2010. Insect infestation and residential property values: a hedonic analysis of the mountain pine beetle epidemic. Forest Pol. Econ. 12, 415–422.

Prieels, H., 1961. Les vices cachés dans les bois sur pied. Bulletins de la Société Royale Forestière de Belgique 7, 323–336.

Priewasser, K., Brang, P., Bachofen, H., Bugmann, H., Wohlgemuth, T., 2013. Impacts of salvage-logging on the status of deadwood after windthrow in Swiss forests. Eur. J. For. Res. 132, 231–240.

Progar, R.A., Eglitis, A., Lundquist, J.E., 2009. Some ecological, economic, and social consequences of bark beetle infestations. In: The Western Bark Beetle Research Group: A Unique Collaboration with Forest Health Protection, 71.Proceedings of a Symposium at the 2007 Society of American Foresters Conference October 23–28, 2007, Portland, Oregon, United States Department of Agriculture Forest Service Pacific Northwest Research Station General Technical Report PNW-GTR-784.

Raffa, K.F., Dahlsten, D.L., 1995. Differential responses among natural enemies and prey to bark beetle pheromones. Oecologia 102, 17–23.

Raffa, K.F., Hobson, K.R., LaFontaine, S., Aukema, B.H., 2007. Can chemical communication be cryptic? Adaptations by herbivores to natural enemies exploiting prey semiochemistry. Oecologia 153, 1009–1019.

Raffa, K.F., Aukema, B.H., Bentz, B.J., Carroll, A.L., Hicke, J.A., Turner, M.G., Romme, W.H., 2008. Cross-scale drivers of natural disturbances prone to anthropogenic amplification: the dynamics of bark beetle eruptions. BioScience 58, 501–517.

Raffa, K.F., Powell, E.N., Townsend, P.A., 2013. Temperature-driven range expansion of an irruptive insect heightened by weakly coevolved plant defenses. Proc. Natl. Acad. Sci. U. S. A. 110, 2193–2198.

Raffa, K.F., Aukema, B.H., Bentz, B.J., Carroll, A.L., Hicke, J.A., Kolb, T. E., in press. Responses of tree-killing bark beetles to a changing climate. In: Björkman, C. Niemelä, P. (Eds.), Climate Change and Insect Pests. CABI, Wallingford.

Ranger, C.M., Reding, M.E., Persad, A.B., Herms, D.A., 2010. Ability of stress-related volatiles to attract and induce attacks by *Xylosandrus germanus* and other ambrosia beetles. Agric. For. Entomol. 12, 177–185.

Ranger, C.M., Reding, M.E., Schultz, P.B., Oliver, J.B., 2013. Influence of flood-stress on ambrosia beetle host-selection and implications for their management in a changing climate. Agric. For. Entomol. 15, 56–64.

Rankin, L.J., Borden, J.H., 1991. Competitive interactions between the mountain pine beetle and the pine engraver in lodgepole pine. Can. J. For. Res. 21, 1029–1036.

Ritchie, C., 2008. Management and challenges of the mountain pine beetle infestation in British Columbia. Alces 44, 127–135.

Ryall, K.L., Fahrig, L., 2005. Habitat loss decreases predator-prey ratios in a pine-bark beetle system. Oikos 110, 265–270.

Saab, V.A., Latif, Q.S., Rowland, M.M., Johnson, T.N., Chalfoun, A.D., Buskirk, S.W., et al., 2014. Ecological consequences of mountain pine beetle outbreaks for wildlife in western North American forests. For. Sci. 60, 539–559.

Safranyik, L., 1985. Infestation incidence and mortality in white spruce stands by *Dendroctonus rufipennis* Kirby (Coleoptera, Scolytidae) in central British Columbia. Z. ang. Entomol. 99, 86–93.

Safranyik, L., 1988. The population biology of the spruce beetle in western Canada and implications for management. In: Payne, T.L., Saarenmaa, H. (Eds.), Integrated Control of Scolytid Bark Beetles. VPI and State University, Blacksburg, Virginia, USA, pp. 3–23.

Safranyik, L., Carroll, A.L., 2006. The biology and epidemiology of the mountain pine beetle in lodgepole pine forests. In: Safranyik, L., Wilson, B. (Eds.), The Mountain Pine Beetle: A Synthesis of Biology, Management, and Impacts on Lodgepole Pine. Natural Resources Canada, Canadian Forest Service, Pacific Forestry Centre, Victoria, B.C., Canada, pp. 3–66.

Safranyik, L., Shore, T.L., Linton, D.A., 1996. Ipsdienol and lanierone increase *Ips pini* Say (Coleoptera: Scolytidae) attack and brood density in lodgepole pine infested by mountain pine beetle. Can. Entomol. 128, 199–207.

Safranyik, L., Carroll, A.L., Régnière, J., Langor, D.W., Riel, W.G., Shore, T.L., et al., 2010. Potential for range expansion of mountain pine beetle into the boreal forest of North America. Can. Entomol. 142, 415–442.

Samalens, J.C., Rossi, J.P., Guyon, D., Van Halder, I., Menassieu, P., Piou, D., Jactel, H., 2007. Adaptive roadside sampling for bark beetle damage assessment. For. Ecol. Manage. 253, 177–187.

Scheepers, D., Eloy, M.C., Briquet, M., 1997. Use of RAPD patterns for clone verification and in studying provenance relationships in Norway spruce (*Picea abies*). Theor. Appl. Genet. 94, 480–485.

Schelhaas, M.J., Nabuurs, G.J., Schuck, A., 2003. Natural disturbances in the European forests in the 19th and 20th centuries. Global Change Biol. 9, 1620–1633.

Schmidt, M., Hanewinkel, M., Kändler, G., Kublin, E., Kohnle, U., 2010. An inventory-based approach for modeling single-tree storm damage experiences with the winter storm of 1999 in southwestern Germany. Can. J. For. Res. 40, 1636–1652.

Schmitz, S., Zini, J., Chandelier, A., 2009. Involvement of *Phytophthora* species in the decline of beech (*Fagus sylvatica*) in the southern part of Belgium. In: *Phytophthoras* in Forests and Natural Ecosystems. Proceedings of the Fourth Meeting of the International Union of Forest Research Organizations (IUFRO) Working Party S07.02.09. United States Department of Agriculture, Forest Service, Pacific Southwest Research Station, pp. 320–323, General Technical Report PSW-GTR-221.

Schroeder, L.M., 2007. Retention or salvage logging of standing trees killed by the spruce bark beetle *Ips typographus*: consequences for dead wood dynamics and biodiversity. Scand. J. For. Res. 22, 524–530.

Schroeder, L.M., Lindelöw, Å., 2002. Attacks on living spruce trees by the bark beetle *Ips typographus* (Col. Scolytidae) following a storm-felling: a comparison between stands with and without removal of wind-felled trees. Agric. For. Entomol. 4, 47–56.

Seidl, R., Rammer, W., Jäger, D., Lexer, M.J., 2008. Impact of bark beetle (*Ips typographus* L.) disturbance on timber production and carbon sequestration in different management strategies under climate change. For. Ecol. Manage. 256, 209–220.

Seidl, R., Schelhaas, M.J., Lexer, M.J., 2011. Unraveling the drivers of intensifying forest disturbance regimes in Europe. Global Change Biol. 17, 2842–2852.

Simard, M., Romme, W.H., Griffin, J.M., Turner, M.G., 2011. Do mountain pine beetle outbreaks change the probability of active crown fire in lodgepole pine forests? Ecol. Monogr. 81, 3–24.

Sims, C., Aadland, D., Finnoff, D., 2010. A dynamic bioeconomic analysis of mountain pine beetle epidemics. J. Econ. Dynam. Contr. 34, 2407–2419.

Six, D.L., Bentz, B.J., 2003. Fungi associated with the North American spruce beetle, *Dendroctonus rufipennis*. Can. J. For. Res. 33, 1815–1820.

Six, D.L., Meer, M.V., DeLuca, T.H., Kolb, P., 2002. Pine engraver (*Ips pini*) colonization of logging residues created using alternative slash management systems in Western Montana. West. J. Appl. For. 17, 96–100.

Six, D.L., Biber, E., Long, E., 2014. Management for mountain pine beetle outbreak suppression: does relevant science support current policy? Forests 5, 103–133.

Stadelmann, G., Bugmann, H., Meier, F., Wermelinger, B., Bigler, C., 2013a. Effects of salvage logging and sanitation felling on bark beetle (*Ips typographus* L.) infestations. For. Ecol. Manage. 305, 273–281.

Stadelmann, G., Bugmann, H., Wermelinger, B., Meier, F., Bigler, C., 2013b. A predictive framework to assess spatio-temporal variability of infestations by the European spruce bark beetle. Ecography 36, 1208–1217.

Stark, D.F., Li, A.H.Y., 2009. Mountain pine beetle: a new aero-allergen. J. Allergy Clin. Immunol. 123, S243.

Stark, R.W., Waters, W.E., 1987. Impacts of forest insects and diseases: significance and measurement. Crit. Rev. Plant Sci. 5, 161–203.

Stednick, J.D., 1996. Monitoring the effects of timber harvest on annual water yield. J. Hydrol. 176, 79–95.

Steed, B.E., Wagner, M.R., 2008. Seasonal pheromone response by *Ips pini* in northern Arizona and western Montana, USA. Agric. For. Entomol. 10, 189–203.

Stennes, B.K., Niquidet, K., van Kooten, C.G., 2010. Implications of expanding bioenergy production from wood in British Columbia: an application of a regional wood fiber allocation model. For. Sci. 56, 366–378.

Steventon, J.D., Daust, D.K., 2009. Management strategies for a large-scale mountain pine beetle outbreak: modelling impacts on American martens. For. Ecol. Manage. 257, 1976–1985.

Steyrer, G., Krehan, H., 2009. Borkenkäfer-Kalamität 2008: ist ein weiterer Rückgang wahrscheinlich? Forstschutz Aktuell. 46, 9–15.

Stinson, G., White, T.M., Blain, D., Kurz, W.A., Smyth, C.E., Neilson, E.T., et al., 2011. An inventory-based analysis of Canada's managed forest carbon dynamics, 1990 to 2008. Glob. Change Biol. 17, 2227–2244.

Sun, J., Lu, M., Gillette, N.E., Wingfield, M.J., 2013. Red turpentine beetle: innocuous native becomes invasive tree killer in China. Annu. Rev. Entomol. 58, 293–311.

Taberlet, P., Fumagalli, L., Wust-Saucy, A.G., Cosson, J.F., 1998. Comparative phylogeography and postglacial colonization routes in Europe. Mol. Ecol. 7, 453–464.

Taylor, S.W., Carroll, A.L., Alfaro, R.I., Safranyik, L., 2006. Forest, climate and mountain pine beetle dynamics. In: Safranyik, L., Wilson, B. (Eds.), The Mountain Pine Beetle: A Synthesis of Biology, Management, and Impacts on Lodgepole Pine. Natural Resources Canada, Canadian Forest Service, Pacific Forestry Centre, Victoria, B.C., Canada, pp. 67–94.

Teale, S.N., Lanier, G.N., 1991. Seasonal variability in response of *Ips pini* (Coleoptera: Scolytidae) to ipsdienol in New York. J. Chem. Ecol. 17, 1145–1158.

Temperli, C., Bugmann, H., Elkin, C., 2013. Cross-scale interactions among bark beetles, climate change, and wind disturbances: a landscape modeling approach. Ecol. Monogr. 83, 383–402.

Wallin, K.F., Raffa, K.F., 2004. Feedback between individual host selection behavior and population dynamics in an eruptive herbivore. Ecol. Monogr. 74, 101–116.

Waring, K.M., Reboletti, D.M., Mork, L.A., Huang, C.-H., Hofstetter, R.W., Garcia, A.M., et al., 2009. Modeling the impacts of two bark beetle species under a warming climate in the southwestern USA: ecological and economic consequences. Environ. Manag. 44, 824–835.

Warzée, N., Gilbert, M., Grégoire, J.-C., 2006. Predator/prey ratios: a measure of bark-beetle population status influenced by stand composition in different French stands after the 1999 storms. Ann. For. Sci. 63, 301–308.

Watson, A.C., Sullivan, J., Amacher, G.S., Asaro, C., 2013. Cost sharing for pre-commercial thinning in southern pine plantations: willingness to participate in Virginia's pine bark beetle prevention program. Forest Pol. Econ. 34, 65–72.

Wermelinger, B., 2004. Ecology and management of the spruce bark beetle *Ips typographus*—a review of recent research. For. Ecol. Manage. 202, 67–82.

Werner, R.A., Holsten, E.H., 1985. Factors influencing generation times of spruce beetles in Alaska. Can. J. For. Res. 15, 438–443.

Werner, R.A., Baker, B.H., Rush, P.A., 1977. The spruce beetle in white spruce forests of Alaska. Gen. Tech. Rep. PNW-61, NW Forest and Range Experimental Station, Portland, OR.

Werner, R.A., Holsten, E.H., Matsuoka, S.M., Burnside, R.E., 2006a. Spruce beetles and forest ecosystems in south-central Alaska: a review of 30 years of research. For. Ecol. Manage. 227, 195–206.

Werner, R.A., Raffa, K.F., Illman, B.L., 2006b. Insect and pathogen dynamics. In: Chapin, I.F.S., Oswood, M., Van Cleve, K., Viereck, L.A., Verbyla, D. (Eds.), Alaska's Changing Boreal Forest. Oxford University Press, Oxford, pp. 133–146.

Westfall, J., Ebata, T., 2014. 2013 Summary of forest health conditions in British Columbia. Pest Management Report Number 15, British Columbia Ministry of Forests, Range and Natural Resources Operations, Victoria, B.C., Canada.

Wichmann, L., Ravn, H.P., 2001. The spread of *Ips typographus* (L.) (Coleoptera, Scolytidae) attacks following heavy windthrow in Denmark, analysed using GIS. For. Ecol. Manage. 148, 31–39.

Williams, D.W., Liebhold, A.M., 2002. Climate change and the outbreak ranges of two North American bark beetles. Agric. For. Entomol. 4, 87–99.

Woo, A., 2012. Softwood-lumber victory unlikely to halt conflict between Canada and U.S. The Globe and Mail, July 18, 2012. Available online, http://www.theglobeandmail.com/report-on-business/industry-news/energy-and-resources/softwood-lumber-victory-unlikely-to-halt-conflict-between-canada-and-us/article4425659/ (Last accessed 18.06.14.).

Wood, S.L., 1982. The bark and ambrosia beetles of North and Central America (Coleoptera: Scolytidae), a taxonomic monograph. Great Basin Nat. Mem. 6, 1–1359.

WSL—Forest Protection Overviews, 2014. Available online: http://www.wsl.ch/fe/walddynamik/waldschutz/wsinfo/fsueb_EN (Last accessed 09.06.14.).

Zhu, J.Y., Luo, X., Tian, S., Gleisner, R., Negrón, J., Horn, E., 2011. Efficient ethanol production from beetle-killed lodgepole pine using SPORL technology and *Saccharomyces cerevisiae* without detoxification. Tappi J. 10, 9–18.

Zimmermann, L., Moritz, K., Kennel, M., Bittersohl, J., 2000. Influence of bark beetle infestation on water quantity and quality in the Grosse Ohe catchment (Bavarian Forest National Park). Silva Gabreta 4, 51–62.

Zolubas, P., Negrón, J., Munson, A.S., 2009. Modelling spruce bark beetle infestation probability. Baltic For. 15, 23–27.

Zycha, H., 1943. Die Buchenrindenfäule. Der Deutsche Forstwirt. 63, 1–2.

Index

Note: Page numbers followed by *f* indicate figures and *t* indicate tables.

Printed and bound by CPI Group (UK) Ltd, Croydon, CR0 4YY

08/05/2025

01865029-0005